ANNALS OF THE NEW YORK ACADEMY OF SCIENCES

Volume 755

EDITORIAL STAFF
Executive Editor
BILL BOLAND
Managing Editor
JUSTINE CULLINAN
Associate Editor
STEFAN MALMOLI

The New York Academy of Sciences
2 East 63rd Street
New York, New York 10021

FUNDAMENTAL PROBLEMS IN QUANTUM THEORY

A CONFERENCE HELD IN HONOR OF PROFESSOR JOHN A. WHEELER

ANNALS OF THE NEW YORK ACADEMY OF SCIENCES
Volume 755

FUNDAMENTAL PROBLEMS IN QUANTUM THEORY

A CONFERENCE HELD IN HONOR OF PROFESSOR JOHN A. WHEELER

Edited by Daniel M. Greenberger and Anton Zeilinger

The New York Academy of Sciences
New York, New York
1995

Library of Congress Cataloging-in-Publication Data

Fundamental problems in quantum theory: a conference held in honor of professor John A. Wheeler/edited by Daniel M. Greenberger and Anton Zeilinger.
p. cm.—(Annals of the New York Academy of Sciences: v. 755).
Includes bibliographical references and index.
ISBN 0-89766-921-5. —ISBN 0-89766-922-3 (pbk.).
1. Quantum theory—Congresses. 2. Quantum optics—Congresses. 3. Wheeler, John Archibald, 1911– . I. Greenberger, Daniel M. II. Zeilinger, Anton. III. Wheeler, John Archibald, 1911– . IV. Series.
Q11.N5 vol. 755
[QC173.96]
500 s–dc20
[530.1′2]

95-3320
CIP

SP
Printed in the United States of America
ISBN 0-89766-921-5 (cloth)
ISBN 0-89766-922-3 (paper)
ISSN 0077-8923

ANNALS OF THE NEW YORK ACADEMY OF SCIENCES

Volume 755
April 7, 1995

FUNDAMENTAL PROBLEMS IN QUANTUM THEORY: A CONFERENCE HELD IN HONOR OF PROFESSOR JOHN A. WHEELER[a]

Editors
DANIEL M. GREENBERGER and ANTON ZEILINGER

Conference Organizer
DANIEL M. GREENBERGER

Conference Organizing Committee
YAN HUA SHIH and CARROLL ALLEY

Scientific Advisory Committee
YAKIR AHARONOV, MICHAEL BERRY, CLAUDE COHEN-TANNOUDJI, DANIEL M. GREENBERGER, DAVID KLYSHKO, ANTHONY LEGGETT, LEONARD MANDEL, N. DAVID MERMIN, JÜRGEN MLYNEK, MIKIO NAMIKI, HELMUT RAUCH, MARLAN SCULLY, ABNER SHIMONY, AKIRA TONOMURA, HERBERT WALTHER, RICHARD WEBB, SAMUEL WERNER, CHEN-NING YANG, and ANTON ZEILINGER

CONTENTS

Part I. Multiphoton Processes and Downconversion

[a]This volume is the result of a conference entitled Fundamental Problems in Quantum Theory: A Conference Held in Honor of Professor John A. Wheeler, which was sponsored by the New York Academy of Sciences and held on June 18–22, 1994, in Baltimore, Maryland.

Part II. Quantum Optics and Micromasers

Part III. Atomic Processes

Part IV. Electron, Neutron, and Atomic Interference

Part V. Geometric Phases

Part VI. Quantum Effects

Part VII. Quantum Mechanical Results

Part VIII. Relativistic and Gravitational Considerations

Part IX. Theoretical Considerations on Measurement

Part X. Quantum Equalities and Inequalities

Part XI. Quantum Entanglement

Part XII. Information Theory and Quantum Communication

Part XIII. Decoherence and Consistent Histories

Part XIV. Randomness in Nature

Part XV. Other Theoretical Considerations

Part XVI. Poster Papers

Financial assistance was received from:

Supporters

- CITY COLLEGE OF NEW YORK
- DEPARTMENT OF ENERGY
- RICHARD LOUNSBERY FOUNDATION
- NATIONAL SCIENCE FOUNDATION
- OFFICE OF NAVAL RESEARCH
- ALFRED P. SLOAN FOUNDATION

Contributors

- HITACHI, LIMITED
- NEC RESEARCH INSTITUTE

JOHN A. WHEELER (photo courtesy of Y. H. Shih)

Preface

DANIEL M. GREENBERGER

Department of Physics
City College of the City University of New York
New York, New York 10031

Quantum theory represents one of the great and most beautiful structures in all of physics. It is the subject that all graduate students must master before going on to their specific area of specialization. Its general principles, like the uncertainty principle, have become metaprinciples that all phenomena must conform to, in all the far-flung branches of the subject. In areas that we understand, such as quantum optics, it provides a basis for predicting the results of as yet unperformed experiments. In areas that we have yet to understand, such as high-T superconductivity, it provides the framework into which new discoveries must be fit and the means for interpreting them. Still, after 70 years, we have no real clue as to how and when the theory will break down (but break down it must—that is the tragedy of the human condition).

Nonetheless, despite its uncontrovertible experimental successes, the theory has a very shaky philosophical foundation. The standard Copenhagen interpretation (whatever that is) requires us to accept so many assumptions that defy common sense that ever since the theory was first developed it has led to enormous debates concerning its interpretation. Most modern physicists accept it without qualification and, indeed, one can develop a creative intuition for using it. The fact that many of its founding fathers turned against the standard interpretation, whereas their followers have tended to accept it without second thoughts can only partly be ascribed to the circumstance that anything tends to grow more familiar with repeated use. Part of the explanation must be related to the fact that those very founders were much more culturally well rounded than most modern physicists. They were philosophically trained and philosophically inclined and did not like what they saw.

In spite of their doubts, the subject grew rapidly and it became fashionable to avoid questions concerning the foundations. This attitude only started to change after Bell's famous theorem in 1964. He showed that one could pose some of one's intuitive doubts experimentally. Since then, a number of alternate interpretations have grown and new experimental tests devised. Over the years, more and more ingenious challenges have been put to the theory and it has passed each of these tests handsomely. Today, we know that the strange predictions of the theory hold up experimentally (even though the foundations remain shaky). We will never go back to classical physics—we must learn to accept and live with the world as it actually is.

What makes quantum mechanics so much fun is that its results run so counter to one's classical intuitions, yet they are always predictable, even if unanticipated. That is why I like to say that quantum mechanics is *magic*, but it is not *black* magic. In recent years, there has been an explosive growth in new experimental techniques that can be adapted to test some of the stranger predictions of quantum theory. Most of the papers of this conference are devoted either to explaining these new techniques or to some of the experiments they have led to. Not only can this serve to sharpen our intuitions concerning this ever fascinating subject, but also it brings us closer to the day when we can determine the limits of the theory. Some of these new techniques include neutron and electron interferometry; experimentally realizable two-dimensional models; photon downconversion in nonlinear crystals; special and ultrafast lasers/micromasers that can create few-photon states; trapping of individual atoms; atomic interferometry, which is the exact opposite of photon interferometry in that atoms are scattered off photon "crystals"; quantum cryptography; SQUIDs; and Josephson tunneling.

The New York Academy of Sciences sponsored a conference similar to this one in 1986, in honor of Eugene P. Wigner. Many of the new techniques mentioned above did not exist or were rudimentary at that time. It was decided that now would be a good time to hold another conference and we feel privileged that John A. Wheeler allowed us to honor him this time. I was surprised to learn at the conference how many participants earned their degree with Wheeler, had a postdoc with him, or otherwise collaborated with him. I suspect that he has probably mentored or supervised as a postdoc more first-rate physicists than anyone in the history of the United States. Of course, even if he had never written a word about fundamental problems in quantum mechanics, he would still be world famous for his major contributions to other fields, but he has had a seminal influence on many ideas in this field; thus, many people participated as a chance to pay homage to him. For my part, although I have never had the chance to work with him directly, I have read many of his books and papers carefully, have listened to him speak many times, have talked with him when I could, and feel as though I have been his student. In fact, few physicists have had a greater influence on my thoughts, indirect as this has been, and I am sure that I speak for many of the participants in the conference when I say this.

The exigencies of the conference made it impossible to schedule all papers on one topic together and so we have arranged the proceedings by topic rather than chronologically. This has allowed us to include a few papers from people who could not attend, but who sent in the papers they would have presented. I also would like to thank each of the session chairmen: Herb Bernstein, Ray Chiao, Serge Haroche, Mike Horne, Rolf Landauer, David Mermin, Mikio Namiki, Herschel Pilloff, Helmut Rauch, Marlan Scully, Abner Shimony, Henry Stapp, Jean-Pierre Vigier, Herbert Walther, Sam Werner, and Anton Zeilinger.

I would like to thank the chief executive of the New York Academy of Sciences, Rodney Nichols, and its board of directors, for letting the conference take place and for providing the experienced use of its staff, and also the conference director, Geraldine Busacco. Also, the conference could not have come to pass without the financial support of the contributing organizations and foundations, to whom we are extremely grateful. Very special thanks go to the two personnel who handled the arduous work and made the conference happen, Renée Wilkerson-Brown and Lynn Serra. The excellent detailed editorial expertise that made the proceedings possible was provided by Stefan Malmoli and the editorial direction by Bill Boland. We are also fully indebted to Carolyn Harriger and her staff at the Baltimore campus of the University of Maryland for handling the conference so smoothly. Finally, I would like to thank all the conference participants for making it a truly memorable experience and for the excellent quality of the papers.

Two-Photon Downconversion Experiments[a]

L. MANDEL

Department of Physics and Astronomy
University of Rochester
Rochester, New York 14627

INTRODUCTION

The process of parametric downconversion in a nonlinear material, in which an incident pump photon of frequency ω_0 effectively splits into two subharmonic photons of frequencies near $\omega_0/2$, has become the basis of numerous experiments for the study of nonclassical and nonlocal effects. The phenomenon appears to have been first investigated theoretically by Klyshko[1] (see also Harris *et al.*[2]) and it was later observed experimentally by Burnham and Weinberg[3] in 1970.

The essential features are illustrated in FIGURE 1. A laser beam of frequency ω_0 falls on the nonlinear crystal, NL, having a $\chi^{(2)}$ nonlinear susceptibility. A parametric interaction that leaves the crystal unchanged causes a pump photon to be absorbed and two downconverted photons of frequencies ω_1 and ω_2 to be emitted simultaneously such that

$$\omega_0 = \omega_1 + \omega_2. \tag{1}$$

For historical reasons, these are known as the signal (s) photon and the idler (i) photon. Because there are many frequencies ω_1 and ω_2 that satisfy equation 1, it follows that both the signal and idler photon may have a substantial bandwidth and appear in the form of a short wave packet, even if the pump is monochromatic. For the quantum state $|\psi\rangle$ of the electromagnetic field after a short interaction time, we may therefore write

$$|\psi\rangle = M|\text{vac}\rangle_{s,i} + \eta V \sum_{\omega} \phi(\omega)|\omega\rangle_s|\omega_0 - \omega\rangle_i. \tag{2}$$

Here V is the complex amplitude of the (classical) pump field of frequency ω_0 such that $|V|^2$ gives the light intensity in photons per second, $|\eta|^2$ is the downconversion efficiency, and $\phi(\omega)$ is a normalized spectral weight function. In equation 2, we have discarded the terms with more than two downconverted photons, which are very improbable, and thus the coefficient M is very close to unity. It is apparent from equation 2 that the signal and idler photons are in a state that is entangled (nonfactorizable) through their Fourier components, and this exhibits many nonclassical features. For example, if a signal photon is detected by a photodetector located in the path of the signal beam, then there has to be an idler photon in the

[a]This work was supported by the National Science Foundation and the Office of Naval Research.

corresponding position at the same time.[4] Moreover, if N signal photons are detected by a photodetector of near 100% efficiency in the signal beam, then N idler photons have to be in the idler beam, and we have a Fock state exhibiting sub-Poisson statistics.[5,6] The probability of simultaneous detections (within the detector resolving time) of a signal and an idler photon pair greatly exceeds the probability of detecting two signal photons in one arm or two idler photons in the other arm. More precisely, the normally ordered cross-correlation between the number $\hat{n}_1$ of signal photons and the number $\hat{n}_2$ of idler photons satisfies the inequality,[7]

$$\langle:\hat{n}_1\hat{n}_2:\rangle > [\langle:\hat{n}_1^2:\rangle\langle:\hat{n}_2^2:\rangle]^{1/2}, \quad (3)$$

which is impossible for a classical electromagnetic field. Recent measurements have shown that classical probability is violated by about 600 standard deviations in coincidence measurements,[8] which may well be the largest observed violation in the optical domain.

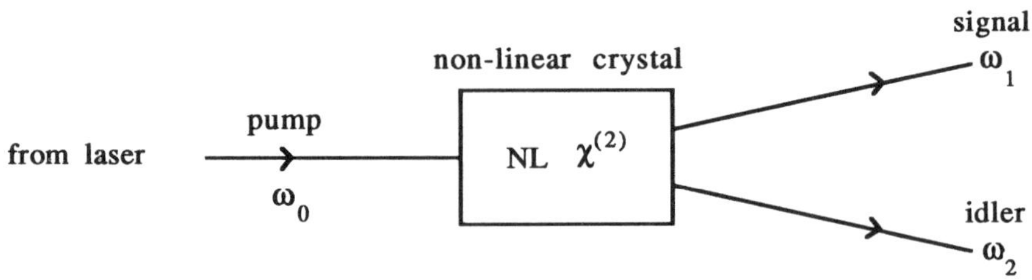

FIGURE 1. Illustration of downconversion of a coherent light beam in a nonlinear crystal.

TWO-PHOTON INTERFERENCE EXPERIMENTS

When the signal and idler beams are allowed to come together, they exhibit no one-photon or second-order interference, but a two-photon or fourth-order interference effect, which is readily revealed by two photodetectors and a coincidence counter.[9–11] FIGURE 2a shows the setup for a two-photon interference experiment in which the signal and idler beams from a parametric downconverter are mixed by a 50%:50% beam-splitter and fall on two detectors, D_a and D_b, at a small angle of inclination relative to each other.[10] Because of the inclinations of the light beams at the detectors, one might expect to find an interference pattern in front of D_a and D_b. Actually, the one-photon counting rates of D_a and D_b reveal no interference, that is, no sinusoidal variations as the distances x_a and x_b are varied. Only the two-photon coincidence counting rate measured with both detectors at once exhibits interference as x_a is varied with x_b held constant, as shown in FIGURE 2b.[10]

An interesting aperiodic two-photon interference effect occurs when the signal and idler photons are mixed at a 50%:50% beam-splitter without relative inclination and the optical path differences are varied by translating the beam-splitter at right angles to its face, as shown in FIGURE 3a.[12] Photodetectors D_1 and D_2 receive the two light beams and the two-photon coincidence rate is measured as a function of the beam-splitter displacement, $\delta x = c\delta\tau$, from the symmetric position. FIGURE 3b shows

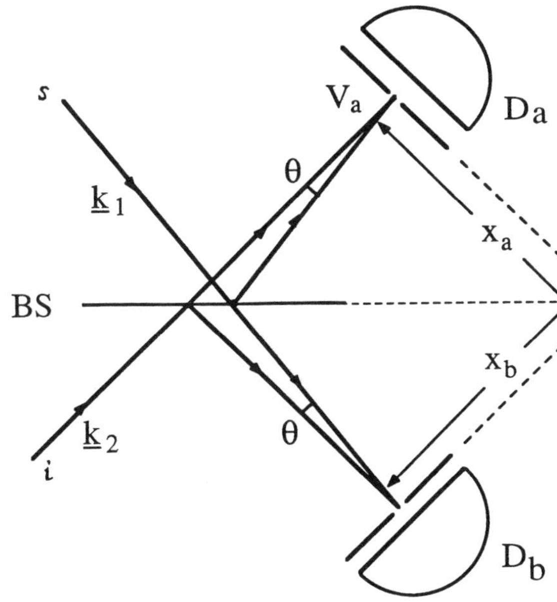

FIGURE 2a. The geometry for a two-photon interference experiment in which signal s and idler i photons are mixed by a 50%:50% beam-splitter, BS, and the mixed beams fall on photodetectors, D_a and D_b. (Adapted from reference 10.)

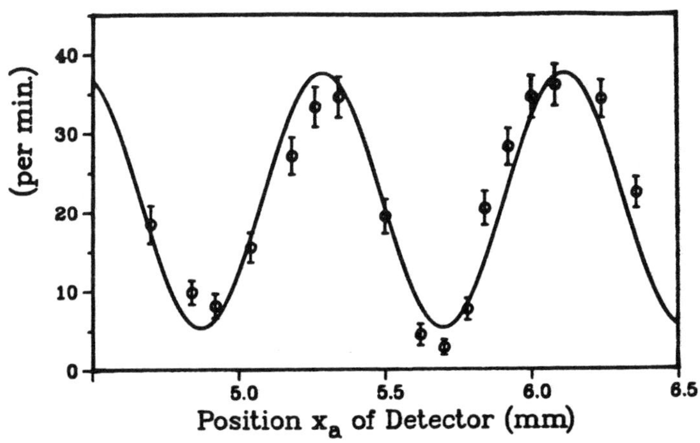

FIGURE 2b. Experimental results for the rate of two-photon coincidence detection as a function of the displacement of detector D_a. The full curve is a sinusoidal function of the expected periodicity. (Adapted from reference 10.)

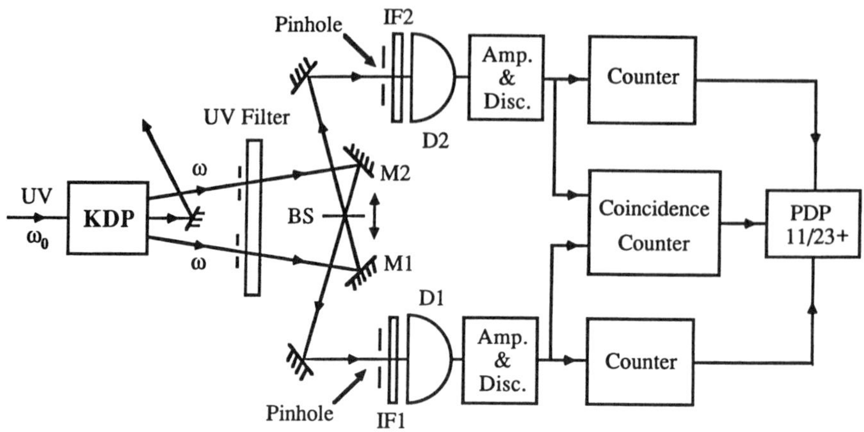

FIGURE 3a. Outline of an experiment to detect mixed signal and idler photons in coincidence as the differential time delay between them is varied by displacement of BS. (Reproduced from reference 12.)

the experimental results superimposed on the theoretical curve. It will be seen that almost no two-photon coincidence detections occur when $\delta\tau = 0$. The coincidence rate rises as $\delta\tau$ increases from zero positively or negatively, and it becomes substantially constant when $\delta\tau$ exceeds the coherence time, which is about 100 fs. This shows that we are dealing with about 100-fs-long photon wave packets, which is the length

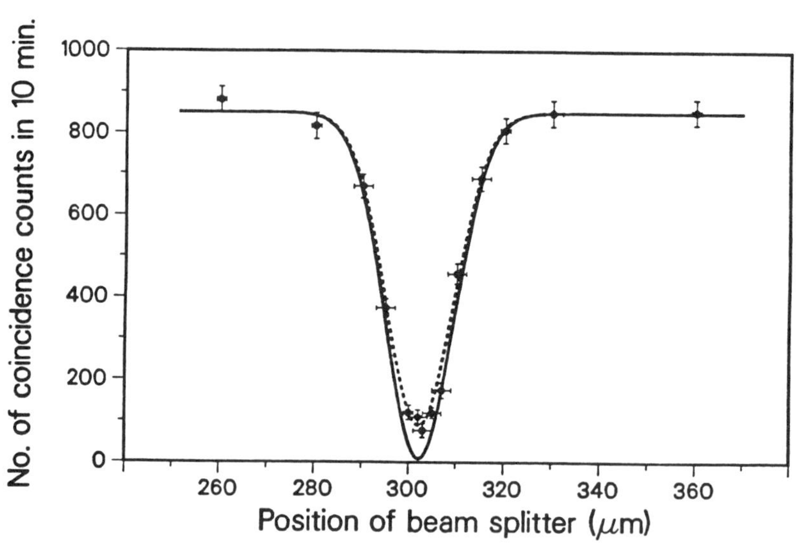

FIGURE 3b. Experimental results for the rate of two-photon coincidence detection as a function of the differential time delay introduced by displacement of the beam-splitter. The full curve is theoretical. (Reproduced from reference 12.)

determined by the 10^{13}-Hz passband of the interference filters in FIGURE 3a. Numerous other experiments based on the same principle have been carried out,[6,13,14] sometimes in order to measure tunneling[15] or propagation times.[16] The technique makes it possible to measure the time separation between two photons with femtosecond time resolution, which is about a million times shorter than the resolving time of the detectors and the electronics. The dip to zero when $\delta\tau = 0$ is a consequence of the destructive interference of the photon pair at the beam-splitter. It can be understood as follows: The detection of one photon in each output beam leaving the beam-splitter can occur in two different ways; either the two incident photons are both transmitted or they are both reflected. Because these possibilities are indistinguishable, we have to add the corresponding probability amplitude before squaring to arrive at the probability. However, because of the phase changes that occur on reflection, the two-photon probability amplitude for two reflections is 180° out of phase with that for two transmissions, and the two amplitudes sum to zero. Of course, this perfect destructive interference only occurs when $\delta\tau = 0$. If one photon wave packet is delayed relative to the other, there is no longer full destructive interference and the coincidence rate rises above zero. The rate becomes constant when the differential delay, $\delta\tau$, is so great that one photon no longer overlaps the other.

The entanglement of signal and idler photons via their Fourier components, as in equation 2, has been demonstrated in a rather striking experiment by Chiao and coworkers.[17] With the help of two-photon coincidence measurements of signal and idler photons, they showed that reducing the bandwidth of the idler photons by a filter also reduces the bandwidth of the signal photons by the same amount. On passing downconverted signal photons through an unbalanced interferometer whose path difference exceeded the coherence length of the signal photons, they found no interference at the interferometer output (see FIGURE 4) as expected. However, after a narrow band filter was inserted in the path of the idler, the signal photons clearly exhibited interference, as shown in FIGURE 4. This is a consequence of the entangled form of the state in equation 2.

A number of two-photon interference experiments have been performed, based on an idea proposed by Franson,[18] with the unusual characteristics that the signal and idler photons never come together and mix. The principle is illustrated in FIGURE 5. Consider a pair of photons, A and B, emitted simultaneously from a common (downconverter) source at some unpredictable time t in two different directions. The two photons never come together, but are directed to two separate photodetectors, D_A and D_B. Some beam-splitters and mirrors are introduced into each channel, as shown, so as to provide both a shorter and a longer path for each photon. Now, suppose that the path difference is similar in each channel and that it is much longer than the coherence or wave-packet length of each photon. Under these conditions, one would not expect to observe any interference in the output of detector D_A as the path difference in channel A is varied, and similarly for detector D_B as the path difference in channel B is adjusted, and this is readily confirmed. However, if the outputs of the two detectors are fed to a coincidence counter that measures the rate of detection of photon pairs, one observes interference when either one of the two path differences is varied.[19–21]

This phenomenon is most readily understandable in terms of the interference of

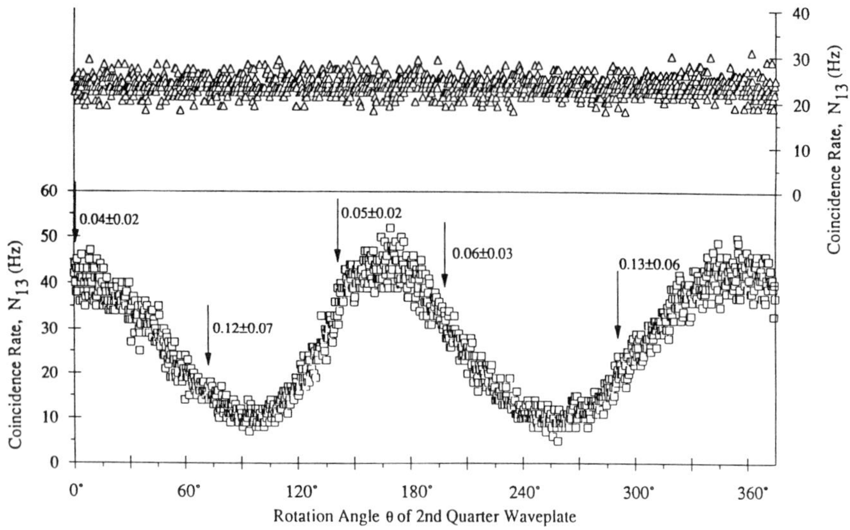

FIGURE 4. Results of an interference experiment in which signal photons are sent through an unbalanced interferometer with unbalance cT and signal and idler photons are detected in coincidence. It is found that the bandwidth or coherence time of the signal photons is determined by the bandwidth of the idlers. The upper trace shows no interference of signal photons when the idler bandwidth exceeds $1/T$, whereas the lower trace exhibits interference when the idler bandwidth is reduced below $1/T$. (Reproduced from reference 17.)

probability amplitudes for the photon pair. A coincidence detection can result from a pair of photons that both follow the short path through the interferometer or from a pair of photons that both follow the longer path (or if the coincidence resolving time exceeds the path-difference time from other combinations). As these are indistinguishable possibilities, we have to add the corresponding probability amplitudes and then square, and this generates interference. Moreover, the interference exhibits nonlocal features to the extent that the output of detector D_A depends not only on path-difference changes made in channel A, but also on those made in channel B. Yet, there need be no causal connection between detector D_A and the interferometer in channel B.

TESTS OF BELL'S INEQUALITIES

The classic experiments of Aspect and coworkers demonstrating locality violations in two-photon correlation measurements made use of the cascade decay of Ca atoms to generate photon pairs in a singlet state.[22–24] It was first noticed by Alley and Shih[25] that the same quantum state could be achieved and the same measurement performed if either circularly or linearly polarized signal and idler photons produced by downconversion were mixed by a 50%:50% beam-splitter. The mixed photon beams are then directed to two photodetectors with polarizers interposed. For example, for a symmetric beam-splitter, if the signal photon is x-polarized and the idler is y-polarized and if subscripts 1 and 2 label the two light beams emerging from the beam-splitter, then the quantum state of the emerging photon pair is given by

$$|\psi\rangle = \left(\frac{1}{2}\right)[|1\rangle_{1x}|1\rangle_{2y} - |1\rangle_{1y}|1\rangle_{2x} + \mathrm{i}|1\rangle_{1x}|1\rangle_{1y} + \mathrm{i}|1\rangle_{2x}|1\rangle_{2y}]. \quad (4)$$

The first two terms correspond to the usual singlet state with respect to polarization. Although terms 3 and 4, corresponding to two photons in beam 1 or two photons in beam 2, represent an undesirable distortion, the singlet-state contribution can be projected out if we always measure coincidences between one photon in beam 1 and one photon in beam 2.

The coincidence detection probability, $\mathscr{P}_{12}(\theta_1, \theta_2)$, as a function of the settings θ_1 and θ_2 of the two linear polarizers is then measured for several different values of θ_1 and θ_2 and also with one or both polarizers removed. If nature is describable by a local realistic theory, then $\mathscr{P}_{12}(\theta_1, \theta_2)$ should satisfy one or more of the Bell inequalities. In practice, it has been found that the Bell inequalities are clearly violated in the course of such two-photon polarization correlation measurements.[25–27] In a recent experiment based on downconversion with type II phase matching, in which the signal and idler photons emerge from the downconverter with orthogonal polarizations, Kiess *et al.* were able to demonstrate the violation of a Bell inequality by 22 standard deviations.[27]

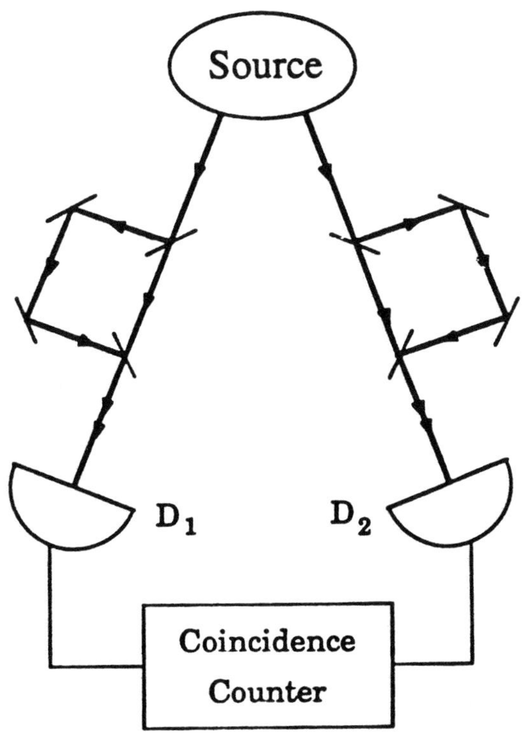

FIGURE 5. Illustration of the principle behind the Franson-type two-photon interference experiment. (Reproduced from reference 19.)

A number of experiments on locality violations based on space-time variables or phases or field quadratures rather than polarizations have also been reported.[28,29]

EXPERIMENTS ON INDUCED COHERENCE WITHOUT INDUCED EMISSION

In the presence of a strong coherent field, the emission from an excited atom or from an optically pumped downconverter can become stimulated or induced as well as being mutually coherent with the inducing field. By contrast, the phenomenon we now describe is of an entirely different kind. Consider the arrangement of two downconverters labeled NL1 and NL2 shown in FIGURE 6. Both are optically pumped by light derived from the same laser and, as a result, downconversion can occur at NL1 with the simultaneous emission of an s_1 and an i_1 photon, or downconversion can occur at NL2 with the simultaneous emission of an s_2 and an i_2 photon. The two downconverters have been aligned so that i_1 passes through NL2 and becomes colinear with i_2. We are interested in the question of whether the two signal beams, s_1 and s_2, are mutually coherent and exhibit interference when they are mixed at the output beam-splitter, BS_0, in which case the photon counting rate of detector D_s varies periodically as BS_0 is translated orthogonally to its face.

Such experiments have been performed.[30,31] So long as i_1 and i_2 are made colinear, an interference pattern is indeed observed, as shown by curve A in FIGURE 7. However, if the i_1 beam is blocked and is prevented from reaching NL2, all interference disappears and the experimental points lie on curve B in FIGURE 7. It is apparent, though, from FIGURE 7 that the mean photon rate is the same in both cases. The rate of photon emission from NL2 remains strictly spontaneous and is not affected by the presence of i_1; only the mutual coherence between s_1 and s_2 is affected. As it is essential for i_1 to reach NL2 for interference to show up, we may describe i_1 as "inducing coherence between s_1 and s_2". However, i_1 induces no

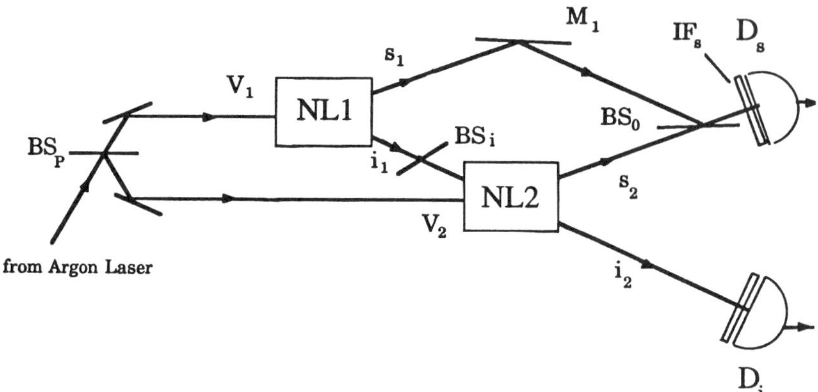

FIGURE 6. Outline of an experiment exhibiting induced coherence between two downconverters, NL1 and NL2, but no induced emission. (Reproduced from reference 30.)

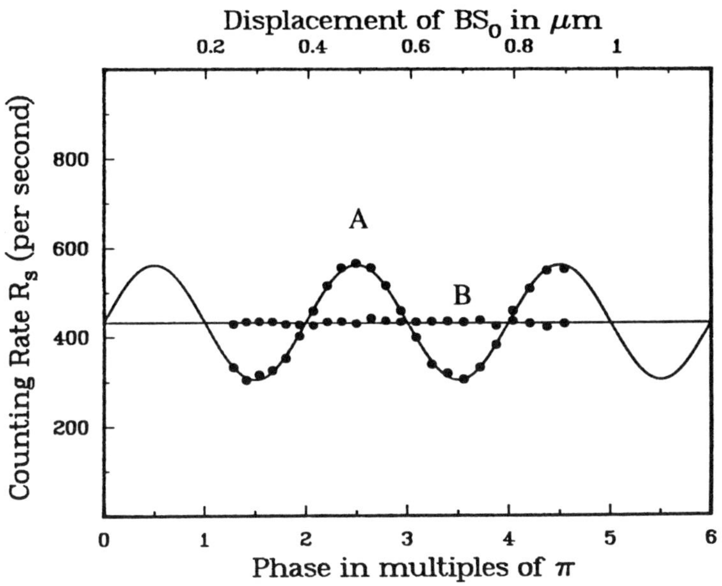

FIGURE 7. Results of an interference experiment obtained with a filter of variable transmissivity placed in the idler i_1 path between NL1 and NL2: 91% filter transmissivity gives result A and zero transmissivity gives result B. (Reproduced from reference 30.)

emission, so this phenomenon is new and quite different from that mentioned at the beginning of this section. In a real sense, there is no interaction between i_1 and NL2.

How is the phenomenon to be understood? One interpretation rests on the quantum mechanical relation between coherence and indistinguishability. If it is impossible, in principle, to determine whether any one signal photon detected by D_s comes from NL1 or from NL2, then the signals will exhibit interference. On the other hand, if there is some scheme that, in principle, allows the source of the detected signal photons to be determined without disturbing the interference experiment, then there will be no interference. It is not necessary for this auxiliary experiment to be actually carried out. The mere possibility of such a measurement is sufficient to wipe out the interference because we are now dealing with a state described by a diagonal density operator, which reflects not so much what is known, but what is knowable in principle.

As an illustration, let us suppose that i_1 is blocked and that an auxiliary, near-perfect detector, D_i, is introduced into the i_2 beam, as shown in FIGURE 6, at the same distance from NL2 as D_s. The introduction of this detector D_i does not physically disturb the s_1, s_2 interference, but the output of D_i provides us with information about the source of the emitted photons. Suppose that when D_s registers a photon, D_i registers a photon simultaneously. Then, this photon pair obviously originated in NL2. On the other hand, suppose that when D_s registers a photon, D_i does not simultaneously register a photon. Then, the detected photon must have come from NL1. We see that the auxiliary device provides us with the information to determine the source of each detected photon, and this is just the quantum prescription for wiping out all interference.

A number of variations on this scheme have been reported that all depend on the same principle.[32–36] A differential time delay, T, inserted into one interferometer arm in general reduces the visibility of the observed interference. If the delay T exceeds the coherence time, T_c, of the interfering light, then the intensity interference is lost altogether, although interference shows up in the spectrum.[32] Because of the quantum entanglement of signal and idler, it is found that the insertion of the delay T into the i_1 beam between NL1 and NL2 has exactly the same consequences as insertion into the s_2 beam.[33] A long delay T makes the sources of i_1 and i_2 distinguishable; therefore, the sources of s_1 and s_2 become distinguishable as well and this wipes out all interference.

FIGURE 8 shows a variation of the former experiment, in which mirrors M_3 and M_4 are introduced to form a resonant cavity for the idlers.[34] Also, a beam-splitter, BS_i, is inserted in the idler i_1 path followed by a detector, D_i. Once again, we look for interference with the help of D_s, but we concentrate on those events in which a

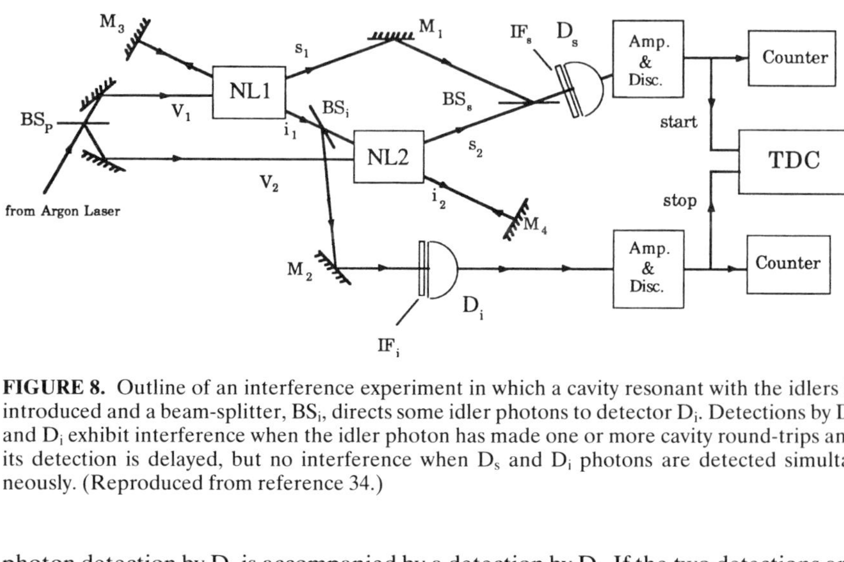

FIGURE 8. Outline of an interference experiment in which a cavity resonant with the idlers is introduced and a beam-splitter, BS_i, directs some idler photons to detector D_i. Detections by D_s and D_i exhibit interference when the idler photon has made one or more cavity round-trips and its detection is delayed, but no interference when D_s and D_i photons are detected simultaneously. (Reproduced from reference 34.)

photon detection by D_s is accompanied by a detection by D_i. If the two detections are simultaneous, they are attributable to an event in which the i_1 photon from NL1 is reflected by BS_i and then detected. As the source of the photon is identifiable, no interference shows up in the coincidence counting rate of D_s and D_i. However, if the idler photon from either NL1 or NL2 makes one or more trips around the cavity before being detected by D_i, the sources of the photons become indistinguishable. Therefore, photon detections by D_s that are accompanied by delayed detections by D_i, as measured with a time-to-digital converter (TDC), are expected to give rise to interference. This is indeed observed.[34]

In this experiment, some signal photons behave like waves and exhibit interference, whereas other signal photons behave like particles that are counted, but exhibit no interference. The "decision" whether to be one or the other is not made by the experimenter nor by the signal photon, but by the associated idler, which is either transmitted or reflected by BS_i. Somewhat similar experiments have also been used to test certain predictions of the de Broglie pilot wave theory.[35,36]

It is apparent that the process of parametric downconversion provides us with entangled photons that are extremely rich in experimental possibilities.

ACKNOWLEDGMENTS

Numerous past and present students and coworkers contributed to the work described in this report, including D. Branning, G. A. Barbosa, E. C. Gage, R. Ghosh, T. P. Grayson, C. K. Hong, B. E. Magill, C. Monken, Z. Y. Ou, J. R. Torgerson, L. J. Wang, and X. Y. Zou. Their help is gratefully acknowledged.

REFERENCES

1. Klyshko, D. N. 1968. Zh. Eksp. Teor. Fiz. **55:** 1006 (1965. Sov. Phys. JETP **28:** 522).
2. Harris, S. E., M. K. Oshman & R. L. Byer. 1967. Phys. Rev. Lett. **18:** 732.
3. Burnham, D. C. & D. L. Weinberg. 1970. Phys. Rev. Lett. **25:** 84.
4. Hong, C. K. & L. Mandel. 1986. Phys. Rev. Lett. **56:** 58.
5. Walker, J. G. & E. Jakeman. 1985. Opt. Acta **32:** 1303.
6. Rarity, J. G. & P. R. Tapster. 1988. *In* Photons and Quantum Fluctuations. E. R. Pike & H. Walther, Eds.: 122. Adam Hilger. Bristol.
7. Graham, R. 1984. Phys. Rev. Lett. **52:** 117.
8. Zou, X. Y., L. J. Wang & L. Mandel. 1991. Opt. Commun. **84:** 351.
9. Gosh, R. & L. Mandel. 1987. Phys. Rev. Lett. **59:** 1903.
10. Ou, Z. Y. & L. Mandel. 1989. Phys. Rev. Lett. **62:** 2941.
11. Rarity, J. G., P. R. Tapster, E. Jakeman, T. Larchuk, R. A. Campos, M. C. Teich & B. E. A. Saleh. 1990. Phys. Rev. Lett. **65:** 1348.
12. Hong, C. K., Z. Y. Ou & L. Mandel. 1987. Phys. Rev. Lett. **59:** 2044.
13. Ou, Z. Y. & L. Mandel. 1988. Phys. Rev. Lett. **61:** 54.
14. Ou, Z. Y., E. C. Gage, B. E. Magill & L. Mandel. 1984. J. Opt. Soc. Am. **B6:** 100.
15. Steinberg, A. M., P. G. Kwiat & R. Y. Chiao. 1993. Phys. Rev. Lett. **71:** 708.
16. Sergienko, A. V., Y. H. Shih & M. H. Rubin. To be published.
17. Chiao, R. Y., P. G. Kwiat & A. M. Steinberg. 1992. *In* Proceedings of the Workshop on Squeezed States and Uncertainty Relations. D. Han, Y. S. Kim & W. W. Zachary, Eds.: 61. NASA CP3135.
18. Franson, D. 1989. Phys. Rev. Lett. **62:** 2205.
19. Ou, Z. Y., X. Y. Zou, L. J. Wang & L. Mandel. 1990. Phys. Rev. Lett. **65:** 321.
20. Kwiat, P. G., W. A. Vareka, C. K. Hong, H. Nathel & R. Y. Chiao. 1990. Phys. Rev. A **41:** 2910.
21. Shih, Y. H., M. H. Rubin & A. V. Sergienko. 1992. *In* Proceedings of the Workshop on Squeezed States and Uncertainty Relations. D. Han, Y. S. Kim & W. W. Zachary, Eds.: 47. NASA CP3135.
22. Aspect, A., P. Grangier & G. Roger. 1981. Phys. Rev. Lett. **47:** 460.
23. Aspect, A., P. Grangier & G. Roger. 1982. Phys. Rev. Lett. **49:** 91.
24. Aspect, A., P. Dalibard & G. Roger. 1982. Phys. Rev. Lett. **49:** 1804.
25. Shih, Y. H. & C. O. Alley. 1988. Phys. Rev. Lett. **61:** 2921.
26. Ou, Z. Y. & L. Mandel. 1988. Phys. Rev. Lett. **61:** 50.
27. Kiess, T. E., Y. H. Shih, A. V. Sergienko & C. O. Alley. 1993. Phys. Rev. Lett. **71:** 3893.
28. Rarity, J. G. & P. R. Tapster. 1990. Phys. Rev. Lett. **64:** 2495.
29. Ou, Z. Y., S. F. Pereira, H. J. Kimble & K. C. Peng. 1992. Phys. Rev. Lett. **68:** 3663.

30. ZOU, X. Y., L. J. WANG & L. MANDEL. 1991. Phys. Rev. Lett. **67:** 318.
31. WANG, L. J., X. Y. ZOU & L. MANDEL. 1991. Phys. Rev. A **44:** 4614.
32. ZOU, X. Y., T. P. GRAYSON & L. MANDEL. 1992. Phys. Rev. Lett. **69:** 3041.
33. ZOU, X. Y., T. P. GRAYSON, G. A. BARBOSA & L. MANDEL. 1993. Phys. Rev. A **47:** 2293.
34. GRAYSON, T. P., X. Y. ZOU, D. BRANNING, J. R. TORGERSON & L. MANDEL. 1993. Phys. Rev. A **48:** 4793.
35. WANG, L. J., X. Y. ZOU & L. MANDEL. 1991. Phys. Rev. Lett. **66:** 1111.
36. ZOU, X. Y., T. P. GRAYSON, L. J. WANG & L. MANDEL. 1992. Phys. Rev. Lett. **68:** 3667.

Quantum Optics

Quantum, Classical, and Metaphysical Aspects[a]

D. N. KLYSHKO

Department of Physics
Moscow State University
Moscow 119899, Russia

INTRODUCTION

Quantum optics (QO) plays a special role in the problem of the interpretation of quantum theory. In optics, we do not in principle need any extra measuring devices for the observation of the quantum object—the electromagnetic field—and thus the interface between the classical and quantum worlds can be put straight into our eyes. It is also essential that the Maxwell equations for classical free-space fields have exactly the same form as the Heisenberg equations for the Bose-type operators of the fields.

One of the mainstreams of the experimental QO for the last two decades has been the observation of the two-photon interference (see references 1–9). The three- and four-photon experiments are also being discussed.[10–14] The usual amplitude or one-photon interference is continuing to attract attention,[15–18] as is the observation of the anticoincidence effect in two-photon counters at the output of a beam-splitter.[19,20]

Usually, these experiments are motivated by the desire to demonstrate once more the quantum nature of light. Many physicists also hope that such investigations would lead to the fulfillment of Einstein's dream—to understand, at last, what the proton really is.

In some types of two-photon experiments, it is possible to introduce a quantitative measure of the nonclassicality. For example, the exceeding of the interference visibility V over a certain level may contradict the classical stochastic models of the light field (the C-language) or more general classical models with dichotomic observables (the C_B-language of John Bell[21]). In other experiments, there is no such measure, but they are frequently understood to be demonstrating the photon structure of light all the same. Sometimes, the optical Berry phase is also considered to be a property of the photons (see the discussion and sources in references 22 and 23).

In this report, I will compare and estimate the three alternative ways of describing the quantum-optical experiments, that is, the Q-, M-, and C-languages (see FIGURE 1).

[a] Financial support was provided by the International Science Foundation.

THE SEMANTICS OF QUANTUM OPTICS

Quantum Language

The Q-language consists of some well-established mathematical models (of the field's state preparation, its evolution, and detection) and of the rule for comparing the mathematical symbols with the experimental data—the projective postulate. We can define also three Q-dialects: the Heisenberg (Q_H) and Schrödinger (Q_S) descriptions of the field's evolution, and the Copenhagen interpretation (Q_C) of the quantum formalism.

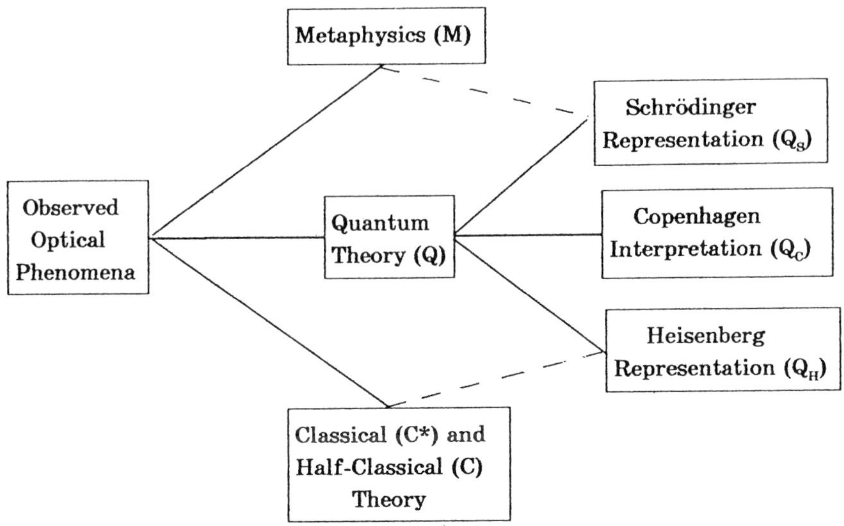

FIGURE 1. The scheme of the main language families used for the description of two-photon experiments.

The Q-language gives excellent descriptions of all known optical phenomena statistics, but it says nearly nothing about the individual counts in the detectors, their immediate causes, and their times of occurrence. According to Q_C, this information does not exist in nature and one simply should not ask improper questions. All we can say a priori about the field is at best its state.

Metaphysical Language

I propose to call metaphysical all terms that are not necessary for the quantitative description of the observable optical phenomena or that have no direct connection with the Q-theory or experiment. I'm risking to be branded such names as positivist or pragmatist, but it seems to me that one should differentiate somehow between terms justified by some theory and nonjustified, redundant terms.

A whole virtual world of quantum metaphysics with its own thesaurus, isolated from experiment and theory, has been built up. We need some criteria to distinguish between it and the traditional physical science. Perhaps, some moderate kind of operationalism and the Popper's principle of falsification should be accepted.

The usual hallmark of an M-language is the absence of exact definitions and of observable predictions (different from the Q-ones) and, at the same time, a pretension to give a more complete description than the Q-language, that is, "to look behind the mirror".

Regretfully, Q-, C-, and M-languages are frequently used together without distinction, even inside one sentence. Some time ago, before the buildup and confirmation of quantum electrodynamics, at the "Heroic Epoch of Quantum Physics", M-terms were unavoidable; however, now there is little hope left that by the way of inventing new fuzzy terms it would be possible to tame Schrödinger's cat and to exorcise the "Devil of Complementarity" and other quantum spooks.

Of course, some useful and traditional M-terms (e.g., the photon) could be used for convenience and consolation, but only with the due reservations about their exact status. The same applies to the reduction of the state vector: if it is only a brief symbolic catchword for a definite situation, which could be described more accurately (but in some length) by exact Q-theory (see Schiff's Q-description of the trace made by an electron in the Wilson camera[24]), then it is a Q-term. Usually, however, the term means something mysterious, inaccessible to experimental confirmation, that is, metaphysical.

A typical example of an unjustified, deceptive term seems to be the nonlocality. It should be understood only as a symbol of the fact that the predictions of Q- and C_B-languages are different. Actually, there is no more nonlocality (in the usual sense) in Q-theory or in experimental optics than negative probabilities there (see references 8 and 25).

Classical and Semiclassical Languages

The semiclassical theory (C-language) considers atoms quantum mechanically and the field classically—as a superposition of some number of wave packets with energies $\hbar\omega$; see reference 26. It should not be taken seriously because it simply serves as a useful auxiliary pictorial model. Most spectacularly, its inadequacy was demonstrated by the crossed-polarizers zero effect in Freedman and Clauser's experiments,[1] depicted in FIGURE 2.

In two-photon optics, the notion of a classical effective field, E_{eff}, whose square gives the joint probability to register two photons, also proves to be useful, especially when supplemented by the picture of fictitious advanced waves, "emitted" by one of the detectors[13,27–30] (see FIGURE 2 and APPENDIX A).

The purely classical C*-language contains the notion of continuous detectors, whose output currents duplicate the intensities of the beams. It permits useful classical analogues of such phenomena as the two-photon anticorrelation and interference.[8,9,13] It gives a trivial explanation of these effects as resulting from the transformation of the phase fluctuations of the input fields into the anticorrelated intensity fluctuations at the output beams (see later sections below).

Only the high visibility of the interference, that is, the practical absence of the background, observed in case of two-photon input light, is an essentially nonclassical effect. The interference phenomenon itself can be observed using the classical squeezed light (CSL) and can be explained in the C*-language.[13,30,31] CSL can be produced by applying strong ($N_0 \gg \frac{1}{2}$) thermal radiation to the input of a parametric downconverter. If in the C*-theory we put $N_0 = \frac{1}{2}$, then the only difference between the output radiation statistics and the true squeezed light would be an extra energy, $\hbar\omega/2$, in each output mode (see APPENDIX B).

Of course, the Q-language is the only true and universal one, but traditionally in physics one should also be using the languages of the lowest possible levels, which still describe the most characteristic features of the phenomenon. Note that even the Davies-Unruh effect[32]—the squeezing of the vacuum by acceleration—has a useful classical counterpart.[33] Regretfully, the C*-language is nearly completely ignored in modern QO.

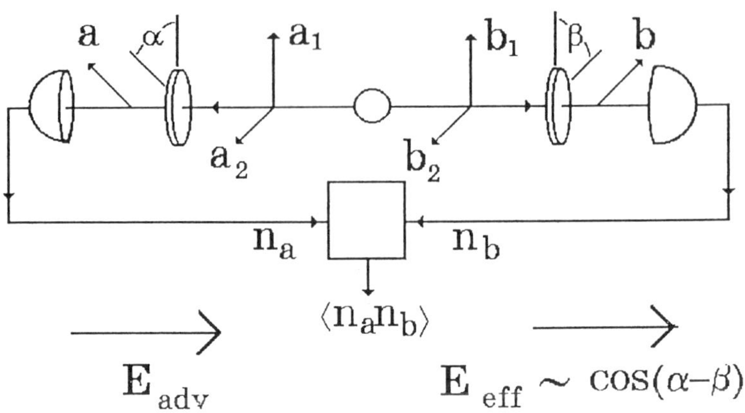

FIGURE 2. The experimental setup for the observation of the two-photon interference of the polarization type.[1] The source emits the photon pairs with correlated polarization states, which are registered by two detectors with analyzers and a coincidence circuit. The coincidence rate is described by the "two-photon Malus law", $\cos^2(\alpha - \beta)$, which cannot be explained in terms of C-photons (wave packets), but which follows at once from the advanced waves interpretation, according to which one detector emits a fictitious unpolarized field, E_{adv}.

The important difference between M- and C-languages is that the latter does not pretend to be giving a better description of quantum phenomena than the Q-language.

What Is a Photon?

Consider the definition of the term "photon" in the Q-language. Usually, it is defined as something that is described by the following state vector, called the one-photon wave packet:

$$\psi_1(t) = \Sigma\, F_k \exp(-i\omega_k t)\,|1\rangle_k, \qquad \Sigma\,|F_k|^2 = 1. \tag{2.1}$$

Note that this is just a very special state vector of the whole field—one of the myriad of possible states. Why are we ascribing to it such a fundamental role? It can be possibly justified only if we consider in detail the measurement process; however, if we are interested in the free-field's a priori state, then all possible states should be treated on equal grounds. The one-photon Fock state $|1\rangle$ is not any better or worse than $|666\rangle$ or than any other state $\psi(t) = \Sigma c_n|n\rangle \exp(-in\omega t)$.

Note also that the one-photon state ψ_1 is very difficult to prepare because one has to first prepare an atom exactly in an excited stationary state. It is possible to do this with a laser π-pulse, but it seems that until now such experiments have not been performed. A little easier way is to prepare a superposition, $\alpha\psi_{\text{vac}} + \beta\psi_1$, of the one-photon state with the vacuum state. However, usually, we have mixed states around us, which are very far away from our ideal ψ_1, and thus we practically never meet Q-photons.

Therefore, if we exclude the exotic one-photon states ψ_1, then we can say that a photon comes to a fleeting existence just at the moment of its absorption in the detector. (Hence, the operators, a and a^+, should swap their names.) Let us rephrase the well-known formulation of N. Bohr: "A photon is a phenomenon only if it is a recorded photon."

Ironically, it is much easier to prepare (using parametric downconversion) a two-photon state (plus a large vacuum component):

$$\psi_2 = |\text{vac}\rangle + \Sigma F_{kl} a_k^+ a_l^+ |\text{vac}\rangle = |\text{vac}\rangle + \Sigma F_{kl} |1\rangle_k |1\rangle_l. \quad \textbf{(2.2)}$$

It is often claimed that the reduction is happening when we detect one photon of the two and, as a result, we prepare the field in the one-photon state. However, this seems to be an M-talk. Actually, in all true two-photon experiments, some kind of coincidence scheme is used and the quantitative Q-description is possible only using the true full state, ψ_2.

Another popular definition of a photon is the following one: it is something living in the space-time and causing this or that individual "click" at the detector's output. This is an M-definition as it has no testable or falsifiable consequences. In Q_C-language, the click means only that the a priori state $|\psi\rangle$ contained a nonzero component, $c_1 \equiv \langle 1|\psi\rangle$, which is possible not only with $|\psi\rangle = |1\rangle$, but with any state with $c_1 \neq 0$. As the detection quantum efficiency, η, is practically always smaller than 1, the stationary a priori states, $|2\rangle$, $|3\rangle$, . . ., are also possible. The quantum retrodiction in case of a single measurement is always uncertain: one cannot reconstruct a vector by its single projection.

Consider as an example a stationary state, prepared by the spontaneous emission in a noncentrosymmetrical molecule[34] (FIGURE 3). There are two possible paths from the excited state to the ground state: a direct path, $1 \Rightarrow 3$, with the emission of one photon, $h\omega_C$, and a two-step path, $1 \Rightarrow 2 \Rightarrow 3$, with the emission of two photons, $h\omega_A$ and $h\omega_B$ ($\omega_A + \omega_B = \omega_C$). As a result, the stationary state with definite energy is prepared:

$$|\psi\rangle = \alpha|0\rangle_A|0\rangle_B|1\rangle_C \exp(-i\omega_C t) + \beta|1\rangle_A|1\rangle_B|0\rangle_C \exp(-i\omega_A t - i\omega_B t). \quad \textbf{(2.3)}$$

It has two peculiarities: the photon number is indefinite (1 or 2) and there is a nonzero constant cube of the field:

$$\langle E^3\rangle \sim 2\,\text{Re}(\langle a_A^+ a_B^+ a_C\rangle) = 2\,\text{Re}(\alpha\beta^*) \quad \textbf{(2.4)}$$

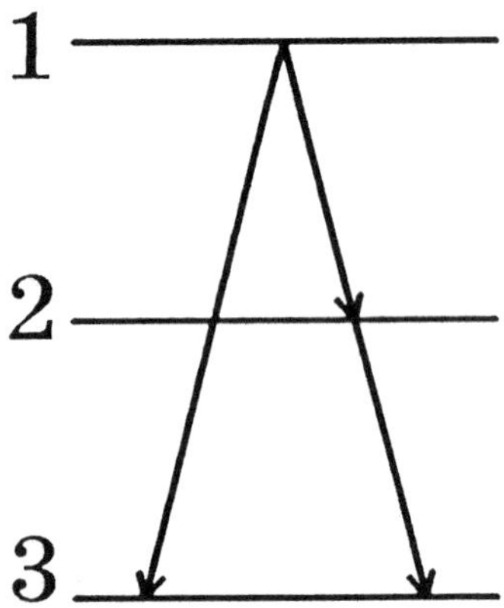

FIGURE 3. Spontaneous emission in a noncentrosymmetrical molecule prepares the stationary state with indefinite photon number and nonzero cube of the electrical field.

(here, α and β depend on the molecule's properties[34]).

The last effect is easily explained in C-language by considering three overlapping wave packets with frequencies ω_A, ω_B, and ω_C:

$$E^3 \sim \cos(\omega_A t + \varphi_A)\cos(\omega_B t + \varphi_B)\cos(\omega_C t + \varphi_C)$$
$$= (1/4)\cos(\varphi_A + \varphi_B - \varphi_C) + f(t). \quad \textbf{(2.5)}$$

However, this model contradicts the first peculiarity: the number of the simultaneously existing wave packets should not be three, but either one or two. This example shows that the usual assertion that "light consists of photons" does not make sense.

The situation is a little easier in case of low-energy Fermi fields, where the number of particles is fixed and only one particle can occupy a coherence volume. As the number of true particles is a fixed integer number, the assertion that they are existing a priori makes sense.

Similar arguments can be applied against many other M-terms. It seems that the textbooks on quantum mechanics should pay more attention to these problems and to the difference between physical theory and its possible interpretations.

ANTICORRELATION EFFECT

Q_H: In the Heisenberg representation, the action of various optical schemes on the input field operators can be described phenomenologically by the scattering matrix D.[34,35] In the lossless case, it is unitary. The scattering matrix can be considered as the Green function (propagator) in the spectral representation. It has identical forms in Q- and C*-theories (for linear optical tracts).

Consider the scheme in FIGURE 4. Two modes with amplitudes a and b are mixed by a beam-splitter (BS), a polarization prism, or a Mach-Zehnder interferometer. A two-mode linear mixer makes the transformation,

$$a' = \mathrm{t}a + \mathrm{r}b, \qquad b' = -\mathrm{r}^*a + \mathrm{t}^*b, \quad \textbf{(3.1)}$$

where t and r are the phenomenological transmission and reflection coefficients with the common phase factor omitted. In the lossless case, $|t|^2 + |r|^2 = T + R = 1$.

In QO, we are usually measuring the normally ordered correlation functions or, in the case of narrow-band radiation, simply the first $N_A = \langle a^+a \rangle$, $N_B = \langle b^+b \rangle$, and the second moments,

$$G_{AA} = \langle a^{+2}a^2 \rangle, \quad G_{BB} = \langle b^{+2}b^2 \rangle, \quad G_{AB} = \langle a^+b^+ba \rangle. \tag{3.2}$$

Let $\langle a^+b \rangle = \langle a^{+2}b^2 \rangle = \langle a^{+2}ab \rangle = \langle b^{+2}ba \rangle = 0$; then, it follows from equation 3.1 that

$$\begin{aligned} G'_{AA} &= T^2G_{AA} + R^2G_{BB} + 4TRG_{AB}, \\ G'_{BB} &= R^2G_{AA} + T^2G_{BB} + 4TRG_{AB}, \\ G'_{AB} &= TR(G_{AA} + G_{BB}) + (T - R)^2G_{AB}. \end{aligned} \tag{3.3}$$

These equations define the output measurable properties of the light field through the input ones and the optical tract parameters, T and R; they can be used in the lossy case as well.[35]

In the lossless case, we have the following invariants: $N_A + N_B$ = constant and $G_{AA} + G_{BB} + 2G_{AB}$ = constant; that is, the energy and the fluctuations plus correlation are conserved.

Let the input be symmetrical: $N_A = N_B = N$, $G_{AA} = G_{BB} = G = g/N^2$; also let $T = R = ½$; then, $N'_A = N'_{\hat{A}} = N$ and

$$g' = g/2 + g_{AB}, \qquad g'_{AB} = g/2 \tag{3.4}$$

(here, $g_{AB} = G_{AB}/N^2$). Thus, if $g_{AB} < g/2$, then the output bunching parameter, g', is smaller than the input one. If $g < 2$ (i.e., the input bunching parameter is smaller than the thermal one), then the output intensities are anticorrelated ($g'_{AB} < 1$)

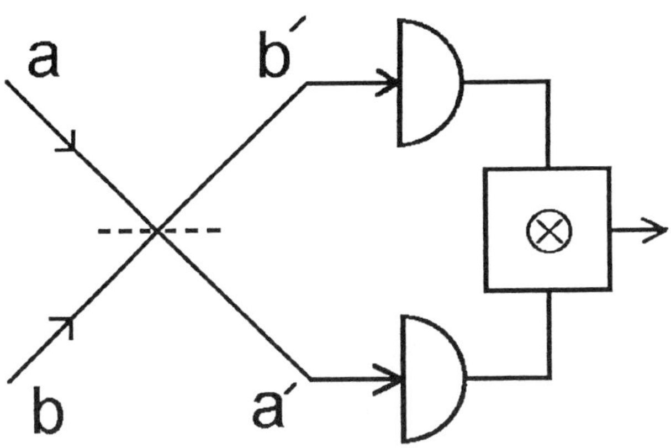

FIGURE 4. The anticorrelation effect. The independent phase fluctuations of two input fields, *a* and *b*, are transformed by a beam-splitter into anticorrelated fluctuations of the output intensities, which are observed by two detectors and the coincidence or multiplication circuit.

independently of the initial correlation. As shown below, this anticorrelation effect has a simple classical explanation.

Consider several input statistics:

1. In case of thermal light, $g = 2$ and thus $g'_{AB} = 1$—the output intensities are uncorrelated independently of any possible initial correlation. This is an example of the general invariance property of thermal light.[13,35]
2. Two independent coherent input beams ($g = g_{AB} = 1$) give some anticorrelation effect: $g'_{AB} = ½$. In the polarization case, this gives the hidden polarization effect.[9]
3. A two-mode squeezed vacuum has thermal fluctuations in each mode and full correlation between the signal and idler modes:

$$N_A = N_B = N = \sinh^2(F), \qquad g_{AA} = g_{BB} = 2,$$

$$g_{AB} = 1 + \coth^2(F) = 2 + (1/N), \qquad K = 1. \tag{3.5}$$

Here, F is the parametric gain, $K = (G_{AB} - N_A N_B)/\sigma_A \sigma_B$ is the correlation coefficient, and $\sigma_A^2 = G_{AA} + N_A - N_A^2 = N_A^2 + N_A$. At the output, the correlation is suppressed: $g'_{AB} = 1, K = 0$. This result should be compared with the strong initial correlation: $g_{AB} = 2 + (1/N)$. Usually, $N \sim 10^{-8}$, so a very large contrast is observed.[4,6]
4. In case of the symmetrical two-photon input state, $|\psi\rangle = |1\rangle_A \otimes |1\rangle_B \equiv |1, 1\rangle$, we have at the input no fluctuations: $g = \sigma = 0$ and $N = g_{AB} = 1$. At the output, $N' = g' = \sigma' = 1$ and $g'_{AB} = 0$; that is, there are some fluctuations and no coincidences.

Q_S: In the Schrödinger approach, we have to find the output state vector, $|\psi\rangle'$, through the input one, $|\psi\rangle$, and the scattering matrix.[22] Consider the two-mode case. Let us define a function $f(x, y)$ by the following equation: $|\psi\rangle = f(a^+, b^+)\ |\text{vac}\rangle$. Reversing the transformation in equation 3.1, we find

$$|\psi\rangle' = f(\text{t}a'^+ - \text{r}^* b'^+, \text{r}a'^+ + \text{t}^* b'^+)|\text{vac}\rangle. \tag{3.6}$$

Consider again the two-photon input state, $|\psi\rangle = |1, 1\rangle$. Now, $f(x, y) = xy$, so

$$|\psi\rangle' = \text{tr}|2, 0\rangle + (\text{T} - \text{R})|1, 1\rangle - \text{t}^*\text{r}^*|0, 2\rangle. \tag{3.7}$$

Thus, the 50% BS makes the following transformation of the state vector:

$$|\psi\rangle = |1, 1\rangle \Rightarrow |\psi\rangle' = (|2, 0\rangle - |0, 2\rangle)/2. \tag{3.8}$$

Q_C: According to the projection postulate, the coincidence probability amplitude equals $c_{11} = \langle 1, 1|\psi\rangle' = \text{T} - \text{R} = 0$; that is, there are no coincidences.

M: In the metaphysical language, the considered transformation, which results from the essentially classical SU(2)-transformation (equation 3.1), is usually explained by the intrinsic properties of the photon. For example, the projections c_{20} and c_{02} in equation 3.7 are considered to reflect the photon's corpuscular component and c_{11} is considered to reflect the wave component.

C: The two-photon anticorrelation effect can be very simply "explained" using the notion of the effective field (two-photon wave packet), built from the advanced wave and "radiated" by one of the detectors:[27,28] there are two paths, which exactly compensate one another.

C*: If we replace in the Q_H-equations the operators a and a^+ by classical dimensionless stochastic amplitudes a and a^* and the quantum-averaging operation $\langle \ldots \rangle$ by the classical one, then we get the classical transformations. Their functional forms are exactly the same. Thus, if there are some nonclassical effects, they are due only to the nonclassicality of the input moments and not to the concrete optical scheme under consideration, which simply performs a linear transformation of the moments. However, in C*-theory, we have a limitation: because $g_{class} \equiv \langle (a^*a)^2 \rangle / \langle a^*a \rangle^2 \geq 1$, $(g'_{AB})_{class} \geq \frac{1}{2}$ according to equation 3.4.

Suppose that there are two classical waves with constant equal amplitudes, $|a| = |b| = 1$, and independently fluctuating phases, $\alpha(t)$ and $\beta(t)$, at the input of a 50% BS. Let $a = \exp[i\alpha(t)]$ and $b = \exp[i\beta(t)]$; then, $a' = (a + b)/\sqrt{2}$ and $b' = (-a + b)/\sqrt{2}$. The output intensities are

$$n'_A = |a'|^2 = 1 + \cos[\varphi(t)], \qquad n'_B = |b'|^2 = 1 - \cos[\varphi(t)], \tag{3.9}$$

where $\varphi = \alpha - \beta$. Thus, the input energy is redistributed between two output beams depending on the transient phase difference, $\varphi(t)$. The total energy, $n_A + n_B = 2$, is conserved and thus $dn'_A/dt = -dn'_B/dt$; that is, the intensities are always changing in the opposite directions and thus are fully anticorrelated, $K = -1$. We see that the scheme in FIGURE 4 serves as a phase detector.

It is clear that this classical anticorrelation effect, which is explained by the transformation of the input phase fluctuations into the anticorrelated output intensity fluctuations, is very similar to the quantum one. Note that in Q-language we can also use the notion of the phase φ by introducing the phase-difference operator (see reference 8).

The given multilingual description of a typical QO effect can be repeated for other more-complicated cases with more modes involved, but the message should already be clear: the most universal, compact, and close to classical notions is the Q_H-language. The M-language cannot add anything testable to it. The C-language gives useful visualizable pictorial analogues. Various linear optical mixing schemes can be considered as parts of the classical detection devices for measuring the input light statistics.

TWO-PHOTON INTERFERENCE

Consider the four-mode scheme in FIGURE 5. We have two amplitudes, a_1 and a_2, with equal frequencies, ω_a, and an adjustable phase delay, $\alpha = k_a(z_{a1} - z_{a2})$, mixed by a 50% BS; analogously, we have two amplitudes, b_1 and b_2, with frequency, ω_b, and delay, $\beta = k_b(z_{b1} - z_{b2})$, mixed by the second BS.

Q_H: The output amplitudes are

$$a = [a_1 \exp(-i\alpha/2) + a_2 \exp(i\alpha/2)]/\sqrt{2},$$

$$b = [b_1 \exp(-i\beta/2) + b_2 \exp(i\beta/2)]/\sqrt{2}. \tag{4.1}$$

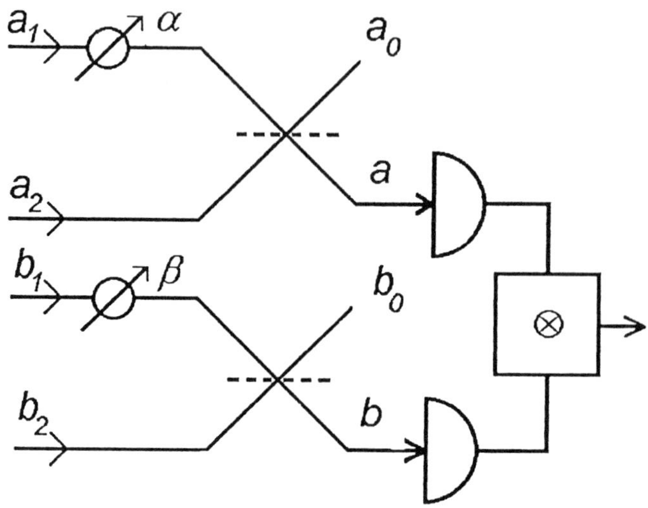

FIGURE 5. The four-mode mixing scheme for the observation of the interference of intensities and the two-photon interference (interference of advanced waves). The output correlator, $G_{ab} = \langle a^+ab^+b\rangle$, depends on the delays, α and β, as $\cos(\alpha - \beta)$ and $\cos(\alpha + \beta)$, depending on the existence of the input correlators, G_- and G_+ (see equation 4.3).

In analogy with earlier discussions, we find

$$G_{ab} = G_0 + G_+ \cos(\alpha + \beta) + G_- \cos(\alpha - \beta), \quad \textbf{(4.2)}$$

where

$$G_0 = (G_{a1b1} + G_{a2b2} + G_{a1b2} + G_{a2b1})/4,$$
$$G_+ = \langle a_1^+ b_1^+ a_2 b_2\rangle/2,$$
$$G_- = \langle a_1^+ b_2^+ a_2 b_1\rangle/2 \quad \textbf{(4.3)}$$

(we take the moments $G_\pm$ to be real).

The terms with $G_\pm$ describe two types of intensity interference with phases $\alpha \pm \beta$ and visibilities $V_\pm = G_\pm/G_0$. The interference with the phase $\alpha - \beta$ is the well-known interference of intensities. The second type with the sum phase can be called the advanced waves interference[28] because the plus sign in $\alpha + \beta$ can be conveniently "explained" in C-language by supposing that fictitious advanced waves are "emitted" by one of the detectors.

It is possible to observe simultaneously both types of intensity interference using a parametric downconverter and three additional beam-splitters (FIGURE 6). A polarization version of this scheme was realized recently by Shih and Sergienko.[36] In this case, the visibilities are

$$V_+ = (G_{cc} - 2G_{cd})/2(G_{cc} + G_{cd}) = -(N + 1)/(4N + 1),$$
$$V_- = G_{cc}/2(G_{cc} + G_{cd}) = N/(4N + 1). \quad \textbf{(4.4)}$$

Here, $N = \langle c^+c\rangle = \langle d^+d\rangle$. In case of weak pumping (small parametric gain F), $N \ll 1$; thus, $V_- = N = 0$ and we have the superclassical 100% visibility, $V_+ = -1$. In case of strong pumping, $N \gg 1$ and hence $V_- = -V_+ = \frac{1}{4}$; thus, the coincidence rate is proportional to $1 + (\frac{1}{2})\sin\alpha \sin\beta$.

Consider next the classical squeezed light, generated by a parametric converter with additional strong ($N_0 \gg \frac{1}{2}$) input signal and idler radiation with thermal statistics (APPENDIX B). Now,

$$N = N_0 \cosh(2F), \qquad G_{cc} = G_{dd} = 2N^2 = N_0^2[1 + \cosh(4F)],$$

$$G_{cd} = N_0^2 \cosh(4F), \qquad g_{cd} = 1 + \tanh^2(2F) = 2 - (N_0/N)^2. \tag{4.5}$$

Hence,

$$V_\pm = [1 \mp \cosh(4F)]/2[1 + 2\cosh(4F)]. \tag{4.6}$$

Thus, in case of small gain ($F \ll 1$, that is, with thermal noise at the converter's output), we have $V_+ = 4F^2/3 \ll 1$ and $V_- = \frac{1}{3}$. In case of strong squeezing, again $V_- = -V_+ = \frac{1}{4}$.

Q_S: In case of weak pumping, the initial state vector with omitted vacuum component is a factored state: $|\psi\rangle = |1\rangle_c|1\rangle_d = c^+d^+|\text{vac}\rangle$. The first BS makes the transformation in equation 3.8. The next two BSs in FIGURE 6 give

$$|\psi_1\rangle = 2^{-3/2}(|2000\rangle - 2|1100\rangle + |0200\rangle - |0020\rangle + 2|0011\rangle - |0002\rangle), \tag{4.7}$$

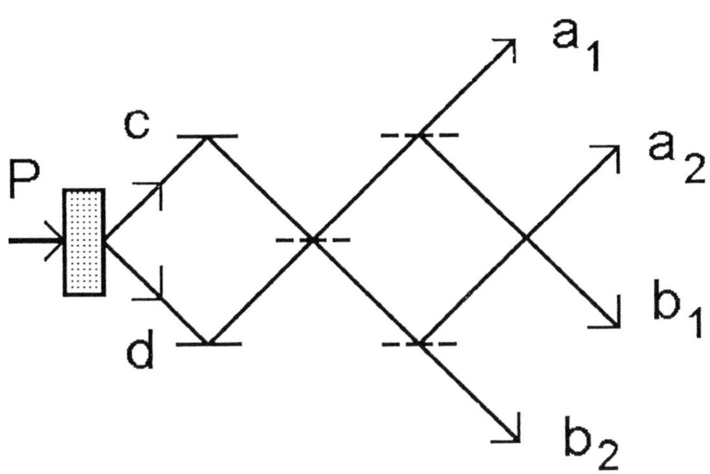

FIGURE 6. A parametric downconverter can be used for the preparation of the four-mode field with nonzero correlators, G_+ and G_- (see FIGURE 5). Terms: P is the pump field; c and d are the signal and idler fields.

where $|klmn\rangle = |k\rangle_{a1}|l\rangle_{b1}|m\rangle_{a2}|n\rangle_{b2}$. The output BSs (see FIGURE 5) give 10 components:

$$|\psi_2\rangle = \{-2i\sin(\alpha)(|2000\rangle + |0020\rangle) - 2i\sin(\beta)(|0200\rangle + |0002\rangle) - 4\cos(\alpha)|1010\rangle + 4\cos(\beta)|0101\rangle + 4i\sin[(\alpha+\beta)/2](|1100\rangle + |0011\rangle) + 4i\cos[(\alpha+\beta)/2](|1001\rangle + |0110\rangle)\}/2^{5/2}. \quad \mathbf{(4.8)}$$

Here, $|klmn\rangle = |k\rangle_a|l\rangle_b|1\rangle_{a0}|1\rangle_{b0}$. Thus, the coincidence probability is $P_{ab} = |c_{1100}|^2 = [1 - \cos(\alpha + \beta)]/4$. We again get the 100% visibility, so the considered scheme can be used for the demonstration of Bell's inequality violation.

M: Usually, for the demonstration of the EPR-Bell-type paradox, it is stressed that the state, describing two or more photons, should be entangled, that is, nonfactorizable. However, all the state vectors, $|\psi\rangle$, $|\psi_1\rangle$, and $|\psi_2\rangle$, considered above are factorizable (they are connected by unitary transformations), so this property is not a necessary one in the optical case.

Because the data, used in constructing Bell's observable, do not depend on the components of the output state vector with two photons in one mode (of the type $|2000\rangle$), one can argue that these components can be omitted. As a result of this "paper" operation, we get from equation 4.7 the desired entangled state, $|\psi_1\rangle_{\text{ent}} = 2^{-1/2}(-|1100\rangle + |0011\rangle)$.

Many other discussions of two-photon interference in M-terms, such as the photon's duality, paths, tunneling, (un)distinguishability, etc., can be found in the literature.

C: Both types of intensity interference have a trivial classical explanation, analogous to that of the anticorrelation effect. Let the classical mode fields in FIGURE 5 have constant amplitudes (=1) and stochastic phases (which play the role of the hidden parameters):

$$a_k(t) = \exp[-ix_k(t)], \quad b_k(t) = \exp[-iy_k(t)], \quad k = 1, 2. \quad \mathbf{(4.9)}$$

The output amplitudes and intensities are (cf. equation 4.1)

$$a(t) = [\exp(-ix_1 - i\alpha/2) + \exp(-ix_2 + i\alpha/2)]/\sqrt{2},$$
$$b(t) = [\exp(-iy_1 - i\beta/2) + \exp(-iy_2 + i\beta/2)]/\sqrt{2},$$
$$n_a = 1 + \cos(x + \alpha), \qquad n_b = 1 + \cos(y + \beta), \quad \mathbf{(4.10)}$$

where $x = x_1 - x_2$ and $y = y_1 - y_2$. Let x_1 and x_2 as well as y_1 and y_2 be independent; then, $\langle n_a\rangle = \langle n_b\rangle = 1$ and

$$G_{ab} = \langle n_a n_b\rangle = 1 + \sum_{\pm} \langle\cos(x + \alpha \pm y \pm \beta)\rangle/2. \quad \mathbf{(4.11)}$$

Thus, if $x(t) \pm y(t) =$ constant, then we have two types of stationary interference with visibilities $V = ½$. The minus sign corresponds to mutually coherent input waves with $a_1 = b_1$ and $a_2 = b_2$ (phase correlation). The plus sign corresponds to the condition, $x_k + y_k =$ constant (phase anticorrelation), which holds in case of the nondegenerate parametric generators or classical squeezed light because the signal and idler frequencies always drift in opposite directions.

CONCLUSIONS

First, Q_H-language (Heisenberg representation) is the most universal, compact, and useful one for the quantitative description of optical experiments. It closely follows the C-description. Moreover, Q_S-language has no advantages over it.

Second, C-language is useful for the qualitative pictorial description of many optical effects in terms of one-photon wave packets. Two-photon wave packets, depending on two time-space points and constructed by means of fictitious advanced waves, give a visualizable description of the effects of two-photon optics.

Third, M-language by definition does not predict any new phenomena and its conclusions cannot be falsified. It has no operationally defined notions. The M-photon has practically no counterpart in Q-language.

Fourth, the interference phenomena themselves are not nonclassical; only the used input light is. All two-photon interference experiments could be repeated using classical squeezed light and continuous detectors. The only difference would be in the smaller visibility.

Fifth and last, the "iron curtain" between Q- and C-worlds, which in QO can be placed right into human eyes, remains impenetrable in spite of all efforts.

ACKNOWLEDGMENTS

I am grateful to Pavel Elyutin, Mikl Horn, Morton Rubin, Alexandr Sergienko, Yanhua Shih, and Anton Zeilinger for many helpful discussions.

REFERENCES

1. FREEDMAN, S. J. & J. F. CLAUSER. 1972. Phys. Rev. Lett. **28:** 938.
2. ASPECT, A., J. DALIBARD & G. ROGER. 1982. Phys. Rev. Lett. **49:** 1804.
3. GHOSH, R. & L. MANDEL. 1987. Phys. Rev. Lett. **59:** 1903.
4. HONG, C. K., Z. Y. OU & L. MANDEL. 1987. Phys. Rev. Lett. **59:** 2044.
5. SHIH, Y. H. & C. O. ALLEY. 1988. Phys. Rev. Lett. **61:** 2921.
6. SHIH, Y. H. & A. V. SERGIENKO. 1994. Phys. Lett. A **186:** 29.
7. CLAUSER, J. F. & A. SHIMONY. 1978. Rep. Prog. Phys. **41:** 1881.
8. BELINSKY, A. V. & D. N. KLYSHKO. 1993. Usp. Fiz. Nauk **163:** 1 (Physics-Uspechi **36:** 653).
9. KLYSHKO, D. N. 1993. Phys. Lett. A **163:** 349.
10. GREENBERGER, D. M., M. HORNE, A. SHIMONY & A. ZEILINGER. 1990. Am. J. Phys. **58:** 1131.
11. MERMIN, N. D. 1990. Phys. Rev. Lett. **65:** 1838.
12. BELINSKY, A. V. & D. N. KLYSHKO. 1993. Phys. Lett. A **176:** 415.
13. BELINSKY, A. V. & D. N. KLYSHKO. 1992. Laser Phys. **2**: 112.
14. SHIH, Y. H. & M. H. RUBIN. 1993. Phys. Lett. A **182:** 16.
15. LAI, M. & J.-C. DIELS. 1992. J. Opt. Soc. Am. **B9:** 2290.
16. GODZINSKI, Z. 1991. Phys. Lett. A **153:** 291.
17. GRANGIER, P., G. ROGER & A. ASPECT. 1986. Europhys. Lett. **1:** 173.
18. OKOSHO, T., A. HIROSE & K. KIMURA. 1989. Opt. Commun. **72:** 7.

19. MIZOBUCHI, Y. & Y. OHTAKE. 1992. Phys. Lett. A **168:** 1.
20. GHOSE, P., D. HOME & G. S. AGARWAL. 1992. Phys. Lett. A **168:** 95.
21. BELL, J. S. 1964. Physics **1:** 195.
22. KLYSHKO, D. N. 1989. Phys. Lett. A **140:** 19.
23. KLYSHKO, D. N. 1993. Usp. Fiz. Nauk **163:** 1 (Physics-Uspechi **36:** 1005).
24. SCHIFF, I. J. 1955. Quantum Mechanics. McGraw–Hill. New York.
25. BELINSKY, A. V. 1994. Usp. Fiz. Nauk **164:** 435 (Physics-Uspechi **37:** in press).
26. KIDD, R., J. ARDINI & A. ANTON. 1989. Am. J. Phys. **57:** 27.
27. KLYSHKO, D. N. 1988. Usp. Fiz. Nauk **154:** 133 (Sov. Phys. Usp. **31:** 74); 1989. Usp. Fiz. Nauk **158:** 327 (Sov. Phys. Usp. **32:** 555).
28. KLYSHKO, D. N. 1988. Phys. Lett. A **128:** 133; **132:** 299.
29. BELINSKY, A. V. & D. N. KLYSHKO. 1994. Sov. Phys. JETP **78:** 259.
30. KLYSHKO, D. N. 1990. Phys. Lett. A **146:** 93.
31. BELINSKY, A. V. & D. N. KLYSHKO. 1993. Phys. Lett. A **166:** 303.
32. UNRUH, W. G. 1976. Phys. Rev. D **14:** 870.
33. KLYSHKO, D. N. 1991. Phys. Lett. A **154:** 433.
34. KLYSHKO, D. N. 1988. Photons and Nonlinear Optics. Gordon & Breach. New York.
35. KLYSHKO, D. N. 1989. Phys. Lett. A **137:** 334.
36. SERGIENKO, A. V. & Y. H. SHIH. 1994. Phys. Rev. Lett. In press.

APPENDIX A

Advanced Waves Interpretation

The advanced waves C-interpretation of two-photon optics[13,27–30] is based on the following correlation function, calculated in the first-order using the effective Hamiltonian, χE^3:

$$P_{12} = \langle\psi_2|E_1^{(-)}E_2^{(-)}E_1^{(+)}E_2^{(+)}|\psi_2\rangle = |E_{12\mathrm{eff}}|^2, \qquad \textbf{(A.1)}$$

where

$$E_k = E(\mathrm{r}_k, \mathrm{t}_k), \quad |\psi_2\rangle = (i\hbar)^{-1}\int d^4x\, \chi E_{\mathrm{p}x}^{(+)}[E_x^{(-)}]^2|\mathrm{vac}\rangle,$$

$$E_{12\mathrm{eff}} = \langle\mathrm{vac}|E_1^{(+)}E_2^{(+)}|\psi_2\rangle = i\hbar\int d^4x\, D(x_1, x)\, \chi E_{\mathrm{p}x}^{(+)} D^*(x, x_2). \qquad \textbf{(A.2)}$$

Here, E_p is the classical pump field and

$$D(x_1, x_2) = -D^*(x_2, x_1) = i[E_1^{(+)}, E_2^{(-)}]/\hbar \qquad \textbf{(A.3)}$$

is the phenomenological Green function, describing the propagation of the field $E^{(+)}$ from (r_1, t_1) to (r_2, t_2) or of the field $E^{(-)}$ in the opposite direction.

APPENDIX B

Classical Squeezed Light

The scattering matrix for any linear scheme, which can include any number of parametric upconverters and downconverters,[31] consists of two parts, u and v. In vector notations, $a' = u^*a + v^*a^+$. In the dissipationless case, $uu^+ = vv^+ + I$ and $v\tilde{u} = u\tilde{v}$.

Let the input field be in a displaced Gaussian state with parameters $N_{kl} = \langle a_k^+ a_l \rangle \delta_{kl}$ and z_k. Then, the output field is described by the following normally ordered characteristic function:[31,34]

$$\chi(\mu, \mu^*) = \exp[-\mu N' \mu^* + (\mu M' \mu + cc)/2 + (\mu z'^* - cc)], \quad \textbf{(B.1)}$$

where

$$N' = uN_{sym}u^+ + vN_{sym}v^+ - I/2 = (2N + I)vv^+ + N, \quad \textbf{(B.2)}$$

$$M' = \langle a'a' \rangle = uN_{sym}\tilde{v} + vN_{sym}\tilde{u} = (2N + I)u\tilde{v}, \quad \textbf{(B.3)}$$

$$z' = u^*z + v^*z^*, \qquad N_{sym} = N + I/2 = (\langle a^+a \rangle + \langle aa^+ \rangle)/2,$$

$$(u^+)_{kl} = u^*_{lk} = \tilde{u}^*_{kl}. \quad \textbf{(B.4)}$$

In case of the squeezed state, $N = 0$; thus,

$$N'_Q = vv^+, \qquad M'_Q = u\tilde{v}. \quad \textbf{(B.5)}$$

In C*-theory, the only difference is that the "spontaneous" terms with $I/2$ in equations B.2–B.4 are missing; that is, $N_{sym} = N$.

If we put $I/2 = 0$ and $N_{sym} = N = I/2$ in equations B.2–B.4, we would get equation B.5, but with N' increased by an extra term, $+I/2$.

If $N \gg 1/2$ in Q-theory or experiment, then we get the classical squeezed light with the same characteristic function of equation B.1, but in which now, in place of equations B.2 and B.3, we have (cf. equation 4.5)

$$N'_C = N(2vv^+ + I) = 2N(N'_Q + I/2), \quad M'_C = 2Nu\tilde{v} = 2NM'_Q. \quad \textbf{(B.6)}$$

Second-Order Photon-Photon Correlations and Atomic Spectroscopy[a]

ULRICH W. RATHE,[b,c,d] MARLAN O. SCULLY,[b,c,e] AND SUSANNE F. YELIN[b,c,d]

[b]*Department of Physics*
Texas A&M University
College Station, Texas 77843

[c]*Max-Planck-Institut für Quantenoptik*
85748 Garching, Germany

[d]*Sektion Physik*
Ludwig-Maximilians-Universität München
80333 München, Germany

[e]*Texas Laser Laboratory*
Houston Advanced Research Center
The Woodlands, Texas 77381

INTRODUCTION

Efforts have been profitably invested into the amelioration of atomic spectroscopy, the goal being the precision measurement of atomic energy differences. In a two-level system, the narrowest line that one can conventionally observe has a width governed by the sum of the decay rates out of the two levels in question. However, in some instances, it is also possible to obtain spectroscopic resolution beyond this "natural line width". We specifically point to "time delay spectroscopy",[1] wherein only the temporal tail of the resonance fluorescence is measured by opening a shutter in front of the detector some time after exciting the atoms. With the help of this method, one observes line widths governed by the *difference* of the decay rates out of the two involved atomic levels, the disadvantage being the exponential decrease of the signal with the time delay.

In time delay spectroscopy, we are interested specifically in single-photon detection, that is, the measurement of first-order correlation functions $G^{(1)}$. However, as we will see, similar techniques can be obtained by considering intensity-intensity correlations measured by using two detectors, that is, second-order correlation functions $G^{(2)}$.[2–6] The pioneering work of Hanbury Brown–Twiss[7] is an example of such a payoff when going from $G^{(1)}$ physics (Michelson stellar interferometry) to $G^{(2)}$ (Hanbury Brown–Twiss interferometry).

In fact, applications of photon correlation techniques extend from astrophysics to the foundations of quantum mechanics. In this latter context, we have in mind the old

[a]This work is dedicated to John Archibald Wheeler, who has given us so many insights into the foundations of quantum physics and in particular the concept of delayed choice, which has been so fruitful. Support was provided by the Office of Naval Research, the Welch Foundation, and the Texas Advanced Research Program.

proposal[8] to investigate complementarity in a Young double-slit experiment, in which two atoms replace the slits; see FIGURE 1. In the case of three-level atoms, which absorb light at ω_{ac} and emit at ω_{ab}, $G^{(1)}$ shows no interference because the atom being left in level $|b\rangle$ provides "which-path" information. However, by extending the story to include two-photon cascade and $G^{(2)}$ physics, interference is restored when we correlate. This is an example of delayed choice[9] and was the initial suggestion of the quantum eraser.[8]

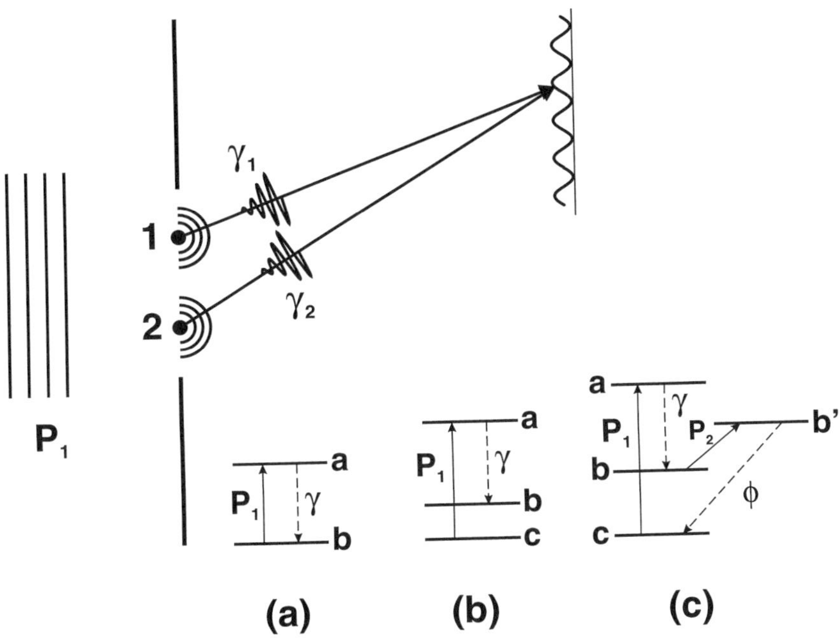

FIGURE 1. Original quantum eraser setup: Light impinges from left on atoms at sites 1 and 2. Scattered photons γ_1 and γ_2 produce an interference pattern on the screen. (a) Two-level atoms excited by laser pulse P_1 emit γ-photons in the $a \to b$ transition. (b) Three-level atoms excited by pulse P_1 from $c \to a$ emit photons in the $a \to b$ transition. (c) Four-level system excited by pulse P_1 from $c \to a$ followed by emission of γ-photons in the $a \to b$ transition. Second pulse P_2 takes atoms from b to b'. Decay from b' to c results in emission of ϕ-photons.

Recently, beautiful experiments verifying these predictions have been carried out by the NIST group[10] involving two trapped ions that scatter light, which is detected by a single detector. These experiments demonstrate the interference expected from two-level atoms and the disappearance of that interference when three levels are used. We emphasize that both cases were single-photon, $G^{(1)}$ experiments.

In this article, we are concerned with "trapped" three-level atoms that radiate spontaneously into two-photon cascades. Interesting results can be obtained that bear resemblance to time delay spectroscopy on the one hand and Hanbury Brown–Twiss correlations on the other, thus deepening our understanding of many-photon processes. In particular, we show in the following calculations of second-order correlation functions that there are possibilities for the measurement of the atomic transition frequencies to a high precision (beyond the natural line width) even if very broad levels are involved.

In the second section, we outline the calculational technique for the determination of second-order correlation functions. This technique is then employed in the third section to analyze an experimental setup with two atoms in a trap, demonstrating the possibility of subnatural spectroscopy. In the fourth section, we put forward another spectroscopic scheme involving only one atom, but utilizing a birefringent medium between this atom and the detectors. Finally, the fifth section contains a discussion of the results.

PHOTON-PHOTON CORRELATIONS FROM SINGLE ATOMS

One way of calculating and understanding Glauber correlation functions utilizes an expression that may be called a "photo-electron detection probability amplitude" Ψ. With Ψ known, the expectation value of light field correlations can be found simply by using $G = |\Psi|^2$; for example, for the present second-order correlation function, we may write

$$G^{(2)}(\vec{r}_1, t_1; \vec{r}_2, t_2) = |\Psi^{(2)}(\vec{r}_1, t_1; \vec{r}_2, t_2)|^2, \tag{1}$$

where the subscripts 1 and 2 label the two detectors involved (see FIGURE 2) and the actual expression for $\Psi^{(2)}$ is derived later in equation 11. We first consider a three-level two-photon cascade in a single atom as depicted in FIGURE 3, governed by the transition frequencies ω_{ab} and ω_{bc} and the inverse lifetimes of the two upper levels γ_a and γ_b. As a first candidate for the two-photon state, one might think of a

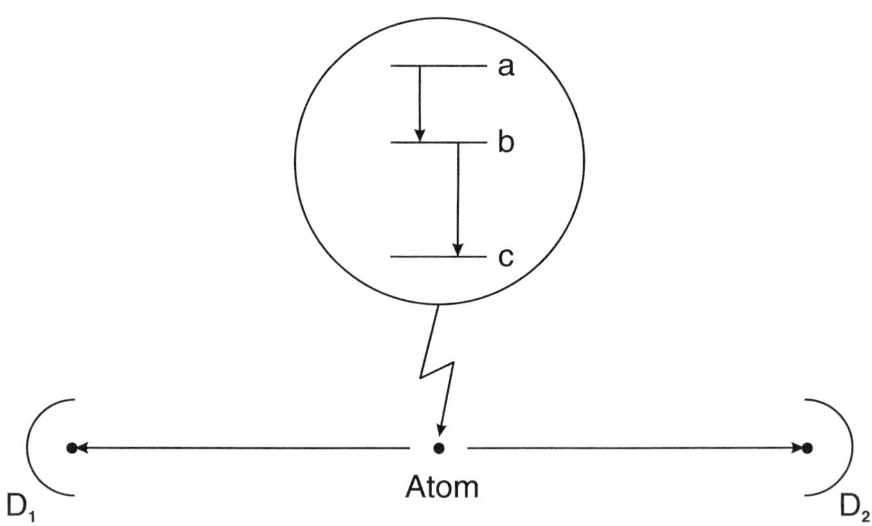

FIGURE 2. The cascade decay of a single excited three-level atom leads to a "Hanbury Brown–Twiss" second-order interference pattern.

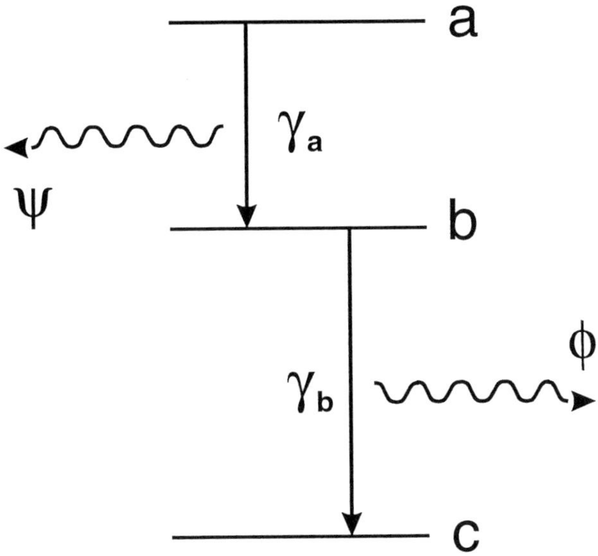

FIGURE 3. The level scheme for the atom in FIGURE 2.

product state of two single-photon states,[8]

$$|\Psi\rangle = |\psi\rangle|\phi\rangle, \tag{2}$$

leading to the expectation value $G^{(2)}(1,2) = |\Psi_{\psi}^{(1)}(1)\Psi_{\phi}^{(1)}(2) + \Psi_{\phi}^{(1)}(1)\Psi_{\psi}^{(1)}(2)|^2$, where we introduced the short-hand notation $(1) = (\vec{r}_1, t_1)$ and $(2) = (\vec{r}_2, t_2)$. However, this result holds only with a rapidly decaying uppermost level and a long-lived intermediate state, as the following actual derivation of $|\Psi\rangle$ shows.

We define the relevant probability amplitudes by writing the atom-field state,

$$|\psi\rangle = A(t)|a, 0\rangle + \sum_{\vec{k}} B_{\vec{k}}(t)|b, 1_{\vec{k}}\rangle + \sum_{\vec{k},\vec{q}} C_{\vec{k},\vec{q}}(t)|c, 1_{\vec{k}}, 1_{\vec{q}}\rangle. \tag{3}$$

This ansatz makes use of the rotating wave approximation. By using the interaction Hamiltonian (in the interaction picture),

$$\mathscr{V} = \hbar \sum_{\vec{k}} [g_{a,\vec{k}}(\vec{r})a_{\vec{k}}|a\rangle\langle b|e^{i(\omega_{ab}-\nu_k)t} + \text{H.c.}] + \hbar \sum_{\vec{q}} [g_{b,\vec{q}}(\vec{r})a_{\vec{q}}|b\rangle\langle c|e^{i(\omega_{bc}-\nu_q)t} + \text{H.c.}], \tag{4}$$

with coupling constants that depend on the atomic position,

$$g_{a,\vec{k}}(\vec{r}) = -\wp_{ab}\sqrt{\frac{\nu_k}{\hbar\epsilon_0 V}}e^{i\vec{k}\cdot\vec{r}} = g_{a,k}e^{i\vec{k}\cdot\vec{r}}, \tag{5}$$

$$g_{b,\vec{q}}(\vec{r}) = -\wp_{bc}\sqrt{\frac{\nu_q}{\hbar\epsilon_0 V}}e^{i\vec{q}\cdot\vec{r}} = g_{b,q}e^{i\vec{q}\cdot\vec{r}}, \tag{6}$$

we obtain the equations of motion for the probability amplitudes of equation 3:

$$\frac{d}{dt}A(t) = -\gamma_a A(t),$$

$$\frac{d}{dt}B_{\vec{k}}(t) = -ig_{a,\vec{k}}(\vec{r})^* A(t)e^{-i(\omega_{ab}-\nu_k)t} - \gamma_b B_{\vec{k}}(t), \tag{7}$$

$$\frac{d}{dt}C_{\vec{k},\vec{q}}(t) = -ig_{b,\vec{q}}(\vec{r})^* B_{\vec{k}}(t)e^{-i(\omega_{bc}-\nu_q)t},$$

where we applied the Wigner-Weisskopf approximation to the decay out of the levels $|a\rangle$ and $|b\rangle$. Solving this system of differential equations for $C_{\vec{k},\vec{q}}$ in the long-time limit, we obtain

$$C_{\vec{k},\vec{q}}(t \to \infty) = \frac{-g^*_{a,k}g^*_{b,q}e^{-i\vec{k}\cdot\vec{r}-i\vec{q}\cdot\vec{r}}}{[i(\omega_{ac} - \nu_k - \nu_q) + \gamma_a][i(\omega_{bc} - \nu_q) + \gamma_b]}. \tag{8}$$

With this result, the state of the radiation field is given by

$$|\Psi\rangle = \sum_{\vec{k},\vec{q}} C_{\vec{k},\vec{q}}|1_{\vec{k}}, 1_{\vec{q}}\rangle. \tag{9}$$

Now, we consider the expectation value of the second-order correlation function:

$$\begin{aligned} G^{(2)}(1,2) &= \langle\Psi|E^{(-)}(1)E^{(-)}(2)E^{(+)}(2)E^{(+)}(1)|\Psi\rangle \\ &= \langle\Psi|E^{(-)}(1)E^{(-)}(2)|0\rangle\langle 0|E^{(+)}(2)E^{(+)}(1)|\Psi\rangle. \end{aligned} \tag{10}$$

With equation 1 in mind, we define

$$\Psi^{(2)}(1,2) = \langle 0|E^{(+)}(2)E^{(+)}(1)|\Psi\rangle \tag{11}$$

and thus have a quantity that can be used as advertised earlier. Using the well-known form of the electric field operator

$$E^{(+)}(i) = \sum_{\vec{k}} \mathscr{E}_k a_{\vec{k}} e^{-i\nu_k t_i + i\vec{k}\cdot\vec{r}_i} \qquad (i = 1, 2) \tag{12}$$

and taking $|\Psi\rangle$ as given by equation 9, we find

$$\begin{aligned} \Psi^{(2)}(1,2) = -\sum_{\vec{k},\vec{q}} &\left(\frac{g^*_{a,k}g^*_{b,q}e^{-i\vec{k}\cdot\vec{r}-i\vec{q}\cdot\vec{r}}}{[i(\omega_{ac} - \nu_k - \nu_q) + \gamma_a][i(\omega_{bc} - \nu_q) + \gamma_b]}\right) \\ &\times \{e^{-i\nu_k t_1 + i\vec{k}\cdot\vec{r}_1}e^{-i\nu_q t_2 + i\vec{q}\cdot\vec{r}_2} + (1 \leftrightarrow 2)\}. \end{aligned} \tag{13}$$

Changing the sums into integrals and choosing the z-axes of the angular integrations along $\vec{k}$ and $\vec{q}$, we obtain

$$\begin{aligned} \Psi^{(2)}(1,2) = &\ \kappa e^{-(i\omega_{ab}+\gamma_a-\gamma_b)(t_1-r_1/c)}\Theta(t_1 - r_1/c) \\ &\times e^{-(i\omega_{bc}+\gamma_b)(t_2-r_2/c)}\Theta[(t_2 - r_2/c) - (t_1 - r_1/c)] \\ &+ (1 \leftrightarrow 2) \end{aligned} \tag{14}$$

as the result for the two-photon probability amplitude, where $r_i \equiv |\vec{r}_i - \vec{r}|$ and κ is an uninteresting constant. Now, we have a useful tool that we will apply in the next section to present a new kind of subnatural spectroscopy not limited by the usual radiative broadening of spectral lines.

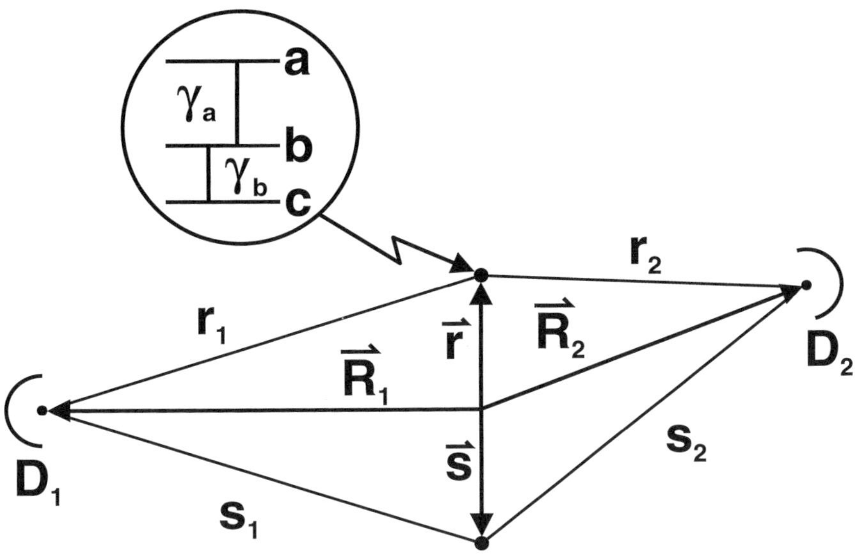

FIGURE 4. Two three-level atoms are stimulated weakly such that either the atom at $\vec{r}$ or the atom at $\vec{s}$ is excited.

CORRELATED EMISSION SPECTROSCOPY

We here consider an experimental setup as depicted in FIGURE 4, with two atoms located at $\vec{r}$ and $\vec{s}$ and two detectors at $\vec{R}_1$ and $\vec{R}_2$. With the cascade scheme of the last section (FIGURE 3), one can determine the energy difference of the levels up to a precision governed by the radiative decay rate γ_a of the upper level $|a\rangle$ independent of γ_b.

As a starting point, we prepare an atomic state such that either one of the atoms is excited to state $|a\rangle$ and the other one is in the ground state $|c\rangle$ with the field in the vacuum state $|0\rangle$. This initial atom-field state may be written as

$$|\Psi(0)\rangle = \left(\frac{1}{\sqrt{2}}\right)(|a, c\rangle + |c, a\rangle) \otimes |0\rangle. \tag{15}$$

For times $t \gg {\gamma_a}^{-1}, {\gamma_b}^{-1}$, this state evolves to

$$|\Psi\rangle = \left(\frac{1}{\sqrt{2}}\right)|c, c\rangle \otimes (|\psi_{\vec{r}}, \phi_{\vec{r}}\rangle + |\psi_{\vec{s}}, \phi_{\vec{s}}\rangle), \tag{16}$$

where $|\psi_{\vec{r}}, \phi_{\vec{r}}\rangle$ and $|\psi_{\vec{s}}, \phi_{\vec{s}}\rangle$ are the field states generated by the atoms at $\vec{r}$ and $\vec{s}$, respectively, and ψ (ϕ) is associated with the $|a\rangle \rightarrow |b\rangle$ ($|b\rangle \rightarrow |c\rangle$) transition; that is, using equation 9,

$$|\psi_{\vec{\rho}}, \phi_{\vec{\rho}}\rangle = \sum_{\vec{k},\vec{q}} \left(\frac{-g^*_{a,\vec{k}} g^*_{b,\vec{q}} e^{-i\vec{k}\cdot\vec{\rho}-i\vec{q}\cdot\vec{\rho}}}{[i(\omega_{ac} - \nu_k - \nu_q) + \gamma_a][i(\omega_{bc} - \nu_q) + \gamma_b]} \right) |1_{\vec{k}}, 1_{\vec{q}}\rangle, \quad \textbf{(17)}$$

in which $\vec{\rho} = \vec{r}$ or $\vec{s}$. Using equation 14, the corresponding two-photon probability amplitude can be written as

$$\Psi^{(2)}(1, 2) = A(1, 2) + A(2, 1) + B(1, 2) + B(2, 1), \quad \textbf{(18)}$$

where

$$A(1, 2) = \kappa e^{-(i\omega_{ab}+\gamma_a-\gamma_b)(t_1-r_1/c)}\Theta(t_1 - r_1/c)$$
$$\times e^{-(i\omega_{bc}+\gamma_b)(t_2-r_2/c)}\Theta[(t_2 - r_2/c) - (t_1 - r_1/c)] \quad \textbf{(19)}$$

and

$$B(1, 2) = A(1, 2)|_{r_i \rightarrow s_i} \quad \textbf{(20)}$$

with $r_i(s_i) \equiv |\vec{R}_i - \vec{r}_i(\vec{s}_i)|$ for $i = 1, 2$. Substituting equation 18 into equation 1 gives

$$G^2(1, 2) = \{[A(1, 2)^*B(1, 2) + A(1, 2)^*B(2, 1) + (1 \leftrightarrow 2)] + \text{c.c.}\}$$
$$+ \{A(1, 2)^*A(1, 2) + B(1, 2)^*B(1, 2) + (1 \leftrightarrow 2)\}, \quad \textbf{(21)}$$

where we used the fact that $A(1, 2)^*A(2, 1)$ and $B(1, 2)^*B(2, 1)$ vanish due to equations 19 and 20. The second curly bracketed term in equation 21 is a dc background term, so we focus on the contribution of the first one. This is the interference cross term (with the factorized state of equation 2 vanishing), as shown in FIGURE 5. Here, $A(1, 2)^*B(1, 2)$ corresponds to the detection of $\psi_{\vec{\rho}}$ at detector D_1 and $\phi_{\vec{\rho}}$ at detector D_2 ($\vec{\rho} = \vec{r}$ or $\vec{s}$) and $A(1, 2)^*B(2, 1)$ corresponds to $\psi_{\vec{r}}$ ($\phi_{\vec{s}}$) at D_1 and $\phi_{\vec{r}}$ ($\psi_{\vec{s}}$) at D_2.

We now define the interesting interference term of the total joint count probability,

$$\tilde{P}^{(2)} = \int_0^\infty dt_1 \int_0^\infty dt_2 \{[A(1, 2)^*B(1, 2) + A(1, 2)^*B(2, 1) + (1 \leftrightarrow 2)] + \text{c.c.}\}, \quad \textbf{(22)}$$

and obtain

$$\tilde{P}^{(2)} = \kappa^2 e^{-\gamma_b\tau}\left[\left(\frac{f(\tau, \Delta)}{2\gamma_a\gamma_b}\right) + \left(\frac{g(\tau, \Delta)}{\Delta^2 + \gamma_a^2}\right)\right], \quad \textbf{(23)}$$

where

$$\tau = \frac{(s_2 - r_2)}{c}, \quad \textbf{(24)}$$

$$\Delta = \omega_{ab} - \omega_{bc}, \quad \textbf{(25)}$$

$$f(\tau, \Delta) = \cos\omega_{bc}\tau + e^{-\gamma_a\tau}\cos\omega_{ab}\tau, \quad \textbf{(26)}$$

$$g(\tau, \Delta) = \cos \omega_{bc}\tau - \left(\frac{\Delta}{\gamma_a}\right) \sin \omega_{bc}\tau - e^{-\gamma_a\tau}\left[\cos \omega_{ab}\tau - \left(\frac{\Delta}{\gamma_a}\right) \sin \omega_{ab}\tau\right]. \quad (27)$$

The term proportional to $(\Delta^2 + \gamma_a^2)^{-1}$ in equation 23 enables us to envision a high-precision measurement of the atomic transition frequencies, ω_{ab} and ω_{bc}, in the following way. As indicated in FIGURE 6, the intermediate level $|b\rangle$ of the atomic

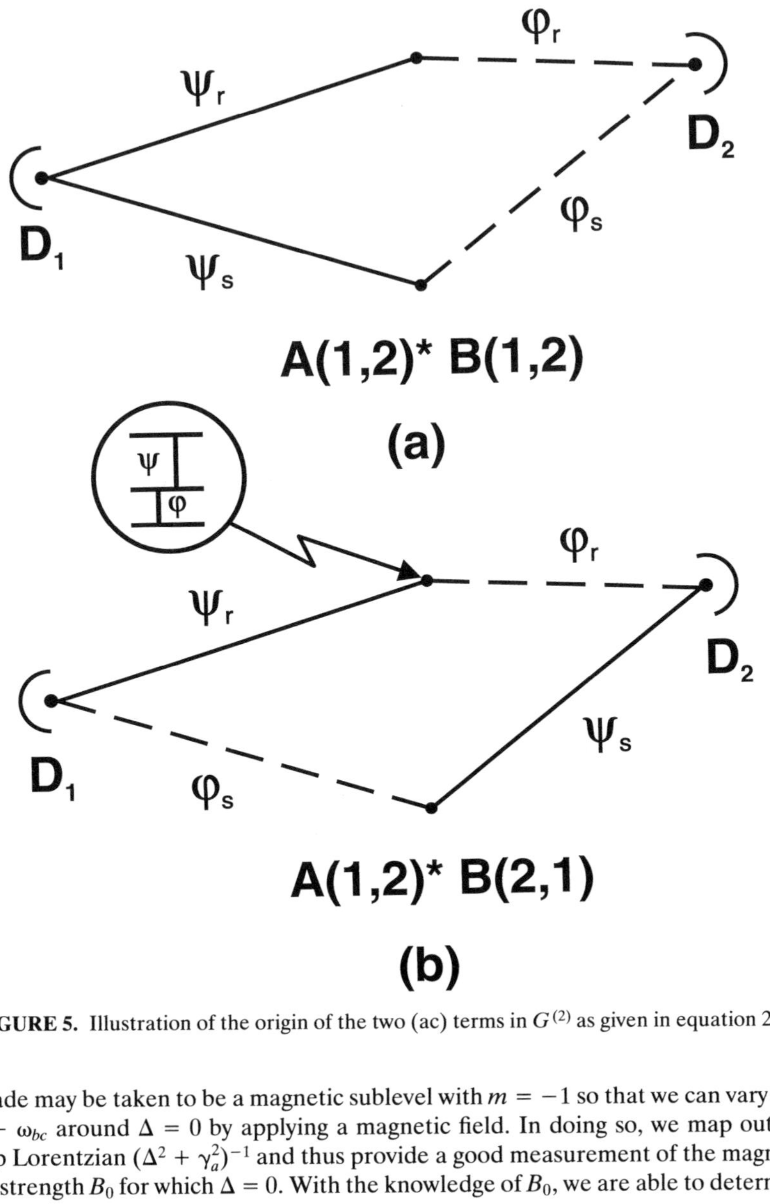

FIGURE 5. Illustration of the origin of the two (ac) terms in $G^{(2)}$ as given in equation 21.

cascade may be taken to be a magnetic sublevel with $m = -1$ so that we can vary $\Delta = \omega_{ab} - \omega_{bc}$ around $\Delta = 0$ by applying a magnetic field. In doing so, we map out the sharp Lorentzian $(\Delta^2 + \gamma_a^2)^{-1}$ and thus provide a good measurement of the magnetic field strength B_0 for which $\Delta = 0$. With the knowledge of B_0, we are able to determine

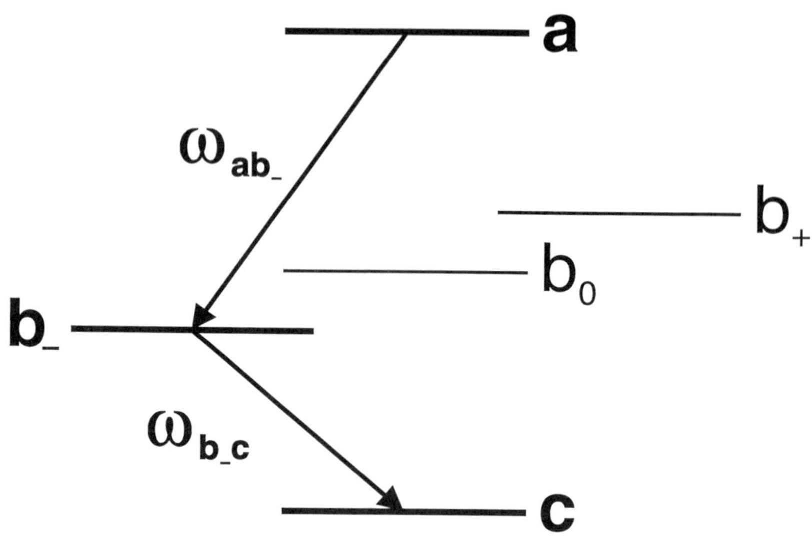

FIGURE 6. Cascade radiation is "tuned" by an external magnetic field such that $\omega_{ab\_} = \omega_{b\_c} \simeq \omega_0$. Then, $\Delta = \omega_{ab\_} - \omega_{b\_c}$ can be varied around zero. (The level b_- plays the role of $|b\rangle$ in the calculation.)

the difference of the energy spacing of the unshifted transition frequencies between $|a\rangle$ and $|b_0\rangle$ ($|b_0\rangle$ being the intermediate state with $m = 0$) and between $|b_0\rangle$ and $|c\rangle$. This procedure thus enables us to measure $\omega_{ab_0} - \omega_{b_0c}$, *limited only by the line width of the atomic level* $|a\rangle$. An additional measurement for $\omega_{ab_0} + \omega_{b_0c} = \omega_{ac}$ could be performed, obviously limited by γ_a alone as well, so that we are finally in a position to determine ω_{ab_0} and ω_{b_0c} to a precision governed only by γ_a.

Note at this point, however, that $g(\tau, \Delta)$ varies with Δ as well as the Lorentzian denominator. This functional dependence is such that, for small $\tau \ll \gamma_a^{-1}$, the signal $g(\tau, \Delta)/(\Delta^2 + \gamma_a^2)$ goes to a constant independent of Δ. Therefore, the time delay τ has to be of order γ_a^{-1}. This, however, leads to an exponential decrease of the signal amplitude by means of the prefactor $\exp(-\gamma_b\tau)$ in equation 23 and eventually to an unwanted oscillatory behavior of the signal by means of the sine and cosine functions in equation 23. Both effects are well known from time delay spectroscopy. As shown in FIGURE 7, we find a signal of width γ_a without oscillations for $\tau \simeq 3\gamma_a^{-1}$; that is, in the optimum case, our signal is damped by the factor $\exp(-3\gamma_b/\gamma_a)$. This places an upper limit on the magnitude of γ_b because of untolerable amplitude loss for $\gamma_b \gg \gamma_a$. However, as stated earlier, the width of the signal is unaffected by γ_b.

ONE-ATOM CORRELATED EMISSION SPECTROSCOPY

We now use one atom as shown in FIGURE 8 involving a three-level two-photon cascade as before. For the sake of simplicity, we choose a symmetric setup with the distances in equation 14: $r_1 = r_2 \equiv r$. The region between the radiating atom and the detectors is filled with a birefringent medium such that photons travel with a polarization-dependent velocity, $c_+ = c_0/n_+$ for right circularly polarized photons and $c_- = c_0/n_-$ for left circularly polarized ones. Here, c_0 is the speed of light in a vacuum and $n_\pm$ is the respective index of refraction.

The two-photon state for this setup can be obtained from equation 14 by noting that the speed of light for $+(-)$-polarized $\psi(\phi)$-photons of frequency $\omega_{ab}(\omega_{bc})$ is now $c_+(c_-)$ and has the form

$$\Psi^{(2)}(r, t_1, t_2) = \kappa e^{-(i\omega_{ab}+\gamma_a-\gamma_b)(t_1-r/c_+)}\Theta(t_1 - r/c_+) \times e^{-(i\omega_{bc}+\gamma_b)(t_2-r/c_-)}\Theta[(t_2 - r/c_-) - (t_1 - r/c_+)] + (1 \leftrightarrow 2). \quad \textbf{(28)}$$

The total joint count probability in this case is given by

$$P^{(2)}(1, 2) = \left(\frac{\kappa^2}{2\gamma_a\gamma_b}\right)\left\{1 + \left[\frac{4\gamma_a\gamma_b e^{-2\gamma_b\tau}}{(\Delta^2 + \gamma_a^2)}\right]\left[1 - e^{-\gamma_a\tau}\left(\cos\Delta\tau - \left[\frac{\Delta}{\gamma_a}\right]\sin\Delta\tau\right)\right]\right\} \quad \textbf{(29)}$$

with $\tau = r(n_+ - n_-)/c_0$. The situation here is similar as in equation 23. We have to trade off signal amplitude against resolution because, for $\tau \ll \gamma_a^{-1}$, the interference term is a constant independent of Δ. From FIGURE 9, we see that, for $\tau \simeq 2\gamma_a^{-1}$, a resolution of order γ_a is obtained without introducing strong oscillations into the signal. Thus, in this case, the damping factor is $\exp(-4\gamma_b/\gamma_a)$.

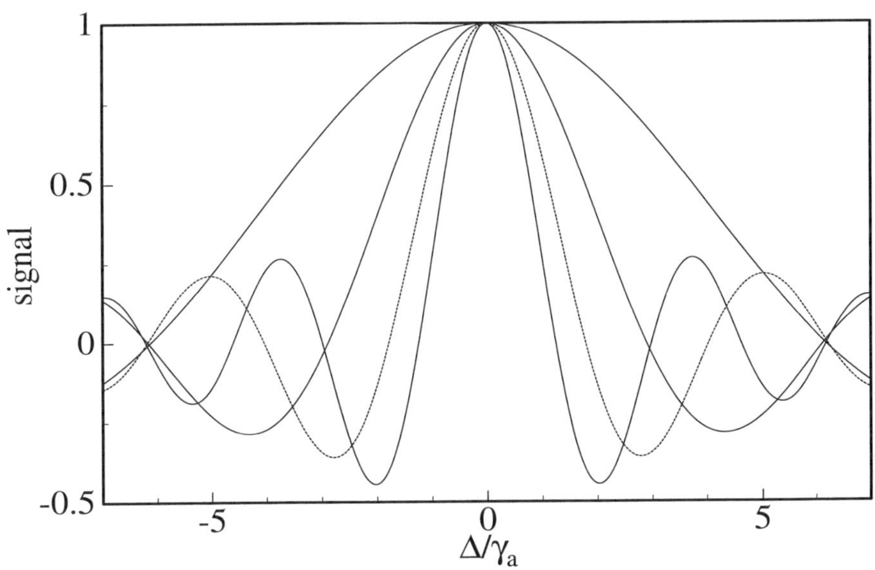

FIGURE 7. Signal in the two-atom setup for different time delays. The time delays are $\tau = \gamma_a^{-1}$, $\tau = 2\gamma_a^{-1}$, $\tau = 3\gamma_a^{-1}$, and $\tau = 4\gamma_a^{-1}$ in the order of decreasing width. The maximum amplitude is normalized to unity for easier comparison.

DISCUSSION

We return to the physical origin of the sharp Lorentzian in equations 23 and 29. As noted in the derivation of equation 23, the interference cross term involves the two possibilities for photon detection as depicted in FIGURE 5. The first term in equation 23 that goes with $f(\tau, \Delta)$ is due to the situation depicted in FIGURE 5a, whereas the interesting term, which is multiplied by the sharp Lorentzian of equation 23, corresponds to the case in FIGURE 5b. We see that this latter term plays a role in equation 23 only when the $\phi_{\vec{r}}$-radiation and the $\psi_{\vec{s}}$-radiation have to traverse different optical path lengths from their respective source to detector D_2 because $g(\tau = 0, \Delta)$ vanishes. In other words, it is the *difference in the time delay* between the superposed probability amplitudes for the detection of ϕ- and ψ-photons at D_2 that forms the essential ingredient for the existence of the sharp Lorentzian.

This idea is confirmed by the results for the one-atom setup including the birefringent medium. Now, we have once more a superposition of probability amplitudes for detection of ϕ- and ψ-radiation in one detector *with a different time delay* because the optical path for +-polarized ψ-radiation is different from the one for −-polarized ϕ-radiation in the birefringent medium. We may therefore say that the possibility for subnatural spectroscopy in both analyzed cases has the same physical origin.

To summarize, we have shown the possibility for subnatural spectroscopy utilizing second-order correlations of spontaneously emitted photons from two atoms that yields results very similar to time delay spectroscopy. Furthermore, we elucidated the underlying physical effect with a single-atom source of polarized radiation involving birefringence.

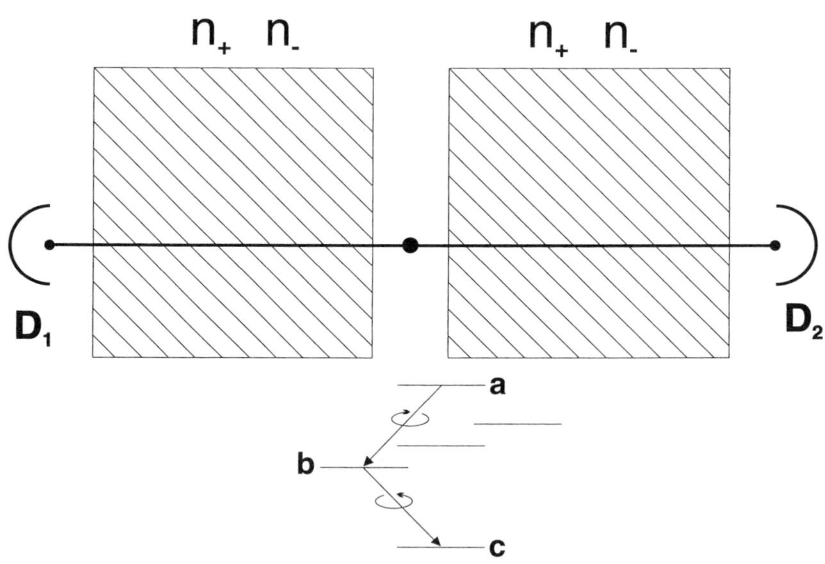

FIGURE 8. One three-level atom in a spectroscopic setup involving a birefringent medium between the atom and the detectors.

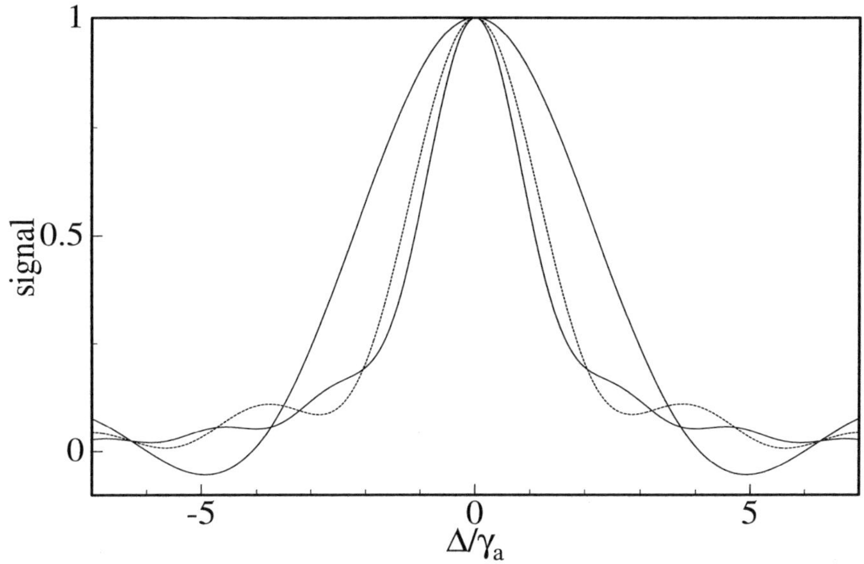

FIGURE 9. Signal in the one-atom setup for different time delays. The time delays are $\tau = \gamma_a^{-1}$, $\tau = 2\gamma_a^{-1}$, and $\tau = 3\gamma_a^{-1}$ in the order of decreasing width. The maximum amplitude is normalized to unity for easier comparison.

REFERENCES

1. WU, C. S., Y. K. LEE, N. BENCZER-KOLLER & P. SIMMS. 1960. Phys. Rev. Lett. **5:** 432; MA, I-J., J. MERTENS, G. ZU PUTLITZ & G. SCHUTTE. 1968. Z. Phys. **208:** 276; ZU PUTLITZ, G. 1969. Comments At. Mol. Phys. **1:** 74; COPLEY, G., B. P. KIBBLE & G. W. SERIES. 1968. J. Phys. B: Proc. Phys. Soc. London **1:** 724; FIGGER, H. & H. WALTHER. 1974. Z. Phys. **267:** 1; HILBORN, R. C. & R. L. DEZAFRA. 1972. J. Opt. Soc. Am. **62:** 1492; SCHENCK, P., R. C. HILBORN & H. METCALF. 1973. Phys. Rev. Lett. **31:** 189; LUK, T. S., L. DIMAURO, M. FELDMAN & H. METCALF. 1981. Phys. Rev. A **24:** 864; MEYSTRE, P., M. O. SCULLY & H. WALTHER. 1980. Opt. Commun. **33:** 153.
2. SHIH, Y. H. & C. O. ALLEY. 1988. Phys. Rev. Lett. **61:** 2921.
3. OU, Z. Y. & L. MANDEL. 1988. Phys. Rev. Lett. **61:** 54; OU, Z. Y., X. Y. ZOU, L. J. WANG & L. MANDEL. 1990. Phys. Rev. Lett. **65:** 321.
4. HORNE, M. A., A. SHIMONY & A. ZEILINGER. 1989. Phys. Rev. Lett. **62:** 2209.
5. FRANSON, J. D. 1989. Phys. Rev. Lett. **62:** 2205; 1991. Phys. Rev. A **44:** 4552.
6. KWIAT, P. G., W. A. VAREKA, C. K. HONG, H. NATHEL & R. Y. CHIAO. 1990. Phys. Rev. A **41:** 2910.
7. HANBURY BROWN, H. & R. Q. TWISS. 1954. Philos. Mag. **45:** 663.
8. SCULLY, M. O. & K. DRÜHL. 1982. Phys. Rev. A **25:** 2208.
9. WHEELER, J. A. 1979. *In* Problems in the Formulation of Physics. G. T. di Francia, Ed. North-Holland. Amsterdam.
10. EICHMANN, U., J. C. BERGQUIST, J. J. BOLLINGER, J. M. GILLIGAN, W. M. ITANO, D. J. WINELAND & M. G. RAIZEN. 1993. Phys. Rev. Lett. **70:** 2359.

EPR and Two-Photon Interference Experiments Using Type-II Parametric Downconversion[a]

Y. H. SHIH, A. V. SERGIENKO, T. B. PITTMAN, AND M. H. RUBIN

Department of Physics
University of Maryland Baltimore County
Baltimore, Maryland 21228

INTRODUCTION

Two-particle entangled states have been known since the early days of quantum mechanics. These states play a particularly important role in the study of the Einstein-Podolsky-Rosen (EPR) paradox[1] and in the test of Bell inequalities.[2] It was Schrödinger who first pointed out that a particular type of two-particle state, which he called an "entangled state", is responsible for the EPR paradox.[3] Entangled states are states of two or more particles that cannot be written as products of single-particle states. An example of a two-particle entangled state was given in EPR's 1935 paper, where the measurement of an observable of either particle determined the value of that observable for the other particle with unit probability.[1] Although two-particle entangled EPR states are predicted by quantum theory, they are not allowed in classical physics. The physical consequences resulting from the EPR states violate classical local realism.[4]

In the past, EPR-type two-particle entanglement has been demonstrated by two types of experiments—(1) two-particle polarization correlation measurements: most of the historical EPR-Bohm experiments[5] and the measurements testing Bell's inequality exhibited a nonlocal two-particle *polarization* correlation;[6–10] these experiments demonstrated the EPR-type two-particle spin entanglement; (2) two-particle interference (fourth-order interference) experiments; recent two-particle nonclassical interference experiments demonstrated two-particle *space-time* entanglement.[11–16]

A typical EPR-Bohm-type two-photon spin entangled state was predicted by Wheeler in the late 1940s and was proved by Wu and Shaknov in the early 1950s:[17]

$$|\Psi\rangle = (1/\sqrt{2})(|R_1\rangle \otimes |R_2\rangle - |L_1\rangle \otimes |L_2\rangle), \tag{1}$$

where $|R_i\rangle$ (or $|L_i\rangle$) stands for a right-hand (or left-hand) circular polarization eigenstate of the superscript photon. If one wants to measure the linear polarization of a single photon, one would find that neither of them has a preferred polarization direction; however, whenever a single photon is measured to be polarized in a certain direction, the other one must be polarized orthogonal to that direction.

[a] This work was supported by the Office of Naval Research (Grant No. N00014-91-J-1430).

A typical EPR-type two-photon space-time entangled state, proposed by Franson recently,[18] is in the form of

$$\Psi_{EPR} = A(L_1, L_2) + A(S_1, S_2), \tag{2}$$

where $A(L_1, L_2)$ [or $A(S_1, S_2)$] is the amplitude in which a pair of photons travel along the long (L) or short (S) optical paths through one or two interferometers. In this state, if one photon is measured to follow the L (S) path, then the other must have followed the L (S) path. One can never predict "which path" for a single photon; however, if one of the photons traveled through the longer (shorter) path, the other must have traveled through the longer (shorter) path. The signature of the state in equation 2 is a cosine sum frequency interference fringe pattern of the coincidence counting rate.

The nonlocal spin or space-time two-particle entanglement phenomenon is striking. It is even more striking that quantum theory also allows *two-particle entanglement both in spin and in space-time simultaneously.* This third type of two-photon entangled state will be discussed in detail by reporting several experiments. In these experiments, it is interesting to see that the measurement of the spin and space-time observables of either particle determines the value of these observables for the other particle with unit probability.

TYPE-II SPONTANEOUS PARAMETRIC DOWNCONVERSION AND THE EFFECTIVE TWO-PHOTON WAVE FUNCTION

Spontaneous parametric downconversion (SPDC) is one of the most effective mechanisms for generating two-photon entangled states. In SPDC, a pump beam is incident on a birefringent crystal. The pump beam is intense enough so that nonlinear effects lead to the spontaneous emission of a pair of entangled light quanta by means of the phase matching condition,[19]

$$\omega_1 + \omega_2 = \omega_p, \qquad \mathbf{k}_1 + \mathbf{k}_2 = \mathbf{k}_p, \tag{3}$$

where ω_i is the frequency and $\mathbf{k}_i$ is the wave number vector, linking pump (p), signal (1), and idler (2). The downconversion is called type-I or type-II, depending on whether the photons in the pair have parallel or orthogonal polarization. The light quanta of the pair that emerges from the nonlinear crystal may propagate in different directions or may propagate collinearly. The frequency and propagation directions are determined by the orientation of the nonlinear crystal and the phase matching relations in equation 3.

In order to understand the two-photon behavior of SPDC, consider the simplified experiment shown in FIGURE 1, which is a simple beam-splitting experiment. Assume that a type-II BBO (β-BaB_2O_4) crystal is used for the SPDC. The collinear downconversion beam is split by a beam-splitter. The beam-splitter is assumed to be polarization-dependent so that the o-ray is transmitted and the e-ray is reflected. Single-photon counting detectors D_1 and D_2 are placed in the transmission and reflection output ports of the beam-splitter for detecting the o-ray and the e-ray, respectively.

It is most convenient to perform the calculation in the Heisenberg picture. An effective wave function $\Psi(t_1, t_2)$ will be developed in the following calculation. The introduction of $\Psi(t_1, t_2)$ is helpful for the understanding of physics.

For collinear type-II SPDC, the two-photon part of the state that exits the downconversion crystal may be calculated from the standard theory of SPDC (first-order perturbation theory) to be[19]

$$|\Psi> = \int d\omega_1 \delta(\omega_1 + \omega_2 - \omega_p)\Psi(k_1 + k_2 - k_p)a_o^\dagger[\omega_1(k_1)]a_e^\dagger[\omega_2(k_2)]|0>, \quad (4)$$

where ω and k represent the frequency and the wave number for signal (1), idler (2), and pump (p). The subscript indices o and e for the creation operators indicate the ordinary and extraordinary rays of the downconversion, traveling along the same direction as the pump, that is, the z-direction. The coordinate axes x and y are chosen

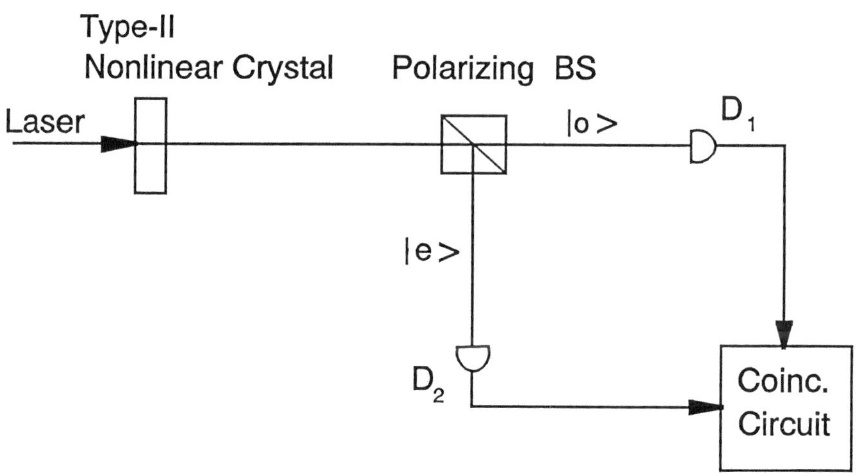

FIGURE 1. Simplified schematic experiment for the study of the type-II SPDC biphoton. BS is a polarization-dependent beam-splitter that transmits the o-ray and reflects the e-ray. D_1 and D_2 are photon counting detectors. A coincidence circuit is used for recording the coincidence rate.

along the polarization direction of the o-ray and the e-ray, respectively. The use of small apertures makes the state of equation 4 a good one-dimensional approximation. The frequency phase matching condition is explicitly displayed by the delta function; the wave number phase matching condition is not of the form $\delta(k_1 + k_2 - k_p)$ because of the finite length of the crystal.[19] The function Ψ determines the natural spectral width of the two-photon state. Taking the origin of the coordinate of z at the output surface of the downconversion crystal,

$$\Psi(\Delta k) = [1 - \exp(-i\Delta k \cdot L)]/(i\Delta k \cdot L), \quad (5)$$

where L is the length of the crystal and $\Delta k = k_1 + k_2 - k_p$.

Suppose that the crystal is oriented so that the perfect phase matching condition (equation 3) is satisfied by a set Ω_o, Ω_e, k_o, and k_e. Because of the finite natural spectral bandwidth of the two-photon state, we may let $\omega_1 = \Omega_o + \nu$ and $\omega_2 = \Omega_e - \nu$, where $|\nu| \ll \Omega_{o,e}$. Now, expand k_1 and k_2 to the first order in ν using the dispersion relations,

$$k_1 = k_o + \nu(dk_o/d\Omega_o) = k_o + \nu/u_o,$$

$$k_2 = k_e - \nu(dk_e/d\Omega_e) = k_e - \nu/u_e,$$

where u_o (u_e) is the group velocity for the ordinary (extraordinary) ray. Equation 5 can be written as

$$\Psi(\nu) = [1 - \exp(-i\nu DL)]/i\nu D, \tag{6}$$

where $D \equiv (1/u_o) - (1/u_e)$ and we assume a negative crystal (BBO), that is, $u_e > u_o$.

The fields at the detectors 1 and 2 are given by

$$E_1^{(+)}(t) = \alpha_t \int d\omega \exp[-i\omega(t - \tau_1)]a_o(\omega),$$

$$E_2^{(+)}(t) = \alpha_r \int d\omega \exp[-i\omega(t - \tau_2)]a_e(\omega), \tag{7}$$

where a_j is the annihilation operator of the photons (j = o, e) and $\tau_i \equiv s_i/c$ (where s_i is the optical path length from the output surface of the BBO crystal to the i-th detector and c is the speed of light). We assume $\tau_1 = \tau_2$ for the following discussion. Also, α_t and α_r are the complex transmission and reflection coefficients of the 50-50 beam-splitter. The average coincidence counting rate is given by

$$R_c = (1/T) \int\int_0^T dT_1\, dT_2 \langle\Psi|E_1^{(-)}E_2^{(-)}E_2^{(+)}E_1^{(+)}|\Psi\rangle$$

$$= (1/T) \int\int_0^T dT_1\, dT_2 |\Psi(t_1, t_2)|^2, \tag{8}$$

where $t_i \equiv T_i - \tau_i$, T_i is the detection time of the i-th detector, and T is the duration time of the measurement. An effective two-photon wave function $\Psi(t_1, t_2)$ is defined by equation 8 as

$$\Psi(t_1, t_2) = \langle 0|E_1^{(+)}E_2^{(+)}|\Psi\rangle. \tag{9}$$

Substituting equations 4 and 7 into equation 9,

$$\Psi(t_1, t_2) = v(t_1 + t_2)u(t_1 - t_2), \tag{10}$$

where

$$v(t) = v_0 \exp(-i\omega_p t/2), \tag{11}$$

$$u(t) = u_0 \exp(-i\omega_d t/2) \int_{-\infty}^{\infty} d\nu[1 - \exp(-\nu DL)]/(i\nu DL) \exp(-i\nu t)$$

$$= \exp(-i\omega_d t/2)\Pi(t). \tag{12}$$

Here,

$$\Pi(t) = \begin{cases} u_0 & DL > t > 0 \\ 0, & \end{cases} \tag{13}$$

v_0 and u_0 are constants (normalization), and $\omega_d \equiv \Omega_o - \Omega_e$. We have approximated the pump to be a plane wave in the calculation. If the pump beam was taken to be a Gaussian with bandwidth σ_p, it is not difficult to show that the constant v_0 will be replaced by a Gaussian function, $v_0 \exp(-\sigma_p^2 t^2/8)$. In this case, we may write the effective wave function in the following form:

$$\Psi(t_1, t_2) = v_0 \exp[-\sigma_p^2(t_1 + t_2)^2/8]\Pi(t_1 - t_2) \exp(-i\Omega_o t_1) \exp(-i\Omega_e t_2), \quad \textbf{(14)}$$

which is a two-dimensional wave packet, referred to as the two-photon effective wave function or, for short, the *biphoton.*[19,20] It is not difficult to understand the physics of the effective two-photon wave function (equation 14). In this simplified setup, the o-ray goes to detector 1 and the e-ray goes to detector 2; the rectangular-shaped function, $\Pi(t_1 - t_2)$, means that if detector 2 is triggered at T_2, detector 1 will be triggered at a later time, but no later than $T_2 + DL$. The "joint triggering" probability at T_1 and T_2 is a constant during this period and is zero otherwise. The $v(t_1 + t_2)$ function has a finite width along the axis of $t_1 + t_2$. This describes the fact that the pair can be produced at any time when the pump wave packet covers the crystal (for a Gaussian pump beam with bandwidth σ_p). This represents a reasonable physical

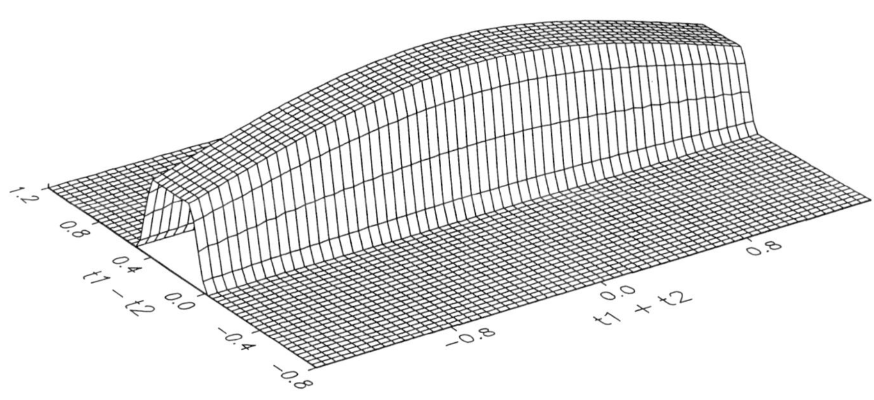

FIGURE 2. A schematic diagram of the biphoton (envelope).

picture for type-II BBO SPDC because each photon pair contains an o-ray and an e-ray. The pair is produced simultaneously in the crystal with equal probability along the z-axis. BBO is a negative uniaxial crystal so that the e-ray exits the BBO crystal first. The maximum possible time delay between the o-ray and the e-ray is $(L/u_o - L/u_e) = DL$, which is the time delay to cross the crystal. It is clear that the biphoton is entangled in space-time because the wave function cannot factor into a function of t_1 times a function of t_2. FIGURE 2 is a schematic diagram of the biphoton wave packet.

The experimental setup for the verification of the Π-shaped biphoton is almost the same as the above simplified experiment, except for the following: (1) the beam-splitter is polarization-independent, so both the o-ray and the e-ray could be transmitted or reflected to trigger D_1 or D_2; (2) a Glan Thompson linear polarization analyzer, oriented at 45° relative to the o-ray and e-ray polarization planes of the BBO crystal, is placed in front of each of the detectors; (3) birefringent material, for example, a set of quartz plates, is introduced into the single incident beam for manipulating the optical delay δ between the o-ray and the e-ray. The fast axes of the quartz plates were carefully aligned to match the o-ray or e-ray polarization planes of the BBO crystal. In order to see the natural spectral bandwidth of the SPDC, no narrow bandwidth spectral filters are used, except UV cutoff filters to cut off the pump scattered light.[21]

We first consider the case in which no quartz plates are used. The coincidence detection measurement realizes the probability amplitudes:

(1) *o-ray transmitted* $\otimes$ *e-ray reflected,*
(2) *e-ray transmitted* $\otimes$ *o-ray reflected.*

The effective wave function is easily calculated as

$$\Psi(t_1, t_2) = \alpha_t\alpha_r v(t_1 + t_2)[u(t_1 - t_2) - u(-t_1 + t_2)]. \tag{15}$$

It is interesting to see that the two terms in the effective wave function (equation 15) do not show any interference. Mathematically, it is easy to see that $\Pi(t_1 - t_2)$ and $\Pi(t_2 - t_1)$ do not overlap. Physically, the o-ray and the e-ray photons are well distinguished in space-time; it is impossible to find any detection time T_1 and T_2, except for $T_1 = T_2$, in which the two terms in equation 15 are both nonzero. If $T_1 > T_2$, the detection only records the amplitude (o-ray to D_1) $\otimes$ (e-ray to D_2); if $T_1 < T_2$, the detection only records the amplitude (e-ray to D_1) $\otimes$ (o-ray to D_2); and the chances of having the o-ray and e-ray photons exit from the BBO at $t_1 = t_2$ (corresponds to $T_1 = T_2$) is infinitesimal.

Now, consider the case of having a quartz plate in the downconversion incident beam. If we align the quartz carefully to match its fast axis to the o-ray polarization direction of the BBO, an optical delay, $\delta \cong (n_o - n_e)l/c$, is introduced between the o-ray and the e-ray of BBO, where n_o and n_e are the indices of refraction of the quartz plates for the o-ray and the e-ray of BBO, and l is the thickness of the quartz plate. The effective wave function becomes (consider that the analyzers are set at 45°)

$$\Psi(t_1, t_2) = \alpha_t\alpha_r v(t_1 + t_2 - \varphi)[u(t_1 - t_2 + \delta) - u(-t_1 + t_2 + \delta)], \tag{16}$$

where $\varphi \cong (n_o + n_e)l/c$ is a phase constant that has no contribution to the coincidence measurement after the normal square of $\Psi(t_1, t_2)$. It is easy to see from equation 16 that there is interference now because the two terms overlap. When $\delta = DL/2$, the two terms completely overlap and therefore cancel each other. This may be considered as a perfect anticorrelation.

If different numbers of quartz plates are introduced into the downconversion incident beam to manipulate the optical delay δ, the coincidence counting rate R_c is calculated by substituting equation 16 into equation 8:

$$R_c = R_{c0}[1 - \rho(\delta)], \tag{17}$$

where ρ is a Λ-shape function of δ (self-convolution of the Π-shape function),

$$\rho = \begin{cases} 0 & -\infty < \delta < 0 \\ \kappa\delta & 0 < \delta \leq DL/2 \\ 1 - \kappa(\delta - DL/2) & DL/2 \leq \delta < DL \\ 0 & DL < \delta < \infty, \end{cases} \qquad (17')$$

and $\kappa \equiv 2/DL$, resulting in a V-shaped R_c. It is easy to see that the width and the shape of the biphoton can be evaluated by the width and the shape of R_c.

FIGURE 3 reports typical observed "V-shape" coincidence rate measurements as a function of the optical delay δ, which verifies the Π-shape effective wave function.[21] The coincidence counts are direct measurements, with no "accidental" subtractions or any other theoretical corrections. Each of the data points corresponds to different numbers of quartz plates remaining in the path of the downconversion incident beam. The negative sign stands for the case in which the slow axis of quartz is aligned with the o-ray polarization direction of the BBO. The solid curve is a calculated curve of equation 17. It is easy to find that the vertex of the V-shape function has a displacement of (72 ± 3) fs from zero, which corresponds to a time delay of $DL/2$ in a (0.56 ± 0.05)-mm BBO crystal.

In the above calculation, we only considered the case of the natural spectral

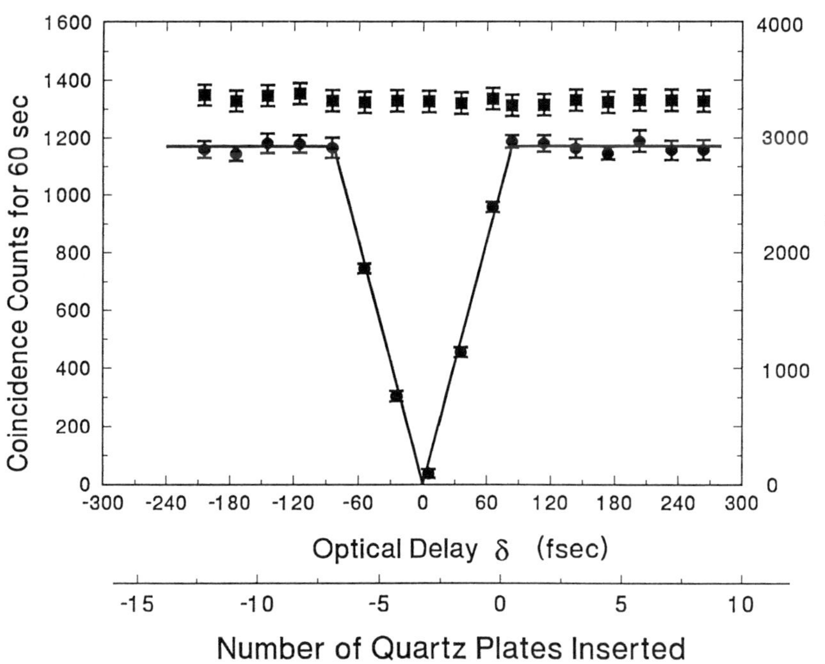

FIGURE 3. Observed "V-shape" coincidence rate measurements as a function of the optical delay δ, which verifies the Π-shape effective wave function. Lower curve: Coincidence counts as a function of the optical delay, which corresponds to a certain number of quartz plates. The solid curve is a fitting curve of equation 17. Upper curve: Single-detector counts.

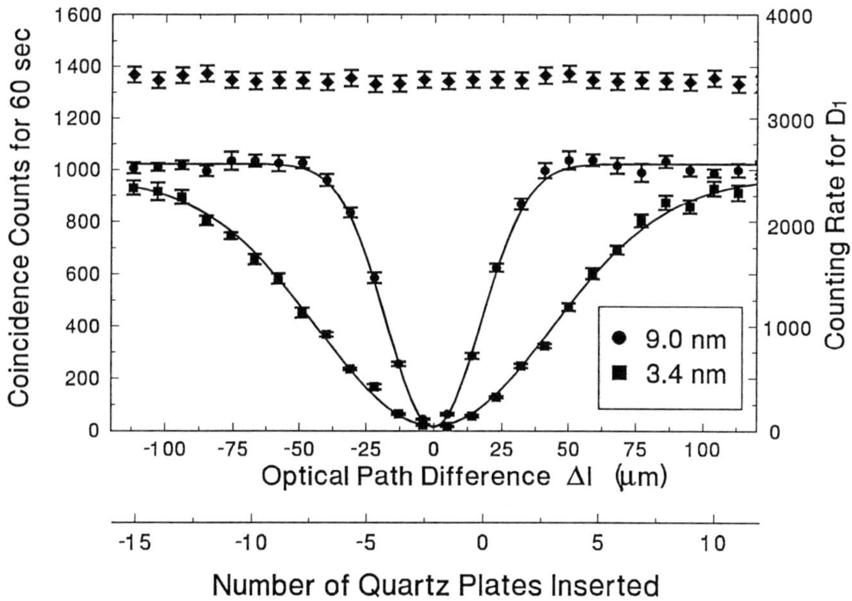

FIGURE 4. A typical "anticorrelation" measurement. Lower curves: Coincidence counts as a function of the optical path difference, which corresponds to a certain number of quartz plates. The solid curves are fitting curves of equation 21 for different bandwidths of the spectral filters. Upper curve: Single-detector counts.

width of the SPDC. If narrow bandwidth spectral filters are used for the detectors in the above experiment, the filter functions must be included in the field operators and $\Pi(t)$ becomes

$$\Pi(t) = \int_{-\infty}^{\infty} d\nu f_1(\Omega_1 + \nu) f_2(\Omega_2 - \nu)[1 - \exp(-\nu DL)]/(i\nu DL) \exp(-i\nu t). \quad \textbf{(18)}$$

The filter function may be taken to be Gaussian with bandwidth σ; then,

$$\Pi(t) = u_0\{\text{erf}(\sigma t/2) - \text{erf}[(\sigma t - DL)/2]\}/2DL, \quad \textbf{(19)}$$

where erf(x) is the error function and u_0 is a normalized constant. This function peaks at $DL/2$ and has a width on the order of $DL + 8/\sigma$. For narrow bandwidth filters, it can be approximated as a Gaussian,

$$\Pi(t) \cong u_0 \exp[-\sigma^2(t - DL/2)^2/4], \quad \textbf{(20)}$$

and the coincidence counting rate becomes

$$R_c = R_{c0}\{1 - \exp[-\sigma^2(\delta - DL/2)^2/2]\}. \quad \textbf{(21)}$$

FIGURE 4 reports typical measurements with the fitting curves of equation 21, which is an "anticorrelation" observation.[22]

The discussion of the effective wave function is important for the understanding of the two-photon double entanglement. The spin and space-time entanglement will be discussed in detail by reporting the following experiments.

EXPERIMENT ONE: TWO-PHOTON QUANTUM BEATS

In this experiment,[23] a pair of orthogonally polarized light quanta with different colors are injected collinearly through a single port of a beam-splitter and are detected by two independent photon counting detectors placed in the two output ports of the beam-splitter with two linear polarization analyzers and two narrow bandwidth spectral filters. The bandwidths of the filters are narrow enough so that each of the detectors only picks up one color. There is no preferred polarization orientation in each of the three (incident, transmitted, and reflected) beams. There is no single-detector counting rate change when the orientation of the polarization analyzers or the optical path difference between the orthogonally polarized components is manipulated. There is not even a coincidence counting rate change when manipulating the optical path difference without having polarizers in each of the transmitted and the reflected beams. However, a beating fringe with 100% modulation shows up in the coincidence counting rate by inserting crystal quartz plates in the incident beam with the help of correctly oriented polarizers in each of the transmitted and reflected beams. It is the spin and space-time entangled state (simultaneously) that makes it possible to demonstrate two-particle nonlocal interference in a simple beam-splitting experiment.

The schematic experimental setup is illustrated in FIGURE 5. A CW argon ion laser line of 351.1 nm is used to pump an 8 mm × 8 mm × (0.56 ± 0.05) mm BBO nonlinear crystal. The BBO is cut at a type-II phase matching angle to generate a pair of orthogonally polarized signal and idler photons collinearly around 702-nm wavelengths. The downconverted beam is separated from the pumping beam by a UV grade fused silica dispersion prism and is then directed collinearly at a near-normal incident angle to a polarization-independent beam-splitter that has 50%-50% reflection and transmission coefficients. A single-photon detector is placed in each transmission and reflection output port of the beam-splitter. The photon detectors are dry-ice-cooled avalanche photodiodes operated in photon counting Geiger mode. A Glan Thompson linear polarization analyzer, followed by a narrow bandwidth interference spectral filter, is placed in front of each of the detectors. The polarization analyzers are oriented at 45° to the o-ray polarization planes of the BBO crystal. The spectral filters f_1 and f_2 have Gaussian-shape transmission functions centered at conjugate wavelengths of the downconversion, $\lambda_1 = 700.7$ nm and $\lambda_2 = 703.7$ nm, respectively. The spectral bandwidths are both 1 nm in full width at half-maximum. The passband overlap between the two filters is negligible. Because of the finite bandwidth of the type-II parametric downconversion, both wavelengths may be o-ray or e-ray; however, if λ_1 is an o-ray, then λ_2 must be an e-ray and vice versa. The output pulses of the detectors are then sent to a coincidence circuit with a 3-ns coincidence time window. The two detectors are separated by about 2 m so that the detection events of the pair are spacelike separated events.

A set of 15 crystal quartz plates is placed in the incident beam for manipulating the optical path difference, Δl, between the orthogonally polarized signal and idler photons. The fast axes of the quartz plates were carefully aligned to match the o-ray or e-ray polarization planes of the BBO crystal during the measurements. Each of the quartz plates is (1 ± 0.1) mm in thickness, resulting in an optical path difference of $(n_e - n_o)l \cong 9$ μm between the o-ray and e-ray of the BBO. The 15 quartz plates were aligned carefully one by one before the taking of data and were moved away one

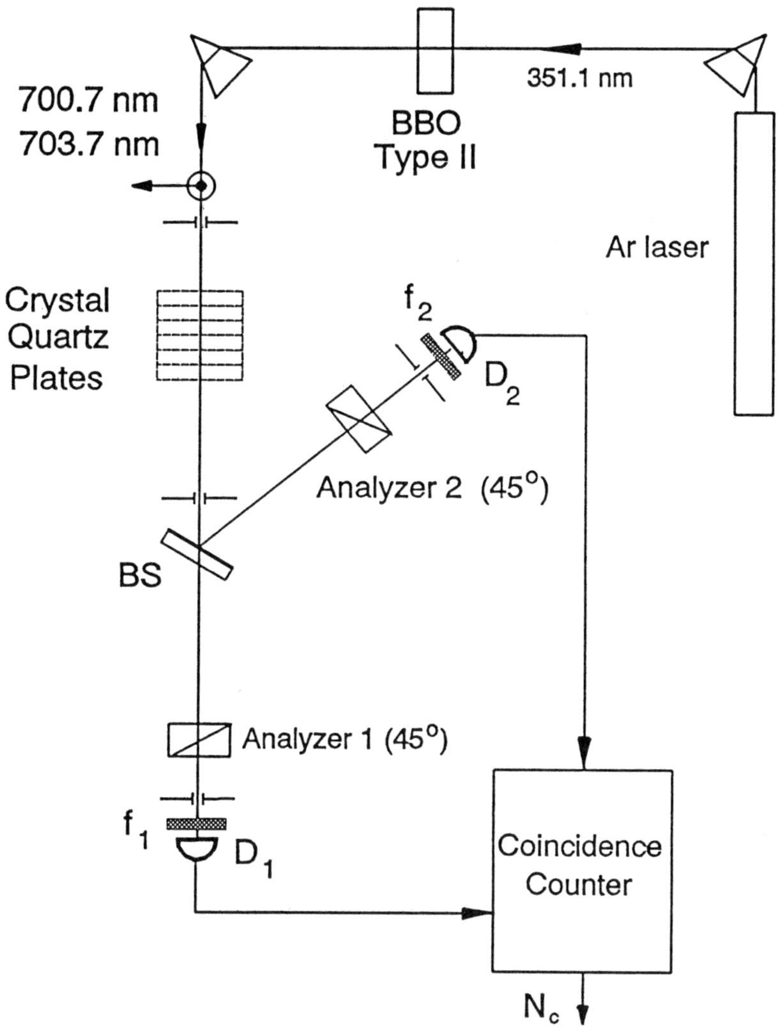

FIGURE 5. Schematic experimental setup for quantum beats.

by one during the measurements. Two sets of measurements were made in order to have 31 experimental points to show the modulation of the interference. In the first (second) set, we aligned the fast axes of the quartz plates to match the o-ray (e-ray) polarization plane of the BBO. The two-photon coincidence counting rate was observed to show a frequency beating fringe when the polarization analyzers were oriented at 45°:

$$R_c \cong R_0[1 - \cos(2\pi/\lambda_1 - 2\pi/\lambda_2)(\Delta l)], \tag{22}$$

where λ_i is the center wavelength of the i-th spectral filter and Δl is the optical path difference introduced by the quartz plates.

FIGURE 6 reports typical observed two-photon coincidence rate measurements as a function of Δl.[23] The coincidence counts are direct measurements, with no "accidental" subtractions or any other theoretical corrections. Each of the data points corresponds to a different number of quartz plates remaining in the incident

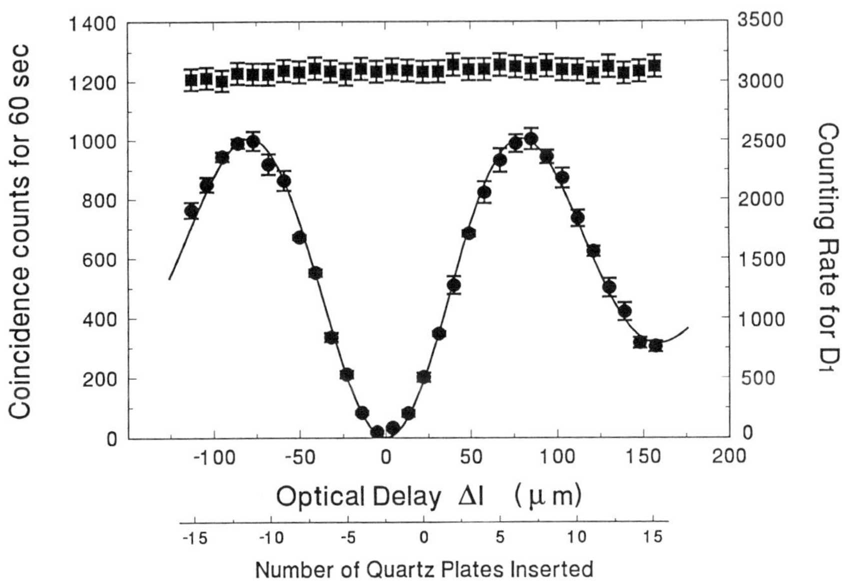

FIGURE 6. Observed beating pattern. Lower curve: Coincidence counts in 60 seconds as a function of Δl, which corresponds to a certain number of quartz plates. The solid curve is a fitting curve of equation 25. The measured period is about 164 μm, which corresponds to the beating frequency, 1.83×10^{12} Hz, of the downconverted beams. The modulation visibility is (97 ± 2)% (no "accidental subtractions"). Upper curve: Single-detector counting rate as a function of Δl.

beam. The left (right) side's 15 points of data, which are indicated by the "−" sign ("+" sign), were taken under the following condition: the fast axes of the quartz plates coincide with the o-ray (e-ray) polarization plane of the BBO crystal. It is clear that the interference pattern has a period of about 164 μm, which corresponds to the beating frequency, 1.83×10^{12} Hz, of the downconverted beams. The modulation visibility is (97 ± 2)%. The single-detector counting rates do not show any modulations, as is reported in the upper part of FIGURE 6.

It is straightforward to calculate the effective wave function $\Psi(t_1, t_2)$:

$$\Psi(t_1, t_2) = v(t_1 + t_2 - \varphi)[\hat{\mathbf{e}}_1 \cdot \hat{\mathbf{e}}_o \hat{\mathbf{e}}_2 \cdot \hat{\mathbf{e}}_e u(t_1 - t_2 + \delta) + \hat{\mathbf{e}}_1 \cdot \hat{\mathbf{e}}_e \hat{\mathbf{e}}_2 \cdot \hat{\mathbf{e}}_o u(-t_1 + t_2 + \delta)], \quad \textbf{(23)}$$

where $\alpha_t \alpha_r$ has been absorbed to the normalization constant. The time delay δ and the phase factor φ have been defined in the previous section. We have used a right-handed natural coordinate system with respect to the **k** vector as the positive z-axis direction. Care has to be taken to follow the rules of the natural coordinate system, especially for the reflected beam.

The effective wave function $\Psi(t_1, t_2)$ can also be written in the following form in order to compare it with the two-photon EPR states (equations 1 and 2):

$$\Psi(t_1, t_2) = [\hat{\mathbf{e}}_1 \cdot \hat{\mathbf{e}}_o \hat{\mathbf{e}}_2 \cdot \hat{\mathbf{e}}_e \mathrm{A}(t_1^o, t_2^e) + \hat{\mathbf{e}}_1 \cdot \hat{\mathbf{e}}_e \hat{\mathbf{e}}_2 \cdot \hat{\mathbf{e}}_o \mathrm{A}(t_1^e, t_2^o)]. \quad \textbf{(23')}$$

The t_i values are given by $t_i^o = t - l_i^o/c$ and $t_i^e = t - l_i^e/c$ $(i = 1, 2)$, where $l_i^{o,e} = \int dz n^{o,e}(z)$ indicates the optical path for the o-ray or e-ray of the i-th beam, with $n^{o,e}(z)$ being the refractive index at position z ($z = 0$ is defined at the output surface of the SPDC crystal). We have approximated $(dn/d\omega)_o - (dn/d\omega)_e \cong 0$ for simplifying the calculation.

The two terms in equations 23 and 23′ correspond to two probability amplitudes:

(1) *o-ray transmitted with long path ⊗ e-ray reflected with short path,*
(2) *e-ray transmitted with short path ⊗ o-ray reflected with long path.*

The effective wave function (equation 23) indicates two-particle entanglement in both spin and space-time.

The coincidence counting rate is calculated as follows (considering the coincidence time window):

$$R_c = (1/T) \int \int_0^T dT_1\, dT_2 |\Psi(t_1, t_2)|^2 S(\tau, \Delta T_c), \quad \textbf{(24)}$$

where $\tau \equiv T_1 - T_2$, T_i is the detection time of the i-th detector, $S(\tau, \Delta T_c)$ is a function that describes the coincidence circuit, and ΔT_c is the time window of the coincidence circuit. For $\tau > \Delta T$, $S(\tau, \Delta T_c) \cong 0$; for $\tau < \Delta T$, $S(\tau, \Delta T_c) \cong 1$. If the coincidence time window is large enough, S can be considered to be equal to 1 at time t. The time integral can be taken to infinity as a good approximation.

Consider the case $\theta_1 = \theta_2 = 45°$; substituting $\Psi(t_1, t_2)$ into equation 24, the coincidence counting rate is calculated as

$$R_c = R_{c0}\{1 - \exp[-\sigma^2(\delta - DL/2)^2/2] \cos(\Omega_1 - \Omega_2)\delta\}. \quad \textbf{(25)}$$

Equation 25 indicates an interference beating of the coincidence counting rate at frequency $\Omega_1 - \Omega_2$ with 100% modulation when δ is near zero. The modulation visibility vanishes exponentially when δ is off from zero. The solid curve in FIGURE 6 is a fitting curve of equation 25.

It is interesting that the zero point of the counting rate is not seen in the case of zero quartz plates. From the experimental data, we conclude that the zero point of the coincidence counting rate happens at a point in which about 2.4 × (1 ± 0.1)–mm quartz plates are placed in the beam path, which corresponds to the time delay of $DL/2$ in a (0.56 ± 0.05)-mm BBO crystal. This is in agreement with the Π-shape effective wave function verification experiment reported in the previous section.

The above quantum beating phenomenon demonstrated a two-photon entangled state in spin and in space-time generated by type-II SPDC. The simultaneous double entanglement may be easily seen from the following argument: neither the transmitted nor the reflected beam has a preferred polarization or a preferred optical path (short or long path in the quartz plates); however, if the photon in the transmitted beam is measured to be polarized at θ_1, the photon in the reflected beam must have been polarized at a specific angle θ_2. At the same time, if the photon that triggered detector 1 is the one that took a shorter optical path in the quartz, then the one that triggered detector 2 must have been the one that took a longer path, and vice versa. The measurement of the observables of either particle (spin and space-time) determines the value of these observables for the other particle with unit probability. The (97 ± 2)% visibility coincidence interference pattern is the signature of this particular double entanglement.

EXPERIMENT TWO: A NEW TYPE OF TWO-PHOTON INTERFEROMETER

Taking advantage of the spin and space-time entanglement of the biphoton, another type of two-photon interference phenomenon can be demonstrated.[24] The experimental setup is similar to the quantum beats experiment, that is, a simple beam-splitting setup. With the addition of a Pockels cell and a reorientation of the quartz plates and polarizers, the coincidence counting rate exhibits interference modulation of the pump frequency when manipulating the voltage across the Pockels cell, regardless of the optical delay by the quartz plates (which is much greater than the coherence length of the signal and idler downconversion fields) and the mutual incoherence between the signal and the idler of the downconversion. This two-photon interference effect is again due to a nonclassical two-photon state that is entangled both in spin and in space-time.

The schematic setup of the experiment is illustrated in FIGURE 7. The type-II SPDC is the same as that in the quantum beats experiment, which is reported in the previous section. The collinear downconversion beam passes through a set of crystal quartz plates before the beam-splitter. The first 3 quartz plates, which sum to 2.4 mm in thickness, are oriented in such a way that the fast polarization plane of the quartz plates coincides with the o-ray polarization plane of the BBO to make the two terms of the Π-shape function completely overlap (see the discussion in the previous two sections). Thirteen more crystal quartz plates follow these 3. The fast axes of these 13 quartz plates are aligned carefully so that they are oriented at 45° relative to the o-ray and the e-ray polarization planes of the BBO. Each of these quartz plates is (1 ± 0.1) mm in thickness, resulting in an optical delay of $\Delta l \cong 9$ μm between the fast and the slow rays of the quartz crystal at wavelengths around 700 nm. The optical delay is about 117 μm after 13 quartz plates, which should be compared with the coherence length of the field, which is about 25 μm. Therefore, the $|X>$ and the $|Y>$ components of the o-ray and e-ray of the downconversion suffer enough optical delay to be nonoverlapping, where $|X>$ and $|Y>$ correspond to the fast and the slow axes of the quartz plates. A Pockels cell with fast and slow axes carefully aligned to match the $|X>$ and the $|Y>$ axes is placed after the quartz plates for fine control of the optical delay between the $|X>$ and the $|Y>$. The quartz plates and the Pockels cell all have antireflection coatings at 702.2 nm. The pair is then injected at a near-normal incident angle to a polarization-independent beam-splitter that has 50%-50% reflection and transmission coefficients, within ±5% accuracy. The detector packages are the same as those used in the quantum beats experiment. The polarization analyzers are oriented at 0° relative to the o-ray polarization planes of the BBO crystal. The spectral filters f_1 and f_2 have Gaussian-shape transmission functions centered at 702.2 nm, with bandwidths of 19 nm (full width at half-maximum). The output pulses of the detectors are then sent to a coincidence circuit with a 3-ns coincidence time window. In order to have spacelike separated detection events, the two detectors are separated by about 2 m.

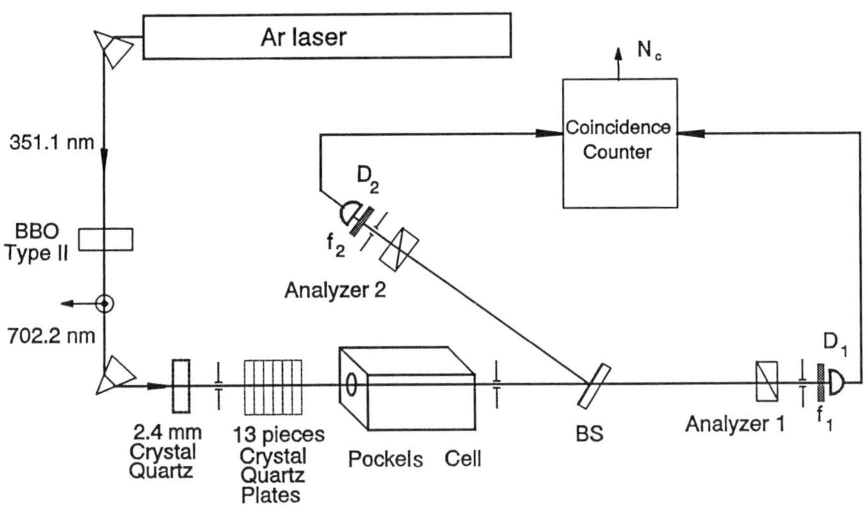

FIGURE 7. Schematic setup for the new type of two-photon interferometer.

FIGURE 8 reports a typical observed interference modulation of the coincidence counting rate as a function of the optical delay Δl_P, where Δl_P is the optical delay (between $|X>$ and $|Y>$) introduced by the Pockels cell.[24] The manipulation of Δl_P is realized by changing the applied voltage of the Pockels cell. The half-wave voltage is calibrated at 702.2 nm. The coincidence counts are direct measurements, with no "accidental" subtractions. Each of the data points corresponds to a different voltage applied to the Pockels cell. The left-side (right-side) data points, which are indicated by a "−" sign ("+" sign), were taken by applying a negative (positive) voltage across the Pockels cell. It is clear that the modulation period corresponds to the pump wavelength, that is, 351.1 nm. The interference visibility is about (88 ± 2)%. Contrary to the coincidence counting rate, the single-detector counting rate remains constant when Δl_P is manipulated, as is reported in the upper part of FIGURE 8.

The downconversion $|o>$ and $|e>$ polarized photons both have certain probabilities to be in the $|X>$ or the $|Y>$ state when passing through the crystal quartz plates and the Pockels cell. The optical delay between the $|X>$ and the $|Y>$ is then introduced by the anisotropic refractive index of the quartz plates and the Pockels cell. When the pair meets the beam-splitter, both of the photons have a 50%-50% chance to be transmitted or to be reflected. The transmitted one is registered by

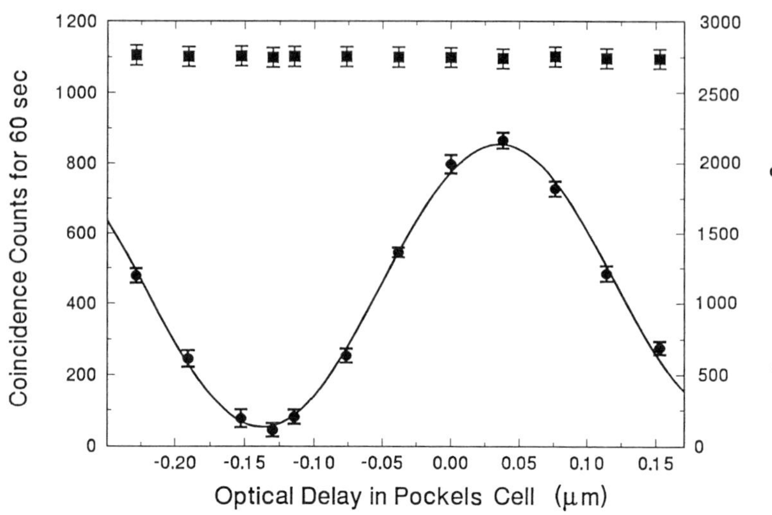

FIGURE 8. Observed sum frequency two-photon interference pattern. Lower part: Coincidence counts in 60 seconds as a function of the optical delay, Δl_P, which corresponds to a certain voltage applied to the Pockels cell. The measured oscillation period is about 351 nm with an (88 ± 2)% visibility. Upper part: Single-detector counting rate remains constant when Δl_P is manipulated.

detector 1 and the reflected one is detected by detector 2. It is interesting to see from the following discussion that the only registered coincidences are the two-photon probability amplitudes in which

(1) ($|o>$ is in $|X>$ and is transmitted) $\otimes$ ($|e>$ is in $|X>$ and is reflected),
(2) ($|e>$ is in $|X>$ and is transmitted) $\otimes$ ($|o>$ is in $|X>$ and is reflected),
(3) ($|o>$ is in $|Y>$ and is transmitted) $\otimes$ ($|e>$ is in $|Y>$ and is reflected),
(4) ($|e>$ is in $|Y>$ and is transmitted) $\otimes$ ($|o>$ is in $|Y>$ and is reflected).

All the other possible two-photon probability amplitudes cancel each other, regardless of the optical delay by 13 pieces of quartz plates (greater than the coherence length of the downconversion field) and the mutual incoherence between the signal and idler. This results in an EPR state. In the following paragraphs, we present a simple quantum mechanical model to explain our experiment, which is essentially the same as that in the previous two sections.

The fields at the detectors 1 and 2 are given by

$$\begin{aligned} E_1^{(+)}(t) = \alpha_t \int d\omega f(\omega)[&\exp(-i\omega t_{1X})(\hat{\mathbf{e}}_o \cdot \hat{\mathbf{e}}_X)(\hat{\mathbf{e}}_X \cdot \hat{\mathbf{e}}_1)a_o(\omega) \\ &+ \exp(-i\omega t_{1X})(\hat{\mathbf{e}}_e \cdot \hat{\mathbf{e}}_X)(\hat{\mathbf{e}}_X \cdot \hat{\mathbf{e}}_1)a_e(\omega) \\ &+ \exp(-i\omega t_{1Y})(\hat{\mathbf{e}}_o \cdot \hat{\mathbf{e}}_Y)(\hat{\mathbf{e}}_Y \cdot \hat{\mathbf{e}}_1)a_o(\omega) \\ &+ \exp(-i\omega t_{1Y})(\hat{\mathbf{e}}_e \cdot \hat{\mathbf{e}}_Y)(\hat{\mathbf{e}}_Y \cdot \hat{\mathbf{e}}_1)a_e(\omega)], \\ E_2^{(+)}(t) = \alpha_r \int d\omega f(\omega)[&\exp(-i\omega t_{2X})(\hat{\mathbf{e}}_o \cdot \hat{\mathbf{e}}_X)(\hat{\mathbf{e}}_X \cdot \hat{\mathbf{e}}_2)a_o(\omega) \\ &+ \exp(-i\omega t_{2X})(\hat{\mathbf{e}}_e \cdot \hat{\mathbf{e}}_X)(\hat{\mathbf{e}}_X \cdot \hat{\mathbf{e}}_2)a_e(\omega) \\ &+ \exp(-i\omega t_{2Y})(\hat{\mathbf{e}}_o \cdot \hat{\mathbf{e}}_Y)(\hat{\mathbf{e}}_Y \cdot \hat{\mathbf{e}}_2)a_o(\omega) \\ &+ \exp(-i\omega t_{2Y})(\hat{\mathbf{e}}_e \cdot \hat{\mathbf{e}}_Y)(\hat{\mathbf{e}}_Y \cdot \hat{\mathbf{e}}_2)a_e(\omega)], \end{aligned} \quad (26)$$

where $\hat{\mathbf{e}}_i$ is in the direction of the i-th linear polarization analyzer axis, $a_o(\omega)$ and $a_e(\omega)$ are the destruction operators for the o-ray and the e-ray of BBO, α_t and α_r are the complex transmission and reflection coefficients of the beam-splitter, and $f(\omega)$ is the spectral transmission function of the filters. The t_{iX} and t_{iY} are given by $t_{iX} = t - l_{iX}/c$ and $t_{iY} = t - l_{iY}/c$ ($i = 1, 2$), where $l_{iX,Y} = \int dz n_{X,Y}(z)$. X and Y, defined by the fast and slow axes of the quartz plates and the Pockels cell, indicate the optical paths for the X and Y components of the i-th beam, and $n_{X,Y}(z)$ is the refractive index at position z for the X or Y components of the beam. We have approximated $(dn/d\omega)_X - (dn/d\omega)_Y \cong 0$ for simplifying the calculation. The use of pinholes allows a good one-dimensional approximation.

It is straightforward to calculate $\Psi(t_1, t_2)$:

$$\begin{aligned} \Psi(t_1, t_2) = \alpha_t\alpha_r[&(\hat{\mathbf{e}}_o \cdot \hat{\mathbf{e}}_X)(\hat{\mathbf{e}}_X \cdot \hat{\mathbf{e}}_1)(\hat{\mathbf{e}}_e \cdot \hat{\mathbf{e}}_X)(\hat{\mathbf{e}}_X \cdot \hat{\mathbf{e}}_2)A(t_{1X}, t_{2X}) \\ &+ (\hat{\mathbf{e}}_e \cdot \hat{\mathbf{e}}_X)(\hat{\mathbf{e}}_X \cdot \hat{\mathbf{e}}_1)(\hat{\mathbf{e}}_o \cdot \hat{\mathbf{e}}_X)(\hat{\mathbf{e}}_X \cdot \hat{\mathbf{e}}_2)A(t_{1X}, t_{2X}) \\ &+ (\hat{\mathbf{e}}_o \cdot \hat{\mathbf{e}}_Y)(\hat{\mathbf{e}}_Y \cdot \hat{\mathbf{e}}_1)(\hat{\mathbf{e}}_e \cdot \hat{\mathbf{e}}_Y)(\hat{\mathbf{e}}_Y \cdot \hat{\mathbf{e}}_2)A(t_{1Y}, t_{2Y}) \\ &+ (\hat{\mathbf{e}}_e \cdot \hat{\mathbf{e}}_Y)(\hat{\mathbf{e}}_Y \cdot \hat{\mathbf{e}}_1)(\hat{\mathbf{e}}_o \cdot \hat{\mathbf{e}}_Y)(\hat{\mathbf{e}}_Y \cdot \hat{\mathbf{e}}_2)A(t_{1Y}, t_{2Y}) \\ &+ (\hat{\mathbf{e}}_o \cdot \hat{\mathbf{e}}_X)(\hat{\mathbf{e}}_X \cdot \hat{\mathbf{e}}_1)(\hat{\mathbf{e}}_e \cdot \hat{\mathbf{e}}_Y)(\hat{\mathbf{e}}_Y \cdot \hat{\mathbf{e}}_2)A(t_{1X}, t_{2Y}) \\ &+ (\hat{\mathbf{e}}_e \cdot \hat{\mathbf{e}}_X)(\hat{\mathbf{e}}_X \cdot \hat{\mathbf{e}}_1)(\hat{\mathbf{e}}_o \cdot \hat{\mathbf{e}}_Y)(\hat{\mathbf{e}}_Y \cdot \hat{\mathbf{e}}_2)A(t_{1X}, t_{2Y}) \\ &+ (\hat{\mathbf{e}}_o \cdot \hat{\mathbf{e}}_Y)(\hat{\mathbf{e}}_Y \cdot \hat{\mathbf{e}}_1)(\hat{\mathbf{e}}_e \cdot \hat{\mathbf{e}}_X)(\hat{\mathbf{e}}_X \cdot \hat{\mathbf{e}}_2)A(t_{1Y}, t_{2X}) \\ &+ (\hat{\mathbf{e}}_e \cdot \hat{\mathbf{e}}_Y)(\hat{\mathbf{e}}_Y \cdot \hat{\mathbf{e}}_1)(\hat{\mathbf{e}}_o \cdot \hat{\mathbf{e}}_X)(\hat{\mathbf{e}}_X \cdot \hat{\mathbf{e}}_2)A(t_{1Y}, t_{2X})]. \end{aligned} \quad (27)$$

To simplify the calculation, $A(t_1, t_2)$ may be taken to be

$$A(t_1, t_2) = A_0 \exp[-\sigma_p^2(t_1 + t_2)^2/8] \exp[-\sigma^2(t_1 - t_2 - DL/2)^2/4] \cdot \exp(-i\Omega_1 t_1) \exp(-i\Omega_2 t_2), \quad (28)$$

where we consider narrow bandwidth Gaussian filters $f_i(\omega)$ with equal bandwidth σ and a Gaussian spectral distribution of the pump field with bandwidth σ_P and where Ω_i is the *i*-th filter's center frequency, which is related to the peak frequency of the pump, Ω_P, by $\Omega_1 + \Omega_2 = \Omega_P$. Considering the phase shift of π due to reflection and the sign of the projections as well as the 45° orientations of the quartz plates and the 0° orientation of the analyzers, it is straightforward to show that the first four terms can be combined into two terms, which correspond to the two-photon probability amplitudes in equation 29, where the Y components (of the o or e) take a longer optical path due to the anisotropic refractive index of the quartz plates and the Pockels cell. It is also straightforward and very interesting to see that other terms in equation 27 are *out of phase* in pairs (terms 5–6 and 7–8) and *cancel each other.* The cancellation makes the interference visibility greater than 50%. We want to remind the reader once again that we have used a right-handed natural coordinate system with respect to the **k** vector as the positive z-axis direction. To realize the state of the type in equation 2, we have taken advantage of the polarization entanglement of the two-photon state for the cancellation of the unwanted amplitudes. An EPR entangled two-photon state is finally realized by the coincidence measurement,

$$\Psi_{EPR} = A(t_{1X}, t_{2X}) - A(t_{1Y}, t_{2Y}), \tag{29}$$

which is similar to the EPR states in equations 1 and 2, with X and Y equivalent to the long and short paths of the interferometer, or the polarizations.

The coincidence counting rate is

$$R_c = R_{c0}\{1 - \exp[-\sigma_P^2(\Delta l/2c)^2] \cos(\omega_P \Delta l/c)\}, \tag{30}$$

which indicates an interference modulation of the pump frequency with 100% visibility when Δl is much smaller than the coherence length of the pump in perfect experimental conditions. The observed visibility is (88 ± 2)%. The reason that we see less than 100% is mainly due to the nonlinear response and the optical quality of our Pockels cell when applying higher voltages.

We have also demonstrated a similar experiment with an orthogonal circularly polarized light quanta pair. The experimental setup is almost the same, except that (1) a quarter-wave plate is placed after the compensation quartz plates; the quarter-wave plate is oriented at 45° relative to the o-ray polarization plane of the BBO to rotate the linear polarization state of the o-ray and the e-ray to an orthogonal circular polarization (left-hand, right-hand) configuration; (2) 13 crystal quartz plates are used to make a large enough delay between the $|x\rangle$ and $|y\rangle$ components of the circular polarized pair, where $|x\rangle$ and $|y\rangle$ are defined by the fast and the slow axes of the quartz plates; $|x\rangle$ and $|y\rangle$ are oriented to coincide with the o-ray and the e-ray polarization planes of the BBO; (3) a Pockels cell is placed after the quartz plates for fine control of the optical delay between $|x\rangle$ and $|y\rangle$; the fast and the slow axes of the Pockels cell are aligned to match $|x\rangle$ and $|y\rangle$; (4) the polarization analyzers are both oriented at 45° relative to $|x\rangle$. The differences of the two configurations are illustrated in FIGURE 9. The coincidence counting rate for the circular configuration exhibits similar interference as that reported in FIGURE 8.

EXPERIMENT THREE: BELL INEQUALITY MEASUREMENTS

In order to have spacelike separated manipulations of the optical path delays, the single Pockels cell before the beam-splitter is replaced by two Pockels cells placed in the transmission and the reflection beams of the beam-splitter; see FIGURE 10.[25] The coincidence time window in this experiment is 1.8 ns, which is much shorter than the distance between the Pockels cells. Bell inequality measurements can be performed

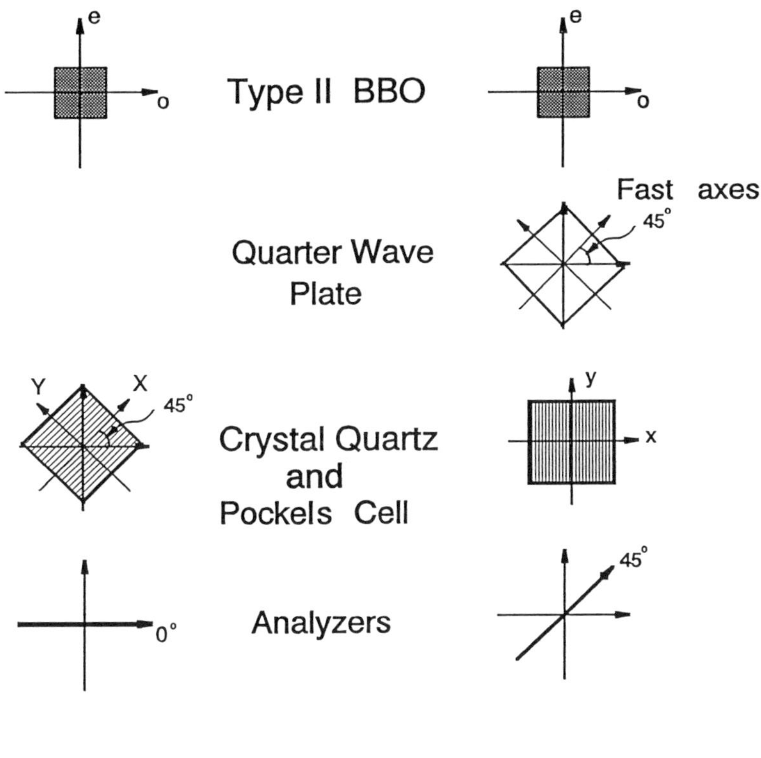

FIGURE 9. Experimental setup differences between the linear and circular configurations.

both for space-time variables and for spin variables in one experiment. Following the calculations in the previous section, for 45°-oriented polarizers, the coincidence counting rate is predicted to be

$$R_c = R_{c0}\{1 - \exp[-\sigma_P^2(\Delta l/2c)^2]\cos(\Omega_1\Delta l_1 + \Omega_2\Delta l_2)/c\}, \tag{31}$$

where $\Delta l_i/c$ is the optical delay introduced by the i-th Pockels cell (to simplify the calculation, we assumed that the optical delays introduced by the Pockels cells are the same).

FIGURE 11 reports a typical measurement of the coincidence rate. The modulation visibility is $(88.2 \pm 1.2)\%$, which violates a Bell inequality for space-time variables (visibility $\leq 70.7\%$) by more than 14 standard deviations.[25]

It is interesting to see that, in the same experiment, a test of a spin variable Bell inequality can be made by manipulating the orientation of the polarizers at a totally constructive or destructive space-time interference point. Because of the symmetries present in the measurement, we are able to study one of the Bell inequalities,[26]

$$\delta = |[R_c(\pi/8) - R_c(3\pi/8)]/R_0| \leq 1/4, \tag{32}$$

and the measured result is $\delta = 0.309 \pm 0.009$, implying a violation of more than 6 standard deviations.[25]

CONCLUSIONS

Experiments starting with type-II downconversion are a very effective mechanism for generating two-photon entangled states (biphoton). The type-II SPDC biphoton is entangled both in space-time and in spin. The double entanglement of the two-photon state makes it possible to perform EPR-type two-photon interference experiments in a simple beam-splitting setup and to test Bell inequalities for space-time variables and spin variables in the same experiment. Two-photon interference visibilities as high as $(97 \pm 2)\%$ have been observed. Experimental tests for the space-time variables and spin variables of Bell inequalities have been measured with violations of 14 and 6 standard deviations, respectively, in one experimental setup.

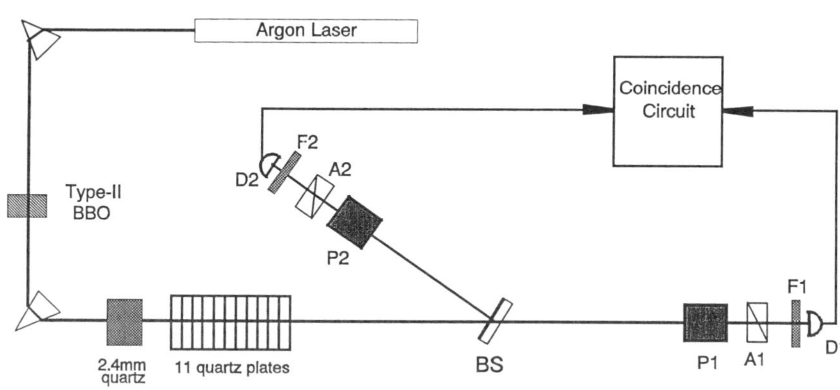

FIGURE 10. Schematic experimental setup for Bell inequality measurements.

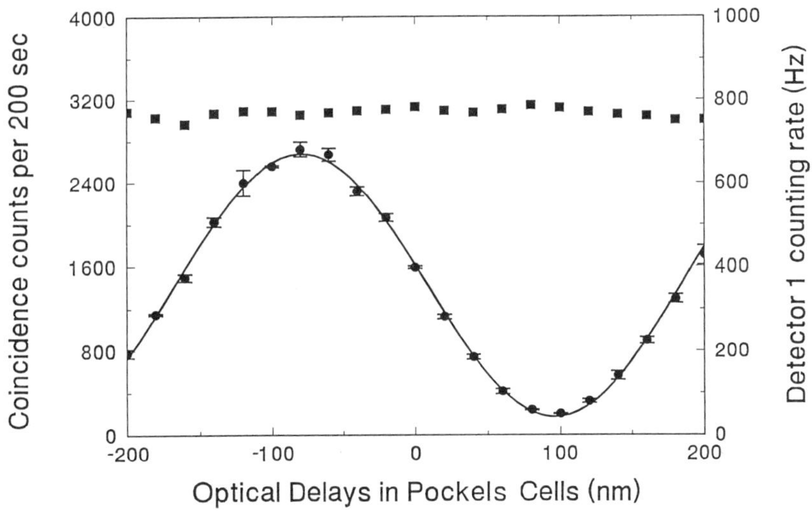

FIGURE 11. A typical observed sum frequency modulation when the cross voltages of the two Pockels cells are manipulated (negative values correspond to negative voltages). The interference visibility is (88.2 ± 1.2)%, which violates a Bell inequality for space-time variables by more than 14 standard deviations.

ACKNOWLEDGMENTS

We wish to thank C. O. Alley and D. N. Klyshko for useful discussions. We thank J. A. Wheeler for many years of encouraging this research.

REFERENCES AND NOTES

1. EINSTEIN, A., B. PODOLSKY & N. ROSEN. 1935. Phys. Rev. **45:** 777.
2. BELL, J. S. 1964. Physics **1:** 195.
3. SCHRÖDINGER, E. 1935. Naturwissenschaften **23:** 807, 823, 844. A translation of these papers appears in: WHEELER, J. A. & W. H. ZUREK, Eds. 1983. Quantum Theory and Measurement. Princeton University Press. Princeton, New Jersey.
4. HORNE, M. A., A. SHIMONY & A. ZEILINGER. 1989. Phys. Rev. Lett. **62:** 2209.
5. BOHM, D. 1951. Quantum Theory. Prentice–Hall. Englewood Cliffs, New Jersey.
6. For a review, see: CLAUSER, J. F. & A. SHIMONY. 1976. Rep. Prog. Phys. **41:** 1981.
7. ASPECT, A., P. GRANGIER & G. ROGER. 1981. Phys. Rev. Lett. **47:** 460; ASPECT, A., P. GRANGIER & G. ROGER. 1982. Phys. Rev. Lett. **49:** 91; ASPECT, A., J. DALIBARD & G. ROGER. 1982. Phys. Rev. Lett. **49:** 1804.
8. SHIH, Y. H. & C. O. ALLEY. 1988. Phys. Rev. Lett. **61:** 2921.
9. OU, Z. Y. & L. MANDEL. 1988. Phys. Rev. Lett. **61:** 50.
10. KIESS, T. E., Y. H. SHIH, A. V. SERGIENKO & C. O. ALLEY. 1993. Phys. Rev. Lett. **71:** 3893.
11. OU, Z. Y. & L. MANDEL. 1988. Phys. Rev. Lett. **61:** 54.
12. KWIAT, P. G., A. M. STEINBERG & R. Y. CHIAO. 1993. Phys. Rev. A **47:** 2472.
13. SHIH, Y. H., A. V. SERGIENKO & M. H. RUBIN. 1993. Phys. Rev. A **47:** 1288.
14. Two pioneer experiments with interference visibilities less than 50% (the long-short and short-long amplitudes were not "cut off" by the coincidence circuits). See: KWIAT, P. G.,

W. A. VAREKA, C. K. HONG, H. NATHEL & R. Y. CHIAO. 1990. Phys. Rev. A **41:** 2910; OU, Z. Y., X. Y. ZOU, L. J. WANG & L. MANDEL. 1990. Phys. Lett. **65:** 321.
15. BRENDEL, J., E. MOHLER & W. MARTIENSSEN. 1991. Phys. Rev. Lett. **66:** 1142.
16. LARCHUK, T. S., R. A. CAMPOS, J. G. RARITY, P. R. TAPSTER, E. JAKEMAN, B. E. A. SALEH & M. C. TEICH. 1993. Phys. Rev. Lett. **70:** 1603.
17. WU, C. S. & I. SHAKNOV. 1950. Phys. Rev. **77:** 136.
18. FRANSON, J. D. 1989. Phys. Rev. Lett. **62:** 2205.
19. KLYSHKO, D. N. 1988. Photons and Nonlinear Optics. Gordon & Breach. New York.
20. For detailed theoretical analysis, see: RUBIN, M. H., D. N. KLYSHKO, Y. H. SHIH & A. V. SERGIENKO. 1994. Phys. Rev. A. To be published.
21. SHIH, Y. H., A. V. SERGIENKO & M. H. RUBIN. 1994. Phys. Rev. To be published.
22. SHIH, Y. H. & A. V. SERGIENKO. 1994. Phys. Lett. A **186:** 29.
23. SHIH, Y. H. & A. V. SERGIENKO. 1994. Phys. Rev. A **50:** 2564.
24. SHIH, Y. H. & A. V. SERGIENKO. 1994. Phys. Lett. A **194:** 201.
25. PITTMAN, T. B., Y. H. SHIH, A. V. SERGIENKO & M. H. RUBIN. 1994. Phys. Rev. Lett. Submitted.
26. FREEDMAN, S. J. & J. F. CLAUSER. 1972. Phys. Rev. Lett. **28:** 938.

Frustrated Downconversion: Virtual or Real Photons?[a]

H. WEINFURTER,[b] T. HERZOG,[b] P. G. KWIAT,[b,c]
J. G. RARITY,[d] A. ZEILINGER,[b,e] AND M. ZUKOWSKI[f]

[b]*Institut für Experimentalphysik*
Universität Innsbruck
A-6020 Innsbruck, Austria

[d]*DRA Malvern*
Malvern, Worcestershire WR14 3PS, United Kingdom

[f]*Instytut Fizyki Teoretycznej i Astrofizyki*
Uniwersytet Gdanski
PL-80-952 Gdansk, Poland

Correlated photon pairs produced by parametric downconversion have become a major tool in demonstrating fundamental properties of quantum mechanics. Nonclassical interference effects, that is, effects not describable by classical Maxwellian electrodynamics, have been used, for example, in experiments on the nonlocality of quantum mechanics,[1] on nonclassical photon statistics,[2] and on superluminal effects in tunneling experiments.[3] However, in nearly all those experiments, only second-order interference can be observed by correlated detection of the photon pair.[4] In contrast, the experiment described here demonstrates the possibility of manipulating downconversion such that one has enhancement or suppression of the spontaneous two-photon emission process.[5] These directly observable intensity variations result from nonclassical first-order interference. However, this interference clearly shows the product-state nature of the photon pair.

Most significantly, our experiment closely resembles the case of a spontaneously decaying atom close to a mirror, where modification of the emission process has already been observed.[6] There, the light emitted by the atom could reach the detectors via two different paths, thus causing constructive (i.e., enhanced emission) or destructive (suppressed emission) interference. For atom-mirror distances on the order of the emitted wavelength, one can achieve perfect inhibition of the emission process into the modes perpendicular to the mirror surface. In the description of this process, one introduces virtual photons, which are understood as being energetically indistinguishable from the vacuum field and therefore not detectable. Similarly, in our experiment, mirrors are used to reflect both the pump beam as well as a pair of downconverted light beams back into the crystal. For the case of destructive

[a]This work was supported by the Austrian "Fond zur Förderung der Wissenschaftlichen Forschung" through Grant No. S6502.

[c]P. G. Kwiat was supported by a "Lise-Meitner" fellowship.

[e]A. Zeilinger was supported by Grant No. PHY92-13964 from the National Science Foundation.

interference, the question arises as to whether any downconversion took place at all, in analogy with the suppressed decay of an atom decoupled from the vacuum field fluctuations due to the possible back-reflection from the mirror. An alternative explanation does not rely on the concept of virtual photons. Instead, it takes into account the fact that a pair of back-reflected photons could be spontaneously upconverted with a probability equal to the one for downconversion, when passing through the crystal. The predictions for the observable interference effects agree with those of the model described before. However, photons would now be immediately detectable in the downconverted modes between the crystal and the mirrors, even in the case of destructive interference.

Our conceptually simple setup allows one to freely manipulate the modes between the crystal and the mirrors, thereby changing the boundary conditions for the interference effect. In the present contribution, we will show how different Welcher-Weg and quantum eraser experiments can now be performed. Also, in contrast to the atom-mirror experiments, one can now try to place a detector (or a switch to a detector) between the crystal and the mirror to test for the existence of photons even in the case of destructive interference.

ENHANCEMENT AND FRUSTRATION OF THE DOWNCONVERSION PROCESS

In the process of downconversion (FIGURE 1a), a UV-pump photon is spontaneously converted into a pair of red photons (not necessarily of equal frequency). Energy and momentum conservation correlates the two photons, represented by the state,

$$|\Psi\rangle = \alpha|s_1\rangle|i_1\rangle + \cdots . \tag{1}$$

Here, α is the amplitude for downconversion (DC) into the idler (i) and signal (s) modes. Higher order terms are neglected because of the very low efficiency of the DC process ($|\alpha| \ll 1$). Also, the coherent pump beam can be treated as essentially unchanged, but still coherent with the photon pair. Therefore, there will be a definite phase relation between DC photons created in the first process and photons created in the process induced if one reflects the pump beam with a phase shift of ϕ_p back through the crystal (FIGURE 1b):

$$|\Psi\rangle = \alpha|s_1\rangle|i_1\rangle + \alpha e^{i\phi_p}|s_2\rangle|i_2\rangle. \tag{2}$$

If one now uses a mirror IM to overlap mode $|i_1\rangle$ (with a phase shift of ϕ_i) with $|i_2\rangle$ to give $|i\rangle$ (FIGURE 1c), a naive picture, neglecting that the phase of each of the DC photons is undetermined, would already predict first-order interference. However, a definite phase relation with the pump phase ϕ_p only exists for the whole product state. In other words, the remaining signal modes could be used for Welcher-Weg detection of the overlapping idler photons; interference effects are therefore excluded.

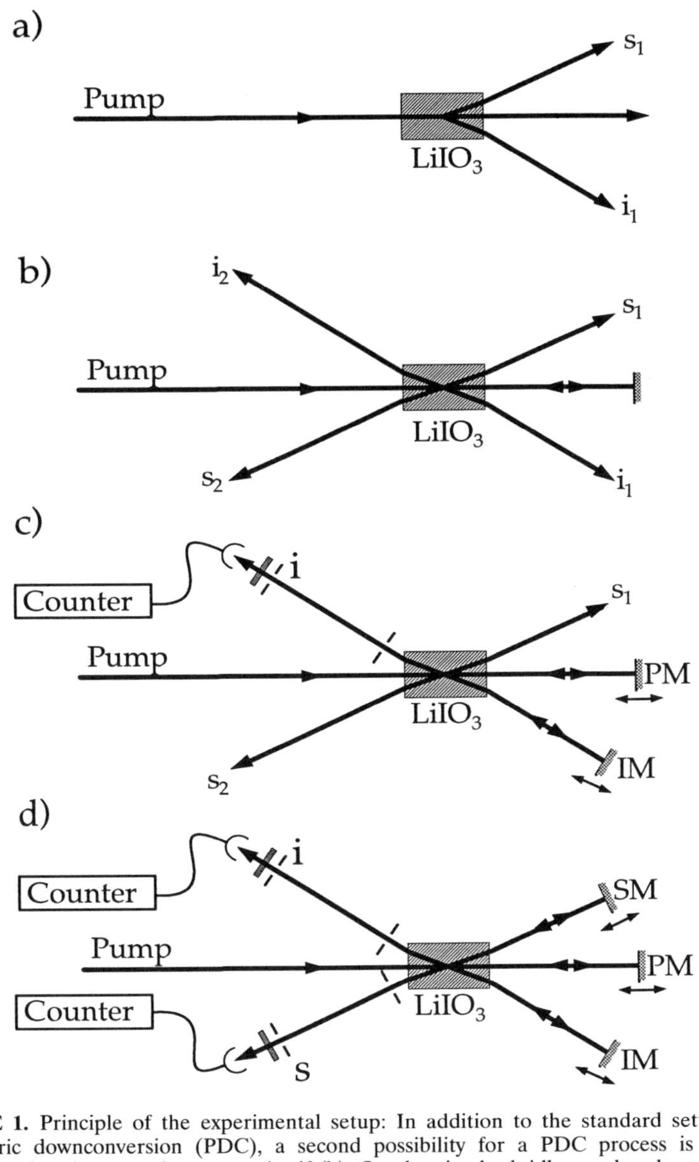

FIGURE 1. Principle of the experimental setup: In addition to the standard setup (a) of parametric downconversion (PDC), a second possibility for a PDC process is given by back-reflecting the pump beam onto itself (b). Overlapping both idler modes alone (c) is not sufficient to obtain interference because the signal photons could still be used for Welcher-Weg analysis. When the signal photons are reflected back as well (d), the two PDC processes become indistinguishable and first-order interference can be observed in both the signal and idler intensities.

Overlapping the signal modes as well as using the mirror SM (i.e., $|s_1\rangle \rightarrow |s_2\rangle = |s\rangle$; see FIGURE 1d) guarantees indistinguishability because no experiment can then give information on when the photon pair was emitted. Consequently, this results in first-order interference for both the signal and idler photons. The two-photon state is now given by

$$|\Psi\rangle = \alpha(e^{i\phi_p} + e^{i(\phi_s+\phi_i)})|s\rangle|i\rangle. \quad (3)$$

Here, ϕ_i and ϕ_s are the phases acquired by the back-reflected signal and idler beams. The observable singles intensity will vary as

$$I_s = I_i = 2I_0\{1 + \cos(\phi_i + \phi_s - \phi_p)\}, \quad (4)$$

where $I_0 = |\alpha|^2 I_p$ is the rate of independent downconversion in either of the two possible processes.

Applying a multimode formalism and thus accounting for the finite coherence length of the light beams involved shows that, as in any standard interferometry, the various path lengths between the crystal and the three mirrors have to agree within the respective coherence lengths. The coherence length of the downconverted radiation is very small for experimental reasons (of the order of 100 μm). The use of filters selecting a narrower frequency band than the one used ($\Delta\lambda = 5$ nm) would lead to an unacceptable loss in intensity. Therefore, a proper path length adjustment is crucial for the mirrors reflecting the downconverted light. The path length requirements are much less stringent for the pump beam mirror because the coherence length of the single-mode UV-line is about 6 m in our experiments. However, only the relative lengths have to be adjusted and, in principle, there is no upper limit for these crystal-mirror distances,[7] in contrast to the atom-mirror case.

Another new feature of this first-order interference effect can be seen from equation 4. The intensity oscillations are modulated by the sum of idler and signal phases, not by the relative phase difference. Therefore, the effects of a movement of one mirror (e.g., the idler mirror) cannot be undone by changing the position of the other (the signal mirror) in the same direction to make both path lengths equal again. Rather, as a consequence of the originally entangled product states (equation 2), one has to move the second mirror in the opposite direction to cancel the first phase change. Also, the nonlocal influence of the change of, say, the signal phase on the idler intensity originates in the product-state properties of the correlated photon pair.

FIRST EXPERIMENTAL DEMONSTRATION

As a source of the downconversion photon pairs, we used a nonlinear crystal ($LiIO_3$) pumped by the 351.1-nm line of a single-mode argon-ion laser (coherence length ≈ 6 m). Interference filters and apertures were used to select photon pairs with wavelengths of 632.8 nm for the signal mode and 788.7 nm for the idler mode. These apertures (≈0.8-mm diameter, 90 cm apart) selected out overlapping and indistinguishable modes behind the crystal (see FIGURE 1d), thus providing high visibility for the first-order interference. The narrow setting of the diaphragms was the limiting factor for the bandwidth of the detected downconversion light: the value of $\Delta\lambda \approx 1.7$ nm corresponds to a coherence length $l_c \approx 260$ μm for the idler photons. Signal and idler photons were detected by silicon-avalanche photodiodes operated in the Geiger mode. After pulse amplification and shaping and an optional coincidence selection, the count rates were recorded by a PC that also controlled the mirror positions via dc motors and piezoelements.

FIGURE 2 shows a first coarse scan of the idler mirror and the narrow region where interference occurs. The coincidence rate was used to find this position because of the much better signal-to-noise ratio. However, its variation is due to the first-order intensity oscillations in the signal and idler beams.

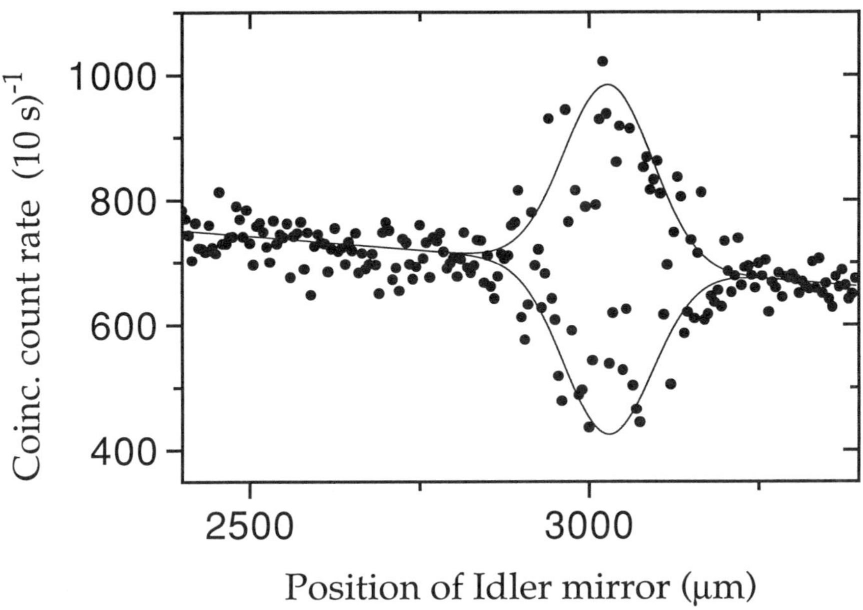

FIGURE 2. Coincidence intensity for a coarse scan of the idler mirror. Interference effects can be observed only where the crystal-mirror distances are equal to within the coherence lengths.

At the position of maximum visibility, we performed fine piezoscans for all three mirrors. FIGURE 3 shows the dependence of the idler count rate on the various mirror translations. One clearly sees the difference in the periodicity of the oscillations when moving mirrors, reflecting different wavelengths. Moreover, this measurement also proves the aforementioned influence of the signal mirror movements on the idler intensity.

One feature of the product state describing a DC-photon pair (equation 2) is strikingly demonstrated in FIGURE 4. Here, both the signal mirror as well as the idler mirror have been moved in the same direction (towards the crystal) in steps proportional to the respective wavelength. Whereas there would be no intensity

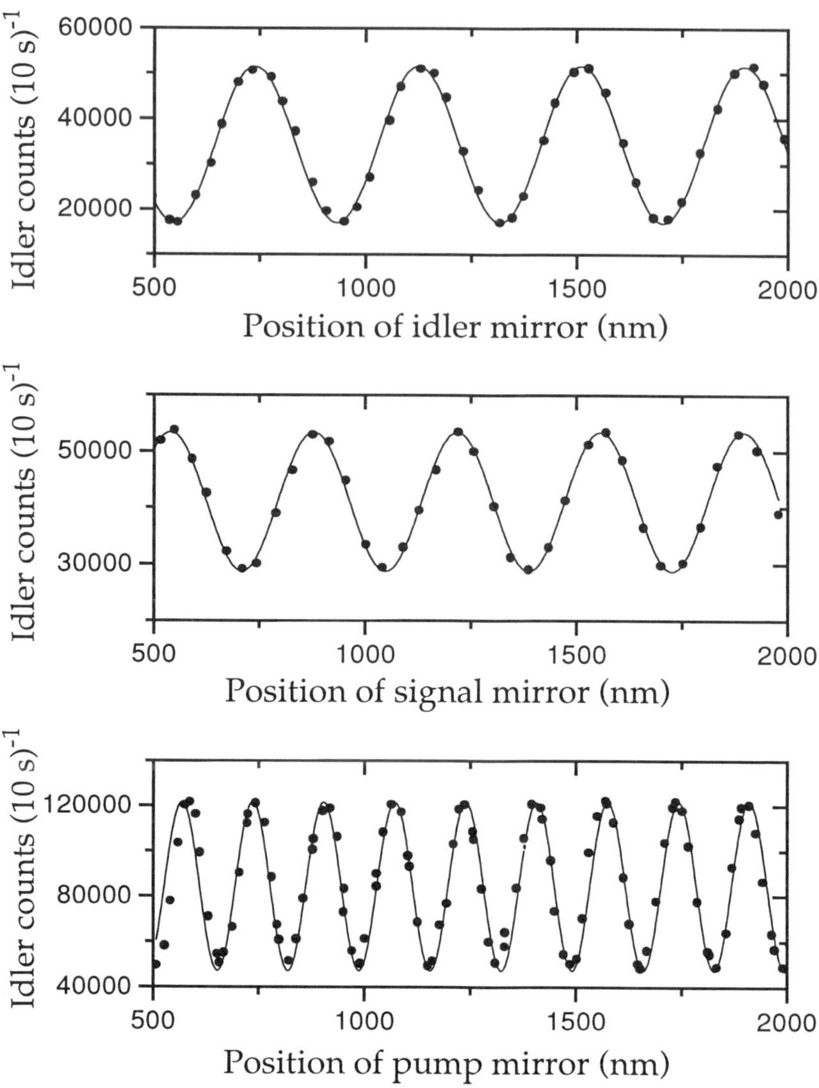

FIGURE 3. Idler count rate as a function of the position of the different mirrors. The oscillation frequency depends on the wavelength of the reflected light.

variation for a standard interferometer, one can now observe oscillations with a period depending on the sum of the phase changes.

JUST A CLASSICAL NONLINEAR EFFECT?

The described experimental setup is topologically very similar to an experiment[8] where two standing-wave infrared modes were upconverted to two UV fields, which were then overlapped. The high visibility interference effects in that case can be described entirely classically in terms of Maxwell's equations.

Applying the theory of parametric amplification to our experiment reproduces many of the observed features. The phase relations of the observed interference phenomena are then described as due to the phase-matching condition implied by this process. However, to also attain visibilities of the first-order intensity variations comparable to those presented here, a very high pump power would be needed.[9] FIGURE 5 shows the dependence of the fringe visibility on the pump power as predicted by such classical considerations. For our experimental parameters (pump intensity of 1 W/mm^2), a visibility of only about 3% is predicted, far below the observed value of up to 51%.

Another unexpected prediction can be deduced from this amplification picture in the case of constructive interference if one considers the very short coherence times of the downconverted radiation ($\Delta\tau_c \approx 100$ fs). It is only during this time that the amplification process can occur, corresponding to an increase of the photon number in the DC modes by a factor of four. The resulting photon bunches in each of the two outgoing downconverted beams should be observable with a Hanbury-Brown and Twiss detection scheme in the two outputs. On the other hand, the Si-avalanche photodiodes used here do not distinguish between one or two photons within the short time of only 100 fs. Therefore, if the increase of DC photons would be only due to this classical amplification process, we could not even have observed the constructive interference shown in FIGURES 2–5.

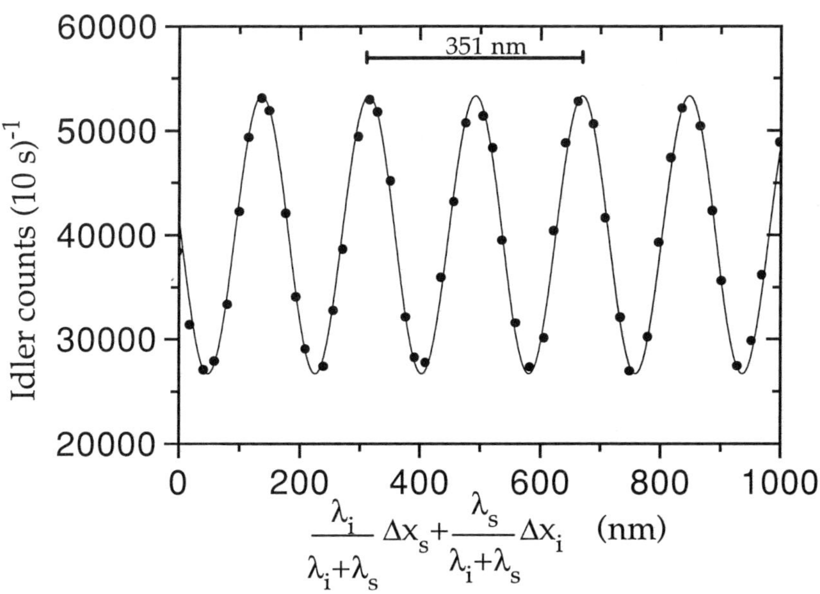

FIGURE 4. Idler count rate for the simultaneous change of the positions of idler mirror and signal mirror in the same direction. Due to the entangled nature of the states, intensity variation with a period corresponding to the pump wavelength can be observed.

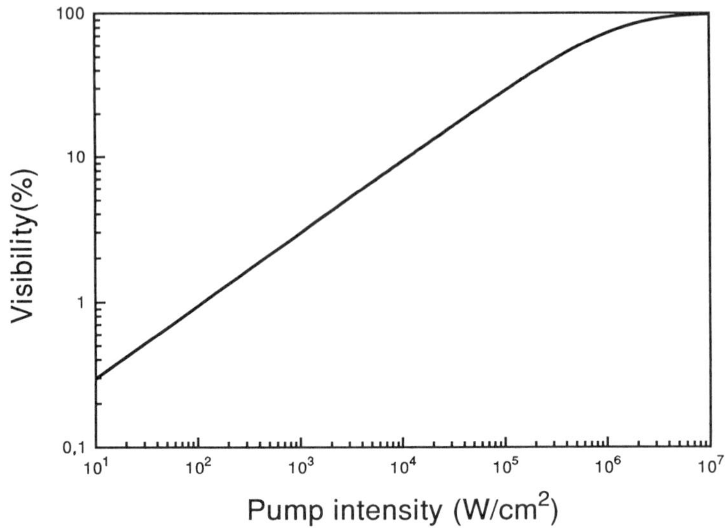

FIGURE 5. Visibility versus pump power in a classical model. A fringe visibility of only 3% is predicted for the pump power of 300 W/mm^2 used in the experiment. This has to be contrasted with the observed value of 51%.

WELCHER-WEG DETECTION AND QUANTUM ERASERS

In the first experimental test of these new interference effects, the distance between crystal and mirror was 13 cm. Although, as mentioned before, there is no upper limit, already these shorter distances suffice to place various optical components in between the crystal and mirrors or to perform path length changes much larger than the coherence length of the DC photons. This allows Welcher-Weg marking that could be subsequently erased after the crystal, restoring the interference.

For example, a path length change of the idler mirror by more than a coherence length provides a means to determine which of the two downconversion processes actually occurred. Therefore, no interference is possible anymore for both the idler and signal photons. Placing a narrow-band filter in front of the idler detector now increases the coherence length and destroys path distinguishability, thus allowing one to recover interference on this beam. However, first-order interference cannot be restored for the signal beam.

Another possibility of manipulating the interference properties is to change the polarization of the back-reflected photons. If, as shown in FIGURE 6a, the polarization of the back-reflected signal beam is changed from vertical (v) to horizontal (h) polarization by means of a quarter wave-plate in front of the mirror, polarization-sensitive detection would allow one to determine the origin of the observed photon pair. The resulting loss of interference is explained by the same Welcher-Weg arguments as before. Again, the signal photons can be made indistinguishable, now by analyzing the polarization at ±45° with respect to the vertical direction, thus allowing interference on the signal beam behind the polarization filter (FIGURE 6b). However, erasing the path information can be done only on photons marked previously. Continuing from equation 2, one obtains, for a polarization change of s_1,

$$|\Psi\rangle = \alpha|s_{1,v}\rangle|i_{1,v}\rangle + \alpha e^{i\phi_P}|s_{2,v}\rangle|i_{2,v}\rangle \xrightarrow[\text{overlap}]{\lambda/4\text{-plate}} |\Psi\rangle = \alpha(|s_h\rangle|i_v\rangle + \alpha e^{i\Delta\phi}|s_v\rangle|i_v\rangle)$$

$$\xrightarrow{\text{polarizer, }45^\circ} |\Psi\rangle = \left(\frac{\alpha}{\sqrt{2}}\right)|i_v\rangle[|s_{45^\circ}\rangle(1 + e^{i\Delta\phi}) + |s_{-45^\circ}\rangle(1 - e^{i\Delta\phi})]. \quad \textbf{(5)}$$

The indistinguishability conditions for interference are not fulfilled at the point of overlap and the idler mode factorizes. As before, there is no way to restore intensity interference for the idler beam.

Changing polarization in both back-reflected beams preserves the entangled nature of the state described by equation 2. Overlap of the signal modes and idler

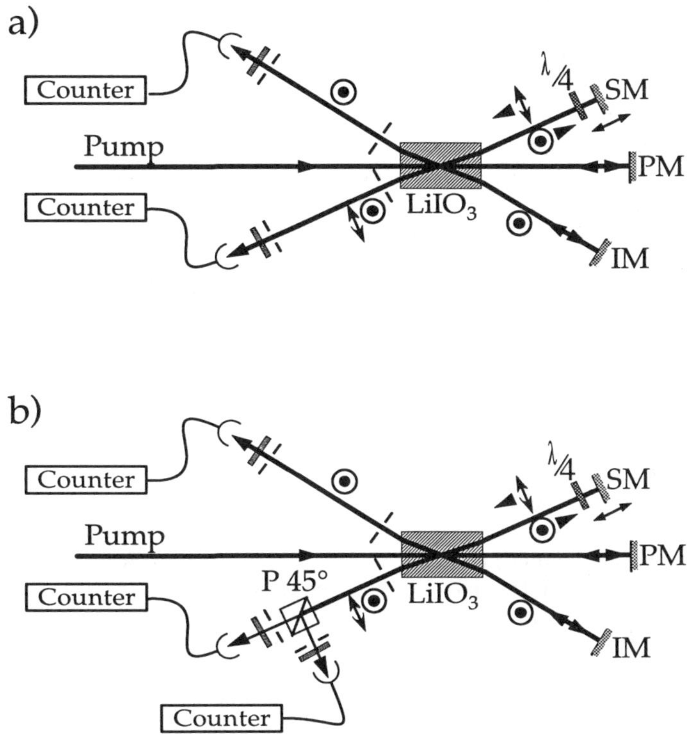

FIGURE 6. The polarization of the signal photons of the first process is rotated by 90° using a retardation plate (λ/4) (a). This destroys the observed interference at the signal detector as well as at the idler detector. (b) Analyzing the signal polarization at 45° can restore the signal interference. This is not possible for the idler beam.

modes, respectively, then results in spin-entanglement:

$$|\Psi\rangle = \left(\frac{\alpha}{\sqrt{2}}\right)(|s_h\rangle|i_h\rangle + e^{i\Delta\phi}|s_v\rangle|i_v\rangle). \tag{6}$$

Attenuation of the back-reflected beams changes the symmetry properties of this entanglement; the resulting incomplete entanglement is required in some classes of loophole-free tests of Bell's inequality.[10]

VIRTUAL OR REAL PHOTONS?

We have shown that the downconversion into the final idler and signal modes can in principle be perfectly suppressed. However, the question arises as to whether or not the downconversion into the modes i_1 and s_1 is then frustrated as well. At first sight, one might conjecture that spontaneous upconversion of a pair of signal and idler photons seems to be too small of an effect ($|\alpha|^2 \approx 10^{-6}$) to explain the observed reduction of the intensities. Two explanations have been given for the process; their main differences are pointed out now.

The close analogy of the experiment described here to the behavior of an atom close to a mirror suggests a similar explanation. When emission is suppressed, the atom sits at a node of the standing wave of the field to be emitted, consequently being described by virtual photons. Changes of the boundary conditions (i.e., mirror position) modify the standing-wave pattern at the atom position after the retardation time, $t = L/c$ (L is the atom-mirror distance), and the atom can start to emit photons. Similarly, replacing a mirror instantaneously by a detector causes emission of photons after the time t and the detector will consequently start to count photons after a total time delay of $2t$.

Application of such a reasoning to our experiment would mean that a detector instantaneously replacing the idler mirror (FIGURE 7) would not detect photons before a time delay of

$$\Delta\tau = 2L/c. \tag{7}$$

In contradiction, a model assuming the same amplitude for upconversion as for downconversion would predict no time delay between insertion of the detector and detection of photons. The description is formally analogous with one of a Michelson interferometer with a very low reflectivity beam-splitter and shall be outlined briefly.

Extending the coarse description of the second section, we start with a pump beam in the state $|p\rangle$, which is downconverted to give

$$|\Psi\rangle = \sqrt{1-|\alpha|^2}|p\rangle + \alpha|i_1\rangle|s_1\rangle + \cdots. \tag{8}$$

Without loss of generality, one can assume that α is a purely imaginary number; higher order terms are neglected again. Reflecting the pump beam back into the nonlinear crystal gives another possibility for downconversion (FIGURE 1b); the state is now described by

$$|\Psi\rangle = \alpha|s_1\rangle|i_1\rangle + \sqrt{1-|\alpha|^2}e^{i\phi_p}(\alpha|s_2\rangle|i_2\rangle + \sqrt{1-|\alpha|^2}|p\rangle). \tag{9}$$

In order to overlap the idler and signal modes, the beams originating in the first process are reflected back into the crystal, as before, with phase shifts ϕ_i and ϕ_s, respectively. If one now also considers the possibility for upconversion of this pair into the pump mode, one obtains

$$|\Psi\rangle = \alpha e^{i\phi_i+\phi_s}(\sqrt{1-|\alpha|^2}|s\rangle|i\rangle + \alpha|p\rangle) + \sqrt{1-|\alpha|^2}\alpha e^{i\phi_p}|s\rangle|i\rangle + (1-|\alpha|^2)e^{i\phi_p}|p\rangle$$

$$= \sqrt{1-|\alpha|^2}\alpha(e^{i\phi_i+\phi_s} + e^{i\phi_p})|s\rangle|i\rangle + [(1-|\alpha|^2)e^{i\phi_p} - |\alpha|^2 e^{i\phi_i+\phi_s}]|p\rangle \quad \textbf{(10)}$$

and, for the intensities,

$$I_s = I_i \propto A[1 + \cos(\phi_i + \phi_s - \phi_p)],$$

$$I_p \propto \{1 - A[1 + \cos(\phi_i + \phi_s - \phi_p)]\}, \quad \textbf{(11)}$$

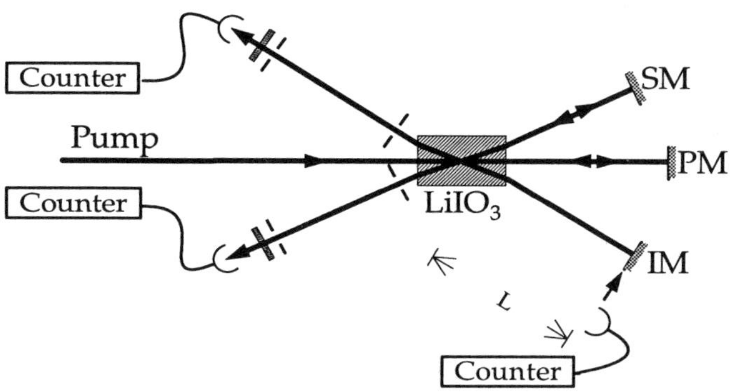

FIGURE 7. Different models predict different behavior for a detector moved rapidly onto the mirror position. An interferometric model predicts immediate detection of photons, whereas the virtual photon picture anticipates a delay of $\Delta\tau = 2L/c$.

with $A = 2|\alpha|^2(1 - |\alpha|^2)$. One sees that the intensity of idler and signal beams behaves the same as before; however, now also the intensity of the backwards-going pump beam depends on the phase settings inside the interferometer. Constructive interference between the amplitude of the back-reflected pump beam and the amplitude resulting from upconversion accompanies the destructive interference of the DC photons. This demonstrates the close analogy to a standard Michelson interferometer if we identify the probability for downconversion (or upconversion, respectively) with the reflection probability. There, too, one obtains, for a very low reflectivity beam-splitter, 100% visibility in one of the two interferometer outputs and very little intensity variation for the other output (the backwards-reflected one). Surely, a detector replacing one of the mirrors of a Michelson interferometer would start counting photons immediately. Then, just as in that case, there would also be in the frustrated downconversion experiment a certain amplitude for detecting photons in the signal or idler arms between the crystal and the mirrors, independent of any phase settings. Therefore, for frustrated downconversion, we infer that the instantaneously inserted detector would start firing without any delay.

The experiment planned to answer which of the two predictions is correct will use a fast switch, consisting of a Pockels cell and a polarizing beam-splitter, in order to change from the reflection at the mirror to the detection of photons. Realizable rise-times of such a switch are about 3–5 ns, implying that the distances between the crystal and the idler and signal mirrors have to be at least 1 m. Spatial filters will be used to guarantee the necessary mode selection without too much intensity loss.

CONCLUSIONS

The experiment described here constitutes a new two-way first-order particle interference effect, which can directly enhance or suppress the downconversion process. Manipulation of the modes involved opens the possibilities for new kinds of Welcher-Weg/quantum eraser experiments. Contrary to well-known examples in standard interferometers, one can restore the interference only in certain modes, not in all. In addition, the possible production of spin-entanglement is a first step towards loophole-free Bell-type experiments.

Different models for this process have been proposed. However, they give significantly different predictions for certain types of experiments. Whether, for example, the picture of virtual photons or an interferometric model as proposed here should be used still has to be answered by experiments.

ACKNOWLEDGMENTS

We thank M. A. Horne, P. W. Milonni, and T. Sleator for helpful discussions.

REFERENCES AND NOTES

1. For a review, see: Greenberger, D. M., M. A. Horne & A. Zeilinger. 1993. Phys. Today **46**(no. 8): 22.
2. Hong, C. K., Z. Y. Ou & L. Mandel. 1987. Phys. Rev. Lett. **59:** 2044.
3. Steinberg, A. M., P. G. Kwiat & R. Y. Chiao. 1993. Phys. Rev. Lett. **71:** 708.
4. There is only a series of experiments on first-order interference of signal modes: Wang, L. J., X. Y. Zou & L. Mandel. 1991. Phys. Rev. A **44:** 4614.
5. Herzog, T., J. G. Rarity, H. Weinfurter & A. Zeilinger. 1994. Phys. Rev. Lett. **72:** 629.
6. Drexhage, K. H. 1974. Prog. Opt. **12:** 163.
7. Grayson, T. P. & G. A. Barbosa. 1994. Phys. Rev. A **49:** 2948.
8. Wu, L. A. & H. J. Kimble. 1985. J. Opt. Soc. Am. B **2:** 697.
9. A thorough discussion of this problem can be found in a comment (by N. N. Senitzky) and a reply (by T. Herzog *et al.*) to appear in Phys. Rev. Lett.
10. Eberhard, P. H. 1993. Phys. Rev. A **47:** R747.

Atoms and Photons in High-Q Cavities

New Tests of Quantum Theory[a]

SERGE HAROCHE

Département de Physique de l'Ecole Normale Supérieure
75231 Paris Cedex 05, France

INTRODUCTION:
CAVITY QED AS A TEST GROUND FOR QUANTUM MECHANICS

John A. Wheeler is one of the leading scientists in quantum theory, who has always pointed to the deep issues and defined clearly the problems to be addressed. Throughout a long and outstanding career, he has had direct interaction with the founding fathers of this marvelously successful, and yet still mysterious, theory. It has been a wonderful experience for me to participate in the conference organized in his honor and to have the opportunity to report on the latest results of the Ecole Normale group in a field so close to his heart and to his mind.

Nonlocal particle correlations[1–3] and macroscopic superpositions of states[4–6] have been given a lot of attention recently. Although restricted for a long time to the realm of "gedankenexperiments", the discussion of these concepts has been revived by the possibility of actually realizing experiments demonstrating some of the most fascinating aspects of quantum theory. "Cavity Quantum Electrodynamics" (Cavity QED)[7] is a new domain of quantum optics where such experiments have become feasible. Cavity QED deals with individual atoms and a few photons coupled together in an electromagnetic resonator. Due to recent progress in the technology of high-Q cavities and in atomic beam manipulation, photons could now be continuously observed in a cavity and could be counted nondestructively, in a way quite similar to the counting and manipulation of material particles.[8–12] New "Einstein-Podolsky-Rosen" (EPR) situations[1] involving atoms correlated at macroscopic distances via their interaction with a cavity field could be studied. Multiple atom correlations of the type discussed by Greenberger, Horne, and Zeilinger (GHZ)[13] have become practically realizable, opening the way to nonlocality tests that are more stringent than the classical Bell's inequality tests performed so far.[2,3] Nonclassical fields could also be generated in high-Q cavities,[9,14–16] which would display some of the properties discussed by Schrödinger in his famous cat paradox.[4] For example, nonlocal superpositions of fields occupying simultaneously two cavities could be prepared and detected.[14,15] Such nonlocal field correlations could be used in a practical scheme[17] of particle teleportation[18] from one cavity to the other.

These experiments can be performed with a very versatile Rydberg atom-superconducting cavity setup, whose simplest version is sketched in FIGURE 1. An

[a]The research reported here was carried out at Ecole Normale Supérieure (Laboratoire Kastler Brossel) in close collaboration with J. M. Raimond, M. Brune, L. Davidovich, and N. Zagury and a team of students (F. Bernardot, P. Nussenzweig, and A. Maali) and guests (F. Schmidt-Kaler and W. Gawlik).

atomic beam effusing from the oven O crosses the cavity C. The atoms are prepared, before entering C, either into a Rydberg state $|e\rangle$ of given energy or into a superposition of Rydberg states of different energies. The preparation into state $|e\rangle$, which involves laser and radio-frequency excitation, takes place inside the preparation box B. If a superposition of Rydberg states is required, a microwave cavity R_1 between B and C is used to apply a resonant pulse on the atoms, admixing the initially prepared Rydberg state $|e\rangle$ with another state $|g\rangle$ of slightly different energy. In this way, initial atomic states of the form $c_e|e\rangle + c_g|g\rangle$ are injected into C. The coefficients c_e and c_g can be adjusted at will by choosing the parameters of the microwave pulse applied in R_1. The atoms then interact with the field in C while crossing it. The cavity sustains a mode whose frequency is resonant or nearly resonant with the transition linking $|e\rangle$ to $|g\rangle$. After leaving C, the atoms are detected

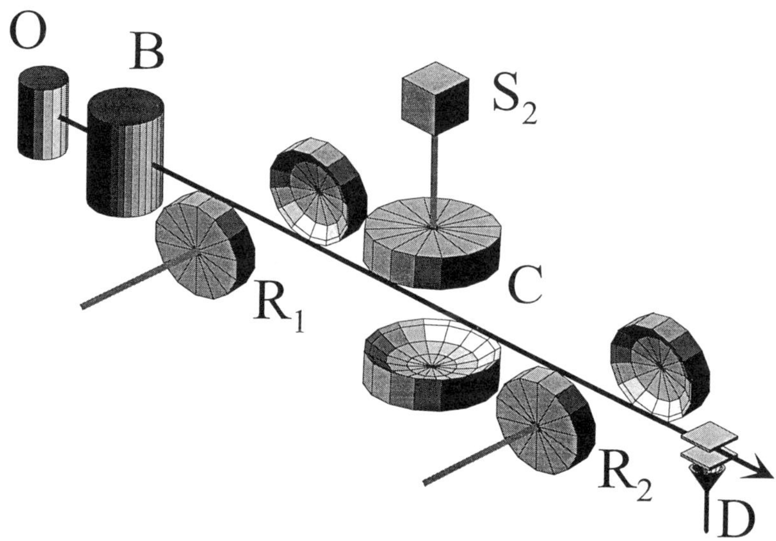

FIGURE 1. General sketch of the circular Rydberg atom-superconducting cavity setup.

downstream by field-ionizing them and counting the resulting ions in the detector D. By adjusting the ionizing electric field, this detection can be made energy-sensitive and one can thus count the atoms in $|e\rangle$ and $|g\rangle$ and determine how the probability of finding them in either level is affected by the coupling with the apparatus.[7] One can also make this detection more subtle by inserting, between C and D, a second microwave zone R_2, admixing again $|e\rangle$ and $|g\rangle$. The combination of R_2 and D acts as a detector sensitive to any linear superposition $c'_e|e\rangle + c'_g|g\rangle$ of the two relevant Rydberg levels.

Let us focus here on the important orders of magnitudes of these experiments, which dictate the choice of atom and of the cavity. The atom-field coupling is characterized by the constant Ω, which represents the rate at which the atom and the empty cavity exchange a photon (vacuum Rabi frequency).[7] The quantum correlations produced by the atom-field interaction are destroyed in a time on the order of the atom and field relaxation times, T_{at} and T_{cav}. Thus, the experiments must be carried out within a time shorter than T_{at} and T_{cav} and the conditions, $\Omega T_{at} \gg 1$ and $\Omega T_{cav} \gg 1$, must be fulfilled. Atoms prepared in circular Rydberg states with a principal quantum number of $n \cong 50$ appear as ideal tools to perform such experiments.[19] Circular states are a special kind of Rydberg states, in which the valence electron precesses in an orbit corresponding to the maximum possible value of the angular momentum projection along the quantization axis.[20] Not only do these atoms correspond to huge electric dipoles associated with very large Ω values, but they also have extremely long damping times, T_{at}, in the 10^{-2}-second range for $n = 50$. The preparation of these atoms, which occurs in the box B (FIGURE 1), is described in detail in reference 19. The resonant frequency between the n and $n - 1$ circular states (which are the e and g states of our present discussion) falls in the 50-GHz range for $n = 50$ (wavelength of about 0.6 cm). The high-Q cavity C coupled to the atom thus has a centimeter size. It is convenient to use a cavity made of separated mirrors facing each other because such a resonator is easy to tune (by displacing the mirrors) and can sustain static electric fields across the mirrors, which is very useful for maintaining the direction and the shape of the circular orbit while the atom crosses the cavity.[21] An intracavity electric field could also be used to tune by Stark effect the atomic transition in and out of resonance with the cavity mode at well-adjusted times. The cavity C, made of superconducting niobium cooled at 1 K or below and submitted to a proper surface treatment, has presently a damping time in the millisecond range ($Q \cong 10^8$). This is longer than the flight time of thermal-velocity atoms over a distance on the order of a few centimeters from C to D, a condition required to keep quantum coherences alive during the experiment. The typical coupling parameter between an $n = 50$ circular atom and the cavity is $\Omega = 10^5$ rad/s, corresponding to $\Omega t = \pi/2$ for an atom crossing a 0.5-cm cavity waist at 300 m/s.

An important feature of this setup is its interferometric character. The first zone R_1 prepares atoms in linear superpositions of atomic states that undergo different "histories" while the atom crosses C, with a probability amplitude for each. The second zone R_2 admixes again the two parts of the atomic wave function before detection occurs. The probability of detecting the atom in $|e\rangle$ or $|g\rangle$ may then exhibit an interference term, which oscillates between 0 and 1 when the frequency ν of the microwave field applied in R_1 and R_2 is varied. This modulation is known as a "Ramsey fringe" signal.[22] The interaction of the atom with a field in C may affect differently the two probability amplitudes associated with levels $|e\rangle$ and $|g\rangle$, resulting in a shift of the fringe pattern whose measurement yields precious information about the atom-field interaction. FIGURE 2 shows, as an example, a typical recording of the fringes obtained with this interferometer when the cavity is empty (lower trace) and when it contains on average one photon (upper trace). In this experiment, the atomic transition and the cavity mode are slightly detuned (frequency mismatch $\delta = 150$ kHz) so that the interaction is dispersive, leading to small energy shifts of the levels $|e\rangle$ and $|g\rangle$. This experiment, described in detail in reference 23, demonstrates that our circular Rydberg atom-cavity system has the sensitivity required to observe a single photon and to carry out the experiments discussed below.

RESONANT AND DISPERSIVE CAVITY QED EXPERIMENTS ANALYZED FROM THE EPR POINT OF VIEW

Let us first consider the situation where the atom and the field in C are exactly resonant. The atom undergoes in this case a Rabi oscillation between levels $|e\rangle$ and $|g\rangle$ at a frequency that depends upon the number of photons initially present in C.[7] This frequency is equal to 2Ω if the cavity field is initially in its vacuum state. The final state of the "atom + field" system depends upon the initial state of this system and upon the time that the atom has spent in C. Assuming, for example, that C is initially

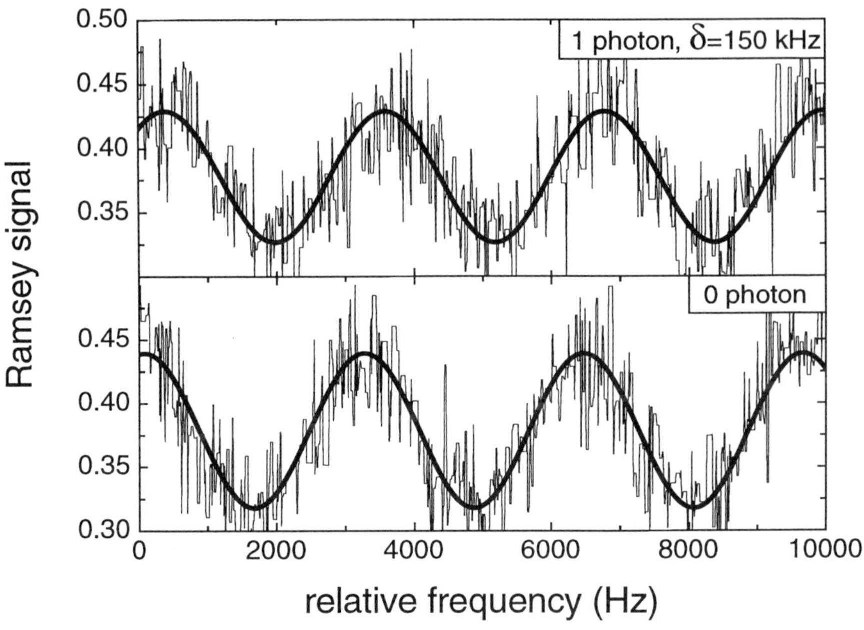

FIGURE 2. Population transfer signal between circular Rydberg levels $|e\rangle$ (n = 51) and $|g\rangle$ (n = 50) as a function of the frequency ν applied in the microwave zones R_1 and R_2. The cavity (detuned by δ = 150 kHz from the atomic transition) contains zero photons (lower trace) or one photon (upper trace) on average. The translation of the Ramsey fringe pattern can reveal the dispersive light shift produced by subphoton fields. (From reference 23.)

empty, we can write the initial state as $|\Psi_1\rangle = c_e|e, 0\rangle + c_g|g, 0\rangle$, where the first symbol in each term refers to the atom's state and the second to the field photon number. During the atom cavity crossing time t, the state $|e, 0\rangle$ evolves into the linear superposition, $\cos(\Omega t)|e, 0\rangle - i\sin(\Omega t)|g, 1\rangle$, whereas the state $|g, 0\rangle$ remains invariant (no photon emission from the lower state of the atomic transition). The state of the global system after the atom has left C is thus $|\Psi_2\rangle = c_e\{\cos(\Omega t)|e, 0\rangle - i\sin(\Omega t)|g, 1\rangle\} + c_g|g, 0\rangle$, which is clearly an entangled state for most values of t. Let us assume for the time being that the microwave zone R_2 is inactive so that the system does not evolve between C and D. Detecting the atom in level $|e\rangle$ means that the field in the cavity is left in its vacuum state, whereas detecting it in level $|g\rangle$ generally projects the field in a linear superposition of zero- and one-photon states. In the special case where $\Omega t = \pi/2$ (probability unity for the excited atom to emit a photon in C), one prepares with certainty the atom in level $|g\rangle$ and the field collapses into the superposition, $-i\,c_e|1\rangle + c_g|0\rangle$, which replicates in the field (with a phase change) the superposition state initially carried by the atom. By properly adjusting the microwave parameters in R_1, any linear photon number superposition can be generated in this way, including the special case of $c_e = 1$ and $c_g = 0$ (one-photon Fock state) and the symmetric superposition, $(1/\sqrt{2})\,(|1\rangle + |0\rangle)$.

If the atom and the field are resonant and $c_e = 1$ and $c_g = 0$, this describes the passage of the first atom through a micromaser cavity fed by excited atoms.[24–26] The analysis can be carried out for successive atoms crossing C. Each atom interacts with the field left in C by the previous ones and a field made of a large number of photons builds up, leading to very interesting nonclassical effects,[26] whose theoretical description, which can be found in various articles,[27–29] is beyond the scope of this review. The atoms are detected by field ionization after they leave C. The field properties are inferred from the statistics of atomic detection events.[28] Note that the micromasers studied so far have involved noncircular, low angular momentum Rydberg atoms, which have a nonnegligible probability of spontaneous decay towards lower states between C and the detector, thus limiting the quantum efficiency of the atomic detection procedure. Micromasers operating with circular Rydberg atoms are now being considered for experiments requiring a very high atomic detection efficiency and hence very long atomic lifetimes. Other interesting resonant cavity QED effects have been observed on two-photon micromasers.[30,31] In this device, the atoms undergo a two-photon transition between two Rydberg states of the same parity and emit photons by pairs in the cavity.

The micromaser situation is complicated (or made richer, depending upon the point of view) by the fact that the atoms play a double role in this device: they are the source of the radiation that they can feed or absorb while they are in C and they also serve as measuring devices for the field, according to the mechanism outlined above. It is, however, also possible to deal with simpler situations, in which the atoms act as mere quantum detectors of the field. If the cavity and the atomic transition are slightly mistuned, with a frequency difference δ, any exchange of energy between atom and field is made impossible. The atom-field coupling then becomes purely dispersive and the atom can be viewed purely as a "measuring device" for the field in C.[8,9] The interaction then produces a mere dephasing of the field (index effect of the atom crossing C) and also dephases the atom's wave function by an angle depending upon the number of photons in the cavity and upon the quantum state of the atom. More precisely, if N photons are present in the cavity, the initial state of the "atom + field" system is now $|\Psi_1\rangle = c_e|e, N\rangle + c_g|g, N\rangle$ and immediately after the atom has crossed C it becomes $|\Psi_2\rangle = c_e \exp[i\epsilon(N + 1)t]|e, N\rangle + c_g \exp[-i\epsilon Nt]|g, N\rangle$, where t is again the atom-cavity crossing time. The Bohr frequency of the atomic transition is shifted by $\epsilon(2n + 1)$, where the quantity $2\epsilon = 2\overline{\Omega^2}/\delta$ is the frequency shift per photon of the atomic transition, averaged over the trajectory of the atom across the cavity. This shift is precisely the quantity measured in the experiment[23] described in the previous section (see FIGURE 2). In addition to the light shift effect (terms proportional to N in equation 2), there is even for $N = 0$ a shift of the upper Rydberg level

$|e\rangle$, which is the "Lamb shift" produced by the vacuum field in the cavity mode. This single-mode Lamb shift has been experimentally observed with our setup.[23]

Assume now that the field is initially in a superposition of different photon number states, for example, in a coherent state, $|\alpha\rangle = \Sigma_N C_N|N\rangle$, with $C_N = \exp(-|\alpha|^2/2)[\alpha^N/\sqrt{N!}]$ (such a field can be produced by coupling C to a classical source of current that is switched off immediately before the atom is sent across the apparatus). The state of the "atom + field" system after the atom leaves the cavity is obtained by mere superposition:

$$|\Psi_2\rangle = c_e e^{i\epsilon t}|e\rangle\,\{\Sigma_N C_N \exp[i\epsilon Nt]\,|N\rangle\} + c_g|g\rangle\,\{\Sigma_N C_N \exp[-i\epsilon Nt]\,|g\rangle\}$$
$$= c_e\, e^{i\epsilon t}|e, \alpha\, e^{i\epsilon t}\rangle + c_g|g, \alpha\, e^{-i\epsilon t}\rangle. \quad (1)$$

This expression shows that the atom in $|e\rangle$ (in $|g\rangle$) dephases the field in the cavity by the angle $+\epsilon t$ $(-\epsilon t)$. This dephasing corresponds to an atomic index effect. If the atom is in a linear superposition of the two levels, this phase shift results in an entanglement of the system after the interaction, with the internal state of the atom being correlated to the phase of the field in the cavity. Note that this entanglement is of a quite different nature than the one discussed above in the resonant case. Here, the number of photons in C cannot be changed by the atom-field interaction and the entanglement results from a purely dispersive phase shift distortion of the wave function, different for each photon number and atomic state.

If the microwave zone R_2 is left inactive, then measuring the atom's state in D "collapses" the phase of the field to a single value, leaving the field in either the state $|\alpha\, e^{i\epsilon t}\rangle$ or the state $|\alpha\, e^{-i\epsilon t}\rangle$, a rather trivial result. However, a very interesting situation arises if a $\pi/2$ microwave pulse mixing levels $|e\rangle$ and $|g\rangle$ is applied on the atoms in R_2. Then, the "atom + field" state immediately after the atom leaves R_2 becomes

$$|\Psi_3\rangle = c_e(e^{i\epsilon t}/\sqrt{2})[\,|e, \alpha\, e^{i\epsilon t}\rangle + |g, \alpha\, e^{i\epsilon t}\rangle] + (c_g/\sqrt{2})[\,|g, \alpha\, e^{-i\epsilon t}\rangle - |e, \alpha\, e^{-i\epsilon t}\rangle] \quad (2)$$

and the subsequent detection of the atom in level $|g\rangle$ or $|e\rangle$ results in the collapse of the field into one of the two states,

$$|\Phi^{\pm}\rangle = c_e\, e^{i\epsilon t}|\alpha\, e^{i\epsilon t}\rangle \pm c_g|\alpha\, e^{-i\epsilon t}\rangle. \quad (3)$$

These are linear superpositions of field states with different classical phases that have been dubbed "Schrödinger cat states" of the field.[6,9]

We recognize here the ingredients of the Einstein-Podolsky-Rosen paradox.[1] The atom and the field in the cavity get entangled by their interaction. This entanglement survives the system separation. One subsystem (here, the field) collapses into a state that depends upon the result of the measurement performed on the other part (here, the atom), even if these two parts are far apart from each other when this measurement is performed. The state into which this collapse occurs depends upon the kind of measurement that one decides to perform (here, by adjusting the microwave parameters in R_2). This decision can even be made after the systems have ceased to interact (one can change these parameters while the atom is flying from C to R_2, thus realizing a "delayed choice" experiment). We must notice, however, a difference at this stage with the discussion of a classical EPR situation. In an EPR experiment, spins are coupled together before flying apart and the experimental data display some kind of correlation between the results of measurements performed on two (or more) such spin particles. Here, we correlate a spinlike particle (the atom) to a harmonic oscillatorlike system (the field). Moreover, we know how to measure the "spin" (by the field ionization detection in D), but we do not have any convenient way of measuring directly the field stored in C. In fact, the only practical way to get information on the field is to couple it to atoms that will be subsequently detected. We are thus naturally led to consider what happens if we send several atoms across the same cavity and perform correlated measurements on them. The atom-field entanglement then translates into easily detectable atom-atom correlations.

PREPARATION OF A STREAM OF ENTANGLED ATOMS AND GHZ EXPERIMENTS

We have shown in the INTRODUCTION that the phase shift produced on the atomic transition by the field in C can be measured by Ramsey interferometry. Let us assume that the field is initially prepared in the state, $|\phi_{\text{field}}\rangle = (1/\sqrt{2})(|1\rangle + |0\rangle)$ (we have indicated in the previous section how such a field can be generated), and let us send nonresonant atoms across the Ramsey interferometer to measure the atomic phase shift. When one photon is stored in C, the atomic coherence between levels $|e\rangle$ and $|g\rangle$ accumulates an extra phase equal to $2\epsilon t$ and the phase of the fringe pattern is shifted by that amount. Let us adjust now the frequency ν of the microwave in the R_1-R_2 zones so that the probability of detecting the atom in $|g\rangle$ when $N = 0$ is equal to unity (this corresponds to a "bright" fringe for this probability). Let us also fix ϵ and t so that $\epsilon t = \pi/2$. In this case, a single photon in C shifts the fringe pattern by half an interfringe and changes a "bright" fringe into a "dark" one. The probability of detecting the atom in $|g\rangle$ then vanishes when $N = 1$. The state of the "atom + field" system immediately before detection of this atom thus must be $|\Psi\rangle = (1/\sqrt{2})(|e, 1\rangle + |g, 0\rangle)$. Our interferometer can thus correlate perfectly the photon number to the state of the atom.[11,12] Detecting this state then provides an ideal measurement of the field photon number and projects the radiation onto a Fock state ($|0\rangle$ or $|1\rangle$). This remarkable dispersive atom-field entanglement is at the basis of a method of quantum nondemolition of photons[8,9] that we have recently proposed and hope to implement soon.

Let us assume now that we do not perform immediately the atomic state measurement, but that we send a second and then a third atom identical to the first one across the cavity. The same argument can be repeated because the state of the field has not been changed by the first (and the second) nonresonant atom. If there is zero photon in C, then necessarily the second (and the third) atom must emerge from R_2 in state $|g\rangle$ with unit probability and, if there is one photon in C, they must emerge in state $|e\rangle$. As a result, the combined system consisting of the three atoms and the cavity field, after all atoms have crossed R_2, is prepared in the state $|\Psi_{\text{total}}\rangle = (1/\sqrt{2})(|e, e, e; 1\rangle + |g, g, g; 0\rangle)$. We have produced in this way a many-atom–field entanglement. If we now want to generate a pure interatomic correlation, we can perform a last operation that "disentangles" the field from the system. We tune back the cavity to resonance and we send in C a last atom (denoted by a prime) in the lower state $|g'\rangle$. The microwave zone R_1 is switched off at this stage. The time t is also adjusted so that $\Omega t = \pi/2$. After this last atom leaves the cavity (and before it enters R_2), the combined state $|g', 1\rangle$ has become $|e', 0\rangle$ (photon absorption in a π pulse), whereas $|g', 0\rangle$ has not evolved. As a result, the total state of the four-atom + field system has become $|\Psi_{\text{total}}\rangle = (1/\sqrt{2})(|e, e, e; e'\rangle + |g, g, g; g'\rangle)|0\rangle$. The cavity is now empty, the atom-field entanglement has been removed, and the state of the three (nonresonant) atoms has been correlated to the state of the last (resonant) atom. This atom finally crosses zone R_2, in which it is subjected to a $\pi/2$ pulse. The state of the total system then becomes $|\Psi_{\text{total}}\rangle = (1/\sqrt{2})[|e, e, e\rangle(|e'\rangle + |g'\rangle) + |g, g, g\rangle(|g'\rangle - |e'\rangle)]|0\rangle$ and the subsequent detection of the last atom (in level $|e'\rangle$ or $|g'\rangle$) results in the "collapse" of the three-nonresonant-atom system into one of the two states:

$$|\Psi_{123}^{\pm}\rangle = (1/\sqrt{2})(|e, e, e\rangle \pm |g, g, g\rangle). \tag{4}$$

The + (−) sign in equation 4 corresponds to the detection of the last atom in $|g'\rangle$ ($|e'\rangle$). Such three-atom entangled states are precisely of the kind required to carry out the GHZ experiment.[13] Translated into a "spin language", the state $|\Psi_{123}^{-}\rangle$ becomes $(1/\sqrt{2})(|+++\rangle - |---\rangle)$, where the symbols "+" and "−" refer to eigenstates of the z component, S_z, of the spin angular momentum. A restricted version of the GHZ argument, presented by Mermin,[32] states that, according to quantum mechanics, a measurement of the transverse component, S_x, of three spins ½ prepared in that entangled state should always yield an odd number (1 or 3) of particles in the $S_x = -½$ state. In contrast, a classical argument based on the existence of "elements of reality" would always predict this number to be even (0 or 2). Coming back to our Rydberg atom system, we remark that the eigenstates of S_x become analogous to the linear superpositions, $(1/\sqrt{2})(|e\rangle \pm |g\rangle)$, which can be produced by "rotating" the states $|e\rangle$ and $|g\rangle$ with the help of yet another $\pi/2$ microwave pulse. Thus, we must extend our apparatus downstream of the atomic beam and add a zone R_3 to perform this pulse on the three entangled atoms, which should not be detected by the detector D (no ionizing field applied while these atoms cross it). Finally, a field ionization detector D′, placed behind R_3, should be used to count the atoms in level $|e\rangle$ or $|g\rangle$. The combination of R_3 and D′ is equivalent to a detector of the "equivalent spin" component along O_x (the R_3-D′ combination is not represented in FIGURE 1). Quantum mechanics predicts the detection in D′ of an odd number of atoms in the $|g\rangle$ state, whereas a classical argument predicts an even number. The strength of the GHZ-Mermin experiment is that it yields, in a single shot, a violation of the prediction of the "element of reality" theory, whereas the Bell's inequality experiments are based on a statistical analysis of a large number of measurements performed on pairs of particles. Here, a single measurement outcome exhibiting two atoms in state $|g\rangle$ would be enough to prove quantum mechanics wrong (provided the detectors are ideal and never yield wrong counts). The argument exposed above for correlating three nonresonant atoms could be generalized to four or more atoms, leading to extensions of the GHZ experiment to many-atom situations.[33]

SCHRÖDINGER CATS AND NONLOCAL FIELD STATES IN CAVITY QED

Another fascinating feature of these Cavity QED experiments is to make possible the preparation and study of nonclassical field superposition states of the "Schrödinger cat" type. We have shown in the second section how this can be achieved by sending a single nonresonant atom through our Ramsey interferometer, when C contains initially a coherent field $|\alpha\rangle$. For the sake of definiteness, let us adjust the atom's parameters to the following values: $c_e = -i/\sqrt{2}$, $c_g = 1/\sqrt{2}$, and $\epsilon t = \pi/2$. Equation 3 then becomes

$$|\phi^{\pm}_{\text{phase}}\rangle = (1/\sqrt{2})(|\beta\rangle \pm |-\beta\rangle) \tag{5}$$

with $\beta = i\alpha$. The field in the cavity is then prepared in a linear superposition of two coherent field states with opposite phases. These particular superpositions are even (odd) photon number states when the sign in equation 5 is + (−). Schrödinger cat states have been studied extensively in theoretical papers.[9,34–37] Cavity QED provides for the first time a practical method to generate and detect them.

We have described elsewhere[16] a possible experiment with these states. It is essential, of course, to perform a measurement probing directly the coherent character of the superposition and enabling us to distinguish between a field described by equation 5 and a mere statistical mixture of the $|\beta\rangle$ and $|-\beta\rangle$ field states. One elegant way to achieve this is to send a second atom after the first one has been detected and to measure the probability of detecting this second atom in $|e\rangle$ or $|g\rangle$. This probability presents an interference term between two probability amplitudes, one associated to each of the $|\beta\rangle$ and $|-\beta\rangle$ states. This interference is constructive for the probability of detecting the second atom in the same state as the first one, making this conditional probability equal to 1, and is destructive for the probability of the second atom to be detected in a state different from the first one, making this conditional probability equal to 0. If the field in C is in a statistical mixture instead, the interference vanishes and both conditional probabilities level off to ½. In fact, we have neglected in the analysis so far the relaxation of the field in the cavity.[9] Dissipative processes have a strong effect on these quantum superpositions. In a time on the order of $T_{\text{cav}}/\overline{N}$, where $\overline{N} = |\alpha|^2$ is the average number of photons in the coherent field, they evolve into a classical statistical mixture. We thus expect the conditional probability of detecting the first and the second atom in the same quantum state to be a function of the delay T between the two atomic detections. For short delays ($T \ll T_{\text{cav}}/\overline{N}$), this probability should be close to 1. For large delays ($T_{\text{cav}} \gg T \gg T_{\text{cav}}/\overline{N}$), it should take the value of ½.

The continuous change of this conditional probability from 1 to ½ as T is increased should be direct evidence of the "Schrödinger cat's" decoherence, a physical process that is at the heart of the quantum measurement process.[38] It is instructive to see the phase of the field in C as a kind of "needle" pointing in two possible directions, each direction being correlated to one of the two Rydberg states $|e\rangle$ or $|g\rangle$ of the first atom crossing C (see equation 5). This "needle" remains for some time in a quantum superposition of its two possible classical positions, but in the end it chooses one or the other (when the quantum superposition has evolved, due to the field dissipative coupling to its environment, into a statistical mixture). For "small needles" ($\overline{N} = |\alpha|^2 \simeq 1$), this decoherence occurs in the relatively long time

T_{cav}. For "long needles" ($\overline{N} = |\alpha|^2 \simeq$ 10–100), the decoherence becomes much faster. Because we can adjust the intensity of the field initially present in the cavity from small to large values of $\overline{N}$, such an experiment would enable us to explore the fuzzy boundary between the quantum world (where "small needles" are, at least for some time, quantum objects existing in several possible states susceptible to the creation of interference effects) and the classical world (where "large needles" decohere into mutually exclusive states much faster than they can be observed).

Other kinds of cats can be generated with simple variants of this Cavity QED setup. Instead of preparing a coherent field inside the cavity prior to the first atom injection, it is possible to employ the atom itself as a kind of "quantum switch" that governs the flow of the field inside the cavity.[15] The cavity must then be connected to a classical source slightly mistuned so that, in the absence of an atom, it cannot feed

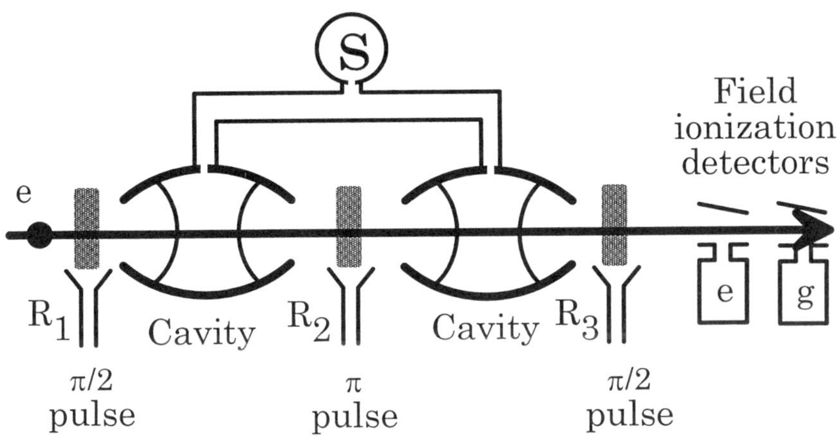

FIGURE 3. Sketch of a setup for the preparation of a field delocalized in two cavities. The source S feeds both cavities through a T-shaped waveguide. An atom sent across both cavities acts as a switch. The state of the switch is controlled by the pulses applied in the zones R_1 and R_2. The nonlocal field is prepared when the atom is detected by the field ionization detectors, following a last $\pi/2$ pulse applied in zone R_3. (Adapted from reference 15.)

any field inside C. The atomic parameters are then adjusted so that an atom crossing C in level $|e\rangle$ provides exactly the mode frequency shift required to tune it into resonance with the source. On the other hand, the atom in level $|g\rangle$ leaves the cavity and the source mistuned. We take again here advantage of the single atom index effect: the atom behaving as a kind of dispersive "plunger" tuning C in and out of resonance with the source. Such a device allows us to prepare "amplitude cat states" of the form, $|\phi^{\pm}_{\text{amplitude}}\rangle = (1/\sqrt{2})(|\alpha\rangle \pm |0\rangle)$, whose coherence can also be tested by sending a second atom across the system and measuring the conditional probability that both atoms end up being detected in the same state.

The quantum switch can also be used to generate nonlocal "Schrödinger cat" states in two identical cavities, C_1 and C_2 (see FIGURE 3).[15] The two cavities are now coupled symmetrically to a slightly mistuned source and a single atom is sent across both cavities. A microwave zone R_1 in front of C_1 prepares again the atom in a linear symmetrical superposition of $|e\rangle$ and $|g\rangle$ states, realizing the quantum switch device. A π microwave pulse applied in zone R_2 turns $|e\rangle$ into $|g\rangle$ between C_1 and C_2 and exchanges the open and closed states of the switch. The two levels $|e\rangle$ and $|g\rangle$ are finally mixed again in the downstream zone R_3 before the atom is detected. In this way, one can generate the superposition state, $|\phi^{\pm}_{\text{nonlocal}}\rangle = (1/\sqrt{2})(|\alpha; 0\rangle \pm |0; \alpha\rangle)$, which represents a nonlocal field, with equal (or opposite) probability amplitudes for the coherent field to be in the first or in the second cavity (the first and the second symbol in each term refer to the field in C_1 and C_2, respectively).

Nonlocal fields can also be generated in a two-cavity system by a resonant atom-field interaction. We present here a simple preparation method for a single photon delocalized between two cavities, which is a variant of a method discussed in reference 14. Let us consider again the setup of FIGURE 3, but assume now that the source is disconnected, as well as the microwave fields in the three zones R_1, R_2, and R_3. A resonant atom, prepared in level $|e\rangle$, is sent across the initially empty cavities. The resonant atom-cavity interaction times with C_1 and C_2, that is, t_1 and t_2, are set so that $\Omega t_1 = \pi/4$ and $\Omega t_2 = \pi/2$. This can be achieved by tuning at conveniently adjusted times the atomic transition in and out of resonance with the cavity mode while the atom is crossing each resonator. The atom crosses C_1 with a probability amplitude of $1/\sqrt{2}$ of leaving a photon in it (and flipping to state $|g\rangle$) and the same probability amplitude of leaving it empty (and remaining in state $|e\rangle$). After the atom leaves C_1, the combined "atom-field" system is thus in the state, $(1/\sqrt{2})(|e; 0, 0\rangle - i|g; 1, 0\rangle)$, where the first and the second figures in each term are the number of photons in C_1 and C_2, respectively. If the atom crosses C_2 in state $|e\rangle$, it deposits a photon in it with unit probability and exits in level $|g\rangle$. Nothing happens, on the other hand, if it crosses C_2 in level $|g\rangle$. Thus, after the atom exits C_2, the atom-field system is in the state, $(1/\sqrt{2})(|g; 0, 1\rangle + |g; 1, 0\rangle)$, and we have prepared the field in the nonlocal one-photon state:

$$|\phi^{\text{one-photon}}_{\text{nonlocal}}\rangle = (1/\sqrt{2})(|1; 0\rangle + |0; 1\rangle). \tag{6}$$

This highly nonclassical field state can be employed in an interesting scheme of particle teleportation between C_1 and C_2 that we briefly analyze in the next and last section.

TELEPORTATION OF A TWO-LEVEL ATOM BETWEEN TWO CAVITIES

It has been recently pointed out by Bennett *et al.*[18] that quantum nonlocality could be used to "teleport" an unknown quantum state from one observer to another. Their scheme involves an EPR pair (1, 2) of two spin-½ particles that are shared by the two observers located at different places. The particle (a) to be teleported is a spin-½ particle in a superposition $c_+|+\rangle + c_-|-\rangle$ of its two quantum states, which is supposed to be unknown to the observers. One of them couples the particle (a) to be replicated to his/her EPR particle (1) and performs a measurement on the combined (a, 1) system. A complete basis for the (a, 1) system is spanned by four "Bell states"[39] in which the (1) and (a) particles are entangled. In the expression of the initial state of the total system made up of the three particles, each of these Bell states is correlated to a particular linear combination of the particle (2) states, whose coefficients are simply related to the c_+ and c_- coefficients. If the first observer can measure an observable having the (a, 1) Bell states as eigenstates, this measurement would amount to projecting the (a, 1) system into one of these Bell states. The quantum correlation between the two EPR particles will then automatically entail the collapse of the (2) particle into a quantum state whose coefficients can be deduced from c_+ and c_- by a unitary transformation that depends upon the outcome of the (a, 1) measurement. If the second observer is informed about the result of the measurement performed by the first one (through a classical communication channel), he/she will then be able to apply the inverse unitary transformation to particle (2) and reconstruct on this particle the quantum state of the (a) particle.

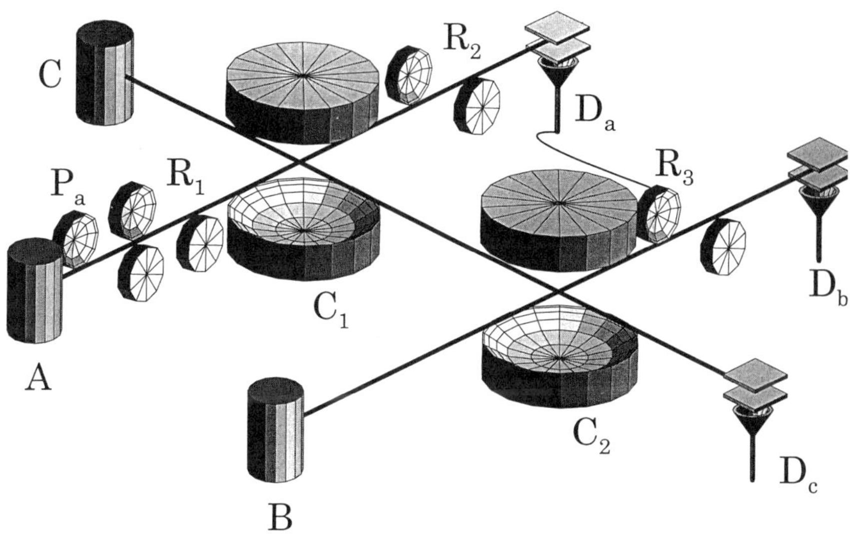

FIGURE 4. General scheme of a two-cavity teleportation machine.

Bennett *et al.* did not discuss any specific way of measuring the Bell states, which is a crucial point of the teleportation process. Translating their spin point of view into the Cavity QED "language", we have recently described a realistic procedure to implement the teleportation program on circular Rydberg atoms crossing superconducting cavities. We present here only the principle of the method, which is described in detail in reference 17. Our experimental scheme is sketched in FIGURE 4. Instead of a pair of EPR particles, our "observers" are attached to two identical cavities C_1 and C_2 in which a nonlocal one-photon state of the form given by equation 6 has been prepared. A single resonant atom, produced by the circular state source C, is used for this purpose, according to the method outlined in the previous section. The detection of this atom by the detector D_c means that the teleportation machine is ready and that the unknown quantum state of an atom A sent across C_1 can be teleported on an atom B sent across C_2. The auxiliary microwave zone P_a is used to prepare the atom A in an arbitrary superposition $c_e|e\rangle + c_g|g\rangle$ of the $|e\rangle$ and $|g\rangle$ levels.

The atom (A) couples dispersively to the field in C_1 and, through a proper manipulation of the atomic states before and after C_1 in the microwave zones R_1 and R_2, one gets upon detection of A by the state-selective detector D_a a bit of information about the (A + C_1) system. This is not sufficient, though, to perform a complete measurement because the (A + C_1) Bell states span a four-dimension space. Another "reference" atom (A′) prepared in state $|g\rangle$ and made resonant with the cavity is then sent across C_1. After a proper manipulation performed in R_2, its detection provides the remaining information required to fully specify the (A + C_1) Bell state. At this time, the field in C_2 "collapses" into a superposition of zero- and one-photon states, which replicates within a unitary transformation the initial superposition of $|e\rangle$ and $|g\rangle$ states in A. The second observer then sends a resonant atom (B) across the second cavity, which replicates on its own quantum state the superposition present in the field, in a "copying process" that is the reverse of the one discussed at the beginning of the second section above. A unitary transformation must then be applied to this state in order to reproduce the initial (A) state. It is performed by applying a microwave pulse in R_3. The parameters of this pulse can be adjusted only if the second observer knows the outcomes of the measurements performed by the observer attached to C_1. Thus, two-bit information has to be transmitted via a classical channel between D_a and R_3 ("wire" in FIGURE 4). Such a teleportation experiment would clearly be a new test of quantum nonlocality. As with the GHZ-type experiments, it would correspond to a new generation of EPR tests based on atomic correlations of higher order, as opposed to the second-order correlations involved in the demonstrations of Bell's inequalities. Note that a similar teleportation scheme based on Cavity QED effects has been independently proposed recently.[40]

This discussion shows the versatility of the circular Rydberg atom-superconducting cavity system for testing quantum mechanics and probing the boundary between the classical and the quantum worlds. Two important conditions are required by these experiments: (i) matter and radiation should be coupled very strongly to make single-photon–single-atom effects observable and (ii) the two coupled systems should be very well protected against relaxation because quantum coherences quickly dissipate into the environment. The Ramsey interferometry experiment mentioned in the first section of this report demonstrates that circular atom-superconducting cavities achieve the strong coupling condition (single-photon and even vacuum effects are observable with these systems). Recent progresses in the technology of open superconducting cavities, in which cavity damping times on the order of 10^{-3} s have been achieved, make us feel confident that the second condition will be met and that some of the fascinating experiments discussed above will soon be realized.

REFERENCES

1. EINSTEIN, A., B. PODOLSKY & N. ROSEN. 1935. Phys. Rev. **47:** 777.
2. BELL, J. S. 1964. Physics (Long Island City, N.Y.) **1:** 195.
3. ASPECT, A., J. DALIBARD & G. ROGER. 1982. Phys. Rev. Lett. **49:** 1804.
4. SCHRÖDINGER, E. 1935. Naturwissenschaften **23:** 807; 823; 844.
5. CALDERA, A. O. & A. J. LEGGETT. 1985. Phys. Rev. **A31:** 1059.
6. YURKE, B. & D. STOLER. 1986. Phys. Rev. Lett. **57:** 13.
7. HAROCHE, S. 1992. Cavity quantum electrodynamics. *In* Systèmes Fondamentaux en

Optique Quantique: Les Houches Session LIII. J. Dalibard, J. M. Raimond & J. Z. Justin, Eds. Elsevier. Amsterdam/New York.
8. BRUNE, M., S. HAROCHE, V. LEFÈVRE, J. M. RAIMOND & N. ZAGURY. 1990. Phys. Rev. Lett. **65:** 976.
9. BRUNE, M., S. HAROCHE, J. M. RAIMOND, L. DAVIDOVICH & N. ZAGURY. 1992. Phys. Rev. **A45:** 5193.
10. HAROCHE, S., M. BRUNE & J. M. RAIMOND. 1992. Appl. Phys. **B54:** 355.
11. HAROCHE, S., M. BRUNE & J. M. RAIMOND. 1992. J. Phys. (Paris) **II:** 659.
12. HAROCHE, S. & J. M. RAIMOND. 1994. Manipulation of nonclassical field states by atom interferometry. *In* Cavity Quantum Electrodynamics (Supplement of Advances in Atomic and Molecular Physics). Academic Press. New York.
13. GREENBERGER, D. M., M. A. HORNE & A. ZEILINGER. 1989. *In* Bell's Theorem, Quantum Theory, and Conceptions of the Universe. M. Kafatos, Ed. Kluwer. Dordrecht.
14. MEYSTRE, P. 1992. *In* Progress in Optics XXX. E. Wolf, Ed. Elsevier. Amsterdam/New York.
15. DAVIDOVICH, L., A. MAALI, M. BRUNE, J. M. RAIMOND & S. HAROCHE. 1993. Phys. Rev. Lett. **71:** 2360.
16. HAROCHE, S., M. BRUNE, J. M. RAIMOND & L. DAVIDOVICH. 1993. Mesoscopic quantum coherences in cavity QED. *In* Fundamentals of Quantum Optics III. Springer-Verlag. New York/Berlin.
17. DAVIDOVICH, L., A. MAALI, M. BRUNE, J. M. RAIMOND & S. HAROCHE. 1994 (August). Phys. Rev. **A50:** R895.
18. BENNETT, C. H., G. BRASSARD, C. CREPEAU, R. JOZSA, A. PERES & W. WOOTTERS. 1993. Phys. Rev. Lett. **70:** 1895.
19. NUSSENZWEIG, P., F. BERNARDOT, M. BRUNE, J. HARE, J. M. RAIMOND, S. HAROCHE & W. GAWLIK. 1993. Phys. Rev. **A48:** 3991.
20. HULET, R. G. & D. KLEPPNER. 1983. Phys. Rev. Lett. **51:** 1430.
21. GROSS, M. & J. LIANG. 1986. Phys. Rev. Lett. **57:** 3160.
22. RAMSEY, N. 1985. Molecular Beams. Oxford University Press. London/New York.
23. BRUNE, M., P. NUSSENZWEIG, F. SCHMIDT-KALER, F. BERNARDOT, A. MAALI, J. M. RAIMOND & S. HAROCHE. 1994. Phys. Rev. Lett. **72:** 3339.
24. MESCHEDE, D., H. WALTHER & G. MÜLLER. 1985. Phys. Rev. Lett. **54:** 551.
25. REMPE, G., H. WALTHER & N. KLEIN. 1987. Phys. Rev. Lett. **58:** 353.
26. REMPE, G., F. SCHMIDT-KALER & H. WALTHER. 1990. Phys. Rev. Lett. **64:** 2783.
27. FILIPOWICZ, P., J. JAVANAINEN & P. MEYSTRE. 1986. Phys. Rev. **A34:** 3077; LUGIATO, L. A., M. O. SCULLY & H. WALTHER. 1987. Phys. Rev. **A36:** 740.
28. REMPE, G. & H. WALTHER. 1990. Phys. Rev. **A42:** 1650.
29. MEYSTRE, P. & E. M. WRIGHT. 1988. Phys. Rev. **A37:** 2524.
30. BRUNE, M., J. M. RAIMOND, P. GOY, L. DAVIDOVICH & S. HAROCHE. 1987. Phys. Rev. Lett. **59:** 1899.
31. DAVIDOVICH, L., J. M. RAIMOND, M. BRUNE & S. HAROCHE. 1987. Phys. Rev. **A36:** 3771.
32. MERMIN, N. D. 1990. Phys. Today **June:** 9.
33. MERMIN, N. D. 1990. Phys. Rev. Lett. **65:** 1838.
34. MALKIN, I. A. & V. I. MAN'KO. 1979. Dynamical Symmetries and Coherent States of Quantum Systems. Nauka. Moscow.
35. JANSKY, J. & A. V. VINOGRADOV. 1990. Phys. Rev. Lett. **64:** 2771.
36. SCHLEICH, W., M. PERNIGO & F. LEKIEN. 1991. Phys. Rev. **A44:** 2172.
37. BUZEK, V., H. MOYA-CESSA, P. L. KNIGHT & S. D. L. PHOENIX. 1992. Phys. Rev. **A45:** 8190.
38. VON NEUMANN, J. 1932. Die Mathematische Grundlagen der Quantenmechanik. Springer-Verlag. Berlin; HEPP, K. 1972. Helv. Phys. Acta **45:** 237; BELL, J. S. 1975. Helv. Phys. Acta **48:** 93; WHEELER, J. A. & W. ZUREK, Eds. 1983. Quantum Theory and Measurement. Princeton University Press. Princeton, New Jersey; ZUREK, W. 1986. Phys. Rev. **D24:** 1516; HAAKE, F. & D. WAALS. 1986. Phys. Rev. **A36:** 730; ZUREK, W. 1991. Phys. Today **44**(No. 10): 36.
39. BRAUNSTEIN, S. L., A. MANN & M. REVZEN. 1992. Phys. Rev. Lett. **68:** 3259.
40. SLEATOR, T. & H. WEINFURTER. 1995. This volume.

Quantum Measurement in Quantum Optics[a]

H. J. KIMBLE,[b] O. CARNAL,[b] Z. HU,[b] H. MABUCHI,[b]
E. S. POLZIK,[c] R. J. THOMPSON,[b]
AND Q. A. TURCHETTE[b]

[b]*Norman Bridge Laboratory of Physics*
California Institute of Technology
Pasadena, California 91125

[c]*Institute of Physics and Astronomy*
University of Aarhus
Aarhus, Denmark

Since the earliest observations of manifestly quantum or nonclassical fields in the 1970s, quantum optics has continued to provide unique opportunities for investigations of fundamental issues in quantum measurement. Indeed, in recent years, exquisite capabilities have been developed for generating a zoology of nonclassical states of the electromagnetic field and for manipulating the field for sensitivity beyond the vacuum-state limit.

Our intent in this abbreviated contribution is to describe some of the activities of the Quantum Optics Group at the California Institute of Technology in the arena of quantum measurement. The themes of the research programs are both (i) retrospective with the laboratory demonstration of what had previously been only gedanken-experiments [such as the realization of the Einstein-Podolsky-Rosen (EPR) experiment] and (ii) prospective with the development of new technical capabilities and the emergence of new theoretical concepts (such as the generation of "arbitrary" states of the electromagnetic field and the implementation of quantum logic).

SQUEEZED STATES OF LIGHT

Parametric downconversion has been employed for the generation of squeezed states of light.[1] This experiment has provided a direct operational verification of the Heisenberg uncertainty principle for light and thereby of the field commutation relation itself.[2] In addition, it has become the prototypical avenue for the production of squeezed light and has led to the largest degree of quadrature-phase squeezing.[3] Our group has employed squeezed light to achieve measurement sensitivity beyond the standard quantum limit.[4–6]

[a]This work was supported by the National Science Foundation, the Office of Naval Research, and the Army Research Office.

QUANTUM CORRELATIONS FOR SPATIALLY SEPARATED BEAMS

The process of nondegenerate parametric downconversion has been utilized in a variety of experiments, including a faithful realization of the original gedankenexperiment of EPR.[7] As opposed to previous work with discrete spin or polarization variables, this is the first experiment for observables with continuous spectra as in the original proposal of EPR. The experiment attempts to confront the important questions of the generalization of the Bell inequalities to continuous variables.

We have also investigated the fundamental noise performance of an optical amplifier experimentally.[8] Here, the standard noise limits from the theorem of Caves are surpassed by "squeezing" the internal modes of the amplifier. The experiment is otherwise significant in that it is an example of a cascaded quantum system, for which the manifestly quantum fields from one system are used in turn to excite a second system.

Finally, back-action evading measurement has been demonstrated for parametric amplification with polarization mixing.[9]

THE STANDARD QUANTUM LIMIT (SQL) FOR THE POSITION OF A FREE MASS

Although in general terms it was a controversial topic throughout the 1980s, Jaekel and Reynaud[10] and others[11,12] have suggested a practical avenue for achieving sensitivity beyond the SQL for the position of a free mass. The basic scheme employs squeezed light for interferometric sensing of position, with the fluctuations of position and momentum correlated in a cunning fashion by proper choice of the orientation of the squeezing ellipse.

Of course, the quantum noise in this scheme is observable only to the extent that it is dominant over other sources of technical noise. Following the work of Saulson,[13] we have shown that it should be possible to find windows in frequency space where in fact the SQL could be experimentally observable.[14] The fundamental ingredients are a large degree of quadrature squeezing,[3] extremely low-loss optical coatings,[15] and materials of exceptionally high mechanical quality factor.[16]

SYNTHESIS OF "ARBITRARY" QUANTUM STATES

In collaboration with Parkins, Marte, and Zoller, we have proposed a new avenue for synthesizing quantum states of the field that should allow for the generation of quite general superpositions of Fock states.[17] The basic scheme employs adiabatic passage and strong coupling in cavity QED. It should be capable of dial-a-state service, whereby the coefficients of the Fock state expansion are preselected; the passage of a single atom then leads with near unit probability to the desired state for the intracavity field. The process is, in principle, reversible (to the extent allowed by cavity dissipation) and hence should also find gainful employment for various applications involving measurement of quantum fields. As well, we and others have suggested processes for the realization of certain quantum logic gates that employ this adiabatic passage scheme.

A necessary condition for the implementation of this scheme is the attainment of strong coupling for individual intracavity atoms. Several years ago, our group demonstrated such capabilities in the optical domain.[18] A comparison to similar developments in the microwave domain is given in reference 19.

ATOM GALLERIES FOR WHISPERING ATOMS

We have proposed coupling atoms to the evanescent fields of whispering gallery modes of quartz microspheres.[20] Because of the small mode volume, the electric field strength per photon (and hence the coherent coupling constant) can be quite large. At the same time, because of the high Q (2×10^9 at 850 nm in our laboratory), the cavity storage time can be long enough to allow for diverse phenomena such as quantum state synthesis by adiabatic passage and even the generation of quantum superpositions for the fields of two different microspheres. The large field strength per photon also should enable investigations that combine the physics of cooling and trapping with that of cavity QED. For example, we have suggested binding atoms in stable orbits around a microsphere resonator and employing the field mode to form a de Broglie resonator for the atomic matter wave.[20]

TRAPPING BY THE NUMBERS

We have recently succeeded in trapping individual cesium atoms in a magneto-optical trap, with resulting localization on the scale of 200 μm.[21] This work is currently being extended to a dipole-force trap, with the objective of achieving spatial localization of several microns. Such capabilities will enable various fundamental experiments in quantum optics, including our work in cavity QED.

REFERENCES

1. Wu, L. A., H. J. Kimble, J. L. Hall & H. Wu. 1986. Phys. Rev. Lett. **57:** 2520.
2. Wu, L. A., M. Xiao & H. J. Kimble. 1987. J. Opt. Soc. Am. B **4:** 1465.
3. Polzik, E. S., J. Carri & H. J. Kimble. 1992. Appl. Phys. B **55:** 279.
4. Xiao, M., L. A. Wu & H. J. Kimble. 1987. Phys. Rev. Lett. **59:** 278.
5. Xiao, M., L. A. Wu & H. J. Kimble. 1988. Opt. Lett. **13:** 476.
6. Polzik, E. S., J. Carri & H. J. Kimble. 1992. Phys. Rev. Lett. **68:** 3020.
7. Ou, Z. Y., S. F. Pereira, H. J. Kimble & K. C. Peng. 1992. Phys. Rev. Lett. **68:** 3663; Appl. Phys. B **55:** 265.
8. Ou, Z. Y., S. F. Pereira & H. J. Kimble. 1993. Phys. Rev. Lett. **70:** 3239.
9. Pereira, S. F., Z. Y. Ou & H. J. Kimble. 1993. Phys. Rev. Lett. **72:** 214.
10. Jaekel, M. T. & S. Reynaud. 1990. Europhys. Lett. **13:** 301.
11. Caves, C. M. 1990. Private communication.
12. Unruh, W. G. 1983. *In* Quantum Optics, Experimental Gravitation, and Measurement Theory. P. Meystre & M. O. Scully, Eds.: 647. Plenum. New York.
13. Saulson, P. 1990. Phys. Rev. D **42:** 2437.
14. Kimble, H. J. 1993. *In* Perspectives in Neutrinos, Atomic Physics, and Gravitation. J. Tran Thanh Van, T. Damour, E. Hinds & J. Wilkerson, Eds.: 349. Editions Frontières. Gif-sur-Yvette, France.
15. Rempe, G., R. J. Thompson, H. J. Kimble & R. Lalezari. 1992. Opt. Lett. **17:** 363.

16. Braginsky, V. B., V. P. Mitrofanov & O. A. Okhrimenko. 1993. Phys. Lett. A **175:** 82; Braginsky, V. B., V. P. Mitrofanov & V. I. Panov. 1985. Systems with Small Dissipation. University of Chicago Press. Chicago.
17. Parkins, A. S., P. Marte, P. Zoller & H. J. Kimble. 1993. Phys. Rev. Lett. **71:** 3095.
18. For a review, see: Kimble, H. J. 1993. *In* Cavity Quantum Electrodynamics: Advances in Atomic, Molecular, and Optical Physics. P. Berman, Ed. Academic Press. New York.
19. Kimble, H. J., O. Carnal, N. Georgiades, H. Mabuchi, E. S. Polzik, R. J. Thompson & Q. A. Turchette. 1994. *In* Proceedings of the 1994 International Conference on Atomic Physics.
20. Mabuchi, H. & H. J. Kimble. 1994. Opt. Lett. **19:** 749.
21. Hu, Z. Y. & H. J. Kimble. 1994. Opt. Lett. In press.

Entangling Photons Radiated by Independent Pulsed Sources[a]

MAREK ZUKOWSKI,[b,c,d] ANTON ZEILINGER,[c]
AND HARALD WEINFURTER[c]

[c]*Institut für Experimentalphysik*
Universität Innsbruck
A-6020 Innsbruck, Austria

[d]*Instytut Fizyki Teoretycznej i Astrofizyki*
Uniwersytet Gdanski
PL-80-952 Gdansk, Poland

INTRODUCTION

By pumping two or more downconversion crystals with a pulsed laser, it is experimentally feasible to obtain with nonnegligible probability one photon pair from each crystal within a pulse. Then, by registering the idlers in such a way that this does not provide path information, we observe that the signal photons emitted together with the idlers are projected onto the desired entangled state. The necessary condition for erasure of idler path information is that the idler filter bandwidth results in an idler coherence time that significantly exceeds the pump pulse length. We show the importance of this technique for photon correlation schemes that would permit one to perform "event-ready" two-photon Bell tests, entanglement swapping, and quantum teleportation and to obtain GHZ states.

Entanglement is at the source of a number of pure quantum phenomena, such as the correlations violating Bell's inequalities,[1] quantum teleportation,[2] GHZ correlations,[3] and various other nonclassical interference phenomena.[4] Entanglement between two or more particles was generally viewed as a consequence of the fact that the particles involved did originate from the same source or at least were interacting at some earlier time. However, it has first been suggested in seminal papers by Yurke and Stoler[5] that the correlations of particle detection events required for a Bell test can even arise for photons, or any kind of particle for that matter, originating from independent sources. What, though, are the operational procedures required to achieve such correlations? If we can observe violations of Bell's inequality for (destructive) registrations of particles coming from independent sources, can we also entangle them in a nondestructive manner? Is this possible for particles that do not interact at all and that share no common past? What would be the requirements for an experimental realization of such a scheme? These questions will be discussed here. In general, the answers could be summed up in the following way:[6] it turns out

[a]This work was supported by the Austrian "Fond zur Förderung der Wissenschaftlichen Forschung" (Project No. S6502) and National Science Foundation Grant No. PHY92-13964.
[b]M. Zukowski was supported by a grant from the University of Gdansk (No. BW/5400-5-0156-4).

that the conditions for obtaining entangled states require specific, not immediately intuitive, choices of coincidence timing that enable one to monitor the emission events of the independent sources as well as to erase the Welcher-Weg information. In this work, we study the possibility of the application of pulsed photon sources to meet these requirements in real experiments.

The initially proposed setup,[6] which serves to correlate independent particles via "entanglement swapping", employs parametric downconversion (PDC) sources pumped by cw lasers. One of the requirements is then narrow filtering of some photons, called the idlers, and detecting them in ultracoincidence, that is, with a coincidence window much narrower than the filter bandwidth time.[6] Commercially available filter bandwidths are of the order of 0.3 nm at best (for 700-nm light), which implies a coherence time of the order of 6 ps. This is at least two orders of magnitude shorter than the time resolution of state-of-the-art single-photon detectors. To overcome this problem, one could use extremely high-Q cavities as filters, a choice that clearly would result in forbiddingly low count rates.

It is the main purpose of the present report to demonstrate that the limitation described above can actually be overcome. We explicitly propose the use of pulsed pump lasers with sufficiently narrow pulse widths to feed the PDC sources. The idlers are now to be observed through filters resulting in a coherence time longer than the pulse width and no stringent requirements for the detection times of the idlers will be necessary.

ENTANGLING INDEPENDENT PARTICLES

Although entanglement of initially independent particles seems to be surprising at first sight, it may easily be understood on the basis of the following observation. Consider, for simplicity, the factorizable two-particle state:

$$|\Psi\rangle = |\alpha\rangle_1 |\beta\rangle_2, \tag{1}$$

where the first ket describes particle 1 and the second ket describes particle 2. Consider also the entangled state,

$$|\Psi^+\rangle = (\sqrt{1/2})(|a\rangle_1 |b\rangle_2 + |c\rangle_1 |d\rangle_2), \tag{2}$$

with, for simplicity, $\langle a|c\rangle = 0$ and $\langle b|d\rangle = 0$. Then, it is always possible to obtain an entangled state of the form of equation 2 out of the state of equation 1 using the projection operator,

$$\hat{P} = |\Psi^+\rangle\langle\Psi^+|, \tag{3}$$

acting on $|\Psi\rangle$:

$$\hat{P}|\Psi\rangle = c|\Psi^+\rangle. \tag{4}$$

Finally, provided that state 1 is not orthogonal to each of both terms in the superposition forming state 2, the relation $|c|^2 = |\langle\Psi^+|\Psi\rangle|^2$ gives the probability for obtaining the entangled state.

We have thus given the general reason of why it is possible to obtain entanglement between independent particles in an initially factorizable state. Obviously, the scheme can be applied to various systems. Slight modifications should account for the specific types of entanglement to be achieved, the number of particles involved, and so on.

In general, the question obviously is how to obtain experimentally a setup realizing the appropriate projection operator.

ENTANGLEMENT SWAPPING

The generic example of the projection method studied here is that of entanglement swapping.[6] Let us first outline its original version that aims at obtaining a two-particle maximally entangled state.

Consider the experimental setup of FIGURE 1. For clarity of the discussion, assume that two independently cw pumped downconversion crystals emit one entangled photon pair each. In a simplified description, the states of the pairs are represented by $(1/\sqrt{2})(|a\rangle_1|b\rangle_2 + |a'\rangle_1|b'\rangle_2)$ and $(1/\sqrt{2})(|c\rangle_3|d\rangle_4 + |c'\rangle_3|d'\rangle_4)$, respectively. Therefore, the initial four-photon state is given by the product,

$$|\Psi\rangle_{\text{in}} = (½)(|a\rangle_1|b\rangle_2 + |a'\rangle_1|b'\rangle_2)(|c\rangle_3|d\rangle_4 + |c'\rangle_3|d'\rangle_4), \tag{5}$$

where, clearly, photon 1 is entangled with photon 2 and photon 3 is entangled with photon 4. Yet, the photons emitted by crystal I are not correlated with the photons emerging from crystal II. Of course, the states written down above should be properly symmetrized. However, as the symmetrization does not change the results discussed here, we drop it for simplicity of the presentation.

For the following, let us call photons 1 and 4 the signals and photons 2 and 3 the idlers. Then, the basic procedure in order to entangle the uncorrelated signal photons and to obtain the state

$$|\Psi^+\rangle_{\text{signal}} = (1/\sqrt{2})(|a\rangle_1|d'\rangle_4 + |a'\rangle_1|d\rangle_4) \tag{6}$$

is to project the two idler photons into an entangled state (cf. equation 2).

This projection can be done (destructively), after overlapping their modes at two beam-splitters and by observing the idler photons in, say, detector i_1 and i_2 each. If the two idlers are indistinguishable, the joint detection projects the state of the initially unentangled idler photons into

$$|\Psi^+\rangle_{\text{idler}} = (1/\sqrt{2})(|b\rangle_2|c'\rangle_3 + |b'\rangle_2|c\rangle_3), \tag{7}$$

giving the maximal entanglement for the signal photons.

One should mention here in passing that, surprisingly, one of the major advantages of the procedure is that the registration acts of particles 2 and 3 can operationally define (at a distance) the ensemble of (now entangled) pairs of 1 and 4. Thus, one can, for example, perform an "event-ready" version of the EPR-Bell experiment.[6]

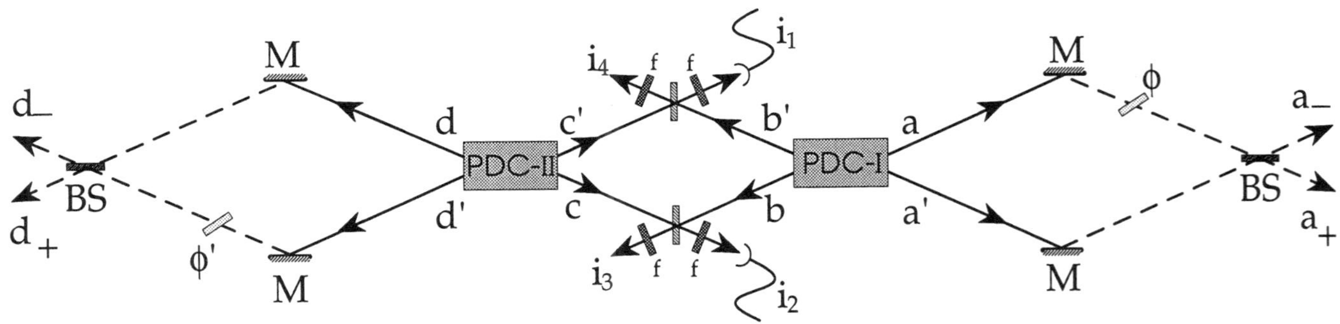

FIGURE 1. Principle of the entanglement swapping (see reference 6). Two cw pumped downconversion sources, PDC-I and PDC-II, emit a photon pair each. The specific geometry of each source is obtainable by a suitable arrangement of mirrors and apertures. The initially independent signal photons are entangled by ultracoincident registration of the idlers at i_1 and i_2. With the dashed setup, one can perform an event-ready version of the EPR-Bell experiment. (Terms: M = mirrors; BS = beam-splitters; f = filters; ϕ,ϕ' = local phase-shifters.)

WELCHER-WEG INFORMATION AND COINCIDENCE REQUIREMENTS

In order to effectively entangle photons emitted from different sources, one has to meet rather stringent—and, in part, not immediately intuitive—conditions on the detection of the photons involved.[6]

The above simplistic single mode description can give only some hints on the possible procedures. However, if the kets (of equations 5–7) are thought of as describing plane waves, one cannot obtain any predictions in the temporal domain. One intuitively feels that the necessary joint detection of the idler photons has to be "in coincidence". Thus, the all-important question is as follows: what are the experimental requirements for the two idler photons to be considered coincident? The conditions for the coincidence time gates can be obtained with the use of a more refined analysis that takes into account the multimode nature of the states involved.

First, one has to clarify the notion of coincidence. Note that even for ideal devices, with perfect time resolution, one always has to impose a certain time gate to *define* two counts as coincident. The idlers at i_1 and i_2 are being registered at completely random times throughout the experiment. We can always select the pairs of idlers out of those counts that are separated in time by less than a certain (ad hoc) imposed gate τ_i.

However, in the case of the coincidence of signals and idlers coming from a *single* downconversion act, the situation is different. It is well known[7] that the actual bandwidth $\Delta\omega$ of the observed PDC radiation determines the relation between the registration times of the twin PDC photons. The detection time of a signal is determined by the registration moment of its idler up to approximately its coherence time of $T_c \approx 1/\Delta\omega$ and vice versa. Because the downconverted light itself is extremely broadband, one defines T_c in a practical experiment through the apertures and filters used. Because of the energy conservation and because of the phase-matching conditions, the filtering of the idlers also defines ("nonlocally") the bandwidth of the coincident signal photons. Thus, if idler and signal counts are coincident (barring retardation effects) to within T_c, they can be thought of as coming from one PDC pair (provided that the rate of production of the signal-idler pairs is low enough so that emergence of two pairs from one source is a very rare event).

Let us return to considering the setup of FIGURE 1. If we select only the idlers detected in "ultracoincidence" (i.e., for $\tau_i \ll T_c$), the Welcher-Weg information is erased.[6] This is because such registrations are too close in time to discriminate which signal shares the source with which idler. Thus, we select signals in a "noninvasive" way. Due to the lack of Welcher-Weg information, the two two-particle processes interfere and we can have high visibility of the fringes. Therefore, we can approximately describe the preselected ensemble of signals by the maximally entangled state (equation 6).

Alternatively, one could also obtain high-visibility fringes by (post-)selection of only ultracoincident signal counts. However, such a procedure would be a selection of a subensemble after actual detection (i.e., destruction) of the particles.

The ultracoincidence condition requires the use of narrow filters in order to make the coherence time as long as possible. Nevertheless, if we consider using state-of-the-art filters, such as those described in the INTRODUCTION, no fast enough detectors exist at present (with a time resolution below 1 ps) as are necessary to select ultracoincident idlers. In the next section, we will show how the use of fast pulsed pump lasers can overcome this limitation.

A minor problem briefly requests our attention: Even if we register two idlers in the way described above, they might actually have come from two downconversion processes within the same crystal. Then, obviously, we do not obtain the desired entangled state. Yet, this case can easily be discarded as irrelevant. These "wrong" events do not lead to coincidence counts at different ends of FIGURE 1 and hence, tautologically, no entanglement arises. A beautiful analysis of the insignificance of such "wrong" events for the Bell theorem has been given by Yurke and Stoler[5] and thus will not be repeated here.

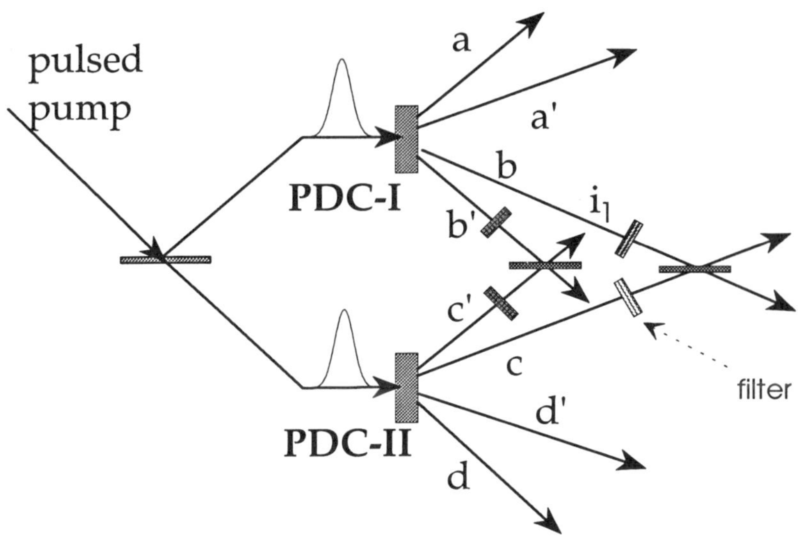

FIGURE 2. The two PDC crystals of the entanglement swapping system are now fed with simultaneously arriving pump pulses (cf. FIGURE 1). The narrow filters make the coherence time of the idler radiation much longer than the pulse width. Thus, detection of the idlers at i_1 and i_2 does not reveal from which crystal they were emitted. The two signal photons emerge in an entangled state via beams a and d' or a' and d.

PULSED PUMP: WELCHER-WEG CONSIDERATIONS

Consider the arrangement of FIGURE 2, where a single pulsed laser pumps two thin downconversion crystals. The crystals are thin so as not to lead to problems related to the property that the downconversion photons inside the crystal might have a different velocity compared to the pump or even compared to each other. As above, we again attempt to project the initially unentangled signal photons onto an entangled state by superposing the idlers and measuring them in coincidence. Then, certainly without any narrow filters in the beams, the tight time correlation within a photon pair emitted by the same crystal permits one again to associate simultaneously detected idlers and signals with each other. This provides path information and hence prohibits the creation of the entangled state.

We thus now insert filters into the idler paths. Using state-of-the-art, commercially available filters, we can achieve a filter coherence time of the order of 5 ps. On the other hand, it is no problem to obtain pulse durations of an order of magnitude shorter. Hence, the time span within which the two photon pairs are emitted can be shorter by more than an order of magnitude than the time interval over which the individual idler photons might emerge from their respective filters. Thus, it follows that the idler photons detected behind the filters carry practically no information anymore on the detection times of their twin signals. This implies that the path information was erased in the process. The signal photons are now projected onto the entangled state. Please note that the time resolution of the detectors plays no role whatsoever here. Of course, it should be better than the pulse period of the laser.

We assumed here that both sources are fed with pulses derived by 50%–50% beam-splitting of the original laser pulses and that they arrive at the crystals simultaneously. Of course, in principle, the pulses may originate from two different lasers, thus making the two sources fully independent. One should then take into consideration only those events where the pulses sufficiently overlap in time.

PULSED PUMP: SKETCH OF A QUANTITATIVE DESCRIPTION

The Welcher-Weg considerations presented above can give the first qualitative predictions and understanding of the basic physics involved. However, it is very important to know the threshold ratios of the pulse width to the time resolution of the idler filters in order to start to speak about real quantum mechanical entanglement. By this, we mean signal states leading to two-particle fringes of high enough visibility as required for the violation of the Bell inequalities.

One can obtain some rough quantitative estimates with the use of the following approximate description. Let us assume that the pulse is weak enough so that we are not troubled by multiple emissions or other effects specific to very strong pulses. Furthermore, one can assume that the pulse is longer than the crystal, as only in this case can one expect that the nonmonochromaticity of the pulse will not blur too much the phase-matching conditions (essentially, momentum conservation). Otherwise, the angular correlation of the emissions could be weakened. Thus, to a good approximation, we can treat the downconversion process as a superposition of processes that would occur for all monochromatic components of the pulse.

Under such assumptions, the PDC photon pair originating from one of the sources can be represented as being emitted in specified, well-defined directions, as dictated by the phase-matching, and the joint wave packet is constructed out of the superpositions of monochromatic modes. Thus, the two-photon state that can be produced by a single pulse at PDC-I (which we consider to incorporate also the idler filters and the aperture arrangement) can be described as

$$|\Psi_{\text{PDC-I}}\rangle = \int d\omega_i \int d\omega_s \int d\omega_0 g(\omega_0)\Delta(\omega_0 - \omega_s - \omega_i) f(\omega_i; \omega_f; \sigma_f)$$

$$\times \left(\frac{1}{\sqrt{2}}\right)(|\omega_s; a\rangle_1 |\omega_i; b\rangle_2 + |\omega_s; a'\rangle_1 |\omega_i; b'\rangle_2). \quad \mathbf{(8)}$$

Here, $|\omega_s; a\rangle_1$ ($|\omega_i; b\rangle_2$) describes the signal (idler) of frequency ω_s (ω_i) in beam a (b), emitted by PDC-I; the function g represents the spectral content of the pulse; and f is the transmission function of the filter with central frequency ω_f and bandwidth σ_f. Finally, function $\Delta(\omega_0 - \omega_s - \omega_i)$, which is highly peaked at the origin, reflects the phase-matching conditions and the energy conservation of the PDC process. For the perfect case, which we assume here, one can replace it by

$$\delta(\omega_0 - \omega_s - \omega_i). \quad \mathbf{(9)}$$

Within this limit, proper normalization of $|\Psi_{\text{PDC-I}}\rangle$ is obtained for both g and f normalized to unity.

The state $|\Psi_{\text{PDC-II}}\rangle$ can be described in a similar way and the two-source emission is described by $|\Psi\rangle = |\Psi_{\text{PDC-I}}\rangle|\Psi_{\text{PDC-II}}\rangle$ (cf. equation 5).

Let us describe in more quantitative terms the results of the wave packet collapse in the experiment. Assume for the moment *gedanken* detectors of infinitely sharp time resolution. Then, the detection of an idler in i_1 at t_1 and of another idler in i_2 at t_2 causes a wave packet collapse of the initial state $|\Psi\rangle = |\Psi_{\text{PDC-I}}\rangle|\Psi_{\text{PDC-II}}\rangle$ into the entangled state of the signals

$$|\Psi; t_1, t_2\rangle = \left(\frac{1}{\sqrt{N}}\right)\langle t_1, t_2, i|\Psi\rangle, \quad \mathbf{(10)}$$

where

$$|t_1, t_2, i\rangle = \left(\frac{1}{\sqrt{2}}\right)(|t_1, b'\rangle_2 |t_2, c\rangle_3 + |t_2, b\rangle_2 |t_1, c'\rangle_3) \quad \mathbf{(11)}$$

with, for example,

$$|t, b\rangle = \left(\frac{1}{\sqrt{2\pi}}\right)\int d\omega e^{i\omega t}|\omega, b\rangle \quad \mathbf{(12)}$$

(compare with the description presented by Fearn and Loudon[8]). The explicit form of the collapsed state is given by

$$|\Psi; t_1, t_2\rangle = \left(\frac{1}{\sqrt{2}}\right)(|g, f; t_1, a'\rangle_1 |g, f; t_2, d\rangle_4 + |g, f; t_2, a\rangle_1 |g, f; t_1, d'\rangle_4). \quad \mathbf{(13)}$$

Here, for example,

$$|g, f; t_1, a\rangle_1 = \left(\frac{1}{\sqrt{N(t_1)}}\right)\int d\omega_0 g(\omega_0)\int d\omega_s e^{-i(\omega_0-\omega_s)t_1} f(\omega_0, \omega_f, \sigma_f)|\omega_s, a\rangle_1, \quad \mathbf{(14)}$$

where $N(t_1)$ is the normalization constant. The above formulas were obtained using a number of simplifying assumptions—for example, that all source-detector optical paths are equal.

Let us now calculate the overall probability of the detection of two signals at a_i and d_j $(i, j = \pm)$ provided that (a) the pulse produced an emission as described by $|\Psi\rangle$ and (b) the two idlers were registered at i_1 and i_2. We shall not impose any time gates on the registration times; the pulse is of a short duration; and we can expect to collect the two signals within the pulse's width T and the idlers within $T_c = 1/\sigma \gg T$. Of course, we are interested in events associated with a single pulse. Thus, if T_c is in turn much smaller than the time separation of two pulses, the time integration can be extended to infinity. The detectors a_i and d_j are behind the interferometers and hence the aforementioned probability can be given by

$$P(i, j|\varphi, \varphi') = [P(i_1; i_2)]^{-1} \int dt \int dt' \int dt_1 \int dt_2 \\ \times P(t_1, t_2)|_1\langle t, a_i|_4\langle t', d_j|U(\varphi, \varphi')|\Psi; t_1, t_2\rangle|^2, \quad (15)$$

where $U(\varphi, \varphi')$ describes the action of the two spatially separated phase-shifters. The term $P(t_1, t_2)$ denotes the temporal distribution of the probability for the two idlers generated by one pulse and to be detected at i_1 and i_2, whereas $P(i_1; i_2)$ is the overall probability of such an event. One can simplify equation 15 by noticing that $P(t_1, t_2) = P(i_1, i_2)N(t_1)N(t_2)$.

Now, it is a straightforward, but otherwise lengthy, manipulation to show that the visibility of the two-signal fringes can be given by

$$V = \int dt \int dt' |G(t)|^2|G(t')|^2| \int dt'' F(t'' - t)F^*(t'' - t')|^2, \quad (16)$$

where $G(t)$, that is, the Fourier transform of $g(\omega)$, is given by

$$G(t) = \left(\frac{1}{\sqrt{2\pi}}\right) \int d\omega_0 g(\omega_0) e^{-i\omega_0 t}. \quad (17)$$

This represents the time-dependence of the pulse; $F(t)$ is the analogous transform of f and it represents the filter response function. Recall that g (and thus also G) is normalized to unity. Next, as f is similarly normalized, one has

$$|\int dt'' F(t'' - t)F^*(t'' - t')| \leq |\int dt'' F(t'' - t)F^*(t'' - t)| = 1. \quad (18)$$

The left-hand side of equation 18 is much smaller than unity if $|t - t'| > 1/\sigma_f$. Thus, the visibility (equation 16) can be close to 100% only if the width T of the pulse function G is much more narrower than $1/\sigma_f$. Only in this case does the region of integration, where the product $|G(t)||G(t')|$ has values that significantly contribute to the overall value of the integral, overlap with the region where the left-hand side of equation 18 is close to 1. Thus, the full integral is close to the square of the normalization value of G, that is, 1.

Let us illustrate this result with two examples. Consider first that both the pulse function and the filter function are Gaussians:

$$g(\omega_0) = \left[\frac{1}{(2\pi\sigma^2)^{1/4}}\right] \exp\left\{-\left(\frac{\omega_0 - \omega_p/2}{2\sigma}\right)^2\right\}, \quad (19)$$

$$f_{\mathrm{Gauss}}(\omega_s, \omega_p/2, \sigma_f) = \left[\frac{1}{(2\pi\sigma_f^2)^{1/4}}\right] \exp\left\{-\left(\frac{\omega_s - \omega_p/2}{2\sigma_f}\right)^2\right\}. \tag{20}$$

The resulting visibility reads

$$V = \frac{\sigma}{\sqrt{\sigma^2 + \sigma_f^2}}. \tag{21}$$

It has the appealing feature that $V \geq 1/\sqrt{2}$ only for $T_{\mathrm{pulse}} = 1/\sigma \leq 1/\sigma_f = T_{\mathrm{coh}}$; that is, we reach the threshold of the visibility of the two-particle fringes to violate the standard Bell inequalities exactly when the time resolution of our filter is equal to the pulse width.

For a rectangular pulse of duration T and for Lorentzian filters of width $\Delta\omega$, that is,

$$f_{\mathrm{Lorentz}}(\omega_s; \omega_p/2; \Delta\omega) \approx \frac{1}{(\omega_s - \omega_p/2) + i(\Delta\omega/2)}, \tag{22}$$

one obtains

$$V = 2\left[\left(\frac{1}{x}\right) + \left(\frac{1}{x^2}\right)(e^{-x} - 1)\right], \tag{23}$$

where $x = T\Delta\omega$. The Bell threshold is now at $x \approx 1.14$. For $x \ll 1$, the formula (equation 23) can be approximated by $V \approx 1 - x/3$.

CONCLUDING REMARKS

The above ideas can be employed in order to obtain high visibility in any other higher-order interference phenomena that involve particles emitted by independent sources.

For example, let us consider the phenomenon of quantum teleportation.[2] In its simplest form (devised for states in a two-dimensional Hilbert space), it involves the following steps. One starts with subsystem A in the arbitrary (in principle, unknown) pure state $|\varphi\rangle_A$ to be teleported. An independently emitted pair of subsystems $B + C$ is EPR-Bell-entangled. One then performs a measurement on $A + B$ that is supposed to result in projecting $A + B$ into one of four possible entangled "Bell" states. Depending on the actual state onto which $A + B$ collapses, one can perform one of four specific unitary transformations on C. This results in the quantum C object ending up in the state $|\varphi\rangle_C$. Thus, the initial state of A is transferred to C. The act of collapse involves a measurement that by its very nature is performed upon the two independently emitted particles $A + B$. If such an experiment is to be performed with photons, then one can devise the actual setup.[9] Following again the ideas explained in previous sections, either one has to impose the ultracoincidence condition for the joint registrations of photons A and B or one has to use the pulsed pumps of sufficiently narrow temporal width. Only then can the Welcher-Weg information be effectively erased, and thus particle C is obtained in the pure state

$|\varphi\rangle_C$. One can check operationally the fidelity of the teleportation process by performing an interference experiment upon C (which will result in high visibility only under the conditions mentioned above).

Another phenomenon that can be experimentally realized with the use of the methods presented above is the production of GHZ states. Let us present here the three-particle case. The scheme of the proposed setup is shown in FIGURE 3. In this case, the three PDC sources independently produce three pairs of EPR-correlated

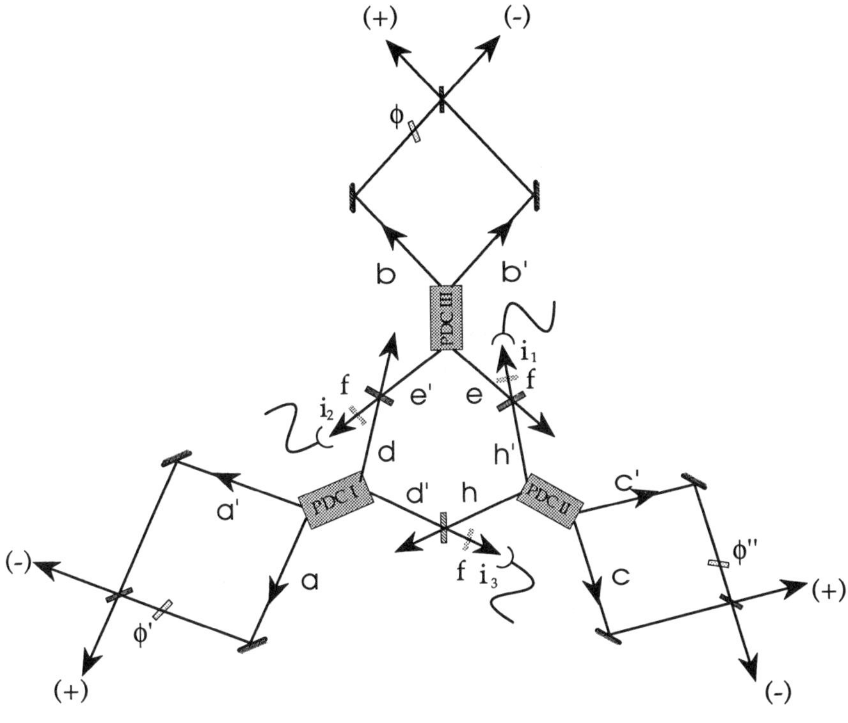

FIGURE 3. Principle of entanglement swapping leading to the GHZ correlations. The initially entangled two-particle emissions of the three PDC sources (equation 24) result in a GHZ state (equation 26) of the outward bound signals, after coincident registration of the idlers at i_1, i_2, and i_3.

particles. The initial state can be written as

$$(\sqrt{1/8})(|a\rangle_1|d\rangle_4 + |a'\rangle_1|d'\rangle_4)(|b\rangle_2|e\rangle_5 + |b'\rangle_2|e'\rangle_5)(|c\rangle_3|h\rangle_6 + |c'\rangle_3|h'\rangle_6). \qquad \textbf{(24)}$$

If we register one photon in each of the three detectors i_1, i_2, and i_3, we collapse (destructively) the idler photons into the GHZ state,

$$(\sqrt{1/2})(|d\rangle_4|e\rangle_5|h\rangle_6 + |d'\rangle_4|e'\rangle_5|h'\rangle_6), \qquad \textbf{(25)}$$

whereas the untouched signals correlated with the three registered idlers continue in the collapsed state,

$$(\sqrt{1/2})(|a\rangle_1|b\rangle_2|c\rangle_3 + |a'\rangle_1|b'\rangle_2|c'\rangle_3), \tag{26}$$

and can be fed into three interferometers to observe GHZ correlations.

The above process can be successfully implemented only if the Welcher-Weg information, contained in the tight time correlation of the individual PDC emissions, is erased. Again, this can be achieved by passing the idlers through sufficiently narrow filters and then either selecting only the ultracoincident idler counts (for cw pumped PDCs) or using pulsed pumps of narrow enough pulse width. Of course, only the latter technique seems to be able to overcome the limitations associated with the poor time resolution of current photodetectors. In fact, numerical estimates using state-of-the-art intensities of pulsed lasers and estimates of downconversion efficiencies suggest that such an experiment is indeed feasible with current technology.

The GHZ states are very interesting as they lead to correlations between three (or more) particles, in clear-cut contradiction with the Einstein-Podolsky-Rosen idea of "elements of reality". However, this holds only for perfect correlations of 100% visibility. In an analysis of a real experiment, one has to resort to some form of Bell's inequalities. Interestingly, the threshold of the visibility of the multiparticle interference pattern that violates such inequalities decreases exponentially with the number of particles involved.[10] In the three-particle case, according to current knowledge, this limit is already achieved at a visibility of 50%,[11,12] as opposed to 71% for two-particle correlations. Thus, the GHZ experiments do not impose sharper coincidence/pump-width requirements than the two-particle EPR-Bell ones (in the "event-ready" mode).

REFERENCES

1. BELL, J. S. 1964. Physics **1:** 195.
2. BENNETT, C. H., G. BRASSARD, C. CREPEAU, D. JOZSA, A. PERES & W. K. WOOTTERS. 1993. Phys. Rev. Lett. **70:** 1895.
3. GREENBERGER, D. M., M. A. HORNE & A. ZEILINGER. 1989. *In* Bell's Theorem, Quantum Theory, and Conception of the Universe. M. Kafatos, Ed. Kluwer. Dordrecht.
4. GREENBERGER, D. M., M. A. HORNE & A. ZEILINGER. 1993. Phys. Today **August:** 22.
5. YURKE, B. & D. STOLER. 1992. Phys. Rev. Lett. **68:** 1251; Phys. Rev. A **46:** 2229.
6. ZUKOWSKI, M., A. ZEILINGER, M. A. HORNE & A. K. EKERT. 1993. Phys. Rev. Lett. **71:** 4287.
7. HONG, C. H. & L. MANDEL. 1986. Phys. Rev. Lett. **56:** 58.
8. FEARN, H. & R. LOUDON. 1989. J. Opt. Soc. Am. B **6:** 917.
9. WEINFURTER, H. 1994. Europhys. Lett. **25:** 559.
10. ZUKOWSKI, M. 1993. Phys. Lett. A **177:** 290.
11. MERMIN, N. D. 1990. Phys. Rev. Lett. **65:** 1838.
12. ARDEHALI, M. 1992. Phys. Rev. A **46:** 5375.

Can We Measure the Wave Function of a Single Wave Packet of Light?

Brownian Motion and Continuous Wave Packet Collapse in Repeated Weak Quantum Nondemolition Measurements

ORLY ALTER AND YOSHIHISA YAMAMOTO

ERATO Quantum Fluctuation Project
Edward L. Ginzton Laboratory
Stanford University
Stanford, California 94305-4085

Two separate structures exist in quantum mechanics: the observables, which are represented by operators of the Hilbert space, and the physical systems, which are described by state vectors or wave functions. The wave function is said to have an epistemological meaning because it contains all the relevant information about the physical system under consideration. The result of a precise measurement on a single quantum system is always one of the eigenvalues of the measured observable. After the measurement, the wave function of the measured system collapses to the corresponding eigenstate, according to the projection postulate. There is no one-to-one correspondence between the result of a single measurement and the state of the system before the measurement. In order to measure the initial wave function of the system, one needs to prepare an ensemble of systems with the same wave function and then measure them all. The wave function is obtained from the statistics of the results of measurements performed on this ensemble. Recently, Aharonov, Anandan, and Vaidman[1,2] suggested that the wave function of a single quantum system could be measured, thereby giving the wave function an ontological significance, that is, physical reality in its own right, in addition to its usual epistemological role. They suggested employing a series of "protective measurements", where an a priori knowledge of the wave function enables one to measure this wave function and to protect it from changing at the same time. However, with this a priori knowledge, one could reproduce the wave function after each measurement for an arbitrarily large number of times and one could then measure the wave function in the conventional manner.

In this report, we investigate the possibility of measuring the wave function of a single quantum system with no a priori knowledge of the wave function in order to explore a real ontological meaning of the wave function. We study the case of repeated weak quantum nondemolition (QND) measurements,[3,4] for which we can assume that the signal and the probe are in pure states before the measurement, without loss of generality. In this case, the signal is left in a pure state after the measurement. The unitary interaction between the probe and the signal does not allow transitions between any two eigenstates of the measured observable, where such transitions cause an appreciable change of the wave function. Also, the QND measurement can be chosen to be as weak as we want. It is possible, therefore, to measure the signal many times, using weak QND measurements, before the wave function of the signal is changed significantly. The measurement results are all generated under some influence of the initial wave function and one may expect the statistics of these results to give at least partial information about this wave function. In this work, we show that this intuitive picture fails and one cannot, in fact, extract any information about the initial wave function of the signal at all. By "information about the wave function", we mean information about both the average and the variance of the measured observable, that is, the center and the width of the wave packet, with finite probability errors. Information about the center position alone corresponds to a measurement of the observable, where information about the variance reveals the wave function.

In our model, a series of photon-number QND measurements is performed on a single wave packet of light.[5–8] A signal wave packet of light, $|\psi_0\rangle_s$, is correlated to a probe wave packet, $|\alpha_0\rangle_p$, in an optical Kerr medium. This process is described by the unitary operator $\hat{U} = \exp(i\mu\hat{n}_s\hat{n}_p)$, where $\hat{n}_s$ and $\hat{n}_p$ are the signal and probe photon-number operators, respectively, and μ is the coupling strength.[9] The photon-number of the signal, $\hat{n}_s$, shifts the phase of the probe, $\Delta\hat{\phi}_p \cong \mu\hat{n}_s$. Then, the second-quadrature amplitude of the probe, $\hat{a}_2$, is measured precisely by a homodyne detection. The inferred signal photon-number, $\tilde{n}$, is obtained from α_2^1, that is, the result of the probe quadrature measurement, $\tilde{n}_1 \cong \alpha_2^1/(|\alpha_0|\mu)$, where $|\alpha_0|$ is the initial excitation of the probe. A back-action noise is imposed on the phase of the signal by the probe photon-number, but this noise does not influence the photon-number distribution of the signal. The probability-amplitude operator, $\hat{Y}_1 = {}_p\langle\alpha_2^1|\hat{U}|\alpha_0\rangle_p$, completely describes the three stages of this QND measurement:[10] the preparation of the probe state, $|\alpha_0\rangle_p$; the interaction of this state with the signal, $\hat{U}$; and the results of the measurement, α_2^1, which corresponds to the state of the probe after the measurement, $|\alpha_2^1\rangle_p$. The probability of obtaining α_2^1 as the readout of the homodyne detection is $P(\alpha_2^1) = \mathrm{Tr}_s[\hat{Y}_1^+\hat{Y}_1\hat{\rho}_0]$, where $\hat{\rho}_0 = |\psi_0\rangle_s\,{}_s\langle\psi_0|$ is the density operator of the signal before the measurement. After a homodyne detection, which results in α_2^1, the signal density operator becomes $\rho_1 = P(\alpha_2^1)^{-1}\hat{Y}_1\hat{\rho}_0\hat{Y}_1^+$ and the corresponding photon-number distribution is $P_1(n) = {}_s\langle n|\hat{\rho}_1|n\rangle_s$.

The same measurement procedure is repeated k times. Each time, the measurement is performed on the output signal of the previous measurement, using a new probe state. We get a series of second-quadrature amplitude readouts, $(\alpha_2^1, \alpha_2^2, \ldots, \alpha_2^k)$, which correspond to a series of inferred photon-number values, $(\tilde{n}_1, \tilde{n}_2, \ldots, \tilde{n}_k)$. It is the statistics of $(\tilde{n}_1, \tilde{n}_2, \ldots, \tilde{n}_k)$, in the limit of weak measurements, that are expected to give the initial photon-number distribution of the signal, $P_0(n) = {}_s\langle n|\hat{\rho}_0|n\rangle_s$. The probability of obtaining a specific series of inferred photon-number values is

$$P(\tilde{n}_1, \tilde{n}_2, \ldots, \tilde{n}_k)\prod_{i=1}^{k} d\tilde{n}_i = P(\alpha_2^1, \alpha_2^2, \ldots, \alpha_2^k)\prod_{i=1}^{k} d\alpha_2^i, \tag{1}$$

where $P(\alpha_2^1, \alpha_2^2, \ldots, \alpha_2^k) = \mathrm{Tr}_s[\hat{Z}_k^+\hat{Z}_k\hat{\rho}_0]$ and where $\hat{Z}_k = \hat{Y}_k\ldots\hat{Y}_2\hat{Y}_1$ is the total probability-amplitude operator that describes the whole process of k repeated QND measurements.[10] The photon-number distribution of the signal wave function after the k-th measurement, $P_k(n) = {}_s\langle n|\hat{\rho}_k|n\rangle_s$, is calculated from the corresponding signal density operator,

$$\hat{\rho}_k = P(\alpha_2^1, \alpha_2^2, \ldots, \alpha_2^k)^{-1}\hat{Z}_k\hat{\rho}_0\hat{Z}_k^+. \tag{2}$$

Note that $\hat{Z}_k$ is symmetric in $(\alpha_2^1, \alpha_2^2, \ldots, \alpha_2^k)$; that is, it is independent of the order in which the results are obtained. This is because the different $\hat{Y}_i$ operators commute with each other. Therefore, the probability of obtaining these results, $P(\alpha_2^1, \alpha_2^2, \ldots, \alpha_2^k)$, and the final photon-number distribution after these results are measured, $P_k(n)$, are both symmetric in $(\alpha_2^1, \alpha_2^2, \ldots, \alpha_2^k)$. The probability of obtaining α_2^2 in the second measurement depends on the result of the first measurement, α_2^1. Yet, the process of measuring α_2^1 first and α_2^2 second has exactly the same probability as the process in which α_2^2 is measured first and α_2^1 is measured second. Also, there is no inherent difference between the changes caused to the wave function by the different consecutive measurements. Because the wave function of the system is slightly different at each measurement, the above observation, namely, that $\hat{Z}_k$ is independent of the order of the measurement results, suggests that no information about the width of the wave function is contained in the statistics of the readouts, $P(\alpha_2^1, \alpha_2^2, \ldots, \alpha_2^k)$.

To confirm this, let us assume that the initial photon-number distribution is a Gaussian, that is, a normal distribution,[11] $P_0(n) = N[n,n_0,\delta_0^{-1}]$. Physically, the photon-number distribution is a discrete distribution, where $n \geq 0$. If the signal is initially in a squeezed state with a large excitation, that is, $n_0 \gg 1$, this Gaussian approximation is valid. The initial distribution of the second-quadrature amplitude of the probe, with the probe being in a squeezed state with a zero phase, is also a Gaussian, centered at zero with the variance $\langle\Delta\hat{a}_2^2\rangle = e^{-2r}/4$, where r is the squeezing parameter. Our model describes a measurement process in which both the signal and the probe have normal distributions. Many other physical schemes are described in the same way—for example, the QND measurement of one of the quadrature amplitudes of a wave packet of light, using a nondegenerate parametric amplification.[12,13] Using $\alpha_2^i \cong |\alpha_0|\mu\tilde{n}_i$ in equations 1 and 2, the probability distribution for inferring a series of photon-number values is

$$P(\tilde{n}_1, \tilde{n}_2, \ldots, \tilde{n}_k) = \int_{-\infty}^{\infty} dn\, N[n, n_0, \delta_0^{-1}] \prod_{i=1}^{k} N[\tilde{n}_i, n, \delta_m^{-1}]. \tag{3}$$

The final photon-number distribution of the signal is

$$P_k(n) = P(\tilde{n}_1, \tilde{n}_2, \ldots, \tilde{n}_k)^{-1} N[n, n_0, \delta_0^{-1}] \prod_{i=1}^{k} N[\tilde{n}_i, n, \delta_m^{-1}]. \tag{4}$$

Here, $\delta_m = |\alpha_0|^2\mu^2/\langle\Delta\hat{a}_2^2\rangle$ is the strength of each consecutive measurement. The measurements are weak when the error associated with each measurement is much larger than the initial width of the photon-number distribution of the signal, that is, when $\delta_m \ll \delta_0$.

Consider the case of one measurement performed on a single wave packet of light. The probability distribution of measuring the inferred photon-number, $\tilde{n}_1$, is

$$P(\tilde{n}_1) = N[\tilde{n}_1, n_0, \delta_0^{-1} + \delta_m^{-1}]. \tag{5}$$

The signal wave function is changed to

$$P_1(n) = N[n, n_0^1, (\delta_0 + \delta_m)^{-1}], \tag{6}$$

$$n_0^1 = (\delta_0 n_0 + \delta_m \tilde{n}_1)(\delta_0 + \delta_m)^{-1}. \tag{7}$$

The center of the wave function, n_0^1, is shifted toward the measurement result, $\tilde{n}_1$, from its original value, n_0, whereas the width of the wave function narrows from δ_0^{-1} to $(\delta_0 + \delta_m)^{-1}$. If the measurement is weak, that is, $\delta_m \ll \delta_0$, both the shift and the narrowing are very small.

Before investigating the case of repeated QND measurements performed on a single wave packet, we analyze the case of one measurement performed on each wave packet in an ensemble of k wave packets, all prepared in the same initial state. In this case, each measurement is independent of the others. The probability of obtaining the inferred photon-number values, $(\tilde{n}_1, \tilde{n}_2, \ldots, \tilde{n}_k)$, is obviously independent of their order, $P(\tilde{n}_1, \tilde{n}_2, \ldots, \tilde{n}_k) = \Pi_{i=1}^k P(\tilde{n}_i)$. It is well known that the statistics of the results of the measurements in this case are analyzed by both the inferred average $\overline{n} = \Sigma_{i=1}^k \tilde{n}_i/k$ and the inferred variance $\overline{\Delta n^2} = \Sigma_{i=1}^k (\tilde{n}_i - \overline{n})^2/(k-1)$, in which all measurement results have the same weight. In terms of $\overline{n}$ and $\overline{\Delta n^2}$, the probability that the measurements performed on the ensemble would result in $(\tilde{n}_1, \tilde{n}_2, \ldots, \tilde{n}_k)$ is

$$\prod_{i=1}^{k} P(\tilde{n}_i)d\tilde{n}_i = [P(\overline{n})dn][P(S)dS]d\Omega_{k-1}, \tag{8}$$

where $S = (k-1)(\delta_0^{-1} + \delta_m^{-1})^{-1}\overline{\Delta n^2}$ and $d\Omega_{k-1}$ is a normalized infinitesimal element of the solid angle in dimension $(k-1)$, that is, $\int d\Omega_{k-1} = 1$. The probability distribution of the inferred average is

$$P(\overline{n}) = N[\overline{n}, n_0, k^{-1}(\delta_0^{-1} + \delta_m^{-1})]. \tag{9}$$

$P(\overline{n})$ is centered at the original center of the wave function, n_0. Therefore, the inferred average, $\overline{n}$, is a statistical measure of n_0. The variance of $P(\overline{n})$ is inversely proportional to the number of measurements, k. The probability error associated with this measurement decreases as the number of measurement results increases. The probability distribution of S is a chi-square distribution,[14] $P(S) = \chi^2[S, (k-1)]$. Therefore, the distribution of the inferred variance, $\overline{\Delta n^2}$, is centered at $\delta_0^{-1} + \delta_m^{-1}$, with the variance $2(k-1)^{-1}(\delta_0^{-1} + \delta_m^{-1})^2$. As k increases, the probability error for $\overline{\Delta n^2}$ to read $\delta_0^{-1} + \delta_m^{-1}$ decreases. By measuring an ensemble of wave packets, all with the same initial wave function, we can conclude that both the center of the wave function and its width can be inferred statistically. This corresponds to a measurement of the wave function.

Next, let us consider the changes in the measured wave function in the process of k repeated measurements performed on a single wave packet. From equation 4, we obtain that the final photon-number distribution after k repeated measurements, which results in $(\tilde{n}_1, \tilde{n}_2, \ldots, \tilde{n}_k)$, is

$$P_k(n) = N[n, n_0^k, (\delta_0 + k\delta_m)^{-1}], \tag{10}$$

$$n_0^k = \left(\delta_0 n_0 + \delta_m \sum_{i=1}^{k} \tilde{n}_i\right)(\delta_0 + k\delta_m)^{-1}. \tag{11}$$

As was noted before, $P_k(n)$ is symmetric in $(\tilde{n}_1, \tilde{n}_2, \ldots, \tilde{n}_k)$. Also, by comparing equations 10 and 11 with equations 6 and 7, the total change in the wave function due to k repeated measurements of strength δ_m is exactly the same as the change due to one measurement of a strength $k\delta_m$, which results in $\tilde{n} = \bar{n}$. After each measurement, the width of the wave packet decreases (continuous wave packet collapse). The center of the wave packet takes a step in a random walk (quantum Brownian motion), which depends on the random result of the measurement, $\tilde{n}_i$. The probability distribution that statistically describes the diffusion of the center position of the wave function after k measurements, n_0^k, is

$$P_k(n_0^k) = N[n_0^k, n_0, k\delta_m\delta_0^{-1}(\delta_0 + k\delta_m)^{-1}]. \tag{12}$$

The average center position is always at the initial center position, n_0. However, the probability of finding the center further away from n_0 increases as the number of measurements increases. As long as the total strength of the measurements is small, that is, $k\delta_m \ll \delta_0$, the variance of n_0^k increases linearly with the number of measurements, $k\delta_m\delta_0^{-1}(\delta_0 + k\delta_m)^{-1} \cong Dk$. In this regime, the movement of the center position is a quantum Brownian motion with a constant diffusion coefficient, $D = \delta_m\delta_0^{-2}$. Here, the time scale is replaced by the discrete scale of the number of measurements. As the wave function narrows, the average step size of this quantum Brownian motion decreases. The statistical variance of the center position saturates and then equals the original variance of the wave function, δ_0^{-1}. At the same time, the wave packet is reduced to a photon-number eigenstate. The measured wave packet, therefore, undergoes a quantum Brownian motion, which is saturated due to the continuous collapse of the wave packet.

Analyzing the statistics of the results of k repeated measurements on a single wave packet, we use the same definitions for the inferred average and variance as for the case of k measurements performed on an ensemble. Both definitions are symmetric in the results of the measurements, $(\tilde{n}_1, \tilde{n}_2, \ldots, \tilde{n}_k)$. In the case of k repeated measurements on a single wave packet, both the final wave function and the probability to obtain a specific series of results are independent of the order in which these results are obtained. Therefore, it is natural to use the same $\bar{n}$ and $\overline{\Delta n^2}$ as before. From equation 3, the probability of obtaining the series $(\tilde{n}_1, \tilde{n}_2, \ldots, \tilde{n}_k)$ as a result of k repeated measurements is

$$P(\tilde{n}_1, \tilde{n}_2, \ldots, \tilde{n}_k) \prod_{i=1}^{k} d\tilde{n}_i = [P(\bar{n})d\bar{n}][P(S)dS]d\Omega_{k-1}, \tag{13}$$

where $S = (k - 1)\delta_m\overline{\Delta n^2}$, and is independent of δ_0. Again, the probability distribution of the inferred average is centered at the original average, n_0, that is,

$$P(\bar{n}) = N[\bar{n}, n_0, \delta_0^{-1} + (k\delta_m)^{-1}]. \quad \textbf{(14)}$$

The variance of the inferred average decreases with an increased number of measurements and, as $k \to \infty$, this variance reaches its minimum value, which equals the original variance of the wave packet, δ_0^{-1}. Therefore, the inferred average has the same probability error in the cases of both an infinite number of repeated weak measurements and one precise measurement. In fact, comparing equation 14 with equation 5, we see that $\bar{n}$ is inferred with equal probabilities by k consecutive measurements of strength δ_m and by one measurement of a strength $k\delta_m$. The probability distribution of S is again a chi-square distribution, $P(S) = \chi^2[S, (k - 1)]$. However, $P(S)$ is now independent of δ_0; therefore, the inferred variance, $\overline{\Delta n^2}$, is not a measure of the original variance, δ_0^{-1}. Indeed, $\overline{\Delta n^2}$ is centered at δ_m^{-1}, with the variance $2(k - 1)^{-1}\delta_m^{-2}$. The statistics of the results of repeated weak QND measurements performed on a single wave packet contain no information about the initial width of the wave packet. In contradiction with our expectations, these statistics do not infer the wave function of the single wave packet.

The mathematical origin of this result is the symmetry of $P(\tilde{n}_1, \tilde{n}_2, \ldots, \tilde{n}_k)$, which appeared already in equation 1. Each time that the wave packet is measured, it is slightly changed. The results of the consecutive measurements are essentially collected from an ensemble of wave packets with different widths. Because all these results have the same weight in $P(\tilde{n}_1, \tilde{n}_2, \ldots, \tilde{n}_k)$, their statistics are independent of the width of the initial wave function. There is no natural way to assign different weights to the different results in the definition of $\overline{\Delta n^2}$ because the changes in the wave function are symmetric in $(\tilde{n}_1, \tilde{n}_2, \ldots, \tilde{n}_k)$ and we cannot overcome the symmetry of $P(\tilde{n}_1, \tilde{n}_2, \ldots, \tilde{n}_k)$.

Physically, it is the exact coordination between the quantum Brownian motion and the continuous collapse of the wave packet that prevents us from distinguishing between two wave packets of large and small widths, both centered at n_0. Probably, the first measurement result obtained from the wide wave packet is further away from n_0 than the result obtained from the narrow wave packet. However, the shift toward the measurement result and the collapse due to the first measurement are more dramatic in the case of the wide wave packet. Therefore, the probability of obtaining the second result in a certain distance from the first result can be the same for both wave packets, regardless of their initial widths.

The above result is consistent with the fundamental theorem of quantum communications,[15] namely, Holevo's theorem. The maximum channel capacity is realized by a photon-number state channel, in which the photon-number state signal is detected by an ideal photon counter.[16,17] The finite capacity of this noiseless channel is due to the discrete spectra of the photon-number; that is, the number of distinguishable states is finite because the photon-number observable has only positive integer eigenvalues. If one could measure the variance as well as the average of a given wave packet, the number of distinguishable states would increase by replacing the photon-number state with other states with the same average and varied variances. The possibility of exceeding the maximum channel capacity is excluded because the measurement of the average is subject to an error, which is determined by the initial variance of the wave packet, and the variance measurement is inhibited.

In conclusion, we have shown that the wave function of a single quantum system cannot be measured by a series of weak QND measurements without an a priori knowledge of the wave function. This is because the statistics of the results of the measurements contain no information about the initial width of the measured wave function. Mathematically, this result originates in the symmetric structure of the probability-amplitude operator. During the measurement process, the wave function undergoes a quantum Brownian motion and continuous collapse. This physical mechanism is responsible for the exact cancellation of the information about the wave function from the statistics of the measurement results.

REFERENCES AND NOTES

1. AHARONOV, Y. & L. VAIDMAN. 1993. Phys. Lett. A **178:** 38.
2. AHARONOV, Y., J. ANANDAN & L. VAIDMAN. 1993. Phys. Rev. A **47:** 4616.
3. BRAGINSKY, V. B., Y. I. VORONSTOV & K. S. THORNE. 1980. Science **209:** 547.
4. CAVES, C. M., K. S. THORNE, W. P. DREVER, V. D. SANDBERG & M. ZIMMERMANN. 1980. Rev. Mod. Phys. **52:** 341.
5. BRAGINSKY, V. B. & S. P. VYATCHANIN. 1981. Sov. Phys. Dokl. **26:** 686.
6. MILBURN, G. J. & D. F. WALLS. 1983. Phys. Rev. A **28:** 2065.
7. IMOTO, N., H. A. HAUS & Y. YAMAMOTO. 1985. Phys. Rev. A **32:** 2287.
8. IMAMOḠLU, A. 1993. Phys. Rev. A **47:** R4577.
9. KITAGAWA, K., N. IMOTO & Y. YAMAMOTO. 1987. Phys. Rev. A **35:** 5270.
10. BRAGINSKY, V. B. & Y. F. KHALILI. 1992. Quantum Measurement. Cambridge University Press. London/New York.
11. In this report, $N[x, x_0, \sigma^2]$ is defined to be a normal distribution of the variable x, which is centered at x_0 with the variance of σ^2: $N[x, x_0, \sigma^2] = (2\pi\sigma^2)^{-1/2} \exp[-(x - x_0)^2/2\sigma^2]$.
12. SHELBY, R. M. & M. D. LEVENSON. 1987. Opt. Commun. **64:** 553.
13. LAPORTA, A., R. E. SLUSHER & B. YURKE. 1989. Phys. Rev. Lett. **62:** 28.
14. In this report, $\chi^2[x, \nu]$ is defined to be a chi-square distribution of the variable x, which is centered at ν with the variance 2ν: $\chi^2[x, \nu] = 2^{\nu/2}\Gamma(\nu/2)^{-1}x^{(\nu-2)/2}\exp(-x/2)$.
15. HOLEVO (KHOLEVO), A. S. 1982. Probabilistic and Statistical Aspects of Quantum Theory. North-Holland. Amsterdam.
16. YAMAMOTO, Y. & H. A. HAUS. 1986. Rev. Mod. Phys. **58:** 1001.
17. CAVES, C. M. & P. D. DRUMMOND. 1994. Rev. Mod. Phys. **66:** 481.

Quantum State Engineering in Pump-coupled High-Q Micromasers[a]

PÁL BOGÁR, JÁNOS A. BERGOU, AND MARK HILLERY

Department of Physics and Astronomy
Hunter College of the City University of New York
New York, New York 10021

INTRODUCTION

The advent of the micromaser[1] has permitted a direct study of the quantum features in the interaction of two-level atoms with a single quantized mode of the electromagnetic field. This device can be used to generate different kinds of nonclassical states of the radiation field,[2–4] for example, macroscopical quantum superpositions as suggested in reference 5. In the present report, we investigate quantum superpositions of nonlocal fields of two lossless micromasers coupled by the common pumping atomic beam[6,7] as depicted in FIGURE 1. Atoms enter the first cavity in their excited states and then proceed to the second one without any time delay between the cavities. We show that a correlation and, ultimately, a steady-state entanglement of the two nonlocal fields arise due to the interference of the two atomic paths that an atom can follow to reach the same final state.[7,8] We assume that there is at most one atom in the cavities at a time to avoid collective effects. The final state of the atoms is detected after the interaction using two different setups. First, we require the atoms to leave in a prescribed sequence of final states, making the fields evolve along the desired trajectory. In particular, we consider the two simplest schemes in the present report where the sequence of final states consists only of the upper or only of the lower state. We do not know prior to the experiment which sequence will be produced, but we can redo the experiment until the desired one is obtained. This is called conditional measurement and is studied in the second section by calculating the pure evolution of the state vector of the fields. In the second case, we do not select a particular result for the final state of the atoms, but average over the two possible outcomes. Thus, the reduced density matrix for the fields is used in this nonselective measurement scheme, the evolution of which is discussed in the third section. The last section is devoted to a summary.

CONDITIONAL MEASUREMENTS OF ATOMS

We consider two kinds of conditional measurement schemes.[7] In the so-called energy-preserving scheme, we require that each atom is detected in its upper state; in the energy-transferring scheme, we require that each atom is detected in its lower

[a]This research was supported by the Office of Naval Research (Grant No. N00014-92-J-1233), the National Science Foundation (Grant No. PHY-9201912), and PSC-CUNY.

state. We assume no cavity losses, zero temperature, 100% detection efficiency, single atom resonant interaction, and no atomic velocity spread.

The state of the fields after the $(k-1)$-th atom left, but before the k-th atom entered the cavity reads

$$|\Psi^{(k-1)}\rangle = \sum_{n_1,n_2} \Psi^{(k-1)}_{n_1,n_2} |n_1, n_2\rangle. \tag{2.1}$$

After the interaction of the k-th atom with the fields, the state of the atom-fields system is

$$|\Phi^{(k)}\rangle = \sum_{n_1,n_2} \Psi^{(k-1)}_{n_1,n_2} [C'_{n_1+1}(C''_{n_2+1}|a, n_1, n_2\rangle - iS''_{n_2+1}|b, n_1, n_2+1\rangle) - iS'_{n_1+1}(C''_{n_2}|b, n_1+1, n_2\rangle - iS''_{n_2}|a, n_1+1, n_2-1\rangle)], \tag{2.2}$$

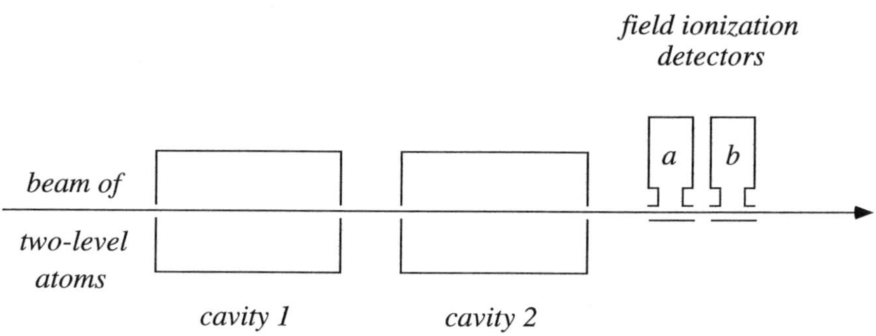

FIGURE 1. Schematic arrangement of two micromasers coupled by a beam of two-level atoms, the state of which is measured after the interaction.

where $S'_{n_1} \equiv \sin(g'\tau'\sqrt{n_1})$ and $S''_{n_2} \equiv \sin(g''\tau''\sqrt{n_2})$, with g', g'' and τ', τ'' being the atom-field coupling constants and the interaction times in the two cavities, respectively, and where C'_{n_1} and C''_{n_2} stand for the cosine functions of the corresponding arguments. Then, the state of the k-th atom is measured, resulting in a reduction of the state of the fields to

$$|\Psi^{(k)}\rangle = N^{(k)} \sum_{n_1,n_2} \Psi^{(k)}_{n_1,n_2} |n_1, n_2\rangle. \tag{2.3}$$

Each measurement means a renormalization of the state vector by $N^{(k)}$. The new amplitudes, $\Psi^{(k)}_{n_1,n_2}$, are functions of the old ones, $\Psi^{(k-1)}_{n_1,n_2}$, and for our two schemes they read as follows:

$$\Psi^{(k)}_{n_1,n_2} = \Psi^{(k-1)}_{n_1,n_2} C'_{n_1+1} C''_{n_2+1} - \Psi^{(k-1)}_{n_1-1,n_2+1} S'_{n_1} S''_{n_2+1} \tag{2.4}$$

for the energy-preserving scheme and

$$\Psi^{(k)}_{n_1,n_2} = \Psi^{(k-1)}_{n_1,n_2-1} C'_{n_1+1} S''_{n_2} + \Psi^{(k-1)}_{n_1-1,n_2} S'_{n_1} C''_{n_2} \tag{2.5}$$

for the energy-transferring scheme. The probability of detecting the k-th atom in the desired state is

$$P^{(k)} = \sum_{n_1,n_2} |\Psi^{(k)}_{n_1,n_2}|^2 = \frac{1}{N^{(k)2}}. \qquad \textbf{(2.6)}$$

The evolution of the two fields is studied by iterating the amplitudes according to equation 2.4 or 2.5 starting from coherent states. Because the state, $|\alpha, \alpha\rangle$, is separable into a product of two coherent states, $|\alpha\rangle_{\text{cavity1}} \otimes |\alpha\rangle_{\text{cavity2}}$, the fields are initially uncorrelated. In the case of an energy-preserving scheme, they settle to a superposition of Fock states at steady state in both micromasers, assuring Rabi angles that are multiples of π. As a result of this, the atoms are in their upper states before, between, and after the cavities. In the example depicted in FIGURE 2, where $\alpha^2 = 30$ and the interaction parameters are $g\tau = \pi$ ($g\tau \equiv g'\tau' = g''\tau''$), the Fock states are located at integer squares minus one, predominantly at 24 and 35. It is easy to see that the two fields are uncorrelated at steady state because the state vector is approximately separable as $(|24\rangle + |35\rangle)_{\text{cavity1}} \otimes (|24\rangle + |35\rangle)_{\text{cavity2}}$ (unnormalized). The probabilities of detecting the atoms in their upper states as required by the conditional measurement scheme are depicted in FIGURE 3.

Let us now consider energy-transferring schemes. FIGURE 4 shows typical evolutions of the fields for interaction parameters $g\tau = 0.3, 0.5$, and 0.8, starting from coherent fields of $\alpha^2 = 10$. It can be seen in the first row of the figure for $g\tau = 0.3$ that the distribution is stretched along a straight line, as it is shown for $k = 20$. The fields are correlated in this regime. The distribution then separates into two regions and finally ends its regular evolution in a rapidly oscillating structure around $k = 40$ that corresponds to the so-called "single-cavity" trapping effect.[7] In the second row for $g\tau = 0.5$, the photon number increases in both cavities in such a way that the distribution localizes around a single point where $n_1 \cong n_2$, showing a balance between the fields. The two micromasers are uncorrelated in this regime. At $n_1 \cong n_2 \cong 40$, the oscillatory structure shows that the system reached a single-cavity trapping point.

For larger interaction parameters, $g\tau = 0.8$, shown in the third and fourth rows of FIGURE 4, the system undergoes a transition between the above two uncorrelated and correlated regimes. On the other hand, the stretched distribution becomes double-peaked around $k = 30$, showing fields with a state vector approximately of the form, $|15,50\rangle + |50,15\rangle$ (unnormalized). This is the result of a new trapping effect called "two-cavity" trapping.[7] Finally, the double-peaked distribution built up by $k = 30$ begins to be destroyed at $k = 35$ by the coexisting single- and two-cavity trapping mechanisms. The probabilities that the atoms are detected in their lower state according to the measurement scheme are depicted in FIGURE 5 for the three examples given in FIGURE 4. It can be seen that the probability drops at the transitions between the uncorrelated and correlated regimes, as well as at the trapping. Defining m-th order correlation by the nonseparability condition,

$$\langle(\hat{a}_1\hat{a}_2^\dagger)^m\rangle \neq \langle\hat{a}_1^m\rangle\langle\hat{a}_2^{\dagger m}\rangle, \qquad \textbf{(2.7)}$$

where $\hat{a}_1$ ($\hat{a}_1^\dagger$) and $\hat{a}_2$ ($\hat{a}_2^\dagger$) are the field operators in micromasers 1 and 2, respectively, it is easy to show that the transient state above, $|15,50\rangle + |50,15\rangle$, exhibits 35th order

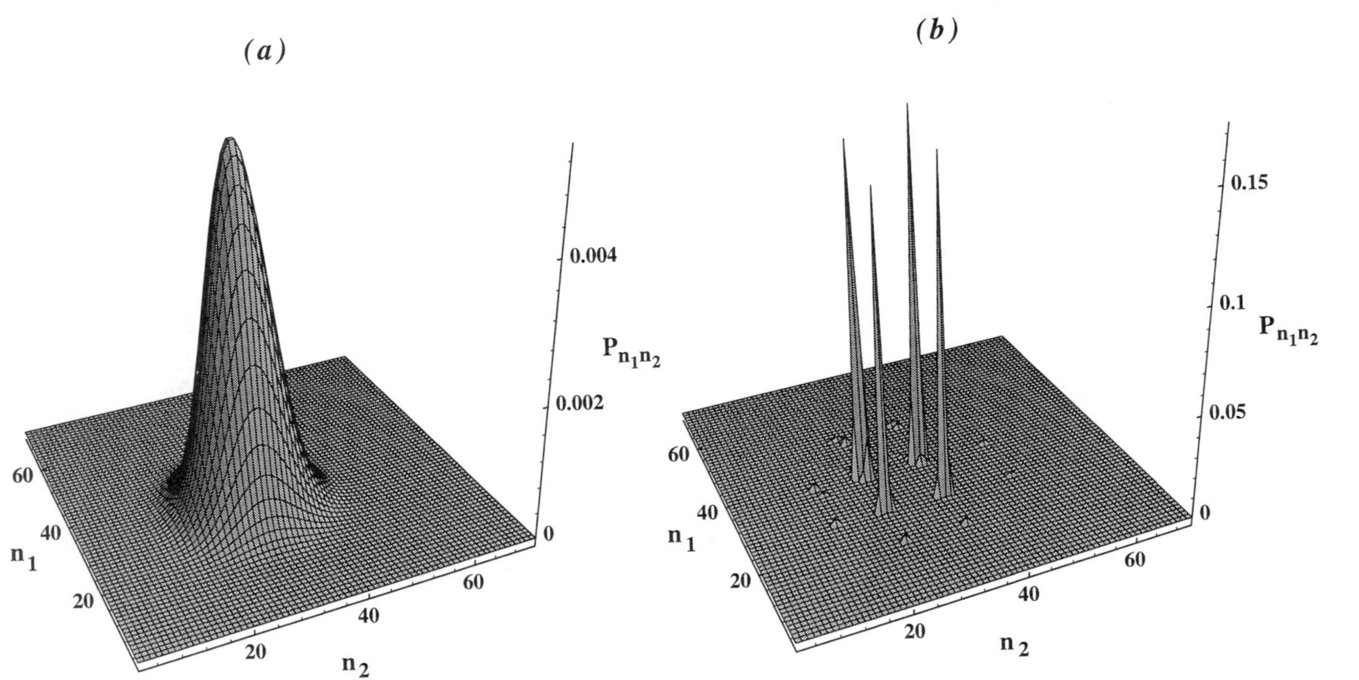

FIGURE 2. Photon statistics (squares of the amplitudes) of the fields—(a) initially in the coherent state of $\alpha^2 = 30$ and (b) after 50 atoms, applying an energy-preserving scheme at $g\tau = \pi$.

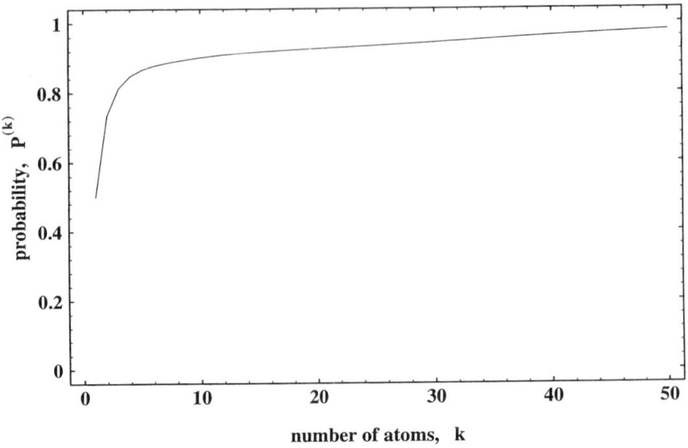

FIGURE 3. The probabilities of detecting the upper state of atom number k in the scheme in FIGURE 2.

correlation. We would like to emphasize that this is a correlation between fields of two spatially separated micromasers, that is, a *nonlocal* entanglement. The state of one of the fields can be inferred from a measurement made on the state of the other field located at a different point in space.

Now, we want to combine these two kinds of schemes, starting with the energy-transferring one and then, at an optimum number of atoms, switching to the energy-preserving one. This way the Fock states generated by the latter scheme will be located under the envelope of the fields generated by the former scheme at steady state. Let us start, for example, the system in the energy-preserving scheme at $g\tau = 0.142$, from uncorrelated coherent fields of $\alpha^2 = 30$. After 100 atoms, the generated fields will exhibit a long stretched distribution as depicted in FIGURE 6a showing strong correlation between the fields. This correlation would be destroyed by the atoms to come due to the appearance of the oscillatory structures if the energy-transferring scheme would be followed any further. We switch our system to the other scheme instead. From the 101st atom, we continue in the energy-preserving scheme using different parameters, $g\tau = \pi/2$. (The interaction time can be changed in an experiment by changing the velocity of the atoms.) After the next 200 atoms, we get a superposition of three Fock states at squares of even integers minus one, depicted in FIGURE 6b, that could be approximated by the state, $|99, 35\rangle + |63, 63\rangle + |35, 99\rangle$ (unnormalized). This is approximately equal to the steady state of the fields. A double-peaked superposition of the fields can be produced if we chose $g\tau = 1.0$ instead of $\pi/2$ as the new parameters. The generated fields are depicted in FIGURE 6c and can be approximated with the state, $|88,38\rangle + |38,88\rangle$ (unnormalized), showing a 50th order correlation at steady state. Although the method is not particularly sensitive to the atom number where the switch is to be made, we show what happens if we switched between the schemes too early in FIGURE 6d: instead of $k = 100$, the switch is at $k = 50$. Because the distribution at the moment of the switch is much broader this time than it was in FIGURE 6a, a peak at $|38, 38\rangle$ can arise. The probabilities of detecting the atoms in the desired states ($|b\rangle$ before and $|a\rangle$ after the switch) are depicted in FIGURE 7. In principle, arbitrary superpositions can be produced by choosing the appropriate initial coherent states and interaction parameters. In the energy-transferring scheme, $g\tau \cdot \alpha = \pi/4$ needs to be satisfied in order to produce a stretched distribution. When it is prepared, we need to switch to new interaction parameters for the energy-preserving scheme that will generate the

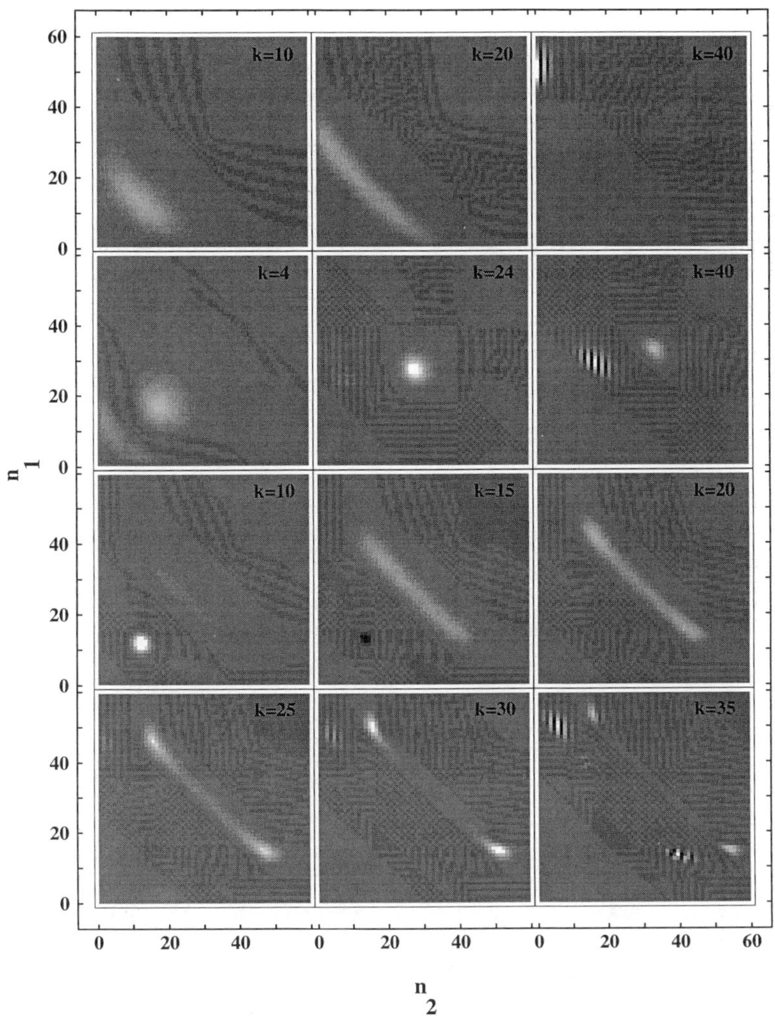

FIGURE 4. Density plots showing the evolution of the amplitude distribution of the fields starting from coherent fields of $\alpha^2 = 10$, using energy-transferring schemes at $g\tau = 0.3$ for the first, 0.5 for the second, and 0.8 for the third and fourth rows. Bright spots correspond to positive values, dark ones correspond to negative values, and the gray level is zero. Atom number k is given in each panel.

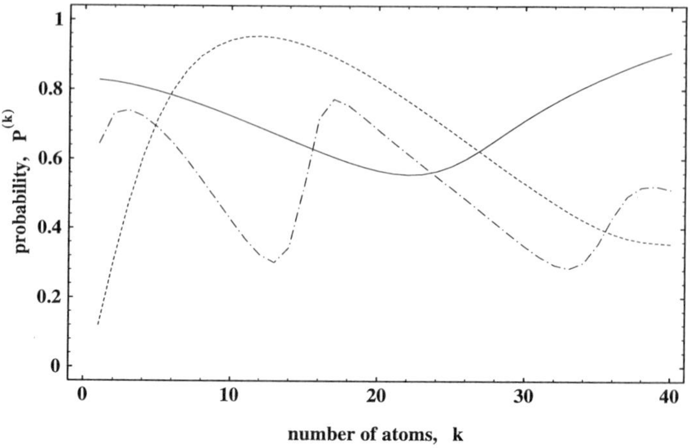

FIGURE 5. The probabilities of detecting the lower state of atom number k in the scheme in FIGURE 4 for $g\tau = 0.3$, 0.5, and 0.8, depicted by the solid, dashed, and dot-dashed lines, respectively.

desired Fock states, assuring Rabi angles of multiples of π under the envelope of the stretched distribution.

NONSELECTIVE MEASUREMENTS OF ATOMS

The state of the fields is described by the reduced density operator obtained by tracing out over the atomic states. This tracing operation is sometimes referred to as a nonselective measurement. The evolution of the system is governed by the Jaynes-Cummings operators U' and U'' in cavities 1 and 2, respectively. At the instant when the k-th atom leaves cavity 2, the field density operator reduces to

$$\rho^{(k)} = \mathrm{Tr}_{\mathrm{atom}}\,[U''U'\rho^{(k-1)}\rho_{\mathrm{atom}}U'^{\dagger}U''^{\dagger}], \tag{3.1}$$

where ρ_{atom} is the atomic- and $\rho^{(k-1)}$ is the field-density operator at the instant when the k-th atom enters cavity 1. It can be seen from equation 3.1 that, for steady states of the fields, one needs

$$C'_{n_1+1}C'_{m_1+1}C''_{n_2+1}C''_{m_2+1} = 1, \tag{3.2}$$

that is, the trapping condition satisfied by any combination of the four numbers n_1, m_1, n_2, and m_2, for all of which $gt\sqrt{n+1} = q\pi$, where q is an integer and n is any of the four numbers above. Thus, the only possible steady states of the system are the superpositions of trapping number states satisfying equation 3.2. They include both mixed and pure quantum states of the fields. It is shown in reference 7 that the most general solution for pure steady states of the fields is given by

$$|\Psi\rangle = \sum_{p,q} \Psi_{N_p,\,M_q}|N_p, M_q\rangle, \tag{3.3}$$

where N_p and M_q satisfy the trapping conditions, $g't'\sqrt{N_p + 1} = p\pi$ and $g''t''\sqrt{M_q + 1} = q\pi$. Any combination of N_p and M_q determines a point in the number space, n_1-n_2, where a nonzero amplitude, $\Psi_{N_p,M_q} \neq 0$, can arise. The initial fields and the parameters of the interactions, $g't'$ and $g''t''$, determine which part of this general state vector the system will evolve into.

We want to show that a state of m-th order correlation, $|\Psi\rangle_N^{(m)}$, exhibiting a structure given by

$$|\Psi\rangle_N^{(m)} = \left(\frac{1}{\sqrt{2}}\right)(|N, N+m\rangle \pm |N+m, N\rangle) \tag{3.4}$$

can be amplified into the trapping state, $|\Psi\rangle_N^{(M)}$, of the same form as above, if the two conditions

$$S'_{N+1} = S''_{N+1} = 0 \qquad \text{and} \qquad S''_N = 0 \tag{3.5}$$

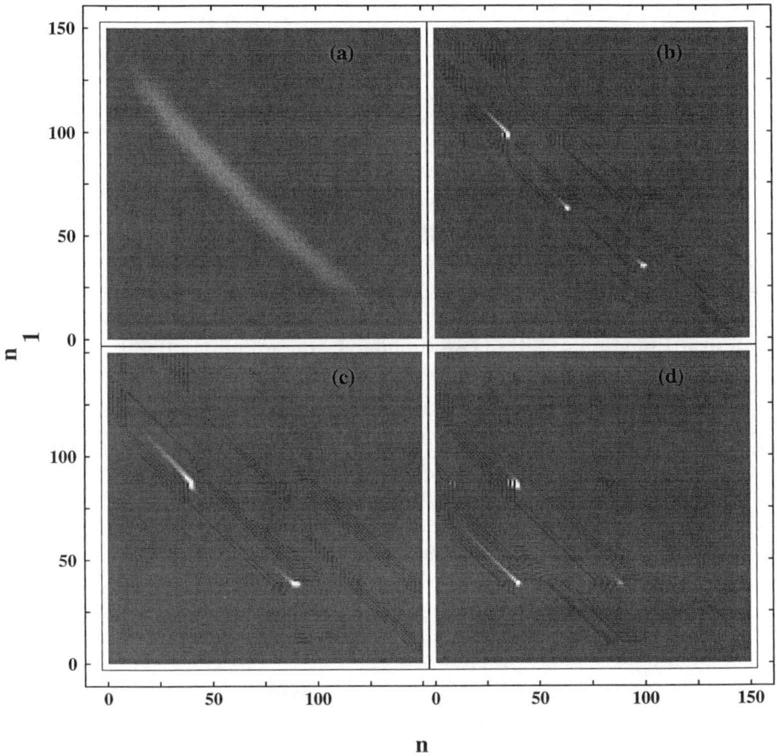

FIGURE 6. Density plots of the amplitude distributions of the fields: (a) at the 100th atom starting from coherent fields of $\alpha^2 = 30$ using an energy-transferring scheme at $g\tau = 0.142$; (b) at the 300th atom after switching from the field produced in part a at the 100th atom to an energy-preserving scheme at $g\tau = \pi/2$; (c) same as part b, but for $g\tau = 1.0$; (d) same as part c, but switching at the 50th atom.

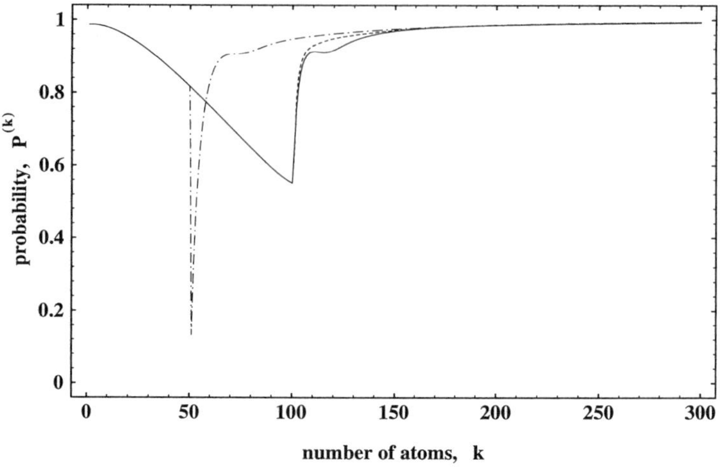

FIGURE 7. The probabilities of detecting the state of atom number k corresponding to the schemes given in FIGURES 6b, 6c, and 6d, depicted by the solid, dashed, and dot-dashed lines, respectively.

were satisfied simultaneously, where N and $N + M$ are trapping numbers. These conditions imply that the interaction parameters must be integer multiples of $\pi/\sqrt{N+1}$ and $\pi/\sqrt{N}$ simultaneously. Although this is not possible exactly, it will be shown that an approximate solution for $g\tau$ works very well and the steady states are very close to the pure entangled trapping states, $|\Psi\rangle_N^{(M)}$. We should mention here that unequal interaction parameters $g't'$ and $g''t''$ result in a loss of correlation and purity of the fields at steady state due to an asymmetrical amplification of the state vector. Therefore, we require $g\tau \equiv g't' = g''t''$ in the amplifying procedure.

It can be seen from equation 2.2 that detecting the first atom emerging from the interactions in the upper state, $|a\rangle$, will not change the initial state of the fields, $|\Psi\rangle_N^{(m)}$, when equation 3.5 is satisfied. On the other hand, detecting the lower state, $|b\rangle$, will result in the state, $|\Psi\rangle_N^{(m+1)}$. The tracing operation averages the two atomic paths out and the state of the fields is a statistical mixture of the two corresponding quantum states, $|\Psi\rangle_N^{(m)}$ and $|\Psi\rangle_N^{(m+1)}$. Similarly, for all the consecutive atoms during the evolution, the states of the fields are statistical mixtures of quantum states of the form of $|\Psi\rangle_N^{(m)}$ only, where N is the same as the initial one and m is any integer below the next trapping number M. This suggests that the system inevitably evolves into the trapping state, $|\Psi\rangle_N^{(M)}$.

Let us start, for example, from the initial state, $|\Psi\rangle_1^{(1)}$, and apply $g\tau = 7\pi/\sqrt{2} = 4.950\pi$. The evolution of the purity factor, $\xi^{(k)} = \mathrm{Tr}[\rho^{(k)2}]$, of the state of the fields as a function of atom number k is depicted in FIGURE 8 by the dot-dashed line. The steady-state field-density matrix elements using the notation (n_1, m_1, n_2, m_2) are the diagonal terms (1, 1, 7, 7) and (7, 7, 1, 1) equal to 0.500, and the off-diagonal ones, (1, 7, 7, 1) and (7, 1, 1, 7) equal to 0.496, resulting in the steady-state purity factor, $\xi = 0.992$. The steady state of the system is approximately equal to the entangled trapping state, $|\Psi\rangle_1^{(6)}$. We have found approximately the same steady states for $g\tau =$

$l\pi/\sqrt{2}$, for $l = 7, 10, 17, \ldots$ (within the $\pm$ sign in equation 3.4), although the time evolution can be very different, as depicted in FIGURE 8 by the dot-dashed, dotted, and solid lines, respectively. The evolution of the system toward its steady state is particularly slow for $l = 17$ (solid line) due to its being attracted by the "pseudotrapping" state, $|1,3\rangle - |3,1\rangle$, along its way around the atom number $k = 9$. The conditions in equation 3.5 ensure that the fields evolve along statistical mixtures of state vectors, $|\Psi\rangle_1^{(m)}$, only ($m \leq 6$). The transient drop in the purity due to the mixture goes back up to 1 as the fields approach the pure trapping state, $|\Psi\rangle_1^{(6)}$. Involving quantum states of different structure into the evolution would result in an irreversible loss of purity and the system would evolve into a mixed quantum state of reduced or vanishing correlation.

In order to prepare the appropriate initial state, $|\Psi\rangle_N^{(1)}$, for amplification starting from the number state, $|N, N\rangle$, we need to have a different pair of conditions satisfied. First, $g't'$ must satisfy the condition, $|C'_{N+1}| = |S'_{N+1}|$. On the other hand, $g''t''$ must approximately be an integer multiple of $\pi/(2\sqrt{N+1})$ and at the same time that of $\pi/\sqrt{N}$ in order to have $|S''_{N+1}| \cong 1$ and $S''_N \cong 0$ simultaneously. In this case, equation 2.2 tells us that, sending one single atom through the fields, the probability of detecting its upper state is close to zero, whereas the other atomic path provides us with fields in a state approximately equal to $|\Psi\rangle_N^{(1)}$. In the case of $N = 1$, choosing for example $g'\tau' = \pi/(4\sqrt{2})$ and $g''\tau'' = 17\pi/(2\sqrt{2}) = 6.01\pi$ gives us the probability of 0.1% to detect the upper atomic state, and the generated fields can be approximated very well by the pure state, $|\Psi\rangle_1^{(1)}$. Once the initial state for amplification, $|\Psi\rangle_N^{(1)}$, is prepared, we can proceed to the amplification procedure by switching the interaction parameters to the appropriate values discussed above.

We have shown in this section that entangled trapping states, $|\Psi\rangle_N^{(M)}$, of arbitrary N and M can be produced from number states, $|N, N\rangle$, using a two-step procedure that is based on conditions regarding the interaction parameters of the two cavities.

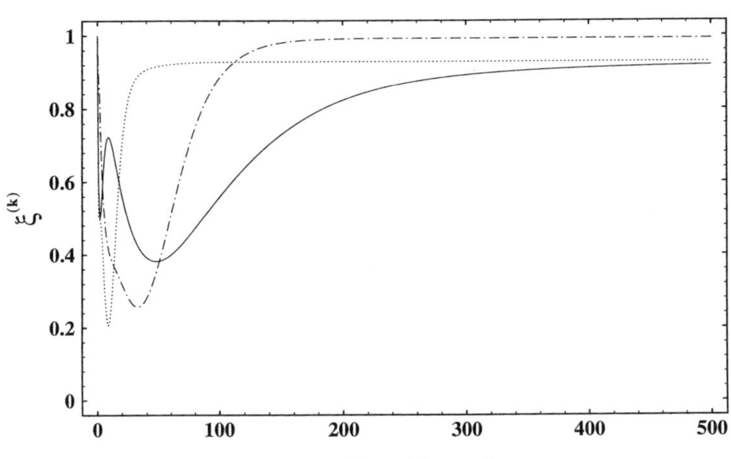

FIGURE 8. The evolution of the purity factor $\xi^{(k)}$ as a function of atom number k applying $g\tau = l\pi/\sqrt{2}$, for $l = 7, 10$, and 17, depicted for the dot-dashed, dotted, and solid lines, respectively. The initial and the final states in each case are $|\Psi\rangle_1^{(1)}$ and $|\Psi\rangle_1^{(6)}$, respectively.

SUMMARY

Two lossless micromasers are coupled by the common pumping beam of excited two-level atoms, the fields of which are studied performing conditional and nonselective measurements of the final state of the atoms. Energy-transferring conditional measurement schemes are shown to produce a transient entanglement of the fields. In the case of energy-preserving schemes, a set of Fock states is generated at steady state under the envelope of the initial fields. If the initial fields were uncorrelated (correlated), the generated ones would be uncorrelated (correlated) too. This suggests that switching from an energy-transferring to an energy-preserving scheme produces a steady-state entanglement. The Fock states generated by the latter scheme will in this case be located under the envelope of the correlated fields generated by the former scheme.

In the case of nonselective measurements, it is shown that entangled trapping states of the form, $|N, N + M\rangle \pm |N + M, N\rangle$, can be produced starting from number states, $|N, N\rangle$, using a two-step procedure. First, we introduce some correlation via generating the above state of $M = 1$; then, we amplify it to a larger M corresponding to the trapping state above. Both steps are based on conditions regarding the interaction parameters, $g't'$ and $g''t''$, of the two cavities. They ensure that the fields evolve along statistical mixtures of state vectors of the above form only. An inclusion of other quantum states or an asymmetrical amplification of the state vector (when $g't' \neq g''t''$) would result in a loss of purity and mixed quantum states of no correlation at steady state.

In the absence of dissipations, both methods can generate steady-state nonlocal quantum superpositions of distinct macroscopical fields, that is, nonlocal "Schrödinger cats".

REFERENCES

1. For a recent review, see: WALTHER, H. 1992. Phys. Rep. **219:** 201; FILIPOWICZ, P., J. JAVANAINEN & P. MEYSTRE. 1986. Phys. Rev. A **34:** 3077.
2. KRAUSE, J., M. O. SCULLY & H. WALTHER. 1987. Phys. Rev. A **36:** 4547; MEYSTRE, P. 1987. Opt. Lett. **12:** 668; PAUL, H. 1989. J. Mod. Opt. **36:** 515.
3. MEYSTRE, P., G. REMPE & H. WALTHER. 1988. Opt. Lett. **13:** 1078.
4. BOGÁR, P., J. A. BERGOU & M. HILLERY. 1994. Phys. Rev. A **50:** 754.
5. SLOSSER, J. J., P. MEYSTRE & S. L. BRAUNSTEIN. 1989. Phys. Rev. Lett. **63:** 934; SLOSSER, J. J. & P. MEYSTRE. 1990. Phys. Rev. A **41:** 3867.
6. BERGOU, J. A. & M. HILLERY. 1991. Phys. Rev. A **44:** 7502.
7. BOGÁR, P., J. A. BERGOU & M. HILLERY. 1995. Phys. Rev. A. To appear. The effect of dissipations on the nonselective measurement scheme is also discussed here.
8. Nonlocal quantum states have been found in a similar system by: DAVIDOVICH, L., A. MAALI, M. BRUNE, J. M. RAIMOND & S. HAROCHE. 1991. Phys. Rev. Lett. **71:** 2360.

Two-Photon "Ghost" Image and Interference-Diffraction[a]

Y. H. SHIH, A. V. SERGIENKO, T. B. PITTMAN, D. V. STREKALOV, AND D. N. KLYSHKO[b]

Department of Physics
University of Maryland Baltimore County
Baltimore, Maryland 21228

INTRODUCTION

One of the most surprising consequences of quantum mechanics is the entanglement of two or more distant particles.[1] The classic example of a two-particle entangled state was given by Einstein, Podolsky, and Rosen in their famous 1935 *gedankenexperiment,*[2] where the measurement of an observable on one of the particles determined the value of that observable for the other particle with unit probability. The progression from *gedanken* to real experiments[3] has been greatly aided by the use of spontaneous parametric downconversion (SPDC),[4] the nonlinear optical process in which a laser pump incident on a crystal leads to the emission of a correlated pair of photons. The distinctively quantum nature of the resulting two-photon state, which has been confirmed in a number of interesting experiments,[5] has allowed us to demonstrate a new type of two-photon phenomenon in the following two experiments.

The first experiment reported in this discussion is a two-photon optical imaging type experiment:[6] the SPDC light beam, which consists of pairs of orthogonally polarized *signal* and *idler* photons, is split into two diverging beams by a polarization beam-splitter (BS) so that coincidence detections may be performed between two distant photon-counting detectors. An aperture (mask) placed in front of one of the detectors, for example, the letters of our institution, is illuminated by the *signal* beam through a convex lens. Surprisingly, an image of this aperture is observed in the coincidence-counting rate by scanning the other detector in the transverse plane of the *idler* beam, even though both detectors' single-counting rates remain constant.

The second experiment demonstrates "ghost" two-photon interference-diffraction patterns.[6] The experimental setup is similar to the image experiment, except that rather than an aperture it is a Young's double slit (or a single slit) inserted into the path of the *signal* beam. Surprisingly, an interference (or diffraction) pattern is observed in the coincidences when scanning the detector in the *idler* beam. This effect is even more striking when one considers that there is not any first-order interference-diffraction pattern behind the slits.

[a]This work was supported by the Office of Naval Research via Grant No. N00014-91-J-1430. Further support was provided by the International Science Foundation for the visit of D. N. Klyshko.

[b]Permanent address: Department of Physics, Moscow State University, Moscow, Russia.

As will be shown in detail, these two experiments demonstrate the EPR phenomena from both a geometrical optics and a physical optics point of view.

TWO-PHOTON "GHOST" IMAGE EXPERIMENT

The experimental setup is shown in FIGURE 1. The 351.1-nm line of an argon ion laser is used to pump a nonlinear BBO (β-BaB_2O_4) crystal that is cut at a degenerate type-II phase-matching angle to produce pairs of orthogonally polarized signal (e-ray of the BBO) and idler (o-ray of the BBO) photons.[4] The pairs emerge from the crystal nearly collinear, with $\omega_s \cong \omega_i \cong \omega_p/2$, where ω_j (j = s, i, p) is the frequency of

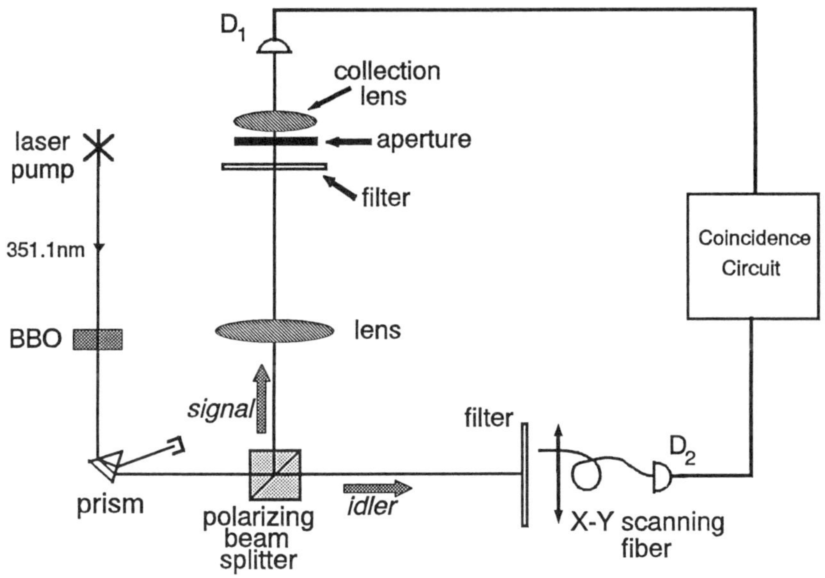

FIGURE 1. A schematic setup for the two-photon "ghost" image experiment.

the signal, idler, and pump, respectively. The pump is then separated from the downconversion beam by a UV-grade fused silica dispersion prism and the remaining signal and idler beams are sent in different directions by a polarization beam-splitting Thompson prism. The signal beam passes through a convex lens with a 400-mm focal length and illuminates a chosen aperture (mask). As an example, one could choose the letters "UMBC" or "UM". Behind the aperture is the detector package D_1, which consists of a 25-mm focal length collection lens in whose focal spot is a 0.8-mm-diameter dry ice–cooled avalanche photodiode. The idler beam is met by detector package D_2, which consists of a 0.5-mm-diameter multimode fiber whose output is mated with another dry ice–cooled avalanche photodiode. The input tip of the fiber is scanned in the transverse plane by two encoder drivers, and the output pulses of each detector, which are operating in the Geiger mode, are sent to a coincidence-counting circuit with a 1.8-ns acceptance window. Both detectors are preceded by 83-nm-bandwidth spectral filters centered at the degenerate wavelength, 702.2 nm.

By recording the coincidence counts as a function of the fiber tip's transverse plane coordinates, we see the image of the chosen aperture (for example, "UMBC" or "UM"), as is reported in FIGURE 2. Moreover, although the size of the "UMBC" aperture inserted in the signal beam is only about 3.5 mm × 7 mm, the observed image measures 7 mm × 14 mm. We have therefore managed linear magnification by a factor of two. Despite the completely different physical situation, the remarkable feature here is that *the relationship between the focal length of the lens f, the aperture's optical distance from the lens S, and the image's optical distance from the lens (lens back through beam-splitter to BBO, then along the idler beam to the image) S′ satisfies the Gaussian thin-lens equation:*

$$\frac{1}{S} + \frac{1}{S'} = \frac{1}{f}. \tag{1}$$

In this experiment, we chose $S = 600$ mm and the twice-magnified clear image was found when the fiber tip was in the plane with $S' = 1200$ mm (see FIGURE 3).

To understand this unusual phenomenon, we examine the quantum nature of the two-photon state produced in SPDC, which is entangled by means of the phase-matching conditions inside the crystal:[4]

$$\omega_s + \omega_i = \omega_p, \qquad \mathbf{k}_s + \mathbf{k}_i = \mathbf{k}_p, \tag{2}$$

where $\mathbf{k}_j$ is the wave vector of the signal, idler, or pump. The angular distribution of the correlated photon pairs, which encourages two-dimensional applications, is the result of the transverse components of the wave vector condition:

$$k_s \sin \alpha_s = k_i \sin \alpha_i, \tag{3}$$

where α_s and α_i are the scattering angles inside the crystal. Upon exiting the crystal, Snell's law thus provides

$$\omega_s \sin \beta_s = \omega_i \sin \beta_i, \tag{4}$$

where β_s and β_i are the exit angles of the signal and idler photons with respect to the $\mathbf{k}_p$ direction. Therefore, near the degenerate case, the photons constituting one pair are emitted at roughly equal, yet opposite, angles relative to the pump, and the measurement of the exit angle of the signal photon determines the exit angle of the idler photon with unit probability and vice versa. This then allows for a simple explanation of the experiment in terms of "usual" geometrical optics in the following manner: Considering the action of the beam-splitter, we envision the crystal as a "hinge point" and "unfold" the schematic of FIGURE 1 into that shown in FIGURE 3. Because of the equal angle requirement of equation 4, we see that all photon pairs that result in a coincidence detection can be represented by *straight lines* (but keep in mind the different propagation directions) and therefore the image is well produced in coincidences when the aperture, lens, and fiber tip are located according to equation 1. In other words, the image is exactly the same as one would observe on a

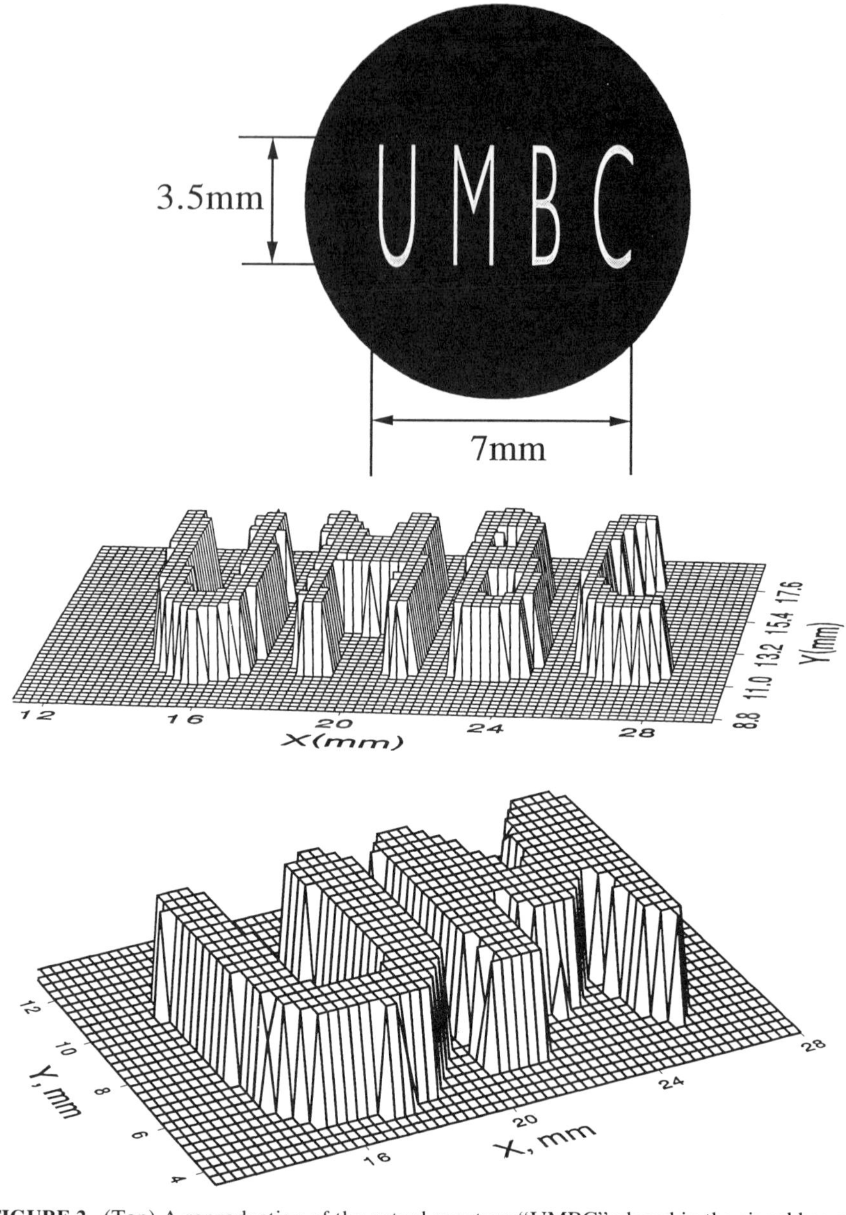

FIGURE 2. (Top) A reproduction of the actual aperture "UMBC" placed in the signal beam. Note that the size of the letters is on the order of standard text. (Middle) The image of "UMBC": coincidence counts as a function of the fiber tip's transverse plane coordinates. The scanning step size is 0.25 mm. The data shown are a "slice" at the half-maximum value, with no image enhancement. (Bottom) The image of "UM".

screen placed at the fiber tip if detector D_1 were replaced by a pointlike light source and the BBO crystal by a reflecting mirror.

The study of fundamental quantum mechanics sometimes leads to interesting and useful applications. In the sense of recent work done on quantum cryptography,[7] the two-dimensional image transfer experiment reported here demonstrates that the possibilities of a quantum "cryptoFAX" are, in principle, quite real. Because the information (i.e., aperture-mask shape) is essentially carried in coincidences, an eavesdropper between detectors would require information from two communication "channels". Therefore, even though the random pulse electric signal from detector D_1 is quite available, the use of indivisible light quanta in the other channel would foil any spying attempts.

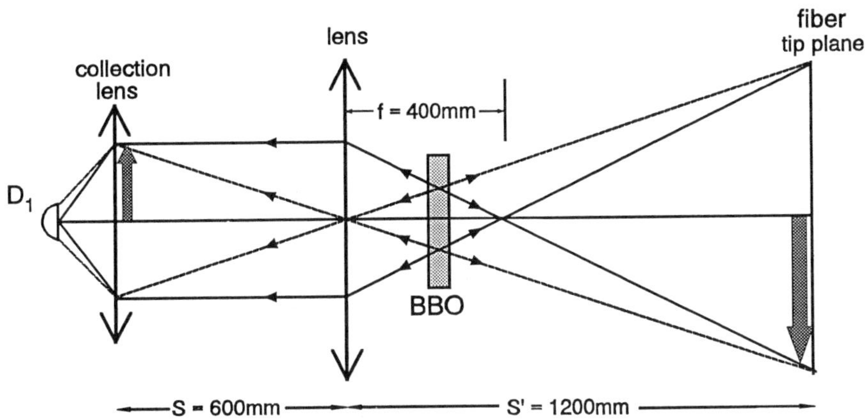

FIGURE 3. A conceptual "unfolded" version of the schematic shown in FIGURE 1, which is helpful for understanding the physics. Although the placement of the lens and the detectors will obey the Gaussian thin-lens equation, it is important to remember that the geometric rays actually represent pairs of SPDC photons that propagate in different directions.

TWO-PHOTON "GHOST" INTERFERENCE-DIFFRACTION

The schematic experimental setup is illustrated in FIGURE 4. It is similar to the first experiment except that, after the separation of signal and idler, the signal photon passes through a double-slit (or single-slit) aperture and then travels about 1 m to be counted by a pointlike photon-counting detector D_1 (0.5 mm in diameter). The idler photon travels a distance about 1.2 m from BS to the input tip of the optical fiber. In this experiment, only the horizontal transverse coordinate, x_2, of the fiber input tip is scanned by an encoder driver.

FIGURE 5 reports a typically observed double-slit interference-diffraction pattern. The coincidence-counting rate is reported as a function of x_2, which is obtained by scanning the detector D_2 (the fiber tip) in the idler beam, whereas the double slit is in the signal beam. The Young's double slit has a slit width of $a = 0.15$ mm and a

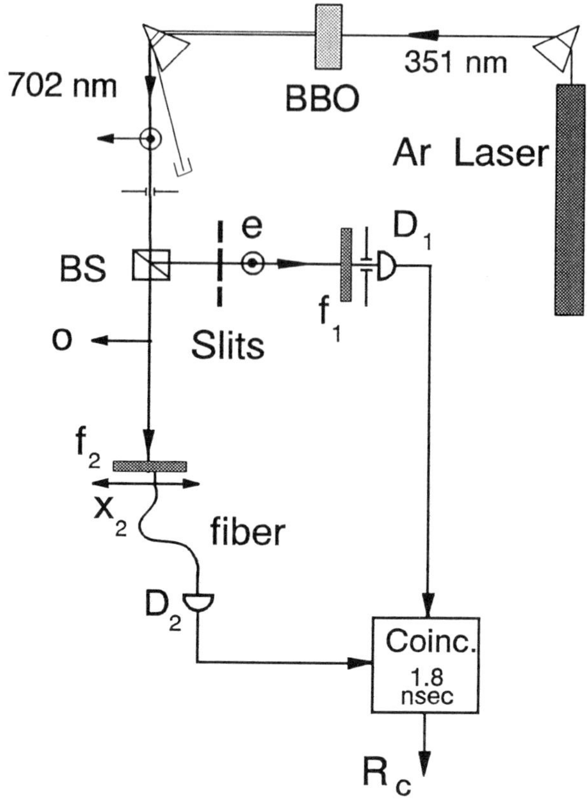

FIGURE 4. A schematic setup for the two-photon "ghost" interference-diffraction experiment. The signal (e-ray of BBO) and idler (o-ray of BBO) photon pair is generated in a nonlinear crystal BBO (β-BaB_2O_4). The ultraviolet pump beam is separated from the downconversion beams by a UV-grade fused silica dispersion prism. BS is a beam-splitting Thompson prism for splitting the signal and idler beams to different directions; f_1 and f_2 are spectral filters with 702.2-nm center wavelength and 10-nm bandwidth. Both photon-counting detectors D_1 and D_2 are dry ice–cooled avalanche photodiodes operated in Geiger mode.

slit distance of $d = 0.47$ mm. The interference period is measured to be 2.7 ± 0.2 mm and the half-width of the envelope is estimated to be about 8 mm. By curve fittings, we conclude that the observation is a standard Young's interference pattern, that is, a sine function oscillation with a sinc function envelope:

$$R_c(x_2) \propto \text{sinc}^2(x_2\pi a / \lambda z_2)\cos^2(x_2\pi d / \lambda z_2). \tag{5}$$

The remarkable feature here is that *z_2 is the distance from the slits' plane, which is in the signal beam, back through BS to the BBO crystal, and then along the idler beam to the scanning fiber tip of detector D_2* (see FIGURE 4). The calculated interference period and the half-width of the sinc function from equation 5 are 2.67 mm and 8.4 mm, respectively.

Even though the interference-diffraction pattern is observed in coincidences, *the single-detector-counting rates are both observed to be constant when scanning detector D_1 or D_2.* Of course, it seems reasonable to not have any first-order interference-diffraction in the single-counting rate of D_2, which is located in the "empty" idler beam. Of interest, however, is that the absence of the first-order interference-diffraction structure in the single-counting rate of D_1, which is behind the double slit, is mainly due to the divergence of the SPDC beam ($\gg \lambda/d$). In other words, the "blurring out" of the first-order interference fringes is due to the considerably large angular propagation uncertainty of a single SPDC photon.

Furthermore, if D_1 is moved to an unsymmetrical point, which results in unequal distances to the two slits, the interference-diffraction pattern is observed to be simply shifted from the current symmetrical position to one side of x_2. This is quite mind-boggling: Imagine that there were a first-order interference pattern behind the double slit and that D_1 were moved to a completely destructive interference point (i.e., zero intensity at that point) and fixed there. Then, we still can observe the same interference pattern in the coincidences (same period, shape, and counting rate), except for a phase shift.

FIGURE 6 reports a typical single-slit diffraction pattern. The slit width equals a = 0.4 mm. The pattern fits to the standard diffraction sinc function, that is, the "envelope" of equation 5, within reasonable experimental error. Here, again, z_2 is the unusual distance described in the above paragraphs.

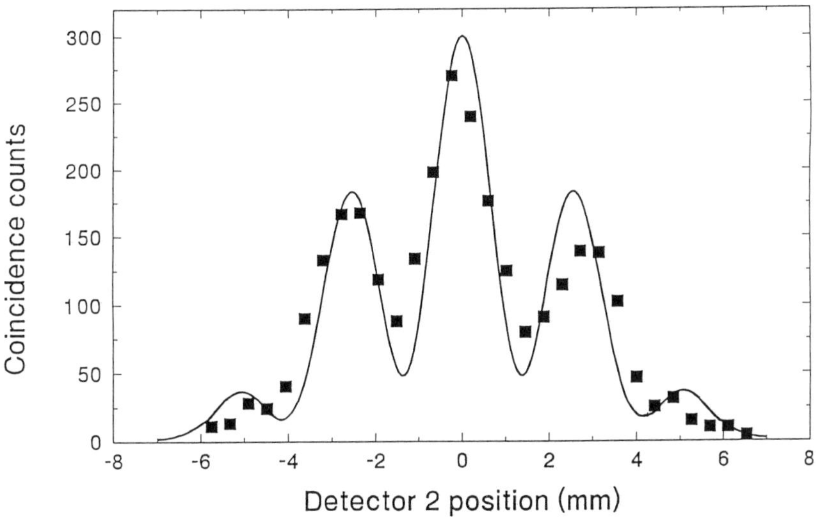

FIGURE 5. Typically observed interference-diffraction pattern: the dependence of the coincidences (per 400 s) on the position of the optical fiber tip of detector D_2, which counts the idler photons, while the signal photons pass through a double slit with a = 0.15 mm and d = 0.47 mm. The solid curve is calculated from equation 5, corrected for the finite size of the detectors and the pump profile. If D_1 is moved to an unsymmetrical point, which results in unequal distances to slits C and D, the interference-diffraction pattern is observed to be simply shifted from the current symmetrical position according to equation 9.

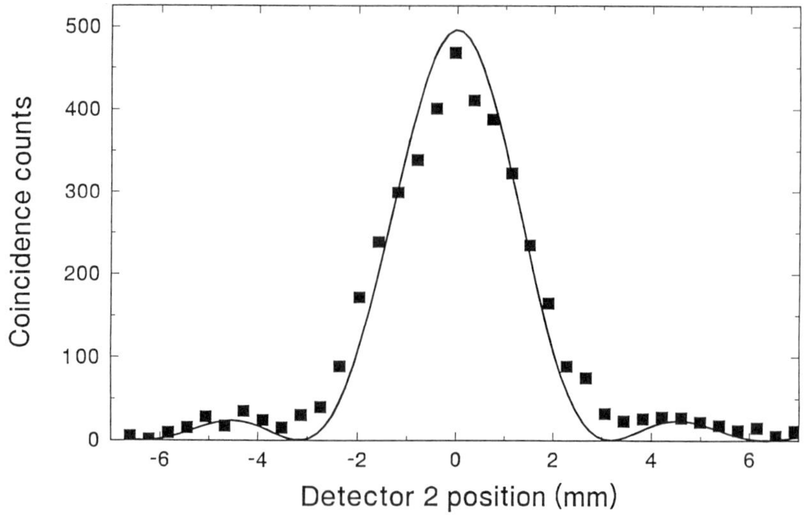

FIGURE 6. Two-photon diffraction: coincidence counts (per 400 s) versus the position of the idler photon-counting detector. A single slit of width $a = 0.4$ mm is in the signal beam. The solid curve is calculated from equation 10.

To explain this unusual phenomenon, we again present a simple quantum model. The quantum entanglement nature of the two-photon state of SPDC has been described in the previous section. Even though the propagation direction for each single photon of the pair has a considerably large uncertainty, the measurement of the exit angle of either photon determines the exit angle of its conjugate twin brother with unit probability. This important peculiarity selects the only possible optical paths in FIGURE 7, when one photon passes through the double-slit aperture while the other gets to D_2. In the near-degenerate case, we can simply treat the crystal as a reflecting mirror.

The coincidence-counting rate R_c is determined by the probability P_{12} of detecting a pair of photons by detectors D_1 and D_2 simultaneously. For SPDC, P_{12} is proportional to the square of the second-order correlation function $\langle E_2{}^{(+)}E_1{}^{(+)}\rangle$ of the fields at points D_1 and D_2 (it thus plays the role of the two-photon effective wave function):

$$P_{12} = \langle E_1^{(-)}E_2^{(-)}E_2^{(+)}E_1^{(+)}\rangle = |\langle E_2^{(+)}E_1^{(+)}\rangle|^2. \qquad \textbf{(6)}$$

In equation 6, $\langle\ldots\rangle \equiv \langle\Psi|\ldots|\Psi\rangle$, with $|\Psi\rangle$ being the four-mode state-vector of the SPDC field:

$$|\Psi\rangle = |\text{vac}\rangle + \epsilon[a_s{\dagger}a_i{\dagger}\exp(i\varphi_A) + b_s{\dagger}b_i{\dagger}\exp(i\varphi_B)]|\,\text{vac}\rangle, \qquad (7)$$

where $\epsilon \ll 1$ is proportional to the pump field (classical) and the nonlinearity of the crystal, φ_A and φ_B are the phases of the pump field at A and B, and $a_j{\dagger}$ ($b_j{\dagger}$) are the photon creation operators for the upper (lower) mode in FIGURE 7b (j = s, i). In terms of the Copenhagen interpretation, one can say that the interference is due to the uncertainty in the birthplace (A or B in FIGURE 7) of a photon pair.

In equation 6, the fields at the detectors are given by

$$E_1^{(+)} = a_s \exp(ikr_{A1}) + b_s \exp(ikr_{B1}),$$

$$E_2^{(+)} = a_i \exp(ikr_{A2}) + b_i \exp(ikr_{B2}), \tag{8}$$

where r_{Ai} (r_{Bi}) are the optical path lengths from region A (B) along the upper (lower) path to the *i*-th detector. Substituting equations 7 and 8 into equation 6,

$$R_c \propto P_{12} = \epsilon^2 |\exp(ikr_A + i\varphi_A) + \exp(ikr_B + i\varphi_B)|^2$$

$$\propto 1 + \cos[k(r_A - r_B)], \tag{9}$$

where we assume $\varphi_A = \varphi_B$ in the second line of equation 9 (although this is not a necessary condition to see the interference pattern, the transverse coherence of the pump beam at A and B is crucial). In equation 9, we defined the overall optical lengths between the detectors D_1 and D_2 along the upper and lower paths (see FIGURE 7b): $r_A \equiv r_{A1} + r_{A2} = r_{C1} + r_{C2}$ and $r_B \equiv r_{B1} + r_{B2} = r_{D1} + r_{D2}$, where r_{Ci} and r_{Di} are the path lengths from the slits C and D to the *i*-th detector.

If the optical paths from the fixed detector D_1 to the two slits are equal, that is, $r_{C1} = r_{D1}$, and if $z_2 \gg d^2/\lambda$, then $r_A - r_B = r_{C2} - r_{D2} \cong x_2 d/z_2$ and equation 9 can be

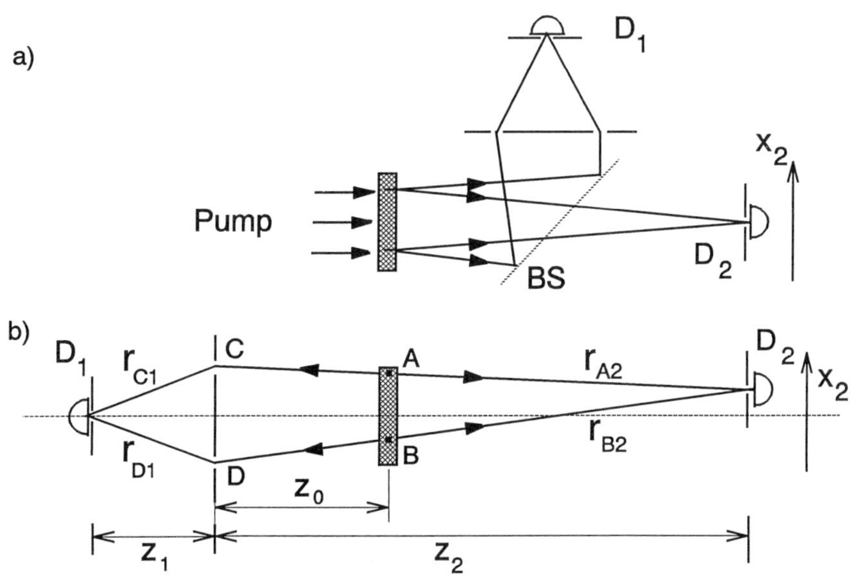

FIGURE 7. Simplified experimental scheme (a) and its "unfolded" version (b). The overall optical path lengths between D_1 and D_2 along the upper (r_A) and lower (r_B) paths, appearing in equation 9, are defined as $r_A \equiv r_{A1} + r_{A2} = r_{C1} + r_{C2}$ and $r_B \equiv r_{B1} + r_{B2} = r_{D1} + r_{D2}$, where r_{Ci} and r_{Di} are the optical path lengths from the slits C and D to the *i*-th detector.

written as

$$R_c(x_2) \propto \cos^2(x_2 \pi d / \lambda z_2). \tag{10}$$

Equation 10 has the form of a standard Young's double-slit interference pattern. Here, again, z_2 *is the unusual distance from the slits' plane, which is in the signal beam, back through BS to the crystal, and then along the idler beam to the scanning fiber tip of detector* D_2.

If the optical paths from the fixed detector D_1 to the two slits are unequal, that is, $r_{C1} \neq r_{D1}$, the interference pattern will be shifted from the symmetrical form of equation 10 according to equation 9. This interesting phenomenon has been observed and reported following the discussion of FIGURE 5.

There are two conclusions that can be drawn from equation 10:

(1) a two-photon interference pattern can be observed in coincidences by scanning D_2 in the transverse direction of one beam, even though the Young's double-slit aperture is in the other beam;
(2) the interference pattern is the same as one would observe on a screen in the plane of D_2, if D_1 is replaced by a pointlike light source and the SPDC crystal by a reflecting mirror.

To calculate the "ghost" diffraction effect of a single slit such as shown in FIGURE 6, we need an integral of the effective two-photon wave function over the slit width:

$$R_c(x_2) \propto \left| \int_{-a/2}^{a/2} dx_0 \exp\left[-ikr(x_0, x_2)\right] \right|^2 \cong \operatorname{sinc}^2(x_2 \pi a / \lambda z_2), \tag{11}$$

where $r(x_0, x_2)$ is the distance between points x_0 and x_2, x_0 belongs to the slit's plane, and the inequality $z_2 \gg a^2/\lambda$ is assumed.

Repeating the above calculations, the combined interference-diffraction coincidence-counting rate for the double-slit case is given by

$$R_c(x_2) \propto \operatorname{sinc}^2(x_2 \pi a / \lambda z_2) \cos^2(x_2 \pi d / \lambda z_2), \tag{12}$$

which is exactly the same as equation 5 obtained from experimental data fittings.

In the above calculations, we assume that the pump beam is a plane wave and the crystal's transverse dimension is infinite. If, instead, a Gaussian pump beam profile is considered, equation 12 (or equation 5) must be multiplied by a Gaussian function $G(x_2, \sigma z_2/z_0)$:

$$R_c(x_2) \propto G(x_2, \sigma z_2/z_0) \operatorname{sinc}^2(x_2 \pi a / \lambda z_2) \cos^2(x_2 \pi d / \lambda z_2), \tag{13}$$

where σ is the Gaussian width of the pump beam and z_0 is the distance between the slit plane and the crystal. Furthermore, if the finite size of the detectors and the divergence of the pump are also taken into account by a convolution, the interference visibility will be reduced. These factors have been taken into account in the theoretical plots in FIGURES 5 and 6. FIGURE 8 reports a study of the diffraction pattern width when the slit width is changed. The theory curve (solid line) takes into

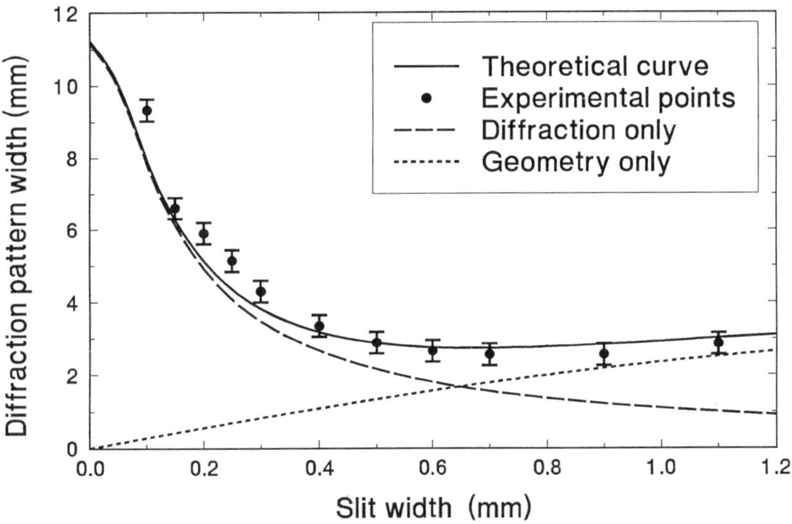

FIGURE 8. A study of the diffraction pattern width when changing the slit width. The calculated solid curve takes into account the corrections for the finite size of the detectors, the Gaussian pump beam profile, and non-far-field corrections. No fitting parameters have been used. The dashed (dotted) line shows the expected purely diffraction (geometrical) width, limited by the pump profile.

account the corrections for the finite size of the detectors, the Gaussian pump profile, and non-far-field corrections, which become important for wider slits.

CONCLUSIONS

A pair of photons is generated in SPDC. The propagation direction of each single photon has a considerably large angular uncertainty. However, if one of them is detected at a certain direction, its conjugate twin brother must have been propagating in a defined certain direction. The above two experiments demonstrated this striking EPR phenomenon from both a geometrical optics and a physical optics point of view. We may consider these experiments as a nonclassical phenomenon of a new variety, in which the two-particle is entangled in *space.*

ACKNOWLEDGMENTS

We gratefully acknowledge useful discussions with C. O. Alley and M. H. Rubin. We thank D. Greenberger for encouraging us to publish these experiments in this volume. After the submission of the second experiment for publication, we learned that an interesting two-photon diffraction experiment was also performed in A. Zeilinger's group by using an optical grating. We would like to thank A. Zeilinger for bringing this to our attention.

REFERENCES

1. SCHRÖDINGER, E. 1935. Naturwissenschaften **23:** 807, 823, and 844. A translation of these papers appears in: WHEELER, J. A. & W. H. ZUREK, Eds. 1983. Quantum Theory and Measurement. Princeton University Press. Princeton, New Jersey.
2. EINSTEIN, A., B. PODOLSKY & N. ROSEN. 1935. Phys. Rev. **47:** 777.
3. For a review, see: CLAUSER, J. F. & A. SHIMONY. 1978. Rep. Prog. Phys. **41:** 1883; ASPECT, A., P. GRANGIER & G. ROGER. 1981. Phys. Rev. Lett. **47:** 460; 1982. Phys. Rev. Lett. **49:** 91 and 1804.
4. KLYSHKO, D. N. 1988. Photons and Nonlinear Optics. Gordon & Breach. New York; YARIV, A. 1989. Quantum Electronics. Wiley. New York.
5. For example, see: ALLEY, C. O. & Y. H. SHIH. 1986. *In* Foundations of Quantum Mechanics in the Light of New Technology. M. Namiki *et al.*, Eds.: 47. Phys. Soc. Japan. Kyoto; HONG, C. K., Z. Y. OU & L. MANDEL. 1987. Phys. Rev. Lett. **59:** 2044; SHIH, Y. H. & C. O. ALLEY. 1988. Phys. Rev. Lett. **61:** 2921; OU, Z. Y. & L. MANDEL. 1988. Phys. Rev. Lett. **61:** 54; RARITY, J. G. & P. R. TAPSTER. 1990. Phys. Rev. Lett. **64:** 2495; BRENDEL, J., E. MOHLER & W. MARTIENSSEN. 1991. Phys. Rev. Lett. **66:** 1142; LARCHUK, T. S., R. A. CAMPOS, J. G. RARITY, P. R. TAPSTER, E. JAKEMAN, B. E. A. SALEH & M. C. TEICH. 1993. Phys. Rev. Lett. **70:** 1603; STEINBERG, A. M., P. G. KWIAT & R. Y. CHIAO. 1993. Phys. Rev. Lett. **71:** 708; KWIAT, P. G., A. M. STEINBERG & R. Y. CHIAO. 1993. Phys. Rev. A **47:** 2472; KIESS, T. E., Y. H. SHIH, A. V. SERGIENKO & C. O. ALLEY. 1993. Phys. Rev. Lett. **71:** 3893; SHIH, Y. H., A. V. SERGIENKO, M. H. RUBIN, T. E. KIESS & C. O. ALLEY. 1994. Phys. Rev. A **50:** 23; SHIH, Y. H. & A. V. SERGIENKO. 1994. Phys. Rev. A **50:** 2564; Phys. Lett. A **191:** 201.
6. For related theory, see: KLYSHKO, D. N. 1988. Phys. Lett. A **132:** 299; Sov. Phys. Usp. **31:** 74; BELINSKII, A. V. & D. N. KLYSHKO. 1994. JETP **78:** 259.
7. EKERT, A. K. 1991. Phys. Rev. Lett. **67:** 661.

Experiments with Single Atoms in Cavities and Traps

H. WALTHER
Sektion Physik der Universität München
and
Max-Planck-Institut für Quantenoptik
D-85748 Garching, Germany

INTRODUCTION

In the last years, the modern techniques of laser spectroscopy have made it possible to investigate quantum phenomena of single atoms. In this report, two groups of these experiments will be reviewed. The first one deals with the one-atom maser and the second one with trapped ions.[1] There is also considerable progress in trapping neutral atoms.[2] These techniques are undergoing rapid development at the moment. However, experiments with single isolated atoms using these techniques have not yet been described.

REVIEW OF THE ONE-ATOM MASER

The most fundamental system to study the generation process of radiation in lasers and masers is to drive a single mode cavity by a single atom. This system, at first glance, seems to be another example of a gedankenexperiment, but such a one-atom maser[3] really exists and also can be used to study the basic principles of radiation-atom interaction. The main features of the setup are as follows:

(1) it is the first maser that sustains oscillations with much less than one atom on the average in the cavity;
(2) the maser allows one to study the dynamics of the energy exchange between an atom and a single mode of the cavity field as treated in the Jaynes-Cummings model;[4]
(3) the setup allows one to study in detail the conditions necessary to obtain nonclassical radiation, especially radiation with sub-Poissonian photon statistics in a maser system directly;
(4) it is possible to study a variety of phenomena of a quantum field, including nonlocal aspects of the quantum measurement process.

What are the tools that make this device work? It is the enormous progress in constructing superconducting cavities with high quality factors together with the laser preparation of highly excited atoms—Rydberg atoms—that have made the realization of such a one-atom maser possible.[3] Rydberg atoms are obtained when one of the outermost electrons of an atom is excited into a level close to the ionization limit. The main quantum number of the electron is then typically of the order of 60–70. These atoms have quite remarkable properties[5] that make them ideal for the maser experiments: The probability of induced transitions between neighboring states of a Rydberg atom scales as n^4, where n denotes the principal quantum number. Consequently, a few photons are enough to saturate the transition between adjacent levels. Moreover, the spontaneous lifetime of a highly excited state is very large. We obtain a maser by injecting these Rydberg atoms into a superconducting cavity with a high quality factor. The injection rate is such that, on the average, there is much less than one atom present inside the resonator.

The experimental setup of the one-atom maser is shown in FIGURE 1. A highly collimated beam of rubidium atoms passes through a Fizeau velocity selector. Before entering the superconducting cavity, the atoms are excited into the upper maser level $63p_{3/2}$ by the frequency-doubled light of a cw ring dye laser. The superconducting niobium maser cavity is cooled down to a temperature of 0.5 K by means of a ^{3}He cryostat. At this temperature, the number of thermal photons in the cavity is about 0.15 at a frequency of 21.5 GHz. The quality factor of the cavity can be as high as 3×10^{10}, corresponding to a photon storage time of about 0.2 s. Two maser transitions from the $63p_{3/2}$ level to the $61d_{3/2}$ and to the $61d_{5/2}$ level are studied. In a new setup equipped with a dilution refrigerator, temperatures in the range of 0.1 K are obtained. Some of the experiments described in this review have been performed with the latter setup.

The Rydberg atoms in the upper and lower maser levels are detected by two separate field ionization detectors. The field strength is adjusted to ensure that, in the first detector, the atoms in the upper level are ionized, but not those in the lower level; the lower-level atoms are then ionized in a second field.

To demonstrate maser operation, the cavity is tuned over the respective transition and the flux of atoms in the excited state is recorded simultaneously. FIGURE 2

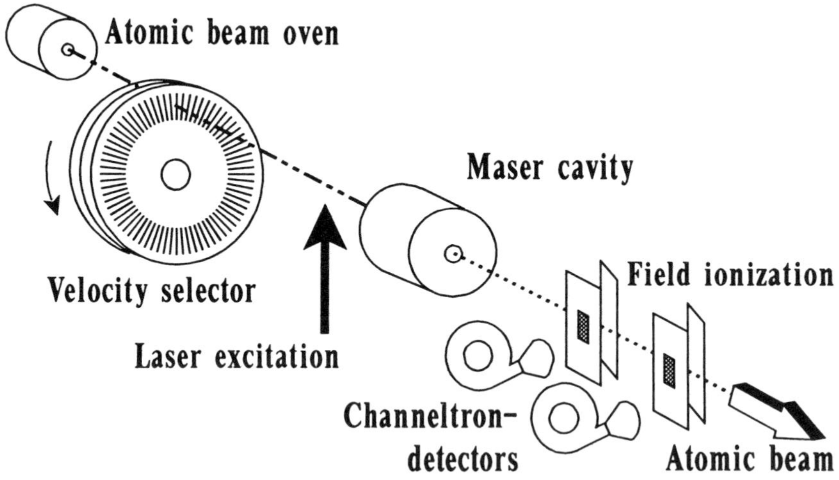

FIGURE 1. Scheme of the one-atom maser. To suppress blackbody-induced transitions to neighboring states, the Rydberg atoms are excited inside the liquid-helium-cooled environment.

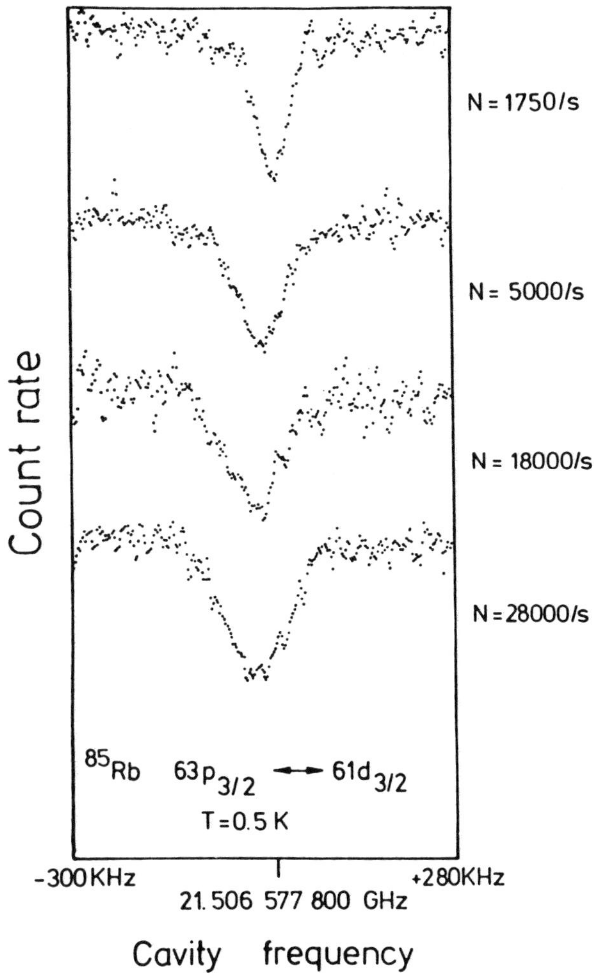

FIGURE 2. A maser transition of the one-atom maser manifests itself in a decrease of atoms in the excited state. The flux of excited atoms N governs the pump intensity. Power broadening of the resonance line demonstrates the multiple exchange of a photon between the cavity field and the atom passing through the resonator.

shows the result for $63p_{3/2}$–$61d_{3/2}$. Transitions from the initially prepared state to the $61d_{3/2}$ level (21.50658 GHz) are detected by a reduction of the electron count rate.

In the case of measurements at a cavity temperature of 0.5 K, shown in FIGURE 2, a reduction of the $63p_{3/2}$ signal can be clearly seen for atomic fluxes as small as 1750 atoms/s. An increase in flux causes power broadening and a small shift. This shift is attributed to the ac Stark effect, caused predominantly by virtual transitions to neighboring Rydberg levels. Over the range from 1750 to 28,000 atoms/s, the field ionization signal at resonance is independent of the particle flux, which indicates that the transition is saturated. This and the observed power broadening show that there is a multiple exchange of photons between Rydberg atoms and the cavity field.

For an average transit time of the Rydberg atoms through the cavity of 50 μs and a flux of 1750 atoms/s, we estimate that approximately 0.09 Rydberg atoms are in the cavity on the average. According to Poisson statistics, this implies that more than 90% of the events are due to single atoms. This clearly demonstrates that single atoms are able to maintain a continuous oscillation of the cavity with a field corresponding to a mean number of photons between unity and several hundred.

Among the studies performed with the one-atom maser are the measurements of the dynamics of the photon exchange between a single atom and a cavity mode.[6,7] Before we discuss some experiments with the one-atom maser, the theory will be briefly reviewed.

THEORY OF THE ONE-ATOM MASER

The simplest form of interaction between a two-level atom and a single quantized mode of the electromagnetic field is described by the Jaynes-Cummings Hamiltonian:[4,8]

$$H = (\tfrac{1}{2})\hbar\omega_0\sigma_z + \hbar\omega a^\dagger a + \hbar(g a^\dagger \sigma_- + \text{adj.}) = H_0 + V,$$

where

$$H_0 = (\tfrac{1}{2})\hbar\omega_0\sigma_z + \hbar\omega a^\dagger a$$

and

$$V = \hbar(g a^\dagger \sigma_- + \text{adj.}).$$

Here, ω_0 is the atomic transition frequency; ω is the field frequency; a and $a^\dagger$ are the photon annihilation and creation operators of the field mode (with $[a, a^\dagger] = 1$); σ_z, σ_-, and σ_+ are atomic pseudospin operators (with $[\sigma_+, \sigma_-] = \sigma_z$); and

$$g = \left(\frac{pE_\omega}{2\hbar}\right) \sin KZ$$

is the electric dipole matrix element at the location Z of the atom, where E_ω is the "electric field per photon", that is, $E_\omega = (\hbar\omega/\epsilon_0 V)^{1/2}$.

The Jaynes-Cummings model plays a central role in quantum optics owing to several reasons: it gives the simplest description of Rabi-flopping in a quantized field and the simplest illustration of spontaneous emission; furthermore, it can be solved exactly and thus describes the true quantum dynamics observed with the one-atom maser, such as collapse and revivals of the atomic inversion. The model describes the situation realized in the one-atom maser and allows a detailed investigation of the complexities of the atom-field dynamics in the simplest of all situations.

In the following, a few results derived from the Jaynes-Cummings model are reviewed that are relevant for the discussions in this report.

The eigenstates of the Jaynes-Cummings Hamiltonian are

$$E_{1n} = \hbar[-(\tfrac{1}{2})\omega_0 + (n + 1)\omega + (\tfrac{1}{2})(\Omega_n + \delta)],$$

$$E_{2n} = \hbar[(\tfrac{1}{2})\omega_0 + n\omega - (\tfrac{1}{2})(\Omega_n + \delta)],$$

where $\delta = \omega_0 - \omega$ is the atom-field frequency detuning. Ω_n is the generalized n-photon Rabi-flopping frequency:

$$\Omega_n = \sqrt{4g^2(n + 1) + \delta^2}.$$

The corresponding eigenvalues are the ones of the dressed atom:

$$|1, n\rangle = \sin\theta_n|a, n\rangle + \cos\theta_n|b, n + 1\rangle,$$

$$|2, n\rangle = \cos\theta_n|a, n\rangle - \sin\theta_n|b, n + 1\rangle,$$

where the states $|a\rangle$ and $|b\rangle$ represent the upper and lower atomic states, respectively, and $|n\rangle$ represents the number state of the field mode with $a\dagger a|n\rangle = n|n\rangle$. The angle θ is defined by means of the following relations:

$$\cos\theta_n = \frac{(\Omega_n - \delta)}{\sqrt{(\Omega_n - \delta)^2 + 4g^2(n + 1)}},$$

$$\sin\theta_n = \frac{2g\sqrt{n + 1}}{\sqrt{(\Omega_n - \delta)^2 + 4g^2(n + 1)}}.$$

Note, in particular, that

$$\cos 2\theta_n = -\delta/\Omega_n$$

and

$$\sin 2\theta_n = 2g\sqrt{n + 1}/\Omega_n.$$

In the vacuum field ($n = 0$) and on resonance ($\omega_0 = \omega$), the dressed states are separated by the frequency $\Omega_0 = 2g$, generally called vacuum Rabi-splitting.

One of the interesting phenomena described by the Jaynes-Cummings model is the dynamical behavior. When we assume that the atom is initially in the upper state $|a\rangle$ and the field is in the number state $|n\rangle$, it follows for the probability of an atom to be in the upper state that

$$|C_{an}(t)|^2 = \cos^2(g\sqrt{n + 1}t).$$

This shows that the upper state population oscillates periodically at the Rabi-flopping frequency, similar to the case of classical fields. If the field is initially described by the photon statistics, p_n, the above results have to be generalized to

$$|C_a(t)|^2 = \sum_n p_n \cos^2(g\sqrt{n + 1}t).$$

In the case where the field mode is initially in a coherent state, it has been shown that $|C_a(t)|^2$ undergoes a collapse followed by a series of revivals.[6] The collapse is due to the destructive interference of quantum Rabi-floppings at different frequencies; a similar phenomenon may also occur with a classical field. However, the revivals are a purely quantum mechanical effect that originates in the discreteness of the quantum field. Collapse and revivals have been observed in a micromaser experiment.[7] In this experiment, the interaction time of the atoms with the cavity was varied and the probability was investigated where the atoms leave the cavity in the excited state. As will be shown below, the photon statistics in the maser cavity changes when the interaction time is varied; therefore, the photon statistics p_n is not a pure distribution. Nevertheless, the revival shows up, as was also confirmed in a computer simulation of the results on the basis of the Jaynes-Cummings model.[7]

In the following, we would like to summarize the results of the quantum theory of the one-atom maser. Because the atom-field interaction takes place in a closed single mode cavity, there is no spontaneous emission rate into free space modes. Owing to the extremely high quality factors achieved in the superconducting cavities, the photon lifetime is extremely long compared to the transit time of the atoms through the resonator. This means that the cavity damping can be practically ignored while an atom interacts with the field. Because the atomic flux is kept small so as to have, at most, one atom present inside the cavity at a time, the cavity is empty most of the time; therefore, cavity damping can be neglected during the rare instances when an atom interacts with the cavity mode.

The one-atom maser theory is therefore based on the following strategy:[9] while an atom is in the cavity, the coupled atom-field system is described by the Jaynes-Cummings Hamiltonian and, for the interval between atoms, the evolution of the field density matrix is described by a master equation considering damping and also the mean number of thermal photons in the cavity.

Besides this microscopic theory, there is also a macroscopic theory based on the quantum theory of the laser.[10] The resulting probability distribution of the photons depends characteristically on the pump rate and on the interaction time t_{int} of the atoms with the cavity field. One obtains the following result for the probability of finding n photons in the maser cavity $[P(n)]$ in steady state:

$$P(n) = P_0\left[\frac{n_b}{(n_b+1)}\right]^n \prod_{m=1}^{n}\left[1+\left(\frac{N}{n_b\gamma}\right)\frac{\sin^2(g\sqrt{m}t_{\text{int}})}{m}\right],$$

with N being the atomic pump rate, n_b being the thermal photon number, and γ being the cavity decay rate. P_0 is determined by the normalization condition, $\Sigma\, P(n) = 1$. One can now evaluate the mean photon number $\langle n\rangle$ and the field variance in the form of the Q_{f} parameter:[9]

$$Q_{\text{f}} = \frac{[\langle n^2\rangle - \langle n\rangle^2 - \langle n\rangle]}{\langle n\rangle}.$$

FIGURE 3 shows the mean photon number as a function of the interaction time of the atoms with the cavity. The photon number is scaled to N_{ex}; here, N_{ex} is the average number of atoms that enter the cavity during the cavity decay time, $N_{\text{ex}} =$

N/γ. The maser thresholds occur at $\Theta = 1$ and regions of sub-Poissonian photon statistics are between $\Theta = \pi$ and 2π as well as between 3π and 4π. The sub-Poissonian character leads to a negative Q_f. A large negative value is obtained close to (before) 2π.[8,9]

To get a more intuitive insight into this effect, we recall that the Fizeau velocity selector preselects the velocity of the atoms. Hence, the interaction time is well defined, which leads to conditions usually not achievable in standard masers. This has a very important consequence when the intensity of the maser field grows as more and more atoms transfer their excitation energy to the field: even in the absence of dissipation, this increase in photon number is halted when the increasing Rabi frequency leads to a situation where the atoms reabsorb the photon and leave the cavity in the upper state. This situation is close to a quantum-nondemolition case. For any photon number, this situation can be achieved by appropriately adjusting the

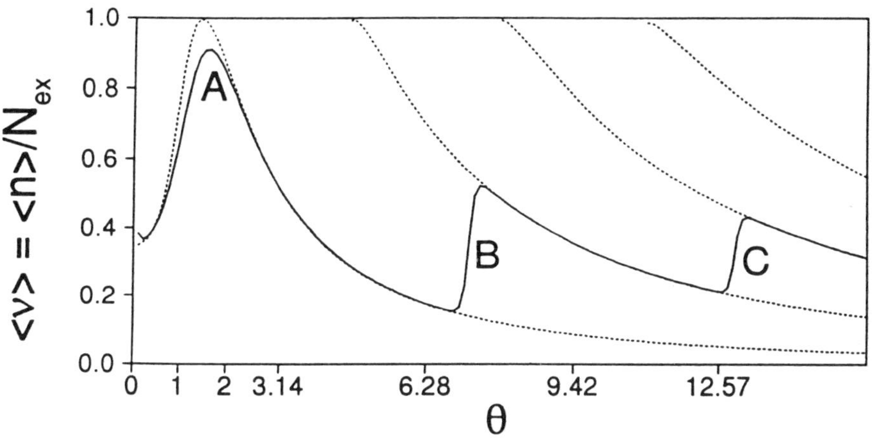

FIGURE 3. Mean value of the photon number $\nu = n/N_{ex}$ versus the pump parameter Θ, where the value of Θ is changed via N_{ex}. The solid line represents the micromaser solution for $\Omega = 36$ kHz, $t_{int} = 35$ μs, and temperature $T = 0.15$ K. The dotted lines are semiclassical steady-state solutions corresponding to fixed stable gain-loss equilibrium photon numbers. The crossing points between a line Θ = constant and the dotted lines correspond to the values where minima in the Fokker-Planck potential $V(\nu)$ occur. For details, see text.

velocity of the atoms. Then, the number distribution of the photons in the cavity is sub-Poissonian.

Unfortunately, the measurement of the nonclassical photon statistics in the cavity is not straightforward. The measurement process of the field involves the coupling to a measuring device, whereby losses lead inevitably to a destruction of the nonclassical properties. The ultimate technique to obtain information about the field employs the Rydberg atoms themselves: for this purpose, the population and the statistics of the atoms in the upper and lower maser levels are probed when they leave the cavity. Accordingly, the atoms play a double role: (i) they pump the cavity and (ii) they are used for the diagnostics. These two roles interfere with one another because the detection of the atom in a known final state leads to a quantum mechanical reduction of the photon state inside the resonator. Frequent detection is accompanied by a quasi-permanent state reduction, which can prevent the cavity field from relaxing to the steady state that would be reached if the atoms were left unobserved. Nevertheless, the steady-state properties determine the statistics of the clicks recorded by the atom detectors.

The theoretical treatment of the one-atom maser produces predictions about both the photon field and the emerging atoms. Only the latter can be tested experimentally and the success of such tests feeds our confidence in the predictions about the quantized radiation field inside the cavity. The pump atoms are statistically independent in the standard one-atom-maser experiments, so their arrival times are subject to Poissonian statistics; therefore, we shall restrict the discussion to this standard Poissonian situation.

Inasmuch as the atoms arrive at random, they are recorded at equally random times; thus, the only reproducible data are statistical. Consequently, one is led to studying the statistics of the detector clicks. Numerical simulations investigating the effect of repeated atomic measurements on the evolution of the cavity field have been performed by Meystre[11] as well as by Meystre and Wright.[12] The relation between the counting statistics of the detected atoms emerging from the resonator and the photon-number statistics of the field inside the cavity has been studied analytically by Rempe and myself[13] as well as by Paul and Richter.[14] In reference 13, the results are also compared with numerical simulations showing good agreement; experimental results are reported by Rempe *et al.*[15]

In a recent paper, a general method for the computation of various statistical properties of the click distribution was presented.[16] This method does not resort to numerical simulations. Naturally, the efficiencies of the detectors (far from the ideal 100%, unfortunately) are taken into account. A central tool used in those calculations is a nonlinear master equation that governs the dynamics of the photon field in periods between detector clicks. The nonlinearity arises from the necessity to distinguish between the notions of observation and detection. When the detectors are active, all emerging atoms are observed, but only a fraction is actually detected. Most of the time, the experimenter is observing, but does not detect; he/she is listening, but does not hear.

In another approach based on the concept of the counting statistics of light beams, the atomic counting probability, the waiting-time distribution, and the "two-atom correlation" function for a Poissonian atomic beam exciting the micromaser cavity are also calculated. In an analytic treatment, it is shown how the waiting-time distribution converges into the atom correlation function for vanishing detection efficiency.[17]

Under steady-state conditions, as mentioned above, the photon number and the photon statistics of the maser field are essentially determined by the dimensionless parameter Θ. The quantity $\langle \nu \rangle = \langle n \rangle / N_{ex}$ shows the following generic behavior (see FIGURE 3): It suddenly increases at the threshold value of $\Theta = 1$ and reaches a maximum for $\Theta \approx 2$. As Θ further increases, $\langle \nu \rangle$ decreases and reaches a minimum at $\Theta \approx 2\pi$ and then abruptly increases to a second maximum. This general type of behavior recurs roughly at integer multiples of 2π, but becomes less pronounced with increasing Θ. The reason for the periodic maxima of $\langle \nu \rangle$ is that, for $\Theta \approx 2\pi$ and multiples thereof, the pump atoms perform an almost integer number of full Rabi-flopping cycles and start to flip over at a slightly larger value of Θ, with this leading to enhanced photon emission. The maser threshold at $\Theta = 1$ shows the characteristics of a continuous phase transition, whereas the subsequent maxima in $\langle \nu \rangle$ can be interpreted as first-order phase transitions. In the intervals between the phase-transition points, the photon statistics is mostly sub-Poissonian. The field is super-Poissonian for all phase transitions,[8,9] with the large photon number fluctuations above $\Theta = 2\pi$ and multiples thereof being caused by the presence of two maxima in the photon number distribution $P(n)$. They result from the fact that atoms in the upper maser level may or may not tip over to the lower level.[8,9]

The phenomenon of the two coexisting neighboring maxima in $P(n)$ was also studied in a semiheuristic Fokker-Planck (FP) approach.[9] There, the photon number distribution $P(n)$ is replaced by a probability function $P(\nu, \tau)$ with continuous variables $\tau = t/\tau_{cav}$ and $\nu(n) = n/N_{ex}$, with the latter replacing the photon number n. The steady-state solution obtained for $P(\nu, \tau)$, $\tau \gg 1$, can be constructed by means of an effective potential, $V(\nu)$, showing minima at positions where maxima of $P(\nu, \tau)$, $\tau \gg 1$, are found. Close to $\Theta = 2\pi$ and multiples thereof, the effective potential $V(\nu)$ exhibits two equally attractive minima located at stable gain-loss equilibrium points of maser operation. The mechanism at the phase transitions mentioned is always the same: a minimum of $V(\nu)$ loses its global character when Θ is increased and is replaced in this role by the next one. This reasoning is a variation of the Landau theory of first-order phase transitions, with $\sqrt{\nu}$ being the order parameter. This analogy actually leads to the notion that, in the limit $N_{ex} \to \infty$, the change of the micromaser field around integer multiples of $\Theta = 2\pi$ can be interpreted as first-order phase transitions.

In the region of the first-order phase transitions, long field evolution time constants τ_{field} are expected. This phenomenon was experimentally demonstrated, as was related phenomena such as spontaneous quantum jumps between equally attractive minima of $V(\nu)$, bistability, and hysteresis.[18] Some of these phenomena are also predicted in the two-photon micromaser.[19]

If there are no thermal photons in the cavity—a condition achievable by cooling the resonator to temperatures below 100 mK—very interesting features such as trapping states show up.[20] The investigation of the trapping states is discussed in detail in a recent review.[21]

In the following, we would like to review two experiments, one on the measurement of the photon statistics of the one-atom maser and another one on the observation of quantum jumps and bistability in the maser field at the first-order phase transition points.

THE PHOTON STATISTICS OF THE ONE-ATOM MASER FIELD

As discussed above in the experiment on the photon statistics of the one-atom maser field,[15] the number N of atoms in the lower maser level is counted for a fixed time interval T roughly equal to the storage time τ_{cav} of the photons in the cavity. By repeating this measurement many times, the probability distribution of finding N atoms in the lower level is obtained. The normalized variance $Q_a = [\langle N^2 \rangle - \langle N \rangle^2 -$

$\langle N \rangle]/\langle N \rangle$ is evaluated and is used to characterize the deviation from Poissonian statistics. A negative (positive) Q_a value indicates sub-Poissonian (super-Poissonian) statistics, whereas $Q_a = 0$ corresponds to a Poisson distribution with $\langle N^2 \rangle - \langle N \rangle^2 = \langle N \rangle$. The atomic Q_a can be related to the normalized variance Q_f of the photon number (for details, see references 13, 15, and 16).

As an example, we discuss one result for the $63p_{3/2}$–$61d_{5/2}$ maser transition with $g = 44$ rd/s; it is shown in FIGURE 4. Velocity-selected atoms with an atom-cavity interaction time of $t_{int} = 35$ μs were used. A very low flux of atoms of $N_{ex} > 1$ is

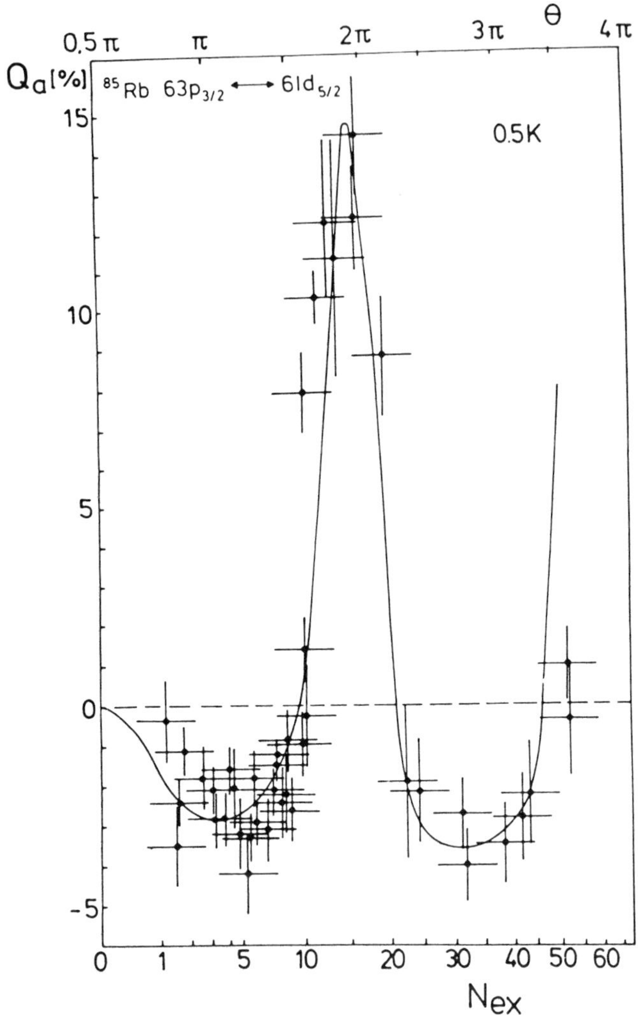

FIGURE 4. Variance Q_a of the atoms in the lower maser level as a function of the flux N_{ex} (see also reference 15).

already sufficient to generate a nonclassical maser field. This is the case because the vacuum field initiates a transition of the atom to the lower maser level, thus driving the maser above threshold.

The sub-Poissonian statistics of the atoms near $N_{ex} = 30$, corresponding to $Q_a = -4\%$, is generated by a photon field with a variance $\langle n^2 \rangle - \langle n \rangle^2 = 0.3\langle n \rangle$, which is 70% below the shot noise level.[15] Again, this result agrees with the prediction of the theory.[9,10] The mean number of photons in the cavity is around 12 and 13 in the regions of $N_{ex} \approx 3$ and $N_{ex} \approx 30$, respectively. Near $N_{ex} \approx 15$, the photon number changes abruptly between these two values. The next maser phase transition with a super-Poissonian photon number distribution occurs above $N_{ex} \approx 50$.

We emphasize that the reason for the sub-Poissonian atomic statistics is the following: A changing flux of atoms changes the Rabi-frequency via the stored photon number in the cavity. By adjusting the interaction time, the phase of the Rabi-nutation cycle can be chosen so that the probability for the atoms leaving the cavity in the upper maser level increases when the flux and therefore the photon number in the cavity are raised. We observe sub-Poissonian atomic statistics in the case where the number of atoms in the lower state is decreasing with increasing flux and photon number in the cavity. This feedback mechanism is also demonstrated when the anticorrelation of atoms leaving the cavity in the lower state is investigated. Measurements of this "antibunching" phenomenon have also been performed (see reference 21 for a detailed review and also reference 14). The fact that anticorrelation is observed shows that the atoms in the lower state are more equally spaced than expected for a Poissonian distribution. Thus, when two atoms enter the cavity close to each other, the second one performs a transition to the lower state with a reduced probability.

The experimental results presented here clearly show the sub-Poissonian photon statistics of the one-atom maser field. In addition, the maser experiment leads to an atomic beam with atoms in the lower maser level showing number fluctuations that are up to 40% below those of a Poissonian distribution found in usual atomic beams. This is interesting because atoms in the lower level have emitted a photon to compensate for cavity losses inevitably present. Although this is a purely dissipative process giving rise to fluctuations, the atoms nevertheless still obey sub-Poissonian statistics.

QUANTUM JUMPS OF THE MICROMASER FIELD

The setup used for these measurements is similar to that described above. As before, ^{85}Rb atoms were used to pump the maser. They are excited from the $5S_{1/2}$, $F = 3$ ground state to $63P_{3/2}$, $m_J = \pm 1/2$ states by linearly polarized light of a frequency-doubled cw ring dye laser. The polarization of the laser light is linear and parallel to the likewise linearly polarized maser field and, therefore, only $\Delta m_J = 0$ transitions are excited. Superconducting niobium cavities resonant with the transition to the $61D_{3/2}$, $m_J = \pm 1/2$ states were used. The experiments were performed in a $^3He/^4He$-dilution refrigerator with cavity temperatures $T \approx 0.15$ K. The cavity Q-values ranged from 4×10^9 to 8×10^9. The velocity of the Rydberg atoms and thus their interaction time t_{int} with the cavity field were preselected by exciting a particular velocity subgroup with the laser. For this purpose, the laser beam irradiated the atomic beam at an angle of approximately 82°. As a consequence, the UV laser light (line width ≈ 2 MHz) is blueshifted by 50–200 MHz by the Doppler effect, depending on the velocity of the atoms.[18]

As before, information on the maser field and interaction of the atoms in the cavity can be obtained solely by state-selective field ionization of the atoms in the upper or lower maser level after they have passed through the cavity. The field ionization detector was recently modified, so there is now a detection efficiency of $\eta = (35 \pm 5)\%$. For different t_{int}, the atomic inversion has been measured as a function of the pump rate; the coupling constant is about $g = 40$ rd/s.

Depending on the parameter range, essentially three regimes of the field evolution time constant τ_{field} can be distinguished.[18] We restrict the discussion here only to the results for intermediate time constants. The maser was operated under steady-state conditions close to the second first-order phase transition (C in FIGURE

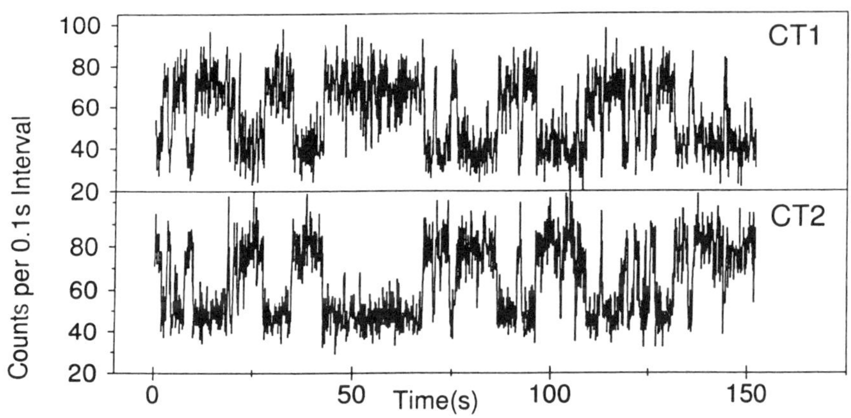

FIGURE 5. Quantum jumps between two equally stable operation points of the maser field. The channeltron counts are plotted versus time (CT1 = upper state, CT2 = lower state).

3). The interaction time was $t_{\mathrm{int}} = 47$ μs and the cavity decay time was $\tau_{\mathrm{cav}} = 60$ ms. The value of N_{ex} necessary to reach the second first-order phase transition was $N_{\mathrm{ex}} \approx 200$. For these parameters, the two maxima in $P(n)$ are manifested in spontaneous jumps of the maser field between the two maxima with a time constant of ≈5 s. This fact and the relatively large pump rate lead to the clearly observable field jumps shown in FIGURE 5. Due to the large cavity field decay time, the average number of atoms in the cavity was still as low as 0.17. The two discrete values for the counting rates correspond to the metastable operating points of the maser, which correspond to ≈70 and ≈140 photons. In the Fokker-Planck description, the two values correspond to two equally attractive minima in the Fokker-Planck potential $V(\nu)$. If one considers, for instance, the counting rate of lower-state atoms (CT2 in FIGURE 5), the lower (higher) plateaus correspond to time intervals in the low (high) field metastable operating point. If the actual photon number distribution is averaged over a time interval containing many spontaneous field jumps, the steady-state result $P(n)$ of the micromaser theory is recovered.

In a parameter range where switching occurs much faster than in the case shown in FIGURE 5, the individual jumps cannot be resolved any longer owing to the reduced N_{ex} necessary in this case. Therefore, a different procedure has to be chosen for the investigation. Those experiments will not be discussed here (details are described in reference 18).

LINE WIDTH AND PHASE DIFFUSION OF THE ONE-ATOM MASER

In the following, we would like to discuss another special feature of the one-atom maser: the spectrum. This is determined by the decay of the expectation value of the electric field:[21]

$$\langle E(t)\rangle \sim \sum_{n=0}^{\infty} (n + 1)^{1/2}\rho_{n,n+1}(t).$$

Hence, the micromaser spectrum is different from the other effects discussed so far because it involves the off-diagonal elements $\rho_{n,n+1}$ of the radiation density matrix rather than, for example, the photon statistics, that is, the diagonal elements $\rho_{n,n}$; moreover, it requires their time dependence rather than their steady-state values. It could also be shown that the line width D of the maser is given by

$$D = 4r \sin^2\left[\frac{gt_{\text{int}}}{4\sqrt{\langle n\rangle}}\right] + \frac{\gamma(2n_b + 1)}{4\langle n\rangle}.$$

For small $g\tau/4\sqrt{\langle n\rangle}$, the sine function can be expanded; this leads to the familiar Schawlow-Townes line width,

$$D = \frac{\alpha + \gamma(2n_b + 1)}{4\langle n\rangle},$$

where

$$\alpha = \gamma(\sqrt{N_{ex}}gt_{\text{int}}).$$

The complicated pattern of the micromaser line width results, in part, from the complicated dependence of $\langle n\rangle$ on the pump parameter, which enters in the denominator.

We emphasize that the one-atom maser line width goes beyond the standard Schawlow-Townes line width. The sine function suggests in the limit of large pumping parameters an oscillatory behavior of the line width confirmed in an exact numerical treatment. (For details, see also references 22 and 23.)

It was pointed out that the maser line width can be measured when two phase-coupled microwave fields are used: one before the atoms enter the microwave cavity and one after corresponding, in principle, to a modified Ramsey setup. The first micromaser field creates a superposition of the two maser states being then probed by the second one. The first field is only applied for an initial period in order to "seed" a phase in the micromaser cavity; in the second field, the phase diffusion is probed. This technique will provide the basis for future experiments to measure the line width. It was tested by computer simulations and analytical calculations that showed that coherent pumping of the maser leads to new and interesting phenomena (not discussed here). The first coherent injection for the one-atom maser was proposed in the paper by Krause *et al.*[24] and was discussed later in connection with the measurement of the phase diffusion line width.[25] Also, the phase dynamics of the maser field in steady-state operation was discussed by Wagner *et al.*;[26] in this latter case, only one microwave field is used for probing after the micromaser cavity. The latter scheme was further pursued in connection with the entanglement of states by Wagner *et al.*[27] The entangled state of the atom-field system occurs because a factorization in the field and atom part is not possible. This fact leads to applications of the micromaser for Einstein-Podolsky-Rosen–type experiments and to nonlocal field correlations in two distant cavities.

A NEW PROBE OF COMPLEMENTARITY IN QUANTUM MECHANICS—THE ONE-ATOM MASER AND ATOMIC INTERFEROMETRY

In one of the preceding sections, we discussed how a nonclassical field inside the maser cavity can be generated. Such a field is extremely fragile because any attenuation causes a considerable broadening of the photon number distribution. Therefore, it is difficult to couple the field out of the cavity while preserving its nonclassical character. However, what is the use of such a field? In the present section, we want to propose a new set of experiments performed inside the maser cavity to test the "wave-particle" duality of nature—or, better said, "complementarity"—in quantum mechanics.

Complementarity lies at the heart of quantum mechanics: Matter sometimes displays wavelike properties manifesting themselves in interference phenomena and, at other times, it displays particlelike behavior, thus providing "which-path" information. No other experiment illustrates this wave-particle duality in a more striking way than the classic Young double-slit experiment. Here, we find it impossible to tell which slit that light went through while observing an interference pattern. In other words, any attempt to gain "which-path" information disturbs the light so as to wash out the interference fringes. This point has been emphasized by Bohr in his rebuttal to Einstein's ingenious proposal of using recoiling slits to obtain "which-path" information while still observing interference. The physical positions of the recoiling slits, Bohr argued, are only known to within the uncertainty principle. This error contributes a random phase shift to the light beams that destroys the interference pattern.

Such random-phase arguments, illustrating in a vivid way how the "which-path" information destroys the coherent wavelike interference aspects of a given experimental setup, are appealing. Unfortunately, they are incomplete: in principle, and in practice, it is possible to design experiments that provide "which-path" information via detectors that do not disturb the system in any noticeable way. Such "Welcher Weg" (German for "which path") detectors have been recently considered within the context of studies involving spin coherence.[28] In this section, we describe a quantum optical experiment[29] that shows that the loss of coherence occasioned by "Welcher Weg" information, that is, by the presence of a "Welcher Weg" detector, is due to the establishment of quantum correlations. It is in no way associated with large random-phase factors as in Einstein's recoiling slits.

The details of this application of the micromaser are discussed in reference 30. Here, only the essential features are given. We consider an atomic interferometer where the two particle beams pass through two maser cavities before they reach the two slits of the Young interferometer. The interference pattern observed is then also determined by the state of the maser cavity. The corresponding interference term is given by

$$\langle \Phi_1^{(f)}, \Phi_2^{(i)} | \Phi_1^{(i)}, \Phi_2^{(f)} \rangle,$$

where $|\Phi_j^{(i)}\rangle$ and $|\Phi_j^{(f)}\rangle$ denote the initial and final states of the maser cavity j.

Let us prepare, for example, both one-atom masers in coherent states $|\Theta_j^{(i)}\rangle = |\alpha_j\rangle$ of large average photon number $\langle n \rangle = |a_j|^2 \gg 1$. The Poissonian photon number distribution of such a coherent state is very broad, $\Delta m \approx \alpha \gg 1$. Hence, the two fields are not changed much by the addition of a single photon associated with the two corresponding transitions. We may therefore write

$$|\Phi_j^{(f)}\rangle \approx |\alpha_j\rangle,$$

which to a very good approximation yields

$$\langle \Phi_1^{(f)}, \Phi_2^{(i)} | \Phi_1^{(i)}, \Phi_2^{(f)} \rangle \approx \langle \alpha_1, \alpha_2 | \alpha_1, \alpha_2 \rangle = 1.$$

Thus, there is a "radiation" interference cross term different from zero.

However, when we prepare both maser fields in number states $|n_j\rangle$, the situation is quite different. Because the number states are orthogonal, the interference term disappears whenever a passing atom deposits a photon in one of the cavities; therefore, no interferences are observed in such a case. (For the case that thermal fields are present, see reference 29.)

At first sight, this result might seem a bit surprising when we recall that, in the case of a coherent state, the transitions did not destroy the coherent cross term, that is, did not affect the interference fringes. However, in the example of number states, we can, by simply "looking" at the one-atom maser state, tell which "path" the atom took.

The atomic interference experiment in connection with one-atom maser cavities is a rather complicated scheme for a "Welcher Weg" detector. There is a much simpler possibility that we will discuss briefly in the following. This is based on the logic of the famous "Ramsey fringe" experiment. In this experiment, two microwave fields are applied to the atoms, one after the other. The interference occurs because the transition from an upper state to a lower one may occur either in the first or in the second interaction region. In order to calculate the transition probability, we must sum the two amplitudes and then square, thus leading to an interference term in the same way as the two-slit experiment. (For details, see reference 31.)

We conclude this section by emphasizing again that this new and potentially experimental example of wave-particle duality and observation in quantum mechanics displays a feature that makes it distinctly different from the Bohr-Einstein recoiling-slit experiment. In the latter, the coherence, that is, the interference, is lost due to a phase disturbance of the light beams. In the present case, however, the loss of coherence is due to the correlation established between the system and the one-atom maser. Random-phase arguments never enter the discussion. We emphasize that the argument of the number state not having a well-defined phase is not relevant here; the important dynamics are due to the atomic transition. It is the fact that "which-path" information is made available that washes out the interference cross terms.[30]

EXPERIMENTS WITH TRAPPED IONS—THE PAUL TRAP

In contrast to neutral atoms, ions can be easily influenced by electromagnetic fields because of their charge. Therefore, it is possible to isolate a single ion quite easily from its surroundings. In most of the experiments performed so far, the Paul trap is used. It consists of a ring electrode and two end-caps as shown in FIGURE 6. Trapping can be achieved if time-varying electric fields[32] are applied between the ring and caps (the two caps are electrically connected). A dc voltage, in addition, changes the relation of the potential depth along the symmetry axis (vertical direction in FIGURE 6) to that in perpendicular direction. The equation of motion of an ion in such a situation is the Mathieu differential equation, well known in classical mechanics, which—depending on the voltages applied to the trap (dc and radio-frequency voltages)—allows stable and unstable solutions. Another way to achieve trapping is the use of a constant magnetic field aligned along the symmetry axis leading to the Penning trap.[1,33] In this case, only a dc voltage has to be applied between the ring and cap electrodes.

In order to produce the ions in the Paul trap, a neutral atomic beam is directed through the trap center and is ionized by electrons. Unfortunately, the resulting trapped ions have a lot of kinetic energy rendering them useless for most applications, such as spectroscopy; therefore, the ions have to be cooled. This is done by laser light. For this purpose, the laser frequency ν is tuned below the resonance frequency, so the energy of the photon is not sufficient to excite the atom. Crudely, the ion can extract the missing energy from its motion and thus reduce its kinetic energy. In other words, the atomic velocity Doppler-shifts the atom into resonance to bridge the detuning gap Δ between laser and resonance frequency and the atom absorbs the photon of momentum $\hbar k = h\nu/c$. After the absorption process, the momentum of the atom in the direction opposite to the laser beam is reduced. This leads to a net cooling effect because the reemission of the energy due to atomic fluorescence is isotropic in space. The lowest temperature achievable is determined by the Doppler limit,[34] which is in the milli-Kelvin region. The low temperatures can be obtained within a fraction of a second.

The results discussed in this report were obtained using a Paul trap with a ring diameter of 5 mm and an end-cap separation of 3.54 mm.[35] This trap is larger than most of the ion traps used in laser experiments.[1,36] The radio frequency of the field used for dynamic trapping is 11 MHz. The trap is mounted inside a stainless-steel, ultrahigh vacuum chamber. We can obtain storage times of hours using a background gas at a pressure of 10^{-10} mbar. The ions are loaded into the trap by means of a thermal beam of neutral atoms (magnesium atoms in our case), which are then ionized close to the center of the trap by an electron beam entering the trap through a small hole in the lower end-cap (see FIGURE 6). In order to not distort the trap potential, the hole is covered by a fine molybdenum mesh. The neutral Mg beam and the laser beam pass through the gaps between the end-cap and the ring electrodes. The laser frequency is shifted by an amount Δ below the $3S_{1/2} \rightarrow 3P_{3/2}$ resonance transition of $^{24}Mg^+$ at 280 nm to extract kinetic energy from the ions by radiation

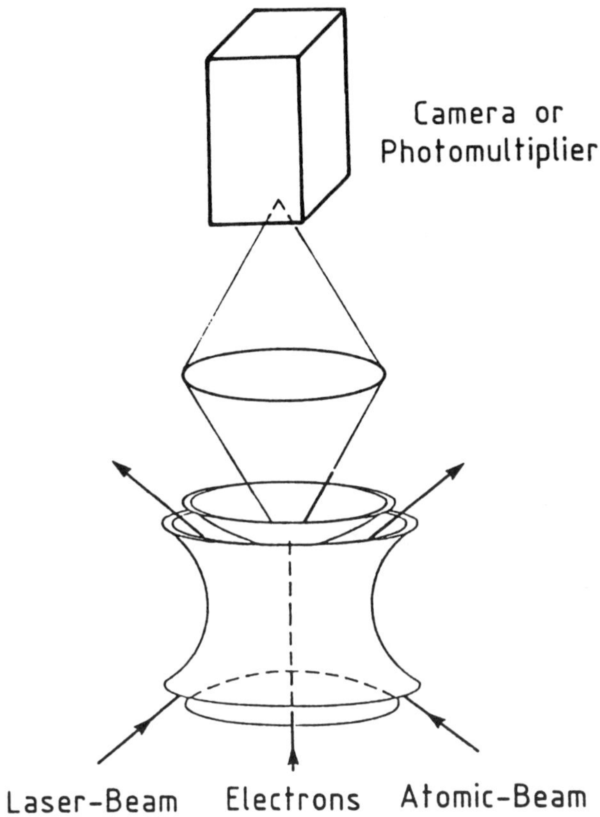

FIGURE 6. Sketch of the Paul trap. The fluorescence light is observed through a hole in the upper electrode.

pressure as discussed above. In this way, a single ion can be cooled to a temperature below 10 mK. The fluorescence from the ions is observed through a hole in the upper end-cap, again covered by a molybdenum mesh. The large size of the trap allows a large solid angle for detecting the fluorescence radiation, either with a photomultiplier or by means of a photon-counting imaging system. To observe the ions, the cathode of the imaging system is placed in the image plane of the microscope objective attached to the trap; in this way, images of the ions could be obtained.[37,38]

PHASE TRANSITIONS OF TRAPPED IONS

The existence of phase transitions in a Paul trap manifests itself by significant jumps in the fluorescence intensity of the ions as a function of the detuning Δ between the laser frequency and the atomic transition frequency.[35] These discontinuities are indicated in FIGURE 7 by vertical arrows and occur between two types of

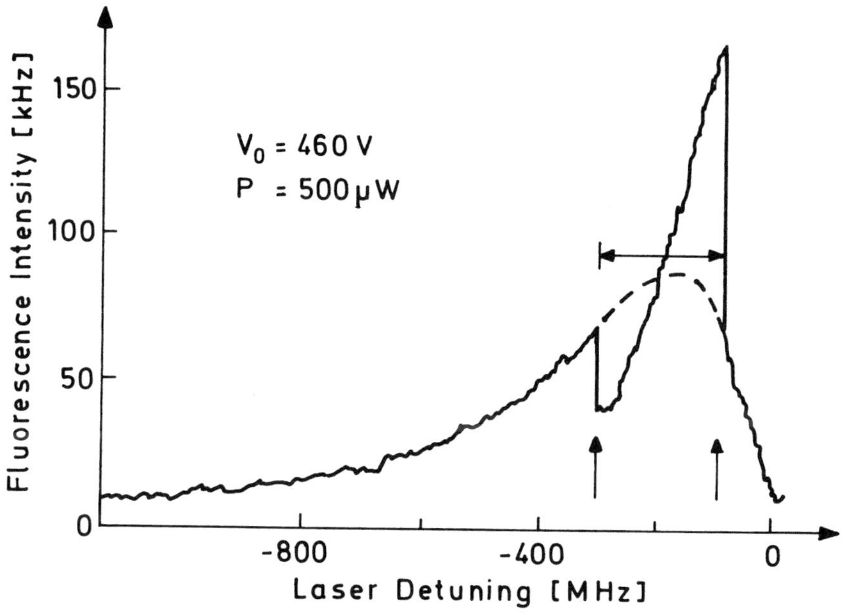

FIGURE 7. Fluorescence intensity, that is, photon counts per second, from five ions as a function of the laser detuning Δ. The vertical arrows indicate the detunings where phase transitions occur. The horizontal arrow shows the range of detunings in which a stable five-ion crystal is observed. The spectrum was scanned from left to right.

spectra: a broad one and a narrow one, analogous to the fluorescence spectrum of a single, cooled ion. We interpret the broad spectrum as a fingerprint of an ion cloud and the narrow spectrum as being characteristic for an ordered many-ion situation with a "single-ion signature". Thus, the jumps clearly indicate a transition from a state of erratic motion within a cloud to a situation where the ions arrange themselves in regular structures. In such a crystalline structure, the mutual ion-Coulomb repulsion is compensated by the external, dynamic trap potential. The regime of detunings in which such crystals exist is depicted in FIGURE 7 by the horizontal arrow. The existence of the two phases—crystal and cloud—can be verified experimentally by direct observation with the help of a highly sensitive imaging system and theoretically by analyzing ion trajectories via Monte Carlo computer simulations.[37,38]

FIGURE 8 shows ion structures as measured with the imaging system. For the measurements, only a radio-frequency voltage was applied to the trap electrodes: in this case, the potential is a factor of two deeper along the symmetry axis rather than perpendicular to it and, therefore, plane ion structures are observed being perpendicular to the symmetry axis (for details, see references 35, 37, and 38).

After we have accepted the existence of the ion crystals and the corresponding phase transitions, how do they actually occur? Would the cooling laser not force any cloud to immediately crystallize? A heating mechanism balancing the cooling effect of the laser must be the answer to this puzzle, but what heating mechanism? Since the early days of Paul traps, this so-called radio-frequency heating has repeatedly been cited.[1] A deeper understanding, though, was missing and was provided only recently by a detailed study of the dependence of the cloud → crystal and crystal → cloud phase transitions on the relevant parameters.[37–39]

The ions are subjected to essentially four different forces: the first one arising from the trapping field, then the Coulomb interaction between the ions, the laser cooling force, and finally a random force, arising from the spontaneously emitted photons. Using these forces, computer simulations of the motion of the ions can be performed.[37] Depending on the external parameters such as the laser power, the laser detuning, and the radio-frequency voltage, the experimentally observed phenomena could be reproduced. Some of the results of the simulations are summarized in FIGURE 9, which plots the radio-frequency heating parameter κ of five ions versus their mean separation.[39] For zero laser power and large r, we did not observe any net heating of the ions. This is confirmed by our experiments, in which, even in the absence of a cooling laser, large clouds of ions can be stored in a Paul trap over several hours without being heated out of the trap. When the ions are far apart, the Coulomb force is small and, on short time scales, the ions behave essentially like independent single stored ions. For this reason, we call this part of the heating diagram the Mathieu regime.[38,39] Turning on a small laser, the root-mean-square radius r reduces drastically, but comes to a halt at about 14 μm. At this distance, the nonlinear Coulomb force between the ions plays an important role and the motion of the ions becomes chaotic. In this situation, the power spectrum of the ions becomes a continuum leading to a radio-frequency heating process.

Increasing the laser power further results in an even smaller cloud. The smaller cloud produces more chaotic radio-frequency heating, as seen clearly by the negative slope of the heating curve in the range of $8 < r < 14$ μm. Finally, in the range of $4 < r < 8$ μm, there is still chaotic heating, but the slope of the heating curve is positive. As a consequence of the resulting triangular shape of the heating curve at a laser power of about $P = 150$ μW, corresponding to $r \approx 8$ μm, the chaotic heating power can no longer balance the cooling power of the laser light and the cloud collapses into the crystalline state located at $r \approx 3.8$ μm. At this point, the amplitude of the ion motion is so small that the nonlinear part of the repulsive Coulomb force is negligible again; therefore, chaotic heating disappears and the phase transition occurs.

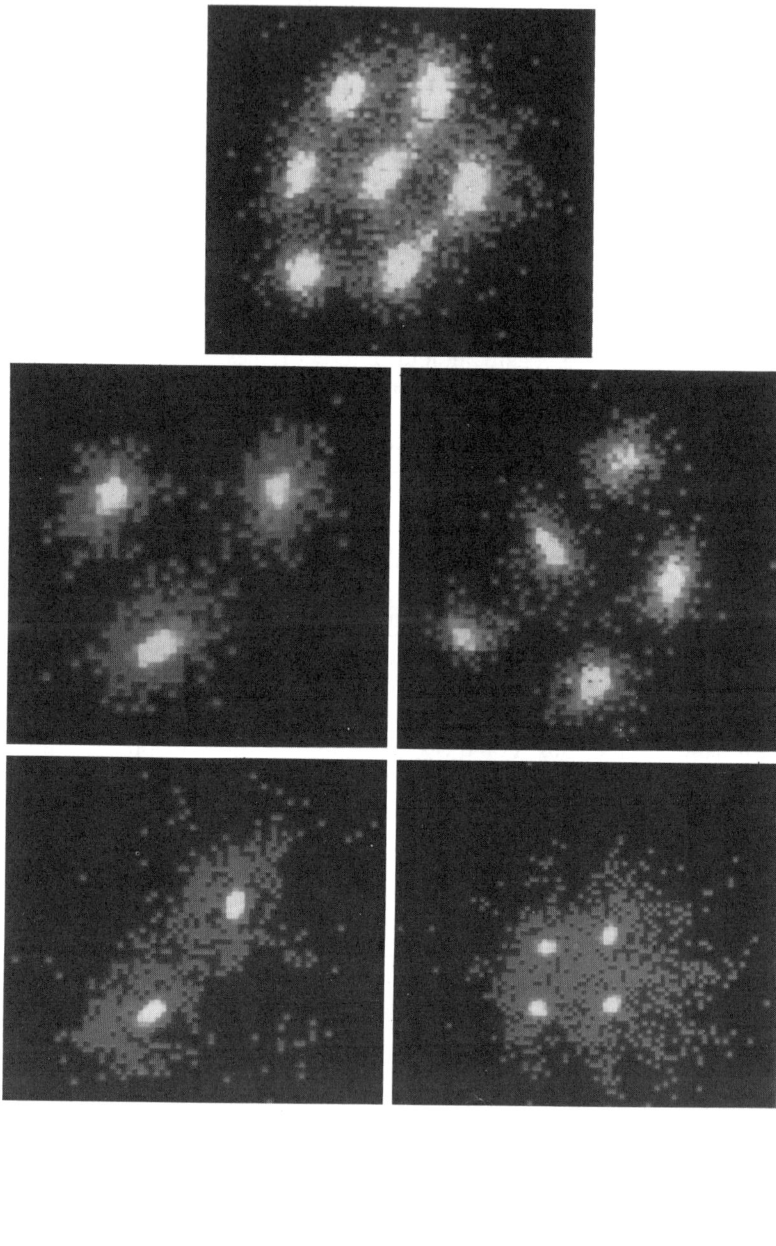

FIGURE 8. Two, three, four, five, and seven trapped ions in the crystallized configuration. The distance between the ions is roughly 20 μm. The scale of the pictures is not completely identical. The trap potential is such that the ions arrange themselves practically in two-dimensional structures. There is a slight distortion of the ion structures by a contact potential caused by a coating on the ring electrode with Mg atoms. This distortion is especially visible for five ions.

THE ION STORAGE RING

A completely new era of accelerator physics could begin if it were possible to produce, store, and accelerate Coulomb crystals in the particle accelerators and storage rings. To work with crystals instead of the usual dilute, weakly coupled, particle clouds has at least one advantage: the luminosity of accelerators (storage rings) could be greatly enhanced and (nuclear) reactions whose cross sections are too small to be investigated in currently existing accelerators would become amenable to experimentation.

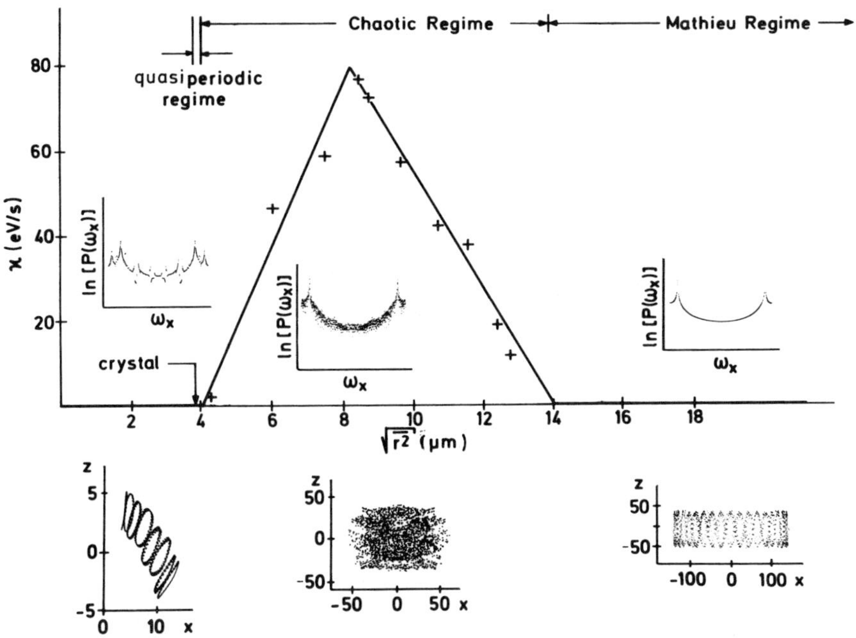

FIGURE 9. Average heating rate κ of five ions in a Paul trap versus mean ion separation. The insets show the power spectrum and the corresponding stroboscopic Poincaré sections (plane perpendicular to the symmetry axis) of the relative separation of two ions in three characteristic domains: the crystal state, the chaotic regime, and the Mathieu regime. The units on the axes are in μm. In order to calculate the power spectrum of the "crystal" shown on the left-hand side, the distance of the two ions was displaced by 1 μm from the equilibrium position. The Mathieu regime shown on the right is dominated by the secular motion.

In the following, we would like to discuss very briefly our recent experiments using a miniature quadrupole storage ring. The storage ring is similar to the one described by Drees and Paul[40] or by Church.[41] We can observe that phase transitions of the stored ions and the observed ordered ion structures are quite similar to the ones expected in relativistic storage rings, although much easier to achieve. The motivation for building this type of small storage ring came from the fact that micromotion perturbs the ion structures in a Paul trap and only a single trapped ion is free of micromotion.[1] The ring trap used consists of a quadrupole field, leading to harmonic binding of the ions in a plane transversal to the electrodes of the quadrupole and with no confinement along the axis (see FIGURE 10). Confinement along the axis is achieved, though, by the Coulomb interaction between the ions when the ring is filled; then, the total number of ions in the ring determines the average distance between them.

A scheme of the ring trap used for our experiment is shown in FIGURE 10.[42,43] It consists of the four electrode rings shown in the insert on the right. The hyperbolic cross section of the electrodes required for an ideal quadrupole field was approximated by a circular one. The experiments were also performed with Mg^+ ions. The ions were produced between the ring electrodes by ionizing the atoms of a weak atomic beam produced in an oven, which injected the atoms tangentially into the trap region. The electrons used for the ionization came from an electron gun, where the electron beam of which was perpendicular to the direction of the atomic beam. A shutter in front of the atomic beam oven allowed the interruption of the atomic flux. The ultrahigh vacuum chamber was pumped by an ion-getter pump. After baking the chamber, a vacuum of 10^{-10} mbar could be reached. Under these conditions, the number of ions stored in the trap stayed practically constant for several hours.

When laser cooling of the ions is started, a sudden change in the fluorescence intensity is observed, resembling very much that seen with stored ions in a Paul trap (FIGURE 7); this indicates a phase transition and the formation of an ordered structure of ions. The ion structure can also be observed using an ultrasensitive imaging system. Pictures of typical ion structures are shown in FIGURE 11. The ions are excited by a frequency-tunable laser beam that enters the storage ring tangen-

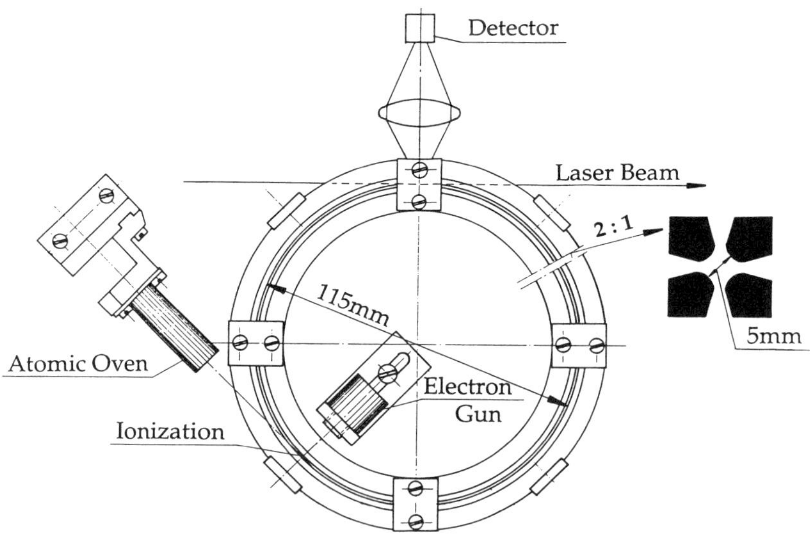

FIGURE 10. Quadrupole storage ring. The cross section of the electrode configuration is shown in the insert on the right-hand side of the figure. The diameter of the ring is 113 mm and the distance between the electrodes is 5 mm.

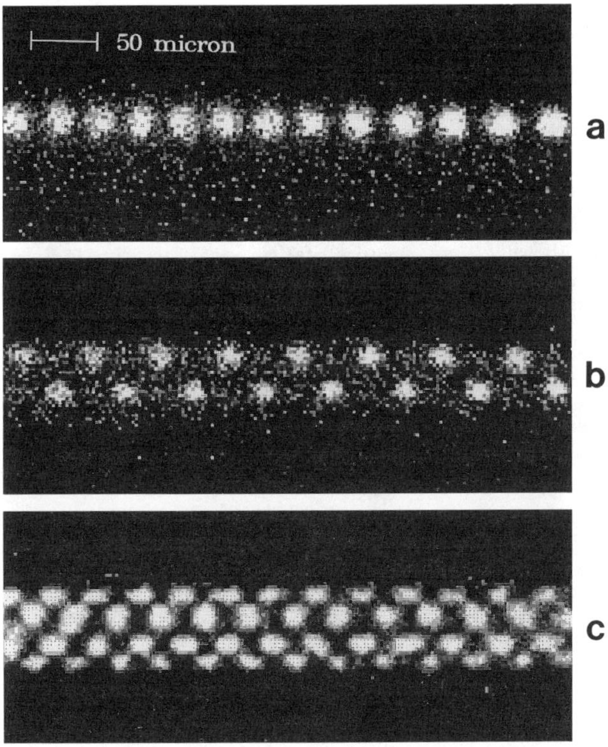

FIGURE 11. Crystalline structures of laser-cooled $^{24}Mg^+$ ions in the quadrupole storage ring. At a low ion density ($\lambda = 0.29$), the ions form a string along the field axis (a, upper). Increasing the ion density transforms the configuration to a zigzag (b, middle; $\lambda = 0.92$). At still higher ion densities, the ions form ordered helical structures on the surface of a cylinder, for example, three interwoven helices at $\lambda = 2.6$ (c, lower). As the fluorescent light is projected onto the plane of observation in this case, the inner spots are each created by two ions seated on opposite sides of the cylindrical surface, resulting in a single bright spot.

tially. In the linear configuration, the ions are all sitting in the center of the quadrupole field; therefore, they do not show micromotion and it is possible to cool them further to temperatures in the micro-Kelvin region. The new cooling methods proposed by Dalibard *et al.*[34] can be applied to the Mg^+ ions so that the single-photon recoil limit can be achieved for the cooling process. At this limit, the kinetic energy of the ions is smaller than the energy resulting from the "zero-energy" motion of the harmonically bound ions; the ion structure then reaches its vibrational ground state—that is, a Mössbauer situation is generated. The observed ion configurations in the quadrupole ring trap are described in a recent paper by Birkl *et al.*[44] We will review the major results reported in this paper in the next section and will compare the observed ordered structure to the results of molecular dynamic calculations.[45]

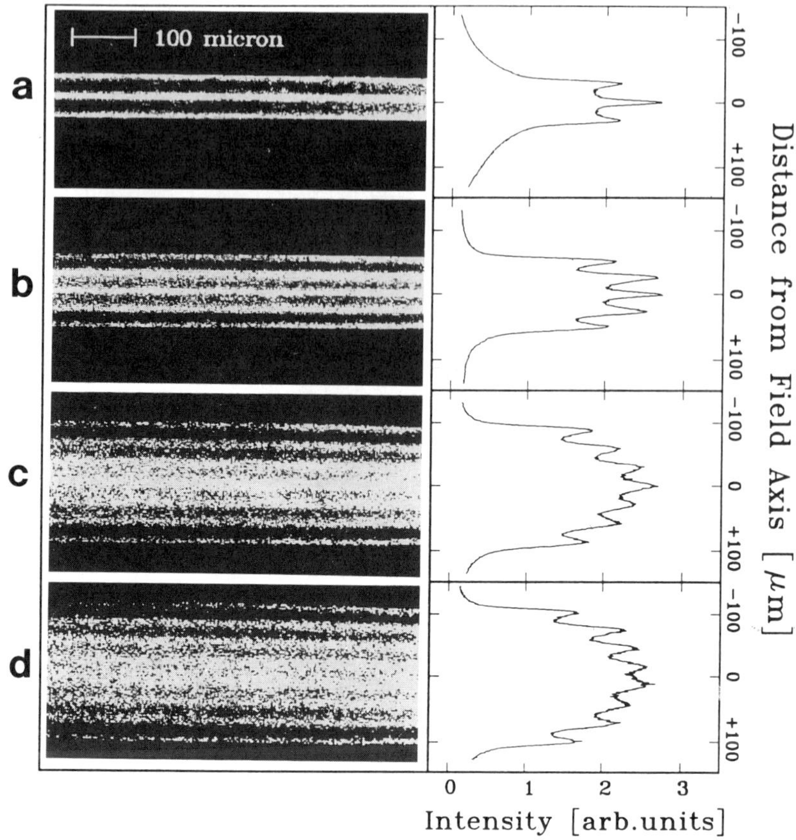

FIGURE 12. Images and intensity profiles of (from the top) one shell plus string (a; $\lambda \approx 4.3$), two shells plus string (b; $\lambda \approx 12.2$), three shells plus string (c; $\lambda \approx 26$), and four shells (d; $\lambda \approx 34$). There are up to $\approx 8 \times 10^5$ ions stored in the ring for the four-shell structure.

ORDERED STRUCTURES IN THE STORAGE RING

In the molecular dynamic calculations,[45] a cylindrically symmetric, static harmonic potential is assumed to describe the confining field. Each particle is subjected to the Coulomb forces of all other particles and to the confining field. The classical equations of motion are integrated for a system of several thousand particles, starting with random positions and velocities, to give the time evolution of the system. Cooling of the stored particles is simulated by scaling down the resulting velocities of the stored particles at defined instants of the integration process. After sufficient cooling, ordered structures such as strings, zigzags, shells, and multiple shells should arise owing to the confining field's harmonic potential. Our experiments are well suited to checking these predictions. To compare the experimental results with theory, we can use the normalized "linear particle density" λ, which is given by the ion density multiplied by the ratio of Coulomb repulsion and confining force of the trap.[45] Low λ values correspond to a deep potential well or to a small number of ions, resulting in an equilibrium structure closely confined to the field axis comprising a string of ions (FIGURE 11a). This is the micromotion-free configuration discussed above and the analogue of the single stored ion in a Paul trap, as in both cases the ions sit in the potential minimum and show no micromotion. For higher values of λ, the structures extend more and more into the off-axis region, giving rise to (in the order of increasing λ) a plane zigzag structure (FIGURE 11b) and cylindrical structures with the ions forming helices on cylindrical surfaces. The structure in FIGURE 11c consists of three interwoven helices with six ions per pitch. The string and the zigzag have also been observed with laser-cooled Hg^+ ions in a linear trap.[46]

Increasing further the number of ions leads to structures with smaller spacings between the ions where we cannot optically resolve individual ions any longer. Images of these structures are presented in FIGURE 12. The radial intensity profiles displayed on the right-hand side of the figure provide information about the structures as they give a measure of the radial distribution of the ions. For increasing λ, it becomes energetically more favorable to create a string inside the first ion shell (FIGURE 12a) to provide space for more particles. This results in a structure that is a three-dimensional analogue of the plane seven-ion crystal for a Paul trap (FIGURE 8). The inner string turns into a second shell at higher densities; a string then develops inside this second shell (FIGURE 12b) and so on. FIGURES 12c and 12d show structures consisting of three shells plus string and four shells, respectively. We have been able to observe all possible structures, from the string up to four shells plus string. The formation of multiple-shell crystalline structures in the quadrupole storage ring contrasts with the observation of shell structures in Penning traps, where the ions do not occupy fixed positions inside the shells.[47]

COMPARISON WITH THEORY

The top panel of FIGURE 13 gives a summary of experimental data for all recorded images in which the ions were individually resolved. The depth of ψ_0 of the potential well and the ion density per unit length are the experimental parameters. The theoretical boundaries between the different shell structures, predicted in reference 45 in terms of the functional dependence of λ on ψ_0 and the ion density, are given by the straight lines with constant λ. String structures are expected for $\lambda < 0.709$, zigzag structures are expected in the range of $0.709 < \lambda < 0.964$, and single shells are expected in the range of $0.964 < \lambda < 3.10$. Many different structures that are degenerate in energy are expected within the single-shell regime. We obtained stable configurations near $\lambda = 1.3$ and $\lambda = 2.0$ (two interwoven helices) and near $\lambda = 3.0$ (three interwoven helices—FIGURE 11c). The observed structures agree with the predicted scheme for a large range of experimental parameters, thus confirming the theoretical results.

A summary of the experimental observations for ordered shell structures with up to four shells plus string and without resolution of individual ions is presented in the bottom panel of FIGURE 13. The depth ψ_0 of the confining potential well is again one of the experimental parameters. As the ion density cannot be determined directly from the images, the radius ρ of the structures is used instead as the second

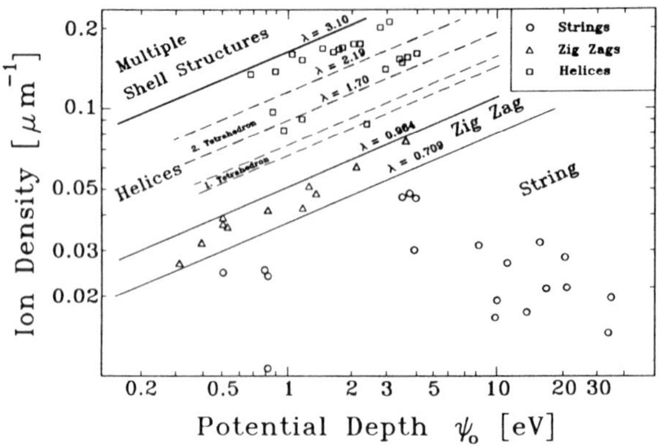

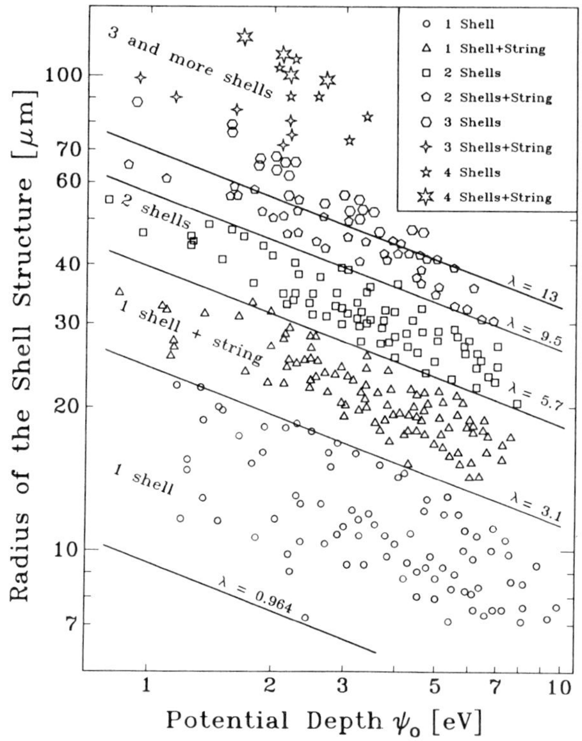

FIGURE 13. Summary of the experimental results. (Top) Individual ions resolved, where the observed structures are characterized by the ion density per unit length and the depth of the potential well ψ_0. These two parameters can be combined to give the normalized linear particle density λ, which fully determines the ion configuration. The straight lines show critical λ values separating the regions corresponding to the various theoretically expected structures. The observed configurations are labeled with different symbols for each structure. (Bottom) Individual ions unresolved, where the observed shell structures with up to four shells plus string are characterized by their radius ρ and the potential depth ψ_0. The various observed structures are again separated by lines of theoretically determined critical λ values.

parameter. The theoretically predicted boundaries between the different shell structures are again given as straight lines of constant λ following reference 45, where the dependence of ρ on λ and ψ_0 was established. The observed ion configurations are seen once more to be determined by λ for a wide range of potential depths and ion densities.

Our results have important implications for two very different fields. Consider first the physics of low-energy particles: an ion string in a quadrupole ring, being free of micromotion, can be cooled to its vibrational ground state in the confining potential using recently proposed laser-cooling techniques.[34] This would place the string in the Lamb-Dicke regime with a vanishing first-order Doppler effect because the spatial amplitude of the motion is smaller than the wavelength of the atomic transition. Furthermore, the second-order Doppler effect, which can only be reduced by further cooling, also disappears, making the stored ions very interesting for frequency standards. The large number of ultracold ions available in the ring will lead to a high signal-to-noise ratio. Finally, cooled ions in the ring represent a quantum object of macroscopic dimensions—a Wigner crystal. In this limit, the photon recoil will disappear because the ion crystal experiences the Mössbauer effect.

REFERENCES

1. Wineland, D. J., W. M. Itano, J. C. Berquist, J. J. Bollinger & J. D. Prestage. 1984. Spectroscopy of stored atomic ions. *In* Atomic Physics. Volume 9. R. S. van Dyck, Jr. & E. N. Fortson, Eds.: 3–27. World Scientific. Singapore.
2. Chu, S., J. E. Bjorkholm, A. Ashkin & A. Cable. 1986. Demonstration and laser cooling and trapping of atoms. *In* Atomic Physics. Volume 10. H. Narumi & S. Shimamura, Eds.: 377–393. North-Holland. Amsterdam; Pritchard, D. E., K. Helmerson & A. G. Martin. 1988. Atom traps. *In* Atomic Physics. Volume 11. S. Haroche, J. C. Gay & G. Grynberg, Eds.: 179–197. World Scientific. Singapore.
3. Meschede, D., H. Walther & G. Müller. 1985. Phys. Rev. Lett. **54:** 551–554.
4. Jaynes, E. T. & F. W. Cummings. 1963. Proc. IEEE **51:** 89–109.
5. For a review, see the articles by: Haroche, S. & J. M. Raimond. 1985. Radiative properties of Rydberg states in resonant cavities. *In* Advances in Atomic and Molecular Physics. Volume 20. D. Bates & B. Bederson, Eds.: 350–411. Academic Press. New York; Gallas, J. A., G. Leuchs, H. Walther & H. Figger. 1985. Rydberg atoms: high resolution spectroscopy and radiation interaction–Rydberg molecules. *In* Advances in Atomic and Molecular Physics. Volume 20. D. Bates & B. Bederson, Eds.: 413–466. Academic Press. New York.
6. See, for example: Eberly, J. H., N. B. Narozhny & J. J. Sanchez-Mondragon. 1980. Phys. Rev. Lett. **44:** 1323–1326 and references therein.
7. Rempe, G., H. Walther & N. Klein. 1987. Phys. Rev. Lett. **58:** 353–356.
8. Meystre, P. 1992. Cavity quantum optics and the quantum measurement process. *In* Progress in Optics. Volume 30. E. Wolf, Ed.: 261–355. North-Holland. Amsterdam.
9. Filipowicz, P., J. Javanainen & P. Meystre. 1986. Opt. Commun. **58:** 327–330; Phys. Rev. **A34:** 3077–3087; J. Opt. Soc. Am. **B3:** 906–910.
10. Lugiato, L., M. O. Scully & H. Walther. 1987. Phys. Rev. **A36:** 740–743.
11. Meystre, P. 1987. Opt. Lett. **12:** 669–671; 1988. Generation and detection of sub-Poissonian fields in micromasers. *In* Squeezed and Nonclassical Light. P. Tombesi & E. R. Pike, Eds.: 115–127. Plenum. New York.
12. Meystre, P. & E. M. Wright. 1988. Phys. Rev. **A37:** 2524–2529.
13. Rempe, G. & H. Walther. 1990. Phys. Rev. **A42:** 1650–1655.
14. Paul, H. & T. Richter. 1991. Opt. Commun. **85:** 508–519.

15. REMPE, G., F. SCHMIDT-KALER & H. WALTHER. 1990. Phys. Rev. Lett. **64:** 2783–2786.
16. BRIEGEL, H. J., B-G. ENGLERT, N. STERPI & H. WALTHER. 1994. Phys. Rev. **A49:** 2962–2984.
17. WAGNER, C., A. SCHENZLE & H. WALTHER. 1994. Opt. Commun. **107:** 318–326.
18. BENSON, O., G. RAITHEL & H. WALTHER. 1994. Phys. Rev. Lett. **72:** 3506–3509.
19. DAVIDOVICH, L., J. M. RAIMOND, M. BRUNE & S. HAROCHE. 1987. Phys. Rev. **A36:** 3771–3787; BRUNE, M., J. M. RAIMOND, P. GOY, L. DAVIDOVICH & S. HAROCHE. 1987. Phys. Rev. Lett. **59:** 1899–1902.
20. MEYSTRE, P., G. REMPE & H. WALTHER. 1988. Opt. Lett. **13:** 1078–1080.
21. RAITHEL, G., CH. WAGNER, H. WALTHER, L. M. NARDUCCI & M. O. SCULLY. 1994. The micromaser: a proving ground for quantum physics. *In* Advances in Atomic, Molecular, and Optical Physics. Supplement 2. P. Berman, Ed.: 57–121. Academic Press. New York.
22. SCULLY, M. O., H. WALTHER, G. S. AGARWAL, T. QUANG & W. SCHLEICH. 1991. Phys. Rev. **A44:** 5992–5996.
23. QUANG, T., G. S. AGARWAL, J. BERGOU, M. O. SCULLY, H. WALTHER, K. VOGEL & W. P. SCHLEICH. 1993. Phys. Rev. **A48:** 803–812; VOGEL, K., W. P. SCHLEICH, M. O. SCULLY & H. WALTHER. 1993. Phys. Rev. **A48:** 813–817.
24. KRAUSE, J., M. O. SCULLY & H. WALTHER. 1986. Phys. Rev. **A34:** 2032–2037.
25. BRECHA, R. J., A. PETERS, C. WAGNER & H. WALTHER. 1992. Phys. Rev. **A46:** 567–577.
26. WAGNER, C., R. J. BRECHA, A. SCHENZLE & H. WALTHER. 1992. Phys. Rev. **A46:** R5350.
27. WAGNER, C., R. J. BRECHA, A. SCHENZLE & H. WALTHER. 1993. Phys. Rev. **A47:** 5068–5079.
28. ENGLERT, B-G., J. SCHWINGER & M. O. SCULLY. 1988. Found. Phys. **18:** 1045–1056; SCHWINGER, J., M. O. SCULLY & B-G. ENGLERT. 1988. Z. Phys. **D10:** 135–144; SCULLY, M. O., B-G. ENGLERT & J. SCHWINGER. 1989. Phys. Rev. **A40:** 1775–1784.
29. SCULLY, M. O. & H. WALTHER. 1989. Phys. Rev. **A39:** 5229–5236.
30. SCULLY, M. O., B-G. ENGLERT & H. WALTHER. 1991. Nature **351:** 111–116.
31. ENGLERT, B-G., H. WALTHER & M. O. SCULLY. 1992. Appl. Phys. **B54:** 366–368.
32. PAUL, W., O. OSBERGHAUS & E. FISCHER. 1958. Forschungsber. Wirtsch. Verkehrminist. Nordrhein-Westfalen **415**; FISCHER, E. 1959. Z. Phys. **156:** 1–26.
33. DEHMELT, H. G. 1967. Radio-frequency spectroscopy of stored ions, I: storage. *In* Advances in Atomic and Molecular Physics. Volume 3. D. R. Bates & I. Estermann, Eds.: 53–72. Academic Press. New York.
34. DALIBARD, J., C. SALOMON, A. ASPECT, E. ARIMONDO, R. KAISER, N. VANSTEENKISTE & C. COHEN-TANNOUDJI. 1988. New schemes in laser cooling. *In* Atomic Physics. Volume 11. S. Haroche, J. C. Gay & G. Grynberg, Eds.: 199–214. World Scientific. Singapore.
35. DIEDRICH, F., E. PEIK, J. M. CHEN, W. QUINT & H. WALTHER. 1987. Phys. Rev. Lett. **59:** 2931–2934.
36. NEUHAUSER, W., M. HOHENSTATT, P. TOSCHEK & H. DEHMELT. 1978. Phys. Rev. Lett. **41:** 233–236; 1980. Phys. Rev. **A22:** 1137–1140.
37. BLÜMEL, R., J. M. CHEN, W. QUINT, W. SCHLEICH, Y. R. SHEN & H. WALTHER. 1988. Nature **334:** 309–313.
38. BLÜMEL, R., J. M. CHEN, F. DIEDRICH, E. PEIK, W. QUINT, W. SCHLEICH, Y. R. SHEN & H. WALTHER. 1988. Phase transitions of stored laser-cooled ions. *In* Atomic Physics. Volume 11. S. Haroche, J. C. Gay & G. Grynberg, Eds.: 243–259. World Scientific. Singapore.
39. BLÜMEL, R., C. KAPPLER, W. QUINT & H. WALTHER. 1989. Phys. Rev. **A40:** 808–823.
40. DREES, J. & W. PAUL. 1964. Z. Phys. **180:** 340–361.
41. CHURCH, D. A. 1969. J. Appl. Phys. **40:** 3127–3134.
42. WALTHER, H. 1991. Chaos and order of laser-cooled ions in a Paul trap. *In* Proceedings of the Workshop on Light-induced Kinetic Effects on Atoms, Ions, and Molecules. L. Moi, S. Gozzini, C. Gabbanini, E. Arimondo & F. Strumia, Eds.: 261–276. ETS Editrice. Pisa.

43. Waki, I., S. Kassner, G. Birkl & H. Walther. 1992. Phys. Rev. Lett. **68:** 2007–2010.
44. Birkl, G., S. Kassner & H. Walther. 1992. Nature **357:** 310–313.
45. Hasse, R. W. & J. P. Schiffer. 1990. Ann. Phys. **203:** 419–448; Rahman, A. & J. P. Schiffer. 1986. Phys. Rev. Lett. **57:** 1133–1136.
46. Raizen, M. G. *et al.* 1992. J. Mod. Opt. **39:** 233–242.
47. Gilbert, S. L., J. J. Bollinger & D. J. Wineland. 1988. Phys. Rev. Lett. **60:** 2022–2025.

A Heisenberg Microscope for Atoms

CH. KURTSIEFER, T. PFAU, S. SPÄLTER,
C. R. EKSTROM, AND J. MLYNEK

Fakultät für Physik
Universität Konstanz
D-78434 Konstanz, Germany

INTRODUCTION

One of the most puzzling aspects of quantum theory is the implementation of the measurement process on a quantum mechanical system and the back-action of this process on the system. A gedankenexperiment from the early days of quantum mechanics dealing with this problem was proposed by Heisenberg in 1927.[1] In order to understand the localization process of an electron in a Wilson cloud chamber, he chose an operational approach of dividing the whole observation of the particle trajectory in a sequence of fundamental quantum processes.

The position of the particle in each fundamental step is described by its center-of-mass wave function, which may have a certain spatial extension. A fundamental step of the measurement is the observation of scattered light from this particle. The uncertainty Δx of the measurable origin of the scattered wave is given from classical optics by the wavelength λ^p of the light used for the illumination of the particle. One expects a localization of the object because a position measurement is carried out that should cause the wave function to collapse to a position distribution with an extension on the order of λ^p.

To analyze the measurement process, the action of the scattered light on the momentum of the particle has to be studied. Because light carries not only energy, but also momentum, one can calculate the momentum transfer from the light field to the object. If the microscopic particle is smaller than the wavelength of the scattered light, the scattered wave may be described in a spherical basis. To evaluate the momentum distribution of such a light field, this mode has to be projected onto a basis of plane waves. Assuming momentum conservation for the scattering process, the momentum distribution from that projection must be transferred to the object under observation. For the most efficient measurement process, the influence on the object for a certain amount of information achieved should be as small as possible. Because the amount of momentum contained in the scattered wave increases with the amount of scattered light, the minimum possible momentum transfer to the particle is given by the minimum detectable amount of light necessary for the localization of the particle, which is a single photon. The momentum uncertainty contained in a spherical wave of one photon is just given by $\Delta p = \hbar k^p$, where $k^p = 2\pi/\lambda^p$ and λ^p is the wavelength of the photon.

This gedankenexperiment is in fact not restricted to a certain combination of quantum objects and scattering waves, but may be extended to a whole variety of pairs. In the following, we use an atom as the quantum object and a spontaneously emitted photon in the place of the scattered photon. This allows the investigation of a single measurement step in the Heisenberg gedankenexperiment.

THE ATOM-PHOTON PAIR

Ideally, an atom in our experiment[2] is prepared into a state where the external motion has to be described by a discrete set of plane waves; the initial position is therefore unknown. Furthermore, the internal structure of the atom is prepared into an excited state, allowing for a single spontaneous emission of a photon. After the decay of this excited state, the atom should show the momentum uncertainty in its center-of-mass motion, reflecting the momentum distribution of the emitted photon (see FIGURE 1). In a first step, we do not look at the photon and observe only the momentum distribution of the atom.

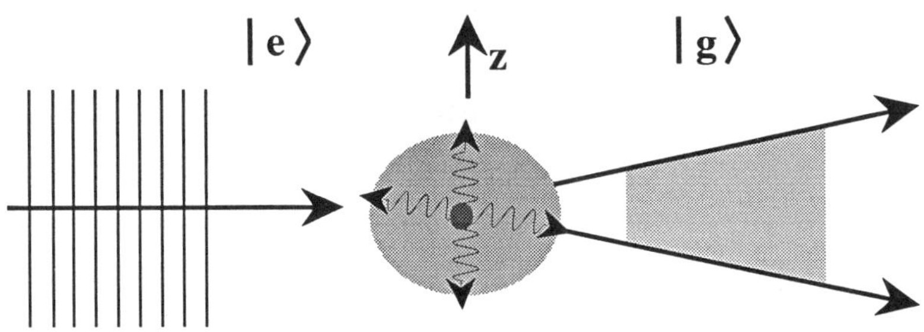

FIGURE 1. An atom prepared in a plane-wave state for its external motion undergoes a spontaneous emission and shows afterwards a transverse momentum distribution. The photon emitted in this process should allow the determination of the position of the atom with an accuracy close to the wavelength of the light.

In the following, we restrict our quantum mechanical treatment to the motion of particles in the transverse direction z. The effect of the spontaneous emission on the wave function may be described as a loss of transverse coherence. A quantity to describe the coherence properties of a wavelike phenomenon is the two-point correlation function of a field. Physically, this function describes the possibility of interference between Huygens waves originating from separated points of the space that is covered by the wave function or a statistical mixture of wave functions. It will be called the transverse coherence function in the following.[3]

If the field state is described by a classical mixture of momentum eigenstates, the coherence of a field $\Psi(z)$ only depends on the separation between two testing points. The coherence function is then identical to the autocorrelation function of the field:[4]

$$g^{(1)}(z) = \int \Psi(z' - z)\Psi^*(z')\, dz'.$$

After the spontaneous emission of a photon from an atom, the initial coherence function has to be multiplied by the Fourier transform of the momentum distribution of the atom, which is complementary to the momentum distribution $I(k_z)$ of the spontaneously emitted light. In an idealized version of our experiment, we would like to begin with a plane-wave state, corresponding to a δ-function-shaped momentum distribution. The initial coherence function would be 1 for all distances z and, after the spontaneous emission of a photon, it will be

$$g^{(1)}(z) = \mathscr{F}[I(k_z)],$$

where $\mathscr{F}$ denotes the Fourier transform. Assuming a spontaneous emission with an isotropic emission probability at a fixed frequency, the one-dimensional momentum distribution is given by a square function ranging from $-\hbar k^p$ to $\hbar k^p$. The extension of the corresponding coherence function is reduced to a small region, with a separation λ^p of the first nodes of $g^{(1)}(z)$. Furthermore, the coherence function shows an oscillatory behavior in the wings (see FIGURE 2).

A proposal for the measurement of this coherence function by Sleator *et al.*[5] uses the visibility of an atomic far-field diffraction pattern from a double slit, where the atom emits a photon. Because this photon allows localization of the atom, the visibility of the diffraction pattern should be reduced according to the amount of position information extractable from the photon. For certain ratios of the slit separation and the wavelength of the emitted photon, there should be an inversion of the contrast of the far-field diffraction pattern (see FIGURE 3a) and, for a large slit separation, the visibility should completely vanish because "which path" information is accessible to an observer. A detailed theoretical treatment of the expected diffraction pattern can be found in reference 6.

An operational definition of the visibility of a double-slit diffraction pattern uses the Fourier component of the intensity distribution at the transverse momentum, h/d, where d is the slit separation. The coherence function, $g^{(1)}(z)$, is simply given by this visibility as a function of the slit separation, d. A negative value of $g^{(1)}(z)$ corresponds to a contrast inversion of the diffraction pattern.

Because it is difficult to realize a double slit with a variable slit separation

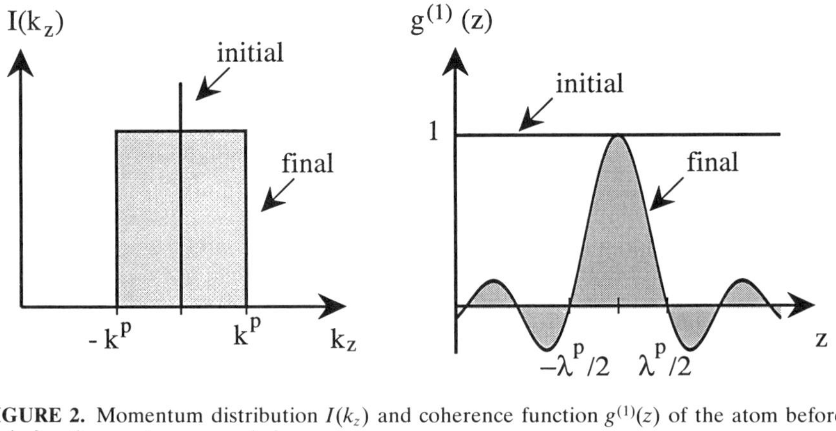

FIGURE 2. Momentum distribution $I(k_z)$ and coherence function $g^{(1)}(z)$ of the atom before and after the spontaneous emission.

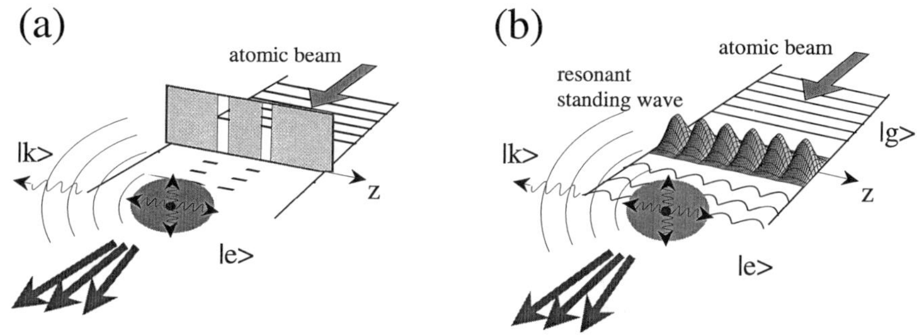

FIGURE 3. Diffraction of atomic matter waves from (a) a double slit and (b) an on-resonant standing light wave, in both cases followed by a spontaneous emission of a photon.

experimentally, we have chosen a different approach to generate a diffraction pattern. The diffractive structure is formed by a resonant standing light wave with a variable period (see FIGURE 3b). This setup allows not only a continuous variation of the diffraction period and thereby a continuous measurement of $g^{(1)}(z)$, but also shows a higher transmission for the atoms than a double slit. Furthermore, the excitation of the internal degrees of freedom of the atom is contained in this interaction region. In addition, the spontaneous emission process is clearly separated from the excitation process, if the interaction time between an atom and the light field is shorter than the natural lifetime of the excited state.

EXPERIMENTAL IMPLEMENTATION

A sketch of the experimental setup is shown in FIGURE 4. The atomic species used are helium atoms excited into the metastable triplet state $^3S_1$ within a gas discharge atomic source with a mean velocity of 2150 m/s. The plane-wave state for the external motion of the atom is prepared by collimation of the atomic beam using two narrow slits (10 μm) with a separation of 110 cm. The variable-period standing light wave was produced by reflecting a Gaussian laser beam under a variable angle α off a mirror close to the second collimation slit. For such a setup, in a region close to the mirror surface, the desired light field configuration with the variable-period standing light field is produced. The light was generated by an LNA-laser tuned to the $2(^3S_1)$–$2(^3P_2)$ transition in helium at 1083.3 nm. During the interaction of the atom with the light field, the probability of a spontaneous emission of a photon is very small because the interaction with the light field takes place within a time of 17 ns, whereas the spontaneous emission should occur on a time scale of the natural lifetime of τ = 100 ns of the excited state. Therefore, the region of spontaneous emission is clearly separated from the standing light field, but still in the near field of the diffraction from the diffracting region. The diffraction pattern itself was detected in the far field of the light grating, so the momentum distribution $I(k_z)$ was converted in a position distribution. The atomic distribution was mapped out using a 5-μm-wide scanning slit and a channeltron detector.

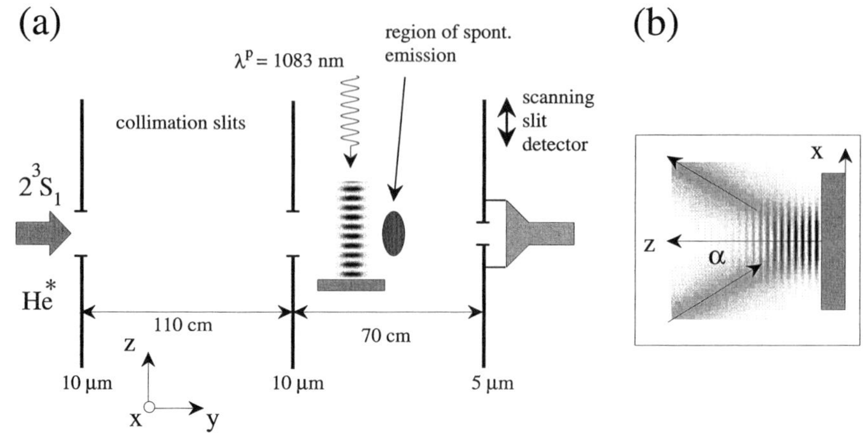

FIGURE 4. Setup for the atomic beam experiment (a). Atoms are diffracted from a standing light wave with a variable period (b).

To extract the coherence function from the diffraction patterns, the excitation process has to be analyzed in more detail. For a light field in resonance with the atom, the interaction can be treated by regarding the local eigenenergies of a two-level atom coupled to a light field as optical potentials[7] for the local energy eigenstates, $|\pm\rangle = (1/\sqrt{2})(|g\rangle \pm |e\rangle)$. The effect of diffraction of the atoms from this light field can be explained by regarding the interaction region as a phase object for the matter wave. The phase acquired by atoms in the two energy eigenstates depends on their transverse position,

$$\Delta\phi_{\pm}(z) = \pm a\,\sin(k^S z),$$

where a is a constant containing the light field strength and the interaction time, and $k^S = k^p/\cos(\alpha)$. Because the atoms are in the ground state of the two-level model system before the interaction, and this ground state does not coincide with an asymptotic energy eigenstate for a light field exactly on resonance, the atomic state has to be projected onto the dressed states. After the interaction, these states with their different acquired phase shifts have to be reexpressed in the bare state basis of the atom. The resulting total transverse wave function of the atom after the interaction takes the form,

$$\Psi(z) = \cos[a \cdot \sin(k^S z)] \cdot |g\rangle + i\,\sin[a \cdot \sin(k^S z)] \cdot |e\rangle.$$

With this procedure, atoms in the ground state leaving the interaction region will be diffracted into orders with a transverse momentum of even multiples of $\hbar k^S$, whereas the atomic component in the excited state will be diffracted into the odd momentum orders. By this formalism, one can see that, for the laser tuned on resonance, half of the atoms leave the interaction region in the ground state and half of the atoms do so in the excited state.

Because only the atoms in the excited state emit a photon, the observed visibility for the diffraction patterns has to be corrected for the contribution of atoms leaving the interaction zone in the ground state. This correction was carried out in a reference diffraction experiment, where the light field was detuned from the atomic resonance. In this case, the asymptotic energy eigenstates of the atom light interaction are the bare states and, for a sufficiently large detuning, the atoms travel adiabatically through the light field. Therefore, the atoms leave the interaction region in the ground state and, for a proper choice of the light field intensity and detuning, the ground state contribution to the diffraction patterns with the resonant laser is reproduced.

MEASUREMENT OF THE LOSS OF COHERENCE

FIGURE 5 shows two pairs of measured atomic diffraction patterns for different standing light wave periods. In both, the visibility for the laser on resonance is

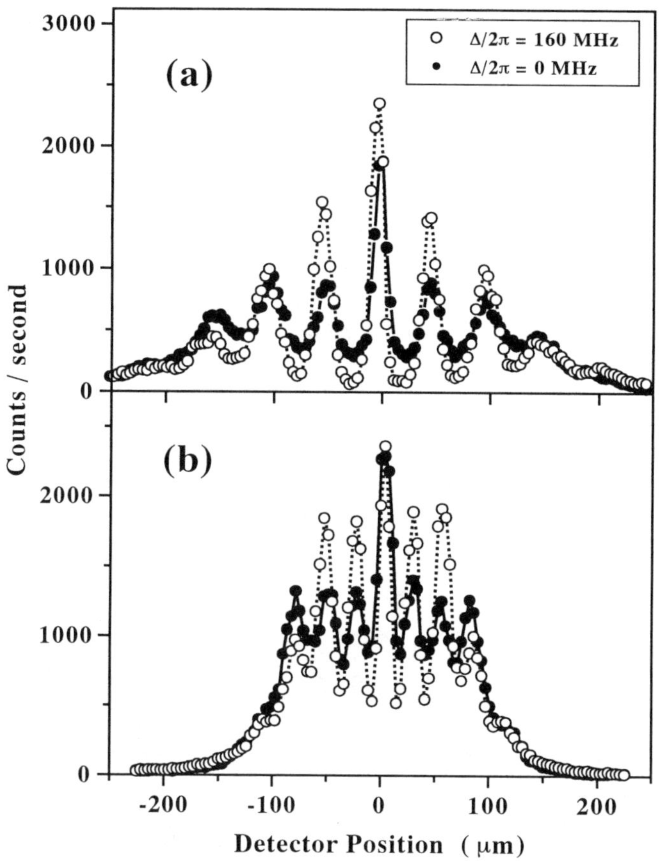

FIGURE 5. Experimental diffraction patterns for a standing wave period of $0.53\lambda^p$ (a) and λ^p (b). The diffraction orders come closer together with increasing period and the visibility is reduced for an on-resonant laser.

smaller due to the loss of transverse coherence by spontaneous emission. For a larger period (FIGURE 5b), the diffraction orders come closer together because the far-field pattern represents the momentum distribution. Also, a reduction of the visibility for the laser off resonance at a large standing wave period can be seen, which is due to the limited momentum resolution in the experiment corresponding to $0.5\hbar k^p$. To correct for that contribution to the visibility and to take into account the finite size of the diffraction pattern, we normalized the visibility of the excited state component to the visibility of patterns with an off-resonant laser. This final visibility value for atoms in the excited state is plotted in FIGURE 6 as a function of the standing wave period. Experimentally, we have access to $g^{(1)}(z)$ for z ranging from $\lambda^p/2$ to $3/2\lambda^p$, corresponding to incident angles α varying from 0° to 70° of the laser beam on the mirror. The solid line in the figure shows the one-dimensional Fourier transform of the angular emission, characteristic for the atomic transition $2(^3P_2)$–$2(^3S_1)$, taking into account the excitation with linearly polarized light. This theoretical description explains the measured visibility within our experimental accuracy.

To our knowledge, this is the first study of the loss of spatial coherence of an atomic wave function, induced by a single spontaneous emission of a photon.

CORRELATION EXPERIMENTS

In the experiment described previously, we have observed the momentum of the atom after a spontaneous emission of a photon and have ignored the information

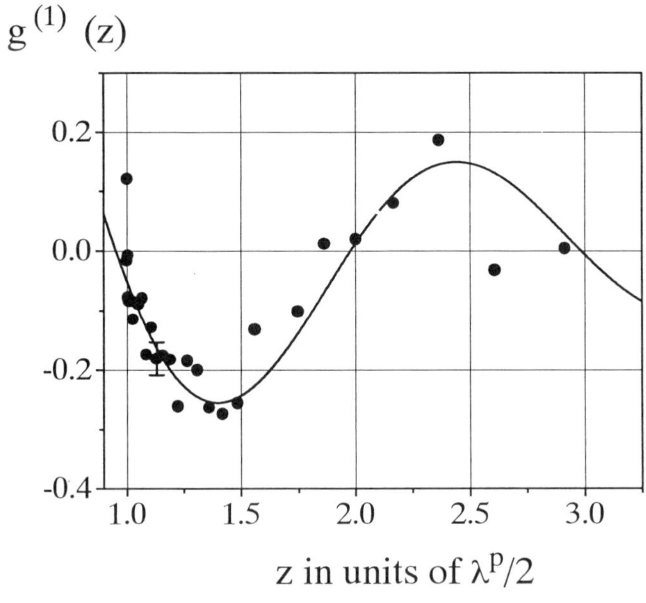

FIGURE 6. Normalized visibility of the diffraction pattern from a standing light wave. The solid line shows the theoretical prediction for the transverse coherence function after one spontaneous emission of a photon and is in good agreement with the experimental data.

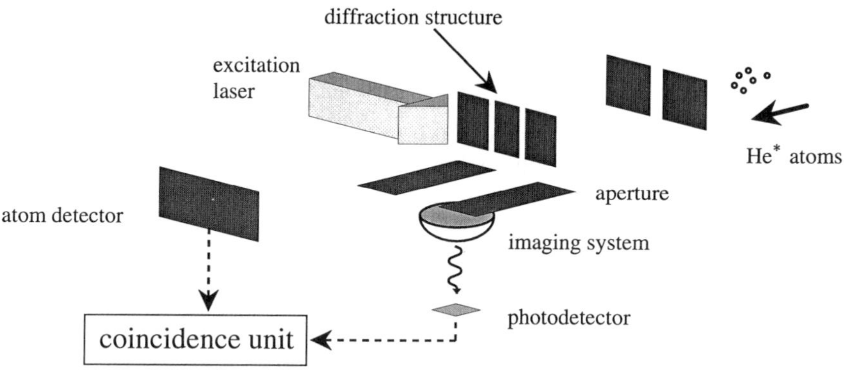

FIGURE 7. Experimental setup to observe conditional diffraction patterns. The photon could be measured either directly in a momentum basis without collection optics or in a position basis using a microscope.

carried away by the photon. The uncertainty in the momentum of the emitted photon was transferred in an uncertainty of the atom, leading to a loss of coherence of the atomic wave function.

Nevertheless, the spontaneous emission process may be described in a slightly larger space, where not only the atomic wave function is regarded, but also the wave function of the photon. To experimentally observe and exploit the entanglement of this two-particle system, one has to carry out correlation experiments between atoms and photons.

The experiments that we have in mind will again be implementations of well-known gedankenexperiments in quantum mechanics. In a first step, we want to look at conditional diffraction patterns if the spontaneously emitted photon is detected in a momentum basis. Analogous to the discussion of "Einstein's recoiling slit" gedankenexperiment, the momentum transfer onto the atom by the spontaneous emission is well known (e.g., zero transverse momentum transfer for the experiment sketched in FIGURE 7). Therefore, the conditional atomic diffraction pattern should show fully visible interference fringes, whereas the total atomic patterns should be washed out completely for certain ratios of slit separation and optical wavelength.

Such an experiment could be extended to a real Heisenberg microscope if, by means of an imaging system, the spontaneously emitted photon is measured in a position basis. Because it could be distinguished whether the atom went through a node or an antinode of a standing light wave, or through one or the other of a pair of mechanical slits, the conditional diffraction pattern is expected to be washed out. By restricting the amount of position information from the spontaneous emission process—for example, by reducing the size of the aperture of the microscope objective—the conditional interference pattern should be reestablished. In the limit of a very "poor" microscope, the aperture is small and this experiment becomes equivalent to an experiment where the momentum of the photon is measured. This experiment may therefore be regarded as a quantum eraser for the atomic wave function. Such experiments have been discussed in the context of micromasers.[8]

Because the combined system of atom and photon forms a clean two-particle entangled state, one could also realize an EPR-type experiment and search for violation of Bell-type inequalities. In order to test local hidden variable theories, the detection of the atom and the choice of the measurement basis for the photons should happen in two spacelike regions.[9] One difficulty in an atom-photon EPR experiment would be the very different propagation velocities of atoms and photons from the origin of the EPR pair to the detection systems. Therefore, the photon must be delayed by nearly the atomic transit time to the detector, for example, in a reasonably long fiber (about 100 km for our current experimental parameters), before the choice of the measurement basis and the detection is carried out (see FIGURE 8).

To perform correlation measurements between atoms and photons in atomic beam experiments, a clear association between an atom and a spontaneous emitted photon has to be ensured. Therefore, either the velocity of the atom has to be known accurately or the density of pairs in time has to be low enough to identify pair events. The way that we plan to achieve the pair identification uses a time-of-flight (t_f) velocity measurement of the atoms by means of a pulsed atomic source and a time-resolved atom detection.

Applying such a technique to our standing light wave diffraction setup, we were able to observe a time-resolved diffraction pattern (see FIGURE 9). This technique allows identification of high diffraction orders, which will be washed out due to chromatic effects for usual atomic beam sources. In the picture shown, the diffraction orders form straight lines because the de Broglie wavelength scales linearly with the time of flight from the source to the atomic detector. Our next experimental step will be the detection of a spontaneously emitted photon and a correlation of the atomic diffraction pattern with the photon momentum.

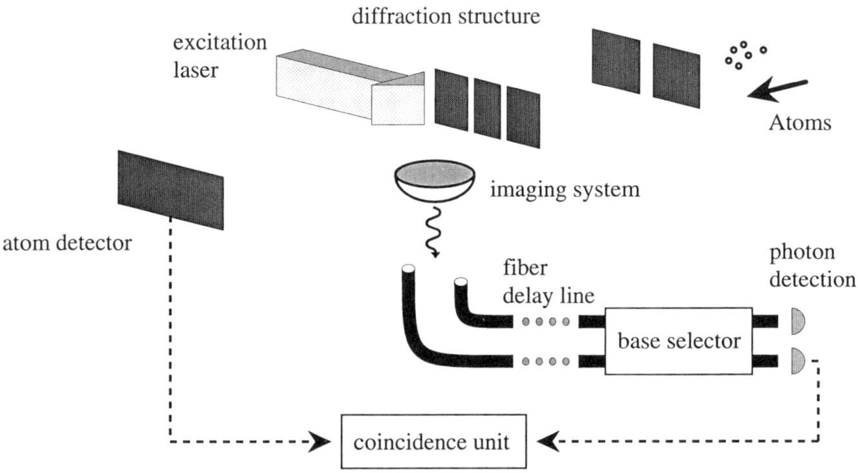

FIGURE 8. A possible experimental setup for an EPR experiment between an atom and a photon. The spontaneously emitted photon is coupled into a pair of fibers to allow the atom to get close to a momentum detector. Then, a measurement basis for the photon is chosen and the particles are detected in two spacelike regions.

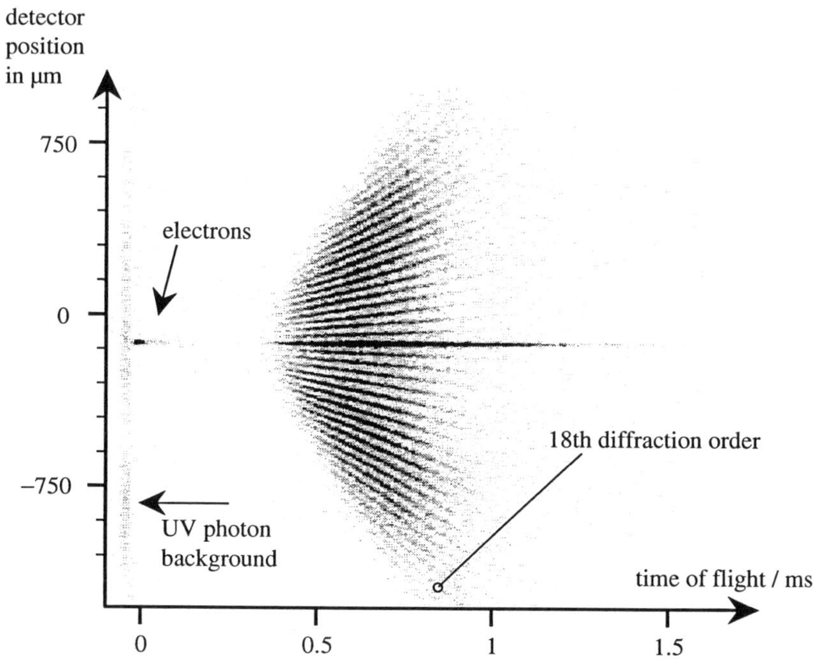

FIGURE 9. Time-resolved diffraction pattern from an off-resonant standing light wave. The additional time information allows us to measure the velocity of each atom to a high accuracy; therefore, the high diffraction orders could clearly be resolved. The line at $t_f = 0$ is caused by UV photons generated together with the atoms in our source. The events at $t_f = 50$ µs may be caused by electrons generated by de-excitation of some atoms at collimating microstructures.

SUMMARY

In conclusion, we were able to prepare a two-particle entangled state between an atom and a single photon. The entanglement was produced by a single spontaneous emission out of an excited internal state. We were able to measure the loss of spatial coherence of the atomic wave function by this emission process. This emission process may be looked at as an elementary process for the localization of an atom with the accuracy of the optical wavelength of the photon. Therefore, the classical gedankenexperiment of a Heisenberg microscope could be realized experimentally with the tools of atom optics. Furthermore, EPR-like experiments could be carried out with one massive and one relativistic particle. We were able to realize a first step towards such correlation experiments by the measurement of a time-resolved diffraction pattern.

REFERENCES

1. HEISENBERG, W. 1927. Z. Phys. **43:** 172.
2. PFAU, T., S. SPÄLTER, CH. KURTSIEFER, C. R. EKSTROM & J. MLYNEK. 1994. Phys. Rev. Lett. **73:** 1223.

3. Born, M. & E. Wolf. 1980. Principles of Optics. Sixth edition. Pergamon. Elmsford, New York.
4. Francon, M. & S. Mallick. 1967. Prog. Opt. **6:** 71.
5. Sleator, T. *et al.* 1992. *In* Proceedings of the Tenth International Conference on Laser Spectroscopy, p. 264.
6. Tan, S. & D. Walls. 1993. Phys. Rev. A **47:** 4663.
7. Dalibard, J. & C. Cohen-Tannoudji. 1985. J. Opt. Soc. Am. **B2:** 1707.
8. Scully, M. O., B-G. Englert & H. Walther. 1991. Nature **351:** 111.
9. Ballentine, L. E. 1970. Rev. Mod. Phys. **42:** 358.

Stern-Gerlach Atomic Interferometry with Space- and Time-dependent Magnetic Fields[a]

J. ROBERT, O. GORCEIX, J. LAWSON-DAKU,[b]
S. NIC CHORMAIC,[c] CH. MINIATURA, J. BAUDON,
F. PERALES, M. EMINYAN,[d] AND K. RUBIN[e]

Laboratoire de Physique des Lasers
Unité Associée du CNRS (URA 282)
Université Paris-Nord
93430 Villetaneuse, France

INTRODUCTION

The basic idea of Stern-Gerlach Interferometry (SGI) is to produce entanglement of the spin variables with the external and internal motion variables. By appropriate spin projections in the preparation and analysis, one gets a signal in which phase information is related to the evolution of the different magnetic levels within the interferometer.[1,2] This allows a transfer of coherence from the spin variable (whose coherence is determined mainly by the direction of the magnetic field) to the external motion variables. As shown in the following section, the extracted phase information is strongly dependent on the way that the entanglement is being produced. No entanglement means pure spin precession and the experimental results give no information on the external motion. On the other hand, strong entanglement yields maximum information on the Fourier component distribution of the external motion wave function. In this report, we do not focus on potential applications of atomic interferometry,[3] but rather limit ourselves to the discussion of the coherence problems that can be studied in the experiments presently performed.

SIGNAL AND ATOMIC BEAM CORRELATION FUNCTION

Principle

Stern-Gerlach Interferometry follows the same pattern as interferences with polarized light. A collimated beam of spin-carrying atoms is first polarized. It passes then through a low-magnetic-field Majorana region where a coherence superposition

[a]This work was supported by EC Contract No. SC1*-CT91-0712 (TSTS).

[b]J. Lawson-Daku was supported by grants from the ALBAR Foundation.

[c]S. Nic Chormaic acknowledges the EC Commission for the provision of a Science Bursary (No. B/SCI*915186).

[d]Present address: Université Paris 7, 75005 Paris, France.

[e]Present address: The City College of the City University of New York, New York, New York 10031.

of spin states is built. The beam then passes through a magnetic field that produces phase shifts in the various spin components. Each state of the coherent superposition undergoes a specific dynamical evolution during its voyage throughout the magnetic field profile. The beam crosses a second Majorana region that produces a second coherent superposition and the beam is analyzed by a second polarizer. Finally, the number of atoms in the selected state is recorded by means of a detector of macroscopic extension.

For an incoming spin polarized wave function of magnetic number ν entering the first Majorana region, that is,

$$\Psi^{\nu}_{\text{in}}(\vec{r}, t) = \int d\vec{k} c(\vec{k}) \exp\left[i\vec{k} \cdot \vec{r} - i\left(\frac{Et}{\hbar}\right)\right], \tag{1}$$

where E stands for the kinetic energy of the atom of mass M, one gets an outgoing spin analyzed wave function of magnetic number μ:

$$\Psi^{\nu\mu}_{\text{out}}(\vec{r}, t) = \sum_{m} \int d\vec{k} c(\vec{k}) \mathscr{D}^{\text{in}}_{m\nu} \mathscr{D}^{\text{out}}_{\mu m} \exp\left[i\vec{k} \cdot \vec{r} - i\left(\frac{Et}{\hbar}\right) + im\varphi(\vec{k}, \vec{u}, \vec{r}, t)\right], \tag{2}$$

where $\varphi(\vec{k}, \vec{u}, \vec{r}, t)$ is the phase shift for an incoming m polarized plane wave scattered by the magnetic field profile; $\vec{u}$ is either the magnetic field direction (when it is kept constant) or the effective one (when there is a change in the magnetic field direction[4,5]); and the $\mathscr{D}$'s are the Wigner rotation matrices describing the Majorana region effect. The Zeeman energy shifts occurring in our experimental conditions are in the neV range and thus are very small compared to the kinetic energy of the atoms (in the eV range). Hence, phase shifts can be determined using Glauber's high-energy approximation.[6] For a ray that passes at point $\vec{r}$ at time t, one has

$$\varphi(\vec{k}, \vec{r}, t) = \left(-\frac{1}{\hbar v}\right) \int_{-\infty}^{\vec{r}.\vec{w}} ds V[X(\vec{r}, s);\ Y(\vec{r}, s);\ Z(\vec{r}, s);\ T(\vec{r}, t, s)], \tag{3}$$

where $\vec{k}$ is directed along $\vec{w}$, V stands for the Zeeman energy, v stands for the group velocity, and s stands for the abscissa along the ray axis. X, Y, Z, and T are space and time coordinates in the fixed reference frame used to compute the magnetic potential energy.

To get the signal S, one has to average $|\Psi^{\nu\mu}_{\text{out}}(\vec{r}, t)|^2$ over the spatial and temporal resolution of the detector, which can be described by a function R of time and position. One gets

$$S = \sum_{m,m'} \int d\vec{k} d\vec{k}' M^{\nu\mu}_{mm'} c(\vec{k}) c(\vec{k}') \int d\vec{r} dt R(\vec{r}, t) \exp[i\Delta\Phi(\vec{k}, \vec{k}', m, m', \vec{r}, t)], \tag{4}$$

where $M^{\nu\mu}_{mm'}$ is the product of the relevant Wigner matrix elements and

$$\Delta\Phi(\vec{k}, \vec{k}', m, m', \vec{r}, t) = (\vec{k} - \vec{k}') \cdot \vec{r} - \left[\frac{(E - E')t}{\hbar}\right] + m\varphi(\vec{k}, \vec{r}, t) - m'\varphi(\vec{k}', \vec{r}, t).$$

Because the detector aperture is macroscopic, one is led to assume that after integration the only nonzero contribution to the signal will be provided by the

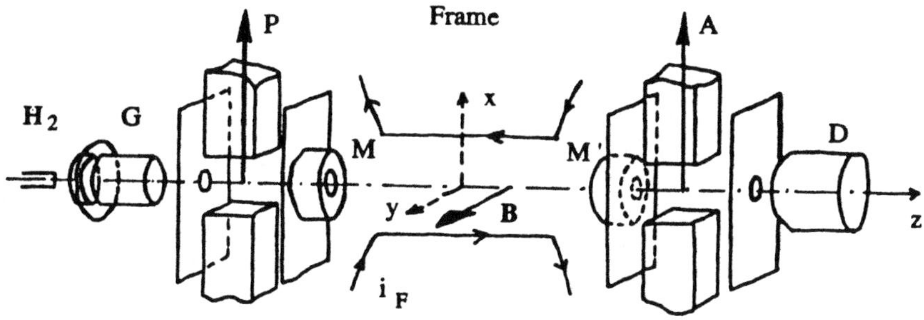

FIGURE 1. Experimental setup.

stationary part of $\Delta\Phi$ with respect to position and time. This condition gives the relations between $\vec{k}$ and $\vec{k}'$ and E and E' and enables us to carry the integration over the primed variables. Indeed, assuming that the detector extension is infinite, one gets

$$\vec{k}' = \vec{k} + \Delta\vec{k} = \vec{k} + \vec{\nabla}_{\vec{r}}[m\varphi(\vec{k}, \vec{r}, t) - m'\varphi(\vec{k}', \vec{r}, t)], \qquad \textbf{(5a)}$$

$$E' = E + \Delta E = E + \partial_t[m\varphi(\vec{k}, \vec{r}, t) - m'\varphi(\vec{k}', \vec{r}, t)]. \qquad \textbf{(5b)}$$

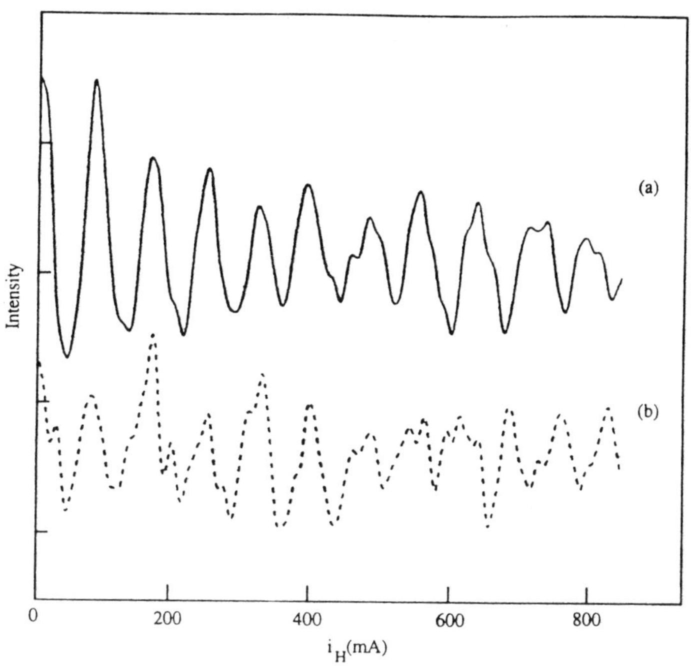

FIGURE 2. Experimental demonstration of the scalar Bohm-Aharonov effect using two different velocity selections [(a) 10 km/s; (b) 5 km/s]. The current height in the Helmholtz coils is scanned from 0 to 910 mA.

As $v \approx v'$ and $\vec{w} \approx \vec{w}'$, one gets the shared phase factor,

$$\varphi(\vec{k}, \vec{r}, t) \approx \varphi(\vec{k}', \vec{r}, t) = \overline{\varphi}(\vec{k}) = \vec{k} \cdot \Delta\vec{r} - \left(\frac{E\Delta t}{\hbar}\right), \tag{6}$$

where the notation of reference 5 is used; the signal expression is thus

$$S = \sum_{m, m'} \int d\vec{k} M^{\nu\mu}_{mm'} c(\vec{k}) c(\vec{k} + \Delta\vec{k}) \exp[i(m - m')\overline{\varphi}(\vec{k})]. \tag{7}$$

Equation 7 shows that by an appropriate choice of the experimental conditions, that is, of the spatial and time dependence of the magnetic field phase object, it is possible to gain information on the beam coherences from the interference patterns obtained by scanning the magnetic field amplitude of the phase object.

Dispersivity Considerations in Continuous versus Pulsed Operation

There is still another issue deserving examination: the wave number (i.e., velocity) dependence of phase shifts and momentum exchange. This is conveniently handled in the framework of Glauber's approximation. The key point is to determine if it is possible to perform the change of variable $\tau = s/v$ in the determination of the

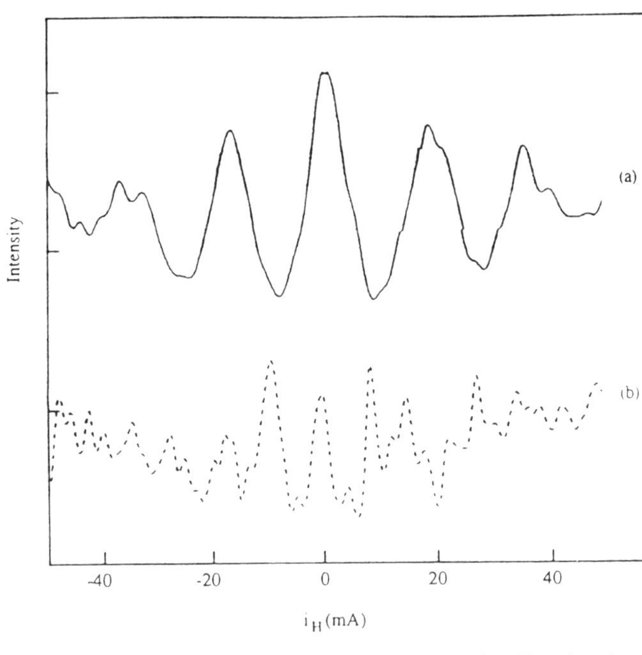

FIGURE 3. Demonstration of the longitudinal Stern-Gerlach effect for the same velocity selections as in FIGURE 2. The current in the Helmholtz coils is scanned from −50 mA to +50 mA.

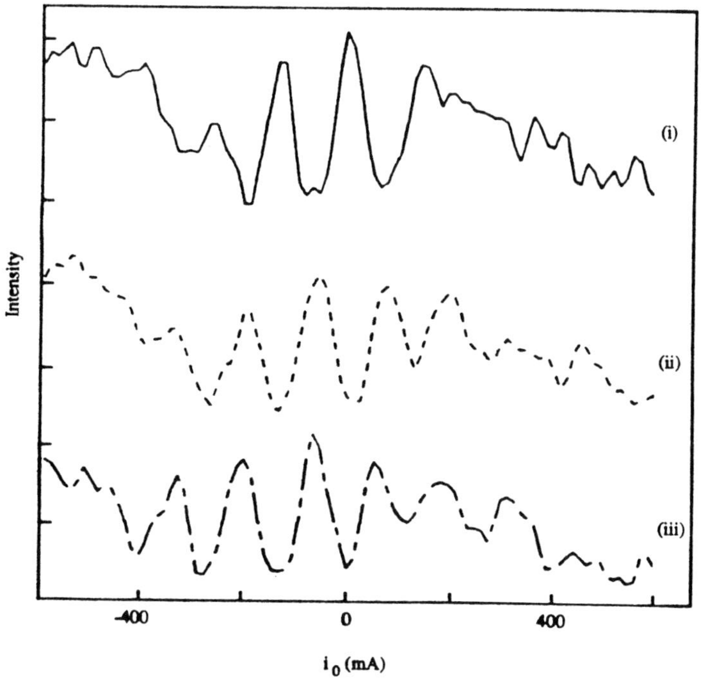

FIGURE 4. SG interference patterns obtained by scanning the continuous current in an additional rectangular frame from −500 mA to +500 mA for fixed values of the pulsed current in the Helmholtz coils. The SAB phase shift is zero (i), π (ii), and 5π (iii).

phase shift and to eliminate the k-dependence in the integration. When one deals with a pulsed magnetic field localized in a specific space region, it is clear that we can disregard the velocity dependence if the pulse duration is less than the transit time through the magnetic profile. The resulting phase shifts and thus the fringe spacing in this type of experiment are velocity-independent and the contrast of the interference patterns is not limited by the beam coherence length.[7] If static fields are used, then the k-dependence remains.

Experimental Setup

The general scheme of our experimental procedure has already been described in detail in previous publications.[2,4,5] FIGURE 1 gives a schematic view of our apparatus. Metastable H* atoms are produced by a 120-eV electron bombardment of a thermal H_2 beam. The time-of-flight (TOF) distribution is roughly Maxwellian with a most probable TOF being a velocity of 10 km/s, corresponding to a de Broglie wavelength of about 40 pm. The polarizer and the analyzer are of the Lamb-Retherford type; that is, they consist of a transverse 600-G magnetic field acting over a 20-mm length that quenches one of the magnetic sublevels. The Majorana regions consist of two

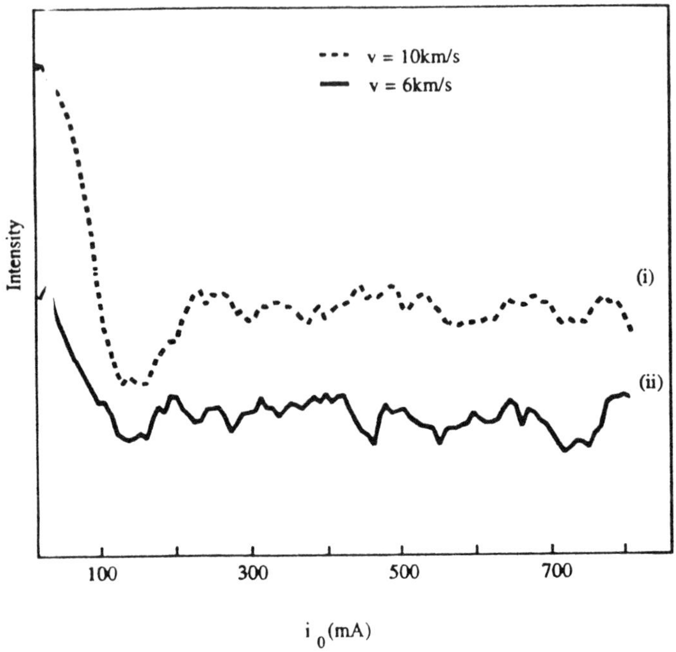

FIGURE 5. Patterns obtained for a pulsed transverse gradient field. The current in the anti-Helmholtz coils is scanned from 0 to 800 mA.

zones, past which the magnetic field has a weak magnitude and a rapidly changing direction. The central region where the phase object lies is protected by a triple Mumetal shielding. The magnetic field inside this region is produced by a pair of electric coils of appropriate geometry. In the pulsed regime, a metallic spring protects the beam from stray electric fields in order to avoid cancellation by eddy current of the high-frequency components of the magnetic field. The H*-sensitive detector consists of a capacitor in which the electric field quenches the metastable hydrogen atoms. The induced Lyman α emission is recorded by means of a channel electron multiplier. Several different experiments can be performed by pulsing, in a definite time sequence, the various elements of the system (source, phase object, detector).

EXPERIMENTAL RESULTS

Magnetic Field Profile Produced by Two Parallel Rectangular Helmholtz Coils

In this configuration, there is neither momentum nor global energy transfer. Furthermore, if the magnetic field is pulsed while the atoms are in the vicinity of the middle of the profile, no momentum transfer occurs at all and one is faced with a scalar Bohm-Aharonov (SAB) effect.[7-9] As shown in FIGURE 2, the phase shift is velocity-independent. As a consequence of this type of pulsed experiment, the number of visible fringes is not related to the coherence length of the beam. On the contrary, if the magnetic field is static, momentum transfers occur at both ends of the profile (although of opposite values); this is a longitudinal Stern-Gerlach effect. The phase shift is inversely proportional to the velocity and thus the number of visible fringes is limited by the velocity dispersion. This is demonstrated in FIGURE 3, from which one can infer that the beam coherence length is about 200 pm under the present experimental condition. If one combines pulsed and static magnetic fields, important information on the finesse of our device can be extracted from the experimental results. For example, when a π or 5π SAB phase shift is superimposed on the Stern-Gerlach dispersive pattern (FIGURE 4), there is an inversion of the fringes and a shift of the envelope. This shift is a signature of the finesse of our apparatus as it can be modeled by adding a random contribution to the phase shift. This shift is not related to the atom wave packet envelope.

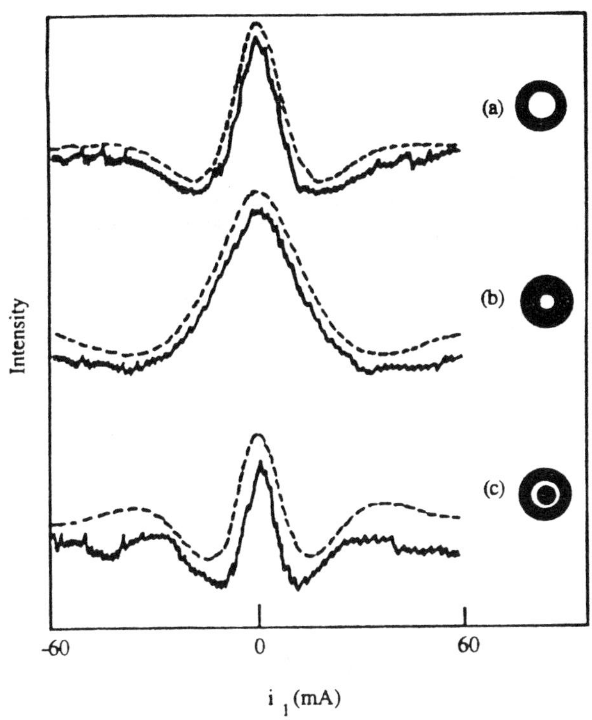

FIGURE 6. Central bright fringe attenuation as a function of the current in the anti-Helmholtz coils for different collimation holes: (a) hole diameter is 4 mm, (b) hole diameter is 2 mm, (c) hole of inner diameter is 3 mm and outer diameter is 4 mm. Dashed lines are theoretical calculations.

Magnetic Field Profile Produced by Two Parallel Rectangular Coils in the Anti-Helmholtz Configuration

The magnetic field produced by this wire configuration is a quadrupole deflector. It has been used in transverse Stern-Gerlach effect studies by Bloom and coworkers.[10] Because the magnetic field gradient is radial, an incoming plane wave is transformed into a conical outgoing wave. Our phase object acts like an axicon. In this configuration, transverse momentum transfer occurs and thus effects related to the beam collimation and to the transverse coherence are observed. If pulsed transverse gradients are used, there is no velocity dependence in the momentum transfer as demonstrated in FIGURE 5. If static transverse gradients are used, a velocity dependence occurs. The sensitivity of the method to the beam collimation is shown in FIGURE 6. In spite of the smallness of the de Broglie wavelength (in the pm range) compared to the collimation-hole diameter (in the mm range), the attenuation of the central fringe visibility as a function of the gradient amplitude is modified by the choice of the macroscopic collimation.

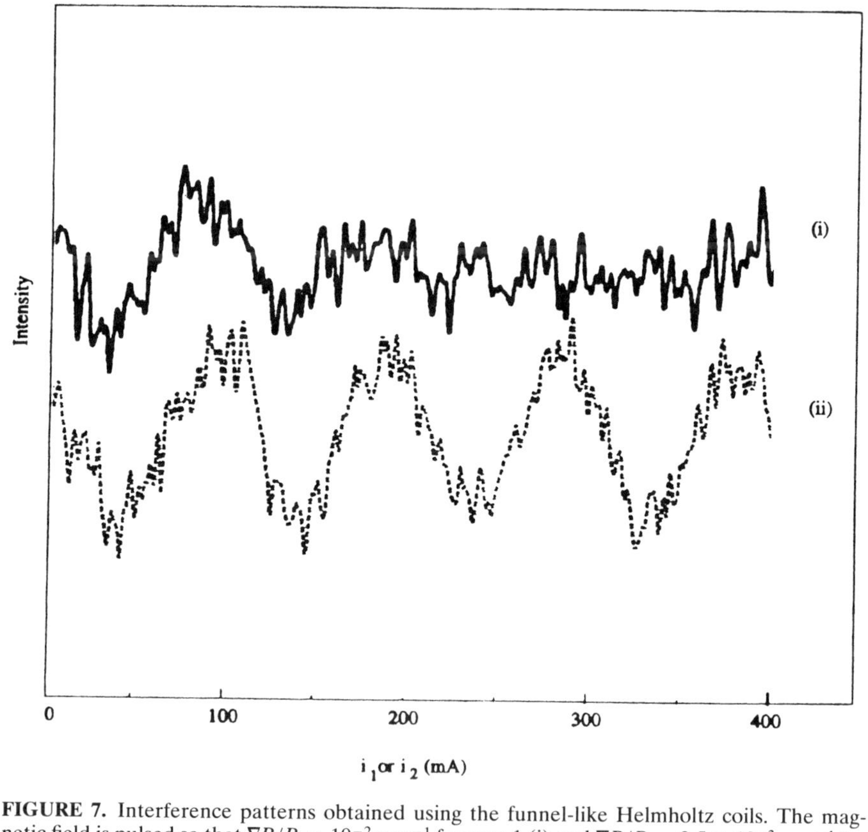

FIGURE 7. Interference patterns obtained using the funnel-like Helmholtz coils. The magnetic field is pulsed so that $\nabla B/B \approx 10^{-2}\,\mathrm{mm}^{-1}$ for case 1 (i) and $\nabla B/B \approx 2.5 \times 10^{-3}\,\mathrm{mm}^{-1}$ for case 2 (ii).

Magnetic Field Profile Produced by Funnel-like Helmholtz Coils

In this configuration, the magnetic field profile in the central part of the phase object indicates that the atoms experience a longitudinal gradient in the pulsed regime. This gives rise to an energy transfer and a longitudinal momentum transfer. The ensuing modifications of the interference patterns displayed in FIGURE 7 can be related to the correlation length of the beam. Under the experimental conditions of case 2 ($\nabla B/B \approx 2.5 \times 10^{-3}$ mm^{-1}), the correlation length is determined to be 0.4 mm (the momentum exchange being 2.5 mm^{-1}).

CONCLUSIONS

We have shown here that Stern-Gerlach Atomic Interferometry is an effective tool for the study of wave function correlation functions. It gives qualitative and quantitative results without involving much technology. The versatility of our method opens a wide range of opportunity for studying quantum phase problems that can be understood in a quite direct way.

ACKNOWLEDGMENTS

G. Lochak, R. Dutheil, and S. Diner of the "Fondation Louis de Broglie" are acknowledged for their help with this work.

REFERENCES

1. MEZEI, F. 1986. Physica **137B:** 295 and references therein.
2. ROBERT, J., CH. MINIATURA, S. LE BOITEUX, J. REINHARDT, V. BOCVARSKI & J. BAUDON. 1991. Europhys. Lett. **16:** 29; MINIATURA, CH., F. PERALES, G. VASSILEV, J. REINHARDT, J. ROBERT & J. BAUDON. 1991. J. Phys. (Paris) II **1:** 425.
3. Special issue on "Optics and Interferometry with Atoms". 1992. Appl. Phys. B **54**.
4. MINIATURA, CH., J. ROBERT, O. GORCEIX, V. LORENT, S. LE BOITEUX, J. REINHARDT & J. BAUDON. 1992. Phys. Rev. Lett. **69:** 261.
5. ROBERT, J., CH. MINIATURA, O. GORCEIX, S. LE BOITEUX, V. LORENT, J. REINHARDT & J. BAUDON. 1992. J. Phys. (Paris) II **2:** 601.
6. GLAUBER, R. J. 1959. *In* Lectures in Theoretical Physics. Vol. 1. W. E. Brittin & L. Dunham, Eds.: 315. Interscience. New York.
7. AHARONOV, Y. & D. BOHM. 1959. Phys. Rev. **115:** 485; ZEILINGER, A. 1984. J. Phys. (Paris) Colloq. **45:** C3–C13; ZEILINGER, A. 1985. *In* Fundamental Aspects of Quantum Theory. NATO ASI Series B. Vol. 144. V. Gorini & A. Frigerio, Eds.: 311. Plenum. New York; ANANDAN, J. 1989. Proceedings of the Third International Symposium on Foundations of Quantum Mechanics (Tokyo), p. 98. Phys. Soc. Japan. Kyoto.
8. BADUREK, G., H. WEINFURTER, R. GÄHLER, A. KOLLMAR, S. WEHINGER & A. ZEILINGER. 1993. Phys. Rev. Lett. **71:** 307.
9. CHORMAIC, S. N., CH. MINIATURA, O. GORCEIX, B. VIARIS DE LESEGNO, J. ROBERT, S. FERON, V. LORENT, J. REINHARDT, J. BAUDON & K. RUBIN. 1994. Phys. Rev. Lett. **72:** 1.
10. BLOOM, M. & K. ERDMAN. 1962. Can. J. Phys. **40:** 179; BLOOM, M., E. ENGA & H. LEW. 1967. Can. J. Phys. **45:** 1481.

Possibility of Quantum Revivals in Long-Time Spin Dynamics of ^{3}He-^{4}He Solutions[a]

DANIEL L. SCHAUER

Department of Physics
University of Illinois at Urbana-Champaign
Urbana, Illinois 61801

INTRODUCTION

For many years, the field of cavity quantum optics has borrowed from the already well established two-level system literature. The Jaynes-Cummings model[1] is essentially an optics generalization of the equations governing an isolated spin evolving in a magnetic field. One can see the influence of this in the names of the equations in cavity QED (the optical Bloch equations and the Bloch sphere are two obvious examples). One of the fascinating recent predictions of the dynamics of an atom interacting with a coherent state field in a cavity is the existence of quantum "collapse" and "revival" effects.[2]

In this report, I would like to discuss the possibility that quantum revivals are being seen in a recent experiment performed by Nunes *et al.*[3,4] investigating the longitudinal spin diffusion coefficient of dilute concentrations of ^{3}He in ^{4}He. In the process of these measurements, they noticed a quite peculiar response of the magnetization to an inverting pi pulse. Immediately following the pi pulse, there is a short free-induction decay that lasts about 100 ms. This is followed by approximately 1.4 seconds of no signal apart from noise. At 1.4 seconds, a rapidly oscillating signal starts to appear. The signal has the appearance of a burst lasting about 0.2 seconds. These bursts, recurring periodically, are separated by intervals of no signal. The width of subsequent bursts increases and the amplitude decreases until neighboring bursts eventually merge together into a long tail, which subsides 10–15 seconds after the initial pi pulse (see FIGURE 1).

Before discussing any mechanisms to describe this phenomenon, it is probably best to describe the apparatus and procedure used. Nunes *et al.* were investigating the longitudinal spin diffusion of spin-polarized ^{3}He-^{4}He solutions. The general procedure used was as follows. A sample cell was constructed that consisted of two chambers (for a diagram of the cell, see reference 4). The upper chamber was about ten times larger than the lower one. The lower chamber was surrounded by a resonant cavity that was attached to the NMR spectrometer for the experiment, and the two chambers were connected by a narrow tube. The experimental procedure was to invert the spins in the lower chamber and to observe how the magnetization in

[a]This research was supported through the MacArthur Professorship endowed by the John D. and Catherine T. MacArthur Foundation.

the lower chamber relaxed back to its original state. It was during this portion of the experiment that the interesting bursting behavior was first observed.

An explanation for this behavior was proposed by Nunes *et al.*[3,4] They hypothesized that this behavior was the result of the inherently nonlinear spin dynamics in the system. In fact, in this system, a purely longitudinal gradient in the magnetization becomes unstable against transverse perturbations. The results of a simple model based on this hypothesis reproduce the initial time delay before the first burst and the long-time scale behavior following the burst, but they do not reproduce the interesting bursting behavior of the experimental signal. Using a simple Jaynes-Cummings type of model, I can reproduce the bursting behavior seen in FIGURE 1. This report will discuss the development of such a model.

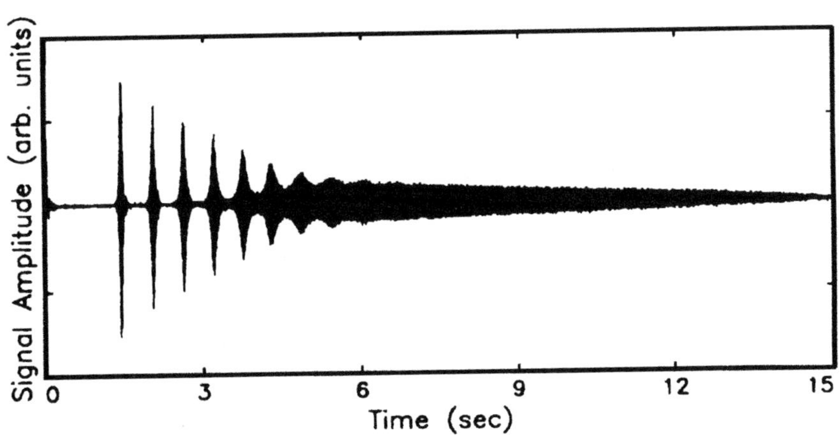

FIGURE 1. Experimental signal amplitude for 350-ppm ^3He in ^4He solution at 10.0 mK. (From reference 3.)

DEVELOPING A MODEL HAMILTONIAN

The pattern of bursts in Nunes' experiment is very reminiscent of quantum revivals in cavity QED. In fact, if one naively plugs some numbers into the standard setup for a quantum revival experiment, one can quite easily generate a signal that closely mimics FIGURE 1. One difficulty in attributing the phenomenon to a simple Jaynes-Cummings revival is that the quantity that undergoes the revival phenomenon is the spin, whereas in this experiment it is the current in the circuit that is measured. It would be nice to formulate the problem in a manner so as to examine the time development of the measured current rather than focus on the behavior of the spins. To do this, I will first derive a Hamiltonian that represents the interaction between the spins and the circuit coupled to the cavity. I will quantize this Hamiltonian in terms of the circuit variables (charge and current). From this quantized Hamiltonian,

I will examine the time development of the expectation value of the current starting with the circuit having a classical-type current and the spins in the cavity being inverted.

The NMR resonator and circuitry prior to signal amplification in Nunes' experiment can be modeled as an LRC circuit with a gas of dipole moments in the inductive element. I would like to consider, at first, a series LC-circuit that has a classical resonance frequency of ω_0. In the inductive element, place a magnetic dipole with moment $\vec{m}$. The total energy of the circuit and the dipole is then

$$E = \left(\frac{1}{2}\right) L\dot{q}^2 + \left(\frac{q^2}{2C}\right) - \vec{m} \cdot \vec{B}, \tag{1}$$

where $\vec{B}$ is the total magnetic field acting on the dipole. The total magnetic field on the dipole arises from two sources. The first is a large static magnetic field, B_0, that is placed perpendicular to the inductor axis ($-z$ direction). The second is the magnetic field (in general, time-varying) generated by the inductor, parallel to the axis of the inductor, and of magnitude (assuming that the inductor is a solid object whose size does not change)

$$B_{\text{ind}} = \frac{LI}{A}, \tag{2}$$

where L is the inductance and I is the current ($\dot{q}$). With this and by using the fact that $\vec{m} = (\hbar\gamma/2)\vec{\sigma}$, we can write the energy as

$$E = \left(\frac{1}{2}\right) L\dot{q}^2 + \left(\frac{q^2}{2C}\right) + \left(\frac{\hbar\gamma}{2}\right) B_0\sigma_z - \left(\frac{\hbar\gamma L}{2A}\right)\dot{q}. \tag{3}$$

Now, I will use the standard identifications to write down a Lagrangian and obtain a canonical momentum conjugate to q. First, the "kinetic energy" is identified with the bilinear terms in $\dot{q}$. The potential energy is then the rest and the Lagrangian is given by

$$\mathscr{L} = \left(\frac{1}{2}\right) L\dot{q}^2 - \left(\frac{q^2}{2C}\right) - \left(\frac{\hbar\gamma}{2}\right) B_0\sigma_z + \left(\frac{\hbar\gamma L}{2A}\right)\dot{q}\sigma_x. \tag{4}$$

From this Lagrangian, the canonical momentum $p = (\partial L/\partial\dot{q})$ is

$$p = L\dot{q} + \left(\frac{\hbar\gamma L}{2A}\right)\sigma_x. \tag{5}$$

We can then use this fact to write down the Hamiltonian and we find that it is very similar to that obtained in the QED case:

$$H = \left[\frac{\left(p - \left[\frac{\hbar\gamma L}{2A}\right]\sigma_x\right)^2}{2L}\right] + \left(\frac{q^2}{2C}\right) + \left(\frac{\hbar\gamma}{2}\right) B_0\sigma_z. \tag{6}$$

The Hamiltonian can be expanded and, after removing all constant terms, it can be written as

$$H = \left(\frac{p^2}{2L}\right) + \left(\frac{q^2}{2C}\right) - \left(\frac{\hbar\gamma}{2A}\right) p\sigma_x + \left(\frac{\hbar\gamma}{2}\right) B_0\sigma_z. \quad (7)$$

It appears that I have started the quantization procedure already as there are some Pauli matrices in the above equation. I could as well replace these by a classical dipole moment and the circuit would then be completely classical. As it stands, it is a classical circuit being driven by a quantum spin. Now, using standard canonical quantization of the circuit variables, we introduce the commutator $[q, p] = i\hbar$ and expand equation 7. The Hamiltonian above is similar to a driven harmonic oscillator; thus, in quantizing it, we introduce the familiar annihilation and creation operators:

$$a = \left(\frac{1}{\sqrt{2\hbar L\omega_0}}\right) (L\omega_0 q + ip), \quad (8)$$

$$a^\dagger = \left(\frac{1}{\sqrt{2\hbar L\omega_0}}\right) (L\omega_0 a - ip). \quad (9)$$

These operators then have the standard commutation relation $[a, a^\dagger] = 1$ and equation 7 can be written, using the rotating wave approximation, as

$$H = \hbar\omega_0 a^\dagger a + \left(\frac{\hbar\gamma}{2}\right) B_0\sigma_z + ig(a\sigma_+ - a^\dagger\sigma_-), \quad (10)$$

where the coupling

$$g = \left(\frac{\hbar\gamma}{2A}\right) \sqrt{\frac{\hbar L\omega_0}{2}}.$$

A GENERAL SOLUTION

At this point, it is just a matter of obtaining a solution for the time-development operator associated with equation 10. Using techniques originally proposed by Walls[5] and used by Yoo and Eberly,[6] we split the Hamiltonian into two pieces. The first piece depends only on constants of the motion and, as such, adds only an overall phase to the wave function. The second contains the nontrivial time development of the system. In the limit of exact resonance, the splitting described above is identical to working in the interaction representation. The Hamiltonian is then

$$H = H_{\mathrm{I}} + H_{\mathrm{II}}, \quad (11)$$

where

$$H_{\mathrm{I}} = \hbar\omega_0\hat{N},$$
$$H_{\mathrm{II}} = \hbar\Delta\sigma_+\sigma_- + ig(a\sigma_+ - a^\dagger\sigma_-), \quad (12)$$

with $\hat{N} = a^\dagger a + \sigma_+\sigma_-$ and $\Delta = \gamma B_0 - \omega_0$.

Because H_{I} is a constant of the motion, we consider a representation where the time-development operator is given in terms of H_{II} alone. Looking at equation 13, it is easy to see that the Hilbert space can be broken down into two-dimensional subspaces with eigenstates

$$|1\rangle^{(n)} = |\downarrow, n\rangle, \tag{13}$$

$$|2\rangle^{(n)} = |\uparrow, n-1\rangle, \tag{14}$$

where the arrows designate the z-component of the spin and n is the number of quanta in the circuit. Because the different n-subspaces are independent, we need only solve for one. Following standard procedures (equation 6), we obtain a time-development operator,

$$U_{\text{II}}^{(n)}(t) = e^{-(i\Delta/2)t}\begin{pmatrix} \cos f_n t + \left(\dfrac{i\Delta}{2f_n}\right)\sin f_n t & -\left(\dfrac{ig_n}{f_n}\right)\sin f_n t \\ \left(\dfrac{ig_n}{f_n}\right)\sin f_n t & \cos f_n t - \left(\dfrac{i\Delta}{2f_n}\right)\sin f_n t \end{pmatrix}, \tag{15}$$

where $f_n = \sqrt{g_n^2 + (\Delta^2/4)}$ and $g_n = \sqrt{n}g$.

THE SOLUTION FOR THE CIRCUIT CURRENT

Let us now examine what the above time-development operator predicts for a gas of N independent spins, initially pointing in the positive z direction and interacting with a circuit initially in a coherent state with respect to the annihilation operator:

$$|\psi(0)\rangle = e^{-(|\alpha|^2/2)}\sum_{n=0}^{\infty}\left(\frac{\alpha^n}{\sqrt{n!}}\right)|\uparrow, n\rangle. \tag{16}$$

Assuming the cavity resonance is tuned to the spin energy splitting, we obtain, from operating equation 15 on the above,

$$|\psi(t)\rangle = e^{-(|\alpha|^2/2)}\sum_{n=0}^{\infty}\left[\frac{\alpha^n}{\sqrt{(n-1)!}}\right](\cos g_n t\,|2\rangle^{(n)} - i\sin g_n t\,|1\rangle^{(n)}). \tag{17}$$

From equation 17, let us calculate the expectation value for the current in the circuit. From equation 5, we see that the current is given by

$$I = \left(\frac{p}{L}\right) - \left(\frac{\hbar\gamma}{2A}\right)\sigma_x. \tag{18}$$

Calculating the expectation values, $\langle p\rangle$ and $\langle\sigma_x\rangle$ are quite straightforward and, after a bit of tedious algebra, we obtain

$$\langle I \rangle = I_0 \left(\frac{e^{-|\alpha|^2}}{|\alpha|} \right) \sum_{n=1}^{\infty} \left[\frac{|\alpha|^{2n}}{\sqrt{n!(n-1)!}} \right] \left\{ \sqrt{n} \cos g_n t \cos g_{n+1} t + \sqrt{n+1} \sin g_n t \sin g_{n+1} t + \left(\frac{\gamma}{A} \right) \sqrt{\frac{\hbar L}{2\omega_0}} \sin g_n t \cos g_{n+1} t \right\}, \quad (19)$$

where $I_0 = \sqrt{2\hbar\omega_0/L}$.

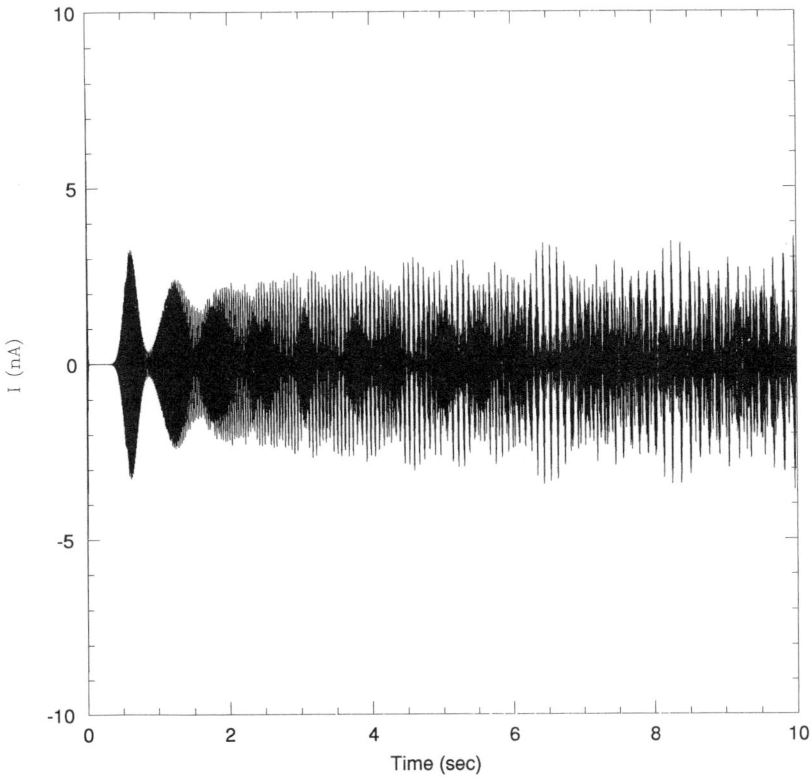

FIGURE 2. Time-dependence of current in a circuit initially in a coherent state with $\bar{n} = 100$.

The above equation can be simplified greatly by introducing the dimensionless parameter, $\xi_0 = (\gamma/A)\sqrt{\hbar L/2\omega_0}$, for then $g_n = N\xi_n(\omega_0/2)$. Furthermore, if we assume that $|\alpha|$ is large, the only terms that contribute significantly to equation 19 will be those near $|\alpha|^2$ and we can approximate $\sqrt{n+1} \sim \sqrt{n} + (1/2\sqrt{n})$. With these identifications, equation 19 becomes

$$\langle I \rangle = I_0 \left(\frac{e^{-|\alpha|^2}}{|\alpha|} \right) \sum_{n=1}^{\infty} \left[\frac{|\alpha|^{2n}}{(n-1)!} \right] \cos \left(\frac{N\xi_0\omega_0 t}{4\sqrt{n}} \right). \tag{20}$$

With equation 20, we can investigate the behavior of the current in this simple model. First, notice that it is a sum of cosines, each of incommensurate period. This sort of sum is what leads to the familiar quantum revivals in the Jaynes-Cummings model. Following along the lines of reference 6, we would expect to find revivals whenever the difference between the two most important terms was a multiple of 2π. The important terms, of course, are those near $\bar{n} \equiv |\alpha|^2$. Again, in the limit of large $\bar{n}$, this criterion becomes

$$\left(\frac{N\xi_0\omega_0 t}{4} \right) \left(\frac{1}{\sqrt{\bar{n}}} - \frac{1}{\sqrt{\bar{n}+1}} \right) = 2\pi. \tag{21}$$

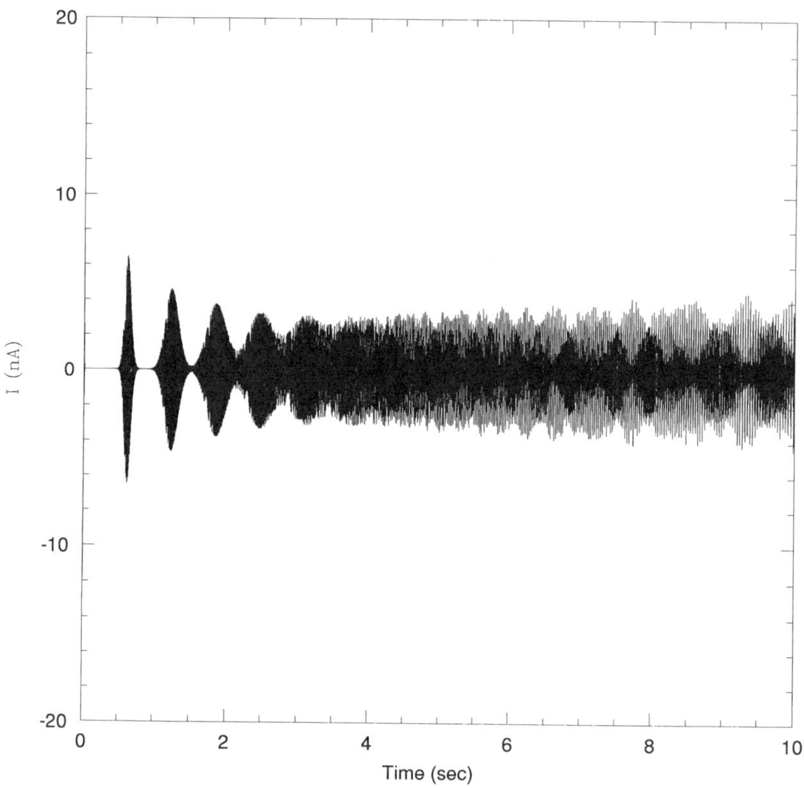

FIGURE 3. Time-dependence of current in a circuit initially in a coherent state with $\bar{n} = 400$.

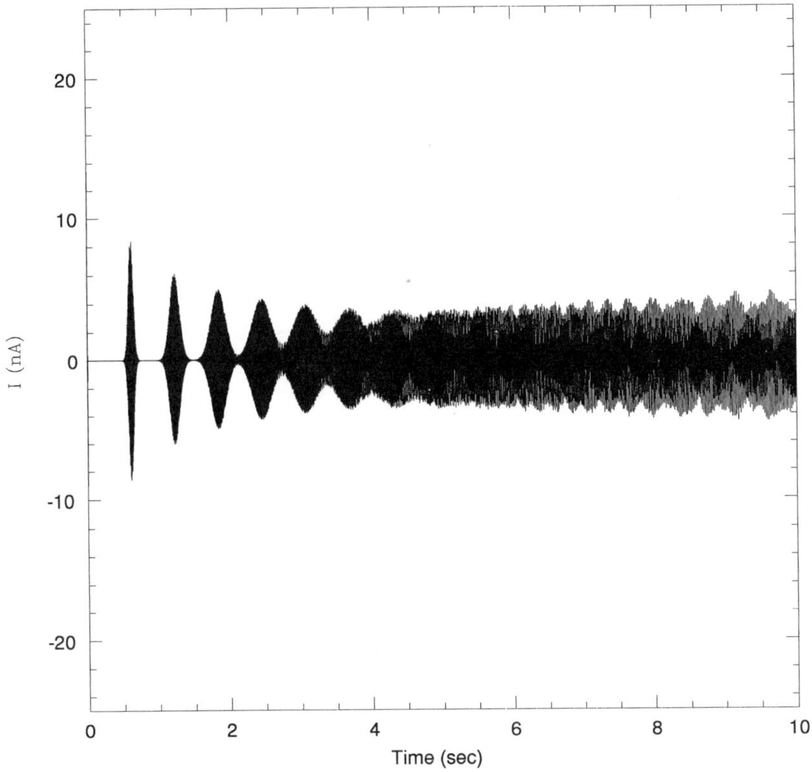

FIGURE 4. Time-dependence of current in a circuit initially in a coherent state with $\bar{n} = 700$.

This leads to the prediction that revivals occur with period

$$t_R = \frac{16\pi\bar{n}^{3/2}}{N\xi_0\omega_0}. \tag{22}$$

In order to get a feel for what this model predicts, let us introduce some experimental parameters. Most of the necessary numbers can be obtained from reference 4. One important number, however, is available only in reference 3. This is the capacitance of the circuit, which was measured to be 0.23 nF. From this and the fact that the resonance frequency was 293 MHz, we find that $I_0 = 1.1$ nA, $\xi_0 = 3.12 \times 10^{-14}$, and $t_R = 0.63$ s. Using these numbers in equation 22, we can obtain a relationship between $\bar{n}$ and N that must be satisfied to match the experimentally observed signal:

$$\frac{\bar{n}^{3/2}}{N} = 1.15 \times 10^{-7}. \tag{23}$$

There is still one free parameter in the model and I use it to match the number of distinct revivals in the experimental trace. FIGURES 2–5 demonstrate how the number of distinct revivals depends on the initial intensity of the current in the circuit ($\bar{n}$). Notice that increasing $\bar{n}$ while keeping t_R fixed increases the number of distinct revivals—specifically, for $\bar{n} = 900$, the number of revivals matches quite well the experimental signal, whereas, for small $\bar{n}$ (FIGURE 2), there is barely one visible revival.

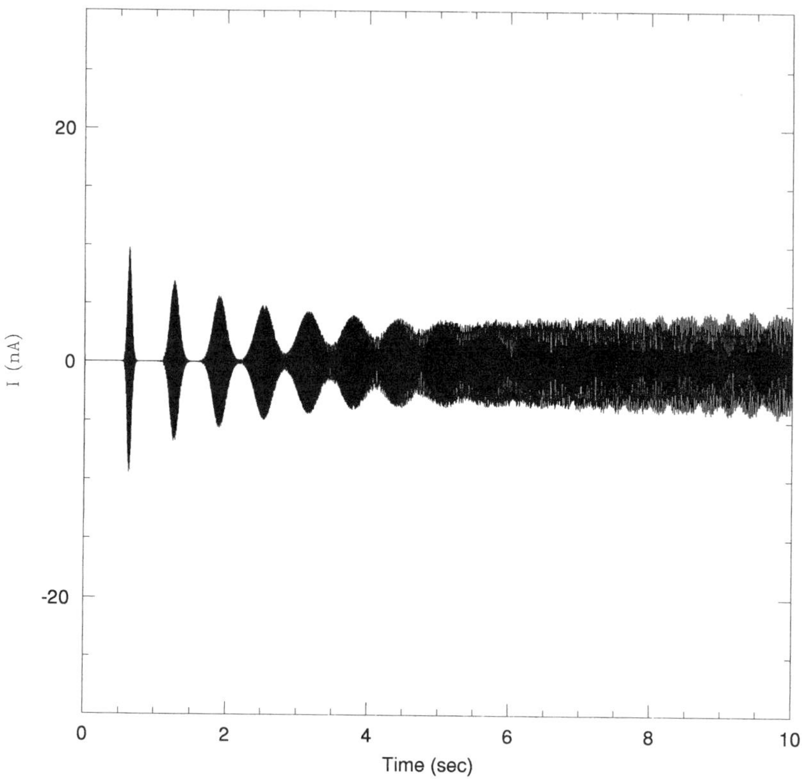

FIGURE 5. Time-dependence of current in a circuit initially in a coherent state with $\bar{n} = 900$.

FIGURE 5 appears to match the experimental signal the best. Using this as a rough estimate, I predict that 2.3×10^{11} ^{3}He atoms are involved in creating this signal. This is to be compared with 1.3×10^{17} atoms in the cavity. Furthermore, $\bar{n} = 900$ implies an initial current in the circuit of approximately 33 nA. One final prediction that can be made, and possibly verified from the raw data, is the base frequency. This is given by the frequency component of the most important term in the sum (equation 20). I predict this to be approximately 17.53 kHz.

DISCUSSION

I would now like to discuss what these results imply for the usefulness of this model in describing the long-time scale oscillations observed in Nunes' experiment. Qualitative comparison of FIGURES 1 and 5 is quite striking. However, direct comparison of the two plots is impossible as Nunes' data have been mixed down to 80 Hz and the base oscillatory frequency has been lost. Also, whereas the revivals in my model are strictly periodic, the first burst in FIGURE 1 is delayed by more than twice t_R. It is not clear if this delay can be reproduced in the simple model that I have described. Furthermore, one must ask the following question: what does resistance do to this model? At first sight, resistance should basically add an overall multiplicative exponential decay to equation 20. The problem that one finds is that one must have a resistance of the order of 5 nΩ in order for the time scale of this model to be 10 seconds. This is clearly unreasonable; thus, in order for this model to be useful, some way must be found to extend the strength of the signal for 10 seconds.

One way to extend the signal lifetime within this model would be to consider a model of correlated spins rather than the one of an independent gas of spins described here. One could do this in such a way that there are not actually N definite spins interacting with the circuit, but rather N effective spins. It is possible that this sort of coherence could counteract the dissipation in the resistor and extend the lifetime of the revivals.

In conclusion, the behavior of the ^{3}He atoms is highly suggestive of quantum revivals. The exact mechanism for these revivals is likely more complicated than the simple model presented in this report, but this simple model seems to capture the essence of the experimental behavior.

ACKNOWLEDGMENTS

I am greatly indebted to Jung Hahm for his help in understanding the details of this experiment.

REFERENCES

1. JAYNES, E. T. & F. W. CUMMINGS. 1963. Proc. IEEE **51:** 89.
2. EBERLY, J. H., N. B. NAROZHNY & J. J. SANCHEZ-MONDRAGON. 1980. Phys. Rev. Lett. **44:** 1323.
3. NUNES, G., JR. 1991. Longitudinal spin diffusion and nonlinear spin dynamics in ^{3}He-^{4}He solutions. Cornell University thesis.
4. NUNES, G., JR., C. JIN, D. L. HAWTHORNE, A. M. PUTNAM & D. M. LEE. 1992. Phys. Rev. B **46**(14): 9082–9102.
5. WALLS, D. F. 1971. J. Phys. A **4:** 813.
6. YOO, H-I. & J. H. EBERLY. 1985. Phys. Rep. **118**(5): 239–247.

Interferometry with Atoms and Molecules[a]

DAVID E. PRITCHARD,[b] MICHAEL S. CHAPMAN,[b]
CHRISTOPHER R. EKSTROM,[b] TROY D. HAMMOND,[b]
JÖRG SCHMIEDMAYER,[c] ALAN LENEF,[b]
RICHARD RUBENSTEIN,[b] AND STEFAN WEHINGER[c]

[b]*Department of Physics*
and
Research Laboratory of Electronics
Massachusetts Institute of Technology
Cambridge, Massachusetts 02139

[c]*Institut für Experimentalphysik*
Universität Innsbruck
A-6020 Innsbruck, Austria

INTRODUCTION

Since the first demonstrations of matter wave interferometry with electrons[1–5] and neutrons,[6–8] matter wave interferometry carried out with these elementary particles has been a central tool for investigating many fundamental aspects of quantum mechanics[9–14] and for making new types of precision measurements.[15] New nanofabrication technology[16,17] and applications of the mechanical forces of light to atoms[18–22] have more recently led to the development of atom optics that manipulate atom de Broglie waves coherently.[16,19,23–41]

Atom interferometers, in particular, are having a significant impact on applications to precision and fundamental measurements. Recent experiments by our group using fabricated gratings exploit a physical separation between interfering matter waves to make precision measurements of absolute energy shifts of individual states. This is in contrast to resonance and spectroscopic techniques that rely on energy differences between states. Using this separated path approach, we have made a measurement of the atomic polarizability of sodium with a precision exceeding 20 times the best previous direct measurements.[42,43] Similar experiments have demonstrated rephasing of different magnetic sublevels from their different Zeeman interactions[44,45] and determinations of the *real* part of the forward scattering amplitude for various gases for the first time.[46] Using Raman pulses to spatially separate

[a]This work was supported by the Army Research Office (Contract Nos. DAAL03-89-K-0082 and ASSERT 29970-PH-AAS), the Office of Naval Research (Contract No. N00014-89-J-1207), and the Joint Services Electronics Program (Contract No. DAAL03-89-C-0001). Support was also provided by the Universität Innsbruck (to J. Schmiedmayer and S. Wehinger), a National Science Foundation graduate fellowship (to T. D. Hammond), and an APART Fellowship of the Austrian Academy of Sciences (to J. Schmiedmayer).

two internal states of the atom, Weiss *et al.*[47] have made measurements of the photon recoil of cesium atoms with the objective of improving the precision of the fine-structure constant. This configuration has also been used to measure fringe shifts from the local gravitational acceleration g.[37] Related experiments using an optical Ramsey fringe configuration have permitted a measurement of the Sagnac phase of atoms.[38] Experimental demonstrations of the scalar Aharonov-Bohm effect have also been performed using atom interferometry[48] and topological effects have been observed in a longitudinal Stern-Gerlach atom interferometer.[49] Optical Ramsey fringe interferometry has also been used to measure the phase shift due to the ac Stark effect.[39,50,51] The shift in fringes from the dc Stark effect has also been observed in a double-slit apparatus using laser-cooled atoms using a gradient electric field,[52] but without the precision inherent in our separated path measurements.

The use of atoms, and now molecules,[53,54] in quantum interferometry experiments raises questions regarding the possible limitations to observing interference with complex particles. We consider the consequences of the particle's size exceeding its de Broglie wavelength, limitations on particle size for observing interference from physical constraints in the interferometer, and the consequences of interfering particles containing a large number of unselected internal states. We performed experiments using our three-grating, achromatic interferometer to investigate some of these limitations imposed by complex interfering particles. High contrast fringes of both atoms and molecules were observed even though the de Broglie wavelength of the atoms ($\lambda \approx 0.16$ Å) was much smaller than the physical size of the sodium atoms (3 Å).[42] Interference fringes with molecules were also observed with contrast nearly identical to that for atoms, showing that a large number of internal states do not reduce contrast.[53] Illuminating the atoms in the interferometer with a laser beam, we observed both loss of fringe contrast and contrast revivals when a single resonant photon was scattered from the two interfering atomic paths.[55] We also discuss applications to atomic and molecular physics and fundamental measurements in quantum mechanics. This includes precision measurements of the polarizability of atomic sodium and the index of refraction of a gas for sodium matter waves. Future experiments are discussed, including accurate tests of predicted topological or geometric phase shifts, such as Berry's,[56] or the Aharonov-Casher (AC) phase.[57]

TABLE 1 summarizes some of the important characteristics of atom interferometry. Most notable for atoms are large source brightness, the wide variety of atomic interactions that can be applied and the corresponding variety of atom optic components, the broad range of atomic parameters, and the ability to apply interferometric approaches to make completely new types of precision measurements in atomic and molecular physics. Difficulties with atoms include complicated fabrication or implementation of optical elements, more complicated interpretation of fundamental experiments, lack of nondiffractive beam-splitting elements, and (currently) small diffraction angles. Although neutron and electron interferometry have provided a wealth of important results,[7,11–13,58–65] neutron sources have lower intensity ($\sim 10^{15}$ s^{-1} sr^{-1}) and both types of interferometry employ effectively elementary particles with only two accessible internal states.

THE MIT ATOM INTERFEROMETER

Our interferometer is a three-grating Mach-Zender type with a diamond path configuration (FIGURE 1) that produces a robust white fringe.[32] We use 200-nm-period silicon nitride transmission gratings that separate the centers of the interfering beams by 60 μm with argon carrier gas. The fabrication of these gratings using nanometer-scale electron beam writing techniques is described in references 24 and 32. A supersonic sodium beam, typically using argon as the carrier gas, provides an intense source of atoms ($>10^{18}$ atoms s^{-1} sr^{-1}, with an effective source diameter of ~ 500 μm) with $v = 1000$ m/s and a corresponding de Broglie wavelength of $\lambda_{dB} \sim 0.16$ Å. The full width at half-maximum (FWHM) of our velocity distribution is

TABLE 1. Characteristics of Atom Interferometers[a]

Sources	Supersonic, effusive, laser-cooled, and trapped atoms
Source Brightness	Up to 10^{20} s^{-1} cm^{-2} sr^{-1} (supersonic)
de Broglie Wavelength	~ 0.1–5.0 Å (effusive and supersonic sources) ~ 100 nm (laser-cooled atoms)
Useful Atomic Properties	
Mass	1–250 amu
Magnetic Moment	$\mu \approx 0$–10^4 μ_N
Polarizabilities	$\alpha \approx 0.5$–100 Å^3
Types	Bosons and fermions (density effects and statistics)
Other Properties/Applications	Rich internal structure Interacts with light and RF Molecules Atomic physics—precision measurements, collisions, fundamental measurements
Approaches with Gratings	Fabricated ($\Delta d/d \sim 10^{-4}$): flexible design, species-independent, ground state interactions Crystal surfaces ($\Delta d/d \sim 10^{-7}$): robust, ground state only (not yet realized) Light gratings ($\Delta d/d \sim 10^{-10}$): species-specific, internal state–dependent, easily integrated with laser-cooled and trapped atoms, high precision, with timed gratings

[a]The phase coherence $\Delta d/d$ is the ratio of deviation of the grating element Δd to the grating period d.

$\Delta v/v < 8\%$ at 3-bar Ar pressure. The mean velocity (de Broglie wavelength) of the beam can be changed by using different seed gases. A rhenium hot-wire detector provides high detection efficiency, a response time of ~ 1 ms, and a background as low as <10 cps. The fringe amplitude of the interferometer is greater than 2000 cps with close to 45% contrast (FIGURE 2), allowing a phase determination of less than 10 milliradians in 1 minute.

As shown in FIGURE 1, an interaction region or septum consisting of a 10-μm-stretched metal foil can be inserted behind the second grating so that portions of the atom wave in the two arms of the interferometer pass on opposite sides of the foil. Insertion of the barrier typically produces less than a 10% reduction in contrast with argon carrier gas. The physical separation of paths in our interferometer is unique

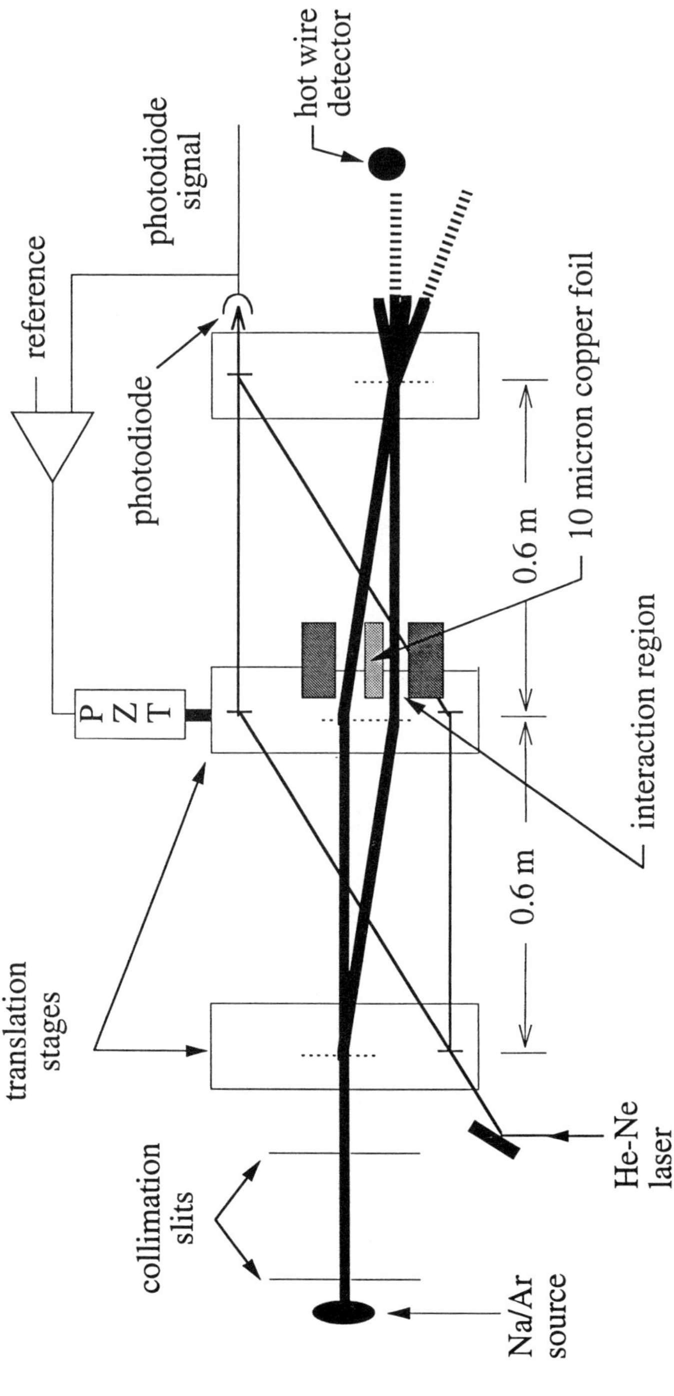

FIGURE 1. Schematic of our interferometer and interaction region. Vertical dashed lines are 200-nm-period diffraction gratings. The interaction region is inserted between the atom waves and consists of a 10-micron copper foil suspended between the side plates.

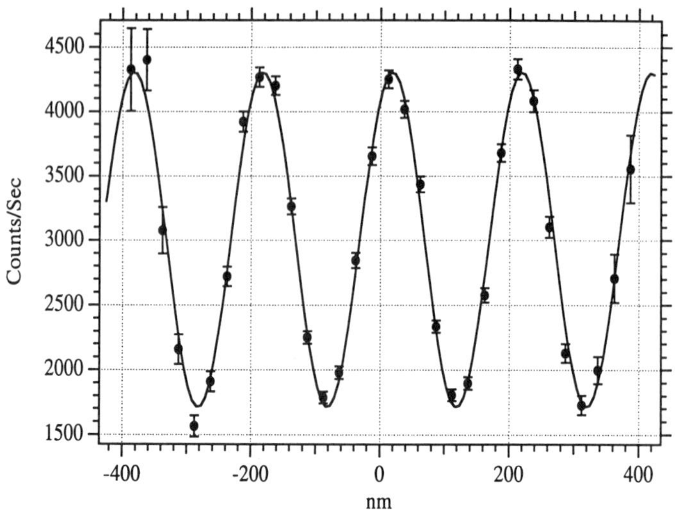

FIGURE 2. Interference pattern from 40 seconds of data (1 second per point). A constant background of 200 cps has been subtracted.

and of central importance to many experiments, as will be discussed in greater detail later.

QUANTUM INTERFERENCE WITH COMPLEX PARTICLES

An important goal of our work is the investigation of fundamental aspects of quantum interference.[66,67] Experimentally, we have already shown that interference can be obtained even when the size of the particle (~ 3 Å for Na atoms) is larger than its de Broglie wavelength (0.16 Å).[32,42] Interference was not degraded by using molecules, in spite of their large numbers of internal states that are closely spaced in energy.[53] The situation is different when we couple internal states to a reservoir. Recent work by our group has shown that the contrast and phase of fringes produced when scattering a photon from interfering matter waves depend upon the mean separation of paths, exhibiting both revivals and loss of contrast as a function of path separation. We discuss these and other possible limitations to producing interference in atom interferometers.

Particle Size and Internal States

Interference with composite particles whose de Broglie wavelength λ_{dB} is smaller than the particle size is expected to occur without degradation of fringe contrast because the wave functions of the bound constituent particles develop a common phase that corresponds to the motion of the center of mass. The formation of fringes depends upon the internal states of the atom along a given path being the same (or at least nonorthogonal in Hilbert space) or upon both paths undergoing the same transition.

However, physical size can limit the ability to obtain interference fringes in a more practical sense. Consider the interferometer in FIGURE 3, in which particles pass through three gratings. Particles of diameter s certainly must be smaller than the grating period d in order to pass between the grating bars. To produce interference fringes with large grating slits, the single-slit diffraction angle $\theta_{dif} \approx \lambda_{dB}/d$ from the first grating must still be large enough to illuminate the adjacent slits in the second grating as shown in FIGURE 3. An interferometer operating in this near-field regime is called a Talbot interferometer and is characterized by having diffraction orders that cannot be spatially separated. If the diffraction angle were reduced further, the pattern at the second grating would be a classical Moiré pattern and we would not observe interference fringes. (In the Moiré case, particles traverse essentially line-of-sight trajectories throughout the interferometer so that one could not recombine different diffracted paths and observe a phase shift.) The condition that the diffraction pattern from a single slit at the first grating extend over at least one pair of adjacent slits at the second grating implies that $L \gg d^2/\lambda_{dB} > s^2/\lambda_{dB}$, where L is the spacing between the first and second gratings (recalling s is the particle diameter and d is the grating period). The quantity $L_{Tal} = d^2/\lambda_{dB}$ is called the Talbot length, which arises in the theory of near-field imaging of periodic structures.[68,69] Using the de Broglie relation $p = h/\lambda_{dB}$ and assuming a spherical particle with mass density ρ, we arrive at the inequality

$$s \ll \left(\frac{8ht}{\pi\rho}\right)^{1/5}, \tag{1}$$

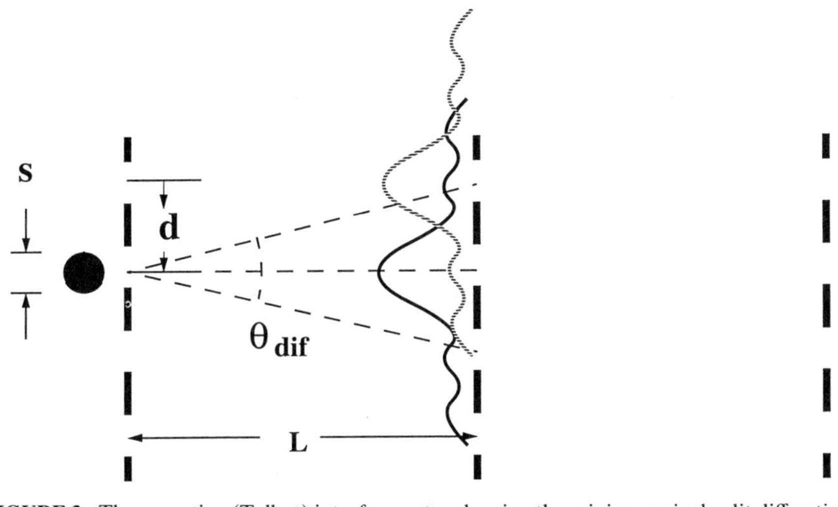

FIGURE 3. Three-grating (Talbot) interferometer showing the minimum single-slit diffraction angle θ_{dif} required to produce fringes as depicted by dashed lines. The particle size s is limited by the slit size.

where t is the traversal time between gratings. The reason that the traversal time and not the grating separation L appears in equation 1 is that the maximum width of the single-slit diffraction pattern that appears at the second grating is limited essentially by the momentum uncertainty, $\Delta p \approx h/d$, which is the grating momentum.

Equation 1 states that it would be impossible to produce interference fringes with a diffractive interferometer using large particles unless the traversal time was long enough. For example, an interferometer with a one-second transit time (between gratings) using sodium particles would require particle diameters smaller than 70 nm or clusters of about 8×10^7 sodium atoms in order to produce interference fringes. Even if we waited one year, we would observe interference from sodium clusters with a diameter of less than about 2 μm or 1.9×10^{12} sodium atoms.

Optics and Interferometry with Molecules

By using mechanical light forces together with our fabricated grating technology, we have obtained molecular diffraction and interference with a contrast essentially equal to that of atoms. These measurements demonstrate that interference is not affected by large numbers of internal states so long as the evolution of the molecular internal states is not dependent upon path in the interferometer.

In these experiments, an intense beam of sodium atoms and dimers is produced in an argon- or krypton-seeded supersonic expansion through a 70-μm nozzle.[53] Pressure and temperature of our source were optimized to yield Na-dimer concentrations of nearly 30% of the detected beam intensity with an FWHM velocity distribution of about 8%. This corresponds to a translational temperature of about 2 K (0.17 meV) and rotational and vibrational temperatures at roughly 30 K (2.6 meV) and ~100 K (8.7 meV), respectively.[70] The much larger 0.8-eV binding energy of Na_2 (see reference 70) implies that the molecules are essentially in their vibrational ground state; however, the population of the (j, m_j) rotational states is on the order of 10^3.

To achieve larger diffraction angles, experiments were performed using Kr-seeded beams that have a mean velocity of about 750 m/s ($\lambda_{Na} = 0.23$ Å, $\lambda_{Na_2} = 0.115$ Å), compared to 1000 m/s ($\lambda_{Na} = 0.16$ Å, $\lambda_{Na_2} = 0.08$ Å) using argon. The Na atoms were removed from the beam with light scattering forces. The atoms were pushed sideways out of the collimated beam with a resonant "pushing laser beam" of circularly polarized light resonant with the $3S_{1/2}, F = 2 \rightarrow 3P_{3/2}, F' = 3$ transition of atomic Na. FIGURE 4a is a diffraction pattern obtained with a microfabricated grating with a 100-nm period, showing well-separated atomic peaks and molecular peaks with half the diffraction angles. FIGURE 4b shows the molecular peaks when the pushing laser is applied.

We have demonstrated a molecular interferometer consisting of three 200-nm-period nanofabricated diffraction gratings using a Mach-Zender geometry employed in our atom interferometer. We have checked that the observed interference is from molecules using two methods: (i) we introduced a laser (decoherence laser) that would destroy any atom interference pattern by scattering resonant light from the split atomic wave function inside the interferometer and observed no change; (ii) we observed that the best contrast of the molecular interference pattern is closer to the

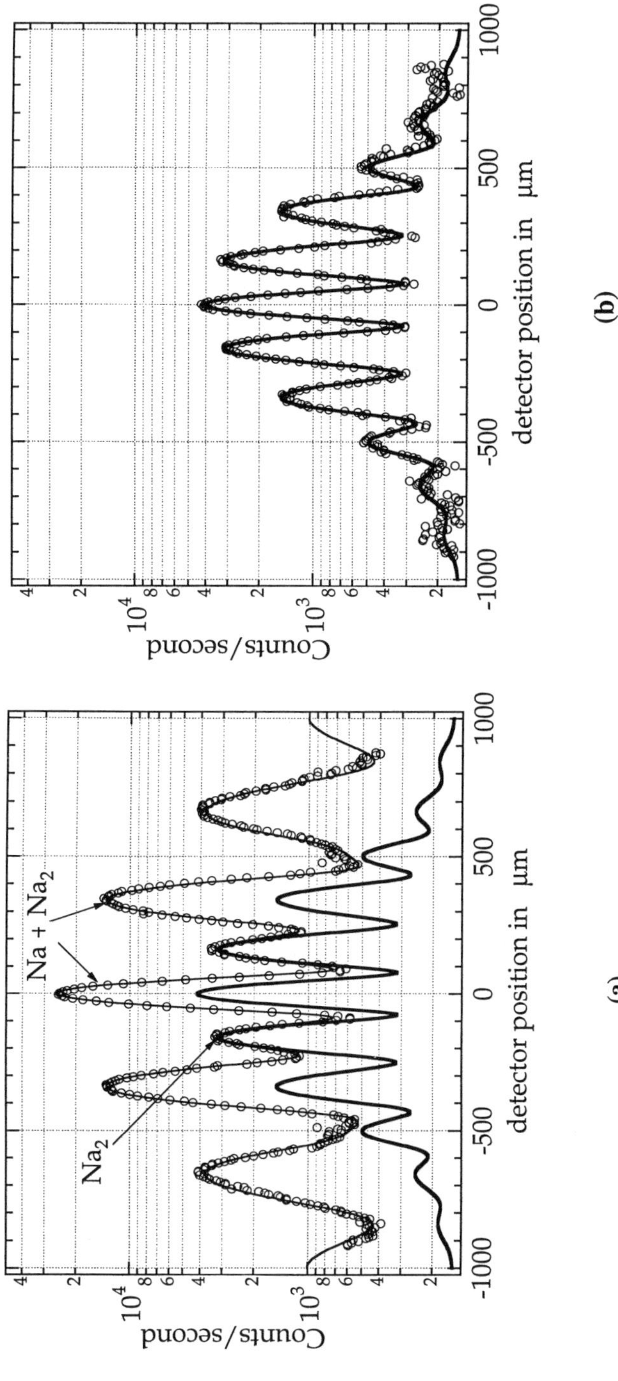

FIGURE 4. Molecular diffraction: (a) Diffraction of the mixed atom-molecule beam from our Kr-seeded source by a 100-nm-period diffraction grating. The thick solid line is the fit to the Na_2 diffraction pattern and also indicates that the diffraction gratings have a 30% open fraction. A fit to the total diffraction pattern (thin solid line) indicates that 16.5% of the intensity is from molecules. (b) Diffraction of 750-m/s sodium dimers compared with the molecular component (Kr as a seed gas). For this measurement, the deflecting laser was on with an intensity of $\sim 1\ \mathrm{W/cm^2}$.

collimation axis, which would be expected from the reduced diffraction angle with molecules.

Coherence Loss by Photon Scattering

We have investigated the loss of coherence between different diffracted wave function components when resonant single photons are scattered from the interfering atoms passing through our three-grating interferometer. In our experiment (FIGURE 5), single photons are scattered from atoms some distance behind the first grating where the path separation is x_0. The fringe contrast and phase are measured as a function of separation. Complementarity would lead to the conclusion that

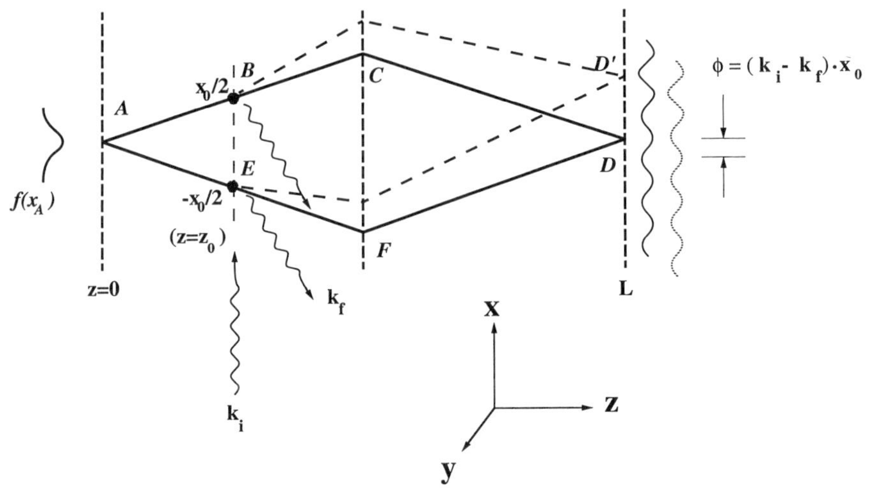

FIGURE 5. Light scattering experiment in a white fringe interferometer: Scattering a photon produces a phase shift in each de Broglie wave path that depends upon the path difference of the photon. The resulting photon-atom state is an entanglement of scattered photons and phase difference between atom paths. The paths ABD' and AED' correspond to the classical paths of the centers of the atom wave packets after receiving a photon recoil kick.

fringe contrast must disappear when the mean separation between the two possible paths is greater than the wavelength of light because, in principle, one could observe from which path the photon was scattered. The physical origin of the loss of coherence is that the dynamics of measuring the position of the atom by scattering a photon of wavelength λ_p produces a corresponding position uncertainty λ_p in the atom. This is simply a consequence of the momentum uncertainty $\sim h/\lambda_p$ produced by the photon recoil as considered originally by Heisenberg.[71] Thus, the wave function becomes uncorrelated on length scales exceeding λ_p, resulting in a loss of fringe contrast when the peak separations are of the same order. Zurek has considered specific cases of quantum measurements, which are closely related to our experiment, in which interactions with the environment produce a loss of coherence or "decoherence" of the measuring apparatus.[67,72]

Our photon scattering experiment shown in FIGURE 5 has two essential features: first is that the effects of displacement of atomic trajectories by the photon-momentum kicks and shifts in the phase of the atom fringes are decoupled because the displacement of the atomic center of mass over the length of the interferometer from the recoil kicks is much larger than the displacement of the fringes at the third grating, as will be shown; second is that the transverse width of the de Broglie waves at the point of scattering can be much larger than λ_p. Experiments in which fringe contrast is reduced by a spontaneously emitted photon after an atom passes through a Young's double slit have been considered theoretically by Sleator[73] and Walls.[74] Clauser[75] and Sterr[39] have also demonstrated complete loss of fringe contrast in an interferometer by light scattering when the path separation was much greater than an optical wavelength. The Konstanz group[76] has recently performed experiments in which atoms initially excited in optical standing waves spontaneously decay, reducing the contrast in their diffraction patterns. However, the loss of contrast in both the experiments and the theoretical analyses could be interpreted as a result of smearing of fringe or diffraction patterns caused by photon recoil displacement of the atomic center of mass.

We illustrate the problem theoretically by evaluating the corresponding atom wave function phase difference between both paths in FIGURE 5 at D when a photon scattered in a particular direction. We later sum over all scattering directions to obtain the total fringe contrast and phase. Scattering a photon from the atom imparts an additional linear phase variation onto the atom wave function. This additional quantum mechanical phase is consistent with momentum conservation and arises from the interaction of the atomic dipole with the (incident and scattered) light field at a particular point in space.

For simplicity, we consider a wave packet incident on the first grating with a well-defined spatial peak and momentum that splits into two packets that propagate along the various $+1$ and -1 grating orders in the interferometer. The symmetric case is considered for simplicity, although the theory developed also holds for our diamond-shaped interferometer in FIGURE 1. The phase difference at D in FIGURE 5 is the difference in phases developed along the paths $ABCD$ and $AEFD$, corresponding to the $+1$ and -1 diffracted orders from the first grating, respectively. Although the center of the atom wave packet would follow the paths to the position D' after scattering, we can still evaluate the phase difference at D, which corresponds to the unperturbed case (no scattering). We consider the amplitudes at B and E after scattering at $z = z_0$. If the initial amplitude at any x along the $z = z_0$ axis were $U_{\text{inc}}(x)$, then after scattering the amplitude would be

$$U_{\text{scat}}(x) = e^{i(\boldsymbol{k}_i - \boldsymbol{k}_f)\cdot \boldsymbol{x}} U_{\text{inc}}(x) \equiv e^{i\Delta k x} U_{\text{inc}}(x), \tag{2}$$

where $\boldsymbol{k}_i$ and $\boldsymbol{k}_f$ are the incident and scattered photon wave vectors and Δk is the difference in the incident and scattered wave vectors in the x-direction. The corresponding amplitudes at B and E are then

$$U(AB)_{\text{scat}} = e^{i\Delta k x_0/2} U_{\text{inc}}(AB) = e^{i\Delta k x_0/2} e^{iS_{\text{cl}}[AB]/\hbar} \tag{3a}$$

and

$$U(AE)_{\text{scat}} = e^{-i\Delta kx_0/2}U_{\text{inc}}(AE) = e^{-i\Delta kx_0/2}e^{iS_{\text{cl}}[AE]/\hbar}, \quad \textbf{(3b)}$$

where $S_{\text{cl}}[AB]$ and $S_{\text{cl}}[AE]$ are the classical actions along the paths AB and AE.

To find the amplitudes at D, we have to sum over all the Feynman paths going through the second grating from B to D and from E to D. Additionally, we would have to sum over the spatial extent of the incident wave packets near B and D. However, we would expect that it is possible to evaluate the phase at the center of the wave packets as expressed by equations 3a and 3b. This is reasonable even if the transverse (x-direction) extent of the wave packets were larger than λ_p because the wave packets are identical, except for being displaced by x_0. Thus, we would expect the average phase difference between them to be Δkx_0. This averaging procedure can be performed more rigorously by summing over the Feynman paths along the transverse extent of the wave packets as shown in reference 55, showing that only the phase difference between the centers of the wave packets is important.

To determine the amplitudes at D, we can therefore use the amplitudes (or propagators) through the second grating only along the unperturbed paths BCD and EFD. The resulting combined amplitude at D is then

$$\begin{aligned} U(D) &= \left(\frac{1}{\sqrt{2}}\right)[e^{i\Delta kx_0/2}U(AB)U(BCD) + e^{-i\Delta kx_0/2}U(AE)U(EFD)] \\ &= \left(\frac{1}{\sqrt{2}}\right)[e^{i\Delta kx_0/2}U(ACD) + e^{-i\Delta kx_0/2}U(AFD)], \end{aligned} \quad \textbf{(4)}$$

where $U(ACD)$ and $U(AFD)$ are the total amplitudes along the unperturbed paths through the second grating when no scattering is present. The unperturbed amplitudes through the second grating $U(BCD)$ and $U(EFD)$ depend upon the classical actions along their respective paths and upon an additional phase factor that comes from the grating.[55]

From the result of equation 4, we see that the phase difference of the interfering de Broglie waves at the third grating is identical to the unperturbed case, except for an additional phase difference imparted by the interaction with the light. Thus, the *phase shift* depends only upon the mean separation between diffracted de Broglie waves and not upon the details of their envelopes as indicated in the previous discussion. Note, however, that the displacement of the peaks of the wave packet *envelopes* near D from the transverse photon recoil kicks is $\Delta x_{\text{recoil}} \approx \hbar\Delta kL/mv_g$, which is much larger than the displacement of the fringes at the third grating, $\Delta x_{\text{fringe}} = \hbar\Delta kx_0/2P_g = \hbar\Delta kz_0/mv_g$ (here, P_g is the grating momentum). This difference in the center-of-mass displacement and the fringe shift is depicted in FIGURE 5 and is a consequence of scattering the photon close to the first grating rather than the third grating. Thus, the effect of the transverse recoil is decoupled from the fringe shift imparted by the light.

To obtain the total fringe contrast and phase shift for the interference pattern at

D, we sum over all possible scattering directions for the photon. This will form an entanglement of the photon and atom states:

$$\psi(D) = \int dk_f U(D, \Delta k) S(\mathbf{k}_i, \mathbf{k}_f) |\mathbf{k}_f\rangle$$
$$= \left(\frac{1}{\sqrt{2}}\right) \int dk_f \{e^{i\Delta kx_0/2} U(ACD) + e^{-i\Delta kx_0/2} U(AFD)\} S(\mathbf{k}_i, \mathbf{k}_f) |\mathbf{k}_f\rangle, \quad (5)$$

where $S(\mathbf{k}_i, \mathbf{k}_f)$ is the scattering matrix for an incident photon with wave vector $\mathbf{k}_i$ and an emitted photon with wave vector $\mathbf{k}_f$ in a state $|\mathbf{k}_f\rangle$. We have explicitly included the Δk dependence in $U(D)$. The magnitude-squared of the scattering matrix, $|S(\mathbf{k}_i, \mathbf{k}_f)|^2$, gives the probability of photon emission in a direction $\hat{\mathbf{k}}_f$ and corresponds to the dipole radiation pattern. Averaging over all scattering directions in equation 5, we obtain the fringe intensity I at D:

$$I = \mathrm{Tr}_{\text{field}}[|\psi(D)\rangle\langle\psi(D)|] = \int dk_f |S(\mathbf{k}_i, \mathbf{k}_f)|^2 |U(D, \Delta k)|^2$$
$$= \left(\frac{1}{2}\right) \int dk_f |S(\mathbf{k}_i, \mathbf{k}_f)|^2 |e^{i\Delta kx_0/2} U(ACD) + e^{-i\Delta kx_0/2} U(AFD)|^2$$
$$= \int dk_f |S(\mathbf{k}_i, \mathbf{k}_f)|^2 [1 + \cos(\phi + \Delta kx_0)]. \quad (6)$$

The phase ϕ is the unperturbed phase of the fringes, where $e^{i\phi} = U^*(ACD)U(AFD)$. Equation 6 explicitly shows that the fringe pattern is an incoherent superposition of fringes shifted by Δkx_0 from the unperturbed phase ϕ. From equations 5 and 6, we see that the loss of coherence is a consequence of the entanglement of the atom and photon states. By detecting only atoms, we lose the two-particle phase relationship established between the photon and the atom during the scattering process.

Destruction of Interference by Scattered Photons

We have investigated the destruction of interference by scattering predominantly single photons in our interferometer using the configuration shown in FIGURE 6. Atomic sodium de Broglie waves are split into two paths by the first grating and are resonantly excited with circular polarized π pulses applied some distance behind the first grating on the Na ($3S_{1/2}, F = 2, m_F = 2 \rightarrow 3P_{3/2}, F' = 3, m'_F = 3$) cycling transition. Varying the z-position of the excitation laser allows the separation x_0 between the diffracted de Broglie wave components to be changed. Optical pumping prepares atoms on the $F = 2, m_F = 2$ stretched state so that excitation and spontaneous decay take place within a closed two-state system. The excitation laser spot that intersects the atom beam has a field amplitude FWHM of about 15 μm. This corresponds to an atom transient time across the laser spot of about 5 ns achieved with He carrier gas. This transient time is much smaller than the 16-ns lifetime of the $3P_{3/2}$ state, so single spontaneous emission events occur with high probability.

A plot of contrast versus separation x_0 between the two interfering beams is shown in FIGURE 7. These data show an overall decay of contrast with increasing separation, falling rapidly when the separation x_0 exceeds more than 1/4 of an optical wavelength. These data also show strong revivals in contrast that have approximately

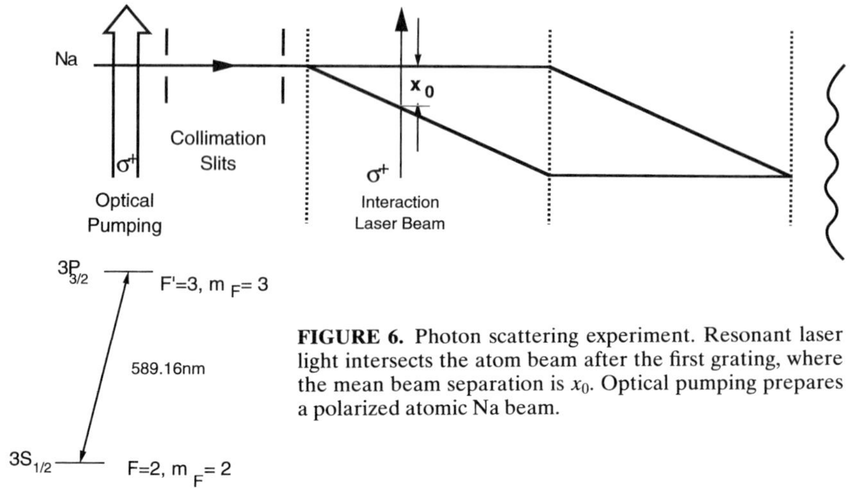

FIGURE 6. Photon scattering experiment. Resonant laser light intersects the atom beam after the first grating, where the mean beam separation is x_0. Optical pumping prepares a polarized atomic Na beam.

periodic maxima. The figure also shows the phases of the fringes, which are nearly periodic with discontinuities approximately at every $x_0 = n\lambda_p/2$.

These data can be understood by considering the phase shift of the fringes produced by scattering a photon in a particular direction. According to equation 6, the phase shift is $\varphi = \Delta k x_0$, where $0 < \hbar\Delta k < 2\hbar k_\mathrm{p}$ is the transverse component (x-direction) of the one-photon recoil momentum of the atom and $\hbar k_\mathrm{p}$ is the photon momentum. The transverse recoil momentum distribution for linear or circularly polarized light is shown in the inset of FIGURE 7. If the distribution were flat in the inset, zeros in contrast would occur at $k_\mathrm{p}x_0 = n\pi$, identical to a single-slit diffraction pattern. The averaged phase near these minima jumps discontinuously from π to zero because we are essentially averaging phasors that are uniformly distributed over just slightly less or slightly more than the entire unit circle. When $k_\mathrm{p}x_0 \gg 2\pi$, the contrast becomes small because the distribution of large phase shifts tends to average out the fringe contributions. The data in FIGURE 7 show these contrast minima close to $k_\mathrm{p}x_0 = n\pi$, slightly shifted by the nonuniform momentum recoil distribution.

A fit to our data from a calculation based on equation 6 is shown in FIGURE 7 and includes contributions from zero- and two-photon scattering processes. The zero-photon processes contribute to a nonzero background contrast at large separations, whereas the two-photon processes further reduce the contrast and smooth the phase discontinuities because of the additional diffusion that occurs with two-photon phase shifts. As emphasized in the theoretical discussion, these results clearly distinguish between the quantum mechanical phase shift induced by the scattering process and the classical recoil momentum that is manifested as a displacement of the envelope of the fringes at the third grating in our experiment. Thus, we have made a direct measurement of the dephasing induced by a spontaneous scattering process.

Delayed Choice

In quantum interferometry, a measurement that distinguishes which path a particle traverses must correspondingly produce a loss of fringe contrast as emphasized by Bohr's discussion of complementarity.[66] The interaction of the path detector with the particle destroys the complementary information needed to produce fringes as demonstrated by Einstein's recoiling slit and Feynman's microscope[77] and also from our photon scattering experiment.[55] Although complementarity forbids the simultaneous observation of path information and interference, Wheeler pointed out that one can wait until the last possible instant to observe either fringes or path information in what he called a delayed-choice experiment.[78] For example, two detectors could be placed after the third grating in FIGURE 1 to intercept the two possible interferometer paths. If the third grating were left in, a fringe pattern would be observed by the detectors from an applied phase shift. On the other hand, the third grating could be removed just before the atom (wave packet) reaches the third grating so that the detectors record the "route" traversed by the atom. The ability to choose between either mode of measurement at the last instant emphasizes that path information always exists in both "routes" simultaneously, whereas the particle itself

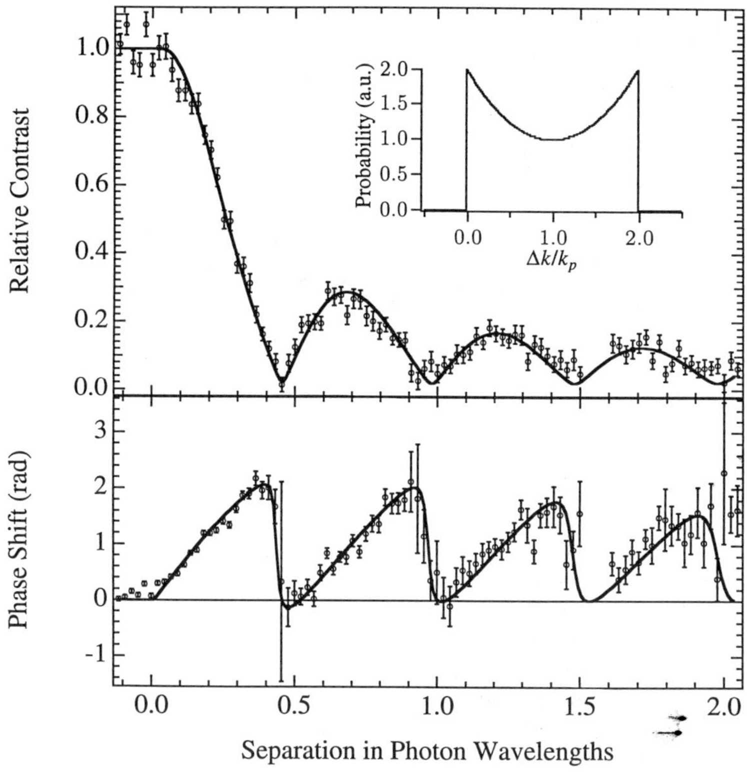

FIGURE 7. (Top) Experimental data showing the ratio of contrast with the scattering laser on to that with the laser off as a function of atom path separation. (Bottom) Experimental data showing the phase difference between fringes with the laser on to that with the laser off as a function of atom path separation. The solid line is a theoretical fit with 83% one-photon, 13% two-photon, and 4% zero-photon processes. (Inset) One-photon transverse (x-direction) recoil momentum distribution for circular polarized light normally incident to the atomic beam.

can be detected in one "route" only. The delayed-choice experiment emphasizes the nonlocal character of quantum mechanics because the detection of the particle in one detector "instantaneously" excludes it from the other.[65]

We consider a type of delayed-choice experiment in which one can *record* path information by entanglement with the internal states of the particle, but the choice can be made later to either *detect* fringes or *detect* path information (FIGURE 8). In accordance with complementarity, we expect interference fringes to disappear when path information is perfectly recorded; loss in contrast is a consequence of orthogonality of the path-dependent internal states. However, the entanglement of the

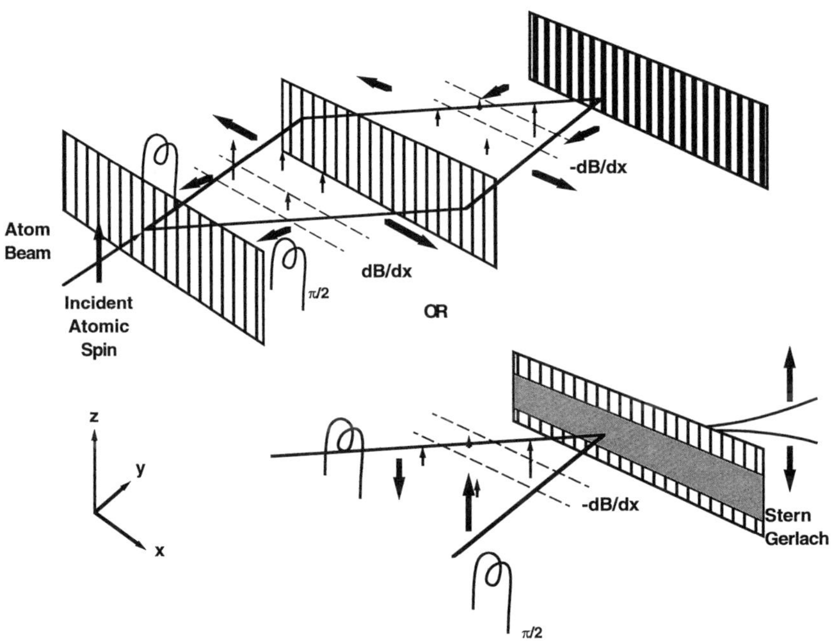

FIGURE 8. Two-state example of a delayed-choice experiment with atoms. Using a combination of gradient fields and $\pi/2$ pulses, the atoms can be prepared in identical spin states or orthogonal spin states before detection. Path information is recorded in both cases, but reversed in the first case. Spins are shown in an unperturbed rotating frame.

particle's internal states with the path can be reversed by applying a second interaction (unitary transformation). In this case, fringes are restored, but with a subsequent loss of path information. An experiment that can regain interference by "erasing" path-distinguishing information has been called a "quantum eraser".[79,80] Closely related experiments using neutron spins to record and undo path information have been performed using *continuous* beams.[64,81] Similar delayed-choice experiments using chopped neutron sources have been proposed by Badurek *et al.*,[61] with some experimental progress being attained in this direction by Rauch *et al.*[82] and Jacobson *et al.*[63]

The necessity to have entanglement with a measuring apparatus *and* entanglement with a macroscopic particle detector to rapidly dephase the correlations between the measuring apparatus (the internal states of the particle) and the path states in a particular preferred basis has been emphasized by Zurek in his theories of measurement.[67,72] The distinction of these two components of the measurement process is a feature of our proposed delayed-choice experiment.

Atoms are particularly useful in this type of experiment because an atom delayed-choice experiment can be generalized to record multiple paths using multiple states. For simplicity, we discuss it in the framework of a spin-1/2 particle (two-state system) experiment, which is illustrated in FIGURE 8. We send a pulsed beam of $S = 1/2$, z-polarized atoms into a three-grating interferometer and apply a resonant $\pi/2$ pulse (x-direction) perpendicular to a uniform magnetic guide field $\boldsymbol{B}_{\text{guide}} = \hat{z}B_0$. This places the atoms in a superposition state whose spin is aligned along the $-y$-axis:

$$\varphi_{\text{I}} = A_{\text{I}}(\varphi_{\text{R}} + \varphi_{\text{L}})(-ie^{-i\omega_a T/2}|\downarrow\rangle + e^{i\omega_b T/2}|\uparrow\rangle), \tag{7}$$

where φ_{R} and φ_{L} are spatial wave functions representing the two paths in the interferometer, $\omega_{ba} = \omega_b - \omega_a$ is the precession frequency, T is the time of interaction, and A_{I} is a normalization factor that depends upon the spatial overlap of φ_{R} and φ_{L}. The spin is oriented in the $-y$-direction in the unperturbed rotating frame at a precession frequency ω_{ba}. Application of a small localized gradient field produces path-dependent Zeeman shifts, $\epsilon_{\text{R}} = \hbar\Delta/2$ and $\epsilon_{\text{L}} = -\hbar\Delta/2$, which momentarily shift the precession rates. By taking the interaction time T' with the gradient field such that $\Delta \cdot T' = \pi/2$, the atoms are now in the entangled state,

$$\varphi_{\text{II}} = A_{\text{II}}[e^{-i\pi/4}(|\downarrow\rangle + e^{-i\phi}|\uparrow\rangle)\varphi_{\text{R}} + e^{i\pi/4}(-|\downarrow\rangle + e^{i\phi}|\uparrow\rangle)\varphi_{\text{L}}], \tag{8}$$

where ϕ is the phase that would be accumulated from the precession with the guide field only. The spins on the left and right sides are now aligned parallel and antiparallel to the x-axis.

At this point, one has "recorded" path information by entangling it with the spins. We now consider two choices: (i) we apply another gradient field, opposite to the first, to align the precessing spins parallel to each other so that they can interfere again or (ii) we apply a second $\pi/2$ pulse at the correct time to align the left and right spins in either the $+z$ or $-z$ directions, respectively, to save the path information in a form useful for detection. In the first case, we apply Zeeman shifts $\epsilon_{\text{R}} = -\hbar\Delta/2$ and $\epsilon_{\text{L}} = \hbar\Delta/2$ just before the third grating, yielding the uncorrelated state with the spins aligned along $-y$ in both paths:

$$\varphi_{\text{III}a} = A_{\text{III}a}(-i|\downarrow\rangle + e^{i\phi'}|\uparrow\rangle)(\varphi_{\text{R}} + \varphi_{\text{L}}). \tag{9a}$$

Here, ϕ' is the new phase from precession due to the guide field alone. In the second case, application of a $\pi/2$ pulse just before the second gradient field yields the entangled state,

$$\varphi_{\text{III}b} = A_{\text{III}b}(|\downarrow\rangle\varphi_{\text{L}} + e^{-i\omega_{ba}T}|\uparrow\rangle\varphi_{\text{R}}). \tag{9b}$$

Here, T and T_{trav} are the $\pi/2$ pulse time and the traversal time between gradient fields, respectively. In order to align the spins along the z-axis following the $\pi/2$ pulse, the phase developed from free precession must satisfy $\omega_{ba}(T + T' + T_{\text{trav}}) = \pi/2$ or it must be adjusted appropriately by varying the RF phase. A Stern-Gerlach apparatus behind the third grating detects the final spin state and corresponding path when a second $\pi/2$ pulse is applied.

This atomic delayed-choice experiment can also be performed with pulsed lasers using Raman pulses rather than RF fields to produce the desired spin flips. This would permit the "choice" of measurement to be made on much shorter time scales. Furthermore, Hammond *et al.*[83] have proposed a scheme for atoms that uses a multiple-peaked velocity distribution to cancel velocity dispersion. This would permit the experiment to be performed with typical atomic beam velocity distributions. In addition to experimental advantages using atoms, an atom delayed-choice experiment begins to address the relationship of information and quantum mechanical behavior of complex particles.

PRECISION MEASUREMENTS OF ATOMIC AND MOLECULAR PROPERTIES

Atom and molecule interferometry with its sensitivity to phase, like the separated oscillatory fields method, will find numerous applications in precision experiments of atomic and molecular properties. Spatially separating the atomic wave function allows a direct measurement of the absolute energy shift of a *single* state with a precision attainable in resonance or spectroscopic methods because of the ability to apply uniform interaction fields to one path only.

Separated Beam Interferometry

Our atom-molecular interferometer is unique in that the two paths are completely separated at the location of the second grating. We have inserted an interaction region or septum consisting of a 10-μm-stretched metal foil behind the second grating to physically isolate the two beams that interfere at the third grating.

With this setup (FIGURE 1), we must consider the effect of an interaction in one path of the interferometer. Applying a potential $U(x)$ to one path changes the unperturbed k-vector from $k_0 = (1/\hbar)\sqrt{2mE_{\text{kin}}}$ to $k(x) = (1/\hbar)\sqrt{2m[E_{\text{kin}} - U(x)]}$. The corresponding phase difference φ between the two paths can be evaluated accurately along the two classical paths in the eikonal approximation ($E_{\text{kin}} \gg U$) to give

$$\varphi(k_0) = \int_{\text{Path 2}} k(x)\,dx - \int_{\text{Path 1}} k_0\,dx \cong \left(\frac{1}{\hbar v}\right) \int_{L_{\text{inter}}} U(x)\,dx. \qquad \textbf{(10)}$$

Here, L_{inter} is the length of the interaction region projected along the path.

The effect of the potential $U(x)$ is equivalent to a wave function traversing a medium with refractive index $n = k/k_0$. Using the eikonal approximation ($U \ll E_{\text{kin}}$), one finds $n = k/k_0 = 1 - (U/2E)$ and the phase shift $\varphi(k_0) = k_0 \int [1 - n(x)]\,dx$. In our interferometer, a phase shift of 1 rad occurs when $|1 - n| \approx 3 \times 10^{-11}$. This gives an energy sensitivity of roughly 7×10^{-14} eV in one minute (using Ar carrier gas, assuming a 10-mrad/$\sqrt{\text{min}}$ phase sensitivity).

The velocity dependence of the phase shift in equation 10 causes a reduction in contrast when large phase shifts are applied because of the averaging over different fringe contributions within the source velocity distribution. The corresponding coherence length is 0.65 Å (1.6 Å FWHM) with argon carrier gas and represents the maximum longitudinal spatial shift between the two paths that can be applied before fringe contrast drops significantly.

Precision Measurements in Atomic and Molecular Physics

Measurement of the Electric Polarizability of the Na Atom

The interaction region allows us to apply different electric or magnetic fields to portions of the atom wave on each side of the interferometer. For an alkali atom such as sodium, an applied electric field $\mathscr{E}$ generates a Stark potential $U = -\alpha\mathscr{E}^2/2$, where α is the scalar electric polarizability. FIGURE 9 shows the measured quadratic dependence of phase shift versus applied voltage for sodium atoms inter-

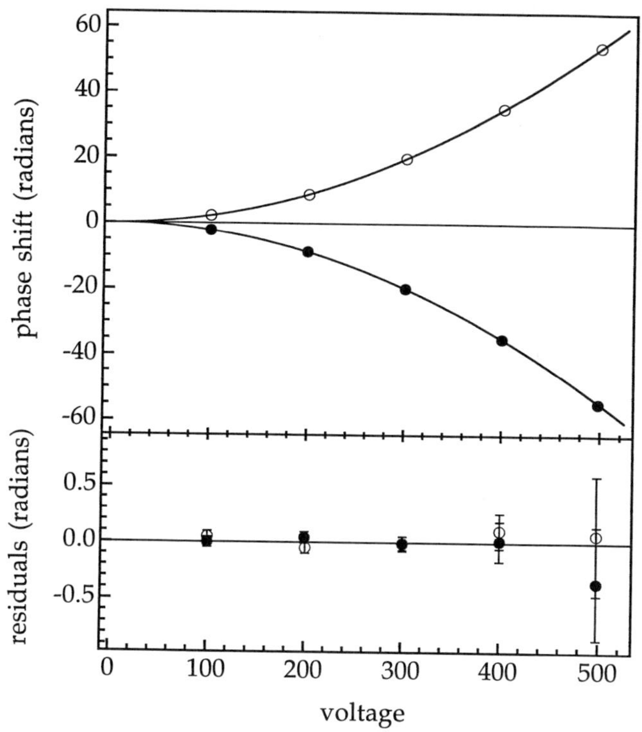

FIGURE 9. Phase shift of the interference pattern as a function of voltage applied to the left (open circles) or right (filled circles) side of the interaction region. The fit is to a quadratic and the residuals are shown in the lower graph.

acting with a very uniform electric field in our interaction region. From these phase-shift measurements, we have determined the ground state polarizability to be $24.11(6)_{\text{statistical}}(6)_{\text{systematic}} \times 10^{-24}$ cm^3. This 0.3% accuracy is about 20 times more accurate than the previous best direct measurement[84] and 6 times more precise than that of the best indirect measurements normalized to the theoretical polarizabilities of metastable helium as a reference.[85] The systematic error is dominated by uncertainties in the geometry of the interaction region and effects related to the 10%-FWHM-wide velocity distribution of our atomic beam. Our statistical error is dominated by uncertainty in the determination of the velocity distribution, by short-term instability of the phase, and to a lesser extent by counting statistics. We recently proposed a velocity multiplexing technique that promises to overcome the problems arising from the finite velocity distribution.[83]

Measuring the Refractive Index for Na Matter Waves

We have applied separated beam interferometry to measure the refractive index of a sodium de Broglie wave in gaseous media. By inserting a gas cell in one arm of an atom interferometer, we have measured both the attenuation and the phase shift of a sodium matter wave that passes through monatomic (He, Ne, Ar, Kr, Xe) (FIGURE 10 and TABLE 2) or molecular gases (N_2, CO_2, NH_3, H_2O) (see TABLE 2).[46] From these measurements, we have determined both the imaginary and *real* parts of the forward scattering amplitude, $f(k, 0)$. In analogy to light and neutron optics, the forward scattering amplitude is related to the complex index of refraction by $n = 1 + (2\pi/k_{lab}k)N \cdot f(k, 0)$, where k_{lab} is the wave vector in the lab frame, k is the wave vector in the center-of-mass frame of the collision, and N is the density of the gaseous medium.

The results of these measurements are summarized in TABLE 2. The complex

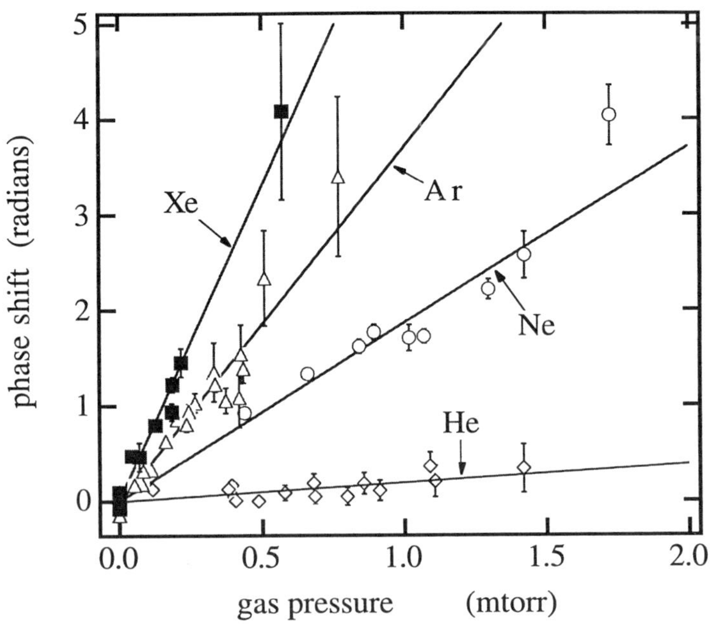

FIGURE 10. Phase shift of Na matter waves passing through He, Ne, Ar, and Xe gas as a function of the gas pressure in a 10-cm-long gas cell.

TABLE 2. Phase Shift $\Delta\varphi$, Refractive Index n, and the Ratio $\mathrm{Re}[f(k,0)]/\mathrm{Im}[f(k,0)]$ for 1000-m/s Na Atoms Passing through Various Gases at 300 K and 1-mtorr Pressure

	$\Delta\varphi$ (mtorr^{-1})	$(n-1)$ (10^{-10} mtorr^{-1})	$\frac{\mathrm{Re}(f)}{\mathrm{Im}(f)}$
He	0.50	$0.14 + 1.18i$	0.12 (2)
Ne	2.0	$0.55 + 0.56i$	0.98 (2)
Ar	3.9	$1.07 + 1.81i$	0.59 (3)
Kr	5.4	$1.51 + 2.45i$	0.62 (6)
Xe	6.5	$1.81 + 2.49i$	0.73 (3)
N_2	4.7	$0.91 + 1.39i$	0.60 (2)
NH_3	3.3	$1.30 + 2.16i$	0.65 (4)
CO_2	5.0	$1.37 + 2.21i$	0.62 (2)
H_2O	6.2	$1.71 + 2.40i$	0.72 (3)

indices of refraction given are based on estimates of gas cell pressure that are subject to large uncertainties. The ratios of the real and imaginary parts of the forward scattering amplitudes are independent of gas cell pressure and are much more accurate. Using Na_2 molecules in the interferometer, we also investigated Na-dimer scattering from neon gas to measure the index of refraction of molecular de Broglie waves passing through a gas sample in one arm of the interferometer. For Na_2 traversing Ne gas, we find a ratio of $\mathrm{Re}[f(k,0)]/\mathrm{Im}[f(k,0)] = 1.4 \pm 0.3$.

This type of experiment provides a fundamentally different probe of scattering processes and interatomic potentials. The real part of the forward scattering amplitude is particularly sensitive to the shape of the interatomic and molecular potentials, especially in long-range potentials, whereas the imaginary part depends on the overall strength of the long-range interaction. Experimentally, we confirm that the ratio of $\mathrm{Re}[f(k,0)]/\mathrm{Im}[f(k,0)]$ depends much more on the form of the interaction potential than conventional total cross-section experiments.

POSSIBLE FUTURE EXPERIMENTS

Aharonov-Casher Phase

Anandan[86] and Aharonov and Casher[57] recently proposed a topological phase, analogous to the Aharonov-Bohm effect, for the magnetic moment of a neutral particle rather than a charge. Like the Aharonov-Bohm effect, this phase shift can occur in the absence of any classical force. This effect has been studied in neutron interferometers[87] and recently in atomic systems by the group at Yale in electronic singlet molecules using a Ramsey fringe technique.[88] These groups have reported phase shifts from nuclear spin contributions on the order of 2–3 mrad. Using ground state Na atoms, with their much larger magnetic dipole moments and greater beam intensities, we will be able to achieve up to 1-radian Aharonov-Casher phase shift in our interferometer. This should greatly reduce the statistical error and will also allow us to study the predicted dependence on the dipole orientation. Varying the velocity and the velocity distribution of our Na beam will allow us to explicitly investigate precisely the nondispersive character of the Aharonov-Casher effect and will therefore give direct evidence of its topological nature.[89]

Geometric Phase

When a quantum system evolves (even adiabatically) around a cyclic path in phase space, it gains an additional phase, as first described by Berry.[90–92] Various demonstrations of this effect have been performed with photons,[93,94] in spin rotation experiments with neutrons,[95,96] in nuclear quadrupole resonance experiments with rotating samples,[97] and with atomic hydrogen.[49] By exploiting the separated beam geometry in our interferometer, one can demonstrate the geometric phase of the wave function directly by measuring the phase difference produced when the atom traverses two different paths in phase space. Such an explicit determination of the geometric phase acquired by the wave function of a massive particle was prohibited in previous experiments because of unknown phase shifts that developed between the two paths in phase space.

For a geometric phase experiment, one ideally wants a magnetic field configuration that gives the same dynamical phase, but transports the angular momentum vector of the particle on different paths in the two interfering beams. This can be accomplished using the configuration shown in FIGURE 11.[44,45] A screw coil produces a helical magnetic field that rotates in the center-of-mass frame of the atom. Opposing magnetic fields oriented in the $\pm z$-direction are produced by passing a vertical current through the metal foil in the interaction region. The resulting magnetic field approximately traces out cones in opposite directions so that the solid angles, and therefore the geometric phases, are different on either side of the foil; however, the dynamic phase contributions that depend only on the field magnitude

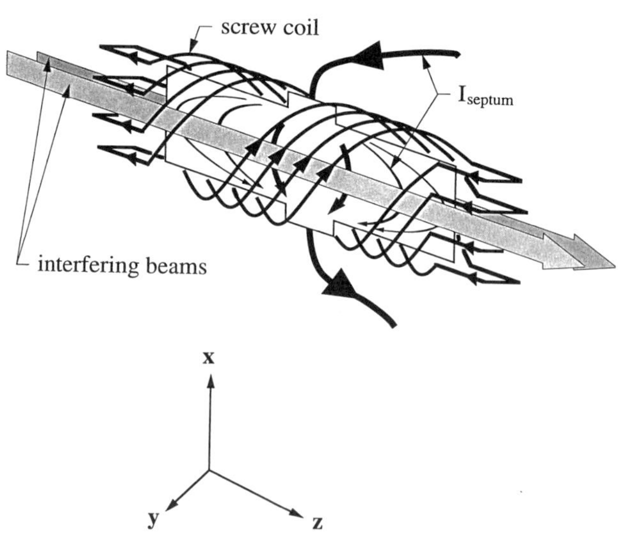

FIGURE 11. Schematic for the magnetic coil configuration in the Berry phase experiment.

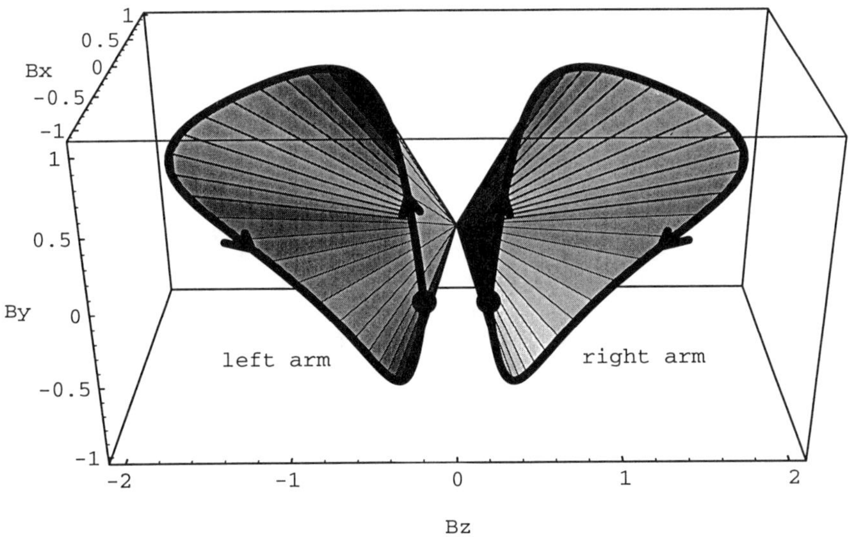

FIGURE 12. Movement of the magnetic field vectors on both sides of the interaction region in the atom frame. The septum field is twice the size of the screw field. The bias field is zero. Entering the interaction region, the atom finds the magnetic field vector at the marked location (●). While the atom moves through the interaction region, the magnetic field vector moves along the two trajectories (—) as given by the arrows and returns to the starting point when the atom leaves the interaction region. The solid angle *between* the two curves is a measure of the geometric phase obtained by the adiabatic evolution of the state.

are the same and cancel. An additional uniform field in the *y*-direction provides a quantization axis and additional field adjustability. The resulting magnetic fields on either side of the interaction region are shown in FIGURE 12 for a particular set of currents. Reference 45 describes this separated path configuration in greater detail. This experiment can also be seen as a demonstration of geometric forces[98,99] because the phase shift has the same relation to the geometric force as the Aharonov-Bohm phase shift has to the Lorentz force.

CONCLUSIONS

Matter wave interferometry using atoms and molecules has tremendous applications to fundamental measurements in quantum mechanics and to new types of precision measurements in atomic and molecular physics. The ability to measure phase shifts produced by an interaction is already allowing the measurement of physical quantities that have previously been inaccessible in intensity-based measurements. Interferometry with particles containing large numbers of populated internal states does not seem fundamentally limited, as demonstrated by our molecular interferometer. However, larger particles may require impractically long traversal times to achieve measurable fringe contrast. Interferometry with complex particles also has significant applications to the theory of quantum measurement because of the relative ease of forming entangled states with a variety of internal states and states with macroscopic degrees of freedom. This, in particular, should permit investigations of the relationship between recording information, entanglement with the measuring apparatus, and the destruction of coherence. The photon scattering experiment, in particular, shows that entanglement with the radiation field does not necessarily destroy coherence.

We have also summarized some of our experimental work on applications of matter wave interferometry to atomic and molecular physics. This includes a number of completely new results such as direct measurements of the ground state polarizability of sodium with an accuracy more than 6 times higher than previous measurements, the demonstration of fringes from molecules, and the determination of the index of refraction of a gas from both incident atoms and molecules.

REFERENCES

1. MARTON, L. 1952. Phys. Rev. **85:** 1057.
2. MARTON, L., J. A. SIMPSON & J. A. SUDDETH. 1954. Rev. Sci. Instrum. **25:** 1099.
3. MARTON, L., J. A. SIMPSON & J. A. SUDDETH. 1954. Phys. Rev. **90:** 490.
4. MOLLENSTEDT, G. & H. DUKER. 1955. Naturwissenschaften **42:** 41.
5. MOLLENSTEDT, G. & H. DUKER. 1956. Z. Phys. **145:** 377.
6. MAIRE-LEIBNITZ, H. & T. SPRINGER. 1962. Z. Phys. **167:** 368.
7. RAUCH, H., W. TREIMER & U. BONSE. 1974. Phys. Lett. A **47:** 369.
8. BAUSPIESS, W., U. BONSE, H. RAUCH & W. TREIMER. 1974. Z. Phys. **271:** 177.
9. COLELLA, R., A. W. OVERHAUSER & S. A. WERNER. 1975. Phys. Rev. Lett. **34:** 1472.
10. STAUDENMANN, J. L., S. A. WERNER, R. COLELLA & A. W. OVERHAUSER. 1980. Phys. Rev. A **21:** 1419.
11. WERNER, S. A., J. L. STAUDENMANN & R. COLELLA. 1979. Phys. Rev. Lett. **42:** 1103–1105.
12. RAUCH, H., A. ZEILINGER, G. BADUREK, A. WILFING, W. BAUSPIESS & U. BONSE. 1975. Phys. Lett. **54A:** 425–427.
13. WERNER, S. A., R. COLELLA, A. W. OVERHAUSER & C. F. EGAN. 1975. Phys. Rev. Lett. **35:** 1053–1055.
14. SHULL, C. G., D. K. ATWOOD, J. ARTHUR & M. A. HORNE. 1980. Phys. Rev. Lett. **44:** 765–768.
15. RAUCH, H. 1979. Neutron Interferometry. U. Bonse & H. Rauch, Eds.: 161–194. Oxford University Press. London/New York.
16. KEITH, D. W., M. L. SHATTENBURG, H. I. SMITH & D. E. PRITCHARD. 1988. Phys. Rev. Lett. **61:** 1580.
17. KEITH, D. W., R. J. SOAVE & M. J. ROOKS. 1991. J. Vac. Sci. Technol. B **9:** 2846.
18. See: SPECIAL ISSUE—LASER COOLING AND TRAPPING OF ATOMS. 1989. J. Opt. Soc. Am. B **6**.
19. See, for example: PRITCHARD, D. E. & W. KETTERLE. 1992. *In* Laser Manipulation of Atoms and Ions (Varenna). E. Arimondo, E. D. Phillips & F. Strumia, Eds. North-Holland. Amsterdam.
20. KASEVICH, M. A., D. S. WEISS & S. CHU. 1990. Opt. Lett. **15:** 607.
21. SLEATOR, T., T. PFAU, V. BALYKIN, O. CARNAL & J. MLYNEK. 1992. Phys. Rev. Lett. **68:** 1996–1999.
22. See: SPECIAL ISSUE—MECHANICAL EFFECTS OF LIGHT. 1985. J. Opt. Soc. Am. B **2**.
23. SHIMIZU, K., H. TAKUMA & F. SHIMIZU. 1991. *In* Laser Spectroscopy X: Proceedings of the 10th International Conference on Laser Spectroscopy. M. Ducloy, E. Giacobino & G. Camy, Eds. World Scientific. Singapore.
24. EKSTROM, C. R., D. W. KEITH & D. E. PRITCHARD. 1992. Appl. Phys. B **54:** 369.
25. See: SPECIAL ISSUE—OPTICS AND INTERFEROMETRY WITH ATOMS. 1994. J. Phys. **4**.
26. CHRIST, M., A. SCHOLZ, M. SCHIFFER, R. DEUTSCHMANN & W. ERTMER. 1994. Opt. Commun. **107:** 211–217.

27. Goldner, L. S., C. Gerz, R. J. C. Spreeuw, S. L. Rolston, C. I. Westbrook & W. D. Phillips. 1994. Phys. Rev. Lett. **72:** 997–1000.
28. Lawall, J. & M. Prentiss. 1994. Phys. Rev. Lett. **72:** 993–996.
29. Moskowitz, P. E., P. L. Gould, S. R. Atlas & D. E. Pritchard. 1983. Phys. Rev. Lett. **51:** 370.
30. Gould, P. L., G. A. Ruff & D. E. Pritchard. 1986. Phys. Rev. Lett. **56:** 827.
31. See the recent review by: Adams, C. S., M. Sigel & J. Mlynek. 1994. Phys. Rep. **240:** 143–210.
32. Keith, D. W., C. R. Ekstrom, Q. A. Turchette & D. E. Pritchard. 1991. Phys. Rev. Lett. **66:** 2693.
33. Pritchard, D. E. 1993. *In* Atomic Physics 13. T. W. Hansch, H. Walther & B. Neizert, Eds. Amer. Inst. Phys. New York.
34. Robert, J., C. Miniatura, S. L. Boiteux, J. Reinhardt, V. Bocvarski & J. Baudon. 1991. Europhys. Lett. **16:** 29.
35. Carnal, O. & J. Mlynek. 1991. Phys. Rev. Lett. **66:** 2689.
36. Clauser, J. F. & S. Li. 1994. Phys. Rev. A **49:** R2213.
37. Kasevich, M. & S. Chu. 1991. Phys. Rev. Lett. **67:** 181.
38. Riehle, F., T. Kisters, A. Witte, J. Helmcke & C. J. Bordé. 1991. Phys. Rev. Lett. **67:** 177.
39. Sterr, U., K. Sengstock, J. H. Muller, D. Bettermann & W. Ertmer. 1992. Appl. Phys. B **54:** 341.
40. For an overview of recent atom interferometer experiments, see: Optics and Interferometry with Atoms. 1992. Special Issue Appl. Phys. B **54**.
41. Shimizu, F., K. Shimizu & H. Takuma. 1992. Phys. Rev. A **46:** 46.
42. Ekstrom, C. R., J. Schmiedmayer, M. S. Chapman, T. D. Hammond & D. E. Pritchard. 1994. Phys. Rev. A. Submitted.
43. Ekstrom, C. R. 1993. Ph.D. thesis. Massachusetts Institute of Technology. Unpublished.
44. Schmiedmayer, J., C. R. Ekstrom, M. S. Chapman, T. D. Hammond & D. E. Pritchard. 1993. *In* Fundamentals of Quantum Optics III (Küthai, Austria). F. Ehlotzky, Ed. Springer-Verlag. Berlin/New York.
45. Schmiedmayer, J., C. R. Ekstrom, M. S. Chapman, T. D. Hammond & D. E. Pritchard. 1994. J. Phys. II **4:** 2029–2042.
46. Schmiedmayer, J., M. S. Chapman, C. R. Ekstrom, T. D. Hammond, S. Wehinger & D. E. Pritchard. 1994. Phys. Rev. Lett. To appear.
47. Weiss, D. S., B. N. Young & S. Chu. 1993. Phys. Rev. Lett. **70:** 2706.
48. Chormaic, S. N. *et al.* 1994. Phys. Rev. Lett. **72:** 1–4.
49. Miniatura, C. *et al.* 1992. Phys. Rev. Lett. **69:** 261–264.
50. Riehle, F., A. Witte, T. Kisters & J. Helmcke. 1992. Appl. Phys. B **54:** 333.
51. Morinaga, A., T. Tako & N. Ito. 1993. Phys. Rev. A **48:** 1346–1368.
52. Shimizu, F., K. Shimizu & H. Takuma. 1992. Jpn. J. Appl. Phys. **31:** L436.
53. Chapman, M. S. *et al.* 1995. To be submitted.
54. Bordé, C. J., N. Courtier, F. de Burck, A. N. Goncharov & M. Gorlicki. 1994. Phys. Lett. A **188:** 187.
55. Chapman, M. S., T. D. Hammond, R. Rubenstein, A. Lenef, J. Schmiedmayer & D. E. Pritchard. 1995. To be published.
56. Berry, M. V. 1984. Proc. R. Soc. London Ser. A **392:** 45.
57. Aharonov, Y. & A. Casher. 1984. Phys. Rev. Lett. **53:** 319.
58. Hasselbach, F. & M. Nicklaus. 1993. Phys. Rev. A **48:** 143–151.
59. Nicklaus, M. & F. Hasselbach. 1993. Phys. Rev. A **48:** 152–160.
60. Klein, A. G. & S. A. Werner. 1983. Rep. Prog. Phys. **46:** 259–335.
61. Badurek, G., H. Rauch & J. Summhammer. 1988. Physica B **151:** 82–92.
62. Werner, S. A., H. Kaiser, M. Arif & R. Clothier. 1988. Physica B **151:** 22–35.
63. Jacobson, D. L., S. A. Werner & H. Rauch. 1994. Phys. Rev. A **49:** 3196–3200.
64. Werner, S. A., R. Clothier, H. Kaiser, H. Rauch & H. Wölwitsch. 1991. Phys. Rev. Lett. **67:** 683.
65. Greenberger, D. M. 1983. Rev. Mod. Phys. **55:** 875.
66. Bohr, N. 1928. Nature **121:** 580.

67. ZUREK, W. H. 1982. Phys. Rev. D **26:** 1862–1880; 1991. Phys. Today **44:** 36–44.
68. TALBOT, H. 1836. Philos. Mag. **9:** 401.
69. CHAPMAN, M. S. *et al.* 1994. Phys. Rev. A **51:** R1.
70. SERRI, J. A. 1974. Ph.D. thesis. Massachusetts Institute of Technology.
71. HEISENBERG, W. 1927. Z. Phys. **43:** 172–198. [English translation by: WHEELER, J. A. & W. H. ZUREK, Eds. 1983. *In* Quantum Theory and Measurement. Princeton University Press. Princeton, New Jersey.]
72. ZUREK, W. H. 1981. Phys. Rev. D **24:** 1516–1525.
73. SLEATOR, T., O. CARNAL, T. PFAU, A. FAULSTICH, H. TAKUMA & J. MLYNEK. 1991. *In* Laser Spectroscopy X: Proceedings of the 10th International Conference on Laser Spectroscopy. M. Dulcoy, E. Giacobino & G. Camy, Eds. World Scientific. Singapore.
74. WALLS, D. F. 1992. *In* Laser Manipulation of Atoms and Ions (Varenna). E. Arimondo, E. D. Phillips & F. Strumia, Eds. North-Holland. Amsterdam.
75. CLAUSER, J. F. & S. LI. 1994. Phys. Rev. A **50:** 2430.
76. PFAU, T., S. SPÄLTER, C. KURTSIEFER, C. R. EKSTROM & J. MLYNEK. 1994. Phys. Rev. Lett. **73:** 1223.
77. JAMMER, M. 1994. The Philosophy of Quantum Mechanics, p. 1–121. Wiley. New York.
78. WHEELER, J. A. 1978. *In* Mathematical Foundations of Quantum Theory. A. R. Marlow, Ed. Academic Press. New York.
79. HILLERY, M. & M. O. SCULLY. 1983. *In* Quantum Optics, Experimental Gravitation, and Measurement Theory. P. Meystre, Ed.: 65–85. Plenum. New York.
80. KWIAT, P. G., A. M. STEINBERG & R. Y. CHIAO. 1992. Phys. Rev. A **45:** 7729.
81. SUMMHAMMER, J., G. BADUREK, H. RAUCH, U. KISCHKO & A. ZEILINGER. 1983. Phys. Rev. A **27:** 2523–2532.
82. RAUCH, H., H. WÖLWITSCH, R. CLOTHIER, H. KAISER & S. A. WERNER. 1992. Phys. Rev. A **46:** 49–57.
83. HAMMOND, T. D., D. E. PRITCHARD, M. S. CHAPMAN, A. LENEF & J. SCHMIEDMAYER. 1995. Appl. Phys. B (Special Issue dedicated to Herbert Walther: Fundamental Systems in Quantum Optics). In press.
84. HALL, W. D. & J. C. ZORN. 1974. Phys. Rev. A **10:** 1141.
85. MOLOF, R. W., H. L. SCHWARTZ, T. M. MILLER & B. BEDERSON. 1974. Phys. Rev. A **10:** 1131.
86. ANANDAN, J. 1982. Phys. Rev. Lett. **48:** 1660.
87. CIMMINO, A. *et al.* 1989. Phys. Rev. Lett. **63:** 198.
88. SANGSTER, K., E. A. HINDS, S. M. BARNETT & E. RILS. 1993. Phys. Rev. Lett. **71:** 3641–3644.
89. ZEILINGER, A. 1986. *In* Fundamental Aspects of Quantum Theory. V. Gorini & A. Frigerio, Eds. Plenum. New York.
90. WILCZEK, F. & A. SHAPERE. 1989. Geometric Phases in Physics. World Scientific. Singapore.
91. BERRY, M. V. 1984. Proc. R. Soc. London Ser. A **392:** 45.
92. BERRY, M. V. 1990. Phys. Today **34:** 34–40.
93. KWIAT, P. G. & R. Y. CHIAO. 1991. Phys. Rev. Lett. **66:** 588.
94. CHIAO, R. Y. & Y. S. WU. 1986. Phys. Rev. Lett. **57:** 933.
95. BITTERS, T. & D. DUBBERS. 1987. Phys. Rev. Lett. **59:** 251.
96. WEINFURTER, J. & G. BADUREK. 1990. Phys. Rev. Lett. **64:** 1318.
97. TYCKO, R. 1987. Phys. Rev. Lett. **58:** 2281–2284.
98. STERN, A. 1992. Phys. Rev. Lett. **68:** 2597.
99. AHARONOV, Y. & A. STERN. 1992. Phys. Rev. Lett. **69:** 3593–3597.

Observations of the Wave Nature of an Ultracold Atom[a]

H. TAKUMA,[b] K. SHIMIZU,[b] AND F. SHIMIZU[c]

[b]*Institute for Laser Science*
University of Electro-Communications
Tokyo 182, Japan

[c]*Department of Applied Physics*
University of Tokyo
Tokyo 113, Japan

INTRODUCTION

The validity of quantum mechanics has been proved in enumerable examples. However, direct observations of the most basic quantum phenomena (such as "quantum jump") have been brought to the range of experimental observation only recently by the progress of laser technology. There still are several hypotheses in the basis of quantum mechanics that have not been tested by direct observation and it should be worth carrying out such tests whenever possible in the light of advanced techniques.

The present report discusses the results of a series of works carried out in order to demonstrate the validity of the concepts of quantum mechanical observations in the case of free motion of an ultralow-energy single atom in a confined space. It shows that a single atom is observed at a point in the space-time coordinates and the accumulation of the observed points reflects directly the wave function obtained as the solution of the Schrödinger equation. It is an experimental demonstration of the validity of the most basic concept that the wave function gives the amplitude of the probability of the atom being observed in each point in the space at a given time. Observation of an atom, the wave function of which is given by a wave-function spread in the space, at a point in the space at a given time is sometimes referred to as a collapse of a wide-spread wave function into a point. As will be discussed later, the experiment is independent of the validity of such a concept as "the collapse of a wave function". The range of the scope of the present work is simply to demonstrate that the waves given as the solution of the Schrödinger equation (which we shall call "de Broglie waves" in the present text without taking the historical definitions and principles given by de Broglie seriously) reflect the probability amplitude of the atom, and the de Broglie waves show excellent analogy with the optical waves propagating in the space having an index-of-refraction distribution corresponding to the potential distribution,[1] including the phase and not only the amplitude.

[a]This work was supported by a grant-in-aid from the Ministry of Education, Science, and Culture of Japan.

The present discussion starts by comparing the Schrödinger equation for a free particle moving in a space having a potential distribution $V(\mathbf{r})$, that is,

$$\nabla^2\phi + \left(\frac{2m}{\hbar^2}\right)[W + V(\mathbf{r})]\phi = 0, \tag{1}$$

with the Maxwell equation for optical waves propagating in a space having an index-of-refraction distribution of $n(\mathbf{r})$, that is,

$$\nabla^2 A + k^2 A = 0, \qquad k = n(\mathbf{r})k_0, \tag{2}$$

where A represents the amplitude of the electromagnetic wave and k and k_0 are the wave vector in the space and that in vacuum, respectively. Similarity between equations 1 and 2 is apparent. If we assume that

$$k^2 = \left(\frac{2m}{\hbar^2}\right)[W + V(\mathbf{r})], \tag{3a}$$

where

$$k_0^2 = \left(\frac{2m}{\hbar^2}\right)W, \tag{3b}$$

the solution of the Schrödinger equation should be equivalent to the optical waves propagating in a space having an index-of-refraction distribution given as

$$n(\mathbf{r}) = \left|\frac{k}{k_0}\right| = \sqrt{1 + \left[\frac{V(\mathbf{r})}{W}\right]}. \tag{4}$$

Such a consideration indicates that we will be able to do on de Broglie waves everything that we can do on optical waves. This is the origin of the motivation of a new field called "atom optics". Although there should be no problem in pursuing such works, the similarity of the de Broglie waves with the optical waves has to be accurately tested by experiments. The purpose of the present work is to carry out such experiments with an accuracy given by the present-day technology. A neon atom in the $1s_5$ metastable state was chosen as a test material because of the following reasons:[2,3] (1) its natural lifetime of about 20 seconds is long enough to assume that it is stable during the course of the experiment, (2) its high internal energy of 16 eV makes it possible to detect one atom with a good efficiency, (3) there is an ideal cooling transition between the $1s_5$ ($J = 2$) and $2p_9$ ($J = 3$) metastable states (transition marked A in FIGURE 1), and (4) atoms in the $1s_5$ metastable state can be efficiently transferred to the nonmagnetic $1s_3$ ($J = 0$) metastable state via the $2p_2$ state by exciting the $1s_5$ state with 598-nm radiation (transition marked B in FIGURE 1).

The kinetic energy of atoms treated in the present report is in such a low range that the excitation of the internal atomic state never occurs by mechanical collision. Among naturally existing isotope species, ^{20}Ne was selected because it is the most abundant isotope, it does not have hyperfine structure, and the lowest energy level above the ground state is the 16-eV-high metastable $1s_5$ state. Such a high internal energy of ^{20}Ne causes a high rate of Penning ionization by collision against all kinds of atoms and molecules. However, collision is considered as a loss of the atom, and those atoms experiencing collision during the course of the experiments are independent of the present observations. In order to make the atomic collision rate as low as possible, all the experiments were carried out in ultrahigh vacuum. The effect of electrons and ions, which were produced by metastable state collisions, together with

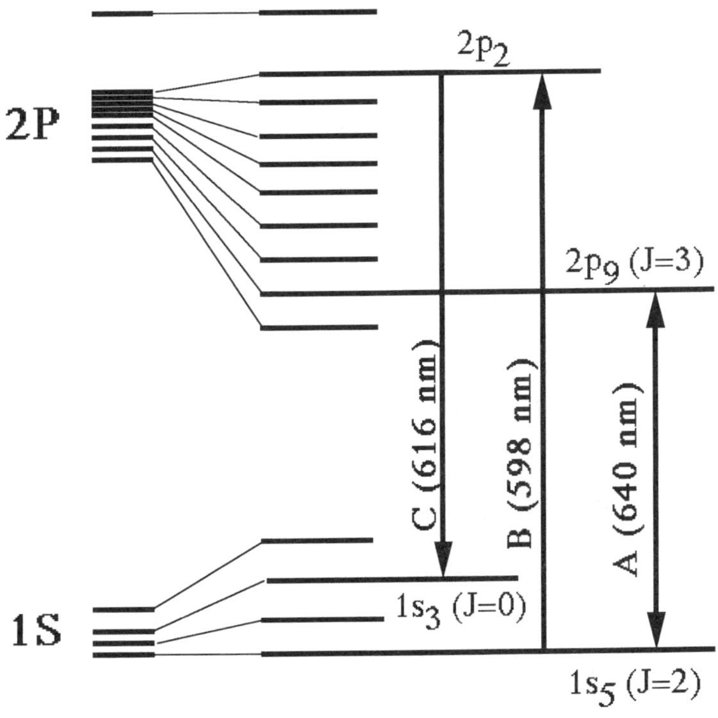

FIGURE 1. Energy level of neon as related to the present work. Each of the allowed transitions is shown by a line, with an arrow showing the direction of the transition used in the present work. The width of the arrow-headed lines roughly represents the transition intensities. Note that the lowest energy level shown here is a metastable state having a lifetime of 21 s and is 13,400 cm^{-1} above the ground state of neon.

the ultraviolet radiations, which were emitted by the transitions induced by the metastable state collisions and the spontaneous emission from the $1p_2$ state in the $1s_5$ to $1s_3$ transition, disturbed the experimental observation strongly by adding excess noises. Reduction of such extra noises is one example of important efforts needed in the present experiments, although such experimental details shall not be described in the present report.

DOUBLE-SLIT INTERFEROMETRY

The first observation of interference fringes of an atom was carried out by Carnal and Mlynek[4] on cold metastable state helium atoms in a supersonic beam. However, they observed only a one-dimensional scan of the fringes because the experiment was carried out at a de Broglie wavelength of only 0.056 nm, and the observation of a two-dimensional fringe pattern by the use of a multichannel plate was impossible because of the limited spatial resolution. Their experiment was an extremely difficult one that could be repeated only with great patience and an extremely stable setup because the de Broglie wavelength was extremely short. It has become much easier to do by the use of a well-developed technique of laser cooling and trapping because

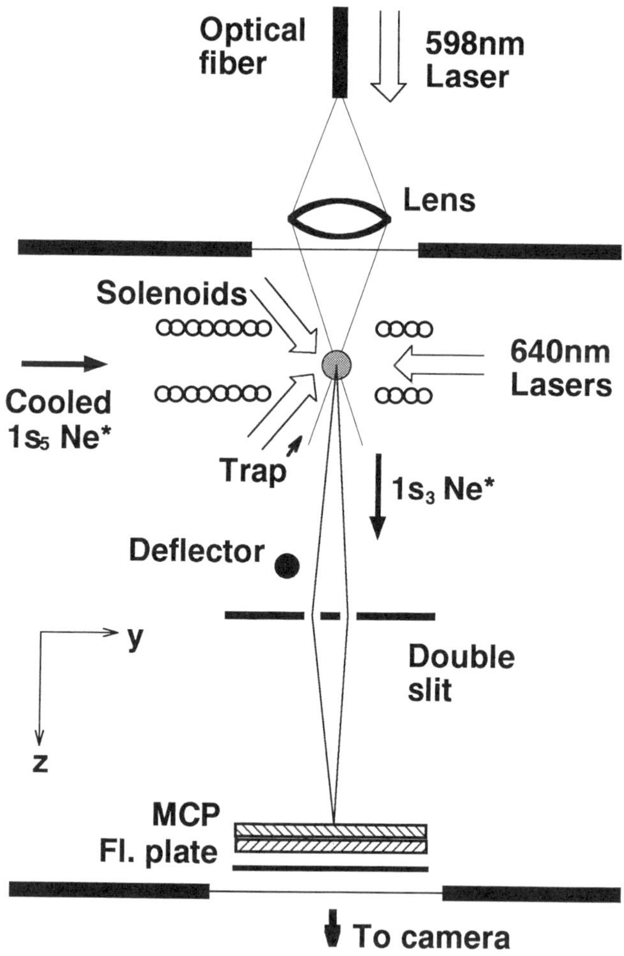

FIGURE 2. Schematic illustration of the setup for Young's interferometer.

FIGURE 3. Observed Young's interference fringes of the metastable state neon atoms.

this allows us to work on a much longer de Broglie wavelength.[5–7] However, most of the observed fringes in the atom interferometry showed only poor visibility, including those cases where the fringes were observed with an excellent signal-to-noise ratio. Therefore, it is imperative to observe a two-dimensional fringe pattern and also to confirm that the fringes show exactly the same high visibility as is described in the textbooks of quantum mechanics.

The present experimental setup of Young's interferometer is schematically shown in FIGURE 2. The magneto-optical trap held metastable state neon atoms at a density of about 10^{10} cm^{-3} cooled to a temperature of 2.5 mK. The pulsed output of a dye laser having a 598-nm wavelength was introduced in vacuum by an optical fiber and was focused in the center of the trap in order to transfer a significant part of the $1s_5$ ($J = 2$) state to the $1s_3$ ($J = 0$) state through the $2p_2$ ($J = 1$) state. Those atoms transferred to the $1s_3$ state are released from the trap, start free-fall motion, and pass through the double-slit if their horizontal velocity component is small enough.

Because the initial speed of the atomic source is extremely low, the axis of symmetry was set in the vertical direction. The double-slit—2-μm opening width, 6-μm separation, and 1 mm long—was made of gold foil and was placed 7.6 cm below the source on a horizontal plane. Because of insufficient mechanical strength, the double-slit was deformed and had only 0.5 mm of usable length. The atoms that passed through the double-slit were detected by a time-resolved two-dimensional detection system consisting of doubled multichannel plates, a fluorescent screen, a CCD camera, and a videocassette recorder.

The clock in the experimental operation was synchronized with the framing pulse of the video recorder and the following series of operations were repeated in a cycle of 16/60 s: The cooling laser of 640 nm was turned on for 4/60 s and the trapped atom grew up with a time constant of 10 ms. The releasing laser pulse of 598 nm was applied at 2/60 s after the turning on of the cooling laser, thereby releasing a significant portion of the trapped atoms in the focal zone. Those atoms having a very small horizontal component of the initial velocity passed through the double-slit and struck the surface of the multichannel plate. An atom was observed as a bright sharp spot and the images were recorded by the videocassette recorder with a rate of 60 frames/s, each frame corresponding to a different range of the initial velocity of atoms. The result was digitally processed afterward and the images were given for each of the different initial velocity groups.

A typical fringe pattern obtained for a group having a 20-nm de Broglie wavelength at the double-slit is shown in FIGURE 3. These atoms correspond to those having a zero initial vertical velocity. In FIGURE 3, the fringes are not parallel in the upper part of the picture. This is due to the deformation of the double-slit in one of the edges. A plot of the fringe pattern scanned in the direction perpendicular to the slit is shown in FIGURE 4. The calculated fringe pattern is shown as the solid curve in FIGURE 4. The calculation is based on the assumption that the contrast ratio is determined by the spatial coherence due to the size of the source having an 80-μm diameter with respect to the slit separation of 6 μm and a 76-mm source for the

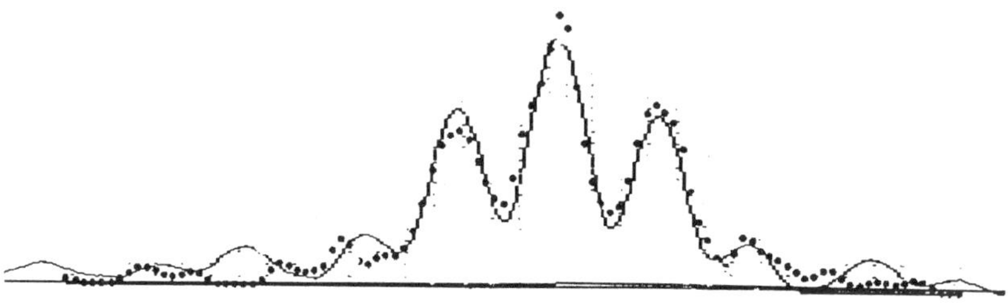

FIGURE 4. Plot of the fringe intensity scanned in the direction perpendicular to the slit (shown by dots) and the calculation based on the classical theory of optics (shown by the solid curve).

double-slit separation. The theoretical curve shows a good agreement with the experimental result, proving that the de Broglie waves behaved exactly like optical waves in Young's double-slit interferometer.

PHASE SHIFT BY ELECTRIC FIELD

The second trial was to observe the phase shift by the potential distribution of electric field distributed in the path of the atom interferometer.[8] The electric field was produced by applying a dc voltage to a copper wire of 0.7-mm diameter placed in parallel to the double-slit, separated by 2 mm from the double-slit surface and 1.5 mm from the vertical plane passing through the slit edge. By applying a dc field E, the internal energy of the metastable state is shifted by the second-order Stark effect and the potential energy $V(\mathbf{r})$ in equation 1 is given by

$$V(\mathbf{r}) = -\chi \cdot E^2, \tag{5}$$

where χ is the polarizability of the neon atom. The value of χ has been determined on the neon $1s_3$ state as[9]

$$\chi \approx 1 \times 10^{-38}\, Fm^2.$$

The effect of the electric field on the neon atom appears in two ways. One is the change of the propagation due to the gradient force and the other is the phase shift. In Young's interferometer, only the difference in the phase of the beam passing through the two slits affects the appearance of the fringes. The difference in the phase integral for the two partial waves passing through each slit is given by

$$\delta = \left(\frac{m}{\hbar^2 k}\right) \int \left[\frac{\partial V(\mathbf{r})}{\partial y}\right] \cdot d \cdot dz, \tag{6}$$

where d is the double-slit separation, the z-axis is taken to be parallel to the vertical direction, and the y-axis is taken to be perpendicular to the slit.

As the path difference due to the nonuniform electric field is rather complicated to calculate in the wave model, we used a classical approximation, which should apply well in this case as geometrical optics does in tracing the wave propagation in optical systems. Because the effect of the electric field should be limited in the vicinity of the double-slit and the force does not apply to the atom passed through the double-slit, which is made of gold and kept at ground potential, the effect of the electric field on the atomic motion is given as the increase in the y component of the atomic velocity v_y, which is given by

$$v_y = \left(\frac{1}{mv_s}\right) \int \left[\frac{\partial V(\mathbf{r})}{\partial y}\right] \cdot dz, \tag{7}$$

where v_s is the atomic speed at the double-slit. The increase in v_y causes a shift in the position of atoms hitting at the target by as much as Δ:

$$\Delta = \left(\frac{\sqrt{1+\alpha} - 1}{mg}\right) \int \left[\frac{\partial V(\mathbf{r})}{\partial y}\right] \cdot dz, \tag{8}$$

where α is a number defined by $2gl/v_s$. On the other hand, the separation between the fringes is given by

$$\epsilon = \left(\frac{\hbar}{mv_s}\right)\left(\frac{L}{d}\right)\left[\frac{2(\sqrt{1+\alpha} - 1)}{\alpha}\right]. \tag{9}$$

Combining equations 6, 8, and 9, we have

$$\delta = 2\pi\Delta/\epsilon. \tag{10}$$

This result means that the fringe pattern is not changed, but only shifted by the electric field. The experimental result is shown in FIGURE 5 for various voltages applied to the pin electrode. It is apparently seen that the interference fringes are shifted by an amount proportional to the applied voltage without changing the shape. This result shows that the shift of the phase integral due to the applied electric field is given by equation 6.

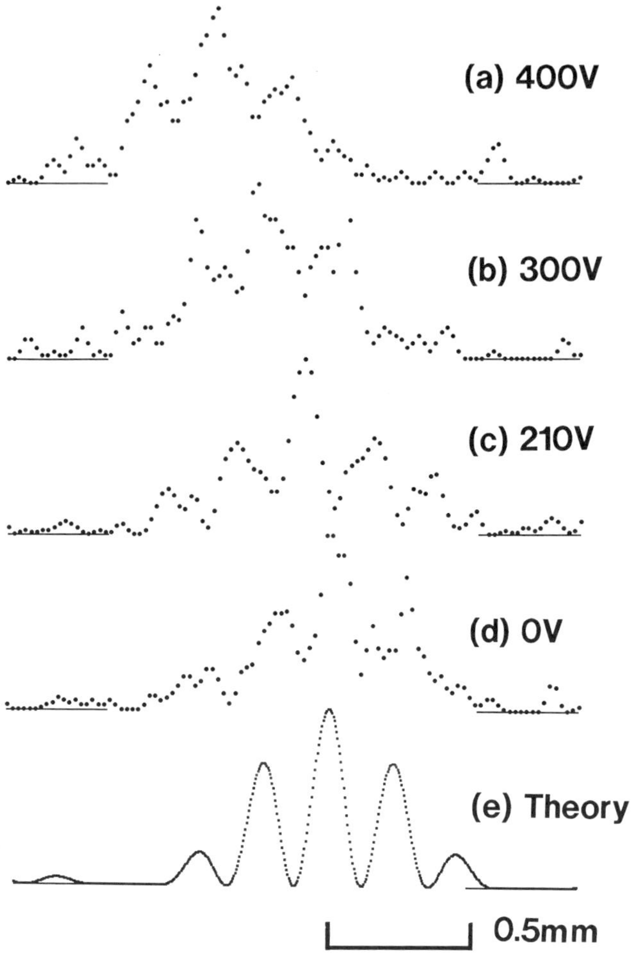

FIGURE 5. Shift of the interference fringes by the application of a dc field.

AN AXIALLY SYMMETRIC CONDENSING LENS FOR ATOMS

The first experiment on an imaging system for atoms was carried out at the University of Constance using a Fresnel zone plate made for X-ray optics.[10] However, because it was necessary to place a wire to stop the incoherent component of atoms, only a one-dimensional image was obtained. A condensing lens with axial symmetry can be realized theoretically by making a field distribution equivalent to the distribution of the index of refraction in a condensing optical lens. However, it is theoretically impossible to make a point having a maximum dc field in a free space.

Because the change of the internal energy of a neon atom by application of an electric field is due to the second-order Stark effect, it is independent of the direction of the field. Therefore, a nonresonant ac field should have the same effect on it as the dc field, and the use of the lowest frequency TM_{01} resonant mode seems to be the most appropriate system for this. In this mode, the distribution of the electric field is represented by $\mathbf{J}_0(k_0 r)$, with the direction being parallel to the axis. It is easily derived that those atoms incident to the cavity in parallel to the axis are focused at a point on the axis separated from the lower edge of the cavity by f, given by the following equation:

$$f = 5.76\left(\frac{mv^2c^4\epsilon_0\pi}{\chi\omega^2 QP}\right), \qquad (11)$$

where Q and P are, respectively, the quality factor and the power dissipated in the cavity.

We have experimentally tested the focusing of an atom by a setup as shown schematically in the left side of FIGURE 6. The procedure is quite similar to Young's interference experiment described in the second section of the present report. A 13-cm-long cylindrical cavity, which is resonant at 17 GHz in the TM_{01} cutoff mode,

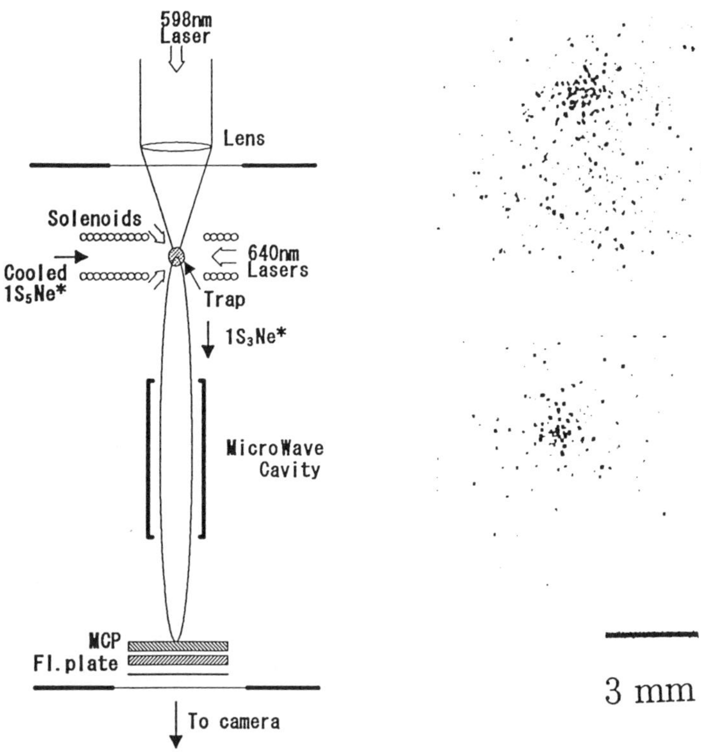

FIGURE 6. Experimental setup (left) and the results (right) of an atomic lens experiment. Upper right: Atoms detected without an electric field. Lower right: Atoms detected with an electric field.

was placed 39 cm below the trap and the multichannel plate was placed 20 cm below the lower edge of the resonator. An example of observed results is shown in the right side of FIGURE 6, where the upper figure corresponds to the observed atoms without any microwave field in the cavity and the lower one corresponds to those with a microwave field. Apparently, atoms are focused toward the center, although the spot size is still not small enough.

Such an atom lens has several intrinsic problems that should be solved in order to obtain a high-quality focusing lens. First of all, it has strong dependence on the atomic velocity. In the present experiment, such a "chromatic aberration" was avoided by a time-resolved measurement and by choosing only the slowest group of atoms. The chromatic aberration should be negligibly small after the atoms have fallen a long distance if the initial velocity spread of the atoms is small enough. It may also be possible to compensate for the effect of changing the velocity by adjusting the microwave power fed to the cavity when the velocity of the atoms is a well-known function of time, as in the present case. Other types of aberration, such as that corresponding to the spherical aberration in conventional optics, are now under investigation.

CONCLUSIONS

Through the three types of experiments reported above, in which the wave nature of an atom is directly observed, it may be concluded that, at least in the limit of present accuracy and for ultralow-energy atoms, the validity of the nonrelativistic Schrödinger equation has been verified, together with a fact that the wave function represents the amplitude of the probability of finding an atom at a given time and space. As the Schrödinger equation indicates, the probability amplitude waves, which we referred to as "de Broglie waves" in the present report, behave exactly similar to the optical waves propagating in a space having an index-of-refraction distribution equivalent to the potential distribution.

REFERENCES

1. TAKUMA, H., K. SHIMIZU & F. SHIMIZU. 1993. *In* Quantum Control and Measurement. H. Ezawa & Y. Murayama, Eds.: 289–293. Elsevier. Amsterdam/New York.
2. SHIMIZU, F., K. SHIMIZU & H. TAKUMA. 1989. Phys. Rev. **A39:** 2758.
3. SHIMIZU, F., K. SHIMIZU & H. TAKUMA. 1987. Jpn. J. Appl. Phys. **26:** L1847.
4. CARNAL, O. & J. MLYNEK. 1991. Phys. Rev. Lett. **66:** 2689.
5. KEITH, D. W., CH. R. EKSTROM, Q. A. TURCHETTE & D. E. PRITCHARD. 1991. Phys. Rev. Lett. **66:** 2693.
6. KASEVICH, M. & S. CHU. 1991. Phys. Rev. Lett. **67:** 181.
7. SHIMIZU, F., K. SHIMIZU & H. TAKUMA. 1992. Phys. Rev. **A46:** R17.
8. SHIMIZU, F., K. SHIMIZU & H. TAKUMA. 1992. Jpn. J. Appl. Phys. **31:** L436.
9. SHIMIZU, F. & M. MORINAGA. 1992. Jpn. J. Appl. Phys. **31:** L1721.
10. CARNAL, O., M. SIEGEL, T. SLEATOR, H. TAKUMA & J. MLYNEK. 1991. Phys. Rev. Lett. **67:** 3231.

Recent Advances in Electron Interferometry

AKIRA TONOMURA

Advanced Research Laboratory
Hitachi, Limited
Hatoyama, Saitama 350-03, Japan

INTRODUCTION

The wavelength of an electron wave is extremely short; for electrons accelerated to 100 kV, for example, it is only 0.04 Å. Microscopic objects and fields can therefore, in principle, be measured or observed with this resolution by electron interferometry. In fact, a few extremely interesting electron-interference experiments were carried out in the 1950s to 1970s (see reference 1). In those days, however, electron-interference experiments required highly skillful techniques. This was partly because there were no convenient optical parts like those in light optics. There is, for example, a convex magnetic electron lens, but not a concave lens, nor are there simple mirrors or half-mirrors in electron optics: the only practical interferometer is the electron biprism.[2] Another reason that full advantage of electron interferometry could not be taken is that there was no "coherent" electron beam like the laser beam used in light optics.

Difficulties in electron-interference experiments in those days can thus be compared to those in optical-interference experiments using high-pressure mercury-arc lamps as a light source. These kinds of experiments were therefore done in only a few laboratories, such as those at Tübingen University in Germany,[3] CNRS Toulouse,[4] Berlin University,[5] Bologna University,[6] and Tohoku University.[7]

The advent of a "coherent" field-emission electron beam[8] in 1979 changed the situation. The maximum number of observable interference fringes increased by an order of magnitude, and interference fringes became observable directly on a fluorescent screen when their number was less than 50.

This coherent beam improved the performance of electron holography[9] to the extent that it can now be used for practical applications. Because electron holography faithfully transforms electron wave fronts into optical wave fronts, versatile optical techniques can be used in the electron optics. Electron holography has made it possible to obtain phase contour maps, which are inaccessible with an electron microscope equipped with an electron biprism. Furthermore, the precision in phase measurement was increased to $2\pi/100$ by using a technique peculiar to holography.[10]

The development of such electron interferometry has made it possible to carry out fundamental physical experiments that had not been feasible: experiments concerning the single-electron buildup of a double-slit interference pattern[11] and experiments confirming the Aharonov-Bohm effect.[12] It has also engendered new methods of measurement and observation, such as measurements of the thickness distribution of a uniform material[10] and of the magnetic field distribution inside a ferromagnetic film[13] and the observation of flux lines[14] in a superconductor.

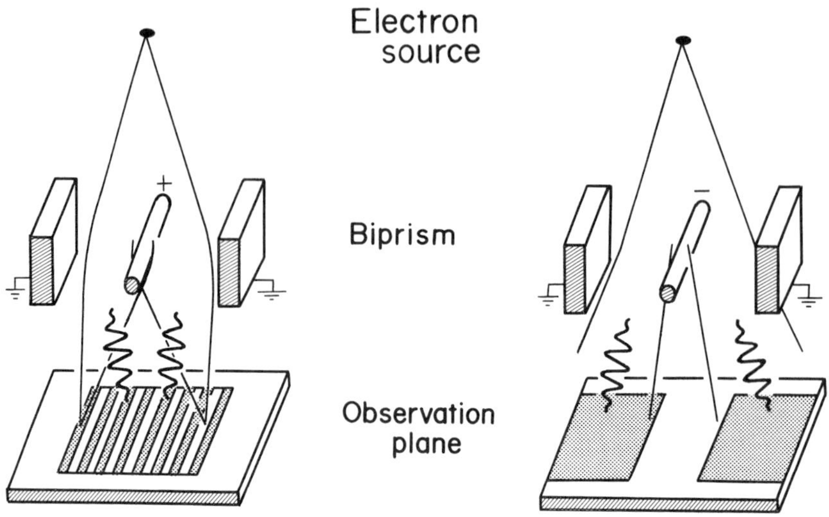

FIGURE 1. Wave packets in the electron biprism: (left) positive potential on the central filament; (right) negative potential on the central filament.

INTERFERENCE OF ELECTRONS

One of the most striking conclusions derived from quantum mechanics can be found in Young's double-slit experiment with electrons when only a single electron exists in the apparatus at one time. Feynman[15] referred to such an experiment as "impossible, absolutely impossible to explain in any classical way, and has in it the heart of quantum mechanics". This experiment has never been done in just this way because the apparatus would have to be made on an impossibly small scale.

Such an experiment is feasible,[11] however, in a field-emission electron microscope equipped with an electron biprism and a two-dimensional position-sensitive electron-counting system.[16] Electrons pass through the biprism, are detected by the electron-counting system, and can then be displayed one by one on a TV monitor (see FIGURE 1, left panel). When there are few electrons, their distribution on the monitor seems quite random (see FIGURE 2a). As their number increases, however, an interference pattern formed by two waves passing through both sides of the biprism becomes recognizable (FIGURES 2b–d). Even when the electron arrival rate is as low as 10 electrons/s over the entire field of view (so that there is at most only one electron in the apparatus at one time), the accumulation of single electrons still forms the interference pattern shown in FIGURE 2d, as if single electrons had passed through both sides of the biprism. In quantum mechanical terms, two partial electron wave functions overlap to interfere on the observation plane, forming a probability distribution of an interference pattern. When detected, however, the two overlapping partial electron waves can be observed only as a single electron, never as two. This phenomenon is interpreted as the result of measurement making the extended wave function instantly collapse into a single point.

This collapse is even more mysterious when a negative instead of positive potential is applied to the central filament of the electron biprism (see FIGURE 1, right panel). In this case, two partial electron waves having passed through both sides of the biprism filament are thoroughly separated as in the right panel of FIGURE 1. When measured, a single electron is detected on either side of the biprism filament. Because two partial wave functions must exist on both sides of the filament until just before the measurement, the partial wave function on one side of the filament must have jumped to the other side to collapse. It should be noted here that the coherence length of the electron wave packet in the traveling direction is only 1 μm, whereas the distance between the biprism and the observation plane is on the order of 10 cm.

This experiment was designed especially for a demonstration and, therefore, the electron frequency in the experiment was made to be much lower than the frequency of the usual hologram formation. However, the single-electron accumulation remains the same for forming any electron hologram. The following experiment was carried out under these single-electron conditions.

AHARONOV-BOHM EFFECT

Another strange phenomenon in quantum mechanics is the Aharonov-Bohm (AB) effect,[17,18] which describes the fundamental interactions of electron waves with electromagnetic fields. The AB effect means that electrons can be physically influenced by a magnetic field without actually entering it. For example, if two electron waves travel in field-free regions on both sides of an infinite solenoid, the electron

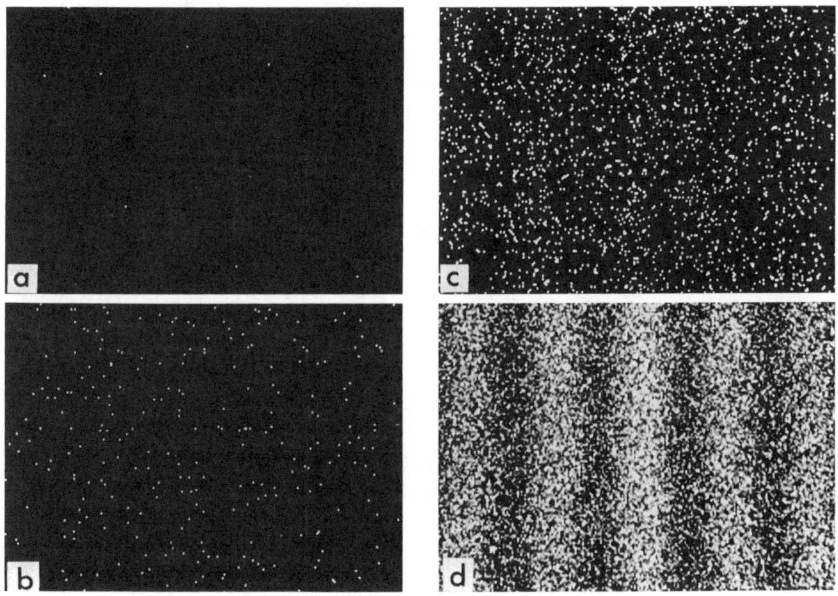

FIGURE 2. Single-electron buildup of the electron interference pattern: (a) $N = 8$, (b) $N = 100$, (c) $N = 3000$, and (d) $N = 100{,}000$.

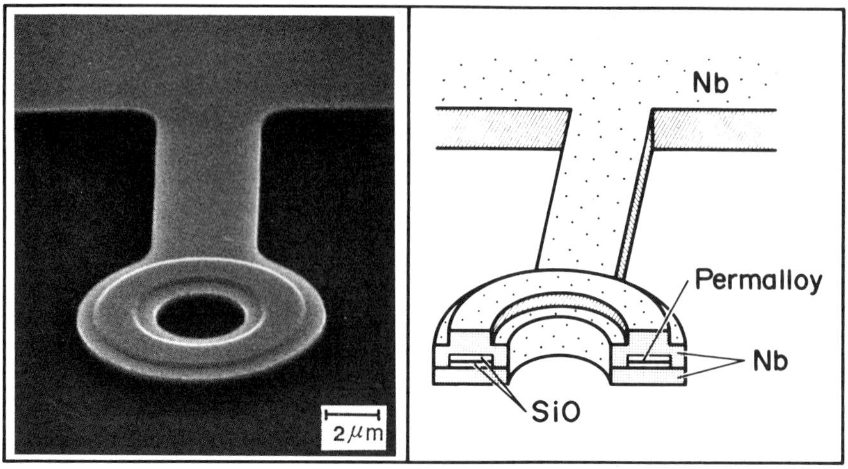

FIGURE 3. Toroidal magnet covered with a superconductor: (left) scanning electron micrograph and (right) schematic cross section.

waves are physically influenced (resulting in a relative phase shift) when a current flows through the solenoid. Aharonov and Bohm attributed this effect to the vector potential surrounding the solenoid, the circulation integral of which does not vanish, but is equal to the magnetic flux inside the solenoid. This effect is purely quantum mechanical because electrons pass through only field-free regions and thus no force is exerted on them. Although the AB effect is a straightforward consequence of the Schrödinger equation, it was long a center of controversy because it contradicts classical electrodynamics. Many conflicting assertions have been made concerning its physical implications and even its existence.

The significance of the AB effect increased[19] in the late 1970s, when the theory of gauge fields was revived as the most probable candidate for the unified theory of all fundamental interactions in nature. In this theory, vector potentials are extended to gauge fields and are regarded as fundamental physical quantities. The AB effect demonstrates the physical reality of gauge fields.

The last in a series of experiments that we made used a toroidal ferromagnet instead of a straight solenoid. An infinitely long solenoid is experimentally impossible, but an ideal geometry can nonetheless be produced by using an actual toroidal magnet.[20] The toroidal magnet was also covered with a superconducting niobium layer to completely confine the magnetic field.

Even with such a complicated structure, the distance across the toroid should be less than the transverse coherence length of an electron beam, say, on the order of 10 μm. Yet, for complete magnetic shielding, the thickness of the superconductor layer should be three times the penetration depth ($\sim$0.1 μm) of Nb. Such samples were thus fabricated by using the most advanced photolithography techniques. The resultant sample is shown in FIGURE 3.

When the sample was cooled to 5 K, the relative phase shift was measured between two electron waves passing through the inside of the hole and outside the toroid.[12] Although phase shifts were measured for many samples with different magnetic flux values, only two phase shifts, 0 or π, were observed (FIGURE 4). The conclusion is obvious—a relative phase shift of π is produced even when the magnetic field is confined within the superconductor and is shielded from the electron beam. This proves that the AB effect exists.

However, why is the phase shift either 0 or π? This quantization of the phase shift provides key evidence for the complete shielding of the magnetic field by the covering superconductor: When a magnetic flux is no other way to be surrounded by a superconductor, the magnetic flux is quantized in $h/2e$ units. Because a magnetic flux of $h/2e$ produces a phase shift of π, whether the relative phase shift is 0 or π depends on whether the number of trapped flux quanta is even or odd.

The AB effect tells us that a phase shift of 2π is produced between two electron beams enclosing a magnetic flux of h/e. It can therefore be concluded that an electron interference micrograph displays the flow of magnetic flux. This can be more clearly understood when we see the result obtained by Aharonov and Bohm. From the Schrödinger equation, they calculated the following phase difference $\Delta S/\hbar$

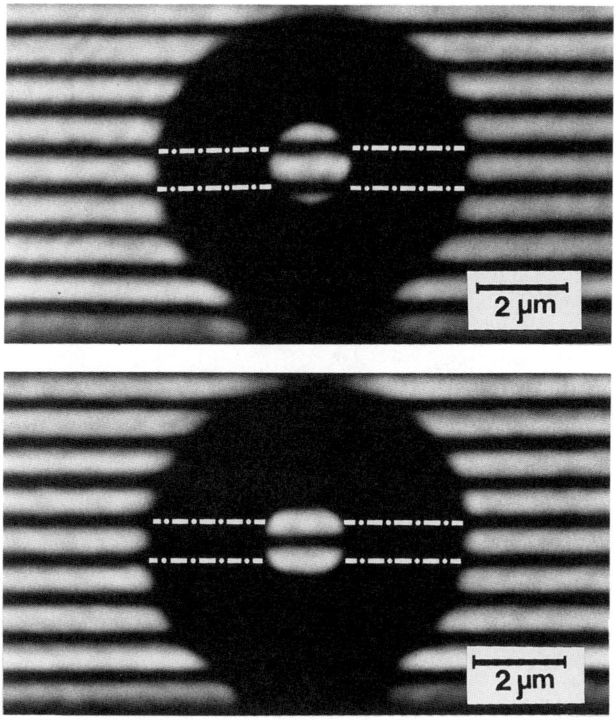

FIGURE 4. Electron interferograms indicating the relative phase shift: (top) phase shift = 0 and (bottom) phase shift = π.

between the two electron beams starting from one point and ending at another point:

$$\Delta S/\hbar = (1/\hbar) \oint (mv - eA) \cdot ds$$
$$= (1/\hbar) \oint (\sqrt{2meV} - e\mathbf{t} \cdot A)ds, \quad (1)$$

where the integral is carried out along the route connecting two electron trajectories and **t** is the unit tangent vector of the electron trajectory.

It can be seen from this equation that electromagnetic potentials (A, V) can be detected by measuring the phase shift of an electron beam, although what we can obtain is not the electromagnetic potentials themselves, but their integrals along the electron beam trajectory.

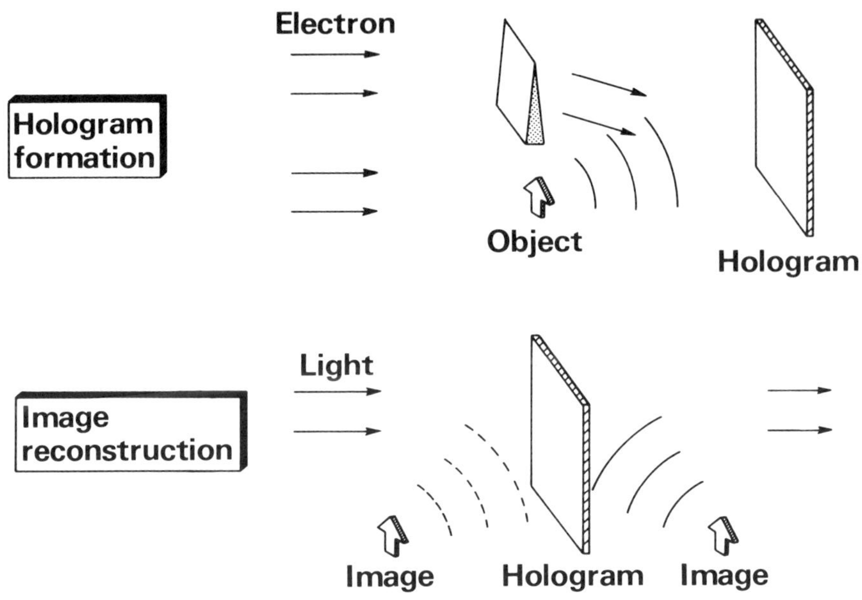

FIGURE 5. Principle behind electron holography.

ELECTRON HOLOGRAPHY

The precise measurement of electron phase shifts became possible by combining electron holography[21] and a "coherent" electron beam.[8] Electron holography, first devised by Gabor for breaking through the resolution limit of electron microscopes,[9] is a two-step imaging technique consisting of hologram formation with an electron wave and image reconstruction with a light beam (FIGURE 5). Its significance, however, was not fully recognized at first because of the lack of coherent waves. Since the advent of the coherent laser beam, holography has blossomed in the field of light optics. Likewise, practical applications of electron holography were opened up by the introduction of a field-emission electron beam.

Once an electron image is transformed into an optical image, versatile optical techniques can be used. For example, the phase distribution of the electron beam having passed through an object can be displayed as a phase-amplified interference micrograph[22] in which phase can be measured with a precision of $2\pi/50$.[10]

The optical reconstruction is simple, but is off-line because of the time required for developing the film. On-line or real-time reconstruction techniques using computers and optical devices are therefore being developed. An image can be numerically reconstructed from a hologram recorded on a charge-coupled device (CCD) attached to an electron microscope, and the amplitude image, the phase image, or the interference image can be displayed. These images can be obtained in a fairly short time (depending on the computer used), but not yet in real time.

A real-time method using a liquid-crystal panel for a phase hologram has been developed recently.[23] The image signal of the hologram, detected with a TV camera attached to the electron microscope, is transmitted to a liquid-crystal panel, where the signal is displayed as the phase distribution in the panel for an incident light beam. Images are reconstructed instantly by shining a laser beam onto this phase hologram. The time resolution depends on the TV system, and dynamic phenomena can be observed in real time.

A two-dimensional phase-shifter having an arbitrary phase distribution can be obtained by applying the appropriate electric signal to the liquid-crystal panel. Therefore, another liquid-crystal panel located on the Fourier plane of the image in the optical reconstruction system acts as a phase-shifter for spatial filtering,[24] compensating for the lens aberrations caused by the electron lens and producing images under arbitrary focusing conditions (FIGURE 6).

APPLICATIONS TO ULTRAFINE MEASUREMENTS

Specimen Thickness Distribution in Atomic Dimensions

The phase distribution of a specimen of uniform material maps the thickness contour of the specimen. A cleaved molybdenite thin film, phase-amplified 24-fold, is shown in FIGURE 7. The phase distribution is displayed here as a deviation from regular fringes (i.e., as an interferogram). Steps A, B, and C in the micrograph correspond to one, three, and five layers of atomic surface steps. The thickness change at step A is only 6.2 Å (one-half of the *c*-axis spacing) and produces a phase shift of $2\pi/50$. This figure shows that a phase shift on the order of $2\pi/100$ can be detected.

In the transmission mode, surface topography can only be inferred from thickness measurements. In the reflection mode, however, topographic features can be measured precisely because surface height differences are directly measured (in units of extremely short electron wavelengths) as geometrical path differences.[25]

Magnetic Field Observation

The phase difference between two electron beams passing through a pure magnetic object is given by

$$\Delta S/\hbar = -(e/\hbar)\oint \boldsymbol{A}\cdot d\boldsymbol{s} = -(e/\hbar)\int \boldsymbol{B}\cdot d\boldsymbol{S}, \qquad (2)$$

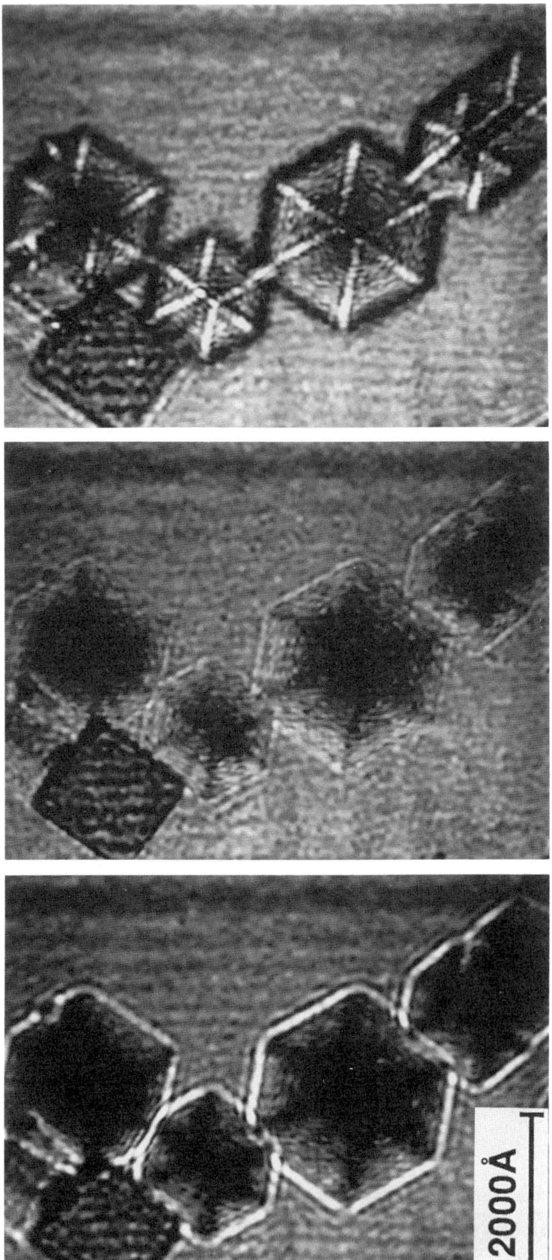

FIGURE 6. Reconstructed images of MgO particles under different focusing conditions: (left) $\Delta f = 10\ \mu m$, (middle) $\Delta f = 0$, and (right) $\Delta f = -10\ \mu m$.

where the first integral is carried out along a closed path along two electron trajectories and the second integral is carried out over the surface determined by the two paths.

From this equation, we can draw the following conclusions:[13]

(1) contour fringes in the interference micrograph indicate magnetic lines of force because the phase difference $\Delta S/\hbar$ vanishes between two beams passing through points along a magnetic line;
(2) a magnetic flux of h/e flows between two adjacent contour fringes.

An interference micrograph of a fine particle of cobalt is shown in FIGURE 8. When the particle is observed with an electron microscope that displays the intensity

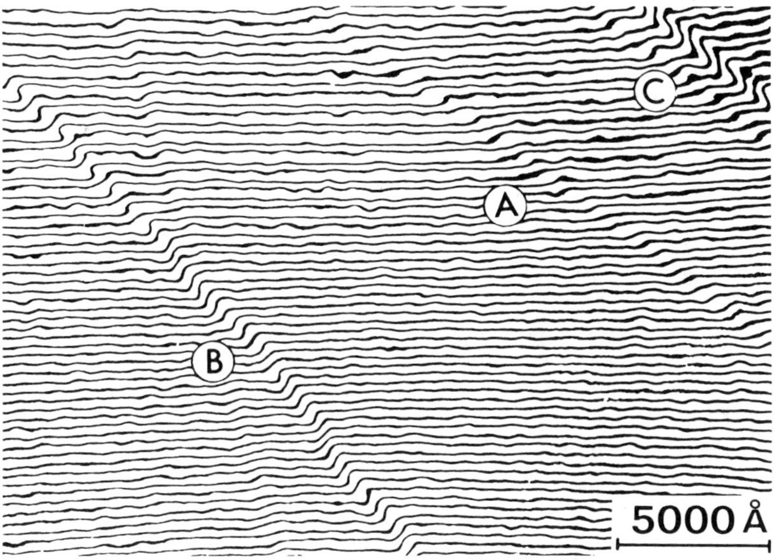

FIGURE 7. Interference micrograph of a molybdenite film (×24).

of the transmitted electron beam, only the triangular outline can be observed. In this interference micrograph, however, two kinds of fringes appear. The fringes parallel to the edges indicate that the thickness increases to 550 Å linearly from the edges, and the fringes in the inner region where the thickness is uniform indicate magnetic lines of force. The smoothly rotating magnetization is observable at a glance even in such a fine particle.

The diameter of this particle is about 3000 Å. When particles are smaller, the magnetization is not closed inside; the particles are uniformly magnetized. A barium-ferrite particle[26] is shown as an example in FIGURE 9. Magnetic fields are leaking outside from the upper north pole of the particle and are sucked up at the south pole below. It can be seen that the particle has a single magnetic domain.

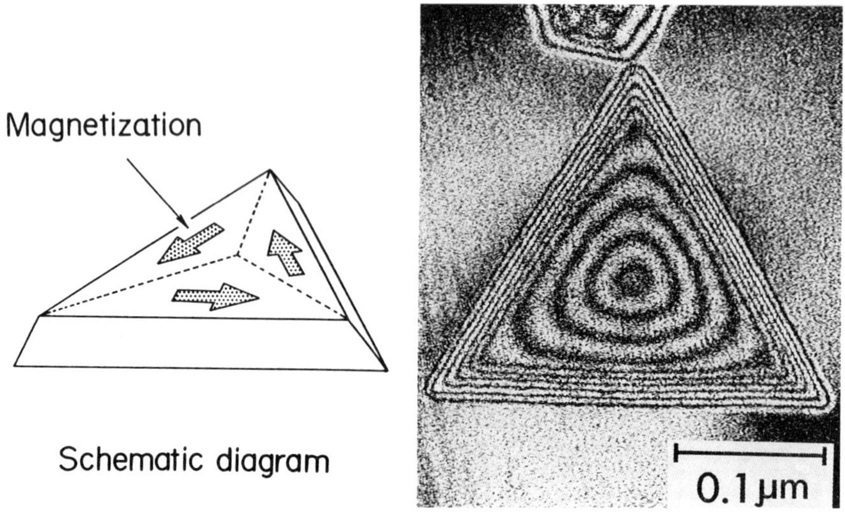

FIGURE 8. Interference micrograph of a Co particle.

Flux Lines in Superconductors

Flux lines in superconductors can be observed quantitatively by interference microscopy[27,28] and Lorentz microscopy[29] with our 350-kV holography electron microscope.[30] In these experiments, we tilted a superconductive thin film with respect to both the electron beam and the magnetic field. The experimental arrangement is shown in FIGURE 10. An Nb thin film set on a low-temperature stage was tilted 45° to an incident beam of 300-kV electrons so that the electrons could be influenced by the flux-line magnetic fields. An external magnetic field of up to 150

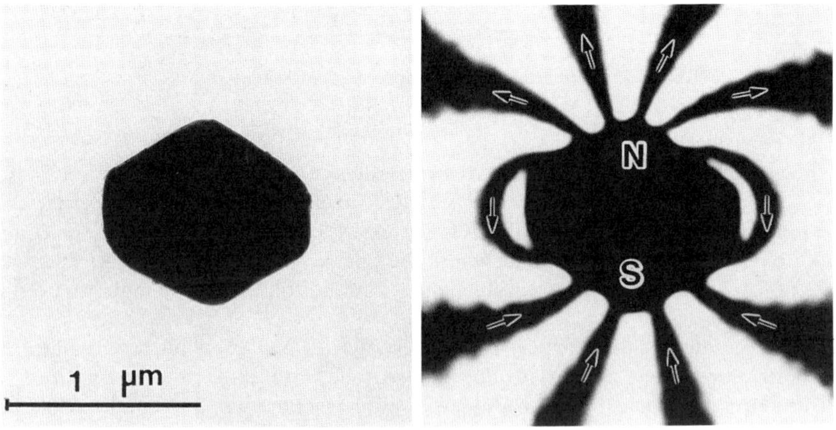

FIGURE 9. Interference micrograph of a barium-ferrite particle.

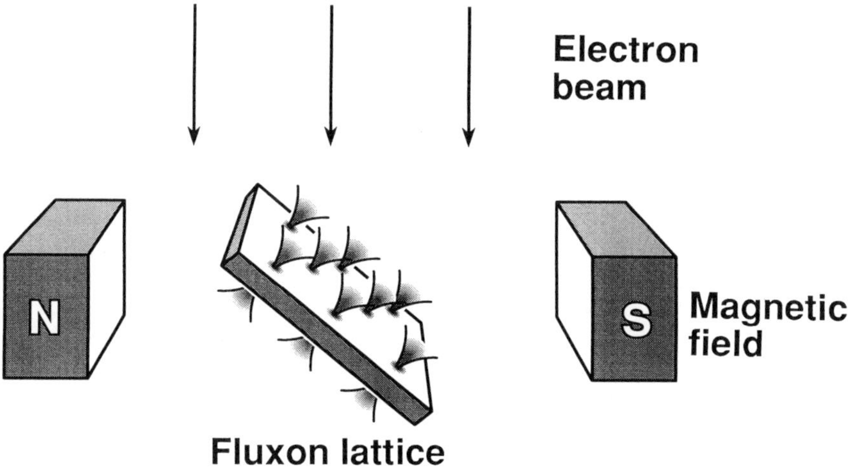

FIGURE 10. Schematic diagram of the flux-line lattice observation.

gauss was applied horizontally. A flux-line array in a single-crystalline Nb thin film[27] is shown in FIGURE 11. Projected magnetic lines of force can be seen in this interference micrograph. They become dense in the localized regions indicated by circles, which correspond to individual flux lines.

Although interference microscopy is a high-resolution and quantitative technique, Lorentz microscopy is more convenient for observing the dynamic behavior of flux lines. After the sample was first cooled to 4.5 K, the magnetic field B was gradually increased. At $B = 32$ gauss, flux lines suddenly began to penetrate the film and the number of flux lines increased as B was increased further. Their dynamic

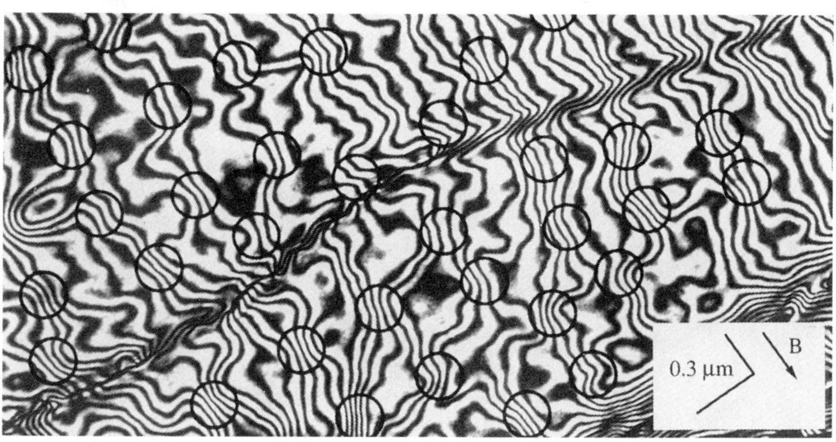

FIGURE 11. Interference micrograph of a superconducting Nb film at $B = 100$ gauss (phase amplification, ×16).

behavior was interesting: At first, only a few flux lines appeared here and there within a (15 × 10)-μm field of view. They oscillated around their pinning centers and occasionally hopped from one center to another. These movements continued as long as the flux lines were not closely packed ($B \leq 100$ gauss).

An equilibrium Lorentz micrograph of a superconducting Nb film at $B = 100$ gauss[29] is shown in FIGURE 12. The film thickness is fairly uniform in the region shown, but the film is bent along the black curves (bend contours), which are caused by Bragg reflections at the atomic planes brought to a favorable angle by bending. Each spot showing a black-and-white contrast is an image of a single flux line. As expected, this contrast reversed when the applied magnetic field was reversed. The tilt direction of the sample can be read from the line dividing the black and white parts of the spots. Because the black part is on the same side of all the spots, the polarities of all the flux lines seen in the region are the same. At a low B (i.e., up to 30–50 gauss), the flux lines are too sparse to form a lattice, even in equilibrium. At $B = 100$ gauss, the flux-line density is so high that they can only form a hexagonal lattice.

A high-T_c superconductor has also been investigated by means of Lorentz microscopy.[31] High-T_c superconductors are difficult to use practically because the critical current vanishes at high temperatures and at high magnetic fields, even when the temperature is well below the critical temperature T_c. This phenomenon most probably arises from the behavior of flux lines, but has not yet been proven concretely. Some researchers believe that these flux lines melt like molecules in a liquid and it is therefore difficult to fix flux lines at some pinning sites.[32] Evidence for flux-line melting was provided by a Bitter BSCCO figure in which the flux-line image was blurred even at 15 K and 20 gauss ($T_c = 85$ K).[33] Accordingly, the practical-use

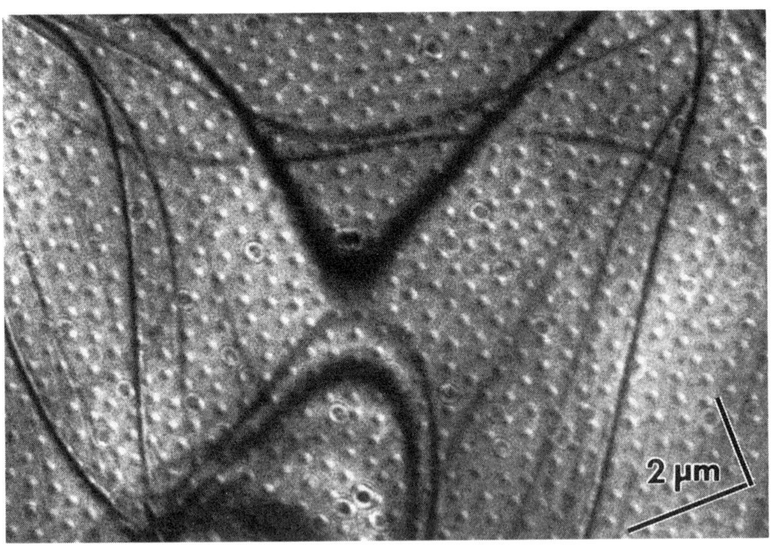

FIGURE 12. Lorentz micrograph of a two-dimensional array of flux lines in a superconducting Nb film.

(a)
(b)
(c)
(d)

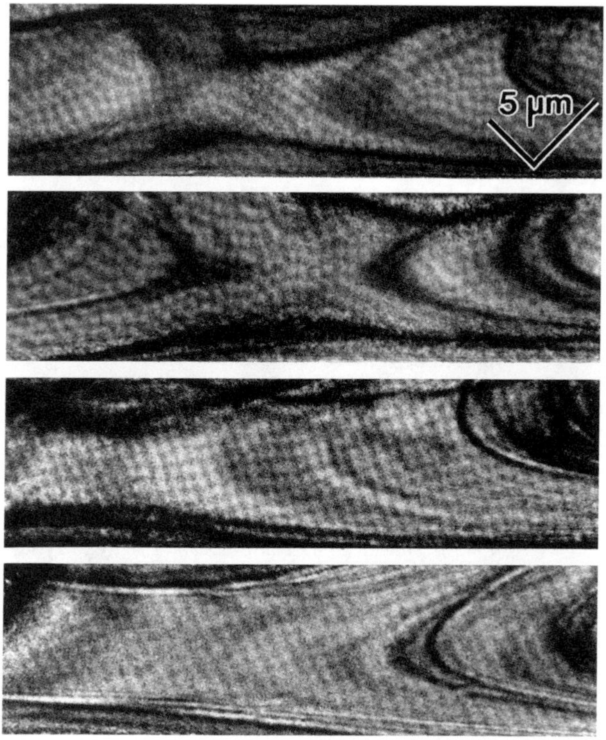

FIGURE 13. Lorentz micrographs of a BSCCO (2212) film: (a) $T = 4.5$ K, (b) $T = 20$ K, (c) $T = 56$ K, and (d) $T = 68$ K.

temperature would not be T_c, but the melting temperature T_m. Other researchers, however, disagree and attribute the low critical current to weak pinning effects.

The flux lines were dynamically observed to find out whether they begin to move under such conditions. They were observed under a fixed magnetic field B while increasing the sample temperature from 4.5 K to above T_c. A Lorentz micrograph at $T = 4.5$ K and $B = 20$ gauss is shown in FIGURE 13a. Flux lines are distributed at random. When the temperature was raised stepwise by a few K, flux lines moved. After a few minutes, they reached an equilibrium state and became still. They did not melt even at 20 K. The flux-line configuration changed between 40 and 50 K and, at temperatures above those of this transition region, the flux lines formed a regular lattice (FIGURE 13c). The flux-line lattice persisted at higher temperatures, although the image contrast gradually decreased and then disappeared above 77 K.

CONCLUSIONS

The advent of the coherent electron beam has opened up a new way to visualize the phase distribution of the electron wave function. This technique has enabled thought experiments in fundamental physics to actually be performed and has provided a new way of observing microscopic objects and fields.

REFERENCES

1. Missiroli, G. F. *et al.* 1981. J. Phys. E **14:** 649.
2. Möllenstedt, G. & H. Düker. 1956. Z. Phys. **145:** 377.
3. Möllenstedt, G. & M. Keller. 1957. Z. Phys. **148:** 34.
4. Faget, J. & C. Fert. 1957. Cah. Phys. **83:** 285.
5. Lischke, R. 1969. Phys. Rev. Lett. **22:** 1366.
6. Pozzi, G. & G. F. Missiroli. 1973. J. Microsc. **18:** 103.
7. Hibi, T. & S. Takahashi. 1963. J. Electron Microsc. **12:** 129.
8. Tonomura, A. *et al.* 1979. J. Electron Microsc. **28:** 1.
9. Gabor, D. 1949. Proc. R. Soc. London **A197:** 454.
10. Tonomura, A. *et al.* 1985. Phys. Rev. Lett. **54:** 60.
11. Tonomura, A. *et al.* 1989. Am. J. Phys. **57:** 117.
12. Tonomura, A. *et al.* 1986. Phys. Rev. Lett. **56:** 792.
13. Tonomura, A. *et al.* 1980. Phys. Rev. Lett. **44:** 1430.
14. Matsuda, T. *et al.* 1989. Phys. Rev. Lett. **62:** 2519.
15. Feynman, R. P. *et al.* 1965. *In* The Feynman Lectures on Physics. Vol. III, p. 1.1–1.5. Addison–Wesley. Reading, Massachusetts.
16. Tsuchiya, Y. *et al.* 1982. *In* Advances in Electronics and Electron Physics. Vol. 64A. P. W. Hawkes, Ed.: 21–31. Academic Press. New York.
17. Aharonov, Y. & D. Bohm. 1959. Phys. Rev. **115:** 485.
18. Peshkin, M. & A. Tonomura. 1989. The Aharonov-Bohm Effect: Lecture Notes in Physics, Vol. 340. Springer-Verlag. Berlin/New York.
19. Wu, T. T. & C. N. Yang. 1975. Phys. Rev. **D12:** 3845.
20. Kuper, C. G. 1980. Phys. Lett. **79A:** 413.
21. Tonomura, A. 1993. Electron Holography. Springer-Verlag. Berlin/New York.
22. Endo, J. *et al.* 1979. Jpn. J. Appl. Phys. **18:** 2291.
23. Chen, J. *et al.* 1993. Opt. Lett. **18:** 1887.
24. Chen, J. *et al.* 1995. Appl. Opt. To be published.
25. Osakabe, N. *et al.* 1989. Phys. Rev. Lett. **62:** 2969.
26. Hirayama, T. *et al.* 1993. Appl. Phys. Lett. **63:** 418.
27. Bonevich, J. *et al.* 1993. Phys. Rev. Lett. **70:** 2952.
28. Bonevich, J. *et al.* 1994. Phys. Rev. **B49:** in press.
29. Harada, K. *et al.* 1993. Nature **360:** 51.
30. Kawasaki, T. *et al.* 1990. Jpn. J. Appl. Phys. **29:** 508.
31. Harada, K. *et al.* 1993. Phys. Rev. Lett. **71:** 3371.
32. For example, see: Bishop, D. J. *et al.* 1992. Science **255:** 165.
33. Kleiman, R. N. *et al.* 1989. Phys. Rev. Lett. **62:** 2331.

Neutron Interferometry Tests of Quantum Theory[a]

SAMUEL A. WERNER

Department of Physics
and
Research Reactor Center
University of Missouri
Columbia, Missouri 65211

HISTORICAL PERSPECTIVES

The neutron, a seemingly simple, yet exquisitely complex particle, was discovered over 60 years ago by Chadwick.[1] Elasser first suggested that the motion of neutrons would be determined by quantum mechanics such that they would be diffracted by crystalline matter.[2] This was soon verified experimentally by Halban and Preiswerk[3] and by Mitchell and Powers[4] using very weak (by today's standards) radium-beryllium neutron sources. With the advent of nuclear reactors after the Second World War and the subsequent construction of high-flux research reactors using highly enriched U-235 fuel during the past 30 years, the scattering of thermal neutrons having a de Broglie wavelength comparable to the interatomic spacings of atoms in solids has become one of the most powerful techniques available in the elucidation of the structure and dynamics of condensed matter.

Diffraction effects at wavelengths of the order of angstroms have been known since Max von Laue's demonstrations of X-ray diffraction by crystals in 1912. However, observing interference effects between well-separated, coherent beams is much more difficult to arrange. In 1965, Bonse and Hart invented the perfect silicon-crystal interferometer. They observed interference effects between two beams of X rays with a wavelength of about 1 Å and separated by about 1 cm.[5] In 1974, Rauch, Treimer, and Bonse were the first to demonstrate that this perfect silicon-crystal interferometer would also work for thermal neutrons in an experiment carried out at the small reactor in Vienna.[6] Thus, this year marks the twentieth anniversary of neutron interferometry. A list of the increasingly sophisticated quantum interference experiments carried out with this marvelous device over the intervening years is given in TABLE 1.

It is very fitting that a conference on Fundamental Problems in Quantum Theory honoring John A. Wheeler should have a session devoted to neutron interferometry. The neutrons that we use in the experiments come from the fission of U-235, which was explained to us in the famous 1939 Bohr-Wheeler paper, "The Mechanism of Nuclear Fission".[7] This, combined with Wheeler's substantial contributions to the applications and clarification of quantum mechanics, makes this series of papers in this session particularly apropos.

[a]This work was supported by the Physics Division of the National Science Foundation through Grant No. PHY-9024608.

TABLE 1. Neutron Interferometry Experiments (1974–1994)

- First Test of Perfect Si-Crystal Interferometer with Neutrons: Vienna (1974)
- Observation of Gravitationally Induced Quantum Interference: Michigan, Missouri (1975, 1980, 1988, 1993)
- Observation of the Change of Sign of the Wave Function of a Fermion due to Precession of 360° in a Magnetic Field: Michigan, Vienna–Grenoble (1975, 1978)
- Observation of the Effect of the Earth's Rotation on the Quantum Mechanical Phase of the Neutron (Sagnac Effect): Missouri (1980)
- Measurement of the Energy-dependent Scattering Length of Sm-149 in the Vicinity of a Thermal Nuclear Resonance: Missouri (1982)
- Charge Dependence of the Four-Body Nuclear Interaction (n-^{3}He versus n-^{3}H): Vienna–Grenoble (1979, 1985)
- Search for Nonlinear Terms in the Schrödinger Equation: MIT (1981)
- Search for the Aharonov-Bohm Effect for Neutrons with a Magnetized-Single-Crystal-of-Fe-inside Interferometer: MIT (1981)
- Measurement of the Longitudinal Coherence Length of a Neutron Beam: Missouri (1983)
- Observation of the Coherent Superposition of Spin States ("Wigner Phenomenon") with Both Static and RF Spin Flippers: Vienna–Grenoble (1983, 1984)
- Neutron Interferometric Search for Quaternions in Quantum Mechanics: Missouri (1984)
- Sagnac Effect Using a Laboratory Turntable—Shows Phase Shift due to Rotation Is Linear in ω: MIT (1984)
- Observation of Acceleration-induced Quantum Interference: Dortmund–Grenoble (1984)
- Experiment on the Null-Fizeau Effect (Stationary Boundaries) for Thermal Neutrons in Moving Matter: Missouri–Melbourne (1985)
- Observation of the Neutron Fizeau Effect with Moving Boundaries of Moving Matter: Dortmund–Grenoble (1985)
- Double-RF Coil Experiment—Analogue of the Magnetic Josephson Experiment: Vienna–Grenoble (1986)
- Precision Measurement of the Bound-Coherent Neutron Scattering Lengths of U-235, U-238, V, Eu, Gd, Th, Kr, H, D, Si, Bi, etc.: Vienna–Grenoble, Missouri (1975–1993)
- Observation of a Motion-induced Phase Shift of Neutron de Broglie Waves Passing through Matter near a Nuclear Resonance (Sm-149): Missouri–Melbourne (1988)
- Observation of Stochastic versus Deterministic Absorption of the Neutron Wave Function: Vienna–Grenoble (1984, 1987, 1990)
- Observation of the Topological Aharonov-Casher Phase Shift: Missouri–Melbourne (1989)
- Test of Possible Nonergodic Memory Effects: Vienna–Grenoble (1989)
- Observation of the Effects of Spectral Filtering in Neutron Interferometry: Missouri–Vienna (1991)
- Counting Statistics Experiments—Particle Number/Phase Uncertainty: Vienna (1990, 1992)
- Observation of the Neutron Phase Echo Effect: Missouri–Vienna (1991)
- Coherence Effects in Time-of-Flight Neutron Interferometry: Missouri–Vienna (1992)
- Observation of the Scalar Aharonov-Bohm Effect: Missouri–Melbourne (1992, 1993)
- Spectral Modulation and Squeezed States in Neutron Interferometry: Missouri–Vienna (1994)

I begin this report with a brief description of the three-crystal *LLL* perfect Si-crystal neutron interferometer. The measurement of the quantum phase shift resulting from the passage of a neutron beam through a slab of matter and interacting with the assembly of nuclei, creating an "optical potential", is then described as an elementary, but important application. The first physics experiment, gravitationally induced quantum interference,[8] carried out with the Si-crystal interferometer has evolved through several stages of improved accuracy and the latest results are described.[9–11] Because this experiment is carried out on the surface of our rotating Earth, a noninertial frame of reference, an additional phase shift, called the Sagnac effect, occurs.[12] Its measurement is described in some detail. I then discuss the observation of the topological Aharonov-Casher effect.[13,14] I conclude by giving an agenda of present and future experiments with some speculations on applications of neutron interferometry.

For overviews of this field, the reader is referred to the proceedings of a conference on Matter Wave Interferometry[15] held in Vienna, Austria, in 1987 and to a review article by Klein and myself in reference 16. A book on Neutron Interferometry by Rauch and myself is in preparation.[17]

PERFECT SILICON-CRYSTAL NEUTRON INTERFEROMETERS

A thermal neutron having a de Broglie wavelength of 2 Å has a kinetic energy of about 20 meV and moves with a velocity of 2000 m/s. This curious dual nature of neutrons—sometimes a particle, sometimes a wave—is wonderfully manifested in the highly nonlocal effects of neutron interferometry. The seemingly incompatible point-by-point motion of particles in space-time as described by relativity considerations and these nonlocal quantum mechanical interference phenomena are brought into close juxtaposition by interference experiments induced by gravity, rotation, and topology.

A schematic diagram of a three-crystal *LLL* interferometer, of the type developed by Bonse and Hart[5] for X rays, is shown in FIGURE 1. It consists of three perfect crystal slabs cut perpendicular to a set of strongly reflecting Bragg planes, typically (220). A collimated, nominally monochromatic beam of thermal neutrons is directed along the line *SA* and is coherently split by Bragg reflection in the first crystal slab. These two coherent beams are split again in the second crystal slab near points *B* and *C*. Two of these beams are directed toward point *D* in the third crystal slab, where they overlap and interfere. The label *LLL* means that this interferometer involves three Laue transmission-geometry crystals. The neutrons are detected in three ^{3}He gas–filled proportional detectors C_1, C_2, and C_3. A photograph of one of our *LLL* interferometers is shown in FIGURE 2.

If the beam traversing path II is phase-shifted relative to the beam traversing path I, then by introducing a slab of material in path II, causing a change in the "optical" path length, the counting rates in detectors C_2 and C_3 will change. On very general grounds, it can be shown that the intensities I_2 and I_3 vary sinusoidally with the phase shift $\Delta\Phi$:

$$I_2 = a_2 - b_2 \cos(\Delta\Phi) \tag{1}$$

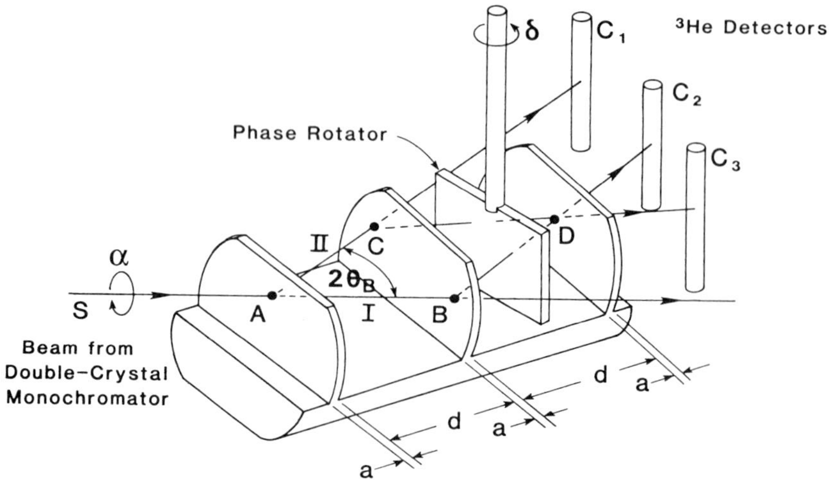

FIGURE 1. Schematic diagram of an *LLL* perfect Si-crystal neutron interferometer used at beam port B at the University of Missouri Research Reactor. The phase rotator allows the neutron "optical" path length on path II relative to path I to be adjusted. The dimensions of this interferometer used in the gravity experiments are $d = 34.518 \pm 0.002$ mm and $a = 2.464 \pm 0.002$ mm.

and

$$I_3 = a_3 + b_3 \cos(\Delta\Phi). \tag{2}$$

The constants a_2, b_2, a_3, and b_3 depend upon the transmission and reflection coefficients of the three Bragg reflecting crystals. As the phase shift $\Delta\Phi$ is varied, the neutron intensity is swapped back and forth between detectors C_2 and C_3 such that $I_2 + I_3$ is constant, requiring $b_2 = b_3$.

For an interferometer of the size shown in FIGURE 2, there are of the order of 10^9 oscillations of the thermal neutron de Broglie wave on each path. Upon recombination in the third crystal slab, stable interference fringes are formed. For this "miracle" to occur, very stringent requirements on microphonic and thermal stability must be met. In our two interferometry setups at the University of Missouri Research Reactor (MURR), the interferometers are positioned inside metallic, isothermal enclosures, which are mounted upon a vibration isolation pad.

The incident neutron beam is not precisely monochromatic. Typically, the wavelength dispersion is $\delta\lambda/\lambda \approx 0.01$. The important feature of this interferometer is that it uses Bragg reflection in perfect crystals, where the Darwin width is about 1.0 arc sec. This means that the Bragg reflection process itself defines the wavelength along a given trajectory (ray) to within about 1 part in 10^6.

The phase accumulated on either path is a line integral over the Lagrangian $\mathscr{L}$ in space-time given by

$$\Phi(x, t) = \left(\frac{1}{\hbar}\right) \int \mathscr{L}\, dt'. \tag{3}$$

The Lagrangian $\mathscr{L}$ is related to the Hamiltonian via a Legendre transformation,

$$\mathscr{L} = \boldsymbol{p} \cdot \boldsymbol{v} - \mathscr{H}, \tag{4}$$

where $\boldsymbol{p}$ is the canonical momentum of the neutron and $\boldsymbol{v}$ is its classical group velocity, that is, $\boldsymbol{v} = d\boldsymbol{s}/dt$. Thus, equation 3 gives the phase at the detector at position $\boldsymbol{x}$ as a function of time t, namely,

$$\begin{aligned} \Phi(\boldsymbol{x}, t) &= \left(\frac{1}{\hbar}\right) \int_{x_0}^{x} \boldsymbol{p} \cdot d\boldsymbol{s} - \left(\frac{1}{\hbar}\right) \int_{t_0}^{t} \mathscr{H}\, dt' \\ &= \int_{x_0}^{x} \boldsymbol{k} \cdot d\boldsymbol{s} - \int_{t_0}^{t} \omega\, dt', \end{aligned} \tag{5}$$

where $k = 2\pi/\lambda$ is the wave vector and ω is the frequency of the wave at any point $(\boldsymbol{x}', t')$ along the trajectory. For each of the two paths in FIGURE 1, we must separately evaluate equation 5, namely,

$$\Phi_{\mathrm{I}}(\boldsymbol{x}, t) = \left(\frac{1}{\hbar}\right) \int_{x_0}^{x} \boldsymbol{p}_{\mathrm{I}} \cdot d\boldsymbol{s} - \left(\frac{1}{\hbar}\right) \int_{t_0}^{t} \mathscr{H}_{\mathrm{I}}\, dt' \tag{6}$$

and

$$\Phi_{\mathrm{II}}(\boldsymbol{x}, t) = \left(\frac{1}{\hbar}\right) \int_{x_0}^{x} \boldsymbol{p}_{\mathrm{II}} \cdot d\boldsymbol{s} - \left(\frac{1}{\hbar}\right) \int_{t_0}^{t} \mathscr{H}_{\mathrm{II}}\, dt'. \tag{7}$$

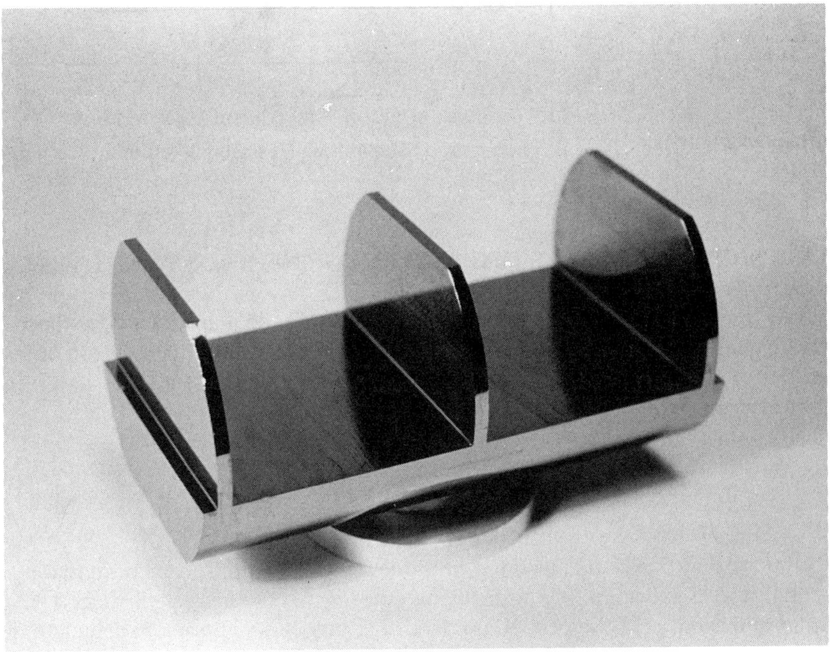

FIGURE 2. A photograph of an *LLL* interferometer.

Thus, the phase difference between the two paths is

$$\Delta\Phi(\boldsymbol{x}, t) = \Phi_{\mathrm{II}}(\boldsymbol{x}, t) - \Phi_{\mathrm{I}}(\boldsymbol{x}, t). \quad \mathbf{(8)}$$

In neutron interferometry, the phase shift $\Delta\Phi_V(\boldsymbol{x}, t)$ caused by a potential $V(\boldsymbol{x}', v, t')$ is the quantity that is measured and is of physical interest, namely,

$$\Delta\Phi_V = \Delta\Phi - \Delta\Phi_0, \quad \mathbf{(9)}$$

where $\Delta\Phi_0$ is the "empty" interferometer phase difference, that is, when $V(\boldsymbol{x}', v, t) = 0$. The phase shift due to the potential is the line integral along the classical trajectories of the neutron around the interferometer.

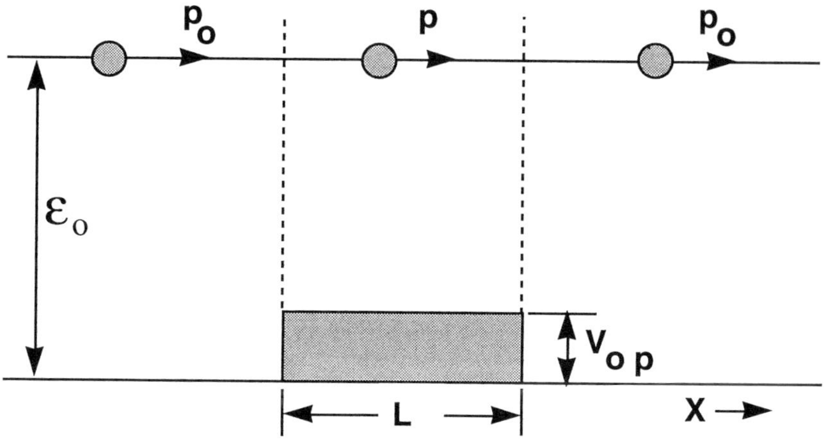

FIGURE 3. A neutron of kinetic energy $\mathscr{E}_0$ passing through a region $\mathscr{R}$ of optical potential V_{op}.

PHASE SHIFT DUE TO THE INTERACTION OF NEUTRONS WITH NUCLEI

Only the *s*-wave component of the scattering amplitude is significant for thermal neutrons scattering from nuclei. It is easy to show from this that the neutron optical potential of a slab of material having an atom density N and a nuclear scattering amplitude b is

$$V_{\mathrm{op}} = \frac{2\pi\hbar^2 bN}{m}. \quad \mathbf{(10)}$$

This potential is typically of the order of 10^{-7} eV.

Suppose that this slab is inserted into one leg of the interferometer, creating a potential step as shown in FIGURE 3. Upon entering $\mathscr{R}$, the neutron is slowed down and, upon leaving $\mathscr{R}$, it is speeded up. Knowing this, it is tempting to calculate the phase shift $\Delta\Phi_V$ caused by V_{op} using equation 3, with limits on the integral given by the time-of-flight across the region $\mathscr{R}$. Surprisingly, this gives the wrong answer. It is important to realize that the energy is conserved here,

$$\mathscr{E}_0 = \frac{p_0^2}{2m} = \left(\frac{p^2}{2m}\right) + V_{\text{op}}, \tag{11}$$

because the Hamiltonian $\mathscr{H}$ is independent of time. Time does not enter the calculation. Because V_{op} is assumed to be small compared to the kinetic energy $p_0^2/2m$, the shift in momentum calculated from equation 11 is

$$\Delta p = p - p_0 \approx \left(-\frac{1}{2}\right)\left(\frac{V_{\text{op}}}{\mathscr{E}_0}\right) \cdot p_0. \tag{12}$$

Then, the phase shift of the neutron wave traversing $\mathscr{R}$ (of length L) due to the potential V_{op} is

$$\Delta\Phi_{\text{nuc}} = \left(\frac{1}{\hbar}\right) \int_0^L \mathbf{\Delta p} \cdot ds = \left(-\frac{1}{2\hbar}\right)\left(\frac{V_{\text{op}}}{\mathscr{E}_0}\right) \cdot p_0 \cdot L = -\lambda_0 NbL, \tag{13}$$

where in the last step we have used equation 10 and the de Broglie relation $p_0 = \hbar k_0 = h/\lambda_0$. For an aluminum slab, 1 mm thick, using 2-Å neutrons, this phase shift is 420 rad $\approx$ 150,000°. It is clear from this that neutron interferometry provides a very sensitive method for accurately measuring neutron-nuclear scattering amplitudes.

GRAVITATIONALLY INDUCED QUANTUM INTERFERENCE

Neutrons, like all matter, are subject to the universal gravitational force. This fact has been demonstrated by verifying that neutrons fall on a parabolic trajectory in the Earth's gravitational field.[18] This is a consequence of classical mechanics and is expected from the principle of equivalence. Is there a quantum phase shift for the neutron accompanying its fall toward the Earth? The answer is yes and it is measurable. Gravity and quantum mechanics do not simultaneously play an important role in most phenomena accessible to terrestrial experimentation. However, a neutron interferometry experiment, for which the outcome necessarily depends upon both the gravitational constant G and Planck's constant h, was first carried out by Colella, Overhauser, and Werner (COW).[8] Subsequently, a series of increasingly precise experiments have been completed by my group at Missouri.[9–11] I now describe briefly the apparatus and geometry of these experiments. Collectively, these experiments are called COW experiments.

A thermal (Maxwellian spectrum) neutron beam is brought out of the 10-MW MURR through a helium gas–filled beam tube and is monochromated by a double-crystal monochromator assembly, which uses either pyrolytic graphite crystals or copper crystals. The monochromatic beam then passes through a series of collimating slits onto the interferometer. The double-crystal monochromator provides a variable-wavelength incident beam directed along the local North-South axis of the Earth, a fact that we will see is important in these experiments.

The experimental procedure involves tilting the interferometer, including the entrance slits and the three ^{3}He detectors C_1, C_2, and C_3, about the incident beam line SAB shown in FIGURE 1. For each angular setting α, neutrons are counted for a preset length of time (actually determined by a monitor detector in the incident beam). Tilting the interferometer allows the path CD to be somewhat higher above the Earth's surface than the beam path AB. The difference in the Earth's gravitational potential between these two levels results in a quantum mechanical phase shift of the neutron wave on trajectory ACD relative to that on the trajectory ABD. The phase accumulated on the rising path AC is equal to the phase accumulated on the opposite rising path BD. The curvature of the trajectories over the distances involved in the interferometer is negligible. If we assume that the gravitational potential is Newtonian, then (as a function of the tilt angle α) we have

$$\Delta V(\alpha) = mgH_0 \sin(\alpha), \tag{14}$$

where m is the neutron mass, g is the local acceleration due to gravity, and H_0 is the perpendicular distance between the lines AB and CD. At Columbia, Missouri, $mg = 1.023 \times 10^{-9}$ eV/cm. The difference in the neutron's momentum between these two levels is

$$\Delta p = p - p_0 = \left(-\frac{1}{2}\right)\left[\frac{mgH_0 \sin(\alpha)}{\mathscr{E}_0}\right] p_0, \tag{15}$$

where the energy $\mathscr{E}_0$ is a constant of the motion; in terms of the initial wave vector k_0, it is

$$\mathscr{E}_0 = \frac{\hbar^2 k_0^2}{2m}. \tag{16}$$

The phase difference for neutron waves on path II relative to path I, induced by the Earth's gravity, is then easily evaluated:

$$\Delta\Phi_{\text{grav}} = \left(\frac{1}{\hbar}\right) \oint \mathbf{\Delta p} \cdot ds = -2\pi m^2 \left(\frac{g}{h^2}\right) \lambda_0 A_0 \sin(\alpha) \equiv q_{\text{COW}} \sin(\alpha). \tag{17}$$

The area A_0 enclosed by the beam paths is

$$A_0 = (2d^2 + 2ad) \tan \theta_B, \tag{18}$$

where d is the distance between the crystal slabs in the interferometer, a is their thickness, and θ_B is the Bragg angle for the Si(220) reflection for neutrons of wavelength λ_0. For $\lambda_0 = 1.4$ Å, this area is about 10 cm^2 for the interferometer used in our experiments. Note that equation 17 contains both the gravitational constant and Planck's constant.

As the interferometer is tilted through various angles α, always maintaining the Bragg condition, equations 1 and 2 predict that we should observe quantum oscillations in the counting rates in detectors C_2 and C_3, induced by the gravitational field of the Earth. An interferogram obtained with 1.407-Å incident neutrons is shown in FIGURE 4.[10] The unexpected loss of contrast was originally interpreted as being due to the fact that the interferometer bends and warps under its own weight (on the scale of angstroms) as it is rotated about the incident beam axis, which is not an axis of elastic symmetry of the Si-crystal interferometer. We have studied these effects with *in situ* X-ray experiments. X rays were directed along the same incident beam path and the interfering X-ray beams were monitored as a function of rotation angle. The effect of gravity on X rays (gravitational redshift) over the distances involved in the interferometer is negligible. The frequency of oscillation of the interferogram due to bending was therefore measured directly with X rays and subtracted from the frequency of oscillation measured with neutrons, leaving only the effects of gravitationally induced quantum interference. Horne[19] has reanalyzed these COW-type experiments, pointing out that the three-crystal *LLL* interferometer is not a simple two-path device, but really an eight-path interferometer. This has the effect that the single interferometer area A_0 appearing in the COW phase shift formula (equation 17) should be replaced by a dynamical diffraction intensity-

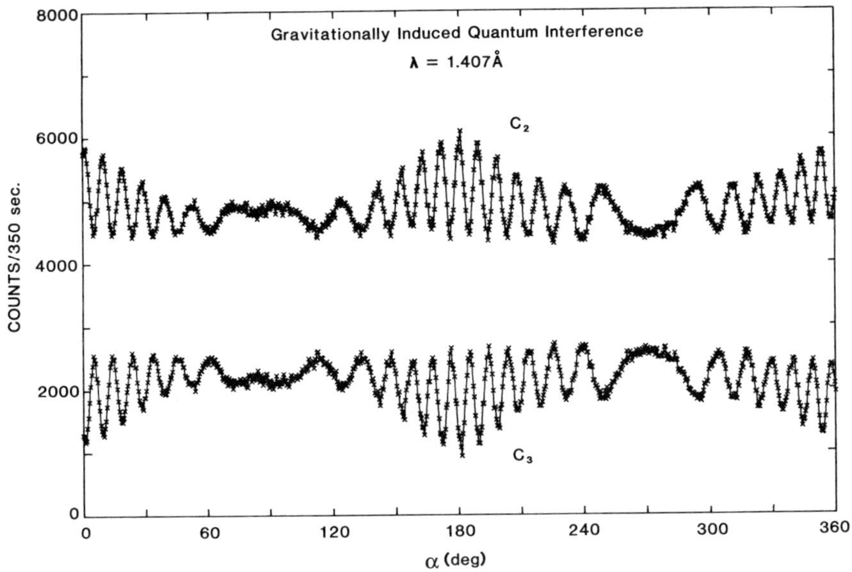

FIGURE 4. Interferogram resulting from tilting the interferometer shown in FIGURE 1 through angles α about the incident beam direction *SAB*. The angle α is zero when the parallelogram *ABDC* defining the beam paths is horizontal. Note that the sum of the counting rates in the two detectors $C_2 + C_3$ is a constant, independent of α, as required by equations 1 and 2. (See reference 10.)

weighted average over three areas, $A_0, A_0 + \delta A$, and $A_0 - \delta A$, where $\delta A/A_0 = \Gamma a/d$. Here, a is the thickness of each of the three slabs, d is the distance between them, and Γ is a factor (less than unity) dependent upon the misset angle $\Delta\theta$ of a given incident ray from the exact Bragg condition. Using Horne's theory, we find that a correction of 4.8% to the experimental frequency of the gravity interferogram should be made before comparison with the COW phase shift formula. The fact that the experiment actually involves more than one interferometer area also explains the loss of contrast with increasing tilt angles α. Similar conclusions, using a complete spherical-wave dynamical diffraction treatment, were reached by Bonse and Wroblewski[20] in analyzing their acceleration-induced interferometry experiment. We have shown that this geometrical correction to the COW formula for the frequency of oscillation is in fact given by[10]

$$q_{\text{grav}} = \left[1 + \left(\frac{2}{3}\right)\left(\frac{a}{d}\right)\right] q_{\text{COW}}. \tag{19}$$

For our interferometer, $a/d = 0.0714$, so

$$q_{\text{grav}} = 1.0476 q_{\text{COW}}. \tag{20}$$

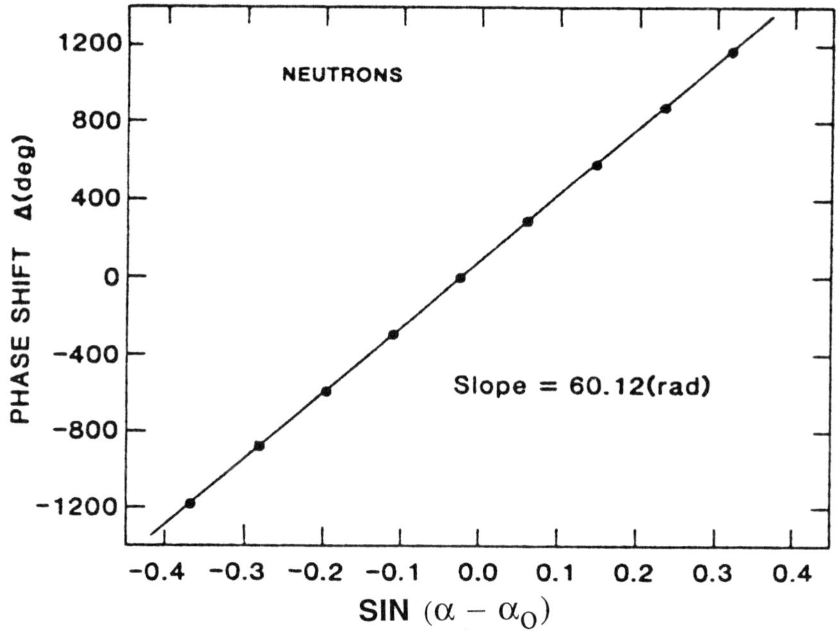

FIGURE 5. For a given tilt angle α, an interferogram is recorded by rotating the phase rotator (a slab of aluminum). The phase of this interferogram is called $\delta(\alpha)$. The functional form of the interferogram is $I_3(\delta, \alpha) = a + b \cos[\Delta\Phi(\delta) + \Delta(\alpha)]$. The tilt angle is then changed and another interferogram is recorded. Note that $\Delta(\alpha) = q_{\text{exp}} \sin(\alpha - \alpha_0)$. See equation 21 and reference 10.

We now turn to comparing this predicted frequency of the gravitationally induced quantum interferogram with our most recent experiments. We have followed a somewhat different procedure than in the original COW experiments. The phase shift $\Delta\Phi_{\text{grav}}(\alpha) = q_{\text{grav}} \sin(\alpha)$ is measured directly by first setting $\alpha = 0$ and rotating the phase rotator through successive angles δ. Due to the neutron-nuclear optical potential of the phase rotator, this results in a sinusoidal interferogram. Then, tilting the interferometer through an angle α and repeating the scan of the phase rotator gives rise to another sinusoidal interferogram. The difference in phase between these two interferograms is $\Delta\Phi_{\text{grav}}(\alpha)$. The results of a series of very accurate measurements, using 1.417-Å neutrons, are summarized in FIGURE 5. The abscissa is $\sin(\alpha - \alpha_0)$, where α_0 is a correction due to the Sagnac effect ($\alpha_0 = 1.41°$), which will be discussed below. The slope of $\Delta\Phi_{grav}$ versus $\sin(\alpha - \alpha_0)$ is the experimental frequency q_{exp} of the gravity interferogram. To compare this frequency with q_{grav}, we must make a correction for bending and for the Sagnac effect, namely,

$$q_{grav} = (q_{exp}^2 - q_{Sagnac}^2)^{1/2} - q_{bend}. \tag{21}$$

Theory gives $q_{Sagnac} = 1.45$ rad and, with X rays, we measure $q_{bend} = 1.41$ rad. Thus, our current experimental result is

$$q_{grav}(\text{observed}) = 58.72 \pm 0.03 \text{ rad}. \tag{22}$$

Equation 20 predicts

$$q_{grav}(\text{theory}) = 59.19 \text{ rad}. \tag{23}$$

Hence, the observed frequency due to gravity is 0.8% lower than theory predicts.

Layer and Greene suggested that this discrepancy might be due to the fact that X rays interrogate a slightly different region of the Si-crystal slabs than the neutron beams;[21] that is, the correction for bending effects using X rays may be in error. Recent X-ray experiments of Arif *et al.*[11] have shown that the bending effects are considerably more complicated than originally thought and corrections using X rays to the neutron frequency of oscillation will require additional analysis.

A way to eliminate the effects of bending and thereby facilitate a more accurate comparison with theory is to float the interferometer in a fluid of density equal to that of Si. This suggestion was first made by Zeilinger (see comments at the end of reference 10). Implementation of this idea presents several experimental challenges.

THE NEUTRON SAGNAC EFFECT

The French scientist, M. G. Sagnac, was the first to demonstrate that optical interferometry is sensitive to angular rotation.[22] This effect is now used in navigation and is the basis for the ring-laser gyroscope. In 1925, Michelson, Gale, and Pearson carried out an heroic experiment.[23] They constructed an interferometer in the form of a rectangle of size 2010 ft × 1113 ft and were able to detect the retardation of light due to the Earth's rotation, corresponding to about ¼ of a fringe, in agreement with relativity theory. Because the coordinate transformation properties of light waves and neutron matter waves are different, it cannot be taken for granted that an analogous quantum mechanical effect will exist for neutrons.

The gravitationally induced quantum interference experiments were carried out on the surface of our rotating Earth, a noninertial frame of reference. The classical Hamiltonian governing the neutron's motion in the gravitational field and frame of our rotating Earth is[24]

$$\mathscr{H} = \left(\frac{p^2}{2m_i}\right) - m_g \boldsymbol{g} \cdot \boldsymbol{r} - \boldsymbol{\omega} \cdot \boldsymbol{L}, \tag{24}$$

where $\boldsymbol{L}$ is the angular momentum of the neutron's motion about the center of the Earth ($\boldsymbol{r} = 0$), namely,

$$\boldsymbol{L} = \boldsymbol{r} \times \boldsymbol{p}, \tag{25}$$

and where $\boldsymbol{p}$ is the canonical momentum and $\boldsymbol{\omega}$ is the angular rotation velocity of the Earth. The third term in equation 24 gives rise to the Coriolis and centripetal accelerations. Over the distances involved in the interferometer, the Coriolis force has a negligible effect on the neutron's trajectory. However, its effect on the neutron's phase is measurable. Using Hamilton's equation $\boldsymbol{v} = \partial\mathscr{H}/\partial\boldsymbol{p}$, we see that the neutron's canonical momentum is

$$\boldsymbol{p} = m_i\boldsymbol{v} + m_i\boldsymbol{\omega} \times \boldsymbol{r}. \qquad \textbf{(26)}$$

We have been more careful here to distinguish between the neutron's inertial mass m_i and its gravitational mass m_g. The Hamiltonian in equation 24 is velocity-dependent, but time-independent; thus, the energy $\mathscr{E}$ is conserved. According to our prescription of the second section for calculating the phase shift, we get two terms from equation 26, namely,

$$\Delta\Phi = \left(\frac{1}{\hbar}\right)\oint \boldsymbol{p} \cdot d\boldsymbol{s} = \Delta\Phi_{\text{grav}} + \Delta\Phi_{\text{Sagnac}}. \qquad \textbf{(27)}$$

Using elementary vector calculus, the Sagnac phase shift is easily found to be

$$\Delta\Phi_{\text{Sagnac}} = \left(\frac{2m_i}{\hbar}\right)\boldsymbol{\omega} \cdot \boldsymbol{A}_0, \qquad \textbf{(28)}$$

where $\boldsymbol{A}_0$ is the normal area vector of the interferometer loop *ABDCA*. When the incident beam is horizontal and directed along the local North-South axis of the Earth, equation 28 gives

$$\Delta\Phi_{\text{Sagnac}} = \left(\frac{4\pi m_i\omega A_0}{h}\right)\cos(\theta_L)\cos(\alpha) = q_{\text{Sagnac}}\cos(\alpha). \qquad \textbf{(29)}$$

The colatitude angle is $\theta_L = 51.376°$ at Columbia, Missouri. Because this phase shift depends upon the cosine of the interferometer tilt angle α and $\Delta\Phi_{\text{grav}}$ depends upon the sine of α, the frequency of oscillation of the interferogram can be written as

$$q = \sqrt{q_{\text{grav}}^2 + q_{\text{Sagnac}}^2} \approx q_{\text{grav}}\left[1 + \left(\frac{1}{2}\right)\frac{q_{\text{Sagnac}}^2}{q_{\text{grav}}^2}\right]. \qquad \textbf{(30)}$$

Numerically, $q_{\text{Sagnac}}/q_{\text{grav}} \approx 0.02$. Thus, the effect of the Earth's rotation adds a small correction of only 2 parts in 10^4 to the interferogram frequency due to the Earth's gravity.

However, in an experiment carried out in collaboration with Staudenmann and Colella,[12] a vertically directed beam was used as shown in FIGURE 6. The Sagnac phase shift was measured (using the phase-rotator technique described earlier for the gravity experiments) as a function of the interferometer orientation angle α about the vertical axis. From symmetry, it is clear that there is no α-dependent gravity-induced phase shift for this geometry. In terms of the colatitude angle Θ_L at the point on the Earth's surface where the experiment was done and of the angle α, the Sagnac phase shift (from equation 28) is given by

$$\Delta\Phi_{\text{Sagnac}} = \left(\frac{2m_i}{\hbar}\right)\omega A_0 \sin(\Theta_L)\sin(\alpha). \qquad \textbf{(31)}$$

The experimental results are shown in FIGURE 7. When the normal area vector A_0 points West or East, the phase shift is zero; when A_0 is directed North or South, it is $+95°$ and $-95°$, respectively. This experimental result is in reasonable agreement with equation 31, which predicts that $\Delta\Phi_{Sagnac}$ should be $+92°$ and $-92°$ for the North and South orientations, respectively.

It is interesting to note that the results of this experiment depend only upon the inertial neutron mass m_i, whereas the results of the gravity experiment depend upon

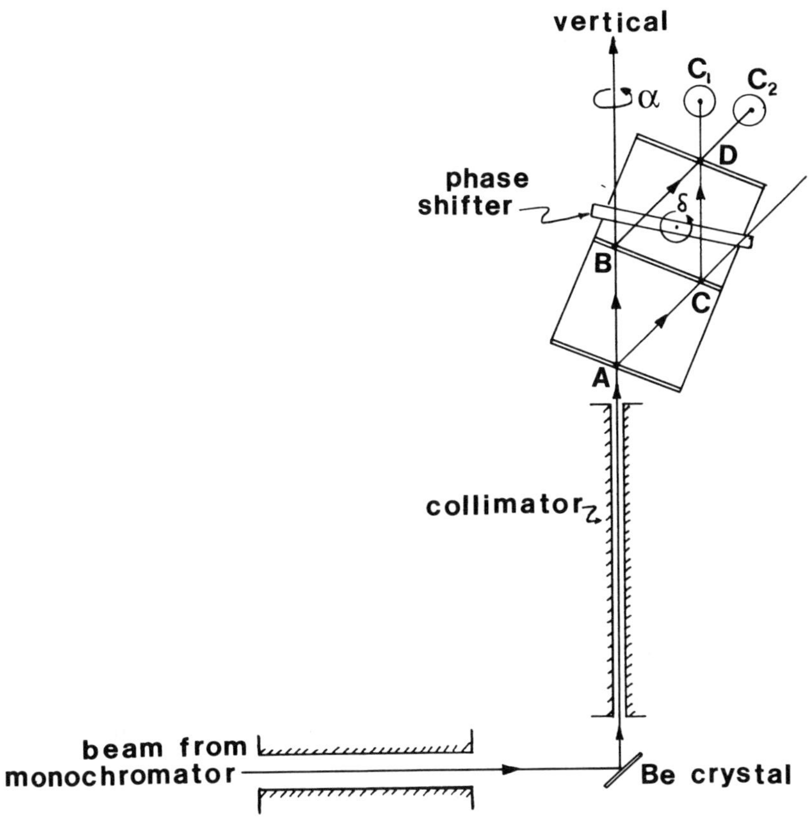

FIGURE 6. Schematic diagram of the vertical beam geometry used in observing the neutron Sagnac effect.[12]

the product of the gravitational mass m_g with the inertial mass m_i. Thus, one can interpret the combination of the two experiments as a quantum mechanical interference measurement of the inertial and gravitational masses. We point out that neutrons are much more sensitive to the Sagnac effect than photons. This can be seen if we write $mv = \hbar k$ and replace $m_i/\hbar$ in equation 28 by k/v. For photons, $v = c$, the velocity of light, which is about 10^5 times larger than for our thermal neutrons.

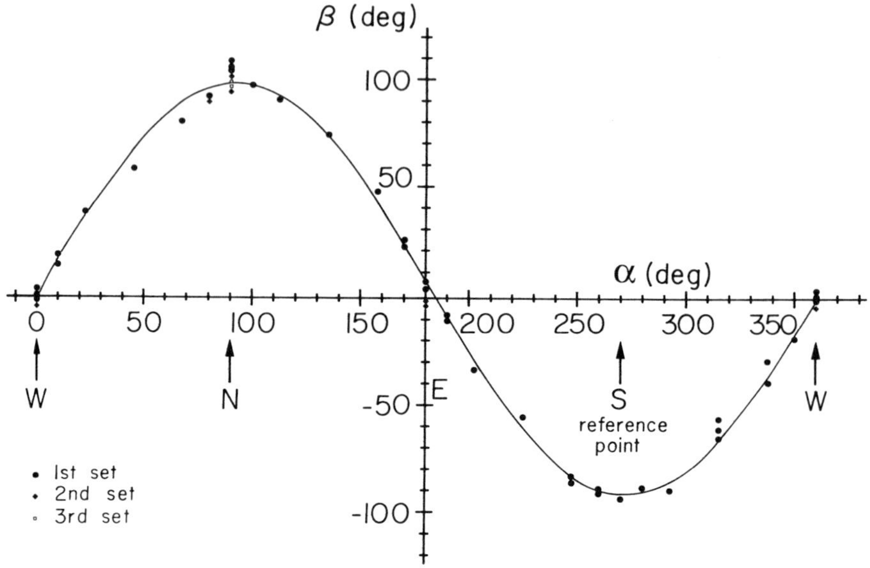

FIGURE 7. Effect of the Earth's rotation on the quantum mechanical phase of the neutron. The angle α specifies the orientation of the interferometer's normal area vector with respect to the local N-S axis of the Earth at Columbia, Missouri.[12] The ordinate, labeled β, is the Sagnac phase shift, $\Delta\Phi_{\text{Sagnac}}$.

THE AHARONOV-CASHER EFFECT

According to the Maxwell theory of electrodynamics, potentials are merely convenient mathematical tools for calculating the electromagnetic fields of force. In quantum mechanics, potentials enter the Schrödinger equation and produce phase shifts, even in geometries where the potential gives rise to no electromagnetic field. These are the situations of interest in the topological interference effects of Aharonov and Bohm.[25] For electrons, the effects are of two types: the magnetic (or vector) AB effect and the electric (or scalar) AB effect. We have observed directly analogous effects with neutrons. We describe the neutron vector AB effect in this section, called the Aharonov-Casher (AC) effect.[13]

In 1984, Aharonov and Casher proposed that a beam of neutral particles with a magnetic dipole moment passing around opposite sides of a line charge will undergo a relative quantum phase shift. This AC effect is considered to be an electrodynamic and quantum mechanical dual of the Aharonov-Bohm effect for charged particles, as can be understood by the schematic diagram shown in FIGURE 8. The AB flux tube has been replaced by a line of electric charge of lineal density Λ and the electron beam of charge e^- has been replaced by a beam of neutrons with magnetic moment μ. If one views the AB flux tube as a line of magnetic dipoles, one sees that the roles of charge and magnetic dipole have been interchanged between the AB effect and the AC effect.

For a neutron of mass m and magnetic moment μ, moving in the region of an electric field $\boldsymbol{E}$, the Hamiltonian is[26]

$$\mathcal{H} = \left(\frac{p^2}{2m}\right) - \left(\frac{1}{mc}\right) \boldsymbol{\mu} \cdot (\boldsymbol{E} \times \boldsymbol{p}). \tag{32}$$

Thus, using Hamilton's equation $\boldsymbol{v} = \partial\mathcal{H}/\partial\boldsymbol{p}$, we see that the canonical momentum is

$$\boldsymbol{p} = m\boldsymbol{v} + \left(\frac{\boldsymbol{\mu}}{c}\right) \times \boldsymbol{E}. \tag{33}$$

It is a straightforward matter to show that there is no force on the neutron if the electric field comes from a line charge along the z-axis, and the neutrons are polarized along z. An expression for the acceleration $\boldsymbol{a}$ is obtained by taking the total time derivative of the velocity $\boldsymbol{v}$ above, using the second of Hamilton's equations $\dot{\boldsymbol{p}} = -\partial\mathcal{H}/\partial\boldsymbol{r}$ and elementary vector algebra. The result is[27]

$$\boldsymbol{a} = \left(-\frac{1}{mc}\right) (\boldsymbol{\mu} \cdot \boldsymbol{\nabla})(\boldsymbol{v} \times \boldsymbol{E}). \tag{34}$$

For neutrons polarized in the z-direction,

$$\boldsymbol{\mu} \cdot \boldsymbol{\nabla} = \mu \frac{\partial}{\partial z}. \tag{35}$$

However, for a line charge, $\boldsymbol{E} = \boldsymbol{E}(x,y)$, independent of z; thus,

$$\boldsymbol{a} = 0 \tag{36}$$

for this line charge geometry.

In other words, the velocity $\boldsymbol{v}$ is a constant of the motion. For a neutron

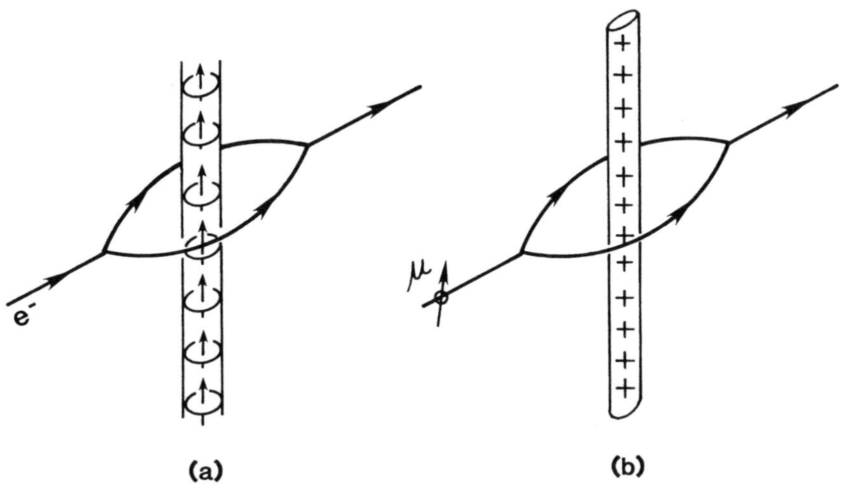

FIGURE 8. Diagram illustrating the duality between (a) the Aharonov-Bohm effect for electrons diffracting around a tube of magnetic flux and (b) the Aharonov-Casher effect for neutrons diffracting around a line charge.

diffracting around a line charge of lineal density Λ, one obtains the AC phase shift by evaluating the line integral of $\boldsymbol{p}$, namely,

$$\Delta\Phi_{AC} = \left(\frac{1}{\hbar}\right)\oint \boldsymbol{p}\cdot d\boldsymbol{s} = \sigma\left(\frac{4\pi\mu\Lambda}{\hbar c}\right), \quad (37)$$

where $\sigma = \pm 1$ depending upon whether the neutron spin is up or down with respect to the plane of the neutron's motion. We have used the fact that the electric field due to a line charge is $\boldsymbol{E} = 2\Lambda\hat{\boldsymbol{r}}/r$. As pointed out by AC, this phase shift depends only

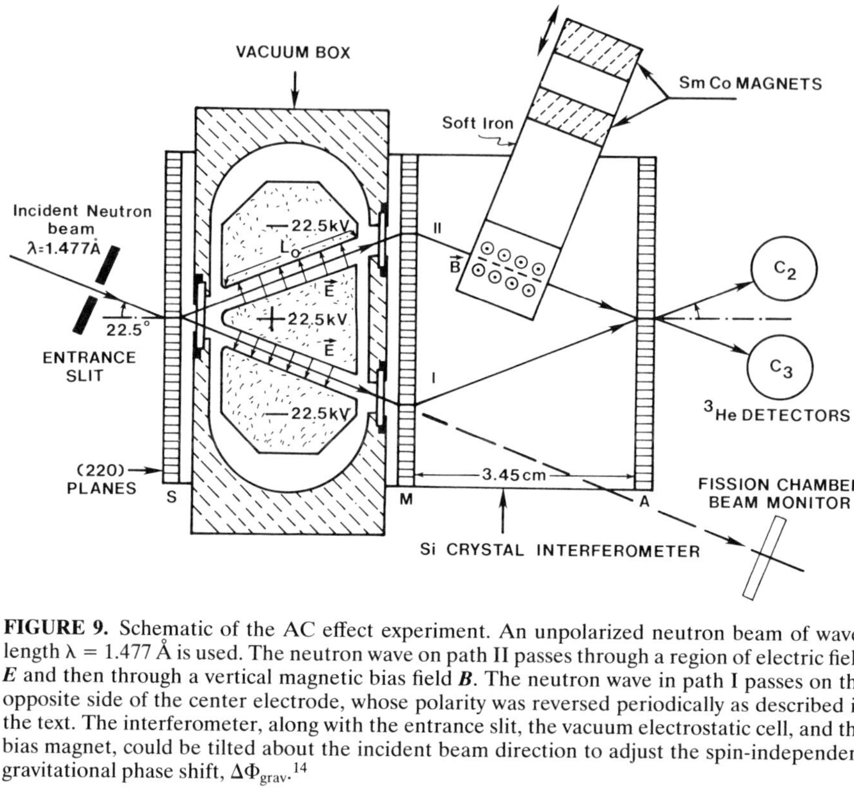

FIGURE 9. Schematic of the AC effect experiment. An unpolarized neutron beam of wavelength $\lambda = 1.477$ Å is used. The neutron wave on path II passes through a region of electric field $\boldsymbol{E}$ and then through a vertical magnetic bias field $\boldsymbol{B}$. The neutron wave in path I passes on the opposite side of the center electrode, whose polarity was reversed periodically as described in the text. The interferometer, along with the entrance slit, the vacuum electrostatic cell, and the bias magnet, could be tilted about the incident beam direction to adjust the spin-independent gravitational phase shift, $\Delta\Phi_{grav}$.[14]

upon the lineal charge density Λ, enclosed by the beam paths, but not on any details of their geometrical shape. In this sense, the effect is topological. Thus, instead of a line charge, a prism-shaped electrode was placed between the splitter (S) and mirror plate (M) of the interferometer as shown in FIGURE 9, thereby enabling a much larger lineal charge density to be obtained. An electrode can be regarded as the superposition of many line charges. The experiment was a collaborative University of Melbourne–University of Missouri project carried out in Columbia, Missouri.

For the electrode assembly of FIGURE 9, Gauss' law allows us to replace Λ by

$2VL/4\pi D$. Here, V is the potential difference between the electrodes, D is their separation, and L is the effective path length as shown in FIGURE 9. In terms of these parameters, for $D = 0.154$ cm, $L = 2.53$ cm, and $V = 45$ kV ($= 150$ stat volts), we find

$$\Delta\Phi_{AC} = 1.50\sigma \text{ milliradians.} \tag{38}$$

It was assumed in the derivation of equation 37 that the neutrons are polarized along an axis parallel to the line charge, that is, the z-axis. However, it is not necessary to use polarized neutrons if an additional spin-independent phase shift is judiciously introduced and fine-tuned. In this experiment, gravity was used for this purpose. The introduction of a further spin-dependent phase shift $\Delta\beta_M$, by means of a magnetic bias field, enabled maximum sensitivity to the AC effect to be accomplished.

An unpolarized beam can be thought of as consisting of two beams, one spin-up and the other spin-down. For the spin-up component, the counting rate in detector C_3 is

$$I_3^{\uparrow} = \left(\frac{1}{2}\right)[a_3 + b_3 \cos(\Delta\alpha + \Delta\beta)], \tag{39}$$

whereas, for spin-down neutrons, it is

$$I_3^{\downarrow} = \left(\frac{1}{2}\right)[a_3 + b_3 \cos(\Delta\alpha - \Delta\beta)]. \tag{40}$$

The spin-independent phase shift is called $\Delta\alpha$ and the spin-dependent phase shift is called $\Delta\beta$. Thus, for an incident beam of unpolarized neutrons, we have

$$I_3 = I_3^{\uparrow} + I_3^{\downarrow} = a_3 + b_3 \cos(\Delta\alpha) \cos(\Delta\beta). \tag{41}$$

We adjust $\Delta\alpha$ by gravitationally induced quantum interference to be zero (mod 2π) and we adjust $\Delta\beta_M$ to be $\pi/2$ or $3\pi/2$, where

$$\Delta\beta = \Delta\beta_M + \Delta\Phi_{AC}. \tag{42}$$

Thus, because $\Delta\Phi_{AC}$ is such a small phase angle, we have

$$I_3(\pm) = a_3 \pm b_3|\Delta\Phi_{AC}| \tag{43}$$

from equation 41, where the + sign is for negative center electrode polarity and the − sign is for positive center electrode polarity. A similar expression applies to the counting rate in detector C_2, namely,

$$I_2(\pm) = a_2 \mp b_2|\Delta\Phi_{AC}|. \tag{44}$$

Thus, the count rates are linearly proportional to $\Delta\Phi_{AC}$.

The magnetic bias field was varied by changing the reluctance of a magnetic circuit. To find the correct operating point for $\Delta\alpha$ and $\Delta\beta_M$, the following procedure was followed: With the magnetic field $\boldsymbol{B}$ set equal to zero, the interferometer was tilted in steps. The first maximum near zero tilt angle sets $\Delta\alpha = 0$ (mod 2π). Leaving the tilt angle fixed at this maximizing value, operating points of negative slope ($\Delta\beta_M = \pi/2$) and positive slope ($\Delta\beta_M = 3\pi/2$) were selected by scanning the magnetic field as shown in FIGURE 10. The magnetic field was then changed in small increments around the value giving $\Delta\beta_M \approx \pi/2$ or $3\pi/2$. Gravity scans were then

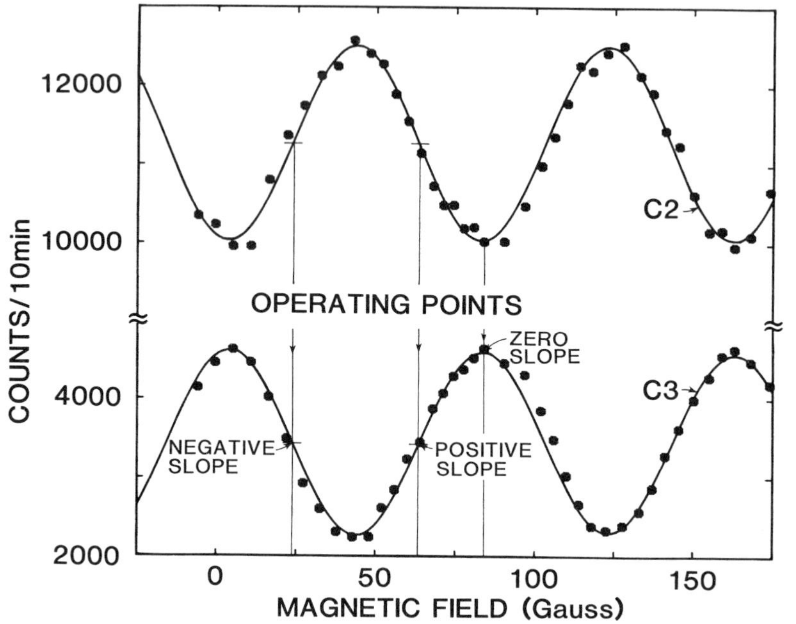

FIGURE 10. Magnetic scan interferogram when the interferometer tilt angle is $\alpha = 1.1°$.[14]

carried out for each value of the magnetic field and the optimum was chosen as that giving minimum oscillation.

A complete sequence of magnetic scans for various tilt angles and of gravity scans for various magnetic fields is shown in FIGURE 11. The procedure for selecting the precise operating points was repeated every week during the course of the experiment.

After the optimum operating conditions were established, the 45-kV high voltage was switched on across the electrode gap and then periodically reversed in polarity. Each switching cycle (positive, zero, negative) took about 30 minutes. The total data accumulation time was about 2 years. A summary of the data is given in TABLE 2. $C_2(+)$ and $C_3(+)$ are the total number of counts accumulated in detectors C_2 and C_3 per cycle for a positive center electrode polarity, whereas $C_2(-)$ and $C_3(-)$ are for a negative center electrode polarity. Only the difference counts, $\Delta C_2 = C_2(+) - C_2(-)$ and $\Delta C_3 = C_3(+) - C_3(-)$, per cycle are given in TABLE 2. The positive slope data apply to when the bias magnetic field is adjusted to give $\Delta\beta_M = 3\pi/2$, whereas the negative slope data apply to the case $\Delta\beta_M = \pi/2$. Notice that the expected symmetry and sign reversals for the difference counts are found in the data of TABLE 2.

By statistically combining the results for the positive and negative slopes according to the rule

$$(\Delta C) = \omega_2 \Delta C_2 + \omega_3 \Delta C_3, \tag{45}$$

where ω_2 (= 0.24) and ω_3 (= 0.76) are the statistical weighting factors (inversely proportional to the square of the standard deviations), we obtain the last column of

TABLE 2. A fit to the magnetic scan of FIGURE 10 gives the slope of the interferogram at the operating points, namely,

$$b = b_2 = b_3 = 1234 \pm 15 \text{ counts/cycle}. \quad \textbf{(46)}$$

The phase shift due to the AC effect is then obtained from the difference counts and this slope, as shown in TABLE 2.

Combining the positive and negative slope data, this experiment yields the first measurement of the AC phase shift:

$$\Delta\Phi_{AC} = 2.11 \pm 0.34 \text{ mrad}. \quad \textbf{(47)}$$

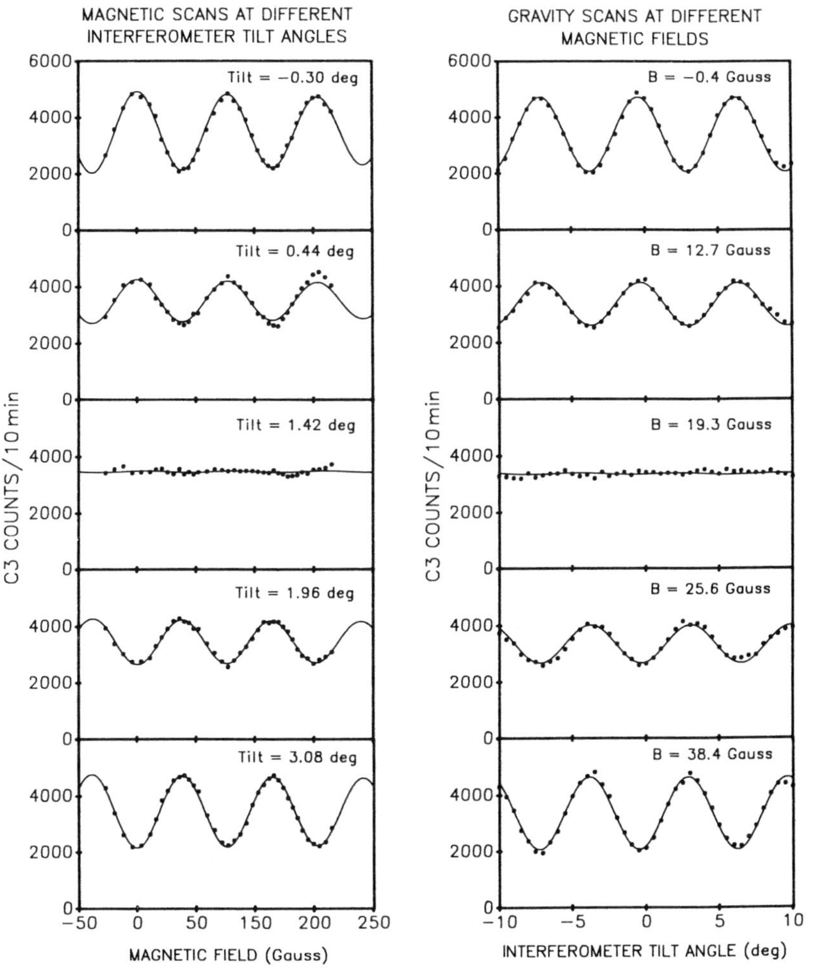

FIGURE 11. A complete sequence of magnetic scans for various tilt angles (left) and a complete sequence of gravity scans at various bias magnetic field values (right).[14]

This result is to be compared to the theoretical prediction of 1.50 mrad. The accuracy of the experiment is limited by available neutron intensity and long-term apparatus stability. Obviously, the much larger interferometers currently being designed and fabricated in our laboratory will be important in the next generation of AC effect experiments. Changing the charged electrode geometry is necessary to elucidate the topological aspects of the AC predictions. An experiment is being developed to pursue this feature of the theory.

The essential necessity of the AB effect for the self-consistency of quantum mechanics was first clearly elucidated by Furry and Ramsey in 1960.[28] Recently, Ramsey has given a similar argument for the AC effect and has discussed the complementarity of two-path neutron interferences with separated oscillatory-field resonances.[29] An experimental observation of the AC phase shift by the Ramsey technique, which can be viewed as interferometry in spin space, has recently been achieved by Sangster *et al.* in a beautiful molecular beam experiment using thallium fluoride molecules.[30] They were able to verify the velocity independence of $\Delta\Phi_{AC}$ and its linear dependence on the electric field $\boldsymbol{E}$ by observing the interference of the two spin states of the fluorine nucleus.

TABLE 2. Difference Counts/Cycle

Experimental Condition	$C_2(+) - C_2(-)$	$C_3(+) - C_3(-)$	$\langle \Delta C \rangle$
positive slope, 3539 cycles, 74 days	$+7.66 \pm 2.48$	-4.80 ± 1.39	5.49 ± 1.21
negative slope, 3654 cycles, 76 days	-5.43 ± 2.36	$+4.66 \pm 1.32$	4.84 ± 1.15

positive slope: $\Delta\Phi_{AC} = \langle \Delta C \rangle / 2b = 2.22 \pm 0.49$ mrad
negative slope: $\Delta\Phi_{AC} = \langle \Delta C \rangle / 2b = 1.96 \pm 0.46$ mrad

CONCLUDING REMARKS

The three experiments discussed in this report (gravitationally induced quantum interference, neutron Sagnac effect, and Aharonov-Casher effect) show the didactic aspects of neutron interferometry and its impact on understanding the nonlocal nature of quantum mechanics. The extensive list (TABLE 1) of experiments carried out with the perfect Si-crystal interferometer encompass a large number of the central ideas and the "magic" of quantum theory. Our current agenda of experiments at Missouri for the immediate future include the following:

(a) A Berry phase experiment along the lines of the proposal of Wagh and Rakhecha, where the nonlocal neutron spin is allowed to precess about magnetic fields $\boldsymbol{B}_1$ on path I and $\boldsymbol{B}_2$ along path II, where there is an angle β between $\boldsymbol{B}_1$ and $\boldsymbol{B}_2$.[31] A geometric phase shift $\Delta\Phi_{geom} = \beta$ is expected, in addition to the dynamical phase shift due to local spin precession.

(b) A multiphoton absorption experiment using polarized neutrons with large amplitude RF flipper coils within the interferometer according to the ideas proposed by Summhammer.[32]

(c) A phase shift due to accelerated matter within one leg of the interferometer. Recently, Kowalski has challenged the validity of the principle of equivalence under conditions where phase shifts due to gravity and matter occur simultaneously.[33] He finds a problem in second order, that is, for terms involving the product of the gravitational potential U_{grav} and the mean nuclear potential of a slab of matter U_{matter} in the index of refraction, $n = (1 - U/E_0)^{1/2}$.

(d) As mentioned in the previous section, the topological aspects of the AC effect have not yet been verified. We intend to pursue this by varying the geometry of the charged electrode assembly within the interferometer.

(e) A cross-correlation Wheeler-delayed-choice experiment, in which He-3 neutron detectors with a 50% transmission are placed in each leg of the interferometer.

Even more challenging experiments have been discussed for some time. They include a neutron version of the Michelson-Morley experiment, a neutron Cavendish experiment where the phase shift due to local moving masses is pursued, the gravitational phase shift due to the Moon and the Sun, and a gravitational version of the AB effect.

ACKNOWLEDGMENTS

I have benefited greatly over the years from working with many fine students and collaborators—in particular, H. Kaiser, J-L. Staudenmann, A. G. Klein, G. I. Opat, H. Rauch, A. W. Overhauser, R. Colella, C. F. Eagen, R. Clothier, M. Arif, and G. L. Greene; and my present students, K. Littrell and D. Jacobson.

REFERENCES

1. CHADWICK, J. 1932. Nature **129:** 312; Proc. R. Soc. London **A136:** 692.
2. ELASSER, W. M. 1936. C. R. Acad. Sci. Paris **202:** 1029.
3. HALBAN, H. V. & P. PREISWERK. 1936. C. R. Acad. Sci. Paris **203:** 73.
4. MITCHELL, D. D. & P. N. POWERS. 1936. Phys. Rev. **50:** 486.
5. BONSE, U. & M. HART. 1965. Appl. Phys. Lett. **6:** 155.
6. RAUCH, H., W. TREIMER & U. BONSE. 1974. Phys. Lett. **A47:** 369.
7. BOHR, N. & J. A. WHEELER. 1939. Phys. Rev. **56:** 426.
8. COLELLA, R., A. W. OVERHAUSER & S. A. WERNER. 1975. Phys. Rev. Lett. **34:** 1472.
9. STAUDENMANN, J-L., S. A. WERNER, R. COLELLA & A. W. OVERHAUSER. 1980. Phys. Rev. Lett. **A21:** 149.
10. WERNER, S. A., H. KAISER, M. ARIF & R. CLOTHIER. 1988. Physica **B151:** 22.
11. ARIF, M., M. S. DEWEY, G. L. GREENE, D. JACOBSON & S. A. WERNER. 1994. Phys. Lett. **A184:** 154.
12. WERNER, S. A., J-L. STAUDENMANN & R. COLELLA. 1979. Phys. Rev. Lett. **42:** 1102.
13. AHARONOV, Y. & A. CASHER. 1984. Phys. Rev. Lett. **53:** 319.
14. CIMMINO, A., G. I. OPAT, A. G. KLEIN, H. KAISER, S. A. WERNER, M. ARIF & R. CLOTHIER. 1989. Phys. Rev. Lett. **63:** 38.
15. BADUREK, G., H. RAUCH & A. ZEILINGER, Eds. 1988. Matter Wave Interferometry: Proceedings of a Workshop held in Vienna, Austria (September 14–16, 1987).
16. WERNER, S. A. & A. G. KLEIN. 1986. Neutron optics. *In* Neutron Scattering (Chapter 4). Volume 23A in the Methods of Experimental Physics series. D. L. Price & K. Sköld, Eds. Academic Press. New York.
17. RAUCH, H. & S. A. WERNER. 1995. Neutron Interferometry. In preparation.

18. DOBBS, J. W. T., J. A. HARVEY, D. DAYA & H. HORSTMANN. 1965. Phys. Rev. **139:** 756.
19. HORNE, M. A. 1986. Physica **B137:** 260.
20. BONSE, U. & T. WROBLEWSKI. 1984. Phys. Rev. **D30:** 1214.
21. LAYER, H. P. & G. L. GREENE. 1991. Phys. Lett. **155:** 450.
22. SAGNAC, M. G. 1913. C. R. Acad. Sci. Paris **157:** 708.
23. MICHELSON, A. A., H. G. GALE & F. PEARSON. 1925. Astrophys. J. **61:** 140.
24. LANDAU, L. D. & E. M. LIFSHITZ. 1969. *In* Mechanics. Second edition, p. 126–129. Pergamon. Elmsford, New York.
25. AHARONOV, Y. & D. BOHM. 1959. Phys. Rev. **115:** 485.
26. FOLDY, L. L. 1958. Rev. Mod. Phys. **30:** 471.
27. AHARONOV, Y., D. PEARLE & L. VAIDMAN. 1988. Phys. Rev. **A37:** 4052.
28. FURRY, W. H. & N. F. RAMSEY. 1960. Phys. Rev. **118:** 623.
29. RAMSEY, N. F. 1993. Phys. Rev. **A48:** 80.
30. SANGSTER, K., E. A. HINDS, S. M. BARNETT & E. RIIS. 1993. Phys. Rev. Lett. **71:** 3641.
31. WAGH, A. G. & V. C. RAKHECHA. 1990. Phys. Lett. **A148:** 17; 1993. Phys. Rev. **A48:** R1729.
32. SUMMHAMMER, J. 1993. Phys. Rev. **A47:** 556.
33. KOWALSKI, F. V. 1993. Phys. Lett. **A182:** 335.

More Quantum Information due to Postselection in Neutron Interferometry

HELMUT RAUCH

Atominstitut der Österreichischen Universitäten
A-1020 Wien, Austria

INTRODUCTION

Different kinds of neutron interferometers have been tested in the past. The slit interferometer is based on wave-front division and provides long beam paths, but only a very small beam separation.[1,2] The perfect-crystal interferometer[3,4] provides the highest intensity and the highest flexibility for beam handling and is now most frequently used due to its wide beam separation and its universal availability for fundamental, nuclear, and solid-state physics research. The interferometer based on grating diffraction is a more recent development and has its main application for very slow neutrons.[5] A schematic comparison is shown in FIGURE 1.

The perfect-crystal interferometer represents a macroscopic quantum device with characteristic dimensions of several centimeters. The basis for this kind of neutron interferometry is provided by the undisturbed arrangement of atoms in a monolithic perfect silicon crystal.[3,6] An incident beam is split coherently at the first crystal plate, reflected at the middle plate, and coherently superposed at the third plate [FIGURE 1 (middle) and FIGURE 2].

From general symmetry considerations, it follows immediately that the wave functions in both beam paths, which compose the beam in the forward direction behind the interferometer, are equal ($\psi_0^{\mathrm{I}} = \psi_0^{\mathrm{II}}$) because they are transmitted-reflected-reflected (TRR) and reflected-reflected-transmitted (RRT), respectively. The de Broglie wavelength of the neutrons diffracted from such crystals is about 1.8 Å and their energy is about 0.025 eV. The theoretical treatment of the diffraction process from the perfect crystal is described by the dynamical diffraction theory, which can also be found in the literature for the neutron case.[7–10] Inside the perfect crystal, two wave fields are excited when the incident beam fulfills the Bragg condition, one of them having its nodes at the position of the atoms and the other in between them. Therefore, their wave vectors are slightly different ($k_1 - k_2 = 10^{-5} \cdot k_0$) and, due to mutual interference processes, a rather complicated interference pattern is built up, which changes substantially over a characteristic length Δ_0—the so-called Pendellösung length, which is of the order of 50 μm for an ordinary silicon reflection. To preserve the interference properties over the length of the interferometer, the dimensions of the monolithic system have to be accurate on a scale comparable to this quantity. The whole interferometer crystal has to be placed on a stable goniometer table under conditions avoiding temperature gradients and vibrations. A phase shift between the two coherent beams can be produced by nuclear, magnetic, or gravitational interactions. In the first case, the phase shift is most easily

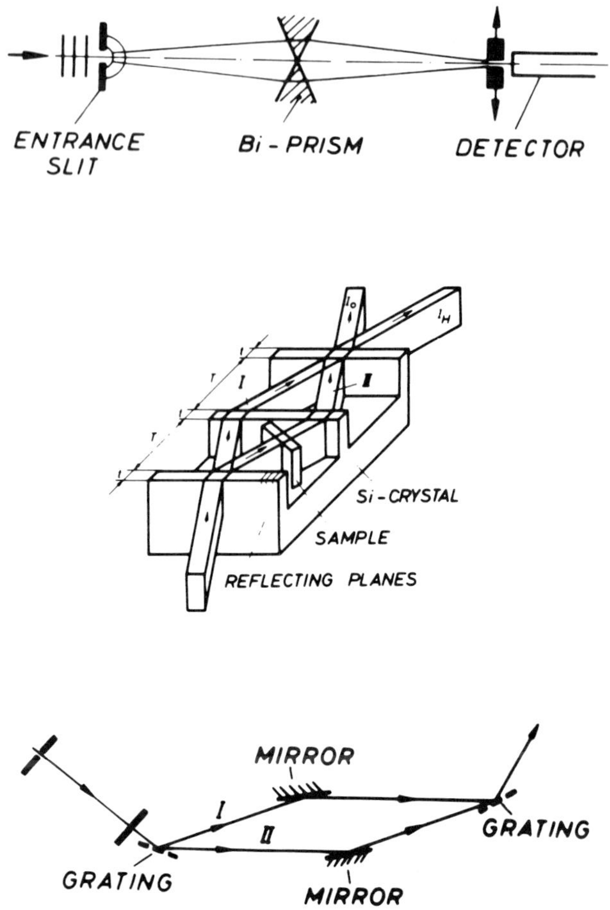

FIGURE 1. Scheme of a slit, a perfect-crystal, and a grating interferometer.

calculated using the index of refraction:[11,12]

$$n = \frac{k}{k_0} = 1 - \left(\frac{\lambda^2 N}{2\pi}\right)\sqrt{b_c^2 - \left(\frac{\sigma_r}{2\lambda}\right)} + i\left(\frac{\sigma_r N \lambda}{4\pi}\right), \tag{1}$$

which simplifies for weakly absorbing materials ($\sigma_r \to 0$) to

$$n = 1 - \lambda^2\left(\frac{N b_c}{2\pi}\right), \tag{2}$$

where b_c is the coherent scattering length and N is the particle density of the phase-shifting material. The different k-vector inside the phase-shifter causes a spatial shift $\vec{\Delta}$ of the wave packet that depends on the orientation of the sample surface $\hat{s}$:

$$\vec{\Delta} = \left(\frac{\vec{k} - k_0}{k}\right) D_0 \qquad \text{with} \qquad \vec{k} - \vec{k}_0 = \frac{(1-n)k\hat{s}}{(\hat{k}\cdot\hat{s})}. \tag{3}$$

As in ordinary light optics, the change of the wave function is obtained as follows:

$$\psi \rightarrow \psi_0 e^{i\vec{\Delta}\cdot\vec{k}} = \psi_0 e^{-iNb_c\lambda D} = \psi_0 e^{i\chi}. \tag{4}$$

Therefore, the intensity behind the interferometer becomes

$$I_0 \propto |\psi_0^{\mathrm{I}} + \psi_0^{\mathrm{II}}|^2 \propto (1 + \cos\chi). \tag{5}$$

The intensity of the beam in the deviated direction follows from particle conservation:

$$I_0 + I_{\mathrm{H}} = \text{const.} \tag{6}$$

Thus, the intensities behind the interferometer vary as a function of the thickness D of the phase-shifter, the particle density N, or the neutron wavelength λ. A wave-packet description has to be used to define the coherence properties in real

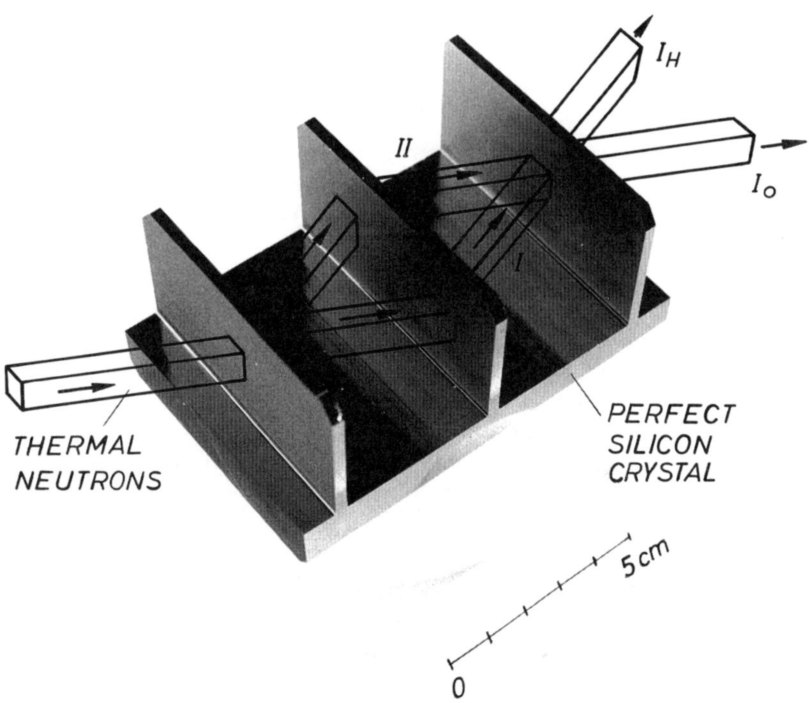

FIGURE 2. A perfect-crystal symmetric neutron interferometer.

experiments:

$$\psi(\vec{r}) \propto \int a(\vec{k})e^{i\vec{k}\cdot\vec{r}}d^3k. \tag{7}$$

Standard quantum mechanics defines the momentum distribution of the beam by

$$g(\vec{k}) = |\psi(\vec{k})|^2 = |a(\vec{k})|^2; \tag{8}$$

therefore, one gets the real part of the coherence function as the Fourier-transform of the momentum distribution,

$$|\Gamma(\vec{\Delta})| \propto \left| \int g(\vec{k})e^{i\vec{k}\cdot\vec{\Delta}}d^3\vec{k} \right|, \tag{9}$$

which simplifies for Gaussian momentum distributions,

$$g(\vec{k}) \propto \exp\left[-(\vec{k} - \vec{k}_0)^2/2\delta\vec{k}^2\right], \tag{10}$$

with characteristic widths δk_i to

$$|\Gamma(\vec{\Delta}_0)| = \prod_{i=x,y,z} \exp\left[-(\Delta_i\delta k_i)^2/2\right]. \tag{11}$$

The mean square distance related to $|\Gamma(\vec{\Delta})|$ defines the coherence length Δ_c, which for Gaussian distribution functions is directly related to the minimum uncertainty relation $[\Delta_c^i = 1/(2\delta k_i)]$.

Any experimental device deviates from the idealized assumptions made by the theory: the perfect crystal can have slight deviations from its perfectness and its dimensions may vary slightly; the phase-shifter contributes to imperfections by variations in its thickness and inhomogeneities; and even the neutron beam itself contributes to a deviation from the idealized situation because of its momentum spread δk. Therefore, the experimental interference patterns have to be described by a generalized relation,

$$I \propto A + B\cos(\chi + \Phi_0), \tag{12}$$

where A, B, and Φ_0 are characteristic parameters of a certain setup. It should be mentioned, however, that the idealized behavior described by equation 5 can nearly be approached by a well-balanced setup.[13] The reduction of the contrast at high order results from the longitudinal coherence length, which is determined by the momentum spread of the neutron beam $[\Delta_x = (2\delta k_x)^{-1}]$. This causes a change of the amplitude factor $\hat{B}$ of equation 12 as $B \rightarrow B\exp[-(\Delta_i\delta k_i)^2/2]$. The wavelength dependence of χ in equation 4 disappears in a special sample position where the surface of the sample is oriented parallel to the reflecting planes, and the path length through the interferometer becomes $D_0/\sin\,\theta_B$; therefore, the phase shift $\chi = -2d_{hkl}Nb_cD_0$ becomes independent of the wavelength, and the damping at high interference orders is strongly reduced and does not appear as in the standard position. Related results showing the interference pattern in the 256th interference order in the dispersive and the nondispersive sample position are shown in FIGURE 3.[14] The much higher visibility of the interferences in the nondispersive sample arrangement is visible and is caused by the much smaller momentum spread perpendicular to the reflecting planes.

Various postselection measurements in neutron interferometry have shown that interference fringes can be restored even in cases when the overall beam does not exhibit any interference fringes due to spatial phase shifts larger than the coherence length of the interfering beams.[15–18] This indicates that the simple picture that predicts interference only when wave packets spatially overlap is untrue. Interference actually occurs no matter how large the optical path difference may be. From classical optics, it has been known for many years that the coherence properties

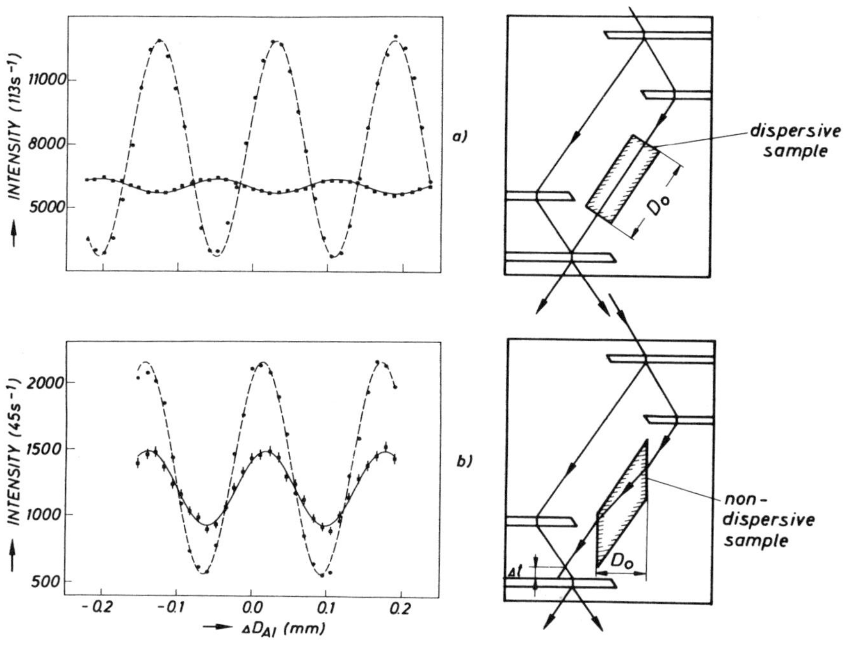

FIGURE 3. Interference pattern observed at high order ($m = 256$) with a dispersively (a) and a nondispersively (b) arranged sample.[14] (Dashed lines correspond to measurements at low order.)

manifest themselves in a spatial intensity variation for phase shifts smaller than the coherence length and in a spectral intensity variation for large phase shifts.[19–23] This phenomenon becomes more apparent for less monochromatic beams and can cause overall spectral shifts[24,25] and even squeezing phenomena.[26,27] The related phenomena for matter waves have been discussed recently[28,29] and will be elucidated in more detail in this report.

All the results of interferometric measurements obtained up until now can be explained well in terms of the wave picture of quantum mechanics and the complementarity principle of standard quantum mechanics. Nevertheless, one should bear in mind that the neutron also carries well-defined particle properties, which have to be transferred through the interferometer. These properties are summarized in TABLE 1 together with a formulation in the wave picture. Both particle and wave properties are well established and, therefore, neutrons seem to be a proper tool for testing quantum mechanics with massive particles, where the wave-particle dualism becomes very obvious.

All neutron interferometric experiments pertain to the case of self-interference, where during a certain time interval only one neutron is inside the interferometer, if at all. Usually, at that time, the next neutron has not yet been born and is still contained in the uranium nuclei of the reactor fuel. Although there is no interaction between different neutrons, they have a certain common history within predetermined limits that are defined, for example, by the neutron moderation process, by their movement along the neutron guide tubes, by the monochromator crystal, and by the special interferometer setup. Therefore, any real interferometer pattern contains single-particle and ensemble properties together.

PARTIAL BEAM PATH DETECTION EXPERIMENTS

A certain beam attenuation can be achieved either by a semitransparent material or by a proper chopper system. The transmission probability in the first case is defined by the absorption cross section σ_a of the material $[a = I/I_0 = \exp(-\sigma_a ND)]$ and the change of the wave function is obtained directly from the complex index of refraction (equation 1):

$$\psi \to \psi_0 e^{i(n-1)kD} = \psi_0 e^{i\chi} e^{-\sigma_a ND/2} = e^{i\chi}\sqrt{a}\,\psi_0. \tag{13}$$

Therefore, the beam modulation behind the interferometer is obtained in the following form:

$$I_0 \propto |\psi_0^{\mathrm{I}} + \psi_0^{\mathrm{II}}|^2 \propto [(1 + a) + 2\sqrt{a}\cos\chi]. \tag{14}$$

On the other hand, the transmission probability of a chopper wheel or another shutter system is given by the open-to-closed ratio, $a = t_{\text{open}}/(t_{\text{open}} + t_{\text{closed}})$, and one obtains (after straightforward calculations)

$$I \propto [(1 - a)|\psi_0^{\mathrm{II}}|^2 + a|\psi_0^{\mathrm{I}} + \psi_0^{\mathrm{II}}|^2] \propto [(1 + a) + 2a\cos\chi]; \tag{15}$$

that is, the contrast of the interference pattern is proportional to $\sqrt{a}$ in the first case and proportional to a in the second case, although the same number of neutrons are observed in both cases. Absorption represents a measuring process in both cases because a compound nucleus is produced with an excitation energy of several MeV, which is usually de-excited by capture gamma rays. These can be easily detected by different means.

FIGURE 4 shows the dependence of the normalized contrast of the measured interference pattern on the transmission probability.[30–32] The different contrast becomes especially obvious for low transmission probabilities where the interfering part of the interference pattern is distinctly larger than the transmission probability through the semitransparent absorber sheet. The discrepancy diverges for $\bar{a} \to 0$,

TABLE 1. Properties of the Neutron

PARTICLE PROPERTIES	CONNECTION	WAVE PROPERTIES
$m = 1.6748220(25) \times 10^{-27}$ kg	de Broglie	$\lambda_c = \frac{h}{mc} = 1.319695(20) \times 10^{-15}$ m
$s = \frac{1}{2}\hbar$	$\lambda_B = \frac{h}{m\text{v}}$	(thermal neutrons: $\lambda = 1.8$ A, v = 2200 m/s)
$\mu = -9.6491783(18) \times 10^{-27}$ J/T		$\lambda_B = \frac{h}{m\text{v}} = 1.8 \times 10^{-10}$ m
$\tau = 882.6\ (2.7)$ s	Schrödinger	$\Delta_c = \frac{1}{2\delta k} \cong 10^{-8}$ m
$R = 0.7$ fm	$H\psi(\bar{r},t) = i\hbar\frac{\delta\psi(\bar{r},t)}{\delta t}$	$\Delta_p = \text{v}\,\Delta t \cong 10^{-2}$ m
$\alpha = 12.0\ (2.5) \times 10^{-4}$ fm^3		$\Delta_d = \text{v}\,\tau = 1.942(5) \times 10^{6}$ m
u - d - d - quark structure		$0 \le \chi \le 2\pi\ (4\pi)$
m ... mass, s ... spin, μ ... magnetic moment, τ ... β-decay lifetime, R ... (magnetic) confinement radius, α ... electric polarizability; all other measured quantities like electric charge, magnetic monopole and magnetic dipole moment are compatible with zero.		λ_c ... Compton wavelength, λ_B ... de Broglie wavelength, Δ_c ... coherence length, Δ_p ... packet length, δk ... momentum width, Δt ... chopper opening time, v ... group velocity, χ ... phase.

but it has been shown that in this regime the variations of the transmission due to variations of the thickness or of the density of the absorber plate have to be taken into account, which shifts the points below the curve.[33] This can most easily be understood if the variation of the beam attenuation due to variations of the thickness or density fluctuations is included, $a = \bar{a} + \Delta a$, which yields after averaging

$$\sqrt{\bar{\bar{a}}} < \sqrt{\bar{a}},$$

indicating that the points fall below the $\sqrt{a}$ curve.

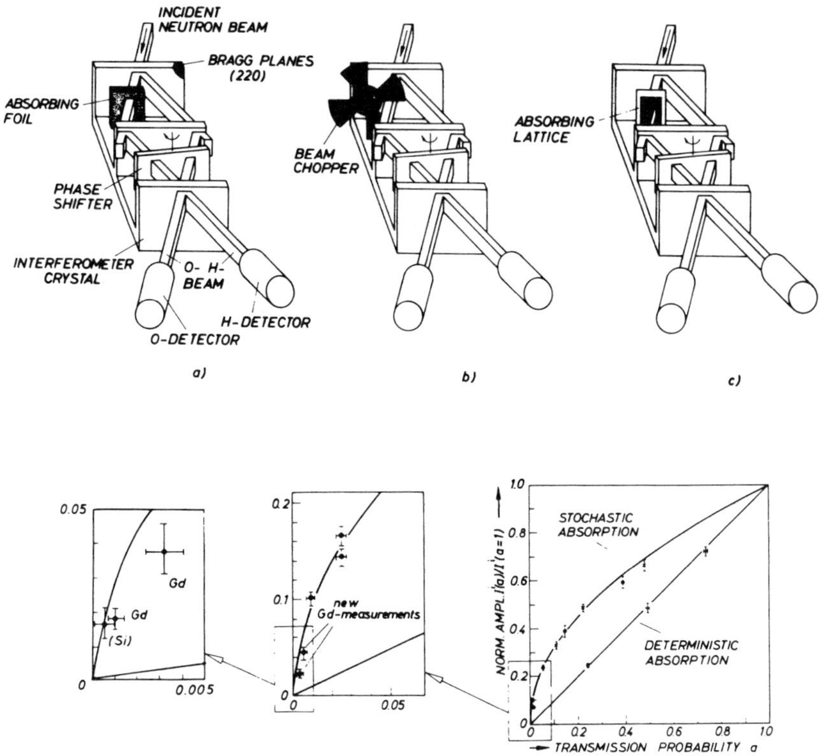

FIGURE 4. (Top) Sketch of the experimental arrangement for absorber measurements: (a) stochastic absorption; (b) deterministic absorption; (c) attenuation by a transmission grating. (Bottom) Reduction of the contrast as a function of beam attenuation for different attenuation methods.[31,32]

The region between the linear and the square-root behavior can be reached by very narrow chopper slits or by narrow transmission lattices, where one starts to lose information concerning which individual slit the neutron went through. This is exactly the region showing the transition between a deterministic and a stochastic situation and, therefore, it can be formulated by a Bell-like[34,35] inequality:

$$\sqrt{a} > x > a. \tag{17}$$

The stochastic limit corresponds to the quantum limit when one does not know anymore through which individual slit the neutron went. Which situation exists depends on how the slit widths l compare to the coherence lengths Δ_i in the related direction. In case that the slit widths become smaller than the coherence lengths, the wave function behind the slits shows distinct diffraction peaks that correspond to new quantum states ($n \neq 0$), which now do not overlap with the undisturbed reference beam. The creation of the new quantum states means that those labeled neutrons carry information about the chosen beam path and hence do not contribute to the interference amplitude.[36] A related experiment has been carried out by rotating an absorption lattice around the beam axis, that is, where one changes from $l \ll \Delta_x$ (vertical slits) to $l \gg \Delta_y$ (horizontal slits) (see FIGURE 5), because the coherence length parallel to the reflecting lattice vector is much larger than in any other direction. Thus, the attenuation on factor $\bar{a}$ has to be generalized, including not only nuclear absorption and scattering processes, but also lattice diffraction effects if they remove neutrons from the original phase space.

A very similar situation exists if a very fast chopper produces beam bursts (packet lengths) shorter than the coherence time of $\Delta t_c = \Delta/v$. In this case, diffraction in time occurs, which also removes neutrons out of the original phase space. This limit is very difficult to reach with a mechanical chopper, but it can probably be tackled with a high-frequency spin flipper.

NEUTRON JOSEPHSON EFFECT

This phenomenon is based on the dipole coupling of the magnetic moment $\vec{\mu}$ of the neutron to a magnetic field $\vec{B}(H = -\vec{\mu}\vec{B})$, which causes the famous 4π-symmetry of spinor wave functions, as measured in early neutron interferometer experiments.[37,38] The change of the wave function reads as

$$\psi \rightarrow \psi_0 e^{-i(Ht/h)} = \psi_0 e^{-i(\vec{\mu}\vec{B}t/\hbar)} = \psi_0 e^{-i\vec{\sigma}\vec{\alpha}/2} = \psi(\alpha), \tag{18}$$

where $\vec{\alpha}$ represents a formal description of the Larmor rotation angle around the field $\vec{B}$ $[\alpha = (2\mu/\hbar) \int Bdt = (2\mu/\hbar v) \int Bds]$. This enabled also the realization of the spin superposition experiments where spin-up ($|\uparrow\rangle$) and spin-down ($|\downarrow\rangle$) states are superposed, producing a final state perpendicular to both initial states.[39,40] It is interesting to mention that, in the case of spin reversal by means of a resonance flipper, the spin term is accompanied by an energy exchange equal to the Zeeman energy, $\hbar\omega_r = 2\mu B$. This provided the basis for the observation of a new quantum beat effect: the magnetic Josephson analogue.

A double-coil arrangement has been used for the observation of this new quantum beat effect. If the frequencies of the two coils are chosen to be slightly different, the energy transfer becomes different too $[\Delta E = \hbar(\omega_{r1} - \omega_{r2})]$. The frequency difference can be made very small, if high quality frequency generators are used for the field generation. The flipping efficiencies for both coils are always very

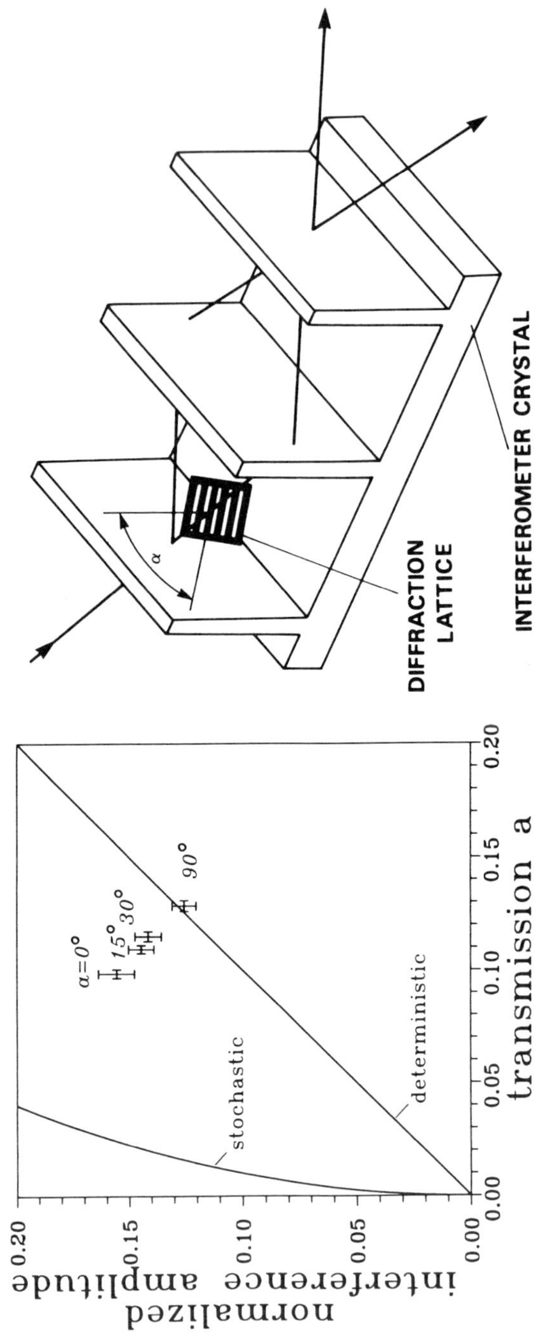

FIGURE 5. Lattice absorber in the interferometer approaching the classical limit when the slits are oriented horizontally and approaching the quantum limit when they are oriented vertically.[31,36]

close to unity (better than 99%). Now, the wave functions change according to

$$\psi \rightarrow e^{i(\omega-\omega_{r1})t}|\downarrow> + e^{i\chi}e^{i(\omega-\omega_{r2})t}|\downarrow>. \tag{19}$$

Therefore, the intensity behind the interferometer exhibits a typical quantum beat effect, given by

$$I \propto 1 + \cos[\chi + (\omega_{r1} - \omega_{r2})t]. \tag{20}$$

Thus, the intensity behind the interferometer oscillates between the forward and the deviated beam without any apparent change inside the interferometer.[41] The time constant of this modulation can reach a macroscopic scale that is correlated to an uncertainty relation $\Delta E \Delta t \leq \hbar/2$. FIGURE 6 shows the result of an experiment where the periodicity of the intensity modulation, $T = 2\pi/(\omega_{r1} - \omega_{r2})$, amounts to $T = (47.90 \pm 0.15)$ s, caused by a frequency difference of about 0.02 Hz. This corresponds to a mean difference (ΔE) of the energy transfer between the two beams of $\Delta E = 8.6 \cdot 10^{-17}$ eV and to an energy sensitivity of $2.7 \cdot 10^{-19}$ eV, which is by many orders of magnitude higher than that of other advanced spectroscopic methods. This high resolution is strongly decoupled from the monochromaticity of the neutron beam, which was $\Delta E = 5.5 \cdot 10^{-4}$ eV around a mean energy of the beam $E_B = 0.023$ eV in this case. It should be mentioned that the result can also be interpreted as being the effect of a slowly varying phase $\Delta(t)$ between the two flipper fields, but the more physical description is based on the argument of a different energy transfer. The extremely high resolution may be used for fundamental, nuclear, and solid-state physics applications.

The quantum beat effect can also be interpreted as a magnetic Josephson effect analogue. In this case, the phase difference is driven by the magnetic energy ($\Delta B_0 = B_{02} - B_{01}$),

$$\left(\frac{\delta}{\delta t}\right)(\Delta_2 - \Delta_1) = \omega_{r2} - \omega_{r1} = \frac{2\mu\Delta B_0}{\hbar}, \tag{21}$$

and therefore

$$\Delta(t) = (\omega_{r2} - \omega_{r1})t = \left(\frac{2\mu\Delta B_0}{\hbar}\right) \cdot t. \tag{22}$$

This yields the observed modulation (compare equation 20):

$$I \propto [1 + \cos\Delta(t)]. \tag{23}$$

This is analogous to the well-known Josephson effect in superconducting tunnel junctions,[42] where the phase of the Cooper pairs in both superconductors is related according to

$$\left(\frac{\delta}{\delta t}\right)(\Phi_2 - \Phi_1) = \left(\frac{1}{\hbar}\right)(E_2 - E_1) = \left(\frac{1}{\hbar}\right)2eV, \tag{24}$$

which is driven by the electric potential V between both superconductors. This gives

$$\Phi(t) = \left(\frac{2eV}{\hbar}\right) \cdot t \tag{25}$$

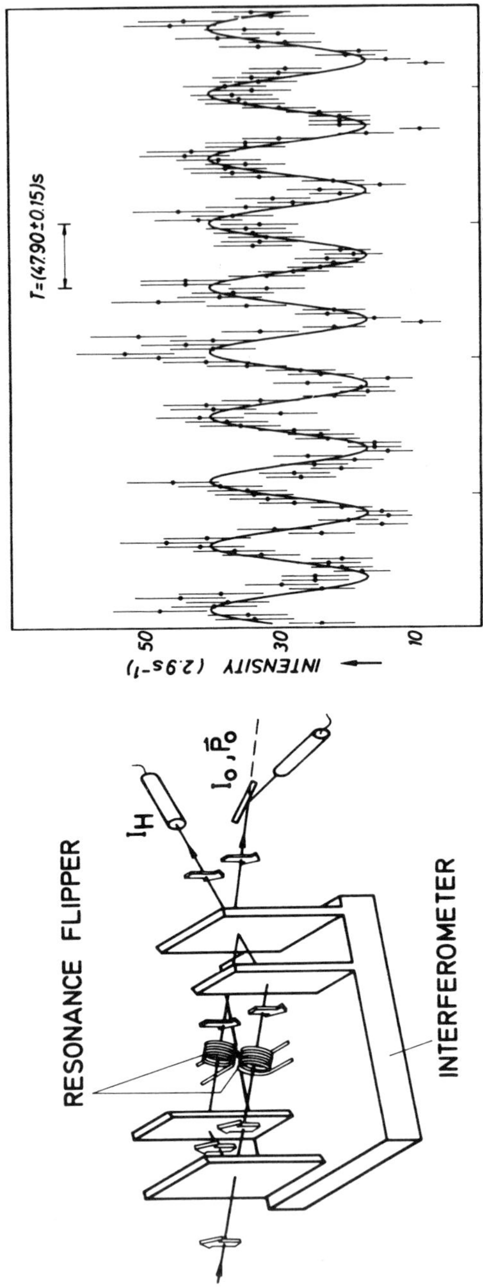

FIGURE 6. Quantum beat effect observed when the frequencies of the two flipper coils differ by about 0.02 Hz around 71.89979 kHz.[41]

and a superconducting Josephson current of

$$I_s = I_{sMax} \sin \Phi(t). \tag{26}$$

POSTSELECTION OF MOMENTUM STATES

In the course of several neutron interferometer experiments,[15–18] it has been established that smoothed-out interference properties at high interference order can be revived even behind the interferometer when a proper spectral filtering is applied. The experimental arrangement with an indication of the wave packets at different parts of the interference experiment is shown in FIGURE 7. An additional monochromatization is applied behind the interferometer by means of various single crystals

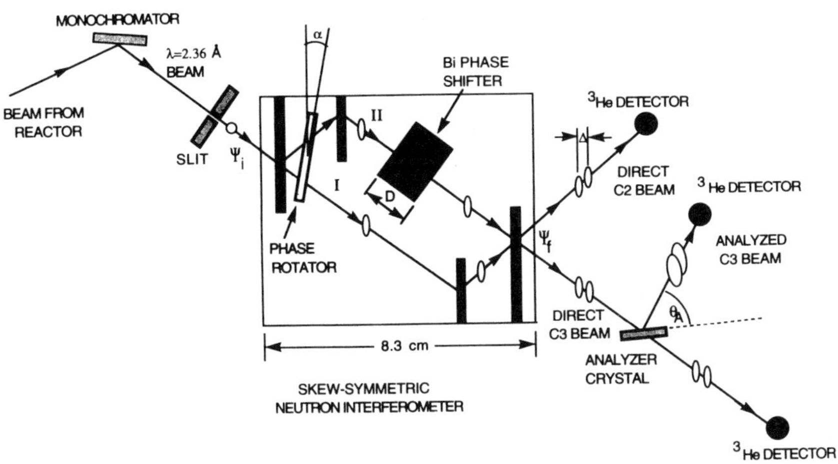

FIGURE 7. Scheme of the experimental arrangement with a skew-symmetrically cut perfect-crystal interferometer and a postselection analyzer crystal.[18,28]

brought into Bragg position. Using equations 5 and 7, the momentum-dependent intensity reads as

$$I_0(\vec{r}, \vec{k}) = |\psi_0^{\mathrm{I}}(\vec{r}, \vec{k}) + \psi_0^{\mathrm{II}}(\vec{r} + \Delta, \vec{k})|^2 \propto |a(\vec{k})|^2\{1 + \cos[\vec{\Delta}(\vec{k}) \cdot \vec{k}]\}, \tag{27}$$

whereas the overall beam reads as

$$I_0(\vec{\Delta}_0) \propto 1 + |\Gamma(\vec{\Delta}_0)| \cos \vec{\Delta}_0 \cdot \vec{k}_0, \tag{28}$$

where $\vec{\Delta}_0$ represents the spatial phase shift for the $\vec{k}_0$ component of the packet. Equation 28 describes the interference fringes when $\vec{\Delta}_0$ is varied. The formula also shows that the overall interference fringes disappear for spatial phase shifts larger than the coherence lengths $[\Delta_i \geq \Delta_c^i = 1/(2\delta k_i)]$ (see equation 11). This behavior is

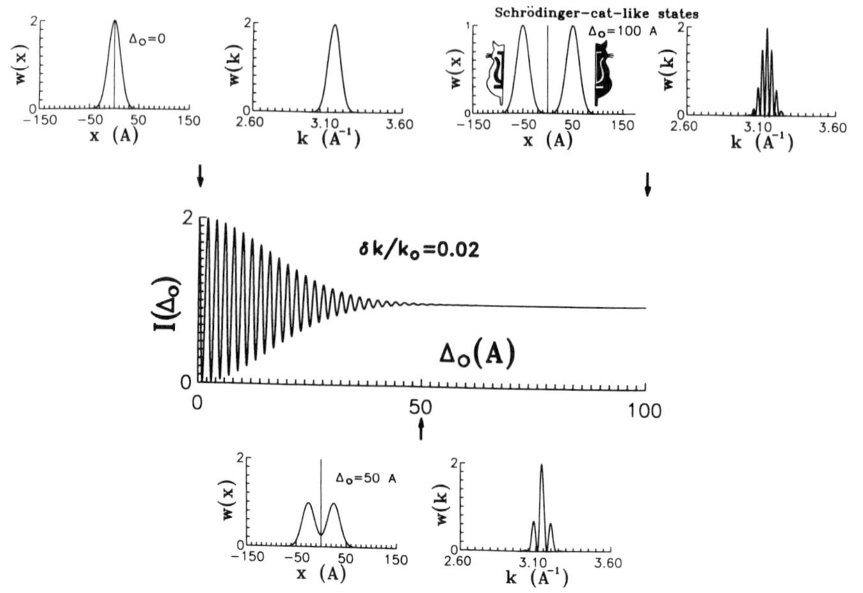

FIGURE 8. Interference pattern as a function of the relative phase shift (middle) and related wave packets and momentum spectra behind the interferometer for different values of the phase shift.[18]

shown in FIGURE 8 and has been verified experimentally by several investigations for Gaussian and non-Gaussian neutron beams.[43–45]

In our experiment, we deal with the coherence properties along the interferometer axis (x), where—according to basic quantum mechanical laws—the (tangential) components of the momentum vectors (and coherence length) do not change due to Bragg diffraction. The related momentum distribution follows from equation 27 and, for Gaussian packets, it can be rewritten in the form,

$$I_0(k) = \exp\left[-(k - k_0)^2/2\delta k^2\right]\left\{1 + \cos\left[\chi_0\left(\frac{k_0}{k}\right)\right]\right\}, \tag{29}$$

where the mean phase shift is introduced ($\chi_0 = k_0\Delta_0 = Nb_c\lambda_0 D_{\text{eff}}$). The surprising feature is that $I_0(k)$ becomes oscillatory for large phase shifts where the interference fringes described by equation 28 disappear (see FIGURE 8). This indicates that interference in phase space has to be considered[46,47] rather than the simple wave function overlap criterion described by the coherence function (equation 10). The second beam behind the interferometer (H) just shows the complementary modulation, $I_H = I_{\text{total}} - I_0$.

The wave function[48] of the packets arising from beam paths I and II determines the spatial shape of the packets behind the interferometer, that is,

$$I_0(x) = |\psi(x) + \psi(x + \Delta)|^2, \tag{30}$$

which separates for large phase shifts into two peaks (FIGURE 8). For Gaussian packets, having a spatial width δx, which corresponds to the coherence length Δ_c, the minimum uncertainty relation $\delta x \delta k = \frac{1}{2}$ is fulfilled. For an appropriately large displacement ($\Delta \gg \Delta_c$), the related state can be interpreted as a superposition state of two macroscopically distinguishable states, that is, a stationary Schrödinger-cat-like state,[26,49,50] but here first for massive particles. These states—separated in ordinary space and oscillating in momentum space—seem to be notoriously fragile and sensitive to dephasing effects.[51–54]

Measurements of the wavelength spectrum were made with a silicon crystal with a rather narrow mosaic spread, which reflects in the parallel position a very narrow band of neutrons only ($\delta k'/k_0 \approx 0.0003$), causing an enhanced visibility at large phase shifts (FIGURE 8). This feature shows that an interference pattern can be restored even behind the interferometer by means of a proper postselection procedure. In this case, the overall beam does not show interference fringes anymore and the wave packets originating from the two different beam paths do not overlap.

The momentum distribution has been measured by scanning the analyzer crystal through the Bragg position. The related results are shown in FIGURE 9 for different phase shifts. These results clearly demonstrate that the predicted spectral modulation (equation 29) appears when the interference fringes of the overall beam disappear. The modulation is somehow smeared out due to averaging processes across the beam due to various imperfections, unavoidably existing in any experimental arrangement. The contrast of the empty interferometer was 60%.

Each peak in the momentum distribution shown in FIGURE 9 corresponds to a

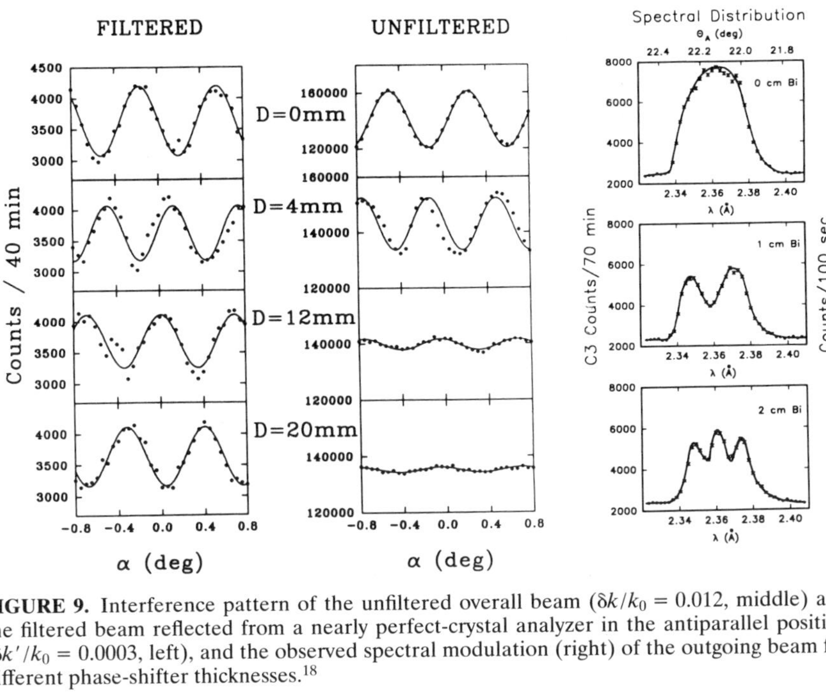

FIGURE 9. Interference pattern of the unfiltered overall beam ($\delta k/k_0 = 0.012$, middle) and the filtered beam reflected from a nearly perfect-crystal analyzer in the antiparallel position ($\delta k'/k_0 = 0.0003$, left), and the observed spectral modulation (right) of the outgoing beam for different phase-shifter thicknesses.[18]

different number of phase shifts experienced by the neutrons of that wavelength band during passage through the interferometer. In that sense, the minimum quantum entity of the incident wave packet becomes a new diverse entity representing different quantum states with distinguishable properties. This kind of labeling shows that constructive interference is restricted to a certain wavelength band only—a situation similar to that where new quantum states have been created due to lattice diffraction inside the interferometer (FIGURE 5).[36]

The new quantum states created behind the interferometer can be analyzed with regard to their uncertainty properties. Analogies between a coherent state behavior and a free, but coherently coupled particle motion inside the interferometer have been addressed previously.[32] In such cases, the dynamical conjugate variables x and p minimize the uncertainty product with identical uncertainties: $(\Delta x)^2 = (\Delta k)^2 = 1/2$

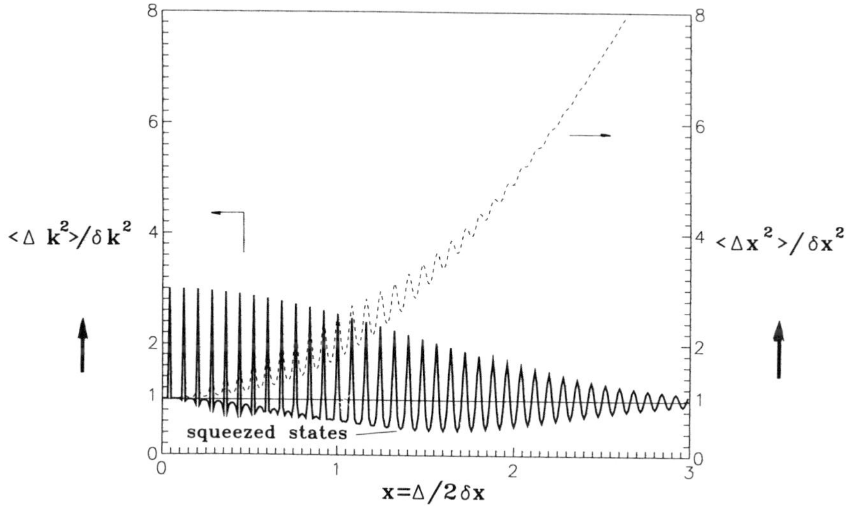

FIGURE 10. Spatial and momentum uncertainties of the outgoing beams with the indication of squeezing in the momentum domain.[18]

(in dimensionless units). Using $I_0(k)$ and $I_0(x)$ (equations 29 and 30) as distribution functions, we get in our case

$$\Gamma(\vec{\Delta}) = \langle \psi^*(0)\psi(\vec{\Delta})\rangle \tag{31}$$

and (for $\delta k/k_0 \ll 1$)

$$\langle(\Delta k)^2\rangle = \langle k^2\rangle - \langle k\rangle^2 = (\delta k)^2\left\{1 - \left(\frac{\Delta_0}{2\delta x}\right)^2 \frac{[e^{-(\Delta_0/2\delta x)^2/2}\cos(\Delta_0 k_0) + e^{-(\Delta_0/2\delta x)^2}]}{[1 + e^{-(\Delta_0/2\delta x)^2/2}\cos(\Delta_0 k_0)]^2}\right\}. \tag{32}$$

These relations are shown in FIGURE 10, indicating that, for $(\Delta k)^2$, a value below the coherent state value can be achieved, which in quantum optic terminology means

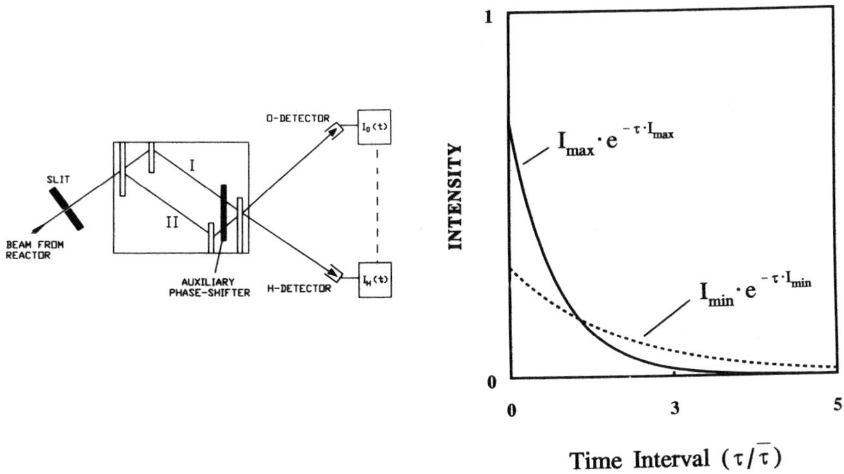

FIGURE 11. Sketch of the experimental setup for time-selection experiments and the pair-correlation function for the maximum and minimum of the interference pattern.[60]

squeezing.[26,55–58] One emphasizes that a single coherent state does not exhibit squeezing, but a state created by the superposition of two coherent states can exhibit a considerable amount of squeezing. Thus, highly nonclassical states are made by the power of the quantum mechanical superposition principle.

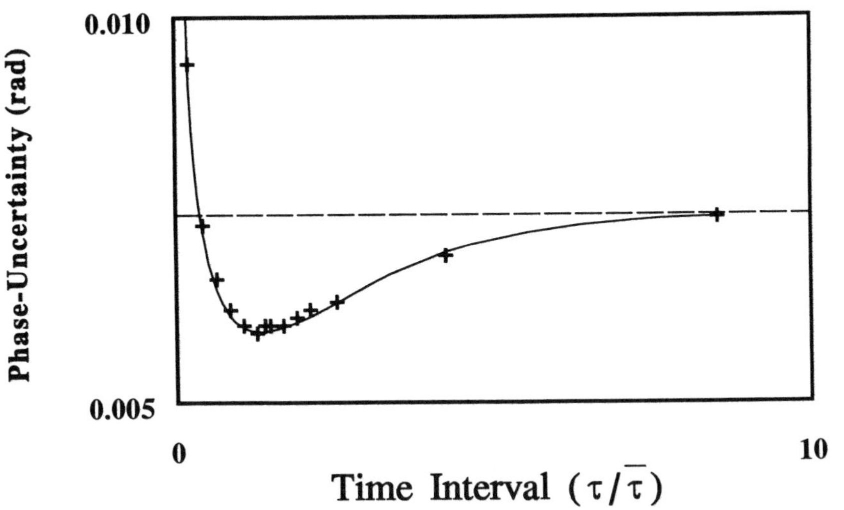

FIGURE 12. Measured phase sensitivity of an interference pattern for the case of proper time-correlation measurements in comparison with the phase sensitivity related to the overall beam (dashed line).[60]

POSTSELECTION IN THE TIME DOMAIN

Instead of measuring the interference pattern by scanning the phase-shifter, one can measure additionally the intensity correlation function,

$$\Gamma^{(2)}(\vec{\Delta}, \tau) = \langle I(0, 0)I(\vec{\Delta}, \tau)\rangle, \tag{33}$$

which defines the probability of registering a neutron at time τ, if there was another one measured at $\tau = 0$. This probability to measure a neutron at a time τ after another neutron has arrived reads for a stationary beam of a thermal (statistical) source as[59]

$$W(\tau) = I(\Delta) \exp[-\tau \cdot I(\Delta)], \tag{34}$$

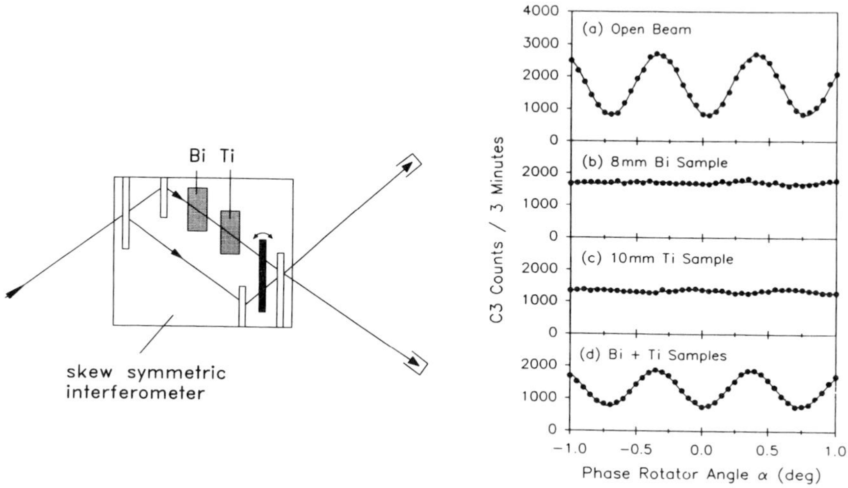

FIGURE 13. Loss of contrast at high interference and its retrieval by an opposite phase-shifter inserted into the same beam.[45]

which exhibits an intensity-dependent "decay time" of $\tau(\Delta) = [I(\Delta)]^{-1}$. FIGURE 11 shows the experimental arrangement and the probability of measuring neutron pairs with a time separation τ for the case when the overall interference pattern is turned to its maximum or minimum, respectively (in our case, the overall contrast was 40%). It becomes visible that the contrast for neutron pairs arriving within short time intervals is higher than the overall contrast. For larger time separations, the contrast vanishes and appears with an opposite sign, reaching values of 100% for timely wide separated pairs. This behavior has been verified experimentally[60] and it has been shown that remarkably higher phase sensitivities can be achieved by using this new measuring technique (FIGURE 12).

These results demonstrate that considerably more information can be deduced even from a statistical beam if the individual arrival times of the neutrons are additionally registered to define the pair-correlation function inherent to the quantum system.

CONTRAST RETRIEVAL BY PHASE ECHO

Phase echo is a similar technique to spin echo, which is routinely used in neutron spectroscopy.[61] A large phase shift ($\Delta > \Delta^c$) can be applied in one arm of the interferometer, which can be compensated by a negative phase shift acting in the same arm or by the same phase shift applied to the second beam path.[44,45,62] According to equation 4, the phase shift is additive and the coherence function depends on the net phase shift only. Thus, the interference pattern can be restored as it is shown schematically and in the form of an experimental example in FIGURE 13. The phase-echo method can also be applied behind the interferometer loop when multiplate interferometers are used.[63] In this case, the situation becomes even more similar to the situation discussed in the previous section.

These results tell us that information first appearing in a spatial phase shift becomes transferred into a momentum modulation, which can be revived to ordinary space modulation effects again. A comment has to be made that it becomes intrinsically more difficult to restore the original contrast when the separation of the wave packets in ordinary space increases.

REQUEST FOR POSTSELECTION IN EPR EXPERIMENTS

The previously discussed neutron experiments have shown us that phase space coupling persists even if the overlap in one parameter space does not exist anymore. The stored information becomes exchanged between parameter spaces and can be measured by a proper experimental method. This has consequences for EPR experiments too. The entangled states (e.g., see reference 64) of two photons produced by an atomic decay cascade (FIGURE 14),

$$\psi \propto |-k\rangle_1 |k\rangle_2 + |-k\rangle_2 |k\rangle_1, \tag{35}$$

are correlated due to the energy conservation of the transition:

$$k_1 + k_2 = k_{01} + k_{02} = \text{const.} \tag{36}$$

This produces a momentum- and space-dependent intensity distribution when the packet structure of the related wave functions is taken into account,[28] that is,

$$I(k_1, k_2, \vec{r}) = |\psi|^2 = 2|a(k_1)|^2 |a(k_2)|^2 \cdot \{1 + \cos[2(k_2 - k_1)r]\}. \tag{37}$$

This shows a characteristic intensity modulation for each photon pair (FIGURE 14) and indicates that individual $|k\rangle$ states remain interacting even at an arbitrarily large spatial separation of the wave packets. For large distances $[r > (2\delta k)^{-1}]$, the appearance of a momentum distribution modulation follows from equation 37 as

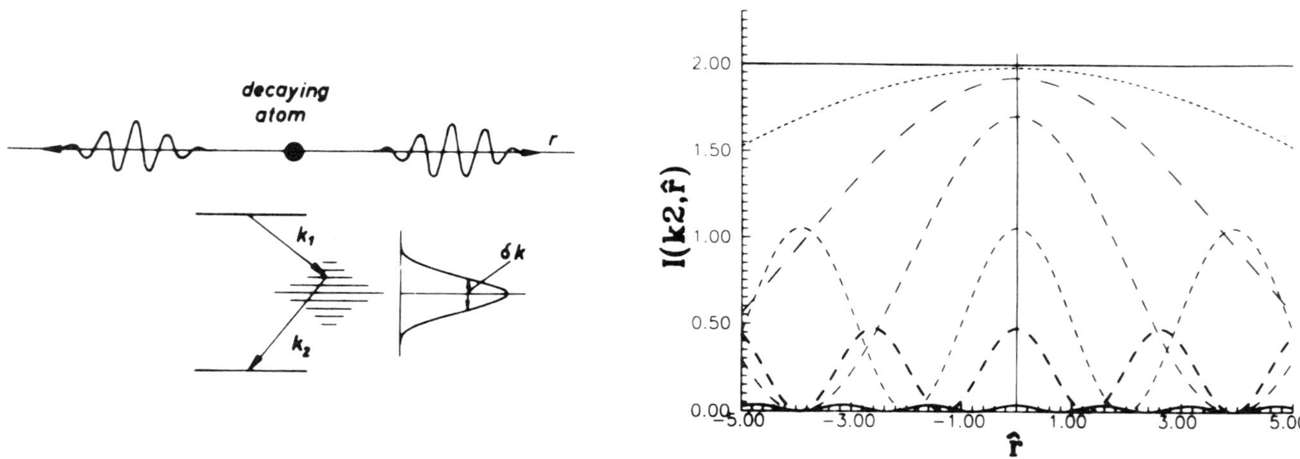

FIGURE 14. Scheme of a typical EPR transition and expected intensity distribution for individual momentum pairs for $|k_1 - k_2|/\delta k = 0, 0.1, 0.2, 0.4, 0.8, 1.2$, and 2 (from top to bottom).[28]

well.[28] If one of these photons are registered on one side, its wave function collapses, which instantaneously changes the wave function on the other side to $||k\rangle_2|^2$. This again shows that much more information can be gained than is usually extracted. Therefore, it is recommended to repeat this experiment with a proper momentum resolution, which would show that the right and the left wave fields of the related momentum band (i.e., the partner photons) remain coupled even at an arbitrarily large spatial separation of the overall wave packets.

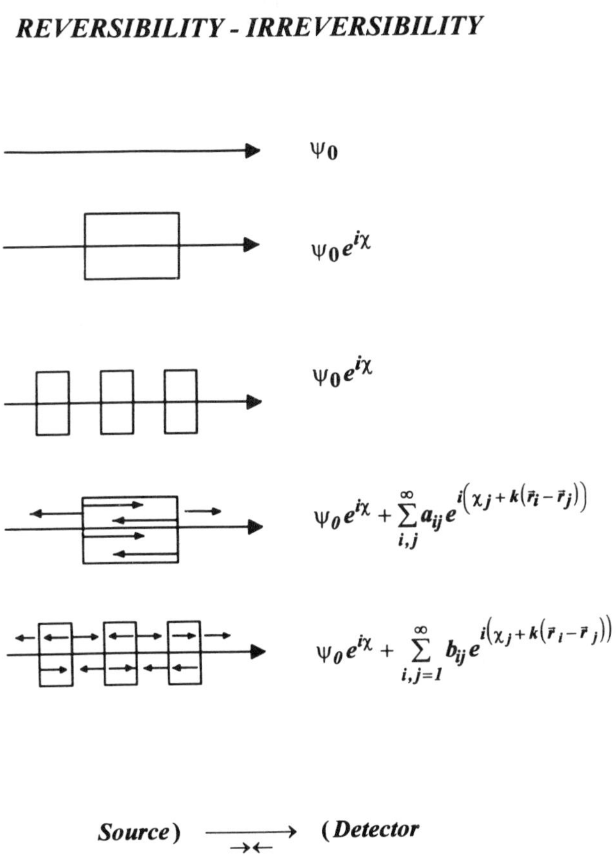

FIGURE 15. Approximate and complete wave functions behind a compact and a split interaction region.

Related experiments will show that this coupling of the partner pairs of photons persists independently from its overall spatial separation. This also indicates that locality should be treated in phase space rather than in ordinary space only. Here, too, the required momentum resolution becomes more stringent when the packets become wider in their separation in ordinary space.

IRREVERSIBILITY AND THE MEASURING PROCESS

In the previous sections, it has been shown that more information about a quantum system can be extracted when more experimentally accessible parameters are measured. It becomes obvious that a system remains coupled in phase space even when it becomes separated in any parameter space. Thus, interference properties can be shifted from one parameter space to another and back again. Related bands of plane wave components that compose the wave packets (equation 7) may be considered as a responsible factor for the understanding of the nonlocality phenomenon in quantum mechanics.

The summaries drawn for the different experimental situations discussed in this article are qualified by a statement that the retrieval of the interference properties by several postselection procedures becomes increasingly more difficult as the width of the separation of the quantum system increases. In fact, a more detailed view shows that a complete retrieval is impossible in principle, as is shown in FIGURE 15 for the case of a phase-echo system. In a more complete and more accurate measurement, more and more parts of the complete (not approximate) wave functions become visible, which contain more and more of the detailed history that the quantum system has experienced between the source and the detector. This indicates a basic irreversibility process not caused by parasitic effects like absorption or incoherent scattering processes, but by the appearance of an infinite number of additional terms in the wave function, indicating that the original state can by no means be restored completely. Unavoidable fluctuations (even zero-point fluctuations) cause an irreversibility effect that becomes more influential for widely separated Schrödinger-cat-like states. All these effects can be described by an increasing entropy inherently associated with any kind of interaction.[65] This also supports the idea that irreversibility is a fundamental property of nature and reversibility is an approximation only, as stated by several authors (e.g., see references 66–68).

This shows that irreversibility and therefore the measurement process start with the first interaction that the quantum system experiences in the experimental setup. The assignment of a source and a detector region defines the direction of increasing entropy.

DISCUSSION

All the results of the neutron interferometric experiments are well described by the formalism of quantum mechanics. According to the complementarity principle of the Copenhagen interpretation, the wave picture has to be used to describe the observed phenomena. The question of how the well-defined particle properties of the neutron are transferred through the interferometer is not a meaningful one within this interpretation, but from the physical point of view it should be an allowed one.

The newly discovered persisting phase space coupling in cases of large spatial shifts of the wave packets may bring some attenuation to the action of plane wave components outside their packets. The shown results clearly demonstrate that a spectral modulation can be observed in neutron interference experiments at high interference order and that interference has to be treated in phase space rather than in ordinary space. It looks like the plane wave components of the wave packets, that is, narrow band width components, interact over a much larger distance than the size of the packets. This interaction guides neutrons of certain momentum bands to the 0- or H-beam, respectively. These phenomena throw a new light on the discussion of Schrödinger-cat-like situations in quantum mechanics and therefore on the discussion about the EPR experiments too.[28,64,69–71] Spatially separated packets remain entangled in phase space and nonlocality appears as a result of this entanglement. The analogy with optical experiments performed in the time-frequency domain is striking.[23] An analogous situation exists in neutron spin-echo systems, where multiple spin rotation plays an equivalent role as the high order interferences discussed here.[61,71]

More complete quantum experiments show that a complete recurrence of all wave components behind an interaction that the quantum system has experienced becomes impossible in principle. This implies—on a high accuracy level—a basic noncommutativity of operators $A \cdot B$ and indicates that the irreversible quantum measuring process starts with the first interaction of the quantum system with the experimental setup.

ACKNOWLEDGMENTS

Most of the experimental results discussed in detail have been obtained by our Dortmund-Grenoble-Vienna interferometer group working at the high flux reactor in Grenoble. Some recent results also stem from our cooperation with the Columbia-Missouri group working at the MURR reactor. The cooperation with these groups and especially the cooperation with colleagues from our Institute, cited in the references, are gratefully acknowledged.

REFERENCES

1. Maier-Leibnitz, H. & T. Springer. 1962. Z. Phys. **167:** 386.
2. Gaehler, R., J. Kalus & W. Mampe. 1980. J. Phys. **E13:** 546.
3. Rauch, H., W. Treimer & U. Bonse. 1974. Phys. Lett. **A47:** 369.
4. Bauspiess, W., U. Bonse, H. Rauch & W. Treimer. 1974. Z. Phys. **271:** 177.
5. Ioffe, A. I., V. S. Zabiyankan & G. M. Drabkin. 1985. Phys. Lett. **111:** 373.
6. Bonse, U. & M. Hart. 1965. Appl. Phys. Lett. **6:** 155.
7. Rauch, H. & D. Petrascheck. 1978. *In* Neutron Diffraction. Chapter 9. H. Dachs, Ed. Springer-Verlag. Berlin/New York.
8. Sears, V. F. 1978. Can. J. Phys. **56:** 1261.
9. Bauspiess, W., U. Bonse & W. Graeff. 1976. J. Appl. Crystallogr. **9:** 68.
10. Petrascheck, D. 1976. Acta Phys. Austriaca **45:** 217.
11. Goldberger, M. L. & F. Seitz. 1974. Phys. Rev. **71:** 294.
12. Sears, V. F. 1982. Phys. Rep. **82:** 1.
13. Bonse, U. & H. Rauch, Eds. 1979. Neutron Interferometry. Oxford University Press (Clarendon). London/New York.
14. Rauch, H., E. Seidl, D. Tuppinger, D. Petrascheck & R. Scherm. 1987. Z. Phys. **B69:** 69.
15. Werner, S. A., R. Clothier, H. Kaiser, H. Rauch & H. Wölwitsch. 1991. Phys. Rev. Lett. **67:** 683.

16. KAISER, H., R. CLOTHIER, S. A. WERNER, H. RAUCH & H. WÖLWITSCH. 1992. Phys. Rev. **A45:** 31.
17. RAUCH, H., H. WÖLWITSCH, R. CLOTHIER, H. KAISER & S. A. WERNER. 1992. Phys. Rev. **A46:** 49.
18. JACOBSON, D. L., S. A. WERNER & H. RAUCH. 1994. Phys. Rev. **A49:** 3196.
19. MANDEL, L. 1962. J. Opt. Soc. Am. **52:** 1335.
20. MANDEL, L. & E. WOLF. 1965. Rev. Mod. Phys. **37:** 231.
21. HEINEGER, F., A. HERDEN & T. TSCHUDI. 1983. Opt. Commun. **48:** 237.
22. JAMES, D. F. V. & E. WOLF. 1991. Phys. Lett. **A157:** 6.
23. ZOU, X. Y., T. P. GRAYSON & L. MANDEL. 1992. Phys. Rev. Lett. **69:** 3041.
24. WOLF, E. 1989. Phys. Rev. Lett. **63:** 2220.
25. FAKTIS, D. & G. M. MORRIS. 1988. Opt. Lett. **13:** 4.
26. SCHLEICH, W., M. PERNIGO & F. LE KIEN. 1991. Phys. Rev. **A44:** 2172.
27. JANSKI, J. & A. V. VINOGRADOV. 1990. Phys. Rev. Lett. **64:** 2771.
28. RAUCH, H. 1993. Phys. Lett. **A173:** 240.
29. RAUCH, H. 1993. *In* Proceedings of Quantum Measurement and Control. E. Ezawa & Y. Murayama, Eds.: 223. North-Holland. Amsterdam.
30. RAUCH, H. & J. SUMMHAMMER. 1984. Phys. Lett. **A104:** 44.
31. SUMMHAMMER, J., H. RAUCH & D. TUPPINGER. 1987. Phys. Rev. **A36:** 4447.
32. RAUCH, H., J. SUMMHAMMER, M. ZAWISKY & E. JERICHA. 1990. Phys. Rev. **A42:** 3726.
33. NAMIKI, M. & S. PASCAZIO. 1990. Phys. Lett. **A147:** 430.
34. BELL, J. 1965. Physics **1:** 195.
35. HOME, D. & F. SELLERI. 1991. Rev. Nuovo Cimento **14:** 1.
36. RAUCH, H. & J. SUMMHAMMER. 1992. Phys. Rev. **A46:** 7284.
37. RAUCH, H., A. ZEILINGER, G. BADUREK, A. WILFING, W. BAUSPIESS & U. BONSE. 1975. Phys. Lett. **A54:** 425.
38. WERNER, S. A., R. COLELLA, A. W. OVERHAUSER & C. F. EAGEN. 1975. Phys. Rev. Lett. **35:** 1053.
39. SUMMHAMMER, J., G. BADUREK, H. RAUCH, U. KISCHKO & A. ZEILINGER. 1983. Phys. Rev. **A27:** 2523.
40. BADUREK, G., H. RAUCH & J. SUMMHAMMER. 1983. Phys. Rev. Lett. **51:** 1015.
41. BADUREK, G., H. RAUCH & D. TUPPINGER. 1986. Phys. Rev. **A34:** 2600.
42. JOSEPHSON, B. D. 1974. Rev. Mod. Phys. **46:** 251.
43. RAUCH, H. 1979. *In* Neutron Interferometry. U. Bonse & H. Rauch, Eds.: 161. Oxford University Press (Clarendon). London/New York.
44. KAISER, H., S. A. WERNER & E. A. GEORGE. 1983. Phys. Rev. Lett. **50:** 563.
45. CLOTHIER, R., H. KAISER, S. A. WERNER, H. RAUCH & H. WÖLWITSCH. 1991. Phys. Rev. **A44:** 5357.
46. SCHLEICH, W. & J. A. WHEELER. 1987. Nature **326:** 574.
47. SCHLEICH, W., D. F. WALLS & J. A. WHEELER. 1988. Phys. Rev. **A38:** 1177.
48. LEVY-LEBLOND, J-M. & F. BALIBAR. 1990. Quantics. North-Holland. Amsterdam.
49. LEGGETT, A. 1984. *In* Proceedings of the Foundations of Quantum Mechanics. S. Kamefuchi, Ed.: 74. Phys. Soc. Japan. Kyoto.
50. YURKE, B., W. SCHLEICH & D. F. WALLS. 1990. Phys. Rev. **A42:** 1703.
51. WALLS, D. F. & G. J. MILBURN. 1985. Phys. Rev. **A31:** 2403.
52. GLAUBER, R. J. 1986. Ann. N.Y. Acad. Sci. **480:** 336.
53. NAMIKI, M. & S. PASCAZIO. 1991. Phys. Rev. **A44:** 39.
54. ZUREK, W. H. 1991. Phys. Today **October:** 36.
55. WALLS, D. F. 1983. Nature **306:** 141.
56. BRAUNSTEIN, S. L. & R. I. MCLACHLAN. 1987. Phys. Rev. **A35:** 1659.
57. LOUDON, R. & P. L. KNIGHT. 1987. J. Mod. Opt. **34:** 709.
58. JANSKI, J. & A. V. VINOGRADOV. 1990. Phys. Rev. Lett. **64:** 2771.
59. GLAUBER, R. J. 1968. *In* Fundamental Problems in Statistical Mechanics. E. G. D. Cohen, Ed.: 140. North-Holland. Amsterdam.
60. ZAWISKY, M., H. RAUCH & Y. HASEGAWA. 1994. Phys. Rev. **A50:** no. 7.
61. MEZEI, F., Ed. 1980. Neutron Spin Echo: Lecture Notes in Physics, Volume 128. Springer Pub. New York.

62. BADUREK, G., H. RAUCH & A. ZEILINGER. 1980. *In* Neutron Spin Echo: Lecture Notes in Physics, Volume 128. F. Mezei, Ed. Springer Pub. New York.
63. HEINRICH, M., D. PETRASCHECK & H. RAUCH. 1988. Z. Phys. **B72:** 357.
64. MERMIN, N. D. 1990. Phys. Rev. Lett. **65:** 1838.
65. LORENTZ, H. A. 1927. Theorie der Strahlung. Akad. Verlagsgesellschaft. Leipzig.
66. PRIGOGINE, I. 1991. Proc. Ecol. Phys. Chem. (Siena), p. 8. Elsevier. Amsterdam/New York.
67. HAAG, F. 1990. Commun. Math. Phys. **123:** 245.
68. BLANCHARD, P. & A. JADCZYK. 1993. Phys. Lett. **A175:** 157.
69. EINSTEIN, A., B. PODOLSKY & N. ROSEN. 1935. Phys. Rev. **47:** 777.
70. GREENBERGER, D. M., M. A. HORNE & A. ZEILINGER. 1989. *In* Bell's Theorem, Quantum Theory, and Conceptions of the Universe. M. Kafatos, Ed.: 69. Kluwer. Dordrecht.
71. BADUREK, G., H. WEINFURTER, R. GÄHLER, A. KOLLMAR, S. WEHINGER & A. ZEILINGER. 1993. Phys. Rev. Lett. **71:** 307.

Experimental Tests of the Foundations of Quantum Mechanics Using Neutrons: The Scalar A-B Effect[a]

A. G. KLEIN

School of Physics
University of Melbourne
Parkville, Victoria 3052, Australia

INTRODUCTION

An experiment using a perfect crystal neutron interferometer, performed by a University of Melbourne/University of Missouri–Columbia collaboration,[1,2] has verified an effect closely analogous to the electric Aharonov-Bohm effect, which shows that scalar potentials, even in the absence of forces, have observable effects in quantum mechanics. The experiment, employing time-dependent interferometry, exhibits certain features that bear upon the famous "delayed choice" proposition by J. A. Wheeler[3] in that, from a classical standpoint, the particles have their motion influenced by a pulse applied after they have left the first beam-splitter.

DISCUSSION

The well-known Aharonov-Bohm (A-B)[4,5] effect demonstrates the fact that electrons suffer additional phase shifts when propagating through regions of space that carry nonzero potentials, even in the absence of forces acting on the particles. The usual magnetic A-B effect, amply verified by beautiful experiments with electron holography,[6] exhibits the influence of magnetic vector potentials. Somewhat less well known is the electric, or scalar, A-B effect, which concerns the phase shift caused by the scalar potential $V = -eU$ in the Schrödinger equation:

$$(H_0 + V) = i\hbar\partial\psi/\partial t. \tag{1}$$

Consider FIGURE 1a, which shows the split halves of an electron wave packet entering conducting cylinders, the interiors of which act as Faraday cages, that is, have field-free interiors no matter what electrostatic potential U they carry. In order to exhibit the scalar A-B effect, the potentials of the cylinders are made to change in a time-dependent manner while the wave packets are in flight through them. Thus, if the cylinders had voltage pulses $U(t)$ applied to them during that time, the electrons would not experience any force; however, a relative phase shift would nevertheless

[a]This work was supported by the Australian Research Council and the United States National Science Foundation (Grant No. NSF-PHY-9024606).

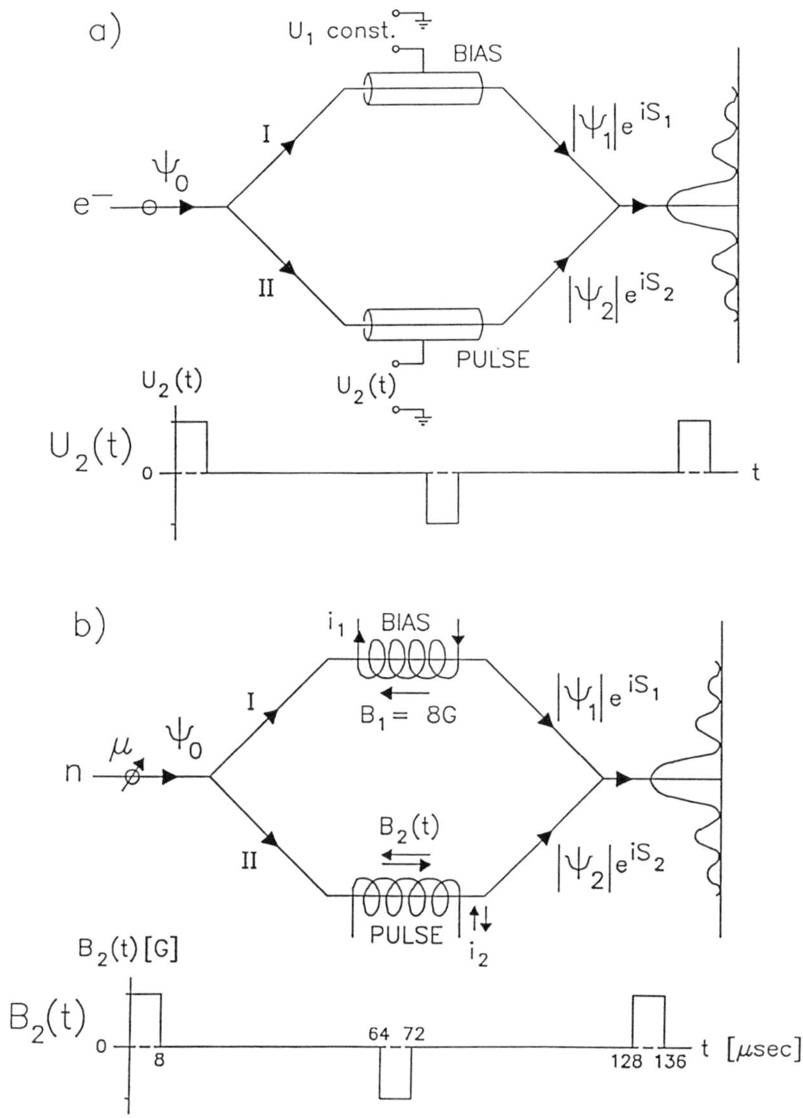

FIGURE 1. Schematic diagram of the scalar Aharonov-Bohm experiment for (a) electrons and (b) neutrons. The waveforms of the applied pulses are also shown.

arise, given by

$$\Delta\phi = (e/\hbar) \int [U_1(t) - U_2(t)]dt. \tag{2}$$

Although few would doubt the correctness of the A-B predictions, the effect is of such fundamental importance to the internal consistency of quantum mechanics that a practical realization of the experiment deserves a serious attempt. However, this simple (in principle) experiment has not yet been carried out because of technical difficulties with existing types of electron interferometers and the lack of suitable interferometers for protons or ions.

Hence, an analogous experiment, with neutrons rather than charged particles, was suggested by Zeilinger[7] and Anandan.[8] We designed and performed such an experiment, making use of a perfect single-crystal neutron interferometer. The experiment is shown in schematic form in FIGURE 1b, which shows the split halves of a neutron wave packet entering solenoids, the interiors of which contain uniform, gradient-free magnetic fields. Thus, neutrons, assumed to be polarized along the direction of B, the axis of quantization, find themselves in force-free regions inside the solenoids, but experience scalar potentials of $V = -\boldsymbol{\mu} \cdot \boldsymbol{B}$, where $\boldsymbol{\mu}$ is the magnetic dipole moment of the neutron. If current pulses, $i(t)$, are applied to the solenoids while the neutrons are in the force-free environment of their interior, they give rise to magnetic fields $\boldsymbol{B}(t)$ and the relative phase shift produced is given by

$$\Delta\phi = (1/\hbar) \int [\boldsymbol{\mu} \cdot \boldsymbol{B}_1(t) - \boldsymbol{\mu} \cdot \boldsymbol{B}_2(t)]dt, \tag{3}$$

in complete analogy with equation 2. In the actual experiment, short-duration current pulses (of a few microseconds) were applied to suitably designed solenoids placed inside a neutron interferometer. The counts detected were gated according to the position of the neutron wave packet relative to the solenoids at the time when the pulses were applied. The phase shift of neutrons that traversed the solenoids when the current was zero may thus be compared with the phase shift of neutrons that traversed the center of the solenoids when the current was at a maximum. For typical solenoid dimensions of ≈ 40 mm and a neutron velocity of ≈ 2 mm per microsecond (de Broglie wavelength $\approx$ 2 Å), magnetic fields of about 20 gauss were required in the solenoids in order to give rise to a relative phase shift of π radians.

The experiments, carried out at the University of Missouri Research Reactor (MURR), are described in detail in the papers of Allman *et al.*[1,2] The results, shown in FIGURE 2, are in excellent agreement with the predictions of equation 3. In what follows, we shall attempt to clarify several aspects of this experiment that have elicited comment.[9,10]

First, regarding the topological nature of the A-B effects, it is easy to see in our experiment that the phase shifts are independent of the exact paths taken by the neutrons through the apparatus, in exactly the same way as in the case of the electrons in the electric A-B experiment of FIGURE 1. The rather artificial definition of what is topological by Peshkin[10] tends to obscure this simple fact. As for his claim that the interiors of the solenoids are not field-free, this is patently true. They are, however, gradient-free and hence force-free, as far as neutrons are concerned, and this, in analogy with the force-free regions in which the electrons travel in FIGURE 1, is the crux of the matter. As further evidence of the fact that there are no forces and hence no dynamical phase shifts involved, it may be shown that the momentum of the particles, and hence their kinetic energy, stays constant when the pulsed magnetic fields are applied, as opposed to the case when the particles enter, traverse, and exit a static magnetic field. This has been demonstrated experimentally in the very beautiful experiment carried out by Badurek *et al.* showing that the A-B phase shifts are nondispersive.[11]

In the actual scalar A-B experiment with neutrons, the prospect of a considerable experimental simplification led us to employ an unpolarized beam of neutrons on the grounds that polarized neutrons would have given exactly the same result as an unpolarized beam, which may be regarded as an incoherent mixture of $\sigma = +1$ and

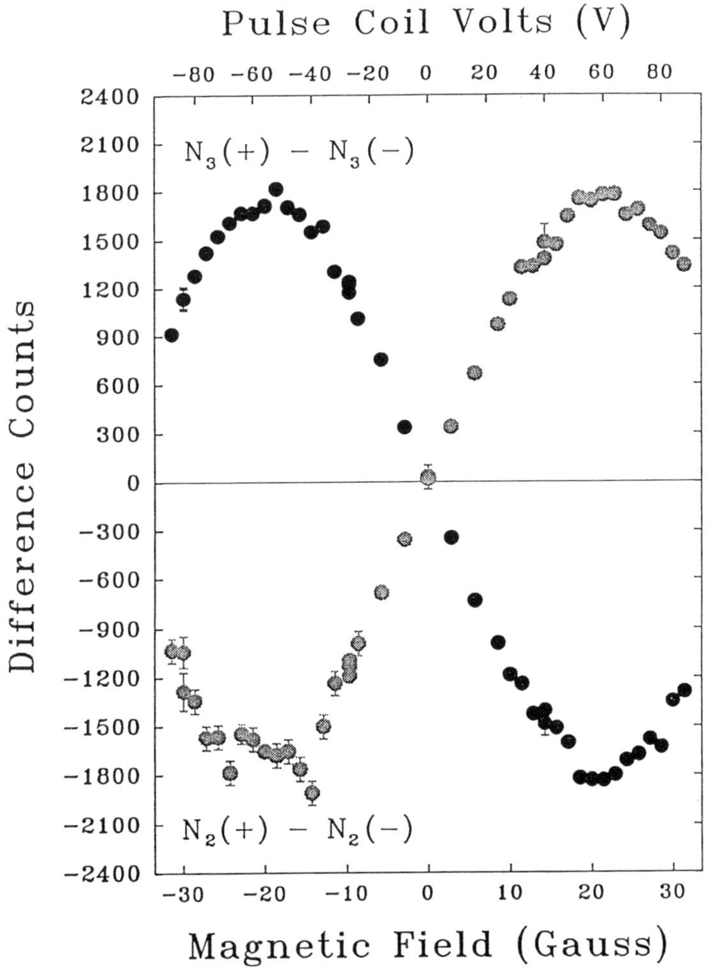

FIGURE 2. Interferometer output signals as a function of pulse coil field strengths.

$\sigma = -1$ neutrons. Because the interferometer signal depends on the cosine of the phase shift, $\Delta\phi$, the two polarization states corresponding to $\sigma = \pm 1$ give rise to identical outputs, that is, counts proportional to $\cos(\pm\Delta\phi)$. The critical comment by Peshkin[9] is based on an alternative interpretation, namely, the precession of neutrons in the B field. This is a perfectly valid alternative viewpoint, but it does not vitiate the analogy on which the A-B interpretation is based.

The whole of the above polemic appears to attack the scalar A-B experiment, as well as the earlier Aharonov-Casher[12] (A-C) experiment,[13] on the grounds that they are not identical to the A-B experiments. Evidently not: the neutron, with a finite dipole moment, but zero charge, is only to be expected to lead to a richer phenomenology that is nevertheless related to the electron effects by simple and highly suggestive analogies. Furthermore, just as the electric and magnetic A-B effects taken jointly are manifestations of the (A, V) 4-vector, the scalar A-B effect with neutrons may be regarded as the fourth component of the Aharonov-Casher effect, as shown by Anandan.[8]

A final observation concerns the time-dependent nature of the scalar A-B experiments: The current pulse applied to the solenoid occurs while the neutron traverses its center, that is, well after the corresponding classical particle had left the first beam-splitter. In other words, the classical particle had already "chosen" the path through the interferometer. Because the output path, behind the interferometer, is determined by the pulse of potential applied after the initial choice of path at the first beam-splitter, this time-dependent interferometry situation bears a close resemblance to the famous "delayed choice" experiment discussed by John Wheeler,[3] to whom this contribution is dedicated.

ACKNOWLEDGMENTS

I wish to thank my Melbourne colleagues, Alberto Cimmino and Geoffrey Opat, and my former student, Brendan Allman, now at Missouri, as well as my collaborators at the University of Missouri–Columbia, Helmut Kaiser and Sam Werner.

REFERENCES

1. Allman, B. E., A. Cimmino, A. G. Klein, G. I. Opat, H. Kaiser & S. A. Werner. 1992. Phys. Rev. Lett. **68:** 2409–2412.
2. Allman, B. E., A. Cimmino, A. G. Klein, G. I. Opat, H. Kaiser & S. A. Werner. 1993. Phys. Rev. A **48:** 1799–1807.
3. Wheeler, J. A. 1978. *In* Mathematical Foundations of Quantum Theory. A. R. Marlow, Ed.: 9–48. Academic Press. New York.
4. Aharonov, Y. & D. Bohm. 1959. Phys. Rev. **115:** 485.
5. Peshkin, M. & A. Tonomura. 1989. The Aharonov-Bohm Effect: Lecture Notes in Physics, Vol. 3450. Springer-Verlag. Berlin/New York.
6. Tonomura, A. 1987. Rev. Mod. Phys. **59:** 639.
7. Zeilinger, A. 1986. *In* Fundamental Aspects of Quantum Theory. NATO ASI Series B144. V. Gorini & A. Frigerio, Eds.: 331. Plenum. New York.
8. Anandan, J. 1990. *In* Proceedings of the Third International Symposium on Foundations of Quantum Mechanics in the Light of New Technology. S. Kobayashi *et al.*, Eds. Phys. Soc. Japan. Kyoto.
9. Peshkin, M. 1992. Phys. Rev. Lett. **69:** 2017.
10. Peshkin, M. 1995. This volume.
11. Badurek, G., H. Weinfurter, R. Gähler, A. Kollmar, S. Wehinger & A. Zeilinger. 1993. Phys. Rev. Lett. **71:** 307.
12. Aharonov, Y. & A. Casher. 1984. Phys. Rev. Lett. **53:** 319.
13. Cimmino, A., G. I. Opat, A. G. Klein, H. Kaiser, S. A. Werner, M. Arif & R. Clothier. 1989. Phys. Rev. Lett. **63:** 3803.

Diffraction in Time and a New Type of Interferometry with Nonseparated Beams[a]

A. I. FRANK[b] AND V. G. NOSOV[c]

[b]*I. M. Frank Laboratory of Neutron Physics*
Joint Institute for Nuclear Research
141980 Dubna, Moscow Region, Russia

[c]*Institute of General and Nuclear Physics*
Russian Scientific Center "Kurchatov Institute"
123182 Moscow, Russia

INTRODUCTION

The problem of nonstationary quantum effects in neutron optics has lately been the subject of extensive discussions. Moshinsky was, apparently, one of the first persons to approach this problem in 1952.[1] He considered the evolution of a neutron wave upon instantaneous extraction of a perfect absorber from a beam of monochromatic neutrons. His result for the wave evolution in the right half-space coincided in form with the familiar pattern of the Fregnel diffraction of light on an abrupt edge. That is why he named the considered phenomenon, "diffraction in time". Ya. B. Zel'dovich analyzed the general problem of periodic processes in 1966 and introduced the concept of quasi-energy.[2] In reference 3, a number of nonstationary quantum phenomena, which, in principle, could be observed in experiments with ultracold neutrons (UCN), were considered. The problem of using a "quantum" chopper as a possible device for performing experiments with cold neutrons was discussed in references 4–6. It is natural to assume that a real chopper operates in a periodic mode and the notion of quasi-energy most closely corresponds to the physical picture of the process. This fact has been noted in references 7 and 8.

NEUTRON QUASI-ENERGY AND THE PROBLEM OF PERIODIC INFLUENCE ON A NEUTRON BEAM

Quasi-energy, as a new physical characteristic, has a clear formulation when the Hamiltonian exhibits an explicit periodic time-dependence. A concrete case for such a phenomenon has been considered in references 7 and 8. The problem considered therein is related to the periodic influence of a perfect chopper, which, when removed from the beam at $t = 0$, put into the beam at time $t = T/2$, and removed again at $t = T$, has a total period of T.

[a]This work has been partially supported by the International Science Foundation (Grant No. ISF RFH 000).

On the basis of the Moshinsky solution[1] and the principle of superposition, the following wave function is found in reference 7 for a periodic chopper:

$$\psi(x, t) = \left(\frac{1}{2}\right) e^{i(kx-\omega t)} + \left(\frac{i}{\pi}\right) \sum_{n=-\infty}^{\infty} \frac{e^{i(k_n x-\omega_n t)}}{(2n - 1)}, \tag{1}$$

$$\omega_n = \omega + \left[\frac{2\pi(2n - 1)}{T}\right], \qquad k_n = \left(\frac{2m\omega_n}{\hbar}\right)^{1/2}. \tag{2}$$

Thus, the state in the right half-space represents a nonstationary superposition of waves, each of which has an energy $\hbar\omega_n$ and a corresponding wave number k_n. These equidistant satellites correspond, together with the nonshifted wave ω, to the same quasi-energy of a particle. It should be noted that the amplitudes of the satellites are Fourier coefficients of a rectangular function characterizing the influence of the chopper.

The relationship with the Fourier transform becomes even more evident when the problem of the diffraction of monochromatic neutrons on a moving grating is being solved.[9–11] Solution of the diffraction problem for a moving reference system connected with a grating reduces to a Fourier transformation of the coordinate part of the wave function. The wave function in the laboratory reference system is found by subsequent application of the Galilean transformation. At the limit, when the grating has a high enough velocity V and its period is $2a$, with the value of $T = 2a/V$ remaining constant, one readily arrives at equations 1 and 2 given above. In the same way, a solution was found for the moving phase "π-grating":

$$\psi(x, t) = \left(\frac{2}{i\pi}\right) \sum_{n=-\infty}^{\infty} \frac{e^{i(k_n x-\omega_n t)}}{(2n - 1)}. \tag{3}$$

To extend Moshinsky's analogy, we may refer to the above results as a "diffraction on a time grating".

THE GENERAL CASE OF NEUTRON WAVE MODULATION

The conclusion concerning the appearance of a discrete energy spectrum with partial amplitudes associated with Fourier coefficients could naturally be made in a more direct way. Let the action of some device placed at the origin of a reference system result in periodic variation of the amplitude or phase of the initial plane wave. Then, at small distances from this device, the wave function will have the form,

$$k^{-1} < x \ll vT, \qquad \psi(x, t) \cong f(t)e^{i(kx-\omega t)}, \tag{4}$$

where $f(t)$ is, generally speaking, a complex function of period T and v is the neutron velocity. Representing $f(t)$ in the form of a Fourier expansion over the frequencies $n\Omega$, one obtains, for $x > 0$,

$$f(t) = \sum_{n=-\infty}^{\infty} C_n e^{-in\Omega t}, \qquad \psi(x, t) = \sum_{n=-\infty}^{\infty} C_n e^{i(k_n x-\omega_n t)}, \tag{5}$$

where

$$\omega_n = \omega + n\Omega, \qquad k_n = k(1 + n\gamma)^{1/2}, \qquad \Omega = \frac{2\pi}{T}, \qquad \gamma = \Omega/\omega \ll 1. \tag{6}$$

The function $f(t)$ is what we will term the function of modulation. For an absorbing chopper and a modulation function of rectangular shape, it can be seen that we arrive at equation 1. In the case of π-phase modulation, we arrive at equation 3.

Note that the requirement of smallness of the modulation region as compared with the size of the spatial element vT in equation 4 is not essential, but is rather of a technical nature. If this region exhibits a significant extension, the main features of the phenomenon are conserved. For a quantitative description, however, somewhat more cumbersome calculations are required. A more detailed analysis of the concrete case of a neutron crossing the region with an oscillating magnetic field is presented in reference 12.

THE WAVE STRUCTURE OF A STATE AND LARGE-SCALE BEATS

Turning to the transformation in equations 5 and 6, we note that its peculiar characteristic is the nonlinear relationship between the quantities k and ω. The same situation is encountered in the propagation of an electromagnetic wave in a dispersing medium.

Passing in equation 6 to the expansion of k_n in $n\gamma$ and substituting the result into equation 5, we obtain, with a precision up to the included quadratic terms,

$$\psi(x, t) = e^{i(kx-\omega t)} \sum_{n=-\infty}^{\infty} C_n e^{i\left\{n\left[\left(\frac{\gamma}{2}\right)kx - \Omega t\right] - \left[\frac{(n\gamma)^2}{8}\right]kx\right\}}. \tag{7}$$

A beam that has passed through the chopper forms a rather complicated wave structure. At points where

$$\left(\frac{\gamma^2}{8}\right)kx = 2p\pi \qquad (p = \text{integer}), \tag{8}$$

it is readily seen that the wave function (equation 7) assumes the form,

$$\psi(x, t) = e^{i(kx-\omega t)} \sum_{n=-\infty}^{\infty} C_n e^{i(t_0 - t)n\Omega} = e^{i(kx-\omega t)} f(t - t_0) \qquad (t_0 = x/v). \tag{9}$$

Precisely the same wave picture is reconstructed in the vicinity of these points that is observed immediately behind the chopper. Thus, the transmitted wave exhibits a structure characteristic of spaced beats. Their large-scale period is[8]

$$L = \frac{16\pi}{\gamma^2 k} = \frac{(vT)^2}{\pi\lambda}, \qquad \lambda = k^{-1}. \tag{10}$$

This pattern of long-scale beating will spread again when the third (cubic) term in the expansion of k_n in $n\gamma$ becomes significant and will then reconstruct again at the distance of

$$L^{(2)} = \frac{32\pi}{\gamma^3 k}, \tag{11}$$

where the phases connected with this term will be equal to 2π. We can see that a characteristic length hierarchy takes place in the case under consideration.

Note that the length, vT, of the reconstructed "primary" structural elements is much smaller than the magnitude of spatial spreading, $\Delta x \simeq \sqrt{x/k}$, of a single neutron bunch. The nature of this reconstruction is the result of interference of the number of waves and occurs only due to the validity of the superposition principle and the linearity of the Schrödinger equation.

THE JOINT ACTION OF TWO CHOPPERS (MODULATORS) AND A TIME INTERFEROMETER

Now, let us see what happens if, at a certain distance l from the first chopper, a second chopper with the same frequency Ω is installed. Clearly, its action also reduces to the Fourier transformation of each partial wave. Thus, we have

$$\psi(x,t) = e^{i(kx-\omega t)} \times \sum_{n=-\infty}^{\infty} \sum_{m=-\infty}^{\infty} C_n e^{i\varphi_n} C_m e^{i\{(n+m)[(\gamma/2)kx - \Omega t] - [(n+m)^2/8]\gamma^2 kx\}}, \tag{12}$$

$$\varphi_n = \left\{ n\left[\left(\frac{\gamma}{2}\right)kl - \Omega\tau\right] - \left[\frac{(n\gamma)^2}{8}\right]kl \right\}. \tag{13}$$

For convenience, the x-coordinate is now read from the second chopper, and the time t is from the moment it is fired. In addition, τ is the time interval between the moments when the choppers are fired and is determined up to an integer number of periods T.

Because the indices n and m in the sum (equation 12) run through the same values from minus infinity to infinity, the energy spectrum of this state may also include a wave of initial energy $\hbar\omega$, for which $n + m = 0$. The amplitude of this wave is

$$A_0 = \sum_{n=-\infty}^{\infty} C_n C_{-n} e^{i\varphi_n}. \tag{14}$$

It is easy to show that in the case of amplitude modulation, when $f(t)$ is real, $|A_0| \leq |C_0|$, where C_0 is the amplitude of the unshifted wave ω behind the first modulator. What this means is that the second modulator can only attenuate this wave.

Another situation occurs in the case of the pure phase modulator. This is easy to show if we restrict ourselves to considering a sole class of symmetrical periodic functions of phase modulation:

$$f(t) = e^{iy(t)}, \qquad y(t + T/2) = -y(t). \tag{15}$$

The Fourier coefficients of such functions satisfy the condition

$$C_n = (-1)^n C^*_{-n}. \tag{16}$$

Thus, if one sets $\varphi_n = n\pi$ in equation 14, the initial wave is totally reconstructed because

$$A_0 = \sum_{n=-\infty}^{\infty} C_n C_{-n} e^{in\pi} = \sum_{n=-\infty}^{\infty} C_n C_{-n}(-1)^n = \sum_{n=-\infty}^{\infty} C_n C_n^* = 1. \tag{17}$$

To meet this requirement, it is necessary that the conditions,

$$\Omega\tau - \left(\frac{\gamma}{2}\right)kl = \pi, \qquad \left(\frac{\gamma^2}{8}\right)kl = 2p\pi \qquad (p = \text{integer}), \tag{18}$$

be satisfied.

Consider our modulators to be initially synchronized and tuned so that the conditions in equation 18 are fulfilled. Imagine that there is a region in the gap between the modulators with a certain potential field or where a refractive substance is present. In either case, refraction results in all the waves acquiring certain phase shifts. To be more definite, we shall deal with a refractive plate of thickness ξ. The material of the plate can be characterized by the potential U. It can be shown that, in this case, the relative phase shift of each wave to an accuracy of the second-power terms of γ and β is

$$\phi = \left(\frac{\beta\gamma}{4}\right) k\xi, \qquad \beta = \frac{U}{\hbar\omega} \ll 1. \tag{19}$$

Setting $\varphi_n = n(\phi + \pi)$ in equation 14, we obtain the expression for the wave function in our case. Now, in the wave picture, all the waves corresponding to harmonics with indices $\mu = n + m$ appear. Their amplitudes are determined by the sums,

$$A_\mu = \sum_{n=-\infty}^{\infty} C_n C_{\mu-n} e^{in(\phi+\pi)} = \sum_{n=-\infty}^{\infty} C_n C_{\mu-n} e^{in\phi}(-1)^n. \tag{20}$$

Substituting the Fourier coefficients $C_{\mu-n}$ into this expression and changing the order of summation and integration (taking into account equation 16), we obtain

$$A_\mu = \left(\frac{1}{2\pi}\right) \int_0^{2\pi} f(\chi) f^*(\chi - \phi) e^{i\mu\chi}\, d\chi \qquad (\chi = \Omega t). \tag{21}$$

In the absence of refraction, when $\phi = 0$, and in accordance with the above reasoning, it is readily seen that the following occurs:

$$A_0 = 1, \qquad A_{\mu\neq 0} = 0.$$

The appearance of the refractive plate, which alters the phases of all the waves, also decreases the amplitude of the reconstructed wave that had an initial energy $\hbar\omega$. Naturally, satellites will again appear here. Because the phase angle ϕ is proportional to the quantity ξ, the amplitudes of all waves exhibit a periodic dependence on the thickness of the plate. Here, the amplitude A_0 of the nonshifted wave changes from unity to $A_0^{\min}$, where

$$A_0^{\min} = \left(\frac{1}{2\pi}\right)\int_0^{2\pi} f^2(\chi)\, d\chi. \tag{22}$$

The value of $A_0^{\min}$ depends on the concrete form of the modulation function and, generally speaking, may equal zero.

The picture obtained is quite reminiscent of a multiray interferometer. Indeed, after the first modulator, instead of the initial wave, a set of coherent waves arises with differing energies and wave numbers. Consequently, it acts like a wave-splitter in the space of momenta and frequencies. The second modulator, if correctly adjusted, results in reconstruction of the initial state; that is, it acts like a recombiner. Introduction into the system of an additional region with a refractive material or potential field leads to a periodic change in amplitude $A_0(\xi)$ of the reconstructed wave with energy $\hbar\omega$. Note that a very similar situation takes place when a neutron with a precessing spin passes through the refractive plate. As has been shown previously, an additional spin rotation angle occurs in this case.[13] The value of this "optical spin rotation" angle coincides in form with the expression of equation 19 if we replace the modulation frequency Ω with the Larmor precession frequency $\omega_L = 2\mu B/\hbar$. This permits us to interpret the spin echo apparatus[14] as a neutron spin interferometer with two nonseparated beams.[15]

CHROMATIC ABERRATIONS AND THE THREE-MODULATOR INTERFEROMETER

Clearly, the phasing condition (equation 18) can be satisfied only for a single wavelength. Therefore, it is important to analyze which are the practical requirements for monochromatization of the initial wave. Such qualitative analysis can be started from the expression for phases (equation 13). Elementary calculations lead to the result,

$$\frac{\partial\varphi_n}{\partial k} \approx \left(-\frac{\gamma}{2}\right) nl.$$

The main contribution to the intensity of the reconstructed wave is, apparently, given only by the lowest harmonics, $n \approx 1$. Therefore, the condition $\delta\varphi_n \ll 1$, necessary for obtaining a satisfactory contrast, leads to the requirement,

$$\frac{\delta k}{k} \ll \frac{T}{t_0}, \qquad t_0 = \frac{l}{v}. \tag{23}$$

If the distance l between the modulators is on the order of L (the period of spaced beats), then equation 23 becomes

$$\frac{\delta k}{k} \ll \gamma \ll 1. \tag{24}$$

The problem of chromatic aberrations may be analyzed analytically as well. It is easy to find that the amplitudes of the partial waves will be expressed by the formulae,

$$A_\mu(\delta k) = \left(\frac{1}{2\pi}\right)\int_0^{2\pi} f(\chi)f^*(\chi - \delta\varphi)e^{i\mu\chi}\,d\chi, \quad \delta\varphi \simeq \left(-\frac{\gamma}{2}\right)l\delta k, \quad f(\chi) = e^{iy(\chi)}, \tag{25}$$

analogous to equation 21.

If the function $y(\chi)$ is analytical, then the value of the chromatic aberration may be calculated. For a change in amplitude δA_0 of the nonshifted wave, one can receive

$$\delta A_0 \cong \left[-\frac{(\delta\varphi)^2}{2\pi}\right]\int_0^{\pi}[y'(\chi)]^2\,d\chi. \tag{26}$$

It is possible to very effectively decrease the chromatic aberration. It is useful to consider, from this point of view, a system consisting of three modulators at identical distances l from each other. Let the first and third modulators be identical and let the depth of modulation of the middle one be twice that of the other two. As before, we shall consider the modulation function $f(t)$ of the extreme modulators to belong to the class given in equation 15 and to satisfy the condition given in equation 16. Then, the modulation function $f^2(t)$ of the middle device is

$$f^2(t) = e^{2iy(t)} = [f(t)]^2. \tag{27}$$

Clearly, it too satisfies the condition given in equation 16. In a general case, there may be waves with arbitrary frequencies $\omega + \nu\Omega$ after neutrons pass through the system. However, for the appropriate tuning of the device, it is possible to obtain,[10] as for a simple interferometer,

$$A_0 = 1, \qquad A_{\nu\neq 0} = 0.$$

Such a device happens to be achromatic in the third order of the wave number variation. Nevertheless, in spite of a quadratic decrease of the chromatic aberration, the requirement of $\delta\varphi \ll 1$ and the estimation given in equation 24 stay formally true in this case as well.

POSSIBLE REALIZATION AND APPLICATION

Obviously, for the successful operation of a time interferometer, one must resolve at least three problems: phase modulation, monochromatization, and analysis of the final state. For the phase modulation, it is natural to seek a solution of the problem on the basis of the possibility of rapidly changing the energy of the neutron magnetic interaction. Transmission of neutrons through an oscillating magnetic field (see reference 12) or magnetic film seems to be most suitable. In this last case, a frequency on the order of 10–20 MHz may be possible to achieve. This corresponds to an energy split of $\hbar\Omega \approx$ 20–40 neV.

Monochromatization of the primary beam and the energy-analyzing of the final state at the level of some neV appear quite within the possibility of UCN spectrometry.[9,16] Unfortunately, a difficulty appears in this case connected with the shortly spaced beats period, L, which is defined by equation 10. In the case under consideration, the beats period is on the order of some tens of microns, whereas the distance between the modulators may be of some millimeters. The value of L is inversely proportional to the square of the modulation frequency, but spectroscopy difficulties increase with a decrease in modulation frequency.

The energy region of very cold neutrons (VCN) with velocities on the order of 20–50 m/s seems to be preferable from this point of view. Then, primary monochromatization may be achieved by the time-of-flight method. For example, it could be possible to construct a VCN source at the IBR-2 pulsed reactor (Dubna, Russia). Then, the duration of the neutron pulse would be on the order of 300 μs. At a flight length of 10 m, the relative velocity resolution would be on the order of 1×10^{-3} for a neutron velocity of 30 m/s, which must be compared with the value of $\gamma \approx 9 \times 10^{-3}$ for the 10-MHz modulation frequency. It is possible to use a relatively wide spectral interval if one changed the modulation frequency periodically to keep the value of the space beats period L constant as neutron energy changed with time.

At the same time, it is necessary to measure the relative intensities of the reconstructed wave with an initial energy $\hbar\omega$ as well as the intensities of the satellite waves. There are several ways to solve this problem, but it seems that the best one is not connected with direct spectroscopy.

It is possible to show[17] for a state of the type of equation 7 that the beam intensity may oscillate. Thus, oscillation occurs in pure phase modulation as well. In this last case, the oscillation takes place "far" from the modulator and is maximal at $x = L/4$. This wholly concerns the state behind the interferometer. When the interferometer is tuned in such a way that the initial wave is fully reconstructed, these intensity oscillations disappear. Therefore, the analysis of the time structure of the beam intensity may be successfully used instead of spectroscopic analysis to detect the presence of satellite waves in the final state.

Finally, we shall briefly discuss the possible application of the proposed device. In our opinion, experimental verification of the above theoretical predictions is very useful in itself, being a direct manifestation of the validity of the nonstationary Schrödinger equation. We stress again that the linearity of the Schrödinger equation is the fundamental principle forming the foundation of this theory. It is possible to roughly estimate the significance of the possible nonlinear term to our problem. Let us put

$$b \approx \left(\frac{\hbar^2}{2m}\right)\left(\frac{\partial^2\psi}{\partial x^2}\right) \approx \left(\frac{\hbar^2}{2m}\right)\left(\frac{1}{(\Delta x)^2}\right). \quad \textbf{(28)}$$

Here, b is a constant with a dimension of energy that characterizes the possible nonlinear term (see, for example, reference 18), limiting wave packet spreading. Using for Δx the length of the spread of the wave packet after the neutron passes along path x, that is,

$$\Delta x = \sqrt{x/k},$$

we obtain $b \approx 4.3 \times 10^{-15}$ eV for $x = 1$ m and $k = 4.7 \times 10^{6}$ cm^{-1} ($v = 30$ m/s). This is very close to its upper limit of 3.3×10^{-15} eV estimated in reference 19. More detailed calculations are needed to more accurately find the sensitivity of the experiment to the possible nonlinear term.

The proposed interferometer may be very useful for performing experiments on measurement effects connected with gravity or with neutron motion in noninertial reference systems that seem very important for the test of neutron gravity and inertial mass equivalence.[20–26] For an estimation of its sensitivity, imagine that a neutron is moving with acceleration a along the interferometer base of l. Then, for the phase difference in equation 21, we obtain

$$\phi = \frac{\Omega a l^2}{v^3}. \tag{29}$$

Substituting $l = 10^2$ cm, $f = (\Omega/2\pi) = 20$ MHz, and $v = 3 \times 10^3$ cm/s, we find $\phi \approx 50a$ [cm/s^2]. Assuming that the accuracy of the phase shift measurement is $\delta\phi \approx 0.05$, the accuracy of the acceleration measurement is

$$\delta a \approx 1 \times 10^{-3}\ \text{cm/s}^2 = (1 \times 10^{-6})g,$$

where g is the acceleration of gravity.

ACKNOWLEDGMENTS

We wish to express our gratitude to R. Gähler, J. Greene, A. M. Kamchatnov, M. V. Kazarnovsky, A. Klein, S. V. Masalovich, H. Rauch, V. P. Smilga, A. Steyerl, J. Summhammer, S. Werner, and A. Zeilinger for useful discussions. Thanks are also given to G. Pontecorvo and A. Schaeffer for their help in the preparation of the English version of this report.

REFERENCES

1. MOSHINSKY, M. 1952. Phys. Rev. **88:** 625.
2. ZEL'DOVICH, YA. B. 1966. JETP **51:** 1492.
3. GERASIMOV, A. S. & M. V. KAZARNOVSKII. 1976. JETP **71:** 1700.
4. GÄHLER, R. & R. GOLUB. 1984. Z. Phys. B **56:** 5.
5. FELBER, J., R. GÄHLER & R. GOLUB. 1988. Physica B **151:** 135.
6. FELBER, J., G. MULLER, R. GÄHLER & R. GOLUB. 1990. Physica B **162:** 191.
7. NOSOV, V. G. & A. I. FRANK. 1990. *In* Proceedings of the Sixth International School on Neutron Physics (Alushta) (in Russian). Volume 1, p. 313.
8. NOSOV, V. G. & A. I. FRANK. 1991. J. Moscow Phys. Soc. **1:** 1.
9. FRANK, A. I. & V. G. NOSOV. 1992. JINR Communication No. E4-92-457.
10. FRANK, A. I. & V. G. NOSOV. 1994. Yad. Fiz. **57:** 1029 (in Russian).
11. FRANK, A. I. & V. G. NOSOV. 1994. Phys. Lett. A **188:** 120.
12. SUMMHAMMER, J. 1993. Phys. Rev. A **47:** 556.
13. FRANK, A. I. 1989. Nucl. Instrum. Methods **A284:** 161.
14. MEZEI, F., Ed. 1980. Neutron Spin Echo: Lecture Notes in Physics 128. Springer-Verlag. Berlin/New York.
15. BARYSHEVSKII, V. G., S. V. CHEREPITSA & A. I. FRANK. 1991. Phys. Lett. A **153:** 299.
16. SCHECKENHOFER, H. & A. STEYERL. 1981. Nucl. Instrum. Methods **179:** 393.
17. FRANK, A. I. & V. G. NOSOV. 1994. Density matrix and the slow neutron beams transformation. To be published.
18. BIALYNICKI-BIRULA, I. & J. MYCIELSKI. 1976. Ann. Phys. (N.Y.) **100:** 62.
19. GÄHLER, R., A. G. KLEIN & A. ZEILINGER. 1981. Phys. Rev. A **23:** 1611.
20. COLELLA, R., A. W. OVERHAUSER & S. A. WERNER. 1975. Phys. Rev. Lett. **34:** 1472.

21. GREENBERGER, D. M. & A. W. OVERHAUSER. 1979. Rev. Mod. Phys. **51:** 43.
22. STAUDENMANN, J-L., S. A. WERNER, R. COLELLA & A. W. OVERHAUSER. 1980. Phys. Rev. A **21:** 1419.
23. BONSE, U. & T. WROBLEWSKI. 1983. Phys. Rev. Lett. **51:** 1401.
24. BONSE, U. & T. WROBLEWSKI. 1984. Phys. Rev. D **30:** 1214.
25. WERNER, S. A., H. KAISER, M. ARIF & R. CLOTHIER. 1988. Physica B **151:** 22.
26. SCHMIEDMAYER, J. 1989. Nucl. Instrum. Methods **A284:** 59.

Two-State Quantum Asymptotics

MICHAEL BERRY

H. H. Wills Physics Laboratory
Bristol BS8 1TL, United Kingdom

Only wimps specialize in the general case. Real scientists pursue examples.
—Adapted from a remark by Beresford Parlett

INTRODUCTION

John A. Wheeler's distinctive intellectual stamp includes the choice of simple examples to demystify difficult and subtle ideas, and I will follow that lead here. In the last decade, we have learned about several general quantum phenomena that share the property that they are emergent as a parameter (ϵ in what follows) vanishes. Remarkably, all can be illustrated by the simplest nonsimple quantum problem, namely, the evolution of two-state systems with a time-dependent Hamiltonian.

Such systems are governed by the Schrödinger equation,

$$i\partial_t|\psi\rangle = \mathbf{R}(\epsilon t)\cdot\mathbf{S}|\psi\rangle. \tag{1}$$

Here, the state is the 2-spinor

$$|\psi\rangle = |\psi(t,\epsilon)\rangle = \begin{pmatrix}\psi_1\\ \psi_2\end{pmatrix}; \tag{2}$$

$\mathbf{S}$ is the vector of spin-½ matrices,

$$\mathbf{S} \equiv \left(\frac{1}{2}\right)\left[\begin{pmatrix}0 & 1\\ 1 & 0\end{pmatrix}, \begin{pmatrix}0 & -i\\ i & 0\end{pmatrix}, \begin{pmatrix}1 & 0\\ 0 & -1\end{pmatrix}\right]; \tag{3}$$

and the vector that drives the system is

$$\mathbf{R}(\epsilon t) \equiv \{X(\epsilon t), Y(\epsilon t), Z(\epsilon t)\}. \tag{4}$$

$\mathbf{R}(\epsilon t)$ will be called the Hamiltonian vector and its track over $-\infty < t < +\infty$ will be called the Hamiltonian curve. The Hamiltonian itself is

$$\mathbf{H}(\epsilon t) = \mathbf{R}(\epsilon t)\cdot\mathbf{S} = \left(\frac{1}{2}\right)\begin{pmatrix}Z(\epsilon t) & X(\epsilon t) - iY(\epsilon t)\\ X(\epsilon t) + iY(\epsilon t) & -Z(\epsilon t)\end{pmatrix}. \tag{5}$$

Vanishing asymptotic parameter ϵ is the adiabatic limit of slow driving (except for the final example in the last section below, where ϵ will have a different interpretation). This two-state formalism has several physical interpretations; one is the spin state of a neutron in a slowly changing magnetic field $\mathbf{R}(\epsilon t)$.

In each of the five examples to follow, I shall first give a brief description of the general phenomenon and then proceed to its two-state illustration.

GEOMETRIC PHASE

A quantum system driven by parameters that are slowly changed round a cycle C will, because of the quantum adiabatic theorem,[1] cling closely to an eigenstate of the instantaneous Hamiltonian and thus return to its original state, apart from a phase factor. This phase factor[2–6] is

$$\langle\psi_n(\text{beginning})|\psi_n(\text{end})\rangle = \exp\left\{\left(-\frac{i}{\epsilon}\right)\int_{\text{cycle}} d\tau E_n(\tau)\right\}\exp\{i\gamma_n(\text{C})\}. \quad (6)$$

Here, n labels the state being transported round C. The first factor, involving the instantaneous energy $E_n(\epsilon t)$, contains the dynamical phase, generalizing the $-i\omega t$ in the evolution of any wave. The second factor contains the geometrical phase, given by the line and surface integrals

$$\gamma_n(\text{C}) = -\text{Im}\oint_{\text{C}}\langle n|dn\rangle = -\text{Im}\int\int_{\partial\text{S}=\text{C}}\langle dn|\wedge|dn\rangle, \quad (7)$$

where $|n\rangle$ is the instantaneous eigenstate, with the phase chosen so that $|n\rangle$ is single-valued round C.

For a two-state system, the parameters are the components (equation 4) of the Hamiltonian vector, C is a loop in **R** space, and n labels the state with energy $(\pm\frac{1}{2})|\mathbf{R}|$ [that is, spin $(\pm\frac{1}{2})\hbar$]. Then, the geometric phase is[2]

$$\gamma_\pm(\text{C}) = (\mp\tfrac{1}{2})\Omega(\text{C}), \quad (8)$$

where Ω is the solid angle subtended by C at $\mathbf{R} = 0$. This version of the geometric phase has been measured with a neutron beam,[7] along which the direction of a magnetic field is varied round a cone. A different implementation was known much earlier[8–11] for light beams whose state of polarization was cycled; in this case, parameter space is the Poincaré sphere of polarizations, that is, $|\mathbf{R}| = 1$.

The geometric phase has been generalized to nonadiabatic cycle evolutions,[12] for which the state returns exactly apart from a phase factor (this can be made to happen even if the Hamiltonian does not depend on time). Then, equation 8 still holds, but now Ω is the solid angle of the loop traversed by the vector $\langle\psi(t)|\mathbf{S}|\psi(t)\rangle$ (which coincides with **R** only adiabatically).

However, the adiabatic case contains surprising richness, in the form of corrections[13] to equation 6 containing higher powers of ϵ. When ϵ is small, but not zero, the state does not return exactly. Instead, we have

$$\langle\psi_n(\text{beginning})|\psi_n(\text{end})\rangle = a(\epsilon)\exp\{i\phi(\epsilon)\}. \quad (9)$$

If the Hamiltonian curve is analytic, the deviation from unity of the modulus $a(\epsilon)$, corresponding to transitions out of the initial state, is exponentially small, that is, beyond all orders, in ϵ. However, the phase $\phi(\epsilon)$ does have a power series in ϵ, whose first two terms (those not vanishing in the adiabatic limit) are given in equation 6. The geometric phase is the term not involving ϵ and can be expressed as an adiabatically emergent phenomenon according to

$$\gamma_n(C) = \lim_{\epsilon \to 0} \left(\frac{d}{d\epsilon}\right) [\epsilon\phi(\epsilon)]. \tag{10}$$

An iteration method[13] for calculating higher orders of the expansion consists of successive ("superadiabatic") transformations to moving frames, aimed at freezing the Hamiltonian vector $\mathbf{R}(t)$. These are time-dependent transformations, which change the Hamiltonian, so the freezing is not quite successful. Instead, C transforms to a sequence of renormalized loops that initially get rapidly smaller and whose geometric phases are the corrections to $\gamma_n(C)$; the r-th such correction is of order ϵ^r. The sequence diverges and the divergence has a universal form[13] related to the exponentially small transition probability.

GEOMETRIC AMPLITUDE

If the adiabatically varied Hamiltonian parameters depend analytically on time—and now the Hamiltonian curve need not be closed—the exponentially small final probability for transitions out of the initial state also contains, in general, a geometric correction,[14,15] independent of ϵ. This was as unexpected as the geometric phase, of which it is an analytic continuation. The geometric amplitude is associated with complex times for which the instantaneous eigenstates are degenerate (it being assumed that there are no real degeneracies); these degeneracies are the source of the weak adiabatic transitions.

For two-state systems, degeneracies arise from complex times τ^* when

$$R(\tau^*) \equiv [X^2(\tau^*) + Y^2(\tau^*) + Z^2(\tau^*)]^{1/2} = 0. \tag{11}$$

The transition probability from (say) the initial state $|\psi(-\infty)\rangle = |+(-\infty)\rangle$, with energy $+R$, to that with $-R$ is[14]

$$|\langle -(+\infty)|\psi(+\infty)\rangle|^2$$

$$= \exp\left\{\left(-\frac{2}{\epsilon}\right)\left|\operatorname{Im}\int_0^{\tau^*} d\tau R(\tau)\right|\right\} \exp\left\{-2\operatorname{Im}\int_0^{\tau^*} d\tau\dot{\phi}(\tau)\cos\theta(\tau)\right\}. \tag{12}$$

Here, $\theta(\tau)$ and $\phi(\tau)$ are the polar angles of the Hamiltonian curve. The first ("dynamical") exponent was familiar[16,17] and is the dominant contribution to the exponentially weak transitions. The second exponent is geometric and can be regarded as a complex generalization of the solid angle in equation 8.

It follows from equation 12 that the geometric exponent is zero if the Hamiltonian curve lies in a plane through $\mathbf{R} = 0$ (as in the familiar Landau-Zener problem[16]) or is reversible in the sense that it can be rigidly rotated about an axis through $\mathbf{R} = 0$ so as to coincide with its time-reverse (this case includes uniform helices). The simplest curve whose geometric exponent is not zero is the "winding-unwinding" helix, for which

$$\mathbf{R}(\tau) = \{\Delta \cos B\tau^2, \Delta \sin B\tau^2, A\tau\}. \tag{13}$$

In this case, the geometric amplitude factor is[14]

$$\exp\left\{-\pi\left(\frac{B\Delta^2}{A^2}\right)\operatorname{sgn}(A)\right\}. \tag{14}$$

The successful observation of the geometric amplitude in a magnetic resonance experiment[18] confirms that it describes real physics in the complex plane and is an asymptotically emergent phenomenon, separating from the dynamical contribution as $\epsilon \to 0$.

WKB PHASES, BOSONS, AND SEMIONS

In the early years of the geometric phase, it was asked whether the phase π responsible for the "½" of the "n + ½" WKB (semiclassical) quantization of oscillators (exact in the harmonic case) is geometric. My first answer was no, but this was wrong. It turns out that this semiclassical π [and related phases in optics, such as Gouy's (1899)[19] $\pi/2$ jump through a focus and Stokes' (1847)[20] similar jump through a caustic] does correspond to a geometric phase.

One way to see this is to cast the one-dimensional stationary Schrödinger equation for a nonrelativistic particle in a potential well $V(z)$, namely,

$$\left(\frac{\hbar^2}{2m}\right)\left(\frac{d^2}{dz^2}\right)\psi(z) + [E - V(z)]\psi(z) = 0, \tag{15}$$

in the form of a two-state problem. Defining

$$\epsilon \equiv \frac{\hbar}{\sqrt{2m}}, \qquad z \equiv \epsilon t, \qquad \psi_1 \equiv \psi, \qquad \psi_2 \equiv \frac{d}{dt}\psi, \tag{16}$$

we find equations 1–4 with

$$X - iY = 2i, \qquad X + iY = 2i[V(\epsilon t) - E], \qquad Z = 0 \tag{17}$$

and with the adiabatic limit now being interpreted as the semiclassical limit $\hbar \to 0$.

The semiclassical quantum condition, determining the allowed energies E, is that the adiabatic eigenstates specified by equations 16 and 17, which correspond to left- and right-moving waves, should be single-valued after a circuit of the two classical turning points defined by $V(z) = E$. From equation 17, these are the degeneracies of the matrix **H** in equation 5. The quantum condition includes two contributions: a dynamical phase, given by the classical action of the circuit divided by $\hbar$, and a geometric phase associated with the degeneracies. An immediate difficulty is that **H** is not Hermitian, but this is resolved by a generalization[21] of the first half of equation 7, namely,

$$\gamma_n(\mathrm{C}) = i \oint_{\mathrm{C}} \frac{\langle \tilde{n} | dn \rangle}{\langle \tilde{n} | n \rangle}, \tag{18}$$

where $\langle \tilde{n} |$ is the left eigenvector of equation 5 (that is, the row vector given by the conjugate transpose of the eigenvector of $\mathbf{H}^\dagger$).

Application of equation 17 shows[22] that equation 18 has two interesting properties. First, it is real and thus does represent a phase. This is surprising because, for general non-Hermitian matrices, equation 18 will give a complex number. Second, it is the same for both of the degenerating states, rather than opposite as in equation 8. This is a consequence of the fundamental property that the eigenvectors of a non-Hermitian matrix become parallel at a degeneracy, rather than remaining orthogonal as in the Hermitian case (the same property has interesting consequences elsewhere in physics—for example, in the optics of absorbing crystals[11]). A calculation shows that the phase is

$$\gamma(C) = \left(-\frac{\pi}{2}\right) \times [\text{number of zeros of } E - V(z) \text{ inside C}]. \tag{19}$$

For an oscillator, there are two turning points, showing that the WKB π phase can indeed be interpreted as geometric.

In the Hermitian case of the second section, a planar circuit of a degeneracy gives the familiar π geometric phase associated with rotation of a spin-½ particle. In the present non-Hermitian case, the $-\pi/2$ phase associated with each degeneracy could also be achieved by planar rotation of a spin-¼ particle, that is, a semion. Given the connection of harmonic oscillators with bosons, we arrive at the curious suggestion that a boson can, in a sense, be regarded as made of two semions.

STOKES' PHENOMENON AND ADIABATIC QUANTUM JUMPS

Asymptotics, that is, the study of divergent series associated with singular limits, is currently enjoying worldwide resurgence. The deepest reason for this is the recognition[23] that relations between physical theories (e.g., quantum and classical mechanics, and statistical mechanics and thermodynamics) take the form of limits (e.g., $\hbar \to 0$ or $N \to \infty$) and these limits are usually singular. A central feature of divergent series is that their divergent tails often yield information about exponentially small terms, which are beyond all orders[24] of the series and whose existence is the cause of the divergence. As variables (not the asymptotic parameter) are changed, the small exponentials can appear and disappear; this is Stokes' phenomenon.[25]

Recent progress in asymptotics[26,27] has enabled Stokes' phenomenon to be understood in detail, with results that will now be described (ignoring several subtleties). Let the function being expanded be $G(\epsilon, \boldsymbol{X})$, where ϵ is the (small) asymptotic parameter and the $\boldsymbol{X} = \{X_1, X_2, \ldots\}$ terms are other variables on which G depends. Let the bare (i.e., not resummed) asymptotic series be

$$G(\epsilon, \boldsymbol{X}) = M_+(\epsilon, \boldsymbol{X}) \exp\left\{\left(\frac{1}{\epsilon}\right)\phi_+(\boldsymbol{X})\right\} \sum_{r=0}^{\infty} \epsilon^r T_r(\boldsymbol{X}), \tag{20}$$

where $T_0 = 1$ and the + indicates that the prefactor and exponent refer to the dominant exponential. As the order r increases, the terms $\epsilon^r T_r$ first get smaller and then diverge (typically factorially,[28] although the results do not depend on this[29]). Let the subdominant exponential (or the largest of these if there are several) have exponent $\phi_-(X)/\epsilon$, where

$$\operatorname{Re} \phi_-(X) < \operatorname{Re} \phi_+(X). \tag{21}$$

If now the series (equation 20) is optimally truncated, that is, summed to its least term r^* (typically, this increases as $1/\epsilon$), the remainder can be resummed and the resummation[26] has a remarkable universality:

$$G(\epsilon, X) \approx M_+(\epsilon, X) \exp\left\{\left(\frac{1}{\epsilon}\right)\phi_+(X)\right\} \sum_{r=0}^{r^*(\epsilon)} \epsilon^r T_r(X) + iS\{F(\epsilon, X)\} M_-(\epsilon, X) \exp\left\{\left(\frac{1}{\epsilon}\right)\phi_-(X)\right\}. \tag{22}$$

This includes the small exponential, which has been born from the divergent tail of the bare series (equation 20). Most important is the Stokes multiplier S, which describes the switching-on of the small exponential. S depends only on the disparity between the two exponents, denoted by F and called the singulant:

$$F(\epsilon, X) \equiv \left(\frac{1}{\epsilon}\right)[\phi_+(X) - \phi_-(X)]. \tag{23}$$

Stokes' phenomenon occurs when the large exponential maximally dominates the small one, that is, when F is real and positive; this occurs on Stokes lines (generally, hypersurfaces). The switching is smooth and is described by the universal function,[26]

$$S(F) = \left(\frac{1}{2}\right)\left[1 + \operatorname{erf}\left(\frac{\operatorname{Im} F}{\sqrt{2 \operatorname{Re} F}}\right)\right], \tag{24}$$

where erf is the error function. Across the Stokes line, this rises from 0 to 1, with a width $\sqrt{\epsilon}$. The universality and compactness of this result is a consequence of the optimal truncation of the original series; if the series is not truncated at or near the least term r^*, these properties are lost.

A physical illustration of this general result is provided by the adiabatic two-state system defined in the INTRODUCTION. Here, the dominant exponential contains the dynamical phase associated with the amplitude for remaining in the initial adiabatic eigenstate and the small exponential describes the transition to the other state. Stokes' phenomenon plays a central part in the *history* of the transition amplitude,[30] that is, in describing how the quantum jump evolves as the amplitude increases from zero to its final exponentially small value (discussed in the third section). A Stokes line issues from the complex-time degeneracy τ^* responsible for the transition and crosses the real-time axis, at (say) $\tau = \tau_0$. It is at this time that the quantum jump can be said to happen.

If the transitions are considered to be between adiabatic states, that is, eigenstates of the instantaneous Hamiltonian, the transition history does not have the universal error function form (equation 24). The reason is that the adiabatic states are merely the first terms in the solution of equation 1 as an adiabatic power series in ϵ. If they are used as a basis to describe the transition, the amplitude increases from zero to $O(\epsilon)$ before falling to the final $O(\exp\{-1/\epsilon\})$, with complicated nonuniversal oscillations en route. However, if the adiabatic series for the solution of equation 1 is optimally truncated and the two solutions are employed as an "optimal superadiabatic" basis for describing the evolution, then the transition amplitude increases smoothly over a time of order $\sqrt{\epsilon}$ and in accordance with the error function (equation 24). Detailed analysis[30] shows that, for transitions from $|+\rangle$ to $|-\rangle$,

$$|\langle -(+\infty)|\psi(t)\rangle| = \left(\frac{1}{2}\right)\left[1 + \text{erf}\left\{\frac{w(\tau)}{\sqrt{2\epsilon|w(\tau^*)|}}\right\}\right]\exp\left\{-\frac{|w(\tau^*)|}{\epsilon}\right\}, \tag{25}$$

where

$$w(\tau) \equiv \int_{\tau_0}^{\tau} d\tau' R(\tau'). \tag{26}$$

This surprising result shows that all adiabatic quantum transitions take place in the same way—that is, as an error function, independent of the form of the Hamiltonian curve—provided that the optimal superadiabatic basis is employed (which differs only slightly from the adiabatic one). Numerical calculations[31] confirm the correctness of the theoretical ideas, even when several degeneracies are involved,[32] which may coalesce.[33] There is a possibility that the Stokes smoothing could be observed directly in a spin experiment or an optical analogue: the spin would be measured at different times in a direction not along the "magnetic field" **R** (which would represent the adiabatic basis), but in a nearby direction, precisely specified by the theory,[30] representing the optimal basis.

DECOHERENCE AND THE QUANTUM ZENO EFFECT

Recent studies have identified two ways in which quantum systems are vulnerable to uncontrolled influences from their environment. The first is decoherence,[34,35] which is intended to explain why superpositions of macroscopically distinct states (as in Schrödinger's cat joke) are not observed. The mechanism is as follows: when the density matrix is represented in a basis of these macroscopically distinct states, averaging over the environmental forces or variables induces rapid vanishing of the off-diagonal elements. Because these elements represent quantum interference between the states, decoherence obliterates such interference, so the occupancies of the states can be described by probabilities rather than by amplitudes. The second way is the Zeno effect[36–39] (also known as the quantum watchdog or quantum watched pot), which is the slowing down, or complete inhibition, of quantum transitions that would occur in the isolated system, again as a result of action by the environment; it could explain, for example, the persistence of chiral molecules that, when isolated, would tunnel to their opposite-handed forms.

These phenomena are very different: in decoherence, the environment acts rapidly (relative to time scales associated with the isolated quantum system), whereas in the Zeno effect the environment induces slowing down. Therefore, it is worth knowing that both can be illustrated by a simple exactly solvable two-state model in the class defined in the INTRODUCTION (but with a different ϵ-dependence), and that is what I will present here. Neither the model nor its solution is new,[40–44] but I wish to emphasize how the two environmental effects are encompassed.

In equation 2, ψ_1 and ψ_2 represent the amplitudes for a quantum particle to inhabit each of the two wells, separated by a high barrier, of a double-bottomed potential. The wells could model, for example, the left- and right-handed forms of a chiral molecule. Forces within the isolated system are invariant under parity, that is, interchange of the two states (ignoring the small effect of weak interactions), so the eigenstates are even and odd superpositions of amplitudes localized in the wells. The energy separation of these eigenstates will be the small parameter ϵ, which is exponentially small in the height and width of the barrier and in $1/\hbar$. A state initially localized in one of the wells will tunnel resonantly between them, with an oscillation time of order $1/\epsilon$. The isolated system satisfies equation 1 with a different ϵ-dependence. The Hamiltonian (cf. equations 4 and 5) is

$$\mathbf{H}_0 = \mathbf{R}_0 \cdot \mathbf{S} = \left(\frac{1}{2}\right)\begin{pmatrix} 0 & \epsilon \\ \epsilon & 0 \end{pmatrix}, \tag{27}$$

where the suffix 0 denotes the isolated system.

A crude, but adequate, representation of the environment is through time-dependent random forces acting differently on the two wells (fuller treatments[45,46] include the environmental coordinates as dynamical variables). The random forces correspond to collisions, for example, with other molecules, or photons from the microwave background. In the asymptotic realm that we are interested in, where the two states are well separated by a high barrier, the environmental forces are large compared with ϵ and vary on time scales short compared with $1/\epsilon$. A Hamiltonian describing this situation is

$$\mathbf{H} = \mathbf{R}(t) \cdot \mathbf{S} = \left(\frac{1}{2}\right)\begin{pmatrix} f(t) & \epsilon \\ \epsilon & -f(t) \end{pmatrix}, \tag{28}$$

where $f(t)$ is white noise. The strength of the noise is normalized by

$$\overline{\left(\int_0^T dt f(t)\right)^2} = T/T_0. \tag{29}$$

Here and hereafter, the overbar denotes ensemble averaging and T_0 is a constant—the noise time—which can be interpreted as the time for the root-mean-square phase drift associated with $f(t)$ to grow to unity.

It is convenient to calculate the evolution of the ensemble-averaged density matrix,

$$\rho(t) \equiv \overline{|\psi(t)\rangle\langle\psi(t)|} = \begin{pmatrix} \overline{|\psi_1(t)|^2} & \overline{\psi_1^*(t)\psi_2(t)} \\ \overline{\psi_2^*(t)\psi_1(t)} & \overline{|\psi_2(t)|^2} \end{pmatrix}$$

$$\equiv \left(\frac{1}{2}\right) + \mathbf{S} \cdot \boldsymbol{r}(t) = \left(\frac{1}{2}\right)\begin{pmatrix} 1 + z(t) & x(t) - iy(t) \\ x(t) + iy(t) & 1 - z(t) \end{pmatrix}, \tag{30}$$

conveniently expressed in terms of the Bloch vector,

$$\boldsymbol{r}(t) = 2\overline{\langle\psi(t)|\mathbf{S}|\psi(t)\rangle}. \tag{31}$$

For a pure state ($\rho^2 = \rho$), $\boldsymbol{r}$ lies on the unit sphere. The north pole corresponds to all the amplitude in well 1 and the south pole to all the amplitude in well 2. The off-diagonal elements of ρ are described by the Bloch components, x and y. For any given $f(t)$ (noisy or otherwise), the unitarity of quantum evolution ensures that $\boldsymbol{r}(t)$ remains on the unit sphere, that is, pure. However, ensemble averaging, reflecting ignorance of the details of the environmental forces (or possibly time averaging—this is a subtle point), turns the pure state into a mixture and makes $\boldsymbol{r}(t)$ flow into the sphere. Then, it is the fate of any density matrix starting on the surface of the sphere to flow to the center, where there is equal probability of finding the particle in the wells and no interference between their amplitudes (for chiral molecules, this represents a racemic mixture of the two handednesses). Decoherence and the Zeno effect are features of the intervening evolution, now to be described.

The main steps in the determination of $\boldsymbol{r}(t)$ are explained in APPENDIX A. The form of the result depends on the parameter u defined by

$$u \equiv 4\epsilon T_0. \tag{32}$$

This is proportional to the ratio of the noise time to the quantum oscillation time. We are interested in small ϵ, that is, small u. For the initial condition of $\boldsymbol{r}(0) = \boldsymbol{r}_0$, the evolution is

$$x(t) = \exp(-t/2T_0),$$

$$\begin{pmatrix} y(t) \\ z(t) \end{pmatrix} = \frac{\exp\{-\alpha_+ t\}}{2\sqrt{1-u^2}} \begin{pmatrix} (1+\sqrt{1-u^2})y_0 + uz_0 \\ -uy_0 - (1-\sqrt{1-u^2})z_0 \end{pmatrix}$$

$$+ \frac{\exp\{-\alpha_- t\}}{2\sqrt{1-u^2}} \begin{pmatrix} -(1-\sqrt{1-u^2})y_0 - uz_0 \\ uy_0 + (1+\sqrt{1-u^2})z_0 \end{pmatrix}, \tag{33}$$

where

$$\alpha_\pm \equiv \left(\frac{1}{4T_0}\right)(1 \pm \sqrt{1-u^2}). \tag{34}$$

From the limits

$$\alpha_+ \to \frac{1}{2T_0}, \qquad \alpha_- \to 2\epsilon^2 T_0, \qquad \text{as} \qquad u \to 0, \tag{35}$$

we arrive at the following picture of the density matrix flow described by equation 34. On the scale of the noise time T_0, which is fast as compared with the quantum oscillation time $1/\epsilon$, the fast exponential, involving α_+, causes the Bloch vector to flow from the surface of the sphere towards the line $x = 0, y = -zu/(1+\sqrt{1-u^2})$, which is close to the z-axis. This fast flow, in which the off-diagonal elements of ρ disappear, is decoherence. Thereafter, the flow down the z-axis towards $\boldsymbol{r} = 0$ is dominated by the slow exponential involving α_-, whose time scale is of order $1/(\epsilon^2 T_0)$ and thus long as compared with the quantum oscillation time. This slowing down of the quantum transitions (tunneling) is the quantum Zeno effect.

Loosely speaking, it is possible to regard the environment, acting through the random force $f(t)$, as monitoring, or measuring, the transitions of the particle between the wells, with the result that a pure state is rapidly reduced to a mixture (decoherence) and transitions are inhibited (Zeno). If $\boldsymbol{r}_0$ is on the line $x = 0$, $y = -zu/(1 + \sqrt{1 - u^2})$, there is no decoherence, but only the Zeno effect in its simplest form; if $\boldsymbol{r}_0$ is in the plane $z = -yu/(1 + \sqrt{1 - u^2})$ (close to the xy plane), there is no Zeno effect, but only decoherence in its simplest form.

It is clear that decoherence and the Zeno effect are phenomena that are asymptotically emergent as the parameter ϵ gets small and the quantum particle becomes increasingly vulnerable to its environment. Then, the three time scales, T_0 (noise and decoherence), $1/\epsilon$ (quantum oscillation), and $1/(\epsilon^2 T_0)$ (Zeno), separate.

The mechanism of the Zeno effect is that the fluctuating environment randomly shifts the wells up and down relative to each other. Then, only a small fraction of the time is available for tunneling because this can occur only when wells are within ϵ of each other and the eigenstates are shared between them (rather than being localized in an individual well). For this mechanism to operate, it is essential for the environment to fluctuate rapidly on the time scale of the transitions; otherwise, as the quantum adiabatic theorem implies, the amplitude will be completely transferred to the other well during each passage. Detailed analysis[47] of a single passage shows how the quantum and classical adiabatic theorems contradict each other in this situation and how this contradiction can be resolved; the same idea has been used in a discussion[48] of the persistence of chirality.

An interesting fact about the Zeno effect is that there is a sense in which its onset is independent of the form of the function $f(t)$, provided this is nonzero. I show in APPENDIX B that, if the initial amplitude is all in one well, the probability of finding the particle in the other well is always smaller with $f(t)$ than for the isolated system, provided t is less than the time π/ϵ for the amplitude to first switch wells in the isolated system.

REFERENCES

1. BORN, M. & V. A. FOCK. 1928. Z. Phys. **51:** 165–169.
2. BERRY, M. V. 1984. Proc. R. Soc. London **A392:** 45–57.
3. SHAPERE, A. & F. WILCZEK, Eds. 1989. Geometric Phases in Physics. World Scientific. Singapore.
4. VINITSKII, S. I., V. L. DERBOV, V. N. DUBOVIK, B. L. MARKOVSKI & YU. P. STEPANOV. 1990. Sov. Phys. Usp. **33:** 403–428.
5. ZWANZIGER, J. W., M. KOENIG & A. PINES. 1990. Adv. Chem. Phys. **41:** 601–646.
6. MEAD, C. A. 1992. Rev. Mod. Phys. **64:** 51–85.
7. BITTER, T. & D. DUBBERS. 1987. Phys. Rev. Lett. **59:** 251–254.
8. PANCHARATNAM, S. 1956. Proc. Indian Acad. Sci. Sect. A **44:** 247–262; 1975. Collected works of S. Pancharatnam. Oxford University Press. London/New York.
9. RAMASESHAN, S. & R. NITYANANDA. 1986. Curr. Sci. **55:** 1225–1226.
10. BERRY, M. V. 1987. J. Mod. Opt. **34:** 1401–1407.
11. BERRY, M. V. 1994. Curr. Sci. **67:** 220–223.
12. AHARONOV, Y. & J. ANANDAN. 1987. Phys. Rev. Lett. **58:** 1593–1596.
13. BERRY, M. V. 1987. Proc. R. Soc. London **A414:** 31–46.
14. BERRY, M. V. 1990. Proc. R. Soc. London **A430:** 405–411.

15. JOYE, A., H. KUNZ & C.-E. PFISTER. 1991. Ann. Phys. **208:** 299–332.
16. ZENER, C. 1932. Proc. R. Soc. London **A317:** 696–702.
17. DYKHNE, A. M. 1962. Sov. Phys. JETP **14:** 941–943.
18. ZWANZIGER, J. W., S. P. RUCKER & G. C. CHINGAS. 1991. Phys. Rev. **A43:** 3233–3240.
19. BORN, M. & E. WOLF. 1959. Principles of Optics, p. 448. Pergamon. Elmsford, New York.
20. STOKES, G. G. 1847. Trans. Cambridge Philos. Soc. **9:** 379–407. *Reprinted in*: STOKES, G. G. 1904. Mathematical and Physical Papers by the Late Sir George Gabriel Stokes. Vol. II, p. 329–357. Cambridge University Press. London/New York.
21. GARRISON, J. C. & E. M. WRIGHT. 1988. Phys. Lett. A **128:** 177–181.
22. BERRY, M. V. 1990. Quantum adiabatic anholonomy. *In* Anomalies, Phases, Defects. U. M. Bregola, G. Marmo & G. Morandi, Eds.: 125–181. Bibliopolis. Naples.
23. BERRY, M. V. 1994. Asymptotics, singularities, and the reduction of theories. *In* Proc. Ninth Int. Congr. Logic Method. Philos. Sci. D. Prawitz, B. Skyrms & D. Westerstahl, Eds. Elsevier. Amsterdam/New York. In press.
24. SEGUR, H. & S. TANVEER, Eds. 1992. Asymptotics beyond All Orders. Plenum. New York.
25. STOKES, G. G. 1864. Trans. Cambridge Philos. Soc. **10:** 106–128. *Reprinted in*: STOKES, G. G. 1904. Mathematical and Physical Papers by the Late Sir George Gabriel Stokes. Vol. IV, p. 77–109. Cambridge University Press. London/New York.
26. BERRY, M. V. 1989. Proc. R. Soc. London **A422:** 7–21.
27. BERRY, M. V. 1989. Publ. Math. Inst. Hautes Études Sci. **68:** 211–221.
28. DINGLE, R. B. 1973. Asymptotic Expansions: Their Derivation and Interpretation. Academic Press. New York/London.
29. BERRY, M. V. 1991. Proc. R. Soc. London **A435:** 437–444.
30. BERRY, M. V. 1990. Proc. R. Soc. London **A429:** 61–72.
31. LIM, R. & M. V. BERRY. 1991. J. Phys. A **24:** 3255–3264.
32. LIM, R. 1993. J. Phys. A **26:** 7615–7635.
33. BERRY, M. V. & R. LIM. 1993. J. Phys. A **26:** 4737–4747.
34. OMNÈS, R. 1992. Rev. Mod. Phys. **64:** 339–382.
35. DOWKER, H. F. & J. J. HALLIWELL. 1992. Phys. Rev. **D46:** 1580–1609.
36. MISRA, B. & E. C. G. SUDARSHAN. 1977. J. Math. Phys. **18:** 756–763.
37. PERES, A. 1980. Am. J. Phys. **48:** 931–932.
38. INAGAKI, S., M. NAMIKI & T. TAJIRI. 1993. Vistas Astron. **37:** 273–276.
39. PASCAZIO, S., M. NAMIKI, G. BADUREK & H. RAUCH. 1993. Phys. Lett. A **179:** 155–160.
40. SIMONIUS, M. 1978. Phys. Rev. Lett. **40:** 980–983.
41. HARRIS, R. A. & L. STODOLSKY. 1981. J. Phys. Chem. **74:** 2145–2155.
42. HARRIS, R. A. & L. STODOLSKY. 1982. Phys. Lett. **116B:** 464–469.
43. HARRIS, R. A. & R. SILBEY. 1983. J. Chem. Phys. **78:** 7330–7333.
44. BLANCHARD, P., G. BOLZ, M. CINI, G. F. DE ANGELIS & M. SERVA. 1993. Localization stabilized by noise. Preprint no. 1022. Department of Physics, University of Rome.
45. PFEIFER, P. 1980. Chiral molecules—a superselection rule induced by the radiation field. Ph.D. thesis. Department of Chemistry, ETH Zürich.
46. BRAY, A. & M. MOORE. 1982. Phys. Rev. Lett. **49:** 1545–1549.
47. BERRY, M. V. 1984. J. Phys. A **17:** 1225–1233.
48. CLAVERIE, P. & G. JONA-LASINIO. 1986. Phys. Rev. **A33:** 2245–2253.

APPENDIX A

Exact Solution of the Two-State System Forced by White Noise

For convenience, we show here how the explicit solution of the model for decoherence and the quantum Zeno effect can be obtained by elementary means.

From equation 1, it follows that the Bloch vector $\boldsymbol{r}(t)$, defined by equation 31 without ensemble averaging, satisfies

$$\partial_t \boldsymbol{r}(t) = \mathbf{R}(t) \wedge \boldsymbol{r}(t), \tag{A1}$$

where

$$\mathbf{R}(t) = \{\epsilon, 0, f(t)\}. \tag{A2}$$

Regarding $\boldsymbol{r}$ as a column vector and introducing the matrix

$$\mathbf{M}(t) = \begin{pmatrix} 0 & -Z(t) & Y(t) \\ Z(t) & 0 & -X(t) \\ -Y(t) & X(t) & 0 \end{pmatrix}, \tag{A3}$$

we obtain the formal solution of equation A1 as the time-ordered product

$$\boldsymbol{r}(t) = T \exp\left\{\int_0^t \mathbf{M}(\tau) d\tau\right\} \boldsymbol{r}_0 \tag{A4}$$

for any $f(t)$. This must be ensemble-averaged. When $f(t)$ is white noise, averaging can be accomplished by dividing the time into small intervals Δ and using the fact that f is independent in different intervals. Then, the ensemble average is

$$\boldsymbol{r}(t) = (\overline{\exp\{\mathbf{M}(\tau)\Delta\}})^{t/\Delta} \boldsymbol{r}_0. \tag{A5}$$

To find the evolution of the density matrix, it is therefore necessary to exponentiate $\mathbf{M}$, average the result, and raise this to the power t/Δ. The exponential is

$$\exp\{\mathbf{M}\Delta\} = \left(\frac{1}{\epsilon^2 + f^2}\right) \left[\begin{pmatrix} 1 & 0 & \epsilon f \\ 0 & 0 & 0 \\ \epsilon f & 0 & f^2 \end{pmatrix} + \mathrm{Re} \exp\{i\Delta\sqrt{\epsilon^2 + f^2}\} \right.$$
$$\left. \times \begin{pmatrix} f^2 & if\sqrt{\epsilon^2 + f^2} & -\epsilon f \\ -if\sqrt{\epsilon^2 + f^2} & 1 + f^2 & i\epsilon\sqrt{\epsilon^2 + f^2} \\ -\epsilon f & -i\epsilon\sqrt{\epsilon^2 + f^2} & \epsilon^2 \end{pmatrix} \right]. \tag{A6}$$

Ensemble averaging eliminates the terms odd in f, and from the terms even in f it is necessary to retain only those contributions that will not vanish in the limit $\Delta \to 0$, after raising to the power t/Δ. To obtain these contributions, we need the variance of f over the interval Δ; from equation 29 (replacing the integral by a sum over intervals Δ), this is

$$\overline{f^2} = \frac{1}{T_0 \Delta}. \tag{A7}$$

Then, the average of equation A6 is

$$\overline{\exp\{\mathbf{M}\Delta\}} \rightarrow \begin{pmatrix} 1-(1/2)\Delta^2\overline{f^2} & 0 & 0 \\ 0 & 1-(1/2)\Delta^2\overline{f^2} & -\epsilon\Delta \\ 0 & \epsilon\Delta & 1 \end{pmatrix}$$

$$= \begin{pmatrix} 1-\left(\dfrac{\Delta}{2T_0}\right) & 0 & 0 \\ 0 & 1-\left(\dfrac{\Delta}{2T_0}\right) & -\epsilon\Delta \\ 0 & \epsilon\Delta & 1 \end{pmatrix}. \quad \textbf{(A8)}$$

Raising this to the power t/Δ gives the x evolution as

$$x(t) = \exp\left\{-\frac{t}{2T_0}\right\} \quad \textbf{(A9)}$$

and the $\{y, z\}$ evolution as

$$\begin{pmatrix} y(t) \\ z(t) \end{pmatrix} = \exp\{-t\mathbf{N}\}\begin{pmatrix} y_0 \\ z_0 \end{pmatrix}, \quad \textbf{(A10)}$$

where (cf. equation 3)

$$\mathbf{N} = \begin{pmatrix} 1/2T_0 & \epsilon \\ -\epsilon & 0 \end{pmatrix} = \left(\frac{1}{4T_0}\right)\mathbf{1} + 2\epsilon\mathbf{S}_x + \left(\frac{1}{2T_0}\right)\mathbf{S}_z. \quad \textbf{(A11)}$$

This is easily exponentiated to give

$$\exp\{-t\mathbf{N}\} = \exp\left\{-\frac{t}{4T_0}\right\}\left[\cosh\left\{\left(\frac{t}{4T_0}\right)\sqrt{1-u^2}\right\}\mathbf{1} - \left(\frac{1}{\sqrt{1-u^2}}\right)\sinh\left\{\left(\frac{t}{4T_0}\right)\sqrt{1-u^2}\right\}\begin{pmatrix} 1 & u \\ -u & -1 \end{pmatrix}\right], \quad \textbf{(A12)}$$

where u is defined by equation 32.

Taken together, equations A9, A10, and A12 yield the claimed evolution (equation 33). It is interesting to examine the case where the noise is not large in comparison with ϵ; then, it is possible to have $u > 1$ and equation A12 is more conveniently written as

$$\exp\{-t\mathbf{N}\} = \exp\left\{-\frac{t}{4T_0}\right\}\left[\cos\left\{\left(\frac{t}{4T_0}\right)\sqrt{u^2-1}\right\}\mathbf{1} - \left(\frac{1}{\sqrt{u^2-1}}\right)\sin\left\{\left(\frac{t}{4T_0}\right)\sqrt{u^2-1}\right\}\begin{pmatrix} 1 & u \\ -u & -1 \end{pmatrix}\right]. \quad \textbf{(A13)}$$

This is the regime of quantum oscillations damped by tunneling friction.[43,46] The quantum Zeno effect corresponds to overdamping. Separating these two cases is critical damping ($u = 1$), for which

$$\exp\{-t\mathbf{N}\} = \exp\left\{-\frac{t}{4T_0}\right\}\left[\mathbf{1} - \left(\frac{t}{4T_0}\right)\begin{pmatrix} 1 & 1 \\ -1 & -1 \end{pmatrix}\right]. \qquad \mathbf{(A14)}$$

(I mention in passing that the matrix here is of the same degenerate non-Hermitian kind as discussed in the fourth section and this fact is responsible for the linear t-dependence.[11]) The limit of the isolated quantum system (no noise) is $T_0 \to \infty$, for which the evolution is unitary and consists of undamped quantum oscillations between the wells:

$$\exp\{-t\mathbf{N}\} = \cos\{\epsilon t\}\mathbf{1} - \sin\{\epsilon t\}\begin{pmatrix} 0 & 1 \\ -1 & 0 \end{pmatrix}. \qquad \mathbf{(A15)}$$

APPENDIX B

Initial Slowing Down of Transitions by Any Environment

For the two-state system (equations 1–4) with Hamiltonian (equation 28), the Schrödinger equation reduces, after the transformation

$$\begin{pmatrix} \psi_1(t) \\ \psi_2(t) \end{pmatrix} \equiv \begin{pmatrix} a_1(t)\exp\{(-\tfrac{1}{2})iw(t)\} \\ a_2(t)\exp\{(+\tfrac{1}{2})iw(t)\} \end{pmatrix}, \qquad \text{where} \qquad w(t) \equiv \int_0^t d\tau f(\tau), \qquad \mathbf{(B1)}$$

to the off-diagonal form

$$\partial_t \begin{pmatrix} a_1(t) \\ a_2(t) \end{pmatrix} \equiv -i\left(\frac{\epsilon}{2}\right)\begin{pmatrix} a_2(t)\exp\{+iw(t)\} \\ a_1(t)\exp\{-iw(t)\} \end{pmatrix}. \qquad \mathbf{(B2)}$$

We start the system in the state 1, that is, $a_1(0) = 1$ and $a_2(0) = 0$. Next, we express the amplitudes a_1 and a_2 in terms of moduli $|a_1|$ and $|a_2|$ and phases χ_1 and χ_2, that is,

$$\begin{pmatrix} a_1(t) \\ a_2(t) \end{pmatrix} \equiv \begin{pmatrix} |a_1(t)|\exp\{i\chi_1(t)\} \\ |a_2(t)|\exp\{i\chi_2(t)\} \end{pmatrix}. \qquad \mathbf{(B3)}$$

Thus, $|a_1|^2$ is the survival probability (for remaining in the initial state) and $|a_2| = \sqrt{(1 - |a_1|^2)}$.

From equation B3, it follows that

$$\partial_t |a_1(t)| = \left(\frac{\epsilon}{2}\right)|a_1(t)|\sin\{\phi(t)\}, \qquad \text{where} \qquad \phi(t) \equiv w(t) + \chi_2(t) - \chi_1(t). \qquad \mathbf{(B4)}$$

This has the solution

$$|a_1(t)|^2 = \cos^2\left\{\left(\frac{\epsilon}{2}\right)\int_0^t \sin[\phi(\tau)]d\tau\right\}. \qquad \mathbf{(B5)}$$

Now, if there is no noise, that is, $f = 0$, the exact solution of equation B2 is

$$a_1(t) = \cos\{(½)\epsilon t\}, \qquad a_2(t) = -i\sin\{(½)\epsilon t\} \qquad \text{if} \qquad f = 0, \tag{B6}$$

so

$$|a_1(t)| = \cos^2\{(½)\epsilon t\} \qquad \text{if} \qquad f = 0, \tag{B7}$$

corresponding to $\phi = \pi/2$ if $f = 0$. For any nonzero $f(t)$, $\sin\phi < 1$, so $|a_1|^2$ decreases from unity more slowly than equation B7, at least until the first zero of equation B7, that is, $t = \pi/\epsilon$. This is the result stated at the end of the last section in the main text: for any environment $f(t)$, the survival probability exceeds that for the isolated system until its amplitude has switched wells for the first time.

Degeneracies and Mirror Symmetries in Polygonal Quantum Billiards[a]

H-M. LAUBER

Physikalisches Institut der Universität Heidelberg
D-69120 Heidelberg, Germany

Recently, quantum billiards with nonintegrable boundary conditions have been the subject of many theoretical and experimental investigations in the context of quantum chaos. The experimental investigations presented here are concerned with wave functions and energy levels of quantum billiards under a perturbation of the boundary conditions. They resemble a simplified model for either the electronic wave function of a molecule that depends on the positions of the nuclei in the Born-Oppenheimer formalism or the wave function of a nucleon depending on the shape of the nucleus.

Mostly for experimental reasons, we used polygonal quantum billiards. Polygonal quantum billiards are not chaotic systems: this means that there is no exponential instability of neighboring trajectories in the phase space of the classical billiards. They are pseudointegrable[1] or presumably ergodic systems, depending on whether the angles of the polygon are rational or irrational multiples of π. Generally, polygonal quantum billiards do not have any symmetry or conserved quantum number besides energy. They are generic quantum systems, in contrast to textbook examples, which are separable and have as many conserved quantum numbers as degrees of freedom.

According to a theorem of von Neumann,[2] generic quantum systems do not have degenerate energy levels. It is not even possible to enforce a degeneracy of two adjacent states by an external perturbation. Typically, two energy levels approaching each other under increasing perturbation will not cross, but will separate again like two branches of a hyperbola (avoided crossing). However, in systems with a real Hamiltonian, like quantum billiards, accidental degeneracies may occur, if two independent perturbations are applied. In case of the quantum billiards investigated here, the boundary conditions are perturbed by shifting one of the corners of the billiards in two dimensions. In the following, the space of corner positions, or different resonator shapes, will be referred to as configuration space.

The planes of energy levels plotted over configuration space do not intersect, but just touch each other at isolated points. These points are called diabolic points because in their vicinity two energy levels split linearly and form a double-cone structure[3] (diabolo).

At very small level spacing, both states can be regarded as a two-level system, described by a linearized 2×2 submatrix of the Hamiltonian. This matrix is equivalent to the Hamiltonian that describes the Zeeman effect of a spin-½ particle

[a]This work has been funded by the German Federal Ministry of Research and Technology (BMFT) under Contract No. 06TM661.

in a magnetic field $\boldsymbol{B}$, which is confined to the xz-plane:

$$\boldsymbol{H} = E_0 + \sigma_x B_x + \sigma_z B_z = \begin{pmatrix} E_0 + B_z & B_x \\ B_x & E_0 - B_z \end{pmatrix}. \tag{1}$$

Rotating B adiabatically by an angle θ will transform an initial state $|\mathrm{up}(0)\rangle$ or $|\mathrm{down}(0)\rangle$ into

$$\begin{pmatrix} |\mathrm{up}(\theta)\rangle \\ |\mathrm{down}(\theta)\rangle \end{pmatrix} = \begin{pmatrix} \cos(\theta/2) & -\sin(\theta/2) \\ \sin(\theta/2) & \cos(\theta/2) \end{pmatrix} \begin{pmatrix} |\mathrm{up}(0)\rangle \\ |\mathrm{down}(0)\rangle \end{pmatrix}. \tag{2}$$

Because the mixing matrix depends on half the rotation angle, the spinor undergoes a sign change, that is, a phase shift of $\pm\pi$. This sign change has to occur for the real wave functions of the quantum billiards too, if they are transported along a loop in configuration space, which encircles the degeneracy.[4] It is the simplest example of a geometric phase.

The most popular concept of a geometric phase, discovered by Berry,[5] is the phase acquired by a pure state transported adiabatically on a closed loop in parameter space. The adiabatic condition ensures that the system will remain in a pure state, whereas the state itself changes during the evolution of the whole system. When the system returns to its initial configuration, the final state can only differ from the initial state by some phase factor, $e^{i\gamma}$. The phase γ consists of a dynamical part, which is the usual time evolution of the state, but additionally there may be a contribution, γ_{geo}, depending only on the topology of the energy levels over the parameter space:

$$\gamma = \left(\frac{-1}{\hbar}\right) \int_0^T \boldsymbol{H}[R_i(t)]t\, dt + \gamma_{\mathrm{geo}}. \tag{3}$$

This geometric phase can be easily calculated for spin states, which are coupled to the three components of an external magnetic field. It depends on the magnetic quantum number m of the state and on the solid angle under which the loop in the space of magnetic fields appears from the origin:[5]

$$\gamma_{\mathrm{geo}} = -m\Omega. \tag{4}$$

The existence of the geometric phase γ_{geo} in spin systems has been experimentally verified using neutrons,[6] nuclei[7] (spin ½), and photons[8] (spin 1).

Strictly speaking, the adiabatic condition is not necessary to determine the geometric phase of a system with a real Hamiltonian. Adiabatic transport of a state is equivalent to parallel transport of a state vector,[9] which means that

$$\mathrm{Im}(\psi \cdot \partial_t \psi) = 0. \tag{5}$$

Therefore, for a small adiabatic change, any real wave function will change into a slightly altered real wave function and will only acquire a dynamic phase in the adiabatic case. Hence, all the information needed to determine the geometric phase is the knowledge of how the wave function rearranges along a closed loop of the system in configuration space. These rearrangements of the wave functions for small steps in configuration space can be easily measured by microwave experiments.

An investigation of the electric field in a microwave resonator is equivalent to a quantum mechanical investigation because the stationary Schrödinger equation for quantum billiards is equivalent to the Helmholtz equation in two dimensions with Dirichlet boundary conditions:

$$\begin{aligned} (\Delta + k^2)\psi(x, y) &= 0 \qquad \text{inside} \\ \psi(x, y) &= 0 \qquad \text{at the boundary.} \end{aligned} \tag{6}$$

For flat microwave resonators, in which the wavelength of the microwaves is larger than twice the height of the resonator, the wave equation for the electric field also reduces to this form.

Our investigations of the mixing of states near a diabolic point and near a triple degeneracy were done with flat microwave resonators of triangular and quadrilateral form. These resonator shapes are easy to change experimentally. A triangular resonator was used for the investigations of diabolic points because triangular quantum billiards were already the subject of a numerical study by Berry and Wilkinson.[10] These researchers discovered several diabolic degeneracies of the lowest 13 energy levels, which could not be related to any symmetry of the triangles.

Our triangular resonator is composed of three brass bars resting on a copper plate. They are covered by a second copper plate with a small antenna. The resonator can be moved underneath the cover so that the coupling antenna can be positioned at any point of the triangle. Using a technique described by Stein and Stöckmann,[11] it was possible to measure standing wave patterns via the reflected microwave intensity at the different coupling points. The shape of our resonator is changed by moving the tip C of the triangle, while keeping its base at a fixed length (cf. FIGURE 1). Each shape is thus defined by two independent parameters: the height H of the triangle and the length P, given by the projection of C on the base of the triangle. Moving C in the whole half-plane above the base line, triangles can be built that are just mirror images of each other or that can be mapped onto each other by a combination of scaling, rotation, and mirroring operations. The spectra and wave functions of these similar triangles scale simply with the size of the triangle and do not contribute any additional information. All unique triangles can be built by restricting the position of C to the gray area in FIGURE 1. Our investigations were confined to a square grid in the HP-plane. H ranged from 13 to 22 cm and P from 8 to 17 cm. This grid is shown in FIGURE 1. For a first survey of the energy surfaces, we measured all resonances in the range from 2 GHz to 7.5 GHz for 100 different shapes, corresponding to the grid points in FIGURE 1.

In order to suppress the effects of different mean level densities, the spectra were normalized to a mean density of one, using area- and perimeter-dependent terms of Weyl's formula[12] for the mean number of levels (N) below frequency (ν), that is,

$$N(\nu) = \left(\frac{\pi A \nu^2}{c^2}\right) + \left(\frac{L\nu}{2c}\right), \tag{7}$$

where A is the area of the billiard, L is its perimeter, and c is the velocity of light.

FIGURE 2 displays the normalized spectra in 10 groups. In each group, P varies from 8 cm to 17 cm, whereas H is kept constant. The series of avoided crossings observed for different values of H can be identified as parallel cuts through double-cone structures, as illustrated in the inset in FIGURE 2.

In order to measure the sign change of the wave function due to the geometric phase, we concentrate on the diabolic degeneracy between levels 13 and 14, plotted as bold lines in FIGURE 2. This degeneracy was not included in the analysis of Berry and Wilkinson.[10]

The position of the degeneracy was determined to an accuracy of 1 mm in the HP-plane. The standing wave patterns of the 13th and 14th resonances were

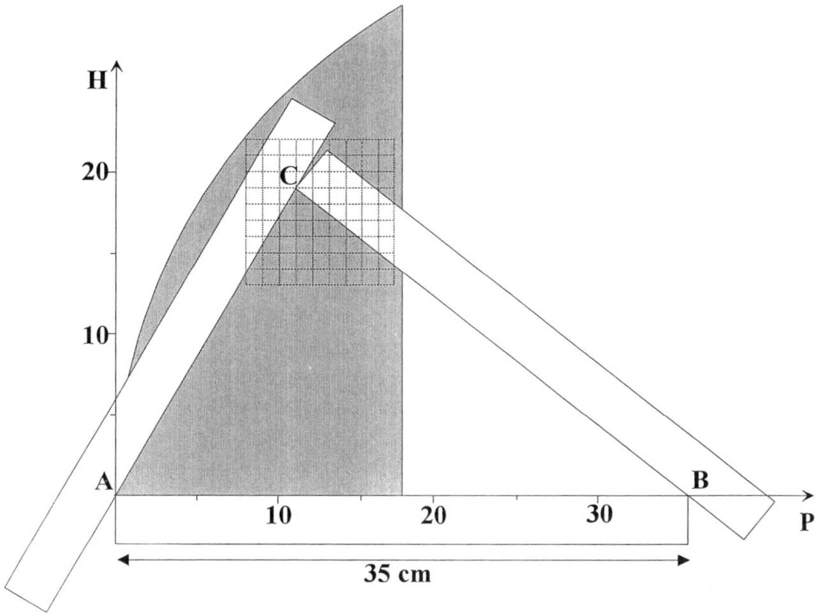

FIGURE 1. Construction and deformation of the triangular microwave resonator. The grid shows the different locations of the upper corner, C. All triangles may be built by moving C inside the gray area.

measured for 12 resonator shapes encircling the diabolic point. Each wave pattern depicted in FIGURE 3 is derived from measurements of the reflected microwave intensity at some 1300 positions within each resonator. Bright regions correspond to high intensities of the standing wave.

The relative phase of the maxima in one mode could be determined by measuring the phase of transmitted microwaves between two antennas positioned at the maxima. However, because the nodal lines of the wave patterns can be clearly identified in each picture, the opposite sign between adjacent maxima can be determined without this measurement. Thus, the wave functions can be determined out of the intensity pattern by tagging adjacent maxima at opposite sides of a nodal

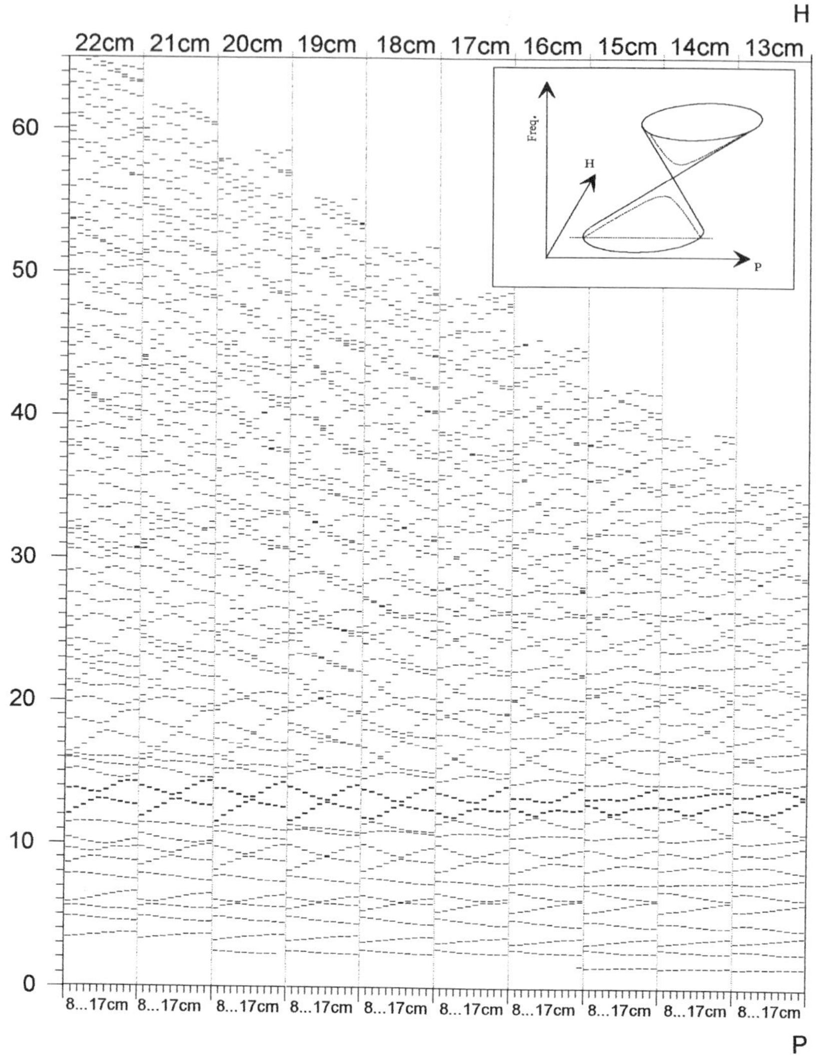

FIGURE 2. Measured spectra of the triangular resonators for 100 different shapes (H, P); levels 13 and 14 (bold lines) trace out the double-cone structure investigated in this article.

line with opposite signs, (+) and (−). For the patterns at resonator shape no. 1, this tagging may be done starting with an arbitrary sign at one maximum (cf. FIGURE 3). At the next resonator shape no. 2, tagging is done according to parallel (or adiabatic) transport of the state. This means that there is a phase relation between states belonging to the same energy surface at different points in configuration space. Because the wave function just changes a little from no. 1 to no. 2, maxima in both patterns can be identified and thus assigned identical phases. As is clearly visible in

FIGURE 3, this procedure cannot be continued after one full revolution around the diabolic point. The topology of wave pattern no. 1 is regained in no. 12, but the positive and negative regions have interchanged, that is, the wave function has changed sign, and $|\gamma_{geo}| = \pi$ as predicted. Also, note that the topologies of the upper and lower states interchange for half a revolution starting at any point; that is, pattern 1 for the upper state coincides with pattern 7 for the lower state, etc., as expected for a spin-½ system.

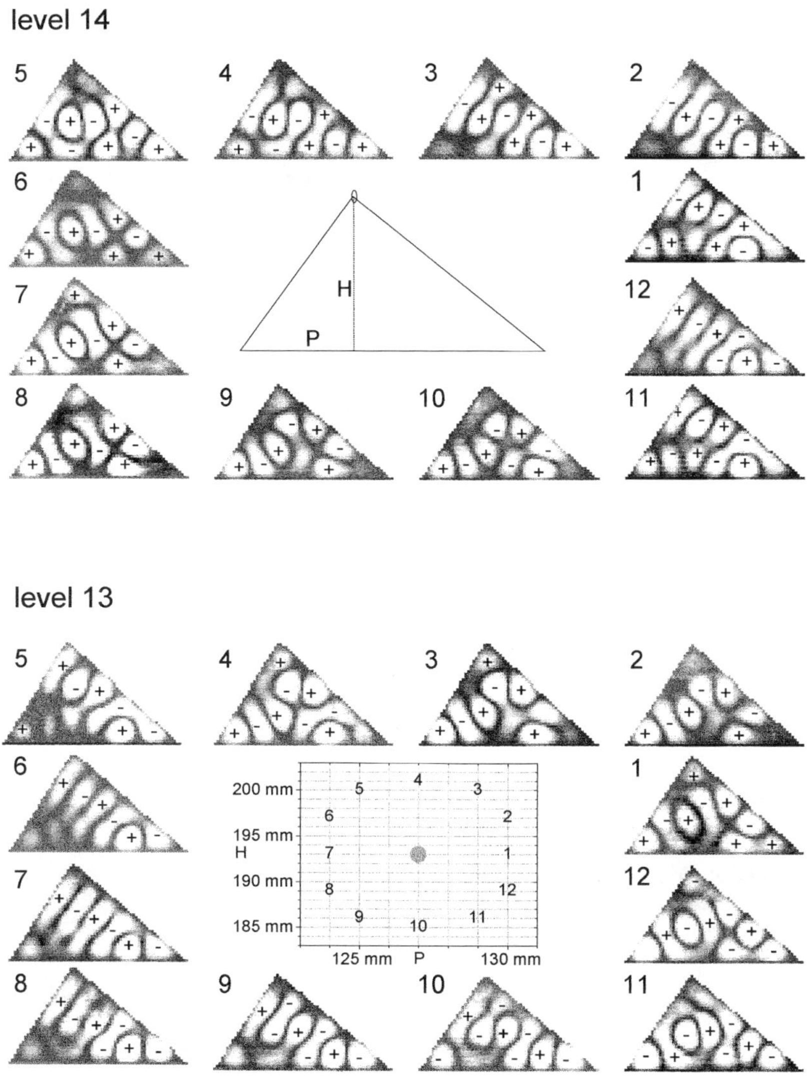

FIGURE 3. Measured wave patterns in the triangular resonator.

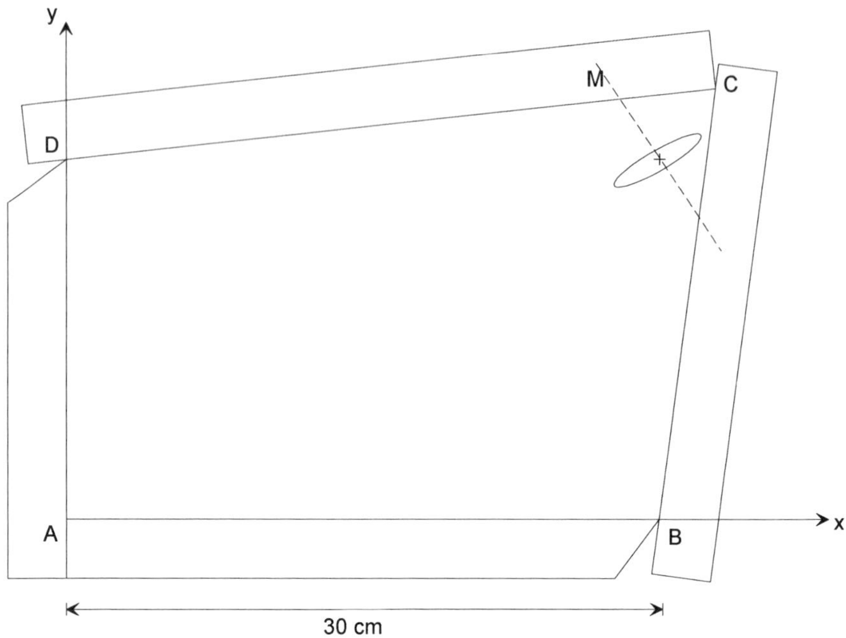

FIGURE 4. Construction and deformation of the rectangular resonator. Corner C was set to 16 positions on the elliptical path around the rectangular configuration marked by +. The mirror plane in configuration space (M) is indicated by a dashed line.

In a second series of experiments, we focused on a pointlike degeneracy of three levels. Because the mixing of states at a diabolic point is similar to the mixing of "fermionic" (spin-½) states under rotation, a question arises as to whether a triple degeneracy resembles a "bosonic" (spin 1–like) system. However, in this case, it was not possible to look for an accidental degeneracy of three levels. According to the theorem of von Neumann, an accidental triple degeneracy can only be found in a configuration space of dimension five or bigger.

In a five-dimensional space, the number of different shapes that must be investigated is too high for the experimental method used here. Hence, we chose a rectangular resonator with a length-to-width side ratio of $\sqrt{3}$. This resonator has many triple degeneracies in its spectrum for number-theoretical reasons.[13] The triplet investigated here consists of the 18th, 19th, and 20th resonance. The shape of the resonator can be changed by moving one of its corners (C) along a loop around the rectangular configuration (cf. FIGURE 4). It should be noted that none of the quadrilaterals at different positions of C can be mapped onto each other by any symmetry or scaling operation.

In order to establish that the system investigated here is topologically equivalent to a spin-1 coupling to an external magnetic field, we verified that the three degenerate states split equidistantly under distortion of the corner position in each direction. An elliptical path was determined along which the splitting of levels was sufficiently high, to resolve all levels experimentally well. For 16 different quadrilaterals on the path, the wave functions of each of the three states were measured.

Subsequently, a similar tagging scheme as in the case of the triangular resonator was applied. FIGURE 5 displays the results of the measurements. During the first half-revolution around the rectangular configuration (pictures 1 through 8), the wave functions of the 18th and 20th states interchange and the wave function of the 19th

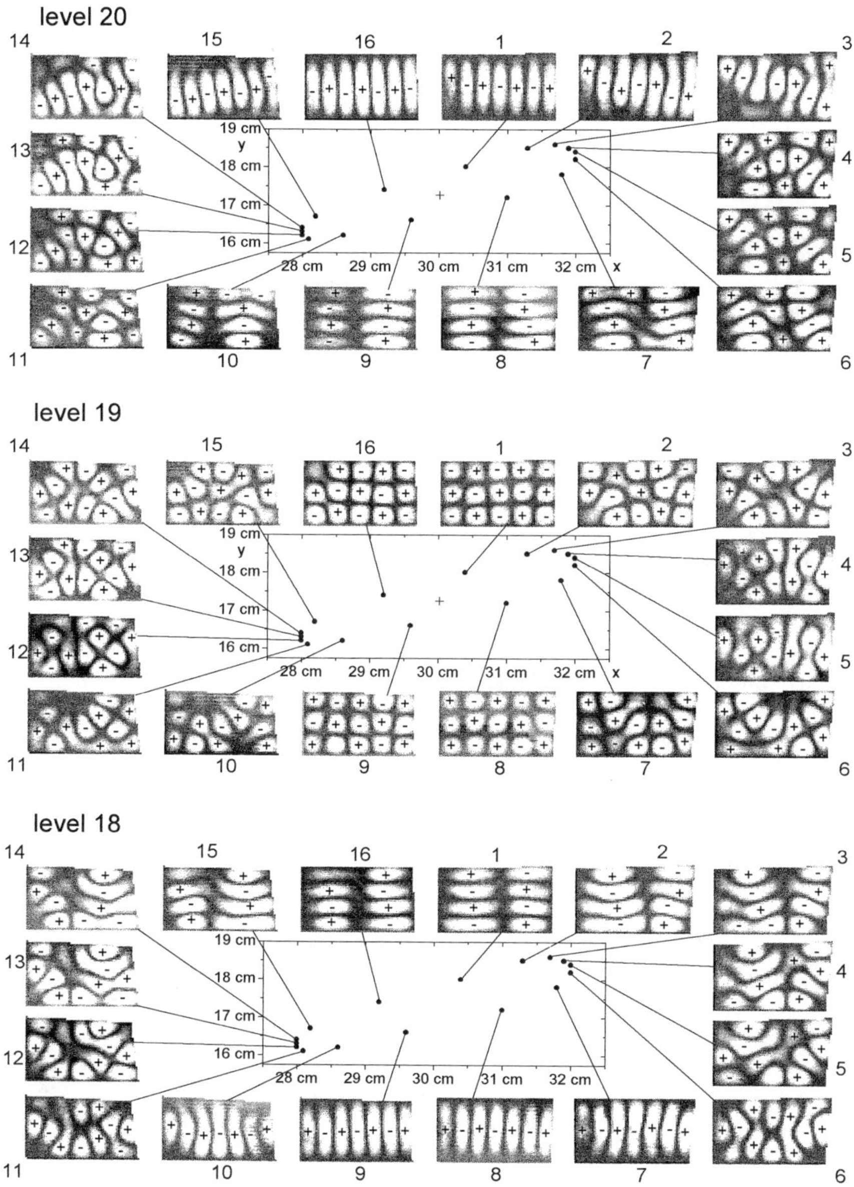

FIGURE 5. Measured wave patterns in the deformed rectangular resonator.

undergoes a sign change, as expected for a spin-1 system under a rotation by π. However, when the loop is continued to the initial configuration (pictures 9 through 16), the outcome is surprising. Again, the wave function of the 19th state changes sign, thus returning to the initial state, and the 18th and 20th states interchange a second time. However, after one full revolution, the wave functions of the 18th and 20th states return to their initial form with opposite signs. This sign change is not expected for a spin-1 system.

Closer investigation reveals another unexpected feature of the wave functions. The wave functions in the left half of FIGURE 5 are nearly mirror images of the wave functions in the right half, although the resonator shapes are not mirror images of each other. (Compare pictures 2 and 15, 3 and 14, and so on.)

The symmetry of the wave patterns indicates that there exists a local mirror symmetry with a mirror plane M in the configuration space, which is shown in FIGURE 4. This symmetry is not obvious when comparing the resonator shapes belonging to points that are mirror images with respect to M. This symmetry is responsible for the unexpected sign change because it occurs in connection with opposite internal parity of the wave functions of states 18 and 20 at the mirror plane. The wave functions of states 18 and 20 interchange after half a revolution from a starting point at M. Hence, the wave functions have to cross the mirror plane two times, but once with odd and once with even parity. For state 20, a continuous crossing of M at the starting point requires that all pairs of wave functions (16–1, 15–2, . . .) are mirror images of each other, including the sign tags of the maxima:

$$\Psi_{1\text{-}8}(x, y) = \Psi_{9\text{-}16}(-x, y) = \Psi_{9\text{-}16}(x, -y). \tag{8}$$

At the second crossing, where the parity of state 20 has become negative, continuity requires that the pair of wave functions are inverse mirror images:

$$\Psi_{1\text{-}8}(x, y) = -\Psi_{9\text{-}16}(-x, y) = -\Psi_{9\text{-}16}(x, -y). \tag{9}$$

Both conditions cannot be fulfilled simultaneously. Thus, a sign change after a full revolution has to occur. For state 18, a similar line of reasoning holds. State 19 crosses M both times with even parity and therefore acquires no sign change after a full revolution.

We then tried to construct a matrix that describes the mixing of the three states. A mixing of three states can always be described by an orthonormal 3×3 matrix like the Kobayashi-Maskawa matrix. This matrix may be parametrized by three angles (ϕ_i) and one phase (δ):

$$\begin{pmatrix} |20;(\theta)\rangle \\ |19;(\theta)\rangle \\ |18;(\theta)\rangle \end{pmatrix} = \begin{pmatrix} c_1c_2c_3 - s_2s_3e^{i\delta} & -s_1c_2 & c_1c_2s_3 + s_2c_3e^{i\delta} \\ s_1c_3 & c_1 & s_1s_3 \\ c_1s_2c_3 + c_2s_3e^{i\delta} & -s_1s_2 & c_1s_2s_3 - c_2c_3e^{i\delta} \end{pmatrix} \begin{pmatrix} |7,1\rangle \\ |5,3\rangle \\ |2,4\rangle \end{pmatrix}, \tag{10}$$

where $c_i = \cos(\phi_i)$, $s_i = \sin(\phi_i)$, and states $|2,4\rangle$, $|5,3\rangle$, and $|7,1\rangle$ are states $|18\rangle$, $|19\rangle$, and $|20\rangle$ at the initial configuration no. 1. We looked for a matrix where the angles ϕ_i should depend only on the rotation angle in configuration space and where the phase δ has the values of 0 or π.

For the mixing matrix that best describes the observed wave functions from no. 1

to no. 8, ϕ_1 increases from π to 2π, whereas the angles ϕ_2 and ϕ_3 have fixed values of $-\pi/4$ and $(3/4)\pi$. The phase δ is always 0. This results in a mixing matrix,

$$K_{0\ldots\pi} = \begin{pmatrix} \frac{(1+c)}{2} & \frac{s}{\sqrt{2}} & \frac{(1-c)}{2} \\ \frac{s}{\sqrt{2}} & -c & \frac{-s}{\sqrt{2}} \\ \frac{(1-c)}{2} & \frac{-s}{\sqrt{2}} & \frac{(1+c)}{2} \end{pmatrix}, \tag{11}$$

where $c = \cos(\theta)$ and $s = \sin(\theta)$; here, θ is the orbiting angle. This matrix is equivalent to the mixing matrix for spin-1 states under rotation. Because it depends on θ instead of $\theta/2$, no sign change of any wave function is expected after a full revolution.

This, though, indicates that $K_{0\ldots\pi}$ is only a good approximation for the right side of the mirror plane. For the left side from no. 9 to no. 16, it does not at all describe the observed mixing. In order to have good agreement for the second half of the revolution, it is necessary to change the angles ϕ_2 and ϕ_3 instantaneously by $\pi/2$ to $\pi/4$ when M is crossed. The mixing matrix for the second half then reads as

$$K_{\pi\ldots 2\pi} = \begin{pmatrix} \frac{(-1-c)}{2} & \frac{s}{\sqrt{2}} & \frac{(1-c)}{2} \\ \frac{-s}{\sqrt{2}} & -c & \frac{-s}{\sqrt{2}} \\ \frac{(1-c)}{2} & \frac{s}{\sqrt{2}} & \frac{(-1-c)}{2} \end{pmatrix}. \tag{12}$$

Notice that this leads to no discontinuity of the mixing scheme at $\theta = \pi$ because the nonzero matrix elements K_{13}, K_{22}, and K_{31} do not suffer any sign change. However, there is a sign change of K_{11} and K_{33}, which leads to the observed sign change after a full revolution.

The wave functions calculated with these matrices are displayed in FIGURE 6. The good agreement with the observed nodal patterns is evident when comparing FIGURES 5 and 6.

In conclusion, using a noninterferometric method, the sign change of a wave function in a resonator was measured. This sign change is due to a geometric phase that builds up whenever an accidental "fermionic" double degeneracy (effective spin ½) is encompassed. Surprisingly, it was also found that "bosonic" triple degeneracies can lead to a sign change of the wave function when an additional and local mirror symmetry is present. This symmetry is not visible in the different boundary conditions, nor does it change the topology of energy surfaces over configuration space. Therefore, it is not possible to predict the geometric phase of multiple degenerate states from the topology of the energy surfaces alone.

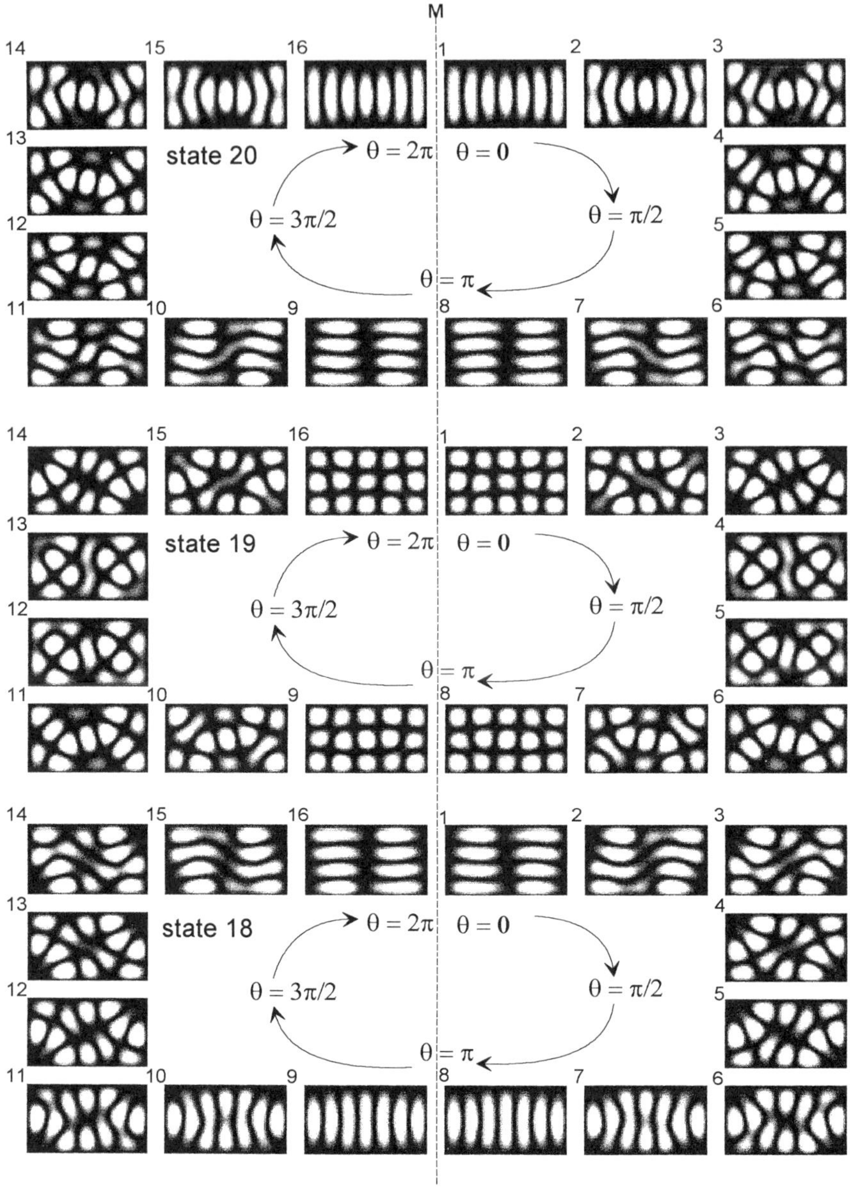

FIGURE 6. Calculated wave functions according to the discontinuous mixing matrices, K.

REFERENCES

1. RICHENS, P. J. & M. V. BERRY. 1981. Physica **2D:** 495.
2. VON NEUMANN, J. & E. J. WIGNER. 1929. Phys. Z. **30:** 467.
3. TELLER, E. 1937. J. Phys. Chem. **41:** 109.
4. HERZBERG, G. & H. C. LONGUET-HIGGINS. 1963. Discuss. Faraday Soc. **35:** 77.
5. BERRY, M. V. 1984. Proc. R. Soc. London Ser. A **392:** 45.
6. BITTER, T. & D. DUBBERS. 1987. Phys. Rev. Lett. **59:** 251.
7. SUTER, D. *et al.* 1987. Mol. Phys. **61:** 1327.
8. TOMITA, A. & R. CHIAO. 1986. Phys. Rev. Lett. **57:** 937.
9. SHAPERE, A. & F. WILCZEK. 1989. Geometric Phases in Physics. World Scientific. Singapore.
10. BERRY, M. V. & M. WILKINSON. 1984. Proc. R. Soc. London Ser. A **392:** 15.
11. STEIN, J. & H-J. STÖCKMANN. 1992. Phys. Rev. Lett. **68:** 2867.
12. BALTES, H. P. & E. R. HILF. 1976. Spectra of Finite Systems. Wissenschaftsverlag. Mannheim.
13. BERRY, M. V. 1981. Ann. Phys. **131:** 163.

On the Aharonov-Bohm Effect with Neutrons[a]

MURRAY PESHKIN

Physics Division
Argonne National Laboratory
Argonne, Illinois 60439-4843

This work was done in collaboration with Harry J. Lipkin. As it has been submitted for publication elsewhere, I will be concise here.

Later in this conference, you will hear of a series of neutron interferometry experiments[1] to demonstrate a new phenomenon that the experimenters have called the "Scalar Aharonov-Bohm effect" (SAB). In an idealization of those experiments,[2] polarized neutrons enter a Mach-Zehnder interferometer, one arm of which contains a region to which an external magnetic field $\boldsymbol{B}$ can be applied. $\boldsymbol{B}$ vanishes except during a time interval of length τ during which a neutron's partial wave packet in that arm is entirely within the field region, and then it has the value B, uniform throughout the support of the wave packet. The neutron has $\sigma_z = +1$ at all times, where z is the direction of the magnetic field. From the Hamiltonian

$$H = \frac{\boldsymbol{p}^2}{2m} - \mu\boldsymbol{\sigma} \cdot \boldsymbol{B}(t), \tag{1}$$

it follows that the partial waves in the two arms of the interferometer will acquire a relative phase

$$\delta\phi = \frac{\mu B}{\hbar}\tau. \tag{2}$$

That relative phase is measured by observing the intensities of the two emergent beams after the partial waves in the two arms are mixed. Thus the motion of the neutrons is influenced by a field in which the neutrons are never subjected to a force, and the experimenters have interpreted their results as demonstrating a new kind of topological effect, similar to the electric version of the usual Aharonov-Bohm (AB) effect (EAB),[3,4] but with the electromagnetic field interacting with the neutron's magnetic moment μ instead of with the electron's charge e.

I will assert here that this interpretation is wrong. SAB is brought about by an ordinary action of the Maxwell field that has all the properties of all other local interactions and shares none of the topological features of EAB. On the other hand, SAB does measure an apparently novel kind of spin correlation that may be of interest.

Let me remind you briefly about EAB. In an adequate idealization, the magnetic moment of the electron can be ignored and we can pretend that spinless electrons

[a]This work was supported by the United States Department of Energy (Nuclear Physics Division) under Contract No. W-31-109-ENG-38.

enter the same Mach-Zehnder interferometer. Instead of having a magnetic field region in one arm, each arm is surrounded by its own conducting cylinder which serves to shield an electron from any external electric field while the electron's wave packet is split into two pieces, each deep inside one of the cylinders. Now potentials V_1 and V_2 are applied to the two cylinders for a time τ while the electron is shielded, and the two beams acquire a relative phase given by

$$\delta\phi = \frac{e(\Delta V)}{\hbar}\tau, \tag{3}$$

where $\Delta V = V_1 - V_2$. The observable effect is the same as in SAB with the substitution $e(\Delta V) \Leftrightarrow \mu B$.

There are of course major differences between SAB and EAB. The SAB Hamiltonian (equation 1) involves the local, contemporaneous Maxwell field $\boldsymbol{B}$, whereas the EAB Hamiltonian

$$H = \frac{\boldsymbol{p}^2}{2m} - eV(t) \tag{4}$$

involves only the gauge field V. Also, the operator equations of motion for the observable spin components σ_x and σ_y in SAB,

$$\frac{\hbar}{2}\dot{\sigma}_x = -\mu B\sigma_y, \tag{5a}$$

$$\frac{\hbar}{2}\dot{\sigma}_y = +\mu B\sigma_x, \tag{5b}$$

exhibit the usual torque in a local magnetic field, whereas the operator equations of motion for the only observables in EAB, the electron's position and velocity, are those of a free particle and involve no electromagnetic fields. However, it is argued that $\boldsymbol{B}$ enters the Hamiltonian as a potential acting on a magnetic moment and that one can replace σ_z by the number $+1$, reducing the SAB Hamiltonian to

$$H = \frac{\boldsymbol{p}^2}{2m} - \mu B(t). \tag{6}$$

Then the Hilbert space is truncated to single-component wave functions with $\sigma_z = +1$, so the two phenomena are described identically in the theory under the substitution $eV \Leftrightarrow \mu B$. SAB appears from that formal argument to be a topological phenomenon in the same sense as is EAB. To complete that argument, equations 5a and 5b have to be written off as questionably involving observable quantities in a state with $\sigma_z = +1$, where

$$\langle\sigma_x(t)\rangle = \langle\sigma_y(t)\rangle = 0. \tag{7}$$

However, the formal similarity between equation 4 and equation 6 is misleading because the spin is a quantum mechanical operator and it cannot simply be replaced by a number, and that changes the physics. The AB effect is nonlocal in that the electron experiences no force and exchanges no momentum, energy, or angular momentum with the electromagnetic field; and in that the Hamiltonian, the equations of motion, and the commutation relations involve no local contemporaneous Maxwell field in the domain of the electron's position. In SAB, the torque

$$\boldsymbol{L} = \mu\boldsymbol{\sigma} \times \boldsymbol{B} \tag{8}$$

has vanishing expectation in a state with $\sigma_z = +1$, but its fluctuations

$$\langle L_x^2 \rangle = \langle L_y^2 \rangle = (\mu B)^2 \tag{9}$$

do not vanish. Angular momentum is exchanged between the electron and the local electromagnetic field. That exchange, as described by the equations of motion of the observables, is local in the sense of Faraday and Maxwell.

This locality is manifested especially clearly by the spin autocorrelation operators:

$$C(t) = \left(\frac{1}{4}\right)[\sigma_x(0)\sigma_x(t) + \sigma_y(0)\sigma_y(t) + \text{h.c.}],$$

$$S(t) = \left(\frac{1}{4}\right)[\sigma_x(0)\sigma_y(t) - \sigma_y(0)\sigma_x(t) + \text{h.c.}]. \tag{10}$$

They obey the equations of motion

$$\dot{C}(t) = \frac{2\mu B}{\hbar} S(t), \tag{11a}$$

$$\dot{S}(t) = -\frac{2\mu B}{\hbar} C(t), \tag{11b}$$

which depend upon the local contemporaneous magnetic field. These spin autocorrelations are Hermitian operators, measurable in principle, and they commute with σ_z, so there is no question of their observability in a state of definite σ_z. The solutions of equations 11a and 11b are given by

$$C(t) = \cos(\omega t),$$

$$S(t) = -\sin(\omega t),$$

$$\omega = \frac{2\mu B}{\hbar}. \tag{12}$$

Thus the spin autocorrelation precesses in the magnetic field under the influence of the local torque according to the usual law.

The autocorrelation has a simple semiclassical meaning that can be seen in the vector model. There, in a state with $\sigma_z = +1$, the spin vector precesses around the magnetic field on a cone, with random starting phase. However, the angle

$$\vartheta(t) = \omega t \tag{13}$$

between the projections of $\boldsymbol{\sigma}(0)$ and $\boldsymbol{\sigma}(t)$ on the xy plane is not random, but grows under the action of the torque. When the two beams merge at the exit from the interferometer, $\boldsymbol{\sigma} = \boldsymbol{\sigma}(\tau)$ in the beam that traversed the magnetic field and $\boldsymbol{\sigma} = \boldsymbol{\sigma}(0)$ in the other arm, so the spin correlation angle is $\vartheta = \omega\tau$, and that influences the interference. In that sense, SAB actually measures what appears to be a new kind of spin correlation.

The SAB interaction is therefore local, unlike that in the EAB effect to which SAB is mathematically analogous. In SAB, the local magnetic field creates a torque that acts on the spin in the ordinary way. The field thereby influences the interference when the neutrons exit the interferometer. The correlation angle is twice the phase shift, as one expects for rotations of spin 1/2.

What about topology? EAB is topological in that the phase shift cannot be ascribed to one particular arm of the interferometer because a gauge transformation can change the individual values of V_1 and V_2. The only relevant gauge-invariant difference between the two arms is that they have different topological winding numbers. (This point is much clearer in the magnetic version of the AB effect.[3,4] There, electrons are diffracted by a cylinder centered on the z-axis. The electrons are excluded from the interior of the cylinder, which contains a magnetic field in the z-direction with flux Φ. Feynman path amplitudes for paths that end on some point on the diffraction screen may go around the cylinder any number of times. Each path amplitude acquires a phase

$$\phi = n\frac{e\Phi}{\hbar c}, \tag{14}$$

where n is the topological winding number of the path, and the diffraction pattern is affected by those phase shifts even though the electron never enters the magnetic field.)

SAB has no analogous topological character. We know exactly in which arm the interaction took place, rotating the spin autocorrelation and exchanging angular momentum with the field.

It has been argued[1] that SAB is in fact a topological effect because the phase shift $\delta\phi$ is independent of the energy of the neutron. However, it has only been shown[5,6] that energy-independence of the phase shift is a necessary condition for a force-free effect, not that it is a sufficient condition for a topological effect. One can imagine an experiment in which one arm of the interferometer contains a phase-shifter consisting of a box whose index of refraction can be varied by pumping some gas in and out. The index of refraction can be made equal to unity except during the time when the neutron wave packet is entirely inside the box, and the energy-dependence of the index of refraction can in principle be such that the phase shift is independent of the neutron's energy over the energy range of the experiment. To accept the energy-independence of the phase shift as proof that SAB is a nonlocal or topological effect, one would also have to accept that the action of this ordinary phase-shifter is also nonlocal or topological.

The same comments apply to the Aharonov-Casher effect (AC)[7] as to SAB. In AC, the neutron moves through an electric field and the torque is $\boldsymbol{\mu} \times \boldsymbol{v} \times \boldsymbol{E}$. Otherwise, the discussion is the same.

Earlier discussions[5,8] of what is now called SAB have dealt with some of the issues presented here from a different perspective.

REFERENCES

1. ALLMAN, B. E. *et al.* 1992. Phys. Rev. Lett. **68:** 2409; 1993. Phys. Rev. A **48:** 1799.
2. For a critique of the experiments as performed with unpolarized neutrons, see: PESHKIN, M. 1992. Phys. Rev. Lett. **69:** 2017.
3. AHARONOV, Y. & D. BOHM. 1959. Phys. Rev. **115:** 485.
4. PESHKIN, M. & A. TONOMURA. 1989. The Aharonov-Bohm Effect: Lecture Notes in Physics 340. Springer-Verlag. New York/Berlin.
5. ZEILINGER, A. 1986. *In* Fundamental Aspects of Quantum Theory. Volume 144. V. Gorini & A. Frigerio, Eds.: 331. NATO ASI Series B.
6. BADUREK, G. *et al.* 1993. Phys. Rev. Lett. **71:** 307.
7. AHARONOV, Y. & A. CASHER. 1984. Phys. Rev. Lett. **53:** 319.
8. ANANDAN, J. 1982. Phys. Rev. Lett. **48:** 1660; 1989. Proceedings of the Third International Symposium on Foundations of Quantum Mechanics in the Light of New Technology. S. Kobayashi, H. Ezawa, Y. Murayama & S. Nomura, Eds.: 98. Phys. Soc. Japan. Kyoto.

Quantum Zeno Effect as a Purely Dynamical Process[a]

SAVERIO PASCAZIO[b,c] AND MIKIO NAMIKI[d,e]

[b]*Dipartimento di Fisica*
Università di Bari
and
Istituto Nazionale di Fisica Nucleare
Sezione di Bari
I-70126 Bari, Italy

[d]*Department of Physics*
Waseda University
Tokyo 169, Japan

INTRODUCTION

The temporal evolution of a quantum mechanical system, initially prepared in an eigenstate of the unperturbed Hamiltonian, is known to be roughly characterized by three distinct regions:[1] a Gaussian behavior at short times, a Breit-Wigner exponential decay at intermediate times, and a power law at long times. It is well known that the asymptotic dominance of the exponential behavior is representative of a purely stochastic evolution and can be derived quantum mechanically[2] in the weak-coupling, macroscopic limit (the so-called van Hove's limit). One may expect a close connection between dissipation and exponential decay.[3] The Gaussian short-time behavior is of particular significance due, in particular, to the so-called quantum Zeno effect.[4,5]

Khalfin[6] and Misra and Sudarshan[4] discovered that, if the short-time behavior is Gaussian and not exponential, it is possible to inhibit the decay of unstable quantum mechanical systems by performing frequent "observations" in rapid succession. This phenomenon was named the quantum Zeno paradox or the quantum Zeno effect (QZE) after the Greek philosopher, Zeno, whose arrows, although darted, did not move. However, the experiment was very difficult to perform.

Recently, Cook[7] proposed an experimental test that made use of a two-level atom. It should be noted, however, that the proposed test did not involve observation of a naturally unstable quantum system, as in the original idea by Misra and Sudarshan. Following Cook's proposal, Itano and his group[8] carried out this experiment and claimed, by obtaining the same result as theoretically predicted by Cook, that the QZE had been proven experimentally. This conclusion has provoked an interesting debate[9–14] on whether this effect is really ascribable to frequent observa-

[a] A major portion of this report will appear in *Physical Review A* (see reference 18) and is here contained by courtesy of the American Physical Society.

[c] S. Pascazio was partially supported by the Italian CNR under the bilateral project, Italy-Japan No. 93.01323.CT02, and acknowledges the High Energy Physics Group of Waseda University for their kind hospitality.

[e] M. Namiki was partially supported by the Ministry of Education, Science, and Culture of the Japanese government.

tions, each of which is described by the *naive* wave-function collapse (i.e., von Neumann's projection rule), or is rather a purely dynamical process. In particular, Petrosky, Tasaki, and Prigogine[9] strongly claimed the latter point of view.

In this context, it is important to notice that, in fact, Itano *et al.* observed only one photon at the final step, but not at every step. Therefore, strictly speaking, they performed a different experiment from the original proposal put forward by Cook.[7] Thus, we are led to an interesting question: why did their experiment (in which one photon was observed only at the final step) yield the same result as the one theoretically derived under the assumption that the naive wave-function collapse (i.e., the simple projection) takes place many times? The main objective of the present report is to answer this question.

In previous papers, we proposed[13,15] to use neutron spin-flip processes, instead of atomic transitions, in order to simplify and clarify the discussion. We drew conclusions essentially similar to those of Petrosky *et al.*

As the whole class of phenomena hinging upon the controversial issue of wave-function collapse,[16,17] the QZE is very interesting from the point of view of quantum measurements. In particular, it seems worth clarifying that the notion of "collapse", as given by von Neumann's projection rule, is not a fundamental requisite for the occurrence of a QZE. Therefore, in our opinion, the widespread belief that the QZE is a clear-cut evidence in support of the wave-function collapse, as given by von Neumann's projection rule, is a misunderstanding. Indeed, we shall endeavor to show that the quantum Zeno phenomenon is a pure dynamical process, always governed by strictly unitary evolutions.

We shall also shortly consider the implications of the QZE on the quantum measurement theory.[16] We shall explain why we believe that the measurement process should be derived as a consequence of the interaction with measuring devices, within the framework of quantum mechanics, and should be ascribed to a dephasing process.[17] This will help us to clarify that the occurrence of the QZE is *no evidence* in support of the Copenhagen interpretation.

More details on the analysis to be shown in this report can be found in reference 18.

MISRA AND SUDARSHAN'S FORMULATION

We shall first formulate the quantum Zeno effect by following the seminal procedure by Misra and Sudarshan,[4] which is entirely based on von Neumann's projection rule. The reader should notice the fundamental role played by the projection operator E in the present section.

Let Q be an unstable quantum system, whose states are vectors in the Hilbert space $\mathscr{H}$ and whose evolution is described by the unitary operator $U(t) = \exp(-iHt)$, where H is a semibounded Hamiltonian. The initial density matrix of system Q is assumed to be an undecayed state, ρ_0, and let E be the projection operator over the subspace of the undecayed states. By definition,

$$\rho_0 = E\rho_0 E, \qquad \mathrm{Tr}[\rho_0 E] = 1. \tag{2.1}$$

Assume that we perform a measurement at time t, denoted by the projection operator E, in order to check whether Q is still undecayed. The measurement is idealized to be *instantaneous.* Accordingly, the state of Q changes into

$$\rho_0 \to \rho(t) = EU(t)\rho_0 U^\dagger(t)E, \tag{2.2}$$

so the probability to find the system undecayed is given by

$$p(t) = \mathrm{Tr}[U(t)\rho_0 U^\dagger(t)E]. \tag{2.3}$$

We shall refer to the process of equation 2.2 as the "naive wave-function collapse". We are now ready to review briefly Misra and Sudarshan's original formulation of what they named the "quantum Zeno paradox". We prepare Q in the initial state ρ_0 at time 0 (this is formally accomplished by performing an initial, "preparatory" measurement of E) and perform a series of observations at times T/N, $2T/N, \ldots,$ $(N-1)T/N$, T. The state $\rho^{(N)}(T)$ of Q after the preparation and the above-mentioned N measurements reads

$$\rho^{(N)}(T) = V_N(T)\rho_0 V_N^\dagger(T), \qquad V_N(T) \equiv [EU(T/N)E]^N \tag{2.4}$$

and the probability to find the system undecayed is given by

$$P^{(N)}(T) = \mathrm{Tr}[V_N(T)\rho_0 V_N^\dagger(T)]. \tag{2.5}$$

Equations 2.4 and 2.5 display what will be referred to as the "quantum Zeno effect": repeated observations in succession modify the dynamics of the quantum system, by slowing down the decay process, as we shall see in a particular example in the following section.

In the $N \to \infty$ limit (continuous observation), Misra and Sudarshan proved under general conditions that

$$\mathscr{P}(T) = \lim_{N\to\infty} P^{(N)}(T) = \mathrm{Tr}[\rho_0 E] = 1. \tag{2.6}$$

If the particle is continuously observed (to check whether it decays or not), it is "frozen" in its initial state and will never be found to decay. This is the essence of the "quantum Zeno paradox".

It is worth stressing the profound difference between the N-finite and the N-infinity case. To perform an experiment with N finite is only a practical problem from the physical point of view. On the other hand, the $N \to \infty$ case is physically unattainable and is rather to be regarded as a mathematical limit (although a very interesting one). In this sense, we shall say that the quantum Zeno *effect*, with N finite, becomes a quantum Zeno *paradox* when $N \to \infty$.

Finally, if the Q system is let to follow its "free" evolution under the action of the Hamiltonian H, its final state at time T reads

$$\rho(T) = U(T)\rho_0 U^\dagger(T) \tag{2.7}$$

and the probability that the system is still undecayed at time T is

$$P(T) = \mathrm{Tr}[U(T)\rho_0 U^\dagger(T)E]. \tag{2.8}$$

We shall now endeavor to show that the consequences of the above theorem are liable to a pure *dynamical* explanation that *does not make use of projection operators.* In this sense, we believe that the quantum Zeno effect is just a consequence (although a peculiar one) of the Schrödinger equation. In the following, we shall show that one needs only a particular quantum dynamics in order to "freeze" the Q system in its initial state.

FIGURE 1A. "Free" evolution of the neutron spin under the action of a magnetic field. An emitter E sends a spin-up neutron through several regions where a magnetic field B is present. The detector D_0 detects a spin-down neutron: no Zeno effect occurs.

It is necessary to stress again that the observations (measurements) schematized via the operator E are *instantaneous.* This is a rather general characteristic of von Neumann–like descriptions of a measurement process: the Q system *instantaneously* makes the transition of equation 2.2 by measurement (naive wave-function collapse). Even though such a picture is often accepted among physicists, it is *misleading.* Indeed, a measurement process, as a physical process, takes place during a *very long time* on a *microscopic* scale, although we can regard it as if it happened *instantaneously* on a *macroscopic* scale.

For a discussion on this point, see reference 19.

QUANTUM ZENO EFFECT WITH NEUTRON SPIN

We shall now discuss a particular solvable example[15] that, in spite of its simplicity, yields rich physical insight and turns out to be very useful for the general analysis found in the next section. We shall show how the same "Zeno-type" evolution can be obtained both by making use of the technique of the previous section and by means of a purely dynamical process.

The example we want to consider is a neutron spin in a magnetic field.[15] (A situation analogous to the one described in this section was outlined by Peres[20] with photons.) We shall consider two different experiments: refer to FIGURES 1A and 1B.

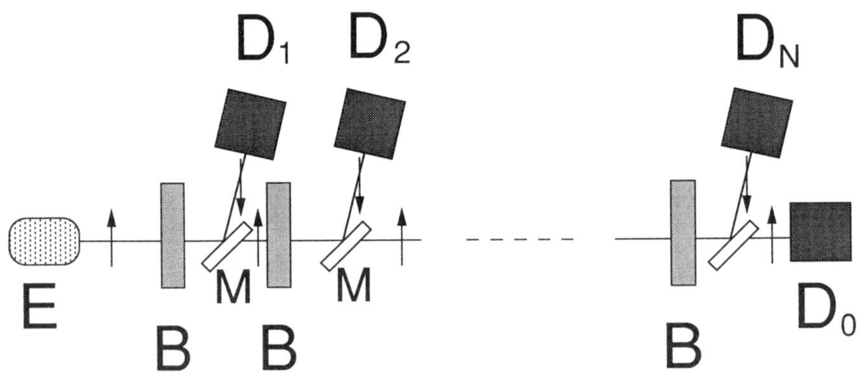

FIGURE 1B. The neutron spin is "monitored" at every step, by selecting and detecting the spin-down component. D_0 detects a spin-up neutron: the Zeno effect takes place.

In the case schematized in FIGURE 1A, the neutron interacts with several identical regions in which there is a static magnetic field, B, oriented along the x-direction. We neglect any losses and assume that the interaction is given by the Hamiltonian,

$$H = \mu B \sigma_1, \tag{3.1}$$

with μ being the (modulus of the) neutron magnetic moment and with σ_i ($i = 1, 2, 3$) being the Pauli matrices. We denote the spin states of the neutron along the z-axis by $|\uparrow\rangle$ and $|\downarrow\rangle$: these can be identified with the undecayed and decayed states of the previous section, respectively.

Let the initial neutron state be $\rho_0 = \rho_{\uparrow\uparrow} \equiv |\uparrow\rangle\langle\uparrow|$. The interaction with the first region simply provokes a rotation of the initial state around the x-direction:

$$\begin{aligned}\rho_0 \to \rho(t) &= e^{-iHt/\hbar}\rho_{\uparrow\uparrow}e^{iHt/\hbar}\\ &= \cos^2\left(\frac{\omega t}{2}\right)\rho_{\uparrow\uparrow} + \sin^2\left(\frac{\omega t}{2}\right)\rho_{\downarrow\downarrow} - i\cos\left(\frac{\omega t}{2}\right)\sin\left(\frac{\omega t}{2}\right)\rho_{\uparrow\downarrow} + \text{h.c.},\end{aligned} \tag{3.2}$$

where $\omega = 2\mu B/\hbar$, $t = \ell/v$ (ℓ is the length of the region where B is present and v is the neutron speed), and the other notation is obvious. (In this section, we do not set $\hbar = 1$.)

We repeat the process N times, as shown in FIGURE 1A. The final density matrix at time $T = Nt$ is

$$\begin{aligned}\rho(T) \equiv e^{-iHT/\hbar}\rho_0 e^{iHT/\hbar} &= \cos^2\left(\frac{\omega T}{2}\right)\rho_{\uparrow\uparrow} + \sin^2\left(\frac{\omega T}{2}\right)\rho_{\downarrow\downarrow}\\ &\quad - i\cos\left(\frac{\omega T}{2}\right)\sin\left(\frac{\omega T}{2}\right)\rho_{\uparrow\downarrow} + \text{h.c.}\end{aligned} \tag{3.3}$$

We call this a "free" evolution, during which the system evolves under the sole influence of H. Note the presence of the off-diagonal terms with respect to the spin states. If T is chosen so as to satisfy the "matching" condition $\cos \omega T/2 = 0$ (notice that this can also be viewed as a fine-tuning requirement, similar to the one experimentally realized by Itano *et al.*[8]), we obtain

$$\rho(T) = \rho_{\downarrow\downarrow} \qquad \left[T = (2m+1)\frac{\pi}{\omega},\, m \in \mathbb{N}\right], \tag{3.4}$$

so the probability that the neutron spin is down at time T is

$$P_{\downarrow}(T) = 1 \qquad \left[T = (2m+1)\frac{\pi}{\omega},\, m \in \mathbb{N}\right]. \tag{3.5}$$

The above two equations correspond to equations 2.7 and 2.8 for a specially chosen T. In our example, H is such that, if the system is initially prepared in the up state, it will evolve to the down state after time T. Notice that, within our approximations, the experimental setup described in FIGURE 1A is equivalent to a single region of length $L = N\ell$ with magnetic field B.

Let us now modify the experiment just described by inserting at every step a device able to select and detect one component [say, the down ($\downarrow$) one] of the neutron spin. This can be accomplished by a magnetic mirror (M) and a detector (D).

The former acts as a "decomposer" by splitting a neutron wave with indefinite spin (a superposed state of up and down spin) into two branch waves, each of which is in a definite spin state (up *or* down) along the z-axis. The down state is then forwarded to a detector, as shown in FIGURE 1B.

The magnetic mirror yields a spectral decomposition[17,21] with respect to the spin states and can be compared to the inhomogeneous magnetic field in a typical Stern-Gerlach experiment. It is very important, in connection with the QZE, to bear in mind that the magnetic mirror does *not* destroy the coherence between the two branch waves. Indeed, the two branch waves corresponding to different spin states can be split coherently and can be brought back to interfere.[22]

We choose the same initial state for Q as in the previous experiment (FIGURE 1A). The interaction with B in the first region still provokes the evolution of equation 3.2. The spectral decomposition and the detection of a spin-down neutron, provoked by M and D, respectively, are (formally) globally represented by the operator $E \equiv \rho_{\uparrow\uparrow}$ (remember that we follow the evolution along the horizontal direction in FIGURE 1B, which corresponds to spin up), so equation 2.2 yields

$$\rho_0 \to \rho(t) = EU(t)\rho_0 U^\dagger(t)E = \left(\cos^2 \frac{\omega t}{2}\right) \rho_{\uparrow\uparrow}, \tag{3.6}$$

where $U = \exp(-iHt/\hbar)$. If the process is repeated N times, as in FIGURE 1B, we obtain

$$\rho^{(N)}(T) = V_N(T)\rho_0 V_N^\dagger(T) = \left(\cos^2 \frac{\omega t}{2}\right)^N \rho_{\uparrow\uparrow} = \left(\cos^2 \frac{\pi}{2N}\right)^N \rho_{\uparrow\uparrow}, \tag{3.7}$$

where $V_N(T)$ has been defined in equation 2.4 and the "matching" condition for $T = Nt$ (see equation 3.4) has been required again. The probability that the neutron spin is up at time T, if N observations have been made at time intervals t ($Nt = T$), is

$$P_\uparrow^{(N)}(T) = \left(\cos^2 \frac{\pi}{2N}\right)^N. \tag{3.8}$$

This discloses the occurrence of a QZE. Indeed, $P_\uparrow^{(N)}(T) > P_\uparrow^{(N-1)}(T)$ for $N \geq 2$, so the evolution is "slowed down" as N increases. Moreover, in the limit of infinitely many observations,

$$\rho^{(N)}(T) \overset{N\to\infty}{\to} \tilde{\rho}(T) = \rho_{\uparrow\uparrow} \tag{3.9}$$

and

$$\mathscr{P}_\uparrow(T) \equiv \lim_{N\to\infty} P_\uparrow^{(N)}(T) = 1. \tag{3.10}$$

Frequent observations "freeze" the neutron spin in its initial state by inhibiting ($N \geq 2$) and eventually hindering ($N \to \infty$) transitions to other states. Notice the difference with equations 3.4 and 3.5: the situation is completely reversed.

The above result, peculiar as it may seem, is a straightforward consequence of the quantum formalism. It is worth stressing that (by setting, for simplicity, $m = 0$ in equation 3.4) the condition, $\omega T = \omega Nt = \pi$, which is to be met at every step in

FIGURE 1B, means that

$$B\ell = \frac{\pi\hbar v}{2\mu N} = O(N^{-1}), \tag{3.11}$$

where ℓ and v were defined after equation 3.2. This implies that the practical realization of the experiment becomes increasingly difficult as N increases in the above equations. We shall come back to this point in the final discussion.

Notice that the final results, equations 3.9 and 3.10, are *automatically properly normalized*: the probability of detecting a spin-down neutron vanishes in the $N \to \infty$ limit. This is the essence and the peculiarity of the Zeno argument. We stress that in the above analysis we have neglected any losses and reflections at the mirrors.

If the state of the total (neutron + detectors) system is taken into account, we obtain the following density matrix in the channel representation:

$$\tilde{\Xi}_{ij} \equiv \begin{pmatrix} c^{2N} & & & & 0 \\ & s^2c^{2N-2} & & & \\ & & s^2c^{2N-4} & & \\ & & & \ddots & \\ 0 & & & & s^2 \end{pmatrix}, \qquad i,j = 0, 1, \cdots, N, \tag{3.12}$$

where $c = \cos(\pi/2N)$ and $s = \sin(\pi/2N)$. This corresponds to the case of frequent observations, in which we confirmed, at every step, the neutron route among the various possibilities $(0, 1, \cdots, N)$. Notice that the $i = j = 0$ component corresponds to detection by D_0, whereas the $i = j = n$ $(n = 1, \cdots, N)$ component corresponds to detection in channel $N - n + 1$. Observe that in the above expression the off-diagonal terms disappeared as a consequence of the "wave-function collapse" by measurement.

It is easy to realize that the *same* result can be obtained without making use of projection operators by simply performing a different analysis involving *only unitary processes*. Assume that $D_1, \cdots, D_N$ are removed in FIGURE 1B. In other words, we make *no* observation of the neutron route, except the final one performed by D_0. The correspondent density matrix is readily shown to be

$$\Xi_{ij} \equiv \begin{pmatrix} c^{2N} & isc^{2N-1} & isc^{2N-2} & \cdots & isc^{N} \\ -isc^{2N-1} & s^2c^{2N-2} & s^2c^{2N-3} & \cdots & s^2c^{N-1} \\ -isc^{2N-2} & s^2c^{2N-3} & s^2c^{2N-4} & \cdots & s^2c^{N-2} \\ \vdots & \vdots & \vdots & \ddots & \vdots \\ -isc^{N} & s^2c^{N-1} & s^2c^{N-2} & \cdots & s^2 \end{pmatrix}, \qquad i,j = 0, 1, \cdots, N. \tag{3.13}$$

Equations 3.12 and 3.13 clearly show that we have the *same* probability $P_{\uparrow}^{(N)} = [\cos^2(\pi/2N)]^N$ of detecting a spin-up neutron at D_0 in both cases with the detectors $D_1, \cdots, D_N$ present and absent in FIGURE 1B. It appears therefore that *no projection rule is necessary* in this context.

Physically, the situation just described corresponds to performing a coincidence experiment between the emitter E and D_0, as was explicitly shown in reference 15. One can also consider this result as a consequence of a negative-result measurement[23] of the neutron spin. See reference 15.

Notice that the *diagonal elements* of the two density matrices (equations 3.12 and 3.13) are the same. Moreover, in both cases, in the $N \to \infty$ limit, only the c^{2N} $(i = j = 0)$ element survives. All other terms disappear because they contain at least an s factor. The correspondent density matrix is

$$\Xi_{ij}^{\infty} \equiv \begin{pmatrix} 1 & 0 & 0 & \cdots \\ 0 & 0 & 0 & \cdots \\ 0 & 0 & 0 & \cdots \\ \vdots & \vdots & \vdots & \ddots \end{pmatrix}, \qquad i, j = 0, 1, \ldots . \tag{3.14}$$

We stress that the $N \to \infty$ limit is only mathematical and is *impossible* to realize *physically* because the elapse of time T/N, even though it can be considered very short on a macroscopic scale, is in fact the time spent by the neutron in each B-region. We have to keep in mind this remark throughout this report.

We shall now generalize this conclusion. We shall also see that the experiment performed by Itano *et al.*,[8] in which a photon was observed *only* at the final step, will appear as a particular case of our analysis. Indeed, if one identifies the effect of the mirrors in FIGURE 1B with the effect provoked by the laser pulses in reference 8, the above discussion unequivocally implies that the same result would be obtained if Itano *et al.* observed the intermediate photons or not. Their experiment is just equivalent to a series of spectral decompositions performed by the laser pulses.

QUANTUM ZENO EFFECT AS A PURELY DYNAMICAL PROCESS

We shall give in this section a general formulation of the QZE as a dynamical process without making use of von Neumann's projection postulate. We shall see when a system evolves under particular conditions and undergoes a "generalized spectral decomposition" (the precise meaning of which will be explained in a while) that its initial state becomes "frozen", in the sense explained in the previous sections.

The evolution we are going to consider is peculiar in that it involves the creation of several quantum correlations in rapid succession. We start (without loss of generality and in line with the formalism of the previous sections) from a two-level quantum system Q, living in a Hilbert space $\mathscr{H}_Q$, and embed the latter in the larger space, $\mathscr{H} \equiv \mathscr{H}_Q \otimes \mathscr{H}_U$, where the subscript U stands for universe. Our "universe" can be anything: the spatial component of the total wave function of Q (as in the previous section), a quantized electromagnetic field interacting with Q (as in the experiment by Itano *et al.*), some detectors surrounding Q, or any atmosphere in which Q exists. Two points are worth noticing. First, it is essential that the universe be treated quantum mechanically: if a "classical" behavior were postulated, we would conceptually go back to von Neumann's projection rule and to Misra and Sudarshan's seminal idea. Second, the only purpose for introducing the universe is to "follow" the quantum correlations engendered by Q, namely, to monitor their "spreading" towards other degrees of freedom of the total wave function ("leakage" and "environment" may be better words than "spreading" and "universe", respectively, but we prefer the former expressions because, as we shall see, the quantum coherence can eventually be recovered).

We start from the initial state

$$|\Psi_I\rangle \equiv |\pm\rangle \bigotimes_{n=0}^{N} \bigotimes_{j=1}^{2^n} |0_j^{(n)}\rangle \equiv |\pm\rangle|0\rangle_N, \tag{4.1}$$

where $|\pm\rangle$ denotes any state of the two-level system Q, $0_j^{(n)}$'s are occupation numbers, and $|0\rangle_N$ is the ground state of the universe. The state + (−) plays the same role of the state ↑ (↓) in the previous section and of the undecayed (decayed) state in the section before that. The structure of $\mathscr{H}_U$ can be understood by looking at FIGURE 2:

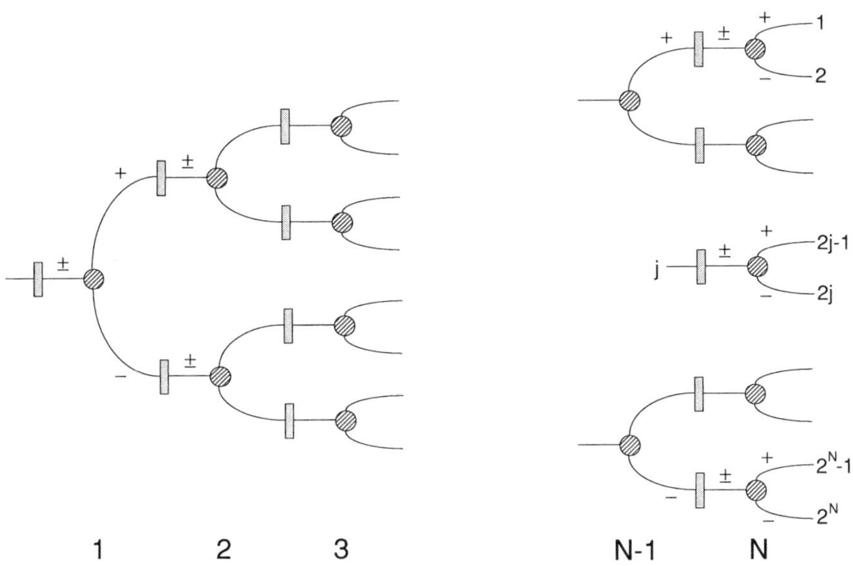

FIGURE 2. The quantum correlation "tree". After N steps, there are 2^N branches. The value of n $(n = 1, \cdots, N)$ is indicated on the bottom.

the "universe" consists of many channels (labeled j, n) that can be in either of the two possible states, $0_j^{(n)}$ and $1_j^{(n)}$. The state $0_j^{(n)}$ ($1_j^{(n)}$) denotes the absence (presence) of an excitation in the corresponding channel. The index n labels the "step" (horizontal direction in FIGURE 2), whereas the index j labels the "branch" (vertical direction) at step n. After n steps, there are 2^n branches.

For the sake of simplicity, the following analysis is performed in terms of wave functions instead of density matrices.

The "Free" Evolution

We assume that the system undergoes two different types of evolution at every step in FIGURE 2. The first evolution is governed by the Hamiltonian H_0 and takes place in the rectangular regions. The second evolution is governed by the Hamiltonian H' and takes place in the circular regions.

The effect of the two evolutions is profoundly different. The evolution engendered by H_0 is identical at every step and will be written as

$$U_0(t)|+\rangle \otimes |\cdots\rangle \equiv e^{-iH_0t}|+\rangle \otimes |\cdots\rangle = [\gamma_+(t)|+\rangle + \epsilon_-(t)|-\rangle] \otimes |\cdots\rangle,$$

$$U_0(t)|-\rangle \otimes |\cdots\rangle \equiv e^{-iH_0t}|+\rangle \otimes |\cdots\rangle = [\epsilon_+(t)|+\rangle + \gamma_-(t)|-\rangle] \otimes |\cdots\rangle, \quad \textbf{(4.2)}$$

where the dots denote any state of the universe. Notice that this leaves the universe unaltered. We shall be interested in the behavior for small t, which is, in general,

$$\epsilon_\pm(t) = O(t), \qquad \gamma_\pm(t) = 1 - O(t^2). \quad \textbf{(4.3)}$$

The Hamiltonian H_0 plays the same role as H in equation 3.1. In some sense, we can consider the above evolution as "free", where "free" simply means that the behavior of the Q system is not "monitored", and Q is let to follow a natural, smooth evolution, under the action of H_0.

The Generalized Spectral Decomposition

The second interaction, governed by H', is a sort of *spectral decomposition*: different states of the Q system become *entangled* with different states of the universe. One can think, for example, of a sort of Stern-Gerlach decomposition of an initial spin state (so that every component of the spin becomes associated with a different wave packet, as in the previous section), of an entanglement of a two-level atomic system with a photonic state (see the analysis of the experiment by Itano *et al.* in the following section and the lucid discussion of Petrosky *et al.* on this point), or (to be more general) of an entanglement of each state of Q with different degrees of freedom of the universe. In this sense, we shall speak of the *generalized spectral decomposition* (GSD) of the Q states.

In order to describe this situation, refer to FIGURE 3 and assume the following Hamiltonian:

$$H'(t) \equiv g(t)[|+\rangle\langle+|\sigma_\beta + |-\rangle\langle-|\sigma_\gamma]\sigma_\alpha \equiv g(t)H', \qquad \int_0^\tau g(t)dt = b \in \mathbb{R}, \quad \textbf{(4.4)}$$

where the interaction is switched on during the time interval $[0, \tau]$, g is a real function, $\sigma_\mu^\dagger = \sigma_\mu$ (the index $\mu = \alpha, \beta, \gamma$ labels the channel), and the effect of σ_μ is defined by

$$\sigma_\mu|0_\mu\rangle = |1_\mu\rangle,$$

$$\sigma_\mu|1_\mu\rangle = |0_\mu\rangle; \quad \textbf{(4.5)}$$

thus, if there is a "particle" in channel μ, the operator σ_μ destroys it, whereas, if there is no particle, σ_μ creates one. The effect of σ_μ $(\forall\mu)$ is therefore identical to that of the first Pauli matrix. (We are implicitly assuming that there cannot be more than one "particle" in channel μ.) We set

$$[\sigma_\mu, \sigma_\nu] = 0 \qquad \forall \mu, \nu. \tag{4.6}$$

It can be shown[18] that H' acts by sending the + (−) state of the Q particle in the upper (lower) channel in FIGURE 3, performing in this way a GSD.

In general, the only effect of a GSD is to set up a correlation between the two states of Q and the different states of the universe. Obviously, for the purpose of our analysis, we are interested in obtaining a perfect GSD (namely, a univocal and unambiguous correspondence between different states of Q and of the universe). This can be easily accomplished by setting $b = \pi/2$ in equation 4.4: this is a sort of fine-tuning and corresponds to the so-called π-pulse condition, widely used for electromagnetic cavities.[24] This condition was experimentally realized in reference 8. As explained in reference 18, $b = \pi/2$ can also be viewed as the requirement that the GSD be "reflectionless". Notice also that all losses are neglected.

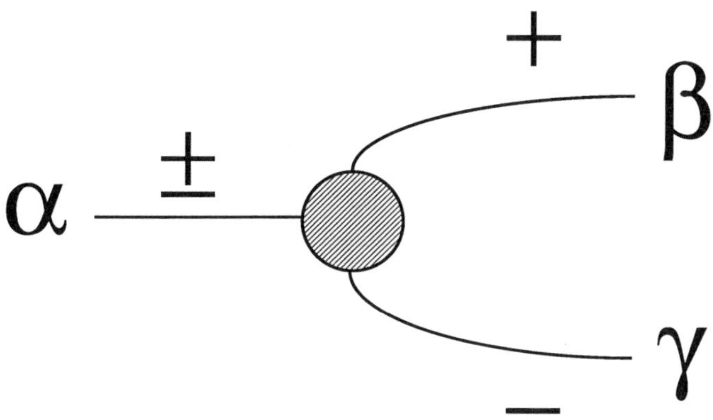

FIGURE 3. The generalized spectral decomposition.

The evolution engendered by H' can be calculated explicitly.[18] The final result is

$$\begin{aligned} U'(t)(c_+|+\rangle + c_-|-\rangle)\otimes|1_\alpha, 0_\beta, 0_\gamma\rangle &\equiv e^{-i\int_0^\tau H'(t')dt'}(c_+|+\rangle + c_-|-\rangle)\otimes|1_\alpha, 0_\beta, 0_\gamma\rangle \\ &= -iH'(c_+|+\rangle + c_-|-\rangle)\otimes|1_\alpha, 0_\beta, 0_\gamma\rangle \\ &= -i(c_+|+\rangle\otimes|0_\alpha, 1_\beta, 0_\gamma\rangle + c_-|-\rangle\otimes|0_\alpha, 0_\beta, 1_\gamma\rangle) \\ &\qquad (t > \tau, b = \pi/2) \end{aligned} \tag{4.7}$$

and yields a genuine GSD.

Our idea is to get a Zeno-type dynamics without making use of nonunitary evolutions (projection operators). In the second section of this report, the operator E represented a measurement that was assumed to be *instantaneous.* As already emphasized, this is clearly an idealized situation that cannot correspond to a real physical process, taking place at a microscopic level. The problem is therefore to understand how we can simulate such an instantaneous and unphysical process in our analysis that makes use of *only* unitary evolutions.

We observe that, in general, a GSD must take place in a very short time. Obviously, the term "very short time" must be understood at a macroscopic level of description because, microscopically, the time required to efficaciously perform a GSD is very long. Therefore, if we restrict our analysis to a macroscopic level of description, we can describe an (almost) instantaneous GSD by means of the so-called impulse approximation,

$$\int_0^\tau g(t)dt = \pi/2, \qquad \tau \to 0^+, \tag{4.8}$$

which roughly amounts to setting $g(t) \to (\pi/2)\delta(t)$ as $\tau \to 0^+$, where δ is the Dirac function $\int_0^\tau \delta(t) = 1$. This is our alternative description of a von Neumann–like instantaneous projection.

The Evolution at the n*-th Step*

We can now tackle the general case. Refer to FIGURE 2 and assume the following Hamiltonian at the n-th step:

$$\begin{aligned} H'_{(n)}(t) &\equiv g(t)H'_{(n)} \\ &\equiv g(t)\sum_{j=1}^{2^{n-1}} \{[|+\rangle\langle+|\sigma_{2j-1}^{(n)} + |-\rangle\langle-|\sigma_{2j}^{(n)}]\sigma_j^{(n-1)}\} \\ &\equiv g(t)\sum_{j=1}^{2^{n-1}} H'_{(n),j}, \qquad \int_0^\tau g(t)dt = b \in \mathbb{R}, \end{aligned} \tag{4.9}$$

where the interaction is switched on during the time interval $[0, \tau]$ and the action of $\sigma_i^{(n)}$ is defined by

$$\begin{aligned} \sigma_i^{(n)}|0_i^{(n)}\rangle &= |1_i^{(n)}\rangle, \\ \sigma_i^{(n)}|1_i^{(n)}\rangle &= |0_i^{(n)}\rangle; \end{aligned} \tag{4.10}$$

thus, if there is a "particle" in the i-th channel at the n-th step, the operator $\sigma_i^{(n)}$ destroys it, whereas, if there is no particle, $\sigma_i^{(n)}$ creates one. We set

$$[\sigma_i^{(n)}, \sigma_k^{(m)}] = 0 \qquad \forall i, k, n, m. \tag{4.11}$$

It can be shown again that $H'_{(n),j}$ acts on the $+$ $[-]$ state of the Q particle in the j-th channel at the $(n-1)$-th step by sending it into the $(2j-1)$-th $[2j$-th$]$ channel at step n, performing in this way a GSD.

If the condition $b = \pi/2$ is used again, one obtains after some algebraic manipulations

$$U'_{(n)}(t)(c_+|+\rangle + c_-|-\rangle) \otimes |\cdots, \otimes_{j=1}^{2^{n-1}} 1_j^{(n-1)}, \cdots\rangle$$
$$= e^{-i\int_0^\tau H'_{(n)}(t')dt'}(c_+|+\rangle + c_-|-\rangle) \otimes |\cdots, \otimes_{j=1}^{2^{n-1}} 1_j^{(n-1)}, \cdots\rangle$$
$$= (-i)^{2^{n-1}}(c_+|+\rangle \otimes |\cdots, \otimes_{j=1}^{2^{n-1}} 1_{2j-1}^{(n)}, \cdots\rangle + c_-|-\rangle \otimes |\cdots, \otimes_{j=1}^{2^{n-1}} 1_{2j}^{(n)}, \cdots\rangle)$$
$$(t > \tau, b = \pi/2), \quad \textbf{(4.12)}$$

where the dots will henceforth signify that all other occupation numbers are 0.

Summarizing, at every step of the process described in FIGURE 2, the system evolves according to

$$U_{(n)}(\tau, t) \equiv U'_{(n)}(\tau)U_0(t), \qquad \textbf{(4.13)}$$

where U_0 was defined in equation 4.2.

Our interest will be focused on the evolution engendered by the limiting operator,

$$\lim_{N\to\infty} U_{TOT}^{[N]}(T) \equiv \lim_{N\to\infty} \left[\prod_{n=1}^{N} U_{(n)}(T_1/N, T_2/N)\right] U_I, \qquad \textbf{(4.14)}$$

where $T = T_1 + T_2$ and the operator U_I creates the initial incoming Q state at $t = 0$. Notice that $T = T_1 + T_2$, that is, the total duration of the "experiment", is kept finite in taking the above limit.

The N $\to \infty$ *Limit*

Let us study the action of the operator $U_{TOT}^{[N]}(T)$ on *any* initial state of the Q system. (The universe is initially taken to be in the ground state. See equation 4.1.) If, for instance, the initial Q state is $|+\rangle$, the final state is readily computed from equations 4.2, 4.3, and 4.12–4.14 as

$$U_{TOT}^{[N]}(T)|+\rangle \otimes |0\rangle_N = (-i)^{2^N}[\gamma_+(T_2/N)]^N|+\rangle \otimes |\cdots, 1_1^{(N)}, \cdots\rangle + O(1/N). \quad \textbf{(4.15)}$$

On the other hand, if the initial state is $|-\rangle$, the final state reads

$$U_{TOT}^{[N]}(T)|-\rangle \otimes |0\rangle_N = (-i)^{2^N}[\gamma_-(T_2/N)]^N|-\rangle \otimes |\cdots, 1_{2^N}^{(N)}\rangle + O(1/N). \quad \textbf{(4.16)}$$

This clearly displays a QZE for N finite and a quantum Zeno paradox in the $N \to \infty$ limit. Indeed, observe that

$$\lim_{N\to\infty} [\gamma_\pm(T_2/N)]^N = \lim_{N\to\infty} [1 - O(1/N^2)]^N = 1, \qquad \textbf{(4.17)}$$

where we made use of equation 4.3. In other words, in the limit of infinitely many GSD's, both states of Q are "frozen", while the universe evolves into the uppermost or lowermost state in FIGURE 2, according to whether the initial Q state is + or −, respectively. Moreover, the final result does not depend on T_1, that is, the total time needed to perform GSD's. In particular, in the impulse approximation (equation 4.8), we can get $T_1 = 0$ and the total duration of the "experiment" is just $T = T_2$. This is interesting and reflects, in our opinion, the essence of the "dynamical QZE": the final state does not depend on the "GSD" time, which can be made arbitrarily small in order to *mimic* the effect of an instantaneous projection E like in the second section. Notice, however, that it is *not* necessary to take the $T_1 \to 0$ limit because we would obtain a dynamical QZE even if $T_1 = N\tau$ is kept finite. On the other hand, if the total duration of the experiment $T = T_1 + T_2$ (with $T_1 = N\tau$) is kept finite, the $N \to \infty$ limit can be taken only in the impulse approximation (equation 4.8).

We stress again that the $\tau \to 0$ limit is unphysical and impossible to realize in practice. Indeed (as already emphasized in the previous section), even though τ can be considered very short in a macroscopic sense, it is in fact a very long time on a microscopic scale. In the case considered in the previous section, for instance, τ is the time elapsed during the interaction between the neutron and a magnetic mirror M, which is of the order of 10^{-7}–10^{-6} s.

However, there is more to this: experimentally, even the requirement that the total time spent in "free" evolutions T_2 be finite appears prohibitive. Indeed, such a total time should be divided into many small time intervals whose duration (T_2/N) vanishes as $N \to \infty$. This additional problem is manifest when one looks at the examples of FIGURES 1A and 1B. There is no conceptual problem related to the experiment in FIGURE 1A because, within our approximation, the experimental arrangement is equivalent to a magnetic field B in a region of length $L = N\ell$ (see definitions after equation 3.2). On the other hand, in the experiment sketched in FIGURE 1B, each single region containing B must have a length $\ell = L/N$ that vanishes in the $N \to \infty$ limit. Consequently, the time T_2/N spent by the neutron in each single region should also vanish. This is clearly impossible to realize in practice.

Observe that the final state at time T is *fully coherent*: the evolution is obviously *unitary* and *no "collapse" of the wave function* has taken place. Needless to say, this result holds true for any possible state of Q. Indeed, application of the superposition principle yields.

$$U^{[N]}_{TOT}(T)\,(c_+|+\rangle + c_-|-\rangle) \otimes |0\rangle_N = c_+(-i)^{2^N}[\gamma_+(T_2/N)]^N|+\rangle \otimes |\cdots, 1^{(N)}_1, \cdots\rangle$$
$$+ c_-(-i)^{2^N}[\gamma_-(T_2/N)]^N|-\rangle \otimes|\cdots, 1^{(N)}_{2^N}\rangle + O(1/N), \quad \textbf{(4.18)}$$

which is still a pure state. At least, the underlying quantum coherence can be simply brought to light by "recombining the two beams" (the uppermost and lowermost states in FIGURE 2) by means of the operator U_F (which plays the same role as U_I in equation 4.14). By defining

$$U^{[N]'}_{TOT}(T) \equiv U_F U^{[N]}_{TOT}(T), \quad \textbf{(4.19)}$$

we obtain

$$U_{TOT}^{[N]'}(T)(c_+|+\rangle + c_-|-\rangle) \otimes |0\rangle_N$$

$$= (-i)^{2^N+1}(c_+[\gamma_+(T_2/N)]^N|+\rangle + c_-[\gamma_-(T_2/N)]^N|-\rangle) \otimes |0\rangle_N + O(1/N). \quad \textbf{(4.20)}$$

In conclusion, by denoting $|0\rangle$ as the ground state of the universe in the $N \to \infty$ limit, we get

$$\lim_{N\to\infty} U_{TOT}^{[N]'}(T)(c_+|+\rangle + c_-|-\rangle) \otimes |0\rangle_N \propto (c_+|+\rangle + c_-|-\rangle) \otimes |0\rangle \quad \textbf{(4.21)}$$

up to a phase factor. This result is stronger than the one obtained in reference 4 and outlined in the second section. Indeed, we have shown that it is possible, by making use of a dynamical process, to freeze *any* initial Q state and not only a suitably chosen initial Q state.

THE EXPERIMENT BY ITANO AND COWORKERS

The recent experiment performed by Itano, Heinzen, Bollinger, and Wineland[8] has provoked a renewed interest and a lively debate on the meaning of the QZE. The above-mentioned investigators claimed to have observed experimentally the quantum Zeno effect by making use of atomic transitions on the basis of Cook's proposal.[7]

This conclusion was challenged by Petrosky, Tasaki, and Prigogine,[9] among others, who proved via a detailed theoretical calculation that the experimental results in reference 8 are liable to a dynamical explanation and therefore need not be ascribed to any collapse of the wave function. Itano *et al.* replied to the above criticisms[12] without anyway withdrawing their original conclusion.

Let us therefore briefly review this experiment and discuss its meaning and implications from the point of view outlined in this report. Itano *et al.* put $^9Be^+$ ions in an rf cavity. The ion energy level configuration was such that $E_1 < E_2 < E_3$, and the resonating rf-field frequency $\omega_2 = (E_2 - E_1)/\hbar$ created a coherent superposition state of the two lower levels. Upon measurement, the ion can be found in level 1 *or* 2, but never in both levels at the same time.

We denote the probability of finding the atom in level 1 [2] at time t by $P_1(t)$ [$P_2(t)$]. If the initial condition $P_1(0) = 1$ is chosen, it is possible, by making use of a "π-pulse",[24] to find a time T such that $P_1(T) = 0$. Notice that the "π-pulse" condition is essentially similar to that described in the third section (see, in particular, equations 3.4 and 3.5) and can be viewed as a fine-tuning condition, as explained in the subsection on the generalized spectral decomposition.

In order to observe the state of the atom, Itano *et al.* irradiated it with very short optical pulses of frequency $\omega_3 = (E_3 - E_1)/\hbar$ and chose the level configuration in such a way that the spontaneous decay $3 \to 1$ was strongly favored, whereas the decay $3 \to 2$ was forbidden. In this way, the atom is known to be in the first level if a photon of frequency ω_3 is observed, whereas it is in the second level if no photon is observed.

According to quantum measurement theory, the wave-function collapse takes place as a consequence of observation and, consequently, the density matrix of the atom loses its off-diagonal components (with respect to the first and second atomic states). If N observations are performed during the time interval $(0, T)$, the probability of finding the atom in the first level is given by

$$P_1^{(N)}(T) = \cos^{2N}\left(\frac{\pi}{2N}\right). \tag{5.1}$$

This displays the quantum Zeno effect because $P_1^{(N)}(T) > P_1^{(N-1)}(T)$ for $N > 2$ and $\lim_{N\to\infty} P_1^{(N)}(T) = 1$. Itano *et al.* experimentally confirmed the above prediction[25] by sending N optical pulses and then claimed to have observed the quantum Zeno effect.

A dynamical explanation, involving no collapse of the wave function, was suggested in reference 9 and is, in our opinion, very convincing.

By making use of the techniques used in the previous section, we propose the following purely dynamical explanation: Let the Q system be the atom, whereas the "universe" is the Fock space of the photons absorbed and then reemitted in the $1 \leftrightarrow 3$ transition. The initial state of the total system is

$$|\Psi_I\rangle \equiv |\phi_1\rangle \otimes |0\rangle, \tag{5.2}$$

where $|\phi_i\rangle$ represents the atomic level i $(i = 1, 2)$ and $|0\rangle$ is the ground state of the Fock space.

The "free" evolution yields simply the Rabi oscillations between the atomic levels 1 and 2 and is obviously in agreement with the general behavior of equations 4.2 and 4.3:

$$|\Psi_I\rangle \to |\Psi_{(1)}\rangle = \left[\cos\left(\frac{\Omega t}{2}\right)|\phi_1\rangle + i\sin\left(\frac{\Omega t}{2}\right)|\phi_2\rangle\right] \otimes |0\rangle, \tag{5.3}$$

where Ω is the frequency of the Rabi oscillations between levels 1 and 2. Notice that we are not mentioning the atomic level 3.

The ω_3 pulse yields in a very short time τ the following evolution:

$$|\Psi_{(1)}\rangle \to |\Psi'_{(1)}\rangle = \cos\left(\frac{\Omega t}{2}\right)|\phi_1\rangle \otimes |1\rangle + i\sin\left(\frac{\Omega t}{2}\right)|\phi_2\rangle \otimes |0\rangle, \tag{5.4}$$

where $|1\rangle$ denotes a one-photon state. Equation 5.4 is a generalized spectral decomposition, in the sense explained in the earlier subsection on that subject (in FIGURE 3, channels α and β coincide and correspond to the vacuum, whereas channel γ represents the one-photon state). The analysis can now proceed along the lines sketched in the previous section.

Observe also that, by repeating the reasoning outlined in the third section, the same result is obtained independently of whether a photon of frequency ω_3 is observed *only* at the final step, after the N-th optical pulse was irradiated (like in reference 8), or after every pulse irradiation. This was discussed at length in reference 15.

DISCUSSION

We have shown that the QZE is liable to a purely dynamical explanation that does not involve any projection operator. Therefore, against widespread belief, we claim that a quantum Zeno-type dynamics is *not* an argument in support of the "collapse of the wave function", provided we observe the same state as the initial one at the final detector D_0. The Schrödinger equation alone can yield a satisfactory explanation of the phenomenon.

Even though we do not question in the least the effectiveness and the practical validity of the projection postulate, we have critically discussed its physical meaning on several occasions.[17,19] We believe that a projection does not correspond to any *physical* operation and thus should be regarded only as a convenient expedient (a "working rule") in order to account for the loss of quantum mechanical coherence (the "collapse" of the wave function). In this sense, von Neumann's projection rule is to be considered as purely mathematical and no physical meaning should be ascribed to it. In our opinion, the projection technique is artificial and extraneous to quantum mechanics as a *physical* theory.

We stress, in this context, that an alternative explanation for the loss of quantum coherence has been proposed,[17,19] in which the decoherence ("collapse") is viewed as a *physical* dephasing process, ascribable to the interaction of the quantum system with a macroscopic object. Notice that, in this approach, the macroscopic system is always treated quantum mechanically and the unitarity of the evolutions is always kept. Dephasing is viewed as a statistical effect, even though it can also be shown to take place (and can be given a definite meaning) for single events.

It is worth stressing that our analysis has been performed under the assumption of "lossless" and "reflectionless" GSD's. In order to realize practically this type of experiment, we have to estimate the effects of such "losses" and "reflections" on the final results. It would also be interesting to understand whether these effects would yield additional phases in the transmitted states. Indeed, in such a case, interesting links with "decoherence" effects[17] might come to light due to (partial) phase randomization. The connection between decoherence, dephasing, and the occurrence of an exponential behavior for very short times has been discussed elsewhere[26] and has interesting spin-offs on the analysis of quantum measurements.[3]

Finally, it is worth noting that there is also another reason for why the $N \to \infty$ limit is physically unattainable. Because of Heisenberg's uncertainty relations, the condition of equation 3.11 must be investigated in great detail.[27] Indeed, it is reasonable to assume that ℓ cannot be made smaller than the spread of the wave packet of the incoming neutron. On the other hand, this bounds the uncertainty on the neutron speed and makes the $N \to \infty$ limit in equation 3.11 impossible to realize from the conceptual point of view.

REFERENCES AND NOTES

1. For old papers, see: NAMIKI, M. & N. MUGIBAYASHI. 1953. Prog. Theor. Phys. **10:** 474; KHALFIN, L. A. 1958. Zh. Eksp. Teor. Fiz. **33:** 1371 (Sov. Phys. JETP **6:** 1053). For a more recent account, see: CHO, G-C., H. KASARI & Y. YAMAGUCHI. 1993. Prog. Theor.

Phys. **90:** 803, in which one can find other references on theories and experiments on the decay of quantum systems.
2. VAN HOVE, L. 1955. Physica **21:** 517.
3. LEGGETT, A. J. 1993. *In* Proceedings of the Fourth International Symposium on Foundations of Quantum Mechanics. M. Tsukada *et al.*, Eds.: 10. Phys. Soc. Japan. Kyoto.
4. MISRA, B. & E. C. G. SUDARSHAN. 1977. J. Math. Phys. **18:** 756. See also: CHIU, C. B., B. MISRA & E. C. G. SUDARSHAN. 1977. Phys. Rev. **D16:** 520; 1982. Phys. Lett. **117B:** 34.
5. KHALFIN, L. A. 1982. Phys. Lett. **112:** 223; HORWITZ, L. P. & E. KATZNELSON. 1983. Phys. Rev. Lett. **50:** 1184; FONDA, L., G. C. GHIRARDI & T. WEBER. 1983. Phys. Lett. **131B:** 309; BERNARDINI, C., L. MAIANI & M. TESTA. 1993. Phys. Rev. Lett. **71:** 2687.
6. KHALFIN, L. A. 1990. Usp. Fiz. Nauk **160:** 185. See also: DEGASPERIS, A., L. FONDA & G. C. GHIRARDI. 1974. Nuovo Cimento **A21:** 471; FONDA, L., G. C. GHIRARDI & A. RIMINI. 1978. Rep. Prog. Phys. **41:** 587.
7. COOK, R. J. 1988. Phys. Scr. **T21:** 49.
8. ITANO, W. H., D. J. HEINZEN, J. J. BOLLINGER & D. J. WINELAND. 1990. Phys. Rev. **A41:** 2295.
9. PETROSKY, T., S. TASAKI & I. PRIGOGINE. 1990. Phys. Lett. **A151:** 109; 1991. Physica **A170:** 306.
10. PERES, A. & A. RON. 1990. Phys. Rev. **A42:** 5720.
11. BALLENTINE, L. E. 1991. Phys. Rev. **A43:** 5165.
12. ITANO, W. H., D. J. HEINZEN, J. J. BOLLINGER & D. J. WINELAND. 1991. Phys. Rev. **A43:** 5168.
13. INAGAKI, S., M. NAMIKI & T. TAJIRI. 1992. Phys. Lett. **A166:** 5.
14. HOME, D. & M. A. B. WHITAKER. 1993. Phys. Lett. **A173:** 327.
15. PASCAZIO, S., M. NAMIKI, G. BADUREK & H. RAUCH. 1993. Phys. Lett. **A179:** 155.
16. VON NEUMANN, J. 1932. Die Mathematische Grundlagen der Quantenmechanik. Springer-Verlag. Berlin.
17. MACHIDA, S. & M. NAMIKI. 1980. Prog. Theor. Phys. **63:** 1457 & 1833; NAMIKI, M. 1988. Found. Phys. **18:** 29; NAMIKI, M. & S. PASCAZIO. 1991. Phys. Rev. **A44:** 39; 1993. Phys. Rep. **232:** 301.
18. PASCAZIO, S. & M. NAMIKI. 1994. Dynamical quantum Zeno effect. Phys. Rev. A. In press.
19. NAMIKI, M., S. PASCAZIO & C. SCHILLER. 1994. Phys. Lett. **A187:** 17.
20. PERES, A. 1980. Am. J. Phys. **48:** 931.
21. WIGNER, E. P. 1963. Am. J. Phys. **31:** 6.
22. SUMMHAMMER, J., G. BADUREK, H. RAUCH, U. KLISCHKO & A. ZEILINGER. 1983. Phys. Rev. **A27:** 2523; BADUREK, G., H. RAUCH & J. SUMMHAMMER. 1983. Phys. Rev. Lett. **51:** 1015.
23. JAUCH, J. M., E. P. WIGNER & M. M. YANASE. 1967. Nuovo Cimento **48B:** 144. Negative-result measurements were used as a counterargument against the so-called ergodic-amplification theory, proposed by: DANERI, A., A. LOINGER & G. M. PROSPERI. 1962. Nucl. Phys. **33:** 297.
24. The meaning and practical realizability of a π (spin-flip) pulse is discussed by: SCULLY, M. O. & H. WALTHER. 1989. Phys. Rev. **A39:** 5229. See also: HAHN, E. 1950. Phys. Rev. **80:** 580.
25. Notice that Itano *et al.* checked the somewhat different formula, $P_1^{(N)}(T) = [1 + \cos^N(\pi/N)]/2$. The two expressions have a similar behavior for large N. See references 7 and 8.
26. NAKAZATO, H., M. NAMIKI & S. PASCAZIO. 1994. Phys. Rev. Lett. **73:** 1063.
27. We thank H. Rauch for discussions on this point. Work is now in progress to clarify the above argument.

Some Studies about Quantum Zeno Effect[a]

YONG-DE ZHANG,[b,c] JIAN-WEI PAN,[c] AND H. RAUCH[d]

[b]*CCAST World Laboratory*
Beijing, 100080, People's Republic of China

[c]*Department of Modern Physics*
University of Science and Technology of China
Hefei, Anhui, 230027, People's Republic of China

[d]*Atominstitut der Österreichischen Universitäten*
A-1020 Vienna, Austria

INTRODUCTION

Quantum Zeno Effect (QZE) was theoretically suggested by Misra and Sudarshan,[1] who proved that the frequent measurements would inhibit the decay of unstable quantum systems. However, as is well known, this effect could never be observed in those quantum systems whose time-dependent laws can strictly obey the negative exponential decay law. Chiu *et al.*[2,3] have pointed out that, for any unstable quantum system, there is a time scale t_1, in the time region $(0, t_1)$, where its decay law has a significant departure from the negative exponential law. By the Flemming rule,[4] Peres estimated that the time of transformation (t_1) is about 10^{-16} δ and 10^{-21} δ, respectively, for the typical unstable systems of atoms and nuclei.[5] However, now neither the resolving time of the apparatus nor the time interval of successive measurements can reach the order of magnitude given above. This situation has brought the experimental difficulty to test this effect. In fact, the conclusion obtained in reference 1 is also suitable for transitions between quantum states; that is, a similar proof process can show that the frequent measurements would "freeze" the quantum system in its initial state and would suppress or inhibit the transitions to other quantum states.[6] Therefore, Cook reexamined QZE from the viewpoint of transitions between atomic states and proposed an actual experimental scheme.[7] In a recent paper, Itano *et al.*[8] claimed to have observed experimentally QZE on the basis of Cook's proposal. This conclusion has been analyzed in references 9–13 from different standpoints; an argument among the above references is that the same result can be obtained by using the notion of "wave function collapse" by measurements[14] or simply regarding the phenomenon as just the natural consequence of a dynamics evolution about the wave function.

In view of the above arguments, it is necessary in order to reveal the genuine meaning of QZE to find new physical examples and design the corresponding experiments. Because the proof of Misra *et al.* is not direct and concise, we give a new proof here, which is more obvious and simple. By references 7, 9, and 11, we easily

[a]This work was supported by a cooperation project of NSF of China and FWF of Austria.

know that, no matter from what standpoint, the disturbance will have a notable influence on those quantum systems whose "survival probability" $P(t)$ has a significant departure from the negative exponential law. Thus, these quantum systems may be used as the proper research objects; by using them, we could find the differences between the two standpoints and could reveal the genuine meaning of QZE.

A THEORETICAL PROOF AND SOME ANALYSES ABOUT QZE

Before starting our analysis, let us introduce the appointment of QZE: "we shall state that a QZE occurs if the 'survival' probability of the initial state increases with N".[15]

Because the usual quantum mechanics could not give a satisfactory description about the reverse effect of the quantum system to the measurement apparatus, it is necessary to use the classical terminology to describe the measurement result; the description of the measurement effect on the quantum system is given by the decomposition law of the probability amplitude.[16–18] According to this law, we could obtain three kinds of meanings for a one-time measurement of some observable quantity: first, measurement means to make a spectra decomposition for the measured wave function with the eigenfunction series of the corresponding mechanical quantity operator; second, the measured wave function will collapse (or transit) to one of these eigenstates (of course, if the measured state has been one of the eigenstates, such collapse will not occur); third, the collapsed or transited state (no matter whether the collapse occurs or not) will independently restart its evolution by taking this moment as the initial moment—that is, this is the preparation of the initial state. It should be emphasized that, in the second and third meanings, the incoherence would be introduced inevitably, that is, there is not any coherence between the different collapses,[19] and the new initial state after the collapse will not inherit the coherence possessed by the original measured state. Just as pointed out in references 11 and 15, the spectra decomposition is not necessary to destroy the coherence of systems, but by the above statement we could say that each time of measurement completed always means the destruction of coherence (and the production of new coherence). Therefore, the incoherence introduced by measurement is the most important character of the quantum measurement effect.

Let us consider an unstable quantum system prepared at moment zero (its initial state is denoted as $|\psi\rangle$), where the evolution of the initial state is given by

$$(i\hbar)\frac{d|\psi(t)\rangle}{dt} = H(t)|\psi(t)\rangle. \quad (1)$$

Here, the Hamiltonian $H(t)$ depends on time; thus, this quantum system will decay or transit to the other states. We define

$$P(t) = |\langle\psi|\psi(t)\rangle|^2 \quad (2)$$

as the "survival probability" with which the system is still in its initial state. From equations 1 and 2, we have

$$\frac{dP(t)}{dt} = \frac{d}{dt}\{\langle\psi\,|\,\psi(t)\rangle\langle\psi(t)\,|\,\psi\rangle\}$$

$$= \langle\psi|\left[\frac{d}{dt}\,|\psi(t)\rangle\right]\langle\psi(t)\,|\,\psi\rangle + \langle\psi\,|\,\psi(t)\rangle\left[\frac{d}{dt}\langle\psi(t)\,|\right]|\psi\rangle$$

$$= \langle\psi\left|\frac{H(t)}{i\hbar}\right|\psi(t)\rangle\langle\psi(t)\,|\,\psi\rangle - \langle\psi\,|\,\psi(t)\rangle\langle\psi(t)\left|\frac{H(t)}{i\hbar}\right|\psi\rangle. \tag{3}$$

As $t \to 0$, we have $|\psi(t)\rangle \to |\psi\rangle$; thus,

$$\left.\frac{dP(t)}{dt}\right|_{t=0} = 0. \tag{4}$$

Equation 4 indicates that the initial transition ratio of an unstable state prepared just now to other states is zero.

However, from the usual negative exponential law of decay, we know

$$\left.\frac{dP(t)}{dt}\right|_{t=0} = -\lambda, \tag{5}$$

where λ is the decay constant. Equation 5 shows that the decay ratio at the initial moment is not zero and this contradicts equation 4. However, we know that equation 4 is deduced directly from the Schrödinger equation without using any other postulations. We have no reason to doubt it. Hence, at least we can say that, for the neighboring region of the initial moment, the decay law of the unstable system will deviate from the negative exponential law. In fact, the negative exponential law originates from the postulation of independence among the decaying particles[5] and from the approximation of the pole point.[2] Thus, the negative exponential law has two meanings: first, after a short-time evolution since the preparation, the coherence among particles has been relieved by various disturbances experienced by the particles during the evolution; therefore, as a statistical law, it is valid for any independent unstable particle ensemble regardless of the decay law of the single unstable particle; second, by the approximation of the pole point, we can say for most of the unstable particles after a very short-time evolution that the decay law of the single unstable particle will be approximately valid as well. These are the reasons why we cannot observe the obvious deviation from the negative exponential law in the usual statistical experiment of radioactive decay.

Now, we consider the following problem: When an unstable system is in state $|\psi\rangle$ at the initial moment, what is the survival probability $P_c(t)$ that the system will still remain undecayed in its initial state $|\psi\rangle$ after carrying out a continuous observation for a period of time t?

In order to calculate $P_c(t)$, we first consider the following case: The system begins to evolve from its initial state $|\psi\rangle$ and undergoes a series of measurements at the moments, t/n, $2t/n$, . . . , $(n-1)t/n$, t. We denote $P_n(t)$ as the probability that the system remains undecayed after these measurements. If the measurement times in the period $(0, t)$ tend to infinity and the limit $P_{n\to\infty}(t)$ exists, then we can naturally identify

$$P_c(t) = \lim_{n\to\infty} P_n(t). \tag{6}$$

From the viewpoint that the measurement certainly prepares the initial state, after being measured at moment t/n, the probability that this system will collapse back to its initial state is $P(t/n)$. Afterward, it will again begin its evolution from the initial state $|\psi\rangle$. Thus, we can obtain the following expression for $P_n(t)$:

$$P_n(t) = [P(t/n)]^n. \tag{7}$$

When n is large enough, we can write $P(t/n)$ as

$$P(t/n) \simeq P(0) + \left.\frac{dP}{dt}\right|_{t=0} \frac{t}{n}. \tag{8}$$

Substituting equations 7 and 8 into equation 6, we have

$$P_c(t) = \lim_{n\to\infty}\left(1 + \left.\frac{dP}{dt}\right|_{t=0} \frac{t}{n}\right)^n. \tag{9}$$

Therefore, we obtain

$$P_c(t) = \exp\left\{\left.\frac{dP}{dt}\right|_{t=0} t\right\}. \tag{10}$$

After considering equation 4 with equation 10, we have

$$P_c(t) = 1. \tag{11}$$

Equation 11 means the following: if an unstable system is undergoing the continuous measurements, it will remain in its initial state all the time. Its decay or transition to other states is inhibited completely. At least, in principle, we can delay the decay or suppress the transitions to other states by means of the frequent measurements. Therefore, we have completed the concise theoretical proof to QZE.

The aforementioned theoretical deduction about QZE shows that, provided the measurement is instantaneously completed (the definite measurement result is given), the QZE will be assured by equation 10, which is derived from the Schrödinger equation and the fact that the measurement is the preparation of the initial state. In fact, the incoherence (i.e., the destruction of the original coherence in the decomposition of the measured state) plays a decisive role in this process. However, from the general viewpoint, it should be emphasized that this is only the sufficient condition for the emergence of QZE. Just as stated above, the spectra decompositions do not necessarily destroy the coherence and, after the frequent spectra decompositions that preserve the coherence, the transitions between quantum states can be suppressed.[11,15] However, it should be pointed out that, after each time of spectra decomposition in the experiment of reference 15, the measurement result about the probability of neutron spin downward is only related with the wave function along the direction of flux and there is not the interference process with the wave functions of other directions; that is, here the so-called coherence does not affect the result of the experiment. Thus, the wave function along the direction of flux can be treated independently and as if this branch underwent a series of preparations of the initial state. Hence, in a generalized sense, the initial state preparation could include two cases for having the wave function collapse and only spectra decomposition in the above case. For this case, we easily find that the above proof process is also suitable.

Here, if an unstable system obeys the exact negative exponential law of decay, by combining equations 5 and 10 we can see that

$$P_c(t) = e^{-M} = P(t). \tag{12}$$

This shows that the measurements (even the continuous measurements) will not influence the negative exponential decay law; that is to say, QZE does not exist or can never be observed in principle. Of course, as it has been pointed out, the negative exponential law of decay is not exact about the neighborhood of the initial moment where this unstable state is prepared.

APPLICATION IN NEUTRON INTERFEROMETRY

Equation 12 shows that, in order to obviously observe QZE, it is necessary to find those quantum systems whose decay laws have significant departure from the negative exponential law. As an example, we consider the experiment of neutron spin state in the magnetic field.

Denoting the spin states of a neutron in the Z direction as $|\uparrow\rangle$ and $|\downarrow\rangle$, let us consider the following experiment where an incident neutron in the initial state $|\downarrow\rangle$ entered a static magnetic field in the direction of the X-axis. At the moment t, the system will evolve into

$$|\psi(t)\rangle = \exp(-i\mu B\sigma_1 t/\hbar)|\downarrow\rangle = \cos(\tfrac{1}{2}\omega t)|\downarrow\rangle - i\sin(\tfrac{1}{2}\omega t)|\uparrow\rangle, \tag{13}$$

where $\omega = 2\mu B/\hbar$, μ is the magnetic moment of the neutron, and σ_1 is the Pauli matrix. Then, the probability that the spin of the neutron remains in its initial state is

$$P(t) = \cos^2\left(\frac{\omega t}{2}\right). \tag{14}$$

Now, let us consider the following experiments where, in the period $(0, T)$ of evolution, the neutron spin state undergoes N times of measurements, which make the neutron spin take a certain direction (upward or downward, respectively) along the Z-axis. Also, assume that each time of measurement has a detection efficiency ξ_0 (≤ 1) and that this almost does not change the momentum of the neutron. Then, the flying direction of the neutron is still along the direction of the Z-axis after each time of measurement. Because the measurement partly destroyed the coherence of the spin superposition states of the neutron, the final state must be a mixed state, instead of the pure state. Now, we introduce a measurement operator ρ:

$$\rho = \sqrt{\xi_0}|\uparrow\rangle\langle\uparrow| \oplus \sqrt{\xi_0}|\downarrow\rangle\langle\downarrow| \oplus \sqrt{1-\xi_0}I. \tag{15}$$

Here, the direct sum symbol $\oplus$ means that, under the action of ρ, the system will collapse to the $|\uparrow\rangle$ or $|\downarrow\rangle$ state by the probability ξ_0, with the respective proportions besides the part of free evolution given by probability $(1-\xi_0)$, and there is not the coherence among the various alternatives.[19] Moreover, if the N points of the measurement positions are arranged with equal distance from each other, the probability amplitude that the neutron spin remains downward at the moment T is given by

$$A^{(N)}_{\downarrow\downarrow}(T) = \langle \downarrow | e^{-iH(T/n+1)/\hbar} \rho e^{-iH(T/n+1)/\hbar} \ldots \rho e^{-iH(T/n+1)/\hbar} | \downarrow \rangle. \tag{16}$$

For example, if $N = 1$, we have

$$A^{(1)}_{\downarrow\downarrow}(T) = \sqrt{\xi_0} A_{\downarrow\uparrow}\left(\frac{T}{2}\right) A_{\uparrow\downarrow}\left(\frac{T}{2}\right) \oplus \sqrt{\xi_0} A_{\downarrow\downarrow}\left(\frac{T}{2}\right) A_{\downarrow\downarrow}\left(\frac{T}{2}\right) \oplus \sqrt{1-\xi_0} A_{\downarrow\downarrow}(T), \tag{17}$$

where $A_{xy}(t) = \langle y | e^{-iHt/\hbar} | x \rangle$ (x, y could be $\downarrow$ or $\uparrow$). Therefore,

$$P_1(T) = \xi_0 i^2_{\downarrow\downarrow}\left(\frac{T}{2}\right) + \xi_0 i_{\downarrow\uparrow}\left(\frac{T}{2}\right) i_{\uparrow\downarrow}\left(\frac{T}{2}\right) + (1-\xi_0) i_{\downarrow\downarrow}(T), \tag{18}$$

where $i_{xy}(t) = |A_{xy}(t)|^2$. For an arbitrary N, we define

$$\vec{i}_{\downarrow,\text{in}}(t) = \begin{pmatrix} \sqrt{\xi_0} i_{\downarrow\downarrow}(t) \\ \sqrt{\xi_0} i_{\downarrow\uparrow}(t) \end{pmatrix},$$

$$\vec{i}_{\downarrow,\text{out}}(t) = \begin{pmatrix} \sqrt{\xi_0} i_{\downarrow\downarrow}(t) \\ \sqrt{\xi_0} i_{\uparrow\downarrow}(t) \end{pmatrix},$$

$$A(t) = \begin{pmatrix} \xi_0 i_{\downarrow\downarrow}(t) & \xi_0 i_{\downarrow\uparrow}(t) \\ \xi_0 i_{\uparrow\downarrow}(t) & \xi_0 i_{\uparrow\uparrow}(t) \end{pmatrix}.$$

Therefore, we have

$$\begin{aligned} P_N(T) &= (1-\xi_0)^N i_{\downarrow\downarrow}(T) + (1-\xi_0)^{N-1} \sum_{K_1=1}^{N} \vec{i}^{+}_{\downarrow,\text{in}}\left(\frac{K_1 T}{N+1}\right) \vec{i}_{\downarrow,\text{out}}\left(\frac{N+1-K_1}{N+1}T\right) \\ &+ (1-\xi_0)^{N-2} \sum_{1=K_1\langle K_2\rangle}^{N} \vec{i}^{+}_{\downarrow,\text{in}}\left(\frac{K_1 T}{N+1}\right) A\left(\frac{K_2-K_1}{N+1}T\right) \vec{i}_{\downarrow,\text{out}}\left(\frac{N+1-K_2}{N+1}T\right) \\ &+ \cdots \\ &+ (1-\xi_0) \sum_{1=K_1\langle\cdots\rangle K_{N-1}}^{N} \vec{i}^{+}_{\downarrow,\text{in}}\left(\frac{K_1 T}{N+1}\right) A\left(\frac{K_2-K_1}{N+1}T\right) \ldots A\left(\frac{K_{N-1}-K_{N-2}}{N+1}T\right) \\ &\cdot \vec{i}_{\downarrow,\text{out}}\left(\frac{N+1-K_{N-1}}{N+1}T\right) + \sum_{1=K_1\langle\cdots\rangle K_N}^{N} \vec{i}^{+}_{\downarrow,\text{in}}\left(\frac{K_1 T}{N+1}\right) \\ &\cdot A\left(\frac{K_2-K_1}{N+1}T\right) \ldots A\left(\frac{K_N-K_{N-1}}{N+1}T\right) \vec{i}_{\downarrow,\text{out}}\left(\frac{N+1-K_N}{N+1}T\right). \end{aligned} \tag{19}$$

If $\xi_0 = 1$, we return to the ideal case:

$$P'_N(T) = \vec{\imath}^{+}_{\downarrow,\text{in}}\left(\frac{T}{N+1}\right)\left[A\left(\frac{T}{N+1}\right)\right]^{N-1}\vec{\imath}_{\downarrow,\text{out}}\left(\frac{T}{N+1}\right). \tag{20}$$

The establishment of equation 19 enables us to solve the difficulty in performing the experiment scheme designed in references 13 and 15; that is, in order to assure the successive spectra decompositions, the intensity of neutron flux will be reduced again and again so that the number of events will markedly decrease. Because the above measurements do not decrease the intensity of neutron flux, the QZE could be easily tested provided that the $P_N(T)$ increases as the number N of measurements increase in this experiment.

DISCUSSION

In proving the existence of QZE, we can see that this effect is based only on two things: (1) the Schrödinger equation, which ensures that the speed of transition to the other states at zero moment is zero; (2) the fact that the measurement or spectra decomposition prepares the initial state as discussed above, which ensures that the system (or some branch of it) evolves from the initial state after every time of measurement (or spectra decomposition). These two conditions are the sufficient and necessary physical basis of QZE. Meanwhile, the analysis in this report shows that those quantum systems whose decay laws departed from the negative exponential law in a larger time region, for example, the $k_0 - \bar{k}_0$ system, the $B_0 - \bar{B}_0$ system, and the neutron spin system in the magnetic field, are proper research objects to test QZE. For further works about the $k_0 - \bar{k}_0$ system, see our discussion in references 20 and 21.

REFERENCES

1. MISRA, B. & E. C. G. SUDARSHAN. 1977. J. Math. Phys. **18:** 756.
2. CHIU, C. B., E. C. G. SUDARSHAN & B. MISRA. 1977. Phys. Rev. **D16:** 520.
3. CHIU, C. B., B. MISRA & E. C. G. SUDARSHAN. 1982. Phys. Lett. **B117:** 34.
4. FLEMMING, G. 1973. Nuovo Cimento **A16:** 232.
5. PERES, A. 1984. Found. Phys. **14:** 1131.
6. INAGAKI, S., M. NAMIKI & T. TAJIRI. 1992. Phys. Lett. **A166:** 5.
7. COOK, R. J. 1988. Phys. Scr. **T21:** 49.
8. ITANO, W. H., D. J. HEINZEN, J. J. BOLLIGER & D. J. WINELAND. 1990. Phys. Rev. **A41:** 2295.
9. PETROSKY, T., S. TASAKI & I. PRIGOGINE. 1990. Phys. Lett. **A151:** 109.
10. PERES, A. & A. RON. 1990. Phys. Rev. **A42:** 5720.
11. BALLENTINE, L. E. 1991. Phys. Rev. **A43:** 5165.
12. ITANO, W. H., D. J. HEINZEN, J. J. BOLLIGER & D. J. WINELAND. 1991. Phys. Rev. **A43:** 5168.
13. HOME, D. & M. A. B. WHITAKER. 1993. Phys. Lett. **A173:** 327.
14. VON NEUMANN, J. 1932. Die Mathematische Grundlagen der Quantenmechanik. Springer-Verlag. Berlin.
15. PASCAZIO, S., M. NAMIKI, G. BADUREK & H. RAUCH. 1993. Phys. Lett. **A179:** 155.
16. LEE, T. D. 1983. Particle Physics and Introduction to Field Theory. Harwood. Chichester.

17. LANDAU, L. D. & E. M. LIFSHITZ. 1977. Quantum Mechanics (Nonrelativistic Theory). Pergamon. Elmsford, New York.
18. BJORKEN, J. D. & S. D. DRELL. 1964. Relativistic Quantum Mechanics. McGraw–Hill. New York.
19. FEYNMAN, R. P. & A. R. HIBBS. 1965. Quantum Mechanics and Path Integrals. McGraw–Hill. New York.
20. ZHANG, Y-D., J-W. PAN & L. MA. 1995. Quantum Zeno effect in k_s regeneration and the data reprocessing of the CP violation experiments. Phys. Rev. A. Submitted.
21. PAN, J-W. & Y-D. ZHANG. 1995. Quantum Zeno effect in oscillation of $k_0 - \bar{k}_0$ system in a medium. Phys. Rev. A. Submitted.

Protective Measurements[a]

YAKIR AHARONOV[b,c] AND LEV VAIDMAN[b]

[b]*School of Physics and Astronomy*
Raymond and Beverly Sackler Faculty of Exact Sciences
Tel Aviv University
Tel Aviv, 69978 Israel

[c]*Department of Physics*
University of South Carolina
Columbia, South Carolina 29208

INTRODUCTION

Recently, we have proposed protective measurements[1,2] that allow the measuring of the Schrödinger wave of a single particle. We have argued that the possibility of such measurements tells us that a quantum state has more physical meaning than is usually assumed; that is, the Schrödinger wave is real in some sense. A quantum state is not only a statistical property of an ensemble, but it is a property of a single system.

Also, in recent years, we developed an approach in which a quantum system is described, at a given time, by two (instead of one) quantum states: the usual one evolving toward the future and the second evolving backwards in time from a future measurement.[3–8] This approach proved itself fruitful at least for describing measurements performed on preselected and postselected ensembles. In this approach, the vector describing a quantum system at a given time consists of two states.

The following questions arise: Is there a contradiction between these two approaches? Which description is appropriate: the standard, single-state or our two-state description? Does the two-state vector have physical meaning for a single system? Is it possible to measure this vector on a single system? In this work, we shall try to answer these questions.

In the following section, we present our method of protective measurements of a single quantum state. This is followed by a brief review of the two-state vector formalism. Then, the main result of this work—the method of protective measurement of a two-state vector—is presented. We conclude with a discussion of the obtained results.

MEASUREMENT OF THE SCHRÖDINGER WAVE OF A SINGLE PARTICLE

At present, the commonly accepted interpretation of the Schrödinger wave is due to Born. He proposed to interpret the wave intensity not as the density of distribution of actual matter, as Schrödinger first imagined, but as a probability density for the presence of a particle. Schrödinger, however, wanted to believe that his wave

[a]This research was supported in part by Grant No. 425/92-1 of the Basic Research Foundation (administered by the Israel Academy of Sciences and Humanities).

represents a single particle: the wave is an extended object really moving in space. Born's interpretation was supported by the fact that nobody knew how to measure the density of the Schrödinger wave on a single system. There was a general belief that the Schrödinger wave could only be tested for an ensemble of particles. We have proposed a new type of measurement: "protective measurements" that allow direct measurement of the Schrödinger wave density on a single particle. We have shown that one can simultaneously measure the density and the current of the Schrödinger wave in many locations. The results of these measurements then allow the reconstruction of the Schrödinger wave.

The simplest protection procedure is introducing a protective potential such that the quantum state of the system will be a nondegenerate eigenstate of the Hamiltonian. In fact, in many important cases, this protection is given by nature: almost isolated systems will eventually decay to their ground state or to some stable excited state.

As an example of a simple protective measurement, let us consider a particle in a discrete nondegenerate energy eigenstate $\Psi(x)$. The standard von Neumann procedure for measuring the value of an observable A in this state involves an interaction Hamiltonian,

$$H = g(t)PA, \tag{1}$$

coupling the system to a measuring device, or pointer, with coordinate and momentum denoted, respectively, by Q and P. The time-dependent coupling $g(t)$ is normalized to $\int g(t)dt = 1$ and the initial state of the pointer is taken to be a Gaussian centered around zero.

In standard impulsive measurements, $g(t) \neq 0$ for only a very short time interval. Thus, the interaction term dominates the rest of the Hamiltonian, and the time evolution $e^{-(i/\hbar)PA}$ leads to a correlated state: eigenstates of A with eigenvalues a_n are correlated to measuring device states in which the pointer is shifted by these values a_n. By contrast, the protective measurements of interest here utilize the opposite limit of extremely slow measurement. We take $g(t) = 1/T$ for most of the time T and assume that $g(t)$ goes to zero gradually before and after the period T. We choose the initial state of the measuring device such that the canonical conjugate P (of the pointer variable Q) is bounded. We also assume that P is a constant of motion not only of the interaction Hamiltonian (equation 1), but of the whole Hamiltonian. For $g(t)$ smooth enough, we obtain an adiabatic process in which the particle cannot make a transition from one energy eigenstate to another and, in the limit $T \to \infty$, the interaction Hamiltonian does not change the energy eigenstate. For any given value of P, the energy of the eigenstate shifts by an infinitesimal amount given by first-order perturbation theory:

$$\delta E = \langle H_{\text{int}} \rangle = \frac{\langle A \rangle P}{T}. \tag{2}$$

The corresponding time evolution $e^{-iP\langle A \rangle/\hbar}$ shifts the pointer by the average value $\langle A \rangle$. This result contrasts with the usual (strong) measurement in which the pointer shifts by one of the eigenvalues of A. By measuring the averages of a sufficiently large number of variables A_n, the full Schrödinger wave $\Psi(x)$ can be reconstructed to any desired precision.

As a specific example, we take the A_n terms to be (normalized) projection operators on small regions V_n having volume v_n:

$$A_n = \begin{cases} \frac{1}{v_n}, & \text{if } x \in V_n, \\ 0, & \text{if } x \notin V_n. \end{cases} \tag{3}$$

The measurement of A_n yields

$$\langle A_n \rangle = \left(\frac{1}{v_n}\right) \int_{V_n} |\Psi|^2 dv = |\Psi_n|^2, \tag{4}$$

where $|\Psi_n|^2$ is the average of the density $\rho(x) = |\Psi(x)|^2$ over the small region V_n. Performing measurements in sufficiently many regions V_n, we can reconstruct $\rho(x)$ everywhere in space. (Simultaneous measurement of all the variables A_n requires slower and weaker interactions and thus takes more time.) For a real state, the density $\rho(x)$ is itself enough to reconstruct the Schrödinger wave; we can fix the sign by flipping it across nodal surfaces.

In the general case, however, we have to measure current density in addition to measurements of the density $\rho(x)$. This time we also adiabatically measure the averages of

$$B_n = \left(\frac{1}{2i}\right)(A_n \nabla + \nabla A_n). \tag{5}$$

Indeed, the $\langle B_n \rangle$ terms are the average values of the current $j = (1/2i) \cdot (\Psi^* \nabla \Psi - \Psi \nabla \Psi^*)$ in the region V_n. Writing $\Psi(x) = r(x)e^{i\theta(x)}$ with $r(x) = \sqrt{\rho(x)}$, we find that

$$\frac{j(x)}{\rho(x)} = \nabla \theta \tag{6}$$

and the phase $\theta(x)$ can be found by integrating j/ρ.

For a charged particle, the density $\rho(x)$ times the charge yields the effective charge density. In particular, it means that an appropriate adiabatic measurement of the Gauss flux out of a certain region must yield the expectation value of the charge inside this region (the integral of the charge density over this region). Likewise, adiabatic measurement of the Ampere contour integral yields the expectation value of the total current flowing through this contour in the stationary case.

Our procedure is not applicable to degenerate energy eigenstates. The simplest way to deal with this case is by adding a potential (as part of the measuring procedure) to lift the degeneracy. This protection does not change the state. However, one can argue that it changes the physical situation. We can bring this change to a minimum by adding strong protection potential for a dense set of very short time intervals. Thus, most of the time, the system has not only the same state, but also the original potential.

We can measure even a superposition of energy eigenstates by a similar procedure. We add a dense set of time-dependent potentials acting for very short periods of time such that the state at all these times is the eigenstate of the Hamiltonian together with the additional potential. Still, most of the time, the system evolves under the free Hamiltonian. The proof of the efficiency of the above strong impulsive potentials is similar to the proof of the Zeno "paradox" in which a quantum system under a dense set of observations evolves in accordance with the evolution tested and not according to the free Hamiltonian. In our case, the two evolutions are identical.

TWO-STATE VECTOR DESCRIPTION OF A QUANTUM SYSTEM

In 1964, Aharonov, Bergmann, and Lebowitz[9] considered measurements performed on a quantum system between two other measurements, results of which were given. They proposed describing the quantum system between two measurements by using two states: the usual one, evolving towards the future from the time of the first measurement, and a second state evolving backwards in time from the time of the second measurement. If a system has been prepared at time t_1 in a state $|\Psi_1\rangle$ and is found at time t_2 in a state $|\Psi_2\rangle$, then at time t, $t_1 < t < t_2$, the system is described by

$$\langle\Psi_2|e^{i\int_{t_2}^{t} H dt} \text{ and } e^{-i\int_{t_1}^{t} H dt}|\Psi_1\rangle. \tag{7}$$

For simplicity, we shall consider the free Hamiltonian to be zero; then, the system at time t is described by the two states $\langle\Psi_2|$ and $|\Psi_1\rangle$. In order to obtain such a system, we prepare an ensemble of systems in the state $|\Psi_1\rangle$, perform a measurement of the desired variable using separate measuring devices for each system in the ensemble, and perform the postselection measurement. If the outcome of the postselection was not the desired result, we discard the system and the corresponding measuring device. We look only at measuring devices corresponding to the systems postselected in the state $\langle\Psi_2|$.

The basic concept of the two-state approach, the weak value of a physical variable A in the time interval between preselection of the state $|\Psi_1\rangle$ and postselection of the state $|\Psi_2\rangle$, is given by

$$A_{\mathrm{w}} \equiv \frac{\langle\Psi_2|A|\Psi_1\rangle}{\langle\Psi_2|\Psi_1\rangle}. \tag{8}$$

Let us present the main idea by way of a simple example. We consider, at time t, a quantum system that was prepared at time t_1 in the state $|B = b\rangle$ and that was found at time t_2 in the state $|C = c\rangle$ ($t_1 < t < t_2$). The measurements at times t_1 and t_2 are complete measurements of, in general, noncommuting variables B and C. The free Hamiltonian is zero and therefore the first quantum state at time t is $|B = b\rangle$. In the two-state approach, we characterize the system at time t by the backward-evolving state $\langle C = c|$ as well. Our motivation for including the future state is as follows: if we know that a measurement of C has been performed at time t, then the outcome is $C = c$ with probability 1. This intermediate measurement, however, destroys our knowledge that $B = b$ because the coupling of the measuring device to the variable C

can change B. The idea of weak measurements is to make the coupling with the measuring device sufficiently weak so that B does not change. In fact, we require that both quantum states do not change, neither the usual one $|B = b\rangle$ evolving towards the future nor $\langle C = c|$ evolving backwards.

During the whole time interval between t_1 and t_2, both $B = b$ and $C = c$ are true (in some sense). However, then $B + C = b + c$ must also be true. The latter statement, though, might not have meaning in the standard quantum formalism because the sum of the eigenvalues $b + c$ might not be an eigenvalue of the operator $B + C$. An attempt to measure $B + C$ using a standard measuring procedure will lead to some change of the two quantum states and thus the outcome will not be $b + c$. A weak measurement, however, will yield $b + c$.

When the "strong" value of an observable is known with certainty, that is, we known the outcome of an ideal (infinitely strong) measurement with probability 1, the weak value is equal to the strong value. Let us analyze the example above. The strong value of B is b, its eigenvalue. The strong value of C is c, as we know from retrodiction. From the definition (equation 1), it immediately follows that $B_w = b$ and $C_w = c$. However, weak values, unlike strong values, are defined not just for B and C, but for all operators. The strong value of the sum $B + C$ when $[B, C] \neq 0$ is not defined, but the weak value of the sum is defined: $(B + C)_w = b + c$.

The system at time t in a preselected and postselected ensemble is defined by two states, the usual one evolving from the time of the preparation and the state evolving backwards in time from the postselection. We may neglect the free Hamiltonian if the time between the preselection and the postselection is very short. Consider a system that has been preselected in a state $|\Psi_1\rangle$ and shortly afterwards postselected in a state $|\Psi_2\rangle$. The weak value of any physical variable A in the time interval between the preselection and the postselection is given by equation 8. Let us show briefly how weak values emerge from a measuring procedure with a sufficiently weak interaction.

We consider a sequence of measurements: a preselection of $|\Psi_1\rangle$, a (weak) measurement interaction of the form of equation 1, and a postselection measurement finding the state $|\Psi_2\rangle$. The state of the measuring device (which was initially in a Gaussian state) after this sequence is given (up to normalization) by

$$\Phi(Q) = \langle\Psi_2|e^{-iPA}|\Psi_1\rangle e^{-Q^2/2\Delta^2}. \tag{9}$$

After simple algebraic manipulation, we can rewrite it (in the P-representation) as

$$\tilde{\Phi}(P) = \langle\Psi_2|\Psi_1\rangle e^{-iA_wP}e^{-\Delta^2P^2/2} + \langle\Psi_2|\Psi_1\rangle \sum_{n=2}^{\infty}\left[\frac{(iP)^n}{n!}\right][(A^n)_w - (A_w)^n]e^{-\Delta^2P^2/2}. \tag{10}$$

If Δ is sufficiently large, then we can neglect the second term of equation 10 when we Fourier-transform back to the Q-representation. Large Δ corresponds to weak measurement in the sense that the interaction Hamiltonian (equation 1) is small. Thus, in the limit of weak measurement, the final state of the measuring device (in the Q-representation) is

$$\Phi(Q) = (\Delta^2\pi)^{-1/4}e^{-(Q-A_w)^2/2\Delta^2}. \tag{11}$$

This state represents a measuring device pointing to the weak value, A_w.

Although we have shown this result for a specific von Neumann model of measurements, the result is completely general: any coupling of a preselected and postselected system to a variable A, provided the coupling is sufficiently weak, results in effective coupling to A_w. This weak coupling between a single system and the measuring device will not, in most cases, lead to a distinguishable shift of the pointer variable, but collecting the results of measurements on an ensemble of preselected and postselected systems will yield the weak values of a measured variable to any desired precision.

When the strength of the coupling to the measuring device goes to zero, the outcomes of the measurement invariably yield the weak value. To be more precise, a measurement yields the real part of the weak value. Indeed, the weak value is, in general, a complex number, but its imaginary part will contribute only a phase to the wave function of the measuring device in the position representation of the pointer. Therefore, the imaginary part will not affect the probability distribution of the pointer position, which is what we see in a usual measurement. However, the imaginary part of the value also has physical meaning. It expresses itself as a change in the conjugate momentum of the pointer variable.[7]

Let us consider a measurement of a spin component of a spin-1/2 particle. We shall consider a particle prepared in the initial state spin "up" in the $\hat{x}$ direction and postselected to be "up" in the $\hat{y}$ direction. At the intermediate time, we measure, weakly, the spin component in the $\hat{\xi}$ direction, which is the bisector of $\hat{x}$ and $\hat{y}$, that is, $\sigma_\xi = (\sigma_x + \sigma_y)/\sqrt{2}$. Thus, $|\Psi_1\rangle = |\uparrow_x\rangle$, $|\Psi_2\rangle = |\uparrow_y\rangle$, and the weak value of σ_ξ in this case is

$$(\sigma_\xi)_w = \frac{\langle \uparrow_y | \sigma_\xi | \uparrow_x \rangle}{\langle \uparrow_y | \uparrow_x \rangle} = \left(\frac{1}{\sqrt{2}}\right) \frac{\langle \uparrow_y | (\sigma_x + \sigma_y) | \uparrow_x \rangle}{\langle \uparrow_y | \uparrow_x \rangle} = \sqrt{2}. \tag{12}$$

This value, of course, is "forbidden" in the standard interpretation where a spin component can obtain the (eigen)values ± 1 only.

The Hamiltonian for measuring σ_ξ is

$$H = g(t) P \sigma_\xi. \tag{13}$$

After the measuring interaction, the quantum state of the system and the pointer of the measuring device is

$$\cos(\pi/8) |\uparrow_\xi\rangle e^{-(Q-1)^2/2\Delta^2} + \sin(\pi/8) |\downarrow_\xi\rangle e^{-(Q+1)^2/2\Delta^2}. \tag{14}$$

The probability distribution of the pointer position, if it is observed now without postselection, is the sum of the distributions for each spin value. It is, up to normalization,

$$\text{prob}(Q) = \cos^2(\pi/8) e^{-(Q-1)^2/\Delta^2} + \sin^2(\pi/8) e^{-(Q+1)^2/\Delta^2}. \tag{15}$$

In the usual strong measurement, $\Delta \ll 1$. In this case, the probability distribution of the pointer is localized around -1 and $+1$ and it is strongly correlated to the values of the spin, $\sigma_z = \pm 1$.

Weak measurement corresponds to a Δ that is much larger than the range of the eigenvalues, that is, $\Delta \gg 1$. The pointer distribution has a large uncertainty and it is peaked between the eigenvalues—more precisely, at the expectation value $\langle \uparrow_x | \sigma_\xi | \uparrow_x \rangle = 1/\sqrt{2}$. An outcome of an individual measurement usually will not be close to this number, but it can be found from an ensemble of such measurements. Note that we have not yet considered the postselection.

In order to simplify the analysis of measurements on the preselected and postselected ensemble, let us assume that we first make the postselection of the spin of the particle and only then look at the pointer of the device that weakly measures σ_ξ. We must get the same result as if we first look at the outcome of the weak measurement, make the postselection, and discard all readings of the weak measurement corresponding to the cases in which the result is not $\sigma_y = 1$. The postselected state of the particle in the σ_ξ representation is $|\uparrow_y\rangle = \cos(\pi/8)|\uparrow_\xi\rangle - \sin(\pi/8)|\downarrow_\xi\rangle$. The state of the measuring device after the postselection of the spin state is obtained by projection of equation 14 onto the postselected state:

$$\Phi(Q) = \mathcal{N}[\cos^2(\pi/8)e^{-(Q-1)^2/2\Delta^2} - \sin^2(\pi/8)e^{-(Q+1)^2/2\Delta^2}], \qquad (16)$$

where $\mathcal{N}$ is a normalization factor. The probability distribution of the pointer variable is given by

$$\text{prob}(Q) = \mathcal{N}^2[\cos^2(\pi/8)e^{-(Q-1)^2/2\Delta^2} - \sin^2(\pi/8)e^{-(Q+1)^2/2\Delta^2}]^2. \qquad (17)$$

If the measuring interaction is strong, that is, $\Delta \ll 1$, then the distribution is localized around the eigenvalues ± 1 (mostly around 1 because the preselected and postselected probability to find $\sigma_\xi = 1$ is more than 85%); see FIGURES 1a and 1b. However, when the strength of the coupling is weakened, that is, Δ is increased, the distribution gradually changes to a single broad peak around $\sqrt{2}$, that is, the weak value; see FIGURES 1c–1e.

The width of the peak is large and therefore each individual reading of the pointer usually will be pretty far from $\sqrt{2}$. The physical meaning of the weak value, in this case, can be associated only with an ensemble of preselected and postselected particles. The accuracy of defining the center of the distribution goes as $1/\sqrt{N}$; thus, by increasing N (the number of particles in the ensemble), we can find the weak value with any desired precision (see FIGURE 1f).

PROTECTION OF A TWO-STATE VECTOR

We are familiar with measurements performed on a single system. In fact, the first work on weak measurements[3] considered such a case. We have shown how a single measurement of the spin component of a spin-N system could yield the "forbidden" value $\sqrt{2}N$ with the uncertainty $\sqrt{N}$. This is the weak value of S_ξ for the two-state vector $\langle S_y = N \| S_x = N\rangle$. Another example that we have investigated is the measurement of the kinetic energy of a tunneling particle.[10] We have shown for any precision of the measurement that we can ensure a negative value reading of the measuring device by an appropriate choice of the postselection state.

However, in these examples, there is no measurement of the two-state vector. If our measuring device for the spin measurement shows $\sqrt{2}N$, we cannot deduce that our two-state vector is $\langle S_y = N \| S_x = N\rangle$. Indeed, there are many other two-state

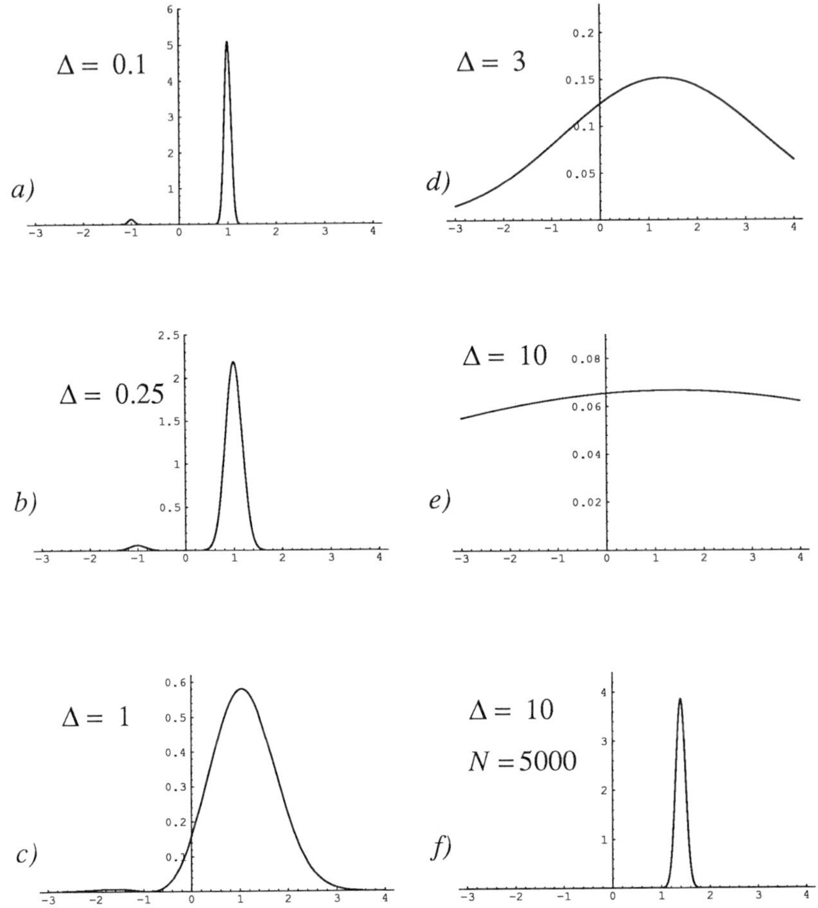

FIGURE 1. Measurement on the preselected and postselected ensemble. Probability distribution of the pointer variable for measurement of σ_ξ when the particle is preselected in the state $|\uparrow_x\rangle$ and postselected in the state $|\uparrow_y\rangle$. The strength of the measurement is parameterized by the width of the distribution, Δ: (*a*) $\Delta = 0.1$; (*b*) $\Delta = 0.25$; (*c*) $\Delta = 1$; (*d*) $\Delta = 3$; (*e*) $\Delta = 10$. (*f*) Weak measurement on the ensemble of 5000 particles; the original width of the peak, $\Delta = 10$, is reduced to $10/\sqrt{5000} \cong 0.14$. In the strong measurements (*a–b*), the pointer is localized around the eigenvalues ± 1; in the weak measurements (*c–e*), the peak of the distribution is located in the weak value $(\sigma_\xi)_w = \langle\uparrow_y|\sigma_\xi|\uparrow_x\rangle/\langle\uparrow_y|\uparrow_x\rangle = \sqrt{2}$. The outcomes of the weak measurement on the ensemble of 5000 preselected and postselected particles (*f*) are clearly outside the range of the eigenvalues $(-1, 1)$.

vectors that yield the same weak value for the spin component, but we cannot even claim that we have one of these vectors because the probability to obtain the "forbidden" outcome $S_\xi = \sqrt{2}N$ due to a statistical error of the measuring device is much higher. The same applies to the measurement of kinetic energy of a tunneling particle. The negative value shown by the measuring device usually is due to a statistical error and only in very rare cases does it correspond to a particle "caught" in the tunneling process.

We could try to use several weak measurements on a single preselected and postselected system in order to specify the two-state vector. However, in that case, these measurements will change the two-state vector. Therefore, as in the case of the measurement of the forward-evolving single-state vector of a single system, we need a protection procedure.

At first look, it seems that protection of a two-state vector is impossible. Indeed, if we add a potential that makes one state to be a nondegenerate eigenstate, then the other state, if it is different, cannot be an eigenstate too (the states of the two-state vector cannot be orthogonal). The Zeno-type protection does not work either: if we test that the system is in one state, then we know that it is not in another state. Nevertheless, protection of the two-state vector is possible, as we will show next.

The procedure for protection of a two-state vector of a given system is carried out by coupling the system to another preselected and postselected system. The protection procedure takes advantage of the fact that weak values might acquire complex values. Thus, the effective Hamiltonian of the protection might not be Hermitian. Non-Hermitian Hamiltonians act in different ways on quantum states evolving forward and backwards in time. This allows simultaneous protection of two different states (evolving in opposite time directions).

Let us start with the description of the protection of a two-state vector of a spin-½ particle considered previously, $\langle \uparrow_y \| \uparrow_x \rangle$. The protection procedure uses an external preselected and postselected system S of a large spin N that is coupled to our spin via the interaction

$$H_{\text{prot}} = -\mathbf{S} \cdot \sigma. \tag{18}$$

The external system is preselected in the state $|S_x = N\rangle$ and postselected in the state $\langle S_y = N|$; that is, it is described by the two-state vector $\langle S_y = N \| S_x = N\rangle$. When N is large and the interaction with our spin-½ particle is not too strong, the latter cannot change significantly the two-state vector of the protective system S, and the spin-½ particle "feels" the effective Hamiltonian in which S is replaced by its weak value,

$$\mathbf{S}_{\text{w}} = \frac{\langle S_y = N|(S_x, S_y, S_z)|S_x = N\rangle}{\langle S_y = N|S_x = N\rangle} = (N, N, iN). \tag{19}$$

Thus, the effective protective Hamiltonian is

$$H_{\text{eff}} = -N(\sigma_x + \sigma_y + i\sigma_z). \tag{20}$$

Straightforward calculations show that this (non-Hermitian) Hamiltonian has two (nonorthogonal) eigenstates: $|\uparrow_x\rangle$ (with eigenvalue $-N$) and $|\downarrow_y\rangle$ (with eigenvalue N). This result provides a certain test of our approach. When we consider the original problem given by the Hamiltonian (equation 18), we can easily see that if we start in the state $|\uparrow_x\rangle$ then all following measurements of σ_x must yield the value 1, whereas if we start with the state $|\downarrow_y\rangle$ then all following measurements of σ_y must yield the value -1.

However, for backward-evolving states, the effective Hamiltonian is the Hermitian conjugate of equation 20 and it has different eigenstates: $\langle \uparrow_y|$ (with eigenvalue

$-N$) and $\langle \downarrow_x|$ (with eigenvalue N). Again, it is easily seen that if the particle is postselected in the state $\langle \uparrow_y|$ then all preceding measurements of σ_y must yield $\sigma_y = 1$, whereas the postselection of $|\downarrow_x\rangle$ ensures $\sigma_x = -1$ for all preceding measurements.

The two-state vectors $\langle \uparrow_y \| \uparrow_x\rangle$ and $\langle \downarrow_x \| \downarrow_y\rangle$ do not change under the action of the Hamiltonian (equation 18). In order to prove that this Hamiltonian indeed provides the protection, we have to show that measuring interactions with the spin components of the particle will not lead to significant changes. For example, we must show that we can measure the weak value of $\sigma_\xi = (\sigma_x + \sigma_y)/\sqrt{2}$, which is $(\sigma_\xi)_w = \sqrt{2}$, on a single particle. [As previously shown,[5] without protection, this weak value is obtained only with an uncertainty that is larger than the observed value; therefore, in order to find the weak value, the preselected and postselected ensemble has to be used (see FIGURE 1).] The effective Hamiltonian during the measuring process is the sum of equation 1 and equation 20:

$$H_{\text{eff}} = -N(\sigma_x + \sigma_y + i\sigma_z) + \left(\frac{P}{\sqrt{2}}\right)(\sigma_x + \sigma_y). \tag{21}$$

For any realistic measurement, P is effectively bounded; thus, for N large enough, the second term will not change significantly the eigenvectors. The two-state vector $\langle \uparrow_y \| \uparrow_x\rangle$ will remain essentially unchanged during the measurement and therefore the measuring device on this single particle will yield $(\sigma_\xi)_w = \sqrt{2}$. This weak value by itself is not enough to establish the two-state vector, but we can perform several weak measurements such as $(\sigma_x)_w = 1$, $(\sigma_y)_w = 1$, and $(\sigma_z)_w = i$ that uniquely define the two-state vector.

We have shown that the Hamiltonian (equation 18), with an external system described by the two-state vector $\langle S_y = N \| S_x = N\rangle$, provides protection for the two-state vector $\langle \uparrow_y \| \uparrow_x\rangle$. It is not difficult to demonstrate that any two-state vector obtained by preselection and postselection of the spin-½ particle can be protected by the Hamiltonian (equation 18). A general form of the two-state vector is $\langle \uparrow_\beta \| \uparrow_\alpha\rangle$, where $\hat{\alpha}$ and $\hat{\beta}$ denote some directions. It can be verified by a straightforward calculation that the two-state vector $\langle \uparrow_\beta \| \uparrow_\alpha\rangle$ is protected when the two-state vector of the protective device is $\langle S_\beta = N \| S_\alpha = N\rangle$.

One can naively suggest the following simple explanation of the above procedure. We preselect the external system in a state $|S_\alpha = N\rangle$. Large N corresponds to the classical limit, so this is equivalent to a "magnetic" field in the $-\hat{\alpha}$ direction. Thus, the quantum states of the system under study evolving to the future "feel" this strong magnetic field. The state of the system, $|\uparrow_\alpha\rangle$, is a ground state and therefore it is protected. Similarly, for the states evolving backwards in time, there is a strong "magnetic" field in the $-\hat{\beta}$ direction, protecting the state $\langle \uparrow_\beta|$. However, this picture is too naive. Based on this argument, one would expect that, in addition to $|\uparrow_\alpha\rangle$, the forward-evolving state $|\downarrow_\alpha\rangle$ is also protected; however, this is not so. There exists another forward-evolving protected state, but it is $|\downarrow_\beta\rangle$. Also, in addition to $\langle \uparrow_\beta|$, there exists another protected backward-evolving state; however, it is $\langle \downarrow_\alpha|$ and not the expected state $\langle \downarrow_\beta|$.

The failure of this naive explanation does not allow a simple protection scheme of the two-state vector of an arbitrary quantum system, $\langle \Psi_2 \| \Psi_1\rangle$. According to this scheme, we construct a coupling of the system under study to the external system such that $|\Psi_1\rangle$ is a ground state when the external system is in a state $|\Phi_1\rangle$ and $\langle\Psi_2|$ is a ground (backward-evolving) state when the external system is in a state $\langle\Phi_2|$. The difficulty here, namely, that the postselection of the state $\langle\Psi_2|$ is impossible because usually in this situation $\langle\Phi_2|\Phi_1\rangle = 0$, cannot be naively solved by adding a tiny component of the preselected state to the postselected one, that is, postselecting $\langle\Phi_2| + \epsilon\langle\Phi_1|$ instead of $\langle\Phi_2|$. Even for ϵ very small, the backward-evolving state is not protected and therefore the two-state vector is not protected either.

The proper way for protecting a two-state vector of an arbitrary system is a generalization of the protection procedure of the two-state vector of a spin-½ particle described above. The task is to protect a two-state vector $\langle\Psi_2\|\Psi_1\rangle$. Let us decompose the postselected state $|\Psi_2\rangle = a|\Psi_1\rangle + b|\Psi_\perp\rangle$. Now, we can define "model spin" states: $|\Psi_1\rangle \equiv |\tilde{\uparrow}_z\rangle$ and $|\Psi_\perp\rangle \equiv |\tilde{\downarrow}_z\rangle$. On the basis of the two orthogonal states, we can obtain all other "model spin" states. For example, $|\tilde{\uparrow}_x\rangle = 1/\sqrt{2}(|\tilde{\uparrow}_z\rangle + |\tilde{\downarrow}_z\rangle)$ and then we can define the "spin model" operator $\tilde{\sigma}$. Now, the protection Hamiltonian, in complete analogy with the spin-½ particle case, is

$$H_{\text{prot}} = -\mathbf{S} \cdot \tilde{\boldsymbol{\sigma}}. \tag{22}$$

In order to protect the state $\langle\Psi_2\|\Psi_1\rangle$, the preselected state of the external system has to be $|S_z = N\rangle$ and the postselected state has to be $\langle S_\chi = N|$, where the direction $\hat{\chi}$ is defined by the "spin model" representation of the state $|\Psi_2\rangle$:

$$|\tilde{\uparrow}_\chi\rangle \equiv |\Psi_2\rangle = \langle\Psi_1|\Psi_2\rangle|\tilde{\uparrow}_z\rangle + \langle\Psi_\perp|\Psi_2\rangle|\tilde{\downarrow}_z\rangle. \tag{23}$$

For general quantum states $|\Psi_1\rangle$ and $\langle\Psi_2|$, the required protection is a gedanken-experiment. In general, the protection Hamiltonian (equation 22) generates nonlocal interactions that can contradict relativistic causality. However, what we investigate here is a conceptual question in the framework of nonrelativistic quantum theory, where any Hamiltonian is allowed.

CONCLUSIONS

We have shown in the framework of nonrelativistic quantum theory that we can measure (or, maybe a better word, "observe") two-state vectors describing preselected and postselected quantum systems. A number of (nonideal) measurements define the two-state vector and we have a procedure to protect the two-state vector from significant change due to these measurements. In order to protect, we have to know the two-state vector. Thus, this procedure is also liable to the criticism[11,12] leveled at our first proposal. Our response to this can be found in reference 13. Although we consider our present proposal as a measurement performed on a single system, it should also be mentioned that in any realistic practical implementation, we will need ensembles of particles, protective systems, and measuring devices. The external system of the protective device has to be not only prepared (preselected) in a certain state, but also postselected in a given state. In all interesting cases, the probability for an appropriate outcome of the postselection measurement is extremely small. Still, there is a nonzero probability that our first run with a single system, a single protective device, and a single set of measuring devices will yield the desired outcomes. In this case, we have a reliable measurement performed on a single system. However, even when we use a preselected ensemble, we actually use only a single preselected and postselected system. After achieving the first successful postselection, we have completed the experiment. For more discussion of this point, see reference 14.

It is interesting to notice that our procedure cannot protect a *generalized two-state vector*,[8] which is a superposition of two-state vectors. The system described by a generalized two-state vector is correlated to some external system. It seems that it is impossible to find any protective procedure of the generalized two-state vector that does not involve coupling to that external system. This feature hints that the generalized vector, although useful as a tool, is not a basic concept. The composite system consisting of the system under study and the system correlated to it is described by the usual, basic two-state vector.

Let us come back to the questions raised in the INTRODUCTION: is there a contradiction between "reality" of the Schrödinger wave, that is, the single-state vector, and "reality" of the two-state vector? Our answer is that the complete reality is described by the two-state vector. The single-state vector gives a partial description when we have only partial information. The apparent paradox of the descriptions is as follows. Consider a spin-½ particle described by the two-state vector $\langle \uparrow_y \| \uparrow_x \rangle$. The value σ_y corresponding to this particle is $\sigma_y = 1$. However, because it is described by the single (preselected) forward-evolving state $| \uparrow_x \rangle$, the value of σ_y is considered as the expectation value, $\langle \uparrow_x | \sigma_y | \uparrow_x \rangle = 0$. According to our claims, both are observable; thus, how can they be different?

In order to observe a quantum state, it has to be protected. When we discussed the protective experiments of single-state vectors, we did not say anything about quantum states evolving backwards in time. (It was not related to the point we wanted to make.) However, the protective procedure that we proposed automatically protects *identical* backward-evolving states. Thus, what we have proposed as an observation of a Schrödinger wave is indeed an observation of a two-state vector with identical forward- and backward-evolving states. For example, the protection of a spin-½ particle state,[2] a strong magnetic field in a given direction, protects the two-state vector with either both states parallel or antiparallel to this direction. This procedure is incompatible with the protection of the forward-evolving state parallel to one direction and the backward-evolving state parallel to another. If the particle is described by $\langle \uparrow_y \| \uparrow_x \rangle$, then the strong magnetic field in the $\hat{x}$ direction will change the backward-evolving spin-state. There exists a protection procedure for $| \uparrow_x \rangle$ that does not change the backward-evolving state as was described in the preceding section. The "observation" of the state protected in such a way will not yield the preselected quantum state, but it will yield the picture defined by the two-state vector.

Thus, the contradiction is resolved by giving a more accurate interpretation of our original protective measurement of the Schrödinger wave. We observed not a single-state vector, but a two-state vector with identical backward- and forward-evolving states.

ACKNOWLEDGMENTS

It is a pleasure to thank Sandu Popescu and Jacob Grunhaus for helpful discussions.

REFERENCES

1. AHARONOV, Y. & L. VAIDMAN. 1993. Phys. Lett. **A178:** 38.
2. AHARONOV, Y., J. ANANDAN & L. VAIDMAN. 1993. Phys. Rev. **A47:** 4616.
3. AHARONOV, Y., D. ALBERT, A. CASHER & L. VAIDMAN. 1987. Phys. Lett. **A124:** 199.
4. VAIDMAN, L., Y. AHARONOV & D. ALBERT. 1987. Phys. Rev. Lett. **58:** 1385.
5. AHARONOV, Y., D. ALBERT & L. VAIDMAN. 1988. Phys. Rev. Lett. **60:** 1351.
6. AHARONOV, Y., J. ANANDAN, S. POPESCU & L. VAIDMAN. 1990. Phys. Rev. Lett. **64:** 2965.
7. AHARONOV, Y. & L. VAIDMAN. 1990. Phys. Rev. **A41:** 11.
8. AHARONOV, Y. & L. VAIDMAN. 1991. J. Phys. **A24:** 2315.
9. AHARONOV, Y., P. G. BERGMANN & J. L. LEBOWITZ. 1964. Phys. Rev. **B134:** 1410.
10. AHARONOV, Y., S. POPESCU, D. ROHRLICH & L. VAIDMAN. 1993. Phys. Rev. **A48:** 4084.
11. UNRUH, W. G. 1994. Phys. Rev. **A50:** 883.
12. ROVELLI, C. 1994. Phys. Rev. **A50:** 2788.
13. AHARONOV, Y., J. ANANDAN & L. VAIDMAN. 1994. The meaning of the protective measurements. Preprint no. TAUP-2195-94.
14. VAIDMAN, L. 1994. *In* Advances in Quantum Phenomena: NATO Advance Science Institute Series. E. Beltrametti & J. M. Levy-Leblond, Eds. Plenum. New York.

Particle Spectroscopy by Application of Michelson's "Light-Wave Analysis" Technique to de Broglie Waves

F. HASSELBACH, A. SCHÄFER, AND
H. WACHENDORFER

Institut für Angewandte Physik
Universität Tübingen
D-72076 Tübingen, Germany

INTRODUCTION

Classical charged particle spectrometers have their basis on the fact that particles with different energies follow different individual trajectories. Different arrival sites in the plane of observation correspond to different energies of the particles. The quantum mechanical counterpart of such a classical measurement is to extract the particle spectrum from the spread of momentum, that is, from the corresponding spread of the de Broglie wavelengths of the ensemble of particles that make up the wave packet. For photons, the spectral decomposition of such a wave packet was proposed and realized for the first time by Michelson at the end of the last century.[1]

MICHELSON'S "LIGHT-WAVE ANALYSIS" TECHNIQUE[a]

In a Michelson interferometer, the path length of one of the two beams is increased continuously and the visibility (≡ contrast, which is used today preferentially) C of the interference fringes is recorded as a function of the path-length difference:

$$C = \frac{(I_1 - I_2)}{(I_1 + I_2)} = G_p \cdot G_q. \tag{1}$$

I_1 denotes the intensity at the center of a bright interference band and I_2 denotes the intensity at the center of the adjoining dark band. The right-hand side of equation 1 shows that the visibility of the interference fringes is the product of the Fourier transformations G_q of the spatial and G_p of the spectral intensity distribution of the light source. This result is obtained under the assumption that the components k of a narrow spectral line are distributed symmetrically around the midfrequency k_0 ($|k - k_0| \ll k_0$). The experiment described below is performed in such a way that $G_q \equiv 1$; that is, from contrast C as a function of the longitudinal shift, the power spectrum can be calculated. The contrast vanishes for longitudinal shifts exceeding

[a] Michelson in his 1892 paper[1] introduces the term "light-wave analysis" (which we venture to call the foregoing method).

the coherence length. By Fourier-analyzing the contrast of such a complete (coherence length–limited) system of interference fringes, the spectrum making up the wave packet is obtained.

Since Michelson's early experiments for electromagnetic waves, this method has been refined into a precision method causing Wiener to write (in 1961) the following about Fourier spectroscopy: "It is indeed the most accurate type of spectrometer known to us."[2] To realize this very promising spectroscopic method also for matter waves was out of reach until recently due to the lack of a phase- or wave packet–shifting device for matter waves.

THE WIEN FILTER: A WAVE PACKET–SHIFTING DEVICE FOR CHARGED PARTICLES

In 1979, Möllenstedt and Wohland discovered in a Wien filter (crossed electric and magnetic fields) a wave packet–shifting device and measured the coherence length of electron waves for the first time.[3] The interferometer they used was not of the Michelson type, but a biprism interferometer because, for electrons, the most versatile beam-splitter has proved to be the electron optical biprism according to Möllenstedt and Düker.

A sketch of the measurement method is given in FIGURES 1a and 1b. FIGURE 1a gives the light optical analogue and FIGURE 1b gives its realization for charged particles. On the left-hand side of FIGURE 1a, the situation for a full overlap of the two wave packets in the interference plane is presented. In the observation plane, maximum fringe contrast is observed. By moving the right-hand-side mirror backwards, an artificial continuously increasing time delay can be inserted into one leg of the interferometer. The contrast of the fringes decreases continuously with increasing delay until the coherence time is exceeded (FIGURE 1a, right). FIGURE 1b shows that the moving mirror, that is, the device introducing a time delay, in our charged particle interferometer is replaced by a Wien filter.

A brief outline of the action of the Wien filter on electron wave packets is given in FIGURE 2. A Wien filter consists of crossed electric and magnetic fields, both perpendicular to the beam path. The Wien filter is said to be in its compensated state when the electric and magnetic forces on the charged particles just cancel each other; that is, the trajectories of the particles are not affected by the electromagnetic fields in the Wien filter. For the case of the compensated Wien filter, it can be shown easily (e.g., see references 4 and 5) that the phase shifts exerted by the electric and magnetic potentials[b] are opposite to each other and are of exactly the same magnitude. To state it differently, the electron optical index of refraction equals 1 (to first order) inside and outside of the Wien filter.[4,5] Therefore, in FIGURE 2, the planes of equal phase (e.g., crests) of the electron waves, represented by the horizontal

[b]Electric and magnetic Aharonov-Bohm phase shifts.[5] The magnetic flux enclosed by the coherent beams creates a certain phase shift that is exactly compensated by the phase shift that arises due to the fact that the waves travel through the Wien filter on paths of different electric potential. In essence, by taking into account the well-known experiments that prove the magnetic Aharonov-Bohm[6] effect, we have here an indirect proof of the existence of the electric (scalar[7]) Aharonov-Bohm effect.

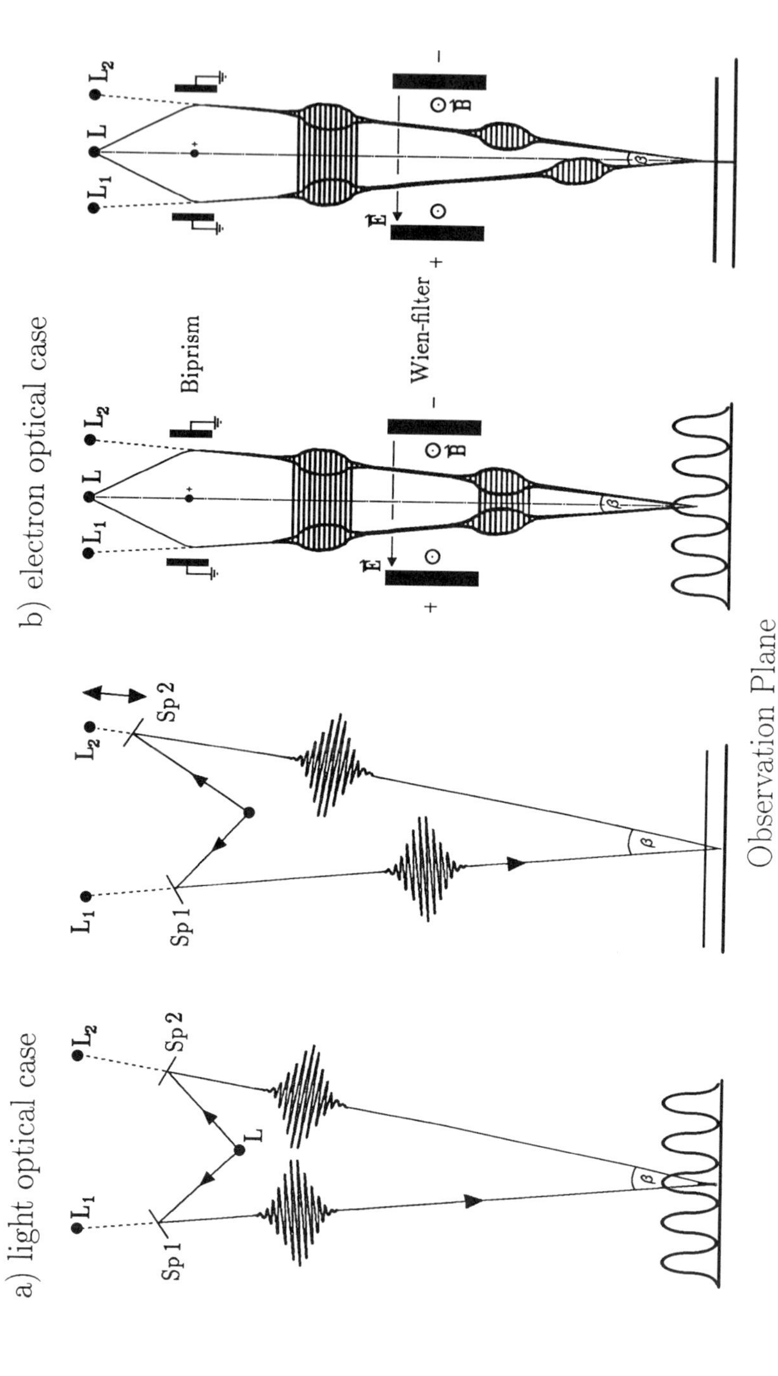

FIGURE 1. Principle of measurement of coherence lengths and interferometric spectroscopy according to Michelson for electromagnetic waves on the left and charged matter waves on the right.

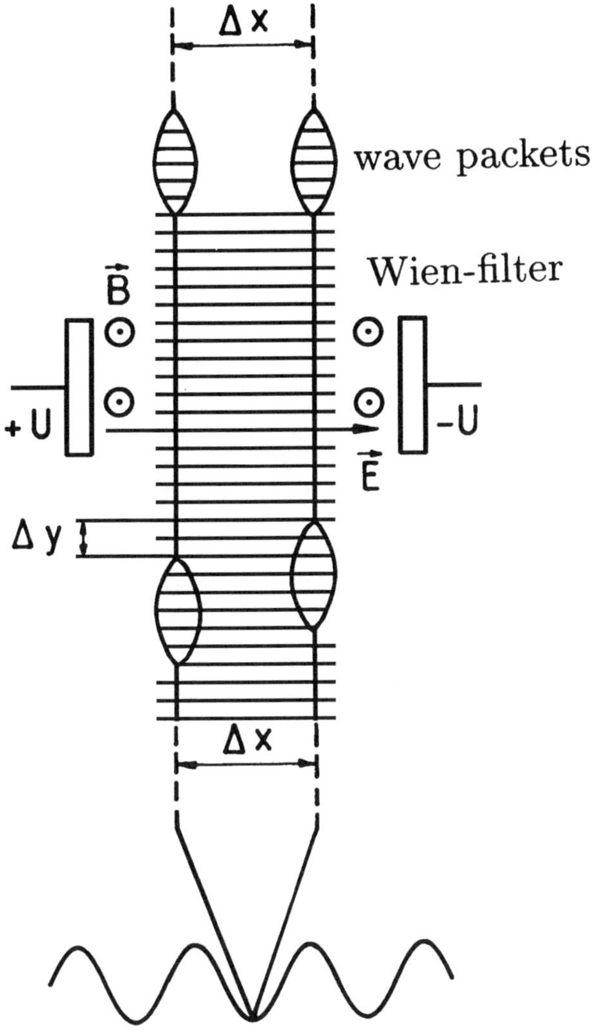

FIGURE 2. The influence of a Wien filter in its compensated state on two spatially separated electron wave packets. The right-hand-side wave packet travels inside the Wien filter on a more negative potential than the left-hand-side one, that is, with a lower group velocity. The acceleration and deceleration of the wave packets takes place in the fringing fields of the Wien filter.

lines, and the phase velocity are not affected at all by the electromagnetic fields of the compensated Wien filter, irrespective of its excitation.

In other words, when we increase the excitation of the Wien filter while always staying in its compensated state, we observe in the observation plane a stationary field of interference fringes; however, with increasing excitation, fringe contrast is decreasing continuously. This is due to the fact that the electron wave packets travel on paths of different electric potentials with different group velocities inside the Wien filter. This leads to a longitudinal shift of the wave packets at the exit plane and consequently to a reduced contrast of the interference fringes.[c] For longitudinal shifts larger than the coherence lengths, fringe contrast vanishes.

It is important to direct attention explicitly to the fact that the Wien filter in its compensated state is not a phase-shifter. The wave packets are shifted longitudinally in a stationary "phase wave sea"; in other words, the internal phase structure of the wave packets is not affected at all when traveling through the electromagnetic fields in a compensated Wien filter.

Refinement of a Wien Filter to a High-Precision Retarding Device for Wave Packets

The delay, that is, longitudinal shift, caused by a Wien filter can be adjusted with a precision of a small fraction of a wavelength when, first, the Wien filter's construction allows the adjustment of the electromagnetic fields in very subtle steps and, second, when it is aligned to its compensated state in a two-step process[8] as follows: At first, the magnetic and electric fields are zero (FIGURE 3a). The full overlap of the wave packets corresponds to maximum fringe contrast. Now, we increase in a first step the electric field only. The Wien condenser works as a deflection element. The interference fringes are deflected, for example, by two fringe widths to the right on the fluorescent screen of the interferometer (FIGURE 3b). This is due to the fact that the wave packet traveling in the more negative region is slower and arrives in the interference plane with a delay. FIGURE 3b demonstrates that a delay of $2\lambda/v$, where v is the phase velocity, corresponds to a shift of the interference fringes of exactly two fringe widths to the right-hand side. We now increase the magnetic field until the deflection due to the electric field is just compensated. This state of the now again compensated Wien filter corresponds to the following physical situation (FIGURE 3c): Both beams travel rectilinearly through the Wien filter, but the right-hand-side wave packet is retarded longitudinally by two wavelengths in the Wien filter. The overlap of the wave packets is reduced by two wavelengths and so too the contrast of the interference fringes correspondingly.

Measurement of Coherence Lengths

In order to measure the coherence length, this procedure is repeated while counting the total number of fringes until the contrast in the fringe field vanishes or (per definitionem) decreases, for example, to $1/e$. The coherence length is then given by twice this number of fringes times the wavelength of the electrons. With the $1/e$ definition, a coherence length of 280 nm was measured[5] for field-emitted electrons of 4 keV with an energy spread of 0.36 eV. The error margin was as small as 3%, compared to 20–30% in Möllenstedt and Wohland's first experiment.[3]

[c] The electric potential difference on the two paths increases with increasing excitation of the Wien filter. The acceleration and deceleration of the wave packets to the value inside the Wien filter occurs in the electric fringing fields of the Wien filter condenser.

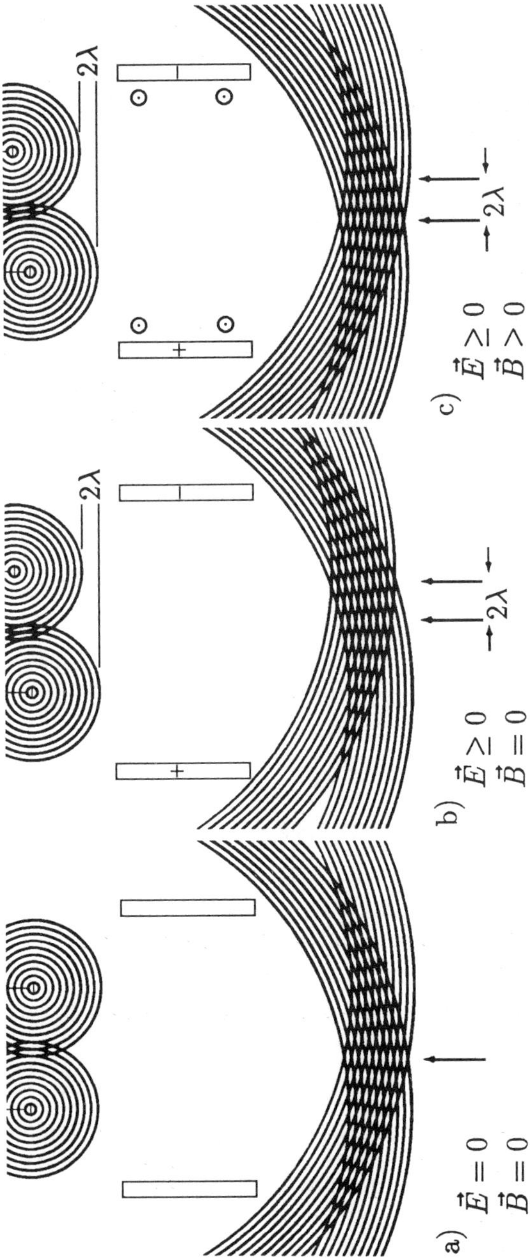

FIGURE 3. The two-step process to reach the compensated state of the Wien filter.

Let us mention here that this measurement method works without any knowledge about the field strengths and (in-)homogeneities of the electromagnetic fields in the Wien filter, not to mention that of the fringing fields. It therefore is intrinsically extremely precise and enables us to measure the longitudinal shift exerted by the Wien filter with a precision on the order of 1% of a wavelength or less. This high precision is a prerequisite for realizing Michelson's visibility technique and Fourier spectroscopy for matter waves.

"Light-Wave Analysis" of Electron Waves

In the following model experiment, the spectrum of a field emission electron gun has been measured. The contrast of the interference fringes must be recorded quantitatively as a function of the longitudinal shift in the whole interference field, consisting of up to 20,000 fringes for the experimental parameters used in our low-voltage interferometer (a few keV of total energy of the field-emitted electrons at an energy spread of about 0.4 eV). This has been done by recording the whole

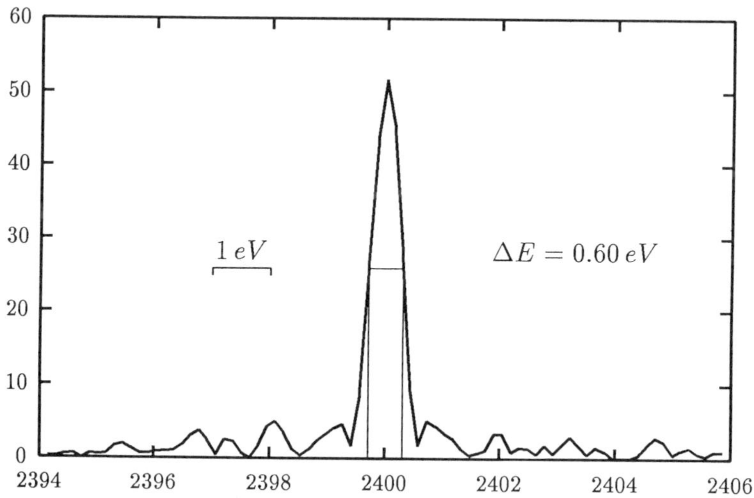

FIGURE 4. Spectrum of a field electron emitter obtained by Fourier analysis. The total energy of the electrons was 2.4 keV.

interference field in sets of, for example, 10 fringes successively with a CCD-camera densitometer. The digitized data sets were corrected for the (small) cylinder lens action of the Wien filter, were put together with matched phases in a personal computer, and were subsequently transferred to a VAX computer for Fourier analysis. In the first experiment,[9] a resolution of about 0.6 eV has been obtained. The state of the art is now about 125 meV.[10] As an example, the energy spectrum of field-emitted electrons measured by this method is given in FIGURE 4.

This result was obtained by taking into account 12,800 interference fringes with a contrast of $C \geq 10\%$. About 2000 low-contrast fringes were neglected, which results in an error of less than 50 meV of the full width at half-maximum of the spectrum given in FIGURE 4. The total error of 125 meV contains 75 meV of error caused by the fact that we recorded 16 sample points per fringe and had an array size of 2^{19} with which we execute the Fourier transformation.

Hence, the present resolving power surpasses that of electron spectrometers used in today's analytical electron microscopes by about an order of magnitude.

CONCLUSIONS

Our Fourier spectrometer is the first spectrometer for particles that fully relies on quantum mechanics, that is, the wave properties of matter. It seems remarkable that the quantum mechanical probability waves exhibit the same features as the "real" classical waves, for example, of an electromagnetic field, in spite of the fact that a field of electrons (fermions) is intrinsically nonclassical. All conventional spectrometers for charged particles are based on the dispersion of particle trajectories in electromagnetic fields and, in turn, their resolution suffers from the well-known aberrations of charged particle optics. The advantages of Michelson's light-wave analysis technique cannot be summarized better than he did in his seminal paper[1] in 1892: "The principal object of the foregoing work is to illustrate the advantages which may be expected from a study of the variations of clearness of interference fringes with increase in difference of path. The fundamental principle by which the 'structure' of a line or a group of lines is determined by this method is not essentially different from that of spectrum analysis by the grating, both depending, in fact, on interference phenomena; but in consequence of the almost complete freedom from errors arising from defects in optical or mechanical parts, the method has extraordinary advantages for this special work."

REFERENCES

1. MICHELSON, A. A. 1891. On the application of interference methods to spectroscopic measurements—I. Philos. Mag. **31:** 338–346; 1892. On the application of interference methods to spectroscopic measurements—II. Philos. Mag. **34:** 280–299.
2. WIENER, N. 1961. Cybernetics or Control and Communication in the Animal and Machine. MIT Press. Cambridge, Massachusetts.
3. MÖLLENSTEDT, G. & G. WOHLAND. 1980. Direct interferometric measurement of the coherence length of an electron wave packet using a Wien filter. *In* Electron Microscopy 1980 (Seventh European Congress on Electron Microscopy Foundation, Leiden). Volume 1. P. Bredoro & G. Boom, Eds.: 28–29.
4. NICKLAUS, M. & F. HASSELBACH. 1993. The Wien filter: a wave packet shifting device for restoring longitudinal coherence in charged matter wave interferometers. Phys. Rev. A **48:** 152–160.
5. HASSELBACH, F. 1993. Experiments with coherent wave packets. *In* Proceedings of the Conference on Fundamental Problems in Quantum Physics (Oviedo, Spain). M. Ferrero & A. van der Merve, Eds. Kluwer. Dordrecht.
6. CHAMBERS, R. G. 1960. Shift of an electron interference pattern by enclosed magnetic flux. Phys. Rev. Lett. **5:** 3–5; BAYH, W. 1962. Messung der kontinuierlichen Phasenschiebung von Elektronenwellen im kraftfeldfreien Raum durch das magnetische

Vektorpotential einer Wolfram-Wendel. Z. Phys. **169:** 492–510; PESHKIN, M. & A. TONOMURA. 1989. The Aharonov-Bohm Effect: Lecture Notes in Physics. Volume 340. Springer Pub. New York.

7. ALLMANN, B. E., A. CIMMINO, A. G. KLEIN, G. I. OPAT, H. KAISER & S. A. WERNER. 1992. The scalar Aharonov-Bohm effect—an experiment with neutrons. Phys. Rev. Lett. **68:** 2409–2412.
8. DABERKOW, I., H. GAUCH & F. HASSELBACH. 1983. Measurement of the longitudinal coherence of electrons from a field emission source. *In* Joint Meeting on Electron Microscopy (Antwerp)—Program and Abstract Books, p. 100.
9. HASSELBACH, F. & A. SCHÄFER. 1990. Interferometric (Fourier-spectroscopic) measurement of electron energy distributions. *In* Proceedings of the Twelfth International Congress for Electron Microscopy (Seattle). Volume 2. L. D. Peachey & D. B. Williams, Eds.: 110–111. San Francisco Press. San Francisco; also: German Patent No. DE 40 24 624 A1.
10. WACHENDORFER, H. 1993. Diploma thesis. Universität Tübingen.

Experimental Realization of Interaction-free Measurements[a]

PAUL KWIAT,[b,c] HARALD WEINFURTER,[c] THOMAS HERZOG,[c] ANTON ZEILINGER,[c] AND MARK KASEVICH[d]

[c]*Institut für Experimentalphysik*
Universität Innsbruck
6020 Innsbruck, Austria

[d]*Department of Physics*
Stanford University
Stanford, California 94305

INTRODUCTION

"No observation can be made with less than one quantum passing through the observed object": so is it stated in an article by D. Gabor, in the first volume of *Progress in Optics.*[1] Recently, Elitzur and Vaidman have pointed out that it is possible to make "interaction-free" quantum mechanical measurements, in which the existence of an object in a given region of space may be ascertained seemingly without interacting with it.[2,3] This possibility is nonclassical, in that it relies on the wave-particle duality. The initial proposal employed a Mach-Zehnder interferometer (see FIGURE 1), aligned so that incident photons (or any other interfering particles) exit to detector D_1 with certainty, in the absence of any object within the interferometer. Thus, detector D_2 *never* fires under this configuration. The presence of an absorbing (or, more generally, a nontransmitting) object in one of the arms changes completely the possible outcomes. For a 50-50 beam-splitter, the photon will encounter the object with 50% probability and will be absorbed. To make the argument more dramatic, Elitzur and Vaidman suggested the notion that this absorption of a single photon would trigger the explosion of an ultrasensitive bomb. There is a 25% probability that the photon will still exit port no. 1, yielding no information. However, there is also a 25% probability that the photon will exit via the *other* exit port, to detector D_2. Detecting this photon, one can conclude that an object was certainly within the interferometer, even though the photon could not have interacted with it. Put differently, one has managed to detect the presence of the ultrasensitive bomb without triggering it, an impossible feat in classical physics.

If one performs the above procedure only once, it is clear that one can have interaction-free verification of the presence of the bomb only 25% of the time. However, by repeating the experiment or recycling the photon in cases where it

[a]This work was supported by the Austria Science Foundation (FWF) under Project No. S065/02.

[b]P. Kwiat was supported by FWF Lise Meitner Postdoctoral Fellowship No. M0077-PHY.

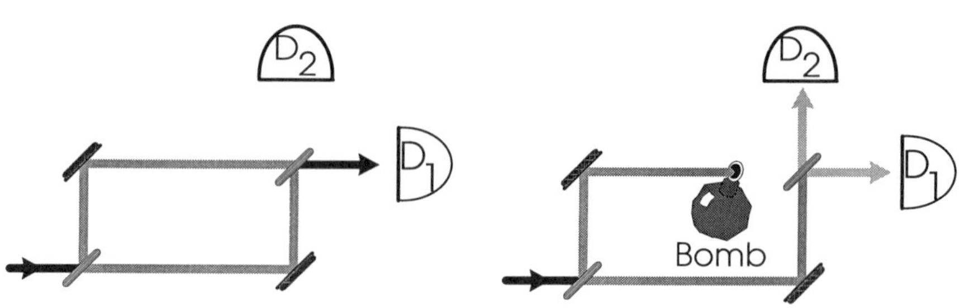

FIGURE 1. Simple Mach-Zehnder interferometer, with and without a sensitive bomb-trigger in one path.

leaves the interferometer by port no. 1, one can increase the total probability of an interaction-free measurement. A useful figure of merit in evaluating any potential scheme is the following:

$$\eta = \frac{P(\text{Det } 2)}{P(\text{Det } 2) + P(\text{boom})}. \tag{1}$$

For a lossless system, this is basically the probability of making an interaction-free measurement (assuming some measurement was made), or the fraction of bombs that can be saved. For the simple case considered above, the fraction of measurements that can be interaction-free is thus ⅓.

It is important to stress why this is a quantum mechanical effect. Using purely classical field calculations, it is certainly true that one arrives at the same probabilities appearing in equation 1. The difference is that, with a classical field, it is possible for both detector 2 *and* the bomb detector to be triggered in a given run. However, if we use a single-photon state incident on the interferometer, this outcome is impossible. Complementarity is essential: in the absence of the object, it is the wavelike nature of the incident quantum that allows us to establish, through destructive interference, a condition in which detector 2 never fires; in the presence of the object, it is the indivisibility of the quantum that enforces the mutual exclusivity of the possible outcomes.

EXPERIMENTAL DESCRIPTION

In our experiment (FIGURE 2), pairs of correlated photons were produced via the process of spontaneous parametric downconversion in a nonlinear crystal ($LiIO_3$). The pump beam at 351 nm originated in an argon-ion laser; using irises and 5-nm (FWHM) interference filters, we selected downconverted photon pairs at 702 nm. One member of each pair was directed to the "trigger" detector D_T, whereas the other one was directed to a Michelson interferometer, adjusted to lie within the "white-light fringe" region so that the difference in path lengths was always less than 3 μm. The detector D_{Ifm} looked at the output port of the interferometer. By means of a translatable mirror, it was possible to direct the photons from one of the arms to the detector D_B, thereby producing the "bomb in" configuration. In the absence of the "bomb mirror", the path difference in the interferometer was adjusted to produce a minimum number of counts at D_{Ifm}; that is, most of the photons exited the interferometer via the entrance port. This position was stabilized by employing an extra helium-neon alignment laser (not shown in FIGURE 2) and a feedback system to piezoelectrically translate one of the interferometer mirrors.

It has been previously shown that by counting the downconversion photons in coincidence one may prepare a very good approximation to a single-photon Fock state.[4] Therefore, our data consist of various coincidence rates between D_T and D_{Ifm} and between D_T and D_B. After locking our interferometer to a minimum of D_{Ifm}, data were recorded, periodically switching from a "bomb out" to a "bomb in" configuration. Typical results are shown in FIGURE 3, where the open and filled symbols refer to the "bomb out" and "bomb in" conditions, respectively. It is immediately seen from the data that, even in the absence of a bomb, one still has counts at D_{Ifm}, falsely indicating the presence of the bomb mirror. These counts constitute the background or noise of our detection scheme, arising from accidental coincidences and nonunity interference fringe visibility.

The beam-splitter in our interferometer was a 1-mm-thick glass plate, with one side antireflection-coated and the other side coated in five sections, each with a different reflectivity. Thus, by horizontally translating the beam-splitter in its plane, we were able to readily choose between reflectivities (measured directly with the downconversion photons) of 54%, 43%, 33%, 19%, and 11% without changing the relative path lengths in the interferometer. The expected figure of merit (equation 1) when the bomb mirror is inserted is a function of beam-splitter reflectivity. In the absence of losses, the probability of an incident photon going towards the bomb is R, whereas the probability of it going towards the interferometer detector is RT. Assuming a lossless beam-splitter, $T = 1 - R$, we have

$$\eta = \frac{RT}{RT + R} = \frac{1 - R}{2 - R}, \tag{2}$$

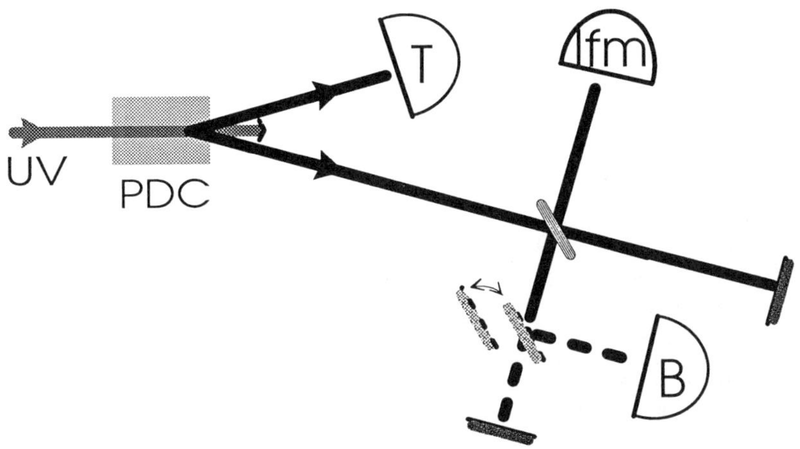

FIGURE 2. Schematic of a downconversion experiment to demonstrate the principle of interaction-free measurement.

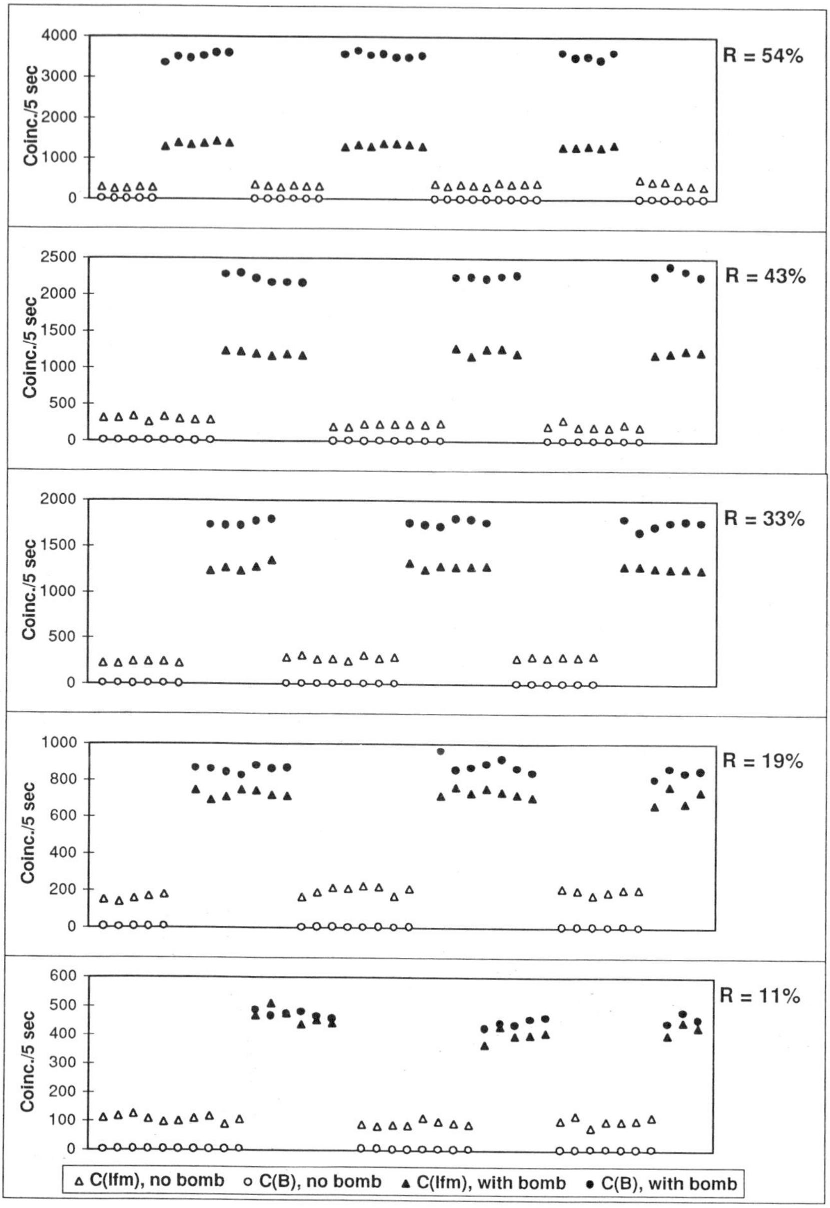

FIGURE 3. Typical experimental results for an interaction-free measurement (from the setup in FIGURE 2). The open symbols refer to the "bomb out" condition, whereas the filled symbols refer to the "bomb in" condition. The five graphs correspond to the five beam-splitter reflectivities examined; the reflectivity R is listed in the upper right corner of each.

a plot of which is given in FIGURE 4. To obtain the figure of merit described by equation 2 from the experimental results, one must account for the finite detector efficiencies. For the purpose of this "proof of principle" experiment, it was convenient to equalize (to within 0.06%) the net detection efficiencies of the bomb detector D_B and the interferometer detector D_{Ifm} by adjusting the detector overbiases (the actual efficiencies were approximately 2%). Under these conditions, the efficiencies cancel out of equation 2 and one may simply use the rates directly: $\eta = C(\mathrm{Ifm})/[C(\mathrm{Ifm}) + C(\mathrm{B})]$. The experimental results for the five regions on our beam-splitter are displayed in FIGURE 4.

HIGH-EFFICIENCY INTERACTION-FREE MEASUREMENTS

Note from equation 2 that the fraction of interaction-free measurements tends to the value of 50% as the reflectivity becomes small. As previously stated by Elitzur and Vaidman,[2] this is the upper limit possible with the interferometer schemes discussed thus far. We have recently discovered[5] that one may exceed this limit by using an application of the quantum Zeno effect.[6,7] Indeed, in principle, the fraction of interaction-free measurements may be made arbitrarily close to unity. Recall that, in the Zeno effect, the evolution of a quantum system is altered by repeated measurements made on the system (see FIGURE 5). Consider now the related scheme in FIGURE 6. Single-photon states are produced using correlated downconversion photons. One of the photons is again used as a trigger, whereas the other photon makes a specified number of cycles in the system before being detected. The number of cycles is determined either from geometry (e.g., the photon spirals up in the loop, until it passes over one of the components) or with fast timing (based on the arrival

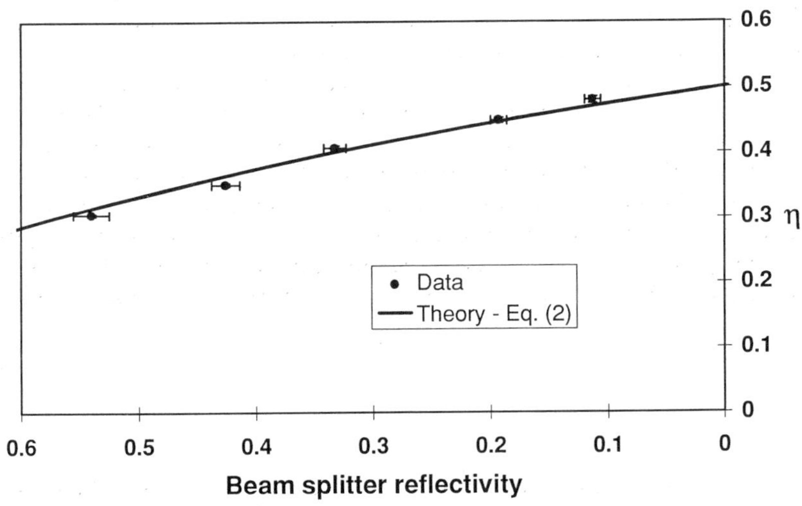

FIGURE 4. Experimental and theoretical values for the figure of merit η in a Michelson-interferometer scheme, as a function of beam-splitter reflectivity.

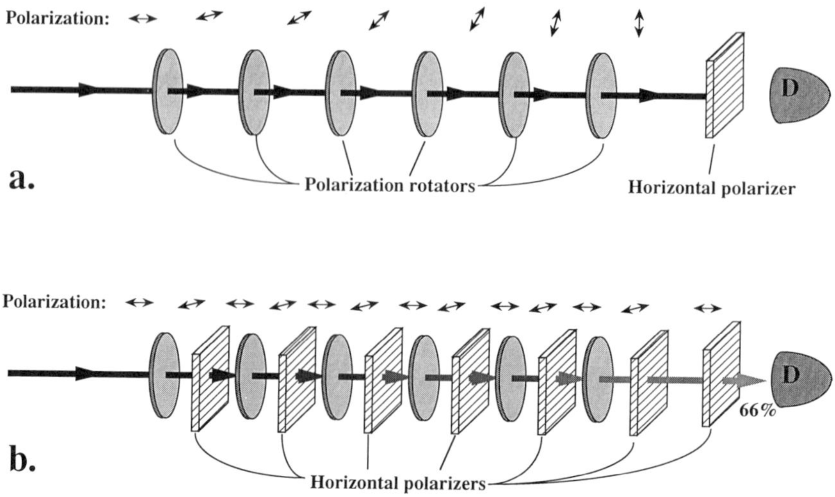

FIGURE 5. Simple optical version of the quantum Zeno effect.[7] (a) A series of polarization rotators is used to rotate the polarization of the input photon from horizontal to vertical. No light is seen at the final detector. (b) When a series of horizontal polarizers is interspersed between the rotators, the light is at every stage projected back onto a state of horizontal polarization, resulting in light at the final detector. If the number of stages is ≥ 4, more than 50% of the input light will be transmitted. In particular, for the case shown ($N = 6$), the chance of transmission is very nearly twice the chance of absorption.

time of the trigger photon, a fast switch [not shown] can be used to direct the other photon out of the system after N cycles). We choose the value of the polarization rotation to be $\pi/2N$ so that, in the absence of any bomb, the initially horizontally polarized photon is found after N cycles to be vertically polarized.[e]

The presence of a "bomb" or any nontransmitting object in the lower path of the internal Mach-Zehnder interferometer changes the situation completely. Now with each cycle there is only a small chance that the photon chooses this lower path and triggers the bomb, and a large probability, $P = \cos^2(\pi/2N)$, that it travels the upper path instead. In this second possibility, the polarization after the recombining polarizing beam-splitter is once again horizontal and the whole process repeats. Clearly, the probability that the photon is found after N cycles to be still horizontally polarized is just the probability for it to have taken the upper path during each cycle:

$$P = \left[\cos^2\left(\frac{\pi}{2N}\right)\right]^N, \tag{3a}$$

[e]To meet this condition, it is important that the path lengths in the internal Mach-Zehnder interferometer (upper right corner, with polarizing beam-splitters) be the same to within 2π; otherwise, incident linear polarization can become elliptical due to the relative phase shift between the vertical and horizontal components.

which in the limit of large N becomes

$$P = 1 - \frac{\pi^2}{4N} + O\left(\frac{1}{N^2}\right). \tag{3b}$$

Of course, the probability that the bomb is triggered is just the complement of P. Equation 3a is plotted in FIGURE 7. One sees immediately that, as long as $N \geq 4$, there exists a greater than 50% probability of making an interaction-free measurement, thereby surpassing the limit of the original Elitzur-Vaidman configurations discussed in the previous sections.

It should be noted that the use of polarization is not essential for these high-efficiency tests. Due to the isomorphism of all two-state systems, it should be possible to use these techniques with any two-level scheme. For example, using two identical cavities that are weakly coupled by a highly reflective beam-splitter, one can realize a practical implementation (see FIGURE 8). A photon is inserted into the left cavity at time $T = 0$. If the beam-splitter reflectivity is given by $\cos^2(\pi/2N)$, then in the absence of any absorber the photon will be located with certainty in the right cavity at time $T_N = N \times$ (round-trip time), so long as the lengths of the two coupled cavities are identical. This is true even if the coherence length of the photon wave packet is much shorter than the cavity length. Therefore, a detector inserted into the left cavity at time T_N would not fire. However, in the presence of an absorber or scatterer (e.g., our ultrasensitive bomb) in the right cavity, the photon wave function is continually projected back onto the left cavity. By making the coupling weaker (and the number N greater), one can arbitrarily reduce the probability that the photon ever leaves the first cavity when the bomb is in the second; now a detector inserted into the left cavity at time T_N will nearly always fire. Consequently, one can

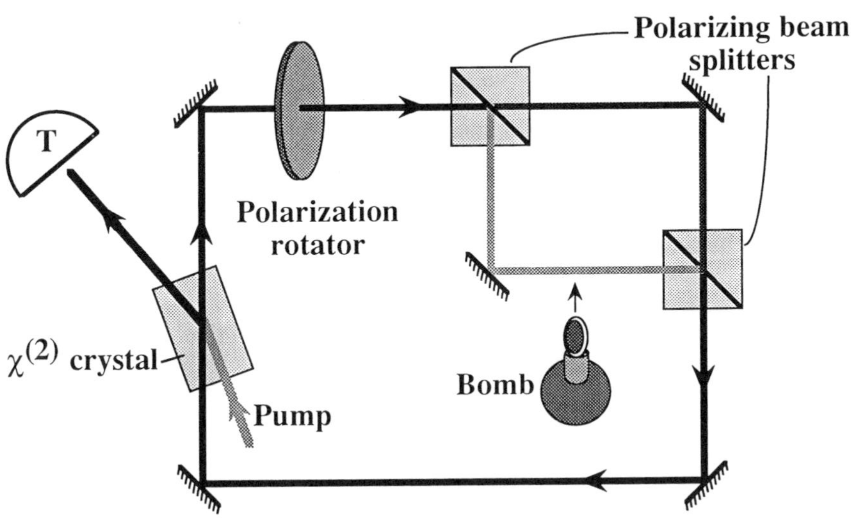

FIGURE 6. Simplified schematic of one method to observe a greater than 50% interaction-free measurement. The downconversion photon makes N cycles before being removed (due to geometry or a fast switch) and its polarization measured.

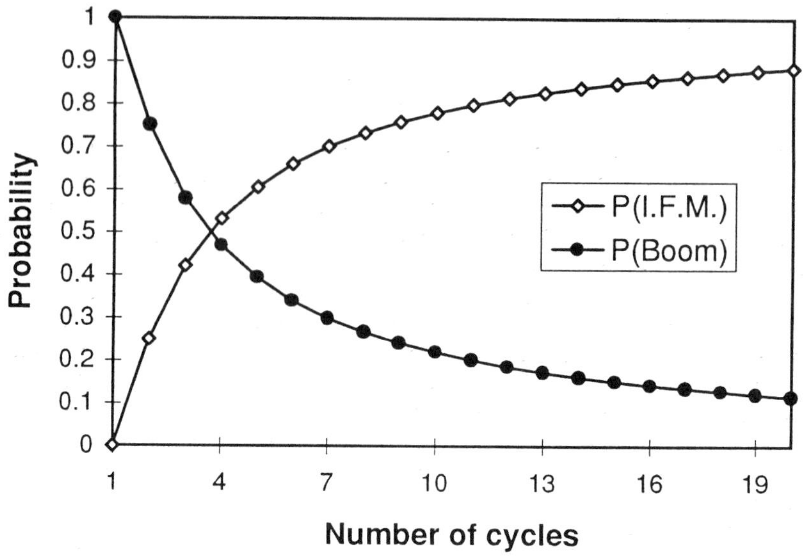

FIGURE 7. The probability P(I.F.M.) of an interaction-free measurement and the probability P(Boom) that the bomb is triggered, as a function of the total number of cycles, assuming ideal optical components.

in principle make the probability of an interaction-free measurement arbitrarily close to 1.

Again, downconverted photons are an appropriate choice because the strong time correlations yield exact information as to when one should "insert" a detector into the cavity. Also, one has the capability of actually creating one of the downconverted photons *inside* the cavity, removing the need for some kind of optical switch at this stage. In practice, losses reduce the effectiveness of such schemes, but it should still be possible in a real setup to have interaction-free measurements more than half of the time.

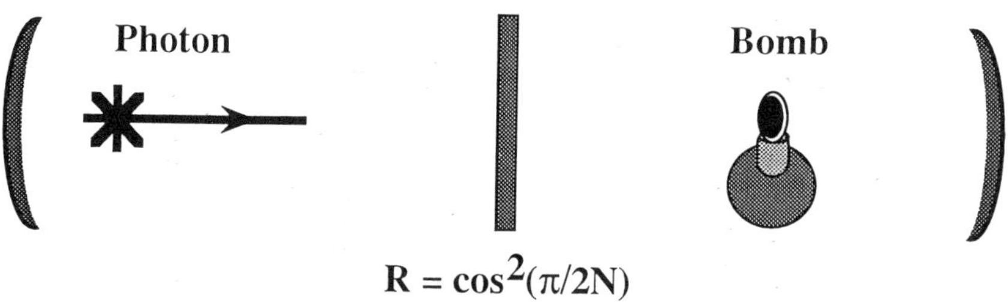

FIGURE 8. A purely interferometric interaction-free measurement system. In the absence of a bomb in the right cavity, a photon initially in the left cavity will be found with certainty in the right cavity after N cycles (where N depends on the precise value of the coupling beam-splitter's reflectivity R). With the bomb, and when $N \geq 4$, the photon will more likely be found in the left cavity on the N-th cycle. It is assumed that the cavity lengths are identical.

EXPERIMENTAL DEMONSTRATION OF REPEATED INTERROGATION

We have performed a simple experiment to illustrate the optical version of the quantum Zeno effect presented in FIGURE 5. It may be interpreted as a limited version of a true high-efficiency interaction-free measurement scheme: one is able to detect a *polarization-sensitive* bomb more than 50% of the time without losing the incident photon. Again, we employ the single-photon states available through parametric downconversion (see FIGURE 9). Photons at 788 nm were used as a trigger (not shown in the figure), whereas the conjugate photons at 633 nm were sent into the quantum Zeno apparatus. The case of six stages was investigated; however, rather than employing six different polarization rotators and six different polarizers, we sent our photons in a spiral through the same components six times. We call this

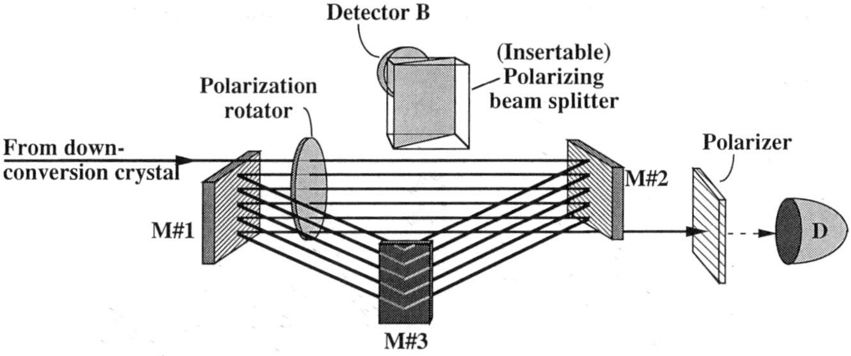

FIGURE 9. Schematic of the experimental setup used to demonstrate the quantum Zeno effect. The photon "cyclotron" consisted of five loops, allowing six interactions with the polarization rotator. In the absence of the polarizing beam-splitter, the polarization is rotated by 15° each cycle. We record the coincidence rates between the trigger detector (not shown) and the D and B detectors.

arrangement a photon "cyclotron". The horizontally polarized photon initially passed (horizontally) over the top of the first mirror, passed through the rotator[f] (and possibly through a removable polarizing beam-splitter), and reflected off mirror no. 2. The photon's trajectory was directed downward by mirror no. 3 so that it could reflect off mirror no. 1, aligned such that the trajectory was once again horizontal. The process was repeated for five loops (six times through the rotator), after which the photon trajectory passed under mirror no. 2 to a detector.

As expected, when the polarization was rotated by 15° each cycle, the photon was vertically polarized at the output and essentially no coincidence counts were seen

[f]For a polarization rotator, it sufficed to use a simple half-wave plate, oriented with its axis at 7.5° from the horizontal. That this functioned was a surprise to us, as we expected that the polarization would simply alternate between horizontal polarization and polarization at +15° with each cycle, yielding no net rotation. The secret to this wonderful surprise is that an *odd* number of mirrors (three in our case) inverts left and right so that, just before encountering the wave plate for the second time, the polarization of the photon is at −15°, which is then rotated to +30° by the wave plate, and so on.

after a horizontal polarizer. (Actually, due to imperfect polarizers, there was a horizontal component at a rate of 5.5 s^{-1}, compared to the vertical component of 441 s^{-1}. The noise figure of this scheme in terms of interaction-free measurements is thus 1.2%.) However, when the polarizing beam-splitter was inserted, we saw that the light reaching the detector was then horizontally polarized, with a rate of 243 s^{-1}. The probability of transmission is then 243/441 = 55.1 ± 0.4%. However, if one corrects for the separately measured transmission factor of the polarizing beam-splitter (= 0.826 ± 0.006, for six passes), this becomes 66.7 ± 0.7%, to be compared with the predicted value of 66.0%.

As indicated in FIGURE 9, we were also able to look directly at the output of the polarizing beam-splitter, using a lens to focus the six parallel output beams onto the surface of another photodiode, suggestively labeled D_B. The temporal distribution of coincidences between this detector and the trigger detector is interesting (see FIGURE 10), as one can see explicitly *when* the photons leave the cyclotron (if they do

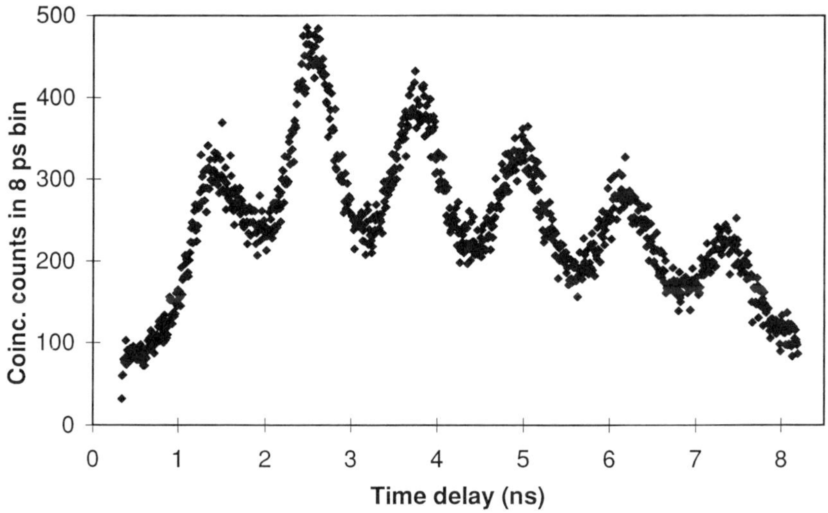

FIGURE 10. The time spectrum of the rate of coincidences between the trigger detector and the detector looking at the output of the polarizing beam-splitter. The time separation between peaks corresponds to the photon time-of-flight through one loop of the "cyclotron". Total accumulation time = 300 s.

not make it to the exit detector). The separation of the peaks is 1.18 ns, precisely what one would expect from the delay of traveling around the equilateral loop with a side of 11.5 cm and through the 2-cm-thick polarizing beam-splitter. That the first peak is actually smaller than the second is due to a problem with our polarizing beam-splitter and is not yet fully understood.

In summary, we have demonstrated in our Michelson-interferometer experiment an interaction-free measurement with a figure of merit η of nearly ½; that is, almost half of our measurements could be interaction-free. In the last experiment discussed, we were able to detect a polarization-sensitive bomb with an η of ⅔. Work is currently in progress to demonstrate the high-efficiency interaction-free measurement of *any* nontransmitting bomb, based on the repeated-interrogation schemes presented here.

REFERENCES

1. Gabor, D. 1961. Light and information. *In* Progress in Optics. Volume 1. E. Wolf, Ed.: 122. North-Holland. Amsterdam.
2. Elitzur, A. & L. Vaidman. 1993. Found. Phys. **23:** 987.
3. Vaidman, L. 1994. Quantum Opt. **6:** 119.
4. Hong, C. K. & L. Mandel. 1986. Phys. Rev. Lett. **56:** 58.
5. Kwiat, P. G., H. Weinfurter, T. Herzog, A. Zeilinger & M. Kasevich. 1994. Phys. Rev. Lett. Submitted.
6. Misra, B. & E. C. G. Sudarshan. 1977. J. Math. Phys. **18:** 756.
7. Peres, A. 1980. Am. J. Phys. **48:** 931.

Negative Kinetic Energy between Past and Future State Vectors[a]

DANIEL ROHRLICH, YAKIR AHARONOV,[b]
SANDU POPESCU, AND LEV VAIDMAN

Raymond and Beverly Sackler Faculty of Exact Sciences
School of Physics and Astronomy
Tel Aviv University
Ramat Aviv, Tel Aviv 69978, Israel

INTRODUCTION

When the word "quantum" first entered the language of physics, it meant a restriction on possible values of energy, and it is still axiomatic that the only observable values of a physical quantity are the eigenvalues of a corresponding quantized operator. When we obtain values that are not eigenvalues, we interpret them as errors. Still, measurements are uncertain in practice and can even yield classically forbidden, "unphysical" values. We have uncovered remarkable regularities in the way that "unphysical" values can appear in sequences of measurements, suggesting that these values may not be unphysical at all. In quantum theory, it seems, not only are physical quantities not restricted: they can take values outside the classically allowed range. Here, we discuss this new effect in the context of barrier penetration by quantum particles.

Barrier penetration, such as tunneling out of a potential well, is a classically forbidden quantum process. Quantum particles can be found in regions where a classical particle could never go: it would have negative kinetic energy. However, the eigenvalues of kinetic energy cannot be negative. How, then, can a quantum particle "tunnel"? The apparent paradox is resolved by noting that the wave function of a tunneling particle only partly overlaps the forbidden region, whereas a particle found within the forbidden region may have taken enough energy from the measuring probe to offset any energy deficit. Nevertheless, actual measurements of kinetic energy can yield negative values. Here, we present a model experiment in which we measure the kinetic energy of a bound particle to any desired precision. We then attempt to localize the particle within the classically forbidden region. The attempt rarely succeeds, but, whenever it does, we find that the kinetic energy measurements gave an "unphysical" negative result; moreover, these results cluster around the appropriate value, that is, the difference between the total and the potential energy. This consistency, which seems to come from nowhere (a background of errors), suggests strongly that the notion of a quantum observable is richer than generally

[a] This research was supported by Grant No. 425/91-1 of the Basic Research Foundation (administered by the Israel Academy of Sciences and Humanities) and by Grant No. PHY-8807812 of the National Science Foundation.

[b] Also at Department of Physics, University of South Carolina, Columbia, South Carolina 29208.

realized. Previous papers making this suggestion have analyzed a measurement of spin[1] and a quantum time machine[2] as well as negative kinetic energy.[3,4]

NEGATIVE KINETIC ENERGY

Our example may be summarized as follows: We prepare a large ensemble of particles bound in a potential well, in an eigenstate of energy, and measure the kinetic energy of each particle to a given precision. Then, we measure the position of each particle and select only those cases where the particle is found within some region "far enough" from the well, with "far enough" depending on how precisely the kinetic energy was measured. In almost all such cases, we find that the measured kinetic energy values are *negative* and cluster around the particular negative value appropriate to particles in the classically forbidden region. Also, the spread of the clustering is the characteristic spread for kinetic energy measurements with this device.

We begin with a particle trapped in a potential well. The Hamiltonian is $H = (p^2/2m) + V(x)$, with $V(x) = -V_0$ for $|x| < a$ and $V(x) = 0$ for $|x| > a$. We prepare an ensemble of particles in the ground state, with energy $E_0 < 0$: $|\Psi_{in}\rangle = |E_0\rangle$. Following von Neumann,[5] we model a measurement of kinetic energy with an interaction Hamiltonian $H_{int} = g(t)P(p^2/2m)$, where P is a canonical momentum conjugate to the position, Q, of a pointer on the measuring device. The time-dependent coupling constant $g(t)$ is nonzero only for a short time interval and is normalized so that $\int g(t)dt = 1$. When the time interval is very short, we call the measurement impulsive. For an impulsive measurement, H_{int} dominates the Hamiltonians of the measured system and the measuring device. Then, since $\dot{Q} = (i/\hbar)\,[H_{int}, Q]$, we obtain

$$Q_{fin} - Q_{in} = \frac{p^2}{2m} \tag{1}$$

for the operator Q.

In an ideal measurement, the position of the pointer is precisely defined and thus we read a precise value of kinetic energy. However, in practice, measurements involve uncertainty. To model a source of uncertainty, we take the initial state of the pointer to be

$$\Phi_{in}(Q) = (\epsilon^2\pi)^{-1/4}e^{-Q^2/2\epsilon^2}. \tag{2}$$

The uncertainty in the initial position of the pointer produces errors of order ϵ; when $\epsilon \to 0$, we recover the ideal measurement. Thus, *any* measured value is possible, although large errors are exponentially suppressed. There is no mystery in such errors; they are expected, given the uncertainty associated with the measuring device. Measurements can even yield negative values. The negative values may be unphysical, but they are part of a distribution representing the measurement of a physical quantity. They should not be thrown out because they give information about the distribution and contribute to the best estimate of the peak value. Given the fact that these errors originate in the measuring device and not in the system under study, it seems that they cannot depend on any property of the system.

However, closer analysis of these errors reveals a pattern that clearly reflects properties of the system under study. The pattern emerges only after selection of a particular final state of the system.

Initially, the particle and device are in a product state $\Psi_{in}(x)\Phi_{in}(Q)$; after the interaction is complete, the state is $e^{-(i/\hbar)Pp^2/2m}\Psi_{in}(x)\Phi_{in}(Q)$, in which the particle and the device are correlated. Now, we consider kinetic energy measurements followed by a final measurement of position, with the particle found far outside the potential well. For the final state, we choose a Gaussian wave packet with its center far from the potential well,

$$\Psi_{fin}(x) = (\delta^2\pi)^{-1/4}e^{-(x-x_0)^2/2\delta^2}, \tag{3}$$

and we require $\delta > \alpha\hbar^2/m\epsilon$. The condition for the particle to be "far enough" from the potential well is

$$\alpha x_0 \gg (\alpha^2\hbar^2/2m\epsilon)^2. \tag{4}$$

Since $\alpha^2\hbar^2/2m = |E_0|$, the expression in the parentheses is the ratio of the magnitude of the effect, $|E_0|$, to the precision of the measurement, ϵ. For more precise measurements of kinetic energy ($\epsilon \to 0$), the final state is selected at increasing distances from the potential well ($x_0 \to \infty$).

The state of the measuring device after the measurement, and after the particle is found in the state $\Psi_{fin}(x)$, is obtained by projecting the correlated state of the particle and measuring device onto the final state of the particle, $\Psi_{fin}(x)$. Apart from normalization, it is $\Phi_{fin}(Q) = \langle\Psi_{fin}|e^{-(i/\hbar)Pp^2/2m}|\Psi_{in}\rangle\Phi_{in}(Q)$. For simplicity, we take $V(x)$ to be a delta-function potential ($a \to 0$). Then, $\Psi_{in}(x)$ is $\sqrt{\alpha}\exp(-\alpha|x|)$. As an integral over x, the final state is

$$\Phi_{fin}(Q) = \int_{-\infty}^{\infty} dx\, e^{-(x-x_0)^2/2\delta^2}e^{-(i/\hbar)Pp^2/2m}e^{-\alpha|x|}\Phi_{in}(Q) \tag{5}$$

up to normalization. Note that the exponential of $-iPp^2/2m\hbar$ acts to translate Q in $\Phi_{in}(Q)$. If we could ignore the part of the integral near $x = 0$, we could replace p^2 with $-\alpha^2$ in equation 5 and the final state of the measuring device would be $\Phi_{fin}(Q) = \Phi_{in}(Q + \alpha^2\hbar^2/2m)$. We cannot ignore this part of the integral, but we can suppress it by choosing x_0 in $\Psi_{fin}(x)$ to be large. If we express $\Psi_{in}(x)$ via its Fourier transform and replace the operator p with its eigenvalue, we obtain (up to a normalizing factor)

$$\Phi_{fin}(Q) = \left(\frac{\pi}{\hbar\alpha}\right) e^{\alpha x_0 - \alpha^2\delta^2/2} \int dp \left[\frac{e^{-p^2\delta^2/2\hbar^2 - ipx_0/\hbar}}{(\alpha^2\hbar^2 + p^2)}\right] \Phi_{in}(Q - p^2/2m). \tag{6}$$

This integral has poles at $p = \pm i\alpha\hbar$; we evaluate it by integration on a contour including a line of p with imaginary part $-ip_0$, for any $p_0 > \hbar\alpha$. The integral in equation 6 then reduces to two terms: a pole term,

$$\Phi_{in}(Q + \alpha^2\hbar^2/2m), \tag{7}$$

and a correction term, the integral in equation 6 with p replaced by $p - ip_0$. The pole term represents the measuring device with its pointer shifted to the negative value of $-\alpha^2\hbar^2/2m$. A short computation (see reference 4) shows that the correction term can be made arbitrarily small by taking x_0 large, as in equation 4. For x_0 large, the final state of the measuring device shows the "unphysical" result of $-\alpha^2\hbar^2/2m$ for the kinetic energy, up to a scatter ϵ characteristic of the device.

We thus obtain a correlation between position measurements and prior kinetic energy measurements: nearly all particles found far outside the potential well yielded negative values of kinetic energy. On the other hand, we could consider all particles that produced negative values of kinetic energy and could ask about their final position. We would find nearly all these particles *inside* the well. The correlation works one way only. Prior kinetic energy measurements on particles found far from the well cluster around a negative value, but position measurements on particles yielding negative values of kinetic energy cluster around zero. How do we interpret this one-way correlation?

INTERPRETATION

Our example suggests that particles in a classically forbidden region have negative kinetic energy. The conventional interpretation of quantum mechanics has no place for negative kinetic energy. However, the conventional interpretation involves an assumption about how measurements are made. The conventional interpretation considers measurements on ensembles of systems prepared in an initial state, without any conditions on the final state of the systems. Such an ensemble, defined by initial conditions only, may be termed a *preselected* ensemble. By contrast, we consider measurements made on *preselected and postselected* ensembles, defined by both initial and final conditions. The experiment of the previous section is an example of a measurement on a preselected and postselected ensemble. It is natural to introduce preselected and postselected ensembles in quantum theory: in the quantum world, unlike the classical world, complete specification of the initial state does not determine the final state.

Also, the measurements that we consider are not *ideal.* Real measurements are subject to error. At the same time, the disturbance they make is bounded. These two aspects of nonideal measurements go together. Suppose our measuring device interacts very weakly with the systems in the ensemble. We pay a price in precision. On the other hand, the measurements hardly disturb the ensemble and therefore they characterize the ensemble during the whole intermediate time. Even noncommuting operators can be measured at the same time if the measurements are imprecise. When such measurements are made on preselected and postselected ensembles, they yield surprising results. An operator yields *weak* values that need not be eigenvalues or even classically allowed.[1,6] The negative kinetic energy of the previous section is an example of a weak value. Another is a measurable value of 100 for a spin component of a spin-1/2 particle.[1]

Let us briefly review how weak values arise. The initial wave function of the measuring device is $\Phi_{in}(Q)$. After an impulsive measurement of an operator C on an initial state $|a\rangle$, and projection onto a final state $|b\rangle$, the final state of the measuring device is $\langle b|e^{-iPC/\hbar}|a\rangle\Phi_{in}(Q) = \Sigma_i\langle b|c_i\rangle\langle c_i|a\rangle\Phi_{in}(Q - c_i)$. If $\Phi_{in}(Q)$ is sharply peaked, then the various terms $\Phi_{in}(Q - c_i)$ will be practically orthogonal. However, suppose $\Phi(Q)$ has a width of ϵ. Its Fourier transform has a width in P of $\hbar/\epsilon$. Small $|P|$ corresponds to a measuring device that is coupled weakly to the measured system. If

ϵ is large, then $|P|$ is small and we can expand the exponential $e^{-iPC/\hbar}$ to first-order in P to obtain $\langle b|e^{-iPC/\hbar}|a\rangle\Phi(Q) \approx \langle b|1 - iPC/\hbar|a\rangle\Phi(Q) \approx \langle b|a\rangle e^{-iPC_w/\hbar}\Phi(Q)$. Here, $C_w \equiv \langle a|C|b\rangle/\langle a|b\rangle$ is the *weak value* of the operator C for the preselected and postselected ensemble defined by $\langle b|$ and $|a\rangle$.

The definition of a weak value provides us with a new and intuitive language for describing quantum processes. In our example, the operators of total energy E, kinetic energy K, and potential energy V do not commute. Therefore, the classical formula $E = K + V$ applies only to their expectation values, and the expectation value of K in any state is positive. However, the formula applies to weak values, $E_w = K_w + V_w$, and the weak value of K is *not* necessarily positive. We know that $E_w = E_0 = -\alpha^2\hbar^2/2m$ because the preselected state is an energy eigenstate, and that V_w vanishes because the postselected state is far from the potential well. Then, $K_w = -\alpha^2\hbar^2/2m$, the "unphysical" result obtained above in our example.

In our example, instead of the condition on the initial state of the measuring device (ϵ large), we had a condition on the final state of the particle (x_0 large and $\delta > \alpha\hbar^2/m\epsilon$). The price is that we must wait for increasingly rare events. As measurements of kinetic energy become more precise ($\epsilon \to 0$), they disturb the particle more. To get negative kinetic energies, we must postselect particles further from the potential well ($x_0 \to \infty$). As the precision of the measurement increases, negative kinetic energies become less and less frequent; in the limit of ideal measurements, the probability vanishes and thus ideal measurements of kinetic energy never yield negative values.

CONCLUSIONS

From the point of view of standard quantum theory, all that we have produced is a game of errors of measurement. Ideal measurements of kinetic energy can yield only positive values because all eigenvalues of the kinetic energy operator are positive. However, in practice, measurements are not exact and, even if their precision is very good, sometimes—rarely—they yield negative values. If particles are subsequently found far from the potential well, we have seen that the measured kinetic energy of these particles comes out negative. Consistently, large measurement "errors" did occur, producing a distribution peaked at the "unphysical" negative value E_0.

What special properties of nonideal measurements led to this result? First, these measurements involve only bounded disturbances of particle position. Second, because their precision is limited, they can supply, "by error", the necessary negative values. These two properties are intimately connected: any measurement of kinetic energy causing only bounded changes of position must occasionally yield negative values for the kinetic energy. The change of x due to the measurement is $\dot{x} = (i/\hbar)g(t)[x, Pp^2/2m]$. P and p are unchanged during the measurement, so $x_{fin} - x_{in} = Pp/m$. From here, it follows that the change of x is bounded only if the pointer is in an initial state with P bounded, that is, if the Fourier transform of $\Phi_{in}(Q)$ has compact support. Then, however, the support of $\Phi_{in}(Q)$ is unbounded,[7] which immediately implies a nonzero probability for the pointer to indicate negative values ($Q < 0$). Indeed, the "game of errors" displays a remarkable consistency, and this consistency allows negative kinetic energies to enter physics in a natural way. The concept of a *weak* value of a quantum operator gives precise meaning to the statement that the kinetic energy of a particle in a classically forbidden region is negative: namely, the weak value of the kinetic energy is negative.

REFERENCES AND NOTES

1. AHARONOV, Y., D. ALBERT & L. VAIDMAN. 1988. Phys. Rev. Lett. **60:** 1351.
2. AHARONOV, Y., J. ANANDAN, S. POPESCU & L. VAIDMAN. 1990. Phys. Rev. Lett. **64:** 2965.
3. AHARONOV, Y., S. POPESCU, D. ROHRLICH & L. VAIDMAN. 1993. *In* Proceedings of the Fourth International Symposium on Foundations of Quantum Mechanics (Tokyo). JJAP Series No. 9, p. 251.
4. AHARONOV, Y., S. POPESCU, D. ROHRLICH & L. VAIDMAN. 1993. Phys. Rev. **A48:** 4084.
5. VON NEUMANN, J. 1955. Mathematical Foundations of Quantum Theory. Princeton University Press. Princeton, New Jersey.
6. AHARONOV, Y. & L. VAIDMAN. 1990. Phys. Rev. **A41:** 11; AHARONOV, Y. & D. ROHRLICH. 1990. *In* Quantum Coherence (Proceedings of the Conference on Fundamental Aspects of Quantum Theory, Columbia, South Carolina). J. S. Anandan, Ed. World Scientific. Singapore.
7. If the Fourier transform of $\Phi_{in}(Q)$ has compact support, then $\Phi_{in}(Q)$ is analytic. The two derivations of our result, via contour integration in the second section and via Taylor expansion of the exponential in the third section, both require $\Phi_{in}(Q)$ to be analytic.

Superluminal (but Causal) Effects in Quantum Physics[a]

RAYMOND Y. CHIAO, JACK BOYCE, AND JOHN C. GARRISON

Department of Physics
University of California
Berkeley, California 94720-7300

It is well known that quantum mechanics implies faster-than-light influences, for example, in the Einstein-Podolsky-Rosen "paradox". These instantaneous actions-at-a-distance, or nonlocal effects, arise in the context of multiparticle systems, where spacelike-separated correlations arise as a result of measurements of an entangled many-body quantum state. However, no genuine signal can be transmitted by these effects. What is less well known is that faster-than-light phenomena can also arise at the single-particle level, for example, in tunneling. We shall review here recent work at Berkeley on this and other closely related superluminal propagation effects. Again, as we shall see, these effects do not permit genuine faster-than-light communication.

At the classical level, Sommerfeld and Brillouin[1] early on treated the problem of propagation of light in dispersive media. They studied the propagation through such media of sinusoidal electromagnetic waves with an abrupt turn-on. When the medium possesses anomalous dispersion, they found it necessary to introduce five different kinds of wave velocities:

(1) the *phase* velocity, at which the zero-crossings of the carrier wave would move;
(2) the *group* velocity, at which the peak of a wave packet would move;
(3) the *energy* velocity, at which energy would be transported by the wave;
(4) the *"signal"* velocity, at which the half-maximum wave amplitude would move;
(5) the *front* velocity, at which the first appearance of a discontinuity would move.

They found that all five velocities differ from each other in the region of anomalous dispersion near an absorption line. The first two kinds of wave velocities, the phase and group velocities, may be superluminal (in fact, they may become infinite or even negative), but the next two kinds, the energy and "signal" velocities, are subluminal under all the circumstances they considered. The last kind, the front velocity, is equal to the vacuum speed of light c. The question naturally arises whether there can be media in which the energy and "signal" velocities, like the phase and group velocities, may become superluminal. Of course, special relativity rules out the

[a]This work was supported by the Office of Naval Research under Grant No. N00014-90-J-1259.

possibility that the front velocity exceeds c, but it is not clear whether or not media with superluminal energy and "signal" velocities can exist. Recently, our group at Berkeley has demonstrated that it is indeed possible to find examples of such media.

Garrett and McCumber[2] have already shown that Gaussian wave packets propagating in the region of anomalous dispersion near an absorption line can in fact propagate at superluminal, infinite, or negative group velocities. Within certain realistic approximations, they showed that an incident Gaussian wave packet will be reshaped by the absorption process (in which later parts of the wave packet are attenuated to a greater extent than the earlier parts) in just such a way as to produce a smaller, but undistorted, Gaussian wave packet at the exit face of the medium. The peak of the wave packet appears to have moved at an abnormal group velocity inside the medium. Chu and Wong[3] verified that this unusual behavior actually occurred for weak picosecond laser pulses propagating near the center of the bound A-exciton line of a GaP:N sample.

Recently, we have verified experimentally that a similar effect occurs in the tunneling of wave packets through a barrier.[4] We showed that single-photon wave packets tunnel superluminally through a 1D photonic bandgap material in the forbidden midgap region. Again, the transmitted wave packets were Gaussian in shape and, although much smaller in amplitude, they had essentially the same shape and width as the incident wave packets. We observed that the peaks of these tunneling wave packets appeared on the far side of the tunnel barrier *earlier* than the peaks of control wave packets, which had propagated through air instead of the barrier. Our results are consistent with the theoretical predictions of MacColl,[5] Eisenbud,[6] and Wigner[7] for the tunneling time based on the stationary-phase method. This theory predicted that the peaks of wave packets would appear superluminally on the far side of the barrier. Again, one can understand this as a pulse-reshaping phenomenon and there is no violation of Einstein causality. Also, recently, superluminal tunneling of classical electromagnetic wave packets was observed in experiments involving microwaves propagating through waveguides beyond cutoff.[8]

One situation in which one might expect superluminal propagation of wave packets is in media with inverted atomic populations. We review here the simplest case of superluminal phase, group, *energy*, and *"signal"* velocities for wave packets propagating in a transparent spectral region far away from the resonance of such an atomic medium, namely, near zero frequency. Sommerfeld and Brillouin's "signal" velocity is rather arbitrarily defined as the propagation velocity of the half-maximum amplitude point of a step-modulated sinusoidal wave. If one adopts this definition for the analytic pulses considered here, then the "signal" velocity for the case of an inverted atomic medium can be equal to the group velocity and can thus also be "superluminal", as we shall see. Quotation marks will henceforth be omitted on "signal" and also on "superluminal", but the reader should remember that there is no genuine violation of Einstein causality because Sommerfeld and Brillouin's *front* velocity in this medium never exceeds the speed of light in the vacuum. Hence, no genuine faster-than-light communication is possible.

In contrast to Sommerfeld and Brillouin, who considered *discontinuous* (step-modulated) signals, we consider here the propagation of *continuous, smooth* wave packets, for example, Gaussians. It is important to note that there is no *information* contained in the peak of an analytic wave packet that is not already contained in its forward tail. Hence, only the *front* velocity is the genuine *information* velocity. The superluminal propagation of wave packets can always be thought of as a pulse-reshaping phenomenon, without any violation of causality. However, in contrast to the previous cases of anomalous dispersion and tunneling, here there is no attenuation of the wave packet after its propagation through the inverted atomic medium because this medium is essentially transparent when it is excited sufficiently far off resonance. One purpose of this report is to dispel the widely held misconception that the group velocity, when it is superluminal, infinite, or negative, is somehow unphysical or unuseful.[9]

Here, we shall consider optical pulse propagation in a wide transparent spectral window far *below* the resonance of a two-level atomic medium with completely inverted populations.[10] Although we shall consider for simplicity only the region near zero frequency for the special case of inverted atoms, it should be noted that the superluminal phenomena discussed here can be found in other transparent spectral regions and also in other media that possess gain, for example, in lasers without inversion.[11] Therefore, the inversion of atomic populations is not an essential prerequisite for superluminality. Nevertheless, let us start by considering a gas of completely inverted two-level atoms in a cell of length L. Treat the electromagnetic field classically. Let a limited-bandwidth (but otherwise arbitrarily shaped) wave packet, whose carrier frequency ω_c and all of its other frequency components lie far below the resonance frequency ω_0 of the two-level atoms, be incident on this medium. The amplitude of this wave packet will be sufficiently small so that only the *linear* response of the medium to this weak perturbation need be considered. The coherent atomic response of the medium to the wave packet, consisting of an ensemble of spatially coherent oscillating atomic dipole moments, gives rise to a macroscopic polarization in the medium, which radiates with such a phase that it *advances* the wave packet. This is in contrast to the more familiar case of an uninverted medium, in which the radiation from the macroscopic polarization *retards* the wave packet, leading to subluminal phase and group velocities. The overall effect of the medium on the wave can be most easily calculated starting from a generalized Lorentz model, in which the atomic oscillator strength f is replaced by its negative $-f$ to characterize an inverted two-level atom. This Lorentz model is a good approximation to the density-matrix equations of motion for a weakly perturbed atomic system[12] and describes the effect of the coherence of the inverted atoms on the propagation of the wave packet. However, note that these results do not depend in any crucial way on the Lorentz model; rather, the crucial assumptions here are (1) that the system responds *linearly* to the classical electromagnetic field and (2) that this response is *causal*, so the Kramers-Kronig relations are valid.

From the generalized Lorentz model, the refractive index of this inverted atomic medium is

$$n(\omega) = \left[1 - \left(\frac{\omega_p^2}{\omega_0^2 - \omega^2 - i\gamma\omega}\right)\right]^{1/2}, \tag{1}$$

where γ is a (small) phenomenological line width and ω_p is the effective plasma frequency,[13]

$$\omega_p = (4\pi N e^2 |f| / m)^{1/2}, \tag{2}$$

where N is the atomic number density, e is the electron charge, and m is the electron mass. The real part of the index calculated from equation 1 is shown in FIGURE 1. In

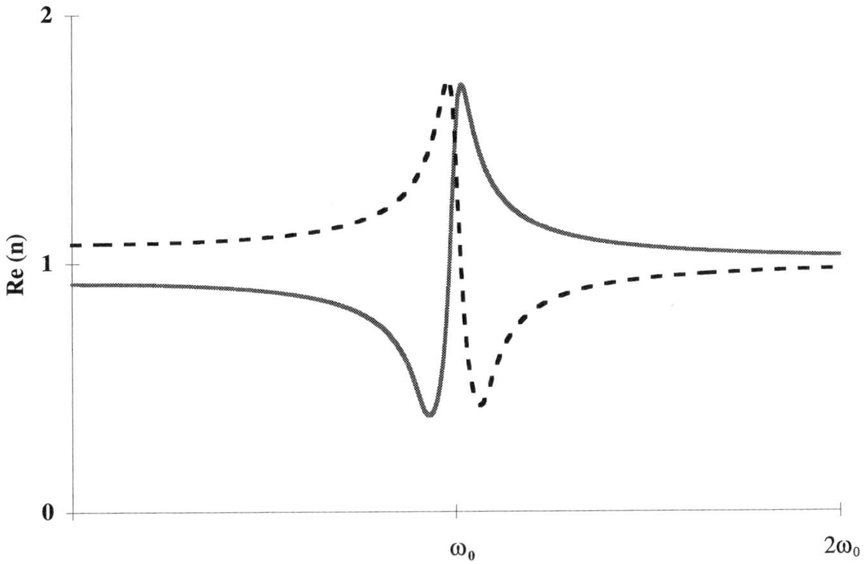

FIGURE 1. The real part of the refractive index for an inverted two-level atomic medium (solid line) calculated from equation 1 as a function of circular frequency ω, compared with that for the same medium with uninverted populations (dashed line). As $\omega \rightarrow 0$, note that the slope of both of these curves vanishes. Therefore, all frequency components sufficiently near zero frequency travel with essentially the same phase velocity. Thus, in a large spectral window below the resonance at ω_0, the medium is essentially transparent and dispersionless for both inverted and uninverted cases. In this window, it follows that the phase, group, energy, and "signal" velocities for the *inverted* atomic medium are all "superluminal". However, a signal with a discontinuous step (or front) possesses high-frequency components extending to infinity. Because $n \rightarrow 1$ as $\omega \rightarrow \infty$, the front velocity is the vacuum speed c and Einstein causality is not violated.

typical situations, the inequalities $\gamma \ll \omega_p \ll \omega_0$ are obeyed. Note that the minus sign in front of the second term on the right-hand side of equation 1 arises from atomic population inversion: it differs from its usual positive sign for an atomic system in its ground state. As a result of this sign change, the index of refraction near zero frequency is less than unity:

$$n(0) = (1 - \omega_p^2/\omega_0^2)^{1/2} < 1. \tag{3}$$

From equation 1, it follows that the slope $d\{\mathrm{Re}\, n(\omega)\}/d\omega$ approaches zero as $\omega \to 0$ (see FIGURE 1). Hence, the medium is essentially dispersionless near DC.

The fact that $n(0) < 1$ implies that the phase velocity,

$$v_p(0) = c/n(0) > c, \tag{4}$$

is greater than the vacuum speed of light c. Near zero frequency, the group velocity,

$$v_g(0) = \frac{c}{\left[n(\omega) + \omega \dfrac{dn}{d\omega}\right]_{\omega_c \to 0}} = \frac{c}{n(0)} = v_p(0) > c, \tag{5}$$

is equal to the phase velocity and is therefore also superluminal; that is, it also exceeds the vacuum speed of light c. Furthermore, in contrast to the *absorptive* media considered by Sommerfeld and Brillouin, here the energy velocity is also superluminal because an *inverted* atomic medium can temporarily give up part of its energy to the forward tail of a signal:

$$v_E(0) = \frac{\langle S \rangle}{\langle u \rangle} = \frac{c}{\sqrt{\epsilon(0)}} = \frac{c}{n(0)} = v_p(0) > c, \tag{6}$$

where $\langle S \rangle$ is the time-averaged Poynting vector, $\langle u \rangle$ is the time-averaged energy density, and $\epsilon(0)$ is the zero-frequency dielectric constant. Also, the half-maximum amplitude velocity (the signal velocity of Sommerfeld and Brillouin) is the same as the group velocity because there is little distortion in the shape of the pulse during its propagation, due to the negligible dispersion in this large spectral window. These velocities are all equal to each other because the inverted atomic medium near zero frequency is essentially transparent and dispersionless. In these respects, it is no different from a normal, uninverted atomic medium near $\omega = 0$. (For details concerning the approximations made in deriving the above wave velocities, see reference 10.)

A more general way to demonstrate these results is to start from the Kramers-Kronig relations for the dielectric constant,[14] from which one can derive the zero-frequency sum rule:

$$\epsilon(0) = 1 + \left(\frac{2c}{\pi}\right) \int_0^\infty \left[\frac{\kappa(\omega')}{(\omega')^2}\right] d\omega', \tag{7}$$

where $\epsilon(0)$ is the zero-frequency dielectric constant and where, given the susceptibility $\chi(\omega)$,

$$\kappa(\omega) = \left(\frac{4\pi\omega}{c}\right) \mathrm{Im}\, \chi(\omega) \tag{8}$$

is the absorption or the gain coefficient of the medium, depending on the sign of $\mathrm{Im}\, \chi(\omega)$. In the case of an inverted two-level atomic medium with a single, isolated gain line,

$$\mathrm{Im}\, \chi(\omega) < 0 \qquad \text{or} \qquad \kappa(\omega) < 0. \tag{9}$$

From the zero-frequency sum rule, it follows that the dielectric constant near zero frequency is less than unity and hence that the index of refraction near zero frequency is also less than unity. Therefore, the phase, group, energy, and signal velocities near zero frequency all exceed the vacuum speed of light, as before. However, because we have now proved this beginning from the Kramers-Kronig relations, it is clear that this result does not depend on the validity of the Lorentz model nor on that of the two-level model. Also, because we have assumed causality from the very beginning, it is clear that causality cannot be violated. The Kramers-Kronig relations imply that any medium with gain must in general give rise to superluminal propagation in a transparent spectral window separate from the region with gain.

To understand how this process of superluminal propagation arises physically, it is helpful to consider the behavior of the electric susceptibility $\chi(0)$ of the medium near DC. In the normal case of an uninverted (absorbing) atomic medium, the susceptibility $\chi(0)$ is always positive, as can be seen from the above zero-frequency sum rule. Thus, if one were to insert such a medium between two charged capacitor plates, energy flows from the electric field into the medium. As a consequence of the fact that energy is initially transferred from the wave to the medium in order to polarize it, the phase shift of the forward scattering amplitude from each atom in this medium near zero frequency has a sign such as to retard the transmitted wave packet. Therefore, the index of refraction of the medium is greater than unity and the propagation speed is subluminal. However, when the medium possesses an inverted atomic population (maintained, for instance, by optical pumping), energy is stored in the medium. It now becomes possible to transfer this energy from the medium to the wave. The DC susceptibility of the medium $\chi(0)$ is now negative. Hence, the sign of the phase shift of the forward scattering amplitude from each atom is reversed. As a result, the transmitted wave packet is advanced rather than retarded and the index of refraction is now less than unity, leading to superluminal propagation.

In analogy with magnetism, where diamagnetism and paramagnetism correspond to the screening and antiscreening, respectively, of the magnetic field by the medium, here we have the phenomena of dielectricity and *parelectricity* corresponding to the screening and antiscreening of the electric field, respectively (see FIGURE 2). *Parelectric* susceptibilities $\chi(0) < 0$ always lead to superluminal phase velocities of low-frequency (e.g., radio) waves through the relation, $n(0) = [1 + 4\pi\chi(0)]^{1/2}$, and thus to superluminal group velocities for wave packets because the medium is essentially dispersionless at low frequencies. Also, just as a small magnet can be levitated above a superconducting surface, here a small ferroelectric sphere can in principle be levitated above a strongly parelectric surface. The origin of this phenomenon lies in the fact that the image charge seen by a real charge outside the parelectric medium has the *same* sign, leading to electrostatic *repulsion.* This levitation phenomenon should occur at room temperature, in contrast to the low temperatures needed for the analogous superconducting phenomenon. However, note that parelectricity is only possible in media *not* in thermodynamic equilibrium.[14]

In the case of normal (uninverted) transparent media, it is possible to view wave packet propagation as a pulse-reshaping process, in which the early part of the incident wave packet is absorbed by the medium in the process of virtual absorption,

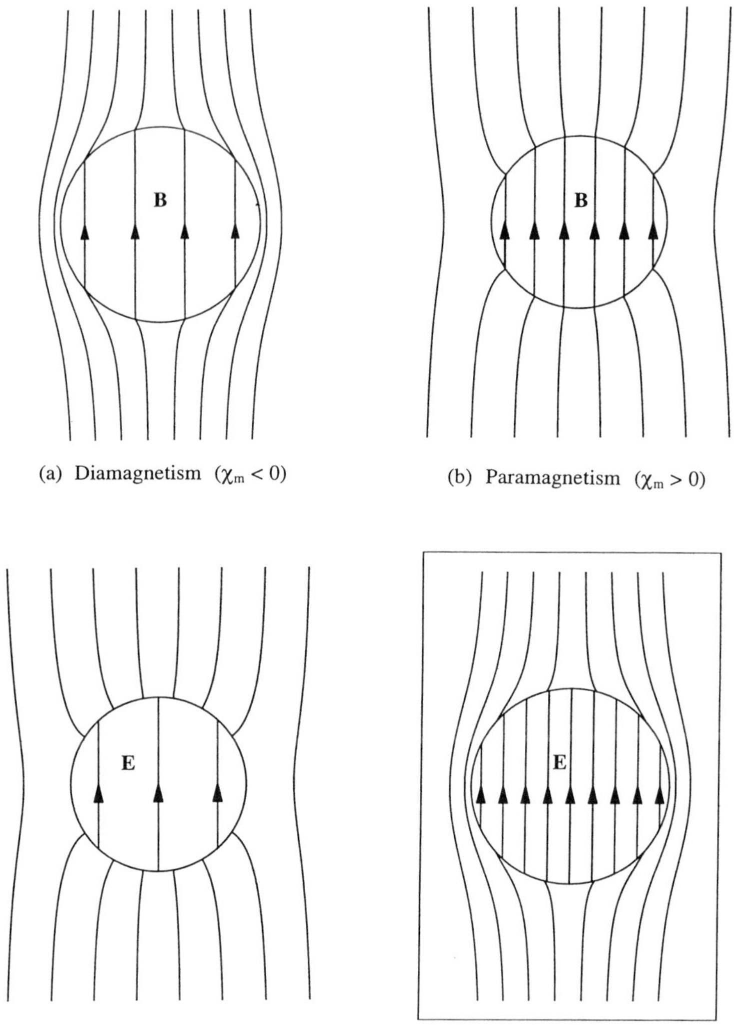

(a) Diamagnetism ($\chi_m < 0$) (b) Paramagnetism ($\chi_m > 0$)

(c) Dielectricity ($\chi_e > 0$) (d) Parelectricity ($\chi_e < 0$)

FIGURE 2. The response of media to an applied static ***B*** or ***E*** field: (a) diamagnetism (the screening of the ***B*** field, as in the Meissner effect), (b) paramagnetism (the antiscreening of the ***B*** field, as in the alignment of electron spins), (c) dielectricity (the screening of the ***E*** field, as in ordinary dielectrics), and (d) parelectricity (the antiscreening of the ***E*** field, as in media with inverted atoms; see FIGURE 1). The last case is always accompanied by superluminal propagation (e.g., of radio-frequency wave packets) because the medium is dispersionless near DC. Note that the *parelectric* medium should not be confused with the *paraelectric* one, which is a ferroelectric medium above its Curie point.

followed by the process of emission during the later part of the incident wave packet, thus producing a retarded transmitted wave packet (see FIGURE 3a). It should be emphasized that no real transitions occur during these processes. In particular, the emission during the later part of the incident pulse, for example, the trailing

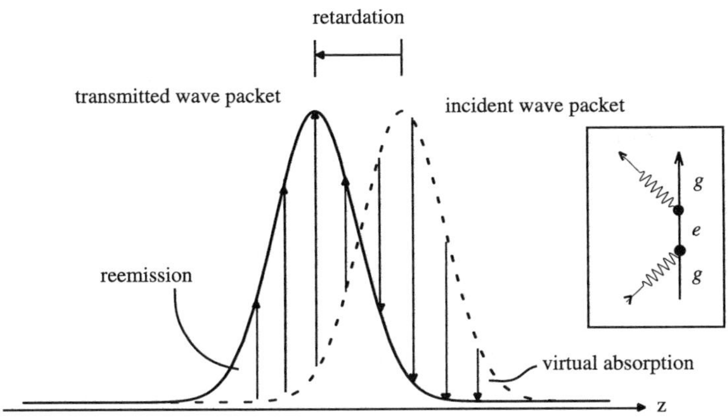

(a) normal, uninverted atomic medium

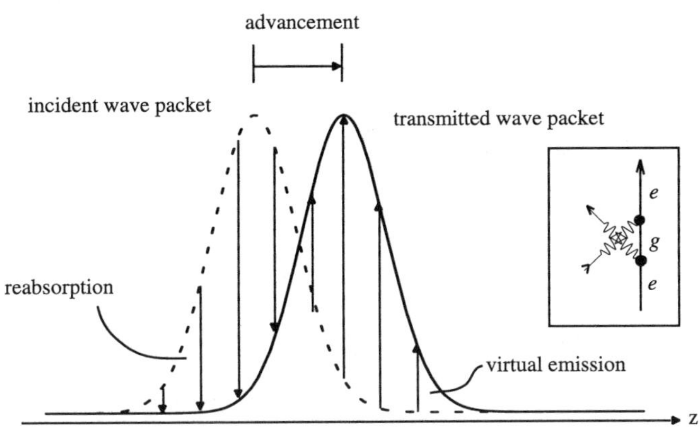

(b) inverted atomic medium

FIGURE 3. (a) A wave packet propagating through an uninverted atomic medium and (b) a wave packet propagating through an inverted atomic medium. The insets depict Feynman diagrams where (a) virtual absorption by an atom in its ground state g is followed by reemission by the atom in its excited state e and (b) virtual emission by an atom in the excited state e precedes reabsorption by the atom in its ground state g, thus leading to an advanced wave packet and hence superluminal propagation.

exponential tail of a Gaussian wave packet, is not accompanied by any real stimulated emission. It therefore follows that there cannot exist spontaneous emission accompanying the emission process during the trailing tail of the incident pulse. Hence, no quantum noise is added to the transmitted wave packets in these transparent media, in spite of the large apparent amplification of the exponentially small tails of the incident wave packet during the pulse-reshaping process. Note that the retardation of the transmitted wave packet due to this medium can be arbitrarily large, so the apparent amplification of the exponentially small tails of the Gaussian must also be arbitrarily large, and yet there is no accompanying addition of noise during this propagation process.

Likewise, in the case of transparent media with inverted atomic populations, it is also possible to view wave packet propagation as a pulse-reshaping process, in which the early part of the incident wave packet is apparently amplified by the medium in the process of virtual emission, followed by the process of absorption of the later part of the incident wave packet, thus producing an advanced transmitted wave packet (see FIGURE 3b). Expressed as a Feynman diagram, this process corresponds to a crossed diagram (see the inset of FIGURE 3b), in which virtual emission *precedes* absorption, and thus leads to an *advanced* wave packet. Again, no real transitions occur during these processes. In particular, the virtual emission during the early part of the incident pulse, for example, the leading exponential tail of a Gaussian wave packet, is not accompanied by any real stimulated emission. Therefore, it follows that there cannot exist any spontaneous emission during the virtual emission process. Hence, virtually no noise is added to the transmitted wave packets in these transparent media. However, in contrast to the case of the uninverted medium, here spontaneous emission from the inverted atoms in the extreme low-frequency tails of the Lorentzian gain line can exist in principle. Nevertheless, spontaneous emission near zero frequency is suppressed due to the two facts that Im $\chi(\omega) \rightarrow 0$ as $\omega \rightarrow 0$ and that the phase space volume for the spontaneously emitted photon rapidly vanishes as $\omega \rightarrow 0$, so an arbitrarily small amount of noise can in principle be added to the leading tail of the transmitted wave packet.

To sum up, the Kramers-Kronig argument demonstrates that, in principle, Sommerfeld and Brillouin's phase, group, energy, and signal velocities can all be superluminal far below a gain line, without violating causality.[10,15] The front velocity, however, is never superluminal because a discontinuity possesses frequency components that extend to infinity and thus propagate at the vacuum speed c (see FIGURE 1). Note that the infinite-frequency (or "f") sum rule implies that any real medium must have spectral regions with absorption as well as gain. Nevertheless, it should be possible in practice to observe superluminal effects outside of a strong gain line. For example, a sufficiently strong, low-frequency gain line can give the dominant contribution to the zero-frequency sum rule, in spite of stronger absorption lines at higher frequencies. As a second example, for the case of two nearby gain lines of comparable strength, Steinberg and Chiao[16] have shown that there exists a point between these lines with a highly superluminal group velocity, but with zero group-velocity dispersion, so nearly dispersionless superluminal propagation is also possible there. As a third example, within a bandwidth of approximately half the effective plasma frequency on either side of a strong gain line, but outside the gain bandwidth γ (assuming that $\omega_p \gg \gamma$), Bolda *et al.*[17] have recently predicted that *negative* group velocities should be observable. A possible practical problem with the above schemes might be the onset of oscillation due to spurious feedback, which will produce laser action on the gain line; however, even in the presence of such lasing, the gain will clamp at a value equal to the loss and thus will still lead to substantial superluminal effects.

These predicted superluminal effects are different from the pulse-reshaping phenomena, which arise from the propagation of laser pulses *inside* the gain band, known as "superluminous" propagation.[2,18] Here, we are *outside* of the gain band, in a spectral window in which the medium is essentially transparent. It is important to differentiate "superluminous" from "superluminal" propagation. In the former, there exists real amplification and hence real transitions associated with stimulated emissions occurring during the pulse reshaping, whereas there are only virtual transitions in the latter. Because spontaneous emissions always accompany stimulated emissions, "superluminous" propagation is always accompanied by noise, whereas "superluminal" propagation is virtually noise-free.

These effects, of course, are also different from the highly subluminal ones in transparent spectral regions next to an *absorption* line.[19–21] (Obviously, the subluminal case does not even pose an apparent conflict with relativity, whereas the same is not true of the superluminal case, which makes it the more important one to understand.) Moreover, these effects are different from the previously observed superluminal group velocities of wave packets propagating inside an absorption band discussed above.[3] Here, the transmitted wave packets are unattenuated; furthermore, Sommerfeld and Brillouin's energy and signal velocities are now superluminal. These effects are different from the recently observed superluminal tunneling of photonic wave packets[4,8] for the same reasons. Also, recently, a superluminal nonlinear pulse-reshaping phenomenon has been observed,[22] but here we shall restrict ourselves only to *linear* phenomena.

The explicit calculations discussed so far have all been carried out using the classical Maxwell equations and this immediately raises the question of their quantal significance. The experimental situations under discussion involve the propagation of weak, off-resonance fields; consequently, they are described by the linear Maxwell equations for a fixed medium. Because this classical model provides no scale of field strengths, the natural scale is set by the vacuum fluctuation fields that occur in the quantum theory. Thus, the question is whether quantum fluctuations can somehow wipe out the superluminal propagation phenomena observed in the classical theory. The answer to this question is no, provided that the classical solutions can be given a Born-like interpretation as "photon wave functions". This approach requires some care because of the nonexistence of a photon position operator or, equivalently, the nonexistence of a local number operator representing the number of photons in a finite volume.[23] Fortunately, this fundamental property of the electromagnetic field does not pose a serious difficulty in the cases of interest. These experimental situations satisfy the conditions for the slowly varying envelope approximation and it has been argued by several researchers[24] that there is a uniquely defined local number operator in that limit. This supports a Born-like interpretation of the classical wave packet; that is, the photodetection rate will be proportional to the square of the field. Thus, the advancement of the classical wave packet corresponding to a superluminal group velocity can be interpreted as an advancement of the time at which the photon counting rate is a maximum. The classical group velocity is the most-probable-photon velocity, that is, the single-photon wave packet group velocity, which can therefore become superluminal. In this context, the fact that the classical front velocity is the vacuum speed of light is closely related to the vanishing of the field commutators for spacelike separations. This fact has been used to show[25] that no information can be transmitted superluminally; this suggests that the arbitrary definition of signal velocity used by Sommerfeld and Brillouin should be replaced by the front velocity as the information velocity in the quantum theory. However, the meaning of information at the quantum level needs further clarification.

As further support for the general arguments adduced above, we have formulated a microscopic, second-quantized model for the quantum propagation of few-photon states through an inverted system of two-level atoms and we are currently beginning to carry out stochastic simulations of the operator equations. We will also investigate an equivalent formulation in which one solves the equations of motion for photon correlation functions. This procedure has the advantage that the correlation functions directly describe the results of counting experiments. Both of these techniques will allow us to investigate the influence of spontaneous emission in the few-photon limit. As pointed out above, there are essentially no real transitions, and hence no stimulated emission, associated with the low-frequency tail of the Lorentzian gain line, but the continuous pumping required to maintain the inverted system can lead to superfluorescence or amplified spontaneous emission near the line center. The effects of these phenomena on the superluminal propagation of off-resonant test wave packets will also be calculated by means of the microscopic model.

We are performing experiments to demonstrate these striking phenomena using a resonantly enhanced stimulated Raman transition in an optically pumped rubidium vapor cell to generate a strong gain line. Pulses propagating in transparent regions with bandwidths on the order of a gigahertz adjacent to this line should exhibit negative group velocities, in which the peak of the transmitted wave packet should leave the exit face of the cell before the peak of the incident wave packet would enter the entrance face of the cell.

Applications of this superluminal effect extend beyond optics. In electronics, one can use this effect to compensate for the retardations of signals (e.g., in computer circuits) caused by ordinary dielectrics, but not beyond the limit set by Einstein causality. In astrophysics, superluminal group velocities may be responsible for certain phenomena observed in radio galaxies. In condensed matter physics, superluminal, indeed infinite, group velocities are also possible. One example may arise in the "incompressible quantum fluid" described by the Laughlin wave function for the quantum Hall system.[26] Because of the scarcity of low-lying excited states, this system is "incompressible"; that is, in its linear response to weak perturbations, it remains *everywhere* adiabatically in its ground state, implying an infinite group velocity for these weak perturbations.

Another example of superluminality may occur in superfluid helium, arising from what London called "the rigidity of the macroscopic wave function", which is associated with the entangled many-body quantum state of the superfluid. Long-range, macroscopic quantum interference in superfluid ^{4}He manifests itself in the quantization of vortex circulation in units of h/m, where m is the mass of the ^{4}He atom.[27] Due to the Bose nature of the ^{4}He atoms, long-range atomic coherence between atoms in the superfluid, no matter how distant they are from each other, becomes possible. Again, due to a scarcity of low-lying excited states, the superfluid system stays adiabatically *everywhere*, and hence "rigidly", in its ground state when it is weakly perturbed on an atomic scale.

Specifically, all the streamlines around a weakly perturbed vortex line, whose velocity remains at all times less than the Landau critical velocity, follow the moving vortex center *instantaneously* in the quantum adiabatic approximation. Note that the spatial scale of the perturbation is determined by the vortex core size, which is on the order of angstroms. From the known dispersion relation for the elementary excitations of the superfluid, it follows that the adiabatic approximation should be valid when the frequency components of the motion of the vortex core are less than the roton gap frequency (180 GHz). (Note that the roton gap frequency does not depend on the spatial extent of the superfluid.) Then, the many-body wave function can adjust itself everywhere adiabatically, and hence instantaneously (within a time scale set by the inverse of the roton gap frequency), to the changing position of the vortex core. This implies an effectively instantaneous response of all streamlines, no matter how far away they are from the vortex center, to the motion of the vortex core. This conclusion also follows from solutions of the Kirchhoff-Onsager equations.[27,28]

The physical origin of this "rigidity" lies in the *discreteness* of phonon emission (i.e., the fact that the vortex core can only emit zero, one, two, etc., phonons, but not a fraction of a phonon). Combined with the restriction that the vortex velocity never exceeds the Landau critical velocity, this implies that phonon emission by the vortex core is forbidden at all times during its motion. Because no sound waves can be emitted at all by the gently perturbed vortex core, the speed of sound cannot be the relevant speed responsible for the retarded response of distant streamlines to these weak perturbations. (If we assume the contrary, then a sound pulse would be produced at the sonic front upon the application of a weak pulsed perturbation to the vortex, contradicting the assumption that no sound is emitted at all by the vortex core.) This is similar to the Mössbauer effect, in which the entire crystal in a zero-phonon state recoils in recoilless nuclear emission of gamma rays: here, the entire superfluid stays adiabatically in the *zero-phonon, zero-roton state* and thus responds instantly everywhere, and hence "rigidly", to the motion of the vortex core. This situation is also similar to that of the Born-Oppenheimer approximation, in which the electronic wave function adjusts itself effectively instantly everywhere inside a molecule to slow changes in the configuration of its nuclei. This *effective* instantaneous-action-at-a-distance (i.e., infinite phase and group velocities) only works for gentle perturbations of the vortex position that are smooth and finite in bandwidth, for example, for a Gaussian pulse whose frequencies are much less than the roton gap frequency, so the adiabatic approximation is valid. Again, as in the above optical case, Einstein causality is not violated.

There results the possibility of long-range mass currents that can oscillate at high frequencies. This leads to a possible application to gravitational antennas: These high-frequency mass currents can generate as well as detect gravitational radiation, in a superfluid helium gravitational wave antenna. Because of the linear, reciprocal nature of this antenna, it can be used as both an emitter and a receiver of gravitational radiation, with equal efficiencies.[29] The efficiency of existing gravitational antennas, such as a Weber bar constructed out of normal matter, is extremely small because the velocity of sound v_s, whose magnitude is a measure of the rigidity of the bar, is typically five orders of magnitude smaller than the velocity of light c. As a result, the length of the Weber bar is far too short compared to the gravitational radiation wavelength for it to be an efficient antenna. Because the Weber bar is a mass-quadrupole antenna, its efficiency[30] is reduced by at least a factor of $(v_s/c)^6 \approx 10^{-30}$ from an analogous electromagnetic antenna. The search for an efficient gravitational antenna is therefore the search for an extremely rigid or incompressible state of matter, in which the effective velocity of sound for some unusual mode is luminal or superluminal. In a rigid system, we can make the individual parts radiate together coherently, whereas, in a typical Weber bar, the velocity of internal mechanical modes is quite small and the individual parts are phased very poorly (destructive interference results). The question arises as to whether or not the "rigidity of the macroscopic wave function" in superfluid helium, in the sense discussed above, may provide a solution to this problem. [Although light can be used in a gravitational antenna, as in LIGO (Laser Interferometer Gravitational wave Observatory), the mass contained in the light is extremely small, so again the antenna efficiency is extremely low.]

One way to construct such an antenna is to attach a charged particle, for example, an electron, to the core of a vortex in superfluid helium. The binding energy[31] of an electron to the vortex core (1040 GHz) is much larger than the roton gap energy (180 GHz), so it is *rigidly* attached at all times (again, in the quantum adiabatic approximation, the electron stays adiabatically in the lowest bound state of the potential well formed by the gently moving vortex core; also, although the electron forms a bubble of a size of 20 Å, it cannot generate any elementary excitations when it moves slower than the Landau critical velocity). Now, let us drive (by means of the Magnus force) high-frequency, small-amplitude oscillatory motions of the vortex by means of an electromagnetic wave, for example, in the few gigahertz range. In this way, macroscopic, high-frequency (but adiabatic) mass currents throughout the superfluid can be driven by electromagnetic radiation. To calculate these currents, we start from Kirchhoff-Onsager equations of vortex motion, modified to include the effect of a tightly bound electron coupled to external electromagnetic fields:

$$\frac{dx_i}{dt} = \left[\frac{Q'E_{0y}}{\rho_m \kappa(1 + R_1)}\right] \sin \Omega t + \left(\frac{\kappa}{4\pi}\right)\left[\frac{(1 + R_2)}{(1 + R_1)}\right] \sum_{j \neq i} \frac{(y_i - y_j)}{r_{ij}^2}, \tag{10}$$

$$\frac{dy_i}{dt} = -\left[\frac{Q'E_{0x}}{\rho_m \kappa(1 + R_1)}\right] \sin \Omega t - \left(\frac{\kappa}{4\pi}\right)\left[\frac{(1 + R_2)}{(1 + R_1)}\right] \sum_{j \neq i} \frac{(x_i - x_j)}{r_{ij}^2}, \tag{11}$$

$$r_{ij}^2 = (x_i - x_j)^2 + (y_i - y_j)^2, \qquad R_1 = \frac{Q'B}{\rho \kappa c}, \qquad \text{and} \qquad R_2 = \frac{4\pi(Q')^2}{\rho \kappa^2}, \tag{12}$$

where ρ_m is the mass fluid density, Q' is the charge per unit length on a vortex, and $\kappa = h/m$ is the unit of vorticity.

In this section, we explicitly calculate the gravitational radiation emitted by an oscillating superfluid vortex. The calculation must be done within the framework of classical general relativity, where space-time curvature is related to massive sources by the Einstein field equation:

$$R_{\mu\nu} - \left(\frac{1}{2}\right) g_{\mu\nu} R = \left(\frac{8\pi}{c^4}\right) G T_{\mu\nu} \qquad (\mu, \nu = 0, 1, 2, 3), \tag{13}$$

where $R_{\mu\nu}$ is the Ricci tensor, $g_{\mu\nu}$ is the metric, $R = R^{\mu}_{\mu}$ is the scalar curvature, G is Newton's constant, and $T_{\mu\nu}$ is the stress-energy tensor. In our case, the sources are quantum mechanical in nature, so we make the assumption that $T_{\mu\nu}$ should be replaced by its quantum expectation value $\langle T_{\mu\nu}\rangle$, just as in semiclassical optics one substitutes $\langle P_i\rangle$ for the polarization P_i as the source in the wave equation $\nabla^2 \boldsymbol{E} - (1/c^2)\partial^2 \boldsymbol{E}/\partial t^2 = (4\pi/c^2)\partial^2 \boldsymbol{P}/\partial t^2$. When the field equations are linearized [$g_{\mu\nu} = \eta_{\mu\nu} + h_{\mu\nu}$, where $\eta_{\mu\nu} = \mathrm{diag}(-1, +1, +1, +1)$ and $|h_{\mu\nu}| \ll 1$], they reduce for nonrelativistic sources to[32]

$$\nabla^2 h_{0i} - \left(\frac{1}{c^2}\right)\left(\frac{\partial^2}{\partial t^2}\right) h_{0i} = -\left(\frac{16\pi G}{c^3}\right) j_i \qquad (i = 1, 2, 3), \tag{14}$$

where j_i is the mass current density, and we have chosen the gauge where $h_{00} = 0$, $h_{ij} = 0$, and $h_{0i} \neq 0$. The problem is formally identical to electromagnetic radiation in the Lorentz gauge, where $\nabla^2 \boldsymbol{A} - (1/c^2)\partial^2 \boldsymbol{A}/\partial t^2 = -(4\pi/c)\boldsymbol{j}$. The solution for the vector potential is

$$\boldsymbol{A}_{\mathrm{EM}}(t) = \left(\frac{1}{c}\right) \int_V \frac{\boldsymbol{j}}{r}\bigg|_{t-r/c} dV, \tag{15}$$

where r is the distance from the observer to a given infinitesimal source volume element and V is the volume of the radiator. The solution to equation 14 can be found by replacing on the right-hand side of equation 15 the electrical current density by the mass current density as the source and by replacing on the left-hand side the vector potential by the components of the metric h_{0i} as the generated fields. In other words, we can calculate the electromagnetic power (treating the mass currents as electrical currents) and then substitute $e^2 \to 4Gm^2$ at the end to find the radiated gravitational power.

The flow of helium around a superfluid line vortex is quantized according to the Feynman-Onsager relation:

$$\oint_C \boldsymbol{v}_{\pm} \cdot d\boldsymbol{l} = \pm \frac{h}{m}, \tag{16}$$

where the oriented path C contains a single vortex of topological charge ± 1 inside. For a single vortex in an infinite fluid, we must have, by symmetry,

$$\boldsymbol{v}_{\pm}(\boldsymbol{r}) = \pm(\hat{z} \times \hat{\boldsymbol{r}})\left(\frac{\hbar}{mr}\right). \tag{17}$$

These velocity fields add linearly when several vortices are present. Also, there is a finite vortex core radius r_0 to prevent divergence near $r = 0$, where $\hbar/mr_0$ is roughly the Landau critical velocity. However, for ^{4}He, this core size is microscopically small and can be ignored for the remainder of the calculation.

The geometry of a possible gravity wave emitter is shown in FIGURE 4. A cylindrical Dewar of radius R contains a single line vortex near the wall. To this vortex, we attach charges, as discussed above, and then apply an oscillating electric field to cause radial harmonic motion of the vortex center. (The Magnus force law causes the vortex to actually move perpendicular to the applied field.) Care must be taken so that the velocity of the vortex never exceeds the Landau critical velocity, roughly 50 m/s. The oppositely charged image vortex at radius R^2/r acts to satisfy the boundary condition that the radial component of the velocity field must vanish at the Dewar wall.

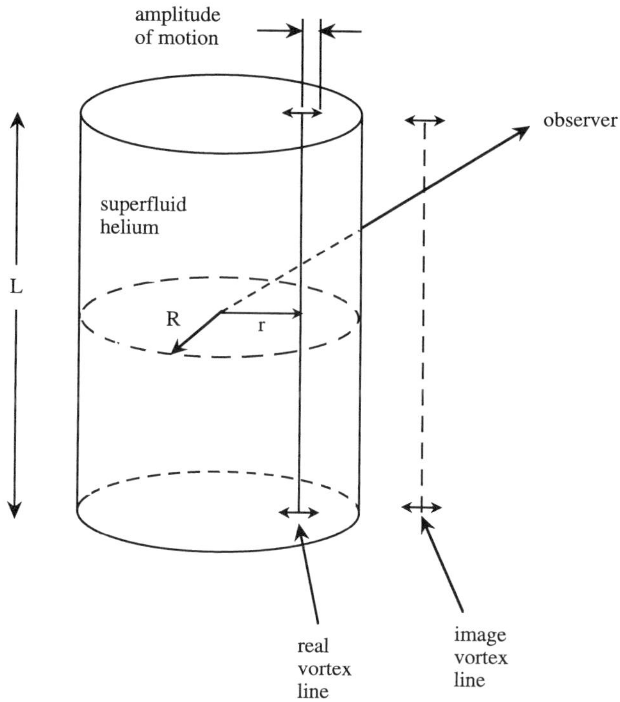

FIGURE 4. Gravitational wave antenna consisting of a vortex line in superfluid helium undergoing simple harmonic motion near a cylindrical Dewar wall.

To complete the calculation, we need to make the following assumptions: (1) the system is changing slowly enough that no elementary excitations can occur; thus, by the quantum adiabatic theorem,[33] the system remains in its instantaneous ground state; (2) the vortex remains parallel to the Dewar's axis at all times, with no pinning at the ends lest longitudinal vortex wave (Kelvin) modes be excited; (3) the Dewar walls are infinitely massive. Assumption 2 requires a somewhat idealized situation, whereas assumption 3 ensures that the fluid center of mass cannot move. These assumptions allow us to use the known fluid flow pattern to calculate A_{EM} for an observer in the radiation zone using equation 15 for the equivalent electromagnetic problem. This solution gives $\boldsymbol{E}$ and $\boldsymbol{B}$ and hence the Poynting vector $\boldsymbol{S}$. As mentioned above, we then substitute $\rho_e^2 \to 4G\rho_m^2$ to get the gravitational power flux. The resulting power is

$$P_{grav} = \left(\frac{\hbar^2 G \rho_m^2 L^2}{m^2 c}\right)\left(\frac{v_0}{c}\right)^2 \int I(\Omega, kR)\, d\Omega, \tag{18}$$

where m is the ^{4}He atomic mass, v_0 is the oscillation speed of the vortex core relative to the Dewar ($v_0 < v_{max} \approx 50$ m/s), L is the length of the Dewar, $kR = (\omega/c)R$ is the size of the system relative to a wavelength, and $I(\Omega, kR)$ is an angular factor of order unity. Using an external electric field to accelerate the vortex, the equations of motion (equations 10–12) yield the oscillation speed, $v_0 = Q'E_0/\kappa\rho_m$, and substitution into equation 18 gives

$$P_{grav} = \left(\frac{GQ^2E_0^2}{c^3}\right) \int I(\Omega, kR)\, d\Omega, \tag{19}$$

where Q is the total charge on the vortex. This leads to the Larmor-like formula for total power radiated gravitationally by an accelerated mass M, whose acceleration is a:

$$P_{grav} \approx (8/3)GM^2a^2/c^3 = (8/3)GF^2/c^3, \tag{20}$$

where $F = Ma$ is the force exerted on this mass. Here, $F = QE_0$ is the force exerted by the external field on the charged vortex line. To get an estimate of the size of this effect, we assume $Q = 300e$ and $E_0 = e/a_0^2 \approx 6 \times 10^9$ V/cm. Then, $P_{grav} \approx 2 \times 10^{-45}$ W, an undetectably small quantity for a single vortex line. In effect, only the matter within a distance d of the vortex line, where d is the amplitude of its oscillation, is radiating appreciably. This power would increase by a factor of N^2 due to coherence for the case of N vortex lines moving synchronously, but it would be difficult to achieve the vortex line density required to obtain a measurable power output. An alternative geometry would be a system of vortex rings, but we have not yet done the explicit calculations.

ACKNOWLEDGMENTS

R. Y. Chiao would like to thank John Archibald Wheeler for early, but lifelong lessons in quantum mechanics and in general relativity. We thank Eric Bolda, Jason Bowie, and Aephraim Steinberg for helpful discussions and Jeff Chamberlain for help in drawing the figures.

REFERENCES

1. BRILLOUIN, L. 1960. Wave Propagation and Group Velocity. Academic Press. New York.
2. GARRETT, C. G. B. & D. E. MCCUMBER. 1970. Phys. Rev. A **1:** 305.

3. CHU, S. & S. WONG. 1982. Phys. Rev. Lett. **48:** 738; Phys. Rev. Lett. **49:** 1293.
4. STEINBERG, A. M., P. G. KWIAT & R. Y. CHIAO. 1993. Phys. Rev. Lett. **72:** 708.
5. MACCOLL, L. A. 1932. Phys. Rev. **40:** 621.
6. EISENBUD, L. 1948. Ph.D. thesis (unpublished). Princeton University. Princeton, New Jersey.
7. WIGNER, E. P. 1955. Phys. Rev. **98:** 145.
8. ENDERS, A. & G. NIMTZ. 1993. J. Phys. I (France) **3:** 1089.
9. STRATTON, J. A. 1941. Electromagnetic Theory, p. 334. McGraw–Hill. New York; BORN, M. & E. WOLF. 1975. Principles of Optics, p. 23. Pergamon. Elmsford, New York; JACKSON, J. D. 1975. Classical Electrodynamics, p. 302. Wiley. New York.
10. CHIAO, R. Y. 1993. Phys. Rev. A **48:** R34.
11. SCULLY, M. O., S. ZHU & A. GAVRIELIEDES. 1989. Phys. Rev. Lett. **62:** 2813.
12. BOYD, R. W. 1992. Nonlinear Optics, p. 209. Academic Press. New York.
13. BURNHAM, D. C. & R. Y. CHIAO. 1969. Phys. Rev. **188:** 667.
14. LANDAU, L. D. & E. M. LIFSHITZ. 1960. Electrodynamics of Continuous Media. Pergamon. Elmsford, New York.
15. BOLDA, E. L., R. Y. CHIAO & J. C. GARRISON. 1993. Phys. Rev. A **48:** 3890.
16. STEINBERG, A. M. & R. Y. CHIAO. 1994. Phys. Rev. A **49:** 2071.
17. BOLDA, E. L., R. Y. CHIAO & J. C. GARRISON. 1994. Phys. Rev. A **49:** 2938.
18. PICHOLLE, E. *et al.* 1991. Phys. Rev. Lett. **66:** 1454; FISHER, D. L. & T. TAJIMA. 1993. Phys. Rev. Lett. **71:** 4338.
19. GRISCHKOWSKY, D. 1973. Phys. Rev. A **7:** 2096; 1974. IEEE J. Quantum Electron. **QE-10:** 723; Appl. Phys. Lett. **25:** 566; WIGMORE, J. K. & D. GRISCHKOWSKY. 1978. IEEE J. Quantum Electron. **QE-14:** 310.
20. CASPERSON, L. & A. YARIV. 1971. Phys. Rev. Lett. **26:** 293; SIEGMAN, A. E. 1986. Lasers (Chapter 9). University Science Books. Mill Valley, California.
21. HARRIS, S. E., J. E. FIELD & A. KASAPI. 1992. Phys. Rev. A **46:** R29.
22. DA SILVA, Y. L., Y. SILBERBERG & J. P. HERITAGE. 1993. Opt. Lett. **18:** 580.
23. NEWTON, T. D. & E. P. WIGNER. 1949. Rev. Mod. Phys. **21:** 400; WIGHTMAN, A. S. 1962. Rev. Mod. Phys. **34:** 400; JAUCH, J. M. & C. PIRON. 1967. Helv. Phys. Acta **40:** 559.
24. MANDEL, L. 1964. Phys. Rev. **136:** 1221; DEUTSCH, I. H. & J. C. GARRISON. 1991. Phys. Rev. A **43:** 2498.
25. EBERHARD, P. H. & R. R. ROSS. 1989. Found. Phys. Lett. **2:** 127.
26. LAUGHLIN, R. B. 1983. Phys. Rev. Lett. **50:** 1395.
27. ONSAGER, L. 1949. Nuovo Cimento Suppl. **6:** 279; FEYNMAN, R. P. 1955. *In* Progress in Low Temperature Physics. C. J. Gorter, Ed. North-Holland. Amsterdam.
28. CRESWICK, R. J. & H. L. MORRISON. 1980. Phys. Lett. A **76:** 267; LUND, F. 1991. Phys. Lett. A **159:** 245.
29. CHIAO, R. Y. 1982. Phys. Rev. B **25:** 1655; ANANDAN, J. & R. Y. CHIAO. 1982. Gen. Rel. Grav. **14:** 515.
30. WEINBERG, S. 1972. Gravitation and Cosmology, p. 271. Wiley. New York.
31. DONNELLY, R. J. 1967. Experimental Superfluidity, p. 50. University of Chicago Press. Chicago.
32. LANDAU, L. D. & E. M. LIFSHITZ. 1951. The Classical Theory of Fields, p. 326. Addison–Wesley. Reading, Massachusetts.
33. SCHIFF, L. I. 1955. Quantum Mechanics, p. 215. McGraw–Hill. New York.

Qualitative View of Quantum Shot Noise

ROLF LANDAUER

IBM Thomas J. Watson Research Center
Yorktown Heights, New York 10598

INTRODUCTION

The last decade has seen an explosion of interest in electron transport in mesoscopic systems.[1,2] In these systems, we can view much of the quantum mechanical world known to us from atoms and molecules, in laboratory structures. Noise in such systems is a subject of growing interest and is more sensitive to the correlation between electrons than steady state transport. The correlations involved include those due to the Pauli exclusion principle, as well as those due to the Coulomb interaction between electrons. Thus, for example, we can calculate the electrical conductance of an elastically scattering sample, in the noninteracting electron approximation, from the transmission of wave packets without regard to the occupation probability of the receiving reservoir.[3] On the other hand, in noise calculations,[4] we have to go beyond that. If an incident wave packet from the left-hand reservoir and one from the right scatter into the same outgoing state, in the noise calculation we have to allow for the relative probability with which none, one, or both of the incoming states are occupied. Let us make this distinction between noise and conductance in an even simpler way. If we calculate the conductivity of a metal from the classical Drude formula, $\sigma = ne^2\tau/m^*$, we get a reasonable answer (n is the electron density, τ is the mean free time, and m^* is the effective mass). On the other hand, in the evaluation of noise currents, the totally occupied energy ranges, way below the Fermi level, do not contribute. A stream of electrons, for example, a succession of wave packets in a given energy range, which is fully occupied, is noiseless. This was understood more than 50 years ago,[5] in contradiction to some implications in the mesoscopic noise literature, suggesting that as modern understanding.

An electron stream is regulated not only by the Pauli principle, but also by the Coulomb interaction between electrons. This was well understood in the analysis of noise in vacuum tubes.[6] In a vacuum diode, where the applied voltage is large enough so that the space charge field due to the emitted electrons is negligible, all thermally emitted electrons are collected at the anode and we observe shot noise. If the voltage is reduced or the emission current increased, then we observe a transition to space charge–limited emission, where the current is determined by the applied voltage, and the Coulomb interaction between electrons becomes significant. Some of the emitted electrons are returned to the cathode by the repulsion of other electrons. In this case, the fluctuations in the current reaching the anode are reduced well below that predicted by the shot-noise equation for the arriving current.

The use of noise as a diagnostic probe, which is more sensitive than simple transport, has been recognized elsewhere.[7,8] On the other hand, this must be balanced, realistically, against the fact that conductance measurements can be accurate. Noise measurements, even under the best conditions, are difficult and inaccurate. This intrinsic difficulty is further aggravated by the fact that mesoscopic samples are typically beset by a number of additional noise sources, less fundamental than those we will emphasize.

Shot noise in diodes in which the applied voltage is large enough to overcome the retarding space charge field in front of the cathode is given by

$$\langle i^2 \rangle_{\Delta\nu} = 2e^2(dn/dt)\Delta\nu. \tag{1.1}$$

Here, $\langle i^2 \rangle_{\Delta\nu}$ is the mean squared noise current in a frequency range $\Delta\nu$; e is the electronic charge, but more generally can be interpreted to be the charge flow in an external circuit due to the elementary stochastic event; and (dn/dt) is the rate at which these events occur. The right-hand side of equation 1.1 is, more frequently, written in the form $2eI\Delta\nu$, where I is the current flow.

DICHOTOMY IN THE LITERATURE

Mesoscopic samples, which scatter elastically, are typically characterized by a transmission matrix t_{ij}, denoting the ratio of the wave transmitted into an outgoing transverse channel i to that incident in channel j. The characteristic values of $tt^\dagger$ are denoted by T_i. In this case, a widely accepted result[9] for the noise current, at zero temperature, is

$$\langle i^2 \rangle_{\Delta\nu} = 2e|V|\left(\frac{e^2}{h}\right)\Delta\nu \sum_N T_i(1 - T_i). \tag{2.1}$$

Under the same conditions, the current flow is given by

$$I = V\left(\frac{e^2}{h}\right) \sum_N T_i, \tag{2.2}$$

where V is the applied voltage. We see that the result in equation 2.1 is less than the result obtained by applying equation 1.1 to equation 2.2. This is manifested by the appearance of the factors $(1 - T_i)$ in equation 2.1. These results apply to the case of independently transmitted electrons; no Coulomb correlation is taken into account. The factors $(1 - T_i)$ are attributed to the Pauli principle and this is most easily understood in terms of a wave packet representation.[10] The Pauli principle prevents the assignment of more than one outgoing electron per successive emerging wave packet and thus regulates that outgoing stream more than equation 1.1, derived for classical particles, would allow. This reduction is now often referred to as shot-noise *suppression*. Beenakker and Büttiker[11] pointed out that the distribution of the eigenvalues T_i is bimodal, in the case of a channel with diffusive carrier motion. There are a few relatively well transmitting channels and many that transmit very poorly,[12] as made particularly clear in reference 13. As a consequence, reference 11 showed

$$\sum_i T_i(1 - T_i) = \frac{1}{3} \sum T_i. \tag{2.3}$$

Thus, the shot noise is cut to one-third of the naive classical expectation, $2eI\Delta\nu$. There are other cases that have been analyzed,[14] not corresponding to diffusive scattering, that behave similarly, but with coefficients in the equivalent of equation 2.3 differing from the exact value of ⅓.

A much more drastic reduction, of order ℓ/L, where ℓ is the mean free path and L is the sample length, has been obtained[11] for the case where there is incoherent scattering along the sample and Coulomb interactions are taken into account. In this particular analysis, the scattering is provided by entry of the electrons into reservoirs tied into the sample, along its length. The reservoirs provide a source of incoherence; the carriers emerging from the reservoirs do not remember their earlier history. The reservoirs absorb carrier energy and the reservoirs are presumed to have no net input current. As a result, fluctuations in current in one link cause the electrochemical potential of adjacent reservoirs to change and hence change the current in the remaining links.

Essentially the same ℓ/L reduction was obtained[15] from a semiclassical model of a metallic conductor. Another example is as follows: Take a set of N identical classical diodes with full shot noise, connected in series, and allow for their shunt capacitances. Then, the mean squared current fluctuations in an external short circuit will be $(1/N)2eI\Delta\nu$, with $1/N$ taking the place of ℓ/L. Note that allowance for capacitive circuit effects is equivalent to some allowance for Coulomb interactions. Shot-noise reduction similar to that for series diodes has been observed for tunneling junctions placed in series, in the case where quantum coherence is disrupted after each tunneling event.[16] This observation is related to a number of theoretical analyses.[8,17–19] The original prototype for such noise calculations, with incoherent tunneling structures in series,[20] addressed the more difficult question of the probability of large noise excursions rather than the small signal noise spectrum of interest here.

A very similar sort of ℓ/L reduction was also invoked in a derivation, by Albert Rose,[21] of thermal equilibrium noise, for nondegenerate conductors, from a shot-noise viewpoint. He takes the elementary stochastic charge transfer event in equation 1.1 to be the result of an electron moving through a single mean free path. Therefore, he replaces e in that equation by $e(\ell/L)$. This approach was also described in section III of reference 15, without awareness of Rose's earlier discussion.

A number of other theoretical calculations end up with results that do not exhibit an ℓ/L reduction. This includes a semiclassical theory[22] where noise is evaluated in terms of the transmitted carriers crossing a surface at the ends of the sample and where no Coulomb interactions of any kind are allowed. Nagaev's theory of noise in metallic point contacts[23] averages carrier flow over a volume and fails to exhibit the ℓ/L reduction. Calculations by de Jong and Beenakker[24] for semiclassical diffusive transport, with no quantum mechanical coherence, but with energy-preserving elastic scattering and with noise evaluated at a plane near the terminating reservoir, also yield $(1/3)(2eI\Delta\nu)$ for the noise current, without an ℓ/L factor. It is our purpose to elucidate this dichotomy and to gain a better understanding of the ℓ/L factor. We do so in a qualitative and suggestive physical way, leaving room for more definitive theories. A much more formal set of discussions, complementary to this one, is presented in reference 25. That is a more explicitly quantum mechanical discussion, following the photon literature. Number and phase noise, as well as squeezing, will be found there, but no reference to electron-electron interactions.

We conclude this section with some simple circuit theory that will typify the cases in which an ℓ/L reduction is observed. FIGURE 1 shows a set of N identical elements in a series, each with a resistance R and a capacitance C. Each element has a noise current generator i_n in parallel; this is the noise current that would flow if the element were by itself and short-circuited. We, instead, now ask what is the noise current through a short circuit for the overall series combination of N elements. (Note that a dc short circuit is not required, only one that is effective at the measurement frequency.) Each noise current generator feeds into the parallel combination of its own RC combination and a series combination of $(N-1)$ elements in parallel with that. Thus, the externally measured current due to one of the noise generators is

$$i = i_n\left[\frac{Y_{N-1}}{(Y_1 + Y_{N-1})}\right], \tag{2.4}$$

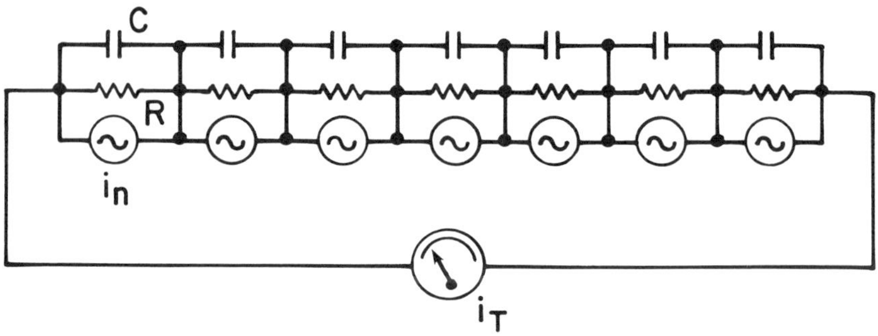

FIGURE 1. Each RC combination is fed by a noise current generator, i_n. The noise current measured in the external circuit is i_T. The circuit supplying the dc current is not shown.

where Y_1 is the admittance of a single element, $\mathscr{G} + i\omega C$, whereas Y_{N-1} is the admittance of the $(N-1)$ series elements, $(N-1)^{-1}(\mathscr{G} + i\omega C)$. Thus,

$$i = i_n/N. \tag{2.5}$$

Now, however, we have N of these contributions, one for each noise generator. These noise sources will be uncorrelated; the mean squared noise contributions will add. Thus, the total measured noise current is

$$\langle i_T\rangle^2 = N\langle i^2\rangle = N\left\langle\frac{i_n^2}{N^2}\right\rangle = \left(\frac{1}{N}\right)\langle i_n\rangle^2. \tag{2.6}$$

Therefore, the total root-mean-squared noise current is reduced by the factor $1/N$, analogous to ℓ/L. Note that the relative values of R and C do not matter. We can, in fact, include an inductor in series with each resistor, to represent electron inertia, and leave equations 2.4–2.6 unchanged. The case of shot-noise diodes in series is a case where $\mathscr{G} = 1/R = 0$. The current through the diodes is determined by thermionic emission and is independent of the voltage across the diode, if that voltage is large enough to collect the total thermionic emission current. The discussion of reference 11 with reservoirs tied in along the length of the sample is at the other extreme. The capacitances are omitted; only resistive equilibration is allowed. In the case of FIGURE 1, where $\mathscr{G}$ and C are both present, it is the measurement frequency that determines whether resistive equilibration or capacitive equilibration is dominant. In the low-frequency limit, it is the resistive circuit that matters.

NOISE MEASUREMENT

Noise theory papers often invoke a quantum mechanical operator and tell us that it is noise. Alternatively, a fictitious plane, unrelated to any measurement apparatus, is used and the electron flow across it is calculated. In actuality, noise is measured in one of two ways. Li *et al.*[26] use a small series resistor and measure the voltage across it; the series resistor is preferably kept small compared to the sample resistance. Alternatively, we can feed a fixed current to the sample, through a large resistor, and can measure the voltage fluctuations across the sample.[27] The voltage fluctuations, V, for an open-circuited sample (at the measurement frequency) must induce a current that exactly cancels the noise current i flowing in the short-circuited sample. Therefore,

$$\langle V^2 \rangle_{\Delta\nu} = |Z^2| \langle i^2 \rangle_{\Delta\nu}, \tag{3.1}$$

where Z is the sample impedance. For conceptual simplicity, we will hereafter concentrate on the current flow as measured via the small series resistor.

The above methods are not necessarily the only way to measure noise. However, in that connection, we need to warn theoreticians fascinated by quantum mechanical measurement. The transport systems under consideration here are dissipative. In the case of purely elastic scattering, the dissipation may occur in the attached reservoirs and not in the sample. This is not a closed Hamiltonian system. Persistent current flow in closed rings without attached leads, in the presence of magnetic flux, has been studied[1] and a number of the papers in reference 1 relate to this topic. That is a Hamiltonian system. Furthermore, the conventional methods that we have described measure noise through an apparatus that is quantum mechanically incoherent with the transport in the sample. It is *not* like a Stern-Gerlach experiment or a spin-flip coupled to a passing electron. It is the total current, electronic flow plus displacement current, that is continuous in a closed circuit. Thus, the current flow in the small series resistor measures that total current in the sample.

Theories that invoke the expectation value of an operator, or the electron flow across a fictitious plane, do not necessarily give incorrect results. However, it is incumbent upon those investigators to relate their methods to experimental reality.

WHERE IS THE IMAGE CHARGE?

The lines of force emanating from a moving carrier in a circuit terminate somewhere. Following that terminating charge is a key ingredient in addressing the issue we raised in the second section.

Let us first consider the case of a charge in a large metallic conductor, a medium densely populated by electrons. Let us also assume that there are terminating electrodes with *magic* properties, still to be defined. What happens if a carrier moves from position A to position B, taken for the moment to be the terminating points between inelastic scattering events? The carrier effectively disappears at A and reappears at B. We can make it disappear by neutralizing it with a compensating charge. Thus, the change in the image charge is that obtained by injecting a compensating positive charge of magnitude $|e|$ at position A and an electronic charge at B. If this seems illegitimate, then break up the path between A and B into many smaller subsegments. At the point separating two such subsegments, we will have an injected electron, followed by an injected compensating positive charge. The two successive compensating events from such an intermediate point will cause a compensating time sequence of charges on the electrodes and elsewhere. Because the two compensating events follow each other immediately, or with a very short lag, no low-frequency charge transients appear elsewhere. Those must come from the charges injected at A and B.

The electrodes are assumed to have a very high density of electrons and very little scattering, so the injected charge first produces an effect in the electrodes before producing one in the conducting sample. After a short transient related to the surface plasma frequency in the electrodes, this charge will be the classical image charge. In the case of plane-parallel electrodes held at fixed potential and large enough to make fringing fields negligible, the two electrodes will bear image charges in inverse proportion to their respective distance from the injected charge.

The injected charge will then produce a response within the conducting medium, which may at first be dominated by plasma effects.[28] In the long term, however, just as if we injected a charge into the network of FIGURE 1, the charge redistribution will be dominated by conductive effects. FIGURE 2 shows a network view of this process. There are capacitive and resistive paths from A to each electrode with respective resistances and capacitances that we denote by R_{A1}, C_{A1}, R_{A2}, and C_{A2}. There are also parallel paths directly between the electrodes, unconnected to A. A much more complex network between the electrodes, and including point A, can always be replaced by the equivalent one shown in FIGURE 2.

A charge injected at A will first generate respective electrode image charges in proportion to C_{A1} and C_{A2}. Subsequently, however, the actual current flow division will be in proportion to $(R_{A1})^{-1}$ and $(R_{A2})^{-1}$. If we apply a voltage V to the electrodes, the voltage across R_{A1} will be $R_{A1}V/(R_{A1} + R_{A2})$ and that across R_{A2} will be $R_{A2}V/(R_{A1} + R_{A2})$. Thus, the conductive current flow to the respective electrodes divides in inverse proportion to the respective voltage drops resulting from an externally applied voltage. Therefore, if we inject a charge at A and then a compensating charge at B, the resulting charge flow in the external circuit will be eV_{AB}/V, where V_{AB} is the voltage difference between A and B resulting from the externally applied voltage V.

If we have a homogeneous medium and can neglect capacitive fringing effects, then $R_{A1}/R_{A2} = C_{A2}/C_{A1}$ and the initial capacitively caused flow in the external short circuit is the only external flow. These image charges then flow in through the conducting medium to neutralize the injected carrier, without further charge flow in the external circuit. In general, however, as a result of sample inhomogeneity or geometry, this will not hold and the long-term (low-frequency) external current flow is determined only by R_{A1} and R_{A2}.

Quantum mechanical particles are, of course, not localizable to a point A as suggested in FIGURE 2. Furthermore, the neutralizing charge flow from the electrodes does not actually reach the injected particle, but reaches only to within a screening length. Nevertheless, we believe that our discussion covers the essential physics. In particular, if the "blemishes" just discussed are identical at A and at B, then our circuit theory results may, in fact, be exact. On the other hand, a conducting sample viewed at a sufficiently microscopic level is inevitably inhomogeneous.

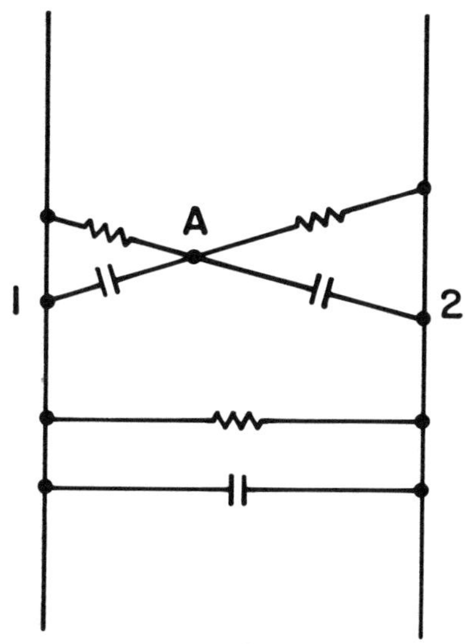

FIGURE 2. A charge injected at A will, at first, induce charge changes in the electrodes according to the electrostatic fields, symbolized by capacitors. Over the long term, it is the conductive effects that matter.

The conductive response discussed above represents the Coulomb interaction between electrons. One carrier moving through a mean free path produced a response from the other carriers. In a homogeneous system, where a voltage applied between electrodes is dropped uniformly, an electron moving through a distance Δz in the direction between electrodes will induce an external current flow $e\Delta z/L$, where L is the electrode separation. Whether homogeneous or not, the sum of all the conductive contributions from a carrier crossing the sample will be e. There may be a temptation then to regard the total current flow as a sequence of uncorrelated electron traversals of the sample, each contributing a charge flow e. However, that would be wrong, the electrons are correlated, and this must serve to reduce the noise. This is the reduction evaluated in section IV of reference 15 and in the closely related earlier discussion of reference 11. (Note that, in the discussion of section IV of reference 15, the conductive equilibration was not invoked explicitly, but would yield the result given there.)

We have left the exact definition of the extent of the journey, between two points A and B, somewhat vague. We alluded to A and B as terminating points between inelastic scattering events. This is an oversimplification. In the detailed analyses of reference 15, care was taken to identify separate uncorrelated charge transport events. The purpose of this section is to justify the $\Delta z/L$ factor, once Δz has been provided from other considerations, and to emphasize that it results from electron-electron interactions. The considerations in this section are an elaboration of reference 29 and are closely related to a more detailed treatment provided by Büttiker.[30]

GATE ELECTRODES

Mesoscopic samples often have nearby electrodes in addition to those actually connected for current flow and for voltage measurements. These additional electrodes are used to define the conducting regions by repelling carriers within regions of a two-dimensional electron gas. If the conducting portion of the two-dimensional electron gas is dense enough so that a carrier is screened within that gas, then these gate electrodes will change the short-term capacitive response discussed in the preceding section, but will not affect the conductive response above and beyond the role of the electrode in defining the carrier density in the sample. After all, it is really the disappearance of the screening charge at an initial point A, through conductive processes, and its reappearance at B that produces the external current flow.

Let us, instead, now consider the case where the moving carrier has its image charge on a gate electrode rather than in the conducting medium. When the carrier leaves the source region and induces an image charge on the gating electrode, there is a capacitive charge flow into that electrode. It is the displacement current that flows into that electrode from the conducting region. If the carrier continues to the other destination electrode (drain), then the inverse sequence takes place there. Nothing happens at the destination electrode, and its external circuit connection, until the carrier leaves the screening of the gate electrode. After that, current flows in the circuit connected to the drain, with the details depending on the electrostatic geometry, the carrier density in the channel, and the scattering in the drain end of the channel. Clearly, while the carrier is screened by the gate, the external circuit sees none of the carrier's details: the Coulomb interaction and the mean free path ℓ involved in the ℓ/L noise reduction cannot be effective. Thus, either the ℓ/L reduction will not be present or there will be a lesser reduction, depending on the details between the screened region and the drain.

Clearly, a myriad of variations will be possible. If the screening by the gate is incomplete, and there is some by carriers in the channel, then the conductive contributions reaching the drain, during the traversal under the gate, will be present to some extent.

Noise measurements in reference 31, in an ordinary field effect transistor configuration, under a high source-drain voltage, yielded $\langle i^2\rangle_{\Delta\nu} = 2eI\Delta\nu$, where I is the actual source-drain current. Neither the ℓ/L reduction nor the factor ⅓ shows up. These measurements were made at a gate voltage below threshold, with a very small channel current. The factor ⅓ is derived from the exclusion principle, whereas these measurements are probably made on nondegenerate electrons. The lack of the ℓ/L reduction is, presumably, due to gate screening in the presence of a very low carrier density. The lack of an ℓ/L reduction observed in reference 27 may also be due to gate screening.

OVERVIEW

We now view the preceding discussion in the light of the matrix shown in FIGURE 3; the left-hand labels refer to the Coulomb interaction between the carriers in the channel:

(I) The upper left-hand corner of the matrix corresponds to the case that yields equation 2.1. It is the best-understood corner of the matrix, with no apparent controversy.

(II) If the carriers do not interact, then at low measurement frequencies noise cannot depend on the fine details of the trip through the sample. All that matters is the extent to which the number of traversing electrons, within a time interval, deviated from the average. Thus, the rate at which carriers are incident on the sample matters; after that, it depends on whether they are transmitted or not. Furthermore, exactly how they emerge after transmission matters, in view of the Pauli principle. Therefore, whether the scattering is coherent or incoherent (for a given resulting average transmission) can only be manifested through the Pauli principle. The so-called "reduction" taught in reference 11, resulting from the Pauli principle, was a result of random matrix theory applied to the transmission matrix t_{ij} and thus seems intimately related to quantum mechanically coherent transmission. Surprisingly, reference 24 shows that the same results as obtained in reference 11 turn up when incoherent elastic scattering in a semiclassical treatment is invoked. Is the distinction between this answer and the ℓ/L

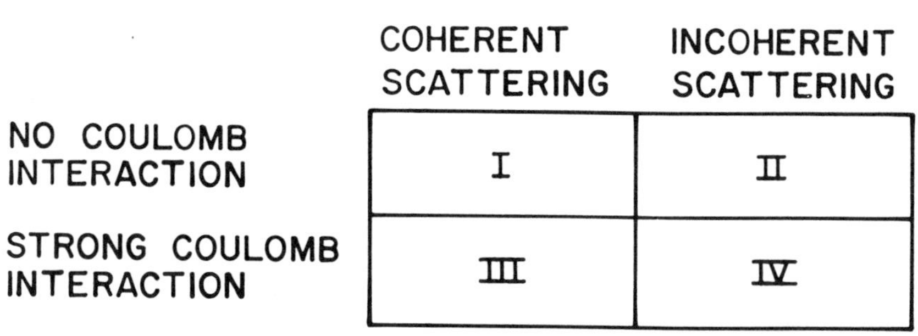

FIGURE 3. Classification scheme for discussion. Coulomb interaction is with other transport carriers within the conducting sample.

reduction for incoherent scattering[11] due to the use of elastic scattering in reference 24? I see no reason for this and presume, instead, that the absence of an ℓ/L reduction results from the absence of Coulomb interactions in reference 24. Nevertheless, it is impressive and intriguing that $\langle T_i(1 - T_i)\rangle = (1/3)\langle T_i\rangle$ can be obtained from an incoherent model.

(III) This corner is best left to the sophisticated many-body theoreticians.[32] I do, however, have some observations. First of all, as long as electrostatic interactions are present, we can expect these to regulate the electron flow and can expect some relative of the ℓ/L reduction to appear. Additionally, we can expect the somewhat complex electrostatic geometry to show up, manifesting some relative of our question: Where is the image charge? All too many sophisticated treatments of Coulomb interactions ignore this very elementary aspect.

(IV) This is the corner that has already been treated via the lateral reservoirs[11] and for bulk samples[15] and where the ℓ/L reduction shows up clearly.

Our discussions leave the role of incoherent scattering as the biggest mystery. It has been invoked in our discussion of class IV in FIGURE 3, but this does not imply that it is a *necessary* condition for that result.

[**Note added in proof:** I have ascribed a reduction factor ℓ/L to reference 11. That is an oversimplification. In the model of reference 11, the inelastic scattering length, the phase coherence length, the length over which the internal electric field is constant, and the length over which currents can fluctuate independently are all the same. If that length, ℓ_{in}, exceeds the mean free path ℓ, then it is (ℓ_{in}/L) that determines the applicable reduction. However, that does not affect the basic dichotomy that we have stressed between reduction factors of order 1/3 versus those that can be much smaller.]

ACKNOWLEDGMENTS

Some of the material discussed here originated in an informal unpublished manuscript of February 9, 1983.[29] That manuscript was a response to stimulating discussions with S. Kivelson, T. D. Schultz, and J. R. Schrieffer and related to the utility of shot-noise experiments in measuring the fractional charge of a soliton in organic one-dimensional conductors. The recent Rencontres de Moriond workshop, *Coulomb and Interference Effects in Small Electronic Structures*, revived my interest, particularly as a result of discussions with M. de Jong and G. Bergmann and amplified later by interaction with M. Büttiker, Th. Martin, and H. Thomas. This report, however, expresses my prejudices, not theirs. I am indebted to R. Sarpeshkar for drawing my attention to reference 21.

As the written contribution to a session honoring John Wheeler, I must apologize for its distance from his work scope. In its style, however, I hope that this paper demonstrates his influence. He has taught us that physics is not just a matter of transforming Hamiltonians. It is an inquiry about nature and one that on occasion requires us to discuss that which we do not yet understand, where we only see questions and possible directions.

REFERENCES

1. GLATTLI, D. C. & M. SANQUER, Eds. 1994. Coulomb and Interference Effects in Small Electronic Structures: Proceedings of the XXIXth Rencontres de Moriond Workshop. Editions Frontières. Gif-sur-Yvette, France. In press.
2. BUOT, F. A. 1993. Phys. Rep. **234**(2 & 3): 73–174; BEENAKKER, C. W. J. & H. VAN HOUTEN. 1991. *In* Solid State Physics. Volume 44. H. Ehrenreich & D. Turnbull, Eds.: 1–228. Academic Press. New York; DATTA, S. 1994. Electronic Transport in Mesoscopic Systems. Cambridge University Press. London/New York. In press; IMRY, Y. 1994. Introduction to mesoscopic physics. To be published.
3. LANDAUER, R. 1992. Phys. Scr. **T42:** 110–114.
4. MARTIN, TH. & R. LANDAUER. 1992. Phys. Rev. B **45:** 1742–1755.
5. BAKKER, C. J. & G. HELLER. 1939. Physica **VI**(3): 262–274.
6. THOMPSON, B. J. 1940. RCA Rev. **IV:** 269–285; NORTH, D. O. 1940. RCA Rev. **IV:** 441–472; RACK, A. J. 1938. Bell Syst. Tech. J. **XVII:** 592–619.
7. BÜTTIKER, M. 1992. Phys. Rev. Lett. **68:** 843–846; LESOVIK, G. B. & L. S. LEVITOV. 1994. Phys. Rev. Lett. **72:** 538–541.
8. WILKINS, J. W., S. HERSHFIELD, J. H. DAVIES, P. HYLGAARD & C. J. STANTON. 1992. Phys. Scr. **T42:** 115–121; HANKE, U., YU. M. GALPERIN, K. A. CHAO & N. ZOU. 1993. Phys. Rev. B **48:** 17209–17216.
9. BÜTTIKER, M. 1990. Phys. Rev. Lett. **65:** 2901–2904.
10. LANDAUER, R. & TH. MARTIN. 1991. Physica B **175:** 167–177.
11. BEENAKKER, C. W. J. & M. BÜTTIKER. 1992. Phys. Rev. B **46:** 1889–1892.
12. DOROKHOV, O. N. 1982. Pis'ma Zh. Eksp. Teor. Fiz. **36:** 259–262 [JETP Lett. **36:** 318–321]; 1984. Solid State Commun. **51:** 381–384; IMRY, Y. 1986. Europhys. Lett. **1:** 249–256; MELLO, P. A., P. PEREYRA & N. KUMAR. 1988. Ann. Phys. (N.Y.) **181:** 290–317.
13. PENDRY, J. B., A. MACKINNON & P. J. ROBERTS. 1988. Proc. R. Soc. London Ser. A **437:** 67–83; NAZAROV, YU. V. 1994. Phys. Rev. Lett. **73:** 134–137.
14. BEENAKKER, C. W. J. & J. A. MELSEN. 1994. Phys. Rev. B **50:** 2450–2457; JALABERT, R. A., J-L. PICHARD & C. W. J. BEENAKKER. 1994. Europhys. Lett. **27:** 255–260.
15. LANDAUER, R. 1993. Phys. Rev. B **47:** 16427–16432.
16. LI, Y. P., A. ZASLAVSKY, D. C. TSUI, M. SANTOS & M. SHAYEGAN. 1990. Phys. Rev. B **41:** 8388–8391.
17. CHEN, L. Y. & C. S. TING. 1992. Phys. Rev. B **46:** 4714–4717.
18. HERSHFIELD, S., J. H. DAVIES, P. HYLGAARD, C. J. STANTON & J. W. WILKINS. 1993. Phys. Rev. B **47:** 1967–1979.
19. KOROTKOV, A. N., D. V. AVERIN, K. K. LIKHAREV & S. A. VASENKO. 1992. *In* Single-Electron Tunneling and Mesoscopic Devices. H. Koch & H. Lübbig, Eds.: 45–59. Springer-Verlag. Berlin/New York.
20. LANDAUER, R. 1962. J. Appl. Phys. **33:** 2209–2216. [*Reprinted*: 1988. *In* Dynamic Patterns in Complex Systems. J. A. S. Kelso, A. J. Mandell & M. F. Shlesinger, Eds.: 104–111. World Scientific. Singapore.]
21. ROSE, A. 1978. Concepts in Photoconductivity and Allied Problems. Chapter 6, p. 97–105. Krieger. Huntington, New York.
22. BEENAKKER, C. W. J. & H. VAN HOUTEN. 1991. Phys. Rev. B **43:** 12066–12069.
23. NAGAEV, K. E. 1992. Phys. Lett. A **169:** 103–107.
24. DE JONG, M. J. M. & C. W. J. BEENAKKER. 1994. *In* Coulomb and Interference Effects in Small Electronic Structures. D. C. Glattli & M. Sanquer, Eds. Editions Frontières. Gif-sur-Yvette, France. In press.
25. LIU, R. C. & Y. YAMAMOTO. 1994. Phys. Rev. B **49:** 10520–10532.
26. LI, Y. P., A. ZASLAVSKY, D. C. TSUI, M. SANTOS & M. SHAYEGAN. 1991. *In* Resonant Tunneling in Semiconductors. L. L. Chang, E. E. Mendez & C. Tejedor, Eds.: 117–125. Plenum. New York.
27. LIEFRINK, F., J. I. DIJKHUIS, M. J. M. DE JONG, L. W. MOLENKAMP & H. VAN HOUTEN. 1994. Phys. Rev. B **49:** 14066–14069.
28. PINES, D. & P. NOZIÈRES. 1966. The Theory of Quantum Liquids: Normal Fermi Liquids. Benjamin. New York.

29. LANDAUER, R. 1983. To my one-dimensional friends. Unpublished.
30. BÜTTIKER, M. 1993. J. Phys. C **5:** 9361–9378.
31. SARPESHKAR, R., T. DELBRÜCK & C. A. MEAD. 1993. IEEE Trans. Circuits Devices **9**(6): 23–29.
32. KRIVE, I. V., R. I. SHEKHTER, S. M. GIRVIN & M. JONSON. 1994. *In* Coulomb and Interference Effects in Small Electronic Structures. D. C. Glattli & M. Sanquer, Eds. Editions Frontières. Gif-sur-Yvette, France. In press; KANE, C. L. & M. P. A. FISHER. 1994. Phys. Rev. Lett. **72:** 724–727; YANG, K., K. MOON, L. ZHENG, A. H. MACDONALD, S. GIRVIN, D. YOSHIOKA & S. C. ZHANG. 1994. Phys. Rev. Lett. **72:** 732–735; LI, Q. P. & X. C. XIE. 1994. Phys. Rev. B **49:** 8273–8276; FABRIZIO, M., A. O. GOGLIN & S. SCHEIDL. 1994. Phys. Rev. Lett. **72:** 2235–2238; OGATA, M. & H. FUKUYAMA. 1994. Phys. Rev. Lett. **73:** 468–471; BERKOVITS, R. & Y. AVISHAI. 1994. Significant interaction induced enhancement of persistent currents in 2D disordered cylinders. To be published.

Proposed New Polarization Correlation Test of Local Realism[a]

SUSANA F. HUELGA,[b] MIGUEL FERRERO,[b] AND EMILIO SANTOS[c]

[b]*Departamento de Física*
Universidad de Oviedo
33007 Oviedo, Spain

[c]*Departamento de Física Moderna*
Universidad de Cantabria
39005 Santander, Spain

LOOPHOLES IN BELL INEQUALITY EXPERIMENTS

Many empirical tests of local-hidden-variables (LHV) theories (or local realism), via Bell's theorem, have been performed in the last two decades.[1–3] In general, the reported results agree with the quantum predictions, but the experiments have not yet refuted the whole family of LHV theories, as loopholes exist in all performed tests. In view of the remarkable implications of an empirical refutation of local realism, the performance of a loophole-free experiment becomes an important task.

Two loophole-free experiments have been recently proposed, one by Fry,[4] involving the spin correlation of atoms, and the other by Chiao *et al.*,[5] involving the polarization correlation of photon pairs produced by parametric downconversion. Here, we propose a new experiment that will consist of measuring the polarization correlation of photon pairs produced in atomic cascades. The novelty with respect to previous experiments of this type[1,2] is the detection of the recoil atom.

There are many gedankenexperiments that have been discussed as tests of local realism, but only a few classes of real experiments have been either performed or seriously proposed. In fact, most of the performed tests of the Bell inequality have used entangled optical photon pairs. In the seventies and early eighties, photon pairs were typically produced in atomic cascades[1,2] and the polarization correlation was measured. In the last decade, the typical sources have been nonlinear crystals, where photon pairs are created by parametric downconversion.[3] All experiments involving optical photons suffer from a loophole due to the low efficiency of single-photon detectors. This loophole is very well known[1,2,6] and we shall not discuss it anymore here. However, there are other loopholes, less well known, that we shall consider below.

Atomic cascade experiments are currently considered to be the most reliable tests of local realism ever made. The setup of a typical experiment[1,2] is sketched in FIGURE 1 (ignore, for the moment, the atomic detector). A pair of photons, emitted from an atomic source placed at the origin of the coordinate system, are collected by

[a]This work was supported financially by DGICYT Project No. PB-92-0507 (Spain) and Universidad de Oviedo Project No. DF-93-214-42 (Spain).

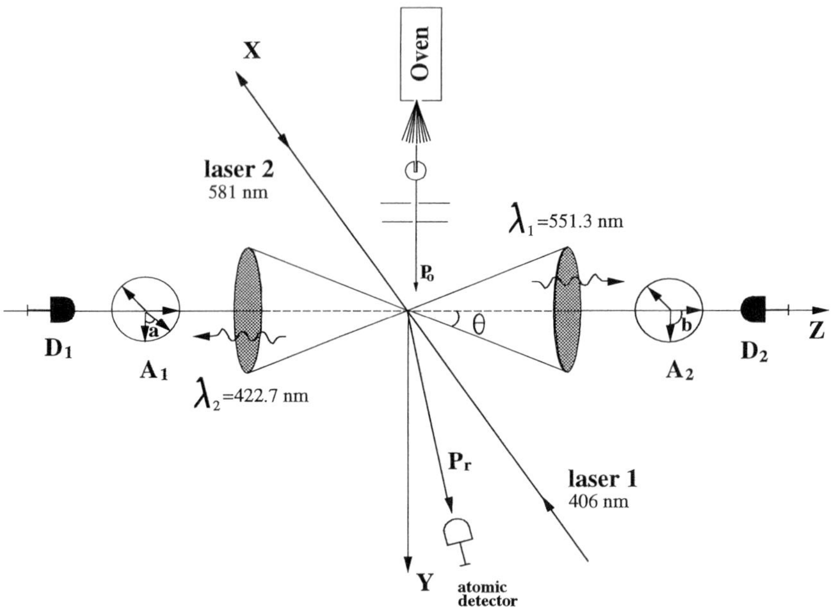

FIGURE 1. Experimental setup for the proposed atomic cascade test.

an appropriate lens system. Each photon in a pair may cross a polarizer and arrive eventually at a detector. The single and coincidence detection probabilities predicted by quantum mechanics are[1]

$$P_{12} = (\tfrac{1}{4})\zeta_{12}[1 + \epsilon \cos(2a - 2b)], \qquad P_1(a) = (\tfrac{1}{2})\zeta_1, \qquad P_2(b) = (\tfrac{1}{2})\zeta_2, \quad \textbf{(1)}$$

where a and b are the angles of the polarizers with respect to a given plane and $\zeta_{12}(\epsilon)$ is an overall efficiency (correlation) parameter, which is a function of the solid angle covered by the lenses and depends also on all practical inefficiencies (of filters, polarizers, detectors, etc.). Similarly, ζ_1 and ζ_2 are single efficiencies.

A test of local realism involves checking whether equation 1 violates the Bell inequality,[6]

$$P_{12}(a, b) - P_{12}(a, b') + P_{12}(a', b) + P_{12}(a', b') - P_1(a') - P_2(b) \leq 0. \quad \textbf{(2)}$$

It is easy to see that equation 2 can be violated only if

$$\zeta(1 + \sqrt{2}\epsilon) > 2, \qquad \zeta := 2\zeta_{12}/(\zeta_1 + \zeta_2); \quad \textbf{(3)}$$

that is, the violation of local realism will occur in the region of the ζ, ϵ plane represented in FIGURE 2 by the hatched triangle area in the upper right corner. However, that region is not accessible in atomic cascade experiments, even if all devices (detectors, polarizers, etc.) were ideal. Indeed, typical experiments lie close to the upper left corner, as shown by a dot in FIGURE 2; that is, the parameter ϵ is fairly close to the ideal value of unity, but ζ is much smaller (less than 0.01). There are two reasons for this: the low efficiency of photon detectors and the low angular correlation of the two photons emitted in a (three-body decay) atomic cascade. Even with perfect detectors, the value of ζ would be about 0.2, still far from the Bell-forbidden region, due to the fact that, most of the time, one of the photons in a pair enters the corresponding aperture, whereas the partner does not. On the other hand, if we attempt to increase ζ by increasing the solid angle covered by the lenses, then the parameter ϵ decreases because photons not emitted in opposite directions are not well correlated in polarization. The net result is that all real experiments lie to the left of the continuous curve in FIGURE 2, always far from the Bell-forbidden region. This is the angular correlation loophole, pointed out 20 years ago by Clauser and Horne,[6] but not well appreciated until the discovery of an LHV model reproducing all quantum predictions, even for ideal measuring devices.[7] Our proposed experiment is able to block this angular correlation loophole by detecting the recoil atom, as explained in the next section.

Before discussing the experiment, we shall explain why photon pairs produced in a nonlinear crystal by parametric downconversion (PDC)[5] or spin entangled atoms[4] may not be better candidates for a loophole-free test of LHV theories. The first reason is that our experiment tolerates a greater amount of noise. In fact, we propose to measure the single and joint probabilities involved in the Bell inequality (equation 2) via the measurement of coincidence rates (atom-photon and atom–two photons, respectively). In contrast, in the other proposals,[4,5] the measurement of single and

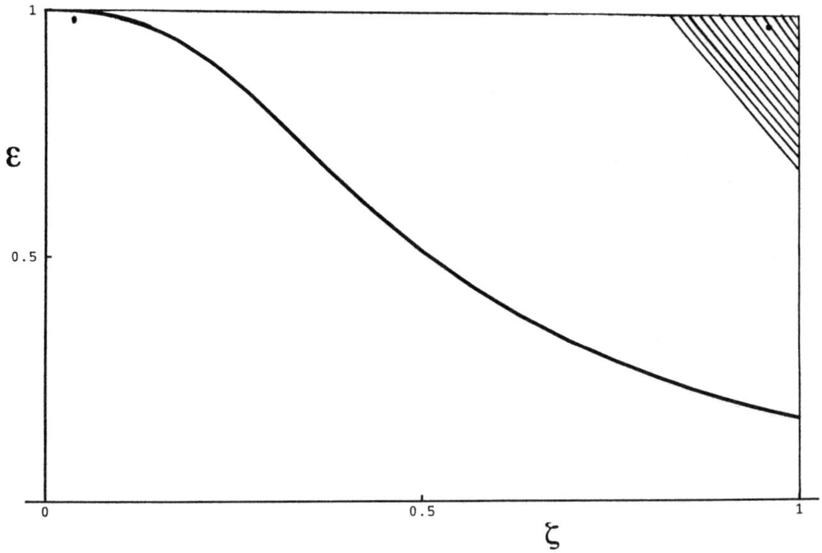

FIGURE 2. Polarization correlation ϵ versus overall detection efficiency parameter ζ in atomic cascade experiments. The hatched region at the upper right corner is forbidden by the Bell inequality,[2] whereas quantum predictions for standard experiments lie to the left of the continuous curve. The small dot in the upper left corner represents the experiment by Aspect, Grangier, and Roger. The small dot in the upper right corner is the quantum prediction for the proposed experiment, with ideal polarizers and detectors.

coincidence rates is required, which demands a strict control of noise. As is well known, any source of noise (e.g., dark rate in photon detectors or accidental counts in atom detectors) should increase single probabilities relative to joint probabilities, thereby preventing or making more difficult the violation of equation 2.

In addition to the noise problem, both PDC and molecular experiments have specific problems that we shall discuss below.

The quantum state of a photon pair produced by PDC may be represented[8] by

$$|\psi\rangle = V^{-1} \sum_{\mathbf{k}_1,\mathbf{k}_2} \phi(\mathbf{k}_1 s_1, \mathbf{k}_2 s_2) \exp\{-i[(\mathbf{k}_1 r_1 + \mathbf{k}_2 r_2) - (\omega_1 + \omega_2)t_0]\} |\mathbf{k}_1 s_1\rangle |\mathbf{k}_2 s_2\rangle, \quad (4)$$

where $|\mathbf{k}_i s_i\rangle$ is a one-photon Fock state of momentum $\mathbf{k}_i$ and spin s_i. The weight function ϕ is appreciably different from zero only when the frequencies and wave vectors fulfill the matching conditions

$$\omega_0 = \omega_1 + \omega_2, \qquad \mathbf{k}_0 \approx \mathbf{k}_1 + \mathbf{k}_2 \qquad (\omega_1 = c|\mathbf{k}_1|). \quad (5)$$

Due to the existence of a sum over modes, state 4 (i.e., equation 4) is entangled, which is a necessary condition for the violation of a Bell inequality. (If there were only two modes, one per photon, the state would be factorable and no violation would be possible.) On the other hand, if we consider all the modes compatible with equation 5, so many photon pairs are produced that any correlation between the two photons of every pair (so-called signal and idler) would be lost. In practice, it is necessary to select, by means of appropriate filters or pinholes, two (or more) narrow ranges of wave vectors for the signal (say, around $\mathbf{k}_s$ and around $\mathbf{k}_{s'}$) and similarly for the idler (around $\mathbf{k}_i = \mathbf{k}_0 - \mathbf{k}_s$ and around $\mathbf{k}_{i'} = \mathbf{k}_0 - \mathbf{k}_{s'}$). All this is very well known and a standard practice, but we recall it in order to emphasize that it is not just a convenient procedure for an easy performance of the experiment, but a fundamental requirement for the test of LHV theories.

If we put two ideal detectors (i.e., 100% efficient at all frequencies) after two appropriate filters, the quantum prediction for the coincidence detection probability is

$$\zeta_{12} = C \int d\omega_1 \int d\omega_2 |f_s(\omega_1 - \omega_s)|^2 |f_i(\omega_2 - \omega_i)|^2 \delta(\omega_1 + \omega_2 - \omega_0), \quad (6)$$

where $\delta(\omega_1 + \omega_2 - \omega_0)$ is the (Dirac's delta-like) function that results after integrating over angles and summing over polarizations the function $|\phi|^2$, with ϕ being the weight function introduced in equation 4. The constant C depends on the setup (e.g., nonlinear crystal, pumping intensity, or detection window) and f_s (f_i) is the transmission function for the signal (idler) filter. Similarly, the single detection probabilities for the signal and idler will be

$$\zeta_1 = C \int d\omega_1 \int d\omega_2 |f_s(\omega_1 - \omega_s)|^2 \delta(\omega_1 + \omega_2 - \omega_0),$$
$$\zeta_2 = C \int d\omega_1 \int d\omega_2 |f_i(\omega_2 - \omega_i)|^2 \delta(\omega_1 + \omega_2 - \omega_0). \quad (7)$$

The relevant question is whether the parameter ζ, defined in equation 3, is high enough to allow the violation of the Bell inequality (equation 2). Despite the assumption of ideal detectors, we see that the parameter may be less than 100% due to the filter functions. It may appear that the value $\zeta = 1$ can be reached with filter functions having a value of 1 within some interval and 0 outside, so chosen that $\zeta_1 = \zeta_{12} = \zeta_2$. (We point out that one of these equalities may be easily obtained and this fact has been used for the absolute measurement of detector efficiencies.[9]) However, $\zeta = 1$ is not possible because the filter function should be causal, that is, complex functions analytic in the upper-half plane of the complex variable ω, in addition to having a modulus bounded by unity. Then, causality of the filter functions introduces a fundamental detection inefficiency and the question is how big it is. This question is equivalent to the mathematical problem of finding the maximum value of ζ, defined by equations 3, 6, and 7, compatible with the causality and boundedness requirements of the filter functions. We have not been able to solve the problem, but it is not difficult to realize that the best-known causal functions, that is, the Lorentzians, give $\zeta \leq ½$, far from the value needed to violate the Bell inequality. If we use pinholes and not filters, the situation is similar due to the strong correlation between direction of emission and frequency in downconverted photons. In summary, the difficulty of guaranteeing that whenever we collect two photons they are a correlated pair is the PDC equivalent to the angular correlation loophole in atomic cascade experiments.

For the sake of completeness, we conclude this section with a discussion of loopholes in molecular experiments,[4,10] where a diatomic molecule in an initial state of zero total angular momentum dissociates into two identical spin-½ atoms in an (entangled) singlet spin state. In contrast with optical photon experiments, there exist highly efficient atom detectors, although the situation with respect to spin discrimination is less clear-cut than in the corresponding case of linear polarization of photons; Stern-Gerlach devices of the necessary efficiency, for example, certainly do not yet exist. However, there is another important loophole. In fact, it is extremely difficult to perform the measurements on the two atoms at two spatially separated regions, that is, in such a way as to prevent the possibility of connection between the measurement regions by signals propagating with the velocity of light. The reason is as follows. Assume that the molecule has an initial velocity dispersion Δv, which corresponds to an energy $E_0 = m\Delta v^2$ for the center-of-mass motion, with m being the mass of an atom. If we use these detectors placed at a distance d from the source, the uncertainty in the time of arrival of the atoms to the detectors will be $\Delta T = d\Delta v/v^2$. Now, the condition of spatial separation of the two detection events puts the constraint $\Delta T < d/c$, with c being the velocity of light. This provides the upper bound to the energy of the center-of-mass motion,

$$E_0 < D^2/mc^2, \tag{8}$$

with $D = mv^2$ being the dissociation energy. For a typical experiment[10] with $D \approx 0.1$ eV and $mc^2 \approx 10^4$ MeV, the bound is as small as 10^{-12} eV, corresponding to a temperature of about 10^{-8} kelvin.

Fry[4] proposed to ensure spacelike separation by measuring the spin projection of the atom by means of a short laser light pulse that selectively excites the atom to one of two hyperfine levels. However, no empirical data about the efficiency of this technique have been reported so far.

ATOMIC CASCADE EXPERIMENTS WITH DETECTION OF THE RECOIL ATOM

In atomic cascade experiments, a good angular correlation could be obtained if the recoil atom is detected. This additional requirement allows one to increase the angular correlation of the photon pairs involved, provides an operational definition for this ensemble, and constitutes a feasible scheme for "event-ready-detectors".[11] Provided that measuring devices have an efficiency above a certain threshold, quantum mechanical (QM) predictions for this ensemble would contradict Bell's inequality, making possible an experiment able to discriminate between local realism and quantum mechanics at high efficiencies.

Consider a typical experimental setup as sketched in FIGURE 1, where a J = 0–1–0 cascade in calcium-40 is selectively excited by two-photon absorption. Fluorescence light is collected by spherical lenses and reaches an analyzer-detector assembly. The arrangement is combined with a detection system for the recoil atoms. The beam's axis has been chosen in the vertical direction Y to avoid gravitational effects. In a standard setup, where the beam's axis is horizontal, the gravitational field will deflect the scattered beam by a significant amount. As a consequence, some difficulties arise, for example, the nonfactorizable form of the distributions that describe the x and z components of the incident velocity and the fact that a static slit will collect only a small percentage of the atoms that would reach the slit without the gravitational field. This could be solved if the acceptance slit moves down with a velocity of approximately 1 mm/s.

In order to evaluate the QM predictions for the ensemble of interest, we assume that each photon has a well-defined linear momentum. Provided that the recoil atom is detected, we can differentiate three kinds of events that could yield single or double counts; the cones subtended by each detector aperture are taken to be equal, with ϑ being the half-angle: R_1 [resp. R_2] will refer to events such that

$$0 \leq \theta_1 \leq \vartheta \quad \text{and} \quad 0 \leq \theta_2 \leq \pi - \vartheta$$

$$[\text{resp. } \vartheta \leq \theta_1 \leq \pi \quad \text{and} \quad \pi - \vartheta \leq \theta_2 \leq \pi] \tag{9}$$

and B will refer to events such that

$$0 \leq \theta_1 \leq \vartheta \quad \text{and} \quad \pi - \vartheta \leq \theta_2 \leq \pi, \tag{10}$$

that is, both emissions are collected for the corresponding lens system. Note that, with the chosen geometry, ϕ_i is always between 0 and 2π.

If there is no absorption in the corresponding polarizer, events R_i $\{i = 1, 2\}$ will contribute as single counts, whereas events B will produce double counts if none of the photons are absorbed and single counts if one of the photons is not absorbed, independently of what happens to the other member of the pair.

If N is the number of atoms that have actually decayed in the source and $N(B)$ and $N(R_i)$ are the number of events of the corresponding class, we can write the quantum mechanical (QM) predictions for the probabilities involved in equation 2 as follows:[6]

$$P_i = \left(\frac{1}{2}\right)\eta\epsilon^i_+\left[\frac{N(B)}{N} + \frac{N(R_i)}{N}\right]\eta_a \qquad (i = 1, 2),$$

$$P_{12}(\Phi) = \left(\frac{1}{4}\right)\eta^2\left[\frac{N(B)}{N}\right][\epsilon^1_+\epsilon^2_+ + \epsilon^1_-\epsilon^2_- F(\vartheta)\cos(2\Phi)]\eta_a, \tag{11}$$

where Φ is the relative angle $|b - a|$; η is the quantum efficiency of the photodetectors; η_a is the efficiency of the atomic detector; $\epsilon_\pm = \epsilon_M \pm \epsilon_m$, where ϵ_M and ϵ_m are the maximum and minimum transmitivities of the analyzers relative to an appropriate orthogonal basis; and $F(\vartheta)$ is a factor that takes into account the fact that the polarization correlation of a photon pair decreases when the angle between their wave vectors departs from π.[6] Substitution of these predictions in equation 2 gives the following: QM predictions will violate Bell's inequality if the following inequality,

$$\eta\epsilon_+\left[1 + \sqrt{2}F\left(\frac{\epsilon_-}{\epsilon_+}\right)^2\right] - 2 \geq t, \qquad t := \frac{[N(R_1) + N(R_2)]}{N(B)}, \tag{12}$$

is satisfied, where we have taken $\epsilon^1_\pm = \epsilon^2_\pm = \epsilon_\pm$ for simplicity. In a more compact form, we can write the condition of equation 12 for incompatibility between the two formalisms as

$$\eta \geq \frac{(t + 2)}{k(\vartheta)}, \qquad k(\vartheta) := \epsilon_+\left[1 + \sqrt{2}F\left(\frac{\epsilon_-}{\epsilon_+}\right)^2\right]. \tag{13}$$

The minimum η to obtain a violation of Bell's inequality requires an arrangement in which the atomic detector is placed in such a way that the ratio t is minimized. We have found that this corresponds to a situation in which the number of events R_i is the lowest possible. This location will be denoted OR (optimal region).

In order to increase the recoil effect, we propose reducing the mean velocity in the beam, typically of order 700 m/s, by one order of magnitude with a velocity filter. As we will show below, a reliable experiment is obtained if we carry out a strong collimation process and if the atomic detector is placed far away from the interaction region.

The OR is estimated by a Monte Carlo simulation. We assume that, after collimation and crossing the velocity filter, the atomic beam is characterized by normal distributions relative to the velocity components, with parameters (μ, σ) equal to $(70, 1.4 \times 10^{-1})$ for the longitudinal component and equal to $(0, 2.1 \times 10^{-4})$ and $(0, 7 \times 10^{-3})$ for the z and x components, respectively. Then, for transverse recoil momenta bounded by

$$P_{rx} \in [-0.415, 0.415] \qquad \text{and} \qquad P_{rz} \in [0.301, 0.319], \tag{14}$$

where all momenta are written in cgs units normalized to 10^{-22}, and taking a detector half-angle $\vartheta = 32°$, we obtain the following results for a simulation with $N_{\text{total}} = 5 \times$

10^6 events:

$N(B)$	$N(R_1)$	$N(R_2)$
4744.79	132.309	118.395

and the mean value for the ratio t is 0.053. The angular correlation has been taken into account, giving different weights to different emissions, according to the value of the angle $(\mathbf{k}_1, \mathbf{k}_2)$. The resulting value for the parameter ζ defined in equation 3 is, for ideal devices, equal to 0.97, whereas the parameter ϵ takes the same value found in previous atomic cascade tests. Then, the proposed experiment would be represented by a point in FIGURE 2 that belongs to the Bell-forbidden region, as expected.

The solid angle subtended by the atomic detector can be calculated as

$$\Omega = \frac{\Delta P_{rx} \Delta P_{rz}}{P_0^2}, \tag{15}$$

where the subscript r refers to the recoil component and P_0 stands for the mean incoming momentum. With bounds given by equation 14, the selected solid angle is equal to 6.92×10^{-10} sr, which corresponds to an area of 0.069 mm^2 for a radial distance of $R = 10$ m from the interaction region. With these values, the detector acceptance slit must have z-edges at 647.6 μm and 686.3 μm, that is, its z-dimension must be 39 μm, with the x-dimension being almost 46 times this value. A vacuum less than 10^{-10} torr is required to reduce the probability of scattering of one calcium atom by background gas to 10^{-4}.

Substitution of the typical values for $\epsilon_+ = 0.99$ and $\epsilon_- = 0.94$ gives the value of 2.2328 for the function $k(\vartheta)$. Inserting the value of t just obtained in the inequality of equation 13, we require that the quantum efficiency of the photodetectors has to satisfy $\eta \geq 92.1\%$ for quantum mechanical predictions to violate Bell's inequality. In these conditions, a reliable experiment can be performed. For a detailed analysis concerning precise experimental requirements, resulting statistics, and production of spurious events, see reference 12.

DISCUSSION

The experiment proposed here blocks the angular correlation loophole in atomic cascade experiments and provides the possibility of discriminating between quantum mechanics and local realism in the high efficiency domain. For the proposed experiment, no LHV model in agreement with quantum mechanics would be tenable with detection efficiencies above a threshold of 0.92 and the previous compatibility for all ranges of efficiency is no longer valid. Only a loophole due to the static character of the experiment would remain open, which might be blocked using a scheme analogous to the one employed by Aspect *et al.*[2] in their third experiment.

Let us finish with a possible improvement of the setup described here. The idea is to manipulate the atomic beam before it reaches the interaction region by means of a laser cooling technique. Preliminary results indicate that a reduction of the mean velocity in the beam to 2 m/s combined with suitable collimation would allow us to reduce drastically the distances involved in the experiment. Moreover, if the residual dispersion in the longitudinal velocity is reduced using a velocity filter, the allowed dark rate for the photodetector could be increased by a factor of 10. A detailed study is in progress.

ACKNOWLEDGMENTS

We are grateful to J. Mlynek and his group in Konstanz, T. W. Marshall, and M. Zukowski for fruitful discussions.

REFERENCES

1. CLAUSER, J. F. & A. SHIMONY. 1978. Rep. Prog. Phys. **41:** 1881–1927. This paper reviews the subject up to 1978. Two examples of more recent experiments are found in references 2 and 3.
2. ASPECT, A., P. GRANGIER & G. ROGER. 1981. Phys. Rev. Lett. **47:** 460–463; ASPECT, A., P. GRANGIER & G. ROGER. 1982. Phys. Rev. Lett. **49:** 91–94; ASPECT, A., J. DALIBARD & G. ROGER. 1982. Phys. Rev. Lett. **49:** 1804–1807.
3. RARITY, J. G. & P. R. TAPSTER. 1990. Phys. Rev. Lett. **64:** 2495.
4. FRY, E. S. 1994. Proceedings of the Third International Workshop on Squeezed States and Uncertainty Relations. D. Han, Y. S. Kim, N. H. Rubin & W. W. Zachary, Eds.: 575–580. NASA Conference Publication Series No. 3270.
5. KWIAT, P. G., P. H. EBERHARD, A. M. STEINBERG & R. Y. CHIAO. 1994. Phys. Rev. **A49:** 3209.
6. CLAUSER, J. F. & M. A. HORNE. 1974. Phys. Rev. **D10:** 526–535.
7. SANTOS, E. 1991. Phys. Rev. Lett. **66:** 1388; 1992. Phys. Rev. Lett. **68:** 894; 1992. Phys. Rev. **A46:** 3646.
8. GHOSH, R., C. K. HONG, Z. Y. OU & L. MANDEL. 1986. Phys. Rev. **A34:** 3962–3968.
9. RARITY, J. G., K. D. RIDLEY & P. R. TAPSTER. 1987. Appl. Opt. **26:** 4616–4619.
10. LO, T. K. & A. SHIMONY. 1981. Phys. Rev. **A23:** 3003.
11. BELL, J. S. 1988. Speakable and Unspeakable in Quantum Mechanics, p. 29 & 105. Cambridge University Press. London/New York.
12. HUELGA, S. F., M. FERRERO & E. SANTOS. 1994. Europhys. Lett. **27:** 181.

An Experiment to Decide between the Causal and the Copenhagen Interpretations of Quantum Mechanics

GERHARD GRÖSSING

Austrian Institute for Nonlinear Studies
A-1030 Vienna, Austria

INTRODUCTION

The recent findings by Rauch, Werner, and others[1] demonstrating that the general criterion for neutron interference to occur is interference in phase space rather than simple wave function overlap have important consequences for the understanding of quantum mechanical nonlocality. The latter can be understood as "the far-reaching action of the plane wave components of the wave function as well. The bandwidth of these plane wave components is determined by the momentum resolution of the measurement with an upper limit defined by the inverse of the source-detector distance."[2] The fact that in EPR experiments individual $|k\rangle$ states of an entangled quantum state, $\psi \propto |-k\rangle_1|k\rangle_2 + |-k\rangle_2|k\rangle_1$, remain interacting at arbitrarily large spatial separation of the wave packets raises the question of how the plane wave components e^{ikr} are acted upon such as to provide the correct quantum mechanical results, namely, the violation of Bell's inequalities.

In the causal interpretation of quantum mechanics (CIQM), nonlocal information transfer is attributed to "some new subquantal superluminal mechanism somehow related to Dirac's ether model."[3] The CIQM is thus not very specific on the nature of this mechanism, apart from the statement that the velocity of the corresponding "phase waves" $e^{ik\cdot\Delta r}$, $u = c^2/v$, is inversely proportional to the particle's velocity v. This, however, cannot explain EPR experiments with photons where $v = u = c$. In a series of papers,[4] I have argued that such phase waves should rather be inversely proportional to the velocity of the alteration of the boundary conditions, and the following discussion with the aid of the standard quantum mechanical formalism is aimed at showing just that.

Any experiment in interferometry is spatially confined to the region between source(s) and detector(s). Thus, instead of the free particle case, where the wave function is defined by the integral

$$\psi(x, t) = \int_{-\infty}^{+\infty} dk\, a(k) e^{-i\Phi},$$

one has to acknowledge that the maximal half-wavelength is given by the distance L_0

between source S and detector D such that the wave function is now defined by

$$\int_{-\infty}^{+\infty} dk\, a(k)e^{-i\Phi} \rightarrow \sum_n a(k_n)e^{-i\Phi_n}. \tag{1}$$

Therefore, in fact, any particle in an interferometer can be treated as a "particle in a box" (limited by practically infinite potential walls next to S and D), where the wave function effectively vanishes for all times outside the said limits. Thus, a treatment of the wave function similar to Greenberger's discussion[5] of a nonlocal effect of moving the walls of a box on the phase of the wave function is required, the only difference to Greenberger's case being that the particle starts next to one wall and ends up next to the other.

"LATE-CHOICE" EXPERIMENTS

Now, considering that the displacement of a wall, say of the detector D, by some amount $\Delta L = L - L_0$ is equivalent to the insertion of a phase-shifter in the region between S and D (FIGURE 1a), this relative displacement can be detected if the phase shift is inserted in one arm of the interferometer only.[6] The crucial point is that an effect can be obtained in so-called "late-choice" experiments[4] even when the phase-shifter is inserted at a location in the interferometer that the main bulk of the wave packet has already passed (FIGURE 1b). As the phase-shifter acts on the plane wave components, this information is (in CIQM language) transported with (superluminal) phase velocity to the last slab of the interferometer. Surprisingly, this statement is in accordance with the standard quantum mechanical formalism, as shall be shown now.

The corresponding Schrödinger equation is given as in reference 5 by

$$-\frac{\hbar^2}{2m}\frac{\partial^2\psi}{\partial x^2} + V\left(\frac{x}{L(t)}\right)\psi = i\hbar\frac{\partial\psi}{\partial t}, \tag{2}$$

with the boundary conditions now being

$$V(y) = \begin{cases} 0 & 0 \pm \delta y \leq y \leq 1 \pm \delta y \\ \infty & \text{elsewhere} \end{cases} \tag{3}$$

where $y = x/L$ and $|\delta y| \ll 1$.

With

$$L(t) = L_0 + \Delta L(t) = L_0 + \alpha t, \tag{4}$$

where α can be positive or negative, and with

$$\frac{\partial^2 L}{\partial t^2} = 0, \tag{5}$$

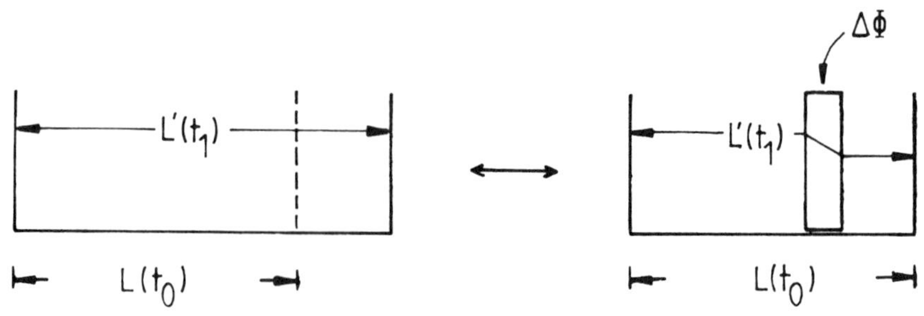

FIGURE 1a. Equivalent altering of boundary conditions: shifting of a wall (left) is equivalent to insertion of a corresponding phase-shifter, ΔΦ (right).

one obtains the exact solutions subject to the boundary conditions:

$$\psi_n(x, t) = e^{-i(m\alpha L/\hbar)y^2} e^{-(i/\hbar)\int E_n dt} e^{ik_n x} \tag{6}$$

with

$$E_n = \frac{\hbar^2 n^2 \pi^2}{2mL^2}. \tag{7}$$

Thus, even though the functions ψ_n are not "stationary" in the usual sense because $L = L(t)$, one has

$$\psi(x, t) = \sum_n a_n \psi_n(x, t) \tag{8}$$

and one obtains the expressions for the dynamical and geometrical phase contributions; that is, the interference term is given by

$$I = 2\cos\Delta\Phi = 2\cos(\Delta\Phi_d + \Delta\Phi_g) := 2\cos\left\{\left[\frac{E_n(L_0)}{\hbar}\frac{\Delta L}{L}t\right] - \frac{m\alpha x^2}{\hbar L(t)}\right\}. \tag{9}$$

Note that the solutions in equation 6 can be rewritten as

$$\psi_n(x, t) = e^{-i[E_n(L_0)/\hbar][1-(\Delta L/L)]t} e^{ik_n[1-(\Delta L/L)]x} \tag{10}$$

such that, relative to the phase of the undisturbed wave function given by

$$\Phi = \frac{E_n(L_0)}{\hbar}t - k_n x, \tag{11}$$

the total phase shift in equation 9 can be rewritten as[7]

$$\Delta\Phi = \frac{E_n(L_0)}{\hbar}\frac{\Delta L}{L}t - k_n\frac{\Delta L}{L}x. \tag{12}$$

This expression vanishes, $\Delta\Phi = 0$, because

$$v = \frac{x}{t} = \frac{d}{dk_n}\left[\frac{E_n(L_0)}{\hbar}\right]. \tag{13}$$

Thus, as long as in our case one continually moves a "wall" by $\Delta L(t)$ (or inserts an increasingly larger phase-shifter, or moves a mirror uniformly, as in double interferometry experiments), dynamical and geometrical phase shifts will always cancel each other exactly. However, once the movement of the wall is stopped (or the phase-shifter is inserted completely), that is, $\alpha = 0$, one obtains a history-dependent net phase shift,

$$\Delta\Phi = \frac{E_n(L_0)}{\hbar}\frac{\Delta L}{L}t', \tag{14}$$

where t' measures the time span of the completely inserted effective phase-shifter. Considering that the phase of the "undisturbed" wave function (equation 11) may be

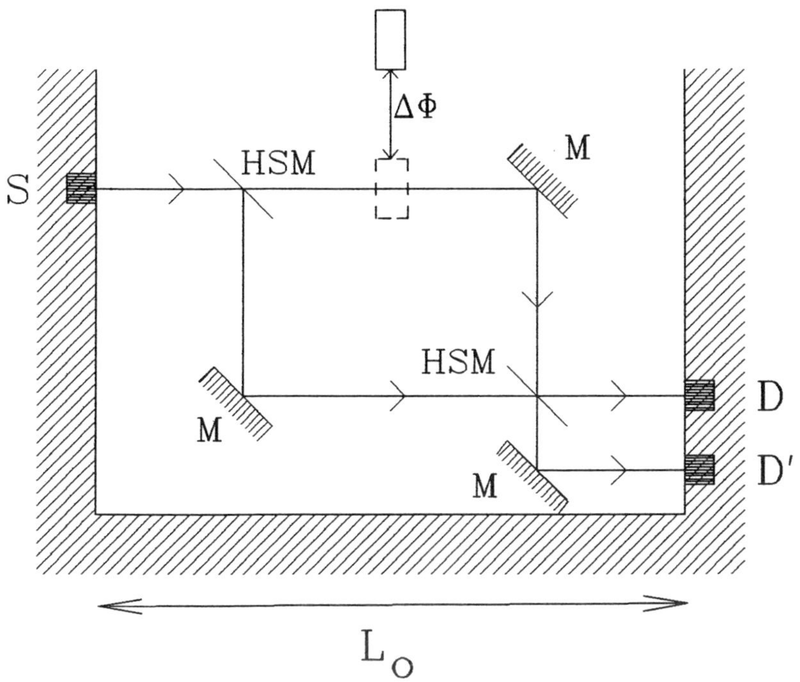

FIGURE 1b. Schematized Mach-Zehnder interferometer within the "walls of a box" of length L_0 limited by source S and detectors D and D' (M, mirror; HSM, half-silvered mirror; $\Delta\Phi$, phase-shifter). The path difference between the second HSM and D or D', respectively, can be easily compensated and is therefore neglected.

defined as

$$\Phi = \frac{E_n(L_0)}{\hbar} t - k_n x := 0, \tag{15}$$

the resulting phase shift can now be written either as

$$\Delta\Phi = \frac{E_n(L_0)}{\hbar} \frac{\Delta L}{L} t' \tag{16}$$

or, equivalently, as

$$\Delta\Phi = k_n \frac{\Delta L}{L} x(t'), \tag{17}$$

where the latter expression reduces to the familiar case of a permanently inserted phase-shifter by choosing $x = L$ in the phase, that is,

$$\Delta\Phi = k_n \Delta L. \tag{18}$$

SUPERLUMINAL SIGNALING

If equations 3, 4, and 5 are reasonable assumptions,[8] it follows that quantum mechanics predicts superluminal signaling: in a very large interferometer, one may let the wave packets travel to a region outside the phase shift's light cone and still have the effect of the inserted phase shift "practically instantaneously" in the interference region. This is due to the very small time lag,

$$\delta t - \delta t' = \frac{\delta x}{u}, \tag{19}$$

where

$$u = \frac{\delta x}{\delta t} \frac{L}{\Delta L} = \frac{\delta x}{\delta t} \frac{L}{t} \frac{t}{\Delta L} = \frac{c^2}{\alpha} \gg c. \tag{20}$$

Hence, if a "no signal" operation of the interferometer were such that the maximum number of particles is registered in the final D-beam due to total constructive interference, an effective "late-choice" relative phase shift of $\Delta\Phi = \pi$ would make the particles arrive in the D'-beam instead (FIGURE 1b). The superluminal switching of the maximum count rate from detector D to detector D' could then be used for signaling.

As has been shown earlier, this would still be completely in agreement with relativity theory. (For example, the resulting net phase shift can always be written covariantly as

$$\Delta\Phi = k_\mu \frac{\Delta L}{L} x^\mu. \tag{21}$$

For more details, see, for example, references 9 and 10.)

Also, with the superluminal phase waves thus introduced, the causality problem usually associated with superluminal signaling would not exist.[10] Concerning the well-known argument by Brillouin and Sommerfeld[11] that the front velocity v_F, at which the first appearance of a discontinuity propagates, never exceeds c and is thus the genuine "information" velocity, their analysis only holds for infinitely long phase waves. However, with the discrete sum over the modes of the wave function subject to the boundary conditions, equation 1, the usual requirement of complex integration for the wave function to vanish, that is, when $k \rightarrow \infty$, is changed into $\psi_n(x, t) = 0$ for $x > v_F t$, where $v_F = \omega_n/k_n = u \geq c$. As to the superluminal signaling itself, it should be possible even if there were a substantial uncertainty $\Delta\tau$ in the timing of the insertion of the phase-shifter, which would only provide a modulation of the pure effect, that is,

$$I = 2 \cos \Delta\Phi \cos\left[\frac{\Delta\Phi(\Delta\tau)}{2}\right]. \tag{22}$$

In effect, the possibility to causally influence real plane waves would thus clearly contradict any version of the Copenhagen interpretation of quantum theory.

Why has this effect not been seen earlier in time-dependent experiments? One can assume that this is basically because of the continuous movements of the objects defining the boundary conditions, which, as we have seen, cancel the effect, or because of too-low time resolutions. The only cases where the effect has been observed implicitly, of course, are EPR-type experiments, which can thus be explained by the action of the plane wave components. However, in principle, no superluminal signaling can be performed in EPR-type experiments unless they are also of a "late-choice" type as presented here.

REFERENCES AND NOTES

1. See, for example: JACOBSON, D. L., S. A. WERNER & H. RAUCH. 1994. Phys. Rev. A **49:** 3196–3200 and further references therein.
2. RAUCH, H. 1993. Phys. Lett. A **173:** 240–242.
3. VIGIER, J. P. 1994. Found. Phys. **24:** 61–83. However, not all versions of the CIQM admit such a mechanism. For example, Dewdney and Lam maintain that "there is no nonlocality in the sense that spacelike separated regions can influence each other." (See: DEWDNEY, C. & M. M. LAM. 1990. *In* Information Dynamics. H. Atmanspacher & H. Scheingraber, Eds. Plenum. New York.)
4. See, for example: GRÖSSING, G. 1986. Phys. Lett. A **118:** 381–386.
5. GREENBERGER, D. M. 1988. Physica B **151:** 374–377.
6. Practically, the phase-shifter could have the shape of a wedge such as to continually increase ΔL with time until it is inserted completely.
7. Note that the factor $(\Delta L/L)$ can, in principle, be attributed to either a modified energy or time, respectively. In other words, one can have either energy conservation with an altered time measure or vice versa. Similar considerations hold for momentum and space coordinates, respectively. In principle, this only demonstrates the genuinely nonlocal character of the quantities involved.
8. As to the idealization by infinite potential walls in equation 3, it is easy to show that eventual wave packets reflected from the walls do not alter our result. In fact, if $\alpha = 0$ before a wave packet can reach the last slab of the interferometer, then with certainty the geometric phase contribution is zero even before any wave packet is reflected back from the "detector wall". This leaves the resulting phase shift of equation 14 un-

changed. Concerning the assumptions of equations 4 and 5, one can see that the smaller that α is, the better the corresponding approximation will hold. See also equations 19 and 20.

9. GRÖSSING, G. 1989. Nuovo Cimento B **103:** 497–510.
10. GRÖSSING, G. 1988. *In* Problems in Quantum Physics. L. Kostro *et al.*, Eds. World Scientific. Singapore.
11. BRILLOUIN, L. & A. SOMMERFELD. 1960. Wave Propagation and Group Velocity. Academic Press. New York; see also, for example: CHIAO, R. *et al.* 1995. This volume.

Relativistic Quantum Measurements[a]

ASHER PERES

Department of Physics
Technion—Israel Institute of Technology
32 000 Haifa, Israel

Relativistic quantum measurements have been a controversial subject since the early days of quantum theory.[1–4] The root of the difficulty is that macroscopic observers, who are inherently localized, cannot perform exhaustive tests on quantum systems, which are inherently nonlocal. As a simple illustration, consider the familiar pair of spin-½ particles, prepared in a singlet state and moving apart from each other toward two distant observers. Assume further that these particles do not interact with each other or with any external agent from the moment they have been produced in the singlet state (their Hamiltonian is $H = 0$). The role of the observers is to ascertain the nature of the preparation state.

Conceptual difficulties arise if you demand that every physical system, such as this pair of particles, has, at every instant, a well-defined quantum state (some authors would even want the entire universe to have a quantum state). To illustrate the difficulty, let our two observers be attached to different Lorentz frames, as shown in FIGURE 1: they recede from each other, with a constant relative velocity. Thus, in each one of the Lorentz frames, the test performed by the observer who is at rest appears to occur earlier than the test performed by the moving observer. If the first observer got a bad education in quantum theory and believes that the pair of particles has, at each instant, a definite wave function, he will say that the singlet state, which existed for $t_1 < 0$, collapsed into an eigenstate of S_{1x} and of S_{2x} for $t_1 > 0$. In the same vein, the second observer may say that the singlet state held for $t_2 < 0$ and thereafter collapsed into an eigenstate of S_{1y} and of S_{2y}, as a result of her test.

Statements like those of our fictitious observers are not only contradictory, but they are utterly meaningless. There is no disagreement about what was actually observed. However, a situation involving several observers cannot be described by a wave function with a relativistic transformation law. No single covariant state history may be defined that properly accounts for all the experimental results.[5]

In this article, I shall examine whether a generalized version of the "collapse" postulate can be formulated in a relativistic framework. This issue was discussed by numerous investigators, who reached conflicting conclusions: depending on the point of view taken, the collapse would occur along past light cones[6,7] or future light cones,[8–10] or both,[11] or on arbitrarily curved spacelike hypersurfaces.[12]

However, we first need to understand how the notion of collapse arises in conventional, nonrelativistic quantum theory. The point of view that I take here is the "orthodox" one—the only one that is actually used by experimental physicists for analyzing the results recorded by their macroscopic instruments.[13] In this orthodox

[a] This work was supported by the Gerard Swope Fund and by the Fund for Encouragement of Research.

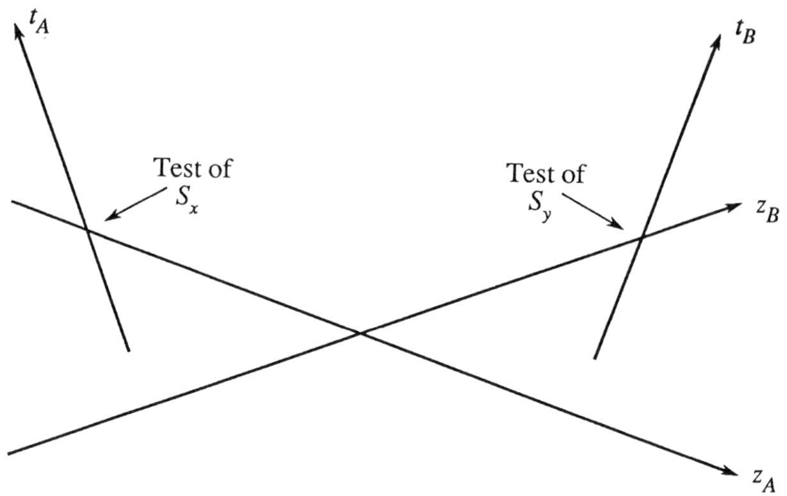

FIGURE 1. In this space-time diagram, the origins of the coordinate systems are the locations of the two tests. The t_A and t_B axes are the world lines of the observers, who are receding from each other. In each Lorentz frame, the z_A and z_B axes are isochronous: $t_A = 0$ and $t_B = 0$, respectively.

(sometimes called "Copenhagen") interpretation, quantum theory does not provide a universal description of nature. It merely is a set of rules for computing the probabilities of occurrence of definite outcomes, in tests that follow definite preparations.[14,15] Anyone who wants to see more than that in quantum theory does so at his own risk.

FIGURE 2a is a symbolic sketch of the domain of existence of a state vector—or a wave function—or, more generally, a density matrix. These notions do not refer to physical objects, akin to classical fields; they are mere mathematical tools for computing probabilities (more precisely, amplitudes). Note that there is no meaning to a quantum state before the preparation of the physical system nor after its final observation (just as there is no "time" before the big bang or after the big crunch).

When a physical system is closed, that is, completely described by its Hamiltonian, the evolution of the state vector is a unitary mapping if we use the Schrödinger picture (or else, if we use the Heisenberg picture, the state vector is constant). The so-called collapse, which is a sudden nonunitary jump of the state vector, occurs only for open systems and is due to temporary external interventions, as sketched in FIGURE 2b, where we see a quantum system with initial state ρ_i interacting with an external object (in this case, a macroscopic apparatus). After that interaction, the quantum states of these two systems are correlated: we no longer have two distinct systems, but a single, indivisible, nonlocal quantum system. No collapse has occurred as yet. However, if we deliberately ignore the quantum correlations (formally, if we force a classical description on the apparatus in spite of these correlations), the new state, ρ_f, is essentially different from the former, ρ_i. It is this "quantum jump" of ρ that is called a collapse and is sketched in FIGURE 2b. There is nothing mysterious here: the jump is solely due to an arbitrary change of the description of the apparatus. This dichotomy in the description of the apparatus was repeatedly emphasized by Bohr:[16]

> However far the [quantum] phenomena transcend the scope of classical physical explanation, the account of all evidence must be expressed in classical terms. The argument is simply that by the word "experiment" we refer to a situation where we can tell others what we have done and what we have learned and that, therefore, the account of the experimental arrangement and the results of the observations must be expressed in unambiguous language with suitable application of the terminology of classical physics.

It is not the quantum system that jumps. It is our description of the apparatus that jumps from the classical language to the quantum language and then back to the classical one.[15] As a consequence of the measuring process, the apparatus has become inseparably entangled with the system under observation, but the language used for describing this event wants to distinguish that quantum system from the observing apparatus. That language also fundamentally distinguishes the past from the future, even if the dynamical laws are basically time-symmetric.

An ideal measurement is repeatable[17] and it is mathematically represented by a complete set of orthogonal projectors.[18] It essentially amounts to the preparation of a new state, identical to the one that has just been observed. In real life, typical measurements are not of this type. They are represented by positive operator valued measures (POVM) with noncommuting elements.[15] If we still want to use a "collapse" description, that collapse must then be considered as incomplete.[19] In the following discussion, I shall consider only ideal measurements.

A typical example is the double Stern-Gerlach experiment illustrated in FIGURE 3. Here, we assume that the angular separation of the beams that leave the first

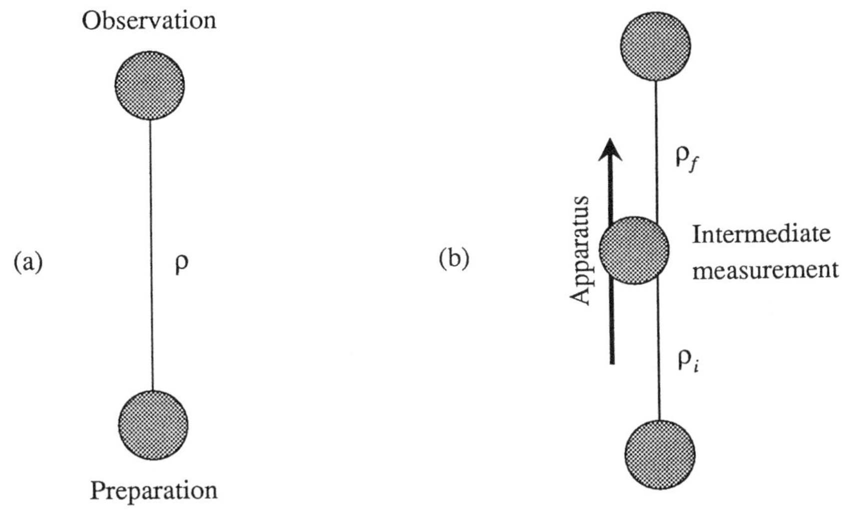

FIGURE 2. (a) The state ρ exists between the preparation of the quantum system and its final observation. (b) An ideal intermediate measurement is equivalent to an observation followed by a preparation and causes a "jump" in the quantum state.

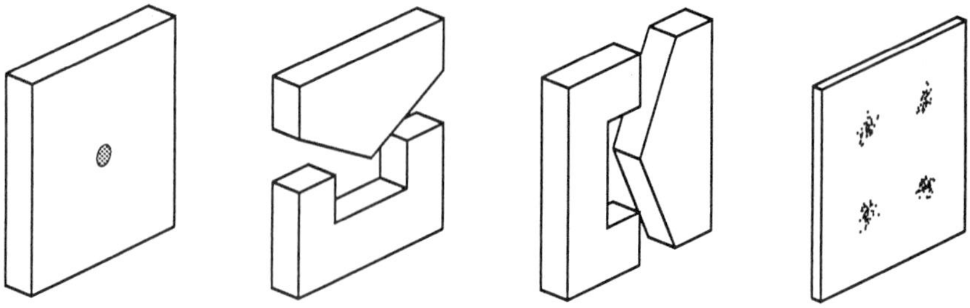

FIGURE 3. Two consecutive Stern-Gerlach experiments. The drawing has been compressed by a factor of 10 in the longitudinal direction. From left to right: the oven from which the atomic beam originates, the inhomogeneous magnets, and a detector plate.

magnet is sufficient so that they do not overlap when they enter the second magnet. If, and only if, this condition is satisfied, it becomes possible to imagine the existence of quasi-classical trajectories. On the other hand, this separation should not be too large so that the second magnet performs essentially the same test for each one of the two beams that impinge on it. We may then consider the setup shown by FIGURE 3 as two consecutive tests rather than a single test with four possible outcomes. We may imagine that each impact on the detector plate is the end point of a trajectory that is not seen, but that can be calculated semiclassically. The location of the impact point reveals not only the outcome of the final test, but also the outcome of the test performed with the first magnet. (Moreover, we assume that if the second magnet had not been there the trajectory through the first magnet would have remained the same. This is a natural, but unverifiable, counterfactual assumption.)

Taking all these assumptions for granted, the four spots on the detector plate can be unambiguously identified as corresponding to definite outcomes of the first test and also definite outcomes of the second test. Note that the first test is repeatable: its different outcomes are preparations of pure states. Therefore, it is admissible to say that the first Stern-Gerlach magnet "collapses" the state of each impinging particle into an eigenstate of the spin component along the magnetic field. In the present case, this terminology is justified, but only provided that we waive the (admittedly difficult) possibility of reuniting in a coherent way the two beams that leave the first magnet before they enter the second one.

How does relativity theory affect these notions? The new feature is that, if observers are localized at different space-time points, the speed of information transfer between these observers is finite so that, when a delocalized quantum system is subject to tests by independent observers, the latter may acquire different types of information. Thus, having different knowledge and predictive power, they will legitimately attribute different state vectors to the same physical system, as was illustrated in FIGURE 1. Note that FIGURE 1 involved both time and space dimensions, whereas in FIGURE 2 there is only a time axis: nonrelativistic time does not mix with the space coordinates.

Under these conditions, is it still possible to describe a partially ordered set of measurements by a suitably modified "collapse" formalism? I sketched an attempt to do so in FIGURE 4, which is the relativistic generalization of FIGURE 2. A single preparation-observation pair of events is easy to represent. Instead of the time segment in FIGURE 2a, we now have a pair of light cones. The region enclosed by these cones is compact even if space-time is curved, provided only that it is globally hyperbolic.[20] The quantum state ρ refers to that compact space-time region and it allows one to compute probabilities for the various preparations and observations that can possibly be performed at its two vertices.

If there also are intermediate measurements by mutually spacelike localized observers and we still want to use a similar language, the best that can be done is shown in FIGURE 4b, which is the relativistic generalization of FIGURE 2b. The vertically hatched regions represent the situation as it could be described by the initial preparer, observer A, and the final observer. The horizontally hatched regions have a similar meaning with respect to observer B. Obviously, there is no universal wave function valid everywhere in space-time or even in the compact region between the two light cones. Likewise, there is no way of defining global quantities, such as the

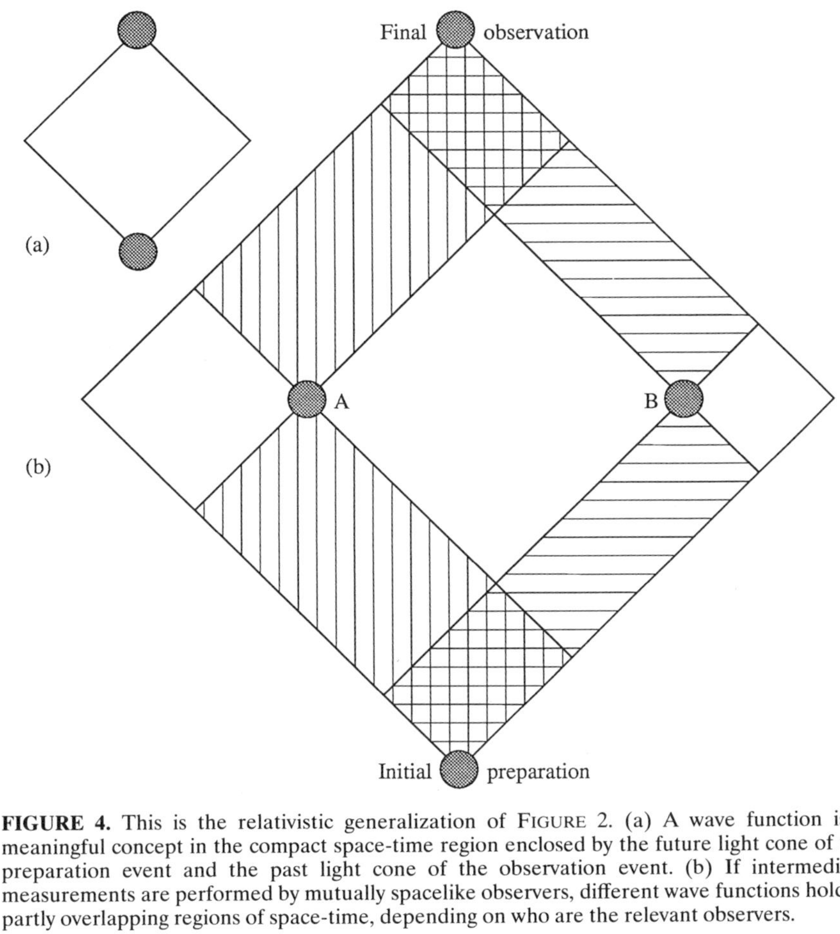

FIGURE 4. This is the relativistic generalization of FIGURE 2. (a) A wave function is a meaningful concept in the compact space-time region enclosed by the future light cone of the preparation event and the past light cone of the observation event. (b) If intermediate measurements are performed by mutually spacelike observers, different wave functions hold in partly overlapping regions of space-time, depending on who are the relevant observers.

total charge of a physical system, nor are there any means for actually preparing and observing such quantities, as this would require having simultaneous access to a complete spacelike hypersurface.

All these conclusions apply only to the case of observers whose measurements are highly localized, when compared with the space-time region under consideration. More sophisticated, nonlocal measuring processes can also be imagined.[21] In particular, quantum teleportation[22] could be used, with the initial preparer sending EPR pairs of test particles—one particle of each pair to each one of the sites being tested and the other particle of that pair to the final observer. The information flow then becomes inherently nonlocal and there is no way that I can imagine for drawing diagrams such as FIGURE 4b to represent these more general processes.

In conclusion, the notion of collapse, which is of dubious value in nonrelativistic quantum mechanics, appears to have no meaning whatsoever in a relativistic context.

REFERENCES

1. LANDAU, L. D. & R. PEIERLS. 1931. Z. Phys. **69:** 56.
2. BOHR, N. & L. ROSENFELD. 1933. Mat. Fys. Medd. Dan. Vidensk. Selsk. **12:** no. 8.
3. BOHR, N. & L. ROSENFELD. 1948. Phys. Rev. **78:** 794.
4. BLOCH, I. 1967. Phys. Rev. **156:** 1377.
5. AHARONOV, Y. & D. Z. ALBERT. 1981. Phys. Rev. D **24:** 359.
6. HELLWIG, K-E. & K. KRAUS. 1970. Phys. Rev. D **1:** 566.
7. MOSLEY, S. N. 1993. Phys. Lett. A **182:** 1.
8. MALIN, S. 1982. Phys. Rev. D **26:** 1330.
9. SQUIRES, E. J. 1990. Phys. Lett. A **145:** 297.
10. FINKELSTEIN, J. 1992. Found. Phys. Lett. **5:** 383.
11. FINKELSTEIN, J. 1994. Phys. Lett. A **188:** 117.
12. AHARONOV, Y. & D. Z. ALBERT. 1984. Phys. Rev. D **29:** 228.
13. VAN KAMPEN, N. 1988. Physica A **153:** 97.
14. STAPP, H. P. 1972. Am. J. Phys. **40:** 1098.
15. PERES, A. 1993. Quantum Theory: Concepts and Methods. Kluwer. Dordrecht.
16. BOHR, N. 1949. *In* Albert Einstein, Philosopher-Scientist. Volume 1. P. A. Schilpp, Ed.: 209. Library of Living Philosophers. Evanston, Illinois.
17. DIRAC, P. A. M. 1947. The Principles of Quantum Mechanics, p. 36. Oxford University Press (Clarendon). London/New York.
18. VON NEUMANN, J. 1955. Mathematical Foundations of Quantum Mechanics. Princeton University Press. Princeton, New Jersey.
19. PERES, A. & A. RON. 1990. Phys. Rev. A **42:** 5720.
20. FULLING, S. A. 1989. Aspects of Quantum Field Theory in Curved Space-Time, p. 21. Cambridge University Press. London/New York.
21. AHARONOV, Y. & D. Z. ALBERT. 1980. Phys. Rev. D **21:** 3316.
22. BENNETT, C. H., G. BRASSARD, C. CRÉPEAU, R. JOZSA, A. PERES & W. K. WOOTTERS. 1993. Phys. Rev. Lett. **70:** 1895.

Quantum Theory, Relativity, and the Bohm Model

EUAN J. SQUIRES

Department of Mathematical Sciences
University of Durham
Durham City DH1 3LE, United Kingdom

INTRODUCTION

I want to discuss some aspects of the relation, or perhaps it would be better to say conflict, between quantum theory and relativity. It seems clear that, in order to understand the nature of this conflict, we need a *complete* quantum theory. Attempts to suppress this need, by using the obscurities of the Copenhagen interpretation, for example, merely hide the conflict and give neither motivation nor guidance towards its solution.

To my knowledge, there are only three methods of providing a "complete" quantum theory that are presently available. These involve hidden variables, in particular as in the Bohm model;[1–3] explicit collapse of the wave function, as in the work of Ghirardi, Rimini, and Weber[4] and of Pearle[5] (see reference 6 for a recent review and further references); or some form of the many-worlds plus consciousness model, as proposed by Albert and Loewer,[7] Lockwood,[8] Stapp,[9] or (in a form that naturally I prefer) Squires.[10] This last method inevitably takes us outside of what we conventionally (at present) regard as physics, so I shall ignore it (which is not to say that I do not believe it might be true). Mainly, I shall concentrate on the Bohm model because there the issues seem more clear (I guess because the concept of trajectories is one with which we are familiar—certainly more so than with stochastic background fields).

THE BOHM MODEL AND LORENTZ-INVARIANCE

In the standard version of the (nonrelativistic) Bohm model, the velocity of particle 1, at time t, is given by the expression,

$$\dot{x}_1(t) = \mathscr{R}\,\frac{\nabla_1\Psi[x_1(t), x_2(t), \ldots]}{m_1\Psi[x_1(t), x_2(t), \ldots]}, \tag{1}$$

where Ψ is the quantum wave function. The nonrelativistic nature of this is apparent from the fact that the velocity of particle 1 at a given time is assumed to depend on the position of the other particles *at the same time.*

A simple (in principle) way to demonstrate the conflict in an experiment is discussed in Hardy and Squires.[11] This paper, based on earlier work by Hardy,[12] also demonstrates that any hidden-variable model (at least within some class defined in the paper) must inevitably violate either Lorentz-invariance or quantum theory. It is important to recognize that this conflict has nothing to do with the fact that particle number cannot be conserved in Lorentz-invariant theories and that it is necessary to use field theory. Indeed, the experiment referred to can be performed with arbitrarily low velocities for all the particles involved, so the field-theoretic aspects should not be important.

Although pre-1905 physicists were, apparently, happy with an absolute concept of simultaneity, as required in equation 1 and in the quantum understanding of the above experiment, it is hard for those of us trained in relativity to accept it. As we shall see below, the idea becomes even harder in more general space-times required in general relativity. Hence, it is natural to suggest that the conflict should be resolved by relaxing the requirement of perfect agreement with the predictions of standard quantum theory. The way to proceed then becomes rather obvious. The problem arises essentially because the wave function, used in calculating the velocities (or the quantum potential, if we write the theory as a Newton-like equation for acceleration), is defined in configuration space; that is, it depends on the position of all the relevant particles. This, however, is normal in nonrelativistic physics; it is true, for example, of the electrostatic potential of several charged particles. There, the conflict with relativity is resolved by going from electrostatics to electrodynamics, which is a fully relativistic theory. What this does, among other things, is to replace the "simultaneous" positions by so-called retarded positions. I have therefore proposed[13] to make a relativistic version of the Bohm model by the same device. Thus, we replace equation 1 by

$$\dot{x}_1(t) = \mathscr{R}\,\frac{\nabla_1\Psi[x_1(t), x_2(t_2), \ldots]}{m_1\Psi[x_1(t), x_2(t_2), \ldots]}, \tag{2}$$

where

$$t_2 = t - |x_1(t) - x_2(t_2)|/c. \tag{3}$$

Of course, the formal similarity between the electrostatics case and quantum theory is probably misleading. For one thing, the wave function usually has an explicit dependence on time, and it is not clear how to treat this. Nevertheless, there is no ambiguity if we apply equation 2 to situations where the particles are far apart, compared to the quantum uncertainty in their positions, and each is described by a wave function that is evolving independently. Then,[13] we use the retarded time t_2 to describe the wave function associated with particle 2, etc.

In order to understand some of the effects of this, we consider an EPR-like measurement of the spins of two correlated particles, for example, in the singlet state. First, we use the standard Bohm model, where it is important to realize that the measured value of the spin is determined by the hidden variables in the *detector*. (Examples of this are given in the work of the Portsmouth group.[14] Exact calculations for a very simple case are given in references 13 and 15.) Suppose, first, that there is only one detector, say, in the path of the L particle. Then, equation 1 shows that it will record $+\frac{1}{2}$ and $-\frac{1}{2}$ with equal probability, with the actual value in a given case depending on the values of the detector hidden variables. Similar results hold if only an R detector is present. When both detectors are included, the value that is recorded depends on the relative values of the hidden variables in the two detectors.

The way that the Bohm model is constructed guarantees that the resulting probabilities are correlated exactly as required by the predictions of quantum theory. This is all illustrated schematically in FIGURE 1.

To see how this changes if we use the retarded model (equation 2), we consider the process depicted in FIGURE 2. The left particle reaches the detector at time t_1, but it is not until the later time t_2 that the formula given in equation 2 can "know about" the presence of the right detector. Clearly then, in the time between t_1 and t_2, the left (L) detector begins to record the two values with equal probability. The same

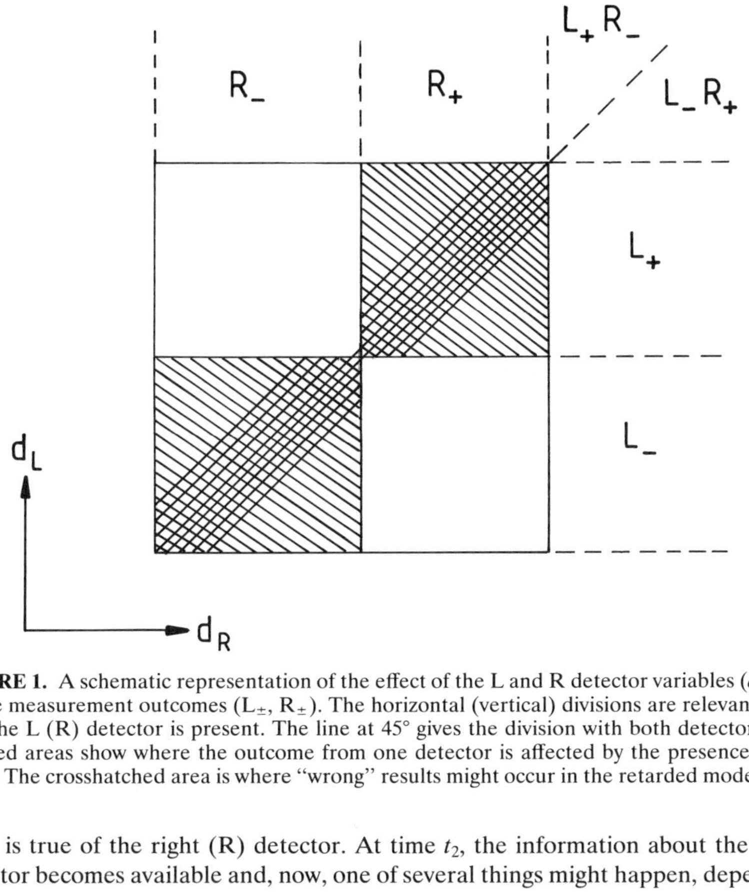

FIGURE 1. A schematic representation of the effect of the L and R detector variables (d_L, d_R) on the measurement outcomes ($L_\pm$, $R_\pm$). The horizontal (vertical) divisions are relevant when only the L (R) detector is present. The line at 45° gives the division with both detectors. The hatched areas show where the outcome from one detector is affected by the presence of the other. The crosshatched area is where "wrong" results might occur in the retarded model.

thing is true of the right (R) detector. At time t_2, the information about the other detector becomes available and, now, one of several things might happen, depending on the detector hidden variables. Consider, for example, the extreme case where the detectors are measuring spins in the same direction. Then, in half of the cases, the initial movement of the detectors will correspond to "wrong" results, that is, both giving the same value of the spin in contrast to the perfect anticorrelation expected in the singlet state. Suppose, for example, that these are + values, which occur if the hidden variables of both detectors are in the regions of space that lead to such values.

When evolution beyond t_2 is considered, the detector whose hidden variable is furthest from the boundary of the region will be unaffected. However, the other will in some cases reverse, to give the "correct" result (−), but in others it will not, and hence the wrong result will be the final reading of the detectors. The condition for

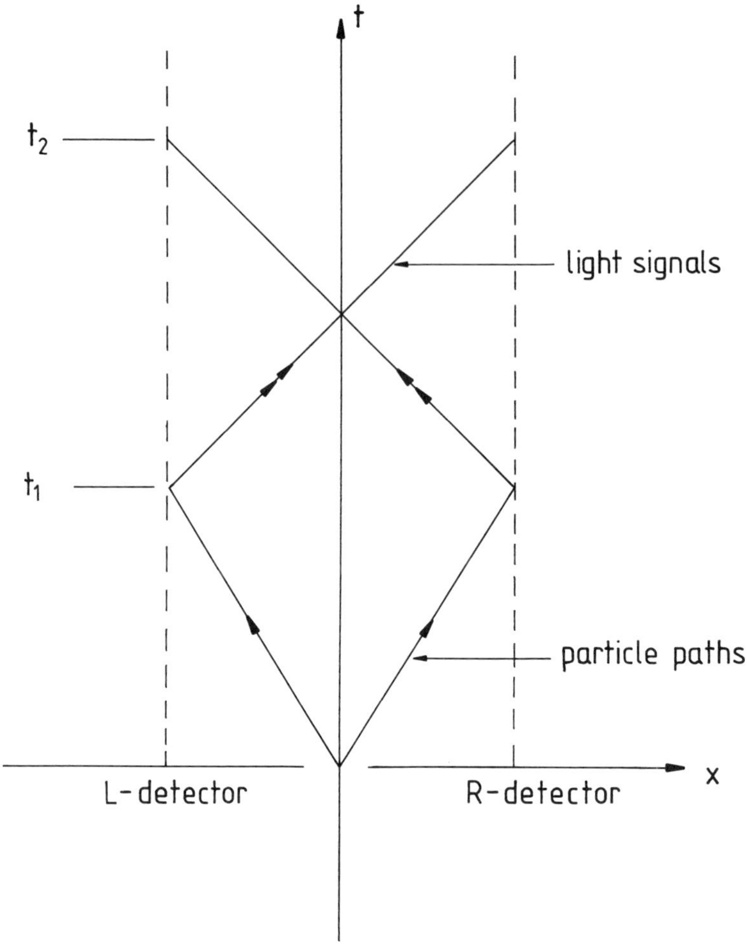

FIGURE 2. The space-time diagram for an EPR-like event in which a correlated pair leaves $x = 0$ at $t = 0$ and arrives at the detectors at t_1. In the retarded Bohm model, each detector, during the time $t_1 < t < t_2$, behaves as though the other were absent.

the wrong results is that the initial hidden variables are sufficiently close to each other, with how close depending on the time difference $t_2 - t_1$.

This is again illustrated schematically in FIGURE 1, where the enormous number of hidden variables in each detector is represented by one variable for each detector.

In this figure, the crosshatched region is the one that gives "wrong" results, and the crucial issue is how large it is in relation to the region available for the hidden variables. We expect (see also reference 13) that the condition for agreement with quantum theory is

$$\frac{T}{T_{\mathrm{M}}} \ll 1, \tag{4}$$

where $T = t_2 - t_1$ is the time for a signal to travel from one detector to another, that is,

$$T = \frac{L}{c} \tag{5}$$

(L being the separation between the detectors) and T_{M} is the time taken for the "measurement" to be complete. Clearly, without a more adequate model of the measurement apparatus, it is difficult to estimate T_{M}; however, if we say that macroscopic systems are unlikely to make significant responses to microscopic stimuli in times less than about 10^{-5} s, then we only expect departures from quantum theory to occur with reasonable frequency if L is of the order of kilometers. I believe that there is a need here for a careful study of actual measurements in order to obtain reliable estimates of what effects might be expected. Without such a study, it is, in my opinion, premature to claim that quantum nonlocality has been experimentally established. (Very similar considerations can be made for explicit collapse models; see reference 16.)

It is worth noting here that we could imagine relativistic models that are more effective in hiding departures from quantum theory (and from angular momentum conservation) than the retarded Bohm model considered here. We could, for example, suppose that the particles carry information about their spin direction; that is, one will be positive in a particular hemisphere, the other negative in the same hemisphere. Then, regardless of any information coming from the other measurement, there will be no violation of strict anticorrelation. Then, the disagreement with quantum theory would arise only because, for sufficiently large separation, the results would have to be consistent with Bell's inequality. Clearly, evidence for this would be harder to find because it would depend on a statistical analysis, whereas, in principle, just *one* result in violation of the anticorrelation would be sufficient to establish a disagreement.

THE BOHM MODEL AND INFORMATION LOSS

Since Hawking demonstrated that black holes radiate a thermal spectrum, there has been growing interest in the apparent loss of information associated with parts of correlated wave functions disappearing down black holes. It seems that an initially pure quantum state turns into a mixed state. However, if the quantum state is supposed to be a statement about "what is", rather than about what we know about it, then this is hard to interpret in a way that makes sense. The statement that the universe is either in state A or state B surely means that it is actually in one or the other; we may be unsure about which, but the universe itself can have no such doubt!

The Bohm model is ideally suited to considering questions of this type because it has a clear ontology. We can ask whether the Bohm model continues to work in its usual deterministic way even in the presence of black holes. To answer such a question requires a formulation of the model that is applicable in nonflat background space-times, and a student of mine (Steve Mackman) is working on this problem here in Durham. Even without this, however, we can discuss the application of the model in flat space-times, which are topologically nontrivial—in particular, which contain closed timelike loops. These allow information loss very similar to that which occurs in black holes.

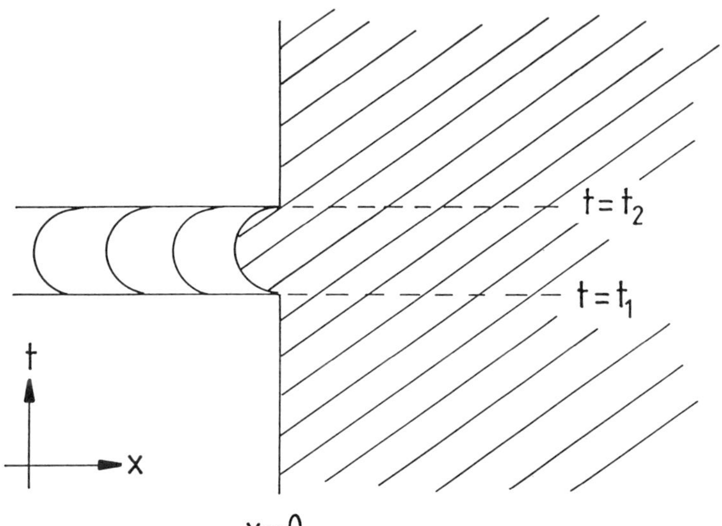

FIGURE 3. The topology of the two-dimensional space used to consider the problem of information loss.

As an example, we consider the two-dimensional space shown in FIGURE 3. This is made from an infinite x, t plane with a cut along $x = 0$ from $t = -\infty$ to $t = t_1 > 0$ and from $t = t_2 > t_1$ to $t = +\infty$, and with the region $x < 0$ being rolled up by the identification of $t = t_3$ with $t = t_1$. We impose the boundary condition that the wave function is zero on the cut.

We take a two-particle wave function chosen by the initial condition that, at $t = 0$, it has the form,

$$\Psi^0(x, y) = \alpha\phi_a^0(x)\chi_a^0(y) + \beta\phi_b^0(x)\chi_b^0(y), \tag{6}$$

where the states $\phi_{a,b}^0$ and $\chi_{a,b}^0$ are normalized to unity and the latter are chosen to be orthogonal,

$$\langle\chi_a^0|\chi_b^0\rangle = 0. \tag{7}$$

Explicitly, we suppose that these states are wave packets approximately localized around $y = a, y = b > a, x = a + c > b$, and $x = b + c$, as shown in FIGURE 4. We assume that the states are solutions of the two-dimensional Dirac equation so that they move as indicated in the figure (see reference 17).

Consider now the trajectory of the x-particle. This is given by the usual Bohm expression,

$$\dot{x} = \mathscr{R} \frac{p_x \Psi}{m\Psi}, \tag{8}$$

which becomes

$$\dot{x} = \mathscr{R} \frac{\alpha\chi_a(y)p_x\phi_a(x) + \beta\chi_b(y)p_x\phi_b(x)}{m\chi_a(y)\phi_a(x) + m\chi_b(y)\phi_b(x)}, \tag{9}$$

an expression that clearly shows how the trajectory for the x-particle depends on the value of y. As long as the two χ wave packets do not overlap, the value of y ensures that the trajectory is determined entirely by either ϕ_a or ϕ_b, exactly as would be the case for a *mixed* state, rather than the pure state of equation 6. This is no longer the case, however, when the χ states overlap, even though the states remain orthogonal. This fact, that in the Bohm model there is a clear distinction between a mixed state and a pure state, was made already in 1980 by John Bell.[18] I am grateful to Shelly Goldstein for telling me about this work.

Now, we suppose, again as shown in FIGURE 4, that the two wave packets ($\chi_{a,b}$) for the y-particle enter the "black hole". In this case, it clearly makes no sense to use the correlated wave function at much larger times, in particular, the shaded region of the figure. It is not reasonable to suppose that anything that happens to the y-particle as it moves along the tube can have any effect on the x-particle outside it. It is this that gives rise to the loss-of-information problem in "orthodox" quantum theory. Because we cannot use the state Ψ to describe the x-particle for large times, we have to use the mixture: *either* ϕ_a or ϕ_b. In the Bohm model, however, there is a natural resolution of this problem. At all stages, the x wave function that is used to calculate the x-trajectory is determined by the value of y; thus, at the moment when the y-particle enters the black hole, with a consequent ending of the correlation, we have a unique wave function as a function of x. This wave function, evolved forward in time by the Dirac equation, determines the trajectory at all future times.

In fact, it seems most natural here to define the time when the correlation ceases as being along the forward light cone from the point $x = 0, t = t_2$ (see FIGURE 4). Thus, we use the Lorentz-invariant modification of the Bohm model, discussed above, to calculate the trajectory from the complete wave function throughout the unshaded region of the figure. On the boundary, that is, along $x = t - t_2$, the x wave function is taken to be

$$\Psi_t(x) = \alpha\chi_a[y(t - |x - y|)]\phi_a(x) + \beta\chi_b[y(t - |x - y|)]\phi_b(x), \tag{10}$$

with y of course being the position of the y-particle. This is a function of x along the boundary and, as such, it can be calculated for all future times using the Dirac equation. The trajectory followed by the particle outside the black hole is now no longer correlated with anything inside and there is no loss of information. Thus, here, as elsewhere, the Bohm model successfully removes any ambiguities in the application of quantum theory.

THE BOHM MODEL AND TIME IN QUANTUM COSMOLOGY

As is well known, the wave function of the universe in canonically quantized gravity is independent of time. Many workers have considered how this fact should be reconciled with a world of change. For a recent review and many other references, see Isham.[19] To some extent, the solutions proposed depend on the attitude adopted

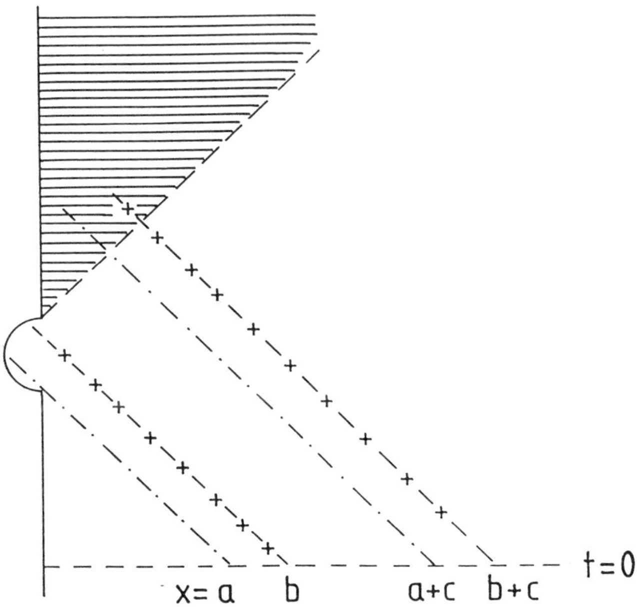

FIGURE 4. Possible paths of two correlated particles: the shaded region is where the possible loss of information might occur.

to the quantum measurement problem. For example, if we follow the explicit collapse route, then there appears to be a natural way of introducing time because, although the uncollapsed wave function might be annihilated by the Hamiltonian, this will not normally be the case with the collapsed wave function (because in general the collapse function will not commute with the Hamiltonian). Thus, "time" begins with the first collapse. In many-worlds quantum theory, on the other hand, there really is no such thing as external time in physics and an endeavor is made to understand our experience of time by equating it to some other variable, that is, something that plays the role of the hands of a clock. Again, I believe that this pushes the issue outside of present physics and into "consciousness".

In the Bohm model, there is, at first sight, a natural solution to the missing time problem because the Bohm formula does not necessarily give zero velocities when the wave function is constant. In other words, a constant wave function need not correspond to a situation where nothing is changing. There are, however, some difficulties with this. The first is conceptual. If we regard the Bohm equation as being the condition that preserves the property that an initial distribution of positions, agreeing with the quantum probability law, will still agree at all later times, then the *natural* solution for a constant wave function will clearly be that with zero velocities. The Bohm formula is just one possible way of satisfying this condition and, at least for a constant wave function, it is not the simplest.

The second difficulty is more practical. Although nonzero velocities are possible, they are not easy to obtain and the obvious (real) solutions of the Wheeler-deWitt equation have zero velocities. For this reason, the Bohm model actually makes the problems arising from a constant wave function harder, not easier, to resolve. A model that does not actually have a radius of the universe, say, is not troubled by its rate of change, but such a thing actually has a value in the Bohm model, so it had better not be a constant! Note also that there seems to be an important difference here between the Bohm model as used in the microscopic world and as used in cosmology (see reference 20). In the former, we are not unduly disturbed by the fact that the model gives zero velocities for particles in systems that are in energy eigenstates, for example, for the electrons in a hydrogen atom. We know from the way that the model is constructed that it will give the correct distribution for *measured* velocities. In other words, measurements of velocities are not "faithful"; that is, values are observed being different from the preexisting values given by the Bohm formula. In quantum cosmology, however, this "escape" is not possible: if everything has zero velocity, according to the Bohm expression, then there simply is no possibility that anything can change with time. There can, indeed, be no "measurements" even.

We can see the relevance of all this if we think about the Bohm description of reflection by a potential barrier. This should be analogous to cosmological models in which an initially expanding universe recollapses after the radius has "bounced" off some potential barrier. We therefore consider a particle, moving in one dimension, under the influence of a square barrier. This is described by the Schrödinger equation:

$$\left[-\frac{\hbar^2}{2m}\frac{\partial^2}{\partial x^2} + V(x)\right]\psi = E\psi, \tag{11}$$

where the barrier potential is given by

$$V(x) = 0, \qquad x < 0,$$

$$V(x) = V_1 > E, \qquad x > 0. \tag{12}$$

The standard method of solution, given in every quantum theory course, leads to the wave function,

$$\Psi = e^{ikx} + Re^{-ikx}, \tag{13}$$

in the region $x < 0$, where $k^2 = 2mE/\hbar^2$. The reflection coefficient, R, is given by

$$R = -\frac{(K + ik)}{(K - ik)}, \tag{14}$$

where $K^2 = [2m(V_1 - E)]/\hbar^2$, so of course $|R|^2 = 1$, corresponding to the fact that all particles are reflected. The "story" that accompanies this in typical quantum theory courses is that particles move from the left and hit the barrier, where, after penetrating for a short distance, they are reflected. This, however, is not what actually happens in the Bohm model, for which the usual formula gives

$$\dot{x} = \frac{(1 - |R|^2)k}{|e^{ikx} + Re^{-ikx}|^2 m} \tag{15}$$

$$= 0. \tag{16}$$

In other words, the particles do not move. There is thus a big difference between the usual story and the actual behavior in the Bohm model.[21] This difference does not, of course, contradict the well-known agreement between the Bohm model and the predictions of orthodox quantum theory. This agreement refers to statistical predictions for position observations and, because in the above situation we have a constant wave function, it is quite in order for the particles to be stationary.

Note that we can recover the usual story if we consider a *time-dependent* wave function with an incident wave packet. This, of course, corresponds more closely to the actual situation in a laboratory experiment, but is not an option that is available for the cosmological case, at least not if we wish to maintain the Wheeler-deWitt equation.

Of perhaps more relevance is the fact that it is possible to obtain nonzero velocities from the Bohm expression even with a stationary wave function. To see this, we write the solution, in $x < 0$, as

$$\psi = Ae^{ikx} + Be^{-ikx}; \tag{17}$$

then, the velocity will be given by

$$\dot{x} = \frac{(|A|^2 - |B|^2)k}{(|A|^2 + |B|^2 + 2\mathscr{R}A^*Be^{-2ikx})m}. \tag{18}$$

The condition that $|A|^2 = |B|^2$ arises from the physical requirement that only the decreasing exponential exists in the "forbidden" region, $x > 0$. If we ignore this requirement, then the Bohm model has nonzero velocities. Of course, such a solution does not show any reflection by the barrier; the particles just move into (or out of) the infinite sink in the region $x > 0$. A model somewhat similar to this is suggested in the work of Kiefer,[22] where it is claimed that "decoherence" may be responsible for effectively removing one of the exponential terms in the allowed region. It is not clear what effect this has on the wave function in the forbidden region, but any solution where this wave function increases exponentially is surely not relevant to physics. Nevertheless, this opens up the possibility that coupling to other systems will in fact allow the static wave function to have nonzero, and even reversing, velocities.

To see how these might occur, we here take a very simple "minisuperspace" model, which has at least some of the desired properties. We write the metric in the form,

$$ds^2 = dt^2 - e^{2\alpha} d^3\Omega, \tag{19}$$

where the last term is the interval on a spatial hypersurface of constant curvature, k. In fact, we take $k = 1$, corresponding to a closed universe. We also introduce a constant scalar field, ϕ. For the "Wheeler-deWitt" equation, we take

$$\left[\frac{\partial^2}{\partial\alpha^2} - \frac{\partial^2}{\partial\phi^2} + A(\alpha)\phi^2 - B(\alpha)\right]\Psi = 0, \tag{20}$$

with

$$\begin{aligned} A(\alpha) &= A_0, \qquad \alpha < 0, \\ A(\alpha) &= A_1, \qquad \alpha > 0, \end{aligned} \tag{21}$$

and

$$\begin{aligned} B(\alpha) &= B_0, \qquad \alpha < 0, \\ B(\alpha) &= B_1, \qquad \alpha > 0. \end{aligned} \tag{22}$$

In the region $\alpha < 0$, we write the solution of equation 20 as

$$\Psi = \sum_n (a_n e^{ik\alpha} + b_n e^{-ik\alpha})\chi_n^{(0)}(\phi), \tag{23}$$

where the $\chi_n^{(0)}$ terms are harmonic oscillator bound states,

$$\chi_n^{(0)}(\phi) = (\pi^{1/2} 2^n n!)^{-1/2} H_n(z_0) e^{-(z_0^2/2)}, \tag{24}$$

with

$$z_0 = A_0^{1/4}\phi, \tag{25}$$

and where

$$k_n^2 = (2n + 1)A_0^{1/2} - B_0. \tag{26}$$

The k_n terms are all real, corresponding to a classically allowed region, provided

$$B_0 < A_0^{1/2}. \tag{27}$$

In restricting the solution to this form, in particular by using equation 24, we have imposed the physical requirement that the wave function should be square-integrable in ϕ. We cannot impose a similar requirement on the α-dependence, but we shall require at least that it should not *increase* exponentially. Thus, in the region $\alpha > 0$, we write

$$\Psi = \sum_n c_n e^{-\kappa_n \alpha} \chi_n^{(1)}(\phi), \tag{28}$$

where the $\chi_n^{(1)}$ terms are defined by a similar equation to equation 24, but with z_0 replaced by z_1, defined by

$$z_1 = A_1^{1/4}\phi. \tag{29}$$

Similarly, the κ_n terms are defined by

$$\kappa_n^2 = B_1 - (2n + 1)A_1^{1/2}. \tag{30}$$

In order that the region $\alpha > 0$ should be classically forbidden, we require the κ_n to be real, which means that B_1 must be many times larger than $A_1^{1/2}$. For sufficiently large n, however, it is inevitable that the right-hand side of equation 30 will change sign, so there will be some "leakage".

To be specific, we choose

$$a_n = \delta_{n0} \tag{31}$$

as the boundary condition, which corresponds to associating the expansion term (with positive $\dot{\alpha}$) with the ground state of the ϕ field. Other modes become excited at the $\alpha = 0$ boundary. To prevent any leakage, we truncate the expansions at the first nontrivial term, that is, that with $n = 2$. The problem is then completely solved by imposing continuity of the wave function and its first derivative at the boundary.

For details, we refer to Hind and Squires.[23] Here, if we ignore the ϕ field, it is not possible for the expansion to reverse unless the wave function is double-valued in α, which in these models it is not. This, of course, follows directly from the Bohm formula where the velocity ($\dot{\alpha}$) is a function only of the value of α (for a wave function that does not have any explicit dependence on time). In the presence of the ϕ field, such a reversal is possible because the velocity can now depend also on the value of ϕ. Indeed, we have a prediction: when the universe recollapses, it does so with *different* values for the fields (only one in our case). In other words, there is not an effective reversal of the direction of time, in the sense of everything returning exactly to its initial state. This would appear to be contrary to the claim of Kiefer and Zeh,[24] who of course do not explicitly use the Bohm model in their discussion.

In actual fact, the velocity given in the above model tends to oscillate (the details depend on the starting points; see reference 23). This occurs because of interference effects between the various terms in the expansion of equation 23. If we were to include more complex states coupling to the α variable, then it seems likely that such oscillation would be suppressed, essentially because there would be less likelihood of the many hidden variables having values that allow both the two exponentials in equation 23 to contribute. This would be analogous to the phenomenon of decoherence,[22] except that here it is not orthogonality of the states that is significant, but rather their limited spatial overlap.

Clearly, much more work is required before reliable conclusions can be drawn from discussions of this nature. In our model, the expansion is associated with low excitation of the ϕ field, whereas in fact it is more likely that it is the de-excitation that causes the bounce (see, for example, equation 30). Also, more complex situations than that involving only a single constant field need to be analyzed. Nevertheless, the results show that there is a real role for the Bohm model in quantum cosmology. In this connection, earlier work by Vink[25] and by Valentini[26] should be mentioned.

REFERENCES

1. BOHM, D. 1985. Phys. Rev. **85:** 166 & 180.
2. HOLLAND, P. R. 1993. The Quantum Theory of Motion. Cambridge University Press. London/New York.
3. BOHM, D. & B. J. HILEY. 1993. The Undivided Universe. Routledge. London.
4. GHIRARDI, G. C., A. RIMINI & T. WEBER. 1986. Phys. Rev. **34D:** 470.
5. PEARLE, P. 1982. Found. Phys. **12:** 249.
6. GHIRARDI, G. C., P. PEARLE & A. RIMINI. 1990. Phys. Rev. **42A:** 78.
7. ALBERT, D. & B. LOEWER. 1988. Synthese **77:** 195.
8. LOCKWOOD, M. 1989. Mind, Brain, and the Quantum. Blackwell. Oxford.
9. STAPP, H. P. 1993. Mind, Matter, and Quantum Mechanics. Springer-Verlag. Berlin/New York.
10. SQUIRES, E. J. 1993. Synthese **97:** 109.
11. HARDY, L. & E. J. SQUIRES. 1992. Phys. Lett. **A168:** 169.
12. HARDY, L. 1992. Phys. Rev. Lett. **68:** 2981.
13. SQUIRES, E. J. 1993. Phys. Lett. **A178:** 22.
14. LAM, M. M. & C. DEWDNEY. 1990. Phys. Lett. **A150:** 127.
15. SQUIRES, E. J. & S. MACKMAN. 1994. Phys. Lett. **A185:** 1.
16. SQUIRES, E. J. 1991. Presentation at the 1991 Cesena Conference; 1994. Found. Phys. Lett. **7:** 353.
17. SQUIRES, E. J. 1994. Phys. Lett. **A190:** 349.
18. BELL, J. S. 1980. Int. J. Quantum Chem. **14:** 155.
19. ISHAM, C. J. 1992. Canonical quantum gravity and the problem of time. Imperial College preprint no. TP/91-92/25.
20. SQUIRES, E. J. 1992. Found. Phys. Lett. **5:** 71.
21. LAM, M. M. 1993. Ph.D. thesis. Portsmouth.
22. KIEFER, C. 1993. Phys. Rev. **47D:** 5414; Freiburg preprint no. THEP-93/27; 1994. The semiclassical approach to quantum gravity. *In* Canonical Gravity—From Classical to Quantum. J. Ehlers & H. Friedrich, Eds. Springer-Verlag. Berlin/New York.
23. HIND, D. & E. J. SQUIRES. 1994. Durham preprint.
24. KIEFER, C. & H. D. ZEH. 1993. Arrow of time in a recollapsing universe. Freiburg preprint no. THEP-93/30 and Heidelberg preprint no. HD-TVP-94-2. To be published in Phys. Lett. A.
25. VINK, J. C. 1992. Nucl. Phys. **B369:** 707.
26. VALENTINI, A. 1991. Phys. Lett. **A156:** 5.

The Yılmaz Theory of Gravity and Its Compatibility with Quantum Theory

CARROLL O. ALLEY

Department of Physics
University of Maryland at College Park
College Park, Maryland 20742

DEDICATION

This paper is dedicated to John Archibald Wheeler, the honoree of this conference, whose search for the understanding of both gravitational and quantum physics has inspired the author for over 45 years, as a listener and learner in his courses and as a seeker of advice and counsel on physics research. Some items in the latter category are the following: The Lunar Laser Ranging (LLR) experiment[1] originated in the research group of Robert Henry Dicke at Princeton for the primary purpose of testing the Brans-Dicke theory[2] of gravity. Wheeler strongly encouraged this investigation. (The result did not confirm the failure of the principle of equivalence for massive bodies predicted by the Brans-Dicke theory.) In my research group at Maryland, we had to learn how to detect and time the arrival of single light quanta in the LLR experiment because of the very weak signals. This technical knowledge provided the basis for the first explicit realization of the Delayed Random Choice gedankenexperiment so strongly emphasized by Wheeler.[3] This experiment led to the first use of parametric downconversion to perform the Einstein-Podolsky-Rosen experiment[4] (Ph.D. thesis of Yan Hua Shih), which has led to so many applications being reported at this conference. Our atomic clock relativity experiments with aircraft[5] were explicitly and effectively endorsed by Wheeler.[6]

In regard to the subject of this paper, Wheeler has urged that the investigations of the differences between the gravity theories of Einstein and of Yılmaz be conducted by calculations, citing the words of Leibniz,[7] that in matters of scientific dispute, "Let us calculate"—"Calculemus".

INTRODUCTION

This report summarizes the physical and mathematical structure of the Yılmaz theory of classical gravity to prepare the reader for the later report of Yılmaz showing how it can be quantized. In these proceedings, Yılmaz has included such summaries in his own paper (see reference 8), which the exposition here may serve to illuminate. We will state and comment on some of the most important physical and conceptual differences between the Einstein and the Yılmaz theories. The most striking concrete results of the "Calculemus" approach, showing explicitly the failure of general relativity to predict the gravitational attraction in a two-slab Cavendish experiment, will be described in some detail. The treatment is similar to that in the written version of my lecture at the Olympia Conference on the Frontiers of Fundamental Physics.[9a]

The compatibility of the new theory of gravity with quantum field theory results from its possession of a field Lagrangian and the resulting canonical field stress-energy tensor of the two-index relativistic generalization of the Newton-Poisson gravitational potential field. This allows the construction of Feynman propagators as in the usual quantum field theories, but would still lead to unrenormalizability because of the dimensional coupling constant. The remarkable approach of Yılmaz, described in print for the first time in these proceedings, gives a physical reinterpretation in terms of local dynamical causality to the loop integrals, avoiding infinite quantities altogether. This eliminates the need for renormalization of any quantum field theory and yields for the first time a viable quantum field theory of gravity.

PHYSICAL AND CONCEPTUAL DIFFERENCES BETWEEN THE THEORIES OF EINSTEIN AND OF YILMAZ

The description of gravitation by curved space-time is a grand concept due to Albert Einstein. The form of his field equations for the determination of the metric coefficients for a given distribution of matter and field stress-energy is an assumption. The source term (right-hand side) of these equations for the Einstein-Hilbert curvature tensor is taken to be the stress-energy tensor τ_μ^ν of all matter and fields, except that of the gravitational field itself:

$$\tfrac{1}{2}G_\mu^\nu = \tau_\mu^\nu. \tag{1}$$

The theory of Hüseyin Yılmaz explicitly includes as an additional source term the stress-energy tensor t_μ^ν of a gauge field, which is a relativistic generalization of the Newton-Poisson potential field in a conservative system:

$$\tfrac{1}{2}G_\mu^\nu = \tau_\mu^\nu + t_\mu^\nu, \tag{2}$$

where G_μ^ν and t_μ^ν have the usual meanings. The expression for t_μ^ν in the low-velocity limit is given in equation 11 below and, more generally, in reference 8.

It is my opinion that the explicit inclusion of the gravitational field stress-energy tensor in equation 2 is as important for our understanding of physics as Maxwell's addition of the displacement current in his equations for the electromagnetic field.

The new theory emphasizes potentials in a successful relativistic generalization of the Newton-Poisson field theory. The metric coefficients $g_{\mu\nu}$ thus become functionals of this gravitational field φ_μ^ν and are relieved of the double burden of serving both as potentials and as space-time metric. The relation is formally exponential:

$$g_{\mu\nu} = (\eta e^{2(\varphi\hat{1} - 2\hat{\varphi})})_{\mu\nu}, \tag{3}$$

where $\eta = \mathrm{diag}(1, -1, -1, -1)$, $\varphi = \mathrm{tr}(\varphi_\mu^\nu)$, $\hat{1}$ is the unit matrix, and $\hat{\varphi}$ is the matrix array φ_μ^ν.

The quantity φ_μ^ν is analogous to the four-potential A^ν in electrodynamics in that the $g_{\mu\nu}$ are determined from it, as the electromagnetic field $F^{\mu\nu}$ is determined from A^ν. With suitable gauge and/or coordinate conditions, φ_μ^ν satisfies the covariant d'Alambert equation with the matter tensor as source:

$$\Box^2 \varphi^{\nu}_{\mu} = \tau^{\nu}_{\mu}. \tag{4}$$

Exact solutions are known in the low-velocity limit where $\varphi^{\nu}_{\mu} \rightarrow \varphi^{0}_{0} \rightarrow \varphi$ and equation 4 becomes the Poisson equation, and for traceless transverse gravity waves.

Many strong physical results are obtained by the new theory, some of which are briefly stated and discussed below.

Newtonian Correspondence in Second Order

This allows N-body interactive solutions including concentrated "point particles". Our calculations show that in general relativity there are no interactive N-body solutions at all.

Local Correspondence to Special Relativity

The metric becomes locally Minkowskian for any observer, even when in accelerated motion. The exponential metric allows the subtraction of a constant from the potential (including the kinematic potential). There is an unambiguous prediction for the local isotropy of the speed of light even for accelerated observers (e.g., an observer on the rotating Earth).

Principle of Equivalence by Local Kinematic Potential Compensation

The transition to a freely falling system is achieved by the addition of a kinematic potential to the gravitational potential. This is the appropriate way of describing the equivalence of gravitational and kinematical accelerations. (Explicit examples are given in reference 9a.)

Localized Stress-Energy of the Gravitational Field

This exists in the new theory as the tensor t^{ν}_{μ} and plays a central role. In general relativity, it is argued[10] that there can be no localized stress-energy tensor since, in that theory, local free-fall is described by a coordinate transformation to a local Riemannian normal coordinate system, where the first derivatives of the metric coefficients vanish. The stress-energy tensor is a quadratic expression in the first derivatives and vanishes in that system, hence vanishes in all coordinate systems. Because the new theory describes the vanishing of t^{ν}_{μ} at a point in a freely falling frame by a compensation, or balancing, of the gravitational potential with a kinematic potential, not by a coordinate transformation, the argument does not apply. Local energy-momentum transfer between matter and gravitational field is consistently treated in the new theory. The serious problems in general relativity with local energy-momentum conservation are well known. (A recent review of the difficulties is given by Carmeli *et al.*[11] See also an early critical comment by Weyl.[12])

Exact Gravity Wave Solutions of Arbitrary Strength

These waves actually carry localized energy-momentum, with an analogous Poynting vector, between quadrupole sources and sinks. In general relativity, gravity wave treatments require the limitation to first-order expansions of the metric tensor. A localized stress-energy tensor for the gravity wave is assumed even though this is inconsistent with the argument against it given in reference 10.

The Strong Principle of Equivalence

This is the name often given to the equivalence of inertial mass, active gravitational mass, and passive gravitational mass. It is readily deduced in the new theory from its second-order Newtonian correspondence. It is not deducible in Einstein's theory.[13]

SOME MATHEMATICAL RESULTS

The above-stated physical differences between the two theories can, of course, only be fully understood and comprehended in mathematical form. A brief exposition of some important results will be given in this section.

Exponential Metric

This is most readily discussed for the low-velocity limit where $\varphi^{\nu}_{\mu} \rightarrow \varphi^{0}_{0} \rightarrow \varphi$. The solution of equation 4 for $t^{\nu}_{\mu} = \sigma u_{\mu} u^{\nu}$, $\sqrt{-g}\sigma = \Sigma\, m_A \delta(\boldsymbol{x} - \boldsymbol{x}_A)$, becomes

$$\varphi(\boldsymbol{x}) = \sum_A \frac{m_A}{|\boldsymbol{x} - \boldsymbol{x}_A|} + \text{constant}; \tag{5}$$

the exponent in equation 3 evaluates as

$$2\begin{pmatrix} \varphi & & & \\ & \varphi & & \\ & & \varphi & \\ & & & \varphi \end{pmatrix} - 4\begin{pmatrix} \varphi & & & \\ & 0 & & \\ & & 0 & \\ & & & 0 \end{pmatrix} \Rightarrow g_{00} = e^{-2\varphi},\ g_{ii} = -e^{2\varphi};$$

and the metric becomes

$$ds^2 = e^{-2\varphi}dt^2 - e^{2\varphi}(dx^2 + dy^2 + dz^2). \tag{6}$$

The very interesting history of the exponential metric is discussed in reference 9a.

Energy-Momentum Conservation

Perhaps the major result of the theory of special relativity is the possibility of transforming rest mass into other forms of energy. The general conservation law that encompasses this possibility is the vanishing divergence of the matter stress-energy tensor $\tau^{\nu}_{\mu} = \sigma u_{\mu} u^{\nu}$:

$$\partial_{\nu}(\sqrt{-g}\sigma u_{\mu} u^{\nu}) = 0. \tag{7}$$

However, in general relativity, this conservation law is replaced by the conservation of rest mass[14] in order to get the equations of motion from the field equations. The argument runs as follows: Take the covariant divergence of the field equations, which is zero by the Bianchi identity,

$$\tfrac{1}{2} D_{\nu} G^{\nu}_{\mu} = D_{\nu}(\sigma u_{\mu} u^{\nu}) = \frac{1}{\sqrt{-g}} \partial_{\nu}(\sqrt{-g}\sigma u_{\mu} u^{\nu}) - \tfrac{1}{2}\partial_{\mu} g_{\alpha\beta} \sigma u^{\alpha} u^{\beta} \equiv 0. \tag{8}$$

Then, differentiate the first term on the right as a product to yield

$$\frac{1}{\sqrt{-g}} \partial_{\nu}(\sqrt{-g}\sigma u^{\nu}) u_{\mu} + \sigma u^{\nu} \partial_{\nu} u_{\mu} - \tfrac{1}{2}\partial_{\mu} g_{\alpha\beta} \sigma u^{\alpha} u^{\beta} \equiv 0. \tag{9}$$

The second term is $\sigma(\partial x^{\nu}/\partial \tau)\partial_{\nu} u_{\mu} = \sigma du_{\mu}/d\tau$, which would produce the geodesic equations of motion if the first term were zero. It is argued that one should require $\partial_{\nu}(\sqrt{-g}\sigma u^{\nu}) = 0$ as a conservation law. However, this is the conservation of rest mass, not the conservation of energy-momentum. This whole procedure is illegitimate in view of the Freud identity, to be discussed below, which requires the identical vanishing of expression 7. If the first term in equation 9 is set equal to zero, the second must also be zero, and one does not get the geodesic equation of motion.

In the new theory, one has

$$\tfrac{1}{2} D_{\nu} G^{\nu}_{\mu} = \frac{1}{\sqrt{-g}} \partial_{\nu}(\sqrt{-g}\sigma u_{\mu} u^{\nu}) - \tfrac{1}{2}\partial_{\mu} g_{\alpha\beta} \sigma u^{\alpha} u^{\beta} + D_{\nu} t^{\nu}_{\mu} \equiv 0. \tag{10}$$

Now, one can require the first term in equation 10 to vanish, expressing the desired conservation law of energy-momentum, and one gets the equation of motion because the divergence of the gravitational field stress-energy tensor t^{ν}_{μ} is the force $\sigma du_{\mu}/d\tau$. In the low-velocity limit, t^{ν}_{μ} is given by

$$t^{\nu}_{\mu} = -\partial_{\mu}\varphi \partial^{\nu}\varphi + \tfrac{1}{2}\delta^{\nu}_{\mu} \partial_{\lambda}\varphi \partial^{\lambda}\varphi. \tag{11}$$

Taking the divergence, one obtains $\nabla^2 \varphi \partial_{\mu}\varphi$, which is just $\sigma_a \partial_{\mu}\varphi$ by the Poisson equation, where σ_a is the active gravitational mass. This is instrumental in establishing the strong principle of equivalence in the new theory.[13]

The Freud Identity

It has become clear recently that there is a mathematical requirement that forces the first term of equation 10 to vanish identically in the most general form with $\sigma u_{\mu} u^{\nu} \Rightarrow \tau^{\nu}_{\mu} = \tfrac{1}{2} G^{\nu}_{\mu}$ in Einstein's theory. In 1939, P. Freud published[15] a decomposition of the Einstein-Hilbert tensor $\tfrac{1}{2} G^{\nu}_{\mu} = U^{\nu}_{\mu} - u^{\nu}_{\mu}$, where u^{ν}_{μ} is the quantity introduced by Einstein to describe the stress-energy of the gravitational field. He transformed the "obstreperous term" in equation 8 (phrase used by Schrödinger[16]) as follows:

$$-\tfrac{1}{2}\partial_\mu g_{\alpha\beta}\sigma u^\alpha u^\beta = \frac{1}{\sqrt{-g}}\,\partial_\nu(\sqrt{-g}u_\mu^\nu). \tag{12}$$

U_μ^ν is an expression satisfying the identity $\partial_\nu(\sqrt{-g}U_\mu^\nu) \equiv 0$. This is an indicial identity resulting from the antisymmetry of the superpotential $H_\mu^{\nu\alpha}$ given by the determinant

$$H_\mu^{\nu\alpha} = \begin{vmatrix} \delta_\mu^\nu & \delta_\mu^\alpha & \delta_\mu^\sigma \\ \sqrt{-g}g^{\nu\rho} & \sqrt{-g}g^{\alpha\rho} & \sqrt{-g}g^{\sigma\rho} \\ \Gamma_{\rho\sigma}^\nu & \Gamma_{\rho\sigma}^\alpha & \Gamma_{\rho\sigma}^\sigma \end{vmatrix} \tag{13}$$

and the relation $\sqrt{-g}U_\mu^\nu = \partial_\alpha H_\mu^{\nu\alpha}$. It has been shown more recently that such an identity is true for a Riemannian geometry of any dimension and arbitrary signature for all symmetric nonsingular metrics.[17] The similarity of the Freud identity as discussed by Pauli[18] to a simpler expression used by Yılmaz to formulate the conservation laws was noted by the author. Their equivalence was established by Yılmaz, who has shown that the Freud identity is nested so that the anharmonic part (the last two terms of U_μ^ν as given in reference 13, appendix B, p. 959; in harmonic coordinates, they themselves vanish) and the coordinate-dependent part separately satisfy the same identity. Freud had actually written $U_\mu^\nu = \tfrac{1}{2}G_\mu^\nu - u_\mu^\nu$ and used the identity to produce the energy-momentum conservation law of general relativity, $\partial_\nu(\sqrt{-g}\tau_\mu^\nu + \sqrt{-g}u_\mu^\nu) \equiv 0$. However, writing the decomposition as $\tfrac{1}{2}G_\mu^\nu = U_\mu^\nu - u_\mu^\nu$ shows that the difference of the two "pseudotensors" must be the true tensor $\tfrac{1}{2}G_\mu^\nu$. There must be a common nontensor part z_μ^ν for each of U_μ^ν and u_μ^ν in addition to their tensor parts, and there must exist coordinate systems in which

$$U_\mu^\nu = \tau_\mu^\nu + z_\mu^\nu \qquad \text{and} \qquad u_\mu^\nu = -t_\mu^\nu + z_\mu^\nu. \tag{14}$$

In Einstein's theory, the field stress-energy tensor t_μ^ν is required to be zero in equation 1 and therefore also in equation 14, forcing the identification of u_μ^ν with the nontensor or coordinate artifact z_μ^ν. This has led to all of the difficulties with the energy-momentum concept in general relativity discussed in references 11 and 12. It was already noted in 1918 by Schrödinger[19] that u_μ^ν evaluated to zero for the Schwarzschild solution when expressed in Cartesian coordinates and by Bauer[20] that u_μ^ν evaluated to nonzero expressions for the flat space Minkowski metric expressed in polar coordinates. Schrödinger's criticism was replied to by Einstein[21] with the additional remark that $u_{\mu\nu} \neq u_{\nu\mu}$, leading to problems with angular momentum conservation. However, he stated that the strange properties of u_μ^ν in the Schwarzschild solution were due to its one-body nature and that u_μ^ν would have appropriate physical properties as soon as the two-body solution to his equations was found. There is no two-body (or N-body) solution for Einstein's equation with interaction between the bodies (see below and references 9b and 13).

In a conservative system, the nontensor z_μ^ν can be identified by starting with a coordinate system in which it is zero and transforming to another coordinate system where it is not. From equation 14, it can be seen that z_μ^ν arises from the nontensor part of the transformation of Christoffel symbols. From the indicial nature of the Freud identity, both τ^{ν}_{μ} and z^{ν}_{μ} have separately vanishing density divergences.[9b,13] Therefore, in Einstein's theory, where $t^{\nu}_{\mu} = 0$ and thus $u^{\nu}_{\mu} = z^{\nu}_{\mu}$, $\partial_{\nu}(\sqrt{-g}u^{\nu}_{\mu}) \equiv 0$.

COMPUTER CALCULATIONS

It is important to provide concrete examples of the foregoing assertions. Using Mathematica (from Wolfram Research) and MathTensor (from Math Solutions) running on Digital Equipment Corporation 5000/240 workstations operating under Ultrix and on a DEC 3000/400AXP workstation under OSF-1, many different metrics for solutions of the new and old theories have been used to evaluate the curvature tensors, Christoffel symbols, etc., emphasizing the study of U^{ν}_{μ}, u^{ν}_{μ}, and their density divergences. There is space for only a very limited presentation of some significant results.

Parameterized Schwarzschild Solution

The following parameterized metric satisfies equation 1 for the values of $\epsilon = \pm 1$ and equation 2 for $\epsilon = 0$. The equations of general relativity give two distinct solutions:

$$ds^2 = \left(1 + \frac{2\epsilon M}{r}\right)^{-1/\epsilon} dt^2 - \left(1 + \frac{2\epsilon M}{r}\right)^{1+(1/\epsilon)} (r^2 d\Theta^2 + r^2 \sin^2\theta d\varphi^2) - \left(1 + \frac{2\epsilon M}{r}\right)^{1/\epsilon} dr^2. \tag{15}$$

If one sets $M/r = \varphi$ in equation 21 below and evaluates the Einstein-Hilbert tensor, one finds

$$\frac{1}{2} G^{\nu}_{\mu} = \begin{matrix} t & \Theta & \varphi & r \end{matrix} \\ \begin{pmatrix} 1+\epsilon & & & \\ & \frac{\epsilon}{2} & & \\ & & \frac{\epsilon}{2} & \\ & & & 0 \end{pmatrix} \nabla^2\varphi + (1 - \epsilon^2) t^{\nu}_{\mu}, \tag{16}$$

where $\nabla^2\varphi$ is the covariant Laplacian and t^{ν}_{μ} is given by equation 11. By correspondence with Newtonian theory, $\nabla^2\varphi$ should be the mass density σ in the matter tensor, $\tau^{\nu}_{\mu} = \sigma u_{\mu} u^{\nu}$, which in the low-velocity limit should be $\tau^0_0 = \sigma$. However, for $\epsilon = -1$, $G^0_0 = 0$ and, for $\epsilon = +1$, $\frac{1}{2}G^0_0 = 2\nabla^2\varphi$. Also, $\nabla^2\varphi$ appears in the G^{Θ}_{Θ} and G^{φ}_{φ} positions. These properties seem physically wrong. In the Yılmaz theory, $\epsilon = 0$ leads to the exponential metric (equation 6) expressed in spherical polar coordinates and the Laplacian occurs as expected only in the G^0_0 position.

The Schwarzschild solution in general relativity clearly seems not to describe a physical mass concentrated at the origin. Does it properly describe any real physical situation in nature?

The metric (equation 15) for $\epsilon = -1$ ($G_0^0 = 0$) is the "standard" form,

$$ds^2 = \left(1 - \frac{2M}{r}\right)dt^2 - \frac{1}{1 - \left(\frac{2M}{r}\right)} dr^2 - r^2 d\Theta^2 - r^2 \sin^2\theta d\varphi^2, \tag{17}$$

which is said to have an "event horizon" at $r = 2M$.

For $\epsilon = +1$ ($\tfrac{1}{2}G_0^0 = 2\nabla^2\varphi$), the metric is

$$ds^2 = \frac{1}{1 + \left(\frac{2M}{r}\right)} dt^2 - \left(1 + \frac{2M}{r}\right)dr^2 - \left(1 + \frac{2M}{r}\right)^2 (r^2 d\Theta^2 + r^2 \sin^2\theta d\varphi^2), \tag{18}$$

giving no "event horizon".

For $\epsilon = 0$, the new theory, equation 15 becomes

$$ds^2 = e^{-2\varphi} dt^2 - e^{2\varphi}(dr^2 + r^2 d\Theta^2 + r^2 \sin^2\theta d\varphi^2) \tag{19}$$

with $\varphi = M/r$.

It is interesting to calculate the square of the Riemann tensor $R^{\alpha\beta\gamma\delta}R_{\alpha\beta\gamma\delta} = K$ (referred to as the Kretschmann invariant and used as an indicator of singular behavior in the current literature). For general relativity[22]

$$K = \frac{48G^2M^2}{r^6} \underset{r\to 0}{\longrightarrow} \infty, \tag{20}$$

whereas for the Yılmaz theory[23]

$$K = \frac{4M^2(7M^2 - 16Mr + 12r^2)}{e^{4M/r}r^8} \underset{r\to 0}{\longrightarrow} 0. \tag{21}$$

The new theory has neither singularities nor "event horizons".

Parameterized Two-Slab Metric

Consider the configuration of two plane-parallel slabs whose separation d in the z direction between their central planes is small compared to their finite transverse x, y dimensions so that one can ignore edge effects—the situation often used in capacitor problems in electrostatics. A realizable experimental arrangement could be a circular disk surrounded by a large annulus of the same thickness, a configuration also often used in electrostatics to achieve a uniform field. Denote the uniform mass densities of the slabs by σ_1 and σ_2 and their thicknesses by w_1 and w_2, as shown in FIGURE 1.

The Newtonian potentials in regions I through V are sketched and their expressions are written beside the drawing. The potentials and their slopes are continuous at the boundaries. The quantities γ_1 and γ_2 are the magnitudes of the acceleration of gravity in the external regions produced by the respective slabs. They are related to the density and thickness by $\gamma = \sigma w/2$.

The metric[24]

$$ds^2 = e^{-2\varphi}dt^2 - e^{2\varphi(1+\epsilon)}(dx^2 + dy^2) - e^{2\varphi(1+\epsilon)}dz^2 \tag{22}$$

has been used to evaluate all of the relevant quantities for each of the five regions.[25] It provides an exact solution of the new theory for $\epsilon = 0$ in each region. However, both $\epsilon = +1$ and $\epsilon = -1$ provide exact solutions for Einstein's theory. Thus, there is no unique solution even for this simple case. The two solutions cannot be related by a coordinate transformation since they have different curvatures. The evaluation of G^{ν}_{μ} yields a general expression in each region that has exactly the same structure as for the parameterized Schwarzschild solution: equation 16 with $(t, \Theta, \varphi, r) \Rightarrow (t, x, y, z)$.

In this two-body case, one can evaluate the right-hand side of the geodesic equation, which by Einstein's introduction of u^{ν}_{μ} is given by equation 12. In reference

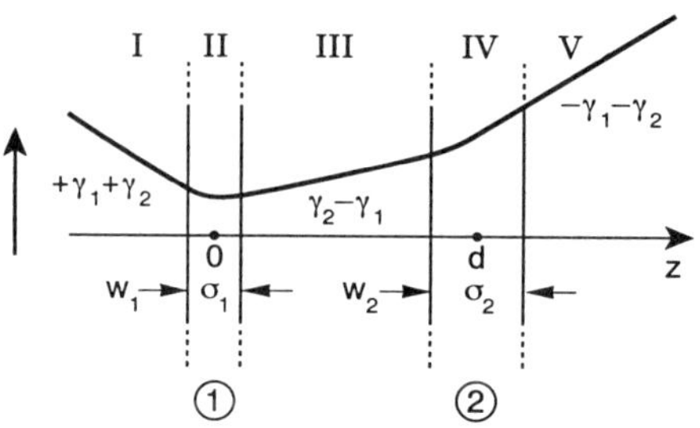

$$\varphi_{I} = -\frac{\sigma_1 w_1}{2} z - \frac{\sigma_2 w_2}{2}(z-d)$$

$$\varphi_{II} = \frac{\sigma_1}{2} z^2 - \frac{\sigma_2 w_2}{2}(z-d) + \sigma_1 w_1^2/8$$

$$\varphi_{III} = \frac{\sigma_1 w_1}{2} z - \frac{\sigma_2 w_2}{2}(z-d)$$

$$\varphi_{IV} = \frac{\sigma_2}{2}(z-d)^2 + \frac{\sigma_1 w_1}{2} z + \sigma_2 w_2^2/8$$

$$\varphi_{V} = \frac{\sigma_1 w_1}{2} z + \frac{\sigma_2 w_2}{2}(z-d)$$

FIGURE 1. Two slabs, neglecting edge effects.

9b, the left-hand side of equation 12 is evaluated for the slabs. Here, we give the results of the computer calculation in region II of $\partial_{\nu}(\sqrt{-g}u^{\nu}_{\mu})$. The only nonzero component is

$$\partial_{\nu}(\sqrt{-g}u^{\nu}_{z}) = (1 - \epsilon^2)[\tfrac{1}{2}\sigma_1\sigma_2 w_2 - \sigma_1^2 z]. \tag{23}$$

The presence of $(1 - \epsilon^2)$ in equation 23 makes it zero for the Einstein case. For the new theory, it has the expected Newtonian value. Integrating over the thickness of the slab (the Jacobian $\sqrt{-g}$ in the integration cancels the $\sqrt{-g}$ in the denominator of the right-hand side of equation 12), the self-force $\sigma_1^2 z$ vanishes and the force per unit area, $\tfrac{1}{2}(\sigma_1 w_1)(\sigma_2 w_2)$, to the right remains. For slab 2, one finds the same expression, but with opposite sign, giving a force to the left.

In this simplest of all gravitational problems, the Einstein theory predicts no interaction: the slabs would remain fixed, contradicting the results of Cavendish-type experiments.[26] The Yılmaz theory, in strong contrast, describes the expected Newtonian interaction and predicts that the slabs will accelerate toward each other.

Einstein often wrote that "above all else" his theory should have the appropriate Newtonian correspondence. In this concrete example, it is shown not to be the case.

The lack of Newtonian correspondence in general relativity holds for any shapes of finite mass (see reference 8) and therefore applies also to binary pulsars. In this respect, a recent remark by J. H. Taylor is particularly significant.[27]

CONCLUDING REMARKS

In the implementation of Einstein's grand design of gravitation as curved space-time, serious problems have appeared. The theory does not possess N-body interactive solutions, nor does it generalize correctly the energy-momentum conservation law of special relativity and relativistic field theories, to mention only two major problems. The computer calculations of simple examples supporting these conclusions seem incontestable. A theory with such defects cannot serve as the foundation on which to build our understanding of the physical universe.

There is a consistent curved space-time theory of gravity that does not have these problems and that is compatible with quantum field theory.[8] I urge that the physics and astronomy communities shift the paradigm from general relativity and concentrate instead on the development and application of the theory of Yılmaz.

ACKNOWLEDGMENTS

I wish to express my gratitude to Hüseyin Yılmaz for the pleasure and privilege of many illuminating and clarifying discussions about physics and its foundations. Ching Yun Ren (now chairman of the Department of Physics, Soochow University, Taipei, Taiwan) developed the computer programs for these investigations and applied them with great knowledge of physics to many different metrics. These programs have been extended with great skill by Robert Kirk Burrows and used to study many additional metrics. Our computer systems have been managed by Steve Sabean.

The endorsement by Willis Lamb of the new theory of gravity and his encouragement of this research is greatly appreciated. Robert Henry Dicke's example of combining experimental and theoretical investigations of gravity has been a continuing inspiration. We thank George Keros for his active interest and encouragement during a critical period of the investigations. The invitation of Daniel Greenberger to present these results at this conference is greatly appreciated.

I have benefited from discussions with many colleagues at College Park, especially James Griffin, George Snow, William MacDonald, Ching Hung Woo, Joseph Weber, Dieter Brill, and Charles Misner.

Finally, I also appreciate the general support by members of my quantum electronics research group at the University of Maryland: P. K. Aschan, R. K. Burrows, T. E. Kiess, A. Krikoriantz, S. W. Sabean, A. V. Sergienko (who also has given very able help in the technical composition of this paper), Y. H. Shih, T. Van Flandern, B. C. Wang, and M. Z. Zhang.

REFERENCES AND NOTES

1. ALLEY, C. O., P. L. BENDER, R. H. DICKE, J. E. FALLER, P. A. FRANKEN, H. H. PLOTKIN & D. T. WILKINSON. 1965. Optical radar using corner reflectors on the moon. J. Geophys. Res. **70**; ALLEY, C. O. 1983. Laser ranging to retro-reflectors on the moon as a test of theories of gravity. *In* Quantum Optics, Experimental Gravitation, and Measurement Theory. P. Meystre & M. Scully, Eds.: 429–495. Plenum. New York; DICKEY, J. O., P. L. BENDER, J. E. FALLER *et al.* 1994. Lunar laser ranging: a continuing legacy of the Apollo program. Science **265**.
2. BRANS, C. & R. H. DICKE. 1961. Mach's principle and a relativistic theory of gravity. Phys. Rev. **124:** 925.
3. ALLEY, C. O., O. G. JAKUBOWICZ & W. C. WICKES. 1986. Results of the delayed random choice quantum mechanics experiment with light quanta. *In* Proceedings of the Second International Symposium on Foundations of Quantum Mechanics in the Light of New Technology. M. Namiki *et al.*, Eds.: 36–47. Phys. Soc. Japan. Kyoto.
4. ALLEY, C. O. & Y. H. SHIH. 1986. *In* Proceedings of the Second International Symposium on Foundations of Quantum Mechanics in the Light of New Technology. M. Namiki *et al.*, Eds.: 47. Phys. Soc. Japan. Kyoto.
5. ALLEY, C. O. 1986. Proper time experiments in gravitational fields with atomic clocks, aircraft, and laser light pulses. *In* Proceedings of the Second International Symposium on Foundations of Quantum Mechanics in the Light of New Technology. M. Namiki *et al.*, Eds.: 363–427. Phys. Soc. Japan. Kyoto; REISSE, R. A. 1976. The effect of gravitational potential on atomic clocks as observed with a laser pulse time transfer system. Ph.D. thesis. University of Maryland; WILLIAMS, R. E. 1976. A direct measurement of the relativistic effects of gravitational potential on the rates of atomic clocks flown in aircraft. Ph.D. thesis. University of Maryland.
6. CALDER, N. 1979. *In* Einstein's Universe, p. 38. Viking. New York. The experiment is also used as an example in: TAYLOR, E. F. & J. A. WHEELER. 1992. Space-Time Physics, p. 133. Freeman. San Francisco.
7. LEIBNIZ, G. W. 1961. *In* Philosophischen Schriften. Volume 7. C. J. Gerhardt, Ed.: 200. Olms Verlagsbuchhandlung. Hildesheim.
8. YILMAZ, H. 1995. This volume.
9. (a) ALLEY, C. O. 1994. Investigations with lasers, atomic clocks, and computer calculations of curved space-time and of the differences between the gravitation theories of Yılmaz and of Einstein. *In* Frontiers of Fundamental Physics. M. Barone & F. Selleri, Eds.: 125–137. Plenum. New York. (b) The above paper was a complement to: YILMAZ, H. 1994. Did the apple fall? *In* Frontiers of Fundamental Physics. M. Barone & F. Selleri, Eds.: 115–124. Plenum. New York.
10. MISNER, C. W., K. S. THORNE & J. A. WHEELER. 1973. Gravitation, p. 466–468. Freeman. San Francisco.
11. CARMELI, M., E. LIEBOWITZ & N. NISSANI. 1990. Gravitation: SL(2,C) Gauge Theory and Conservation Laws. Chapter 4. World Scientific. Singapore.
12. WEYL, H. 1950. Space, Time, Matter. Fourth edition, p. 270. Dover. New York.
13. YILMAZ, H. 1992. Towards a field theory of gravity. Nuovo Cimento **107B:** 941.
14. YILMAZ, H. 1973. Correspondence paradox in general relativity. Nuovo Cimento Lett. **7:** 337.
15. FREUD, P. 1939. On the expression of the total energy and the total momentum of a material system in the theory of general relativity. Ann. Math. **40:** 417.
16. SCHRÖDINGER, E. 1950. Space-Time Structure, p. 102. Cambridge University Press. London/New York.
17. RUND, H. 1991. The Freud identity in Riemannian geometry. Algebras, Groups, and Geometries **8:** 267.
18. PAULI, W. 1958. Theory of Relativity. Supplementary note 8a, p. 215. Pergamon, Elmsford, New York. Pauli acknowledges being informed of Freud's result by V. Bargmann. The present author attended lectures by Bargmann at Princeton in the 1950s on general relativity that included the Freud work. Notes on these lectures were made and retained by D. Brill.

19. SCHRÖDINGER, E. 1918. The energy components of the gravitational field. Phys. Z. **19:** 4.
20. BAUER, H. 1918. On the energy components of the gravitational field. Phys. Z. **19:** 163.
21. EINSTEIN, A. 1918. Note on E. Schrödinger's paper. Phys. Z. **19:** 115.
22. BAHDER, T. B. 1995. Mathematica for Scientists and Engineers, p. 540. Addison–Wesley. Reading, Massachusetts.
23. VIRBHADRA, K. S. Private communication.
24. YILMAZ, H. 1979. Einstein, the exponential metric, and a proposed gravitational Michelson-Morley experiment. Hadronic J. **2:** 997.
25. BURROWS, R. K. 1994. A brief study of the metric theory of Hüseyin Yılmaz using symbolic computer calculations. Scholarly paper to fulfill a requirement of the Graduate School of the University of Maryland at College Park. The calculations were performed under the direction of H. Yılmaz and the author. The paper gives the complete details of the two-slab geometry, but is only a small fraction of the many different metrics evaluated by R. K. Burrows.
26. CAVENDISH, H. 1798. Experiments to determine the density of the earth. Philos. Trans. R. Soc. London **MDCCXCVIII**(part 1): 469–526. It is ironic that Cavendish's experiment was originally constructed by John Mitchell, who also suggested the concept of "black hole". A modern account of such experiments is given by: CHEN, Y. T. & A. COOK. 1993. Gravitational Experiments in the Laboratory. Cambridge University Press. London/New York.
27. TAYLOR, J. H., JR. 1994. Binary pulsars and relativistic gravity. Rev. Mod. Phys. **66**(no. 3): 711–719. In the last paragraph of this paper, he states: "I do not believe that general relativity necessarily contains the last valid words to be written about the nature of gravity. The theory is not, of course, a quantum theory, and at its most fundamental level the universe appears to obey quantum-mechanical results."

Gravity and Quantum Field Theory

A Modern Synthesis

HÜSEYİN YILMAZ

Hamamatsu Photonics K.K.
Hamamatsu City, 430 Japan
and
Electro-Optics Technology Center
Tufts University
Medford, Massachusetts 20155

INTRODUCTION

In this article, we consider the problem of the perturbative quantization of gauge fields in general and that of gravity in particular. To achieve this goal, the conventional quantum field theory and the conventional theory of gravity had to be revised so that they became internally consistent and mutually compatible. The revised system is then conceived as a general quantum field theory within a geometrically based gauge principle so as to recover all known experiments concerning gravity and the standard model of elementary particles at the respective correspondence limits.

Despite the formidable appearance of this setting, what transpires at the end turns out to be surprisingly simple. The whole work can be condensed into three simple rules amending the conventional Feynman method of quantization, which then becomes finite and applicable to the perturbative calculation of any effect within the system and, in principle, to any desired order.

For maximum efficiency of the communication, we shall present the three rules at the beginning (with examples of application) because they are simpler to grasp than the arguments that led to them and because they get us immediately into the calculations. We leave their possible proofs or justifications to a later discussion; if proofs or justifications are forthcoming, they will, of course, become consequences of higher principles; if not, they will be considered as postulates on which perturbative quantization may be based.

RULE 1: The Lagrangian of a dynamical system of fields is the sum of the Lagrangians of its individual members plus their mutual interactions:

$$\mathscr{L} = \mathscr{L}_A + \mathscr{L}_B + \mathscr{L}_{AB}.$$

The rule is utterly obvious, but it must never be forgotten. (We shall see an example of what goes wrong if one does forget it.) It owes its importance to the fact that the Feynman propagators, by which the perturbative quantization is formulated, are obtained from the Lagrangians of the individual fields. In other words, if there is no Lagrangian for a field, then there is no propagator for that field and hence no quantization. The word "dynamical" means the fields possess energy-momentum and can transfer such to other fields and vice versa via the mutual interaction terms, $\mathscr{L}_{AB}$. The whole theory is to be relativistic; hence, when localized sufficiently away from each other, the individual fields are representations of local relativity groups, satisfy positive energy conditions, etc., as usual. Field Lagrangians contain the first derivatives of the fields so that $\mathscr{L}_A \Rightarrow \mathscr{L}_A(\partial_\mu\phi, \phi\ldots)$, which, for simplicity, may be denoted as $\mathscr{L}_A \Rightarrow \mathscr{L}_A(\partial_\mu)$.

RULE 2: The mutual interactions of a dynamical system of fields are absorbable into local gauge-covariant derivatives in terms of which the system appears formally free:

$$\mathscr{L} = \mathscr{L}_A(D_\mu) + \mathscr{L}_B(D_\mu),$$
$$D_\mu = P^\alpha_\mu\partial_\alpha + Q_\mu.$$

This rule, which determines the form of the interaction terms, arises from the fact that $-i\partial_\mu$ is essentially the quantum mechanical energy-momentum operator for individual modes; however, in the presence of other fields, the energy-momentum differs.[1] For example, the energy of a charged particle at rest in a static electromagnetic field is $E \Rightarrow mc^2 + eA_0$ (additive change $dE \Rightarrow edA_0$) and the energy of a neutral particle at rest in a static gravitational field is $E \Rightarrow mc^2\exp(-\phi/c^2)$ [multiplicative change $dE \Rightarrow -md\phi \Rightarrow -p_\nu d(2\phi^\nu_0 - \delta^\nu_0\phi) \Rightarrow -(E/c^2)d\phi$]. Rule 2 is a statement that leads to these as classical correspondence limits. The multiplicative gauge has a two-index potential ϕ^ν_μ and, when appropriately iterated, leads to a functional exponential for the metric (see the second and sixth sections of this article).

Thus, the interaction terms are inseparably connected to the fields via the underlying gauges and cannot be turned off. An important aspect of the gauge fields is that they can be interpreted geometrically (metric connections, fiber bundles, torsions, curvatures). By this property, rule 2, on the one hand, restricts the kinds of fields to be considered physical and, on the other, unifies the description into a geometrically based general field theory.[2] Local Lorentz gauge means that, along the local tangent space, the coordinates are as in special relativity (time measured with an atomic clock and distances by light, with the speed of light being locally c) and satisfy the Lorentz condition (harmonicity) $\partial_\nu g^{\mu\nu} = \partial_\nu\phi^\nu_\mu = 0$. The geometric concept acts as a selection principle, but allows gravity. The usual condition of renormalizability does not; hence, it must be rejected.

RULE 3: The divergent Feynman loop amplitudes $\mathscr{A}_\mathrm{F}(s, a)$ are not physical. The physical (finite and unambiguous) loop amplitudes $\mathscr{A}_\mathrm{ph}(s, a)$ are obtainable from the corresponding Feynman amplitudes by replacing the divergent momentum integrals as

$$\int d^4l[A_1x_1 + \cdots + A_nx_n]^{-n} \Rightarrow \pi^2/\Gamma(n)\left(\int_{a_0}^{a} da\right)^{S+1} a^{-1},$$

where the loop defines an average $l \Rightarrow p_{\mathrm{av.}}$ so that $p \Rightarrow p - p_{\mathrm{av.}}$ is the effective loop momentum. $G \Rightarrow a^{-1}, a = \{m^2 - (p - p_{\mathrm{av.}})^2\}_{\mathrm{av.}}$ is the effective loop propagator, n is the number of internal lines, t is the degree (power) of the trace, and $s = 2 - n + t/2 \geq 0$ is the loop index. The averages are over normalized ($\Sigma_i x_i = 1$) Feynman parameters.

It will turn out that the key phrase is the "normalized Feynman parameters". The rule seems to arise from the fact that, at the loops, the character of the energy-momentum transfer changes. Physically, the loops are different from the trees because they come with a factor $\hbar$ for each; hence, they are truly quantum mechanical. It is possible to give a new physical interpretation to them: namely, a loop defines an average $l \Rightarrow p_{\text{av.}}$ so that it operates relative to this average as $p \Rightarrow p - p_{\text{av.}}$. Then, the inverse propagator becomes

$$G^{-1} \Rightarrow a = \{m^2 - (p - p_{\text{av.}})^2\}_{\text{av.}} = m^2_{\text{av.}} - p^2_{\text{av.}} + (p_{\text{av.}})^2,$$

on which the amplitude depends. This suggests that the loop propagator conveys something like a quantum fluctuation in momentum rather than the usual interpretation of a definite (prescribed) contribution. For $s = -1$, the new and the old amplitudes become the same and, apart from a numerical factor, are equal to a^{-1}, which provides a correspondence argument for the new amplitude. For $s \geq 0$, the two differ: $\mathscr{A}_{\text{F}}(s, a)$ becomes infinite, whereas $\mathscr{A}_{\text{ph}}(s, a)$ is always finite; thus, the two are no longer the same. Note that the function

$$K(s, a) = \left(\int_{a_0}^{a} da\right)^{s+1} a^{-1}$$

is a universal function, although a is different for different loops. A few of its members and the general form are given below:

$$K(0, a) = \ln(a/a_0)$$

$$K(1, a) = a \ln(a/a_0) - (a - a_0)$$

$$K(2, a) = (a^2/2) \ln(a/a_0) - (3/4)(a^2 - a_0^2) + a_0(a - a_0)$$

$$\vdots$$

$$K(s, a) = (a^s/s!)\ln(a/a_0) + \Sigma_k B_k a_0^{s-k}(a^k - a_0^k).$$

$K(s, a)$ satisfies a generalized homogeneity property (called self-similarity):[3] namely, if a is scaled as $a \Rightarrow \xi a$ (the same being, of course, for a_0), then $K(s, a)$ scales as $K(s, a) \Rightarrow \xi^s K(s, a)$. With these and other properties to be mentioned, the final form of a physical amplitude can essentially be written down by inspection as soon as the corresponding diagram is drawn. Note that $K(s, a)$ is short for $K(s, a, a_0)$.

APPLICATIONS OF THE RULES

In this section, we refer to a number of applications of the three simple rules. This is done essentially to get a feel of their working in actual physical situations and to get an appreciation of what can go wrong if they are not carefully followed. For convenience, we shall repeat the statement of the rule in question before the examples of its application. (Also, see APPENDIX A and APPENDIX B.)

RULE 1: The Lagrangian of a dynamical system of fields is the sum of the Lagrangians of its individual members plus their mutual interactions.

Example 1. Consider the electron-photon system (QED): The Lagrangian of the system is

$$\mathscr{L} = -\bar{\psi}(\gamma^{\mu}\partial_{\mu} + m)\psi - (1/4)(\partial_{\mu}A_{\nu} - \partial_{\nu}A_{\mu})^2 + \mathscr{L}_{\psi A}.$$

The rule is obviously satisfied by QED. Consequently, the Feynman propagators, by which the quantization of ψ and A is implemented, are obtainable from the Lagrangians of the individual fields. ($\mathscr{L}_{\psi A}$ is the interaction term to be considered later under rule 2.) If there is no Lagrangian for one of the fields, there would be no propagator—hence, no quantization for that field. Propagators convey energy-momentum and transfer such to other fields at the interaction vertices. To do so, the field itself must have a nonzero stress-energy derivable from its Lagrangian.[4] It is the field stress-energy of a field that is being quantized into discrete quanta of energy-momentum. The Lagrangian itself can be numerically zero as here (the electromagnetic case), but it must have a functional form from which a nonzero stress-energy tensor must be derivable. The present example satisfies all these because both the Dirac field and the Maxwell field possess field stress-energies t^{ν}_{μ}:

$$t^{\nu}_{\mu}(\text{Dir.}) = -\bar{\psi}(\gamma^{\nu}\partial_{\mu}\psi) + \delta^{\nu}_{\mu}\bar{\psi}(\gamma^{\mu}\partial_{\mu} + m)\psi,$$

$$t^{\nu}_{\mu}(\text{Maxw.}) = -(\partial_{\mu}A_{\lambda} - \partial_{\lambda}A_{\mu})(\partial^{\lambda}A^{\nu} - \partial^{\nu}A^{\lambda}) + (1/4)\delta^{\nu}_{\mu}(\partial_{\mu}A_{\nu} - \partial_{\nu}A_{\mu})^2.$$

This pattern must repeat for all fields if we are to have a consistent theory. Let us now see what happens in the next example.

Example 2. Consider the electron-graviton system (QGD): According to rule 1, the Lagrangian of the system must be of the form,

$$\mathscr{L} = -\bar{\psi}(\gamma^{\mu}\partial_{\mu} + m)\psi + \mathscr{L}_g + \mathscr{L}_{\psi g},$$

where $\mathscr{L}_g$ is the Lagrangian of the gravitational field that we seek to find. This $\mathscr{L}_g$ leads, by variation, to the field stress-energy of gravity, but we know that in the conventional theory of gravity the stress-energy of the gravitational field is zero because it is (arbitrarily) set to zero by hand. Hence, accordingly, there is no field stress-energy and no energy-momentum in the gravitational field of the conventional theory. In other words, unlike the electromagnetic theory, the conventional theory of gravity does not satisfy rule 1. It is thus obvious that such a theory cannot be quantized because it does not have the field stress-energy-momentum to quantize. It is also clear that, by the act of setting the field stress-energy to zero by hand, the conventional theory of gravity sets a severe restriction on the Lagrangian so that it cannot lead to a nonzero stress-energy. Such a theory cannot have a meaningful dynamics and cannot lead to a meaningful propagator. Consequently, it also cannot be quantized. These statements are essentially equivalent to another statement given in reference 5 stating that the conventional theory of gravity is mathematically overdetermined because the existing field stress-energy is set to zero by fiat, imposing a most severe restriction.

This problem is solved as follows: The right-hand side of the conventional Einstein theory of gravity is modified as

$$(1/2)G^{\nu}_{\mu} = \tau^{\nu}_{\mu} + t^{\nu}_{\mu},$$

where t^{ν}_{μ} is the stress-energy tensor of the gravitational field. The more reliable left-hand side in fact permits such a field stress-energy t^{ν}_{μ}. This t^{ν}_{μ} must be obtained from the gravitational field Lagrangian by variation (and it is); hence, in the modified theory, there must be a field Lagrangian $\mathscr{L}_g$. Using two geometric identities (see the sixth section) and the equation above, it is then found that the gravitational field Lagrangian and the gravitational field stress-energy are as follows (note that they are nonzero):[5]

$$\mathscr{L}_g = -(\partial_\mu \phi^{\alpha}_{\beta})^2 + (1/2)(\partial_\mu \phi)^2 + 2\partial_\alpha \phi^{\nu}_{\beta} \partial^{\beta} \phi^{\alpha}_{\nu}$$

and

$$t^{\nu}_{\mu}(\text{grav.}) = -2[\partial_\mu \phi^{\alpha}_{\beta} \partial^{\nu} \phi^{\beta}_{\alpha} - (1/2)\delta^{\nu}_{\mu} \partial^{\lambda} \phi^{\alpha}_{\beta} \partial_\lambda \phi^{\beta}_{\alpha}] + \partial_\mu \phi \partial^{\nu} \phi - (1/2)\delta^{\nu}_{\mu} \partial^{\lambda} \phi \partial_\lambda \phi,$$

where ϕ = trace ϕ^{ν}_{μ} and $\partial_\nu \phi^{\nu}_{\mu} = \partial^{\mu} \phi^{\nu}_{\mu} = 0$. The gravitational field $\phi^{\nu}_{\mu} = \hat{\phi}$ is defined as[5,6]

$$\partial_\lambda \phi^{\nu}_{\mu} = -(1/4)g^{\rho\nu}[\Gamma_{\mu\rho\lambda} + \Gamma_{\rho\mu\lambda}] + (1/4)\delta^{\nu}_{\mu}\Gamma^{\alpha}_{\alpha\lambda} = (1/4)\mathbf{g}_{\mu\rho}\partial_\lambda \mathbf{g}^{\rho\nu}$$

or by its general formal solution (all members used here, however, are exact and are in closed form),

$$g_{\mu\nu} = (\eta e^{2(\phi - \hat{\phi} - \hat{\phi}^T)})_{\mu\nu},$$

where $\mathbf{g}^{\mu\nu} = \sqrt{-g}g^{\mu\nu}$, $\mathbf{g}_{\mu\nu} = g_{\mu\nu}/\sqrt{-g}$, and $\Box^2 = (\sqrt{-g})^{-1}\partial_\alpha(\sqrt{-g}\partial^\alpha)$. Note that $\partial_\nu \mathbf{g}^{\mu\nu} = \partial_\nu \phi^{\nu}_{\mu} = 0$—hence, the theory satisfies the harmonic gauge. Surprisingly (or not very surprisingly, for this is the consistent theory), the new theory has exact interactive N-body solutions and exact multimode gravity waves carrying energy-momentum. (The usual theory does not have these because the field stress-energy is forced to zero; the equations are overdetermined.) The Lagrangian $\mathscr{L}_g$ turns out to be exactly $\mathscr{L}_g = (1/4)R$, the source-free curvature scalar calculated in the new theory. Thus, in this new theory, the propagator of the gravitational field can be constructed; hence, applying the Feynman procedure, the gravitational field becomes quantized. That the Lagrangian in the usual theory is also taken to be $\mathscr{L}_g = (1/4)R$ is of no consequence because, there, its chances to be physical are killed by arbitrarily setting its field stress-energy to zero (that is, by rendering $\mathscr{L}_g$ ineffective). The above modification of the conventional theory of gravity was found a long time ago via other arguments. Here, we see another and, in fact, a most compelling argument of its necessity directly from the requirement of its quantizability.

Example 3. Consider the standard model: This covers electromagnetic, electroweak, and strong interaction processes, where the gauge fields are not necessarily Abelian. For simplicity, take the quark-gluon plus the electroweak system. The system Lagrangian is

$$\mathscr{L} = -\bar{\psi}^{\alpha}(\gamma^{\mu}\partial_\mu + m)\psi^{\alpha} - (1/4)(\partial_\mu A^{\alpha}_{\nu} - \partial_\nu A^{\alpha}_{\mu})^2 + \mathscr{L}^{\alpha}_{\psi A}.$$

The situation is very similar to electrodynamics, except that ψ^{α} and A^{α}_{μ} have multiplets denoted by an internal group index, a. In QCD, these are multiplets of SU(3). They are quarks and gluons. There are many details, but the Lagrangian is still of the above form. It is the same in electroweak theory plus the combination of the two, so the formalism can be taken over. In other words, these theories satisfy rule 1 because they follow the pattern of quantum electrodynamics.

RULE 2: The mutual interactions of a dynamical system of fields are absorbable into local gauge-covariant derivatives in terms of which the system appears formally free.

Example 1. Consider again the electron-photon system (QED): According to rule 2, the system Lagrangian can be written, as

$$\mathscr{L} = -\bar{\psi}(\gamma^\mu D_\mu + m)\psi - (\tfrac{1}{4})(D_\mu A_\nu - D_\nu A_\mu)^2,$$
$$D_\mu = \partial_\mu + eA_\mu,$$

where the second line is the (linear) gauge of the electromagnetic theory. Substituting and comparing with rule 1, we find the interaction term as

$$\mathscr{L} = -\bar{\psi}(\gamma^\mu \partial_\mu + m)\psi - (\tfrac{1}{4})(\partial_\mu A_\nu - \partial_\nu A_\mu)^2 + \mathscr{L}_{\psi A},$$
$$\mathscr{L}_{\psi A} = -ie\bar{\psi}\,\gamma^\nu A_\nu \psi.$$

Thus, rule 2 provides a way of finding the interaction term when the individual member fields (and the properties of the potentials) are known.

Example 2. Consider the electron-graviton system: Similar to the above, we write

$$\mathscr{L} = -\bar{\psi}(\gamma^\mu \partial_\mu + m)\psi + \mathscr{L}_g + \mathscr{L}_{\psi g},$$
$$D_\mu = P^\alpha_\mu \partial_\alpha + Q_\mu,$$

where P^α_ν and Q_ν are likewise local gauge substitutions. Such a generalization is both possible and necessary because of the following: whereas in the electromagnetic case the potential affects the energy momentum linearly ($D_\nu = \partial_\nu + eA_\nu \rightarrow E = mc^2 + eA_0$), the relation in the gravitational case is multiplicative ($D_\nu \rightarrow E = mc^2 \exp(-2\phi^0_0 + \phi)$. The latter is due to the principle of equivalence. Thus, via the mass-energy relation $E = mc^2$, one can determine P^α_μ as

$$P^\alpha_\mu = l^\alpha_\mu,$$

where l^α_μ is a tetrad defined by

$$g_{\mu\nu} = l^\alpha_\mu \eta_{\alpha\beta} l^\beta_\nu.$$

According to the new theory's solutions (note that it has exact closed-form solutions),

$$l^\alpha_\mu = (\eta e^{\phi - \hat{\phi} - \hat{\phi}^T})^\alpha_\mu;$$

hence (ignoring torsion), the system Lagrangian can be written as

$$\mathscr{L} = -\bar{\psi}(\gamma^\mu D_\mu + m)\psi - (\partial_\mu \phi^\alpha_\beta)^2 + (\tfrac{1}{2})(\partial_\mu \phi)^2 + 2\partial_\alpha \phi^\nu_\beta \partial^\beta \phi^\alpha_\nu,$$

where we take only the gauge-covariant solutions (see the sixth section) of the field equations. They are the ones mentioned in connection with rule 1 and satisfy $\partial_\nu \phi^\nu_\mu = \partial^\mu \phi^\nu_\mu = 0$. Within quantum field theory, no other solution is needed. (All others needed can be generated via group transformations, as is also the case in electrodynamics.) The interaction term then is

$$\mathscr{L}_{\psi g} = -\bar{\psi}[\gamma^\mu(D_\mu - \partial_\mu)]\psi$$

or, in its form expanded into the local Lorentz frame,

$$\mathscr{L}_{\psi g} = -\bar{\psi}\{\gamma^\alpha[(\phi - \hat{\phi} - \hat{\phi}^T) + (\tfrac{1}{2})(\phi - \hat{\phi} - \hat{\phi}^T)^2 + \cdots]^\mu_\alpha \partial_\mu\}\psi.$$

Note that the static limit $\phi - \hat{\phi} - \hat{\phi}^T \Rightarrow -\phi$ is similar to the coulomb case in first order. In higher orders, it differs because it is exponential (principle of equivalence). Thus, in this case, the Lagrangian is nonpolynomial. We shall see, however, that there is no objection to this in the new theory because (via rule 3) it is finite. In the hydrodynamic limit, $t^\nu_\mu(\text{Dir.}) + t^\nu_\mu(\text{Maxw.}) = \tau^\nu_\mu \Rightarrow -\sigma u_\mu u^\nu$. In general, there is a minute amount of torsion that will be neglected. When torsion is absent, $\phi^T = \phi$. (In any case, torsion does not propagate; hence, the theory may be viewed as essentially a curvature theory.)[7] For the three exact solutions to be used, the first covariant derivatives of the fields reduce to ∂_μ; hence, we can substitute for ∂_μ so that rule 2 is satisfied.

We now summarize what has transpired: If we reject the right-hand side of the Einstein-Hilbert field equations and work with the (more reliable) left-hand side, we have $\mathscr{L}_g = (\tfrac{1}{4})R$ (quadratic part of the curvature scalar) and its canonical stress-energy is $t^\nu_\mu = t^\nu_\mu(\text{grav.})$, which we take as the gravitational field stress-energy. The new right-hand side becomes

$$(\tfrac{1}{2})G^\nu_\mu = \tau^\nu_\mu + t^\nu_\mu,$$

where t^ν_μ is not zero, nor can it be set to zero for it is needed for quantization. In fact, setting $t^\nu_\mu = 0$ (or $t^\nu_\mu \Rightarrow \lambda t^\nu_\mu$, $\lambda = 0$) forces the theory into a mathematical overdetermination.[5]

Example 3. Generalization to the quark-gluon and lepton vector-boson systems: These cases work similarly to the electromagnetic case. Only the group indices for SU(2) and SU(3) are to be added consistently. Because of symmetry-breaking, these cases are more intricate because a scalar field is also involved, although gauged in a similar manner. Despite complexity, no new idea is involved. This extension, called the standard model, is found to be highly successful.

All these theories, although experimentally viable, are nevertheless unsatisfactory (including QED) from a fundamental theoretical point of view: namely, they lead to ambiguous and infinite results that have to be removed by equally ambiguous and unacceptable mathematical manipulations. Rule 3, to which we now turn, seems to solve this problem.

RULE 3: The divergent Feynman loop amplitudes $\mathscr{A}_F(s, a)$ are not physical. The physical (finite and unambiguous) loop amplitudes $\mathscr{A}_{ph}(s, a)$ are obtainable from the corresponding Feynman amplitudes by replacing the divergent momentum integrals as

$$\int d^4l[A_1x_1 + \cdots + A_nx_n]^{-n} \Rightarrow \pi^2/\Gamma(n)\left(\int_{a_0}^{a} da\right)^{S+1} a^{-1},$$

where the loop defines an average $l \Rightarrow p_{av.}$ so that $p \Rightarrow p - p_{av.}$ is the effective loop momentum. $G \Rightarrow a^{-1}$, $a = \{m^2 - (p - p_{av.})^2\}_{av.}$ is the effective loop propagator, n is the number of internal lines, t is the degree (power) of the trace, and $s = 2 - n + t/2 \geq 0$ is the loop index. The averages are over normalized $(\Sigma_i x_i = 1)$ Feynman parameters.

Note that, in this interpretation, no regularization, renormalization, or ambiguity is present. (In fact, the divergent momentum integrals are physically wrong and that is why we are replacing them.) The inverse propagator gives $a = m^2_{av.} - p^2{}_{av.} + p_{av.}{}^2$ and the calculation of a 1-loop amplitude reduces to the mnemonic (polarization sums P may be subsumed in the trace),

$$\mathscr{A}_{ph}(s, a) = \{S\}\{g^2/(4\pi)^2\}\{\textstyle\int dx \ldots\}\{\text{trace}\}K(s, a),$$

$$a = m^2_{av.} - p^2{}_{av.} + p_{av.}{}^2,$$

$$\left\{\int dx \ldots\right\} = \int_0^1 dx_1 \ldots \int_0^1 dx_n \delta(x_1 + \ldots + x_n - 1),$$

where S is the symmetry factor and g is the coupling constant. [π^2 is absorbed into the g block and $\Gamma(n)$ is canceled with a $\Gamma(n)$ in the $\int dx \ldots$ block.] A few examples will show their simplicity.

Example 1. The electron self-energy in QED (FIGURE 1): From the figure, $S = 1$, $s = 0$, $g = e$, etc. On the other hand, the trace is $\text{tr}[\gamma_\mu(m + p - p_{av.})\gamma_\mu] = 4m - 2p(1 - x)$ because $p_{av.} = px$, $\gamma_\mu m \gamma_\mu = 4m$, and $\gamma_\mu p \gamma_\mu = -2p$. Thus, the final expression is

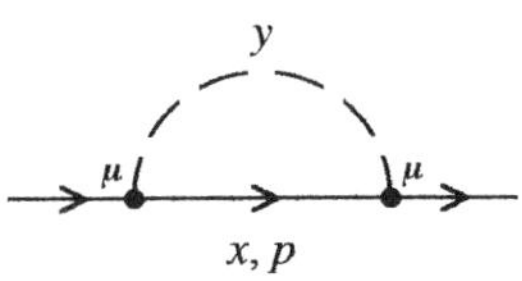

Fig. 1

$$\Sigma_{ph}(0,p) = e^2/(4\pi)^2 \int_0^1 dx\,\{4m - 2p(1-x)\}\,ln(a/a_0)$$

$$a = m^2x - p^2x(1-x) + \lambda^2 y.$$

This is a good model of a loop calculation because it shows the main components of the calculation, including the trace, in a simple and compact manner.

Example 2. Vacuum polarization in QED (FIGURE 2): Applying the mnemonic above and noting that in this case the quadratic part cancels, one has the gauge-invariant part:

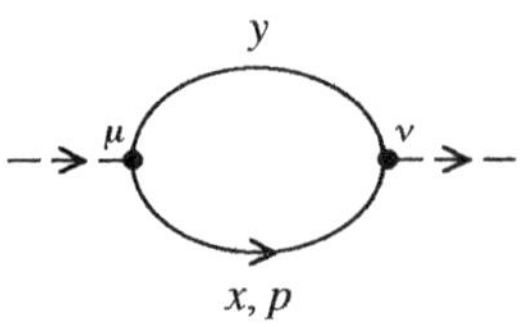

$$\Pi_{\mu\nu\, ph}(0, p) = -e^2/(4\pi)^2 \int_0^1 dx\, \{8x(1-x)Z_{\mu\nu}\}\, ln(a/a_0)$$

$$a = m^2 - p^2x(1-x)$$

Fig. 2

where $Z_{\mu\nu} = p_\mu p_\nu - \delta_{\mu\nu}p^2$. Here, one can set $a_0 = m^2$ because there is no infrared problem for this diagram. In general, a_0 can be thought of as the mass-shell value, but it must be done together with other contributing diagrams as a physical condition. If the original Lagrangian starts at mass shell, the amplitudes are all zero at mass shell, recovering the original Lagrangian. Note that a multiple of the trace of $\Pi_{\text{ph}\,\mu\nu}$, namely, $\Pi_{\text{ph}} = -\text{trace}\ \Pi_{\mu\nu}/3p^2$, enters into the physical calculations:

$$\Pi_{\text{ph}}(0, p) = -e^2/(4\pi)^2 \int_0^1 dx\{8x(1-x)\} \ln(a/a_0),$$

$$a = m^2 - p^2x(1-x).$$

When $k^2 \ll m^2$, we have [setting $\alpha = e^2/(4\pi)$]

$$\Pi_{\text{ph}}(0, p) \simeq (\alpha/3\pi)(p^2/m^2)(-1/5).$$

We shall use this result when calculating the Lamb-shift.

Example 3. Vertex-function in QED (FIGURE 3): For the divergent part $s = 0$, we have

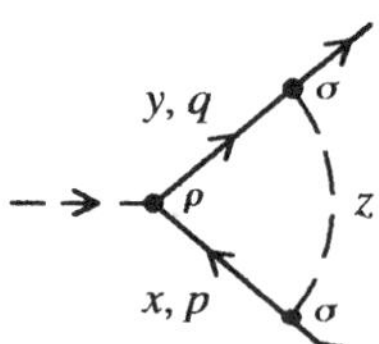

$$\Lambda_{\rho\, ph}(0, p, q) = e^3/(4\pi)^3 \int_0^1 dx \int_0^{1-x} dy\, \{trace\}_\rho\, ln(a/a_0)$$

$$a = m^2(x+y) - (p^2x + q^2y) + (px + qy)^2 + \lambda^2 z$$

Fig. 3

where p and q are vectors. This is a complicated expression, but for small p and q, that is, for low energy scattering, and for $k^2 \ll m^2$, it is calculated to be as follows (λ is a temporary mass assigned to the photon for later convenience of fitting with the soft photon emissions where it cancels out):

$$\Lambda_{\text{ph}}(p) \simeq (\alpha/3\pi)(p^2/m^2)[\ln(m/\lambda) - 3/8] - (\alpha/2\pi)\sigma \cdot \boldsymbol{k}/2m,$$

$$\Lambda_{\text{mag. m.}}(k) \simeq -(\alpha/2\pi)\sigma \cdot \boldsymbol{k}/2m.$$

Example 4. Scalar field $\phi^3/3!$ in 6 dimensions (FIGURE 4): This is an example other than in 4 dimensions. From the figure, $S = \frac{1}{2}$, $s = d/2 - n = 1$, and {trace} = 1. Thus, the result is

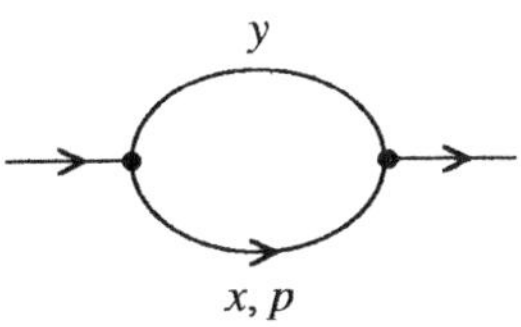

$$\Sigma_{ph}(1, p) = \tfrac{1}{2}\, g^2/(4\pi)^3 \int_0^1 dx\, [a \ln(a/a_0) - (a - a_0)]$$

$$a = m^2 - p^2 x(1 - x).$$

Fig. 4

In the usual theory of $\phi^4/4!$, note that amplitudes in 6, 8, 10, etc., dimensions would be useless because they would be unrenormalizable. Here, there is no problem. Results are all finite and recover the original Lagrangian at mass shell. For dimensions other than 4, one simply substitutes $s = D/2 - n + f/2$. If the external momenta are of the form $p + q = r$ as in the case of $\phi^4/4!$ in $D = 4$ theory, one would use r^2 (a Mandelstam variable) in place of p^2, etc. In the new theory, all powers of ϕ—in fact, even a nonpolynomial function of ϕ—would be allowable. Note that the arrows in these diagrams indicate momentum flows.

Example 5. Gravitational vacuum polarization (QGD) (FIGURE 5): We can again use our method and evaluate, for example, the vacuum diagram in the slow motion (static) limit as

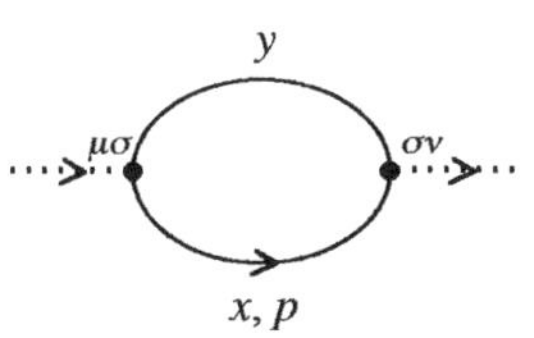

$$\mathcal{A}_{\mu\nu\, ph}(0, a) = -f^2/(4\pi)^2 \int_0^1 dx\, P\{8x(1-x)Z_{\mu\nu}\}\, K(0,a)$$

$$a = m^2 - p^2 x(1 - x)$$

Fig. 5

where $f = \sqrt{(4\pi G)}$. It is similar to the electron-positron loop in electrodynamics.

Overlapping and nested divergences do not pose problems as infinities do not occur. The case of QED is so simple that (after cancellations) only the $K(0, k) = \ln(a/a_0)$ occurs. One may worry about the relativistic invariance or the gauge covariance, but there is no reason because the final results are identical to renormalized electrodynamics, which we know are consistent with them. What we are seeing seems to be the first consistent formulation of quantum field theory without infinities and ambiguities.

What happens if in a diagram there is a particle-antiparticle loop? Can part of the momentum go via the other line? Yes, but probabilistically; one can prove a theorem that it makes no difference to the answer. Let α, β be the respective probabilities with

$\alpha + \beta = 1$. Then,

$$a = m^2x + m^2y - (p^2\alpha^2x + p^2\beta^2y) + (p\alpha x - p\beta y)^2,$$

where the $(-)$ sign exists because the other channel is the antiparticle channel. Substituting, we have

$$a = m^2 - p^2[\alpha^2x + \beta^2y - \alpha^2x^2 - \beta^2y^2 + 2\alpha x\beta y],$$

$$a = m^2 - p^2[\alpha^2xy + \beta^2yx + 2\alpha x\beta y] = m^2 - p^2x(1 - x),$$

the same as if the momentum is in one channel only. This simplifies the evaluation of the diagrams.

A general feature of these calculations seems to be that we may have to take the virtual particle idea more seriously. Originally, this idea was introduced as a purely mathematical aid to describe perturbation expansions. We may be seeing the viability of an interpretation whereby (as Feynman already emphasized) they can be thought of as short-lived physical particles. If this is so, the situation is similar to Planck's resolution of the ultraviolet problem in 1900. The $K(s, a)$ functions, and therefore the amplitudes, are analytic due to the analyticity of the Feynman propagators of which they are composed. The unitarity is subtle as it depends on the measure and in general requires ghost states.[8,9] However, the problem does not arise if the gauge is fixed unitary (axial) and if the theory is considered (at least temporarily) a preferred-gauge theory.

ON THE NATURE OF THE THREE RULES

If, without worrying about their origins or justifications, the three simple rules are consistently applied in conjunction with Feynman quantization, one will get finite and unambiguous (and experimentally viable) loop amplitudes. Because there are no infinities or related ambiguities, there are also no renormalizations. By this result, gravity is relieved from its previous bind of unrenormalizability and can now be quantized when it is properly formulated. What is behind the three simple rules that makes this development possible?

In this section, we shall assume that rule 1 and rule 2 are more or less obvious (although, as we have seen, they were not always carefully applied in the past) and shall concentrate on the more intricate rule 3. This rule was originally arrived at in two different ways based on the principle of causality. As is well known, the principle of causality is an ancient principle having many physical implications. In its most primitive and, even today, most popular form, it may be stated as follows:

(1) every effect must have a cause;
(2) no effect can be its own cause;
(3) cause always precedes its effect.

These statements, however, invite two further considerations:

(4) how fast does the cause travel?
(5) what exactly is a physical cause?

The first three statements are characteristic to Aristotelian physics. Newtonian physics amended them by saying that there are some motions that do not have to have a cause, for example, uniform motion with constant velocity. These are called states of the motion. Only the deviations from uniform motion need to have a cause. This cause is called force and its effect is a change of the state of the motion. However, in some cases, cause and effect may become coexistent or, as in the case of mechanical equilibrium, force may exist, even store energy, but does not have to cause acceleration. Furthermore, the Newtonian idea of force is conceived in relation to motion along a trajectory, which is not very useful in quantum mechanics.

The fourth question was answered only at the beginning of this century by the theory of relativity—namely, cause travels, at most, with the speed of light. This statement is implemented in quantum field theory by the condition that physical observables commute at spacelike separations. Note, however, that this is only a kinematical condition and does not say anything about the dynamical aspects of the cause-and-effect relationship.

Curiously, the fifth question, namely, what is it that constitutes an actual physical cause, seems never to be carefully considered, especially in relation to the loop processes of quantum field theory. (Drake makes a similar observation on the classical principle of causality—namely, philosophers talked about causality for centuries, but never carefully defined what actually constitutes a physical cause.[10]) We consider here the physical cause to be the energy-momentum transfer between physical systems or between the internal states of a physical system. For example, if a proton scatters an electron in a coulomb field, the momentum transferred by the intermediate photon is the cause of the scattering. However, if, on its way to the electron, the photon creates an electron-positron pair that subsequently annihilates, recreating the photon, then what is the cause-and-effect relationship with the scattering of the electron? In the case of trees, the internal momenta are completely determined by the external momenta, but, in the case of loops, there can be circulating (should we say undetermined or spontaneous) momenta that, according to the rules of quantum mechanics, must be integrated (summed) over. This integration ranges from zero to infinity for reasons of relativistic covariance and gives a result as a function of the momenta p external to the loop. In certain cases, the result contains ambiguous and infinite terms.[11]

However, from a causal point of view, the result ought to be *not* just a function of p, but a cumulative function of p. In other words, in the loop effect, p ought to have a lower value where the effect starts and an upper value where it describes the contribution to the prevailing physical situation; that is, the result ought to be a definite integral of p. The problem with Feynman's amplitude is that, although it is not wrong from a relativistic point of view, it gives the loop amplitude as a function of p and not as a cumulative function of p. This ignores the dynamic nature of the causal connection and allows ambiguities and infinities to come into play. Mathematically, it is a form of underdetermination because a necessary physical condition is missing.

If this is so, we can easily correct the deficiency by first differentiating the amplitude and then reintegrating with the desired integration limits. In the simplest case, this would be of the form

$$\int_{p_0}^{p} dp \partial_p.$$

This is actually sufficient for logarithmically divergent amplitudes. If the degree of divergence is higher, the operations would be repeated with the same integration limits because the unified Feynman denominators are integrals of their lower powers. Also, because odd-integer divergences vanish, the variable can be changed to p^2 or, in fact, to a propagator $G^{-1} \Rightarrow a = Am^2 - Bp^2$ (where A and B are functions of Feynman parameters). One can then rewrite the physical condition as

$$\mathscr{A}_{\mathrm{ph}}(s, a) = \left(\int_{a_0}^{a} da\right)^{s+1} (\partial/\partial a)^{s+1} \mathscr{A}_{\mathrm{F}}(s, a);$$

that is, the finite and unambiguous physical amplitudes are obtainable from the divergent Feynman amplitudes by this operation. For $s = -1$, the two coincide, providing a correspondence argument. An interesting result is that, up to a numerical factor, this correspondence turns out to be of the form

$$\mathscr{A}_{\mathrm{ph}}(-1, a) = \mathscr{A}_{\mathrm{F}}(-1, a) = a^{-1},$$

$$a = m^2_{\mathrm{av.}} - p^2{}_{\mathrm{av.}} + p_{\mathrm{av.}}{}^2.$$

This version of rule 3 is sometimes very convenient because it can be automated if the complete integrated form of the Feynman amplitude is given. However, it depends on the original infinite integrals and does not explain why the loop propagator has the above form, which is the same as if causally the loop contributes not via a more or less determinate circulating momentum, but more like a fluctuating one. This leads to a different and independent derivation, where the intermediate momentum defines an average $l \Rightarrow p_{\mathrm{av.}}$ relative to which ($p \Rightarrow p - p_{\mathrm{av.}}$) the loop operates. We then arrive, via the function $K(s, a)$, precisely at the statement of rule 3.

There is an interesting way to see why the conventional integrals are unacceptable. If we analyze the mathematical process leading up to the conventional momentum integrals, we see that they crucially depend on a parameter, $\zeta = a + b + \ldots$, being a variable. The Feynman parameters if defined as $x = a/\zeta, y = b/\zeta, \ldots$, would be normalized ($x + y + \cdots = 1$), but $a, b, \ldots$, are not. What leads to the infinities in the conventional theory (see, for example, reference 12) is that ζ varies from 0 to ∞ and this produces a Γ function that diverges[12] (note here that $t = 0$):

$$\int_0^\infty d\zeta \zeta^{n-1-D/2} e^{-\zeta} = \Gamma(n - D/2) = \Gamma(-s) \Rightarrow \infty.$$

The simplest way to prevent this disaster would be to normalize $a, b, \ldots$, by setting $\zeta = 1$, that is, assuming they become normalized Feynman parameters. In that case a new interpretation consistent with the normalized Feynman parameters becomes possible. It is then obvious that we again fall on our rule 3. As a bonus, this interpretation allows a simpler and quicker way of writing down the expression of an amplitude as soon as the diagram is drawn. Diagrams are drawn with minimum clutter, avoiding redundant information. For example, (m), (λ) and channel associations are obvious; hence, they are not explicitly shown.

QUANTUM ELECTRODYNAMICS (QED)

In this section, we treat quantum electrodynamics both as a model of a quantum gauge-field theory to illustrate how the new situation differs from the old and as a theory rendered consistent by our revision. The Lagrangian of quantum electrodynamics is

$$\mathscr{L} = -\bar{\psi}(\gamma^{\mu}\partial_{\mu} + m)\psi - (1/4)(\partial_{\mu}A_{\nu} - \partial_{\nu}A_{\mu})^2 - ie\bar{\psi}\gamma_{\mu}A^{\mu}\psi,$$

where A_{μ} is the electromagnetic potential, ψ is the Dirac field, and e is the electric charge. Two more terms are usually added to this expression, namely, a gauge-fixing term $(1/2)\xi^2(\partial_{\nu}A^{\nu})^2$ and a photon-mass term $(1/2)\lambda^2(A_{\nu})^2$. For simplicity, we ignore these in the Lagrangian and remember only that, in the final expression, the former of these will imply gauge factors $k_{\mu}k_{\nu} - \delta_{\mu\nu}k^2$ for the observable quantities and the latter will imply a photon mass with λ for the purpose of dealing with the infrared problems. We also omit the $-i\epsilon$ in the propagator, but remember it whenever needed. In the present theory, these procedures are the same as in the usual theory.

The Feynman method of quantization, which we adopt, works directly with the Lagrangian. As soon as the Lagrangian is given, we may assume it is quantized via Feynman's procedure. Also, the Feynman method is so worked out that we can start calculating as soon as the Lagrangian is given. These features are still valid within our three rules. Applying our rules, the three most important 1-loop integrals, namely, the electron self-energy Σ_{ph} (FIGURE 1), the photon self-energy Π_{ph} (FIGURE 2), and the vertex-function Λ_{ph} (FIGURE 3), are already obtained in the previous section and they are finite. The vertex-function has a part describing a correction to the magnetic moment of the electron, which can be measured separately. It is finite and its expression is already given.

As an example of application, we shall calculate below the Lamb-shift via the revised theory. The Lamb-shift depends on the loop diagrams $\Sigma_{\text{ph}}(k)$, $\Pi_{\text{ph}}(k)$, and $\Lambda_{\text{ph}}(k)$, which in the present theory are convergent and unique. The Lamb-shift also gets a contribution from the low-energy incoherent soft photon emission, which is finite and is denoted as $\Omega_{\text{ph}}(k)$. Three contributions are shown below in the low-energy limit, $k^2/m^2 \ll 1$, as

$$\Pi_{\text{ph}}(k) \simeq (\alpha/3\pi)(k^2/m^2)(-1/5),$$

$$\Lambda_{\text{ph}}(k) \simeq (\alpha/3\pi)(k^2/m^2)[\ln(m/\lambda) - (3/8)] - (\alpha/2\pi)\sigma \cdot \mathbf{k}/2m,$$

$$\Omega_{\text{ph}}(k) \simeq (\alpha/3\pi)(k^2/m^2)[\ln(\lambda/2k_0) + (5/6)].$$

The electron self-energy Σ_{ph} (FIGURE 1) is not shown because, as we have said, in the 1-loop level it just cancels a small term in $\Lambda_{\text{ph}}(k)$. Remember in the conventional theory that Π, Σ, and Λ are, strictly speaking, not calculable because, there, they are divergent and ambiguous.[11]

The Lamb-Shift

There are three very important consequences of quantum electrodynamics. The connection between spin and statistics, the Lamb-shift, and the anomalous magnetic moment. The spin-statistic connection is the foundation of many facts (Planck distribution, correlated n-particle states, etc.). The Lamb-shift is probably the subtlest and also the most celebrated effect. It arises from the fact that, in Dirac's theory of the hydrogen atom, the $2s_{1/2}$ and $2p_{1/2}$ levels are equal, whereas the quantum field theory contributions arising from the integrals Π, Λ, and Ω contribute to each differently. In fact, the contribution to the $2s_{1/2}$ state is much larger than to the $2p_{1/2}$ state. We can therefore get a good approximate value of this shift by calculating the contribution to the $2s_{1/2}$ state. To do this, let us first remember that the potential energy is $U(x) = -e^2Z/r_{12}$ in Dirac's hydrogen atom. The effect of our integrals is to cause a perturbation such that $U \rightarrow U + \delta U$:

$$\delta E = (\Pi + \Lambda + \Omega)U.$$

Because the shift is very small compared to U, it can be written simply as the expectation value of δU:

$$\delta U = \bar{\psi}_{2s_{1/2}}(\delta U)\psi_{2s_{1/2}}.$$

Then, noting that $|\psi_{2s_{1/2}}|^2 = (Z\alpha m/n)^3/\pi$, the expectation value of $(\alpha/2\pi)\sigma \cdot \boldsymbol{k}/2m$ in the $\psi_{2s_{1/2}}$ state is $-3/8$, and $k^2U = -\Delta U = 4\pi\delta(r)$, we can immediately write down the contribution to the $2s_{1/2}$ state as

$$\delta E = (4/3)m(Z\alpha^5/\pi n^3)[(-1/5) + \ln(m/2k_0) + (5/6)],$$

where $n = 2$, $k_0 = 225$ eV (fluctuation energy), and $\ln(m/2k_0) = 7.0348$. The overall coefficient is 135.64 MHz; hence, one finds $\delta E \simeq 1040$ MHz, which is close to the experimental value of

$$1057.845\ (09)\ \text{MHz} \qquad \text{(Exp.)}.$$

However, the theoretical value can be calculated more accurately by taking into account the $2p_{1/2}$ state and including higher-order contributions. One then gets the more accurate value of

$$1057.857\ (12)\ \text{MHz} \qquad \text{(Theor.)},$$

which is in remarkable agreement with experiment. Of course, this is also obtained in the conventional theory, but not with the assurance of consistency. We can say that the above is probably the first consistent calculation of the Lamb-shift.

Another triumph of quantum electrodynamics is, of course, the anomalous magnetic moment of the electron. Its prediction is verified to a fantastic accuracy of 11 significant digits. Such numbers are sometimes used to intimidate one to accept the conventional quantum electrodynamics without reservation (but we now seem to know a little better). Feynman, both in his original article and many years later in his beautiful little book *QED*, placed great emphasis on the problems of the physical basis and consistency of quantum electrodynamics.[13] It seems that at long last these problems have found an acceptable solution based on a physical principle.

QUANTUM CHROMODYNAMICS (QCD) AND THE STANDARD MODEL

Quantum electrodynamics is a good theory. In innumerable experimental data, no deviation from it is ever observed. However, the value of quantum electrodynamics does not stop there. Perhaps its greatest value for the progress of physics has been to serve as a model for other gauge-field theories and to inspire confidence in them. As we all know, through the application of similar principles, great strides have been made in electromagnetic and weak interactions and also in the formulation of strong interactions. The standard model, which is a gauge theory model, seems to be able to explain essentially all known effects in elementary particle physics. There have even been some attempts to unify all interactions via gauge-field theory. Although such attempts did not incorporate gravity and at present are not in agreement with experiment, they indicate how strong their influence has been on the thinking of physicists. We shall not be able to go into detail on the standard model. It is a rather complicated theory, but its essence is the gauge-field theory conception. Its treatment is (despite symmetry-breaking, etc.) analogous to quantum electrodynamics. In the following, we shall also model the quantum theory of gravity on quantum electrodynamics.

QUANTUM GRAVIDYNAMICS (QGD)

Quantum gravity is formulated in close analogy to quantum electrodynamics. There are many compelling parallels between the two fields that facilitate this undertaking. For example, the conservation laws of gravity (Freud identity) in local Lorentz metric,[5]

$$\tau^{\nu}_{\mu} = \Box^2\phi^{\nu}_{\mu} - \partial_{\alpha}(\sqrt{-g}\partial^{\nu}\phi^{\alpha}_{\mu})/\sqrt{-g},$$

$$\partial_{\nu}(\sqrt{-g}\tau^{\nu}_{\mu}) \equiv 0,$$

and the electromagnetic counterparts in the same gauge, that is, the conservation of charge,

$$j^{\nu} = \Box^2 A^{\nu} - \partial_{\alpha}(\sqrt{-g}\partial^{\nu}A^{\alpha})/\sqrt{-g},$$

$$\partial_{\nu}(\sqrt{-g}\,j^{\nu}) \equiv 0,$$

are identical, except for the extra index in gravity.[13] If we let $A^{\nu} \Rightarrow \partial^{\nu}\Lambda(x)$, we get $j^{\nu} \equiv 0$. Similarly, $\phi^{\nu}_{\mu} \Rightarrow \partial^{\nu}\Lambda_{\mu}(x)$ leads to $\tau^{\nu}_{\mu} \equiv 0$: namely, the fields $\partial^{\nu}\Lambda(x)$ and $\partial^{\nu}\Lambda_{\mu}(x)$ are sourceless—hence, unobservable. They have no effect on the matrix elements.

The new theory of gravity has exact interactive N-body solutions and has exact multimode gravity waves carrying energy-momentum. These are members of the general exponential solution mentioned in the second section. For example, if only $\phi^0_0 = \phi = \phi(x, y, z)$ is present (coulomb analogue), we have the N-body slow motion limit (usually called time-independent or, inappropriately, static):

$$g_{00} = e^{-2\phi}, \qquad -g_{ik} = \delta_{ik} e^{2\phi},$$
$$\phi(x, y, z) = \sum_A m_A / |x - x_A|.$$

If the only components present are $\phi_1^1 = -\phi_2^2 = \zeta(t, z)$, then

$$-g_{11} = e^{4\zeta}, \qquad -g_{22} = e^{-4\zeta},$$
$$\zeta(t, z) = \sum_k (a e^{ikx} + a^* e^{-ikx}), \qquad \sqrt{-g} = 1$$

(other $g_{\mu\nu}$ terms are as in $\eta_{\mu\nu}$). Similarly, if the only components present are $\phi_1^2 = \phi_2^1 = \xi(t, z)$, then

$$-g_{22} = -g_{11} = \cosh 4\xi, \qquad -g_{12} = -g_{21} = \sinh 4\xi,$$
$$\xi(t, z) = \sum_k (b e^{ikx} + b^* e^{-ikx}), \qquad \sqrt{-g} = 1$$

(other $g_{\mu\nu}$ terms are as in $\eta_{\mu\nu}$). The last two solutions represent T-T (transverse traceless) gravity waves. By quantization, they become graviton emission and absorption operators.

All these solutions adhere to the following: (1) they are exact and are members of the general exponential metric; (2) they are in the local Lorentz gauge[14] (i.e., have no pseudoparts); (3) they possess nonzero field stress-energies t_μ^ν; and (4) they are sufficient for purposes of quantization. In the new theory, it is the time-independent N-body solution that is used for the laboratory and solar-system predictions and it is the two transverse traceless (T-T) gravity wave solutions (spin ± 2) that are used for calculating gravity radiation.

Given these, the construction of the Feynman diagrams is a well-defined technical procedure. One needs only the photon analogue of the normalization for emission (+) and absorption (−), that is,

$$\phi_\mu^\nu \Rightarrow \sqrt{(2/\omega)} \epsilon_\mu^\nu e^{\pm ikx}, \qquad k_0 = \omega = |\boldsymbol{k}|,$$

with $4\pi\hbar c^2 = G = 1$, and the consequent transverse graviton propagator is

$$D_{\mu\nu}^{\alpha\beta}(x - x') = \langle | T\phi(x)_\mu^\alpha \phi(x')_\nu^\beta | \rangle = (½)(\eta_\mu^\alpha \eta_\nu^\beta + \eta_\mu^\beta \eta_\nu^\alpha - \eta_{\mu\nu}\eta^{\alpha\beta}) D(x - x').$$

Here, $D(x - x')$ is the massless scalar propagator. By comparing the above with the photon propagator in the Lorentz gauge, $D_{\mu\nu}(x - x') = -\eta_{\mu\nu} D(x - x')$, it is clear that for any loop process in the local Lorentz gauge the only difference (excepting the self-interactions of gravitons, which are similar to $\phi^n/n!$ theory) is in the polarization sum (a factor of 2, with the longitudinal modes indefinite and metric, ghost, etc., all being similar). As Feynman said of quantum electrodynamics,[15] the quantum theory of gravity is "made to appear more difficult than it actually is." If we analyze the works on the same problem in general relativity,[8,16] we find that they are formally similar to the present theory. The differences are only two: (a) they try to quantize the conventional theory, which is, as we have seen, unquantizable; (b) they work within the usual Feynman loop amplitudes, which are, as we have seen, divergent and ambiguous. The value of the present approach is that both of these problems are overcome. However, most of the tree amplitudes obtained, for example, by Feynman,[8] by DeWitt,[8] by Isham, Salam, and Strathdee,[16] and also by Mandelstam,[17] can be used with some reinterpretations. Further details of the quantum theory of gravity and its relations to stellar collapse and the early universe will be discussed elsewhere.

DISCUSSION OF THE RESULTS

The following statements can now be made:

(1) Through rule 1, we find that the conventional theory of gravity is unquantizable. A new theory of gravity is then formulated by requiring quantizability. The resulting theory is identical to our theory that was found a long time ago using other physical arguments.

(2) Through rule 2, a criterion is provided for admissible theories in a geometrically based gauge principle, in the sense that connections and covariant derivatives are geometrical concepts. This rule restricts the number of admissible fields, but allows gravity.

(3) Through rule 3, a new "dynamical principle of causality" is introduced that makes all quantum field theories obeying the first two rules finite, unambiguous, and calculable to any desired order. Gravity is thereby quantized without the impediment of renormalizability. If infinities persisted, any theory of gravity, including our new theory, would be unrenormalizable.

(4) In the new theory, the chiral anomalies are absent and Ward identities are not violated. The $\pi^0 \rightarrow 2\gamma$ problem is soluble by hadron physics independently of anomaly cancellations. Any result depending on the infinities of the old theory is suspect and must be reinvestigated.

(5) The consistency of quantum field theory in general and that of QED in particular was not provable before because of the ambiguities and infinities. Now, for the first time, the consistency seems to be achieved—hence, the first consistent derivation of many results provided.

(6) Coming to gravity proper, the effects depending on the graviton as a quantum are too small for present-day laboratory experiments. They may, however, have applications in the study of the early universe, collapsing stars, and extreme high-energy elementary particles.

(7) The tree (no loop) limit of quantum gravity can be calculated and tested. It is possible to count some of these limits as tests of quantized theory because they depend for their existence on being a correspondence limit to the quantized theory. If the limit is unviable, it would certainly reflect on the covering theory.

(8) The present implementation is generally covariant in the sense that one can go to more general coordinates by transformations, but not necessarily in the sense of being set up generally covariant from the beginning. If the gauge condition $\partial_\rho \mathfrak{g}^{\rho\nu} = 0$ is relaxed, τ^ν_μ and $-t^\nu_\mu$ change by $z^\nu_\mu =$

$\partial_\alpha[\delta^\nu_\mu \partial_\rho g^{\rho\alpha} - \delta^\alpha_\mu \partial_\rho g^{\rho\nu}]/4\sqrt{-g}$. Because $\partial_\nu(\sqrt{-g}z^\nu_\mu) \equiv 0$, it can be removed or ignored. More generally, $z^\nu_\mu = (U^\nu_\mu + u^\nu_\mu)/2$, subject to $G^\nu_\mu = 0$.

(9) The new theory passes all available tests with flying colors, but it is embarrassing to say that the conventional theory can account for none of them. At the most elementary level, it is shown by Alley[18] and myself[19] that a Cavendish experiment performed with two parallel plates or two spherical balls is predicted by general relativity to have no mutual attraction. This is because, in the conventional theory, t^ν_μ is absent; hence, force $f_\mu \Rightarrow \partial_\nu(\sqrt{-g}t^\nu_\mu)$ is zero.

(10) Similarly, in general relativity, the period decay of the binary pulsar cannot be calculated consistently because of the following: (1) there are no interactive 2-body solutions,[18–20] so a bound system—hence also the mass quadrupole—cannot be formed; (2) there are no gravity waves carrying energy-momentum, so (whatever the rest) there can be no radiation; and (3) the theory is not quantizable and hence cannot be a serious contender in the first place.

In the new theory, all these problems are overcome. Energy loss by the waves is expressible by the quadrupole formula. The rate of period decay is $-dT/T \simeq (192\pi\sqrt{2/5})\sqrt{(Gm/c^2r)}/(1-e^2)^5$, which is very nearly 2.4×10^{-12} as observed. (We have used $m_1 \simeq m_2 \simeq m \simeq 1.41$ solar mass and have approximated the effect of the eccentricity.) In other words, the observed period decay is actually a prediction of the new theory. This is true also for other predictions. The N-body equation of motion of the new theory in the slow motion limit is $m_k d^2\mathbf{r}_k/dt^2 = -\Sigma'_{jk}[1 + 6\phi]m_k m_j \mathbf{r}_k/|\mathbf{r}_{jk}|^3$, where $\phi = m_1/r_1 + m_2/r_2 + \cdots + m_N/r_N$. It replaces the Einstein-Infeld-Hoffmann N-body equations of motion and gives the observed periastron advance of $\bar{\omega} \simeq 4.226°$ per year.

It may appear strange that workers in general relativity claim to get the same results. If all the things we have said are true, how can the two be the same? Very easily, as follows: (1) they implicitly assume that general relativity has interactive N-body solutions; (2) they again implicitly assume that general relativity has energy-carrying multimode gravity wave solutions; and (3) although, in general relativity, they are rendered empty by suppressing the t^ν_μ, the Lagrangians coming from the left-hand side of the field equations are formally the same. Using these implicit assumptions, which are actually true in the new theory, what results will be obtained? Of course, they will be the same as in the new theory. We describe such a situation by the statement, "physicists are smarter than their theories"—that is, the workers intuitively know what assumptions are needed and they supply them implicitly, although the theory itself cannot.

Of course, we are not trying to criticize our colleagues working in general relativity. The problem is that, with the t^ν_μ absent, the usual theory is at most a test-particle theory, which for a small mass near a large one mimics the same limit of the Newtonian N-body theory and gives the illusion that things are working out nicely. Close examination, however, shows that the two bodies (the test body and the central body) are not symmetric (one has a field and the other does not), that is, M_i (central body) $= \infty$, $M_a =$ finite $\neq M_i$, m_i (test body) $=$ finite, and $m_a = 0$; hence, the universal interparticle symmetry of gravity (principle of equivalence) is violated and therein lies the hidden flaw that invalidates the general theory of relativity for any realistic physical application.[21] It was fortunate that this was not noticed in 1915 because the theory might have been discarded and forgotten, making the present development unlikely. I would like to end the discussion with the following quote from Francis Crick:[22]

> *The hallmark of a successful theory is that it predicts facts that were not known when the theory was presented or, better still, which were then known incorrectly.*

[**Note added in proof:** A. S. Wightman of Princeton University and C. H. Woo of the University of Maryland kindly suggested that possible relations of this work to general meromorphic functions and to the method used by T. Applequist[23] and M. C. Bergère and B. J. Zuber[24] (ABZ) be considered. A necessarily incomplete investigation seems to reveal the following: (a) Both methods (ours and ABZ) are applicable to general (ordinary and transcendental) meromorphic functions, but our formula,

$$f(u) - f(u_0) - \cdots - (1/n!)(u - u_0)^n f^{(n)}(u_0) = \left(\int_{u_0}^{u} du\right)^{n+1} (\partial/\partial u)^{n+1} f(u),$$

is more general than that of ABZ,

$$f(u) - f(0) - \cdots - (1/n!)u^n f^{(n)}(0) = \int_0^1 d\xi (1/n!)(1 - \xi)^n (\partial/\partial\xi)^{n+1} f(\xi u),$$

as the latter can be obtained from the former by setting $u_0 = 0$. This is already comforting because by using our formula in the same limit and in the same way we will get the same renormalized results and there will be no contradiction, although in the general case unitarity must be investigated. (b) Our aim is, however, to bypass the complications arising from the process of renormalization (regularization, renormalization, and the subsequent need for prescriptions) and recover the original Lagrangian as a limit in a single step. To this end, we first note that in the case of Feynman diagrams one may take $(\partial/\partial a)^{n+1} f(a) = a^{-1}$, $a = m^2_{\text{av.}} - p^2_{\text{av.}} + (p_{\text{av.}})^2$; hence,

$$K(s, a, a_0) = \left(\int_{a_0}^{a} da\right)^{s+1} a^{-1}$$

as a universal function so as to write the end result directly in terms of $K(s, a, a_0)$. This is not necessarily equivalent to a renormalization procedure because we may interpret the loop contribution not as a result of a parametric integration over a prescribed loop momentum, but as a contribution due to the fluctuating momenta of emitted and reabsorbed virtual quanta. This means that the causal interpretation of how the loop contributes is changed. (c) We see here an interesting analogy to Planck's resolution of cavity divergence by changing the rule of calculation via the quantum principle. Here, we resolve the loop divergence by changing the rule of calculation via the principle of causality. What seems to be physically new is the extension of the principle of causality beyond the mere kinematical condition of commutativity for space-like separations to the dynamics of the loops and the formulation of a consistent and quantizable theory of gravity that now joins the repertoire of standard quantum field theories of physics.]

ACKNOWLEDGMENTS

Thanks are due to Carroll O. Alley for help and many illuminating discussions, to Arthur S. Wightman for discussions and advice, and to Ching H. Woo and James J. Griffin for a critical reading of the paper and valuable suggestions. Special thanks are due to Teruo Hiruma for discussions and encouragement. I also thank Ching Y. Ren for early discussions, Robert K. Barrows for computer calculations, and Abdu R. Yılmaz for computer maintenance and artwork.

REFERENCES AND NOTES

1. We use ∂_μ in the sense of $-i\partial_\mu$ on the Dirac ψ nowhere else. The sign of $\mathscr{L}$ is a matter of tradition.
2. FADDEEV, L. D. & A. A. SLAVNOV. 1989. Gauge Fields, p. 1–11. Addison–Wesley. Reading, Massachusetts.
3. SHIRKOV, D. V. 1993. *In* Renormalization. L. M. Brown, Ed.: 169–186. Springer-Verlag. New York/Berlin. Shirkov mentions the property desired, but does not provide the explicit function. This property seems to be the essence of the "renormalization group" considerations.
4. JAUCH, J. M. & F. ROHRLICH. 1989. Theory of Photons and Electrons, p. 59. Addison–Wesley. Reading, Massachusetts.
5. YILMAZ, H. 1992. Nuovo Cimento **107B:** 941; 1982. Int. J. Theor. Phys. **31:** 871. See also: YILMAZ, H. 1975. Gravitation and source theory. Ann. Phys. (N.Y.) **90:** 256. This is an early work on the quantization of gravity in the context of Schwinger's Source Theory.
6. WEYL, H. 1952. Space, Time, Matter, p. 133. Dover. New York. Weyl had an identity, but did not make use of it for a definition of the exact physical field, ϕ^ν_μ. See also p. 270 for a criticism of z^ν_μ.
7. GOCKELER, M. & T. SCHUCKLER. 1987. Differential Geometry, Gauge Theories, and Gravity, p. 71 (and similar expositions). Cambridge University Press. London/New York.
8. DEWITT, B. S. 1967. Phys. Rev. **162:** 1239; FEYNMAN, R. P. 1963. Acta Pol. **24:** 697; FEYNMAN, R. P. 1963. Lectures on Gravitation. Cal. Tech. Notes (typewritten), p. 78. See the last section. (An Addison–Wesley edition has been recently announced. To be published in March 1995.)
9. FEYNMAN, R. P. 1963. Acta Pol. **24:** 697.
10. DRAKE, S. 1981. Cause, Experiment, and Science, p. xv. University of Chicago Press. Chicago.
11. KROLL, K. & W. E. LAMB. 1949. Phys. Rev. **75:** 388. As in this paper, I was concerned more with the problem of ambiguities than infinities. The latter came as a bonus.
12. COLLINS, J. C. 1984. Renormalization, p. 53–55. Cambridge University Press. London/New York.
13. FEYNMAN, R. P. 1949. Phys. Rev. **76:** 770. Feynman always felt that the post–World War II (1947–51) work in quantum electrodynamics did not solve the divergence problem. See, for example: FEYNMAN, R. P. 1985. QED, p. 128. Princeton University Press. Princeton, New Jersey; SCHWEBER, S. 1994. QED and the Men Who Made It, p. 457. Princeton University Press. Princeton, New Jersey.
14. For any given direction, the T-T gauge may be made axial as well as local Lorentz.
15. FEYNMAN, R. P. 1961. Quantum Electrodynamics, p. 3. Addison–Wesley. Reading,

Massachusetts. Feynman pinpointed the underlying simplicity in the jungle of complexity. He used essentially only plane waves to built up the quantized theory and thus we do so here for gravity. Feynman believed in general relativity and wished to retain it as he did not realize it had fundamental deficiencies.
16. ISHAM, C. J., A. SALAM & J. STRATHDEE. 1971. Phys. Rev. D **3:** 1808.
17. MANDELSTAM, S. 1968. Phys. Rev. **175:** 1604.
18. ALLEY, C. O. 1994. Frontiers of Fundamental Physics. M. Baroni & F. Selleri, Eds. Plenum. New York.
19. YILMAZ, H. 1994. Frontiers of Fundamental Physics. M. Baroni & F. Selleri, Eds. Plenum. New York.
20. BURROWS, R. K. 1994. Master's paper. Department of Physics, University of Maryland. This thesis contains some, but not all, of the calculations conducted by C. O. Alley and myself with the help of other students and colleagues. (Note that what Einstein calls t^{ν}_{μ} is here z^{ν}_{μ}.)
21. YILMAZ, H. 1986. Marcel Grossmann Meeting, p. 1766–1790. R. Ruffini, Ed. North-Holland. Amsterdam.
22. CRICK, F. 1981. Life Itself. Simon & Schuster. New York.
23. APPLEQUIST, T. 1969. Ann. Phys. (N.Y.) **54:** 27.
24. BERGÈRE, M. C. & B. J. ZUBER. 1974. Commun. Math. Phys. **35:** 113.

APPENDIX A

How the New Theory of Gravity Works

We start with the definition of the metric in terms of its connections,

$$\partial_\lambda g_{\mu\nu} = g_{\mu\alpha}\Gamma^{\alpha}_{\nu\lambda} + g_{\alpha\nu}\Gamma^{\alpha}_{\mu\lambda},$$

and introduce the gauge potentials, $\hat{\phi} = \phi^{\nu}_{\mu}$, $\phi = \text{trace } \phi^{\nu}_{\mu}$, as[5,6]

$$\partial_\lambda \phi^{\nu}_{\mu} = (-1/4)g^{\rho\nu}[\Gamma_{\mu\rho\lambda} + \Gamma_{\rho\mu\lambda}] + (1/4)\delta^{\nu}_{\mu}\Gamma^{\alpha}_{\alpha\lambda} = (1/4)g_{\mu\rho}\partial_\lambda g^{\rho\nu}.$$

The metric becomes (at least formally) an exponential of the gauge potentials, that is,

$$g_{\mu\nu} = (\eta e^{2(\phi - \hat{\phi} - \hat{\phi}^T)})_{\mu\nu},$$

because $dg_{\mu\nu} = 2(g_{\mu\nu}d\phi - g_{\mu\alpha}\,d\phi^{\alpha}_{\nu} - g_{\alpha\nu}\,d\phi^{\alpha}_{\mu})$. The integration constants, $\phi^{\nu}_{\mu} \Rightarrow \phi^{\nu}_{\mu} - C^{\nu}_{\mu}$, are omitted, but can be used to make the kinematics locally Lorentz.

Now, consider the following three identities:[5]

$$(1/2)G^{\nu}_{\mu} \equiv U^{\nu}_{\mu} - u^{\nu}_{\mu},$$

$$D_\nu G^{\nu}_{\mu} \equiv 0,$$

$$\partial_\nu(\sqrt{-g}U^{\nu}_{\mu}) \equiv 0,$$

where D is covariant derivative. In general coordinates, the expressions of U^{ν}_{μ} and u^{ν}_{μ} are very complicated and have a pseudotensor z^{ν}_{μ}, $\partial_\nu(\sqrt{-g}z^{\nu}_{\mu}) \equiv 0$, additive to both ($U^{\nu}_{\mu} = \tau^{\nu}_{\mu} + z^{\nu}_{\mu}$ and $u^{\nu}_{\mu} = -t^{\nu}_{\mu} + z^{\nu}_{\mu}$ so that, in G^{ν}_{μ}, it cancels). However, if the coordinates are chosen locally Lorentz and harmonic (local Lorentz gauge), the expressions simplify enormously and the pseudotensor z^ν_μ vanishes. One then gets

$$\tau^\nu_\mu = \Box^2\phi^\nu_\mu - (\sqrt{-g})^{-1}\partial_\alpha(\sqrt{-g}\partial^\nu\phi^\alpha_\mu),$$

$$t^\nu_\mu = -2[\partial_\mu\phi^\alpha_\beta\partial^\nu\phi^\beta_\alpha - (\tfrac{1}{2})\delta^\nu_\mu\partial^\lambda\phi^\alpha_\beta\partial_\lambda\phi^\beta_\alpha] + \partial_\mu\phi\partial^\nu\phi - (\tfrac{1}{2})\delta^\nu_\mu\partial^\lambda\phi\partial_\lambda\phi.$$

The original equations then become[5]

$$(\tfrac{1}{2})G^\nu_\mu \equiv \tau^\nu_\mu + t^\nu_\mu,$$

$$D_\nu G^\nu_\mu \equiv 0,$$

$$\partial_\nu(\sqrt{-g}\tau^\nu_\mu) \equiv 0,$$

where all quantities entering are true tensors. Note that, when $\phi^\nu_\mu \Rightarrow 0$, they reduce to special relativity and, when $\tau^\nu_\mu = -\sigma u_\mu u^\nu$, $u_\mu u^\nu \Rightarrow u_0 u^0 \Rightarrow 1$, they reduce to the Newtonian theory of gravitation.[5] Also, the two identities have a consistency condition, $(\tfrac{1}{2})\partial_\mu g_{\alpha\beta}(\tau^{\alpha\beta} + t^{\alpha\beta}) \equiv (\sqrt{-g})^{-1}\partial_\nu(\sqrt{-g}t^\nu_\mu)$, which leads to the equations of motion, $-\sigma du_\mu/ds = (\tfrac{1}{2})\partial_\mu g_{\alpha\beta}(\tau^{\alpha\beta} + t^{\alpha\beta}) \equiv (\sqrt{-g})^{-1}\partial_\nu(\sqrt{-g}t^\nu_\mu)$, via a variation in the Lagrangian. By just looking at these results, it is hardly possible to avoid the interpretation that these equations represent a space-time theory of gravitation. Einstein's theory of gravity ($t^\nu_\mu = 0$) is unacceptable because it leads to a null acceleration, $\sigma du_\mu/ds = (\tfrac{1}{2})\partial_\mu g_{\alpha\beta}\tau^{\alpha\beta} = 0$.[5,18–20]

APPENDIX B

How the New Theory May Have Been Overlooked

All evil seems to be hiding in a simple coincidence: namely, if one writes the field equations as

$$(\tfrac{1}{2})G^\nu_\mu = \tau^\nu_\mu + \lambda t^\nu_\mu,$$

then, in first and second order, a 1-body solution with $\phi = M_a/r$ is found and it has the form,

$$g_{00} = 1 - 2\alpha\phi + 2\beta\phi, \qquad -g_{ik} = \delta_{ik}(1 + 2\gamma\phi + 2\delta\phi),$$

$$\alpha = \beta = \gamma = 1, \qquad \delta = (3 + \lambda)/4;$$

that is, only δ is affected by λ. Because the three test-particle predictions—the gravitational redshift ($\alpha = 1$), the deflection of light ($\alpha + \beta = 2$), and the advance of the perihelion of Mercury [$2\alpha(\alpha + \gamma) - \beta = 3$]—are independent of δ, one can, if everything else is all right, set $\lambda = 0$ and write the simpler (Einstein) equation, $(\tfrac{1}{2})G^\nu_\mu = \tau^\nu_\mu$, which, with his characteristic love of simplicity, is what Einstein seems to have done.

Unfortunately, everything else is not all right. The explanations above are made on the basis of a test-particle theory (one big body M in the solution plus a number of noninteracting test particles not in the solution) and this is unacceptable for several reasons:

(1) A test-body theory ($\lambda = 0$) cannot predict the 532″ per century planetary perturbative part of the total perihelion advance of Mercury because the test bodies do not interact.
(2) A test-body theory ($\lambda = 0$) predicts infinite inertial mass for the central body, which we know is wrong. Moreover, its active mass is finite—hence, violates the equality $M_a = M_i$.
(3) For $\lambda = 0$, there are no interacting ($N \geq 2$)-body solutions; thus, for two finite bodies of comparable mass (binary pulsar), we cannot form a bound system and a mass quadrupole.[5]
(4) For $\lambda = 0$, there are no gravity waves carrying energy-momentum ($t^{\nu}_{\mu} = 0$); thus, even if everything else were all right, one cannot have gravity radiation.[5]
(5) From Appendix A, we have, for $\lambda = 0$, that acceleration (force) is zero, so any two objects (parallel plates, point particles, spheres) do not attract each other.[18–20]
(6) The $\lambda = 0$ theory does not satisfy the Newtonian correspondence in the N-body interactive sense, so even the simple Cavendish experiment cannot be explained.[18,19]
(7) It is now also found that the $\lambda = 0$ theory cannot be quantized because the necessary Feynman propagator for the gravitational field cannot be constructed.

In view of all these difficulties of the conventional theory (all of which are actually the consequences of the missing t^{ν}_{μ}), it is a great relief that the $\lambda = 1$ theory has none of these deficiencies and meets all classical demands of theory and experiment successfully. In fact, it is even a greater relief that the same theory is now found to be compatible with and obtainable from the quantum theory of fields through the requirement of consistency.

Gravitational Fluctuations: Decoherence and Stochastic Jumps

J. L. SANCHEZ-GOMEZ AND J. L. ROSALES

Departamento de Física Teórica
Universidad Autónoma de Madrid
Cantoblanco, 28049 Madrid, Spain

INTRODUCTION

There are, essentially, two ways of solving the measurement problem of quantum mechanics (QM), that is, of finding a procedure that allows QM to consistently describe the measurement process that, at least in principle, belongs to the classical province. These two procedures can be called (in Stapp's[1] terminology) "Heisenberg" and "von Neumann" reductions, in the respective cases that a particular result is picked up from all outcomes potentially present in the wave function ("Heisenberg") or where the coherence between off-diagonal elements of the density matrix, in some suitable representation, is destroyed ("von Neumann"). Let us point out that the former, which is actually endorsed by the orthodox ("Copenhagen") interpretation, involves either the projection postulate (with all the metaphysics carried along with it) or the existence of a suitable stochastic mechanism that operates beyond QM, being then responsible for the quantum jumps. This is the philosophy of the stochastic models to be discussed at the end of this report. On the other hand, in the latter—"von Neumann"—one need not go beyond QM as far as a statistical (i.e., "ensemble") interpretation of QM is concerned.

In this communication, the relevant features of a recent model[2] are discussed. The fundamental assumption in the model is that the space-time metric is not strictly defined because it undergoes very tiny fluctuations. These stochastic fluctuations have a conformal nature and, in spite of their smallness, are able to produce an effective decoherence in the density matrix of the CM (center of mass) of a "macroscopic" body, without affecting at all the quantum mechanical time evolution of "microscopic" systems (i.e., they preserve microscopic unitary evolution).

The general ideas of the model are introduced in the second section, whereas the third section is devoted to analyzing some important features of it, especially from a cosmological point of view. Finally, in the fourth section, a discussion is presented about the possible connection between this model and the aforementioned stochastic reduction models.

MAIN FEATURES OF THE MODEL

As said in the INTRODUCTION, the main idea of the model[2] is the existence of universal fluctuations of the gravitational field (of the vacuum) that have a stochastic nature. Nevertheless, such fluctuations are not completely chaotic as they preserve the vacuum symmetries—the assumption is then that they are conformal. Hence, one has

$$g_{\mu\nu}(x) = e^{\phi(x)}\langle g_{\mu\nu}(x)\rangle, \tag{1}$$

with ϕ being a stochastic field ($< >$ stands for the stochastic mean and the background metric is taken as Minkowskian for simplicity).

From the above equation, one derives the following:

$$\langle e^{\phi(x)}\rangle = 1, \qquad \langle e^{\phi(x)}\phi_\mu(x)\rangle = 0, \tag{2}$$

where $\phi_\mu(x) = \partial_\mu\phi(x)$. Assuming the probability distribution to be Riemannian implies[3]

$$\langle \phi_\mu(x)\rangle = 0, \qquad \langle \phi_\mu\phi\rangle = 0. \tag{3}$$

The preceding properties enable us to think of the field ϕ as an (approximately) Gaussian stochastic process whose mean, without any loss of generality, can be set equal to zero and having a variance

$$\langle \phi^2(x)\rangle = \sigma^2. \tag{4}$$

We will also assume the following correlation function:

$$\langle \phi(\boldsymbol{x}, t), \phi(\boldsymbol{x}', t')\rangle = \sigma^2 e^{-(t-t')^2/\tau^2} e^{-(\boldsymbol{x}-\boldsymbol{x}')^2/L^2}, \tag{5}$$

where $\tau = L/c$ and L is a free (by now) parameter representing the correlation length. Note that time has been distinguished somehow from space in equation 5; this is really a classic problem in stochastic processes (one very difficult to deal with in the relativistic case). It will not be discussed here any longer; however, we should just recall that we are in fact dealing with nonrelativistic ("low velocity") processes, so equation 5 is probably an adequate expression for our current purposes.

Now, we only need to use dimensional arguments to obtain an approximate expression relating the parameters L and σ to each other:

$$\sigma^2 \sim \hbar G/(L^2c^3), \tag{6}$$

with G being the Newtonian gravitational constant. We are then left with a single free parameter: the correlation length L. It can be easily seen that $\sigma^2 \ll 1$ for all reasonable values of L, so we are in the weak field case; all gravitational effects are then properly accounted for by using a stochastic Newtonian potential:

$$V(\boldsymbol{x}, t) = (½)mc^2\phi(\boldsymbol{x}, t). \tag{7}$$

Now, we can use the well-known Feynman and Vernon's method[4] to obtain the following equation for the time evolution of the density matrix of a particle in the coordinate representation (technical details are given in reference 5):

$$\left(\frac{\partial}{\partial t}\right)\rho(\boldsymbol{x}, \boldsymbol{x}'; t) = \mathscr{L}[\rho] - g(\boldsymbol{x}, \boldsymbol{x}')\rho(\boldsymbol{x}, \boldsymbol{x}'; t), \tag{8}$$

where the operator $\mathscr{L}$ representing purely quantum (unitary) evolution is given by

(also in coordinate representation)

$$\mathscr{L}[\rho] = -\left(\frac{i\hbar}{2m}\right)[\Delta_x - \Delta_{x'}]\rho - \left(\frac{i}{\hbar}\right)[U(x, t) - U(x', t)]\rho. \tag{9}$$

Here, U is any external, "deterministic" potential (of course, for a free particle, $U = 0$) and the "decoherence part" in equation 8 is

$$g(x, x') = \left(\frac{\tau}{\hbar^2}\right)[\langle V^2\rangle - \langle V(x, t)V(x', t)\rangle]. \tag{10}$$

To derive equation 10, one has to assume that the correlation time τ is much smaller than the relevant quantum times of the system. As we have already commented, and as will be seen in more detail below, L should lie in the range of 10^{-5}–10^{-3} cm. Now, if, for definiteness, we take $L = 10^{-4}$ cm, then $\tau \simeq 3 \times 10^{-15}$ s; in contrast, the corresponding quantum time is $\simeq mL^2/\hbar$, which for a proton, say, is $\simeq 10^{-9}$ s. Hence, the above condition clearly holds.

An explicit expression for g can be obtained by inserting the correlation function given in equation 5 into equation 10 (taking equation 6 into account). In order to obtain simple—but accurate enough—expressions, we approximate the space part as follows:

$$\begin{array}{ll} \sigma^2[1 - (x - x')^2/L^2], & |x - x'| < L \\ 0, & |x - x'| > L. \end{array} \tag{11}$$

This must be a good approximation, except, perhaps, for $|x - x'| \simeq L$.

By using equation 11, equation 8 can be written as

$$\begin{aligned} \left(\frac{\partial}{\partial t}\right)\rho(x, x'; t) &= \mathscr{L}[\rho] - \left(\frac{Gm^2}{4\hbar L^3}\right)(x - x')^2\rho(x, x'; t), \qquad |x - x'| < L \\ &= \mathscr{L}[\rho] - \left(\frac{Gm^2}{4\hbar L}\right)\rho(x, x'; t), \qquad |x - x'| > L. \end{aligned} \tag{12}$$

One can see that the nonunitary term in equation 8 is negligible in the "microscopic" case. For instance, with $L = 10^{-4}$ cm and by taking such a system to be a proton, we would obtain

$$\lambda_m = \frac{Gm^2}{4\hbar L^3} \simeq 10^{-17}\ \text{cm}^{-2}\ \text{s}^{-1}. \tag{13}$$

Hence, the relative coherence between parts of the proton wave packet that are separated by up to 1 cm would survive up to $\sim 10^{24}$ s. This means that the time evolution is practically unitary (quantum). Let us now go to the more interesting, in the present context, macroscopic case.

First of all, it should be pointed out that this work will be concerned with macroscopic systems with "solid" structure; that is, the relative distance between any pair of constituents will not be changed by the quantum fluctuation; also—and this is for the sake of technical simplicity—the corresponding body is taken to be homogeneous and spherical. Now, the treatment of the microscopic case can be straightforwardly extended to the present one simply by noting that the stochastic potential will be given by

$$V(\boldsymbol{x}_1, \ldots, \boldsymbol{x}_N; t) = (½)mc^2 \sum_i \phi(\boldsymbol{x}_i, t). \tag{14}$$

In the "very" macroscopic case, that is, when $R \gg L$, with R being the radius of the body, the CM can be factored out; then, after a long, but easy, calculation, one obtains the following equation for the density matrix representing the CM quantum state, ρ_0:

$$\left(\frac{\partial}{\partial t}\right) \rho_0(\boldsymbol{x}_0, \boldsymbol{x}_0'; t) = \begin{cases} \mathscr{L}[\rho_0] - \lambda_0(\boldsymbol{x}_0 - \boldsymbol{x}_0')^2\rho_0(\boldsymbol{x}_0, \boldsymbol{x}_0'; t), & |\boldsymbol{x}_0 - \boldsymbol{x}_0'| < L \\ \mathscr{L}[\rho_0] - \gamma_0\rho_0(\boldsymbol{x}_0, \boldsymbol{x}_0'; t), & |\boldsymbol{x}_0 - \boldsymbol{x}_0'| > L \end{cases} \tag{15}$$

where

$$\lambda_0 \simeq GMD/\hbar, \qquad \gamma_0 = \lambda_0 L^2, \tag{16}$$

with M and D being, respectively, the mass and density of the body. Now, equation 15 effectively shows the decoherence between the off-diagonal terms of the CM density matrix of a macroscopic system.

Consider, for instance, a body of $M = 1$ g and $D = 1$ g/cm^3. From equations 15 and 16, we see that correlations between macroscopic quantum states separated by more than L (10^{-4} cm, say) would decay in about 10^{-12} s. Notice that this decoherence is produced just by stochastic fluctuations of the gravitational vacuum, which are supposed to be universal, thereby in some sense giving rise to an "effective" environment for any body, even in the case of it being ideally isolated.

FEATURES OF THE CONFORMAL STOCHASTIC FLUCTUATIONS

First, we shall discuss the energy conservation. Obviously, the nonunitary term in equation 15 is responsible for the change with time of the energy of the system (even if it is isolated). In the present model, such a change is negligible; nevertheless, it has the unpleasant feature of always being positive. From equation 15, one easily obtains

$$\frac{d}{dt}\langle E_{\mathrm{CM}}\rangle = 2GD\hbar \simeq 10^{-41}\ \mathrm{W}, \tag{17}$$

which is independent of the mass.

Such a constant increase of the energy can be understood by taking into account that the vacuum fluctuations "shake" the particle in a nondissipative way, with the particle's reaction on the field being disregarded. Precisely this reaction could be rather relevant in the "mesoscopic" case; however, its treatment in a correct mathematical way is quite complicated (some work has already been reported[6] and a more refined treatment is in progress). We think the mesoscopic case is worth dealing with because some kind of behavior is expected in between the quantum and classical ones, which perhaps could give rise to some experimental tests of the present model (notice that the model predicts the aforementioned mesoscopic region to lie within the range of 10^{-16} g $< M < 10^{-8}$ g).

Let us now discuss some relevant cosmological points that could affect the consistency of this model.

It has been pointed out by Diósi and Lukács[7] that the so-called K-model[8] has the unpleasant feature (perhaps critical for the reliability of that model) of predicting a very large mean value of the space-time curvature—such a curvature would imply a mean energy density of the universe much larger than that of a neutron star. This unpleasant fact does not occur in the present model due mainly to the assumed conformal nature of the stochastic metric fluctuations. It is easy to see that the mean value of the curvature is given by ($c = 1$)

$$\tilde{R} = (-\tfrac{1}{2})\langle \exp(-2\phi)\{6\Box\phi + 3[(\partial\phi/\partial t)^2 - (\nabla\phi)^2]\}\rangle. \tag{18}$$

Now, conformality plus the form of the correlation function (equation 5) implies

$$\langle \Box\phi \rangle = 0,$$

$$\langle (\partial\phi/\partial t)^2 \rangle = \langle (\nabla\phi)^2 \rangle = 4\sigma^2/L^2,$$

so $\tilde{R} = 0 + O(\phi^4)$ (recall that the background metric is Minkowskian). No problem then appears in the present model, in regard to the mean curvature of the universe. One can also compute the 00-component of the energy-momentum tensor, which of course is not a Minkowski scalar. Note, however, that the relative *dynamic* velocity (i.e., cosmic expansion not included) between any pair of *physically relevant* systems seems to be, as a matter of fact, much smaller than the velocity of light; hence, one is (probably) allowed to think that $<T_{00}>$ [computed in the reference frame (ether?) wherein equation 5 holds] gives a reliable estimate of the mean energy density of the universe stored in the fluctuating vacuum. By using standard techniques, one obtains

$$\langle T_{00}/c^2 \rangle = c^2\sigma^2/(32\pi G L^2). \tag{19}$$

If it is now imposed that $<T_{00}>/c^2$ be less than 10^{-29} g/cm^3 (as it seems reasonable that vacuum fluctuations alone cannot make the universe closed) and if, according to equation 6, one takes $\sigma = \Lambda/L$, where Λ is the Planck length, one can obtain a lower bound to L: $L > 10^{-3}$ cm. Of course, this is a very rough bound, but it should give a hint (at least) of the order of magnitude of the parameter L.

Finally, as to the physical origin of the hypothetical metric fluctuations introduced here, very little can be said at present. Perhaps one should relate such fluctuations to primordial gravity waves, originating at the end of the "Planck era"; however, as we do not have a consistent theory of quantum gravity, this is at present a mere speculation.

GRAVITATIONAL FLUCTUATIONS AND STOCHASTIC REDUCTION

As a good example of a fairly elaborate stochastic reduction model (from a mathematical point of view, at least), one should mention the CSL (Continuous Spontaneous Localization) model of Ghirardi, Pearle, and Rimini.[9] In this model, one assumes a stochastic evolution equation of the Ito form, which, after one imposes the conservation of the norm, reads

$$d|\phi\rangle = [-(1/2)\gamma(\boldsymbol{A} - \boldsymbol{R})^2 dt + (\boldsymbol{A} - \boldsymbol{R})d\boldsymbol{B}]|\phi\rangle, \tag{20}$$

where $\boldsymbol{A}$ is a set of commuting self-adjoint operators (the non-Schrödinger terms in equations 21a and 21b induce, for large times, the reduction of the state vector on the common eigenspaces of the operators $\boldsymbol{A}$), $\boldsymbol{R} = <\phi|\boldsymbol{A}|\phi>$, and $d\boldsymbol{B}$ is a (vector) Wiener process such that

$$\overline{dB_i} = 0, \tag{21a}$$

$$\overline{dB_i dB_j} = \delta_{ij}\gamma dt. \tag{21b}$$

In fact, the modification of "orthodox" QM represented by equations 21a and 21b consists of superimposing a Markov process in Hilbert space upon the ordinary Schrödinger (i.e., Hamiltonian) evolution, giving rise[9] to a continuous spontaneous localization of the wave function. Furthermore, as also shown in reference 9, decoherence in the density matrix of "classical" systems is obtained. (However, we stress that the reduction obtained in models of this sort is certainly more "complete" than that of the simple decoherence models.) Now, the question is how to obtain a sensible physical process that could give rise to the aforementioned Markov process. When one analyzes the CSL model, he or she realizes that there are, in principle, two stochastic processes in it, one linear and one nonlinear. The linear one, to some extent, can be understood in terms of metric fluctuations;[10] however, this is not so in the case of the nonlinear terms. Our conjecture[10] is that perhaps the nonlinearity apparently present once the particle's gravitational field is taken into account[11] could have something to do with such a strange stochastic term. Thus, we are studying such a possibility at present.

REFERENCES

1. STAPP, H. P. 1992. Phys. Rev. **A46:** 78.
2. SANCHEZ-GOMEZ, J. L. 1993. *In* Fundamental Questions in Quantum Physics and Relativity. F. Selleri, Ed. Hadronic Press. Palm Harbor, Florida.
3. ROY, S. & M. SINHA. 1980. Einstein equation for the gravitational field and stochastic fluctuations (Calcutta preprint); see also: ROSEN, G. 1980. Nuovo Cimento **57B:** 125.
4. FEYNMAN, R. P. & F. L. VERNON. 1963. Ann. Phys. **24:** 118.
5. GALLIS, M. R. 1992. Phys. Rev. **A45:** 47.
6. UNTURBE, J. & J. L. SANCHEZ-GOMEZ. 1992. Nuovo Cimento **107B:** 211.
7. DIÓSI, L. & B. LUKÁCS. 1993. Budapest preprint.
8. KÁROLYHÁZY, F. 1990. *In* Sixty-two Years of Uncertainty (and references therein). A. Miller, Ed. Plenum. New York.
9. GHIRARDI, G. C., P. PEARLE & A. RIMINI. 1990. Phys. Rev. **A42:** 78.
10. ROSALES, J. L. & J. L. SANCHEZ-GOMEZ. 1995. To be published.
11. ROSALES, J. L. & J. L. SANCHEZ-GOMEZ. 1992. Phys. Lett. **A166:** 111.

Spontaneous Wave Packet Reduction[a]

GIANCARLO GHIRARDI

Department of Theoretical Physics
University of Trieste
and
International Center for Theoretical Physics
Trieste, Italy

INTRODUCTION

With his research activity, with his deep investigations, and with his stimulating suggestions of new experiments, J. A. Wheeler has given extremely relevant contributions to the debate on the foundational problems of quantum mechanics. Some of his penetrating remarks focusing in a lucid and concise way on the challenge that quantum mechanics represents for scientific enterprise constitute an ideal starting point to elucidate the motivations for the so-called spontaneous localization program.

Let us consider one of his preferred sentences:[1] "No elementary phenomenon is a phenomenon until it is a registered (observed) phenomenon." This phrase, which concludes the presentation of the delayed choice experiments, is accompanied by the following remarks: "It makes no sense to talk of the phenomenon until it has been brought to a close by an irreversible act of amplification." The argument acquires a greater strength when combined with the brilliant and picturesque way of stressing the enigma we have to face by resorting[3] to the image of the "Great Smoky Dragon", with a sharply defined tail and mouth and a fog of uncertainty in between.

We consider it appropriate to compare the above sentences with the desiderata put forward[4] by J. S. Bell for an "exact theory": ". . . it should allow electrons to enjoy the cloudiness of waves, while allowing tables and chairs, and ourselves, and black marks on photographs, to be rather definitely in one place rather than another and to be described in 'classical terms'." The cloudiness of waves parallels the fog of uncertainty of the Smoky Dragon and the definiteness, the particularity of the world of experience, parallels the sharp definiteness of its tail and mouth, which is due to the irreversible acts of amplification involved in the preparation and detection procedures.

With reference to the above, we consider it appropriate to stress that the fundamental attitude behind the spontaneous localization approach that we are going to analyze in this report (contrary to the position taken by, for example, the supporters of hidden-variable theories) is to accept that, in general, microphenomena are basically "foggy" and "cloudy", demanding however that such fogginess disappears as a consequence of specific events whose occurrence and effects are clearly identified and described in a mathematically precise way.

Although it is very seldom mentioned, we would like to recall that even Einstein

[a] This work has been supported in part by the Sezione di Trieste of the INFN.

was, in a sense, prepared to accept the foggy nature of microscopic phenomena. In his "Reply to Criticisms",[5] he remarks that, in the case of the decay of a radioactive atom, the orthodox quantum theorist would claim that "it is not reasonable even to posit the existence of a definite point of time for the transformation of a single atom" (i.e., the microworld is cloudy). He states clearly that such a position is sensible. However, he also stresses that Schrödinger taught us that the question about the time of decay can be straightforwardly transformed (by simply coupling the atom to an appropriate apparatus) into the question of the presence of a macroscopic registration mark on a paper strip. He states, "the location of the mark is a fact which belongs entirely within the sphere of macroscopic concepts. . .", and there is hardly likely to be anyone inclined to accept that the existence of that location is essentially dependent upon the carrying out of an observation. In fact, "in the macroscopic sphere it is simply considered certain that one must adhere to the program of a realistic description in space and time [i.e., the macroworld is definite], whereas in the sphere of microscopic situations one is more readily inclined to give up or at least to modify this program [!]." Some pages later, he comes back to this point and stresses that his realistic attitude does not derive from a philosophical prejudice: "The real in physics is to be taken as a type of program, to which we are, however, not forced to cling a priori. No one is likely to be inclined to attempt to give up this program within the realm of the 'macroscopic' (location of the mark on the paper strip 'real')." Then, he draws his conclusion: "But the 'macroscopic' and the 'microscopic' are so interrelated that it appears impracticable to give up this program in the realm of the 'microscopic' alone."

We have chosen to devote so much space to this quoted excerpt to call attention to the fact that, within the context of our analysis, the crucial problem of our best theory is to succeed in accounting for the smooth merging of the cloudy microworld with the definite macroworld—stated differently, to make precise where and how acts occur that can be unambiguously identified as "irreversible acts of amplification".[2] Quite appropriately, in our opinion, Wheeler held, with Bohr, that locating the shifty split cannot be related to acts of empirical assertions. On the contrary, the occurrence of the irreversible acts of amplification must be part of the physical description and one should not ascribe to them any exceptional position above the rest.

Nowadays, this position is shared by almost all scientists seriously involved in the foundational problems of quantum mechanics. Many interesting attempts along this line have appeared recently, ranging from detailed critical investigations about the possibility of a deterministic completion of the theory, to the consideration of the so-called environment-induced superselection rules, to the interesting features of the so-called Quantum Histories Approach. All such lines, as well as many others deserving attention, are represented in this conference; we refer the reader to the pertinent papers in these proceedings for a general view of the status of the matter. From now on, we deal with the line of research that is the subject of this report.

The Spontaneous Wave Packet Reduction program represents an attempt to answer the puzzling questions raised above by identifying in a precise way within a strictly quantum context (i.e., by assuming completeness of the Hilbert space description of the states of individual physical systems) the shifty split between micro and macro, and between reversible and irreversible. The very existence of precise dynamical reduction models shows that, contrary to Einstein's expectations, the program of building up a consistent and unified theory allowing the microsystems to be foggy and nevertheless implying the definiteness of the macroscopic world is viable. It is important to stress that the emergence of such definiteness is implied and precisely described by the formalism so that, to account for it, there is no need to invoke (as many other attempts do) the practical impossibility of revealing interference effects at the macroscopic level. We leave it to the reader to evaluate the potential relevance of the approach and to identify its advantages and drawbacks with respect to the many interesting recent proposals. At any rate, it seems to us that having the explicit proof that a theory exhibiting such features is possible has, by itself, a certain relevance for the investigations about the foundations of quantum mechanics.

GENERAL REMARKS ABOUT SPONTANEOUS LOCALIZATIONS

As stressed above, the motivation for this line of research derives from the desire of forbidding the embarrassing superpositions of macroscopically different states without requiring the consideration of measuring apparatuses and, more generally, macroscopic objects as peculiar systems differing from all other physical systems. Because the most characteristic differences between standard quantum evolution and wave packet reduction consist of the fact that the first is linear and deterministic, whereas the second is nonlinear and stochastic, one entertains the idea of modifying the standard quantum dynamics by adding stochastic and nonlinear terms to the evolution equation. Such modifications are assumed to describe universal mechanisms governing all physical processes.

Several attempts have been made,[6–10] but they did not lead (up to recent times) to a real breakthrough due to some crucial problems that had been left unsolved. The first one is that of the choice of the preferred basis: which specific properties of individual physical systems should one require to be dynamically and spontaneously objectified? Second, how can the dynamical modifications satisfy the two diverging desiderata of having a practically negligible effect for all microsystems (a necessary requirement due to the extremely high degree of accuracy of tested predictions of quantum theory) and simultaneously being able to induce an extremely rapid suppression (the amplification mechanism) of the superpositions of macroscopically distinguishable states? The solution[11,12] came from the identification of the appropriate preferred basis, that is, the one associated with positions.

Before proceeding, with reference to the celebrated sentence by Wheeler quoted previously and calling attention to the crucial role of the irreversible act of amplification in making it legitimate to speak of a phenomenon, we point out that the stochastic and nonlinear features of the theory introduce irreversibility at the very fundamental level. Thus, as we will see, "irreversible acts" making an elementary phenomenon become a phenomenon can occur and actually do occur (even with extremely low probability) already at the genuinely microscopic level. To elucidate this point, let us consider the famous cosmological example[2] of the gravitational lens effect pointed out by Wheeler. If the elementary particle coming from the quasar and following two faraway routes would be a massive particle rather than a photon, then, due to the extremely long time (billions of years) that it has to spend traveling, it would surely suffer at least one spontaneous localization process (see below) before reaching us, thus being compelled "to choose a specific route". However, it is appropriate to remark that within the model, although a microphenomenon takes hundreds of millions of years to become a phenomenon in Wheeler's sense, the elementary and very tiny irreversibility built in it is such as to imply a tremendous "amplification" when such a microscopic system triggers macroscopic changes. Thus, irreversibility and amplification are two distinct elements of the formalism, but they combine just in such a way as to make appropriate Wheeler's statements about microphenomena and his sharing of Bohr's view that the central point is not the observer's consciousness, but the experimental device, bringing the phenomenon to a close. What is relevant about the models under discussion is that, although to account for this process standard quantum mechanics has to embody the (inconsistent) postulate of wave packet reduction, within such models the universal dynamical laws governing all natural processes do the desired job.

A CONCISE REVIEW OF DYNAMICAL REDUCTION MODELS

As already stated, a satisfactory elaboration of the spontaneous localization program at the nonrelativistic level requires various steps.[11–15] In order to grasp the conceptually relevant points as well as to understand precisely how the new dynamics works, it is useful to start by discussing the simplest model of this type: Quantum Mechanics with Spontaneous Localization (QMSL).

The QMSL Model

The first model of spontaneous reduction,[11,12] QMSL, is based on the assumption that, besides the standard evolution, physical systems are subjected to spontaneous localizations occurring at random times and affecting their elementary constituents. Such processes, which we will call "hittings", are formally described in the following way. When the i-th constituent of the system suffers a hitting, the wave function changes according to

$$\Psi(\boldsymbol{r}_1, \ldots, \boldsymbol{r}_N) \rightarrow \Psi_{\boldsymbol{x}}(\boldsymbol{r}_1, \ldots, \boldsymbol{r}_N) = \Phi_{\boldsymbol{x}}(\boldsymbol{r}_1, \ldots, \boldsymbol{r}_N)/\|\Phi_{\boldsymbol{x}}\|,$$

$$\Phi_{\boldsymbol{x}}(\boldsymbol{r}_1, \ldots, \boldsymbol{r}_N) = (\alpha/\pi)^{3/4} \exp[-(\alpha/2)(\boldsymbol{r}_i - \boldsymbol{x})^2]\Psi(\boldsymbol{r}_1, \ldots, \boldsymbol{r}_N). \tag{3.1}$$

The localization processes occur at randomly distributed times with a mean frequency of $\lambda = 10^{-16}\ \mathrm{s}^{-1}$. The probability density of the process occurring at point $\boldsymbol{x}$ is given by $\|\Phi_{\boldsymbol{x}}\|^2$. The localization parameter $1/\sqrt{\alpha}$ is assumed to take the value of 10^{-5} cm.

To understand the physical relevance of the hitting processes as well as the reasons for which QMSL meets the requirements imposed on it, let us consider first the case of a single particle in one dimension in the superposition of two faraway states (see FIGURES 1a–c). The separation between the two regions L and R in which the configuration space wave function is appreciably different from zero is assumed

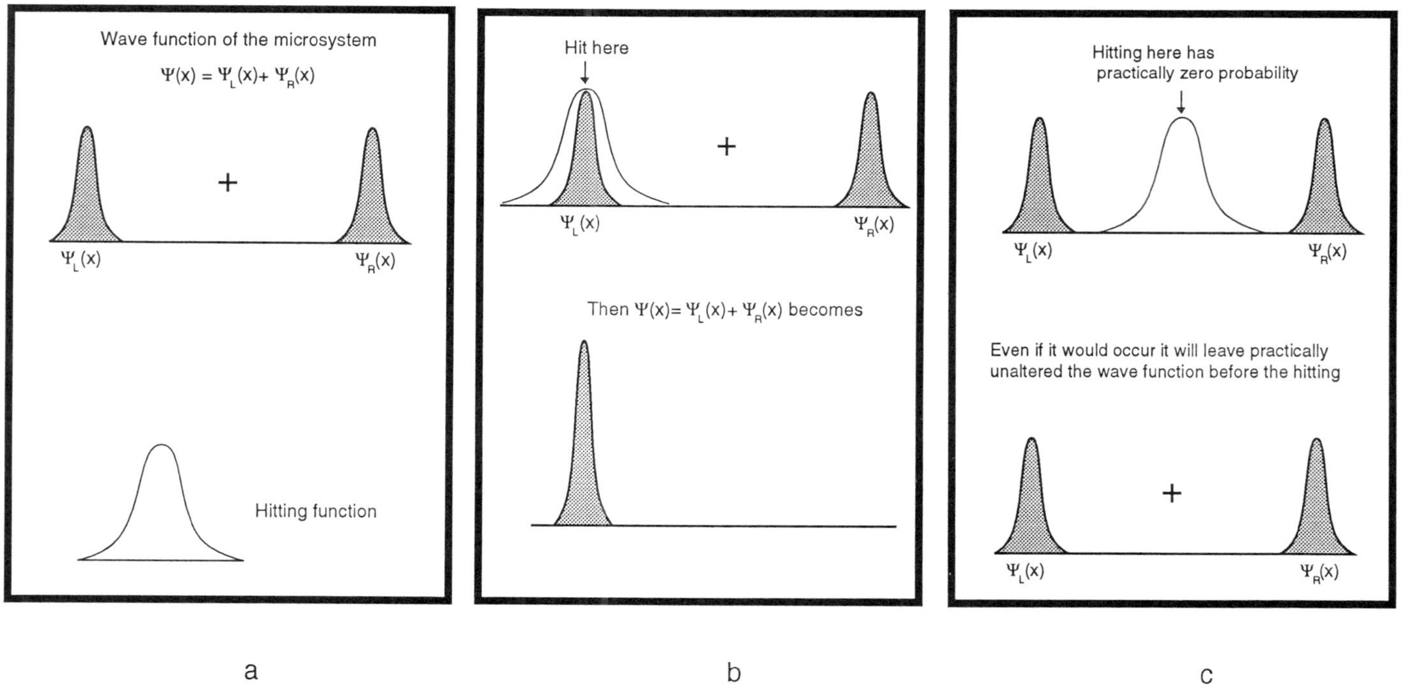

FIGURE 1. A typical spontaneous localization process of the QMSL model for a microsystem in a superposition of faraway states: (a) initial state and hitting function; (b) a localization around L leads to the suppression of the R component of the state vector; (c) a very improbable localization has practically no effect.

to be much larger than the characteristic localization parameter $1/\sqrt{\alpha}$. In FIGURE 1b, we have supposed the localization to occur[b] around the point L and we have shown how such a process leads to a state that is well localized around L. This is an example of the irreversible acts that take place according to the model. Obviously, in the case of a single elementary particle, even when the separation L-R is of thousands of light-years, a spontaneous localization has an appreciable probability of occurring only after about 10^8 years. At any rate, if the process occurs, the microsystem has to choose in which space region it is.

To understand why all those measurementlike processes that we are compelled to recognize as occurring almost all the time, almost everywhere in the universe, lead to the definiteness of the world of our experience, that is, to allow the reader to grasp the basic role of the amplification mechanism, we have considered, in FIGURE 2, the

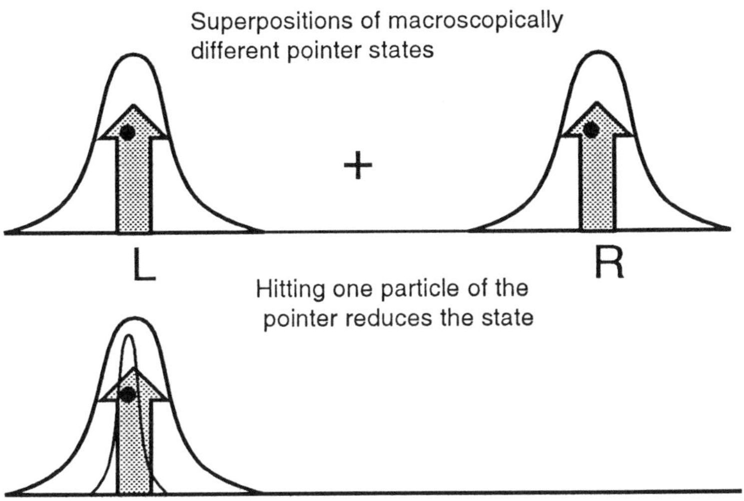

FIGURE 2. The amplification mechanism of the QMSL model: any localization of one of the constituents of the pointer amounts to a localization of the whole pointer.

case of the superposition of two faraway states of a macroscopic pointer. If such a state would occur, as is evident both from equations 3.1 and from the figure, a localization process affecting just one of the particles of the pointer would lead to a suppression of the linear superposition. In fact, in the state corresponding to the pointer being around L, all its constituent particles are around L, whereas, in the state at R, they are all around R. Hitting one of the particles at left means to multiply the whole wave function by a Gaussian centered around L in the position variable of the particle suffering the localization. It is obvious that, after normalization of such a

[b]We have also considered, in FIGURE 1c, the effect of the occurrence of one of the extremely improbable localizations, that is, at a point where the wave function practically vanishes. One sees from the figure (as one can easily understand from the precise rules governing localization processes) that such a process, even if it occurs, leaves the state practically unaffected.

function, the R term in the superposition (practically) disappears. One then simply remembers that the pointer contains about 10^{24} particles, so one of them will suffer a hitting in about 10^{-7} s. This is the way in which the elementary irreversible processes are amplified whenever a measurementlike process takes place.

The QMSL mechanism does not respect the symmetry properties of the wave function in the case of identical constituents. Its generalization satisfying such a requirement, the Continuous Spontaneous Localization model (CSL), has been presented and discussed in various papers.[13–15]

The CSL Model

The model is based on a linear stochastic evolution equation for the state vector. The evolution does not preserve the norm, but only the average value of the square norm. The equation, in the Stratonovich version, is

$$\frac{d\,|\Psi_w(t)\rangle}{dt} = \left[-\left(\frac{i}{\hbar}\right)H + \sum_i A_i w_i(t) - \gamma \sum_i A_i^2\right]|\Psi_w(t)\rangle. \tag{3.2}$$

In equation 3.2, the quantities A_i are commuting self-adjoint operators, whereas the quantities $w_i(t)$ are c-number stochastic processes with a probability of occurrence satisfying

$$P_{\text{Cook}}[w(t)] = P_{\text{Raw}}[w(t)]\,|\,\|\Psi_w(t)\rangle\|^2. \tag{3.3}$$

Here, $P_{\text{Raw}}[w(t)]$ is equal to

$$P_{\text{Raw}}[w(t)] = \left(\frac{1}{\mathcal{N}}\right)e^{(-1/2\gamma)\Sigma_i\int_0^t d\tau w_i^2(\tau)} \tag{3.4}$$

($\mathcal{N}$ being a normalization factor), that is, equal to the probability density of a white-noise process satisfying

$$\langle\langle w_i(t)\rangle\rangle = 0, \qquad \langle\langle w_i(t)w_j(t')\rangle\rangle = \gamma\delta_{ij}\delta(t-t'). \tag{3.5}$$

To clarify the physical meaning of the model, let us assume, for the moment, that the operators A_i have a purely discrete spectrum and let us denote by P_σ the projection operators on their common eigenmanifolds.

Now, we make the following precise assumption: if a homogeneous ensemble (pure case) at the initial time $t = 0$ is associated with the state vector $|\Psi(0)\rangle$, then the ensemble at time t is the union of homogeneous ensembles associated with the normalized vectors $|\Psi_w(t)\rangle / \,|\,\|\Psi_w(t)\rangle\|$, where $|\Psi_w(t)\rangle$ is the solution of equation 3.2 with the assigned initial conditions and for the specific stochastic process $w(\tau)$ that occurred in the interval $(0, t)$. The probability density for such a subensemble is that given by equation 3.3.

One can prove[14,15] that the map from the initial ensemble to the final ensemble obeys the forward-time translation semigroup composition law (once more, we see that irreversibility is built in the model from the very beginning). It is also easy to prove that the evolution, at the ensemble level, is governed by the dynamical equation for the statistical operator:

$$\frac{d\rho(t)}{dt} = \left(-\frac{i}{\hbar}\right)[H, \rho(t)] + \gamma \sum_i A_i\rho(t)A_i - \left(\frac{\gamma}{2}\right)\left\{\sum_i A_i^2, \rho(t)\right\}; \tag{3.6}$$

if one disregards the Hamiltonian evolution, one immediately sees that the off-diagonal elements $P_\sigma\rho(t)P_\tau$ ($\sigma \neq \tau$) are exponentially damped.

For our concerns, the relevant feature of the dynamical process (equation 3.2) with the prescription (equation 3.3) is that it drives the state vector of each individual member of the ensemble into one of the common eigenmanifolds of the operators A_i, with the appropriate probability. To clarify this, we consider a simplified case[13] in which only one operator A appears in equation 3.2. The solution of this equation corresponding to the particular initial condition (involving only two eigenmanifolds of A with eigenvalues α, β),

$$|\Psi(0)\rangle = P_\alpha|\Psi(0)\rangle + P_\beta|\Psi(0)\rangle, \tag{3.7}$$

when the Hamiltonian is disregarded,[c] is

$$|\Psi_B(t)\rangle = e^{\alpha B(t) - \alpha^2\gamma t}P_\alpha|\Psi(0)\rangle + e^{\beta B(t) - \beta^2\gamma t}P_\beta|\Psi(0)\rangle. \tag{3.8}$$

Here, $B(t)$ is the Brownian process:

$$B(t) = \int_0^t d\tau w(\tau). \tag{3.9}$$

Taking into account equation 3.8 and the "cooking" prescription, one gets the "cooked" probability density for the value $B(t)$ of the Brownian process at time t:

$$P_{\text{Cook}}[B(t)] = \|P_\alpha|\Psi(0)\rangle\|^2\left(\frac{1}{\sqrt{2\pi\gamma t}}\right)e^{(-1/2\gamma t)[B(t)-2\alpha\gamma t]^2} + \|P_\beta|\Psi(0)\rangle\|^2\left(\frac{1}{\sqrt{2\pi\gamma t}}\right)e^{(-1/2\gamma t)[B(t)-2\beta\gamma t]^2}. \tag{3.10}$$

From equation 3.10, it is evident that, for $t \to \infty$, the Brownian process $B(t)$ can assume only values belonging to an interval[d] of width $\sqrt{\gamma t}$ around either the value $2\alpha\gamma t$ or the value $2\beta\gamma t$. The corresponding probabilities are $\|P_\alpha|\Psi(0)\rangle\|^2$ and $\|P_\beta|\Psi(0)\rangle\|^2$, respectively. The occurrence of a value "near" to $2\alpha\gamma t$ for the random variable $B(t)$ leads (according to equation 3.8) to a state vector that, for $t \to \infty$, lies in

[c]In equation 3.8 and the following, we have changed the notation for the state vector from the one labeled by the white-noise symbol w as in equation 3.2 to the one labeled by the Brownian motion symbol B to stress the fact that, under our assumptions, the state at time t does not depend on the specific sample function $w(\tau)$ in the interval $(0, t)$, but only on its integral (equation 3.9).

[d]Even though the spread $\sqrt{\gamma t}$ tends to ∞ for $t \to \infty$, its ratio to the distance $2(\alpha - \beta)\gamma t$ between the two considered peaks of the distribution tends to zero.

the eigenmanifold corresponding to the eigenvalue α of A. In fact, one gets

$$\frac{\|P_\beta|\Psi_B(t)\rangle\|^2}{\|P_\alpha|\Psi_B(t)\rangle\|^2} \cong e^{-2\gamma t(\alpha-\beta)^2}\left[\frac{\|P_\beta|\Psi(0)\rangle\|^2}{\|P_\alpha|\Psi(0)\rangle\|^2}\right] \underset{t\to\infty}{\to} 0. \tag{3.11}$$

Analogously, when the random variable $B(t)$ takes a value "near" to $2\beta\gamma t$, for $t \to \infty$, the state vector is driven into the eigenmanifold corresponding to the eigenvalue β of A.

It is then clear that the model establishes a one-to-one correspondence between the "outcome" (the final "preferred" eigenmanifold into which an individual state vector is driven) and the specific value (among the only ones having an appreciable probability) taken by $B(t)$ for $t \to \infty$, a correspondence irrespective of what $|\Psi(0)\rangle$ is.[e] In the general case of several operators A_i, a similar conclusion holds for the "outcomes" α_i of A_i and the corresponding Brownian processes $B_i(t)$.

This concludes the exposition of the general structure of the CSL model. Obviously, to give a physical content to the theory, one must choose the so-called preferred basis, that is, the eigenmanifolds on which reduction takes place or, equivalently, the set of commuting operators A_i. The specific form that has been presented and shown to possess all the desired features is obtained[13–15] by identifying the discrete index i and the operators A_i of the above formulae with the continuous and discrete indices $(\boldsymbol{r}, k)$ and the operators

$$N^{(k)}(\boldsymbol{r}) = \left(\frac{\alpha}{2\pi}\right)^{3/2} \sum_s \int d\boldsymbol{q} e^{(-\alpha/2)(\boldsymbol{q}-\boldsymbol{r})^2} a_k^+(\boldsymbol{q}, s) a_k(\boldsymbol{q}, s). \tag{3.12}$$

Here, $a_k^+(\boldsymbol{q}, s)$ and $a_k(\boldsymbol{q}, s)$ are the creation and annihilation operators of a particle of type k (e.g., k = electron, proton, . . .) at point $\boldsymbol{q}$ with spin component s, satisfying the canonical commutation or anticommutation relations. Correspondingly, one has a continuous family of stochastic Gaussian processes satisfying

$$\langle\langle w_k(\boldsymbol{r}, t)\rangle\rangle = 0, \qquad \langle\langle w_k(\boldsymbol{r}, t) w_j(\boldsymbol{r}', t')\rangle\rangle = \gamma\delta_{kj}\delta(\boldsymbol{r} - \boldsymbol{r}')\delta(t - t'). \tag{3.13}$$

The parameter α is assumed to take the same value (10^{10} cm^{-2}) as in the case of QMSL, whereas γ is related to the frequency $\lambda = 10^{-16}$ s^{-1} of that model according to $\gamma = \lambda(4\pi/\alpha)^{3/2}$.

How Does Dynamical Reduction Work?

Due to the choice of the parameters for QMSL and the corresponding ones for CSL, the considered dynamics has the following nice features:[15]

(1) In the case of microscopic systems, the non-Hamiltonian terms have negligible effects.

(2) On the other hand, in the macroscopic case, the reduction mechanism is extremely effective in suppressing linear superpositions of states in which a macroscopic number of particles are displaced by more than the characteris-

[e] Obviously, $|\Psi(0)\rangle$ plays a crucial role in determining the probability of occurrence of the Brownian processes $B(t)$.

tic localization length. Such a suppression occurs at the individual level, so any individual macroscopic system acquires[16] definite macroscopic properties in a split second.

These features have already been analyzed in great detail with reference to the QMSL model. To discuss the decoherence properties of CSL ensuing from the choice (equation 3.12), even though the reduction processes occur at the individual level, one can limit his or her considerations to the evolution equation for the statistical operator:

$$\frac{d\rho(t)}{dt} = -\left(\frac{i}{\hbar}\right)[H, \rho(t)] + \gamma \sum_k \int d\mathbf{r} N^{(k)}(\mathbf{r})\rho(t)N^{(k)}(\mathbf{r}) - \left(\frac{\gamma}{2}\right) \sum_k \left\{ \int d\mathbf{r} N^{(k)2}(\mathbf{r}), \rho(t) \right\}. \tag{3.14}$$

For the sake of simplicity here, we will further restrict ourselves to a simplified version of CSL obtained by disregarding the Hamiltonian term and discretizing the space. This allows us to derive in a straightforward way[14,15] the main consequences that are of interest for the subsequent discussion.

We divide the space into cells of volume $(\alpha/4\pi)^{-3/2}$ and we denote by $N_i^{(k)}$ the number operator counting the particles of type k in the i-th cell. As follows from the discussion of the preceding subsection, the dynamical evolution in the considered case drives the state vector into a manifold such that the number of particles present in any cell is definite. The simplified equation for the statistical operator reads

$$\frac{d\rho(t)}{dt} = \gamma\left(\frac{\alpha}{4\pi}\right)^{-3/2} \sum_k \left[\sum_i N_i^{(k)}\rho(t)N_i^{(k)} - \left(\frac{1}{2}\right)\left\{ \sum_i N_i^{(k)2}, \rho(t) \right\} \right]. \tag{3.15}$$

In accordance with the relation of the previous subsection, we will often use the QMSL frequency parameter λ in place of the expression $\gamma(\alpha/4\pi)^{-3/2}$. If we denote by $|n_1^{(k)}, n_2^{(k)}, \ldots, n_i^{(k)}, \ldots\rangle$ the state with the corresponding occupation numbers for the various types of particles and for the various cells, the solution of equation 3.15 reads, in the considered basis,

$$\langle n_1^{(k)}, n_2^{(k)}, \ldots |\rho(t)| m_1^{(k)}, m_2^{(k)}, \ldots \rangle$$
$$= e^{(-\lambda/2)\Sigma_k \Sigma_i (n_i^{(k)} - m_i^{(k)})^2 t} \langle n_1^{(k)}, n_2^{(k)}, \ldots |\rho(0)| m_1^{(k)}, m_2^{(k)}, \ldots \rangle. \tag{3.16}$$

Equation 3.16 shows that linear superpositions of states containing a different number of particles in the various cells are dynamically reduced to one of the superposed states with an exponential time rate depending on

$$\left(\frac{\lambda}{2}\right) \sum_k \sum_i (n_i^{(k)} - m_i^{(k)})^2.$$

The amplification process in going from the microscopic to the macroscopic case and the preferred role assigned to position make it clear how also CSL, like QMSL, overcomes the difficulties of quantum measurement theory. This is easily understood by remarking once more that, in measurement processes, different eigenstates of the measured microobservable trigger (through the system-apparatus interaction) different displacements of a macroscopic pointer from its "ready" position. The unique dynamical principle of CSL leads then, in extremely short times, to the dynamical suppression, with the appropriate probability, of all but one of the terms in the superposition, that is, to the emergence of an outcome.

The above analysis should have made it evident that Dynamical Reduction Models describe in mathematically precise terms when, where, and how a microphenomenon becomes a phenomenon as a consequence of the occurrence of an irreversible act and how it is brought to a close by an amplification process.

GENERAL IMPLICATIONS OF DYNAMICAL REDUCTION

In the previous section, we concisely presented the precise formal aspects of QMSL and CSL, and we clarified how such models overcome the difficulties of the formalism. We stress that the considered models represent truly rival theories of standard quantum mechanics (in fact, they can in principle be tested against it), but they exhibit, at the microscopic level, empirical divergences from it that are so small that they can claim all the same experimental support. Said differently, they meet the two divergent desiderata for dynamical reduction theories that have been mentioned previously; that is, they imply no appreciable changes in the behavior of such systems and, at the same time, they induce an extremely rapid suppression of the unwanted superpositions at the macroscopic level.

Due to the (present) impossibility of performing *experimenta crucis* allowing one to discriminate between dynamical reduction models and quantum mechanics, to accept or to refuse the dynamical reduction philosophy is, to a large extent, a matter of taste. At any rate, we consider it to be of some conceptual interest to have shown that one can follow such a line of thought to overcome the "difficulties" met by quantum mechanics. In a recent interesting paper,[17] on the basis of a comparison of the effects of the CSL dynamics with those due to the environment, it has been stated that, in the absence of possible experimental tests, the consideration of new physical principles that are not motivated by the necessity of explaining new phenomena makes the dynamical reduction program vulnerable to Ockham's razor.

However, we disagree on this point for two reasons. One has been sharply expressed by Bell:[18] "I think that theoretical physics owes much to insisting on more than agreement with experiment." The second emerges naturally when one takes into account the specific context, that is, the conceptual problems that motivate the reduction program. The central issue is the conceptual implications of quantum formalism for what concerns the possibility of adopting a macrorealistic position about nature.

A quite natural way to grasp the core of the question is to consider the hidden-variable models that have played and that continue to play (quite appropriately in our opinion) an important role for the debate on the conceptual foundations of quantum mechanics. Within such approaches, one requires the theory to be in complete agreement with quantum mechanics from the very beginning. In spite of this, it would be quite unappropriate to deny the conceptual relevance of having been able to prove that a deterministic completion of quantum mechanics is possible and of having identified the price (i.e., contextuality) that one has to pay for this.

Having stated this, we will, in this section, call attention to recent developments of the dynamical reduction program. As we will see, the investigations furthering the considered line have led to some interesting general results about the requests that one has to respect when attempting to modify quantum mechanics and have required a reconsideration of nonlocality and of the criteria for property attribution to individual physical systems. We will also review briefly some experimental aspects of the new dynamics and we will call attention to some open problems on which active research is still going on.

Some General Results Concerning the Role of Nonlinearity and Stochasticity

As is well known,[19] standard quantum mechanics exhibits nonlocal features that, however, allow[20] its peaceful coexistence with relativity because quantum nonlocality is of the uncontrollable type. It can also be proved[15,16] that QMSL and CSL satisfy the no-faster-than-light signaling constraint.

In connection with this problem, it is appropriate to call attention to a quite general result by Gisin.[21] It can be summarized by stating that the inclusion of nonlinear elements in the Schrödinger equation unavoidably leads to violations of the above constraint.

On the other hand, even though the inclusion of stochastic features in the dynamics may lead to ensemble reduction (i.e., to diagonalization of the statistical operator in the preferred basis), it is worth mentioning that one can prove[15,22] that it cannot induce, by itself, individual reductions (i.e., the fact that the state vector of each individual system is driven into one of the preferred eigenmanifolds). It goes without saying that the very reason for considering the dynamical reduction program is the desire of accounting for individual reductions.

The conclusion of this subsection is that it is just the combined interplay of nonlinearity and stochasticity that makes it possible for CSL to peacefully coexist with relativity and, at the same time, to satisfy the fundamental desideratum that individual macroprocesses have outcomes.

Locality from the Dynamical Reduction Point of View

Because CSL reproduces the quantum correlations about measurement outcomes at the two wings of the apparatus in an EPR-like situation, it goes without saying that it also exhibits nonlocal features, just as standard quantum mechanics does. Thus, one is led to raise the problem of the precise nature of the nonlocal aspects of the theory. Immediately after the formulation of QMSL, Bell felt the necessity of discussing this problem and reached the conclusion[16] that the model "is as Lorentz invariant as it could be in the nonrelativistic version."

This problem has been reconsidered in great detail in two recent papers.[23,24] In them, the two different kinds of nonlocality that can characterize a theory and that are usually denoted as parameter dependence (PD) and outcome dependence (OD) have been taken into account. For those who are not familiar with this problem, one can state that a theory violates locality by exhibiting PD when the outcome at one wing can depend on the settings at the other wing, whereas, in theories exhibiting OD, the outcome at one wing can depend only on the outcome at the other wing.

Although quantum mechanics exhibits OD, it is appropriate to recall that[23] all deterministic hidden-variable theories exhibit PD. In references 23 and 24, it has been shown that both QMSL and CSL violate locality by exhibiting only OD. This is of some relevance because, as extensively discussed in reference 18 and in the papers under consideration, the two just-mentioned nonlocal features have a completely different status from the point of view of the possibility of getting relativistic generalizations of the theory itself. In fact, it is easily proved[23] that all theories exhibiting PD admit, at most, what Bell has called nongenuinely relativistic generalizations;[18] note that this specification makes reference to the fact that, although physics turns out to be the same for all observers, there is nevertheless a hidden preferred reference frame. On the other hand, the violation of locality by OD does not preclude the possibility of a relativistic generalization in the true Lorentz sense.[18]

Are Experimental Tests of Dynamical Reduction Possible?

In spite of the fact that we have repeatedly stressed the extreme difficulty in devising experimental tests of CSL against quantum mechanics, it is appropriate to mention that there have been various investigations aimed at identifying possible ways of tackling such a problem.

Let us list some of the effects of the modified dynamics that deserve to be discussed:

(1) The theory is fundamentally irreversible; as such, it implies a continuous increase of the energy with the elapsing of time. This is easily understood by taking into account the fact that localizing a system implies inducing high-momentum components. The corresponding energy increase can be explicitly evaluated[15] and turns out to be quite negligible and well below experimental testability.

(2) A quite natural area to search for effects of the modified dynamics is the area of so-called macroscopic quantum effects, typically superconductivity and the like. There have been various interesting investigations[25–27] about this point, the conclusion being that the theory actually implies a change in the resistivity of a superconductor with respect to the quantum mechanical value. Once more, however, testing such an effect is not possible with the present technology.

(3) Another effect is particularly interesting and its investigation has recently led to some relevant conclusions about CSL. Consider a bound system like, for example, a hydrogen atom and suppose that its electron suffers a localization process. One can easily evaluate[15,28,29] the probability that such a process leads to the excitation or to the dissociation of the atom. Due to the fact that the localization accuracy is orders of magnitude larger than the dimensions of the atom, even if such a process occurs, the corresponding probability turns out to be extremely small. When one takes into account the extremely low probability that a microscopic system suffers a localization, one reaches

> the conclusion that, on the average, one atom per second per mole will be excited or dissociated as a consequence of the modifications of the dynamics. This seems to be a too improbable event to be tested. In spite of this, investigations[30] on this effect have brought a new understanding of the phenomenon and have suggested a modification of CSL along the lines that we are going to discuss.

The argument of reference 30 goes as follows. Let us suppose that the CSL mechanism is operative at the level of quarks and that it makes sense to apply it to the quark model for nucleons (both of these assumptions are by no means obvious). Then, one can go through a calculation strictly similar to the previous one for the atom. In spite of the fact that the probability of dissociation is much smaller than before (due to the extremely small dimensions of a nucleon), nevertheless, if reduction is governed by the number of displaced particles in accordance with the CSL parameters λ and α, one gets a lifetime for the proton appreciably shorter than the one already confirmed by experiments. To avoid this problem, one should change the parameters of the model. This in turn would imply violating the other requests that it must meet (typically, the suppression of the superpositions of macroscopically different states would require an unacceptably long time to occur).

The authors of reference 30 then make an important remark. If one replaces the number density operators of the CSL model by the corresponding mass density operators and if one assumes the reduction rates to be those of CSL for nucleons, then the excitation and/or dissociation probabilities are depressed by large factors. The advantages are remarkable. First of all, the dissociation rate for the proton turns out to have a value well below the experimental bound, whereas the reduction rates for macroscopic objects coincide practically with those of CSL, with the decoherence being governed by the nucleons in ordinary matter and the contribution from electrons becoming negligible. Moreover, because of this, one relates reduction to gravity, an interesting possibility that has been suggested by various authors.[6,31–34] Actually, a model exhibiting this feature and having the further advantage of replacing one of the two parameters of CSL with Newton's gravitational constant had been presented[35] some years ago. Other advantages of taking such a position have been discussed in reference 27.

Relativistic Generalizations and Property Attribution

In spite of the fact that, as pointed out in the subsection on locality, there is no reason of principle forbidding a relativistic generalization of CSL due to its nonlocality being of the OD type, it turns out to not be an easy task to reach such a goal. Several interesting attempts have been made in recent years,[36–39] but they have not led to a satisfactory solution of the problem. Trying to embody stochastic elements in a quantum field theory context leads to intractable divergences. The considered investigations, however, have led to a better understanding of some crucial points and have thrown some light on relevant conceptual issues.

First, they have led to a completely general and rigorous formulation[37,38] of the concept of stochastic invariance. Second, they have stimulated a critical revisitation of the problem of the criteria for the attribution of objective local properties to physical systems. A way to do this, having the following implications, has been proposed. In specific situations, one cannot attribute any local property to a microsystem; any attempt to do so gives rise to ambiguities. However, in the case of macroscopic systems, the impossibility of attributing to them local properties (or, equivalently, the ambiguities about such properties) lasts only for time intervals of the order of those that are necessary for the dynamical reduction to take place.

The above picture has stimulated a deeper investigation of the problem and a critical reconsideration (taking appropriately into account the role of nonlocality within a relativistic context) of the logical structure of the EPR argument.[40] The conclusion is that, when the appropriate criterion is adopted, no objective property corresponding to a local observable can emerge, even for microsystems, as a consequence of a measurementlike event occurring in a spacelike separated region. Such properties emerge only in the future light cone of the considered macroscopic event. Correspondingly, it turns out to be impossible to establish, even conceptually, cause-effect relations between spacelike events.

To conclude, if a way to circumvent the present difficulties (i.e., the intractable divergences) can eventually be found, we can anticipate that, even in the relativistic version, dynamical reduction models will allow microsystems to be foggy, whereas requiring macrosystems to always have definite macroproperties.

Closing the Circle within the Spontaneous Reduction Program

The last point that we consider worth mentioning is that the spontaneous localization models represent theoretical constructions allowing one to close the circle in A. Shimony's sense, that is, to elaborate a worldview based on a genuinely quantum formalism (i.e., on the Hilbert space description of physical systems) that can accommodate our knowledge about microscopic phenomena and, at the same time, can account for our definite perceptions. This program has been proven to be viable in two recent papers.[41,42]

Reference 41 gives an answer to a criticism[43] that had been raised, completely in general, against the dynamical reduction program. It is based on the remark that one can easily imagine situations leading to definite perceptions and that nevertheless do not involve the displacement of a large number of particles. Typically, consideration has been given to a "measurementlike" process in which the two paths followed by a microsystem going through a Stern-Gerlach setup end on two different regions of a fluorescent screen and that then excite a small number of atoms that decay by emitting a small number of photons. Then, one is dealing with a superposition of two states corresponding to photons emerging from two different points. However, the process involves such a small number of particles that the CSL dynamics cannot lead to its suppression. On the other hand, because the visual perception threshold is quite low (about seven photons), the naked eye of a human observer is sufficient to detect the point from which the luminous spot originates. The conclusion is obvious: in the considered example, no dynamical reduction can take place and thus the measurement is not over—the outcome is not definite—up to the moment in which a conscious observer perceives the signal.

This criticism is inappropriate. It is perfectly true in the considered case that the superposition persists for long times (actually it must do so because, due to the fact that the system under consideration is microscopic, one could perform on it interference experiments that everybody would expect to confirm standard quantum predictions); however, if one takes seriously the above remark, one cannot avoid considering the whole system that brings about the definite "outcome", that is, the unambiguous perception. One must then give a simple estimate of the number of ions that are involved in the visual perception mechanism. Such an analysis makes it perfectly plausible[41] that, in the process, a sufficient number of particles are displaced by a sufficient spatial amount to satisfy the conditions that are necessary, according to CSL, for the suppression of the superposition of the two nervous signals to take place within the perception time.

It has to be stressed that this analysis, even though resorting to the mechanism of visual perception, does by no means attribute a special role to the conscious observer. The observer's brain is simply the only system that enters into the game in which a superposition of two states involving different locations of a large number of particles occurs. As such, it is the only place where the amplification act bringing to a close the microscopic phenomenon can occur. However, if in place of the eye of a human being one puts in front of the photon beam a spark chamber or any device leading to the displacement of a macroscopic pointer or producing ink spots on a computer output, reduction will take place. Once more, this example shows the appropriateness, within the considered theoretical models, of Wheeler's position that the central point is not the observer's consciousness, but the experimental device bringing the phenomenon to a close.

Before concluding, we would like to mention that, in reference 42, it has been shown how a reinterpretation of the considered model theory allows one to make a further relevant step towards closing the circle within a nonrelativistic framework. One starts by defining at each fixed time t an average mass density function $M(\boldsymbol{r}, t)$ in the real three-dimensional space. Such a function is simply the expectation value of the mass operator of a cell of volume 10^{-15} cm^3 centered at $\boldsymbol{r}$, evaluated on the state vector $|\Psi, t\rangle$ describing the physical system that represents "our universe". It is obvious within standard quantum mechanics that such a function cannot be endowed with an objective physical meaning precisely due to the occurrence of linear superpositions of macroscopically different mass distributions. One then considers a CSL model relating reduction to the mass density. The theory dynamically suppresses in extremely short times the embarrassing linear superpositions. Limiting his or her considerations to the set of states that are allowed (i.e., are dynamically stable) by the model, one can give a description of the world in terms of the considered function $M(\boldsymbol{r}, t)$. Second, one can define an appropriate topology on such a set that allows a quite natural specification of macroscopic similarity of allowed Hilbert space states. In turn, this topology can be taken as a basis to define a sensible principle of psychophysical correspondence for the theory.

CONCLUSIONS

In this report, we have taken into account the main conceptual difficulties met by standard quantum mechanics in dealing with physical processes involving a macroscopic system. We have stressed how J. A. Wheeler's remarks and lucid analysis have been relevant for pinpointing and for bringing to its extreme consequences the puzzling aspects of quantum phenomena. We hope to have made plausible how the recently proposed models of spontaneous dynamical reduction represent a consistent way to overcome the conceptual difficulties of the standard theory. Obviously, many nontrivial problems remain open, the first and most relevant one being that of generalizing the model theories considered here to the relativistic case. This is the challenge of the dynamical reduction program.

ACKNOWLEDGMENTS

We acknowledge illuminating discussions with R. Grassi.

REFERENCES

1. Bohr, N. 1928. Naturwissenschaften **16:** 245–257.
2. Wheeler, J. A. 1983. *In* Quantum Theory and Measurement. J. A. Wheeler & W. H. Zurek, Eds.: 182–200. Princeton University Press. Princeton, New Jersey.
3. Miller, W. A. & J. A. Wheeler. 1983. *In* Proceedings of the International Symposium of Quantum Mechanics. S. Kamefuchi, Ed.: 140–151. Physical Society of Japan. Kyoto.
4. Bell, J. S. 1986. *In* Proceedings of the Nobel Symposium. Volume 65: Possible Worlds in Arts and Sciences. Stockholm.
5. Einstein, A. 1949. *In* Albert Einstein: Philosopher Scientist. P. A. Schlipp, Ed.: 665–688. Tudor. New York.
6. Karolyhazy, F. 1966. Nuovo Cimento **A42:** 390–402.
7. Pearle, P. 1976. Phys. Rev. **D13:** 857–868.
8. Pearle, P. 1982. Found. Phys. **12:** 249–263.
9. Pearle, P. 1984. Phys. Rev. **D29:** 235–240.
10. Gisin, N. 1984. Phys. Rev. Lett. **52:** 1657–1660.
11. Ghirardi, G. C., A. Rimini & T. Weber. 1985. *In* Quantum Probability and Applications. L. Accardi & W. von Waldenfels, Eds.: 223–232. Springer-Verlag. Berlin/New York.
12. Ghirardi, G. C., A. Rimini & T. Weber. 1986. Phys. Rev. **D34:** 470–491.
13. Pearle, P. 1989. Phys. Rev. **A39:** 2277–2289.
14. Ghirardi, G. C., P. Pearle & A. Rimini. 1990. Phys. Rev. **A42:** 78–89.
15. Ghirardi, G. C. & A. Rimini. 1990. *In* Sixty-two Years of Uncertainty. A. Miller, Ed.: 167–191. Plenum. New York.
16. Bell, J. S. 1987. *In* Schrödinger-Centenary Celebration of a Polymath. C. W. Kilmister, Ed.: 41–52. Cambridge University Press. London/New York.
17. Tegmark, M. 1993. Found. Phys. Lett. **6:** 571–590.
18. Bell, J. S. 1989. *In* Themes in Contemporary Physics II. S. Deser & R. J. Finkelstein, Eds.: 1–26. World Scientific. Singapore.
19. Redhead, M. 1987. Incompleteness, Nonlocality, and Realism. Oxford University Press (Clarendon). London/New York.
20. Shimony, A. 1978. Int. Philos. Q. **18:** 3–17.
21. Gisin, N. 1990. Helv. Phys. Acta **62:** 363–371.
22. Ghirardi, G. C. & R. Grassi. 1990. *In* Nuovi Problemi della Logica e della Filosofia della Scienza. D. Costantini & M. C. Galavotti, Eds.: 83–95. Clueb. Bologna.
23. Ghirardi, G. C., R. Grassi, J. Butterfield & G. N. Fleming. 1993. Found. Phys. **23:** 341–364.
24. Butterfield, J., G. N. Fleming, G. C. Ghirardi & R. Grassi. 1993. Int. J. Theor. Phys. **32:** 2287–2304.
25. Gallis, M. R. & G. N. Fleming. 1990. Phys. Rev. **A42:** 38–48.
26. Rae, A. I. M. 1990. J. Phys. **A23:** 57–60.

27. Rimini, A. 1994. *In* International Course on Advances in Quantum Phenomena. E. Beltrametti & J. M. Levy-Leblond, Eds. Plenum. New York.
28. Benatti, F., G. C. Ghirardi, A. Rimini & T. Weber. 1988. Nuovo Cimento **B101:** 333–355.
29. Squires, E. 1991. Phys. Lett. **A158:** 431–432.
30. Pearle, P. & E. Squires. 1994. Phys. Rev. Lett. To appear.
31. Komar, A. B. 1969. Int. J. Theor. Phys. **2:** 157–160.
32. Penrose, R. 1986. *In* Quantum Concepts in Space and Time. R. Penrose & C. J. Isham, Eds.: 129–146. Oxford University Press (Clarendon). London/New York.
33. Diosi, L. 1989. Phys. Rev. **A40:** 1165–1174.
34. Frenkel, A. 1990. Found. Phys. **20:** 159–188.
35. Ghirardi, G. C., R. Grassi & A. Rimini. 1990. Phys. Rev. **A42:** 1057–1064.
36. Pearle, P. 1990. *In* Sixty-two Years of Uncertainty. A. Miller, Ed.: 193–214. Plenum. New York.
37. Ghirardi, G. C., R. Grassi & P. Pearle. 1990. Found. Phys. **20:** 1271–1316.
38. Ghirardi, G. C., R. Grassi & P. Pearle. 1991. *In* Symposium on the Foundations of Modern Physics 1990. P. Lahti & P. Mittelstaedt, Eds.: 109–123. World Scientific. Singapore.
39. Pearle, P. 1992. *In* Quantum Chaos–Quantum Measurement. P. Cvitanovic, I. Percival & A. Witzba, Eds.: 283–297. Kluwer. Dordrecht.
40. Ghirardi, G. C. & R. Grassi. 1994. Stud. Hist. Philos. Sci. **25:** 97–121.
41. Aicardi, F., A. Borsellino, G. C. Ghirardi & R. Grassi. 1991. Found. Phys. Lett. **4:** 116–128.
42. Ghirardi, G. C. & R. Grassi. 1994. ICTP preprint no. IC/107/94. To appear in Found. Phys.
43. Albert, D. & L. Vaidman. 1988. Phys. Lett. **A139:** 1–4.

The Quantum State Diffusion Model of Open Systems

N. GISIN

Group of Applied Physics
University of Geneva
1211 Geneva 4, Switzerland

INTRODUCTION

The Quantum State Diffusion (QSD) model[1–7] can be considered from two quite different points of view.

From the first point of view, it provides a tool for the description of individual open quantum systems. With this tool, one can compute numbers that cannot be computed with the traditional quantum mechanical description based on the density matrix formalism. These numbers are directly related to experimental data. The power of the QSD model is due, formally, to the efficient numerical algorithm it provides and, more deeply, to the fact that it represents an actual laboratory experiment in a way more faithful than what density matrices can do. In particular, individual runs of experiments are described explicitly.

From the second point of view, the QSD model aroused from problems related to the so-called "quantum measurement problem". More precisely, it follows from the question, "why should a theory well known for its stochasticity be based on a deterministic (Schrödinger) equation?"[8] and from the idea that, because classical (i.e., Kolmogorov) probabilities are "seen" (frequencies are measured), why not introduce such probabilities in the basic equations.

These two aspects of QSD are treated in the second and fourth sections, respectively. In the third section, quantum Brownian motion illustrates the QSD model. However, before these more technical sections, let us try to motivate the reader by some general statements about the QSD model:

(1) Ordinary quantum mechanics, based on the Schrödinger equation, is no longer the best for all practical purposes. Although the new model depends on ordinary quantum mechanics for its formulation, it provides very different pictures of what happens to an open quantum system and much better means of computation for all, but the simplest, open quantum systems.
(2) In this model, the changing state of a single quantum system is represented directly by a stochastically evolving pure state vector, as for a single run of a laboratory experiment. The state vector is then not merely a device for computing statistical expectations in an ensemble of systems.
(3) Localization or reduction appears dynamically, as a result of the diffusion process (in contrast to the stochastic dynamics of the quantum jump or trajectory methods[9–11]).
(4) Quantum measurement is nothing special. It is just one example of the interaction of a system with its environment. This interaction of systems with

measuring apparatus can be treated in the same way and, using the same basic equations, as the interaction with any other environment such as a heat bath or reservoir.

(5) Heat baths or reservoirs produce localization or reduction, in the absence of anything resembling a measurement.

(6) The state diffusion model provides graphic illustrations of quantum fluctuations that mimic experimental fluctuations. In particular, the time dependence of the mean photon number for induced atomic transitions bears a remarkable similarity to the corresponding experimental plots.[12]

(7) The path from an open quantum system to a closed classical Hamiltonian system involves two noncommuting limits. The correct order is (first) "quantum → classical" and (next) "open → close", as illustrated in FIGURE 1 and discussed in the fourth section.

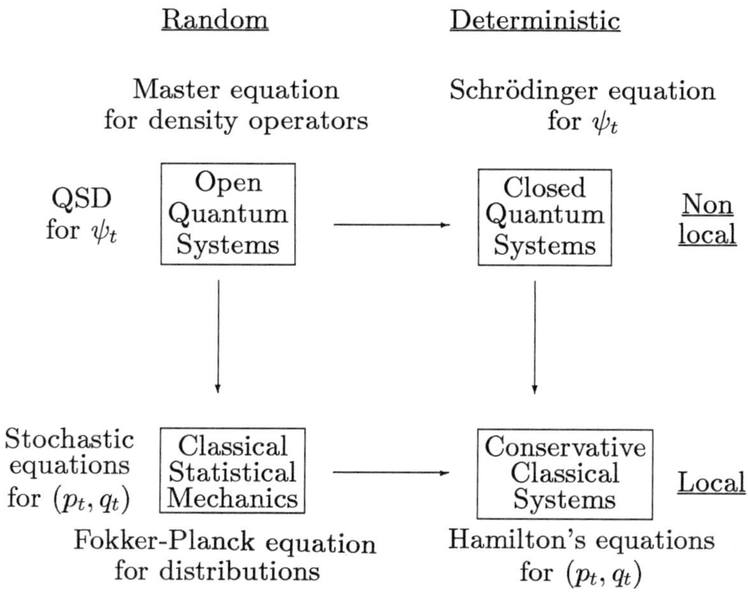

FIGURE 1. Noncommuting diagram of the "classical limit" from open quantum systems to closed classical systems.

QSD AS A PRACTICAL TOOL

The evolution equation for the density matrix ρ_t of an open system contains nonunitary terms characterized by "environment operators" L_m:

$$\dot{\rho}_t = \left(-\frac{i}{\hbar}\right)[H, \rho_t] + \sum_m \left[L_m \rho_t L_m^\dagger - \left(\frac{1}{2}\right)L_m^\dagger L_m \rho_t - \left(\frac{1}{2}\right)\rho_t L_m^\dagger L_m\right]. \quad (1)$$

The QSD model associates to each such dissipative evolution a unique[1,13] (Itô) stochastic evolution for the normalized state vector ψ_t:

$$d|\psi_t\rangle = \left(-\frac{i}{\hbar}\right)H|\psi_t\rangle\, dt + \sum_m \left[\langle L_m^\dagger\rangle_\psi L_m - \left(\frac{1}{2}\right)L_m^\dagger L_m - \left(\frac{1}{2}\right)\langle L_m^\dagger\rangle_\psi\langle L_m\rangle_\psi\right]|\psi_t\rangle dt + \sum_m (L_m - \langle L_m\rangle_\psi)|\psi_t\rangle d\xi_m, \quad \textbf{(2)}$$

where the ξ_m terms are complex-valued Wiener processes of zero mean and correlations,

$$Md\xi_m = 0, \qquad M(d\xi_n d\xi_m) = 0, \qquad M(d\xi_n^* d\xi_m) = \delta_{nm}dt. \quad \textbf{(3)}$$

Equations 1 and 2 are related by averaging the pure states ψ_t over the noises ξ_m:

$$\rho_t = M(|\psi_t\rangle\langle\psi_t|). \quad \textbf{(4)}$$

The relations 2 and 4 are the basic equations of the QSD model.

In reference 1, we emphasized the use of equation 2 for practical computations based on a Monte Carlo–type algorithm, especially in quantum optics. In reference 2, we proved general localization theorems and, in reference 3, we emphasized the physical picture and insight provided by the state diffusion model. Finally, let us mention some applications and tests of the model:

(1) in reference 12, QSD is used to describe a quantum jump experiment;
(2) in reference 14, QSD is applied to some nonlinear optical processes;
(3) in reference 15, QSD and quantum jump simulations are compared for two-photon processes;
(4) in reference 16, quantum jumps and the "phase-space picture" of QSD are compared;
(5) in reference 17, an approach to thermal equilibrium of harmonic oscillators is numerically investigated;
(6) in reference 18, QSD is applied to an open angular system, such as a quantum capacitor or rotor;
(7) in reference 19, some explicit solutions are presented;
(8) in reference 20, the Heisenberg picture is investigated and algorithms for multitime expectation values and correlation functions are presented.

The phenomenon of localization of QSD trajectories has been investigated in references 2, 3, 5, 13, 21, and 22. In the remainder of this section, we shall simply try to convey some intuition for the generality of this phenomenon. Due to the random diffusion, the solutions of equation 2 concentrate on regions (in the Hilbert space) where the fluctuations are minimal. These regions are characterized by

$$\text{Min}\left[\sum_k |(L_k - \langle L_k\rangle)|\psi\rangle d\xi_k|^2\right] = \text{Min}\sum_k (\Delta L_k)^2. \quad \textbf{(5)}$$

Accordingly, localization takes place with respect to the L_k variables. Because interactions are local in space, the position operator often appears in the L_k terms. Consequently, position localization is common in QSD (and in nature).

If the environment operator L is self-adjoint, the minima (equation 5) correspond to the eigenstates. In this way, one can model idealized measurements, reproducing the projection postulate as asymptotic solutions: $\psi_0 \to \phi_n$ as $t \to \infty$ with probability $|\langle\phi_n|\psi_0\rangle|^2$, where ϕ_n is an eigenstate of L.[1,5]

If $L = a$ or $a^\dagger$, that is, the annihilation or creation operator of the harmonic oscillator, the minima (equation 5) correspond to the coherent states. This example is developed in the next section.

Note that these states of minimum fluctuations are also those satisfying the Zurek conditions[23] in that they remain maximally pure under the evolution described by equation 2:

$$\text{Min} \sum_k (\Delta L_k)^2 = \text{Max}\left[\text{Tr}\left(P_{\psi_t} \frac{d}{dt} P_{\psi_t}\right)\right].$$

The QSD model thus provides explicit equations for the "narrow wave packets" discussed, for instance, by Zurek[24] and by Joos and Zeh.[25,26] If the states minimizing $\Sigma_k(\Delta L_k)^2$ are not preserved by the Hamiltonian, then the asymptotic solutions of the QSD model (equation 2) are no longer given by equation 5, but instead result from a competition between the Hamiltonian and dissipative part of the evolution equation, as described by equation 2. This contrasts with the Zurek condition because any Hamiltonian preserves pure states.

QUANTUM BROWNIAN MOTION

In this section, we consider a pumped and damped harmonic oscillator. We study its approach to thermal equilibrium, as described by the QSD model.[17] All expectation values agree with standard quantum mechanics, as always with QSD. However, this example also illustrates some specific aspects of QSD, like localization for instance.

Other simple examples of quantum Brownian motion can be found in references 13 and 19.

The master equation describing the approach to thermal equilibrium of a harmonic oscillator is

$$\dot{\rho}_t = -i\omega[a^\dagger a, \rho_t] + \bar{n}\gamma\left[a^\dagger\rho_t a - \left(\frac{1}{2}\right)\{aa^\dagger, \rho_t\}\right] + (\bar{n}+1)\gamma\left[a\rho_t a^\dagger - \left(\frac{1}{2}\right)\{a^\dagger a, \rho_t\}\right], \quad \textbf{(6)}$$

where $\bar{n}$ is the equilibrium mean photon number, $\bar{n} = \langle a^\dagger a\rangle_{\text{equ.}}$, and γ is the inverse of the relaxation time.

The corresponding QSD equation can be readily written down by substituting the Hamiltonian $H = \omega a^\dagger a$ and environment operators $L_1 = \sqrt{\bar{n}\gamma}a^\dagger$ and $L_2 = \sqrt{(\bar{n}+1)\gamma}a$ into equation 2. The environment operators have the tendency of driving the system towards states of minimum ΔL_1 and ΔL_2, as explained in the previous section. In the present case, these states are the coherent states $|\alpha\rangle$, labeled by the complex number $\alpha = (\langle q\rangle + i\langle p\rangle)/\sqrt{2}$. Because the coherent states are preserved by the Hamiltonian, any initial state evolves asymptotically under the QSD equation (equation 2) towards a coherent state. Then, they are preserved and the complex number α follows the well-known (classical) diffusion equation:

$$d\alpha_t = -i\omega\alpha_t dt - \left(\frac{\gamma}{2}\right)\alpha_t dt + \sqrt{\bar{n}\gamma}\,d\xi_t. \tag{7}$$

Note that, at zero temperature, $\bar{n} = 0$ and the fluctuations disappear.

FIGURE 2 illustrates this example. After an initial transient time, the state tends first to a Gaussian wave function; next, after some oscillations of the standard deviations Δp and Δq, it evolves into a coherent state. The mean position and momenta continue to oscillate, as should happen for an oscillator, with random fluctuations due to the finite temperature.

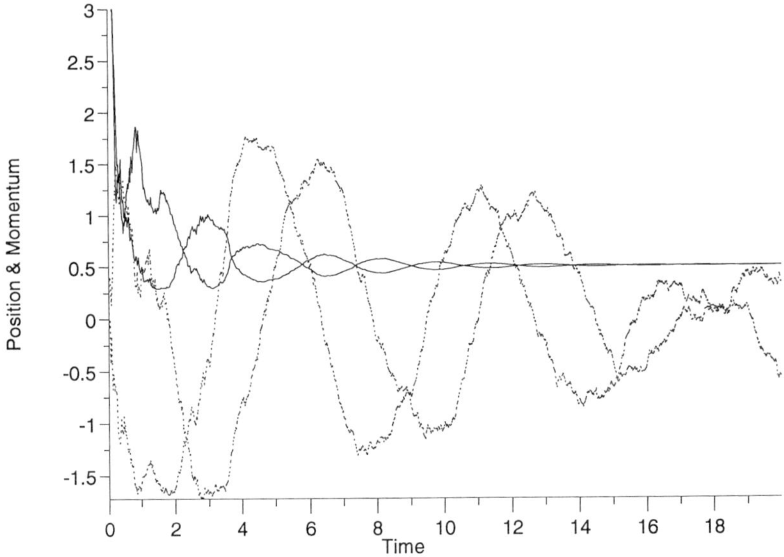

FIGURE 2. Example of a realization of QSD illustrating the approach to thermal equilibrium of a harmonic oscillator; see the third section. The initial state is the pure number state $|3\rangle$. The two dotted lines represent the mean position and momentum, $\langle p\rangle$ and $\langle q\rangle$. The two full lines represent the standard deviations, $(\Delta p)^2$ and $(\Delta q)^2$. This example illustrates how an arbitrary initial state tends asymptotically to a coherent state. The center of these coherent states will follow the classical stochastic process (equation 7).

This example illustrates two general features of QSD:

(1) Any initial state localizes: In this example, the localization is in phase space, but in general it could be on energy eigenstates or in position, depending on the environment operators. Most interactions, if not all, are local; hence, the environment operators usually involve the position operator. Thus, position localization is the most common. Combined with a Hamiltonian evolution, phase-space localization is also often realized.

(2) After localization, the wave packet spread is the minimum compatible with the Heisenberg uncertainty (quantum fluctuations) and the wave packet center follows classical trajectories with thermal fluctuations. Hence, to classical eyes, the quantum state solutions of the QSD equation (equation 2) look like classical points in phase space evolving under the classical diffusion equation (equation 7).

Before concluding this section, let us note that the master equation (equation 6) can be equivalently written in terms of the following three environment operators:

$$L_1 = \sqrt{\gamma}a, \qquad L_2 = \sqrt{\bar{n}\gamma}q, \qquad L_3 = \sqrt{\bar{n}\gamma}p. \tag{8}$$

Although these environment operators are different from the ones in equation 6, the corresponding master equations are identical. Consequently, any physically meaningful unraveling of that master equation must be independent of the choice of environment operators. This is indeed the case with QSD,[1,13] contrary to the stochastic jump process studied in references 9 and 10.

QSD AND THE CLASSICAL LIMIT OF QUANTUM MECHANICS

In reference 22, Ian Percival showed that, because of the generic localization phenomena in QSD, the corresponding states "appear to classical eyes like phase-space points". This localization has been explored in references 2 and 3 and is well illustrated in the previous section. This, together with the presence in QSD of classical probabilities and of distributions of pure states, makes it natural to investigate the classical limit starting from the QSD equation (equation 2) rather than from the density operator formalism.

The intuitive idea, as depicted in FIGURE 1, is the following: the two limits, "quantum → classical" and "open → close", do not commute. In cases where $\hbar$ is relatively small, some effects of the environment cannot be neglected, like localization and randomness. Accordingly, the first limit should be "quantum → classical", after which classical statistical physics will be obtained. It is only after the second limit "open → close" that the classical Hamilton equations are obtained.

Conservative classical mechanics is local and deterministic. Open quantum systems are nonlocal and nondeterministic. If one tries first the limit "quantum → classical", one obtains a theory (centered on the Schrödinger equation) that is deterministic, but nonlocal. No limit can remove smoothly this nonlocality. The Wigner function and related functions, for instance, reintroduce "probability" distributions in the classical state space, but these "probabilities" are difficult to interpret: first, because of the negative values; next and mainly, because there are no probabilities in the Schrödinger equation. If, on the contrary, one tries first the limit "open → close", locality comes out naturally. Indeed, although QSD is still nonlocal, the nonlocal states evolve in general to local states. The mean values of these local states then undergo classical trajectories with noise. In the second limit "quantum → classical", this noise is smoothly removed and classical Hamilton equations are obtained as illustrated in the previous section.

In FIGURE 3, another example is illustrated: a free particle with a friction "à la

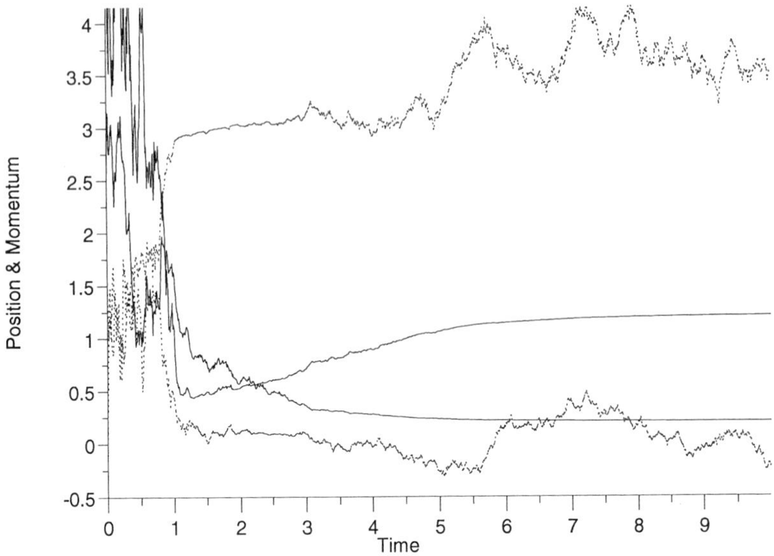

FIGURE 3. Friction "à la Caldeira-Leggett-Diósi"[27] in the QSD model. Any initial state tends asymptotically toward a Gaussian wave function of fixed shape, as illustrated on this realization. The Hamiltonian $H = (\lambda^2/4)\{q,p\}$ stretches the q coordinate and damps the conjugate coordinate p. The environment operator $L = \lambda a$ damps both the p and q coordinates in such a way that, together, the Hamiltonian and environment operator produce, on average, $\dot{p} = -\lambda^2 p$ and $\dot{q} = 0$. In this example, $\lambda = \sqrt{0.5}$ and the initial state is the following superposition of number states: $\psi_0 = (|4\rangle + i|5\rangle)/\sqrt{2}$. The two dotted lines represent the mean position and momentum, $\langle p\rangle$ and $\langle q\rangle$. The two full lines represent the standard deviations, $(\Delta p)^2$ and $(\Delta q)^2$. They tend asymptotically to $(\sqrt{2} - 1)/2$ and $[2(\sqrt{2} - 1)]^{-1}$, respectively.

Caldeira-Leggett-Diósi".[27] Any initial state tends asymptotically toward a Gaussian wave function of fixed shape (i.e., fixed Δp and Δq). FIGURE 4 illustrates an open nonlinear oscillator. The classical counterpart exhibits chaos for certain values of the parameters (see, for instance, reference 28). This figure illustrates again that the wave function tends to be close to a Gaussian (although in this case Gaussian states are not preserved by the nonlinear Hamiltonian). In this example, the wave function breathes: the standard deviations of position and momentum oscillate. The amplitude of this oscillation is irregular and lasts forever.

In QSD, the randomness is not arbitrary, but follows strict laws. The relation to the density matrix evolution is unique (under the assumptions of Markovicity and continuity): all other stochastic equations unraveling the density matrix evolution do depend on an arbitrary choice of the explicit form of the density matrix evolution equation.[1,13] Although the individual evolutions are not local, the statistics makes it compatible with relativity (no arbitrary fast signaling).[5,29]

The necessity to consider open systems has already been realized and emphasized by Zurek and Zeh, among others.[24,25] This alone, however, is insufficient[30] because a density operator mixes classical and quantum probabilities in such a way that, in general, many (classical) distributions of pure (quantum) states do correspond to a given density matrix. Moreover, classical mechanics deals with pure states (points in phase space) or with distributions of pure states. There is no classical analogue to the density matrix.

The consistent history approach[31] is another attempt to introduce classical probabilities into a quantum evolution. In reference 32, we showed that this approach is in fact very similar to the QSD model, at least for open systems. Indeed, the consistent histories (also called decoherent histories) are precisely the solutions of the QSD equation (equation 2). Let us make this connection plausible by showing that two different realizations ψ_{ξ_1} and ψ_{ξ_2} of the stochastic process (equation 2) tend to decohere. The mean of the square of their scalar product provides a measure of the mean "distance" between arbitrary trajectories in QSD:

$$M_{\xi_1} M_{\xi_2} |\langle \psi_{\xi_1} | \psi_{\xi_2} \rangle|^2 = \mathrm{Tr}(\rho^2). \quad \mathbf{(9)}$$

In general, the trace $\mathrm{Tr}(\rho^2)$ tends to zero (maximum entropy) for long times. In such cases, two arbitrary QSD trajectories tend to become orthogonal, that is, to decohere, for long times.

CONCLUSIONS

The two aspects of Quantum State Diffusion have been presented: first, QSD as a practical tool; next, the emergence of classical physics out of QSD. Is it to say that

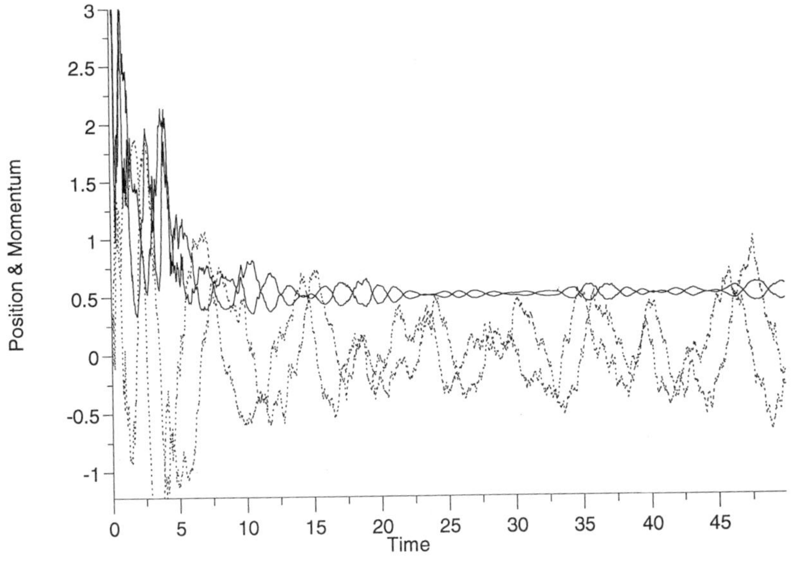

FIGURE 4. Open nonlinear oscillator: $H = a^\dagger a + (1/10)a^\dagger a^\dagger aa$, $L_1 = \sqrt{(\bar{n}+1)\gamma}a$, and $L_2 = \sqrt{\bar{n}\gamma}a^\dagger$, with $\bar{n} = \gamma = 1/8$. The initial state is a pure number state: $\psi_0 = |3\rangle$. Only one realization is shown. After a short transient time, the state is close to a Gaussian. However, the Gaussian state is not preserved by the nonlinear Hamiltonian. The pseudo-Gaussian state breathes, with an irregular amplitude. The two dotted lines represent the mean position and momentum, $\langle p\rangle$ and $\langle q\rangle$. The two full lines represent the standard deviations, $(\Delta p)^2$ and $(\Delta q)^2$.

QSD is a new fundamental theory? The answer is negative because of the following two—related—limitations of QSD:

(1) QSD applies only in the Markovian limit. This is often an excellent approximation. However, it is an approximation and a fundamental theory cannot be based on approximations.
(2) The distinction system/environment is in part arbitrary. The question of "what is a system" in quantum physics is a very difficult one, if not the most difficult one.

However, QSD could be considered as providing a first approximation to the next theory: a kind of "semi-quantum-mechanical" approximation.[33]

To conclude, let us mention a recent article from the Sussex group working on quantum experiments:[18]

> There is thus clearly a very good case for adopting an individual system approach, like QSD, in tandem with the usual statistical approach when discussing such devices theoretically. For example, it should be an invaluable tool for providing experimenters with a feel for the situation in which individual events will be revealing, and those in which the ensemble average will be more so. Thus, when single runs are seen to exhibit highly characteristic features, it should provide considerable insight into the environment interactions present.... With all this in mind, we therefore think that the individual system view of open quantum mechanics should prove to be most useful, both from a fundamental and from a practical aspect, for future quantum device research.

ACKNOWLEDGMENTS

It is a pleasure to thank Lajos Diósi, Jonathan Halliwell, and Ian Percival for invaluable discussions.

REFERENCES

1. GISIN, N. & I. C. PERCIVAL. 1992. J. Phys. A **25:** 5677–5691.
2. GISIN, N. & I. C. PERCIVAL. 1993. J. Phys. A **26:** 2233–2244.
3. GISIN, N. & I. C. PERCIVAL. 1993. J. Phys. A **26:** 2245–2260.
4. GISIN, N. 1984. Phys. Rev. Lett. **52:** 1657–1660.
5. GISIN, N. 1989. Helv. Phys. Acta **62:** 363–371.
6. GISIN, N. & M. CIBILS. 1992. J. Phys. A **25:** 5165–5176.
7. DIÓSI, L. 1988. J. Phys. A **21:** 2885–2898.
8. GISIN, N. 1993. Am. J. Phys. **61:** 86.
9. DALIBARD, J. Y., Y. CASTIN & K. MOLMER. 1992. Phys. Rev. Lett. **68:** 580–583; see also: 1993. J. Opt. Soc. Am. **10:** 524.
10. CARMICHAEL, H. 1993. An Open Systems Approach to Quantum Optics: Lecture Notes in Physics (m18). Springer Pub. New York.
11. COHEN-TANNOUDGI, C., B. ZAMBON & E. ARIMONDO. 1993. J. Opt. Soc. Am. **B10:** 2107.
12. GISIN, N., P. L. KNIGHT, I. C. PERCIVAL, R. C. THOMPSON & D. C. WILSON. 1993. J. Mod. Opt. **40:** 1663–1671.
13. GISIN, N. 1993. Symposium on the Foundations of Modern Physics, p. 178. World Scientific. Singapore.
14. GOETSCH, P. & R. GRAHAM. 1993. Ann. Phys. **2:** 706.
15. GARRAWAY, B. M. & P. L. KNIGHT. 1994. Phys. Rev. A **49:** 1266.

16. Garraway, B. M. & P. L. Knight. 1994. Dissipative environments: quantum jumps or phase space localization? *In* Quantum Optics VI. J. D. Harvey & D. F. Walls, Eds. In press.
17. Spiller, T. P., B. M. Garraway & I. C. Percival. 1993. Phys. Lett. A **179:** 63–66.
18. Spiller, T. P., J. F. Ralph, T. D. Clark, R. J. Prance & H. Prance. 1994. QSD applied to an angular system. Preprint. University of Sussex.
19. Salama, Y. & N. Gisin. 1993. Phys. Lett. A **181:** 269.
20. Gisin, N. 1993. J. Mod. Opt. **40:** 2313.
21. Gisin, N. 1994. *In* Proceedings of the Workshop on Stochastic Evolution of Quantum States in Open Systems and in Measurement Processes (Budapest). L. Diósi & B. Lukacs, Eds.: 40. World Scientific. Singapore.
22. Percival, I. C. 1994. J. Phys. A **27:** 1003.
23. Zurek, W. H. 1993. Phys. Rev. Lett. **70:** 1187.
24. Zurek, W. H. 1991. Phys. Today **October:** 36; 1993. Phys. Today **April:** 13.
25. Joos, E. & H. D. Zeh. 1985. Z. Phys. B **59:** 223–243; Zeh, H. D. 1993. Phys. Lett. A **172:** 189–192.
26. Zeh, H. D. 1994. *In* Proceedings of the Workshop on Stochastic Evolution of Quantum States in Open Systems and in Measurement Processes (Budapest). L. Diósi & B. Lukacs, Eds. World Scientific. Singapore.
27. Diósi, L. 1993. Physica A **199:** 517.
28. Gerry, Ch. 1990. Phys. Lett. A **146:** 363.
29. Gisin, N. 1984. Phys. Rev. Lett. **53:** 1776.
30. Gisin, N. & I. C. Percival. 1993. Phys. Lett. A **175:** 144–145.
31. Gell-Mann, M. & J. B. Hartle. 1993. Phys. Rev. D **47:** 3345; Griffiths, R. 1984. J. Stat. Phys. **36:** 219; Omnés, R. 1992. Rev. Mod. Phys. **64:** 339; Dowker, H. F. & J. J. Halliwell. 1992. Phys. Rev. D **46:** 1580.
32. Diósi, L., N. Gisin, J. Halliwell & I. C. Percival. 1994. Decoherent histories and QSD. Preprint.
33. Gisin, N. 1993 (September). Proceedings of the Third Max Born Symposium: Stochasticity and Quantum Chaos (Wrocław).

Order Parameter for Quantum Measurements and Related Topics

MIKIO NAMIKI[a] AND SAVERIO PASCAZIO[b]

[a]*Department of Physics*
Waseda University
Tokyo 169, Japan

[b]*Department of Physics*
University of Bari
I-70126 Bari, Italy

WHAT IS THE WAVE-FUNCTION COLLAPSE?

We shall start by reformulating the wave-function collapse along the line of thought of the many-Hilbert-space approach.[1]

Consider the measurement problem of an observable F performed by a detector D on a quantum system Q in a normalized superposed state:

$$\psi^Q = \sum_i c_i |u_i> \tag{1}$$

or

$$\rho^Q \equiv |\psi^Q><\psi^Q| = \sum_{ij} c_i c_j^* |u_i><u_j| \tag{2}$$

with $\Sigma_i |c_i|^2 = 1$, where we have assumed that $F|u_i> = \lambda_i|u_i>$ and $<u_i|u_j> = \delta_{ij}$.

When we find an eigenvalue λ_k in a measurement, we often consider the measurement process within the scope of the naive Copenhagen view.

Naive Copenhagen view—Projection rule: The measurement should instantaneously change the Q-system as

$$|\psi^Q> \rightarrow |u_k> \qquad \text{or} \qquad \rho^Q \rightarrow \xi_k = |u_k><u_k|, \tag{3}$$

$$\text{all other branch waves disappear.} \tag{4}$$

This type of transition is often called the wave-function collapse (wfc) and is widely used to describe the measurement process. This wfc is usually introduced in theoretical calculations by hand or as a postulate, but not through a physical process.

As for the above naive view, we should state several unsatisfactory points as follows: We know that the above transition is an acausal and probabilistic process to be found with probability $|c_k|^2$, but it does not describe the appearance of probabilities in a measurement process. For this reason, one can say that the naive view is not satisfactory. Moreover, we know that quantum mechanics can never give a definite prediction for a single experiment on a single system, but only the theoretical probability distribution, which is to be compared with the accumulation of many independent experiments performed under the same conditions.

For this reason, we often use an improved expression called the system-only wave-function collapse:

$$\rho_I^Q = \sum_{ij} c_i c_j^* |u_i\rangle\langle u_j| \rightarrow \rho_F^Q = \sum_k |c_k|^2 \xi_k, \qquad \xi_k \equiv |u_k\rangle\langle u_k|. \tag{5}$$

However, this is still not satisfactory for the following reason. In the special case with $c_1 = c_2 = 1/\sqrt{2}$ for a dichotomic observable F, the system-only description of the wfc becomes

$$\rho_I^Q \rightarrow \rho_F^Q = (½)(\xi_1 + \xi_2) = (½)(\xi_+ + \xi_-), \tag{6}$$

where $u_\pm = (u_1 \pm u_2)/\sqrt{2}$ and $\xi_\pm \equiv |u_\pm\rangle\langle u_\pm|$. This describes a measurement of F via the expression in the middle right-hand term, whereas it describes a measurement of another observable G, with eigenstates $u_\pm$, via the expression in the far right-hand term. In general, $[F, G] \neq 0$, so we are led to the contradiction that the above expression describes the incompatible measurements of two uncommutable observables at the same time.

Within the scope of the naive view, it is very difficult to explain the negative-result measurement and we can hardly discuss the possibility of having imperfect measurements (see later discussions on the decoherence parameter).

Remember that all measurement processes take place during a very long time on a microscopic scale, but not instantaneously, even though we can regard them as if they happened instantaneously on a macroscopic scale.

The widespread idea that observation of the so-called quantum Zeno effect offers experimental evidence of the above naive view is misleading, as we discussed elsewhere in this volume.[2]

Thus, we are led to the system-detector description of the wave-function collapse,

$$\Xi_I^{\text{tot}} = \rho_I^Q \otimes \sigma_I^D = \sum_{ij} c_i c_j^* |u_i\rangle\langle u_j| \otimes \sigma_I^D \rightarrow \Xi_F^{\text{tot}} = \sum_k |c_k|^2 \xi_{F(k)t}^Q \otimes \sigma_{F(k)t}^D, \tag{7}$$

as the "final goal" of the measurement theory. We should derive this type of wave-function collapse by applying quantum mechanics to the total (Q + D) system. Here, Ξ^{tot} stands for the density matrix of the total system, σ_I^D stands for the initial density matrix of D, $\xi_{F(k)t}^Q$ stands for the final Q-state corresponding to the k-th eigenvalue, and $\sigma_{F(k)t}^D$ stands for the final D-state. Note that $\sigma_{F(k)t}^D$ describes secondary processes that have the purpose of generating a signal (such as counterdischarge) to display the observation of the k-th eigenstate of F. As was repeatedly mentioned,[1] the wave-function collapse itself should be strictly distinguished from the thermal irreversible processes displaying the result of the measurement.

Note that the right-hand side of this expression is the sum of the probabilities of finding Q in mutually exclusive states. Therefore, once a particular event takes place, all other results cannot occur. This is nothing other than the very original idea of wave-function collapse for a single measurement, as expected from the original probabilistic interpretation.

Consequently, this expression exactly describes not only the statistical distribution for accumulation of many independent measurements, but also a single measurement on a single system.

Remember that the final goal of the measurement theory is not to derive the naive Copenhagen view for individual events, but to derive the above system-detector description of the wfc.

We should, in addition, remark the following facts:

(a) The wave-function collapse represented by the above transition is a perfectly dephasing process provoked by the interaction of Q with D.
(b) In order to experimentally observe the whole transition scheme, we have to repeat the same experiment many times and then accumulate many independent results. In this context, the wfc is a statistical process.
(c) We have to repeat that quantum mechanics can never give a definite prediction for a single experiment on a single system, but only the theoretical probability distribution to be compared with the experimentally accumulated one.
(d) Note that the presence of the D-states in the system-detector description of the wfc is very important because equation 7 goes back to the unsatisfactory system-only description of the wfc if we remove the D-states from equation 7.

DECOHERENCE PARAMETER—AN "ORDER PARAMETER" FOR THE WAVE-FUNCTION COLLAPSE

Let us restart our discussion from von Neumann's process:

$$|\Psi_I\rangle = \sum_i c_i|u_i\rangle \otimes |\Phi_0\rangle \rightarrow |\tilde{\Psi}_F\rangle = \sum_i c_i|u_i\rangle \otimes |\Phi_i\rangle \tag{8}$$

or

$$\Xi_I^{\text{tot}} \equiv |\Psi_I\rangle\langle\Psi_I| = \sum_{ij} c_i c_j^*|u_i\rangle\langle u_j| \otimes |\Phi_0\rangle\langle\Phi_0|$$

$$\rightarrow \sum_{ij} c_i c_j^*|u_i\rangle\langle u_j| \otimes |\Phi_i\rangle\langle\Phi_j| = |\tilde{\Psi}_F\rangle\langle\tilde{\Psi}_F| \equiv \tilde{\Xi}_F^{\text{tot}}, \tag{9}$$

where Φ_0 and Φ_i are the initial and final normalized detector-environment states (DE-states), subject to the asymptotic orthogonality,

$$\langle\Phi_i|\Phi_j\rangle = \delta_{ij} + O(\eta), \tag{10}$$

in which η is a very small number tending to zero as the number of degrees of freedom of the detector goes to infinity.

Remember the following:

(1) von Neumann's process is characterized by a unitary temporal evolution, provoking no dephasing among different branch waves.
(2) We often accept the simple idea that von Neumann's process with the orthogonality or the nonoverlapping of different Φ_k's is equivalent to the wfc because we cannot observe interference among them. However, note that the

apparent disappearance of interference does not always mean the wfc because it is possible that coherence is still kept.

(3) One of the simplest ways to destroy the unitarity by hand is to take the partial trace with respect to the DE-states:

$$\mathrm{Tr}_{\mathrm{DE}}\tilde{\Xi}_F^{\mathrm{tot}} = \sum_k |c_k|^2 \xi_k, \tag{11}$$

which is derived by the above orthogonality. The result is nothing other than the unsatisfactory system-only description of the wfc. Note that partial tracing is a human-dependent manipulation.

(4) The off-diagonal part of $\tilde{\Xi}_F^{\mathrm{tot}}$ never vanishes, even when taking the partial trace with respect to the DE-states, because

$$\mathrm{Tr}_{\mathrm{DE}}\tilde{\Xi}_{F,\mathrm{off}}^{\mathrm{tot}} = 0, \qquad \text{but} \qquad \mathrm{Tr}_{\mathrm{DE}}[\tilde{\Xi}_{F,\mathrm{off}}^{\mathrm{tot}}]^2 \neq 0, \tag{12}$$

as is easily derived by the asymptotic orthogonality. The vanishing of interference terms of the partial inner product with respect to DE-states does not imply the lack of the off-diagonal part of the whole density matrix. This implies that partial tracing or the partial inner product should not be considered responsible for the exact wave-function collapse.

Here, let us recall that different incoming Q-particles in an experimental run will meet, particle by particle, the D-system (particularly, one of its local systems) in microscopically different states because of its open-system character and internal motions, although we can look at it as being in the same state from a macroscopic point of view (see FIGURE 1).

For this reason, we should label different repetitions of the same experiment with ℓ ($\ell = 1, \ldots, N_p$), with N_p being the total number of repetitions. Accordingly, the final D-state wave function and the final total density matrix for von Neumann's process should read

$$|\Phi_i^{(\ell)}> \text{ and } \tilde{\Xi}^{\mathrm{tot}(\ell)}, \tag{13}$$

respectively. The resultant total density matrix to describe the probability distribution should naturally be given through its average over many events:

$$\overline{\Xi^{\mathrm{tot}}} = \left(\frac{1}{N_p}\right) \sum_{\ell=1}^{N_p} \Xi^{\mathrm{tot}(\ell)}. \tag{14}$$

The most important and fundamental average in the present theory is the bar-average, but we can usually or sometimes replace it with the statistical ensemble-average ($\ll \Xi^{\mathrm{tot}} \gg$ for the total density matrix), under the assumption of ergodicity. Of course, the validity of ergodicity should depend on the internal state of D. Remember that the Gaussian distribution law that is used in our numerical simulation is finally introduced only as a simplified substitute for the resultant complicated internal states of D.

If we can use the statistical ensemble-average, then we can put

$$\overline{\Xi^{\mathrm{tot}}} = \ll \Xi^{\mathrm{tot}} \gg. \tag{15}$$

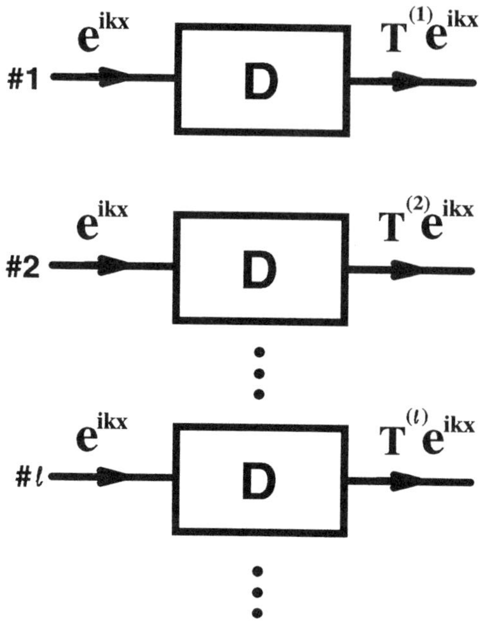

FIGURE 1. Repetitions of the same experiment in an experimental run: detector D put in the i-th channel will yield, in the ℓ-th experiment, each transmission coefficient, $T^{(\ell)i} = e^{i\gamma^{(\ell)i}}$, for the i-th branch wave, different from others due to different microscopic internal states of D.

In this case, we are led to the following description of the measurement process:

$$\Xi_I^{\rm tot} \to \Xi_F^{\rm tot} \equiv \overline{\tilde{\Xi}_F^{\rm tot}} = \ll \tilde{\Xi}_F^{\rm tot} \gg . \tag{16}$$

For simplicity, and without a big loss of generality, we can write

$$|\Phi_i^{(\ell)}> \; = e^{i\gamma_i^{(\ell)}} |\Phi_i>, \tag{17}$$

where $\gamma_i^{(\ell)}$ is in general a complex number with a positive imaginary part. Defining

$$\Delta_{ij} \equiv \overline{e^{i(\gamma_i - \gamma_i^*)}} = \ll e^{i(\gamma_i - \gamma^*_i)} \gg \tag{18}$$

$$\equiv d_i[\delta_{ij} + \epsilon_{ij}(1 - \delta_{ij})], \tag{19}$$

we obtain the final-state total density matrix:

$$\Xi_F^{\rm tot} = \sum_{ij} \Delta_{ij} c_i c_j^* |u_i><u_j| \otimes |\Phi_i>< \Phi_j|. \tag{20}$$

In the case of no losses and no reflection, that is, $d_i = 1$, a perfect measurement provokes a complete loss of coherence, yielding

$$\Delta_{ij} = \delta_{ij}, \tag{21}$$

so we obtain

$$\Xi_I^{\rm tot} \to \Xi_F^{\rm tot} = \sum_k |c_k|^2 \xi_k \otimes |\Phi_k><\Phi_k|. \tag{22}$$

Thus, we can define, via equation 19, the decoherence parameter ϵ_{ij}, which is an order parameter for quantum measurements.

Summarizing, we can conclude that

(1) $\epsilon_{ij} = 0$: for perfect measurement;
(2) $\epsilon_{ij} = \delta_{ij}$: for perfect coherence;
(3) $1 > |\epsilon_{ij}| > 0$: for imperfect measurement.

DETECTOR MODEL AND NUMERICAL SIMULATIONS

Consider the measurement of a dichotomic observable, for which we can use a measuring apparatus of the Stern-Gerlach type as schematized in FIGURE 2. The branch waves $\psi'_a = T\psi_a$ and ψ_b running in channels A and B at the first step are combined in the same channel of the second step, where T stands for the transmission coefficient at D. Labeling the relevant quantities with (ℓ) and introducing the bar-averages, we obtain the detection probability by D_0, proportional to

$$P = \overline{|\psi'_a + \psi_b|^2} = (½)(1 + \overline{|T|^2} + 2\mathrm{Re}\,\overline{T}), \tag{23}$$

where we have assumed that $|\psi_a|^2 = |\psi_b|^2 = |\psi_b^*\psi_a| = ½$. For simplicity, we have suppressed the D-states.

We can compare $T^{(\ell)}$ with $e^{i\gamma_i^{(\ell)}}$ in the above general discussion; thus, we obtain

$$\epsilon = 1 - \frac{|\overline{T}|^2}{\overline{|T|^2}} \tag{24}$$

for the decoherence parameter. Note that $1 > \epsilon > 0$. This ideal is also applied to analyses of neutron interference experiments.[1]

By making use of a detector model as schematized in FIGURE 3, we perform numerical simulations of $T^{(\ell)}$ (FIGURES 4a–c) and ϵ (FIGURE 5).[3] Now, assume that

$$\frac{\delta d}{\langle d\rangle} = \frac{\delta\Lambda}{\langle\Lambda\rangle} \equiv \Delta, \tag{25}$$

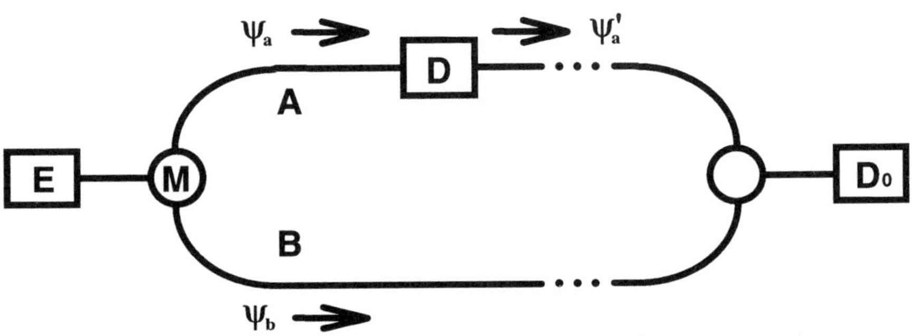

FIGURE 2. The first step on the left is a measurement of the Stern-Gerlach type. The second step of the experiment on the right is put in for the purpose of observing the physical effects of detector D in the first step.

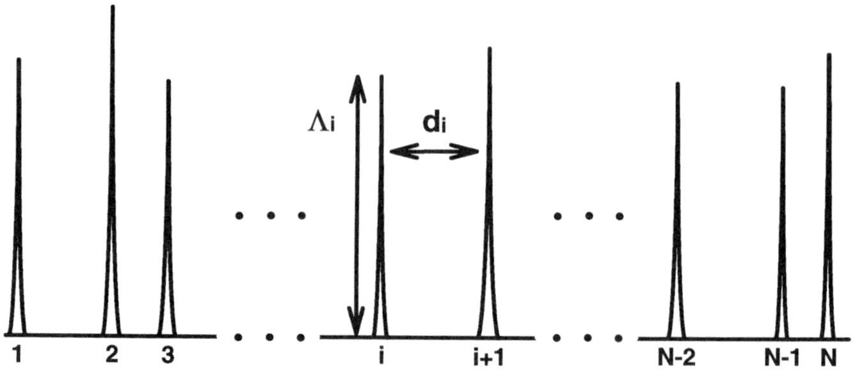

FIGURE 3. Modeling of the detector: $V(x) = \Sigma_{i=1}^{N} \Lambda_i \delta(x - d_i)$, with N standing for the total number of potentials, Λ_i standing for the potential strength of the i-th potential, and $d_i = b_{i+1} - b_i$ standing for the spacing between the i-th and the $(i + 1)$-th ones.

$k\langle d \rangle = 4.5\pi$, and $\langle \Lambda \rangle / \hbar v = 0.01$ (v being the particle speed) and suppose that an experimental run includes $N_p = 5000$ repetitions of the same experiment. We have assumed Gaussian distributions of the spacings and the strengths as conventional substitutes for the internal motions.

For the above detector model with very weak potentials, we can easily derive the following formula,

$$\epsilon \cong 1 - \exp[-\{k\langle d \rangle\}^2 \cdot (N - 1)\Delta^2], \tag{26}$$

where $k\langle d \rangle = 4.5\pi$, for the decoherence parameter (see the last paper in reference 3). This formula is excellently fitted to the results of the numerical simulation, as was seen in FIGURE 5.

POSSIBLE RELATION BETWEEN DEPHASING AND CHAOS

FIGURE 5 indicates that we can describe ϵ in terms of almost the same function of a single variable $N\Delta^2$ for various N and Δ^2. Therefore, we have the possibility of obtaining the wfc even for small N if Δ^2 is large enough.

We may realize such a case by making use of a chaotic system with small degrees of freedom, as D in FIGURE 2, in order to give the wave-function collapse.[4] We are now performing numerical simulations for a few D's with some chaotic behavior, for example, as given by a (classical and quantum) kicked rotator.[4] In this case, we have to use the bar-average, $\overline{\Xi^{\rm tot}}$, but not the statistical ensemble-average, $\ll \Xi^{\rm tot} \gg$. Suppose now that we have the second moment, fourth moment, etc., for the momentum transfers of the incoming Q-particle in collisions with the rotator, as well as the rotator position in terms of the bar-averages. Assume further that we can approximate their resultant distributions (of the momentum transfers $\mathbf{K}$ and the rotator position $\mathbf{x}$) by the Gaussian ones with the mean square deviations of them, $\overline{K_{\rm L}^2}$, $\overline{K_{\rm T}^2}$, $(\Delta x_{\rm L})^2$, and $(\Delta x_{\rm T})^2$. Here, the subscripts L and T stand for the longitudinal and transverse components, respectively. Under this assumption, we easily obtain

$$\epsilon \cong 1 - \exp[-\{\overline{K_L^2}(\Delta x_L)^2 + \overline{K_T^2}(\Delta x_T)^2\}] \tag{27}$$

as a rough approximation. This work will be reported in the near future.

CONCLUSIONS

In this report, we first reformulated the system-detector description of the wave-function collapse as a dephasing process, which should be considered to be the final goal of quantum measurement theory. Starting from von Neumann's description of the measurement process, we next introduced the decoherence parameter, which enables us to quantitatively judge whether an instrument can work well or not

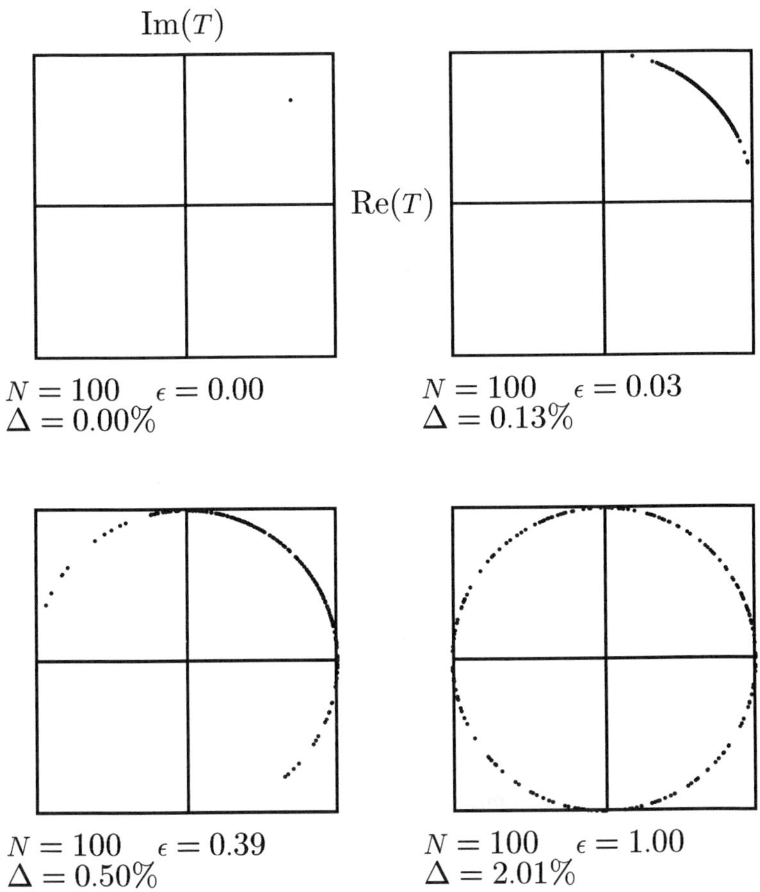

FIGURE 4a. T-diagrams and ϵ for various N and Δ: effects of fluctuations for fixed N (= 100).

as a detector. The dephasing process or the wave-function collapse must be provoked by the fluctuations of the constituents of the detector. In the present theory, we can take the effect of fluctuations into account by taking the bar-average (given by equation 14), which was introduced by faithfully following an experimental procedure. This is the most important one and can be replaced with the statistical average

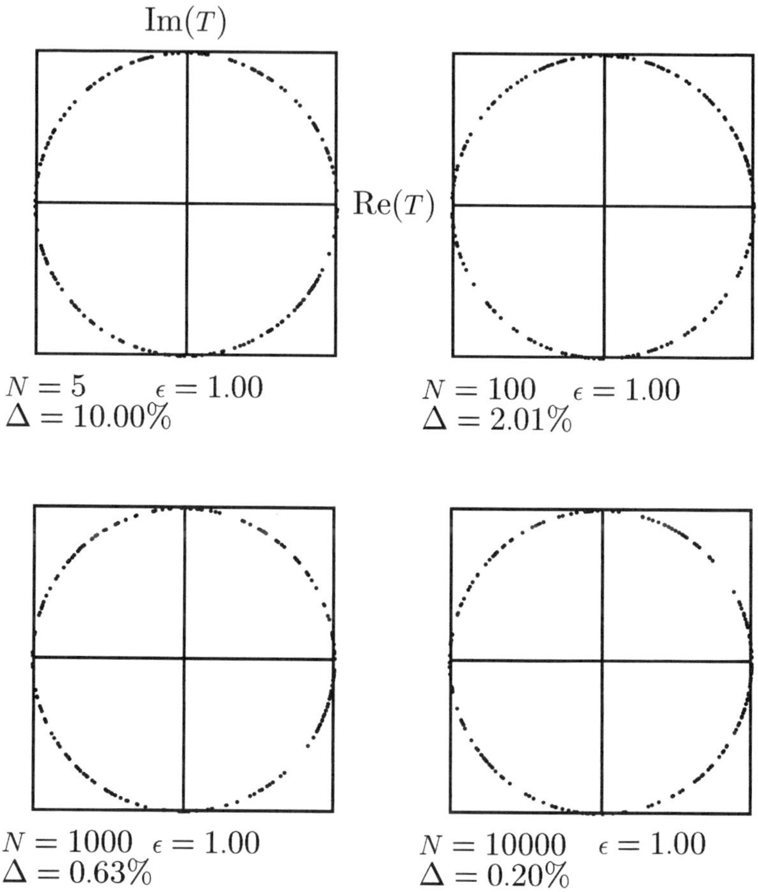

FIGURE 4b. T-diagrams and ϵ for various N and Δ: the wave-function collapse for N = 5, 100, 1000, and 10,000.

given by equation 15, depending on the internal state of the detector, under the assumption of ergodicity. In the case of numerical simulations based on the Dirac-comb detector model, we further simplified the statistical average by making use of the Gaussian distributions of potential strengths and spacings. In this case, we found that the decoherence parameter ϵ can be described by a simple function of a single

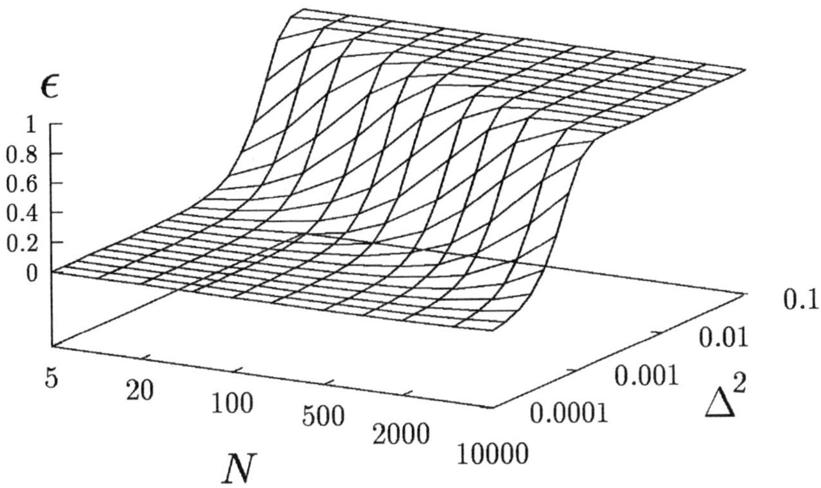

FIGURE 4c. Dependence of ϵ on N and Δ^2.

variable $N\Delta^2$, where N stands for the total number of constituents and Δ^2 stands for the mean square deviation of potential strengths or spacings. Finally, we discussed the possibility of provoking the wave-function collapse, in the case of a small system, through its chaotic behavior, but the problem is still open.

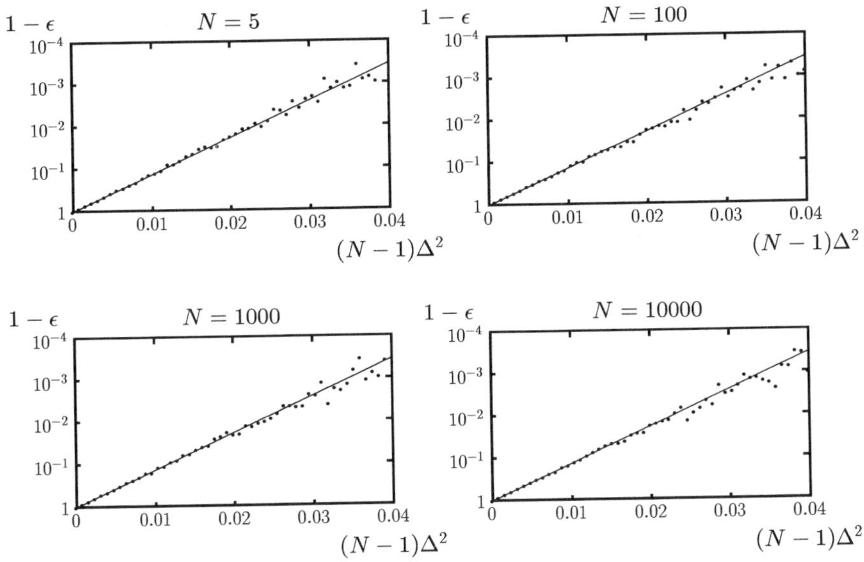

FIGURE 5. ϵ versus $N\Delta^2$: the numerical results are well fitted by the theoretical formula (equation 26) (see the last paper in reference 3).

ACKNOWLEDGMENTS

We are very much indebted to M. Kamimura and N. Kohno for their collaborations on numerical simulations.

REFERENCES

1. MACHIDA, S. & M. NAMIKI. 1980. Prog. Theor. Phys. **63:** 1457 & 1833; NAMIKI, M. 1988. Found. Phys. **18:** 29; NAMIKI, M. & S. PASCAZIO. 1992. Found. Phys. **22:** 451; 1991. Phys. Rev. **A44:** 39; 1993. Phys. Rep. **232:** 301; NAMIKI, M., S. PASCAZIO & C. SCHILLER. 1994. Phys. Lett. **A182:** 17.
2. PASCAZIO, S. & M. NAMIKI. 1995. This volume; also see: PASCAZIO, S., M. NAMIKI, G. BADUREK & H. RAUCH. 1993. Phys. Lett. **A179:** 155.
3. See the fifth and sixth papers of reference 1; also see: NAMIKI, M., S. PASCAZIO, M. KAMIMURA & N. KOHNO. 1995. In preparation.
4. SAITO, N. Private communication; also see the last paper in reference 3.

Quantum Interference, State Engineering, and Quantum Eraser

D. S. KRÄHMER, K. VOGEL, V. M. AKULIN,[a]
AND W. P. SCHLEICH[b]

Abteilung für Quantenphysik
Universität Ulm
D-89069 Ulm, Germany

INTRODUCTION

To summarize in one pregnant sentence the most recent result has always been and still is the requirement every scientist has to fulfill when he or she wants to explain his or her research to John A. Wheeler, the great scientist, deep philosopher, and extraordinary human being we honor at this conference. The central lesson of quantum mechanics, that is, interfering transition probability amplitudes rather than probabilities, stands out nowhere clearer than in the double-slit experiment, the center of the long-standing debate between Bohr and Einstein.[1] Nobody has identified this simple, but far-reaching difference between classical and quantum physics as the origin of so many different physical phenomena ranging from nuclear physics[2] via rainbow scattering[3,4] to squeezed state physics[5] as John A. Wheeler. Nothing is more appropriate at a conference celebrating this great man than to present yet another illustration of this one-sentence summary—path information implies probabilities, whereas no path information implies interfering probability amplitudes.

In this article, we transfer the double-slit configuration into the rather abstract space of atomic excitation and show how by erasing "*which path*" information[6–11] we can generate quantum coherence and even do state engineering,[12–26] that is, manipulate quantum states in a controlled way. Our *quantum eraser* is spontaneous emission together with reabsorption in a detector.

The article is organized as follows: In the second section, we send two excited two-level atoms, one at a time, through a cavity and find noninterfering contributions to the photon statistics of the cavity field because we still have the information of which atom has deposited a photon in the cavity. We erase this information via spontaneous emission and reabsorption of the emitted photon and obtain interfering parts in the photon statistics of the cavity field. In the third section, we first extend our investigations from two two-level atoms to two three-level atoms and then to many three-level atoms. In the fourth section, we show that we can use the scheme of the third section in order to manipulate the state of the cavity field in a controlled way; that is, we can do quantum state engineering. In the fifth section, we give a brief summary of our investigations.

[a] Also at Moscow Institute of Physics and Technology, Dolgoprudny, Moscovskaja Obl., Russia.

[b] Also at Max-Planck-Institut für Quantenoptik, D-85748 Garching bei München, Germany.

MOTIVATION AND BASIC CONSIDERATIONS

We consider the interaction of two two-level atoms with a quantized field in a high-Q cavity. The spatial separation of the atoms is such that there is at most *one* atom inside the cavity. We describe the interaction of a single two-level atom with the radiation field by the well-known Jaynes-Cummings Hamiltonian:

$$H_{\text{int}} = \hbar g(a^\dagger\sigma + a\sigma^\dagger),$$

where $a, a^\dagger$ and $\sigma, \sigma^\dagger$ are the usual lowering and raising operators for the field and the atom, respectively, and g is the coupling strength. The atoms are resonant with the radiation field. Before they enter the cavity, we excite them by an appropriate laser pulse into the upper state $|a\rangle$. The quantized field in the resonator is assumed to be in an arbitrary state $|\psi_f\rangle = \Sigma_{n=0}^{\infty} w_n|n\rangle$. After an interaction time τ, the first atom leaves the cavity and the complete state of the system consisting of two atoms and the field reads in the interaction picture

$$|\psi\rangle = \sum_{n=0}^{\infty} w_n(C_n|n; a, a\rangle - iS_n|n+1; b, a\rangle).$$

Here, $|n + 1; b, a\rangle$ describes a state where the field is in the number state $|n + 1\rangle$, the first atom is in the ground state $|b\rangle$, and the second atom is still in the excited state $|a\rangle$. In addition, we have introduced the abbreviations $C_n = \cos(g\tau\sqrt{n+1})$ and $S_n = \sin(g\tau\sqrt{n+1})$. After the second atom has left the cavity, the state of the system is

$$|\psi\rangle = \sum_{n=0}^{\infty} w_n[C_n^2|n; a, a\rangle - iC_nS_n|n+1; a, b\rangle - iS_nC_{n+1}|n+1; b, a\rangle - S_nS_{n+1}|n+2; b, b\rangle]. \quad \mathbf{(1)}$$

This state is the starting point for the considerations in the remainder of this report. The interesting feature about state 1 (i.e., equation 1) is that there are two contributions that contain exactly one atomic excitation. The probability of finding n photons in the cavity and *one* atomic excitation reads

$$\begin{pmatrix}\text{probability for } n \text{ photons}\\ +\ 1 \text{ atomic excitation}\end{pmatrix} \equiv W_n\begin{pmatrix}|a, b\rangle\\ |b, a\rangle\end{pmatrix} = |w_{n-1}|^2(C_{n-1}^2S_{n-1}^2 + S_{n-1}^2C_n^2).$$

In FIGURE 1, we depict this probability for the case of a coherent state with a mean photon number of 64 and a scaled interaction time of $g\tau = 50$. We emphasize that this expression does not contain interference terms because we had to sum up two probabilities rather than probability amplitudes. The reason for this is that the two states that contribute, namely, $|n + 1; a, b\rangle$ and $|n + 1; b, a\rangle$, are orthogonal.

At this point, one might ask the question of whether it is possible to figure out a specific mechanism that leads to interference between the two contributions. For this purpose, we take into account the spontaneous decay of the excited state $|a\rangle$ of the atoms outside the cavity.

According to APPENDIX A, we find for times much larger than the decay time $1/\Gamma$ of the atom that an initially excited two-level atom with no photons present, that is, $|a\rangle \otimes |0\rangle$, transforms into the state $|b\rangle \otimes |\gamma_1\rangle$, where $|\gamma_1\rangle$ is a superposition of one-photon states of different wave vectors; see equation 12 in APPENDIX A.

Now, we return to our problem of retrieving interference between the two contributions in equation 1 that contain the state $|n+1\rangle$. Let us first consider state 1 (i.e., equation 1) for a time much larger than $1/\Gamma$. Then, atoms in the excited state $|a\rangle$ will have decayed to the ground state $|b\rangle$ and state 1 reduces to

$$|\psi(t \gg 1/\Gamma)\rangle = \sum_{n=0}^{\infty} w_n[\ldots - i(C_nS_n|\gamma_1\rangle + S_nC_{n+1}|\gamma_2\rangle)|n+1; b, b\rangle + \ldots]. \quad (2)$$

Here, we have focused on the contributions containing one spontaneously emitted photon and have indicated the terms with two or no spontaneously emitted photons by dots. As shown in APPENDIX A, the field state $|\gamma_1\rangle$ depends on the location of the atom. Therefore, the field states $|\gamma_1\rangle$ and $|\gamma_2\rangle$ created by the first and the second atom are different because the location of the atoms is different.

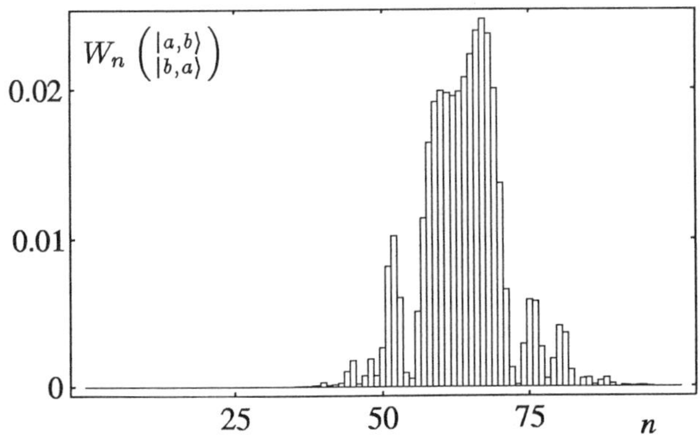

FIGURE 1. Probability distribution for finding n photons in the cavity field and exactly one atomic excitation. The initial field state is a coherent state with a mean photon number of 64. The scaled interaction time is $g\tau = 50$.

Now, both atoms are in the ground state, but still we do not get interference. Indeed, the states $|\gamma_1\rangle$ and $|\gamma_2\rangle$ are approximately orthogonal[7] provided $|\vec{r}_2 - \vec{r}_1| \gg c/\omega = \lambda\!\!\!^{-}$. Therefore, we are forced to consider more complicated schemes in order to regain interference. Suppose we place another two-level atom in its ground state as a detector atom in the region behind the cavity as shown in FIGURE 2. Then, the radiation emitted by the two atoms can be absorbed by this detector atom. Using the field states

$$|\pm\rangle \equiv \left(\frac{1}{\sqrt{2}}\right)(|\gamma_1\rangle \pm |\gamma_2\rangle),$$

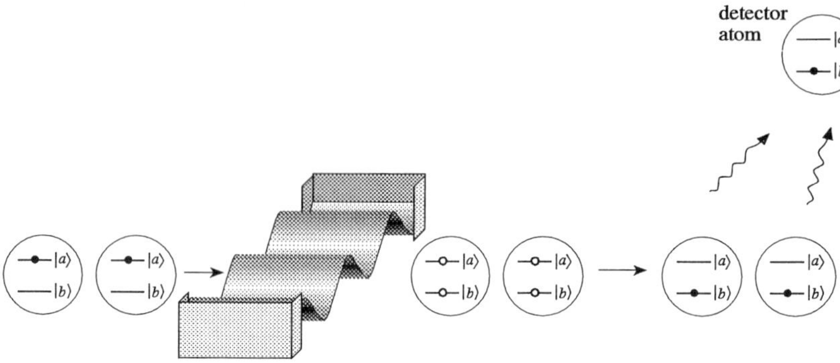

FIGURE 2. Two two-level atoms interact with a resonant field and leave the cavity in a superposition of upper and lower level. An atom in the excited state will decay to the ground state. The emitted radiation can be absorbed by another two-level atom placed far away from the others.

we can write state 2 (i.e., equation 2) as

$$|\psi\rangle = \sum_{n=0}^{\infty} w_n \left\{ \dots - \left(\frac{i}{\sqrt{2}}\right) S_n[(C_n + C_{n+1})|+\rangle + (C_n - C_{n+1})|-\rangle]|b\rangle_{\text{det}} \otimes |n+1; b, b\rangle + \dots \right\},$$

where $|b\rangle_{\text{det}}$ denotes the ground state of the detector atom. If we place the detector atom such that the distance to the first atom and the distance to the second atom are equal, the radiation described by the state $|-\rangle$ cannot excite the detector atom; see APPENDIX B. Therefore, only radiation described by the state $|+\rangle$ can be absorbed.

Hence, the probability of finding n photons in the cavity and an excited detector atom reads

$$\begin{pmatrix}\text{probability for } n \text{ photons}\\ +\text{ excited detector atom}\end{pmatrix} \equiv W_n(|b, b\rangle; |a\rangle_{\text{det}}) = \mathcal{N}\{|w_{n-1}|^2 |C_{n-1}S_{n-1} + S_{n-1}C_n|^2 + \dots\}, \quad \textbf{(3)}$$

where $\mathcal{N}$ is a normalization constant. The dots indicate another contribution to the photon statistics coming from the part of state 2 with two spontaneously emitted photons. This expression clearly contains interference—we first add probability amplitudes and then take the square. The physical reason for this interference is that there exist two different paths (in an abstract space) that in our approach cannot be distinguished, but they both lead to the same final state. One of these paths corresponds to the situation where the first atom made a transition from $|a\rangle$ to $|b\rangle$ in the resonator and the second atom did not. For the other path, the roles of atoms 1 and 2 are just interchanged. In FIGURE 3, we depict the one-photon part of the probability in equation 3 for the same parameters as in FIGURE 1. We note, however, a structure completely different from the one in FIGURE 1.

We conclude this section by noting that, in order to get interference terms in the probability describing a joint measurement on the field and atomic part of the system, we are forced to *erase* the information[11] about which path the system actually took. The "*which path*" information about the system is hidden in the states of the two atoms. Therefore, as a first possibility to erase this information, we let the atoms decay to their ground state by spontaneous emission and absorb the emitted radiation by a detector atom. This very process of absorption serves as an eraser for the "*which path*" information. The central lesson is that we need a detector system that tells us that both atoms are in the ground state without providing us the information of how this happened, that is, via which path they got into the ground state. In the further sections of this report, we present a more sophisticated way to erase the "*which path*" information and show how the cavity field state can be manipulated by erasing that information.

MULTICHANNEL INTERFERENCE BY SPONTANEOUS EMISSION

Up to now, we have learned how, in principle, we could get interference by erasing "*which path*" information. In the present section, we extend our considerations from a two-level atom to a three-level system with a level scheme as shown in FIGURE 4. In particular, we assume that our atoms have a ground state $|g\rangle$ and we allow transitions $|a\rangle \rightarrow |g\rangle$ or $|b\rangle \rightarrow |g\rangle$ in the optical regime and transitions $|a\rangle \rightarrow |b\rangle$ in the microwave domain.[27]

The first part of our scheme, namely, the interaction of the atoms with this cavity field, is the same as before because the resonant cavity field couples only the levels $|a\rangle$ and $|b\rangle$ very strongly and the third level $|g\rangle$ can be neglected completely.

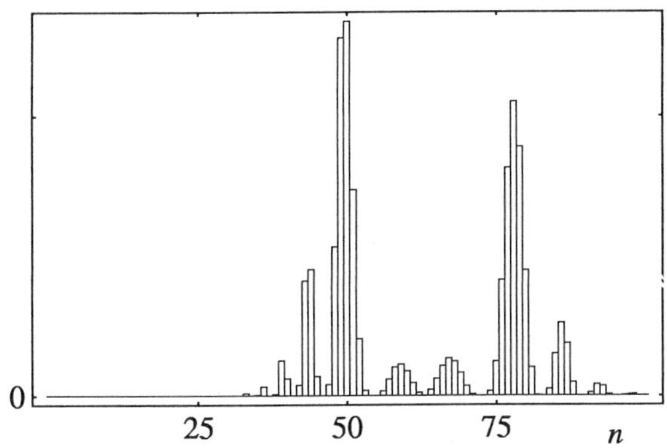

FIGURE 3. One-photon part of the probability distribution (arbitrary units) for finding n photons in the cavity field and the detector atom in its excited state. The initial field state is a coherent state with a mean photon number of 64. The scaled interaction time is $g\tau = 50$.

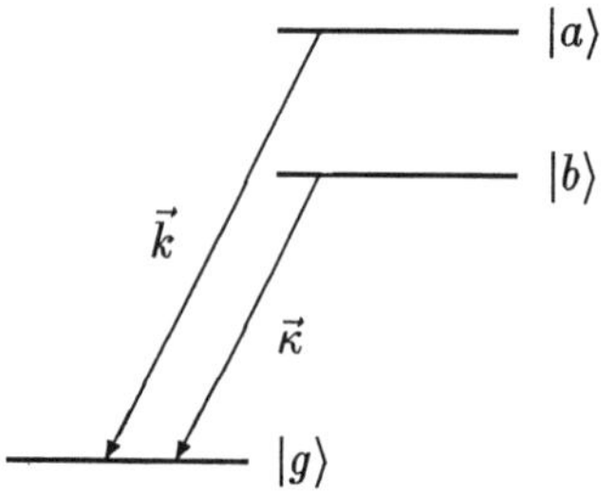

FIGURE 4. Extended three-level system. State $|a\rangle$ decays to the ground state $|g\rangle$ and emits simultaneously a photon with wave vector $\vec{k}$, whereas the decay of state $|b\rangle$ is accompanied by the emission of a photon with wave vector $\vec{\kappa}$.

However, after the interaction with the resonator field, we now allow for spontaneous decay of the levels $|a\rangle$ and $|b\rangle$ to the ground state $|g\rangle$. The emitted photons are measured by two photodetectors as illustrated schematically in FIGURE 5.

Because the decay constant Γ (see equation 13 in APPENDIX A) contains the frequency to the third power, a transition with a large energy difference is very likely compared to one with a small difference. The transition $|a\rangle \to |b\rangle$ involves frequencies that are 4–5 orders of magnitude lower than those corresponding to transitions to $|g\rangle$. Therefore, we can neglect transitions from $|a\rangle$ to $|b\rangle$ during the evolution in the free-field region. We describe the complete system by a state vector consisting of three main parts:

- the cavity field mode,
- the two atoms,
- the quantized free field outside the resonator that induces spontaneous emission.

We start our investigation when both atoms have left the cavity, but have not yet decayed to the ground state. The state vector consists of the state 1 describing the resonator field and the atoms and, in addition, the vacuum state of the free field

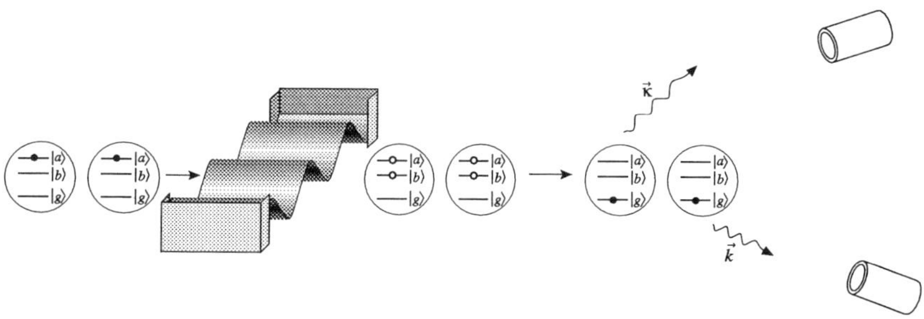

FIGURE 5. Schematic illustration of the model. Two three-level atoms interact with a quantized electromagnetic field inside a resonator. Only levels $|a\rangle$ and $|b\rangle$ are important because the atom is resonant with the field. After the atoms have left the cavity, they decay to the ground state $|g\rangle$ and the simultaneously emitted photons are measured by two photodetectors.

outside the resonator, that is,

$$|\psi\rangle = \sum_{n=0}^{\infty} w_n[C_n^2|n; a, a\rangle - iC_nS_n|n+1; a, b\rangle - iS_nC_{n+1}|n+1; b, a\rangle - S_nS_{n+1}|n+2; b, b\rangle] \otimes |0\rangle. \quad \textbf{(4)}$$

After some time, the atoms will spontaneously emit optical photons. Although the emitted radiation has a whole spectrum of different wavelengths, there are essentially only two wavelengths that are important. They correspond to the energy differences between the levels $|a\rangle$, $|g\rangle$ and $|b\rangle$, $|g\rangle$ and the corresponding wave numbers are denoted by k and κ, respectively (see FIGURE 4). We place frequency-sensitive and direction-sensitive detectors—one sensitive for k-photons and the other sensitive for κ-photons—in the far-field regime. This setup selects the modes on which we want to perform a measurement. The detection of one photon of each kind corresponds to a projection of the state vector of the complete system onto the state $|1_{\vec{k}}, 1_{\vec{\kappa}}\rangle$. Which parts of our initial state 4 (i.e., equation 4) can contribute to such an event? Because we concentrate on those cases where one measures one photon of each type, it is clear that only those terms that contain the combination $|a, b\rangle$ or $|b, a\rangle$ can contribute. This shows that our new scheme again serves as an eraser because, by only measuring the two photons, we cannot decide whether it was $|a, b\rangle$ or $|b, a\rangle$ that led to the final result. In other words, we do not know whether atom 1 emitted photon k and atom 2 emitted photon κ or vice versa. Consequently, both paths are used by the system and we end up with interference between both contributions.

After the detection of the two photons, both atoms are in the ground state $|g\rangle$; hence, we are left with a pure field state of the resonator field. The central question we now want to address is as follows: What is this pure state of the cavity field after photon detection? A rigorous calculation of the time evolution of the system taking into account the spontaneous emission and also the measurement of the photons at different times is rather complicated and lengthy; therefore, it will be published elsewhere. In the present report, we only give a brief outline of the derivation. For this purpose, we modify the asymptotic photon states $|\gamma\rangle$ discussed in APPENDIX A to fit our three-level problem. We introduce the states[c]

$$|\gamma_i^j\rangle = \sum_{\vec{k}} \frac{[g_{\vec{k}}^* e^{-i\vec{k}\cdot\vec{r}_i}/\hbar]}{\left[\Delta_{\vec{k}} + \left(\frac{1}{2}\right)i\Gamma_j\right]} |1_{\vec{k}}\rangle$$

produced by the transition from $|j = a, b\rangle$ to $|g\rangle$. Here, $i = 1, 2$ denotes the first and the second atom at position $\vec{r}_1$ and $\vec{r}_2$. The decay constants Γ_a and Γ_b are defined analogous to equation 13 and $g_{\vec{k}}$ is the coupling constant to the mode $\vec{k}$. For a time much larger than the lifetimes $1/\Gamma_a$ and $1/\Gamma_b$ of states $|a\rangle$ and $|b\rangle$, respectively, state 4

[c]The label $\vec{k}$ contains also the polarization.

reads

$$|\psi\rangle = \sum_{n=0}^{\infty} w_n[\ldots - i(C_nS_n|\gamma_1^a, \gamma_2^b\rangle + S_nC_{n+1}|\gamma_1^b, \gamma_2^a\rangle)|n+1; g, g\rangle + \ldots]$$

or

$$|\psi\rangle = \sum_{n=0}^{\infty} w_n\left[\ldots - i\sum_{\vec{k}',\vec{\kappa}'}(C_nS_ne^{-i\vec{k}'\cdot\vec{r}_1 - i\vec{\kappa}'\cdot\vec{r}_2} + S_nC_{n+1}e^{-i\vec{k}'\cdot\vec{r}_2 - i\vec{\kappa}'\cdot\vec{r}_1})\right.$$

$$\left.\times\frac{[g_{\vec{k}'}^{*}g_{\vec{\kappa}'}^{*}/\hbar^2]}{\left[\left(\Delta_{\vec{k}'} + \left[\frac{1}{2}\right]i\Gamma_a\right)\left(\Delta_{\vec{\kappa}'} + \left[\frac{1}{2}\right]i\Gamma_b\right)\right]}|n+1; g, g; 1_{\vec{k}'}, 1_{\vec{\kappa}'}\rangle + \ldots\right].$$

Here, we have assumed that the decay of the atoms is independent from one another because the atoms are separated by a distance larger than the wavelength corresponding to the k or κ transition. After measuring two photons in the specific modes $\vec{k}$ and $\vec{\kappa}$, the state of the cavity field and the atoms is governed by

$$\langle 1_{\vec{k}}, 1_{\vec{\kappa}}|\psi\rangle \propto \sum_{n=0}^{\infty} w_n(C_nS_ne^{-i\vec{k}\cdot\vec{r}_1 - i\vec{\kappa}\cdot\vec{r}_2} + S_nC_{n+1}e^{-i\vec{k}\cdot\vec{r}_2 - i\vec{\kappa}\cdot\vec{r}_1})|n+1\rangle \otimes |g, g\rangle$$

$$\propto \sum_{n=0}^{\infty} w_n(C_nS_n + e^{i(\vec{\kappa}-\vec{k})\cdot(\vec{r}_2-\vec{r}_1)}S_nC_{n+1})|n+1\rangle \otimes |g, g\rangle.$$

Hence, the field in the cavity is in the pure state

$$|\psi_{\text{cavity}}\rangle = \mathcal{N}\sum_{n=0}^{\infty} w_n(C_nS_n + e^{i(\vec{\kappa}-\vec{k})\cdot(\vec{r}_2-\vec{r}_1)}S_nC_{n+1})|n+1\rangle, \tag{5}$$

where $\mathcal{N}$ is a normalization constant. In particular, the probability to find n photons in this state,

$$\begin{pmatrix}\text{probability for } n \text{ photons after}\\ \text{detection of photons } \vec{k} \text{ and } \vec{\kappa}\end{pmatrix} = |\mathcal{N}|^2|w_{n-1}|^2|C_{n-1}S_{n-1} + e^{i(\vec{\kappa}-\vec{k})\cdot(\vec{r}_2-\vec{r}_1)}S_{n-1}C_n|^2, \tag{6}$$

clearly displays interference coming from the two indistinguishable paths that the system takes in order to arrive at the final state.

In contrast to equation 3, equation 6 depends via the phase factor $e^{i(\vec{\kappa}-\vec{k})\cdot(\vec{r}_2-\vec{r}_1)}$ on the particular experimental arrangement, namely, on the mutual distance of the atoms and on the wave vectors of the modes we decide to measure. This allows us to control the final state in the resonator by changing these experimental parameters.

The method described above can be generalized to an arbitrary number N of atoms. However, there exist many different ways of how to do this. We could, for example, repeat the experiment until we measure a fixed number, say l $(1 \leq l \leq N)$, of photons of type k and $(N - l)$ photons of type κ. In order to keep things as simple as possible, we take $l = 1$, that is, we concentrate on the case where we measure one κ-photon and $(N - 1)$ k-photons, denoted by $\vec{k}_2, \ldots, \vec{k}_N$. It can be shown[27] that the field state after the detection of the photons reads

$$|\psi_{\text{cavity}}\rangle = \mathcal{N} \sum_{n=0}^{\infty} \text{per}\, M_n |n + 1\rangle,$$

where the matrix

$$M_n = \begin{pmatrix} c_n^{(1)} & c_n^{(2)} & \ldots & c_n^{(N)} \\ 1 & \epsilon_{22} & \ldots & \epsilon_{2N} \\ \vdots & \vdots & \ddots & \vdots \\ 1 & \epsilon_{N2} & \ldots & \epsilon_{NN} \end{pmatrix}$$

has the elements

$$\epsilon_{ml} = e^{i(\vec{\kappa} - \vec{k}_m)\cdot(\vec{R}_l - \vec{R}_1)}$$

and

$$c_n^{(l)} = w_n C_n^{l-1} S_n C_{n+1}^{N-l}.$$

Here, per A denotes the permanent of the matrix $A = (a_{ij})$ defined as

$$\text{per}\, A \equiv \sum_{\pi \in P_N} a_{1\pi_1} a_{2\pi_2} \ldots a_{N\pi_N}.$$

The sum runs over all permutations of length N. The permanent is very similar to a determinant; the only difference is that the familiar factor $(-1)^{\text{sign}\pi}$ is missing in the case of the permanent. Note that the case $N = 2$ immediately reproduces equation 5.

The appearance of the permanent reflects the fact that the final state contains all contributions originating from different decay paths; that is, each atom could have emitted each of the detected photons. Each combination thus leads to one term of the permanent.

Where are the parameters that determine the final field state and how many are there? First of all, there is the interaction time that fixes the first row of the matrix M_n. Then, we have the $(N - 1) \times (N - 1)$ submatrix ϵ_{ml} at our disposal. Each of its $(N - 1)^2$ elements is a complex number with modulus unity. The corresponding phase is determined by the relative distance of the atoms and by the wave vectors of the emitted photons.

Let us summarize the results of this section: By letting N three-level atoms interact with a cavity field, we create a highly entangled state. Outside of the resonator, we place detectors that can distinguish between k- and κ-photons. Some time after the atoms have left the cavity, they decay to the ground state and emit optical photons that are detected. Just by measuring the photons, we cannot infer which atom emitted which photon or, in other words, we cannot infer the decay channel that the system took. Therefore, all possible decay channels interfere—we have multichannel interference. After the detection of the photons, we are left with a pure field state in the resonator. Although the detection process is separated in space and time from the interaction with the resonator field, the mentioned field state depends on parameters like the direction of the wave vectors of the detected photons. This offers the possibility to manipulate the field state just by changing the position of the detectors. The method presented in this section relies on the entanglement between the cavity field, the atoms, and the field outside the cavity.

QUANTUM STATE ENGINEERING USING SPONTANEOUS EMISSION

At this stage, we want to discuss the kind of field states that we can create by the method described above. Because the number of parameters grows approximately as N^2, it is very difficult to keep track of the influence of all these parameters. Therefore, we consider the special case where (i) all atoms are equidistant, that is, $\vec{R}_l = l\vec{a}$ with a fixed distance $\vec{a}$ between consecutive atoms, and (ii) all k-photons are assumed to be emitted in the same direction, that is, $\vec{k}_2 = \ldots = \vec{k}_N \equiv \vec{k}$. This reduces the set of $(N-1)^2$ independent parameters of the matrix ϵ_{ml} to a single parameter, namely,

$$\epsilon_{ml} = e^{-il\beta} \qquad \text{for all } m,$$

where $\beta = (\vec{k} - \vec{\kappa}) \cdot \vec{a}$. For this special case, it is possible to calculate the permanent of M_n explicitly[27] and we find for the field state

$$|\psi_{\text{cavity}}\rangle = \mathcal{N} \sum_{n=0}^{\infty} \sum_{l=1}^{N} w_n e^{i(l-1)\beta} C_n^{l-1} S_n C_{n+1}^{N-l} |n+1\rangle. \tag{7}$$

Let us now take a coherent state with a real amplitude $\alpha_0 \gg 1$ as the initial field state. We get a good approximation for the coefficients in equation 7 by expanding the square root $\sqrt{n+1}$ up to first-order around the maximum $n = \alpha_0^2$ of the Poisson distribution:

$$|w_n|^2 = e^{-\alpha_0^2} \left(\frac{\alpha_0^{2n}}{n!}\right).$$

In this case, it is possible to perform the sum over n and we can represent the field state as a sum of $(N+1)$ coherent states located on a circle of radius α_0, that is,

$$|\psi_{\text{cavity}}\rangle \cong \mathcal{N} \sum_{\lambda=0}^{N} G_\lambda |\alpha_0 e^{i\varphi_\lambda}\rangle, \tag{8}$$

where

$$\varphi_\lambda = (N - 2\lambda)\theta$$

and

$$\theta = \frac{g\tau}{2\alpha_0}.$$

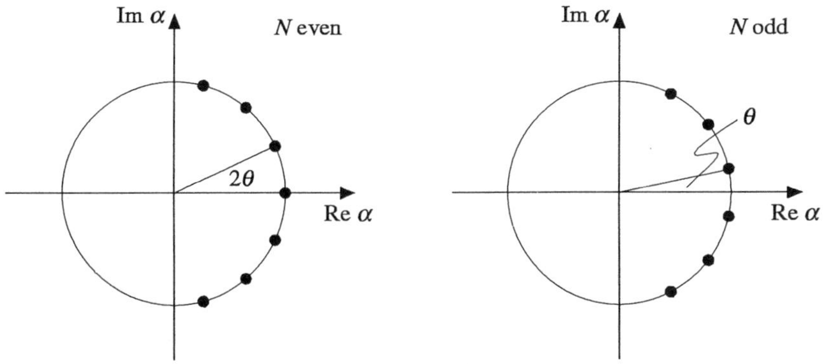

FIGURE 6. Illustration of state 8 (i.e., equation 8). It consists of $(N + 1)$ coherent states lying on a circle. The angle between neighboring peaks is always 2θ. In the case of even N, one of the peaks is located at α_0.

Here, the weight function G_λ is a known, but complicated, function of θ, α_0, N, and (most importantly) our parameter β.

FIGURE 6 illustrates the structure of state 8 (i.e., equation 8) in Q-function phase space. As is already known,[28,29] the interaction of a single atom with a coherent state leads to a splitting of the peak in the Q-function into two counterrotating peaks. If we take a whole sequence of N atoms instead of 1 atom, each of them will lead to a further splitting of the Q-function. In the case of identical interaction times, some of these peaks overlap and effectively we get $(N + 1)$ peaks lying on a circle. The weight

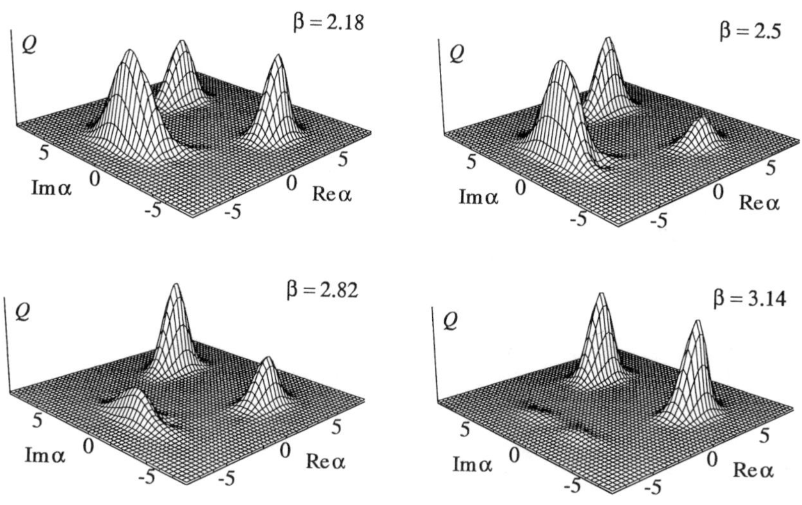

FIGURE 7. Influence of the parameter β. By changing β appropriately, we can suppress the third peak almost completely. The field is initially in a coherent state with a mean photon number of 25 and we have used 5 atoms.

of each peak depends, among other parameters, on our choice of β. By changing β, we are able to change the weight of the different peaks. We demonstrate this in FIGURE 7. Here, we use 5 atoms and the initial field state is a coherent state with a mean photon number of 25. The scaled interaction time is $g\tau = 10$. From our former considerations, we would expect to find 11 peaks. However, the weight of the outermost peaks is already exponentially small and 2 of the visible peaks overlap strongly at $\alpha = -\alpha_0$. By changing the parameter β, we can make one of the peaks vanish almost completely.

This example demonstrates state engineering using spontaneous emission in a most vivid way.

SUMMARY

We have used a system consisting of two- or three-level atoms, one mode of a cavity field, and the free radiation field outside the cavity as an example to demonstrate some fundamental laws of quantum mechanics: (i) for distinguishable paths, we must add *probabilities*; (ii) for indistinguishable paths, we must add *probability amplitudes.* We have used the absorption of spontaneously emitted photons by detectors as *quantum erasers* in order to erase the "*which path*" information. Using a *quantum eraser*, we were able to retrieve interference terms in the photon statistics of the cavity field. Furthermore, we have shown that we can use our erasers in order to manipulate the cavity field state in a controlled way—that is the idea of quantum state engineering. The way that we manipulate the field state in the cavity is based on another fundamental idea of quantum mechanics: a measurement on a subsystem of an entangled state can drastically change the state of another subsystem without any "physical interaction" between the two subsystems, that is, without an interaction described by a Hamiltonian.

REFERENCES

1. WHEELER, J. A. & W. H. ZUREK. 1983. Quantum Theory and Measurement. Princeton University Press. Princeton, New Jersey.
2. HILL, D. L. & J. A. WHEELER. 1953. Phys. Rev. **89:** 1102–1145.
3. FORD, K. W. & J. A. WHEELER. 1959. Ann. Phys. **7:** 259–286.
4. FORD, K. W. & J. A. WHEELER. 1959. Ann. Phys. **7:** 287–322.
5. SCHLEICH, W. P. & J. A. WHEELER. 1987. Nature **326:** 574–577.
6. SCULLY, M. O. & K. DRÜHL. 1982. Phys. Rev. A **25:** 2208–2213.
7. HILLERY, M. & M. O. SCULLY. 1983. *In* Quantum Optics, Experimental Gravitation, and Measurement Theory. P. Meystre & M. O. Scully, Eds.: 65–85. Plenum. New York.
8. SCULLY, M. O. & H. WALTHER. 1989. Phys. Rev. A **39:** 5229–5236.
9. SCULLY, M. O., B-G. ENGLERT & J. SCHWINGER. 1989. Phys. Rev. A **40:** 1775–1784.
10. ENGLERT, B-G. & M. O. SCULLY. 1990. *In* New Frontiers in Quantum Electrodynamics and Quantum Optics. A. O. Barut, Ed.: 507–512. Plenum. New York.
11. SCULLY, M. O., B-G. ENGLERT & H. WALTHER. 1991. Nature **351:** 111–116.
12. LAMB, W. E. 1969. Phys. Today **22**(4): 23–28.
13. MEYSTRE, P. 1987. Opt. Lett. **12:** 669–671.
14. KRAUSE, J., M. O. SCULLY & H. WALTHER. 1989. Phys. Rev. A **36:** 4547–4550.

15. KRAUSE, J., M. O. SCULLY, T. WALTHER & H. WALTHER. 1989. Phys. Rev. A **39:** 1915–1921.
16. PAUL, H. 1989. J. Mod. Opt. **38:** 515.
17. BRUNE, M., S. HAROCHE, V. LEFEVRE, J. M. RAIMOND & N. ZAGURY. 1990. Phys. Rev. Lett. **65:** 976–979.
18. MEYSTRE, P., J. J. SLOSSER & M. WILKENS. 1990. Opt. Commun. **79:** 300–304.
19. HOLLAND, M. J., D. F. WALLS & P. ZOLLER. 1991. Phys. Rev. Lett. **67:** 1716–1719.
20. BRUNE, M., S. HAROCHE, J. M. RAIMOND, L. DAVIDOVICH & N. ZAGURY. 1992. Phys. Rev. A **45:** 5193–5214.
21. KNIGHT, P. L. & B. W. SHORE. 1993. Phys. Rev. A **48:** 642–655.
22. VOGEL, K., V. M. AKULIN & W. P. SCHLEICH. 1993. Phys. Rev. Lett. **71:** 1816–1820.
23. VOGEL, K., V. M. AKULIN & W. P. SCHLEICH. 1993. Phys. Bl. **49**(12): 1111–1112.
24. PARKINS, A. S., P. MARTE, P. ZOLLER & H. J. KIMBLE. 1993. Phys. Rev. Lett. **71:** 3095–3098.
25. GARRAWAY, B. M., B. SHERMAN, H. MOYA-CESSA, P. L. KNIGHT & G. KURIZKI. 1994. Phys. Rev. A **49:** 535–547.
26. BUŽEK, V. 1994. Acta Phys. Slovaca **44:** 1–7.
27. KRÄHMER, D. S. 1994. Diploma thesis. University of Ulm.
28. EISELT, J. & H. RISKEN. 1989. Opt. Commun. **72:** 351–355.
29. EISELT, J. & H. RISKEN. 1991. Phys. Rev. A **43:** 346–360.

APPENDIX A

Asymptotic Field State of Spontaneous Emission

In the present appendix, we consider the interaction of an initially excited two-level atom with a multimode field that does not contain any photons at the beginning. The Hamiltonian for this system reads[d]

$$H = \left(\frac{1}{2}\right)\hbar\omega\sigma_z + \sum_{\vec{k}} \hbar\Omega_{\vec{k}} a_{\vec{k}}^\dagger a_{\vec{k}} + \sum_{\vec{k}} [g_{\vec{k}}(\vec{r}_1)\sigma^\dagger a_{\vec{k}} + g_{\vec{k}}^*(\vec{r}_1)\sigma a_{\vec{k}}^\dagger],$$

where ω is the atomic transition frequency, σ_z is the usual Pauli spin matrix, and

$$g_{\vec{k}}(\vec{r}_1) = g_{\vec{k}} e^{i\vec{k}\cdot\vec{r}_1} = \left(-\sqrt{\frac{\hbar\Omega_{\vec{k}}}{2\epsilon_0 V}}\right)\vec{e}_{\vec{k}} \cdot \vec{p} e^{i\vec{k}\cdot\vec{r}_1}$$

is the coupling strength between an atom at position $\vec{r}_1$ with dipole moment $\vec{p}$ and the field mode $\vec{k}$ with polarization vector $\vec{e}_{\vec{k}}$. We expand the state vector of the system,

$$|\phi\rangle = \alpha(t)e^{-(1/2)i\omega t}|a, 0\rangle + \sum_{\vec{k}} \beta_{\vec{k}}(t)e^{(1/2)i\omega t - i\Omega_{\vec{k}} t}|b, 1_{\vec{k}}\rangle, \tag{9}$$

where the initial conditions are $\alpha(0) = 1$ and $\beta_{\vec{k}}(0) = 0$. When we substitute this ansatz into the Schrödinger equation

$$(i\hbar)\frac{d}{dt}|\phi\rangle = H|\phi\rangle,$$

[d]The label $\vec{k}$ contains also the polarization.

we find the equations of motion

$$\dot{\alpha} = \left(\frac{1}{i\hbar}\right) \sum_{\vec{k}} \beta_{\vec{k}} g_{\vec{k}}(\vec{r}_1) e^{-i\Delta_{\vec{k}} t} \tag{10}$$

and

$$\dot{\beta}_{\vec{k}} = \left(\frac{1}{i\hbar}\right) \alpha g^*_{\vec{k}}(\vec{r}_1) e^{i\Delta_{\vec{k}} t} \tag{11}$$

for the coefficients α and $\beta_{\vec{k}}$. Here, we have introduced the detuning $\Delta_{\vec{k}} = \Omega_{\vec{k}} - \omega$. Within the Wigner-Weisskopf approximation, we can solve these differential equations (equations 10 and 11) and obtain, for times much larger than the atomic decay time $1/\Gamma$, the state[7]

$$|\phi(t \gg 1/\Gamma)\rangle \cong \sum_{\vec{k}} \left[\frac{g^*_{\vec{k}}(\vec{r}_1)/\hbar}{\left[\Delta_{\vec{k}} + \left(\frac{1}{2}\right) i\Gamma\right]} \right] |b, 1_{\vec{k}}\rangle = |b\rangle \otimes |\gamma_1\rangle,$$

where we have introduced the asymptotic photon state

$$|\gamma_1\rangle = \sum_{\vec{k}} \left[\frac{g^*_{\vec{k}}(\vec{r}_1)/\hbar}{\left[\Delta_{\vec{k}} + \left(\frac{1}{2}\right) i\Gamma\right]} \right] |1_{\vec{k}}\rangle \tag{12}$$

and the damping constant

$$\Gamma = \frac{\omega^3 p^2}{3\pi\hbar\epsilon_0 c^3}. \tag{13}$$

APPENDIX B

Reabsorption of Spontaneously Emitted Photons

In this appendix, we investigate the absorption of a photon by a detector atom after it has been spontaneously emitted by another atom of the same kind. After the first two-level atom at position $\vec{r}_1$ has spontaneously emitted a photon, the state of the radiation field is given by equation 12. Because the detector atom at position $\vec{x}$ is of the same kind as the atom that has emitted the photon, the interaction of the detector atom with the radiation field is described in the same way as the interaction of the emitting atom with the radiation field. We therefore make the ansatz of equation 9 and find equations 10 and 11 with $\vec{r}_1$ replaced by $\vec{x}$ as the equations of motion. Furthermore, because the system is initially in the state $|b\rangle_{\text{det}} \otimes |\gamma_1\rangle$, we now have the initial conditions

$$\alpha(0) = 0$$

and

$$\beta_{\vec{k}}(0) = \frac{[g_{\vec{k}}^{*}(\vec{r}_1)/\hbar]}{\left[\Delta_{\vec{k}} + \left(\frac{1}{2}\right)i\Gamma\right]}.$$

For $\alpha(t)$, we find

$$\dot{\alpha}(t) = \left(-\frac{1}{\hbar^2}\right)\sum_{\vec{k}} |g_{\vec{k}}|^2 \int_0^t e^{-i\Delta_{\vec{k}}(t-t')}\alpha(t')dt' - \left(\frac{i}{\hbar^2}\right)\sum_{\vec{k}}\left(\frac{|g_{\vec{k}}|^2 e^{i\vec{k}\cdot(\vec{x}-\vec{r}_1)}}{\left[\Delta_{\vec{k}} + \left(\frac{1}{2}\right)i\Gamma\right]}\right)e^{-i\Delta_{\vec{k}}t}.$$

The first part does not depend on the position of the atoms and is usually approximated by $(-\Gamma/2)\alpha(t)$, that is, the Wigner-Weisskopf approximation for a two-level atom. As shown by Hillery and Scully[7] for a similar expression, the second part depends only on the distance $|\vec{x} - \vec{r}_1|$ between the absorbing and the emitting atom and not on the actual positions of the atoms. Therefore, $\alpha(t)$ depends only on $|\vec{x} - \vec{r}_1|$.

We now consider the radiation field described by the state

$$|-\rangle = \left(\frac{1}{\sqrt{2}}\right)(|\gamma_1\rangle - |\gamma_2\rangle),$$

that is, a superposition between the radiation field generated by a photon emitted by atom 1 at position $\vec{r}_1$ and the radiation field generated by a photon emitted by atom 2 at position $\vec{r}_2$. We place the detector atom such that the distance to atom 1 is equal to the distance to atom 2 and thus find two identical contributions to $\alpha(t)$ with different signs. Because these two contributions cancel, the detector atom cannot absorb radiation from the field state $|-\rangle$.

Varieties of Quantum Measurement[a]

W. G. UNRUH

CIAR Cosmology Program
Department of Physics
University of British Columbia
Vancouver, British Columbia, Canada V6T 1Z1

The problem of quantum measurement has been with us since the foundations of the theory were laid in the mid-1920s. It has generated much discussion, with little resolution of the questions raised. I will argue here that this situation has arisen in part because of the confusion brought about by giving two very different concepts the same name, with the expected result that the valid questions related to the two concepts have become entangled. It furthermore has led to a restriction on the types of measurements considered within the theory. In this report, I am not going to propose any radical or even very new interpretations of the theory of quantum mechanics. I am rather going to engage in an ancient philosophical pastime, namely, to propose that we use distinct terms for distinct concepts. I am then going to review some of the novel insights that have been obtained recently (especially by the group around Aharonov) regarding some novel types of measurement.

MEASUREMENT, DETERMINATION, AND KNOWLEDGE

The concept of measurement in quantum mechanics has had a long and confused history. There are essentially two separate concepts that have been conflated under the same title, concepts with a very different status in the theory a priori. In part, the intense confusion surrounding the word results from the attempt to reconcile these two different concepts, or rather to apply the properties of one concept to the other.

The first concept subsumed under the term "measurement" is an axiomatic concept. Quantum mechanics, as with all of our theories in physics, is based on a set of mathematical structures. In the case of quantum mechanics, these structures are complex Hilbert spaces and operators on those Hilbert spaces. In addition to such mathematical structures, the theory must also make contact with the physical world. Structures in the theory must be correlated by structures in our experience of the world itself. As with all theories, quantum mechanics is a means of answering questions about our experiences of the world. Furthermore, they are questions related to the particulars and peculiarities of the actual world we live in. The theory requires a mapping of the mathematical structure onto our experiences. As in all physical theories, this takes the form of a general map, both of general structures of the world that we expect to have a broad range of validity and of structures that reflect the particulars and peculiarities of our experiences.

[a] This work was performed under a fellowship and other support from the Canadian Institute for Advanced Research and under a grant from the Natural Science and Engineering Research Council of Canada.

In classical physics, the former constitute what is called the dynamical theory, whereas the latter are called the initial conditions. The theory encompasses the identification of dynamical variables and equations of motion, whereas the "initial conditions" encompass those aspects of our experiences that are felt to be peculiar to the individual time and place of those experiences.

Quantum mechanics contains both of these aspects as well, but in a very different form from that of classical physics. The dynamics is represented by the operators, whereas, in the simplest case, the particulars of the situation are represented by the vector in the Hilbert space, that is, the wave function. I will denote these particulars by the term "knowledge" or "conditions" rather than the term "initial conditions" because, as we will see, conditions need not be initial nor are they equivalent in general to initial conditions (as they are in classical physics).

In addition to explanations, the theory must produce answers, that is, answers to questions that we have about the physical situations that we are interested in. It is here that the theory actually makes contact with the physical world. In quantum mechanics, these answers are in terms of probabilities. The usual phraseology goes something like this: "When one measures a quantity, and the system is in the state $|\phi\rangle$, the outcome of that measurement is one of the eigenvalues, say a, of the operator, say **A**, representing the physical variable measured, and the probability is given by the usual expression $|\langle a|\phi\rangle|^2$."

However, the word "measure" brings with it the image of a physical process. Measurements are performed by means of measuring apparatuses. As aspects of the physical world, such measuring apparatuses should themselves be describable by quantum mechanics itself. However, it is difficult to have a system in which, at the same time, a concept is an axiomatic feature of the theory and one describable by the theory. Therefore, I would suggest that the word "determine" be used for this axiomatic feature of the theory. Thus, I would rephrase the above sentence as follows: "When one determines a quantity, and the knowledge (or conditions) under which one wishes to determine that quantity is represented by the vector $|\phi\rangle$, then the determination of a quantity represented by **A** gives one of the eigenvalues of **A**, say a, with probability $|\langle a|\phi\rangle|^2$."

Determination, in this axiomatic sense, says nothing about how the determination was made. It is simply a statement of a mapping from the theory to our experience, in which some knowledge sets the conditions on the questions we wish to ask and some knowledge represents the answers to the questions we want to ask.

What then is a measurement? I will reserve the term "measurement" for a physical process, that is, a process describable in terms of quantum theory itself. A measurement is a process in which one has two separate physical systems, represented by two separate sets of dynamical operators. Furthermore, the dynamical evolution is such that, given certain conditions on the measuring apparatus, a determination of some quantity associated with the measuring apparatus will give information about the system of interest.

Under certain conditions, as von Neumann[1] showed, a measurement on a system could be treated as a determination of that system; that is, certain types of measurement (in which one makes a determination of some aspect of the measuring apparatus only) acted in all ways as though one had instead made a determination of the system itself. There is a consistency in quantum mechanics such that the axiomatic concept that I call determination is closely related to the physical process that I call measurement. However, in von Neumann's analysis, notice that one has not done away with the concept of determination. One still must apply the axiomatic concept of determination to the measuring apparatus before one can draw any conclusions at all from the theory. It is just that such measurements allow us to reduce a complicated system (apparatus plus system of interest) to a simple system (the system of interest alone) under certain conditions. This mapping of a complex system onto a simpler system does not, however, in any way change the requirement for the axiomatic concept of "determination". It simply changes the system to which we need to apply the concept.

At least, in part, the measurement problem in quantum mechanics is the disquiet that physicists feel for the concept of "determination". It feels like an extra and extraneous concept, a nonphysical concept. In classical physics, one can imagine that the theory and reality are in complete correspondence. The position of a particle really is a number and our experience of that position is simply the experience of that number. The physical map from experience to theory is just an identification of those numbers in the theory with the experience. (That some fairly sophisticated manipulations of experience are necessary to extract that number is a technical detail.) In quantum mechanics, on the other hand, there seems to be no direct map from our experience to the theory. The operators themselves have far too much structure for experience. The state, or the Hilbert space vector itself, has the wrong properties to map onto our experience. The only map is the rather indirect and seemingly unnatural one of "determination". One would like either to subsume determination under some physical concept of the theory (but that would lose the only relation between experience and theory that the theory contains) or to introduce some other relation between the theory and experience from which one could derive "determination" in a natural way. That neither of these objectives has ever been achieved is a large part of the "problem of measurement" in quantum mechanics.

However, I do not want to spend any more of my time on this issue; rather, I want to point out that the concern about this problem has warped our thinking about quantum mechanics and about the types of measurement possible in the theory. Because the von Neumann type of measurement creates the possibility of reduction of a complex system to a simpler system, the idea has become implanted that all measurement must be of the same sort. It seems to be the thinking that, because determination has a certain form, measurement must have the same form. However, it is becoming clear, especially through the work of the group around Aharonov, that this is too restrictive.

Measurement is a physical process by which one has two systems interacting and, by making a determination on one system, one can obtain information about the other system. In certain cases, the information obtained is the same as a determination; however, in other cases, it can differ significantly. Furthermore, because of the similarity of wave mechanics to classical wave theory, the impression has also arisen that conditions in quantum mechanics are entirely equivalent to conditions in classical mechanics, namely, initial conditions. Let us now look at the last case first.

It has long been known to some (but ignored or resisted by most) that the conditions in quantum mechanics differ significantly from those of classical physics.[2] In classical physics, all conditions can be mapped onto initial conditions by use of the equations of motion. Whether one measures the position now and the momentum two days hence, or measures them both now, is really irrelevant. For any condition, imposed at a time, one can always, by use of the equations of motion, produce initial conditions that are entirely equivalent in all of their predictions to those general conditions. However, as Aharonov *et al.*[2] already showed about 30 years ago (and as has been independently rediscovered often since; e.g., see reference 3), setting conditions at different times may not be equivalent to any initial conditions. The simplest example is that of a spin-½ particle whose x-component of spin is known at 9 A.M. and whose y-component is known at 11 A.M. Now, say both are known to have a value of +½. The probability that the answer will be +½ if one determines the component $\cos(\theta)S_x + \sin(\theta)S_y$ at 10 A.M. is

$$P_{S_\theta=1/2} = \frac{[1 + \cos(\theta)][1 + \sin(\theta)]}{2[1 + \sin(\theta)\cos(\theta)]}. \tag{1}$$

Note that there exists no initial condition—wave function nor density matrix—for the spin-½ particle that would give this answer. It is unity for both $\theta = 0$ and $\theta = \pi/2$. The conclusion drawn from this simple example is true in general—conditions in quantum mechanics are not equivalent to initial conditions.

Already at the last New York Academy of Sciences meeting in 1986, Aharonov[4] mentioned a surprising new effect that combines the inequivalence of conditions to initial conditions with what he calls "weak" measurements. If we set both initial and final conditions and, at an intermediate time, perform a particular type of inexact measurement of a quantity, the outcome of that measurement can be very counterintuitive. Although the measuring apparatus and the interaction are designed so that, if the initial state is an eigenstate of the measured quantity, the outcome will be approximately given by that value for the measured quantity, the expected value for the measurements in this pre- and post-conditioned experiment is impossible according to all the usual tenets of quantum mechanics.

Let us make this clear by an example. Our measuring apparatus is a trivial infinite-mass free particle. It is coupled to a spin-s particle (in our example, $s = 20$). The coupling is of the form,

$$H_I = \left[\frac{\epsilon}{\sqrt{(2)}}\right](S_x + S_y)p\delta(t - t_0); \tag{2}$$

that is, the interaction is such that, if the initial state of the free particle is $\psi(x)$ and the state of the spin is in an eigenstate of the operator $S_{//}$, say with eigenvalue σ, then the final state of the free particle is $\psi(x - \epsilon\sigma)$. Thus, by measuring the displacement of the free particle due to the interaction, one can estimate σ and hence measure $S_{//}$. If the particle begins in an eigenstate (or almost an eigenstate) of X (i.e., ψ is sharply peaked about some value x_0 with an uncertainty much less than ϵ), then the displacement during the interaction can be measured precisely by determining the value x of X after the interaction and $\sigma = (x - x_0)/\epsilon$ will be a measurement of S. On the other hand, if the initial ψ has a spread of Δx, then the final determination of X will give the displacement only to $\pm\Delta x$; that is, we will have $\sigma = [(x - x_0)/\epsilon] \pm (\Delta x/\epsilon)$. This is the sense in which the measuring apparatus is inexact. The determina-

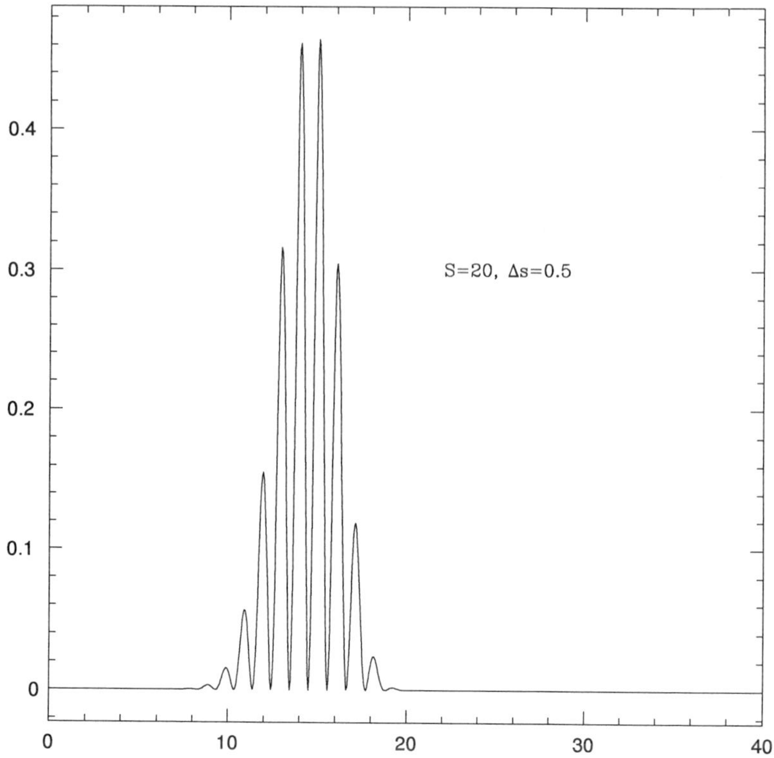

FIGURE 1. The probability distribution for the pointer of the measuring apparatus with pre- and post-conditions with a small error (≈ 0.5) for the measurement of the spin. The maximum value of the spin is 20.

tion of some variable of the measuring apparatus only gives an inexact estimate of the value of some dynamic variable of the system.

Now consider the following situation. Set conditions before the interaction with the measuring apparatus such that the value of S_x is known to be its maximum possible value, s. Furthermore, after the interaction, the value of the y-component, S_y, is known to be the maximum possible value, s. What is the distribution of possible outcomes for the measuring apparatus? One would expect this to be something like the probability distribution over the various possible values for $S_{//}$ convoluted with the initial probability distribution for the position of the free particle; that is, one would expect something like $\Sigma_\sigma P_\sigma |\psi(x - \epsilon\sigma)|^2$, where P_σ is a probability for the spin to have a value $-s \leq \sigma \leq s$. In particular, the average value (expectation value) for X should lie somewhere between $x_0 - \epsilon s$ and $x_0 + \epsilon s$. If the measurement is sufficiently accurate, this expectation is fulfilled. FIGURE 1 plots the probability distribution for the location of the particle ($x_0 = 0$ and $\epsilon = 1$) in the case where the initial spread of the wave function for the particle is small. However, FIGURE 2 is the plot of the distribution for the value of the position of the particle in the case where the initial spread for X is large [of order $\pm\epsilon\sqrt{(s)}$]. Note that the center of the probability distribution is at $x \approx 28$, and the probability that x would lie between -20 and 20 (the naively expected values) is very small. Using the determined value of X to infer the value of $S_{//}$ gives a value at all times larger than the maximum eigenvalue of $S_{//}$.

If we regard S as a classical vector spin and if we know that S_x and S_y both have value s, then $S_{//}$ will have value $\sqrt{S_x^2 + S_y^2} = \sqrt{2}s \approx 28$. Note also that this works in this way only if the initial state $\psi(x)$ is sufficiently smooth. (In my case, I have chosen it to be a Gaussian.) In particular, sharp features in ψ will destroy this property.

One reaction to this example is that it is not a real measurement. However, it meets all of the criteria of a measuring apparatus in that, if the state of the spin is an eigenstate, the measuring apparatus produces the value to the accuracy to which the apparatus is designed. What we have here is a strange result that arises from the combination of an inexact measuring apparatus with the inequivalence of conditions in quantum mechanics to be equivalent to initial conditions. (For any initial conditions, the expectations that the result would simply have been the sum of the probabilities of the result for the eigenstates would have been true.) Note that this is

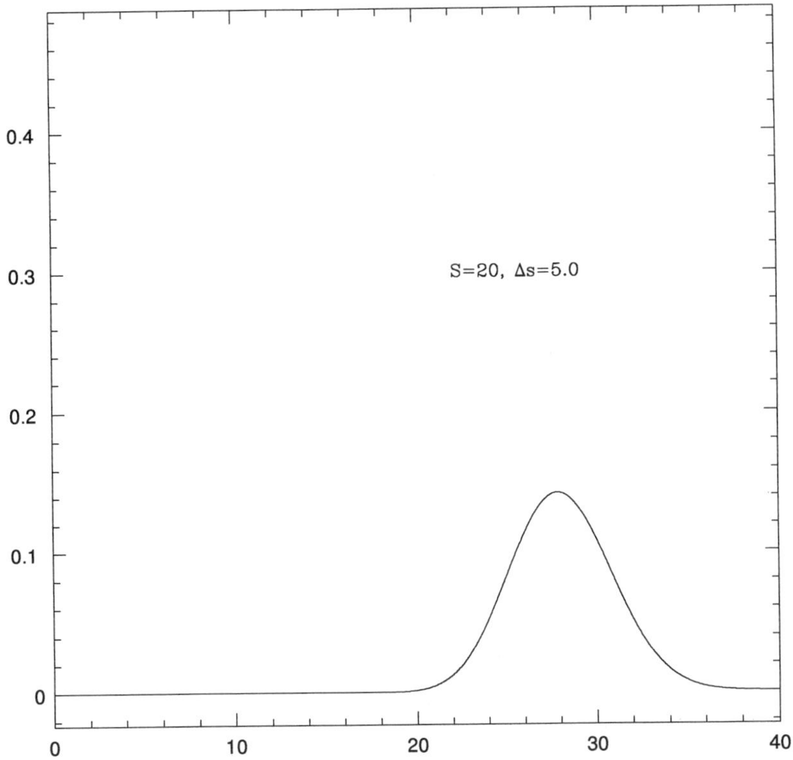

FIGURE 2. The probability distribution for the pointer with the same conditions as in FIGURE 1, but with a large error (5.0) for the inferred value of the spin. Note that the distribution centers around the value of 28 and has only a very small probability of lying between 20 and -20.

a measurement situation in which the measurement is not equivalent to any determination on the spin system itself.

There is another measurement situation that leads to results in conflict with the von Neumann equivalence of measurements and determination. This is a situation that I call adiabatic measurement. It arises out of another situation noted by the group around Aharonov.[5] (They call it "protected" measurements, a term I feel to be highly misleading. They furthermore use it to argue that the wave function is "real" in some sense, a conclusion I also have great difficulty with.[6]) This is a situation in which the measuring apparatus is coupled to the system sufficiently weakly, and the system's evolution during the interaction with the measuring apparatus is dominated by a Hamiltonian with sufficiently widely spaced energy levels that the interaction with the apparatus can be treated throughout as an adiabatic perturbation.

Consider a system whose Hamiltonian during the course of the interaction is given by H_0. Consider couplings to a set of measuring apparatuses (which, for simplicity, we will take as free infinitely massive particles again, although nothing changes if we use more complex measuring apparatuses):

$$H = H_0 + \sum_i \epsilon_i(t) A_i P_i, \tag{3}$$

where the A_i terms are a variety of operators associated with the system (in general, noncommuting) and the P_i terms are the momenta of a set of free infinitely massive particles. We can solve this assuming that the measuring apparatuses are in the momentum eigenstates $|p_i\rangle$ to obtain the adiabatic approximation to the Schrödinger equation for the system:

$$|\psi(t)\rangle = \sum_E a_E(t)|E\rangle, \tag{4}$$

$$i\dot{a}_E = E(t)a_E(t) - i\sum_{E'} a_{E'}\langle E(t)\left|\frac{d}{dt}\right|E'(t)\rangle \approx E(t)a_E, \tag{5}$$

where

$$E(t) \approx E_0 + \sum_i \epsilon(t)\langle E_0|A_i|E_0\rangle p_i, \tag{6}$$

with $|E_0\rangle$ being the eigenstates of H_0 and E_0 being their eigenvalues. Thus, the equation of motion for the state of the system plus measuring apparatus can be written as

$$|\Psi(t)\rangle = \sum_E a_E(0)e^{i\int^t[E_0 + \Sigma_i\epsilon(t)\langle E_0|A_i|E_0\rangle P_i dt]}|E(t)\rangle \prod_i |\phi_i\rangle \tag{7}$$

$$= \sum_E e^{iE_0t}|E(t)\rangle \prod_i e^{i\int\epsilon(t)dt\langle E_0|A_i|E_0\rangle P_i}\phi_i(x_i) \tag{8}$$

$$= \sum_E e^{iE_0t}|E(t)\rangle \prod_i \phi_i\left[x_i - \int \epsilon_i(t)dt\,\langle E_0|A_i|E_0\rangle\right]. \tag{9}$$

After the interaction with the apparatus is finished, the state is

$$|\Psi(t)\rangle = \sum_{E_0} a_E(0)e^{iE_0t}|E_0\rangle \prod_i |\phi\left(x_i - \int \epsilon_i dt\, \langle E_0|A_i|E_0\rangle\right)\rangle. \tag{10}$$

Each of the measuring apparatuses has been displaced by an amount $\int \epsilon_i(t)dt \cdot \langle E_0|A_i|E_0\rangle$, that is, by an amount proportional to the expectation value of the measured operator A_i in the state $|E_0\rangle$. Now, if we assume that the states $\phi(x_i)$ are sufficiently narrow that there is at least one A_i such that $\int \epsilon_i(t)dt(\langle E_0|A_i|E_0\rangle - \langle E_0'|A_i|E_0'\rangle)$ is larger than the initial uncertainty in $\phi_i(x)$, then the various energy eigenvalues will decohere. The measuring apparatuses will point to a value of $\langle E_0|A_i|E_0\rangle$, that is, an expectation value, for some value of E_0, with the probability of that E_0 given by $|a_i(0)|^2 = |\langle E_0|\psi\rangle|^2$.

There are a number of strange features of this result. In the first place, the value to which the measuring apparatus points is not that corresponding to one of the eigenvalues of A_i. The measuring apparatus measures A_i, but the pointer does not give one of A_i's eigenvalues; rather, it gives an expectation value, $\langle E_0|A_i|E_0\rangle$, in any single measurement. Furthermore, if we repeat the experiment, we will, as expected, get a variety of answers that the pointer points to, namely, each of the various expectation values for the various possible values of E_0. Over a large number of trials, we can expect to get a number of trials in which we get a specific value $\langle E_0|A_i|E_0\rangle$ a number of times, given by $N|\langle E_0|\psi\rangle|^2$. Thus, the statistical expectation value for the measurement of A_i is

$$\langle A_i\rangle_{\text{stat}} = \sum_{E_0} |\langle\psi|E_0\rangle\langle E_0|A_i|E_0\rangle\langle E_0|\psi\rangle. \tag{11}$$

However, the quantum mechanical expectation value of A_i is given by

$$\langle A_i\rangle_{\text{QM}} = \sum_{E_0}\sum_{E_0'} |\langle\psi|E_0\rangle\langle E_0|A_i|E_0'\rangle\langle E_0'|\psi\rangle. \tag{12}$$

In general, these two expressions are the same only if the vectors E_0 are also eigenvectors of A_i; that is, the statistical expectation value of A_i obtained by performing a large number of such adiabatic measurements is not the quantum expectation of A_i in the state of the system.

We thus have a situation that violates almost all of the standard lore about measurement. Because the A_i terms are not necessarily commuting (there is nothing in the above derivation that demands that they commute), we can, in a single measurement, measure noncommuting variables. Furthermore, if the initial state is an eigenstate of H_0, then every measurement in any ensemble of measurements will give exactly the same value for the measurement of those noncommuting variables. There will be no statistical uncertainty in the result. Moreover, the outcome of the measurement is not an eigenvalue of the operator corresponding to the measured quantity A_i. It is rather an expectation value of that quantity. The statistical distribution of the results does not depend on the quantities A_i being measured. Instead, the statistical distribution depends on the eigenvectors of the Hamiltonian H_0, which is not coupled to any measuring apparatus at all.

It is interesting to note that the standard von Neumann measurement falls into exactly this class as well. In the von Neumann measurement, the interaction with the measuring apparatus is such that the coupling to the apparatus dominates the dynamics during the measurement; for example, the Hamiltonian is of the form,

$$H = H_{\text{free}} + \epsilon\delta(t)AP. \tag{13}$$

In this case, the dominant Hamiltonian during the interaction is A because $\delta(0)$ is infinite. The coupling to the measuring apparatus A clearly commutes with the dominant Hamiltonian A and, thus, the interaction is adiabatic for an arbitrary time-dependence of $\epsilon(t) = \epsilon\delta(t)$. According to our adiabatic analysis, the measurement will give us various expectation values $\langle E_0|A|E_0\rangle$, where the E_0 terms are the eigenvalues of the dominant Hamiltonian A; that is, the E_0 terms are just the eigenvalues a of A. Hence, the measured quantities will be $\langle a|A|a\rangle = a$, that is, the eigenvalues of A. The probability of obtaining the value of a in the measurement is $|\langle E_0|\psi\rangle|^2 = |\langle a|\psi\rangle|^2$ and the statistical expectation value of A is

$$\langle A\rangle_{\text{stat}} = |\langle\psi|a\rangle|^2 a = \langle A\rangle_{\text{QM}}. \tag{14}$$

We thus see that the usual rules on measurement are simply a special case of the results obtained for adiabatic measurements.

Note, though, that the general adiabatic measurement is not equivalent to a determination. This, however, does not make the measurements any less interesting. In fact, the archetypal quantum measurement example, the Stern-Gerlach experiment, used in almost all the textbooks as an example of the von Neumann measurement is actually an adiabatic measurement, in which noncommuting observables (the spin in both of the transverse directions) are adiabatically measured. For details, see reference 6.

CONCLUSIONS

The key points of this report have been as follows:

(1) In the standard formulation of quantum mechanics, the term "measurement" is used to denote two distinct concepts. In order to clarify the problem, I have suggested that it would be useful to use separate terms to denote separate concepts and have proposed that we use "determination" for the axiomatic concept and reserve "measurement" for the physical notion of using changes induced into one system to deduce properties of another system.

(2) I have pointed out the old, but little understood, feature of quantum mechanics that the conditions in quantum mechanics are not equivalent to initial conditions. A couple of examples have emphasized this unexpected nature of the results obtained in quantum mechanics when conditions span the time during which one wants to ask questions of the quantum system.

(3) If we liberate the notion of measurement from determination, the varieties of measurements are in fact much larger than simply those that are equivalent to a determination. Although this has been well known for a long time in the case of inexact measurements, the example of adiabatic measurements shows that many of

the features of the measurements of the von Neumann type are features restricted to that type of measurement alone.

ACKNOWLEDGMENTS

I would like to thank Y. Aharonov and L. Vaidman for helpful discussions.

REFERENCES

1. VON NEUMANN, J. 1955. Mathematical Foundations of Quantum Mechanics. Princeton University Press. Princeton, New Jersey.
2. AHARONOV, Y., P. G. BERGMANN & J. L. LEBOWITZ. 1964. Phys. Rev. B **134:** 1410.
3. UNRUH, W. G. 1986. Ann. N.Y. Acad. Sci. **480:** 242–249.
4. AHARONOV, Y. *et al.* 1986. Ann. N.Y. Acad. Sci. **480:** 417–421.
5. AHARONOV, Y., J. ANANDAN & L. VAIDMAN. 1993. Phys. Rev. A **47:** 4616.
6. UNRUH, W. G. 1994. Phys. Rev. A **50:** 882–886.

Making Sense of the Kochen-Dieks "No-Collapse" Interpretation of Quantum Mechanics Independent of the Measurement Problem

ROB CLIFTON

Department of Philosophy
University of Western Ontario
London, Ontario N6A 3K7, Canada

VON NEUMANN VALUE DEFINITENESS AND SCHRÖDINGER'S CAT

The quantum measurement problem is really the by-product of adopting a particular, rather conservative, view about the conditions necessary for an observable to possess a definite value. Because this view seems to have become entrenched by von Neumann's classic analysis of quantum measurement,[1] I shall call it vN-definiteness. It asserts that an observable possesses a definite value at any given time only if one of its eigenvalues can be predicted with certainty using the state vector that applies to the system at that time. One can do no better to see the awkward consequences that vN-definiteness has for quantum measurement than to recall Schrödinger's simple, but striking, example.[2]

Think of a device that detects whether or not a radioactive atom has decayed. If the answer is vN-definitely **Yes**, the decay products trigger off a chain of events rapidly culminating in the death of a cat:

$$|\text{Decay}\rangle_{\text{Atom}} \otimes |\text{Alive}\rangle_{\text{Cat}} \rightarrow |\text{Decay}\rangle_{\text{Atom}} \otimes |\text{Dead}\rangle_{\text{Cat}}. \tag{1}$$

If the answer is vN-definitely **No**, no such chain of events is triggered and the cat remains unscathed:

$$|\text{No Decay}\rangle_{\text{Atom}} \otimes |\text{Alive}\rangle_{\text{Cat}} \rightarrow |\text{No Decay}\rangle_{\text{Atom}} \otimes |\text{Alive}\rangle_{\text{Cat}}. \tag{2}$$

Thus, the unitary evolutions given by equations 1 and 2 define the cat's "life observable" to be the pointer observable for detecting the atom's state of decay.

Now, suppose that we use this preposterous measurement device to look at the atom at a moment in time when its state of decay is vN-indefinite:

$$\sqrt{w_1}\,|\text{Decay}\rangle_{\text{Atom}} + \sqrt{w_2}\,|\text{No Decay}\rangle_{\text{Atom}}, \tag{3}$$

where $w_1 + w_2 = 1$ and $0 < w_1, w_2 < 1$. Equations 1 and 2, together with the linearity of the Schrödinger equation, then force the atom-cat system into the following state:

$$\rightarrow \sqrt{w_1}\,|\text{Decay}\rangle_{\text{Atom}} \otimes |\text{Dead}\rangle_{\text{Cat}} + \sqrt{w_2}\,|\text{No Decay}\rangle_{\text{Atom}} \otimes |\text{Alive}\rangle_{\text{Cat}}. \tag{4}$$

Thus, the cat, although at first vN-definitely alive, soon gets into an entangled state in which its life observable is vN-indefinite. The striking thing is that, once one accepts quantum theory as a fundamental theory with universal applicability, vN-indefiniteness knows no bounds and makes our perceptions of everyday macroscopic objects, like cats, inexplicable.

Fortunately, the proponent of vN-definiteness has an easy way out championed by von Neumann himself. Faced with horrific entangled states like that of equation 4, one introduces another kind of state evolution according to which each member of an ensemble of atom-cat systems in that state stochastically "collapses" into a new quantum state given by either the first or second term of equation 4, with a frequency given by the squares of their coefficients:

$$\overset{w_1}{\mapsto} |\text{Decay}\rangle_{\text{Atom}} \otimes |\text{Dead}\rangle_{\text{Cat}}, \qquad \overset{w_2}{\mapsto} |\text{No Decay}\rangle_{\text{Atom}} \otimes |\text{Alive}\rangle_{\text{Cat}}. \tag{5}$$

Collapse ensures that, at the end of the measurement, there will be two subensembles of the original ensemble, in each of which the cat's life observable can once again be said to be vN-definite.

However, talk of collapse is vague and unphysical. One may try to demystify it by locating the moment of collapse as far along the chain of events leading to the cat—or our perception of the cat—as to be effectively undetectable by any practical interference experiment. Better, one might seek a generalized law of evolution of quantum states that effectively reduces to Schrödinger evolution or stochastic collapse in appropriate microscopic or macroscopic limits. However, for Kochen[3] and Dieks[4]—whom I will henceforth refer to as just KD—all of this adds nothing more than insult to injury. It is the criterion for the definiteness of values embodied in vN-definiteness that must go.

In the next section, I will define KD-definiteness and show how it sidesteps collapse in securing a definite life state for Schrödinger's cat in spite of entanglement. However, because there are a number of controversial questions surrounding this particular resolution of the measurement problem, my main aim, pursued in the final section, will be to give an analysis of the underpinnings of KD-definiteness without making any explicit reference to that problem.

Specifically, I shall show that KD's new criterion for when an observable possesses a definite value is the only criterion that satisfies six natural conditions on what set of observables one should regard as having definite values—conditions that I believe have an intuitive appeal quite apart from Schrödinger-cat-type considerations. I shall also indicate how this result is meant to strengthen an earlier related result that I have proved elsewhere.[5]

KOCHEN-DIEKS VALUE DEFINITENESS AND SCHRÖDINGER'S CAT

Most presentations of KD's alternative to vN-definiteness employ the Schmidt decomposition of a quantum state. However, the essential idea for our purposes can be extracted more efficiently using the familiar concept of the statistical operator of a system.

Due to the possibility of entanglement, a quantum system in general cannot be ascribed a state vector of its own. Nevertheless, we can represent the statistics of all measurements that one might conduct on that system alone using a trace-one, Hermitian operator W, that is, its statistical operator. This operator will be uniquely fixed at any time by the Schrödinger-evolving state vector of the composite system of which the given system is a part, so it seems an appropriate mathematical object in the theory from which to define that system's definite-valued observables.

Here, in effect, is how KD do it. Every statistical operator W possesses a unique spectral decomposition:

$$W = \sum_k w_k \cdot P_k + 0 \cdot P_0, \tag{6}$$

where I have distinguished the 0 eigenvalue, if W has one, from its other eigenvalues $\{w_k\}$ for reasons shortly to become clear. By way of reminder, we have

$$\mathrm{Tr}(W) = \sum_k w_k \cdot \dim(P_k) = 1, \quad [P_i, P_j] = \delta_{ij}, \quad \text{and} \quad \sum_k P_k + P_0 = I. \tag{7}$$

Henceforth, let $\mathrm{Def}(W)$ denote the set of definite-valued observables of a given system having the statistical operator W.

KD's proposal for this set is

$$\mathrm{Def}_{\mathrm{KD}}(W) = \left\{ O \,|\, O = \sum_k o_k \cdot P_k + \sum_m o_m \cdot P_m, \text{ with } \sum_m P_m = P_0 \right\}. \tag{8}$$

There is no demand that the (necessarily real) coefficients in the above expansion for an observable O in the set $\mathrm{Def}_{\mathrm{KD}}(W)$ be distinct for distinct nontrivial (i.e., nonzero eigenvalue) spectral projectors of W. Thus, an equivalent way to say the above, which will be of use to us later, is as follows: an observable O is KD-definite exactly when, for each of W's nontrivial spectral projectors, O has an eigenspace that contains the range of that projector. Note how this differs from vN-definiteness, which corresponds to making the following much more straightforward proposal for the set $\mathrm{Def}(W)$:

$$\mathrm{Def}_{\mathrm{vN}}(W) = \{O \,|\, OW = o_i W, \text{ for some eigenvalue } o_i \text{ of } O\}. \tag{9}$$

Having settled which observables get definite values, KD then take the various values that they can possess to be statistically distributed in accordance with the usual formula:

$$\text{for } O \in \mathrm{Def}_{\mathrm{KD}}(W),\ \mathrm{Prob}_W(O = o_i) = \mathrm{Tr}(P_{o_i} W) = \sum_{\{k | o_k = o_i\}} w_k \cdot \dim(P_k). \tag{10}$$

Thus, in the most typical case, where $W^2 \neq W$ and the system at issue is in a mixed statistical state, equation 10 allows us to see even more clearly that having a value for KD is more than just having a value predictable with certainty in that state, like it is for von Neumann.

It should now be pretty obvious how KD-definiteness manages to avoid collapse in resolving the measurement problem. The statistical operator for Schrödinger's cat, given the composite atom-cat entangled state of equation 4, is

$$W^{\text{Cat}} = w_1 \cdot P_{|\text{Dead}\rangle} + w_2 \cdot P_{|\text{Alive}\rangle}. \tag{11}$$

Supposing $w_1 \neq w_2$, that is, the most typical case, equation 11 gives the unique spectral decomposition for W^{Cat}. Thus, the cat's life observable, whose spectral projectors coincide with those of W^{Cat}, will get counted by $\text{Def}_{\text{KD}}(W^{\text{Cat}})$ as having a definite value, entanglement notwithstanding. For exactly the same reason and due to the fact that KD-definiteness recognizes no boundary between the macroscopic and microscopic, the state of decay of the atom will also be KD-definite at the end of the interaction because

$$W^{\text{Atom}} = w_1 \cdot P_{|\text{Decay}\rangle} + w_2 \cdot P_{|\text{No Decay}\rangle}. \tag{12}$$

However, unlike the cat's life observable, which is KD-definite both before and after the measurement, the atom's decay only achieves KD-definiteness afterwards. Prior to measurement, its state is given by equation 3, yielding the statistical operator:

$$W^{\text{Atom}} = P_{\sqrt{w_1}|\text{Decay}\rangle + \sqrt{w_2}|\text{No Decay}\rangle}, \tag{13}$$

which manifestly fails to commute with either $P_{|\text{Decay}\rangle}$ or $P_{|\text{No Decay}\rangle}$, excluding these projections from the set $\text{Def}_{\text{KD}}(W^{\text{Atom}})$ by definition. Hence, there is a sense in which KD-definiteness, although eschewing collapse, still retains Bohr's insight that measurements make definite the properties of the systems that they measure.

Now, quite apart from qualms that one might have about whether KD-definiteness is always sufficient to account for the definiteness in our everyday perceptions of macroscopic objects, one is compelled to pose the following question: why should we accept what KD say about the definite values of observables? Surely, KD's shift to a new criterion of definiteness just takes by theft the solution to a problem that should really be achieved through honest toil! What we want is some sort of analysis of the underpinnings of KD-definiteness, independent of the measurement problem, that shows it to be a reasonable criterion to adopt in general.

Providing an analysis of this sort will be the main aim of the next (final) section. However, let me be clear. Through this analysis, I do not pretend to be "justifying" KD's definiteness criterion once and for all. Rather, I simply wish to broaden the debate surrounding the criterion and try to remove the sense that KD-definiteness represents merely an ad hoc modification of vN-definiteness designed specifically with the thought of avoiding collapse.

AN ANALYSIS OF KOCHEN-DIEKS VALUE DEFINITENESS

Apart from the measurement problem, the project of attributing definite values to observables faces another long-standing obstacle: the "no-hidden-variable" theorems of Bell[6] and Kochen and Specker.[7] This suggests using the limitations that these theorems impose to tighten up the reasons one might give for accepting KD-definiteness, that is, for accepting KD's new proposal for the set $\text{Def}(W)$ over others that one might entertain, such as $\text{Def}(W) = \{O \,|\, OW = WO\}$.

Of course, things are not quite that easy and we need to impose further conditions on the set Def(W) before there is any hope of being forced to adopt $\mathrm{Def}_{\mathrm{KD}}(W)$. What I have come up with is six conditions in all, each introduced below formally, followed by an informal discussion of its motivation. As promised, none of these conditions refer explicitly to the measurement problem, unlike those of my previous paper.[5] I do not claim that these conditions are not debatable; I only assume that they can be reasonably entertained as constraints on Def(W).

To check one's understanding of each condition below, it is a useful exercise to satisfy oneself that $\mathrm{Def}_{\mathrm{KD}}(W)$ conforms to it (in some cases, trivially). (The conditions labeled **Invariance** and **Ignorance** are slight strengthenings of conditions used in my previous paper.[5])

Certainty: $\mathrm{Def}(W) \supseteq \mathrm{Def}_{\mathrm{vN}}(W)$.

When an observable has a value certain to be found on measurement, it is quite natural to suppose that it had that value all along—that much about vN-definiteness seems right and is worth preserving in any proposal for what gets definite values. However, to say that this is the *only* way that an observable could possess a definite value is to dogmatically rule out the possibility that values might come to be possessed by observables contingently following some indeterministic process. Thus, there is no reason to demand equality above. **Certainty** is reminiscent of EPR's well-known reality criterion.[8] However, because **Certainty** only refers to predictions with certainty obtained from the system's statistical operator, which contains no record of correlations with other systems, the condition does not allow such predictions to be obtained at the expense of conditionalizing on the results of measurements on other correlated systems. Hence, Bohr's specific objections[9] to EPR's reality criterion do not apply here.

Function: $O = f(\{O_i\})$, where $\{O_i\} \subseteq \mathrm{Def}(W) \Rightarrow O \in \mathrm{Def}(W)$.

It seems intuitively plausible that any observable that is a function of (compatible) definite-valued observables should itself be regarded as having a definite value. For example, if we believe that it makes sense to entertain ideas like "the particle is 3 cm to the left or right of the origin", attributing a definite value to its position X, then that should also license us to speak of such things as "the particle is to the left or right of the origin", which attributes a definite value to the observable $|X|/X$. **Function** has nothing to say about how the particular values of functionally related observables are themselves correspondingly related, so it brings no immediate threat of Kochen-Specker contradiction.[10]

Locality.

There is no elegant way to formalize this condition without introducing new, cumbersome notation. But the basic idea of **Locality** is simple: that definite-valued observables pertaining to spacelike separated systems must not sustain Bell inequality–violating correlations. Of course, it was Bell's great lesson to teach us that *any* interpretation of quantum mechanics, whether it be von Neumann's, KD's, or somebody else's, needs to be nonlocal in *some* sense. The sense that I want to exclude with the above condition dubbed **Locality** is what Shimony[11] has termed "parameter dependence" because it, in contrast to "outcome dependence", could *in principle* be exploited to send a superluminal signal, making it all the more difficult to reconcile with relativity theory (although opinions about this vary). Gisin[12] has shown that, for any entangled state of two spacelike separated systems, there exist observables on each system that bear correlations violating a Bell inequality. Thus, what **Locality** is meant to exclude is the attribution of definite values to precisely such observables; otherwise, values would then need to change as a result of spacelike separated measurement events to enforce the violation of parameter dependence.

Ortholattice: $P, P' \in \mathrm{Def}(W) \Rightarrow P^{\perp}, P \oplus P', P \cap P' \in \mathrm{Def}(W).$

If we view the projection operators of a system that have definite values as representing the propositions that may be said to be determinately true or false of the system, then it is reasonable to require that negations, disjunctions, and conjunctions of such propositions should also be determinately true or false. This requires that the set of definite-valued projections—which are in 1:1 correspondence with the subspaces onto which they project—be closed under orthocomplementation ("negation"), span ("disjunction"), and intersection ("conjunction").

Invariance: $O \in \mathrm{Def}(W) \Leftrightarrow UOU^{-1} \in \mathrm{Def}(UWU^{-1}).$

Following the general approach to picking out definite-valued observables as set out in the previous section means committing oneself to the idea of defining this set using the system's statistical operator and perhaps also the structure of the Hilbert space that represents the system. Thus, if we consider any nontrivial isomorphism of the system's Hilbert space that preserves its statistical operator, the definite-valued set that they jointly pick out should remain invariant. Moreover, if the isomorphism is a mapping between Hilbert spaces of the same dimension that changes the statistical operator to a new isomorphic one, the definite-valued set should change accordingly to a new isomorphic "copy" of itself.

Ignorance: $\mathrm{Def}(W) = \bigcap_{\{k|w_k \neq 0\}} \bigcap_{\{P|P_kP=P\}} \mathrm{Def}(P).$

This final condition is perhaps the most difficult to motivate because it tangles with the thorny issue of how to interpret a mixed statistical state (the condition is vacuous when $W^2 = W$). Consider first the case where the system, although in a mixed state, is not entangled with any other system. Then, in many circumstances, we want to say that each member of an ensemble of similar systems prepared in the mixture has really been prepared in one of the pure states lying in the range of one of the statistical operator's nontrivial spectral projectors (i.e., the pure states having nonzero probability in the mixture) and we are simply ignorant as to which pure state obtains in each individual case. Given this, what we say has a definite value in the mixed state should not be inconsistent with what we say has definite values in all those possible pure states that the system may be in. When the system is in the given mixed state, **Ignorance** demands that we say as much as we can say about what gets definite values, short of such inconsistency. What if the system *is* entangled with another so that this sort of ignorance interpretation of its mixture no longer applies? Still, the **Ignorance** condition must hold, *at least formally*, for the given system's statistical operator. For whatever that operator is, it could just as well have arisen in circumstances where there is no entanglement with other systems because a system's statistical operator contains no record of any possible entanglement with other systems. Thus, because we have committed ourselves to using it alone (in conjunction with the system's Hilbert space) to pick out the definite-valued observables, we are thereby compelled to impose **Ignorance** regardless of whether entanglement is present.

Having said about as much as one *can* say to cast the above conditions in a plausible light, let me state the precise result, using H to refer to the Hilbert space describing the system:

THEOREM: When $\dim(H) > 2$,

$$\mathrm{Def}(W) = \mathrm{Def}_{\mathrm{KD}}(W) = \left\{ O \mid O = \sum_k o_k \cdot P_k + \sum_m o_m \cdot P_m, \text{ with } \sum_m P_m = P_0 \right\}$$

is the *unique* choice for $\mathrm{Def}(W)$ satisfying the above six conditions.

The proof of this theorem relies on a little result proved in my earlier paper,[5] which was later strengthened by Bub.[13] I shall not take the trouble to reproduce the proof of that result here (which, however, only relies on some elementary Hilbert space geometry), but instead just state the result in the form of a lemma. It is this lemma that forces the dimensionality restriction (and one can easily see why).

LEMMA: When $W^2 = W = P$ (i.e., W is pure) and $\dim(H) > 2$, **Certainty** and **Ortholattice** jointly imply that either $\mathrm{Def}(P)$ contains only those projections on H with values predictable with certainty in state P or $\mathrm{Def}(P)$ contains *all* projections on H (i.e., there is no middle ground between these two extremes).

Proof: Fix an arbitrary statistical operator W and Hilbert space H of dimension greater than 2. Combining **Certainty** with **Ignorance** (no irony intended), it must be the case that $\mathrm{Def}(W)$ should at least include all those observables whose values are predictable with certainty in each of the pure states that lie in the range of one of W's nontrivial spectral projectors. Call that set of observables $S \subseteq \mathrm{Def}(W)$. Obviously, $\mathrm{Def}_{\mathrm{KD}}(W) \subseteq S$, but it is also useful (mainly for the next paragraph) to note that $\mathrm{Def}_{\mathrm{KD}}(W) \supseteq S$. Thus, if some observable O has a value predictable with certainty in all pure states contained in the range of a particular nontrivial spectral projector of W, those pure states must all be eigenstates of O with the same eigenvalue because distinct eigenvalues require orthogonal eigenspaces. Moreover, because this point applies for every nontrivial spectral projector of W, O must have, for each such projector, an eigenspace that contains its range, which is the equivalent characterization of the observables in the set $\mathrm{Def}_{\mathrm{KD}}(W)$ I gave following equation 8. Thus, we have $\mathrm{Def}(W) \supseteq S = \mathrm{Def}_{\mathrm{KD}}(W)$. What we are left to show is that $\mathrm{Def}(W) \subseteq \mathrm{Def}_{\mathrm{KD}}(W)$, that is, that we cannot enlarge the set $\mathrm{Def}(W)$ beyond $\mathrm{Def}_{\mathrm{KD}}(W)$ without destroying the satisfaction of the six conditions. This is where the limitations imposed by the no-hidden-variable theorems come in.

Suppose that we add any observable to the set $\mathrm{Def}(W)$ that it not already in that set by virtue of being in $\mathrm{Def}_{\mathrm{KD}}(W)$. Thus, because $\mathrm{Def}_{\mathrm{KD}}(W) = S$, the observable we add, call it O, must fail to have a value predictable with certainty in at least one of the pure states, call it P, lying in one of W's nontrivial eigenspaces. By **Ignorance,** this O must also be in $\mathrm{Def}(P)$. Indeed, so must all of O's spectral projectors, by **Function**, because each can be written as a characteristic function of O. However, at least one of these spectral projectors in Def(P), say P_{o_i}, must be such that $P_{o_i}P \neq P$ and $\neq 0$ because O's value (by hypothesis) is not predictable with certainty in state P. Thus, using **Certainty**, **Ortholattice**, and the lemma [whose conclusion follows whenever W is pure and dim(H) > 2], Def(P) must include *all* projections on H. Furthermore, by **Function**, Def(P) must include all observables on H because the spectral theorem tells us that every observable is some linear combination of compatible projections on H. **Invariance** then forces us to say the same for any pure state P on H because any two pure states can be transformed into one another by an isomorphism. Moreover, now that Def(P) must include all observables regardless of the pure state P on H that we consider, **Ignorance** (trivially) implies that Def(W) must include all observables for any mixed state W on H. Finally, take any mixed state W of the system at issue and consider another system described by a Hilbert space H' (of the same dimension as H), but in mixed state $UWU^{-1} = W'$ for some isomorphism U. Because Def(W) includes all observables on H, Def(W') must similarly include all observables on H' by **Invariance**. However, clearly, there is an entangled state of these two systems that yields W and W' as their statistical operators, so we are now forced (by Gisin's theorem[12]) into violating **Locality**. □

The second paragraph above gives what one might call a "slippery slope" argument: if we want to assert that more observables are definite than what KD allow, then our conditions force us in the end to admit so many definite-valued observables that we must violate **Locality**. (Clearly, the lemma is the main thing in the proof that makes the slope slippery!) In my previous paper,[5] I had an assumption of **Noncontextualism** playing the role of **Locality** because, by the end of the proof above, one clearly has enough observables of the system with definite values that Kochen-Specker contradictions pose a threat too. However, I now feel that motivating KD-definiteness with the requirement that it avoid the need for contextual value assignments is perhaps not the most convincing way to do things because interpretations like Bohm's[14] can quite naturally and unobjectionably incorporate contextualism (another great lesson from Bell[15]). On the other hand, although the final verdict is not yet in on the matter, it is not so easy to swallow an interpretation that violates the condition that I have called **Locality** (i.e., parameter independence), even given that Bohm's interpretation does (see reference 15).

At the very least, I claim that the above analysis of KD-definiteness, which uniquely characterizes it, "tidies up" an otherwise unmotivated and ad hoc interpretation initially designed to solve the measurement problem without collapse. One may also feel that my analysis opens up a whole new can of worms about KD's interpretation, depending on how plausible or implausible the theorem's six conditions are taken to be. However, if that is all that one takes away from this result, I will still have achieved my aim of shifting the debate about KD-definiteness onto whether the criterion for definiteness it embodies is even a reasonable one.

REFERENCES

1. VON NEUMANN, J. 1932. Mathematische Grundlagen der Quantenmechanik. Springer-Verlag. Berlin.
2. SCHRÖDINGER, E. 1935. Die gegenwärtige situation in der quantenmechanik. Naturwissenschaften **23:** 807–812, 823–828, and 844–849.

3. KOCHEN, S. 1985. A new interpretation of quantum mechanics. *In* Symposium on the Foundations of Modern Physics. P. Lahti & P. Mittelstaedt, Eds. World Scientific. Singapore.
4. DIEKS, D. 1989. Resolution of the measurement problem through decoherence of the quantum state. Phys. Lett. **A142:** 439–446.
5. CLIFTON, R. 1994. Independently motivating the Kochen-Dieks modal interpretation of quantum mechanics. Br. J. Philos. Sci. In press.
6. BELL, J. 1964. On the Einstein-Podolsky-Rosen paradox. Physics **1:** 195–200.
7. KOCHEN, S. & E. SPECKER. 1967. The problem of hidden variables in quantum mechanics. J. Math. Mech. **17:** 59–87.
8. EINSTEIN, A., B. PODOLSKY & N. ROSEN. 1935. Can quantum mechanical description of physical reality be considered complete? Phys. Rev. **47:** 777–780.
9. BOHR, N. 1935. Can quantum mechanical description of physical reality be considered complete? Phys. Rev. **48:** 696–702.
10. REDHEAD, M. 1987. Incompleteness, Nonlocality, and Realism. Oxford University Press (Clarendon). London/New York.
11. SHIMONY, A. 1986. Events and processes in the quantum world. *In* Quantum Concepts of Space and Time. R. Penrose & C. Isham, Eds. Oxford University Press (Clarendon). London/New York.
12. GISIN, N. 1991. Bell's inequality holds for all nonproduct states. Phys. Lett. **A154:** 201–202.
13. BUB, J. 1994. On the structure of quantal proposition systems. Found. Phys. In press.
14. BOHM, D. 1952. A suggested interpretation of the quantum theory in terms of "hidden variables": parts I and II. Phys. Rev. **85:** 166–179 and 180–193.
15. BELL, J. 1966. On the problem of hidden variables in quantum mechanics. Rev. Mod. Phys. **38:** 447–475.

Limitation to Quantum Measurements of Space-Time Distances[a]

Y. JACK NG[b,c,d] AND H. VAN DAM[d]

[b]*School of Natural Sciences*
Institute for Advanced Study
Princeton, New Jersey 08540

[c]*Center for Theoretical Physics*
Massachusetts Institute of Technology
Cambridge, Massachusetts 02139

[d]*Institute of Field Physics*
Department of Physics and Astronomy
University of North Carolina
Chapel Hill, North Carolina 27599-3255

Anything I have written on the topic is primarily testimony as to how important I consider it.
—J. A. Wheeler (private correspondence)

We start by recalling the fundamental nature of space-time distance measurements.[1–4] In quantum mechanics, one specifies a space-time point by its coordinates, without bothering to give a prescription as to how these coordinates are to be measured. However, general relativity ordains that coordinates do not have any meaning independent of observations; in other words, according to relativity, a coordinate system is defined only by explicitly carrying out space-time distance measurements. We will pay heed to this dictum of general relativity and will follow Salecker and Wigner[1] to use clocks and light signals to measure distances (see reference 5).[e]

Suppose we want to measure the length between two spatially separated points A and B. The measurement can be carried out in the following way. A clock is put at point A. Set the clock to read zero when a light signal is sent from A towards B, where a mirror is stationed to reflect the light signal back to A. From the reading of the clock, to be denoted by t, when the light signal arrives at A, one deduces that the length AB is given by $\ell = ct/2$, where c denotes the speed of light. There are two major sources of errors in the length measurement: one comes from the uncertainty principle of quantum mechanics and the other is due to space-time curvature effects.

First, we note that the clock is not absolutely stationary, its spread in speed being given by the uncertainty principle of quantum mechanics:

$$\delta v = \frac{\delta p}{m} \gtrsim \left(\frac{1}{2}\right)\frac{\hbar}{m\delta\ell_{QM}}, \tag{1}$$

where m is the mass of the clock. Because the clock is the agent in measuring the

[a]This work was supported in part by the United States Department of Energy under Grant No. DE-FG05-85ER40219 and by the Z. Smith Reynolds Fund of the University of North Carolina.

[e]Our work has some overlap with that of reference 6.

length, the fact that it is not stationary implies an uncertainty in the length measurement given by, at time t,

$$\delta\ell_{QM}(t) = t\delta v \gtrsim \left(\frac{1}{2}\right)\left(\frac{\hbar}{m\delta\ell_{QM}(0)}\right)t = \left(\frac{1}{m}\right)\left(\frac{\hbar}{\delta\ell_{QM}(0)}\right)\left(\frac{\ell}{c}\right), \tag{2}$$

where in the last two steps we have used equation 1 and $t = 2\ell/c$, respectively. Next, we minimize $\delta\ell_{QM}(0) + \delta\ell_{QM}(t)$ so that the uncertainty in the length measurement due to quantum mechanical effects is given by

$$(\delta\ell_{QM})^2 \gtrsim \frac{\hbar\ell}{mc}. \tag{3}$$

However, the uncertainty in the position of the clock cannot be made arbitrarily small by making the clock very massive as that would disturb the space-time curvature.[7] If one assumes the clock to be spherically symmetric and to have a radius (to be denoted a) larger than the Schwarzschild radius $r^* = (2Gm/c^2)$, where G is the gravitational constant, then $\delta\ell_c$, the error in the length measurement caused by the curvature, may be calculated from the Schwarzschild solution. The result is r^* multiplied by a logarithm involving r^*/a and $r^*/(a + \ell)$. For $a \gg r^*$, one finds $\delta\ell_c = (½)r^*\ln[(a + \ell)/a]$; hence,

$$\frac{Gm}{c^2} \lesssim \delta\ell_c. \tag{4}$$

The combined error in the length measurement, $\delta\ell = \delta\ell_{QM} + \delta\ell_c$, due to quantum mechanical and curvature effects, can be minimized by adjusting m. Using equations 3 and 4, we find

$$\delta\ell \gtrsim (\ell\ell_P^2)^{1/3}, \tag{5}$$

where $\ell_P = (\hbar G/c^3)^{1/2}$ is the Planck length.[f] We expect the presence of the mirror at point B to contribute an error of comparable magnitude. We can also deduce the minimum error in time interval measurements by using the same experimental setup:

$$\delta t \gtrsim (tt_P^2)^{1/3}, \tag{6}$$

where $t_P = \ell_P/c$ is the Planck time. These errors in space-time measurements induce an uncertainty in the space-time metric. Noting that $\delta\ell^2 = \ell^2\delta g$ and using equations 5 and 6, we readily get

$$\delta g_{\mu\nu} \gtrsim \left(\frac{\ell_P}{\ell}\right)^{2/3} \sim \left(\frac{t_P}{t}\right)^{2/3}. \tag{7}$$

Our results (equations 5–7) should be contrasted with those according to the canonical[8] viewpoint. The derivation of the latter goes as follows. The vacuum

[f] Carrying out the measurement at nonzero temperature results in an additional error.

functional for the theory of pure gravity has roughly the form

$$Z \sim \int \mathscr{D}[g] \exp\left\{\left(\frac{ic^3}{\hbar G}\right)\int\left(\frac{\partial g}{\partial x}\right)^2 d^4x\right\}. \tag{8}$$

Thus, if one is making measurements in a space-time region of volume ℓ^4, contributions to the Feynman integral are more or less in phase until variations in the gravitational field amplitudes from their classical values become as large as

$$\delta g \gtrsim (\hbar G/c^3)^{1/2}/\ell = \ell_P/\ell. \tag{9}$$

These represent the quantum fluctuations of the gravitational fields. They give rise to errors in space-time measurements that are constants:

$$\delta\ell \gtrsim \ell_P; \qquad \delta t \gtrsim t_P. \tag{10}$$

Let us see how reasonable the canonical results (equations 9 and 10) are in light of the analysis above for our actual experimental setup involving a clock, a mirror, and light signals. There, to obtain the canonical results instead of equations 5–7, all one has to do is to replace equation 4 by the requirement $Gm/c^2 \lesssim \ell$. However, this requirement is trivially true because otherwise the mirror would be required to be inside the Schwarzschild radius of the clock, a nonsensical situation. It should be possible to impose a more restrictive (but still physically sensible) requirement to arrive at a more useful and better bound for the minimum errors in space-time measurements (such as those given by equations 5–7).

Two more comments on the results (equations 5–7) are as follows:

(1) Mathematically, it is not surprising that the uncertainty in the length ℓ involves the two lengths in the problem: ℓ and ℓ_P. There is an analogous result that is actually relevant for long thin rulers. A quantum mechanical calculation for a one-dimensional chain of N ions with a spring of constant k between successive ions gives, in the high-temperature limit,

$$\delta\ell = \left(\frac{1}{\pi}\right)\sqrt{N\overline{\Delta x_i^2}} = \left(\frac{1}{\pi}\right)\sqrt{Nb\frac{\overline{\Delta x_i^2}}{b}} = \left(\frac{1}{\pi}\right)\sqrt{\ell\frac{\overline{\Delta x_i^2}}{b}}$$

for the uncertainty in the length,[g] where b is the distance between two successive ions when the spring is relaxed and where $\overline{\Delta x_i^2} = (½)k_BT/k$, that is, the mean square displacement of a mass on a spring of force constant k. Thus, for a long thin ruler, the uncertainty in the length depends on both the length ℓ itself and the lattice constant b. Note that $\delta\ell$ is proportional to $\ell^{1/2}$. [In the zero-temperature limit, one finds $\delta\ell$ to be proportional to $\sqrt{\log \ell/b}$.] This is one of the reasons that long thin rulers are not the best tools for measuring distances. In addition, rulers, being macroscopic objects, will influence other objects in the measurement process through their gravitational attraction. The Lorentz contraction of rulers will also complicate matters.

[g]This result is originally due to E. P. Wigner (in response to a question raised by H. van Dam [private communication]). Details of the calculation will appear elsewhere.

(2) Because the Planck length is so small, even for the universe,[h] for which $\ell = 10^{10}$ light-years, $\delta\ell$ is only 10^{-13} cm long, the size of a nucleus, which is quite small, but in principle is there.

As a simple application of equations 5–7, we deduce the minimum error in momentum-energy measurements. Imagine sending a particle of momentum p to probe a certain structure of spatial extent ℓ so that

$$p \sim \frac{\hbar}{\ell}. \tag{11}$$

Consider the coupling of the metric to the energy-momentum tensor of the particle,

$$(g_{\mu\nu} + \delta g_{\mu\nu})t^{\mu\nu} = g_{\mu\nu}(t^{\mu\nu} + \delta t^{\mu\nu}), \tag{12}$$

where we have noted that the uncertainty in $g_{\mu\nu}$ can be translated into an uncertainty in $t_{\mu\nu}$. Equations 7 and 12 can now be employed to give

$$\delta p \gtrsim p\left(\frac{\ell_P}{\ell}\right)^{2/3}, \tag{13}$$

which, with the aid of equation 11, yields

$$\delta p \gtrsim p\left(\frac{p}{m_P c}\right)^{2/3}, \tag{14}$$

where $m_P = (\hbar c/G)^{1/2}$ is the Planck mass.

[An alternative derivation of equation 14 is provided by considering δp, the "uncertainty" of the momentum operator, $p = (\hbar/i)(\partial/\partial x)$, associated with $\delta x = (x\ell_P^2)^{1/3}$ (equation 5). For any function $f(x)$, $(\delta p)f$ is given by

$$(\delta p)f = \left(\frac{\hbar}{i}\right)\left[\delta x\left(\frac{\partial^2 f}{\partial x^2}\right) + \left(\frac{\partial f}{\partial x}\right)\left(\frac{\partial \delta x}{\partial x}\right)\right].$$

Taking the function $f(x)$ to be the linear momentum eigenstate $f = \exp(ipx/\hbar)$, one gets

$$|(\delta p)e^{ipx/\hbar}| = \hbar\left|\left[i\left(\frac{p^2}{\hbar^2}\right)\ell_P^{2/3}x^{1/3} + \left(\frac{1}{3}\right)\left(\frac{p}{\hbar}\right)\ell_P^{2/3}x^{-2/3}\right]e^{ipx/\hbar}\right|.$$

The minimum value of $|\delta p|$ is attained at $x \sim (2/3)(\hbar/p)$, yielding equation 14.]

The analogous statement for the minimum error in energy measurements is

$$\delta E \gtrsim E\left(\frac{E}{m_P c^2}\right)^{2/3}. \tag{15}$$

For energy-momentum near the Planck scale, the error is not negligible; for example, at the grand unification scale $\sim 10^{16}$ GeV, the error is of order 1%. In analyzing a

[h] Since this conference is held in honor of J. A. Wheeler, it may not be out of place here at the mentioning of the word "universe" to quote him (private correspondence): "[I recall] the well-known statement of Rutherford, 'When a student of mine uses the word "universe", I tell him it is time for him to leave.' However, maybe that is why so many of us live in America!"

high-energy experiment, an experimentalist should keep in mind that energy-momentum conservation holds up only to the errors given by equations 14 and 15. In passing, we mention that the minimum errors in measurements given by equations 6, 7, 14, and 15 are fixed by dimensional analyses once the minimum error in spatial distance measurements is found, given by equation 5.

As another application of the above results (equations 5–7), let us consider the quantum (de)coherence phenomenon for a scalar particle of mass m moving slowly. Let us assume that the particle satisfies the Schrödinger-Klein-Gordon type equation,

$$i\hbar\left(\frac{\partial}{\partial t}\right)\psi = \left[\left(-\frac{\hbar}{2m}\right)\left(\frac{\partial^2}{\partial x^2}\right) + V(x,t) + g^{00}mc^2\right]\psi, \tag{16}$$

where we have kept the most important term involving $g^{\mu\nu}$. We are interested here in the effects caused by the uncertainty in the metric given by equation 7. Obviously, δg induces a multiplicative phase factor in the wave function

$$\psi \to e^{i\delta\varphi}\psi, \tag{17}$$

where

$$\delta\varphi = \left(\frac{1}{\hbar}\right)\int^t mc^2\delta g^{00}\,dt'. \tag{18}$$

For consistency, the integral should be fairly insensitive to the lower integration limit as long as $t \gg t_P$. If one is making measurements in a short time interval, contributions to the phase of the wave function from the different time elements in this time interval will be more or less in phase (i.e., $\delta\varphi \ll 1$) until the time interval reaches the decoherence time t_D when $\delta\varphi$ becomes sizable, that is,

$$1 \sim \left(\frac{1}{\hbar}\right)\int_0^{t_D} mc^2\left(\frac{t_P}{t'}\right)^{2/3} dt' \sim \left(\frac{mc^2}{\hbar}\right)t_P^{2/3}t_D^{1/3} = \ell_P^{2/3}\ell_D^{1/3}/\lambda_c, \tag{19}$$

where $\lambda_c = (\hbar/mc)$ is the Compton wavelength of the particle of mass m and $\ell_D = ct_D$ is what we will call the decoherence length. This system can be treated classically if the decoherence length is less than the Compton wavelength—in other words, if

$$\ell_D \leq \lambda_c$$

or, via equation 19,

$$m \gtrsim m_P. \tag{20}$$

Therefore, due to the uncertainty of the space-time metric, it suffices to give a particle heavier than the Planck mass a classical treatment.

ACKNOWLEDGMENTS

We thank J. A. Wheeler for an enjoyable correspondence and G. Domokos for calling our attention to reference 6. Y. J. Ng thanks the faculty (especially S. Adler) at the School of Natural Sciences at the Institute for Advanced Study and the faculty (especially A. Guth) at the Center for Theoretical Physics (Massachusetts Institute of Technology) for their kind hospitality.

REFERENCES

1. SALECKER, H. & E. P. WIGNER. 1958. Phys. Rev. **109:** 571–577.
2. DEWITT, B. S. 1962. *In* Gravitation: An Introduction to Current Research. L. Witten, Ed.: 266–381. Wiley. New York.
3. MARZKE, R. F. & J. A. WHEELER. 1964. *In* Gravitation and Relativity. H. Y. Chiu & W. F. Hoffmann, Eds.: 40–64. Benjamin. New York.
4. WHEELER, J. A. 1984. *In* Quantum Theory of Gravity: Essays in Honor of the Sixtieth Birthday of Bryce S. DeWitt. S. M. Christensen, Ed.: 223–233. Adam Hilger. Bristol.
5. NG, Y. J. & H. VAN DAM. 1994. Mod. Phys. Lett. **A9:** 335–340.
6. KAROLYHAZY, F. 1966. Nuovo Cimento **A42:** 390–402.
7. WIGNER, E. P. 1957. Rev. Mod. Phys. **29:** 255–268.
8. MISNER, C. W., K. S. THORNE & J. A. WHEELER. 1973. Gravitation, p. 1190–1194. Freeman. San Francisco.

Two-Particle versus Three-Particle EPR Experiments[a]

DANIEL M. GREENBERGER

Department of Physics
City College of the City University of New York
New York, New York 10031

There is a class of three-particle EPR[1] experiments that depend on the detector angles only through their sum. Such an experiment was used to prove the GHZ theorem.[2] One can produce a classical, realistic, local model to reproduce this dependence, but the restriction constrains such models to behave differently from what one would expect from two-particle experiments. Their expectation values show *greater* correlation than the quantum case and their detection efficiencies are necessarily low. For the two-particle case, the expectation value showed a necessarily *lower* coherence than the quantum case and one could have 100% counter efficiency.

INTRODUCTION

When the GHZ paper was being written in 1988, a model was developed in order to help us think through the problems of constructing a three-particle local, realistic, classical model to explain EPR experiments. It was included in some unpublished notes we made at that time, but was never published, and there were no plans to publish it because it seemed unnecessary.

However, recently, there have been some attempts to consider the possibility of performing a three-particle experiment and also there have been some attempts to derive inequalities for such experiments in the spirit of such inequalities for two-particle experiments. It turns out that our model shows that there are important differences between the two- and three-particle cases and, if one wants to derive realistic inequalities for such experiments, one had better be aware of these differences.

The first limitation comes from the fact that the Bell theorem in its original form[3] treated counters with 100% efficiency. In other words, it was assumed that all emitted particles were to be counted. Later, in order to derive results that could be related to actual experiments, inefficient counters had to be considered. However, most of the models I am aware of treat the efficiencies of the counters as some constant percentage of the total particles entering the counter; that is, the counting rate is independent of the special values of the hidden variables of the emitted particles. One then makes seemingly plausible assumptions about the model one will

[a]This work was supported in part by a grant from the National Science Foundation.

encounter, one arrives at upper limits for correlations that can be tested experimentally, and a generation of such experiments have been done.[4]

In the two-particle case, the two particles enter separated detectors, each of which has an internal variable (such as the spin orientation to which each is sensitive), which we shall call α and β for the two detectors. The quantum mechanical correlation between these two variables, for 100% efficient counters, is $-\cos\theta$, where $\theta = (\alpha - \beta)$. In the special cases $\theta = 0$ or π, when one has maximum correlation ($= \pm 1$), or the case $\theta = \pi/2$, when there is no correlation, one can find a classical model that reproduces the result. However, for all other values of θ, the quantum mechanical correlation is greater than any possible classical correlation, whose upper bound is given by a linear function of θ, according to the bounds set by Bell's theorem. This situation is shown in FIGURE 1, where the quantum correlation and the greatest classical correlation are drawn and labeled "Quantum" and "Classical", respectively. (The other two curves will be explained later.)

For convenience, we have removed the negative sign and have plotted the correlation as $+\cos\theta$. We shall do this for convenience throughout this report and it has no theoretical significance. Also, note that we have only plotted the correlation between $\theta = 0$ and $\theta = \pi/2$ because, by symmetry, the correlation will be the same (except for sign) for $\theta' = \pi - \theta$ as it is for θ. The important thing to learn from FIGURE 1 is that the quantum mechanical correlation is greater than the maximum classical one everywhere, except for those special angles where they are equal. This is the basis for the experiments that have been done to show that nature violates the upper bounds set by Bell's theorem.

The second limitation is that, according to the GHZ theorem, it is impossible to find any model for three particles that will reproduce the results of quantum theory, even at the maximum correlation angles $\theta = 0$ and π, for counters that are 100% efficient. Hence, in this case, it is very important to consider models that allow the counters to not count in certain configurations. Because the quantum correlation in the three-particle case will be $\cos\theta$, where $\theta = (\alpha + \beta + \gamma)$ (and again we forget the negative sign), then for a particular value of θ only two of the variables α, β, and γ can be independent. Thus, the three variables are constrained by this fact and only two of them are independent.

We will seek a classical three-particle model where the correlation depends only on θ so that we can compare the classical and quantum results easily. Because this constrains the system as just mentioned, our counter efficiency will be a complicated function of the three angles, with only two of them really independent. The three counters will intrinsically behave the same, but in any configuration the efficiency will depend on θ. The variables α and β will be allowed to vary independently over their range, but γ will be fixed by the value of θ; thus, there is really only a two-dimensional phase space over which the three variables are free. This makes inequalities derived on the basis of fixed efficiencies for the separate counters unrealistic in the three-particle case.

As an example of how different the results for the three-particle case are from those of the two-particle case, we will state the conclusion from our model. For a given value of θ, it is difficult to make any consistent fit of the model and the efficiency of the counters will be low, much lower than the nominal efficiency of the counters themselves, because only a small percentage of the two-dimensional phase space is compatible with there being any counts at all. In fact, it is sufficiently difficult to make a situation where all the detectors count—even in the case $\theta = 0$ and there is a perfect correlation—that when $\theta \neq 0$ we find that it is difficult to reduce the

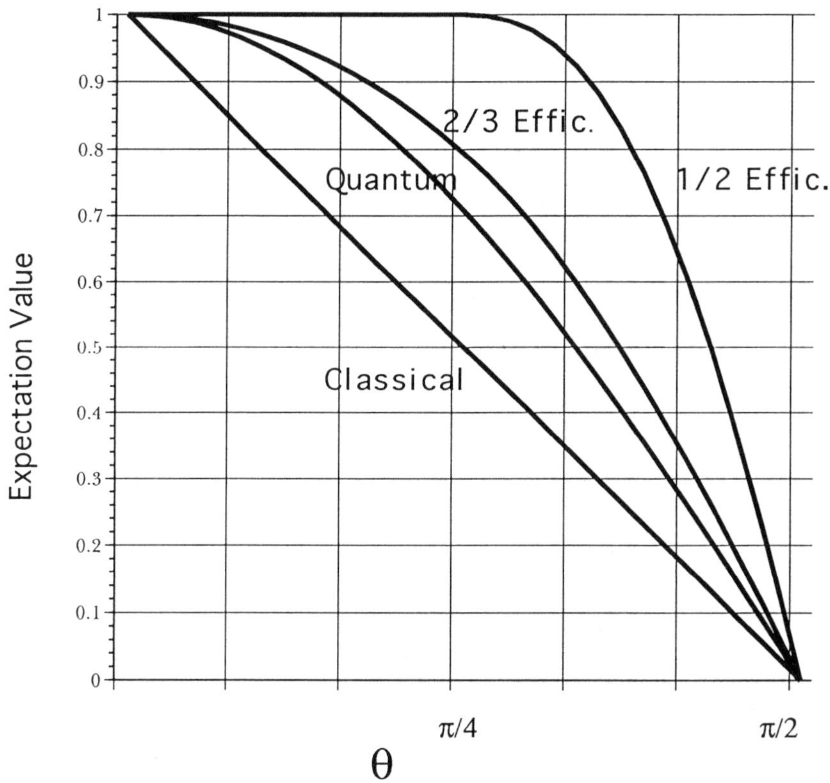

FIGURE 1. Expectation values for various models versus quantum theory. In the two-particle case, Bell's model gives the maximum possible correlation of the two spins, yielding a straight line, the curve marked "Classical". The counters can be 100% efficient in this model (i.e., they count at all angles and for all values of the hidden variables). This is to be contrasted with the quantum case, marked "Quantum", which has a higher expectation value of the spin correlation. In the three-particle case, to be described, the expectation values are higher than the quantum ones, but the counter efficiencies are necessarily lower, because of the GHZ theorem. The maximum efficiency for our type of model is ⅔ and the expectation value is plotted as "⅔ Effic.". For a less efficient model, where the counters fire only one-half (½) of the time, marked "½ Effic.", the expectation value is very high.

correlation. As a result, in this model, the classical results are *always* more highly correlated than the quantum mechanical ones. Of course, one must remember that this is for inefficient counters because there can be no classical model when the counters are 100% efficient, according to the GHZ theorem.

BELL'S CLASSICAL TWO-PARTICLE MODEL

Bell, in his early papers, gave a two-particle model that is very simple, but it gives a linear correlation, which is as good as one can get with a local, realistic model. In any Bell-type model, a single particle can decay into two particles, each of which heads toward a different detector. The detectors are each labeled by a setting—α for detector 1, which catches particle 1, and β for detector 2, which catches particle 2. Each detector can respond to the arriving particle by jumping into one of two states. We call these states $+1$ and -1. Thus, particle 1, on arriving at detector 1, which is set to a particular value of α, can lead to the detector ending up in either state $+1$ or -1. We introduce a function $A(\alpha, \lambda)$ to describe the situation. The function A can take on the two values ± 1 to denote the final state of the detector. A can depend not only on α, but on any internal (hidden) variables given to the particle in the decay. We describe these by the catchall label λ. Similarly, for particle 2, we introduce $B(\beta, \lambda)$, which also takes on the values ± 1. The locality condition is expressed by the fact that A cannot depend on β (the setting of B) and B cannot depend on α (the setting of A).

The experiment is described by a correlation function, $E(\alpha, \beta)$, which is the classical expectation value:

$$E(\alpha, \beta) = \int d\lambda \rho(\lambda) A(\alpha, \lambda) B(\beta, \lambda).$$

This means that the original particle decays into particle 1 and particle 2, and each heads toward its respective counter; then, what is counted is the product of the outputs of the two detectors, $AB = \pm 1$. The values of λ are not known and, over a large number of counts, one may include the various possibilities by some weight function $\rho(\lambda)$, where $0 \leq \rho \leq 1$ and $\int d\lambda\, \rho(\lambda) = 1$.

Now, quantum mechanically, as we have noted, the correlation corresponding to E will be $E_{qm} = \cos\theta = \cos(\alpha - \beta)$. (Here, as afterward, we are neglecting a trivial negative sign.) In order for a classical model to be useful for comparison purposes, it should be a function of θ alone. One may wonder how a product of two functions can become a function of their difference, but, remember, we are averaging over a large number of events. The way that any correlation between α and β comes about is by having α correlated with λ_1, a hidden variable for particle 1, and β correlated with λ_2, from particle 2, and then having λ_1 and λ_2 correlated. This is reasonable because particles 1 and 2 were produced in the same decay. Thus, we will consider λ_1 and λ_2 as hidden internal angles, which we take (as also with α and β) to vary between $-\pi$ and π, and we write

$$E(\alpha, \beta) = \left(\frac{1}{2\pi}\right) \int d\lambda_1 d\lambda_2 \delta(\lambda_1 - \lambda_2) A(\alpha - \lambda_1) B(\beta - \lambda_2) = E(\alpha - \beta) = E(\theta).$$

Here, the function ρ has become a Dirac δ function.

In order to produce the linear result that we want, we can take A and B to be

$$A(\varphi) = B(\varphi) = \begin{cases} +1, & -\pi/2 \leq \varphi \leq +\pi/2, \\ -1, & |\varphi| > \pi/2. \end{cases}$$

This situation is depicted in FIGURE 2. In the case where $\alpha = \beta$ or $\theta = 0$, the two disks, one for each particle, can be thought of as being superposed exactly so that the + and − regions coincide. This comes about in the model because, according to the δ function, $\lambda_1 = \lambda_2$. Hence, the angles on the two disks will coincide and, when one is +, the other will also be. Thus, the product will always be $(+) \cdot (+)$ or $(-) \cdot (-)$ and therefore always +. Then, one will always have a perfect correlation for $\theta = 0$, as in the quantum case. However, if $\theta \neq 0$, then one of the two disks will be rotated with respect to the other and there will be two border regions of angle θ where the product will have a negative sign. Because this region varies linearly with θ, the expectation value E will also vary linearly.

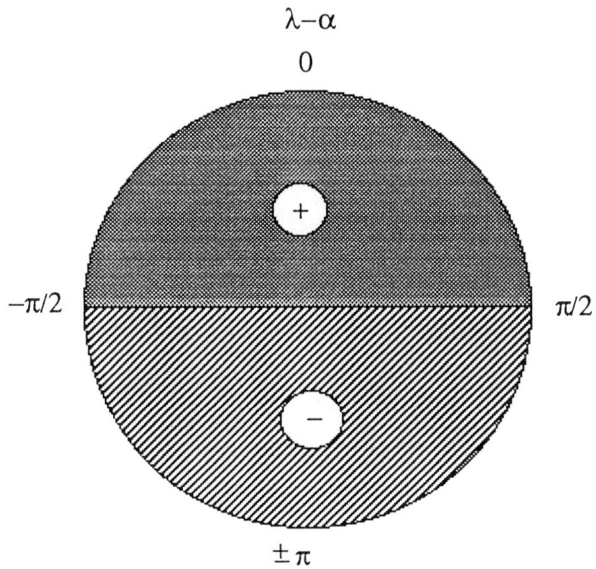

FIGURE 2. Detector behavior in Bell's two-particle model. In this and all subsequent figures, the stippled regions represent the value +1, whereas the hatched regions represent the value −1. When the internal hidden variable λ is within π/2 of the detector angle (α for detector 1), the detector will measure spin +1 along the direction α; when λ differs from α by more than π/2, the detector will measure spin −1 along α.

THE THREE-PARTICLE MODEL

In order to make a classical, realistic, local model like the above, which is good for three particles, there is no hope in having 100% efficient detectors as in Bell's model. That would be self-inconsistent, according to the GHZ theorem. Thus, when we try to construct a model that is as similar as possible to the above one, we have to take this into account. In this case, we assume a particle that decays into three particles, 1, 2, and 3, each of which heads toward a different detector. Again, each detector has a setting parameter, labeled by α, β, and γ. Also, there will be three separate functions that determine whether each detector will fire or not. Here, the functions are

$A(\lambda_1, \alpha)$, $B(\lambda_2, \beta)$, and $C(\lambda_3, \gamma)$. Once again, locality restricts A to depend only on α and not on the settings of the other detectors, which may be far away, and similarly for B and C. We are again assuming that the general hidden variable λ will reduce to one for each particle, which can be interdependent, as they are all determined at the initial decay, and we again assume that they can be in the form of angles so that, again, all the λ_i will vary between $-\pi$ and π.

The main difference between this model and the previous one is that now we assume that the functions A, B, and C can have the values $+1$, -1, and 0. Again, the ± 1 values represent the two firing states, but now 0 indicates that the counter did not fire when the particle hit it. The counter may not fire because of some experimental inefficiency, but this is not what interests us. We must build in the possibility that the counters will not fire for certain values of the λ_i because, even if the counters are experimentally perfect, the GHZ theorem guarantees that there is no combination of the λ_i that is self-consistent for 100% efficient counters.

In the quantum mechanical cases that have been considered, we have

$$E_{\mathrm{qm}} = \cos(\alpha + \beta + \gamma) = \cos\theta,$$

$$\theta = \alpha + \beta + \gamma.$$

(Again, we are ignoring a trivial negative sign.) Thus, again, we would like our model to depend on the one variable θ. Note that the perfect correlation cases are for $\theta = 0$ and $\theta = \pi$. However, our model, like the previous one, will also have the symmetries that $E(\pi - \theta) = -E(\theta)$ and $E(-\theta) = +E(\theta)$, so we will only be concerned with $0 \leq \theta \leq \pi/2$.

In order to have the results depend only on θ, we resort to the same trick as before and write

$$E(\alpha, \beta, \gamma) = \left(\frac{1}{N}\right) \int d\lambda_1 d\lambda_2 d\lambda_3 \delta(\lambda_1 + \lambda_2 + \lambda_3) A(\lambda_1 - \alpha) B(\lambda_2 - \beta) C(\lambda_3 - \gamma) = E(\theta).$$

The last equality is true because A, B, and C are periodic functions and, in fact, we shall take $A = B = C$, so all the detectors are identical. Thus, although it appears that we are varying parameters within a three-dimensional space of the λ_i, the restriction $\Sigma\, \lambda_i = 0$ nonetheless reduces it to a two-dimensional space. In our analysis, we shall consider λ_1 and λ_2 as varying freely, whereas λ_3 will be determined by the constraint. The quantity N is a normalizing factor, which is needed because, for most values of λ_i, at least one of the functions A, B, C will be 0 and the experimental event will not be completed. Thus, the definition of E above will have to be normalized to only include those cases where all of the counters fire. Otherwise, E will never equal ± 1. Therefore, we take N to include all those cases in which all three counters fire:

$$N = N(\alpha, \beta, \gamma) = \int d\lambda_1 d\lambda_2 d\lambda_3 \delta(\lambda_1 + \lambda_2 + \lambda_3) |A(\lambda_1 - \alpha) B(\lambda_2 - \beta) C(\lambda_3 - \gamma)| = N(\theta).$$

We define the experimental efficiency of the experiment by

$$\eta(\theta) = [N(\theta)/(2\pi)^2]^{1/2}.$$

This is defined over the two-dimensional subspace in which the variables are free to roam and η^2 gives the probability of choosing a λ_1 and a λ_2 such that all three counters will fire. The one-half exponent is used to give an equivalent "one-particle" effective detection rate.

All we must do now is to choose a function $A = B = C$ for the detectors. We do this so that at $\theta = 0$ we will have $E = 1$. For this to be so, if we choose $\alpha = \beta = \gamma = 0$, and each $\lambda_i = 0$, and find $A = B = C = 1$, then we may expect one or more of the A, B, C terms to become zero when we vary the λ_i subject to $\Sigma\, \lambda_i = 0$, but we should never get the product $ABC = -1$. (However, two of the functions can simultaneously become -1.) For this to be so, no region where $A = -1$ can abut a region where $A =$

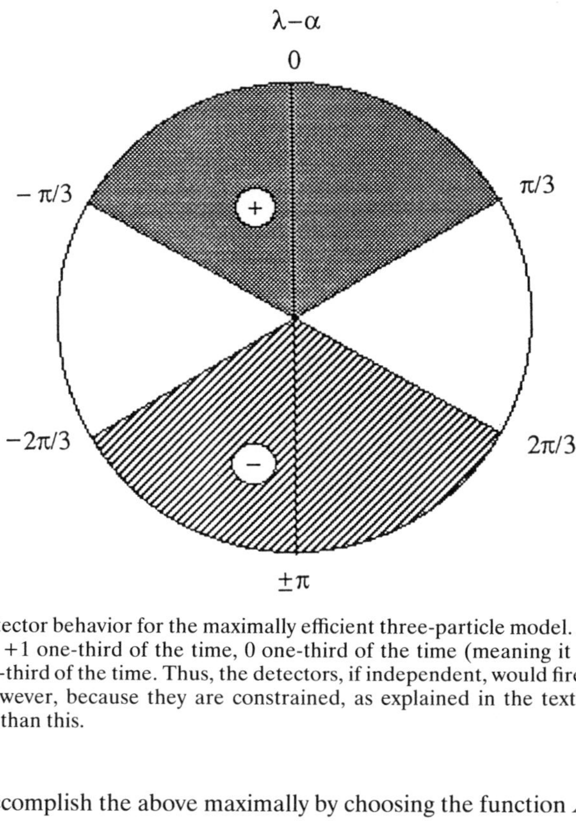

FIGURE 3. Detector behavior for the maximally efficient three-particle model. In this case, the detector counts +1 one-third of the time, 0 one-third of the time (meaning it does not fire at all), and −1 one-third of the time. Thus, the detectors, if independent, would fire two-thirds (⅔) of the time. However, because they are constrained, as explained in the text, their effective efficiency is less than this.

$+1$. We can accomplish the above maximally by choosing the function A as

$$A(\varphi) = \begin{cases} +1, & 0 \le |\varphi| < \pi/3, \\ 0, & \pi/3 \le |\varphi| < 2\pi/3, \\ -1, & 2\pi/3 \le |\varphi|. \end{cases}$$

This is shown graphically in FIGURE 3. The "nominal" efficiency of this counter, and thus also the counters B and C, is ⅔ because it will count two-thirds (⅔) of the time for random values of λ_1. However, the restriction $\Sigma\, \lambda_i = 0$ makes them not

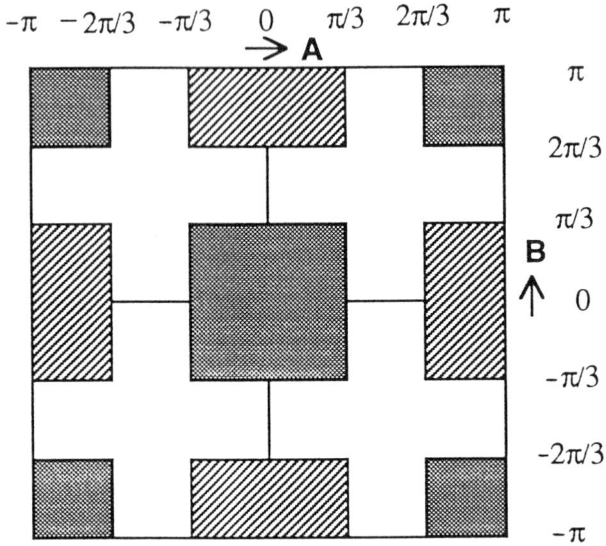

FIGURE 4. Independent firing of two detectors. Because the detectors A and B can fire independently, each two-thirds (⅔) of the time, the product of the two will show that they both fire together four-ninths (4/9) of the time. The four combinations $(\pm 1)(\pm 1)$ occur with equal frequency, so there are as many regions where the product is $+1$ (stippled) as there are where the product is -1 (hatched).

independent of each other and this will reduce the effective experimental efficiency of a three-particle event to much lower than this.

There is one further simplification that we shall make in order to make the model easier to graph. We make the simple substitutions,

$$\lambda_1 - \alpha = \mu_1, \qquad \lambda_2 - \beta = \mu_2, \qquad \lambda_3 - \gamma = \mu_3,$$

which converts us from the restriction $\Sigma\, \lambda_i = 0$ to $\Sigma\, \mu_i = \theta$. Now, α, β, and γ no longer appear separately, but only in the combination θ. Hence, the equation for E becomes

$$E(\theta) = \left[\frac{1}{N(\theta)}\right] \iint_{\Sigma\mu_i=\theta} d\mu_1 d\mu_2 A(\mu_1)B(\mu_2)C(\mu_3),$$

whereas $N(\theta)$ is given by

$$N(\theta) = \iint_{\Sigma\mu_i=\theta} d\mu_1 d\mu_2 |A(\mu_1)B(\mu_2)C(\mu_3)|.$$

We shall start off with the case $\theta = 0$. If we vary μ_1 and μ_2 independently in our model (according to the values of A and B assigned in FIGURE 3, subject only to $-\pi \leq \mu_1,\mu_2 \leq \pi$) and if we ignore μ_3 altogether, we shall get for the product AB the values indicated in FIGURE 4. However, each value of (μ_1, μ_2) uniquely determines the value of μ_3 (via the constraint $\Sigma\mu_i = 0$) and therefore C, and so too the value of ABC. The value of ABC is shown in FIGURE 5 for each value of (μ_1,μ_2). Note that one only has $ABC = +1,0$ for $\theta = 0$, as must be the case. By looking at FIGURE 5, one can

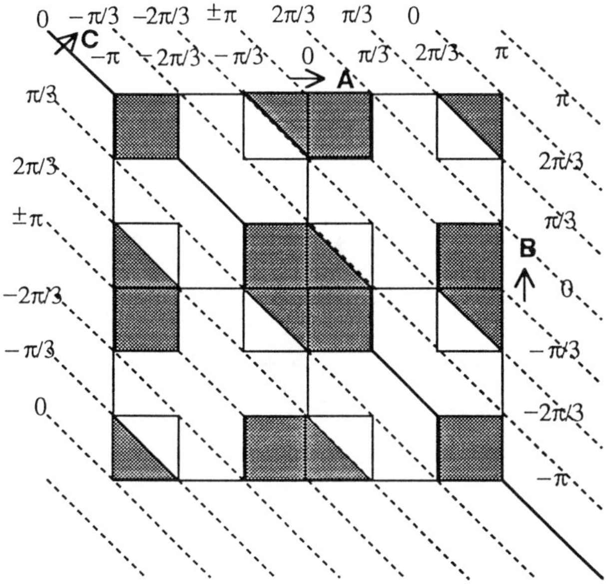

FIGURE 5. Firing of the three detectors when $\theta = 0$. When the third detector is added, the three detectors are constrained by the restraint $\Sigma\mu_i = 0$. The angle μ_3 for the third detector is constant along the 45° diagonals and is labeled along the left-hand side and in the very upper line. For $\theta = 0$, there can be no negative regions for the product ABC because $E(0) = +1$.

see that the fraction of the two-dimensional space for which $ABC \neq 0$, or η^2, is ⅓. This gives $\eta = 0.577$, the effective efficiency per detector. This is the maximum possible efficiency because, when θ varies, η decreases, reaching a minimum at $\theta = \pi/2$, where $\eta = 0.5$.

Next, we can calculate $E(\theta)$ for $0 \leq \theta \leq \pi/3$. The situation here is shown in FIGURE 6. In order to calculate $E(\theta)$, we will have to be able to calculate the area in the partial squares, the four types of which are shown in FIGURE 7. If we normalize the area of each square to 1, the area of each type of square is

$$A_a = \left(\frac{9}{2}\right)\left(\frac{\theta}{\pi}\right)^2, \qquad A_b = 1 - A_a,$$

$$A_c = \left(\frac{1}{2}\right)\left(1 - \frac{3\theta}{\pi}\right)^2, \qquad A_d = 1 - A_c.$$

In this case, some of the areas have $ABC = +1$, whereas some of them have $ABC = -1$, so $E < 1$.

In FIGURE 8, we show $E(\theta)$ for $\pi/3 \leq \theta \leq 2\pi/3$. In this case, the area in the four types of partial squares (again as shown in FIGURE 7) can be given by

$$A_a = \left(\frac{1}{2}\right)\left(\frac{3\theta}{\pi} - 1\right)^2, \qquad A_b = 1 - A_a,$$

$$A_c = \left(\frac{1}{2}\right)\left(2 - \frac{3\theta}{\pi}\right)^2, \qquad A_d = 1 - A_c.$$

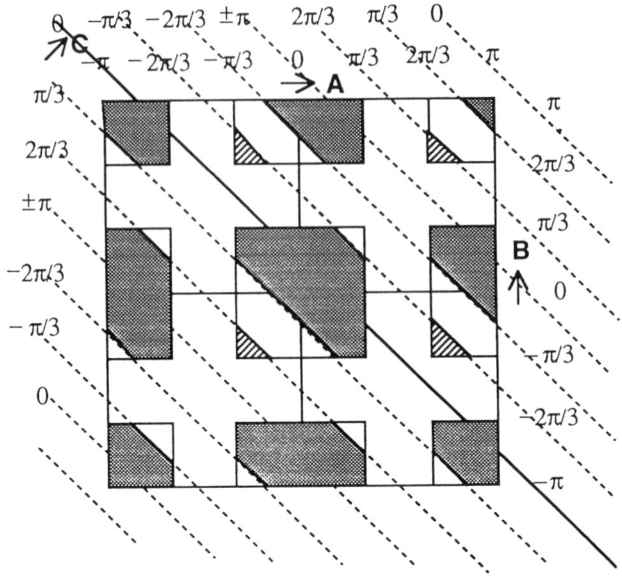

FIGURE 6. Firing of the three detectors when $\theta \leq \pi/3$. In this case, the restraint is $\Sigma\mu_i = \theta$ and thus the third detector is displaced by less than the side of one of the small squares. This introduces some negative values into the product ABC, which makes $E(\theta) < 1$.

We now present a table (TABLE 1) of the values of $E(\theta)$ (the expectation value) and $\eta^2(\theta)$ (the efficiency) as determined from FIGURES 5 and 7 for angles between $\theta = 0$ and $2\pi/3$. (Of course, we only need them up to $\pi/2$.) The average is determined by subtracting the negative areas (striped) from the positive areas (dotted) and then dividing by the total area (36 squares). The efficiency η^2 is determined by adding the

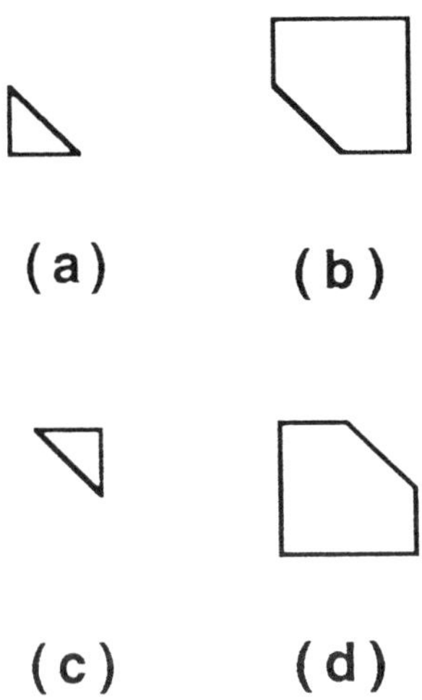

FIGURE 7. The four types of incomplete squares. In calculating $E(\theta)$ in FIGURES 6 and 8 and in modified form in FIGURES 10 and 11, one needs to know the areas of the four types of incomplete squares that appear. For each case, the results are given in the text.

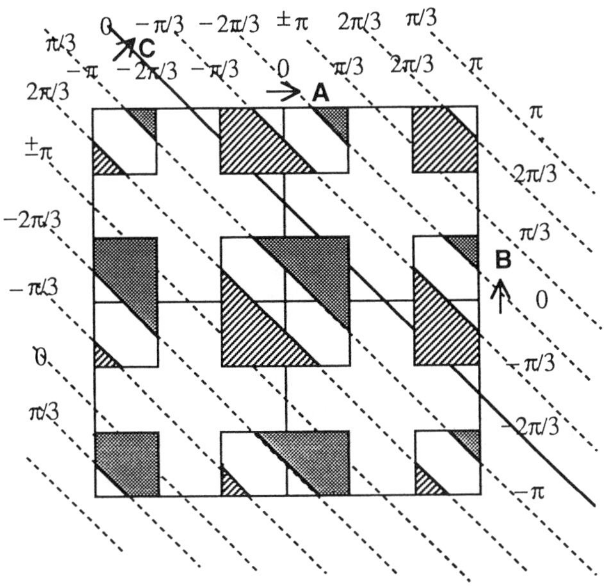

FIGURE 8. Firing of the three detectors when $\pi/3 \leq \theta \leq 2\pi/3$. In this case, the third detector is displaced horizontally by between one and two lengths of one of the square sides, yielding more negative areas than in FIGURE 6.

positive and negative areas and then dividing by the total area. This is the percentage of area in which all three detectors fire. Here $x = \theta/\pi$.

$E(\theta)$ from TABLE 1 is plotted in FIGURE 1, labeled as "⅔ Effic." because each of the individual detectors would be ⅔ efficient if they were not correlated. $E(\theta)$ does not differ very much from the quantum mechanical answer; however, note that it is *greater* than the quantum mechanical one for all θ, in contrast to the case for two particles, where it is always smaller than the quantum mechanical one. Therefore, it does not seem that the way to try to detect the difference between this case and the quantum one is to measure averages. Rather, the thing to measure is efficiencies. The quantum $E(\theta)$ is plotted in FIGURE 1 for a detector efficiency of 100%. However, our three-particle model has a maximum value of η^2 of ⅓ at $\theta = 0$, which decreases to $\eta^2 = ¼$ at $\theta = \pi/2$. Hence, a model such as ours can be eliminated through its low efficiency. This is made possible, of course, by the GHZ theorem, which guarantees

TABLE 1. Average and Efficiency of the Model

$x = \theta/\pi$	$E(x)$	$\eta^2(x)$
$0 \leq x \leq 1/3$	$\dfrac{(2 - 9x^2)}{(2 - 3x^2)}$	$\left(\dfrac{1}{3}\right) - \left(\dfrac{x^2}{2}\right)$
$1/3 \leq x \leq 2/3$	$\dfrac{[(1 - x)^2 - x^2]}{[(1 - x)^2 + x^2]}$	$\dfrac{[(1 - x)^2 + x^2]}{2}$

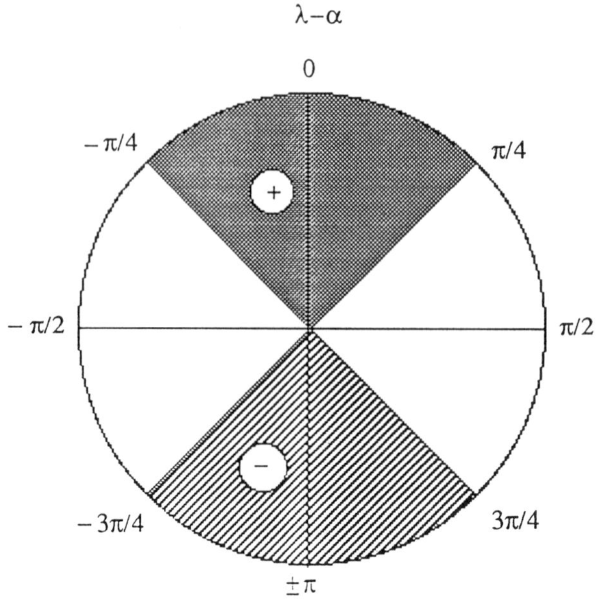

FIGURE 9. Detector behavior for a 50% efficient detector. This case is similar to that of FIGURE 3, except that the detector only counts half the time, excluding the region $\pi/4 \leq |\theta| \leq 3\pi/4$.

that no deterministic three-particle model will have 100% efficiency. If the intrinsic efficiency of our detectors were less than ⅔, then the overall efficiency of the correlated three-particle system would be even less than for the model above. We will now give an example of this.

AN INEFFICIENT THREE-PARTICLE MODEL

In order to show that one will get even more highly correlated values for $E(\theta)$ by further decreasing the efficiency of the individual counters, and at the same time get lower effective values of η^2, we will examine one further model. In this model, the intrinsic efficiency of each of the counters will be ½. The detailed operation of each of the counters is shown in FIGURE 9. Here, each of the counters will count according to the prescription,

$$A(\varphi) = \begin{cases} +1, & -\pi/4 \leq \varphi \leq +\pi/4, \\ 0, & +\pi/4 < |\varphi| \leq 3\pi/4, \\ -1, & 3\pi/4 < |\varphi| \leq \pi. \end{cases}$$

Similar to our previous case, one can give a graphic demonstration of which possible regions the three counters will count in simultaneously. In this case, the angles that separate the various cases are $0 \leq \theta \leq \pi/4$ and $\pi/4 < \theta \leq \pi/2$. The two cases are shown in FIGURES 10 and 11. Notice that, for angles $\theta \leq \pi/4$, there are *no* negative counts at all, so $E(\theta) = 1$. From $\theta = \pi/4$ to $\theta = \pi/2$, $E(\theta)$ must drop sharply from 1 to 0. This case is also plotted in FIGURE 1, labeled as "½ Effic." because, if the detectors were not coupled, each of them would have efficiency ½. In FIGURE 10, each of the four incomplete square sections shown in FIGURE 7 has area

$$A_a = \frac{8\theta^2}{\pi^2}, \qquad A_b = 1 - A_a,$$

$$A_c = \left(\frac{1}{2}\right)\left(1 - \frac{4\theta}{\pi}\right)^2, \qquad A_d = 1 - A_c.$$

In FIGURE 11, each of these four incomplete square sections has area

$$A_a = \left(\frac{1}{2}\right)\left(\frac{4\theta}{\pi} - 1\right)^2, \qquad A_b = 1 - A_a,$$

$$A_c = \left(\frac{1}{2}\right)\left(2 - \frac{4\theta}{\pi}\right)^2, \qquad A_d = 1 - A_c.$$

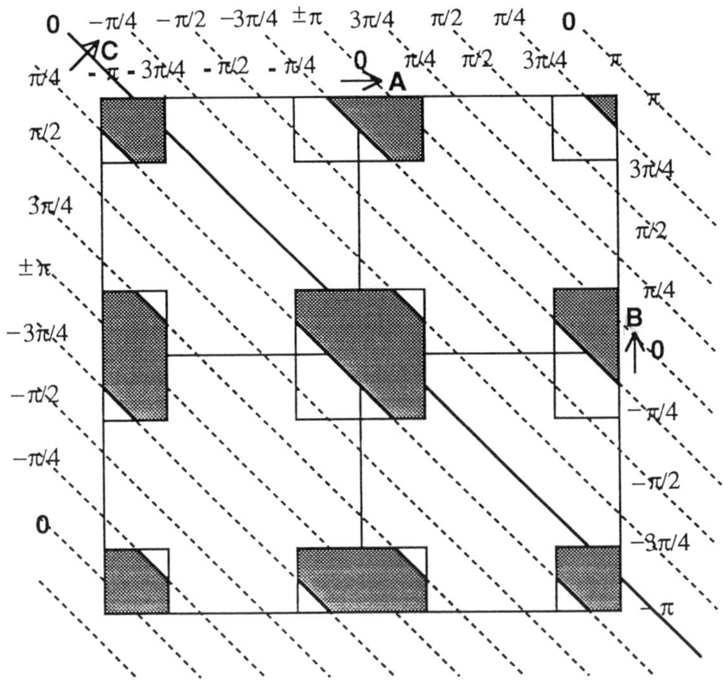

FIGURE 10. Firing of detectors for $\theta \leq \pi/4$ in the ½ efficiency model. This case is determined similarly to that in FIGURE 6, except that the squares represent only a length of $\pi/4$; however, for $\theta \leq \pi/4$, there are no regions where the product $ABC = -1$, so $E(\theta) = +1$, although the efficiency of the counting process is very low.

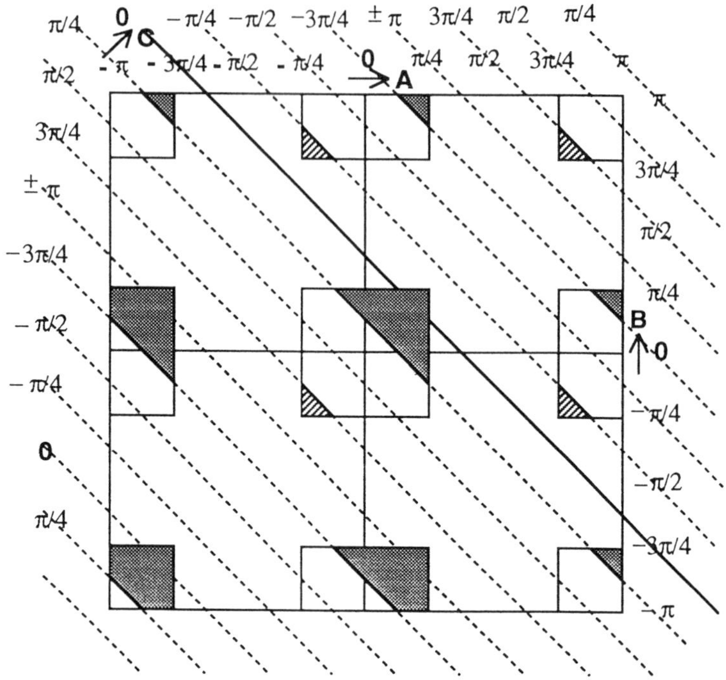

FIGURE 11. Firing of detectors for $\pi/4 < \theta \leq \pi/2$ in the ½ efficiency model. This case is similar to that of FIGURE 8, but with smaller squares. Now, there are some negative regions, but the counting efficiency is still low.

We can also give the functions $E(\theta)$ and $\eta^2(\theta)$ (see TABLE 2), determined in the same way as for our former model. (Note again that $x = \theta/\pi$.) The expectation value E in this case is much higher than the quantum mechanical model, but the efficiency is even lower than in our previous model, varying from $\eta^2(0) = 3/16$ to $\eta^2(\pi/2) = 1/16$.

CONCLUSIONS

The major lesson from this exercise is as follows: although one can always derive Bell-type inequalities for inefficient counters in the three-particle case, just as in the

TABLE 2. Average and Efficiency of the New Model

$x = \theta/\pi$	$E(x)$	$\eta^2(x)$
$0 \leq x \leq 1/4$	1	$\left(\frac{3}{16}\right) - x^2$
$1/4 \leq x \leq 1/2$	$\frac{(4 - 8x)}{(5 - 16x + 16x^2)}$	$\left(\frac{5}{16}\right) - x + x^2$

two-particle case, if one wants to preserve the overall dependence on a single variable, as occurs in quantum theory, then the three detectors cannot possibly be independent of each other. We have formed a simple model that has the maximum efficiency allowed by the GHZ theorem and we find that the expectation value of the results closely approximates that of quantum theory, although with low efficiencies for the counting rates of the detectors. Although the results do come close to quantum theory, the expectation value is nonetheless always greater than the quantum case, in contrast with what one expects for two particles. With a model that is less efficient than the maximum allowed, we found that the expectation value became even higher, giving a large discrepancy from the quantum results, whereas the overall efficiency became even lower. Thus, the types of experiments that one will have to perform in the three-particle case are not at all the same types of experiments that one performed in the two-particle case. Of course, these are simple models and the results very well may not be general.

ACKNOWLEDGMENTS

I would like to thank Mike Horne, Anton Zeilinger, and Abner Shimony for valuable conversations.

REFERENCES AND NOTES

1. Einstein, A., B. Podolsky & N. Rosen. 1935. Phys. Rev. **47:** 777. This paper will be referred to as EPR.
2. A good summary of this work is contained in: Greenberger, D. M., M. A. Horne, A. Shimony & A. Zeilinger. 1990. Am. J. Phys. **58:** 1131. This shall be referred to as GHZ.
3. Bell, J. S. 1964. Physics (N.Y.) **1:** 195. All of Bell's papers on the subject are collected in: Bell, J. S. 1987. Speakable and Unspeakable in Quantum Mechanics. Cambridge University Press. London/New York.
4. A good summary of experimental and theoretical work on the subject can be found in: Selleri, F. 1989. Quantum Paradoxes and Physical Reality. Kluwer. Dordrecht.

The EPR Argument and Nonlocality without Inequalities for a Single Photon[a]

LUCIEN HARDY[b]

Institut für Experimentalphysik
Universität Innsbruck
A-6020 Innsbruck, Austria

INTRODUCTION

The Einstein, Podolsky, and Rosen (EPR) paper[1] of 1935 is often presented as the starting point of the discussion regarding locality in quantum mechanics and leading to Bell's proof that quantum mechanics must necessarily be nonlocal. However, we will see that the essential points in the EPR argument had already been made by Einstein some eight years earlier at the Fifth Solvay Congress. Einstein illustrated his points by considering a single particle rather than two particles as used later in the EPR discussion and subsequently in most discussions of locality. However, he was still able to argue that the wave function provides an incomplete description of a system or that there must be some kind of "peculiar action-at-a-distance", this being the same as the conclusion of the EPR argument. In order to see more clearly that a single particle can be used to run an argument against the completeness of the wave function description of a system, Einstein's 1927 argumentation is cast into a similar form to that in the EPR paper. One natural question that then arises is whether it is possible to also find a version of Bell's theorem[2] (demonstrating that quantum mechanics is nonlocal) using a single particle. In fact, some attempts at this have already been made. The first, using a single photon, was due to Tan, Walls, and Collett[3] (see also Oliver and Stroud[4]) and it is their work that has provoked recent interest on this matter. However, their proof makes use of an untestable supplementary assumption and therefore rules out only those local interpretations that are consistent with this assumption.[5–8] A more recent proof due to Czachor,[9] although interesting, is cast in a form quite different to that usually considered in discussions of Bell's theorem. In particular, the local classical variable (of which there is only one rather than two) is necessarily in the past light cone of the region in which its variation is supposed to have no effect. It is desirable to find a proof of the nonlocality of a single particle that is as strong as Bell's original proof for two particles. In this report, we will consider a source that never emits more than a single photon at a time and, on average, emits less than a single photon at a time. We will see how, with this source, it is possible to find measurements that allow a demonstration of nonlocality without using inequalities. The form of the contradiction when locality is assumed is particularly striking. The type of proof of nonlocality without inequalities considered here, originally motivated by the work of Elitzur and

[a]This work was supported by the Royal Society (London).

[b]Present address: Department of Mathematical Sciences, University of Durham, Durham DH1 3LE, United Kingdom.

Vaidman,[10] was first presented in reference 11 and developed in reference 12 (see also sources cited in these papers). However, in references 11 and 12, two-particle situations were considered.

EINSTEIN'S REMARKS IN 1927

A very readable account (and the source of the material in this section) of Einstein's remarks at the Fifth Solvay Congress is given by Jammer.[13] Having apologized for "not having gone deeply into quantum mechanics", Einstein proceeded to consider a system in which a single particle is allowed to impinge on a hole in a diaphragm so that it is diffracted and then impinges on a hemispherical scintillation-screen placed behind the hole. The quantum mechanical description of this process says that the wave function, ψ, associated with the particle is spread at the hole to form something approximating a spherical wave that propagates towards the screen. Then, according to the Born probability interpretation, the probability (density) for the particle being detected at some particular point on the screen is given by the square magnitude of ψ at that point. According to Einstein, there are two possible viewpoints that one can take of this process.

Viewpoint I says that ψ does not represent an individual particle, but an ensemble of particles. According to this viewpoint, $|\psi(\boldsymbol{r})|^2$ represents the probability (density) that some particle of the ensemble exists at the point $\boldsymbol{r}$.

Viewpoint II says that the wave function is considered as a complete description of an individual process. Thus, just before causing a scintillation on the screen, the individual particle must be considered to be potentially present at all points where $|\psi(\boldsymbol{r})|^2$ is nonzero. Then, when the particle is detected at one point on the screen, there is a "peculiar action-at-a-distance" that prevents the particle from being detected at another point on the screen.

It is clear that Einstein had seen that one must make a choice: either one could accept locality, in which case ψ could not be considered to provide a complete description of an individual process (viewpoint I), or one could consider ψ to provide a complete description of an individual process, but then one must abandon locality (viewpoint II). Although the issue of locality was not emphasized much in the EPR paper, it is clear that the conclusion of EPR is the same as Einstein's conclusion here. We will simplify this single-particle example by considering a single particle incident on a beam-splitter because then there are only two paths that the particle can take. First, however, we will introduce some basic tools to be used in this example and also in the nonlocality example.

SOME BASICS

In the remaining sections, we will be a little more specific and take the particle under discussion to be a photon. So far as the EPR discussion in the fourth section below is concerned, this amounts only to a convenience. However, in the nonlocality discussion in the sixth section, it is not the case that any type of particle could be used to run an analogous argument.

Beam-Splitters

Typically, a beam-splitter is a partially silvered mirror that transmits some fraction of the light. We will only consider 50:50 beam-splitters that transmit half the light and that reflect the other half. Let the beam-splitter have inputs a and b and outputs c and d. The annihilation operators for each of these modes are related by a transformation equation:

$$\begin{pmatrix} \hat{c} \\ \hat{d} \end{pmatrix} = \left(\frac{1}{\sqrt{2}}\right)\begin{pmatrix} 1 & i \\ i & 1 \end{pmatrix}\begin{pmatrix} \hat{a} \\ \hat{b} \end{pmatrix}. \tag{1}$$

Here, we have made a particular choice for the transformation matrix. More generally, this matrix need only be a member of the group SU(2) (although, of course, additional constraints are imposed for a 50:50 beam-splitter), but the above choice is quite sufficient for our purposes. Using this matrix, it is possible to calculate the resulting output state from any input state. We require just two results. The first says that the vacuum transforms to the vacuum:

$$|0\rangle_a|0\rangle_b \rightarrow |0\rangle_c|0\rangle_d. \tag{2}$$

Second, if a single photon is incident on one of the two inputs, a say, with the vacuum incident on the other input, then there is equal probability amplitude associated with the photon emerging in either output:

$$|1\rangle_a|0\rangle_b \rightarrow \left(\frac{1}{\sqrt{2}}\right)(|1\rangle_c|0\rangle_d + i|0\rangle_c|1\rangle_d). \tag{3}$$

This result follows because $|1\rangle_a|0\rangle_b = \hat{a}^\dagger|0\rangle_a|0\rangle_b$ and, from equation 1, we obtain $\hat{a}^\dagger \rightarrow (1/\sqrt{2})(\hat{c}^\dagger + i\hat{d}^\dagger)$.

Coherent States

The state produced by a laser is a coherent state. Coherent states[14] are eigenstates of the annihilation operator $\hat{a}$:

$$\hat{a}|\alpha\rangle = \alpha|\alpha\rangle,$$

where α is the complex amplitude ($|\alpha|^2$ is the average number of photons). The coherent state can be written as a sum over number states:

$$|\alpha\rangle = e^{-(1/2)|\alpha|^2}\sum_n \frac{\alpha^n}{\sqrt{n}}|n\rangle = e^{-(1/2)|\alpha|^2}\left(|0\rangle + \alpha|1\rangle + \frac{\alpha^2}{\sqrt{2}}|2\rangle + \cdots\right). \tag{4}$$

Parametric Downconversion

Certain crystals (referred to as nonlinear crystals), when pumped by a laser, exhibit parametric downconversion.[15,16] This means, essentially, that a photon in the pump laser beam is sometimes converted into two photons, one called the signal and one called the idler (these names are just conventional). If the frequency and wave vector of the pump, signal, and idler are $(\omega_p, \boldsymbol{k}_p)$, $(\omega_s, \boldsymbol{k}_s)$, and $(\omega_i, \boldsymbol{k}_i)$, respectively, then the following conditions (called phase-matching conditions) hold to a good approximation:

$$\omega_s + \omega_i = \omega_p, \tag{5a}$$

$$\boldsymbol{k}_s + \boldsymbol{k}_i = \boldsymbol{k}_p. \tag{5b}$$

The second condition implies that the pairs have some directional correlation and, therefore, it is possible to pick out pairs of photons of given colors by placing diaphragms appropriately. If we restrict out attention to a pair of modes picked out in this way, then the interaction Hamiltonian between the pump beam (treated classically and represented by the complex amplitude V) and the signal and idler modes (with annihilation operators $\hat{a}_s$ and $\hat{a}_i$) is

$$\hat{H}_I = i\hbar g V \hat{a}_s{\dagger}\hat{a}_i{\dagger} + hc, \tag{6}$$

where g is a coupling constant. With the initial states of the signal and idler modes as the vacuum, after a short interaction time t (equal to the time of passing of the pump photons through the interaction region of the crystal), the state becomes

$$|\psi\rangle = \exp\left(-iH_I t/\hbar\right)|0\rangle_s|0\rangle_i = |0\rangle_s|0\rangle_i + gtV|1\rangle_s|1\rangle_i + \cdots. \tag{7}$$

This can be used as a single-photon source because, if we detect a photon in the idler mode, for example, then we know that there is a photon in the signal mode.

THE SINGLE-PHOTON EPR ARGUMENT

A very careful discussion of the EPR argument is given by Redhead[17] and we will follow his argumentation here. In particular, we will make use of the fact, noticed by Redhead, that it is not necessary to consider two possible measurements on system 1 (e.g., σ_{1x} and σ_{1y} in the singlet state example considered by Bohm[18]). One measurement is sufficient. EPR introduced a necessary condition for a theory to be considered complete—"Necessary condition for completeness: every element of the physical reality must have a counterpart in the physical theory." Then, in order to be able to discuss such elements of physical reality, at least in some situations, EPR gave a sufficient condition for their existence. We give it in the form given by Redhead—"Sufficient condition for existence of elements of physical reality: if we can predict with certainty, or at any rate with probability equal to one, the result of measuring a physical quantity at time t, then at the time t there exists an element of physical reality corresponding to the physical quantity and having a value equal to the predicted measurement result." In addition, the EPR paper contained an implicit locality assumption that was stated explicitly by Redhead—"Locality principle: elements of physical reality pertaining to one system cannot be affected by measurements performed 'at-a-distance' on another system."

We will now apply the above two conditions and one principle to the single-photon experiment shown in FIGURE 1. A nonlinear crystal is used as a single-photon source in the way described above. The single photon in the signal mode is incident

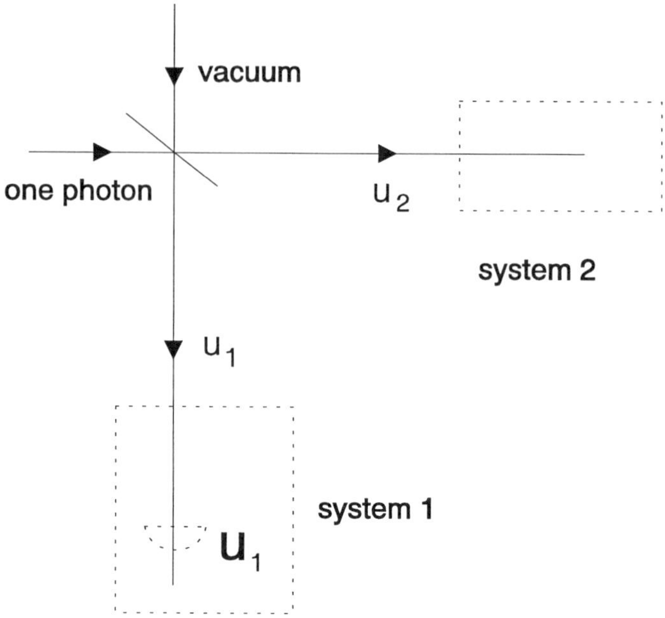

FIGURE 1. Apparatus used to illustrate the single-particle EPR argument.

on one input of a 50:50 beam-splitter with the vacuum incident on the other input. Then, the two output modes, u_1 and u_2, separate to two distant regions where they form system 1 and system 2, respectively. When the detector in the idler mode has clicked, we know that a photon has impinged on the beam-splitter and therefore, from equation 3 above, the state of the system becomes

$$|\psi\rangle = \left(\frac{1}{\sqrt{2}}\right)(i|1\rangle_{u_1}|0\rangle_{u_2} + |0\rangle_{u_1}|1\rangle_{u_2}) \quad \textbf{(8)}$$

(note that the coefficient i is associated with the term in which the photon has been reflected). Now, if we wanted, we could insert a detector U_1 in path u_1. If we did this, then there would be two possibilities: (1) the detector would fire, in which case we could predict with certainty that a detector U_2, if it were placed in path u_2, would not fire; in this case, we could apply the reality condition and deduce that there exists an element of reality corresponding to U_2 that we can write as $[U_2] = 0$; or (2) detector U_1 would not fire, in which case we could predict with certainty that a detector U_2, if it were placed in path u_2, would fire; thus, there would again exist an element of reality, this time given by $[U_2] = 1$. For an individual run, one of these two possibilities would have to happen. Furthermore, although we could not predict which one in advance, there would always be a definite element of reality associated with system 2. Now, according to the locality principle, any element of reality at system 2 cannot be affected by whether or not a detector is placed in path u_1. This means that, even in the case where the detector U_1 is not used, there must still exist an element of reality at system 2 (whether it is $[U_2] = 0$ or $[U_2] = 1$) because, were this detector used, we would then be able to deduce the existence of the element of reality. However, the state vector describing the system in the absence of any measurements is $|\psi\rangle$, given in equation 8. This state vector is not an eigenstate of the projection operator, $\hat{U}_2$, onto the state $|1\rangle_{u_2}$. Hence, there is nothing in $|\psi\rangle$ that can correspond to the element of reality $[U_2]$ that we have deduced must exist and therefore, according to the completeness condition above, this state vector does not provide a complete description of reality.

As is now well known, the locality assumption is not consistent with quantum mechanics as was shown by Bell. Before giving a single-photon demonstration of the violation of locality, we need to consider how to produce a superposition of the vacuum with a single-photon state.

HOW TO PREPARE $e|0\rangle + f|1\rangle$

A coherent state is a superposition of many number states. However, we want to produce a superposition of only the vacuum and single-photon number states. This can be prepared using the apparatus shown in FIGURE 2. A nonlinear crystal is pumped by a strong laser and a low intensity state $|\beta\rangle$, of the same frequency as the idler mode, is incident on the crystal behind the idler mode. Ou, Wang, Zou, and Mandel[19] considered a similar setup to this (although they were interested in the high intensity limit of the laser behind the idler mode) and some of the working below is taken from their work. A detector is placed in the idler mode. There are two ways in which one photon can come to be detected at this detector (we assume that the detector can measure photon number, at least to the extent that it can distinguish between one and more photons). Either it comes from a downconversion process, in which case there will be an accompanying signal photon, $|1\rangle_s$, in the signal mode, or it comes from the low intensity coherent state source behind the idler, in which case there will be only the vacuum, $|0\rangle_s$, in the signal mode. As these two processes are indistinguishable, the resulting state in the signal mode will be a superposition

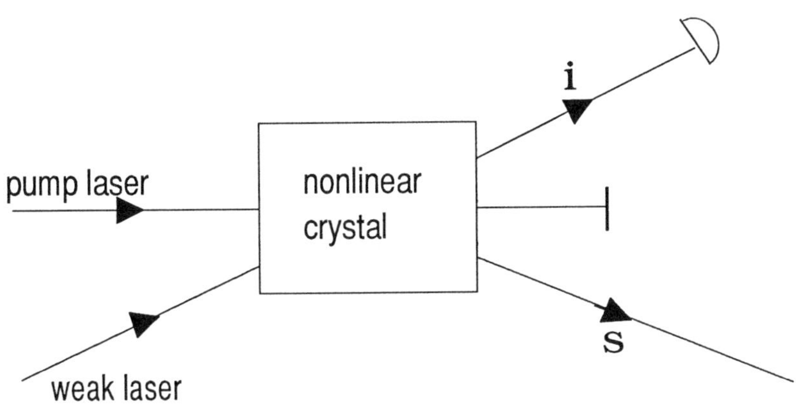

FIGURE 2. Apparatus used to prepare the state $e|0\rangle + f|1\rangle$.

$e|0\rangle_s + f|1\rangle_s$, where e and f are the relevant amplitudes. We can give a more rigorous treatment of this process by using some of the results from above. The initial state is $|\beta\rangle_i|0\rangle_s$. After some short interaction time t, the state becomes

$$|\psi\rangle = \exp(-iH_I t/\hbar)|\beta\rangle_i|0\rangle_s.$$

This can be expanded as

$$|\psi\rangle = |\beta\rangle_i|0\rangle_s + gtV\hat{a}_i{}^\dagger|\beta\rangle_i|1\rangle_s + \cdots.$$

Substituting in the number state expansion for the coherent state (like that given in equation 4) and collecting terms, this becomes

$$|\psi\rangle = \exp[(-\tfrac{1}{2})|\beta|^2][|0\rangle_i|0\rangle_s + |1\rangle_i(\beta|0\rangle_s + gtV|1\rangle_s) + |\varphi\rangle], \quad \mathbf{(9)}$$

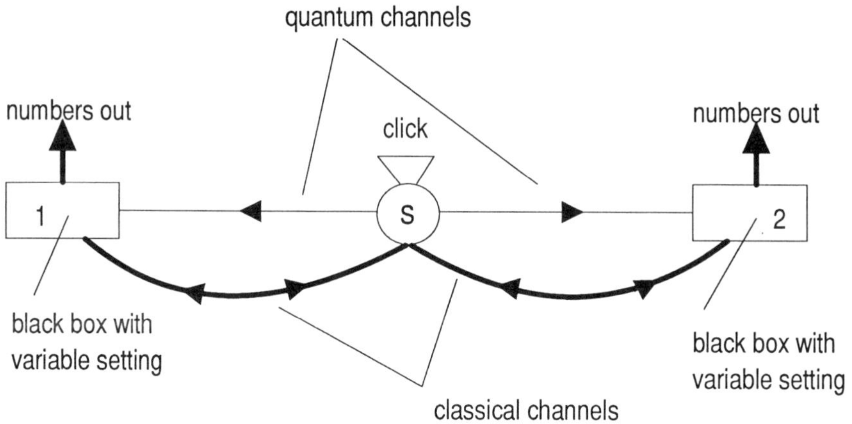

FIGURE 3. Configuration required of an experiment designed to demonstrate the nonlocality of a single photon.

where the state $|\varphi\rangle$ contains only terms with more than one photon in the idler mode. If we consider those times when one photon is detected in the idler mode, then we see from equation 9 that the required state, $\beta|0\rangle_s + gtV|1\rangle_s$, is produced in the signal mode.

SINGLE-PHOTON NONLOCALITY

The Rules of the Game

First, we will set down the rules of the game—those conditions that must be satisfied before we can say that we have the nonlocality of a single photon.

Rule 1: The experimental configuration must be as shown in FIGURE 3. From the source, there are two quantum channels (that is, channels that are allowed to carry a quantum state from the source) that impinge on two distant "black boxes". The source has an event-ready device that goes "click" when a state is emitted. There are classical channels that are allowed to carry classical information (that is, information that can be put into binary code) between the source and the two ends (at speeds less than or equal to c). Each black box has a variable setting that can take at least two values. The outputs from the black boxes are some numbers (these are essentially the measurement results, although in an operational description like this there is no need to use the word "measurement").

Rule 2: If the source clicks at time t and a detector is placed into each of the quantum channels at a distance cT from the source, then, during the time interval $t + T - \tau/2$ till $t + T + \tau/2$, at most only one of the two detectors should click (and that detector should only detect one photon). Here, τ is the time taken for the quantum state to be emitted from the source. This rule ensures that we only consider single photons.

Rule 3: If the quantum channels are blocked so that the quantum state does not reach the black boxes, then there should be no possible way of getting a violation of locality between the two black boxes.

Rule 4: The predictions must be such that the entire class of local hidden variable theories is ruled out (so that no supplementary assumptions are used).

Here, we envisage a scheme in which two output quantum channels are taken from a single-photon source. In the case where three, four, or even an infinite number of such channels are taken, the above rules could be revised appropriately. These rules could also be revised to apply to particles other than photons. The main purpose of the classical channels allowed for in rule 1 would be to transmit the information that the source had clicked so that it is known when to take measurement results from the black boxes. We could have instead stated rule 2 more generally in the following way: if two observers, Alice and Bob say, make measurements in the quantum channels, Alice on the left and Bob on the right, then at most only one of them is able to establish on the basis of this measurement that the source has clicked. Such a formulation of the rule is more operational than that above and captures the essential point (it is because of this property that nonlocality of a single particle is more surprising than the nonlocality of two particles). This formulation does not rule out the possibility of a source that emits, for example, two photons, but always such that they both go into the same quantum channel; still, though, such a source could be regarded as a single-particle source, the two photons being a kind of "molecule". The purpose of rule 3 is to ensure that any nonlocality is really due to the state emitted from the source and not because of some preexisting entanglement between the two black boxes. For example, already entangled photons could be stockpiled. The scheme of Tan, Walls, and Collett satisfies the first three rules, but not rule 4.

Four Experiments

The apparatus to be considered is shown in FIGURE 4. The state $q|0\rangle_s + r|1\rangle_s$, prepared in the above way, is incident on the s input of a beam-splitter, with the vacuum incident on the other input, t. The initial state is

$$(q|0\rangle_s + r|1\rangle_s)|0\rangle_t = q|0\rangle_s|0\rangle_t + r|1\rangle_s|0\rangle_t.$$

Upon evolving through the beam-splitter, this becomes, using equations 2 and 3,

$$|\Psi\rangle = q|0\rangle_{u_1}|0\rangle_{u_2} + \left(\frac{ir}{\sqrt{2}}\right)|1\rangle_{u_1}|0\rangle_{u_2} + \left(\frac{r}{\sqrt{2}}\right)|0\rangle_{u_1}|1\rangle_{u_2}. \quad \textbf{(10)}$$

We will now consider four different possible experiments that could be performed on this state.

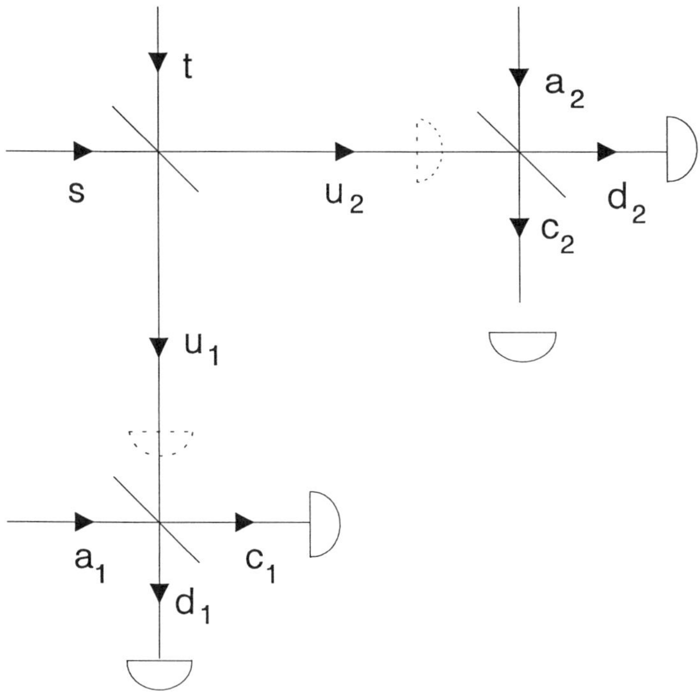

FIGURE 4. Experimental setup used to demonstrate the nonlocality of a single photon. The state $q|0\rangle + r|1\rangle$ is incident on the s mode, the vacuum is incident on the t mode, and the coherent states $|\alpha_k\rangle$ are incident on the a_k modes ($k = 1, 2$).

Experiment 1: Detectors U_1 and U_2 are placed in modes u_1 and u_2, respectively. We put $U_1 = n$, where n is the number of photons detected at detector U_1, and similarly for detector U_2. Because there is never more than one photon at a time, it is clear that

$$U_1 = 1 \text{ and } U_2 = 1 \text{ never happens.} \quad \textbf{(11)}$$

More formally, we can see that this is true because there is no $|1\rangle_{u_1}|1\rangle_{u_2}$ term in equation 10.

Experiment 2: Detector U_2 is placed in mode u_2 as before, but detector U_1 is removed and mode u_1 is allowed to impinge onto a beam-splitter. The coherent state $|\alpha_1\rangle_{a_1}$ impinges onto the other input of this beam-splitter. Detectors C_1 and D_1 are placed in the output modes c_1 and d_1 of this beam-splitter. The state of the system before measurements are made is given in equation 10 and can be written

$$|\Psi\rangle = \left[q|0\rangle_{u_1} + \left(\frac{ir}{\sqrt{2}}\right)|1\rangle_{u_1}\right]|0\rangle_{u_2} + \left[\left(\frac{r}{\sqrt{2}}\right)|0\rangle_{u_1}\right]|1\rangle_{u_2}. \quad \textbf{(12)}$$

Now, concentrate on the case where no photon is detected at detector U_2 so that $U_2 = 0$. When this happens, it follows that the state in mode u_1 becomes $N[q|0\rangle_{u_1} + (ir/\sqrt{2})|1\rangle_{u_1}]$, where N is a normalization factor. Thus, the state impinging on the beam-splitter is

$$|\alpha_1\rangle_{a_1}N\left[q|0\rangle_{u_1} + \left(\frac{ir}{\sqrt{2}}\right)|1\rangle_{u_1}\right].$$

Expanding the coherent state (using equation 4) and collecting terms, we get

$$Ne^{-(1/2)|\alpha_1|^2}\left[q|0\rangle_{a_1}|0\rangle_{u_1} + \left(\frac{ir}{\sqrt{2}}\right)|0\rangle_{a_1}|1\rangle_{u_1} + \alpha_1 q|1\rangle_{a_1}|0\rangle_{u_1} + |\varphi'\rangle\right],$$

where $|\varphi'\rangle$ contains only terms with a total of two or more photons (as we shall only be interested in those cases in which a total of one photon is detected, we need not pay any special attention to the evolution of these terms). Upon evolving through the beam-splitter, the state becomes, using equations 2 and 3,

$$Ne^{-(1/2)|\alpha_1|^2}\left\{q|0\rangle_{c_1}|0\rangle_{d_1} + \left[\alpha_1 q - \left(\frac{r}{\sqrt{2}}\right)\right]|1\rangle_{c_1}|0\rangle_{d_1} + i\left[\alpha_1 q + \left(\frac{r}{\sqrt{2}}\right)\right]|0\rangle_{c_1}|1\rangle_{d_1} + |\varphi''\rangle\right\}, \quad \textbf{(13)}$$

where the state $|\varphi'\rangle$ has evolved to $|\varphi''\rangle$. There are two terms corresponding to the case in which a total of one photon is detected, either at C_1 or at D_1. Consider the case where one photon is detected at D_1 and zero photons are detected at C_1. There are two possible ways in which this could have happened. Either the photon came from the source along the u_1 path and the vacuum came from the coherent state, or vice versa. The amplitudes corresponding to these two possibilities can be made to interfere destructively such that it never happens that one photon is detected in D_1 and no photons in C_1 (remember that we are here considering the case where $U_2 = 0$). It is clear from equation 13 that this condition is met when $\alpha_1 q + (r/\sqrt{2}) = 0$, that is, when

$$\alpha_1 = -\frac{r}{\sqrt{2}q}. \quad \textbf{(14)}$$

Assuming from now on that α_1 has been set according to equation 14, we have the prediction

$$\text{if } U_2 = 0, \qquad \text{then } \{D_1 = 1 \;\&\; C_1 = 0\} \text{ never happens.} \quad \textbf{(15)}$$

It follows logically from this that

$$\text{if } \{D_1 = 1 \;\&\; C_1 = 0\}, \qquad \text{then } U_2 = 1 \tag{16}$$

because U_2 can only take values of 0 or 1. This is the prediction of interest for experiment 2.

Experiment 3: This is the same as experiment 2, but with 1 and 2 interchanged. By similar reasoning to that above, if we set

$$\alpha_2 = \frac{ir}{\sqrt{2}q}, \tag{17}$$

then

$$\text{if } \{D_2 = 1 \;\&\; C_2 = 0\}, \qquad \text{then } U_1 = 1. \tag{18}$$

Experiment 4: Now, detectors U_1 and U_2 are both removed so that mode u_1 impinges onto one input of a beam-splitter with the coherent state $|\alpha_1\rangle$ incident on the other input and, similarly, mode u_2 impinges onto one input of another beam-splitter with the coherent state $|\alpha_2\rangle$ incident on the other input. The settings of α_1 and α_2 are taken to be the same as above. The result of interest is that

$$\{D_1 = 1 \;\&\; C_1 = 0\} \text{ and } \{D_2 = 1 \;\&\; C_2 = 0\} \text{ happens sometimes.} \tag{19}$$

To see this, consider the evolution of the state in this case. After the state from the source has passed the first beam-splitter, the total state of the system, including the coherent states, is $|\Psi\rangle|\alpha_1\rangle_{a_1}|\alpha_2\rangle_{a_2}$, where $|\Psi\rangle$ is given in equation 10. Expanding this out using equations 4 and 10, we get

$$e^{-(1/2)(|\alpha_1|^2+|\alpha_2|^2)}\left[\alpha_1\alpha_2 q|1\rangle_{a_1}|0\rangle_{u_1}|1\rangle_{a_2}|0\rangle_{u_2} + \left(\frac{\alpha_1 r}{\sqrt{2}}\right)|1\rangle_{a_1}|0\rangle_{u_1}|0\rangle_{a_2}|1\rangle_{u_2} + \left(\frac{i\alpha_2 r}{\sqrt{2}}\right)|0\rangle_{a_1}|1\rangle_{u_1}|1\rangle_{a_2}|0\rangle_{u_2} + |\varphi'''\rangle\right], \tag{20}$$

where $|\varphi'''\rangle$ contains all those terms that do not have a total of one photon at each end of the apparatus. The state now evolves through the beam-splitters. We can concentrate only on the term relevant to the prediction in equation 19. The state becomes

$$e^{-(1/2)(|\alpha_1|^2+|\alpha_2|^2)}\left\{\left[\alpha_1\alpha_2 q\left(\frac{i}{\sqrt{2}}\right)^2 + \left(\frac{\alpha_1 r}{\sqrt{2}}\right)\left(\frac{i}{\sqrt{2}}\right)\left(\frac{1}{\sqrt{2}}\right) + \left(\frac{i\alpha_2 r}{\sqrt{2}}\right)\left(\frac{1}{\sqrt{2}}\right)\left(\frac{i}{\sqrt{2}}\right)\right] \times |0\rangle_{c_1}|1\rangle_{d_1}|0\rangle_{c_2}|1\rangle_{d_2} + |\vartheta\rangle\right\}, \tag{21}$$

where all the terms that we are not interested in are in $|\vartheta\rangle$. Using equations 14 and 17, we find that the square modulus of the coefficient in front of the $|0\rangle_{c_1}|1\rangle_{d_1}|0\rangle_{c_2}|1\rangle_{d_2}$ term is equal to $(|r|^4/16|q|^2)e^{-(|\alpha_1|^2+|\alpha_2|^2)}$. Provided that q and r

are nonzero, this probability is also nonzero and, hence, the prediction of equation 19 above follows.

The Contradiction

Consider two observers, Alice and Bob. Alice stands at end 1 of the apparatus and has a choice of whether to put detector U_1 in place and measure U_1 or to remove detector U_1 and measure D_1 and C_1. Bob stands at end 2 of the apparatus and has a choice of whether to put detector U_2 in place and measure U_2 or to remove detector U_2 and measure D_2 and C_2. Suppose that they make this choice randomly, not deciding until after the state has been emitted from the source. Now, consider a run of the experiment in which Alice chooses to measure D_1 and C_1 and Bob chooses to measure D_2 and C_2 (i.e., the U_1 and U_2 detectors are removed). Furthermore, suppose that Alice obtains the result $D_1 = 1$ and $C_1 = 0$ and Bob obtains the result $D_2 = 1$ and $C_2 = 0$. That this can happen follows from prediction 19. From her results of $D_1 = 1$ and $C_1 = 0$ and prediction 16, Alice knows that if Bob has chosen to measure U_2 then he will certainly detect a photon there (in path u_2) and hence she may reason that the photon from the source has gone towards Bob. From his results of $D_2 = 1$ and $C_2 = 0$ and prediction 18, Bob knows that if Alice has chosen to measure U_1 then she will certainly detect a photon there (in path u_1) and hence he may reason that the photon from the source has gone towards Alice. However, there is at most only one photon emitted from the source at a time and it cannot have gone both towards Alice with certainty and towards Bob with certainty (this would violate prediction 11). It is not possible that Alice and Bob are both right. Thus, we have a contradiction. This contradiction is slightly worrying because it is not clear that there are any assumptions in the above argument. If there really were no hidden assumptions, then the only resolution to the contradiction would be that quantum theory is wrong. However, under closer scrutiny, it is clear that both Alice and Bob are making locality assumptions in their reasoning. Consider Alice. She deduces that the photon is going towards Bob, which must mean that if he had put his U_2 detector in place then he would have detected the photon there. However, in the above scenario, Bob did not put his U_2 detector in place. Therefore, Alice is arguing counterfactually. When one argues counterfactually, it is important to consider properly all the respects in which the world might have been different had the other choice been made. In this case, Alice must consider how the world might have been had Bob instead chosen to measure U_2. In such a case, it is possible that there might have been a nonlocal effect (or an absence of a nonlocal effect that had actually happened in the world where Bob did not measure U_2) with the consequence that Alice might not then have obtained the results $D_1 = 1$ and $C_1 = 0$. Consequently, Alice cannot argue that Bob would have obtained $U_2 = 1$ had he measured U_2 because, in the world where he made that choice, she might not have obtained the results on which she based the deduction. Similar remarks apply to Bob. Thus, by allowing nonlocality, we remove the contradiction.

Actually, the situation is a little more subtle than was revealed above because of the possibility of indeterminism. To see this, first consider the following philosophical conundrum (see reference 17, pages 92–93). A roulette wheel is spun by Alice. The number 6 comes up. Bob is sitting some miles away and we assume that no nonlocal influence can pass between the two places. After some time, Bob finds out that the result was 6. Then, he asks himself the following question: would the result have been different if I had raised my hand while the wheel was spinning? If the laws that govern the motion of the roulette wheel are entirely deterministic, then it is clear in the world where Bob had raised his hand that the number 6 would still have come up because the initial conditions for the wheel that determine the outcome would not have been different if Bob had raised his hand. However, if the laws that govern the motion of the roulette wheel are indeterministic, then the answer is not so clear. Indeed, there is no conclusive argument against the view that, had Bob raised his hand, Alice might have then got some other number, even in the absence of nonlocal effects. Consider again the photon experiment. If we allow the possibility of indeterminism, then it is possible that in the world where Bob had measured U_2 (analogous to raising his hand) Alice might not have got $D_1 = 1$ and $C_1 = 0$ (analogous to not getting a 6), even when there are no nonlocal influences. Hence, even assuming locality, it seems that Alice cannot infer that Bob would have got $U_2 = 1$ for the reasons expressed above and thus it now seems that we can avoid the above contradiction even without allowing nonlocality. However, and this is the subtlety, this reasoning is wrong. For the particular run of the experiment that we are considering, Alice did actually obtain $D_1 = 1$ and $C_1 = 0$, although Bob was not measuring U_2. Alice could not have obtained the result $D_1 = 1$ and $C_1 = 0$ had the probability for this particular outcome been equal to zero for this run of the experiment. If we assume locality, then the stochastic process at end 1 cannot depend on whether U_2 is measured or not. It follows that the probability of Alice getting $D_1 = 1$ and $C_1 = 0$ in the counterfactual world in which Bob does measure U_2 must still be nonzero for this run of the experiment. If Bob had measured U_2, then the system (or "local reality") at end 2 would have had to come up with a choice between $U_2 = 0$ or $U_2 = 1$. However, because we are assuming locality, there can exist no information at end 2 that reveals what the outcome of any indeterministic process at end 1 would be. Therefore, the system at end 2 would have to opt for $U_2 = 1$ just in case the outcome at end 1 is still $D_1 = 1$ and $C_1 = 0$ (because otherwise there is a chance of a violation of prediction 16). This means that, so long as she assumes locality, Alice is justified in deducing that the photon went towards Bob (meaning that Bob would detect it if he used his U_2 detector). Similarly, if he assumes locality, Bob is justified in deducing that the photon went towards Alice. Hence, we cannot avoid the contradiction above by invoking indeterminism and therefore abandoning locality remains the only way out.

Some Inequalities

The above discussion, involving counterfactual reasoning, may seem a little involved. To put matters on more familiar ground, we will show that the predictions of quantum mechanics violate the Clauser-Horne-Bell inequalities.[20] These inequalities can be put into the form,

$$-1 \leq \text{Prob}\,(F_1 = 1\ \&\ F_2 = 1) - \text{Prob}\,(F_1 = 1\ \&\ U_2 = 0) - \text{Prob}\,(U_1 = 0\ \&\ F_2 = 1) - \text{Prob}\,(U_1 = 1\ \&\ U_2 = 1) \leq 0, \quad \textbf{(22)}$$

where F_k and U_k are two measurements that can be made on side k (where $k = 1, 2$).

By putting $F_k = 1$ when $D_k = 1$ and $C_k = 0$, we see from prediction 19 that the first probability in equation 22 is nonzero [equal to $(|r|^4/16|q|^2)e^{-(|\alpha_1|^2 + |\alpha_2|^2)}$] and, from predictions 11, 16, and 18, we see that the other three probabilities are equal to zero. Hence, the upper limit of the inequalities is violated, which means that locality is violated.

Multimode Treatment

We know that the state $q|0\rangle_s + r|1\rangle_s$ has been created in the signal mode when the detector in the idler mode detects one photon. In order for the discussion of nonlocality to make sense, we require that this state is emitted within some time interval $\tau \ll L/c$, where L is the dimension of the apparatus (as would indeed be the case because the idler and signal photons produced in parametric downconversion are very strongly correlated in time). Thus, if the idler detector clicks at time t, then the state must be emitted during a time interval $t - \tau/2$ to $t + \tau/2$. It follows from the energy-time uncertainty principle that there must be a spread of frequency in the photons emitted from the source and, consequently, a proper treatment requires a full multimode calculation. The input state is actually

$$q|0\rangle_s + \sum_{\omega} r_\omega |1\rangle_{s_\omega}, \tag{23}$$

where ω is the frequency. The sum over frequencies will tend to an integral as the mode spacing tends to zero. The notation in equation 23 is shorthand: $|0\rangle_s$ denotes that all frequency modes in path s are in the vacuum mode (i.e., $\Pi_\omega |0\rangle_{s_\omega}$) and $|1\rangle_{s_\omega}$ denotes that the mode with frequency ω in path s has one photon and all the other modes in this path are in the vacuum state (i.e., $|1\rangle_{s_\omega} \Pi_{\omega' \neq \omega} |0\rangle_{s'_\omega}$). The photons from the source must be indistinguishable from the photons in the coherent states used at ends 1 and 2. Hence, these coherent states should also have a spread of frequencies. Thus, we must use multimode coherent states[14] $|\{\alpha_k\}\rangle_{a_k} = \Pi_\omega |\alpha_k^\omega\rangle_{a_{k\omega}}$, where $|\alpha_k^\omega\rangle$ is a single mode coherent state of frequency ω and with the form given in equation 4. The first few terms in this product can be expanded out to yield

$$|\{\alpha_k\}\rangle_{a_k} = e^{[-(1/2)\Sigma_{\omega'}|\alpha_k^{\omega'}|]}\left(|0\rangle_{\alpha_k} + \sum_{\omega} \alpha_k^\omega |1\rangle_{a_{k\omega}} + \cdots\right), \tag{24}$$

where we are using similar shorthand as before. The above calculations can be easily repeated using the multimode states of equations 23 and 24, and predictions 11, 16, 18, and 19 are recovered if we set the complex amplitudes of each frequency mode according to

$$\alpha_1^\omega = i\alpha_2^\omega = -\frac{r_\omega}{\sqrt{2}q}. \tag{25}$$

CONCLUSIONS

We have seen that the essential reasoning behind the EPR argument had already been given by Einstein in 1927 for a single-particle rather than two-particle situation.

Furthermore, we saw above that Einstein's reasoning in 1927 can be put into a rigorous form similar to that of the EPR argument. Next, a demonstration of the nonlocality of a single photon was given without using inequalities (although it was shown that some Bell inequalities are violated). The four rules listed in the first subsection of the sixth section are all satisfied. A number of questions are raised by this proof. First, it is the case that, on average, less than a single photon was used. The above proof does not go through in the case where $q = 0$, that is, in the case where there is no vacuum term. This raises the question of whether it would be possible to construct a proof where there was exactly one photon each time. One such proof has been constructed[8] in which two atoms sitting in the paths from the source go into an entangled state as a consequence of the interaction and some postselection. This method is rather cumbersome. It seems likely that this could be done, instead, by taking a similar approach to that above. However, this comes down to a matter of calculation and will be the subject of future work. The second question is as follows: how significant is the role of the vacuum in this nonlocal effect? Could these predictions be recovered without having to incorporate the vacuum as a quantum object in the calculations? The answer to this is not clear. It is, however, interesting to note that a mixture, rather than a superposition, of $|0\rangle$ and $|1\rangle$ would lead to quite different predictions. The third question, already hinted at above, is as follows: could a proof of a single particle be run for any type of particle? The answer to this is almost certainly no. The above proof makes use of a superposition of the vacuum and the single-photon state in the source and also in the measurement process. As just mentioned, it is possible that this superposition is not necessary in the source. However, it is difficult to see how it could be avoided in the measurement process. Nonlocality in quantum mechanics is always a second or higher order interference effect. To obtain such an effect when only a single photon comes from the source, it is necessary that we take advantage of the quantum uncertainty pertaining to which direction the photon went at the beam-splitter. To do this, we must consider measurements that destroy this information. Without making use of the indistinguishability of photons, and using a superposition of different number states, this could not be done. Indistinguishability is a property of any given class of quantum particles. However, there exist superselection rules for a vast range of particles that rule out the possibility of forming a superposition of different number states (even a superposition of the vacuum and the single-particle state) and, in such cases, it seems that no proof for a single particle could be given. Boson statistics for the photons, however, appear to play no significant role and a proof like that above could be constructed for a fermionic particle if it were possible to prepare a superposition of a vacuum and a single-particle state.

ACKNOWLEDGMENTS

I would like to thank Rob Clifton, Mike Horne, Paul Kwiat, Sandu Popescu, Euan Squires, Anton Zeilinger, and Marek Zukowski for many stimulating discussions on this topic.

REFERENCES

1. EINSTEIN, A., B. PODOLSKY & N. ROSEN. 1935. Phys. Rev. **47:** 138.
2. BELL, J. S. 1964. Physics **1:** 195.
3. TAN, S. M., D. F. WALLS & M. J. COLLETT. 1991. Phys. Rev. Lett. **66:** 252.
4. OLIVER, B. J. & C. STROUD, JR. 1989. Phys. Lett. A **135:** 407.
5. SANTOS, E. 1992. Phys. Rev. Lett. **68:** 894.
6. TAN, S. M., D. F. WALLS & M. J. COLLETT. 1992. Phys. Rev. Lett. **68:** 895.
7. GRANGIER, P., M. J. POTASEK & B. YURKE. 1988. Phys. Rev. A **38:** 3132.
8. HARDY, L. 1992. Ph.D. thesis. Durham University.
9. CZACHOR, M. 1994. Phys. Rev. A **49:** 2231.
10. ELITZUR, A. C. & L. VAIDMAN. 1992. Found. Phys. **23:** 987.
11. HARDY, L. 1992. Phys. Rev. Lett. **68:** 2981.
12. HARDY, L. 1993. Phys. Rev. Lett. **71:** 1665.
13. JAMMER, M. 1974. The Philosophy of Quantum Mechanics, p. 115–116. Wiley–Interscience. New York.
14. LOUDON, R. 1983. The Quantum Theory of Light (second edition). Oxford University Press. London/New York.
15. MOLLOW, B. R. 1973. Phys. Rev. A **8:** 2684.
16. HONG, C. K. & L. MANDEL. 1985. Phys. Rev. A **31:** 2409.
17. REDHEAD, M. 1987. Incompleteness, Nonlocality, and Realism. Oxford University Press (Clarendon). London/New York.
18. BOHM, D. 1951. Quantum Theory. Prentice–Hall. Englewood Cliffs, New Jersey.
19. OU, Z. Y., L. J. WANG, X. Y. ZOU & L. MANDEL. 1990. Phys. Rev. A **41:** 1597.
20. CLAUSER, J. F. & M. A. HORNE. 1974. Phys. Rev. D **10:** 526.

The Best Version of Bell's Theorem[a]

N. DAVID MERMIN

Laboratory of Atomic and Solid State Physics
Cornell University
Ithaca, New York 14853-2501

LUCIEN HARDY'S GEDANKENEXPERIMENT

Consider two far-apart black boxes, each with a switch that can be set in one of two positions. The boxes act as detectors for a pair of particles emanating from a distant source. Each box responds to its particle by flashing a red or green light. There are no connections between the detectors or between the source and the detectors, beyond those mediated by the particles. We press a button at the source to send the particles off to their detectors. After they have left the source, but before they arrive at and trigger the detectors, we randomly and independently set the switch on each detector to either of its two positions (labeled 1 and 2) by tossing a coin at each detector.

The data collected in many runs exhibit the following features:

(a) in runs in which the detectors end up with different settings (21 or 12), the lights never both flash green;
(b) in runs in which both detectors end up set to 1, the lights never both flash red;
(c) in runs in which both detectors end up set to 2, the lights sometimes both flash green.

Because there are no direct connections between detectors, the explanation for their coordinated behavior must come from the fact that both are triggered by particles that originated at a single source. Something in the common origin of the particles must be responsible for the correlations. Since the switches on the detectors are not set until after the particles have left their source, the features of the particles that produce these correlations cannot depend on the setting of the switches. And since each detector is triggered by only one of the two particles, it can only respond to features residing in that particle and not to features residing in the particle that went off to the other faraway detector. We can then reason as follows: Because any run has an even chance of ending up as a 12 or 21 run, whenever one of the particles is of a type that *allows* a type 2 detector to flash green, its partner must be of a type that *requires* a type 1 detector to flash red, for both detectors never flash green in a 12 or 21 run. Therefore, in any of those 22 runs in which both detectors flash green, because each particle has revealed itself to be of a type that does allow its type 2 detector to flash green, each must also be of a type that requires a type 1 detector to flash red. Thus, if the random setting of the detectors resulting in any such 22 run had resulted instead in a 11 run, both detectors would have had to flash red. Such behavior, however, is never observed in a 11 run.

[a]This work was supported by the National Science Foundation (Grant No. PHY9320821).

Something is therefore wrong with the reasoning in the last paragraph. This gedankenexperiment, discovered by Lucien Hardy,[1] offers by far the simplest and cleanest case for exploring where the error in this kind of reasoning might lie. I shall not indulge in such explorations here, except to define the central issue by adding three more lines to one of John Wheeler's favorite Bohr quotations:

To be, to be. What does it mean to be?
Or is the problem nonlocality?
That is the question, vexing many souls:
What's the idea for which Bell's theorem tolls?

The quantum mechanics underlying Hardy's wonderful trick is remarkably simple and general.[2] Pick any nontrivial one-particle observable you like for each detector to measure when its switch is set to 1—spin components, polarizations, position, momentum, or anything else. Divide the spectrum of the observable into any two sets you wish and fix the detector so a flash of red or green indicates which set the measured value was found in. The source produces a pair of particles in a two-particle state $|\Psi\rangle$ constructed as follows: Pick, for the left particle, two superpositions of eigenstates of the observable measured in mode 1 on the left, taken entirely from the red and entirely from the green part of its spectrum. Call these two orthogonal states $|1R\rangle_l$ and $|1G\rangle_l$. Pick two states $|1R\rangle_r$ and $|1G\rangle_r$ for the particle on the right, similarly defined in terms of its own mode-1 observable. The two-particle state $|\Psi\rangle$ is a superposition of the three two-particle states:

$$\begin{aligned} |1R, 1G\rangle &= |1R\rangle_l |1G\rangle_r, \\ |1G, 1R\rangle &= |1G\rangle_l |1R\rangle_r, \\ |1G, 1G\rangle &= |1G\rangle_l |1G\rangle_r. \end{aligned} \qquad (1)$$

That both detectors never flash red when both detectors are in mode 1 is guaranteed by the absence of $|1R, 1R\rangle$ from the superposition, which it is simplest to take in a symmetric form with real coefficients:[3]

$$|\Psi\rangle = a|1R, 1G\rangle + a|1G, 1R\rangle - b|1G, 1G\rangle. \qquad (2)$$

The choices for the observables measured by the detectors in mode 2 are also quite flexible. All we require is that the observable measured on the left have an eigenstate $|2G\rangle_l$ with nonzero components along both $|1G\rangle_l$ and $|1R\rangle_l$. A green flash on the left indicates that the measurement in mode 2 did indeed find the particle in the state $|2G\rangle_l$. Anything else results in a red flash. A similar arrangement is made on the right with a state $|2G\rangle_r$. As an explicit example, we take the state $|2G\rangle$ on each side to be the same superposition with real nonzero coefficients of the states $|1R\rangle$ and $|1G\rangle$ on that side:[3]

$$|2G\rangle = c|1G\rangle + d|1R\rangle. \qquad (3)$$

The absence of two green flashes when one detector is in mode 1 and the other is in mode 2 is guaranteed by picking c and d so that $|1G, 2G\rangle$ and $|2G, 1G\rangle$ are both orthogonal to $|\Psi\rangle$. This requires

$$ad = bc. \qquad (4)$$

The probability of two green flashes when both detectors are in mode 2 is thus

$$p(22GG) = |\langle 2G, 2G|\Psi\rangle|^2 = (2acd - bc^2)^2 = b^2c^4, \tag{5}$$

so we require b and c not to be zero and then equation 4 requires a and d not to be zero.

To make the probability p for the "impossible" 22GG events as large as possible, note that equations 4 and 5 and the normalization of the states defined in equations 2 and 3 combine to give

$$p(22GG) = \left(\frac{1 - c^2}{1 + c^2}\right) c^4. \tag{6}$$

This is maximum when $c^2 = 1/\tau$, where τ is the golden mean, $(\frac{1}{2})(\sqrt{5} + 1)$, which results in a maximum probability,

$$p(22GG) = 1/\tau^5 = 9.017\%. \tag{7}$$

[The rational approximation to the golden mean $c^2 = 3/5$ does almost as well, yielding a $p(22GG)$ that is exactly 9%.]

How did something this simple and powerful go undiscovered during the 30 years since John Bell proved his theorem? My guess is that nobody noticed it before Hardy because everybody was too strongly focused on states that exhibited Einstein-Podolsky-Rosen (EPR) correlations.

The Hardy state is one in which the full EPR argument cannot be made. According to EPR, an "element of reality"—preexisting information carried by a particle that specifies the color to be flashed by its detector—must exist if it is possible to predict in advance the result of a localized measurement of that observable by other localized experiments done far away. In the Hardy experiment, this condition of predictability is not invariably met. Sometimes you can predict with certainty the result of a distant experiment, but you can only do so in an uncontrollable fraction of the runs. Suppose, for example, you try to learn what the particle on the left will do at a type 2 detector by measurements on the faraway particle on the right. If you subject the particle on the right to a type 1 detector, then if that detector flashes green you can indeed predict with certainty that the particle on the left will cause its detector to flash red because two green flashes are never observed when the switches have different settings. But if the detector on the right flashes red for these settings, you cannot predict what will happen on the left. Because the outcome on the right is not under your control, your ability to predict with certainty the behavior on the left is a matter of chance. If you're lucky, you can do it; otherwise, you can't. The same difficulty besets any of the other ways one might try to exploit the correlations (a) and (b) to make an invariably successful prediction of what will happen at one detector, based on what happens at the other.

Thus Hardy's data do not demand that each particle always carries a set of instructions telling its detector what color to flash for both settings of its switch. That is why the argument I gave above had to start by considering particles that "allow" a detector to flash green. "Allow" is as much as one can say in the Hardy experiment and "allow" is, in fact, much more in accord with the inherent randomness in the response of a detector to a particle that quantum mechanics has taught us to accept. Nevertheless, that one softening of an EPR "require" to a Hardy "allow" does not prevent the behavior of the Hardy device from refuting the apparently obvious explanation for the correlations just as effectively as Bell's theorem does.

As Alain Aspect emphasized in a discussion remark early in this conference, beautiful as the Hardy experiment is as a gedankenexperiment, in a real experiment "never" can never be established, if only because the number of runs is necessarily finite. Hardy's argument can, however, be cast in the form of an inequality that bounds the probability $p(22GG)$ by the sum of the probabilities of the three events forbidden in the ideal gedankenexperiment:[4]

$$p(22GG) \leq p(21GG) + p(11RR) + p(12GG). \qquad \mathbf{(8)}$$

This inequality can be derived from various sets of assumptions, none of them as compelling as the simple argument that led us to conclude that the *vanishing* of the probabilities on the right required the *vanishing* of the one on the left.[5] Furthermore, while the Hardy state violates inequality 8 by making the left side 9% and the right side zero, the version of Bell's theorem subject to actual experimental tests gives a significantly bigger disparity, making the left side 43% and the right side 22%. Thus, Hardy's argument, although of enormous conceptual interest as a gedankenexperiment, will not lead to a more definitive experiment than the ones already performed.

THE QUESTION OF POPESCU AND ROHRLICH

One can use the Hardy state to examine a question recently raised by Sandu Popescu and Daniel Rohrlich.[6] A maximum violation of the Clauser-Horne (CH) inequality 8 occurs when $p(22GG) = 1$ and the other three probabilities are zero. Although quantum mechanics violates inequality 8, it cannot give a maximum violation, for the analysis I have given above (more precisely, the slight generalization of it that I give in reference 3) establishes that, when the right side of inequality 8 is zero, the largest that quantum mechanics allows the left side to be is 9.017%. Could there be a simple reason why the distributions allowed by quantum mechanics cannot give a larger violation? Popescu and Rohrlich asked whether such a reason might be provided by a requirement that I would call physical locality—that the distribution of results at a single detector should be independent of the setting of the other.

Physical locality does indeed prohibit a maximal violation of inequality 8. For if $p(22GG) = 1$ and $p(21GG)$, $p(11RR)$, and $p(12GG)$ are all zero, then we can reason as follows (now appealing only to physical locality): both detectors always flash green when both are set to 2. Physical locality then requires a detector set to 2 to always flash green even when the other detector is set to 1. (Otherwise, one could determine the setting of the faraway detector merely by observing the one nearby.) Because the probabilities $p(21GG)$ and $p(12GG)$ are zero, this requires in turn that a detector set to 1 must always flash red when the other detector is set to 2. Physical locality then requires a detector set to 1 to always flash red when the other detector is set to 1. But far from happening always, this never happens because $p(11RR)$ is zero. So indeed, a maximal violation of inequality 8 contradicts the straightforward, unambiguous, and experimentally testable notion of physical locality—not merely the plausible, but surprisingly flawed reasoning that I set forth at the beginning of this essay, based on what one might call metaphysical locality.

Avoiding metaphysical locality and reasoning only from physical locality, one can in fact derive a new inequality for the four probabilities appearing in inequality 8 that is less restrictive than inequality 8 itself, but strong enough to rule out a maximal violation of inequality 8 (see APPENDIX A for the derivation):

$$2p(22GG) - 1 \leq p(21GG) + p(11RR) + p(12GG). \tag{9}$$

Inequality 9 constrains the size of $p(22GG)$ not to exceed 50% when $p(21GG)$, $p(11RR)$, and $p(12GG)$ are all zero. This is less than the 100% maximal violation, but considerably larger than the 9% limit imposed by quantum mechanics.

Although I am not sure how meaningful it is to ask such questions about these kinds of inequalities, I nevertheless found this particular feature of the Hardy state puzzling for the following reason: Popescu and Rohrlich phrase their question in terms of the inequalities of Clauser, Horne, Shimony, and Holt (CHSH),[7] which have a different form from inequality 8. Define a variable m that is 1 if a detector flashes red and -1 if it flashes green, and define the correlation coefficient $\langle mm' \rangle_{ij}$ to be the mean of the product mm' in those runs in which the right detector is set to i and the left to j. The CHSH inequalities are the requirement that

$$-2 \leq \langle mm' \rangle_{11} - \langle mm' \rangle_{12} - \langle mm' \rangle_{21} - \langle mm' \rangle_{22} \leq 2 \tag{10}$$

(and the three inequalities in which the + sign goes with the 12, 21, or 22 correlation coefficient). Although quantum mechanics can violate inequality 10, it does not permit the magnitude that inequality 10 bounds by 2 to exceed $2\sqrt{2}$, even though one can construct general pair distributions that make this quantity as large as 4. Popescu and Rohrlich show, however, that physical locality is compatible with this quantity having this maximum magnitude.

In view of the impossibility of maximally violating the CH inequality with physically local pair distributions, which follows from thinking about Hardy states, this conclusion of Popescu and Rohrlich puzzled me. I have always viewed the CH and CHSH inequalities as completely equivalent and have been criticized more than once for forgetting which was which. I therefore could not understand how an inequality from the CHSH set could suffer a maximal violation with a physically local set of pair distributions, while one from the completely equivalent CH set could not. After brooding about this paradox for longer than it should have taken me, I realized what was going on: the proof of equivalence between the CH and CHSH inequalities (see APPENDIX B) makes an implicit assumption of physical locality, which it is easy to overlook. Those distributions for which the CH inequalities violate inequality 9, but not maximally, necessarily violate physical locality and therefore do not correspond to any combination of correlations appearing in a CH inequality.

Physical locality is so fundamental a property and so readily confirmable by direct experimental test that it is almost invariably taken completely for granted in manipulating either CH or CHSH inequalities from one form to another. Existing relations between and among such inequalities are infected by unstated assumptions of physical locality. This is true even for the relation between the form 8 in which I have written the CH inequalities and the form (involving single-particle distributions as well as joints) used by Clauser and Horne. It is important to be aware of this when considering how such relations might be violated in the absence of physical locality. In the context of the question of Popescu and Rohrlich, the Hardy states make the importance of such an awareness quite explicit.

REFERENCES AND NOTES

1. HARDY, L. 1993. Non-locality for two particles without inequalities for almost all entangled states. Phys. Rev. Lett. **71:** 1665–1668. Hardy gives an earlier version of his gedankenexperiment in: 1992. Quantum mechanics, local realistic theories, and Lorentz-invariant realistic theories. Phys. Rev. Lett. **68:** 2981–2984.
2. This simplicity and generality was pointed out by: GOLDSTEIN, S. 1994. Nonlocality without inequalities for almost all entangled states for two particles. Phys. Rev. Lett. **72:** 1951. A very simple discussion of Hardy's gedankenexperiment has been given by: STAPP, H. 1993. Mind, Matter, and Quantum Mechanics, p. 5–9. Springer-Verlag. New York/Berlin. I have described the version of Hardy's gedanken-demonstration given here along with my own version of the underlying quantum mechanical analysis in: MERMIN, N. D. 1994. What's wrong with this temptation? Phys. Today **June:** 9–11; 1994. Quantum mysteries refined. Am. J. Phys. **62:** 880–887.
3. In the sources cited in reference 2 above, I take general superpositions, which reduce to this slightly simpler form when the probability of two green flashes with both detectors in mode 2 is maximum.
4. This is a form of one of the Clauser-Horne inequalities. See: CLAUSER, J. F. & M. A. HORNE. 1974. Experimental consequences of objective local theories. Phys. Rev. D **10:** 526–535.
5. For example, inequality 8 follows directly from the assumption that one can simulate the data for all four choices of detector settings by a single statistical ensemble of particle pairs in which each member of a pair specifies what its detector will do for either setting of its switch. One merely notes that every member of the subensemble of pairs contributing to the probability on the left of inequality 8 (i.e., those pairs in which both members specify green for setting 2) must belong to at least one of the three subensembles contributing to the three probabilities on the right. The conventionally stated assumptions are less blatantly at odds with the spirit of quantum mechanics, but they invariably have the existence of just such an ensemble as a fairly straightforward consequence.
6. POPESCU, S. & D. ROHRLICH. 1994. Quantum nonlocality as an axiom. Found. Phys. **24:** 379–385.
7. CLAUSER, J. F., M. A. HORNE, A. SHIMONY & R. A. HOLT. 1969. Proposed experiment to test local hidden-variable theories. Phys. Rev. Lett. **23:** 880–884.

APPENDIX A

An Inequality of the CH Form Based Entirely on Physical Locality

Define $p(i[j]X)$ to be the probability of the light on the left flashing X in mode i when the detector on the right operates in mode j:

$$p(i[j]X) = p(ijXR) + p(ijXG); \tag{11}$$

similarly define

$$p([i]jX) = p(ijRX) + p(ijGX). \tag{12}$$

Physical locality is the requirement that each of these one-detector probabilities is independent of the choice of what to measure at the other detector:

$$p(i[1]X) = p(i[2]X), \qquad p([1]jX) = p([2]jX). \tag{13}$$

To derive inequality 9 using only this condition of physical locality, note first that quite generally

$$p(11RR) \geq p(1[1]R) + p([1]1R) - 1. \tag{14}$$

This simply states that the fraction of 11 runs in which both detectors flash red cannot be less than the excess over unity of the sum of the fractions of runs in which one or the other detector flashes red.

Note next the quite general pair of inequalities,

$$p(1[2]R) + p(12GG) \geq p([1]2G), \qquad p([2]2G) \geq p(22GG), \tag{15}$$

which, with the condition of physical locality given in equation 13, give

$$p(1[2]R) + p(12GG) \geq p(22GG). \tag{16}$$

In the same way, physical locality extracts from the pair,

$$p([2]1R) + p(21GG) \geq p(2[1]G), \qquad p(2[2]G) \geq p(22GG), \tag{17}$$

the inequality

$$p([2]1R) + p(21GG) \geq p(22GG). \tag{18}$$

The sum of the inequalities 14, 16, and 18 gives inequality 9 when we use physical locality to identify $p(1[2]R)$ with $p(1[1]R)$ and $p([2]1R)$ with $p([1]1R)$.

Note that equality can actually be attained in inequality 9 with physically local distributions by taking, for example, the 22 joint distributions to be 1/2 for equal colors and zero for different colors, and the 21, 11, and 12 distributions to be zero for equal colors and 1/2 for different colors. (This is precisely the physically local distribution used by Popescu and Rohrlich to violate maximally the CHSH inequalities.)

APPENDIX B

Equivalence of the CH and CHSH Inequalities Provided Physical Locality Holds

When physical locality holds, the inequalities given by Clauser and Horne in reference 4 are easily shown to be equivalent to the set of all possible versions of inequality 8:

$$p(ijmm') \leq p(ij'mm''') + p(i'j'\overline{m''}\ \overline{m'''}) + p(i'jm''m') \tag{19}$$

(where $\overline{m} = -m$, etc.).

The equivalence follows from the fact that all these simple two-particle distributions can be represented in terms of correlation coefficients and mean values by

$$p(ijmm') = (\tfrac{1}{4})(1 + m\langle m\rangle_{i[j]} + m'\langle m\rangle_{[i]j} + mm'\langle mm'\rangle_{ij}). \tag{20}$$

Here, $\langle m \rangle_{i[j]}$ is the mean of the result in mode i on the left when the detector on the right is in mode j, and similarly for $\langle m \rangle_{[i]j}$. If all such means are independent of the setting of the detector in the other wing, then they drop out of inequality 19, which reduces to the various forms of inequality 10 when examined for all 16 possible values of m, m', m'', m'''. However, if physical locality is violated and the means are not independent of the faraway choice of what to measure, then there need be no equivalence.

Interference of Single Photons from Separate Sources[a]

J. G. RARITY

DRA Malvern

Malvern, Worcestershire WR14 3PS, United Kingdom

INTRODUCTION

The fact that pairs of photodetections can occur in coincidence is itself a manifestation of the nonclassical nature of parametric downconversion. Furthermore, two-photon interference experiments[1–3] have shown that the pairs are coincident to within their coherence length, which can be much less than 100 fs. This time is several orders of magnitude less than the fastest resolving time of typical photon-counting detectors. The entanglement arising from the momentum and energy conservation inherent in the downconversion process has provided strong support for the nonlocal nature of quantum mechanics[4–6] extending now out to 4-km separation.[7] In these experiments, the coincident nature of the photon pairs is used to postselect the pair detection events and to define the ensemble to which tests of local realism can be applied. As these postselected events often form a very small proportion of the total number of events, this type of experiment can in principle be explained by hidden-variable theories and thus does not provide a watertight test for quantum nonlocality.

The fact that the photons occur in coincidence can also be used to define a 1-photon state simply by time gating[8] and has been used to demonstrate true 1-photon interference.[9] Several recent papers also suggested ways of entangling a number (from 1 to 3) of 1-photon states by mixing them in simple interferometers.[10,11] These papers have not addressed the problem of creating these states and ensuring that they overlap to show high visibility interference effects. As stated above, the time gating afforded by conventional detectors is several orders of magnitude longer than the typical coherence length of filtered downconverted photons.

Here, we propose using a mode-locked short pulse laser as a pump for parametric downconversion. Gated 1-photon states from such a source are located in a time window governed by the length of the pump pulse. With suitable narrow band filters, we can arrange a coherence time that is comparable to or longer than this "gate" time. The "length" of the photons is now the inverse of their bandwidth, as a direct result of Heisenberg's uncertainty principle. Taking two such sources and feeding one-states simultaneously into opposite ports of a beam-splitter as shown in FIGURE 1 will demonstrate the boson statistics first highlighted by Fearn and Loudon.[12]

[a]This work was supported in part by NATO Grant No. CRG.910571.

THEORY

In a simple particle picture, we expect there to be no correlation in the output taken by each photon of the pair. A quantum mechanical picture includes the 1-photon probability amplitudes $|1\rangle_1$ and $|1\rangle_2$ and the phase shift between transmission amplitude t and reflection ir. If the input photons are indistinguishable from measurements made at the beam-splitter outputs, the beam-splitter output state can be represented by

$$|\Psi\rangle = \left(\frac{1}{\sqrt{2}}\right)[t^2|1\rangle_3|1\rangle_4 - r^2|1\rangle_3|1\rangle_4 + irt|2\rangle_3 + irt|2\rangle_4]. \quad (1)$$

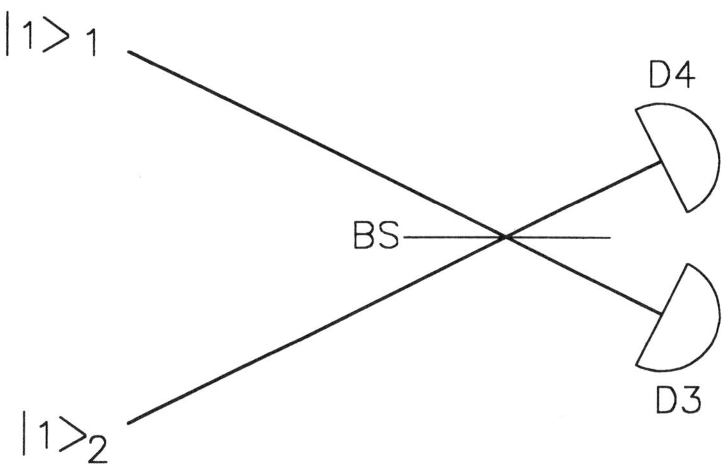

FIGURE 1. The interference of two 1-photon states $|1\rangle_1$ and $|1\rangle_2$ at a beam-splitter (BS). Coincidences are measured between photon-counting detectors D3 and D4.

We see that a destructive interference effect can occur in the probability for seeing photons simultaneously in both beam-splitter outputs:

$$P_{34} = t^4 + r^4 - 2r^2t^2. \quad (2)$$

In fact, when $r^2 = t^2 = ½$ (a 50/50 beam-splitter), $P_{34} = 0$ and the photons always appear as pairs in random outputs of the beam-splitter.

Originally, experiments of this type used the indistinguishable, but correlated photon pairs obtained from parametric downconversion.[1,2] Essentially, the coincidence rate drops to near zero when the pairs overlap at the beam-splitter. By scanning the path length difference from the crystal to the beam-splitter inputs, one can show that the pairs are coincident to within their coherence length (see FIGURES 2A and 2B).

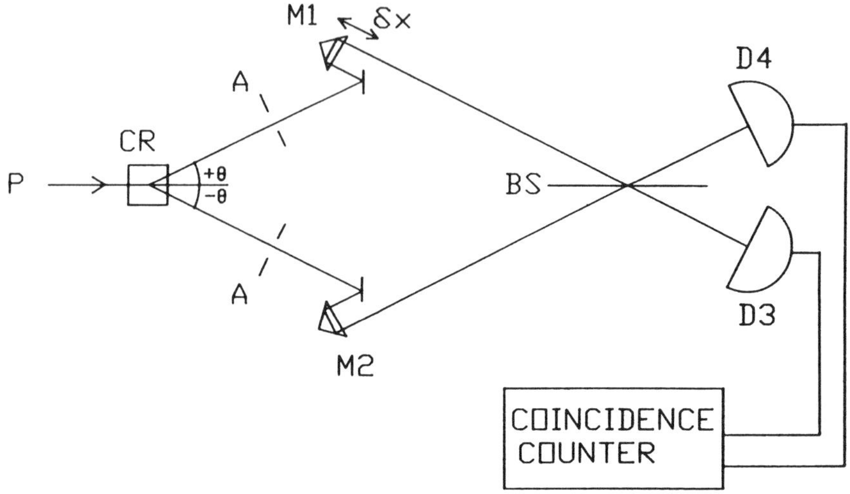

FIGURE 2A. Experiment showing the suppression of coincidences when correlated photon pairs are mixed at a beam-splitter BS. Pump beam P illuminates a downconversion crystal CR. Apertures A select degenerate photon pairs that are directed via mirror arrangements M1 and M2 onto opposite sides of beam-splitter BS. Coincident detections are measured in D3 and D4 as a function of the path length difference δx varied by moving the prism in M1.

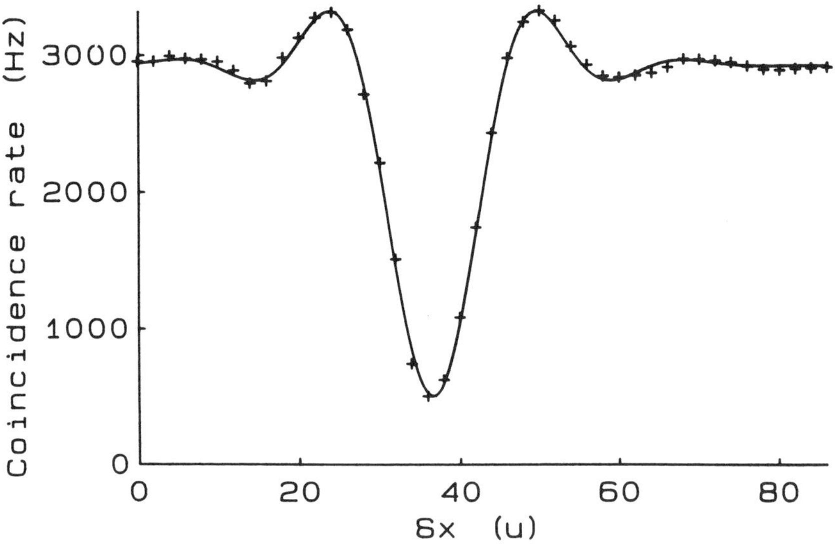

FIGURE 2B. Coincidence rate as a function of path length difference δx (in microns) showing the reduction of coincidences when the pairs overlap. The width of the coincidence reduction reflects the coherence time of the light, which is less than 40 fs here.

Here, we propose to use uncorrelated pairs that, in principle, could have no common history. We can gate out single photons from a parametric downconversion source,[8] but we cannot guarantee overlap of the photons until their inverse bandwidth (therefore, time uncertainty) is longer than the detector resolution time. With present technology, this bandwidth is about 1 GHz, which can be achieved using small etalons, but leads to low single-photon rates. Here, we choose to time gate using a short laser pump pulse. The proposed experiment is shown in FIGURE 3. A mode-locked laser pumps two parametric downconversion crystals cut for nondegenerate operation. Signal and idler photons satisfying energy conservation are emitted in a broad band cone behind each (thin) crystal and two sets of pairs are selected by

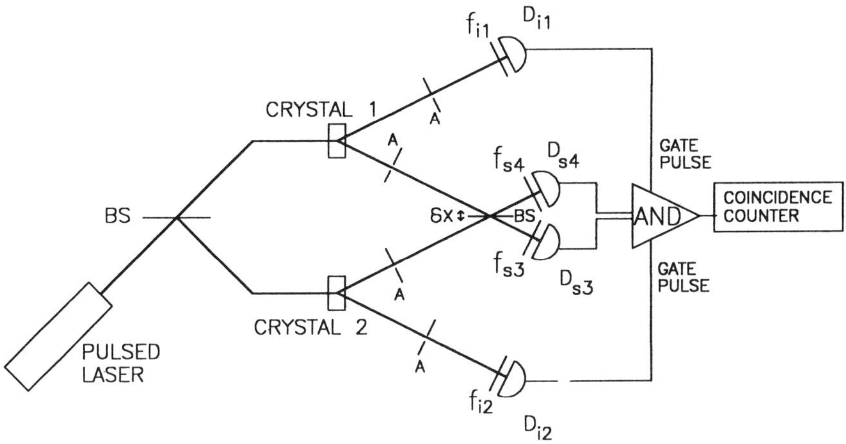

FIGURE 3. Schematic experiment where a pulsed laser pumps two crystals (1 and 2). Apertures A select the phase-matched pairs, which are further selected by energy-matched filters $f_{i1,i2,s3,s4}$. Coincident detections in the idler detectors D_{i1} and D_{i2} are used to gate out instances when there are 1-photon states in each signal beam at the input ports of beam-splitter BS. When the signal photons are indistinguishable at detectors D_{s3} and D_{s4}, we see the coincidence rate dip when the path length difference is adjusted to zero by moving the beam-splitter. Note that the δX shown corresponds to the δX in the text when multiplied by sin θ, where θ is the angle subtended by the beam and beam-splitter, and we assume that the total distance moved does not change the mode overlap significantly. In any experiment, we would arrange for $\theta > 80°$.

placing four apertures as shown. At low pump powers, the signal photons from each crystal are entirely uncorrelated and could in principle come from separate (synchronized) pump lasers. The state Ψ at crystal 1 is obtained by integrating the interaction Hamiltonian[13–15] over the interaction time t_{int}:

$$\Psi(t_0) = |\text{vac}\rangle + \frac{\alpha}{2\pi}|1\rangle_{i1}|1\rangle_{s1}\int_{t_0-t_{int}}^{t_0} d\tau \int d\omega_P d\omega_{s1} d\omega_{i1} \cdot f_P(\omega_P) f_{s1}(\omega_{s1}) f_{i1}(\omega_{i1}) e^{i[\omega_{s1}+\omega_{i1}-\omega_P]\tau}, \quad (3)$$

where $|\text{vac}\rangle$, $|1\rangle_{i1}$, $|1\rangle_{s1}$ represent the vacuum and one-states in idler (i) and signal (s) modes; ω_P, ω_{s1}, ω_{i1} represent the pump, signal, and idler frequencies; and we assume that pairs are selected to satisfy momentum conservation and have their bandwidth limited by external filters $f_{s1}(\omega_{s1})$ and $f_{i1}(\omega_{i1})$. A similar equation with subscripts 1 replaced by 2 describes the state at crystal 2. Note that α is an amplitude arising from the incident pump power, volume, and nonlinearity in the crystal, and frequency spectra are normalized such that $|\alpha^2| \ll 1$ is the probability of seeing a photon pair in each pump pulse. Also, $f_P(\omega_P)$ describes the pump frequency spectrum, which is that of a CW mode-locked laser. It is assumed that the laser is turned on at an earlier time, which we can approximate as $-\infty$, and runs until well after time $\tau = t_0$. We thus can express the pump beam as a sum of infinite-duration single modes with weights defined by f_P and can allow the interaction time t_{int} to tend to infinity. Performing the time integral leads to the energy conservation expressed as $\delta(\omega_{sj} + \omega_{ij} - \omega_P)$ $[j = 1, 2]$. Following Shih and Rubin,[16] we can define an effective 2-photon wave function at the detectors by

$$\Psi(\bar{t}_s, \bar{t}_i) = \langle \text{vac}|\hat{A}_s(\bar{t}_s)\hat{A}_i(\bar{t}_i)|\Psi\rangle, \tag{4}$$

where $\hat{A}_{s,i}$ are (multimode) annihilation operators (or electric field operators),

$$\hat{A}_i(\bar{t}) = \int d\omega f(\omega)e^{i\omega\bar{t}}, \tag{5}$$

and reduced time $\bar{t} = t - X/c$ includes the propagation time from crystal to detectors. Here (see FIGURE 3),

$$\Psi_{j,k}(\bar{t}_{sj}, \bar{t}_{ik}) = \alpha \int d\omega_P d\omega_{sj} d\omega_{ik} f_P(\omega_P) f_{sj}(\omega_s) f_{ik}(\omega_i)\delta(\omega_{sj} + \omega_{ik} - \omega_P)e^{i[\omega_{sj}\bar{t}_{sj} + \omega_{ik}\bar{t}_{ik}]}, \tag{6}$$

with $j = 3, 4$ and $k = 1, 2$ and with $f_{sj,ik}$ now being the limiting filters placed at the detectors.

Referring again to FIGURE 3, we can now express the effective wave function at detectors D_{i1}, D_{i2}, D_{s3}, D_{s4} as

$$\Psi(\text{i1, i2, s3, s4}) = t^2\Psi_{14}(\bar{t}_{s14}, \bar{t}_{i1})\Psi_{23}(\bar{t}_{s23}, \bar{t}_{i2}) - r^2\Psi_{13}(\bar{t}_{s13}, \bar{t}_{i1})\Psi_{24}(\bar{t}_{s24}, \bar{t}_{i2}), \tag{7}$$

where the subscripts define the source and detector and we ignore the *irt* terms appearing in equation 1 as they correspond to detection of pairs in either detector D_3 or D_4. Higher order terms when more than one pair are produced per pump pulse are of order α^3 and can also lead to coincident detections in D_3 and D_4, but are negligible here as we restrict ourselves to the situation where $|\alpha^2| \ll 1$. The probability $P(\text{i1, i2, s3, s4})$ of seeing a fourfold coincidence detection in the idler and signal detectors gated by an electronic timing pulse from the laser at time t_0 can thus be calculated from

$$P(\text{i1, i2, s3, s4}) = \eta_{i1}\eta_{i2}\eta_{s3}\eta_{s4} \int dt_{i1}dt_{i2}dt_{s3}dt_{s4}\, H(\bar{t}_{i1} - t_0, \Delta T)H(\bar{t}_{i2} - t_0, \Delta T)$$
$$\cdot H(\bar{t}_{s3} - t_0, \Delta T)H(\bar{t}_{s4} - t_0, \Delta T)|\Psi(\text{i1, i2, s3, s4})|^2, \tag{8}$$

where $\eta_{i1,i2,s3,s4}$ are effective detector efficiencies and $H(t_x - t_0, \Delta T)$ is a normalized detector response function centered on t_0 that falls to zero when $t_x - t_0 > \Delta T$. Typically, $\Delta T \simeq 1$ ns and the pump pulse separation is ~ 10 ns; thus, by arranging all propagation times including electronic delays to be equal within 1 ns, the above probability is confined to events arising from individual pump pulses. This allows us to model the pump spectrum as a continuous function of frequency (given by the gain envelope of the laser) rather than as the normal discrete sum of modes. We also note that the propagation times from the crystals to the beam-splitter must be equal to within the pump pulse length with small adjustments ΔX made by moving the beam-splitter closer to one or the other of the crystals. We thus write

$$\bar{t}_{s13} = \bar{t}_{s3} - 2\Delta X/c, \qquad \bar{t}_{s24} = \bar{t}_{s3} + 2\Delta X/c, \qquad \bar{t}_{s14} = \bar{t}_{s4}, \qquad \bar{t}_{s23} = \bar{t}_{s3}, \tag{9}$$

noting that no change in path length occurs on transmission. Equation 8 will in general involve eight independent frequency integrals, but we are using narrow band detectors with a long integration time, $\Delta T \gg \sigma_P^{-1}, \sigma_s^{-1}, \sigma_i^{-1}$, where $\sigma_{P,i,s}$ are the pump, idler, and signal bandwidths. Allowing $\Delta T \to \infty$ and performing the time integral, we constrain frequency pairs at each detector to be identical, thus reducing the number of integrals to four. In general, we obtain

$$P(\text{i1, i2, s3, s4}) = |\alpha|^4 \eta_{i1}\eta_{i2}\eta_{s3}\eta_{s4}\{r^4 F_{13}F_{24} + t^4 F_{14}F_{23} - r^2t^2[F_{\text{int}}(\Delta X) + F^*_{\text{int}}(\Delta X)]\}, \tag{10}$$

where F_{jk} ($j = 1, 2; k = 3, 4$) is

$$F_{jk} = (2\pi)^2 \int d\omega_{sj} d\omega_{ij} |f_P(\omega_{sj} + \omega_{ij})|^2 |f_{sk}(\omega_{sj})|^2 |f_{ij}(\omega_{ij})|^2 \tag{11}$$

(here, the subscripts j label the crystal source and k the detector). By contrast, the cross terms give rise to $\delta(\omega_{s1} - \omega'_{s2})\delta(\omega_{s2} - \omega'_{s1})$, where the primes denote association with Ψ^*. This essentially characterizes the indistinguishability of signal photons at the detectors, thus leading to the interference effect. F_{int} then has the form,

$$F_{\text{int}} = (2\pi)^4 \int d\omega_{s1} d\omega_{i1} d\omega_{s2} d\omega_{i2} f_P(\omega_{s1} + \omega_{i1}) f^*_P(\omega_{s2} + \omega_{i1}) f_P(\omega_{s2} + \omega_{i2}) f^*_P(\omega_{s1} + \omega_{i2}) \cdot |f_{s4}(\omega_{s1})|^2 |f_{i1}(\omega_{i1})|^2 |f_{s3}(\omega_{s2})|^2 |f_{i2}(\omega_{i2})|^2 e^{i(\omega_{s1} - \omega_{s2})\cdot 2\Delta X}. \tag{12}$$

In the simplest case, we specialize to energy-matched Gaussian filter functions,

$$f_P(\omega_P) = \left(\frac{1}{N}\right) \exp\left[\frac{(\omega_{P0} - \omega_P)^2}{\sigma_P^2}\right],$$

$$f_{s3}(\omega) = \left(\frac{1}{N'}\right) \exp\left[\frac{(\omega_{P0}/2 - \omega)^2}{\sigma^2}\right] = f_{s4}(\omega) = f_{i1}(\omega) = f_{i2}(\omega), \tag{13}$$

and set the normalization $F_{jk} = 1$ for $\sigma \gg \sigma_P$; thus, we obtain

$$P(\text{i1, i2, s3, s4}) = \left[\frac{|\alpha|^4 \eta^4}{(1 + \sigma_P^2/2\sigma^2)}\right]\left\{r^4 + t^4 - 2Vr^2t^2 \exp\left[\frac{\Delta X^2 \sigma^2}{2c^2(1 + \sigma^2/\sigma_P^2)}\right]\right\}, \tag{14}$$

where η^4 implies roughly equal detector efficiencies and includes now losses due to filter and aperture edge effects.[17] The visibility function,

$$V = \frac{(1 + 2\sigma^2/\sigma_P^2)^{1/2}}{(1 + \sigma^2/\sigma_P^2)}, \tag{15}$$

is unity for $\sigma \ll \sigma_P$, showing that overlap is ensured when the time uncertainty due to detection filtering ($\sim \sigma^{-1}$) is larger than the time duration of the pump pulse ($\sim \sigma_P^{-1}$). The fourfold coincidence probability shows a Gaussian dip to near zero (assuming $r^2 \simeq t^2 \simeq 0.5$) when $\Delta X = 0$ (similar to that shown in FIGURE 2B), indicating that we are gating out single signal photons and seeing the nonclassical effect predicted in our naive picture (equation 2). Although a classical model can reproduce this dip, the visibility would be limited to 50%.[1] The high visibility is balanced by a reduction in the probability of seeing fourfold coincidences, which falls off as σ^2/σ_P^2 in this region. When $\sigma \geq \sigma_P$, the width of the dip changes to reflect the pump bandwidth, but the visibility falls as σ_P/σ because we can no longer guarantee full overlap of the photons and hence they are, in principle, distinguishable.

In the above analysis, we have not mentioned the effect of dispersion, which can be significant in short pulse experiments. Dispersion can be included in the above by adding frequency-dependent corrections to t_{s1} and t_{s2} of the form,

$$t_{sj} = t_s + A_j(\omega_{P0}/2 - \omega_{sj}) + B_j(\omega_{P0}/2 - \omega_{sj})^2 \qquad (j = 1, 2). \tag{16}$$

When the paths s1 and s2 contain identical amounts of dispersing material, $A_1 = A_2$ and $B_1 = B_2$. On adjusting equation 9 appropriately and substituting into equations 7 and 8, we find that all dispersion terms cancel, indicating simply that this is a zero-order interference effect.

PROSPECTS

The fourfold coincidence rate (FCR) is given by the product of the laser repetition rate (100 MHz) and $P(\text{i1, i2, s3, s4})$. If we assume $|\alpha|^2 \sim 0.01 \ll 1$ and detector effective efficiencies of order 10%, we obtain an FCR of 0.5 counts per second, which, although low, is a similar rate to those achieved in the early experiments with correlated photons (see reference 1, for example).

The above estimates of efficiency are low considering that modern photon-counting detectors can achieve greater than 70% efficiency[18] in a suitably arranged downconversion apparatus. In interferometry measurements, however, we use small apertures and narrow band filters, and edge effects reduce the maximum effective efficiency. In interference experiments performed to date, effective efficiencies have not exceeded 10%. Similarly, the value of $|\alpha|^2$ could be increased to 0.1 without drastically compromising visibility. In an optimistic analysis of the experiment, we may assume 31% efficiencies and $|\alpha|^2 = 0.1$ and expect a coincidence rate of 500 cps.

The experiment can be easily extended to investigate 3-, 4-, and N-photon interference effects, but we find that the coincidence rate is proportional to

$$C \propto |\alpha|^{2N}\eta^{2N},$$

which gives 5 cps for $N = 3$ and 0.05 cps for $N = 4$. Obviously, to begin studies of higher order entanglement effects, we need a more efficient source of 1-photon states.

Another experiment that may be more viable with pulsed sources is the demonstration of nonlocality in a single photon as suggested by Tan and Walls.[19] Detector efficiency constraints are reduced in this type of experiment and we are at present working on a similar simplified scheme.[20]

DISCUSSION

The experiment described here has a lot in common with an earlier experiment carried out by Pfleegor and Mandel[21] where the interference between separate laser sources was investigated in the limit of extremely low power levels when only one photon could be in the apparatus at one time. In an experiment with a single beam-splitter as in our FIGURE 1 (but with classical sources), they saw interference effects in the detected intensity even at the lowest power levels. As the phases of the separate sources drifted, the fringes moved and fringe visibility dropped when the resolving time was long compared to the drift time. With detection times shorter than the fringe drift time, high visibility fringes could be seen, leading to an anticorrelation in the intensities in detectors D_3 and D_4. This effect would lead to a coincidence dip analogous to that discussed here. A similar effect should be seen with independent pulsed classical sources such as two mode-locked lasers producing time-bandwidth-product-limited pulses. However, a simple classical analysis averaging over the random relative phases shows that the coincidence dip with classical sources cannot exceed 50% visibility. The Pfleegor and Mandel experiment limits itself to considering interference of single photons, supporting Dirac's viewpoint. Here, we see the interference is not confined to be a single-photon effect, but arises because of indistinguishability of two (or more) possible paths to a measurement.

REFERENCES AND NOTES

1. HONG, C. K., Z. Y. OU & L. MANDEL. 1987. Phys. Rev. Lett. **59:** 2044.
2. RARITY, J. G. & P. R. TAPSTER. 1988. *In* Photons and Quantum Fluctuations. E. R. Pike & H. Walther, Eds.: 122. Adam Hilger. Bristol; 1989. J. Opt. Soc. Am. B **6:** 1221.
3. RARITY, J. G. & P. R. TAPSTER. 1990. Phys. Rev. A **41:** 5139.
4. RARITY, J. G. & P. R. TAPSTER. 1990. Phys. Rev. Lett. **64:** 2495.
5. BRENDEL, J., E. MOHLER & W. MARTIENNSEN. 1992. Europhys. Lett. **20:** 575.
6. KWIAT, P. G., A. M. STEINBERG & R. Y. CHIAO. 1993. Phys. Rev. A **47:** R2472.
7. TAPSTER, P. R., J. G. RARITY & P. C. M. OWENS. 1994. Phys. Rev. Lett. **73:** 1923–1926.
8. RARITY, J. G., P. R. TAPSTER & E. JAKEMAN. 1987. Opt. Commun. **62:** 201.
9. GRANGIER, P., G. ROGER & A. ASPECT. 1986. Europhys. Lett. **1:** 173.
10. YURKE, B. & D. STOLER. 1992. Phys. Rev. Lett. **68:** 1251.
11. ZEILINGER, A., A. EKERT & H. WEINFURTER. 1994. Phys. Rev. Lett. **71:** 4287.
12. FEARN, H. & R. LOUDON. 1987. Opt. Commun. **64:** 485–490.
13. OU, Z. Y., C. K. HONG & L. MANDEL. 1987. Opt. Commun. **63:** 118–122.
14. MOLLOW, B. R. 1973. Phys. Rev. A **8:** 2684.
15. HONG, C. K. & L. MANDEL. 1985. Phys. Rev. A **31:** 2409.
16. SHIH, Y. H & M. H. RUBIN. 1993. Phys. Lett. A **182:** 16–22.
17. Filter edge effects in the case of identical Gaussian filter functions as here lead to an effective efficiency of $\eta/\sqrt{2}$. This can be avoided by using broad band filters in the idler detectors. This then changes the effective filter bandwidth to $\sqrt{2}\sigma$, thus altering the form of equations 14 and 15.
18. KWIAT, P. G., A. M. STEINBERG & R. Y. CHIAO. 1993. Phys. Rev. A **48:** R867–R870.
19. TAN, S. M., D. F. WALLS & M. J. COLLETT. 1991. Phys. Rev. Lett. **66:** 252–255.
20. RARITY, J. G., P. R. TAPSTER & R. LOUDON. 1995. In preparation.
21. PFLEEGOR, R. L. & L. MANDEL. 1967. Phys. Rev. **169:** 1084.

Parametric Downconversion Photon Sources, Beam-Splitters, and Bell's Inequality[a]

A. GARUCCIO[b]

Department of Physics and Astronomy
University of Rochester
Rochester, New York 14627

INTRODUCTION

After the first generation of experiments on the EPR paradox based on atomic cascade sources,[1] different sources have been used in the past years in order to produce pairs of correlated photons. All these new sources are based on type-I spontaneous parametric downconversion.

In this process, a photon beam with energy ω and wave number $\mathbf{k}$ impinges on a nonlinear crystal. The impinging photons are polarized along the extraordinary direction (i.e., in the plane containing the direction of propagation of the impinging beam and the optical axis of the crystal). Inside the crystal, the pump photon is split into two photons, called signal and idler, which satisfy the phase-matching condition, $\omega = \omega_s + \omega_i$ and $\mathbf{k} = \mathbf{k}_s + \mathbf{k}_i$. Both of the emerging photons are polarized along the ordinary direction (FIGURE 1).

In general, $\mathbf{k}_s$ and $\mathbf{k}_i$ make an angle θ_s and θ_i with the direction of the impinging beam and are distributed along two coaxial cones. The condition $\mathbf{k} = \mathbf{k}_s + \mathbf{k}_i$ implies that the two photons are strictly correlated in momentum; thus, if the signal photon is detected along the direction defined by $\mathbf{k}_s$, we are sure that the idler photon travels, along $\mathbf{k}_i = \mathbf{k} - \mathbf{k}_s$, in the plane containing the direction of the impinging beam and of the signal photon, at angle θ_i with the photon axis and on the opposite side with respect to $\mathbf{k}_s$.

Using this source, it is possible to perform different types of tests of Einstein locality, using either the original Bell's inequality with polarization correlation functions[2] or the more recently proposed Bell-type inequalities concerning momentum-phase[3] and energy-time[4] measurements.

In all these experiments, the number of detected photon pairs is less than or equal to one-half of the number of total emitted pairs, even in the case of perfect photon detection. We will prove in this report that this fact makes the experiments, in principle, unable to discriminate between quantum mechanics (QM) and Einstein locality (EL). We will discuss in detail in the next section one of these classes of experiments, namely, the polarization correlation measurements, but our reasoning can be extended to all the other experiments.

[a]The research work at the University of Rochester was supported by the Administration Council of the University of Bari and the INFN.

[b]Present address: Dipartimento di Fisica, Università di Bari, I-70126 Bari, Italy.

POLARIZATION CORRELATION EXPERIMENTS

In the experiments on Bell's inequality concerning the polarization, with a suitable choice of the angle between the impinging beam and the optical axis of the crystal, it is possible to obtain the condition $\omega_s = \omega_i$ or, equivalently, $\theta_s = \theta_i$. One of the two photons passes through a 90° polarization rotator, whereas the other crosses a compensating glass plate. The two photons are then reflected from two mirrors and impinge onto a beam-splitter from opposite sides (FIGURE 2).

If the beams emerging from the beam-splitter pass through linear dichotomic polarizers, set at angles θ_1 and θ_2, respectively, the correlation function between the results of the measurement of polarization is

$$E(\theta_1; \theta_2) = P(\theta_{1+}, \theta_{2+}) - P(\theta_{1+}, \theta_{2-}) - P(\theta_{1-}, \theta_{2+}) + P(\theta_{1-}, \theta_{2-}), \qquad (1)$$

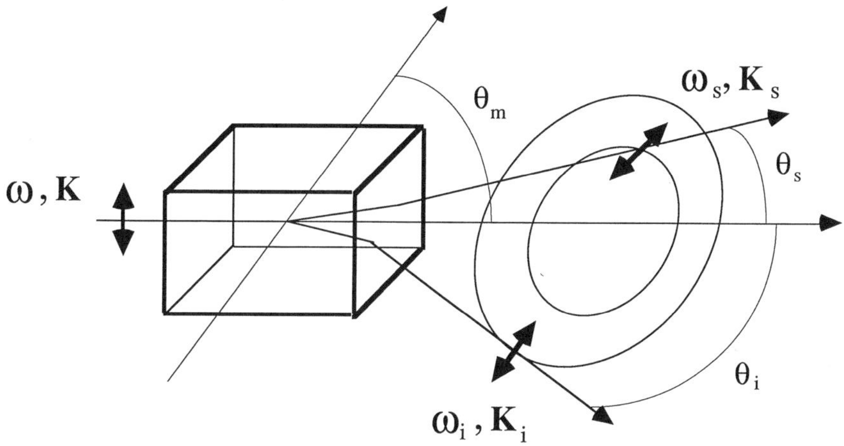

FIGURE 1. Type-I parametric downconversion. The pump beam is polarized along the extraordinary direction, whereas the two emerging photons are polarized in the ordinary direction.

where the $P(\theta_{1i}, \theta_{2j})$ values are the probabilities that the photon that travels along channel 1 is transmitted through polarizer 1 along the i-channel ($i = +$ for the ordinary channel, $i = -$ for the extraordinary one) and the photon that travels along channel 2 is transmitted through polarizer 2 along the j-channel.

In 1964, Bell[5] proved that for a realistic and local theory the expectation value (equation 1) must obey the inequality,

$$|E(\theta_1; \theta_2) - E(\theta_1; \theta'_2)| + |E(\theta'_1; \theta_2) + E(\theta'_1; \theta'_2)| \leq 2, \qquad (2)$$

where θ_1, θ'_1, θ_2, θ'_2 are four different directions of the polarizers.

In most of the performed experiments, only the photomultipliers monitoring the ordinary channel of the polarizers are used and therefore only the joint probabilities $P(\theta_{1+}, \theta_{2+})$ have been measured; consequently, a different inequality has been used.

Two different approaches have been developed to deduce this new inequality. The first was proposed by Clauser, Horne, Shimony, and Holt[6] (CHSH) in 1969 and the second was proposed by Clauser and Horne[7] (CH) in 1974. Both approaches start from Einstein locality and, with some supplementary assumptions, arrive at the following inequality:

$$B(\theta_1, \theta'_1; \theta_2, \theta'_2) = P(\theta_1; \theta_2) - P(\theta_1; \theta'_2) + P(\theta'_1; \theta_2) + P(\theta'_1; \theta'_2) - P(\theta'_1; \infty) - P(\infty; \theta_2) \leq 0, \quad (3)$$

where we have dropped out the index + for simplicity and where $P(\theta'_1; \infty)$ and $P(\infty; \theta_2)$ are the corresponding probabilities when either one of the linear polarizers is removed.

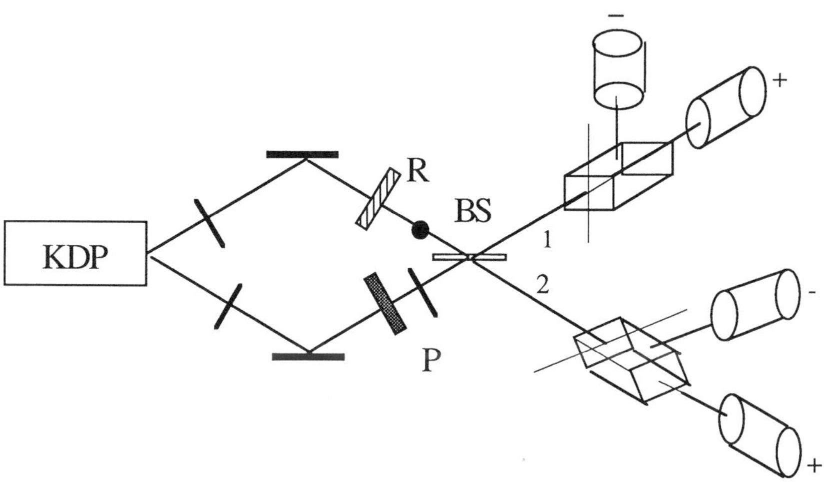

FIGURE 2. Setup of an experiment of the correlation in polarization with a parametric downconversion source.

We will prove now that the CHSH approach leads to an inequality different from equation 3 if it is correctly applied to the parametric downconversion source.[8] Moreover, it is possible to prove that the approach of CH is not applicable *in principle* to this class of experiment.[9]

In general, the following relations hold:

$$P(\theta_{1+}, \theta_{2+}) + P(\theta_{1+}, \theta_{2-}) + P(\theta_{1-}, \theta_{2+}) + P(\theta_{1-}, \theta_{2-}) = P(\infty, \infty),$$

$$P(\theta_{1+}, \theta_{2+}) + P(\theta_{1+}, \theta_{2-}) = P(\theta_{1+}, \infty) \equiv P(\theta_1, \infty),$$

$$P(\theta_{1+}, \theta_{2+}) + P(\theta_{1-}, \theta_{2+}) = P(\infty, \theta_{2+}) \equiv P(\infty, \theta_2), \quad (4)$$

where $P(\theta_1, \infty)$, $P(\infty, \theta_2)$, and $P(\infty, \infty)$ are the corresponding measurable probabilities with one, the other, or both linear polarizers removed, respectively.

In the case of dichotomic observables, $P(\infty, \infty)$ is equal to 1 and equations 1 and 4 allow us to transform the Bell's inequality of equation 2 into the inequality of equation 3.

However, in the case of parametric downconversion, the state of the emerging pair is given by

$$|\psi> = (\tfrac{1}{2})|x_1>|y_2> + (\tfrac{1}{2})|y_1>|x_2> - i(\tfrac{1}{2})|x_1>|y_1> + i(\tfrac{1}{2})|x_2>|y_2>, \quad \textbf{(5)}$$

where $|x_i>$ $[|y_i>]$ is the polarization state along the x-direction [y-direction] for the photon in the i-th output channel of the beam-splitter and it is assumed (in order to simplify the reasoning) that the beam-splitter reflectivities and transmissivities are all equal to 0.5 for both the impinging polarizations.

The state (equation 5) is composed of four terms: the first two give the contributions to the probability of detecting one photon in channel 1 and the other in channel 2; the last two terms give the contributions to the probability of finding both photons in the same channel.

Consequently, the joint detection probability with both of the polarizers removed, $P(\infty, \infty)$, is less than or equal to one-half (even in the case of perfect photon detection) and this causes the inequality of equation 3 to be replaced by

$$(-\tfrac{3}{4}) \equiv (-\tfrac{1}{2})[1 + P(\infty, \infty)] \le B(\theta_1, \theta_1'; \theta_2, \theta_2') \le (\tfrac{1}{2})[1 - P(\infty, \infty)] \equiv \tfrac{1}{4}. \quad \textbf{(6)}$$

The inequality given by equation 6 cannot be violated by the quantum mechanical joint transmission probabilities for the correlated photon pairs described by equation 5. For ideal polarizers and detectors, the latter are given by

$$P(\theta_1; \theta_2) = (\tfrac{1}{4}) \sin^2(\theta_1 + \theta_2), \quad \textbf{(7a)}$$

$$P(\theta_1; \infty) = \tfrac{1}{4}, \quad \textbf{(7b)}$$

and the maximum value of observable B according to the quantum mechanical predictions expressed by equations 7a and 7b is

$$B_{QM} = (\tfrac{1}{4})(\sqrt{2} - 1) < \tfrac{1}{4}.$$

Therefore, the parametric downconversion photon source cannot give quantum states with which to test quantum mechanics versus local realism, *not even in the case of ideal behavior of polarizers and detectors.*

This result is in contradiction with the claimed locality violation in this class of experiments,[2] but it is in complete agreement with a deeper theoretical analysis of the subject.

THEORETICAL ANALYSIS OF THE EXPERIMENTAL CONFIGURATION

Equation 1 can be easily written in the factorized form,

$$|\psi> = \left[\left(\frac{1}{\sqrt{2}}\right)|x>_1 + i\left(\frac{1}{\sqrt{2}}\right)|x>_2\right]\left[\left(\frac{1}{\sqrt{2}}\right)|y>_2 - i\left(\frac{1}{\sqrt{2}}\right)|y>_1\right], \quad \textbf{(8)}$$

and it has been proved that factorable states always satisfy Einstein locality or, equivalently, Bell-type inequalities.[10]

Furthermore, it has been observed that in Bell-type inequality only joint probabilities are present. Hence, to overcome the problems related to the use of equation 5 or 8, it may seem "reasonable" to cut away the last two terms that do not give contribution to the joint detection probability. If this procedure is applied, the result will be the entangled (but not normalized) state:

$$|\psi> = (\tfrac{1}{2})(|x>_1|y>_2 + |y>_1|x>_2). \tag{9}$$

Imposing the normalization of the quantum state to *the subset of photon pairs traveling in both channels*, the wave function becomes

$$|\psi> = \left(\frac{1}{\sqrt{2}}\right)(|x>_1|y>_2 + |y>_1|x>_2), \tag{10}$$

which leads to violation of the Bell-type inequality.

This is a crucial point. The problem of the renormalization of the state to the subset of "detectable" photons has been posed by several investigators. For example, Santos[11] discussed in detail this problem in the context of atomic cascade experiments, denying the possibility to realize physically a singlet state due to the depolarization introduced by the finite collecting solid angle. Moreover, Lepore and Selleri[12] discussed the role of low quantum efficiency in decreasing the measured correlation function to such a small value so as to verify Bell's inequality in all the cases.

We will underline that, in this particular case of parametric downconversion, in the reasoning previously drawn in reducing the state of equation 5 to the state of equation 10, there is a chain of incorrect steps either in QM or in EL. It is incorrect to consider the ensemble as consisting of two subensembles: one in which one photon is in channel 1 and the other is in channel 2, and a second in which both photons are in the same channel. In QM, each pair described by the state of equation 5 has the photons traveling at the same time along the two different channels and in the same channel. A set consisting of the two previous subsets is described by a state that is a mixture of

$$\begin{cases} 50\% & \left(\frac{1}{\sqrt{2}}\right)(|x>_1|y>_2 + |y>_1|x>_2) \\ 50\% & \left(\frac{1}{\sqrt{2}}\right)(|x>_1|y>_1 - |y>_2|x>_2). \end{cases} \tag{11}$$

The states of equations 5 and 11 are physically distinguishable (for example, overlapping the two beams onto another beam-splitter); moreover, there is no unitary evolution from the state of equation 5 to the state of equation 11. Hence, "discarding" the last two terms requires a state preparation or a measuring process.

It is worth noting that, in this case, the conceptual operation of the selection of the subensemble of the coincidence pairs is neither a measuring process nor a state preparation. It is not a measuring process because the measurements are made behind the polarizers and thus give a factorized state of two photons in a well-defined polarization state. It is not a state preparation because other pairs of photons are traveling along the two channels and these contribute, for example, to the measured single-photon detection rates.

Moreover, the procedure of identifying an entangled state with a violation of the locality can lead to error. In fact, in a previous paper,[13] we argued that the entangled state (equation 9) can be reproduced in a hidden-variable local realistic model for physical correlated systems. If the quantum operation of the normalization of the probability amplitudes is imposed on the state of equation 9, then the state of equation 10 can be obtained, but any possibility of physical and local interpretation of the model will be lost. This result shows the critical importance of the probability interpretation of the wave function. In fact, quantum mechanics can be obtained starting from local realism, but only after the normalization step, which allows the raising of correlation function values and, consequently, the violation of Bell's inequality. Then, the entangled state is a necessary, but not a sufficient, condition for violating the locality.

Even from the point of view of Einstein locality, the application to the subset of pairs traveling in the opposite direction presents some difficulties. The problem is that of the extension of the theory from the dichotomic case, in which the results are $+1$ or -1, to the three-valued observable (trichotomic) case, in which even the event of "no detection" or "0" occurs. The problem was posed in 1971 by Bell[14] and later by CH in 1974, but up to now it has not been completely solved.

In the appendix of their famous paper, CH tried to develop a probabilistic hidden-variable theory based on a trichotomic observable. They deduced a new inequality,

$$-1 + Q \leq B(\theta_1, \theta'_1; \theta_2, \theta'_2) \leq Q, \tag{12}$$

where Q is a linear combination of some unmeasurable quantities, and they tried to prove that $Q \leq 0$, but were unsuccessful.

Recently, De Caro and myself,[15] starting from Bell's original approach to the problem, deduced the following value for Q:

$$Q = \frac{[1 - P(\infty, \infty) - \mu_0(\theta_2, \theta'_2)]}{2}, \tag{13}$$

where $\mu_0(\theta_2, \theta'_2)$ is the measure of the intersection of the two subsets $\Lambda_0(\theta_2)$ and $\Lambda_0(\theta'_2)$ of the set Λ of hidden variables in which we have no detection in channel 2 when the polarizations along θ_2 and θ'_2 are measured.

Obviously, Q is model-dependent, that is, it depends in general on the model or on the class of models that we are dealing with. This is a limit from one point of view, but an improvement from another: now, we are able to insert in the hidden-variable theory the *same assumptions* that we make in the quantum mechanical treatment of the problem. Moreover, the quantities $P(\infty, \infty)$ and $\mu_0(\theta_2, \theta'_2)$ are either measurable or, in principle, deducible from our assumptions.

In general, it is possible to determine the minimum and the maximum value of Q.

Let us suppose, for simplicity, that the probability of no detection of a photon is equal in both channels and for all settings of the polarizers and that its value is β. If we assume that the no-detection process is a random process, then the probability of a double no-detection is

$$\mu_0(\theta_2, \theta'_2) = \beta^2,$$

whereas

$$P(\infty, \infty) = (1 - \beta)^2.$$

Then,

$$Q = \beta(1 - \beta).$$

If we suppose that the sets of no-detection events are *identical* for each measurement, then

$$\mu_0(\theta_2, \theta'_2) = \beta,$$

$$P(\infty, \infty) = (1 - \beta),$$

and

$$Q = 0.$$

In conclusion, the upper limit of the inequality of equation 3 becomes 0 only when β is 0 (i.e., in the dichotomic case) or when the set Λ of hidden variables is the sum of two disjoint sets, Λ' and Λ_0, the first giving signals (and coincidences) regardless of the setting of the polarizing angles and the second giving no signal regardless of the value of the polarizing angles. In this case, and only in this case, we can take into account the photon pairs with $\lambda \in \Lambda'$, redefine a new density distribution, and reduce the original three-valued observable to a dichotomic observable.

In the particular case of type-I parametric downconversion,

$$P(\infty, \infty) = \tfrac{1}{2}$$

and

$$\mu_0(\theta_2, \theta'_2) = \tfrac{1}{4}$$

because the probability of having no counts in channel 2, either when we measure θ_2 or θ'_2, is equal to the probability of having both photons in channel 1. Then, $Q = \tfrac{1}{8} = 0.125$ and the maximum value of B for QM (0.103) cannot violate this limit.

SOURCES OF PHOTON COUPLES BASED ON TYPE-II PARAMETRIC DOWNCONVERSION

In order to avoid the previous criticisms of type-I sources, we are studying a new type of source of correlated photon pairs based on type-II parametric downconversion. Let us describe briefly this new source.

In type-II phase-matching, the pump beam has ordinary polarization and the output photons have orthogonal polarization. More precisely, the signal photons travel along the cone with angle θ_s and ordinary polarization, whereas the idlers travel along the cone with angular aperture θ_i and extraordinary polarization (FIGURE 3).

Let us suppose now that we choose the phase-matching angle so that the two cones overlap. In this configuration, the state of the pairs traveling in the plane containing the intersection points of the two cones is the superposition of signal photon in direction $\mathbf{k}_1$ and idler photon in direction $\mathbf{k}_2$ *plus* idler photon in direction $\mathbf{k}_1$ and signal photon in direction $\mathbf{k}_2$, or equivalently

$$|\psi> = \left(\frac{1}{\sqrt{2}}\right)\{|s>_1|i>_2 + |i>_1|s>_2\}, \tag{14}$$

which is an entangled and normalized state.

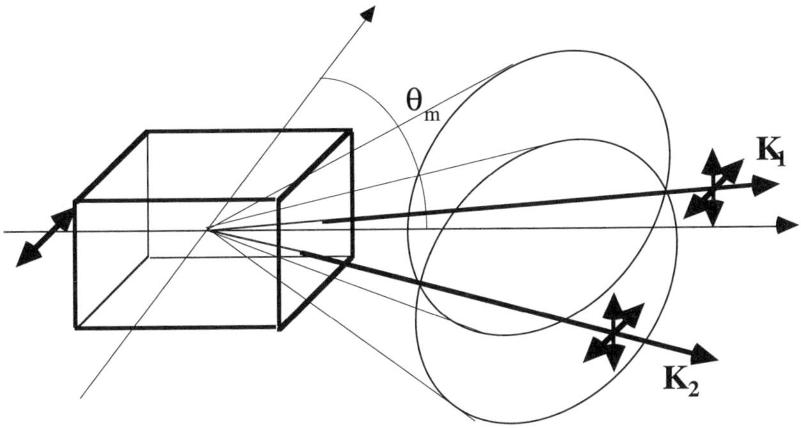

FIGURE 3. Type-II parametric downconversion. The two photons have orthogonal polarization. In the points in which the cones overlap, the photon pair is described by an entangled state.

Unfortunately, in type-II parametric downconversion, if we impose the overlapping of the cones, we cannot verify the condition of $\omega_s = \omega_i$. If it is possible to find a commercial crystal in which the energy difference of $\omega_s - \omega_i$ is so narrow as to lie inside the bandwidth of the filters usually used in these experiments, the two photons will be indistinguishable and the source will produce pairs totally correlated in momentum.

A similar experiment has been proposed recently by Kwiat *et al.*[16] In their apparatus, two parametric downconverters are used and the output beams are overlapped using a cube polarizer.

REFERENCES

1. FREEDMAN, S. J. & J. F. CLAUSER. 1972. Phys. Rev. Lett. **28:** 938; CLAUSER, J. F. 1976. Phys. Rev. Lett. **36:** 1223; FRY, E. S. & R. C. THOMPSON. 1972. Phys. Rev. Lett. **27:** 465;

ASPECT, A., P. GRANGIER & G. ROGER. 1981. Phys. Rev. Lett. **47:** 460; ASPECT, A., P. GRANGIER & G. ROGER. 1982. Phys. Rev. Lett. **49:** 91; ASPECT, A., J. DALIBARD & G. ROGER. 1982. Phys. Rev. Lett. **49:** 1804; DUNCAN, A. J., W. PERRIE, H. J. BEYER & H. KLEINPOPPEN. 1985. *In* Fundamental Processes in Atomic Collision Physics, p. 555. Plenum. New York.

2. OU, Z. Y. & L. MANDEL. 1988. Phys. Rev. Lett. **61:** 50; SHIH, Y. H. & C. O. ALLEY. 1988. Phys. Rev. Lett. **61:** 2921; OU, Z. Y., C. K. HONG & L. MANDEL. 1988. Opt. Commun. **67:** 159; TAN, S. M. & D. F. WALLS. 1989. Opt. Commun. **71:** 235.
3. FRANSON, J. D. 1989. Phys. Rev. Lett. **62:** 2200; KWIAT, P. G., W. A. VAREKA, C. K. HONG, H. NATHAL & R. Y. CHIAO. 1990. Phys. Rev. **A41:** 2910; OU, Z. Y., X. Y. ZOU, L. J. WANG & L. MANDEL. 1990. Phys. Rev. Lett. **65:** 321; RARITY, J. G. & P. R. TAPSTER. 1990. Phys. Rev. Lett. **64:** 2495; RUBIN, M. H. & Y. H. SHIH. 1992. Phys. Rev. **A45:** 8138.
4. KWIAT, P. G., A. M. STEINBERG & R. Y. CHIAO. 1993. Phys. Rev. **A47:** 2472; BRENDEL, J., E. MOLLER & W. MARTIENSSEN. 1992. Europhys. Lett. **20:** 572.
5. BELL, J. S. 1965. Physics **1:** 195.
6. CLAUSER, J. F., M. A. HORNE, A. SHIMONY & R. A. HOLT. 1969. Phys. Rev. Lett. **23:** 880.
7. CLAUSER, J. F. & M. A. HORNE. 1974. Phys. Rev. **D23:** 526.
8. DE CARO, L. & A. GARUCCIO. 1994. The reliability of Bell's inequality measurement using parametric down-conversion photon sources. Phys. Rev. A. In press.
9. GARUCCIO, A. 1993. On the validity of Clauser and Horne factorizability. Preprint.
10. CAPASSO, V., D. FORTUNATO & F. SELLERI. 1973. Int. J. Theor. Phys. **7:** 319.
11. SANTOS, E. 1991. Phys. Rev. Lett. **66:** 1388.
12. LEPORE, V. L. & F. SELLERI. 1990. Found. Phys. Lett. **3:** 203.
13. DE CARO, L. & A. GARUCCIO. 1992. Found. Phys. Lett. **5:** 393.
14. BELL, J. S. 1971. *In* Foundations of Quantum Mechanics: Proceedings of the International School of Physics "Enrico Fermi", Course XLIX. Academic Press. New York.
15. DE CARO, L. & A. GARUCCIO. 1994. *In* Frontiers of Fundamental Physics—Proceedings of the International Conference (September 1993). M. Barone & F. Selleri, Eds. Plenum. New York.
16. KWIAT, P. G., P. H. EBERHARD, A. M. STEINBERG & R. Y. CHIAO. 1994. Phys. Rev. **A49:** 3209.

Three-Particle Bell Inequalities

T. E. KIESS

Department of Physics
University of Maryland
College Park, Maryland 20742

INTRODUCTION

Recent work[1–4] in N-particle Bell inequalities focuses on the entangled[5] three-particle GHZ state,[6] and N-particle analogues,[7] to deduce quantum mechanical violations of local reality predictions (based on this state) that grow with N.

This work is not the only possible approach to the issue of N-particle quantum violations of local reality predictions. Indeed, it lacks the flavor of some of the two-particle Bell inequality work, which is laden with specific choices of detection variables ($\vec{a}$, $\vec{b}$). These variables are angles of polarizers (for photon polarization measurements),[8] orientation of spin direction (for spin-½ particle measurements),[9] or phases of interferometers (for Franson-type[10] interferometry).

Another approach[11] is more similar to the traditional $N = 2$ program,[12,13] in that expectation values dependent on detection variables ($\vec{a}$, $\vec{b}$, $\vec{c}$) and on a hidden variable (λ) are written prior to specialization to a particular quantum state. Then, specific quantum states are considered that violate the deduced inequalities.

Our work represents another approach to the $N = 3$ case, more similar to reference 11 than to references 1–4. Our method differs from these and can be extended to arbitrary N. We generalize the method[14] of Clauser, Horne, Shimony, and Holt (CHSH) to write down Bell-type inequalities for expectation values and probabilities in a three-particle situation.

One outcome of this approach is that violations are, as in the two-particle case, linked to choices of as yet unspecified quantum states and detection variables. With a rich dependence on angle variables, we recover the statement of the $N = 2$ violation, which is absent from reference 1. We also derive an $N = 3$ violation differing in degree from that of references 1–4 and 11. It is a weaker inequality, and a weaker violation, and the associated detector efficiency minimum is correspondingly higher.

However, it is straightforward to show that a violation exists with the third particle unentangled with the other two. This simplification of the requisite three-particle state is important in achieving a nongedankenexperimental test of a three-particle Bell-type inequality.

REALITY DEFINITIONS AND BELL LOCALITY

We start with a two-particle source in view of two detectors, where the detection variable $\vec{a}$ is set in front of detector 1 and $\vec{b}$ is in front of detector 2. The standard definition of the expectation value, $E_{12}(\vec{a}, \vec{b})$, is

$$E_{12}(\vec{a}, \vec{b}) = \lim_{N\to\infty} \left(\frac{1}{N}\right) \sum_{i=1}^{N} A_i B_i,$$

with $A_i = \pm 1, 0$ and $B_i = \pm 1, 0$ denoting the outcome of the i-th measurement of $\vec{a}$ and $\vec{b}$, respectively. Here, an outcome of $+1$ denotes an alignment (or an "up") with $\vec{a}$, -1 denotes an antialignment (or a "down"), and 0 signifies a nondetection (a no show or no count recorded). For this expectation value,

$$-1 \leq E_{12}(\vec{a}, \vec{b}) \leq +1.$$

Other quantities of interest are the probabilities,

$$p^+ = \frac{\text{no. of } \mathit{ups}}{N},$$

$$p^- = \frac{\text{no. of } \mathit{downs}}{N},$$

$$p^0 = \frac{\text{no. of } \mathit{nondetections}}{N},$$

which sum to 1. In terms of these probabilities, we may write the average value of A_i as

$$\overline{A}(\vec{a}, \lambda) = p^+(\vec{a}, \lambda) - p^-(\vec{a}, \lambda);$$

similarly for $\overline{B}(\vec{b}, \lambda)$, that is, the average value of B_i:

$$\overline{B}(\vec{b}, \lambda) = p^+(\vec{b}, \lambda) - p^-(\vec{b}, \lambda).$$

Here, the probabilities and expectation values for recording an event at each detector depend on the setting of the analyzer in front of the detector and on an as yet unspecified parameter λ, a "hidden variable" that may characterize the source or may even characterize the passage through the analyzer and the detection process. We have denoted only one such hidden variable, although there may be several. In this sense, the variable λ symbolizes the presence of all of them and it is straightforward to generalize the notation to include multiple hidden variables explicitly. This treatment and interpretation of the role of λ follows standard practice.

We can now write the mathematical statement of Bell locality for the expectation value of coincidence detections:

$$E_{12}(\vec{a}, \vec{b}) = \int d\lambda \rho(\lambda) \overline{A}_1(\vec{a}, \lambda) \overline{B}_2(\vec{b}, \lambda). \quad \textbf{(1)}$$

The key feature of locality is that the dependence of $E_{12}(\vec{a}, \vec{b})$ on $\vec{a}$ and $\vec{b}$ is factorizable.

By writing $\overline{A}$ and $\overline{B}$ in terms of the probabilities $p^{\pm}$, this expectation value can be rewritten as

$$E_{12}(\vec{a},\vec{b}) = \int d\lambda\rho(\lambda)[p^+(\vec{a},\lambda) - p^-(\vec{a},\lambda)][p^+(\vec{b},\lambda) - p^-(\vec{b},\lambda)]$$
$$= 4P_{12}^{++}(\vec{a},\vec{b}) + 1 - 2P_1^+(\vec{a}) - 2P_2^+(\vec{b}), \quad \textbf{(2)}$$

where in the second line we have used the identity $p^+ - p^- = 2p^+ - 1$ and the following definitions:

$$P_1^+(\vec{a}) \equiv \int d\lambda\rho(\lambda)p^+(\vec{a},\lambda),$$

$$P_2^+(\vec{b}) \equiv \int d\lambda\rho(\lambda)p^+(\vec{b},\lambda),$$

$$P_{12}^{++}(\vec{a},\vec{b}) \equiv \int d\lambda\rho(\lambda)p^+(\vec{a},\lambda)p^+(\vec{b},\lambda).$$

THREE-PARTICLE GENERALIZATION

Our three-particle generalization starts with the expression,

$$E_{123}(a,b,c) - E_{123}(a,b',c) + E_{123}(a,b,c') - E_{123}(a,b',c'). \quad \textbf{(3)}$$

Using a procedure analogous to that of reference 14, the inequality for expectation values is

$$|E_{123}(a,b,c) - E_{123}(a,b',c) + E_{123}(a,b,c') - E_{123}(a,b',c') + E_{123}(a',b',c') + E_{123}(a',b,c') + E_{123}(a',b',c) + E_{123}(a',b,c)| \le 4 \quad \textbf{(4)}$$

and the inequality for probabilities is

$$0 \le (\textit{triples}) - (\textit{doubles}) + (\textit{singles}) \le 1, \quad \textbf{(5)}$$

where

$$\textit{triples} = P_{123}(a,b,c) - P_{123}(a,b',c) + P_{123}(a',b,c) + P_{123}(a',b',c) + P_{123}(a,b,c') - P_{123}(a,b',c') + P_{123}(a',b,c') + P_{123}(a',b',c'),$$

$$\textit{doubles} = P_{12}(a,b) + P_{12}(b,c) - P_{12}(a,b') + P_{12}(a',b') + P_{12}(a',c) + P_{12}(a',b) + P_{12}(b,c') + P_{12}(a',c'),$$

$$\textit{singles} = P_1(a) + P_2(b) + (1/2)P_3(c) + (1/2)P_3(c').$$

These are our three-particle Bell inequalities.

SPECIFIC VIOLATIONS

For a three-particle state in which the third particle is unentangled with the other two, the three-particle inequality of equation 5 reduces to the two-particle form of reference 14.

We shall explore this violation in more detail. For the state

$$|\Psi\rangle = \left(\frac{1}{\sqrt{2}}\right)[|X\rangle_1|Y\rangle_2 - |Y\rangle_1|X\rangle_2]|X\rangle_3 \tag{6}$$

and the choices $\vec{c} = \vec{c}\,' = 0°$, $\vec{a} \equiv \theta_1$, $\vec{a}\,' \equiv \theta_1'$, $\vec{b} \equiv \theta_2$, and $\vec{b}\,' \equiv \theta_2'$, equation 5 reduces to

$$\sin^2(\theta_1 - \theta_2) - \sin^2(\theta_1 - \theta_2') + \sin^2(\theta_1' - \theta_2') + \sin^2(\theta_1' - \theta_2) - 1 \le 1.$$

This is the familiar two-particle CHSH situation, with multiple solutions. One solution is given by the choice of angles, $\theta_1 = 22.5°$, $\theta_1' = 67.5°$, $\theta_2 = -45°$, and $\theta_2' = 0°$. For these settings, the above inequality reads $\sqrt{2} \le 1$. This is the violation of equation 5 due to the quantum state of equation 6.

Equation 6 also conveniently generates detector efficiency requirements. With an efficiency ϵ for each detector explicitly included, the above inequality becomes a cubic equation in ϵ, with violations for an efficiency bound of $\approx$91.7%. This is competitive with, but larger than, the bounds in other three-particle work: 90.9% in reference 15 and 79.3% in reference 4.

DISCUSSION AND CONCLUSIONS

We have generalized the CHSH method[14] to write inequalities for expectation values and for probabilities based only on the two assumptions of locality and reality. Although we have recorded here only the $N = 3$ results, our approach can be extended in a straightforward way to generate similar inequalities for arbitrary N.

This is a different approach to N-particle Bell-type inequalities than other work[1–4] based on arguments using the GHZ state. Here, we recover a rich dependence on the N detection variables to describe the degree of violation. We recover the $N = 2$ inequality and an $N = 3$ violation. As in $N = 2$, the settings of the detection variables control the degree of the $N = 3$ violation.

In an application of this work, demonstrations of violations of the Bell-type inequalities shown here represent a rigorous and plausible experimental program. We emphasize that these inequalities contain no additional assumptions other than Bell reality and locality. The Clauser-Horne no-enhancement assumption,[16] for example, has not yet been applied.

This program can be compared to experimental realizations[17,18] of the GHZ gedanken.[6] The GHZ-type experiments, although originally advocated[19] to be free from the statistics of accumulating many detection events over time, have upon closer scrutiny suffered from detector inefficiency loophole criteria.[4,15] The result is that inequalities are necessarily involved.

Although the violation of the inequality derived here is weaker than the violations of the inequalities in references 1–4 and 11, the advantage of our approach is that a less entangled state is needed. This can be appreciated in experiments involving two-photon decays. The atom is caught and detected to form an "event-ready" situation. Atoms, unlike optical photons, can be detected with nearly 100% probability. Hence, the state of equation 6 is precisely that of a cascade decay (photons being particles 1 and 2) from an atom (particle 3). With $\epsilon_3 = 1$, the detector efficiency cubic equation reduces to the standard two-particle quadratic,[20] with solution $\epsilon \approx 82.84\%$ for the common efficiency of detectors 1 and 2.

Our view is that this situation is then competitive with a scheme involving the 79.3% efficiency criterion in reference 4, especially because such a scheme involves the difficult manufacture of a triple entangled state, such as the GHZ state; that is, we would claim that the ease of using a state entangled in only two particles is a favorable trade-off compared to a few percent in detection efficiency. The test of a three-particle Bell-type inequality of the kind considered here is then a worthy competitor in its degree of experimental difficulty to an experimental realization of a GHZ-type gedanken based on the GHZ state. This comparison shows that there exist different and competing approaches to the goal of exhibiting violations to local realism for $N = 3$.

ACKNOWLEDGMENTS

Useful discussions with A. Stairs are gratefully acknowledged.

REFERENCES

1. Mermin, N. D. 1990. Phys. Rev. Lett. **65:** 1838–1840.
2. Ardehali, M. 1992. Phys. Rev. **A46:** 5375–5378.
3. Braunstein, S. L., A. Mann & M. Revzen. 1992. Phys. Rev. Lett. **68:** 3259–3261.
4. Braunstein, S. L. & A. Mann. 1993. Phys. Rev. **A47:** R2427–R2430.
5. Horne, M. A., A. Shimony & A. Zeilinger. 1989. Phys. Rev. Lett. **62:** 2209–2212.
6. Greenberger, D. M., M. A. Horne & A. Zeilinger. 1989. *In* Bell's Theorem, Quantum Theory, and Conceptions of the Universe. M. Kafatos, Ed.: 69–74. Kluwer. Dordrecht.
7. Mermin, N. D. 1990. Phys. Rev. Lett. **65:** 3373–3376.
8. Kiess, T. E., Y. H. Shih, A. V. Sergienko & C. O. Alley. 1993. Phys. Rev. Lett. **71:** 3893–3897 and references therein.
9. Clauser, J. F. & A. Shimony. 1978. Rep. Prog. Phys. **41:** 1882–1927 and references therein.
10. Franson, J. D. 1989. Phys. Rev. Lett. **62:** 2205–2208.
11. Roy, S. M. & V. Singh. 1991. Phys. Rev. Lett. **67:** 2761–2764.
12. Bell, J. S. 1965. Physics **1:** 195–200.
13. Bell, J. S. 1971. Introduction to the hidden variable question. *In* Foundations of Quantum Mechanics. B. d'Espagnat, Ed.: 171–181. Academic Press. New York.
14. Clauser, J. F., M. A. Horne, A. Shimony & R. A. Holt. 1969. Phys. Rev. Lett. **23:** 880–884.
15. Greenberger, D. M., M. A. Horne, A. Shimony & A. Zeilinger. 1990. Am. J. Phys. **58:** 1131–1143.
16. Clauser, J. F. & M. A. Horne. 1974. Phys. Rev. **D10:** 526–535.
17. Yurke, B. & D. Stoler. 1992. Phys. Rev. Lett. **68:** 1251–1254.
18. Reid, M. D. & W. J. Munro. 1992. Phys. Rev. Lett. **69:** 997–1001.
19. Mermin, N. D. 1990. Phys. Today **43**(no. 6): 9–11.
20. Garg, A. & N. D. Mermin. 1987. Phys. Rev. **D35:** 3831–3835.

A GHZ Argument for a Single Spinless Particle

GORDON N. FLEMING

Davey Laboratory
Pennsylvania State University
University Park, Pennsylvania 16802

INTRODUCTION

The original argument of Einstein, Podolsky, and Rosen[1] (EPR) concerned itself with the position and momentum of a two-particle system. To keep the argument simple, they employed the idealized notion of a simultaneous eigenstate of the total momentum and relative position, idealized because of the continuous spectra of those observables. In a move designed to make the argument more realistic and the quantum correlations involved more susceptible to experimental test, David Bohm[2] reformulated the EPR argument in terms of measurements of the spin components of two spatially separated particles. From that day to the present, the observables of choice, in arguments demonstrating the incompatibility of quantum theory with noncontextual possessed values, have usually been spin/polarization or spinlike observables of multiparticle systems.[3] In this report, I indulge my own curiosity as to whether versions of the simplest[4] of the most recent forms of these arguments, discovered by Greenberger, Horne, and Zeilinger or GHZ[5,6] and generalized by Clifton, Redhead, and Butterfield,[7] have variants employing only "simple" functions of position and momentum observables for a single particle. I actually construct two arguments, of exactly the same algebraic structure as the Mermin simplification of the GHZ argument, the first involving three particles, as does GHZ, and the second employing only one particle. In Greenberger *et al.*,[6] a three-particle argument involving only the momenta of the particles was given and my three-particle argument (which can be rephrased to focus on momenta, although in fact it focuses on position) is novel only in exploiting the algebraic form recognized by Mermin.[4]

The reasons for the traditional preference for spin/polarization-like observables, in the arguments of interest heretofore, include and go beyond the original motivations of David Bohm. First, discrete spectrum observables make possible a very clean and simple statement of the requirements of quantum theory (because normalizable eigenstates of the observables exist) and of the possessed value assumptions (because determination of a definite value is implied by the value lying in a sufficiently small interval). This consideration will be respected in the arguments constructed here. Second, the spin/polarization-like observables can live in finite-dimensional vector spaces that, being factor spaces of the infinite-dimensional state spaces of real physical systems, keep the mathematics simple without making the argument unrealistic. To maintain these virtues, we might note that any argument that can be carried out in a six- or eight-dimensional factor space, say (as the Mermin simplifications of GHZ are), can, in principle, be replicated in any six- or eight-dimensional subspace, respectively, of the infinite-dimensional state spaces in which position and momentum observables live. However, in almost all, if not all such subspaces, the relevant observables for the arguments would be such exotic functions of the irreducible sets of position and momentum observables as to disembowel the conclusions reached of all intuitive physical impact. Instead, by diverting the search for simplicity from a focus on lower-dimensional vector spaces to a focus on simple algebraic relationships between the relevant observables (regardless of the dimensionality of the supporting vector space), we will see here that we can find versions of the GHZ argument that arguably involve only "simple" functions of position and momentum observables.

These versions also lend a novel perspective on the still-controversial issue of the role that locality considerations play in either mounting or assessing arguments of the GHZ type. On the one hand, my first argument involving three particles, widely separated spatially, yields conflicting conclusions, coming from quantum theory and noncontextual possessed values, respectively, which are themselves conclusions about just where the particles are located in space. A champion of the view that a nonlocal structure of quantum theory follows from GHZ-type arguments, as well as their predecessors,[8] might regard this result as grist for his mill. On the other hand, my second argument, of exactly the same algebraic structure, involving only a single particle, yields conflicting conclusions concerning where that one particle is located in space. A champion of the view that locality considerations are insufficient to ground noncontextuality and are something of a red herring in GHZ-type arguments, as well as their predecessors,[9,10] might regard the existence of such a one-particle argument as grist for her mill (I have heard such rumblings). We now move to the arguments themselves.

THE FORMAL STRUCTURE OF THE ARGUMENTS

We need three mutually commuting pairs of self-adjoint operators such that, within each pair, the operators anticommute, that is,

$$(A_i, B_i) \qquad i = 1, 2, 3,$$

where

$$[A_i, A_k] = [A_i, B_k] = [B_i, B_k] = 0 \text{ for } i \neq k \tag{1a}$$

and

$$\{A_i, B_i\} = 0 \text{ for each } i. \tag{1b}$$

It then follows that the operators

$$C_i \equiv A_i B_k B_l \qquad (i \neq k \neq l \neq i \text{ and } i = 1, 2, 3) \tag{2a}$$

and

$$C_0 \equiv A_1 A_2 A_3 \tag{2b}$$

are all mutually commuting and therefore possess simultaneous eigenvectors. Furthermore, each C operator is itself composed of mutually commuting operators. Finally, the C operators themselves satisfy

$$C_0C_1C_2C_3 = A_1A_2A_3A_1B_2B_3A_2B_3B_1A_3B_1B_2 = -A_1^2A_2^2A_3^2B_1^2B_2^2B_3^2 \le 0; \quad \textbf{(3)}$$

that is, the eigenvalues and expectation values of the product of the C operators must always be nonpositive numbers.

Contradiction is immediate if we now assume noncontextual possessed values for the observables represented by the A, B, and C operators and all products of them that have been listed above. Denoting the possessed values by correspondingly subscripted lowercase letters, a_i, b_k, etc., the composition of the C operators out of mutually commuting A's and/or B's guarantees for the observables represented by the C's the possessed values given by

$$c_i \equiv a_ib_kb_l \qquad (i \neq k \neq l \neq i \text{ and } i = 1, 2, 3) \quad \textbf{(4a)}$$

and

$$c_0 \equiv a_1a_2a_3. \quad \textbf{(4b)}$$

Furthermore, the product of the mutually commuting C operators represents an observable, and noncontextual possessed values for it require that value to also be given by the product of the possessed values of the C factors. However,

$$c_0c_1c_2c_3 = a_1a_2a_3a_1b_2b_3a_2b_3b_1a_3b_1b_2 = +a_1^2a_2^2a_3^2b_1^2b_2^2b_3^2 \ge 0 \quad (5)$$

and the possessed values of the product of the C's must always be nonnegative. With possessed values for any observable restricted to the eigenvalues of the representing operator, our conclusions are consistent iff the possessed values of the product of the C's are always zero. However, this requires the possessed values of at least one of the A's or B's to be zero. Any instance of A's and B's satisfying the assumed commutation relations in which none of them has a zero eigenvalue provides a contradiction.

In the original presentations of GHZ and Mermin, the eigenvalues of the A's and the B's were restricted to ± 1. Furthermore, the need for an assumption of noncontextual possessed values for the C's and their product was eliminated by applying the operators to a simultaneous eigenstate of the C's, with eigenvalues ± 1. As Mermin[8] emphasizes, the application of the argument to a specific simultaneous eigenstate of the composite C's, in a context in which the subscripts of the A's and B's refer to distinct, spatially separated particles, allows the noncontextuality of the possessed value assignments to be supported by a locality assumption.

A PHASE-SPACE VERSION OF THE ARGUMENT

By the term "phase-space", I mean to imply a version of the argument, described abstractly above, that employs "simple" functions of the operators representing the position and momentum of particles only. In this first example, the three mutually commuting pairs of operators will, as was the case in the original GHZ and Mermin arguments, refer to three distinct particles. Let X_1, X_2, and X_3 denote the operators for the x-components of position of the three particles. Then, playing the role of the

A's in the abstract account are the operators

$$S_i \equiv X_i / |X_i|, \tag{6}$$

that is, the algebraic signs of the X_i. These operators are simultaneously well defined only over the (nonclosed) subspace of states with position-representation state functions that vanish for $x_1x_2x_3 = 0$. In that subspace, they each have infinitely degenerate eigenvalues of ± 1 only. To play the role of the B's of the abstract account, we need mutually commuting self-adjoint operators that selectively anticommute with the S_i. These are provided by the rotations through π radians, about the y-axis, generated, respectively, by the operator for the y-component of the orbital angular momentum of the i-th particle:

$$L_{yi} \equiv Z_i P_{xi} - X_i P_{zi}; \tag{7}$$

that is, the B's are given by

$$R_i \equiv \exp\,[(i/\hbar)L_{yi}\pi] = \cos\,[L_{yi}(\pi/\hbar)]. \tag{8}$$

Because obviously

$$R_i X_k R_i^{-1} = (1 - 2\delta_{ik})X_k \qquad \text{(no sum on } i \text{ or } k) \tag{9}$$

and $R_i^{-1} = R_i = R_i^{+}$, it follows that

$$R_i S_k = (1 - 2\delta_{ik}) S_k R_i \qquad \text{(no sum on } i \text{ or } k). \tag{10}$$

These are just the commutation/anticommutation relations that we need.

Following the lines of the abstract argument, we now introduce the composite operators,

$$\Sigma_i \equiv S_i R_k R_l \qquad (i \neq k \neq l \neq i,\ i = 1, 2, 3) \tag{11a}$$

and

$$\Sigma_0 \equiv S_1 S_2 S_3, \tag{11b}$$

which are mutually commuting and are each composed of mutually commuting factors. Multiplying the Σ's together, we find

$$\Sigma_0\, \Sigma_1\, \Sigma_2\, \Sigma_3 = -1 \tag{12}$$

as a consequence of the commutation/anticommutation relations among the S's and R's and the relations

$$S_i^2 = R_i^2 = 1 \qquad (i = 1, 2, 3). \tag{13}$$

However, the corresponding presumed noncontextual possessed values of the observables that these operators represent must satisfy, from equation 11,

$$\sigma_i = s_i r_k r_l \qquad (i \neq k \neq l \neq i,\ i = 1, 2, 3), \tag{14a}$$

$$\sigma_0 = s_1 s_2 s_3 \tag{14b}$$

and, from equation 13,

$$s_i^2 = r_i^2 = 1 \qquad (i = 1, 2, 3). \tag{15}$$

Consequently, we have

$$\sigma_0\sigma_1\sigma_2\sigma_3 = 1, \tag{16}$$

in conflict with equation 12.

If, following GHZ and Mermin, we think that the assumptions of noncontextual possessed values are rendered more secure when made only for observables with widely separated spatial associations, then we can make the argument more secure by eliminating the assumption for the composite observables represented by the Σ's. This is done by applying the argument to specific states that give the Σ's their values by being simultaneous eigenstates of them. (Note that the particles can be guaranteed to be far from each other in space by having their y-coordinates differ greatly; that is, the three-particle position-representation state function can have its support limited to three widely separated intervals of the y-coordinates of the three particles, respectively. Both the S's and the R's commute with the y-coordinates.) There are many different choices for the simultaneous eigenfunctions. All of them will lead to contradiction with the presumed noncontextual possessed values of the spatially separated observables, the S's and the R's. Within the subspace in which the S's are all well defined, a state function is a simultaneous eigenstate of the Σ_i $(i = 1, 2, 3)$ with eigenvalues $\sigma_i = 1$ iff it is of the form,

$$\begin{aligned}\Psi(x_1, x_2, x_3) = {} & \theta(-x_1)\theta(-x_2)\theta(-x_3)\Phi(|x_1|, y_1, z_1; |x_2|, y_2, z_2; |x_3|, y_3, z_3) \\ & - \theta(-x_1)\theta(x_2)\theta(x_3)\Phi(|x_1|, y_1, z_1; |x_2|, y_2, -z_2; |x_3|, y_3, -z_3) \\ & - \theta(x_1)\theta(-x_2)\theta(x_3)\Phi(|x_1|, y_1, -z_1; |x_2|, y_2, z_2; |x_3|, y_3, -z_3) \\ & - \theta(x_1)\theta(x_2)\theta(-x_3)\Phi(|x_1|, y_1, -z_1; |x_2|, y_2, -z_2; |x_3|, y_3, z_3),\end{aligned} \tag{17}$$

where θ is the standard unit step function with positive support and Φ is an arbitrary function except for the requirement that it vanish when $x_1x_2x_3 = 0$. For these state functions, which span an infinite-dimensional subspace, the possibilities are that all three particles will be found in the negative-x half-space or any one of them, but just one. Noncontextual possessed values for the S's and the R's allow either no particles in the negative-x half-space or any two of them, but just two—the strongest possible contradiction.

A PHASE-SPACE GHZ ARGUMENT FOR A SINGLE PARTICLE

This time around, I will not belabor the details so much. The role of the A operators is played by the algebraic signs of the three Cartesian coordinates of the one-particle's position:

$$S_i \equiv X_i/|X_i| \qquad (i = 1, 2, 3). \tag{18}$$

Unlike the previous case, we cannot use rotations through π radians to play the role of the B's here because such rotations reverse both coordinates orthogonal to the axis of rotation and not just one, as we require. Instead, the role of the B's is played by the operators for reflection, R_i ($i = 1, 2, 3$), through the three planes defined separately by the equations,

$$x_i = 0 \qquad (i = 1, 2, 3). \tag{19}$$

As before, the role of the C's is played by the products,

$$\Sigma_i \equiv S_i R_k R_l \qquad (i \neq k \neq l \neq i,\ i = 1, 2, 3) \tag{20a}$$

and

$$\Sigma_0 \equiv S_1 S_2 S_3. \tag{20b}$$

Finally, as before, the required assumptions of noncontextual possessed values can be minimized by considering these operators in the presence of the simultaneous eigenstates of the Σ's with $\sigma_i = 1$ for $i = 1, 2, 3$. A state function for the particle is such an eigenfunction iff it has the form,

$$\Psi(x_1, x_2, x_3) = \{\theta(-x_1)\theta(-x_2)\theta(-x_3) - \theta(-x_1)\theta(x_2)\theta(x_3) \\ - \theta(x_1)\theta(-x_2)\theta(x_3) - \theta(x_1)\theta(x_2)\theta(-x_3)\}\Phi(|x_1|, |x_2|, |x_3|), \tag{21}$$

where, to guarantee that all the S_i are well defined on Ψ, we require that Φ vanish when $x_1x_2x_3 = 0$. Otherwise, Φ need only be square-integrable.

In this state, quantum theory restricts the particle's position to the four octants in which either all three Cartesian coordinates are negative or one and only one of them is. In contradiction, the assumption of noncontextual possessed values for the signs of the coordinates and the planar reflection observables restricts the position of the particle to exactly the remaining octants.

THE STATUS OF THE REFLECTION OPERATORS

Among all the observables of elementary quantum theory, perhaps none seem more likely a candidate for noncontextual possessed values than the positions of the particles of the system. However, what of these reflection operators? Although it would go against the grain of orthodox quantum theory, one might argue that the reflection operators do not represent any property of the particle at all, but merely an operation on the state function for the particle. Because a champion of noncontextual possessed values is going against the grain of orthodox quantum theory anyhow, she might very well choose to avoid the contradiction by refusing to recognize the reflection operators as representing a particle property in need of a possessed value at all. Well, the position and momentum operators do represent properties of the particle and, for a spinless particle, as we are considering, the reflection operators can be expressed as functions of the position and momentum operators. How "simple" are those functions? Can they force a particle-property interpretation of reflection on a die-hard champion of noncontextual possessed values for such properties?

For a particle in a one-dimensional position space, the reflection operator is given by

$$\begin{aligned} R \equiv \int dx\,|-x\rangle\langle x| &= \int dx \exp[(i/\hbar)Px]|0\rangle\langle 0| \exp[(i/\hbar)Px] \\ &= \int dx \exp[(i/\hbar)Px]\,\delta(X) \exp[(i/\hbar)Px] \\ &= (1/2\pi\hbar) \int dx\,dp \exp[(i/\hbar)Px] \exp[(i/\hbar)Xp] \exp[(i/\hbar)Px] \\ &= (1/4\pi\hbar) \int dx\,dp \exp[(i/\hbar)(Px + Xp)] \\ &= (1/4\pi\hbar) \int dx\,dp \cos[(1/\hbar)(Px + Xp)], \end{aligned} \tag{22}$$

where the next to last line follows from the Baker-Hausdorf theorem and a change of integration variables. For our R_i in three dimensions, we need only put the subscript i on the X and the P operators. The assessment of the status of these operators, vis-à-vis the assignment of noncontextual possessed values, is left to the reader.

A WORD ON CONTEXTUAL POSSESSED VALUES

At the risk of belaboring the obvious, I will make a few comments on the nature of contextual possessed value assignments that can avoid the contradictions derived here and in previous GHZ arguments. The perhaps obvious point is that it is not enough that the possessed value assignments to the relevant observables merely depend upon the experimental environment (pace Bell) of the system. They must depend upon the environment in such a way as to violate some or all of the product relations between possessed values that have been employed in the argument. For example, the environment may be such as to not perform any measurement at all, but if the possessed values, for that environment, are assigned so that equations 14 and 15 are satisfied in the three-particle argument, or the analogues of equations 14 and 15 are satisfied in the one-particle argument, then the contradiction still holds. The same is true for any other environment that one might consider. For an environment suitable for measuring the relevant position variables, equation 14b or its one-particle analogue must hold, as well as equation 15, if all possessed values must be possible eigenvalues. Consequently, all three or any one of the possibilities of equation 14a must be violated, but not just two of them. Similar considerations hold for the other possible measuring environments.

ACKNOWLEDGMENTS

I wish to thank Abner Shimony for pointing out an oversight and Robert Clifton for pointing out an algebraic error in an earlier version of this paper.

REFERENCES

1. EINSTEIN, A., B. PODOLSKY & N. ROSEN. 1935. Phys. Rev. **47:** 777.
2. BOHM, D. 1951. Quantum Theory. Prentice–Hall. Englewood Cliffs, New Jersey.

3. BELL, J. 1987. Speakable and Unspeakable in Quantum Mechanics. Cambridge University Press. London/New York.
4. MERMIN, D. 1990. Phys. Today **43**(6): 9.
5. GREENBERGER, D., M. HORNE & A. ZEILINGER. 1989. *In* Bell's Theorem, Quantum Theory, and Conceptions of the Universe. M. Kafatos, Ed.: 73–76. Kluwer. Dordrecht.
6. GREENBERGER, D., M. HORNE, A. SHIMONY & A. ZEILINGER. 1990. Am. J. Phys. **58:** 1131–1143.
7. CLIFTON, R., M. REDHEAD & J. BUTTERFIELD. 1991. Found. Phys. **21:** 149.
8. MERMIN, D. 1993. Rev. Mod. Phys. **64:** 803–815.
9. FINE, A. 1989. *In* Philosophical Consequences of Quantum Theory. J. Cushing & E. McMullin, Eds.: 175–194. University of Notre Dame Press. Notre Dame, Indiana.
10. WESSELS, L. 1989. *In* Philosophical Consequences of Quantum Theory. J. Cushing & E. McMullin, Eds.: 80–96. University of Notre Dame Press. Notre Dame, Indiana.

The Entanglement of Virtual Photons[a]

J. D. FRANSON

Applied Physics Laboratory
The Johns Hopkins University
Laurel, Maryland 20723

Real photons exhibit a number of interesting nonlocal effects. This report will instead consider some of the nonlocal properties of virtual photons. They play an important role in many systems and, in particular, are responsible for the macroscopic electric and magnetic fields of bar magnets, solenoids, transformers, etc. One might ask whether or not there might be a Bell inequality associated with bar magnets, for example.

Violations of Bell's inequality involving virtual photons do exist but will not be discussed here. Instead, a new phase associated with the electromagnetic field that is dynamic in the sense that it vanishes in the quasi-static limit of slowly varying currents will be described. This dynamic phase is observable only through the entanglement of virtual photons with an electron, in which case it can produce a fractional Aharonov-Bohm effect.[1]

Some of the results described here have been submitted for publication elsewhere.[2] Rather than repeat those derivations, this report will review the earlier results while adding some intuitive comments and figures to further illustrate their meaning. In addition, the geometric or topological[3,4] nature of the dynamic phase will be briefly discussed.

Consider a known current distribution $\vec{j}(\vec{r}, t)$ and suppose that it can propagate the field from an initial state to the same final state along two or more different paths in some parameter space. The final state is assumed to be the same aside from a path-dependent phase factor that will be written in the form

$$|\psi_P\rangle = e^{i\phi_D} e^{-i\int\langle H\rangle dt/\hbar} |\psi_f\rangle. \tag{1}$$

Here the energy-dependent phase shift defined by

$$\phi_E \equiv -\int \langle H\rangle \, dt/\hbar \tag{2}$$

is the same for all paths if energy is conserved and is of no further interest. Any remaining phase shift is contained in ϕ_D, which will be shown to vanish in the quasi-static limit and will therefore be referred to as the dynamic phase. Mathematically, ϕ_D is a geometric phase, but there are some physical difficulties in that interpretation, as will be discussed below. It should also be noted that ϕ_E is often referred to as "dynamic" by other authors.[3]

In the Coulomb gauge the Hamiltonian is given by

$$H = \frac{\epsilon_0}{2} \int [\vec{E}_\perp^2(\vec{r}) + c^2 \vec{B}^2(\vec{r})] \, d^3\vec{r} - \int \vec{j}(\vec{r}, t) \cdot \vec{A}(\vec{r}) \, d^3\vec{r} \equiv H_0 + H' \tag{3}$$

[a]This work was supported by the Office of Naval Research.

or

$$H = \sum_i \left[\left(a_i^\dagger a_i + \frac{1}{2} \right) \hbar\omega_i - \left(\frac{\hbar}{2\epsilon_0 L^3 \omega_i} \right)^{1/2} (a_i^\dagger j_i + a_i j_i^*) \right], \tag{4}$$

where j_i is the Fourier transform of the current distribution. One might expect the field to be essentially classical since the current distribution is classical, which suggests that the state of the field may be a coherent state,[5] aside from an overall phase factor:

$$|\psi(t)\rangle = e^{i\phi(t)} |\alpha(t)\rangle. \tag{5}$$

Here the multimode coherent state is defined as usual by

$$|\alpha(t)\rangle = \prod_i e^{-\alpha_i^*(t)\alpha_i(t)/2} e^{\alpha_i(t) a_i^\dagger} |0\rangle. \tag{6}$$

Inserting this state vector into Schrödinger's equation gives the following requirements:

$$\dot{\alpha}_i + i\omega_i \alpha_i = \frac{i}{\sqrt{2\epsilon_0 L^3 \hbar\omega_i}} j_i(t), \tag{7}$$

$$\frac{d\phi}{dt} = \sum_i \left[-\frac{\omega_i}{2} + \frac{\alpha_i j_i^*}{\sqrt{2\epsilon_0 L^3 \hbar\omega_i}} - \frac{i}{2} \frac{d}{dt} (\alpha_i^* \alpha_i) \right]. \tag{8}$$

Equation 7 is identical in form to the corresponding classical equation for the normal-mode expansion of the field, while equation 8 governs the nonclassical phase factor. Neglecting the zero-point energy (which is the same for any path) and taking the inverse Fourier transform allows this result to be rewritten as

$$\frac{d\phi}{dt} = \frac{1}{2} \frac{\int \vec{j}(\vec{r}, t) \cdot \langle \vec{A}(\vec{r}) \rangle \, d^3\vec{r}}{\hbar} = -\frac{1}{2} \frac{\langle H' \rangle}{\hbar}. \tag{9}$$

Equation 5 is an exact solution to Schrödinger's equation regardless of how fast the current may vary and despite the appearance of the expectation value in equation 9.

Glauber[6] previously derived a formula for $\phi(t)$ in the interaction picture that involves the commutation relations for the field operators, which are quite complicated in the Coulomb gauge. $\phi(t)$ is a different function in the Schrödinger picture used here and can be seen to be proportional to the expectation value of the interaction Hamiltonian. Glauber also noted that ϕ has no effect on the density matrix of the field, which suggests that it is unobservable—this illustrates the need for the entanglement of the virtual photons with some other particle.

The Aharonov-Bohm (AB) effect[1] depends on the line integral of the vector potential and is the same as the expression in equation 9, aside from the rather surprising factor of 1/2, which will play an important role in the AB effect for quantized fields. The origin of this factor of 1/2 can be understood by considering a single mode of the field, for which the vector potential operator is proportional to the displacement x of a simple harmonic oscillator. First consider a classical oscillator with potential $U_0 = x^2$ to which an external force corresponding to $U'(x) = j_i x$ is

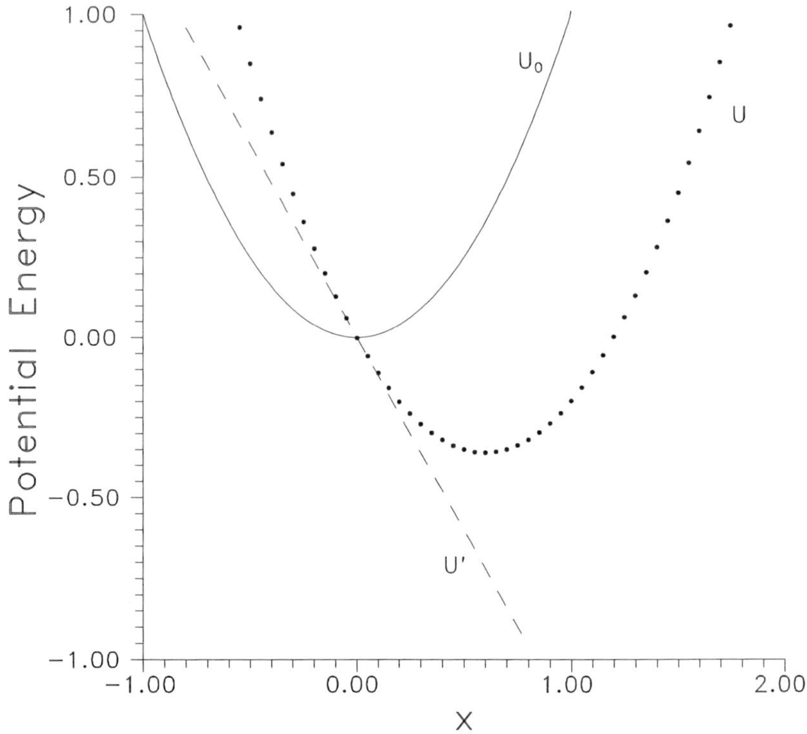

FIGURE 1. Change in the energy of a classical harmonic oscillator subjected to a slowly varying external force, which illustrates the origin of the factor of ½.

slowly applied. The total energy changes by $-U'(\delta x)/2$, where δx is the displacement induced by the external force, as illustrated in FIGURE 1. The total energy change is only half of the externally applied potential due to the increase in the internal energy of the harmonic oscillator.

The same results are obtained quantum mechanically in the quasi-static limit of a slowly varying current, in which case $\dot{\alpha}_i$ of equation 7 becomes negligible compared to the current and equation 7 reduces to

$$\alpha_i = \frac{1}{\sqrt{2\epsilon_0 L^3 \hbar \omega_i^3}} j_i(t). \tag{10}$$

Substitution of this result into the Hamiltonian (and once again neglecting the zero-point energy) gives

$$\langle H \rangle = \langle H' \rangle / 2,$$
$$\phi = -\int \langle H \rangle \, dt/\hbar = \phi_E. \tag{11}$$

The expectation value of the energy is half that of the interaction energy, just as in the classical case, and the total phase shift reduces to the energy-dependent phase.

This demonstrates that the dynamic phase ϕ_D vanishes in the quasi-static limit. (The quasi-static limit should not be confused with the adiabatic limit, as will be discussed shortly.)

Now consider the opposite limit in which an impulsive current is applied over a time interval δt that is much smaller than the period of the relevant modes of the field. Here the $\alpha_i \delta t$ term is negligible compared to the current and equation 7 reduces to

$$\delta\alpha_i = \frac{i}{\sqrt{2\epsilon_0 L^3 \hbar\omega_i}} j_i \delta t. \quad \textbf{(12)}$$

If we further suppose that mode k had previously been excited to a large value of α_k, then to lowest order in δt

$$\delta\phi_D = -\frac{1}{2}\frac{\int \vec{j}_{\text{imp}}(\vec{r}) \cdot \langle \vec{A}(\vec{r}, t_0)\rangle \, d^3\vec{r}}{\hbar}\delta t \quad \textbf{(13)}$$

because $\langle H_0 \rangle \delta t$ is negligible compared to the interaction Hamiltonian. It can be seen that the effect of an impulsive current is to generate a phase shift reminiscent of the AB effect, aside, once again, from the factor of 1/2.

It is instructive to consider the generation of a dynamic phase in a different manner, starting instead from the vacuum state. Suppose that an impulsive current coupled to mode k is applied at some initial time in such a way as to move the value of α_k along the real axis, as shown in FIGURE 2. The system will subsequently precess around a circle in the complex-α plane until a second impulsive current is used to

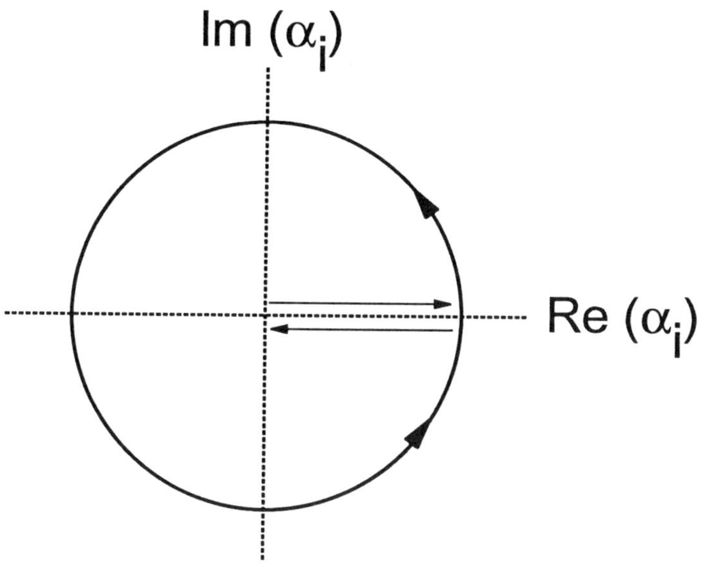

FIGURE 2. Generation of a dynamic phase starting from the vacuum state.

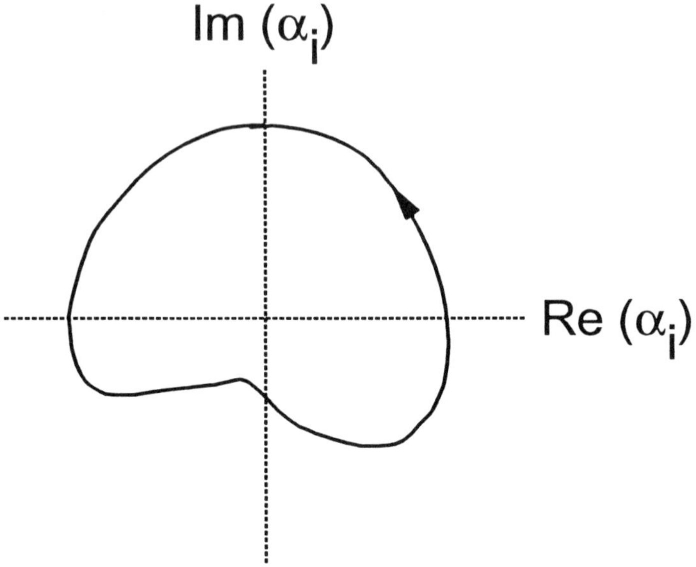

FIGURE 3. Geometric properties of ϕ_D, which is equal to minus twice the area enclosed in the complex-α plane.

restore the field to the vacuum state. The energy-dependent phase will have precessed by an amount equal to $-\langle n_k \rangle \omega_k$ during the time interval between the impulses, where n_k is the number of photons created by the impulse. But the total phase would not have changed during this time because the current was zero, so we must have that $\phi_D = -\phi_E$. Intuitively, those excited states with large values of n did precess through a large phase angle, but the system essentially loses all "memory" of that when the amplitude of those modes is reduced to zero at the end.

ϕ_D is a geometric phase as defined by Aharonov and Anandan,[4] at least mathematically. In their classic paper, they showed that some Hamiltonian can always be found to propagate a system at any desired rate along a path in a projective Hilbert space and that the geometric phase is a function only of this path and not the rate of propagation. However, several difficulties arise if these ideas are applied to the dynamic phase. The Hamiltonian required to propagate an electromagnetic wave adiabatically does exist, but it is nonphysical in nature. (Imaginary currents are required, for example.) In addition, the electric field (conjugate momentum) depends on the rate of change of the vector potential and thus depends on the rate of propagation in the projective Hilbert space, which suggests that the physical state of the system is not the same for different rates of propagation. In fact, the electric field of a radiative state vanishes in the adiabatic limit.

Aside from these issues of interpretation, ϕ_D does possess the mathematical properties of a geometric phase. As illustrated in FIGURE 3, the dynamic phase is equal to minus twice the area enclosed by any closed path in the complex-α plane. Its value can indeed be shown to be independent of the rate of traversal of the path and it can be computed using Berry's formulas[3] in the adiabatic limit.

The geometric phase of a generalized simple harmonic oscillator has been considered by a number of authors,[7] where the Hamiltonian was taken to be

$$\frac{1}{2}[Xq^2 + Y(qp + pq) + Zp^2]. \quad \textbf{(14)}$$

This Hamiltonian does not contain any terms linear in q or p. In contrast, the Hamiltonian considered here can be rewritten (for a single mode) as

$$H = c_1(q - q_0)^2 + c_2(p - p_0)^2 + \text{constant}, \quad \textbf{(15)}$$

which does contain terms linear in both q and p. In particular, the offset p_0 in the momentum is essential in order to obtain the geometric properties illustrated in FIGURE 3. It is interesting to note that the "generalized" simple harmonic oscillator is not sufficiently general to describe the electromagnetic field. (I also recently learned that the geometric properties associated with the Hamiltonian of equation 15 have been described elsewhere.[8])

As mentioned earlier, the adiabatic limit and the quasi-static limit correspond to two different limiting procedures. The quasi-static limit as used here consists of restricting ourselves to the Hamiltonian and currents that actually occur, considering some finite time interval, and then taking the limit of slowly varying currents. The adiabatic limit consists of choosing whatever Hamiltonian is required to propagate the system along a chosen path given an indefinitely large time interval. Both limits are mathematically valid and the only question is which limit is more appropriate to any particular physical situation. Actually, the dynamic phase can be shown to vanish in both limits if the cancellation between modes corresponding to wave vectors $\vec{k}$ and $-\vec{k}$ is taken into account.

An overall phase factor like ϕ_D can only be observed in an interference experiment such as the electron interferometer illustrated in FIGURE 4, which encloses a static current source S. An electron will become entangled with the virtual photons from S because a different dynamic phase will be induced in the field depending on which path the electron takes through the interferometer. The dynamic phase associated with the state of the field is in addition to the usual AB phase shift associated with the state of the electron.

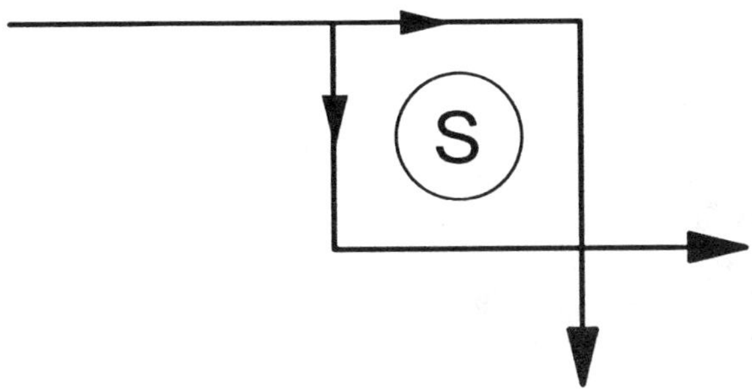

FIGURE 4. Observability of the dynamic phase in an electron interferometer enclosing a static current distribution S.

Consider the case in which the electron is described by a highly localized wave packet, in which case the uncertainty in the position of the electron (in a given path) is negligible and its current is essentially known. The self-interaction terms can be neglected because they are the same for both paths in a symmetrical interferometer. Under these conditions, an approximate solution to the Schrödinger equation can be shown to be

$$|\Psi\rangle = |\psi_{e1}(\vec{r}_e, t)\rangle|e^{i\phi_1(t)}\alpha_1(t)\rangle + |\psi_{e2}(\vec{r}_e, t)\rangle|e^{i\phi_2(t)}\alpha_2(t)\rangle, \tag{16}$$

which clearly shows the entangled nature of the system. The approximation consists of neglecting the A^2 terms, which is valid for a weak field, as well as

$$(\vec{A} - \langle\vec{A}\rangle)\cdot(\vec{j} - \langle\vec{j}\rangle), \tag{17}$$

which corresponds to the correlation between the fluctuations in the field and the electron current. Both of these terms are negligible for a localized wave packet when ignoring any self-field effects.

Subject to these approximations, Schrödinger's equation is satisfied with

$$\frac{d\phi}{dt} = -\frac{1}{2}\frac{\int\langle\vec{j}_e(\vec{r}_e, t)\rangle\cdot\langle\vec{A}_S(\vec{r}_e)\rangle\, d^3\vec{r}_e}{\hbar} + \frac{1}{2}\frac{\int\vec{j}_S(\vec{r}_S)\cdot\langle\vec{A}_e(\vec{r}_S, t)\rangle\, d^3\vec{r}_S}{\hbar} \tag{18}$$

for each path. The nonretarded contributions to the field are negligible in the limit of a distant source, even in the Coulomb gauge, in which case

$$\langle\vec{A}_e(\vec{r}_S, t)\rangle = \frac{1}{4\pi\epsilon_0 c^2}\int\frac{\langle\tilde{j}_e(\vec{r}_e, t_{\rm ret})\rangle}{|\vec{r}_e - \vec{r}_S|}\, d^3\vec{r}_e, \tag{19}$$

where $\tilde{j}$ is the component of the current perpendicular to $\vec{r}_e - \vec{r}_S$.

First consider the quasi-static limit in which the currents are slowly varying on the time scale of the transit time across the apparatus at the speed of light so that

$$t_{\rm ret} = t. \tag{20}$$

It can then be shown that the two integrals in equation 20 are equal and

$$\frac{d\phi}{dt} = -\frac{1}{2}\frac{\int\langle\vec{j}_e(\vec{r}_e, t)\rangle\cdot\langle\vec{A}_S(\vec{r}_e)\rangle\, d^3\vec{r}_e}{\hbar} + \frac{1}{2}\frac{\int\vec{j}_S(\vec{r}_S)\cdot\langle\vec{A}_e(\vec{r}_S, t)\rangle\, d^3\vec{r}_S}{\hbar} = 0. \tag{21}$$

Thus, the dynamic phase is zero and the usual AB effect is obtained in the quasi-static limit.

Retardation effects are significant, however, for an electron interferometer located such a large distance from S that the wave packet will pass through the interferometer in less time than it would take for light to propagate from the interferometer to S, as illustrated in FIGURE 5. Then, the field A_e at S must be evaluated at retarded times for which the electron would not yet have reached the first beam-splitter. A_e and the second term of equation 21 are thus the same for both paths and can be neglected, in which case equation 18 reduces to

$$\frac{d\phi}{dt} = -\frac{1}{2}\frac{\int \langle \vec{j}_e(\vec{r}_e, t)\rangle \cdot \langle \vec{A}_S(\vec{r}_e)\rangle \, d^3\vec{r}_e}{\hbar}. \tag{22}$$

When combined with the usual AB phase from the electron's wave function, this gives a total geometric phase of

$$\phi_{AB} = \frac{1}{2}\frac{q}{\hbar}\oint \langle \vec{A}_S\rangle \cdot d\vec{l} = \frac{1}{2}\left(2\pi\frac{\Phi}{\Phi_0}\right), \tag{23}$$

which is half the usual AB effect.

That is not the end of the story, however, because we have neglected the fact that the electron current will generate a slightly different state of the field, $|\alpha_1\rangle$ or $|\alpha_2\rangle$,

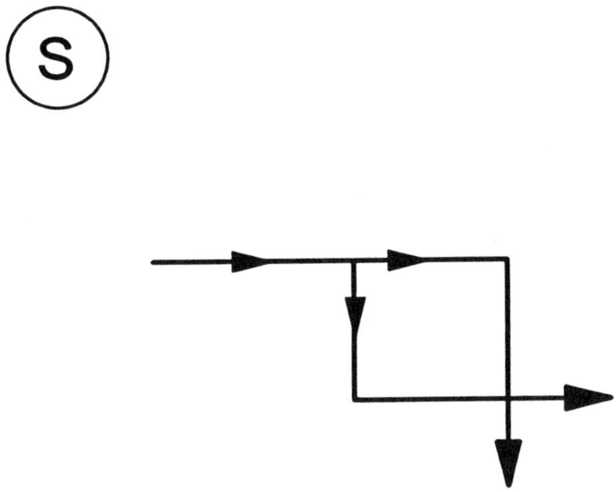

FIGURE 5. An electron interferometer where the retarded nature of the potentials has a major effect on the dynamic phase.

depending on which path it took. The difference in the phases of the parameters α_1 and α_2 is not a Berry phase, but is related instead to the classical phase of the field. The total phase must be determined from the inner product,

$$\langle e^{i\phi_2}\alpha_2 | e^{i\phi_1}\alpha_1\rangle = e^{i\gamma_R}e^{-\gamma_I}, \tag{24}$$

where

$$\dot{\gamma}_R = \frac{1}{2}\frac{\int (\vec{j}_1 - \vec{j}_2)\cdot(\langle\vec{A}_1\rangle + \langle\vec{A}_2\rangle)\, d^3\vec{r}}{\hbar},$$

$$\dot{\gamma}_I = \frac{\mathrm{Im}\sum_i [(\alpha_{1i} - \alpha_{2i})(j^*_{1i} - j^*_{2i})]}{\sqrt{2\epsilon_0 L^3 \hbar\omega_i}}. \tag{25}$$

The diagonal terms involving $\vec{j}_1 \cdot \vec{A}_1$ and $\vec{j}_2 \cdot \vec{A}_2$ correspond to the geometric phase, whereas the cross terms involving $\vec{j}_1 \cdot \vec{A}_2$, etc., are due to the difference in the parameters α_1 and α_2. This restores the usual AB effect but only half of the observed phase is "geometric" in the dynamic limit.

The fractional phase shift can be directly observed in an experiment of the type shown in FIGURE 6. Here there is no static, external field and, instead, one path of an electron interferometer passes through a cavity where an electromagnetic field is generated by the electron. This induced field in the cavity subsequently interacts with the electron to produce a dynamic phase shift as well as the usual AB phase. Now $\langle\vec{A}_2\rangle$ is zero in equation 25 and the total phase shift is purely geometric and retains the factor of 1/2.

Most inherently quantum mechanical effects in quantum optics are due to energy conservation or the entanglement of two photons. However, such effects may be consistent with a theory by Barut and his colleagues,[9] in which the field variables have been eliminated from the equations of motion in favor of the particle coordinates at

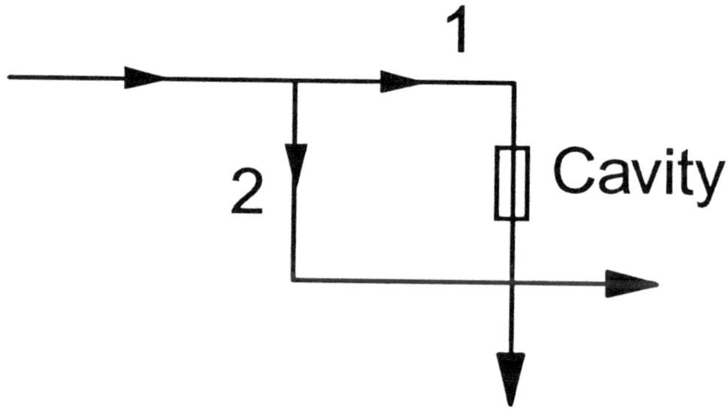

FIGURE 6. An experiment suitable for direct observation of the fractional phase shift.

earlier times. What is normally viewed as the entanglement of two photons, for example, would then be a consequence of the previous entanglement of two atomic states. The dynamic phase, on the other hand, is a result of the quantum mechanical evolution of the internal degrees of freedom of the field and would appear to be missing in an alternative theory where those degrees of freedom have been eliminated. In that sense, one could say that the dynamic phase is more dependent on the quantization of the field than are other effects such as EPR correlations. In any event, the dynamic phase allows an experimental investigation of new and fundamental properties of the field.

In summary, the electromagnetic field exhibits a variety of phase phenomena, including the classical phase (and its quantum mechanical generalizations[10]), various Berry phases[11] (actually, Hannay angles), and the dynamic phase described here. The dynamic phase vanishes in the quasi-static limit and is only observable if the virtual photons become entangled with an electron or some other particle. The dynamic phase allows an experimental probe of the quantum mechanical evolution of the harmonic oscillators comprising the field.

REFERENCES

1. AHARONOV, Y. & D. BOHM. 1959. Phys. Rev. **115:** 485.
2. FRANSON, J. D. 1994. Phys. Rev. Lett. Submitted.
3. BERRY, M. V. 1984. Proc. R. Soc. London Ser. A **392:** 45.
4. AHARONOV, Y. & J. ANANDAN. 1987. Phys. Rev. Lett. **58:** 1593.
5. GLAUBER, R. J. 1963. Phys. Rev. **131:** 2766.
6. GLAUBER, R. J. 1965. *In* Quantum Optics and Electronics: 1964 Les Houches Lectures. C. DeWitt, A. Blandin & C. Cohen-Tannoudji, Eds. Gordon & Breach. New York.
7. BERRY, M. V. 1985. J. Phys. A **18:** 15–27; HANNAY, J. H. 1985. J. Phys. A **18:** 221–230; KOBE, D. H. 1990. J. Phys. A **23:** 4249–4268; BOSE, S. K. & B. DUTTA-ROY. 1991. Phys. Rev. A **43:** 3217–3220.
8. CHATURVEDI, S., M. S. SRIRAM & V. SRINIVASAN. 1987. J. Phys. A **20:** 1071–1075; SIMON, R. & N. KUMAR. 1988. J. Phys. A **21:** 1725–1727.
9. BARUT, A. O. & J. P. DOWLING. 1987. Phys. Rev. A **36:** 649; BARUT, A. O., J. KRAUS, Y. SALAMIN & N. UNAL. 1992. Phys. Rev. A **45:** 7740–7745.
10. SUSSKIND, L. & J. GLOGOWER. 1964. Physics (N.Y.) **1:** 49; PEGG, D. T. & S. M. BARNETT. 1988. Europhys. Lett. **6:** 483; NOH, J. W., A. FOUGERES & L. MANDEL. 1991. Phys. Rev. Lett. **67:** 1426; 1992. Phys. Rev. A **45:** 424; 1992. Phys. Rev. A **46:** 2840; FRANSON, J. D. 1994. Phys. Rev. A **49:** 3221–3228.
11. PANCHARATNAM, S. 1956. Proc. Indian Acad. Sci. Sect. A **44:** 247; CHIAO, R. Y. & Y. S. WU. 1986. Phys. Rev. Lett. **57:** 933; TOMITA, A. & R. Y. CHIAO. 1986. Phys. Rev. Lett. **57:** 937; CHIAO, R. Y. & T. F. JORDAN. 1988. Phys. Lett. A **132:** 77–81.

Multipath Interferometry of the Biphoton[a]

MICHAEL HORNE[b] AND ABNER SHIMONY[c]

[b]*Department of Physics*
Stonehill College
North Easton, Massachusetts 02357

[c]*Departments of Physics and Philosophy*
Boston University
Boston, Massachusetts 02215

INTRODUCTION

The purpose of this report is to suggest some extensions of two-particle interferometry. Generically, two-particle interferometry considers ensembles of particle pairs, with each pair produced in the same quantum mechanical entangled state and propagating in several different two-particle paths to a pair of detectors. Probabilities of joint detection are calculated after adding the amplitudes of all the contributing two-particle paths. The phases along different paths can be manipulated experimentally and one can observe "two-particle fringes", which essentially are the variation of joint detection probabilities.

Both theoretically and experimentally, the generic field of two-particle interferometry has been quite constrained. The particle pair is usually (indeed, in experiments so far, always) taken to be a pair of photons, aptly called a biphoton by Klyshko.[1] Usually, there are two paths for each photon, which gives a maximum of four two-particle paths to the pair of detectors, but correlations due to the entangled state typically reduce the number of two-particle paths to two, which implies sinusoidal fringes. Typically, the detections are also coincident counts, occurring almost at the same time, within a time window whose duration is set by the experimenter.

There have already been some interesting theoretical investigations that have relaxed some of the foregoing constraints, for example, by using a new device called the multiport,[2] which generalizes the ordinary beam-splitter by yielding three or more output paths. We shall show that further unusual phenomena will result from interferometric arrangements in which each particle can take more than two paths of unequal length and the times of joint detection are allowed to be well separated as well as nearly simultaneous. Two arrangements, both using biphotons, will be analyzed in detail: (i) one in which each photon propagates in three paths, giving nine biphoton paths, and (ii) a double Fabry-Perot interferometer, in which each photon propagates in infinitely many paths.

The second section will briefly state the results of a theoretical analysis of the

[a]This research was supported by the National Science Foundation (Grant Nos. PHY92-13964 and PHY93-21992).

propagation of a biphoton in a simple basic arrangement. The third section will apply these basic results to the double three-path arrangement (i), whereas the last section will apply them to the double Fabry-Perot arrangement (ii).

We note that an entirely different direction of generalization is to go from two-particle to n-particle interferometry ($n > 2$), but this is outside the scope of this report.

THE BIPHOTON IN A SIMPLE ARRANGEMENT

Consider a source S that emits two beams of radiation, one along axis x_1 and the other along x_2, as shown in FIGURE 1.[3] More specifically, let the radiation consist of

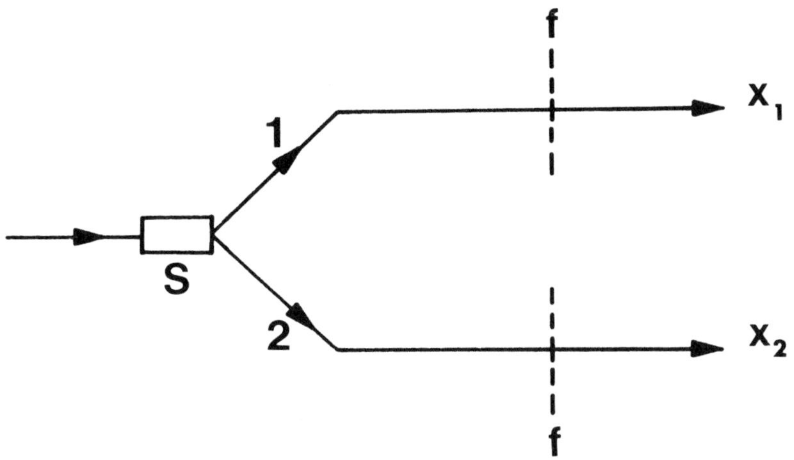

FIGURE 1. Schematic representation of a standard biphoton preparation. A photon of wave number $2k_0$ illuminates the source crystal S, generating by downconversion a biphoton, which propagates in beams 1 and 2 with coordinates x_1 and x_2. Gaussian wave number filters, labeled f, each with center k_0 and width σ, prepare the biphoton in the state given by equations 1a and 1b.

pairs of photons such that, if one is in beam 1, the other is in beam 2. One way of achieving this direction correlation is via momentum conservation, if the pair is produced by the decay of a parent particle having negligible momentum in the vertical direction of FIGURE 1. In addition, we assume for simplicity that the parent particle had a specific energy and hence a specific wave number, which will be taken to be $2k_0$ if the parent is massless (notably, a photon, in the parametric downconversion process). Finally, we imagine that each beam is filtered by a Gaussian wave number filter, f, with center k_0 and width σ. Then, the wave function of the photon pair, neglecting spin, is

$$\phi(k_1, k_2, t_1, t_2) = \delta(k_1 + k_2 - 2k_0)e^{-i(ck_1t_1+ck_2t_2)}e^{-[(k_1-k_0)^2+(k_2-k_0)^2]/2\sigma}. \tag{1a}$$

The delta function ensures energy conservation and the Gaussian factors describe the transmitted amplitude beyond the filters. Double Fourier transforming yields the equivalent configuration space wave function:

$$\psi(x_1, x_2, t_1, t_2) = e^{ik_0(x_1-ct_1)}e^{ik_0(x_2-ct_2)}e^{-\sigma^2[(x_2-x_1)-c(t_2-t_1)]^2/4}. \tag{1b}$$

It is somewhat surprising that equation 1b has monochromatic one-photon phase factors, $\exp ik_0(x_j - ct_j)$ $(j = 1, 2)$, whereas the entanglement of the two photons and the actual spread of the wave numbers of each, indicated explicitly in equation 1a, are buried in the real Gaussian factor,

$$g(x_1, x_2, t_1, t_2) = e^{-\sigma^2[(x_2-x_1)-c(t_2-t_1)]^2/4}, \tag{2}$$

but Fourier analysis is rich in surprises.

The absolute square of ψ is the (unnormalized) probability density of finding the two photons at x_1, x_2 at specified times t_1, t_2. The single-photon probability density,

$$p_1(x_1, t_1) = \int |\psi(x_1, x_2, t_1, t_2)|^2\, dx_2, \tag{3}$$

is obviously constant for all x_1 and t_1, and likewise for $p_2(x_2, t_2)$. The lack of localization of each photon singly is consistent with our assumption that the parent particle is monochromatic—hence, completely unlocalized. However, each photon is localized relative to the other. If t_1 and t_2 are specified, then the probability density of $x_2 - x_1$ is Gaussian, with a center, $c(t_2 - t_1)$, and a standard deviation, $1/\sigma$. Alternatively, each x_j has a Gaussian distribution when t_1, t_2, and the other position coordinate are specified. One can also say that, when x_1 and x_2 are fixed, there is a Gaussian distribution for the time difference, $\tau \equiv t_2 - t_1$, with a center at $(x_2 - x_1)/c$ and a standard deviation of $1/(c\sigma)$.[4]

We shall now consider an idealized experimental arrangement in which perfectly efficient detectors are placed in beams 1 and 2 of FIGURE 1, equidistant from the source. The origins of the two coordinates x_1 and x_2 may both be placed at the respective detectors. Equation 1b, evaluated at $x_1 = x_2 = 0$, determines the probability amplitude of joint detection at times t_1 and t_2, respectively:

$$\psi(0, 0, t_1, t_2) = e^{ik_0(t_1+t_2)}e^{-(1/4)(c\sigma)^2(t_2-t_1)^2}. \tag{4}$$

Imagine that the detections are registered by dots in the left and right halves of a long tape on which t_1 and t_2 are respectively represented by the vertical distance upward from a common origin. If we assume that the mean temporal interval between a biphoton and the next emitted from S is very large compared to $1/c\sigma$, then the pattern of dots is determined by equation 4. The dots will occur in nearly simultaneous pairs, where the meaning of "nearly simultaneous" is implicit in the Gaussian factor of equation 4. If there is a dot in one half of the tape, then with virtual certainty there will be a nearly simultaneous dot in the other half. Of course, equation 4 contains more detailed information about the temporal correlation of the members of a biphoton than "nearly simultaneous" and this information would be experimentally interesting if coincidence time windows on the order of $1/c\sigma$ or less were employed,[3] as is not the case in the experimental arrangements of the following sections.

A DOUBLE THREE-PATH INTERFEROMETER

FIGURE 2 represents schematically a double three-path interferometer for the biphoton. The left-hand part of the apparatus (to the left of the ladders in FIGURE 2) is the same as FIGURE 1. We focus first on photon 1, which impinges on the upper ladder, containing three "rungs" labeled l, m, and u (for "lower", "middle", and "upper"). Beam-splitters at the ends of rungs l and m, of intensity reflectivities 2/3 and 1/2, respectively, permit photon 1 to "choose" among possible paths with definite amplitudes. Three detectors, similarly labeled L, M, and U, are located to

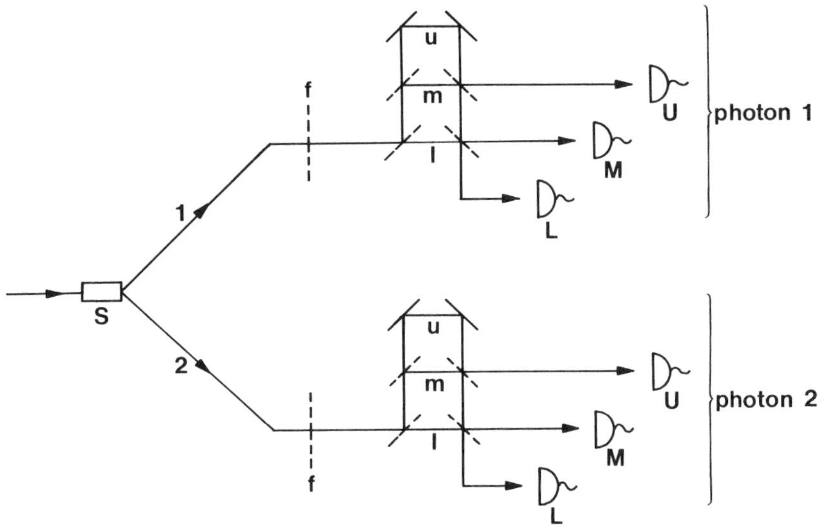

FIGURE 2. The biphoton of FIGURE 1 is here fed into a double three-path interferometer. The beams on these paths, labeled l, m, and u, are first formed and are then recombined by beam-splitters at the left and right ends of the rungs l and m; the recombined beams feed detectors L, M, and U. Each horizontal and vertical step of the ladders has length $\Delta/2$, and detectors L, M, and U are located $\Delta/2$, Δ, and $3\Delta/2$, respectively, to the right of the ladders. When the beam-splitters on rungs l and m reflect intensities 2/3 and 1/2, respectively, the probabilities of joint detection by various pairs of detectors at various time separations can be given as in TABLE 2.

the right of the ladder: U on the same horizontal as m, M on the same horizontal as l, and L on a horizontal below the ladder. Beam-splitters and mirrors on the right-hand side and on top of the ladder permit photon 1 to reach detector U via either m or u and to reach L or M via l, m, or u. Each rung and each vertical step of the ladder have length $\Delta/2$, as do the vertical and horizontal steps from the lower right corner of the ladder to detector L. Detectors M and U are, respectively, Δ and $3\Delta/2$ to the right of the ladder. Consequently, for a fixed rung, the paths to all the accessible detectors have the same total length, and the paths via rung u are Δ longer than the paths via rung m, which in turn are Δ longer than the paths via rung l. TABLE 1 gives the

TABLE 1. Single-Photon Amplitudes for Each Specification of Detector and Rung

$a_{\mathrm{Ll}} = -(2/9)^{1/2}$
$a_{\mathrm{Ml}} = 1/3$
$a_{\mathrm{Ul}} = 0$
$a_{\mathrm{Lm}} = (1/18)^{1/2}$
$a_{\mathrm{Mm}} = 1/3$
$a_{\mathrm{Um}} = -(1/6)^{1/2}$
$a_{\mathrm{Lu}} = (1/18)^{1/2}$
$a_{\mathrm{Mu}} = 1/3$
$a_{\mathrm{Uu}} = (1/6)^{1/2}$

amplitudes of all nine two-photon paths and also the phases due to a factor i for each reflection, but not the phases due to path length (which will be taken into account separately). The first subscript of the amplitude labels the detector and the second labels the rung; for instance, a_{Um} is the amplitude for photon 1 to pass to detector U via rung m. Photon 2 impinges on the lower ladder, which has exactly the same features as the upper. Consequently, TABLE 1 also gives the amplitudes of all nine paths of photon 2.

In biphoton interferometry, we must consider the amplitudes of paths in two-particle configuration space. Except for additional phase contributions due to the path length, these are given simply by the products of two items of TABLE 1. For example, the amplitude for photon 1 to impinge on its detector U via rung m while photon 2 impinges on its detector L via rung u is $a_{\mathrm{Um}}a_{\mathrm{Lu}}$, multiplied by a phase factor. We now combine these amplitude calculations with appropriate adaptations of the biphoton wave function of the previous section. For specificity, focus on detectors M for both photons 1 and 2 and locate both detectors at the origins of the coordinates x_1 and x_2 as before. Then, the total amplitude $a_{\mathrm{MM}}(0, 0, t_1, t_2)$ for the respective photons to reach the detectors at t_1, t_2 is the sum of nine terms adapted from equation 1b:

$$\begin{aligned} a_{\mathrm{MM}}(0, 0, t_1, t_2) = {} & a_{\mathrm{Ml}}a_{\mathrm{Ml}}\psi(0, 0, t_1, t_2) + a_{\mathrm{Mm}}a_{\mathrm{Ml}}\psi(\Delta, 0, t_1, t_2) \\ & + a_{\mathrm{Mu}}a_{\mathrm{Ml}}\psi(2\Delta, 0, t_1, t_2) + a_{\mathrm{Ml}}a_{\mathrm{Mm}}\psi(0, \Delta, t_1, t_2) \\ & + a_{\mathrm{Mm}}a_{\mathrm{Mm}}\psi(\Delta, \Delta, t_1, t_2) + a_{\mathrm{Mu}}a_{\mathrm{Mm}}\psi(2\Delta, \Delta, t_1, t_2) \\ & + a_{\mathrm{Ml}}a_{\mathrm{Mu}}\psi(0, 2\Delta, t_1, t_2) + a_{\mathrm{Mm}}a_{\mathrm{Mu}}\psi(\Delta, 2\Delta, t_1, t_2) \\ & + a_{\mathrm{Mu}}a_{\mathrm{Mu}}\psi(2\Delta, 2\Delta, t_1, t_2). \end{aligned} \tag{5}$$

Here, the substitutions of 0, Δ, and 2Δ for x_1 and x_2 arise from the path lengths: 0 for paths via rung l, Δ for paths via rung m, and 2Δ for paths via rung u, from either ladder. Because of the real Gaussian factor $g(x_1, x_2, t_1, t_2)$ of equation 2, a_{MM} is negligible, except when $\tau = t_2 - t_1$ has approximately one of the values, $s\Delta/c$, where s = 0, 1, −1, 2, or −2; as in the previous section, the meaning of "approximately" is determined by the Gaussian. To avoid the accidental confusion of members from two different biphotons, it is useful (though not essential) that the mean temporal interval of biphoton emissions from the source be greater than $2\Delta/c$; it is essential, however, that both Δ/c and the mean temporal interval be greater than $1/c\sigma$. When s is specified, only three of the nine terms in equation 5 are nonnegligible. For example, if $s = 0$ (approximately simultaneous detection), only the terms with rungs l for both photons, rungs m for both photons, and rungs u for both photons contribute. Hence, for $s = 0$, the amplitude $a_{\mathrm{MM}}(0, 0, t_1, t_2)$ can be rewritten by normalizing and disregarding an overall phase factor as

$$a_{\mathrm{MM}}(0) = a_{\mathrm{Ml}}a_{\mathrm{Ml}} + a_{\mathrm{Mm}}a_{\mathrm{Mm}}e^{2ik_0\Delta} + a_{\mathrm{Mu}}a_{\mathrm{Mu}}e^{4ik_0\Delta}, \tag{6}$$

where the argument 0 refers to the value of s. There are, in fact, 45 amplitudes of the form of $a_{\mathrm{IJ}}(s)$, where I is L, M, or U and J is L, M, or U. The normalization is imposed by requiring

$$\sum_{\mathrm{I,J},s} |a_{\mathrm{IJ}}(s)|^2 \equiv \sum_{\mathrm{I,J},s} p_{\mathrm{IJ}}(s) = 1, \tag{7}$$

where $p_{\mathrm{IJ}}(s)$ is the probability that the two photons will be detected respectively by detectors I and J, with a time difference of approximately $s\Delta/c$. A complete compilation of the probabilities $p_{\mathrm{IJ}}(s)$ is presented in TABLE 2.

Note that the amplitude $a_{\mathrm{MM}}(0)$ of nearly simultaneous detection at the two detectors labeled M is, according to equation 6, a superposition of three terms corresponding to three different path lengths through the double interferometer and hence to three different transit times and hence to three different times of birth. This superposition is possible only because the two photons were born together, but at an indefinite time, and the time of birth cannot be determined without changing the experimental arrangement. The probability of nearly simultaneous detection at both

TABLE 2. Probabilities of Joint Detections with Various Time Separations

$p_{\mathrm{LL}}(0) = 324^{-1}(18 + 10\cos 2k_0\Delta + 8\cos 4k_0\Delta)$
$p_{\mathrm{LM}}(0) = p_{\mathrm{ML}}(0) = 324^{-1}(12 - 4\cos 2k_0\Delta - 8\cos 4k_0\Delta)$
$p_{\mathrm{LU}}(0) = p_{\mathrm{UL}}(0) = 324^{-1}(6 - 6\cos 2k_0\Delta)$
$p_{\mathrm{MM}}(0) = 324^{-1}(12 + 16\cos 2k_0\Delta + 8\cos 4k_0\Delta)$
$p_{\mathrm{MU}}(0) = p_{\mathrm{UM}}(0) = 324^{-1}(12 - 12\cos 2k_0\Delta)$
$p_{\mathrm{UU}}(0) = 324^{-1}(18 + 18\cos 2k_0\Delta)$
$p_{\mathrm{LL}}(1) = p_{\mathrm{LL}}(-1) = 324^{-1}(5 - 4\cos 2k_0\Delta)$
$p_{\mathrm{LM}}(1) = p_{\mathrm{ML}}(-1) = 324^{-1}(10 - 8\cos 2k_0\Delta)$
$p_{\mathrm{LU}}(1) = p_{\mathrm{UL}}(-1) = 324^{-1}(15 + 12\cos 2k_0\Delta)$
$p_{\mathrm{ML}}(1) = p_{\mathrm{LM}}(-1) = 324^{-1}(4 + 4\cos 2k_0\Delta)$
$p_{\mathrm{MM}}(1) = p_{\mathrm{MM}}(-1) = 324^{-1}(8 + 8\cos 2k_0\Delta)$
$p_{\mathrm{MU}}(1) = p_{\mathrm{UM}}(-1) = 324^{-1}(12 - 12\cos 2k_0\Delta)$
$p_{\mathrm{UL}}(1) = p_{\mathrm{LU}}(-1) = 3/324$
$p_{\mathrm{UM}}(1) = p_{\mathrm{MU}}(-1) = 6/324$
$p_{\mathrm{UU}}(1) = p_{\mathrm{UU}}(-1) = 9/324$
$p_{\mathrm{LL}}(2) = p_{\mathrm{LL}}(-2) = 4/324$
$p_{\mathrm{LM}}(2) = p_{\mathrm{ML}}(-2) = 8/324$
$p_{\mathrm{LU}}(2) = p_{\mathrm{UL}}(-2) = 12/324$
$p_{\mathrm{ML}}(2) = p_{\mathrm{LM}}(-2) = 2/324$
$p_{\mathrm{MM}}(2) = p_{\mathrm{MM}}(-2) = 4/324$
$p_{\mathrm{MU}}(2) = p_{\mathrm{UM}}(-2) = 6/324$
$p_{\mathrm{UL}}(2) = p_{\mathrm{LU}}(-2) = 0$
$p_{\mathrm{UM}}(2) = p_{\mathrm{MU}}(-2) = 0$
$p_{\mathrm{UU}}(2) = p_{\mathrm{UU}}(-2) = 0$

M's is the absolute square of the amplitude $a_{\mathrm{MM}}(0)$:

$$p_{\mathrm{MM}}(0) = |a_{\mathrm{MM}}(0)|^2 = 81^{-1}(3 + 4\cos 2k_0\Delta + 2\cos 4k_0\Delta), \tag{8}$$

showing that the contribution of the three terms in equation 6 gives rise to a nonsinusoidal interference fringe pattern. By contrast, many of the probabilities presented in TABLE 2 do exhibit a sinusoidal fringe pattern [namely, $p_{\mathrm{LU}}(0)$, $p_{\mathrm{MU}}(0)$, and $p_{\mathrm{IJ}}(\pm 1)$ for many values of I and J]; for each of these, exactly two paths with different phases contribute to the total amplitude. Finally, a few of the $p_{\mathrm{IJ}}(\pm 1)$ and all of the $p_{\mathrm{IJ}}(\pm 2)$ exhibit no fringes as Δ varies because, in each of these cases, there is only one contributing path in the two-photon configuration space. It is interesting that this variety of fringe pattern occurs in a single experimental arrangement.

TABLE 2 shows a symmetry whose explanation is obvious:

$$p_{\mathrm{IJ}}(s) = p_{\mathrm{JI}}(-s). \tag{9}$$

It also shows that the probability of single detection at any detector of either photon, at a fixed s, exhibits no fringes as Δ is varied, that is,

$$\begin{aligned} p_{\mathrm{I}}(s) &\equiv p_{\mathrm{IL}}(s) + p_{\mathrm{IM}}(s) + p_{\mathrm{IU}}(s) = p_{\mathrm{LI}}(s) + p_{\mathrm{MI}}(s) + p_{\mathrm{UI}}(s) \\ &= \text{a constant dependent on I, but independent of } \Delta. \end{aligned} \tag{10}$$

There is a simple physical explanation. For specificity, let $s = 0$. Then, $p_{\mathrm{I}}(0)$ is the probability that one of the photons, for example, the first, goes to the detector I, whereas the other photon goes to any of its three detectors. However, the paths of the two photons are correlated: $s = 0$ implies that both go via rung l, both go via rung m, or both go via rung u. Now, if the arrangement of FIGURE 2 is modified by substituting double-sided full mirrors for the beam-splitters on the right side of the ladder for photon 2, then clearly $p_{\mathrm{I}}(0)$ will not change, even though there will be a change of $p_{\mathrm{IL}}(0)$, $p_{\mathrm{IM}}(0)$, and $p_{\mathrm{IU}}(0)$. In the original arrangement, photons traversing a specific rung could go to more than one detector, whereas in the modified arrangement l feeds only L, m feeds only M, and u feeds only U. In the modified arrangement, one can determine with certainty which path photon 1 takes to detector I, by observing which detector is triggered by the partner photon 2. Therefore, by the classical rule of Bohr, Dirac, and Feynman, the three amplitudes contributing to the modified $p_{\mathrm{I}}(0)$ are absolute-squared and added, rather than added prior to absolute-squaring. Hence, $p_{\mathrm{I}}(0)$ is independent of Δ in the modified arrangement and, by the invariance of this probability under the change of arrangement, the unmodified $p_{\mathrm{I}}(0)$ is independent of Δ.

We conclude our discussion of the double three-path interferometer with two remarks. First, the correlations between the various paths of photon 1 and those of photon 2 are an extension of the long-long and short-short path length correlations exhibited by Franson in a double two-path interferometer.[5] Second, the entries in TABLES 1 and 2 could be made much more uniform if the pair of beam-splitters on the left of the ladder of photon 1 and the pair of beam-splitters on the right were each replaced by a "sixport"[5]—a generalization of the ordinary beam-splitter in which each of three input beams produces three output beams with equal intensities;

likewise for photon 2. With the sixports, $p_{IJ}(0)$ is nonsinusoidal for all pairs of detectors I and J, $p_{IJ}(\pm 1)$ is sinusoidal for all pairs, and $p_{IJ}(\pm 2)$ is flat for all pairs.

A DOUBLE FABRY-PEROT INTERFEROMETER

FIGURE 3 represents schematically an arrangement that presents each biphoton with infinitely many path options. These options arise because each beam emerging from the source as in FIGURE 1 is fed at normal incidence into a Fabry-Perot interferometer, where each photon separately may enjoy an arbitrary number of reflections between the two beam-splitters, $\Delta/2$ apart, before proceeding to its detector. As in the previous section, imagine that detections are recorded on the left and right halves of a tape for photons 1 and 2, respectively, of a biphoton. The

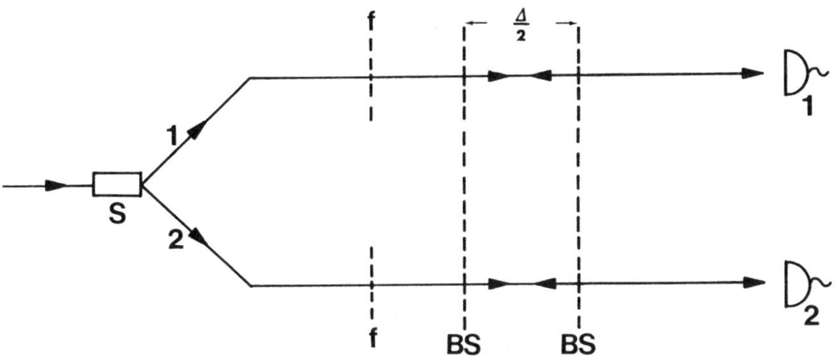

FIGURE 3. The biphoton prepared in FIGURE 1 is here fed into a Fabry-Perot interferometer, which is formed by beam-splitters (BS) separated by distance $\Delta/2$. The amplitudes of transmission of the beam-splitters are t and t', respectively, for external and internal incidence, and r is the inside amplitude of reflection for both beam-splitters. The resulting probabilities of joint detection by D_1 and D_2 with various delay times are given in equation 12.

round-trip time, Δ/c, across the Fabry-Perot interferometer is taken to be substantially larger than the temporal width, $1/c\sigma$, of the biphoton Gaussian, and the intensity of the source is sufficiently low that members of two different biphotons are seldom confused. For any biphoton, the detection dots may be "nearly simultaneous", indicating that the photons reflected the same, but indefinite, number of times inside the Fabry-Perot; or the detection of photon 1 may precede [follow] that of photon 2 by approximately Δ/c, indicating that particle 1 made round-trips in the Fabry-Perot one less [more] time than particle 2; or the detections may differ by $2\Delta/c$, indicating that one photon made two more round-trips than the other; etc.

Let $a(s)$, where $s = 0, \pm 1, \pm 2, \ldots$, denote the amplitude that the pair will be detected with a time difference of approximately $t_2 - t_1 = s\Delta/c$. Clearly, $a(0)$ is the superposition of an infinite number of terms corresponding to both particles reflecting zero times, once, twice, etc.; in general, any $a(s)$ is a superposition of infinitely many terms. By a straightforward generalization of the reasoning leading to equations 5 and 6, we obtain, for nonnegative s,

$$a(s) = t^2(t')^2 \sum_{n=0} r^{2n} r^{2(n+s)} \psi[n\Delta, (n+s)\Delta, 0, s\Delta/c]; \tag{11a}$$

to cover negative values of s, we use the obvious relation

$$a(-s) = a(s). \tag{11b}$$

In equation 11a, t and t' are the amplitudes of transmission through either of the two beam-splitters forming the Fabry-Perot interferometer for a photon incident from the exterior or from the interior of the interferometer, r is the reflection amplitude on the "inner" surfaces of the two beam-splitters (assumed the same for both), n $[n + s]$ is the number of round-trips of photon 1 [2] in the Fabry-Perot, and ψ is the biphoton wave function of equation 1b. Then, for any integer s (positive, negative, or zero),

$$p(s) = |a(s)|^2 = t^4(t')^4 r^{4|s|}(1 - 2r^4 \cos 2k_0\Delta + r^8)^{-1}. \tag{12}$$

This probability of joint detection, at a temporal separation of approximately $s\Delta/c$, when studied as a function of $k_0\Delta$ is a two-particle fringe pattern and it is essentially the same kind of function that occurs in single-particle Fabry-Perot interferometry.[6] Note, however, that for the biphoton a factor r^2 occurs wherever r occurs in the single-photon case and that the period in Δ is twice as large as for one-photon fringes in incident light of the same color.

A number of comments are now appropriate.

First, equations 11a and 12 were derived in the ideal case of a monochromatic, hence everlasting, pump beam. As a result, there is a superposition of infinitely many terms in $a(s)$ for any s, each corresponding to a different transit time through the Fabry-Perot for each photon, and hence a superposition of infinitely many birth times of the biphoton. As a result, $p(s)$ has the same form for all s, except for the factor $r^{4|s|}$. Hence, the two-particle fringe visibility is exactly the same for each value of s, even though the probability $p(s)$ falls off as the $4|s|$-th power of r. By contrast, in the double three-path interferometer (see previous section), the fringe pattern of $p_{IJ}(s)$ varied with s (whether one used the arrangement of FIGURE 2 or a sixport arrangement) because the number of two-photon paths contributing to the total amplitude $a_{IJ}(s)$ depended upon s.

Second, in a realistic situation, the Fabry-Perot fringe pattern $p(s)$ will vary with s because the pump beam is not monochromatic and everlasting, so the number of superposed birth times contributing to $a(s)$ falls off with increasing $|s|$. The spatial characteristics of the downconversion process may also affect the coherence time of the biphoton and could contribute to the variation of the fringe pattern of $p(s)$. One may, in fact, be able to make inferences from the variation of the fringe pattern of $p(s)$ about the spectrum and coherence time of the pump beam and about the downconversion process.

Third, the Fabry-Perot arrangement of FIGURE 3 is impractical because the factor t^2 in equation 11a drastically diminishes the biphoton detection rate. Anton Zeilinger (personal communication) has suggested the arrangement of FIGURE 4, in which each beam 1 and 2 enjoys its own separate Fabry-Perot and, most significantly, the source crystal S is now within the Fabry-Perot cavities. A similar arrangement, but without the multireflection aspect, has already been employed experimentally.[7] Assuming for simplicity that the lengths of the two Fabry-Perot cavities are the same and neglecting reflection and refraction at the crystal interfaces, one finds that equations 11a and 12 still hold with a few changes: (a) the length L and refractive index n_0 of the crystal produce an effective cavity length, $\Delta \rightarrow \Delta + 2L(n_0 - 1)$; (b) the full mirrors at the left end of FIGURE 4 halve the number of powers of r; and (c) because the biphoton is born within the cavities, the troublesome entrance factor t^2 does not appear. If the optional mirror of FIGURE 4 is present to reflect the pump beam, the pump photon may downconvert "earlier" while going right or "later" while going left, as in the superpositions reported in reference 7; equations 11a and 12 would have to be modified accordingly.

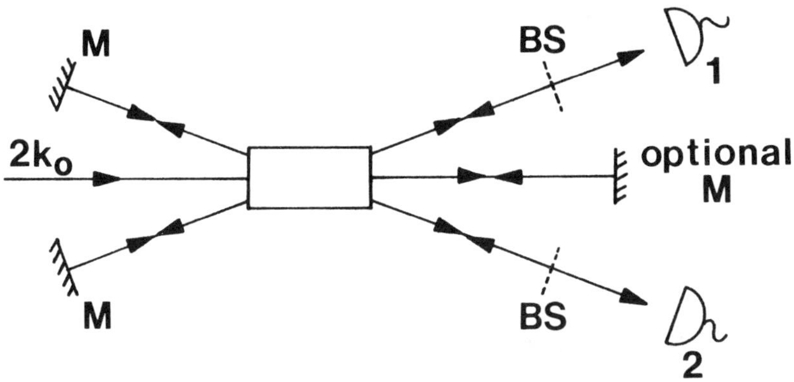

FIGURE 4. The filtered biphoton beams of FIGURE 1, although still proceeding directly away from the crystal, are here reflected by two beam-splitters (BS) back through the crystal to the full mirrors (M) on the left, where they reflect again, etc. Thus, two separate Fabry-Perot cavities are formed, one for each member of the biphoton, with the crystal source in the center. The optional full mirror on the right reflects the pump beam back through the crystal, thereby providing a second opportunity for downconversion to occur.

Fourth and last, when each member of the biphoton enjoys its own Fabry-Perot, as for example in the preceding paragraph and in FIGURE 4, then two cavity length parameters Δ_1 and Δ_2 can be varied independently. The evaluation of amplitudes $a(\tau)$ for joint detection of the two photons with a time separation of $\tau = t_2 - t_1$ becomes much more intricate than in equation 11a. Especially if Δ_1 and Δ_2 are incommensurable, there is no simple rule determining which integral numbers n_1 and n_2 of round-trips of 1 and 2 in their respective interferometers will make the Gaussian of equation 2 nonnegligible. An analytic solution is likely to be difficult, but a computational solution may be revealing.

ACKNOWLEDGMENTS

We wish to thank Anton Zeilinger for valuable discussions.

REFERENCES AND NOTES

1. KLYSHKO, D. N. 1988. Photons and Nonlinear Optics. Gordon & Breach. New York.
2. ZEILINGER, A., H. J. BERNSTEIN, D. M. GREENBERGER, M. A. HORNE & M. ZUKOWSKI. 1993. *In* Quantum Control and Measurement. H. Ezawa & Y. Murayama, Eds.: 9–22. North-Holland. Amsterdam.
3. A more detailed analysis of this arrangement is given in: HORNE, M. A., A. SHIMONY & A. ZEILINGER. 1990. *In* Quantum Coherence. J. Anandan, Ed.: 356–372. World Scientific. Singapore.
4. This statement treats time as if it is an observable, despite well-known objections against doing so. However, in very simple situations, like the one under consideration, in which both particles propagate in the same direction and at a constant speed, there is an effective equivalence between taking time as a parameter and position as an observable and vice versa. See: WIGNER, E. P. 1972. *In* Aspects of Quantum Theory. A. Salam & E. P. Wigner, Eds.: 237. Cambridge University Press. London/New York.
5. FRANSON, J. D. 1989. Phys. Rev. Lett. **62:** 2205–2208.
6. BORN, M. & E. WOLF. 1980. Principles of Optics. Sixth edition, p. 325. Pergamon. Elmsford, New York.
7. HERZOG, T. J., J. G. RARITY, H. WEINFURTER & A. ZEILINGER. 1994. Phys. Rev. Lett. **72:** 629.

Degree of Entanglement[a]

ABNER SHIMONY

Departments of Philosophy and Physics
Boston University
Boston, Massachusetts 02215

INTRODUCTION

As Schrödinger discovered, the general principles of quantum mechanics imply that a quantum state of a many-particle system may be "entangled", in the sense of not being a product of single-particle states. The concept of entanglement enters some deep investigations of the foundations of quantum mechanics, including the argument of Einstein-Podolsky-Rosen and the theorem of Bell. It is basic to the program of two-particle (or n-particle) interferometry. Of course, entanglement is exhibited by almost all of the interesting quantum states of atoms, molecules, nuclei, and condensed matter systems.

The present report exhibits some mathematical relations among three concepts associated with entanglement. The second section proposes a general quantitative definition of the degree of entanglement $E(\phi)$ of any n-particle quantum state and when $n = 2$ evaluates $E(\phi)$. (Note that the same symbol ϕ is used for the quantum state and for the normalized vector in a Hilbert space that represents this state.) The third section is restricted to the case of two particles, each associated with a two-dimensional Hilbert space, called "the two-by-two case". In this case, it is possible to calculate the two-particle fringe visibility $V_{12}(\phi)$ and also the quantity $B(\phi)$, which is the maximum deviation achievable, when the state is ϕ, from the limit allowed by one of Bell's inequalities. It is shown that $V_{12}(\phi)$ and $B(\phi)$ are monotonic increasing functions of $E(\phi)$.

PROPOSED DEFINITION OF THE DEGREE OF ENTANGLEMENT

The Hilbert space for representing the states of any n-particle system $1 + 2 + \ldots$ (consisting of a finite number or a denumerably infinite set of particles) is the direct product space,

$$H = H_1 \otimes H_2 \otimes \ldots, \tag{1}$$

where H_k is the Hilbert space associated with the k-th particle. The dimension of H_k is denoted by d_k, which may be denumerable infinity or an integer greater than zero. By a theorem of Schmidt,[1] any $\phi \in H$ can be represented without a loss of generality

[a]This research was supported in part by the National Science Foundation (Grant No. PHY-9321992).

by a sum with a single index if $n = 2$,

$$\phi = \sum_{j=1}^{d} \alpha_j (u_{1j} \otimes u_{2j}), \tag{2}$$

where u_{kj} is a normalized vector in H_k, the α_j factors are positive real numbers satisfying

$$\alpha_{j+1} \leq \alpha_j, \tag{3}$$

and

$$d = \min(d_k), \qquad k = 1, 2. \tag{4}$$

It is convenient to define the degree of entanglement of ϕ only in the case of normalized ϕ,

$$\|\phi\| = 1. \tag{5}$$

With this restriction, we propose the following definition:

$$E(\phi) = (½)\min\|\phi - \chi\|^2, \tag{6}$$

where χ is a normalized product state in H and the minimum is taken over the set of normalized product states. There is a considerable amount of conventionality in this definition. The norm of $\phi - \chi$ is a reasonable ingredient in the concept of the degree of entanglement, but there is no compelling reason for preferring to use the square of the norm rather than some other positive power. Clearly, any monotonic increasing function of $E(\phi)$ gives the same ordering of normalized vectors ϕ. The proposed definition of $E(\phi)$ has the obvious virtue of being zero when ϕ is a product state and the less-obvious virtue that the least upper bound of $E(\phi)$ is unity.

To evaluate $E(\phi)$ for $n = 2$, we take a basis u_{kj} in the Hilbert space H_k, using the $u_{k1}, \ldots, u_{kd}$ terms that appear in equation 2 as the first d basis vectors. If χ is a normalized vector in H, then it can be written as

$$\chi = \sum_j c_{1j} u_{1j} \otimes \sum_j c_{2j} u_{2j}, \tag{7}$$

where

$$\sum_j |c_{kj}|^2 = 1 \tag{8}$$

for $k = 1, 2$. Then,

$$\begin{aligned}\|\phi - \chi\|^2 &= \|\phi\|^2 + \|\chi\|^2 - (\phi, \chi) - (\chi, \phi) \\ &= 2 - 2\sum_{j=1} \alpha_j \left(\prod_k |c_{kj}|\right) \cos\left(\sum_k \theta_{kj}\right),\end{aligned} \tag{9}$$

where

$$c_{kj} = |c_{kj}| e^{i\theta_{kj}} \tag{10}$$

and

$$\alpha_j = 0 \qquad \text{for} \qquad j > d. \tag{11}$$

By equations 3 and 11, the right-hand side of equation 9 is minimized by choosing θ_{kj} and c_{kj} such that

$$\sum_k \theta_{kj} = 0, \tag{12}$$

$$|c_{k1}| = 1, \tag{13}$$

and

$$|c_{kj}| = 0 \qquad \text{for} \qquad j > 1. \tag{14}$$

The conclusion is that

$$E(\phi) = 1 - \alpha_1. \tag{15}$$

This result is remarkable not only for its generality, but for its independence of the values of α_j for $j > 1$. Clearly,

$$\text{glb}\, E(\phi) = 0, \tag{16a}$$

and this value is attained when $\alpha_1 = 1$, that is, when ϕ is a product state. For a fixed finite d,

$$\text{lub}\, E(\phi) = 1 - (1/\sqrt{d}), \tag{16b}$$

which is attained when all the coefficients α_j $(j = 1, \ldots, d)$ are equal. If d is infinite,

$$\text{lub}\, E(\phi) = 1, \tag{16c}$$

but this limit is not attained for any ϕ.

DEVIATION FROM THE BELL LIMIT AND TWO-PARTICLE FRINGE VISIBILITY

In this section, we restrict our attention to the two-by-two case,

$$\dim H_1 = \dim H_2 = n = 2. \tag{17}$$

In this case, Gisin[2] considered the Bell inequality,

$$-2 \le E(a,b) + E(a,b') + E(a',b) - E(a',b') \le 2, \tag{18}$$

where a is an abbreviation for the bivalent observable represented by the operator A_a on H_1,

$$A_a = 2|u_a\rangle\langle u_a| - 1, \tag{19}$$

with a being a normalized vector in H_1; b is an abbreviation for an analogous observable on H_2; $E(a,b)$ is the expectation value of these two observables in a specified ensemble; and $E(a, b')$, etc., have similar meanings. Gisin showed that if ϕ is entangled, then there exist a, a', b, b' such that the expectation values computed using the quantum state ϕ will violate inequality 18. Furthermore, he showed that the maximum violation of the inequality for fixed ϕ, but variable a, a', b, b', is[3]

$$B(\phi) = 2\{[1 + 4\alpha_1^2(1 - \alpha_1^2)]^{1/2} - 1\}. \tag{20}$$

Comparison of equations 15 and 20 shows that each is a monotonic increasing function of the other in the relevant range of α_1, that is, $\alpha_1 \geq 1/\sqrt{2}$.

Another quantity associated with entanglement is the two-particle interferometric fringe visibility, investigated by Jaeger, Horne, and Shimony[4] and by Jaeger, Shimony, and Vaidman.[5] They considered a quantum state ϕ in the two-by-two case and an experimental arrangement in which a pair of transducers (mathematically represented by unitary unimodular matrices T_1 and T_2) connect the initial states with two pairs of output beams. If one focuses attention upon one specific output beam for particle 1 and one specific output beam for particle 2, then one can consider the probability of joint detection by ideal detectors placed in the specified output beams, denoted by $P_{12}(T_1T_2\phi)$. One can also consider the probabilities $P_i(T_1T_2\phi)$ $(i = 1, 2)$ of single detection in the specified beams. In order to extract correlations, one can define a "corrected joint probability" as

$$\overline{P}_{12}(T_1T_2\phi) = P_{12}(T_1T_2\phi) - P_1(T_1T_2\phi)P_2(T_1T_2\phi) + (1/4). \tag{21}$$

There is a standard expression for the one-particle interferometric fringe visibility,

$$V_i(\phi) = \frac{[P_i(T_1T_2\phi)]_{\max} - [P_i(T_1T_2\phi)]_{\min}}{[P_i(T_1T_2\phi)]_{\max} + [P_i(T_1T_2\phi)]_{\min}} \qquad (i = 1, 2), \tag{22}$$

but the two-particle interferometric fringe visibility is most appropriately defined in terms of the "corrected" probabilities:

$$V_{12}(\phi) = \frac{[\overline{P}_{12}(T_1T_2\phi)]_{\max} - [\overline{P}_{12}(T_1T_2\phi)]_{\min}}{[\overline{P}_{12}(T_1T_2\phi)]_{\max} + [\overline{P}_{12}(T_1T_2\phi)]_{\min}}. \tag{23}$$

In both of these expressions, the maximum and the minimum are computed by varying T_1 and T_2 with ϕ fixed.[6] In reference 5, it is found that

$$V_{12}(\phi) = 2\alpha_1(1 - \alpha_1^2)^{1/2}.$$

It is obvious that $V_{12}(\phi)$ is a monotonic increasing function of $E(\phi)$ in the relevant range of α_1.

It is satisfying that, in the two-by-two case, three quantities connected with entanglement, but defined in entirely different ways, turn out to be monotonic increasing functions of each other. This result may be of heuristic value in future research. As shown in the previous section, $E(\phi)$ can be defined for any quantum state of a many-particle system and this quantity may be a guide to generalizations of $B(\phi)$ and $V_{12}(\phi)$ beyond the two-by-two case.

ACKNOWLEDGMENTS

I am grateful to Asher Peres for correcting an error.

REFERENCES AND NOTES

1. SCHMIDT, E. 1907. Math. Ann. **63:** 433; **64:** 161.
2. GISIN, N. 1991. Phys. Lett. A **154:** 201.
3. This expression has also been derived by Alonso Botero, using a geometrical construction, and by: POPESCU, S. & D. ROHRLICH. 1992. Phys. Lett. A **166:** 293.
4. JAEGER, G., M. A. HORNE & A. SHIMONY. 1993. Phys. Rev. A **48:** 1023.
5. JAEGER, G., A. SHIMONY & L. VAIDMAN. 1994. Phys. Rev. A. To appear.
6. In reference 4, it is shown that $V_{12}^2 + V_i^2 \leq 1$ and, in reference 5, this inequality is strengthened to an equality. The earlier result is weaker because only a proper subset of the transducers T_1 and T_2 is used in evaluating the max and min.

Statistics of Entangled-Photon Coincidences in Parametric Downconversion[a]

T. S. LARCHUK,[b] M. C. TEICH,[b,c,d] AND B. E. A. SALEH[e]

[b]*Department of Electrical Engineering*
[c]*Department of Applied Physics*
Columbia University
New York, New York 10027

[e]*Department of Electrical and Computer Engineering*
University of Wisconsin
Madison, Wisconsin 53706

INTRODUCTION

We present experimental counting distributions and interevent-time histograms for photon detections from spontaneous parametric downconversion,[1] both marginally and as coincidences. The experiments were conducted with a lithium-iodate downconverter pumped by 413-nm krypton-ion laser light. The data are consistent with Poisson statistics; a model leading to this result is presented.

EXPERIMENT

A block diagram of the experimental setup is shown in FIGURE 1. Light from a single-mode Kr^+-ion laser, operated at 413 nm and attenuated to 0.75 mW, is directed by a lens onto an l = 10-mm-long 42.8°-cut lithium iodate ($LiIO_3$) nonlinear optical crystal, oriented for type-I (ooe) phase matching. Unconverted pump photons pass straight through the crystal and enter a beam dump. Downconverted photons emerge at angles to the pump beam determined by energy- and phase-matching, with degenerate photons emerging symmetrically in a cone of full angle ~ 15°. By using apertures for the downconverted beams of about 2-mm diameter, we selected out desired degenerate photon pairs[2] with center wavelengths of ~826 nm. The entangled photon pairs were directed to two passively quenched avalanche-photodiode (APD) counting modules. A pulse counter and a time-interval counter recorded events from the detectors. Coincidences were generated by passing the sequence of standardized pulses from the two detectors through a 10-ns AND gate. The coincidence events were also fed to the pulse and time-interval counters for statistical analysis.

[a]This work was supported by the Office of Naval Research under Grant No. N00014-93-1-0547, by the Joint Services Electronics Program through the Columbia Radiation Laboratory, and by NATO Collaborative Research Grant No. CGR-910571.

[d]To whom all correspondence should be addressed.

In the upper portion of FIGURE 2, we present the marginal photon-counting distribution $p(n, T)$ obtained from one of the twin beams (detector A) using a counting time of $T = 1$ ms (solid dots). Using a total of $N = 50{,}004$ samples, the mean number of photocounts was found to be $\bar{n} = 20.238$ and the Fano factor was $F = \mathrm{Var}(n)/\bar{n} = 0.974$.

The solid curve in the upper portion of FIGURE 2 is a Poisson distribution with the same mean as the data. The standard deviation of the Fano factor estimate for a Poisson distribution based on N samples is known to be $\approx\sqrt{2/N}$ so that, for $N = 50{,}004$, F is expected to lie between 0.994 and 1.006. The Poisson distribution fits the data quite well, but the observed value of F (= 0.974) is too low.

The discrepancy is readily resolved by making use of the nonparalyzable dead-time-modified Poisson (DTMP) counting distribution.[3,4] This distribution is, in fact, appropriate because the electronic pulses produced by the APD-module circuitry incapacitate the counting system for a duration $\tau_d \approx 1$ μs following the registration of a photon. The theoretical expression for the DTMP Fano factor is[5]

$$F \approx (1 - \bar{n}\tau_d/T)^2 \leq 1. \tag{1}$$

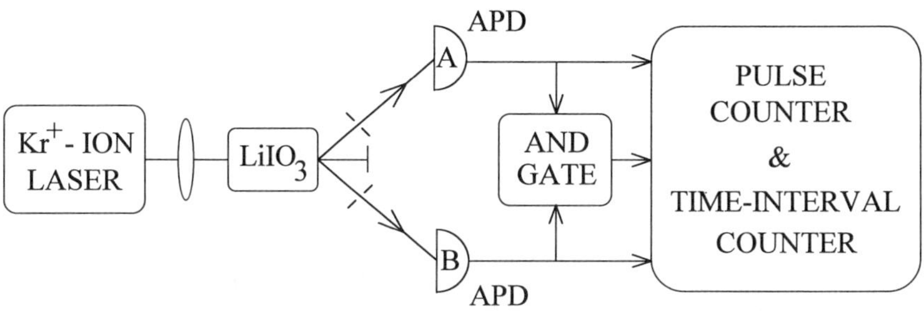

FIGURE 1. Block diagram of the experimental arrangement. The pulse-counter electronics produces a pulse of finite duration that acts as a dead time τ_d.

Using the observed value for F, we obtain $\tau_d \approx 0.64$ μs, which is in reasonable accord with the APD-module pulse duration. Photon-counting distributions were obtained using counting times T ranging from 5×10^{-7} to 2×10^{-3} s, and the DTMP counting distribution was found to describe properly the data in all cases.

The marginal interevent-time probability density function $p(t)$ from detector A is shown in the lower portion of FIGURE 2 (solid dots). The histogram bins are 1 μs in duration and the number of samples is $N = 100{,}000$. The data are well fit by a straight line that, on this semilogarithmic plot, represents an exponential distribution with the same mean as the data. The only exception is the initial data point, which lies below the exponential curve, and this is a result of dead time. The observed mean interevent time of $\bar{t} = 50.336$ μs accords well with the expected value of

$$\bar{t} \approx \tau_d + T/\bar{n} \tag{2}$$

using the parameters determined from the counting distribution.

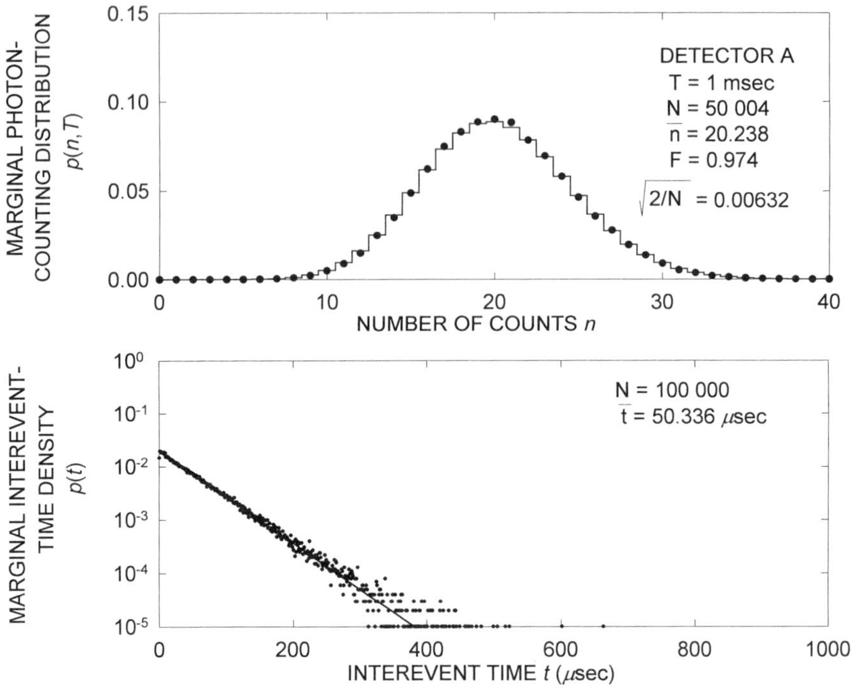

FIGURE 2. Marginal photon-counting distribution $p(n, T)$ versus number of counts n at the output of detector A (upper panel, solid dots): the solid curve is the Poisson distribution with the same mean as the data. Marginal interevent-time probability density function $p(t)$ versus interevent time t at the output of detector A (lower panel, solid dots): the solid curve is the exponential distribution with the same mean as the data. Both results are consistent with a photon stream at the input to the detector that obeys a Poisson point process, observed by a detector with dead time.

We conclude that the photocount occurrences observed at the output of detector A are well modeled by a dead-time-modified Poisson point process. This is consistent with a photon stream at the input to the detector described by a Poisson point process.

The marginal photon-counting distribution and interevent-time probability density function for detector B are shown in FIGURE 3, again using a counting time of $T = 1$ ms (solid dots). In this case, the mean number of photocounts was $\bar{n} = 9.501$, about half that in detector A (probably as a result of the different quantum efficiencies of the two detectors). The solid curve in the upper portion of FIGURE 3 represents a Poisson distribution with the same mean as the data. Again, it fits well, but the observation that $F = 0.983 < 1$ suggests the use of the DTMP counting distribution. Using this value for F in equation 1 leads to $\tau_d \approx 0.88$ μs, which is again reasonable. The observed mean interevent time $\bar{t} = 102.32$ μs from detector B also accords well with the value expected from equation 2. Thus, the photon stream impinging on detector B also appears to be describable by a Poisson point process.

The statistics of the coincidences are displayed in FIGURE 4. Using a counting time of $T = 2$ ms (solid dots), the coincidence-counting distribution is shown in the upper panel. Using a total of $N = 25{,}002$ samples, the mean number of coincidences was $\bar{n} = 3.206$ and the Fano factor was $F = 0.988$. The solid curve is a Poisson distribution with the same mean as the data. Similar results were obtained using counting times T ranging from 5×10^{-7} to 2×10^{-3} s. We can therefore also invoke the DTMP counting distribution to describe the coincidence data. Using the observed value for F, we infer that $\tau_d \approx 3.8$ μs, which is somewhat greater than the values obtained for the rates of singles. The reason for the increased dead time in the coincidence case is unclear. Except for the initial point, the coincidence interevent-time probability density function is also well fit by an exponential distribution. The mean interevent time $\bar{t} \approx 623.77$ μs is in accord with equation 2. Thus, the coincidences are also well modeled as a DTMP, which is consistent with the photon pairs arriving at the detectors as a joint Poisson point process.

THEORY

The theory developed by Yurke and Potasek[6] for the photon statistics of parametrically downconverted light is based on the assumption of a classical pump

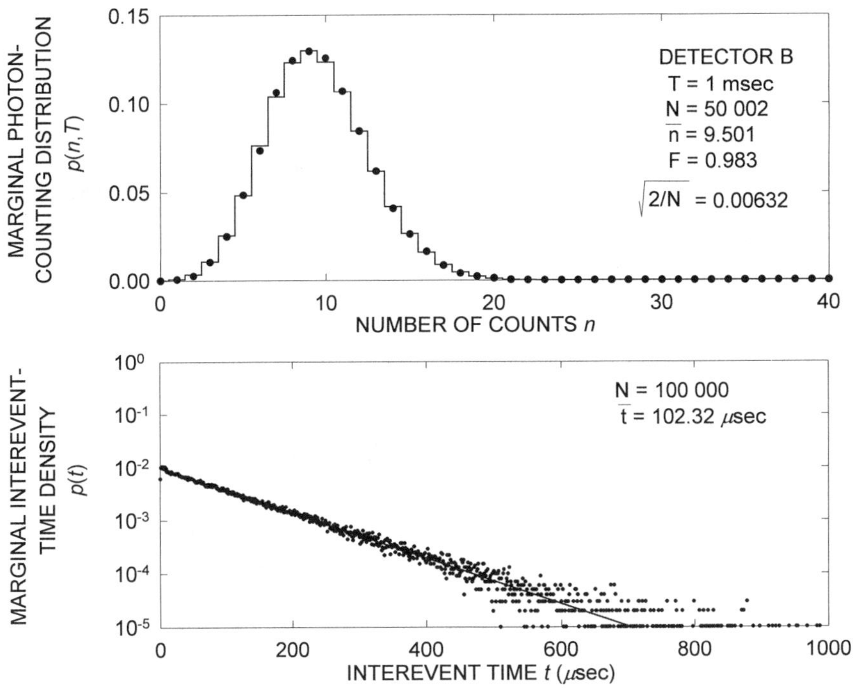

FIGURE 3. Same as FIGURE 2, but at the output of detector B.

applied for a time interval between 0 and t and of single signal and idler modes of a cavity that are initially in the vacuum state. Writing the Hamiltonian for the process, they showed that the final (pure) state is one for which the marginal photon-counting distributions for each of these modes constitute the Bose-Einstein distribution, and the numbers of photons in the two modes are always the same. In the framework of this model, the mean number of photons is a monotonically increasing function of time because the emitted photons continue to build up in the cavity.

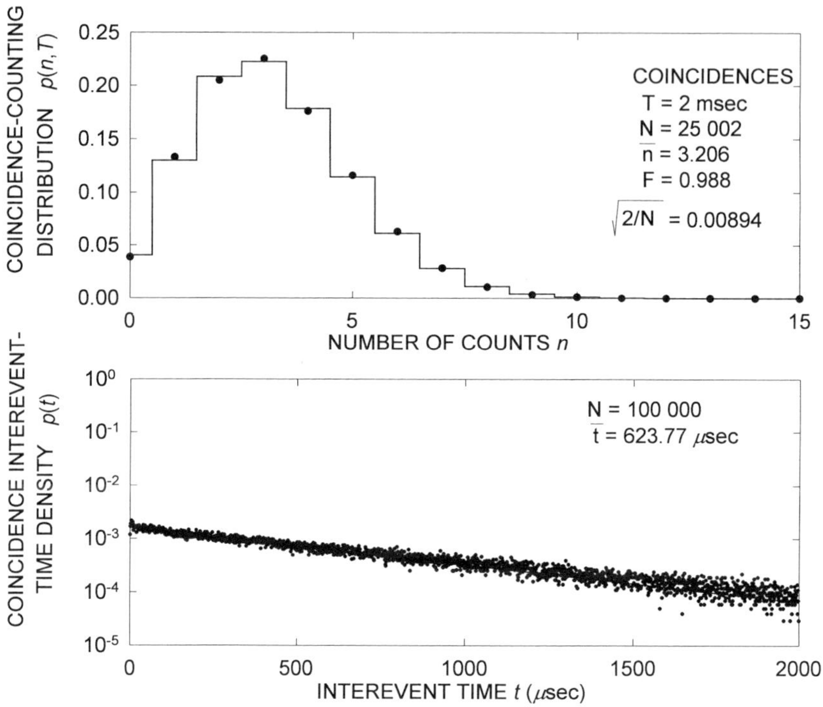

FIGURE 4. Coincidence-counting distribution $p(n, T)$ versus number of coincidences n at the output of the AND gate (upper panel, solid dots): the solid curve is the Poisson distribution with the same mean as the data. Coincidence interevent-time probability density function $p(t)$ versus interevent time t at the output of the AND gate (lower panel, solid dots): the solid curve is the exponential distribution with the same mean as the data. These results are consistent with the photon pairs arriving at the detectors as a joint Poisson point process.

These results can be adapted to a cavityless open system in which the pump travels continuously through the downconversion crystal and the emitted photons propagate away from the interaction region. We consider the continuous classical pump wave as divided into $N = T/\tau_r$ contiguous segments, each of which has a duration given by the transit time τ_r through the crystal. Because each transit-time segment may be described by the Yurke-Potasek cavity formulation with $t = \tau_r$, it will give rise to a Bose-Einstein-distributed number of photons in each of the downconverted beams. Random deletion, associated with loss and finite detector quantum efficiency, leaves the form of the Bose-Einstein distribution intact.[7] The total number of photons n generated in each beam, in the counting time T, is therefore a sum of N independent Bose-Einstein random variables. The result is the well-known negative-binomial distribution[8] with a mean $\bar{n} = \lambda T$ that is the sum of the individual means and with a variance given by

$$\mathrm{Var}(n) = \bar{n} + \left(\frac{\bar{n}^2}{N}\right) = \bar{n}(1 + \lambda\tau_r). \tag{3}$$

The quantity $\lambda\tau_r$ represents the mean number of photons emitted in a single transit time. In our experiments, $\lambda = \bar{n}/T \approx 20{,}000$ and $\tau_r = l/c \approx 30$ ps, so $\lambda\tau_r \approx 6 \times 10^{-7}$. The fact that $\lambda\tau_r$ is small means that $\mathrm{Var}(n) \approx \bar{n}$ in equation 3, as for Poisson statistics. Stated differently, the mean number of photons emitted into each downconversion beam in a transit time is sufficiently small such that the probability of generating more than one photon is negligible; thus, the overall sequence of photon emissions is consistent with a Poisson point process. If it is desired to observe the bunching accompanying Bose-Einstein emissions, $\lambda\tau_r$ would have to be increased significantly.

The results described above assume that the pump is classical; that is, it exhibits no fluctuations. It also applies if the pump exhibits phase fluctuations with coherence time τ_c, but is effectively sinusoidal during any transit time ($\tau_r \ll \tau_c$).

Using the foregoing results, we conclude that the coincidence events must arise from the intersection of two initially identical point processes that have been independently randomly deleted. Because the relevant point processes all maintain their form under random deletion,[7] the surviving pairs of events comprising the coincidences follow the same statistics as the marginals, albeit with lower mean. For the parameters of our experiment, joint Poisson coincidences ensue, as confirmed by our observations.

More comprehensive results for the statistical behavior of the downconverted light require the use of a multimode approach. Joobeur *et al.*[9] constructed a theory of downconversion based on an interaction Hamiltonian that couples a pump wave of finite spectral width with a nonlinear crystal of finite spatial width l. The downconverted light is not cross-spectrally pure. Marginal signal and idler spectral densities, denoted $S_{\theta_s}(\omega_s)$ and $S_{\theta_i}(\omega_i)$, respectively, have been calculated for different observation directions. These can be used to obtain a coherence volume V_c and a degrees-of-freedom parameter $M_V = V/V_c$ characterizing the marginal count variances.

The signal-idler cross-spectral density $\tilde{S}_{\theta_s,\theta_i}(\omega_s)$ has also been obtained.[9] It is characterized by entanglement volumes V_s and V_i and associated spatiotemporal entanglement degrees of freedom M_s and M_i that determine the joint (and coincidence) statistics. One important result has already emerged from using this approach: the normalized coincidence rate R_{si} has been found to be suppressed substantially below unity unless M_s and M_i are both $\gg 1$.

All of this notwithstanding, for the range of parameters used in our experiments, the data are consistent with a simple model in which the pump may be regarded as a sequence of primary photons obeying Poisson statistics, each of which splits into a pair of downconverted photons that are created essentially simultaneously. Models of this type have been considered previously in other contexts.[10–12]

ACKNOWLEDGMENT

We are grateful to B. Yurke for valuable discussions.

REFERENCES

1. BURNHAM, D. C. & D. L. WEINBERG. 1970. Observation of simultaneity in parametric production of optical photon pairs. Phys. Rev. Lett. **25:** 84–87.
2. RARITY, J. G., P. R. TAPSTER, E. JAKEMAN, T. LARCHUK, R. A. CAMPOS, M. C. TEICH & B. E. A. SALEH. 1990. Two-photon interference in a Mach-Zehnder interferometer. Phys. Rev. Lett. **65:** 1348–1351.
3. CANTOR, B. I. & M. C. TEICH. 1975. Dead-time-corrected photocounting distributions for laser radiation. J. Opt. Soc. Am. **65:** 786–791.
4. TEICH, M. C. & G. VANNUCCI. 1978. Observation of dead-time-modified photocounting distributions for modulated laser radiation. J. Opt. Soc. Am. **68:** 1338–1342.
5. TEICH, M. C. 1985. Normalizing transformations for dead-time-modified Poisson counting systems. Biol. Cybern. **53:** 121–124.
6. YURKE, B. & M. POTASEK. 1987. Obtainment of thermal noise from a pure quantum state. Phys. Rev. A **36:** 3464–3466.
7. TEICH, M. C. & B. E. A. SALEH. 1982. Effects of random deletion and additive noise on bunched and antibunched photon-counting statistics. Opt. Lett. **7:** 365–367.
8. SALEH, B. E. A. 1978. Photoelectron Statistics, p. 13 & 81. Springer-Verlag. New York/Berlin.
9. JOOBEUR, A., B. E. A. SALEH & M. C. TEICH. 1994. Spatio-temporal coherence properties of entangled light beams generated by parametric downconversion. Phys. Rev. A **50:** 3349–3361.
10. SALEH, B. E. A., D. STOLER & M. C. TEICH. 1983. Coherence and photon statistics for optical fields generated by Poisson random emissions. Phys. Rev. A **27:** 360–374.
11. TEICH, M. C., B. E. A. SALEH & J. PEŘINA. 1984. Role of primary excitation statistics in the generation of antibunched and sub-Poisson light. J. Opt. Soc. Am. B **1:** 366–389.
12. TEICH, M. C. & B. E. A. SALEH. 1988. Photon bunching and antibunching. *In* Progress in Optics. Volume 26. E. Wolf, Ed.: 1–104. Elsevier. Amsterdam/New York.

The Lattice Dynamics of Completely Entangled States and Its Application to Communication Schemes

DANIEL I. FIVEL

Department of Physics
University of Maryland
College Park, Maryland 20742-4111

This report is organized as follows: First, I will make precise the concept of completely entangled states, which will be referred to as "EPR states" because of their relevance to discussions of the Einstein-Podolsky-Rosen problem.[1] I will derive some interesting physical consequences of the definition and then show that orthonormal sets of EPR states can be chosen in such a way that they also have a remarkable finite Heisenberg group structure that makes it possible to interpret them as points of a lattice phase space. A by-product will be an improvement of recently suggested cryptographic schemes.[2,3] I will then use the lattice phase space structure to show that certain quadratic Hamiltonians can induce EPR states to "hop" from one site to another at discrete time intervals and that the solution of the Schrödinger equation for these Hamiltonians exhibits peculiar phase oscillations that are expressed by the number-theoretic Legendre symbol. The Legendre symbol has a celebrated symmetry property given by Gauss' law of quadratic reciprocity, which now manifests itself as a symmetry of EPR states whose physical significance we do not yet understand. Indeed, as we examine the phases that appear for more general quadratic Hamiltonians, there appears to be a fascinating relationship between the structure of EPR states and the theory of quadratic Diophantine equations in number theory, of which the law of quadratic reciprocity is the precursor.

Consider a state $|u\rangle$ of two spin-J particles so that each partner inhabits a Hilbert space of dimension $N = 2J + 1$. Let $|\alpha\rangle$ be the one-particle states. We use the notation $|j\rangle$ ($j = 1, 2, \cdots, N$) to indicate an arbitrary one-particle basis with an extra 1 or 2 index when it is necessary to specify the particle in question. To carry through the familiar EPR argument, one requires the following properties of the state $|u\rangle$: First, there is a probability of $1/N$ to find particle-1 in any state $|\alpha\rangle$. Second, there is a one-to-one correspondence $|\alpha\rangle \to |\tilde{\alpha}\rangle$ such that the conditional probability $\mathscr{P}(\alpha|\tilde{\alpha})$ is unity for finding particle-1 in state $|\alpha\rangle$ when it is given that its partner is in state $|\tilde{\alpha}\rangle$. States with this property are said to be completely entangled states or "EPR states". The structure of EPR states is determined by the following theorem:

THEOREM 1: A necessary and sufficient condition that two spin-J particles form an EPR state $|u\rangle$ is that there exists an antiunitary one-particle transformation $\mathscr{U}$ on particle-2 such that

$$|u\rangle = N^{-1/2} \sum_{j=1}^{N} |j, 1\rangle \mathscr{U} |j, 2\rangle. \tag{1}$$

The proof of sufficiency depends on first establishing that the expression on the right of equation 1 is independent of the basis. To prove basis-independence, note that if one makes a change of basis from $|j\rangle$ to $|j'\rangle$ there will be a unitary matrix C_{jk} such that

$$|j\rangle = \sum_k C_{jk}|k'\rangle. \tag{2}$$

However, from antiunitarity of $\mathscr{U}$, we have $\mathscr{U}|j\rangle = \Sigma_l C^*_{jl}\mathscr{U}|l\rangle$, in which the "*" signifies complex conjugation. Thus, summing on j and using the unitarity of C, it follows that the right side of equation 1 is unchanged if the j in both kets is replaced with j'. Next, one deduces the EPR property as follows: Let $\pi(\alpha) = |\alpha\rangle\langle\alpha|$ be the projector on the state $|\alpha\rangle$. Then, the probability of finding particle-1 in the state $|\alpha, 1\rangle$ when the two-particle system is in the state $|u\rangle$ will be $\mathrm{Tr}[\pi(\alpha, 1)\pi(u)]$ and, to evaluate this, we can change the basis in equation 1 so that $|\alpha\rangle$ is one of the basis elements. Hence, only the α term survives in the trace, which evaluates to $1/N$ as required. Then, if we choose

$$|\tilde{\alpha}\rangle = \mathscr{U}|\alpha\rangle, \tag{3}$$

the conditional probability $\mathscr{P}(\alpha|\tilde{\alpha})$ is given by

$$\mathscr{P}(\alpha|\tilde{\alpha}) = \frac{\mathrm{Tr}[\pi(\alpha, 1)\pi(\tilde{\alpha}, 2)\pi(u)]}{\mathrm{Tr}[\pi(\tilde{\alpha}, 2)\pi(u)]} = 1, \tag{4}$$

where we again used the basis-independence and chose a basis containing $|\tilde{\alpha}\rangle$ to compute the trace, thereby verifying the second required property for EPR states and completing the proof of sufficiency in theorem 1. The proof of necessity is a bit more involved and is given in APPENDIX A.

Let us now explore some consequences of this theorem. First of all, we observe that every antiunitary operator $\mathscr{U}$ can be written in the form,

$$\mathscr{U} = u\mathscr{T}, \tag{5}$$

where $\mathscr{T}$ is some fixed antiunitary operator and u ranges over the set of unitary operators. Thus, we may arbitrarily select $\mathscr{T}$ to be the time-reversal operator and then let u range over the group U_N of unitary transformations on particle-2. This justifies using u as a label for the EPR states as we did on the left side of equation 1. Hence, the EPR states form a manifold in the two-particle Hilbert space whose geometry is that of the set of one-particle unitary operators. There will be N^2 linearly independent EPR states that thus span the two-particle spin space. However, it is important to keep in mind that the EPR states do not form a linear manifold because, in general, linear combinations of entangled states will not be of this form. Indeed, the sum of two EPR states can be a factorized state that clearly does not exhibit EPR correlations. The fact that two-particle states can be labeled by one-particle operators was observed long ago by von Neumann[4] and was extensively studied by Herbut and Vujĉiĉ.[5,6] We shall use their term correlation operator to refer to u. Computations involving states of the form of equation 1 can be done using the representation of the state by its correlation operator because of the readily verified identity:

$$\langle u | v \rangle = N^{-1}\mathrm{Tr}(u^\dagger v). \quad \textbf{(6)}$$

Moreover, we can describe the transformation of an EPR state $|u\rangle$ by the action of a unitary transformation v acting on particle-2 through the relation,

$$v|u\rangle = |vu\rangle. \quad \textbf{(7)}$$

Thus, one makes new EPR states from old ones, not by adding vectors, but by multiplying one-particle unitary operators, thereby changing the form of the correspondence expressed by equation 3. When no confusion will result, we shall drop the ket notation and adopt the following convention: u will indicate either an EPR state or the operator that produces that state from some arbitrarily selected fiducial EPR state through a transformation on particle-2.

Before going further, it may be helpful to see how the above general results are expressed in the familiar case of $N = 2$, that is, for spin-½ particles. Consider the four orthogonal entangled states:

$$\Psi^{(1)} = 2^{-1/2}\{|\alpha, 1\rangle|\alpha', 2\rangle - |\alpha', 1\rangle|\alpha, 2\rangle\}, \quad \Psi^{(2)} = 2^{-1/2}\{|\alpha, 1\rangle|\alpha', 2\rangle + |\alpha', 1\rangle|\alpha, 2\rangle\},$$

$$\Psi^{(3)} = 2^{-1/2}\{|\alpha, 1\rangle|\alpha, 2\rangle - |\alpha', 1\rangle|\alpha', 2\rangle\}, \quad \Psi^{(4)} = 2^{-1/2}\{|\alpha, 1\rangle|\alpha, 2\rangle + |\alpha', 1\rangle|\alpha', 2\rangle\},$$

$$\textbf{(8)}$$

in which α and α' label an arbitrarily selected pair of orthogonal spins. (Observe that the sum of the first two is a factorized state.) For spin-½ particles, one can define the time-reversal operator $\mathscr{T}$ so that in a particular basis we have

$$\mathscr{T}|\downarrow\rangle = |\uparrow\rangle, \qquad \mathscr{T}|\uparrow\rangle = -|\downarrow\rangle. \quad \textbf{(9)}$$

Thus, $\Psi^{(1)}$ (the familiar spin-0 state) is the special case of equation 1 where u is the unit operator in equation 5. Observe how the four states $\Psi^{(j)}$ $(j = 1, 2, 3, 4)$ are obtained from $\Psi^{(1)}$ by applying, respectively, the four unitary transformations, I, σ, τ, $\sigma\tau$, to particle-2, where I is the unit operator and

$$\sigma = \begin{pmatrix} -1 & 0 \\ 0 & 1 \end{pmatrix}, \qquad \tau = \begin{pmatrix} 0 & 1 \\ 1 & 0 \end{pmatrix}. \quad \textbf{(10)}$$

The rules (equation 3) relating $|\tilde{\alpha}\rangle$ to $|\alpha\rangle$ in the four cases are

$$\mathscr{U} = \xi\mathscr{T}, \qquad \xi = I, \sigma, \tau, \sigma\tau. \quad \textbf{(11)}$$

Although it is not the principal thrust of the present discussion, it is perhaps worth remarking here that there is another important fact that emerges from theorem 1 that has been pointed out elsewhere:[7] Suppose we prepare a particle-1 beam in the state $|\alpha\rangle$ noninvasively, that is, by doing what one does in the EPR argument, admitting to the beam only the particles in an EPR state whose partners have passed a filter for the correlated state $|\tilde{\alpha}\rangle$. Let us then compute the attenuation of such an $|\alpha\rangle$ beam when it is passed through a filter for the state $|\beta\rangle$. If one uses equation 1 to compute this, the result is $|\langle\beta|\alpha\rangle|^2$; that is, it is precisely the same result that one would have obtained if the $|\alpha\rangle$ beam had been prepared invasively by first passing it through a filter for this state. Thus, a corollary of theorem 1 is that quantum mechanics has a "Markovian" property that it shares with classical mechanics, by which I mean that it makes the same prediction for different methods of initial state preparation. In the absence of such a property, the notion of a "state" would be ambiguous. On the other hand, it can be shown[7] that this Markovian property cannot hold for any locally realistic theory that reproduces the observed one-particle beam result (Malus' law). In particular, the "toy" hidden-variable model given by Bell,[8] which he used to argue for the necessity of two-particle experiments, is actually non-Markovian. The importance of this observation is that it supports the conclusion drawn by Griffiths[9] from a consistent history analysis showing that the counterintuitive aspect of quantum mechanics is already revealed in single-particle beam experiments. Thus, if one believes Malus' law and that nature is Markovian (as both classical and quantum mechanics agree it must be), then two-photon correlation measurements[10,11] are unnecessary for the ruling out of locally realistic theories. To put the matter positively, one may say that what these experiments do accomplish is to verify that nature is indeed Markovian.

Let us now return to our main theme: the analysis of EPR states in the general case with arbitrary N. It is the ability to transform EPR states by action on just one of the two particles that makes the cryptographic schemes possible. For this purpose, "Alice" and "Bob" agree on an orthonormal set of EPR states or, equivalently, upon the set of N^2 operations on particle-2 that will transform the fiducial state into one of this set. Alice prepares one of the states and sends particle-2 to Bob while retaining particle-1. Bob applies one of the agreed unitary operators to particle-2 and returns it to Alice. Because she possesses both particles, she can determine the state and hence what Bob did by a generalized Stern-Gerlach measurement. No eavesdropper can get information because it is encoded into the relative quantum mechanical phase.

Because the operators in question belong to the group U_N, we now ask the following question: can one select N^2 operators from U_N so that they not only generate the required N^2 orthogonal states, but also give the ray representation of a finite subgroup $\mathscr{G}$ within U_N. If this is possible, the communication scheme gains flexibility by virtue of the fact that Bob (or Alice) can *repeatedly* apply the operators without causing the state to leave the set of N^2 states that are recognizable by Alice. I shall next demonstrate that not only does such a possibility exist, but that the species of finite group generated is uniquely determined and has an elegant, physically interpretable structure.

One must first observe that the qualifier "ray" is essential; that is, there will be no such group $\mathscr{G}$ for which we obtain a "true" representation for the following reason: clearly, $\mathscr{G}$ cannot be Abelian because, with $N \times N$ matrices, at most N linearly independent ones can commute. Now, suppose that $N = p$, where p is prime. It is known[12] that the only groups of order p^2 are Abelian, namely, the cyclic group of order p^2 and the direct product of two cyclic groups, each of order p. The first possibility does not give a non-Abelian result even when we extend to a ray representation by allowing arbitrary phases. However, the second one does. To see why, one need only recall how the coordinate and momentum operators X, P of a particle each generate Abelian groups $e^{i\alpha X}$, $e^{i\beta P}$ that also commute with one another except for a phase multiple. The resulting group is the so-called Heisenberg group over the real numbers. This immediately suggests how to construct the solution in the case where N is prime and leads us to expect a finite Heisenberg group over the integers mod(p). The analysis for the case of nonprime N is complicated by the fact that in such cases there exist EPR correlations between subspaces of more than one dimension (nontrivial divisors of N) as well as the familiar type of correlation between pure states. The theory of higher manifold entanglement is under investigation by myself, but is not yet complete. Hence, I shall confine myself to the case of prime N for the remainder of this report and shall replace N by p to keep this restriction in mind. It is to be noted that the integers mod(p) form a field $\mathscr{Z}_p$, which means that all matrix manipulations can be done just as for the reals, remembering only that one may not divide by multiples of p. (We shall not have to compute eigenvalues, so the fact that $\mathscr{Z}_p$ is not algebraically closed will not cause problems.)

From the analogy with the $e^{i\alpha X}$, $e^{i\beta P}$ operators, let σ, τ be unitary operators in the p-dimensional particle-2 Hilbert space $\mathscr{H}_p^{(2)}$ satisfying

$$\sigma\tau = \omega\tau\sigma, \qquad \omega = e^{2\pi i/p}. \tag{12}$$

These can be realized by introducing a basis $|j\rangle$ [$j = 0, 1, \ldots, (\bmod\, p)$] in $\mathscr{H}_p^{(2)}$ and letting

$$\sigma|j\rangle = \omega^j|j\rangle, \qquad \tau|j\rangle = |j+1\rangle. \tag{13}$$

(Note that $\tau|p-1\rangle = |0\rangle$.) We now define the p^2 u-matrices of our EPR basis to be $u(\mathbf{j})$, where $\mathbf{j} = (j, k)$ with $j, k = 0, 1, \ldots, (p-1)$ and

$$u(\mathbf{j}) = e^{-i\pi jk/p}\sigma^j\tau^k. \tag{14}$$

With this choice of phases, we have the simple group multiplication law:

$$u(\mathbf{j})u(\mathbf{j}') = e^{i\pi\mathbf{j}\wedge\mathbf{j}'/p}u(\mathbf{j}+\mathbf{j}'), \qquad \mathbf{j}\wedge\mathbf{j}' \equiv jk' - kj', \tag{15}$$

which is the analogue mod(p) of the quantum mechanical rule for multiplying operators of the form $u(\alpha, \beta) \equiv e^{i(\alpha X + \beta P)}$. Note that the phase factor on the right is such that the operators on the left commute whenever $\mathbf{j}$ and $\mathbf{j}'$ are linearly dependent, whence in particular

$$u(n\mathbf{j}) = [u(\mathbf{j})]^n, \qquad n\in\mathscr{Z}. \tag{16}$$

Using the familiar character identity

$$\sum_{n=0}^{p-1} e^{2\pi in/p} = \delta_{n0}, \tag{17}$$

where the Kronecker symbol is understood as mod(p), one obtains

$$\mathrm{Tr}[u(\mathbf{j})] = p\delta_{\mathbf{j},\mathbf{o}}, \tag{18}$$

where $\mathbf{o}$ is the zero vector. This gives the trace-orthonormality of the $u(\mathbf{j})$ terms.

The Heisenberg group structure that we have obtained shows that the states of the lattice constitute a set of coherent states associated with a phase space on the field $\mathscr{Z}_p$. It is worth noting that there is a striking difference between this and the phase space over the real numbers $\mathscr{R}$ in that the former are mutually orthogonal, whereas the latter are not. Thus, the coherent states on $\mathscr{Z}_p$ give a complete rather than an overcomplete characterization of compatible measurements. We shall denote this lattice as $\mathscr{L}_p$.

The states of $\mathscr{L}_p$ are generated by the actions of the $u(\mathbf{j})$ on a fiducial state and thus we need only consider what happens to these operators when the two particles evolve under some specified time evolution. If unitary transformations V_1, V_2 are applied to the two particles, respectively, then it is shown in APPENDIX B that $u(\mathbf{j})$ is transformed by

$$u(\mathbf{j}) \rightarrow V_2 u(\mathbf{j}) V_1. \tag{19}$$

Note that the trace-orthogonality of the u terms is preserved for any choice of V_1, V_2; however, to preserve the Heisenberg group structure, we must require

$$V_2 = V_1^{-1}. \tag{20}$$

Thus, we shall restrict to dynamical laws of the form,

$$V_2 = U(t), \qquad V_1 = U^{-1}(t) = U(-t), \tag{21}$$

which means that the two particles evolve as they would if they were antiparticles of one another.

The phase space analogy now suggests that we look for dynamical processes that will cause the states on the lattice $\mathscr{L}_p$ to hop from one point to another at discrete time intervals. This is a difficult constraint to implement for it requires

$$U(t)u(\mathbf{j})U^{-1}(t) = u[\mathbf{j}(t)], \qquad t = 0, 1, 2, \ldots, \tag{22}$$

so $\mathbf{j}(t)$ defines an orbit on $\mathscr{L}_p$. Indeed, it is not difficult to see that this can only be achieved by a linear transformation on $\mathbf{j}$. For if one puts $u(\mathbf{j}') = Uu(\mathbf{j})U^{-1}$, one may check from equations 15 and 16 that, for any integers n_1, n_2, we must have

$$(n_1\mathbf{j}_1 + n_2\mathbf{j}_2)' = n_1\mathbf{j}_1' + n_2\mathbf{j}_2', \qquad \mathbf{j}_1' \wedge \mathbf{j}_2' = \mathbf{j}_1 \wedge \mathbf{j}_2. \tag{23}$$

In other words, we must have the following theorem:

THEOREM 2:

$$Uu(\mathbf{j})U^{-1} = u(\mathscr{M}\mathbf{j}), \tag{24}$$

in which $\mathscr{M}$ is a linear transformation with integer coefficients mod(p) that leaves the cross product invariant mod(p). Thus, $\mathscr{M}$ is a member of the group $SL_2(\mathscr{Z}_p)$.

We then see that the quantum mechanical evolution operator U will be associated with a linear transformation $\mathscr{M}$ on the finite two-dimensional lattice $\mathscr{L}_p$ of integers mod(p) and we note that $SL_2(\mathscr{Z}_p)$ is just the $\mathscr{L}_p$ analogue of a symplectic (canonical) transformation defining a hopping orbit at $t = 0, 1, 2, \ldots$, that is,

$$\mathbf{j} \rightarrow \mathbf{j}(t) = \mathscr{M}(t)\mathbf{j}, \tag{25}$$

on the two-dimensional phase space lattice $\mathscr{L}_p$.

An elegant relationship between the quantum evolution generated by $U_{\mathscr{M}}$ and the $\mathscr{L}_p$ hopping described by the corresponding $\mathscr{M}$ will now be obtained: We can guess at the explicit form of $U = U_{\mathscr{M}}$ that satisfies equation 24 for various choices of $\mathscr{M}$ from our experience with X, P in quantum mechanics. There, it is quadratic Hamiltonians $\mathscr{H}$ in X, P that generate the linear transformations of X, P, that is,

$$e^{i\mathscr{H}}Xe^{-i\mathscr{H}} = aX + bP, \qquad e^{i\mathscr{H}}Pe^{-i\mathscr{H}} = a'X + b'P,$$

$$\mathscr{H} = AP^2 + B(PX + XP) + CX^2. \tag{26}$$

Moreover, $e^{i\mathscr{H}}$ can be written as an integral over $e^{i(\alpha X+\beta P)}$ in which the Fourier coefficient is the exponential of a quadratic form in α, β. Because the $u(\mathbf{j})$ terms that are the analogues of $e^{i(\alpha X+\beta P)}$ are also complete over the p^2-dimensional vector space of $p \times p$ matrices, we know that $U_{\mathscr{M}}$ can be similarly expressed as a linear combination of them. We naturally guess that the coefficient will once again be the exponential of a quadratic form $\tilde{\mathbf{j}}\mathscr{Q}\mathbf{j}$ in which $\mathscr{Q}$ is a two-by-two matrix and the tilde indicates a row vector. Indeed, we find that for each $\mathscr{M}$ there will be a two-by-two matrix $\mathscr{Q}_{\mathscr{M}}$ such that

$$U_{\mathscr{M}} = \sum_{\mathbf{j}} \exp\{(\pi i/2p)\tilde{\mathbf{j}}\mathscr{Q}_{\mathscr{M}}\mathbf{j}\}u(\mathbf{j}) \tag{27}$$

or possibly a degenerate form of this in which the double sum reduces to a simple sum. A straightforward approach to the determination of the relationship between $\mathscr{Q}_{\mathscr{M}}$ and $\mathscr{M}$ exploits the fact that the group $SL_2(\mathscr{Z}_p)$ is generated[13] by ρ, χ with

$$\rho = \begin{pmatrix} 1 & 1 \\ 0 & 1 \end{pmatrix}, \qquad \chi = \begin{pmatrix} 1 & 0 \\ 1 & 1 \end{pmatrix}. \tag{28}$$

One readily verifies that the corresponding U_ρ and U_χ will be degenerate forms of equation 27. Specifically, one verifies using equation 17 that the following formula holds for any powers of the generators:

$$U_{\rho^m} = \sum_{j=0}^{p-1} e^{-\pi imj^2/p}u(j, 0), \qquad U_{\chi^m} = \sum_{k=0}^{p-1} e^{\pi imk^2/p}u(0, k), \qquad m = 1, 2, \ldots. \tag{29}$$

Hence, one might, in principle, find a $U_{\mathscr{M}}$ for any $\mathscr{M}$ by decomposing it into a product of powers of ρ and χ and then one can use equation 29. However, simpler and more instructive methods are available as indicated in APPENDIX C.

Because our primary purpose here is to clarify the relationship between the quantum dynamics determined by U and the associated classical motion on the lattice determined by $\mathscr{M}$, it will suffice to focus on the simplest example, namely, the dynamical process obtained by iterating one of the group generators (equation 28). The corresponding quadratic form gives us the lattice analogue of free-particle motion because the associated quadratic form is one that behaves like the free-particle Hamiltonian P^2. It will turn out to simplify matters if we consider a sequence of even numbers and examine

$$U(2t) = (U_\rho)^{2t}, \qquad t = 0, 1, \ldots. \tag{30}$$

In quantum mechanics, the solution of the Schrödinger equation entails the explicit computation of the time-evolution operator. In the case of equation 30, this means that we must explicitly compute the right side for any choice of t. Now, we note that it follows from equation 24 that U_{ρ^m} must coincide with $(U_\rho)^m$ up to a phase. Hence, equation 30 can be written as

$$U(2t) = e^{i\psi(t)}U_{\rho^{2t}}, \tag{31}$$

in which we have an explicit formula for $U_{\rho^{2t}}$ from equation 29 with $m = 2t$; thus, "solving the Schrödinger equation" reduces to determining the phase $e^{i\psi(t)}$. As we shall now see, this phase turns out to be extraordinarily interesting. First, from equation 29, note that the taking of traces on both sides of equation 31 will give

$$e^{i\psi(t)} = p^{-1}\text{Tr}[(U_\rho)^{2t}]. \tag{32}$$

If one inserts the left side of equation 29 for U_ρ (with $m = 1$), there will be a product of sums indexed by $j_1, j_2, \ldots, j_{2t}$ containing $u(j_1 + \cdots + j_{2t}, 0)$ that has zero trace unless the sum of the j terms is 0 mod(p). One can then use equation 17 to pick out this term (a standard trick) and one obtains

$$\text{Tr}[(U_\rho)^{2t}] = \sum_n [F(n)]^{2t}, \tag{33}$$

where

$$F(n) = \sum_j e^{-i\pi j^2/p}e^{2\pi i n j/p} = S(-2p)e^{i\pi n^2/p}, \qquad S(p) = \sum_n e^{2\pi i n^2/p}, \tag{34}$$

in which the second step results from completion of the square. Thus,

$$e^{i\psi(t)} = p^{-1}[S(-2p)]^{2t}S(p/t). \tag{35}$$

The function $S(x)$ is known as a Gaussian sum and such sums are fundamental in the solution of quadratic Diophantine equations. Fortunately, thanks to Gauss (who is reputed to have worked for five years to prove it),[14] we have a beautiful formula for $S(p/t)$ when p is an odd prime, namely,

$$S(p/t) = \left(\frac{t}{p}\right)\sqrt{\left(\frac{-1}{p}\right)p}, \tag{36}$$

where the Legendre symbol is defined by

$$\left(\frac{t}{p}\right) = \begin{cases} +1, & \text{for } t \text{ a square mod}(p) \\ -1, & \text{for } t \text{ not a square mod}(p). \end{cases} \tag{37}$$

Now suppose that our $U(t)$ is generated by a Hamiltonian. If we add an arbitrary constant E to that Hamiltonian, it will introduce an extra factor e^{-2iEt} in equation 31. Hence, the argument of the factor $[S(-2p)]^{2t}$ can be "gauged" away along with the t-independent factor in equation 36, which is removed by changing the time origin. Thus, due to the gauge, we have established the extremely surprising fact of theorem 3:

THEOREM 3:

$$e^{i\psi(t)} = \left(\frac{t}{p}\right), \tag{38}$$

that is, the time-evolving phase of the "free" EPR state follows a pattern of +1's and −1's in a manner with basic number-theoretic significance. Although the notion of "sign" in the usual sense does not exist in the field $\mathscr{Z}_p$, we can give it meaning if, as for real numbers, we define a positive number as one that is the square of something, whereas a negative number is one that is not. Thus, the Legendre symbol extends the notion of "sign" to $\mathscr{Z}_p$. It can also be shown that for $p > 2$ there are just as many squares (quadratic residues) as nonsquares. Therefore, we have obtained the quite pleasing result that the analogue of free-particle motion in the EPR lattice is characterized by a wave function with the $\mathscr{Z}_p$ counterpart of a sign-alternating phase.

The computation of the Legendre symbol is facilitated by a factorization law that reduces it to a product of Legendre symbols whose upper members are the prime factors of t. These in turn obey the celebrated and profound Gaussian law of quadratic reciprocity:[14,15]

$$\left(\frac{p^*}{q}\right) = \left(\frac{q}{p}\right), \qquad p^* = (-1)^{(p-1)/2}p. \tag{39}$$

One cannot avoid the obvious question here as to what this symmetry means physically when the Legendre symbol appears in determining the dynamical phase evolution of EPR states. I do not know the answer.

It is clear from the above and from the discussion in the appendices that, when we come to investigate and classify more general lattice Hamiltonians, we will encounter generalized Gaussian sums (theta-series)[13,15] and will have to invoke the general theory of quadratic Diophantine equations. It thus appears that we have just scratched the surface of fruitful connections between the lattice dynamics of EPR states and one of the richest areas of contemporary mathematics.

REFERENCES

1. EINSTEIN, A., B. PODOLSKY & N. ROSEN. 1935. Phys. Rev. **47:** 777.
2. BENNETT, C. H. & S. J. WIESNER. 1992. Phys. Rev. Lett. **69:** 2881.
3. BENNETT, C. H., G. BRASSARD, C. CRÊPEAU, R. JOSZA, A. PERES & W. WOOTTERS. 1993. Phys. Rev. Lett. **70:** 1895.
4. VON NEUMANN J. 1955. Mathematical Foundations of Quantum Mechanics. Princeton University Press. Princeton, New Jersey.
5. HERBUT, F. & M. VUJĈIĈ. 1971. Scuola Internationale Enrico Fermi. B. d'Espagnat, Ed.
6. HERBUT, F. & M. VUJĈIĈ. 1967. J. Math. Phys. **8:** 1345.
7. FIVEL, D. I. 1992. University of Maryland preprint no. UMD-PP-132.
8. BELL, J. S. 1964. Physics **1:** 195.
9. GRIFFITHS, R. B. 1987. Am. J. Phys. **55:** 11.
10. SHIH, Y. H. & C. O. ALLEY. 1988. Phys. Rev. Lett. **61:** 2921.
11. ASPECT, A., J. DALIBARD & G. ROGER. 1982. Phys. Rev. Lett. **49:** 1804.
12. HALL, M. 1959. Theory of Groups, p. 51. Macmillan Co. New York.
13. HUA, L. K. 1982. Introduction to Number Theory, p. 367. Springer Pub. New York.

14. FLATH, D. E. 1989. Introduction to Number Theory. Wiley. New York.
15. SCHOENEBERG, B. 1974. Elliptic Modular Functions, p. 218. Springer Pub. New York.

APPENDIX A

Proof of Necessity in Theorem 1

Let $|S\rangle$ be an EPR state with each particle having spin J. Here, x, y are used as labels for one-particle states and $|x\rangle \to |\tilde{x}\rangle$ is the map associated with the EPR correlation that is presumed to exist. Thus, $|S_x, 1\rangle \equiv \langle\tilde{x}, 2|S\rangle$, $|S_x, 2\rangle \equiv \langle x, 1|S\rangle$ (which make sense notationally because $|S\rangle$ is a two-particle state). Then, $\langle x, 1|S_x, 1\rangle = \langle\tilde{x}, 2|S_x, 2\rangle$. One checks that the perfect correlation condition can be expressed in the form,

$$\langle S_x, 1|x, 1\rangle\langle x, 1|S_x, 1\rangle = \langle S_x, 1|S_x, 1\rangle \neq 0, \tag{40}$$

and is equal to the same expression with 1 replaced by 2. Hence, one deduces

$$|S_x, 1\rangle = \gamma(x)|x, 1\rangle \qquad \text{and} \qquad |S_x, 2\rangle = \gamma(x)|\tilde{x}, 2\rangle, \tag{41}$$

where $\gamma(x)$ is a nonvanishing complex number. Then,

$$\gamma(x)\langle y, 1|x, 1\rangle = \langle y, 1|S_x, 1\rangle = \langle\tilde{x}, 2|y, 1|S\rangle = \gamma(y)\langle\tilde{x}, 2|\tilde{y}, 2\rangle. \tag{42}$$

Multiplying the left and right members by the complex conjugate $\gamma(x)^*$ and noting that

$$\gamma(x)^*\langle\tilde{x}, 2|\tilde{y}, 2\rangle = \gamma(x)\langle\tilde{y}, 2|\tilde{x}, 2\rangle^* = \gamma(y)^*\langle y, 1|x, 1\rangle, \tag{43}$$

it follows that

$$|\gamma(x)| = |\gamma(y)| \qquad \text{if} \qquad \langle x|y\rangle \equiv 0, \tag{44}$$

so $\gamma(x)$ is unimodular up to a constant factor. Hence, by redefining $|\tilde{x}, 2\rangle$ to absorb the unimodular factor $\gamma(x)$, it follows from equation 42 that the map $|x\rangle \to |\tilde{x}\rangle$ is antiunitary. One may then select any basis to express $|S\rangle$ in the form,

$$|S\rangle = \sum_{i,j=1}^{N} \alpha_{ij}|i, 1\rangle|\tilde{\mathbf{j}}, 2\rangle, \tag{45}$$

and can use equation 41 to show that $\alpha_{ij} = \delta_{ij}/\sqrt{N}$, thereby giving the theorem. □

APPENDIX B

Behavior of **u***-Operators under Dynamical Transformation of Particles*

If particles 1 and 2 are transformed by unitary operators V_1 and V_2, respectively, one sees from the fact that the expression in equation 1 is invariant to a change of basis that the effect is the same as transforming the correlation operator by

$$u \to V_2 u V_1^*, \qquad \text{where} \qquad V^* \equiv \mathcal{T} V^{-1} \mathcal{T}^{-1}. \tag{46}$$

If V is a unitary time-evolution operator associated with a time-reversal invariant Hamiltonian, then one sees that

$$V^*(t) = V(t) \tag{47}$$

as asserted.

APPENDIX C

Relation of $\mathscr{Q}_{\mathscr{M}}$ to $\mathscr{M}$

Suppose that in equation 27 we put

$$\mathscr{Q}_{\mathscr{M}} = \begin{pmatrix} 2a & b \\ b & 2c \end{pmatrix}, \tag{48}$$

with integers a, b, c and with the following restrictions on the discriminant $\Delta \equiv b^2 - 4ac$:

$$\text{(i)} \quad \Delta \not\equiv 1 \ (\mathrm{mod}\, p), \qquad \text{(ii)} \quad \Delta \equiv 1 \ (\mathrm{mod}\ 4), \tag{49}$$

that is, b odd. Then, it can be shown that $\mathscr{Q}_{\mathscr{M}}$ is related to $\mathscr{M}$ by a Cayley transform:

$$\mathscr{M} = \frac{(\nu \mathscr{Q}_{\mathscr{M}} + I)}{(\nu \mathscr{Q}_{\mathscr{M}} - I)}, \qquad \nu = \begin{pmatrix} 0 & 1 \\ -1 & 0 \end{pmatrix}, \tag{50}$$

where the computations in equation 50 are in $\mathscr{Z}_p$. One sees from equation 50 how the analogue of the canonical structure of quantum mechanics is expressed in the finite lattice phase space of EPR states: if $\mathscr{R}$ is a unimodular matrix, one verifies that $\mathscr{R}^{-1}\nu = \nu\tilde{\mathscr{R}}$; thus, equation 50 continues to hold under the transformation,

$$\mathscr{M} \rightarrow \mathscr{R}^{-1}\mathscr{M}\mathscr{R}, \qquad \mathscr{Q}_{\mathscr{M}} \rightarrow \tilde{\mathscr{R}}\mathscr{Q}_{\mathscr{M}}\mathscr{R}. \tag{51}$$

Therefore, each $U_{\mathscr{M}}$ producing a "hopping" of EPR states from one lattice site to another will have a counterpart under the canonical transformation $\mathscr{R}$ that, as one sees from its relation to ν, is a finite symplectic transformation of the lattice. Note that $\mathscr{R}$ preserves the discriminant and thus the two conditions used in deriving equation 50.

Sending Classical Bits via Quantum Its

PAUL HAUSLADEN,[a] BENJAMIN SCHUMACHER,[b]
MICHAEL WESTMORELAND,[c] AND
WILLIAM K. WOOTTERS[d]

[a]*Department of Physics*
University of Pennsylvania
Philadelphia, Pennsylvania 19104

[b]*Department of Physics and Astronomy*
Kenyon College
Gambier, Ohio 43022

[c]*Department of Mathematics*
Denison University
Granville, Ohio 43023

[d]*Department of Physics*
Williams College
Williamstown, Massachusetts 01267

Two of us have had the privilege of being students of John Wheeler. One of us is a grandstudent. We have all been close enough to this great teacher to have been strongly influenced not only by his ideas, but also by his passionate drive to understand the world. It is a pleasure to present this report in his honor.

In recent years, John has argued persuasively that physics stands to learn a great deal about the world by looking at it in terms of information.[1] Information occupies a wonderfully ambiguous place somewhere between the concrete and the subjective. It does not exist without observers, but it is concrete enough to be measured and manipulated. In this report, we look at one aspect of the measurement and manipulation of information, namely, the transmission of classical information via quantum objects.

Our focus is on the following communication problem. Alice wants to send to Bob a classical message, which we picture as a long sequence of 0's and 1's. We suppose that Alice has already compressed her classical message to the minimum number of bits, in which case the 0's and 1's occur with roughly equal frequency. Unfortunately for Alice, she is required to transmit this information in a sequence of photons, using only the photons' polarizations to encode the information. Moreover, she is allowed to use only certain states of polarization and, finally, she is required to use them in certain specified proportions. Our question is this: Under those conditions, how many classical bits can Alice convey per photon? (The answer will always be less than or equal to 1.)

We can state the question again in slightly more technical terms. Alice is given an *ensemble* with which she is required to perform her transmission, an ensemble being a set of quantum states with associated probabilities:

$$\mathscr{E} = \{(\rho_i, p_i)\}, \qquad i = 1, \ldots, m. \tag{1}$$

We have written ρ_i for the states of the ensemble because in general these states may be mixed. However, for the time being, we will consider only pure-state ensembles. Our question can be rephrased as follows: What is the classical information capacity of an ensemble $\mathscr{E}$?

In order to get a feel for the problem, it may help to consider various examples. As the first example, suppose that Alice is allowed to use only vertical and horizontal polarizations, and she is required to use them equally often. This is actually the best situation she could be in because all she has to do in that case is to let "vertical" represent 0 and "horizontal" represent 1. At the receiving end, Bob can distinguish perfectly between these two orthogonal states, so each photon carries a full bit of classical information. On the other hand, if Alice is allowed to use only "vertical" and "45° to the right of vertical", then Bob will not be able to distinguish the states perfectly. In that case, Alice will have to use an error-correcting code—that is, she will have to use some redundancy—to make up for Bob's unavoidable errors and, as a result, she will have to use more than one photon for each classical bit. If we continue to allow Alice only two states, but make them closer and closer to each other, then it is plausible that she will have to send more and more photons per bit in order to get her message across with negligible error. In this report, we aim to find the exact quantitative relation that goes along with this intuition.

Immediately one may object: What a stupid way to send information! Why would Alice ever be required to use nonorthogonal states?

There are two answers to this objection. First, in quantum cryptography, one intentionally uses nonorthogonal signal states to avoid undetected eavesdropping.[2] It is therefore interesting to ask at what rate information can be transmitted via such states. Second, and more generally, one would like to develop a set of concepts and theorems that can be used to speak rigorously and quantitatively about information in quantum mechanics in a variety of contexts. Much work has already been done toward that end, some of which we mention below. The question we have posed here can be thought of as one piece in a puzzle that, once it is all assembled, will constitute a complete theory of quantum information.

To begin to answer our question, let us first turn our attention to Bob's end of the communication channel. Bob knows what states Alice will be sending and he is faced with the problem of figuring out what measurement to make. Some measurements will clearly be better than others, but it is by no means immediately obvious how we are to *rate* his possible measurements in a way that is relevant to our question. How does a particular measurement strategy at Bob's end translate into some definite number of photons that Alice needs to send per bit?

As we said before, the fact that Bob cannot perfectly distinguish the states forces our communicators to use an error-correcting code. Does this mean that we need to study up on error-correcting codes before we can make progress? Fortunately, the answer is no. There is a wonderful, simple theorem due to Shannon[3] that fully characterizes the capabilities of optimal error-correcting codes. Applied to our context, the theorem reads as follows: If Bob gets I bits of information on average from each photon, then in order to convey n classical bits with negligible error Alice will need to send n/I photons (as long as n is large enough). Here, the quantity I is defined in terms of Bob's probabilities of the possible states. Before he makes his measurement, the various possible states have certain *a priori* probabilities. After he makes his measurement, he will assign new *a posteriori* probabilities to the same states because he will have learned something. Both before and after his measurement, one can speak of the *entropy* H of Bob's probability distribution:

$$H = -\sum_{i=1}^{s} p_i \log p_i, \tag{2}$$

where s is the number of states he is trying to distinguish and the logarithm is base 2. H measures the amount of information that Bob lacks about the state; for example, H is zero if Bob knows the state with certainty. If all s states are equally likely, then $H = \log s$. The *information* I that Bob gains from his measurement is simply the difference between his initial entropy and his final entropy:

$$I = H_{\text{initial}} - H_{\text{final}}. \tag{3}$$

Once we know the average value of I for a particular measurement strategy—the average is over the possible outcomes of the measurement—we immediately know, by virtue of Shannon's theorem, how many photons Alice will have to send per classical bit.

Let us now apply this tool, focusing for the next several paragraphs on a particular ensemble of photon polarization states. The ensemble consists of three states of linear polarization with equal *a priori* probabilities: vertical, 60° to the right of vertical, and 60° to the left of vertical. We will consider a number of strategies that Alice and Bob could use to send classical information via this ensemble and we will rate these strategies according to how much information is conveyed per problem.

To formulate the first strategy, we imagine Bob reasoning as follows: "There are three possible states of each photon and they are all equally likely. There must be, then, some measurement I can make that will give me, on average, the most possible information about a given photon's state under the circumstances." Indeed, there is an optimal measurement for distinguishing the three given states and one can even find out what it is.[4] It happens not to be an orthogonal measurement, but rather a three-outcome generalized measurement, represented by a probability-operator-valued measure (POM).[5] This particular measurement, applied to the given ensemble, always has the effect of eliminating one of the three possibilities and leaving the other two equally likely. The amount of information that Bob gains from this measurement is thus

$$I = H_{\text{initial}} - H_{\text{final}} = \log 3 - \log 2 = 0.585 \text{ bits}. \tag{4}$$

One can show that there is no measurement that Bob can make on a single photon in this ensemble that yields more information on average than this one. If Bob uses this strategy and if Alice wants to send 100 classical bits, then she must use approximately $100/0.585 = 171$ photons.

It turns out that Alice and Bob can do better than that, as we will see shortly. First, though, let us pause for a moment and mention an important and relevant theorem from over twenty years ago. The theorem was stated by Levitin[6] and the proof was given by Kholevo.[7] It says that the amount of information that one can extract, on average, from an ensemble of pure quantum states cannot exceed the von

Neumann entropy of the ensemble,[8] defined as

$$S = -\mathrm{tr}\,\rho \log \rho, \tag{5}$$

where ρ is the ensemble's density matrix: $\rho = \Sigma_{i=1}^{m} p_i|s_i\rangle\langle s_i|$. For our three-state ensemble, one finds that the von Neumann entropy is 1 bit. Hence, the theorem states that the amount of information that one can extract from a photon in that ensemble cannot exceed 1 bit. In fact, as we have just said, the actual amount of information that one can extract in this case is 0.585 bits, which is considerably less than 1 bit. Thus, the Kholevo-Levitin theorem provides only an upper bound on the accessible information and, at least in this case, a rather weak one.

One of our motivations in considering the communication problem was to see whether the von Neumann entropy had a greater information-theoretic significance than merely being an upper bound on the accessible information. Somehow, one feels that it ought to have a greater significance, if only because in a thermodynamic context the von Neumann entropy is *the* entropy of the system. As long as one is dealing with single photons, though, the von Neumann entropy is indeed no more than an upper bound, at least as far as information theory is concerned. However, in our communication problem, Bob is not restricted to measuring the photons one at a time. He can, in principle, make a coherent measurement on several photons at once. We now consider strategies involving such multiphoton measurements to see whether the information transmitted per photon can be made greater than the amount achievable with single-photon measurements and, if so, whether it can be made equal to the von Neumann entropy of the ensemble.

In order to make use of Bob's ability to perform multiphoton measurements, it turns out that Alice must introduce classical correlations among the states of the photons she sends. She is not allowed to introduce quantum *entanglements* between photons because to do so would amount to using a different ensemble of states. However, she certainly is allowed to introduce classical correlations. Indeed, an error-correcting code depends on such correlations. In our three-state ensemble, let us call the states a, b, and c (representing vertical, $+60°$, and $-60°$, respectively). We now consider all the possible states of a *pair* of photons; they are aa, ab, ac, ba, bb, bc, ca, cb, cc. Now, instead of using all nine of these states in her transmission, Alice can choose to use only a subset, as long as the three single-particle states are still used with equal frequency. In particular, let us suppose she chooses to use only the three states ab, bc, and ca; that is, she is choosing special two-letter code words rather than coding in individual letters. These particular code words are farther apart in their two-photon Hilbert space than the original letters are in their Hilbert space—76° apart instead of 60° apart—and they are therefore easier for Bob to distinguish. On the other hand, Alice and Bob are using up two photons for each code word, so it is not immediately clear whether the information transmitted per photon will be greater than it was before.

As it happens, it is not hard to construct on paper a good measurement for Bob to use to try to distinguish among the three two-photon states ab, bc, and ca.[9] These three states can legitimately be represented as unit vectors in a real three-dimensional space, all separated from each other by 76°. The measurement that we have in mind is the orthogonal measurement whose eigenvectors straddle symmetrically the three states. Not surprisingly, this measurement is not one that can be realized by measuring each photon separately; it requires either that the two photons interact with each other as part of the measurement or that they both interact with a third quantum system. One finds[9] that this particular measurement allows Bob to extract 1.369 bits of information from each pair of photons, which translates to 0.685 bits per photon. This amount is larger than the 0.585 bits per photon that Bob got earlier using a one-photon-at-a-time strategy. Thus, there is indeed an advantage in using multiphoton code words and multiphoton measurements. Our question now is how large this advantage can become.

We jump now from two-photon code words to n-photon code words, where n is large. For the same ensemble of three polarization states, a string of n photons will have 3^n allowed states. Of these 3^n states, let Alice choose at random $2^{(1-\epsilon)n}$ states that she will actually use in her transmission. These are her code words. One finds that these $2^{(1-\epsilon)n}$ states typically become very nearly orthogonal as n becomes very large, so they become almost perfectly distinguishable. To put it more precisely: For any given ϵ, one can find an n large enough to make Bob's probability of error as small as one likes. Therefore, the amount of information transmitted per code word becomes arbitrarily close to $\log[2^{(1-\epsilon)n}] = (1 - \epsilon)n$ bits. Each code word contains n photons, so the information transmitted per photon becomes arbitrarily close to $1 - \epsilon$ bits. We state this result here without proof; the proof will be published later in a longer paper.[10]

This result is precisely what we were hoping for, at least for our three-state ensemble. It shows that the classical information capacity of a photon in this ensemble is 1 bit because Alice can transmit up to 1 bit per photon. However, 1 bit is precisely the von Neumann entropy of the ensemble, so the classical information capacity, at least for this ensemble, is not merely bounded above by the von Neumann entropy; it is equal to the von Neumann entropy. The key to changing the inequality to an equality was to consider an arbitrarily long message rather than focusing on a single photon.

Is this result general? In other words, is it always the case that the classical information capacity of an ensemble is equal to the ensemble's von Neumann entropy? Our answer is part theorem and part conjecture. We have been able to show that this equality holds for all ensembles of pure states in a two-dimensional Hilbert space. Thus, it always works for photon polarization, for example. For that case, we can say with confidence that *information capacity equals von Neumann entropy*. We have not yet generalized the theorem to higher-dimensional Hilbert spaces, but we are quite willing to conjecture that it does hold in arbitrary dimension.

In the scenario that we have been imagining, Alice has been forced not only to use certain signals, but also to use them in the proportions given in the specification of the ensemble. It is likely, however, that in many applications only the *set* of allowed signals will be forced on Alice, and not the probabilities. For example, she may be constrained to use a fixed set of nonorthogonal coherent states of an electromagnetic mode, while remaining free to use these states in any proportions that allow her to transmit the most information. For a given set of quantum states, without any specified probabilities, we can define the *channel capacity of the set* as the optimal amount of classical information that one can convey per signal using only those states. It follows from the above theorem/conjecture that the channel capacity of a set of states is the maximum von Neumann entropy over all ensembles constructed from the given states. For example, consider the set consisting of three linear polarization states: vertical, horizontal, and 45° to the right of vertical. The best way to use these states is to send the vertical and horizontal states equally often and never to send the diagonal state. This particular weighting of the signals maximizes the von Neumann entropy for that set of states. The channel capacity of this set is thus 1 bit, that is, the entropy of an equal mixture of two orthogonal states. This definition of quantum channel capacity is analogous to Shannon's definition of the capacity of a noisy classical channel, in which one likewise looks for the optimal probability distribution over a fixed set of signals.[3]

We have now presented our main results, but there are two more topics that we would like to consider briefly: (i) a possible generalization of our capacity theorem to ensembles containing mixed states and (ii) a comparison between the capacity theorem of this report and another coding theorem recently published by one of us.[11]

So far, we have been considering only ensembles of pure states. One can, though, also talk about ensembles of mixed states—that is, Alice may be required to use as signals states that are degraded in some way; for example, she may be required to use partially polarized photons. It may seem that whoever is making these rules for Alice is now becoming unnecessarily mean. However, in real life, Alice will often have to use mixed states as signals. Suppose she sends a perfectly pure polarization state through an optical fiber. By the time it gets to Bob's end, the photon will typically have been degraded to some extent by unpredictable fluctuations in the fiber. As far as the transmission of information is concerned, it is as if she had sent the degraded signal; that is, it is as if we were using mixed states as signals. Thus, it is of considerable interest to determine the classical information capacity of an ensemble of mixed states.

We have not yet made a serious attempt to determine the capacity of such an ensemble, but it is not hard to make a guess at the answer. Consider by way of analogy a purely classical communication problem. Alice sends a string of 0's and 1's to Bob, but there is noise in the channel so that a 0 has a small probability of becoming a 1 and vice versa. One can say that, in effect, Alice is using two mixed states as her signals. One of the states—call it x—consists mostly of 0 with a small admixture of 1, and the other state, y, is mostly 1 with a small admixture of 0. One knows from Shannon's theorem what the information capacity is of this classical ensemble. We write the result here suggestively in terms of von Neumann entropy because no one can stop us from regarding these classical states as special cases of quantum states and from computing their von Neumann entropies. The capacity, as given by Shannon's theorem, is the quantity χ defined by

$$\chi = S(\rho) - [p_x S(\rho_x) + p_y S(\rho_y)]. \tag{6}$$

Here, ρ_x and ρ_y are the density matrices of the individual signals, p_x and p_y are their probabilities, and ρ is the density matrix of the ensemble as a whole: $\rho = p_x\rho_x + p_y\rho_y$. To put it in words, χ is the difference between the von Neumann entropy of the whole ensemble and the average von Neumann entropy of the individual signals.

It is plausible that the quantity χ will continue to be the correct answer even when we consider truly quantum mechanical ensembles of mixed states. It certainly reduces correctly to our above result when all the states are pure because in that case each signal has zero entropy and χ becomes $S(\rho)$. Interestingly, Levitin called attention some time ago to the quantity we have called χ, referring to it as the "entropy deficit" of the ensemble.[6] This quantity also appears in the Kholevo-Levitin theorem as it applies to ensembles containing mixed states.[7]

We turn now to our comparison between the capacity theorem that we have described in this report and the quantum coding theorem of Schumacher.[11] Both of these theorems have to do with the coding of long strings of signals, both are quantum mechanical, and both are restricted (at present) to pure states. It may therefore be easy to confuse the two theorems and our intention here is to head off any such confusion. For simplicity, let us state Schumacher's theorem nonrigorously and only for photon polarization, even though it is actually quite general. It answers the following question: Suppose you start with a sequence of photons, each prepared independently from a given pure-state ensemble. To what extent can you compress the information contained in this sequence with negligible loss? (The test of a successful compression is that the original sequence can be recovered with negligible error.) The answer is very simple: You can compress it by the factor S, the von Neumann entropy of the ensemble measured in bits. For example, an ensemble consisting of two equally likely linear polarization states separated by 45° has a von Neumann entropy of 0.6 bits. If you start with a sequence of 1000 photons prepared from this ensemble, you can compress all that information into 600 photons and later be able to reconstruct the original sequence with very small error. Moreover, you cannot compress it into fewer than 600 photons.

In Schumacher's theorem, one is taking a quantum message into another quantum message, compressing it by a factor of S. In our capacity theorem, one is taking a classical message into a quantum message, *expanding* it by a factor of $1/S$. (Here, we are still assuming that the system is photon polarization.) There is evidently some sort of reciprocity between the two scenarios, and the interesting fact is that, in both cases, the von Neumann entropy is the determining factor. In this respect, the von Neumann entropy plays a role in quantum communication theory analogous to that played by the Shannon entropy in classical communication theory.[12]

To put the two quantum theorems in other words, one can say that Schumacher's theorem gives the *information content* of a quantum system, whereas the theorem of this report gives the *classical information capacity* of a quantum ensemble. Both are equal to S, but we advocate using different units of information for the two concepts. The classical information capacity is clearly to be measured in bits (or nats, or some other unit of classical information) because it is classical information that is being measured—how many 0's and 1's can be transmitted per photon? On the other hand, there is nothing classical in Schumacher's theorem. The units in which the compressed information is to be stored are not 0's and 1's, but photons, whose quantum character is crucial to the proof of the theorem. Thus, *information content* of a quantum system is to be measured in *qubits*, that is, quantum binary objects. Entropy measures both the *cost* of quantum information (i.e., the number of qubits necessary for its representation) and the *value* of quantum information (in bits) for transmitting classical messages.

John Wheeler has raised the very interesting and difficult question: How can intangible bits of information be understood as giving rise to tangible entities, that is, to real "its"? In this report, we have begun to answer an easier question: How many bits can you fit in a quantum mechanical it?

REFERENCES AND NOTES

1. WHEELER, J. A. 1990. It from bit. *In* Complexity, Entropy, and the Physics of Information. W. H. Zurek, Ed. Addison–Wesley. Reading, Massachusetts.
2. WIESNER, S. 1983. Sigact News **15**(no. 1): 78; BENNETT, C. H. & G. BRASSARD. 1985. *In* Proceedings of the IEEE International Conference on Computers, Systems, and Signal Processing (Bangalore, India), p. 175–179; EKERT, A. K. 1991. Phys. Rev. Lett. **67:** 661; BENNETT, C. H., G. BRASSARD & N. D. MERMIN. 1992. Phys. Rev. Lett. **68:** 557–559; BENNETT, C. H. 1992. Phys. Rev. Lett. **68:** 3121; EKERT, A. K., J. G. RARITY, P. R. TAPSTER & G. M. PALMA. 1992. Phys. Rev. Lett. **69:** 1293; BENNETT, C. H., G. BRASSARD, C. CRÉPEAU & M-H. SKUBISZEWSKA. 1991. *In* Advances in Cryptology—Crypto '91 Proceedings, p. 351–366; BRASSARD, G. & C. CRÉPEAU. 1990. *In* Advances in Cryptology—Crypto '90 Proceedings, p. 49–61.
3. SHANNON, C. E. 1948. Bell Syst. Tech. J. **27:** 379 & 623.
4. This example was given in: KHOLEVO, A. S. 1973. Probl. Inf. Transm. **9:** 110 (translated from Probl. Peredachi Inf.). The determination of the optimal measurement is based on the methods of: DAVIES, E. B. 1978. IEEE Trans. Inf. Theory **IT24:** 596.
5. PERES, A. 1993. Quantum Theory: Concepts and Methods. Chapter 9. Kluwer. Dordrecht; 1990. Found. Phys. **20:** 1441–1453; HELSTROM, C. W. 1976. Quantum Detection and Estimation Theory, p. 74–83. Academic Press. New York.
6. LEVITIN, L. B. 1969. On the quantum measure of the amount of information. *In* Proceedings of the IVth National Conference on Information Theory (Tashkent), p. 111–115 (in Russian); 1987. Information theory for quantum systems. *In* Information, Complexity, and Control in Quantum Physics. A. Blaquière, S. Diner & G. Lochak, Eds. Springer-Verlag. Berlin/New York.
7. KHOLEVO, A. S. 1973. Probl. Inf. Transm. **9:** 1977 (translated from Probl. Peredachi Inf.); YUEN, H. P. & M. OZAWA. 1993. Phys. Rev. Lett. **70:** 363–366.
8. VON NEUMANN, J. 1932. Mathematische Grundlagen der Quantenmechanik. Springer-Verlag. Berlin. [Translated by: BEYER, E. T. 1955. Mathematical Foundations of Quantum Mechanics. Princeton University Press. Princeton, New Jersey.]
9. PERES, A. & W. K. WOOTTERS. 1991. Phys. Rev. Lett. **66:** 1119–1122.
10. Many elements of the proof can be found in: HAUSLADEN, P. 1993. On the quantum mechanical channel capacity as a function of the density matrix. B.A. thesis. Williams College, Williamstown, Massachusetts.
11. SCHUMACHER, B. 1995. On quantum coding. Phys. Rev. A. To appear.
12. Schumacher's theorem is the quantum analogue of Shannon's noiseless coding theorem having to do with the compressibility of classical information.[3] The theorem of the present report resembles Shannon's channel capacity theorem for noisy channels, which we used earlier to figure out the expansion factor for Alice's error-correcting codes. However, in our channel, the only noise is the noise inherent in quantum mechanics (for the case of pure-state ensembles).

Bounds for Accessible Information in Quantum Mechanics[a]

CHRISTOPHER A. FUCHS AND CARLTON M. CAVES

Center for Advanced Studies
Department of Physics and Astronomy
University of New Mexico
Albuquerque, New Mexico 87131-1156

Long before anyone else, John A. Wheeler saw a role for information theory in tackling the still-unresolved foundational questions of quantum mechanics. To pursue this grand vision, Wheeler exhorts us to begin building a structure of concrete technical results. We offer one such example here in honor of Wheeler's eighty-third birthday.

A quantum communication channel is defined by the action of sending one of n possible messages, with prior probabilities $p_1, \ldots, p_n$, to a specified receiver in the form of one of n distinct (possibly mixed) density operators $\hat{\rho}_1, \ldots, \hat{\rho}_n$ on an N-dimensional Hilbert space. The message states $\hat{\rho}_i$, together with their probabilities p_i, constitute the *message ensemble.* At the end of the transmission, the receiver can perform any *generalized* quantum measurement described by a positive-operator-valued measure (POVM) in an attempt to discern which message was actually sent. The fundamental question of quantum communication theory[1] is this: Which measurements maximize the Shannon mutual information about the actual message and just how much information is that maximal amount $I_{\rm acc}$? Previous results on $I_{\rm acc}$ include an upper bound due to Holevo,[2–4] a lower bound recently exhibited by Jozsa, Robb, and Wootters,[5] and the classification of a few examples where $I_{\rm acc}$ can be calculated exactly.[5–8] Of the two bounds, both are uniquely distinguished in the sense that they are the best bounds expressible solely in terms of the total density operator $\hat{\rho} = \Sigma p_i \hat{\rho}_i$ whenever all the $\hat{\rho}_i$ are pure states. For this reason, however, both bounds are fairly loose for many message ensembles. Here, we derive ensemble-dependent bounds (an upper bound for the general channel and a lower bound for the binary channel) by way of reexamining and simplifying Holevo's original derivation.

The quantum communication problem[1] is made precise through a formalization of the most general measurements allowed by quantum theory, the POVM.[9] A POVM is a set of nonnegative, Hermitian operators $\hat{E}_b$, which are complete in the sense that $\Sigma_b \hat{E}_b = \hat{\mathbb{1}} =$ (N-dimensional unit operator). The subscript b indexes the possible outcomes of the measurement; the conditions on the $\hat{E}_b$ are just those necessary and sufficient for the quantity $\mathrm{tr}(\hat{\rho}\hat{E}_b)$ to be a valid probability distribution.

[a] This work was supported in part by the Office of Naval Research (Grant No. N00014-93-1-0116).

The Shannon mutual information[10] with respect to a measurement $\{\hat{E}_b\}$ is defined by

$$I = H(\hat{\rho}) - \sum_{i=1}^{n} p_i H(\hat{\rho}_i), \tag{1}$$

where $H(\hat{\rho}) = -\Sigma_b[\mathrm{tr}(\hat{\rho}\hat{E}_b)] \ln [\mathrm{tr}(\hat{\rho}\hat{E}_b)]$ is the average information gain upon finding outcome b, when the density operator is assumed to be $\hat{\rho}$.

The *accessible information* I_{acc} of the channel is defined to be the maximum of the mutual information I over all measurements $\{\hat{E}_b\}$. The Holevo upper bound to I_{acc} is

$$I_{\mathrm{acc}} \leq S(\hat{\rho}) - \sum_{i=1}^{n} p_i S(\hat{\rho}_i), \tag{2}$$

where $S(\hat{\rho}) = -\mathrm{tr}(\hat{\rho} \ln \hat{\rho}) = -\Sigma_j \lambda_j \ln \lambda_j$ is the von Neumann entropy of the density operator $\hat{\rho}$, whose eigenvalues are λ_j. Because $\epsilon \ln \epsilon \to 0$ as $\epsilon \to 0$, $S(\hat{\rho}_i) = 0$ whenever $\hat{\rho}_i$ is a pure state. Therefore, the upper bound (equation 2) reduces to $I_{\mathrm{acc}} \leq S(\hat{\rho})$ when all the input states $\hat{\rho}_i$ are pure.

Holevo's derivation of inequality 2 can be summarized as follows. Let $\{\hat{E}_b\}$ be an arbitrary POVM. The total mutual information (equation 1) can be written as a sum of binary-channel mutual informations,

$$I = \sum_{k=2}^{n} s_k \left[H(\hat{\tau}_k) - \left(\frac{p_k}{s_k}\right) H(\hat{\rho}_k) - \left(\frac{s_{k-1}}{s_k}\right) H(\hat{\tau}_{k-1}) \right], \tag{3}$$

where $s_k = \Sigma_{i=1}^{k} p_i$ is the probability for the first k messages and $\hat{\tau}_k = s_k^{-1} \Sigma_{i=1}^{k} p_i \hat{\rho}_i$ is their associated density operator. (All the terms in this sum containing an $H(\hat{\tau}_k)$ cancel, except the contributions from $\hat{\tau}_n = \hat{\rho}$ and $\hat{\tau}_1 = \hat{\rho}_1$.) Thus, if one can find an upper bound $B(\hat{\tau}_k, \hat{\rho}_k; p_k/s_k)$ for the mutual information in each of these imagined binary channels, one can immediately build an upper bound for the general case:

$$I_{\mathrm{acc}} \leq \sum_{k=2}^{n} s_k B\left(\hat{\tau}_k, \hat{\rho}_k; \frac{p_k}{s_k}\right). \tag{4}$$

The bound (equation 2) can, in fact, be built in just this way; the derivation thereby reduces to a study of the binary case.

For a binary channel specified by density operators $\hat{\rho}_0$ and $\hat{\rho}_1$ with probabilities $1 - t$ and t, the mutual information $I \equiv I(t)$ can be considered as a function of the parameter t; so, too, in this case can the Holevo bound $S(t) \equiv S(\hat{\rho}) - (1 - t)S(\hat{\rho}_0) - tS(\hat{\rho}_1)$, where $\hat{\rho} = (1 - t)\hat{\rho}_0 + t\hat{\rho}_1 = \hat{\rho}_0 + t\hat{\Delta} = \hat{\rho}_1 - (1 - t)\hat{\Delta}$ and $\hat{\Delta} = \hat{\rho}_1 - \hat{\rho}_0$. The key to deriving the bound $I_{\mathrm{acc}}(t) \leq S(t)$ is in realizing the importance of properties of $I(t)$ and $S(t)$ as functions of t. Note first that $I(0) = I(1) = S(0) = S(1) = 0$. Moreover, both $I(t)$ and $S(t)$ are downwardly convex, as can be seen by working out their second derivatives. For $I(t)$, this is

$$I''(t) = -\sum_{b} \frac{[\mathrm{tr}(\hat{\Delta}\hat{E}_b)]^2}{\mathrm{tr}(\hat{\rho}\hat{E}_b)}. \tag{5}$$

$S''(t)$ is most easily found by representing $S(\hat{\rho})$ as a contour integral:[11]

$$S(\hat{\rho}) = \left(-\frac{1}{2\pi i}\right)\oint_C (z \ln z)\mathrm{tr}[(z\hat{\mathbb{1}} - \hat{\rho})^{-1}]dz, \qquad (6)$$

where the contour C encloses all the *nonzero* eigenvalues of $\hat{\rho}$; by differentiating within the integral and using the operator identity $(\hat{A}^{-1})' = -\hat{A}^{-1}\hat{A}'\hat{A}^{-1}$, one finds that

$$S''(t) = -\sum_{\{j,k|\lambda_j+\lambda_k\neq 0\}} \Phi(\lambda_j, \lambda_k)|\Delta_{jk}|^2, \qquad (7)$$

where $\Phi(x,y) = (\ln x - \ln y)/(x - y)$ if $x \neq y$, $\Phi(x,x) = 1/x$, $\Delta_{jk} = \langle j|\hat{\Delta}|k\rangle$, and $|j\rangle$ is the eigenvector of $\hat{\rho}$ with eigenvalue λ_j. Equations 5 and 7 are clearly nonpositive.

The conjecture that $S(t)$ is an upper bound to $I(t)$ for *any* t is equivalent to the property that, when plotted versus t, the curve for $S(t)$ has a more negative curvature than the curve for $I(t)$ (regardless of which POVM $\{\hat{E}_b\}$ is used in its definition); that is,

$$S''(t) \leq I''(t) \leq 0 \text{ for any POVM } \{\hat{E}_b\}. \qquad (8)$$

The meat of the derivation is in showing this inequality. Holevo does this by demonstrating the existence of a function $L''(t)$, independent of $\{\hat{E}_b\}$, such that $S''(t) \leq L''(t)$ and $L''(t) \leq I''(t)$. From this, it follows, upon enforcing the boundary condition $L(0) = L(1) = 0$, that $I_{\mathrm{acc}}(t) \leq L(t) \leq S(t)$ (although, in reference 2, $L(t)$ is never explicitly computed).

It is at this juncture that a fairly drastic simplification can be made to the original proof. The easiest way to get at such a function $L''(t)$ is simply to minimize $I''(t)$ over all POVMs $\{\hat{E}_b\}$ and thereafter to show that $S''(t) \leq L''(t)$. This is distinctly more tractable than maximizing the mutual information $I(t)$ itself because no logarithms appear in $I''(t)$. It turns out that this approach generates exactly the same function $L''(t)$ as used by Holevo, although the two derivations appear to have little to do with each other. The difference in substance—aside from a deeper understanding—is that this approach pinpoints a measurement that actually minimizes $I''(t)$. This measurement in turn, although it generally does not maximize $I(t)$ itself, necessarily provides a *lower* bound $M(t)$ to the accessible information $I_{\mathrm{acc}}(t)$.

The difficulty that crops up in extremizing quantities such as equation 5 is that, so far at least, there seems to be no way to make the problem amenable to a variational approach. Nevertheless, there is an easy way to minimize $I''(t)$ through a clever application of the Schwarz inequality, as shown in a recent paper by Braunstein and Caves.[12] The problem considered here, in fact, is formally the one considered there: the expression for $-I''(t)$ is well known as the Fisher information,[13] a quantity used in statistical parameter estimation. The steps are as follows. The Schwarz inequality for the operator inner product $\mathrm{tr}(\hat{A}^\dagger\hat{B})$ is given by $|\mathrm{tr}(\hat{A}^\dagger\hat{B})|^2 \leq \mathrm{tr}(\hat{A}^\dagger\hat{A})\,\mathrm{tr}(\hat{B}^\dagger\hat{B})$, where equality is achieved if and only if $\hat{A} = \mu\hat{B}$ for some constant μ. The idea now is to think of the numerator within the sum (equation 5) as analogous to the left-hand side of this Schwarz inequality. One would like to use the Schwarz inequality in such a way that the $\mathrm{tr}(\hat{\rho}\hat{E}_b)$ term in the denominator is canceled and only an expression linear in $\hat{E}_b$ is left; for then, upon summing over the index b, the completeness property for POVMs will leave the final expression independent of the given measurement. This is done by introducing the "lowering" linear superoperator $\mathscr{L}_{\hat\rho}$, whose action on the operator $\hat{A}$ is defined by

$$(½)[\hat\rho\mathscr{L}_{\hat\rho}(\hat A) + \mathscr{L}_{\hat\rho}(\hat A)\hat\rho] = \hat A. \tag{9}$$

In a basis $|j\rangle$ that diagonalizes $\hat\rho$, $\mathscr{L}_{\hat\rho}(\hat\Delta)$ becomes

$$\mathscr{L}_{\hat\rho}(\hat\Delta) \equiv \sum_{\{j,k|\lambda_j+\lambda_k\neq 0\}} \left[\frac{2}{(\lambda_j + \lambda_k)}\right]\Delta_{jk}|j\rangle\langle k|, \tag{10}$$

which depends on the fact that $\Delta_{jk} = 0$ if $\lambda_j + \lambda_k = 0$. (For further discussion of why equation 10 is the appropriate extension of $\mathscr{L}_{\hat\rho}(\hat A)$ to the zero-eigenvalue subspaces of $\hat\rho$, see reference 12; note that $\mathscr{L}_{\hat\rho}$ is denoted there by $\mathscr{R}_{\hat\rho}^{-1}$.) This superoperator has the property that it acts much like "division by $\hat\rho$" when used within a trace; in particular, using equation 9, one easily derives the identity that $\mathrm{tr}(\hat B\hat\Delta) = \mathrm{Re}\{\mathrm{tr}[\hat\rho\hat B\mathscr{L}_{\hat\rho}(\hat\Delta)]\}$ for Hermitian $\hat B$.

The desired optimization now follows in short order:

$$\begin{aligned} I''(t) &= -\sum_b (\mathrm{Re}\{\mathrm{tr}[\hat\rho\hat E_b\mathscr{L}_{\hat\rho}(\hat\Delta)]\})^2[\mathrm{tr}(\hat\rho\hat E_b)]^{-1} \\ &\geq -\sum_b |\mathrm{tr}[\hat\rho\hat E_b\mathscr{L}_{\hat\rho}(\hat\Delta)]|^2[\mathrm{tr}(\hat\rho\hat E_b)]^{-1} && \text{(A)} \\ &= -\sum_b |\mathrm{tr}\{(\sqrt{\hat E_b}\sqrt{\hat\rho})^\dagger[\sqrt{\hat E_b}\mathscr{L}_{\hat\rho}(\hat\Delta)\sqrt{\hat\rho}]\}|^2[\mathrm{tr}(\hat\rho\hat E_b)]^{-1} \\ &\geq -\sum_b \mathrm{tr}[\hat E_b\mathscr{L}_{\hat\rho}(\hat\Delta)\hat\rho\mathscr{L}_{\hat\rho}(\hat\Delta)] && \text{(B)} \\ &= -\mathrm{tr}[\mathscr{L}_{\hat\rho}(\hat\Delta)\hat\rho\mathscr{L}_{\hat\rho}(\hat\Delta)] = -\mathrm{tr}[\hat\Delta\mathscr{L}_{\hat\rho}(\hat\Delta)]. && \text{(11)} \end{aligned}$$

The conditions for equality in equation 11—that is, for the achievability of the lower bound—arise from steps A and B: $\mathrm{Im}\{\mathrm{tr}[\hat\rho\hat E_b\mathscr{L}_{\hat\rho}(\hat\Delta)]\} = 0$ for all b and (with a little rearrangement)

$$\sqrt{\hat E_b}[\hat{\mathbb{1}} - \mu_b\mathscr{L}_{\hat\rho}(\hat\Delta)]\sqrt{\hat\rho} = 0 \text{ for all } b. \tag{12}$$

As discussed more fully in reference 12, these conditions can always be met by choosing the operators $\hat E_b$ to be one-dimensional projectors onto a basis that diagonalizes $\mathscr{L}_{\hat\rho}(\hat\Delta)$ and by choosing the constants μ_b to be the inverse eigenvalues of $\mathscr{L}_{\hat\rho}(\hat\Delta)$. Note that this set of projectors is ensemble-dependent in that it depends not only on $\hat\rho$, but also on $\hat\Delta = \hat\rho_1 - \hat\rho_0$.

The function $L''(t)$ can now be defined as

$$L''(t) = -\mathrm{tr}[\hat\Delta\mathscr{L}_{\hat\rho}(\hat\Delta)] = -\sum_{\{j,k|\lambda_j+\lambda_k\neq 0\}} \left[\frac{2}{(\lambda_j + \lambda_k)}\right]|\Delta_{jk}|^2. \tag{13}$$

This, as stated above, is exactly the function $L''(t)$ used by Holevo, but obtained there by other means. The remainder of the derivation of equation 2, to show that $S''(t) \leq$

$L''(t)$, consists of demonstrating the arithmetic inequality $\Phi(x, y) \geq 2/(x + y)$ (see reference 2). The bound (equation 2) then follows from equation 4 by the particular structure of the bound $S(t)$. This completes our discussion of the Holevo upper bound.

Now, we focus on deriving explicit expressions for the binary-channel ensemble-dependent upper and lower bounds $L(t)$ and $M(t)$. The lower bound $M(t)$, in particular, can be written in a surprisingly simple form. We start with the generic formula for the binary-channel mutual information expressed in a slightly different guise, as the weighted sum of two Kullback-Leibler informations:[14]

$$I(t) = (1 - t) \sum_b \mathrm{tr}(\hat{\rho}_0 \hat{E}_b) \ln\left[\frac{\mathrm{tr}(\hat{\rho}_0 \hat{E}_b)}{\mathrm{tr}(\hat{\rho} \hat{E}_b)}\right] + t \sum_b \mathrm{tr}(\hat{\rho}_1 \hat{E}_b) \ln\left[\frac{\mathrm{tr}(\hat{\rho}_1 \hat{E}_b)}{\mathrm{tr}(\hat{\rho} \hat{E}_b)}\right]. \tag{14}$$

Using the linearity of the trace operation, this becomes

$$I(t) = \mathrm{tr}\left[(1 - t)\hat{\rho}_0 \sum_b (\ln \alpha_b)\hat{E}_b + t\hat{\rho}_1 \sum_b (\ln \beta_b)\hat{E}_b\right], \tag{15}$$

where $\alpha_b = \mathrm{tr}(\hat{\rho}_0 \hat{E}_b)/\mathrm{tr}(\hat{\rho}\hat{E}_b)$ and $\beta_b = \mathrm{tr}(\hat{\rho}_1 \hat{E}_b)/\mathrm{tr}(\hat{\rho}\hat{E}_b)$. The lower bound $M(t)$ is defined by inserting the projectors $\hat{E}_b$ onto a basis that diagonalizes $\mathscr{L}_{\hat{\rho}}(\hat{\Delta})$ into this formula. Now, a curious fact can be used: even though $\hat{\rho}_0$ and $\hat{\rho}_1$ need not commute, $\mathscr{L}_{\hat{\rho}}(\hat{\Delta})$, $\mathscr{L}_{\hat{\rho}}(\hat{\rho}_0)$, and $\mathscr{L}_{\hat{\rho}}(\hat{\rho}_1)$ do all commute. This follows from the linearity of the $\mathscr{L}_{\hat{\rho}}$ superoperator: $\mathscr{L}_{\hat{\rho}}(\hat{\rho}_0) = \mathscr{L}_{\hat{\rho}}(\hat{\rho} - t\hat{\Delta}) = \hat{\mathbb{1}} - t\mathscr{L}_{\hat{\rho}}(\hat{\Delta})$ and $\mathscr{L}_{\hat{\rho}}(\hat{\rho}_1) = \mathscr{L}_{\hat{\rho}}[\hat{\rho} + (1 - t)\hat{\Delta}] = \hat{\mathbb{1}} + (1 - t)\mathscr{L}_{\hat{\rho}}(\hat{\Delta})$. Thus, the same projectors $\hat{E}_b$ that diagonalize $\mathscr{L}_{\hat{\rho}}(\hat{\Delta})$ also diagonalize $\mathscr{L}_{\hat{\rho}}(\hat{\rho}_0)$ and $\mathscr{L}_{\hat{\rho}}(\hat{\rho}_1)$. With this, it immediately follows from equation 9 that α_b and β_b are the respective eigenvalues of $\mathscr{L}_{\hat{\rho}}(\hat{\rho}_0)$ and $\mathscr{L}_{\hat{\rho}}(\hat{\rho}_1)$ corresponding to the projector $\hat{E}_b$. Hence, $M(t)$ takes the form,

$$M(t) = \mathrm{tr}\{(1 - t)\hat{\rho}_0 \ln[\mathscr{L}_{\hat{\rho}}(\hat{\rho}_0)] + t\hat{\rho}_1 \ln[\mathscr{L}_{\hat{\rho}}(\hat{\rho}_1)]\}. \tag{16}$$

The trace of the operators $\hat{\rho}_0 \ln[\mathscr{L}_{\hat{\rho}}(\hat{\rho}_0)]$ and $\hat{\rho}_1 \ln[\mathscr{L}_{\hat{\rho}}(\hat{\rho}_1)]$ in this expression can be interpreted, to an extent, as quantum analogues of the Kullback-Leibler informations in equation 14.

The upper bound $L(t)$ has not so far yielded such a simple form. Of course, all that need be done in principle is to integrate equation 13 twice, applying the boundary conditions of $L(0) = L(1) = 0$; the problem lies in finding a tractable representation for $\mathscr{L}_{\hat{\rho}}(\hat{\Delta})$. At least *when $\hat{\rho}$ is invertible*, $\mathscr{L}_{\hat{\rho}}(\hat{A})$ can be written as a contour integral:

$$\mathscr{L}_{\hat{\rho}}(\hat{A}) = \left(\frac{2}{2\pi i}\right)\oint (z\hat{\mathbb{1}} - \hat{\rho})^{-1}\hat{A}(z\hat{\mathbb{1}} + \hat{\rho})^{-1} dz, \tag{17}$$

where the contour contains the pole at $z = \lambda_j$ for all eigenvalues λ_j of $\hat{\rho}$, but does not contain the pole at $z = -\lambda_j$ for any j. This contour representation leads to a Fourier series expansion for $L(t)$. It is not difficult, again using the operator-inverse differentiation formula, to work out that

$$\hat{\Delta}\,\frac{d^n}{dt^n}\mathscr{L}_{\hat{\rho}}(\hat{\Delta}) = \left[\frac{2(n!)}{2\pi i}\right]\sum_{k=0}^{n}(-1)^k D_{\hat{\rho}}(n;k), \tag{18}$$

where

$$D_{\hat{\rho}}(n;k) = \oint [\hat{\Delta}(z\hat{\mathbb{1}} - \hat{\rho})^{-1}]^{n+1-k}[\hat{\Delta}(z\hat{\mathbb{1}} + \hat{\rho})^{-1}]^{k+1}dz. \tag{19}$$

With this, one can derive a Taylor series expansion for $L''(t)$ and then use the standard algorithm for Fourier expansions to obtain

$$L(t) = \sum_{m=1}^{\infty} b_m \sin(m\pi t), \tag{20}$$

where

$$b_m = \left(\frac{1}{m^3\pi^4 i}\right)\sum_{n=0}^{\infty}(n!)\left[b(n;m) - (-1)^m\sum_{j=0}^{n}\left(\frac{1}{(n-j)!}\right)b(j;m)\right]\sum_{k=0}^{n}(-1)^k \mathrm{tr}[D_{\hat{\rho}_0}(n;k)] \tag{21}$$

and $b(j;m) = (-1)^{j/2}[1 + (-1)^j](m\pi)^{-j}$. The advantages of this representation are that it automatically satisfies the boundary conditions and only the first few terms in equation 20 are significant.

Finally, we consider a special case of some practical interest—binary communication channels on two-dimensional Hilbert spaces. Here, the new bounds are readily expressible in terms of elementary functions and, moreover, the optimal orthogonal projection-valued measurement can be found via a variational calculation. With this case, one can gain a feel for how tightly the new bounds delimit the true accessible information $I_{\mathrm{acc}}(t)$.

Let the signal states $\hat{\rho}_0$ and $\hat{\rho}_1$ be represented by two vectors within the Bloch sphere, that is, $\hat{\rho}_0 = (\frac{1}{2})(\hat{\mathbb{1}} + \boldsymbol{a}\cdot\boldsymbol{\sigma})$ and $\hat{\rho}_1 = (\frac{1}{2})(\hat{\mathbb{1}} + \boldsymbol{b}\cdot\boldsymbol{\sigma})$, where $a^2 \leq 1$, $b^2 \leq 1$, and $\boldsymbol{\sigma}$ is the Pauli spin vector. (In this representation, the signal states are pure if $\boldsymbol{a}$ and $\boldsymbol{b}$ have unit modulus.) The total density matrix for the channel can then be written as $\hat{\rho} = (\frac{1}{2})(\hat{\mathbb{1}} + \boldsymbol{c}\cdot\boldsymbol{\sigma})$, where $\boldsymbol{c} = (1-t)\boldsymbol{a} + t\boldsymbol{b} = \boldsymbol{a} + t\boldsymbol{d}$ and $\boldsymbol{d} = \boldsymbol{b} - \boldsymbol{a}$. For an orthogonal projection-valued measurement in the direction specified by the unit vector $\boldsymbol{n}$, the mutual information $I(t)$ takes the form,

$$I(t) = (1-t)K(\hat{\rho}/\hat{\rho}_0) + tK(\hat{\rho}/\hat{\rho}_1), \tag{22}$$

where

$$K(\hat{\rho}/\hat{\rho}_0) = \left(\frac{1}{2}\right)\left[(1 + \boldsymbol{a}\cdot\boldsymbol{n})\ln\left(\frac{1+\boldsymbol{a}\cdot\boldsymbol{n}}{1+\boldsymbol{c}\cdot\boldsymbol{n}}\right) + (1 - \boldsymbol{a}\cdot\boldsymbol{n})\ln\left(\frac{1-\boldsymbol{a}\cdot\boldsymbol{n}}{1-\boldsymbol{c}\cdot\boldsymbol{n}}\right)\right] \tag{23}$$

and similarly for $K(\hat{\rho}/\hat{\rho}_1)$. The optimal projector can be found by varying equation 22 over all *unit* vectors $\boldsymbol{n}$ (i.e., the variation satisfies $\delta\boldsymbol{n}\cdot\boldsymbol{n} = 0$). The resulting equation for $\boldsymbol{n}$ is

$$0 = (1 - t)\ln\left(\frac{(1 + \boldsymbol{c}\cdot\boldsymbol{n})(1 - \boldsymbol{a}\cdot\boldsymbol{n})}{(1 - \boldsymbol{c}\cdot\boldsymbol{n})(1 + \boldsymbol{a}\cdot\boldsymbol{n})}\right)\boldsymbol{a}_\perp + t\ln\left(\frac{(1 + \boldsymbol{c}\cdot\boldsymbol{n})(1 - \boldsymbol{b}\cdot\boldsymbol{n})}{(1 - \boldsymbol{c}\cdot\boldsymbol{n})(1 + \boldsymbol{b}\cdot\boldsymbol{n})}\right)\boldsymbol{b}_\perp, \tag{24}$$

where $\boldsymbol{a}_\perp = \boldsymbol{a} - (\boldsymbol{a}\cdot\boldsymbol{n})\boldsymbol{n}$ and $\boldsymbol{b}_\perp = \boldsymbol{b} - (\boldsymbol{b}\cdot\boldsymbol{n})\boldsymbol{n}$ are vectors perpendicular to $\boldsymbol{n}$. Equation 24 is, of course, a transcendental equation and, as such, generally has no explicit solution. Nevertheless, there are four nontrivial situations where it can be

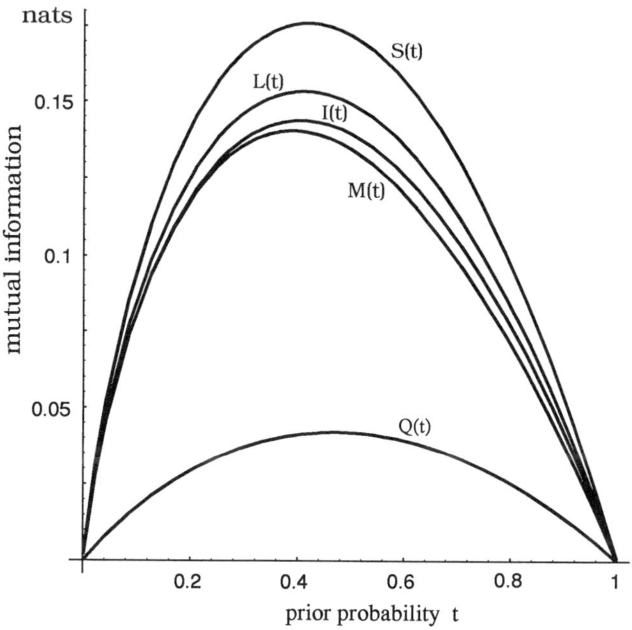

FIGURE 1. The signal states in this example are a pure-state Bloch vector $\boldsymbol{a}$ aligned along the z-axis and a mixed state $\boldsymbol{b}$, with $b = 2/3$, aligned at an angle of $\pi/3$ away from the z-axis. The respective prior probabilities for the messages are $(1 - t)$ and t. Pictured are the Holevo upper bound $S(t)$, the ensemble-dependent upper bound $L(t)$, the information $I(t)$ extractable by the optimal orthogonal projection-valued measurement, the ensemble-dependent lower bound $M(t)$, and the Jozsa-Robb-Wootters lower bound $Q(t)$.

solved exactly: (i) a classical channel, where $\hat\rho_0$ and $\hat\rho_1$ commute ($\boldsymbol{a}$ and $\boldsymbol{b}$ parallel); (ii) $\hat\rho_0$ and $\hat\rho_1$ both pure states ($a = b = 1$); (iii) $a = b$ and $t = 1/2$; and (iv) $t = [1 + \sqrt{(1 - b^2)/(1 - a^2)}]^{-1}$. When these conditions are fulfilled, equation 24 can be solved by choosing $\boldsymbol{n}$ such that $(1 - t)\boldsymbol{a}_\perp = t\boldsymbol{b}_\perp$ and by requiring that the arguments of the logarithms be multiplicative inverses.

In case (iv), the optimal $\boldsymbol{n}$ (unnormalized) is given by

$$\boldsymbol{n} \propto \left[\frac{\boldsymbol{a}\cdot(\boldsymbol{a} - \boldsymbol{c})}{\boldsymbol{a}\cdot(\boldsymbol{a} - \boldsymbol{b})}\right]\boldsymbol{b} - \left[\frac{\boldsymbol{b}\cdot(\boldsymbol{b} - \boldsymbol{c})}{\boldsymbol{b}\cdot(\boldsymbol{b} - \boldsymbol{a})}\right]\boldsymbol{a}. \tag{25}$$

In cases (i)–(iii), the optimal $\boldsymbol{n}$ is

$$\boldsymbol{n} = \left(\frac{1}{D}\right)[(1 - \boldsymbol{a}\cdot\boldsymbol{c})\boldsymbol{b} - (1 - \boldsymbol{b}\cdot\boldsymbol{c})\boldsymbol{a}] = \left(\frac{1}{D}\right)[\boldsymbol{d} + \boldsymbol{c}\times(\boldsymbol{c}\times\boldsymbol{d})], \tag{26}$$

where $\boldsymbol{c}\cdot\boldsymbol{n} = \boldsymbol{c}\cdot\boldsymbol{d}/D$, $D = \sqrt{\delta^2(1-c^2) + (\boldsymbol{c}\cdot\boldsymbol{d})^2}$, and

$$\delta = \sqrt{D\boldsymbol{d}\cdot\boldsymbol{n}} = \sqrt{d^2 - |\boldsymbol{c}\times\boldsymbol{d}|^2} = \sqrt{d^2 - |\boldsymbol{a}\times\boldsymbol{b}|^2}. \tag{27}$$

Cases (ii) and (iii) of these have been reported by Levitin.[8] Case (ii), in particular, is of special interest because two pure states in a Hilbert space of any dimension still span only a two-dimensional subspace; hence, it remains valid as the optimal orthogonal projection-valued measurement for a pure-state binary channel in all dimensions.

Under the assumption of an orthogonal projection-valued measurement in the direction $\boldsymbol{n}$, $I''(t)$ takes the form,

$$I''(t) = -\frac{(\boldsymbol{d}\cdot\boldsymbol{n})^2}{[1 - (\boldsymbol{c}\cdot\boldsymbol{n})^2]}. \tag{28}$$

The vector $\boldsymbol{n}$ that minimizes this is given easily enough by a variational calculation; it, too, is given by equation 26. Inserting this vector into equation 23 produces the lower bound $M(t)$, which as we now see sometimes coincides with the maximum information attainable from an orthogonal projection-valued measurement. The upper bound $L(t)$, found by integrating $I''(t)$ back up with this measurement in place, is given by

$$L(t) = \left(-\frac{\delta}{2d^2}\right)[(\delta - \boldsymbol{c}\cdot\boldsymbol{d})\ln(\delta - \boldsymbol{c}\cdot\boldsymbol{d}) + (\delta + \boldsymbol{c}\cdot\boldsymbol{d})\ln(\delta + \boldsymbol{c}\cdot\boldsymbol{d})] + \beta_1 t + \beta_2, \tag{29}$$

where β_1 and β_2 are easily determined by the $L(0) = L(1) = 0$ boundary condition. The extent to which these bounds are tighter than the Holevo upper bound and the Jozsa-Robb-Wootters lower (ensemble-independent) bound is illustrated by a typical example in FIGURE 1.

ACKNOWLEDGMENTS

We thank S. L. Braunstein and C. Chandler for helpful discussions.

REFERENCES

1. CAVES, C. M. & P. D. DRUMMOND. 1994. Rev. Mod. Phys. **66:** 481.
2. HOLEVO, A. S. 1973. Probl. Peredachi Inf. **9**(3): 3. [1973. Probl. Inf. Transm. **9:** 177.]
3. YUEN, H. P. & M. OZAWA. 1993. Phys. Rev. Lett. **70:** 363.
4. HALL, M. J. W. & M. J. O'ROURKE. 1993. Quantum Opt. **5:** 161.
5. JOZSA, R., D. ROBB & W. K. WOOTTERS. 1994. Phys. Rev. A **49:** 668.
6. HOLEVO, A. S. 1973. Probl. Peredachi Inf. **9**(2): 31. [1973. Probl. Inf. Transm. **9:** 110.]
7. DAVIES, E. B. 1978. IEEE Trans. Inf. Theory **IT-24:** 596.

8. LEVITIN, L. B. 1993. *In* Workshop on Physics and Computation: Phys. Comp. '92. D. Matzke, Ed. IEEE Computer Society Press. Los Alamitos, California.
9. KRAUS, K. 1983. States, Effects, and Operations: Fundamental Notions of Quantum Theory. Springer-Verlag. Berlin/New York.
10. SHANNON, C. E. 1948. Bell Syst. Tech. J. **27:** 379 & 623.
11. POINCARÉ, H. 1899. Trans. Cambridge Philos. Soc. **18:** 220.
12. BRAUNSTEIN, S. L. & C. M. CAVES. 1994. Phys. Rev. Lett. **72:** 3439.
13. FISHER, R. A. 1992. Proc. R. Soc. Edinburgh **42:** 321.
14. KULLBACK, S. & R. A. LEIBLER. 1951. Ann. Math. Stat. **22:** 79.

Quantum Teleportation and Quantum Computation Based on Cavity QED

TYCHO SLEATOR[a] AND HARALD WEINFURTER[b,c]

[a]*Department of Physics*
New York University
New York, New York 10003

[b]*Institute of Experimental Physics*
University of Innsbruck
A-6020 Innsbruck, Austria

INTRODUCTION

Recently, Bennett *et al.*[1] have proposed a scheme for "teleporting" an unknown quantum state by splitting it up into a purely classical channel and a purely quantum channel. The quantum channel consists of a pair of particles with perfectly correlated spins, such as a spin singlet state of two spin-½ particles (henceforth referred to as an EPR pair[1]). One of the two EPR particles is sent to the sender of the state (Alice) and the other is sent to the receiver (Bob). To send the unknown state (labeled $|u\rangle$), which is also carried by a spin-½ particle, Alice makes a measurement in an entangled state basis (made up of the four orthogonal two-particle states of perfectly correlated spins) on the two-particle system consisting of the particle in the state $|u\rangle$ and her EPR particle. She sends the result of her measurement (one of four possible outcomes) via a classical channel to Bob, who then applies one of four corresponding unitary transformations to his EPR particle, leaving it in the state $|u\rangle$. It is important to note that, in this scheme, Alice need not have any knowledge about the state $|u\rangle$, nor does she learn anything about $|u\rangle$ in the teleportation process.

The crucial point in the teleportation scheme is the measurement of the two-particle system in the basis of entangled states (also referred to as "Bell states"[2]). This type of measurement is also important for a related scheme, in which one can transmit two bits of information with a single two-level system, effectively doubling the channel capacity.[3] In references 1 and 3, little discussion was devoted to how one could carry out the measurement of the two-particle system in the Bell-state basis. A Bell-state measurement of spin entangled particles using upconversion of a photon pair in a nonlinear crystal[3] does not seem feasible due to its very low efficiency. On the other hand, a measurement by linear optical elements only cannot be performed unambiguously.[4]

In this report, we describe how one can use cavity QED techniques to produce EPR correlations and to make measurements in the Bell-state basis. This will form a basis for a specific implementation of the teleportation scheme of Bennett *et al.*,[3] as well as that of encoding two bits of information on a single two-level system.[3] In these

[c]H. Weinfurter was supported by the Austrian "Fonds zur Förderung der Wissenschaftl. Forschung" (Project No. S6502).

implementations,[5] one of the EPR particles is replaced by a cavity mode and the other particles consist of two-level atoms, at or near resonance with the cavity. Recent experiments[6,7] involving the interaction of spherical Rydberg atoms with a microwave cavity demonstrate that the atom-cavity interaction is sufficiently strong and that the atomic and cavity lifetimes are sufficiently long that an experimental implementation of the proposed schemes should be feasible.

We also discuss in this report how one could produce perfectly correlated states of n two-level systems, as well as demonstrate some applications of cavity QED to quantum computation.

ELEMENTS OF THE SYSTEM

Our proposed implementation for teleporting quantum states, which is shown in FIGURES 1 and 2, consists of a microwave cavity, two Ramsey zones, and a set of two-level atoms. There is never more than one photon in the cavity, so the state of the cavity field can be described in terms of the basis states $|0\rangle$ and $|1\rangle$, corresponding to the vacuum and one-photon states, respectively. The atoms are described by the basis states $|g\rangle$ and $|e\rangle$.

Two different kinds of atom-cavity interactions are used in our schemes; an "on-resonant" and an "off-resonant" interaction. The on-resonant atom-cavity interaction can be described by the Hamiltonian,

$$H_{\text{on}} = i\hbar\Omega_1(\sigma_- a^\dagger - \sigma_+ a), \tag{1}$$

where $\sigma_+ = |e\rangle\langle g|$ and $\sigma_- = |g\rangle\langle e|$ are the raising and lowering operators for the atom, and Ω_1 is the one-photon Rabi frequency. In our schemes, the purpose of the on-resonant interaction is to prepare the cavity in the desired initial state and to read out the final state of the cavity.

In the off-resonant interaction, the $|g\rangle$ to $|e\rangle$ transition is sufficiently detuned from the cavity frequency that there are no transitions between these two levels during the interaction. We can model the corresponding interaction Hamiltonian by[7]

$$H_{\text{off}} = \hbar\Omega_2 a^\dagger a|e\rangle\langle e|, \tag{2}$$

where Ω_2 is the change in atomic level spacing per photon in the cavity. The two types of interactions described above could be produced from a single atomic species by appropriate Stark shifting of the atomic levels.

The Ramsey zone consists of a classical rf-field, whose interaction with the atoms can produce a general unitary transformation on the atomic two-level systems when appropriate values of the frequency and amplitude of the field are used. In the special case where a resonant field produces a $\pi/2$ rotation about the y-axis in spin-space, the effect of the Ramsey zone can be described by the operator,

$$R = (I + i\sigma_y)/\sqrt{2}, \tag{3}$$

where $\sigma_y = i(|e\rangle\langle g| - |g\rangle\langle e|)$.

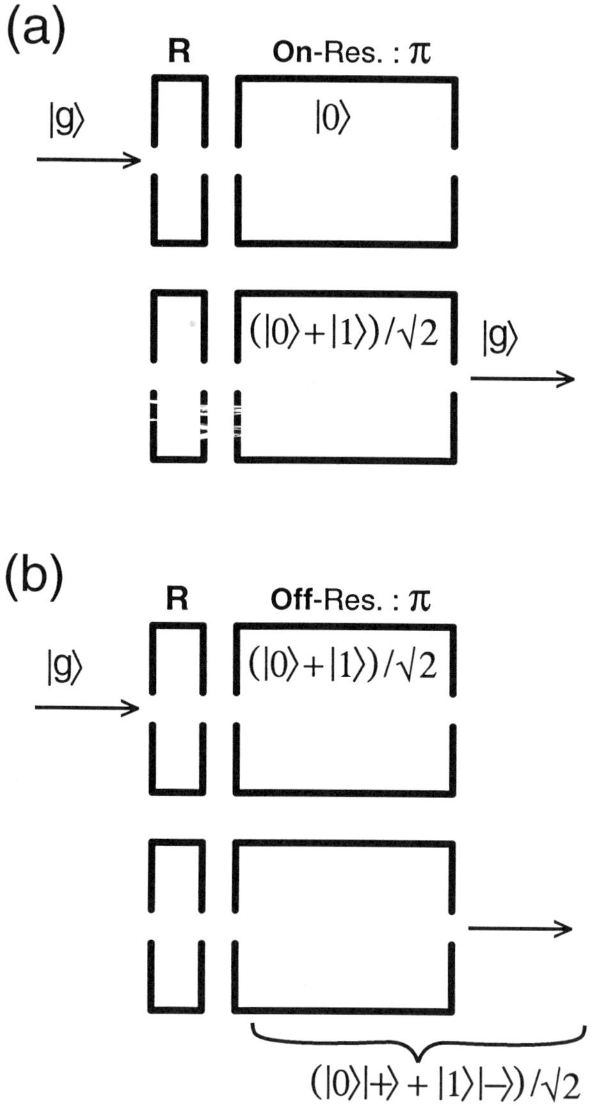

FIGURE 1. Cavity preparation (a) and generation of entanglement (b) used in the teleportation scheme. In both cases (a and b), the figure shows the state of the atom and cavity before the interaction (upper part) and after the interaction (lower part). The arrow represents the atom and the larger rectangular box depicts the cavity. The smaller rectangular box labeled "R" represents the Ramsey zone. "On-Res." indicates that the interaction is on-resonant and "Off-Res." indicates that it is off-resonant. The "π" indicates that the interaction strength satisfies $\Omega\tau = \pi$, as described in the text.

DESCRIPTION OF THE CAVITY QED TELEPORTATION SCHEME

The first step of the scheme—the production of an entangled atom-cavity EPR pair—starts by preparing the cavity in the state $(|0\rangle + |1\rangle)/\sqrt{2}$ (a superposition of the ground and one-photon states). This is done by passing an atom prepared in the state $|g\rangle$ through the Ramsey zone and then resonantly (equation 1) through the cavity, as shown in FIGURE 1a. After passing through the Ramsey zone, the atom is in the state $(|g\rangle + |e\rangle)/\sqrt{2}$. The atom-cavity interaction time τ_1 is chosen to satisfy $\Omega_1\tau_1 = \pi$ so that, after the interaction, the atom is left in state $|g\rangle$ (independent of its initial state) and the cavity in the state $(|0\rangle + |1\rangle)/\sqrt{2}$. The effect of this resonant atom-cavity interaction is to transfer the state of the atom to that of the cavity. The atom can now be discarded.

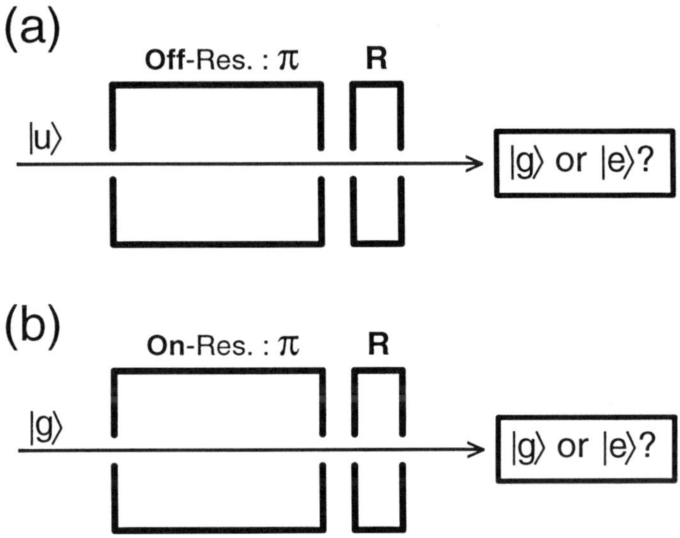

FIGURE 2. Setup used for Bell-state detection in the teleportation scheme: (a) Unknown atom passes first through the cavity and then through the Ramsey zone. The box on the right measures the atomic state in the $\{|g\rangle, |e\rangle\}$ basis. (b) Cavity state is "read out" when an atom prepared in the ground state passes through the cavity and Ramsey zone and the final atomic state is measured. Other symbols are defined in the caption to FIGURE 1.

After the cavity is prepared in the state $(|0\rangle + |1\rangle)/\sqrt{2}$, a second atom is sent first through the Ramsey zone and then nonresonantly through the cavity, as shown in FIGURE 1b. After passing through the Ramsey zone, the atom is in the state $|+\rangle \equiv (|g\rangle + |e\rangle)/\sqrt{2}$, where the definition of $|+\rangle$ is used for notational convenience. The atom-cavity interaction time τ_2 for this nonresonant interaction is chosen to satisfy $\Omega_2\tau_2 = \pi$. With this interaction strength, we find that the interaction results in a transformation defined by the unitary operator,

$$U = \exp(i\Omega_2\tau_2 a^\dagger a|e\rangle\langle e|) = (|+\rangle\langle -| + |-\rangle\langle +|)|1\rangle\langle 1| + |0\rangle\langle 0|, \tag{4}$$

where we have defined $|-\rangle \equiv (|g\rangle - |e\rangle)/\sqrt{2}$. Equation 4 can be interpreted in the following way. If there is no photon in the cavity, then the atom emerges from the cavity in the same state as it entered. If there is a photon in the cavity, the atomic excited state undergoes a π phase shift and the atomic transformations $|+\rangle \rightarrow |-\rangle$ and $|-\rangle \rightarrow |+\rangle$ take place. For an initial state $|\Psi_{in}\rangle = (1/\sqrt{2})(|0\rangle + |1\rangle)|+\rangle$, the atom-cavity interaction leaves the system in the state

$$|\Psi^{(+)}\rangle = (\sqrt{1/2})\,(|0\rangle|+\rangle + |1\rangle|-\rangle). \tag{5}$$

We have generated an EPR pair consisting of the cavity field and the nonresonant atom. The atom is then sent to the receiving part of the apparatus (Bob).

In the next phase of the experiment, a measurement is made on the system consisting of the cavity field and the unknown state to be teleported. According to the scheme of Bennett *et al.*, the measurement is made in the Bell operator basis[2] defined by the states

$$|\Psi^{(\pm)}\rangle = (\sqrt{1/2})\,(|0\rangle|+\rangle \pm |1\rangle|-\rangle) \tag{6}$$

and

$$|\Phi^{(\pm)}\rangle = (\sqrt{1/2})\,(|0\rangle|-\rangle \pm |1\rangle|+\rangle). \tag{7}$$

We now outline the method by which such a measurement can be made. We consider the result of an off-resonant interaction (equation 4) between an atom and the cavity field, where the atom-cavity system is initially in one of the states defined by equations 6 and 7. One can show from equation 4 that the interaction results in the following transformations:

$$\begin{aligned} |\Psi^{(\pm)}\rangle &\rightarrow (\sqrt{1/2})\,(|0\rangle \pm |1\rangle)|+\rangle, \\ |\Phi^{(\pm)}\rangle &\rightarrow (\sqrt{1/2})\,(|0\rangle \pm |1\rangle)|-\rangle. \end{aligned} \tag{8}$$

The atom-cavity interaction has removed the entanglement. Thus, a pair of single-particle measurements in the appropriate single-particle bases *after the interaction* is equivalent to a measurement *before the interaction* in the Bell operator basis of equations 6 and 7. The single-particle measurements are made as follows. After the atom in the unknown state has interacted with the cavity, it passes through a Ramsey zone (FIGURE 2a). The effect of the Ramsey zone, defined by equation 3, is to produce the transformations $|+\rangle \rightarrow |g\rangle$ and $|-\rangle \rightarrow -|e\rangle$. The atomic state is then measured in the $\{|g\rangle, |e\rangle\}$ basis by an ionizing detector.[7]

To complete the measurement, one must measure the state of the cavity in the basis $\{(|0\rangle \pm |1\rangle)/\sqrt{2}\}$. The cavity measurement is achieved by passing an atom, initially in the state $|g\rangle$, through the cavity with the *on-resonant* interaction (equation 1) and then through the Ramsey zone (FIGURE 2b). The effect of this atom-cavity interaction is to transfer the state of the cavity to the atom (the cavity is left in the vacuum state). Examination of equations 1 and 3 shows that, after the atom interacts with the cavity and Ramsey zone, it will emerge in the state $|g\rangle$ if the initial cavity state is $(|0\rangle + |1\rangle)/\sqrt{2}$ and will emerge in the state $-|e\rangle$ if the initial cavity state is

$(|0\rangle - |1\rangle)/\sqrt{2}$. Thus, a measurement of the atom in the basis $\{|g\rangle, |e\rangle\}$ constitutes a measurement of the cavity in the basis $\{(|0\rangle \pm |1\rangle)/\sqrt{2}\}$.

Before Alice's measurement, the three-particle system (consisting of the cavity, the atom to be teleported, and Bob's EPR particle) is in the state

$$|\Psi_{123}\rangle = (\sqrt{1/2})|u\rangle_1(|0\rangle|+\rangle_3 - |1\rangle|-\rangle_3), \tag{9}$$

where we have labeled the particles to distinguish the particle in the unknown state (1) from Bob's EPR particle (3). Following reference 1, we express the unknown state as $|u\rangle_1 = c_+|+\rangle_1 + c_-|-\rangle_1$ and write the state of the three-particle system as

$$|\Psi_{123}\rangle = (1/2)[|\Psi^{(-)}\rangle_{12}(-c_+|+\rangle_3 - c_-|-\rangle_3) + |\Psi^{(+)}\rangle_{12}(-c_+|+\rangle_3 + c_-|-\rangle_3) + |\Phi^{(-)}\rangle_{12}(c_+|+\rangle_3 + c_-|-\rangle_3) + |\Phi^{(+)}\rangle_{12}(c_+|+\rangle_3 - c_-|-\rangle_3)]. \tag{10}$$

Thus, Alice's measurement projects the state of Bob's EPR particle into one of four states. These states differ from the state $|u\rangle$ by one of four unitary transformations, depending on the outcome of Alice's measurement. In our proposed implementation, one can transform the state of Bob's particle into $|u\rangle$ by applying an appropriate signal to a second Ramsey zone, through which Bob's particle passes after Alice's measurement is made.

This completes our description of the implementation. To test the scheme, one would use it to teleport a *known state* $|u\rangle$, which could be prepared by an additional Ramsey zone. Bob's particle could then be analyzed in a basis consisting of the state $|u\rangle$ and an orthogonal state. The necessary measurements can be made by using a Ramsey zone and an ionizing detector. One would then repeat the experiment using a variety of initially prepared states and measure the fidelity of transmission. More detailed discussions of transmission fidelity in teleportation have been made in references 8 and 9.

RELATED CAVITY QED SCHEMES

In a closely related paper, Bennett and Wiesner[3] have suggested a scheme by which two bits of information could be transmitted by a single two-state particle. In this scheme, an EPR pair is created, one particle going to Alice and the other to Bob. Bob is now able to send two bits of information to Alice by applying one of four transformations on his EPR particle and sending that particle to Alice. Alice can unambiguously determine which transformation Bob applied by making a measurement in the Bell operator basis. An experimentally realizable implementation of this scheme could be made with a slight modification to our implementation of the teleportation scheme. After the cavity-atom EPR pair is created (equation 5), the atom (Bob's particle) would pass through a Ramsey zone and would undergo one of four transformations. This atom would then be sent back through the cavity for the measurement in the Bell operator basis (by Alice). To facilitate this scheme, the state of Bob's particle could be stored in a second cavity. Some time later, another particle, traveling in the opposite direction, would pass through this cavity and would pick up the cavity state. It would then go through the first cavity, where the measurement is made.

Another modification to our scheme could be used to generate perfectly correlated states of n particles ($n \geq 3$). Such n-particle correlated states have been of great interest recently[10,11] because of their potential to provide a strong test of nonlocality in quantum mechanics. Very little discussion, however, has been given to their production.[12] In our scheme, such states could be produced by first creating the EPR state of equation 5. One could then send an arbitrary number ($n - 2$) of atoms (prepared in the state $|g\rangle$), first through the Ramsey coil and then through the cavity with the off-resonant interaction. It can be shown that the resulting state of the system will be

$$|\Psi\rangle = (\sqrt{1/2})\,(|0\rangle|+\rangle_1 \cdots |+\rangle_{n-1} - |1\rangle|-\rangle_1 \cdots |-\rangle_{n-1}), \tag{11}$$

which leaves the system of ($n - 1$) atoms and the cavity in an entangled state. One could read out the state of the cavity by sending through a resonant atom prepared in the ground state to obtain the n-atom entangled state. An additional feature of this method of producing n-particle entangled states is that one could use a very similar technique for detecting such an n-particle state. This would be done essentially by running the above procedure in reverse.

APPLICATION TO QUANTUM COMPUTATION

It has recently been shown that for certain applications, such as factoring, quantum computers are potentially much faster than "classical" computers.[13] In this section, we briefly outline some applications of cavity QED to quantum computation. Specifically, we indicate how one may produce quantum gates. A quantum gate is a generalization of a logic gate, with the additional property that superpositions of input states will yield superpositions of output states. We discuss how one can produce a reversible quantum "exclusive-or" (or XOR) gate, as well as a quantum "Toffoli" gate.

A reversible XOR gate has two input channels (which we denote a_i and b_i) and two output channels (a_o and b_o). The values on the output channels are determined by $a_o = a_i$ and $b_o = a_i \oplus b_i$, where the XOR operation $\oplus$ is defined to be 0 if the two inputs are the same and 1 if the two inputs are different.[14]

For a cavity QED scheme, the logical states 0 and 1 can be encoded as the ground and excited states, respectively, of the atoms or of the cavity. The XOR gate can be realized in a similar manner to the quantum teleportation scheme (see the third section of this report) by first transferring the state of atom a to the cavity field (using the on-resonant interaction), as shown in FIGURE 3a. Atom b then passes through the cavity (FIGURE 3b) and the interaction is chosen to be off-resonant with a strength determined by $\Omega_2\tau_2 = \pi$ (see equations 2 and 4). Equation 4 indicates that, if the cavity is in the state $|1\rangle$ (logical "1"), the state of atom b is switched from $|+\rangle$ to $|-\rangle$ or from $|-\rangle$ to $|+\rangle$. By placing a Ramsey zone R (equation 3) before the cavity and its inverse R^{-1} after the cavity, we find that a logical "1" of atom a results in the logical transformations $0 \rightarrow 1$ and $1 \rightarrow 0$ of atom b and that a logical "0" of atom a leaves the state of atom b unchanged. This is just the action of the XOR operation. Afterwards, the state of the cavity can be transferred again to that of an atom, thereby accomplishing the operation of the quantum XOR gate.

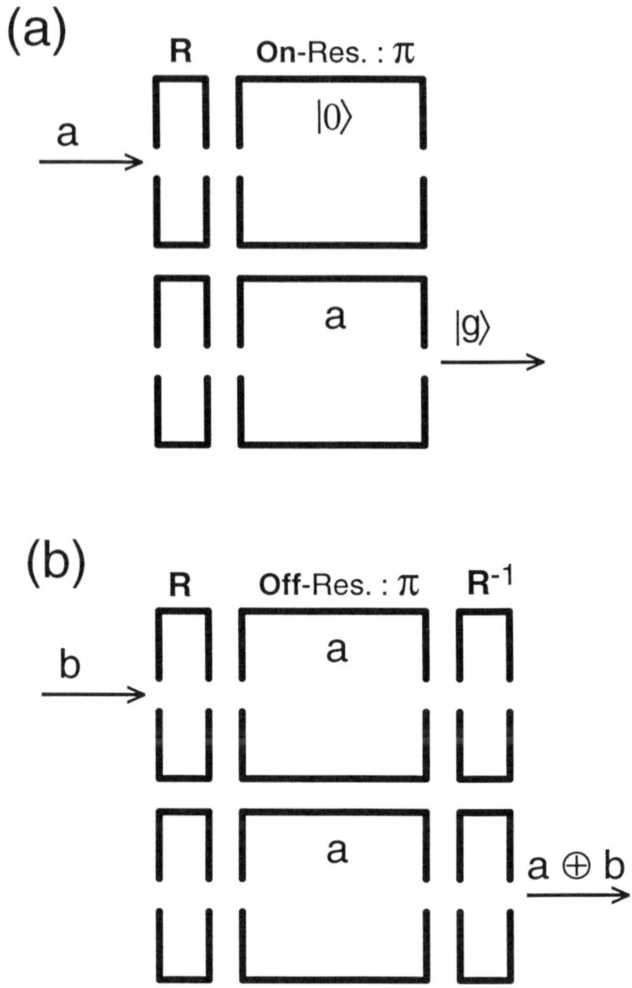

FIGURE 3. The operation of a reversible quantum XOR gate: (a) The state of atom a is transferred to the cavity. (b) Interaction between the cavity and atom b yields $a \oplus b$. Interaction parameters are defined in the caption to FIGURE 1.

The outputs of the reversible quantum Toffoli gate are defined as $a_o = a_i$, $b_o = b_i$, and $c_o = (a_i \cdot b_i) \oplus c_i$. An additional AND operation of inputs a and b $(a \cdot b)$ is needed in combination with the XOR operation. This universal gate can then be used for all necessary quantum logic operations.[14] The operation of the Toffoli gate is achieved by the following steps: (i) the states of atom a and atom b are transferred to the fields in two cavities (labeled C_1 and C_2); (ii) a third cavity (C_3) is put into the state $a_i \oplus b_i$ (this can be accomplished by preparing C_3 in state $|0\rangle$ and then sending an atom through C_1, C_2, and C_3, with appropriate interaction parameters); (iii) atom c passes

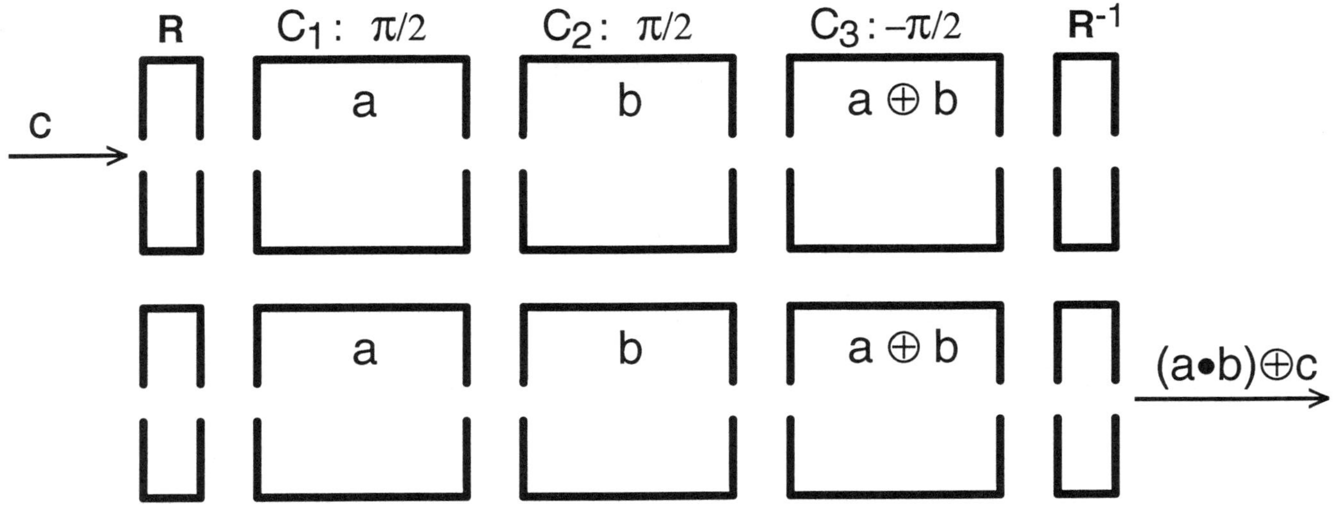

FIGURE 4. One step in the operation of a reversible quantum Toffoli gate. The atom interacts off-resonantly with all three cavities (C_1, C_2, and C_3). Notice the phase shifts of $\pm\pi/2$ in this setup, in contrast to the π phase shift in the XOR gate shown in FIGURE 3.

through the three cavities (and two Ramsey zones) in the manner shown in FIGURE 4. The atom-cavity interactions are all off-resonant and satisfy the conditions $\Omega_2\tau_2 = \pi/2$ for both C_1 and C_2 and $\Omega_2\tau_2 = -\pi/2$ for C_3.[15] Under these conditions, the total phase shift experienced by the excited state of atom c after interacting with the three cavities is 0 unless both a and b are initially in the excited state, in which case the phase shift of atom c is π. By taking into account the action of the two Ramsey zones, we find that we have produced the logical transformation $c_o = (a_i \cdot b_i) \oplus c_i$. Appropriately resetting C_3 (to the ground state) and transferring the states of C_1 and C_2 back to atomic states assures the reversibility of the operation. Because the atoms can always be in any superposition of ground and excited states, this scheme enables one to perform a fully quantum logic computation for the first time.

It is interesting to note that we have constructed the three-bit Toffoli gate with the use of only two-particle interactions. In doing so, we have demonstrated that, in quantum mechanics, the Toffoli gate can be reduced to a finite number of smaller (two-bit) quantum gates.[16]

CONCLUSIONS

In conclusion, we have presented an implementation of the scheme of Bennett *et al.*[1] for teleportation of an unknown quantum state based on cavity quantum electrodynamics. We have shown that the high coupling strength of the atom-cavity interaction allows one to disentangle a system of two two-state particles and to make a measurement in a basis of entangled states. This so-called Bell-state measurement is also essential to a dense coding quantum communication scheme, which could be implemented in a very similar way. In addition, it was shown how cavity QED interactions could be used for the production of n-particle entangled states, necessary for new tests of basic features of quantum mechanics. Finally, the application of cavity QED to the construction of universal quantum gates was outlined.

ACKNOWLEDGMENTS

We acknowledge useful discussions with C. H. Bennett. Tycho Sleator thanks Anton Zeilinger for his hospitality during a stay in Innsbruck.

REFERENCES AND NOTES

1. BENNETT, C. H., G. BRASSARD, C. CRÉPEAU, R. JOZSA, A. PERES & W. K. WOOTTERS. 1993. Phys. Rev. Lett. **70:** 1895.
2. BRAUNSTEIN, S. L., A. MANN & M. REVZEN. 1992. Phys. Rev. Lett. **68:** 3259.
3. BENNETT, C. H. & S. J. WIESNER. 1992. Phys. Rev. Lett. **69:** 2881.
4. WEINFURTER, H. 1994. Europhys. Lett. **25:** 559.
5. A brief description of our scheme is presented in: SLEATOR, T. & H. WEINFURTER. 1994. *In* International Quantum Electronics Conference. Vol. 9, p. 140. OSA Technical Digest Series. A similar scheme has been proposed by Davidovich *et al.* in reference 9.
6. BERNARDOT, F., P. NUSSENZVEIG, F. SCHMIDT-KALER, M. BRUNE, J. M. RAIMOND & S. HAROCHE. 1994. *In* International Quantum Electronics Conference. Vol. 9, p. 223. OSA Technical Digest Series.

7. BRUNE, M., S. HAROCHE, V. LEFEVRE, J. M. RAIMOND & N. ZAGURY. 1990. Phys. Rev. Lett. **65:** 976; BRUNE, M., S. HAROCHE, J. M. RAIMOND, L. DAVIDOVICH & N. ZAGURY. 1992. Phys. Rev. A **45:** 5193.
8. POPESCU, S. 1994. Phys. Rev. Lett. **72:** 797.
9. DAVIDOVICH, L., N. ZAGURY, M. BRUNE, J. M. RAIMOND & S. HAROCHE. 1994. Submitted.
10. GREENBERGER, D. M., M. HORNE & A. ZEILINGER. 1989. *In* Bell's Theorem, Quantum Theory, and Conceptions of the Universe. M. Kafatos, Ed.: 107. Kluwer. Dordrecht; GREENBERGER, D. M., M. HORNE, A. SHIMONY & A. ZEILINGER. 1990. Am. J. Phys. **58:** 1131.
11. MERMIN, N. D. 1990. Am. J. Phys. **58:** 731; Phys. Rev. Lett. **65:** 1838.
12. WÓDKIEWICZ, K., L. WANG & J. H. EBERLY. 1993. Phys. Rev. A **47:** 3280. These investigators have proposed a scheme for generating three-particle correlated states using cavity QED; KUDRYAVTSEV, I. K. & P. L. KNIGHT. 1993. J. Mod. Opt. **40:** 1673.
13. SHOR, P. 1994. Proceedings of the Twenty-sixth Annual ACM Symposium on the Theory of Computing. To be published.
14. DEUTSCH, D. 1989. Proc. R. Soc. London Ser. A **425:** 73.
15. The interaction of the atom with each cavity results in an operation that can be expressed as "a square root of XOR". Such an operation is possible only for quantum logic gates and is an extension of the idea of the "square root of NOT", introduced in reference 14.
16. A similar result (using an infinite number of two-bit gates) has been obtained by: DIVINCENZO, A. 1994. Two-bit gates are universal for quantum computation. Submitted.

A Review of the Decoherent Histories Approach to Quantum Mechanics[a]

J. J. HALLIWELL

Theory Group
Blackett Laboratory
Imperial College
London SW7 2BZ, United Kingdom

INTRODUCTION

Quantum mechanics was originally developed to account for a number of unexplained phenomena on the atomic scale. The theory was not thought to be applicable to physics at larger scales, nor was there felt any need to do so. Indeed, it was only by reference to an external, classical, macroscopic world that the theory could be properly understood. This view of quantum mechanics, the Copenhagen interpretation, has persisted for a very long time with not one shred of experimental evidence against it.[1]

Today, however, more ambitious views of quantum mechanics are entertained. Experiments have been contemplated (e.g., involving SQUIDS) that may probe domains traditionally thought of as macroscopic.[2] Even in the absence of such experiments, the Copenhagen interpretation rests on unsatisfactory foundations. Macroscopic classical objects are made from microscopic quantum ones. The dualist view of the Copenhagen interpretation may therefore be internally inconsistent and is at best approximate. Most significantly, there has been a considerable amount of recent interest in the subject of quantum cosmology in which the notion of an external classical domain is completely inappropriate.[3] Generalizations of conventional quantum theory are required to meet these new challenges.

John A. Wheeler was one of the very first people to be so bold as to even talk about "the wave function of the universe".[4] He has contributed extensively to our understanding of quantum mechanics and quantum cosmology, both through his own work and through his inspiration to many others in the field. It is a great pleasure to contribute to this conference organized in his honor.

The Histories Approach

The object of this report is to review one particular approach to quantum mechanics that was specifically designed to overcome some of the problems of the orthodox approach. This is the decoherent (or "consistent") histories approach due to Griffiths,[5–9] Gell-Mann and Hartle,[10–19] and Omnès.[20–26] It is, in particular, a predictive formulation of quantum mechanics for genuinely closed quantum systems

[a]This work was supported by a University Research Fellowship from the Royal Society.

that is sufficiently general to cope with the needs of quantum cosmology. In brief, its aims are as follows:

(1) To understand the emergence of an approximately classical universe from an underlying quantum one, without becoming embroiled in the details of observers, measuring devices, or collapse of the wave function. Prediction of a classical domain similar to the one in which we live will generally depend on the initial condition of the universe and, moreover, could be one of many possibilities predicted by quantum mechanics. Accommodation, rather than absolute prediction, of our particular classical universe may be as much as can be expected.

(2) To supply a quantum mechanical framework for reasoning about the properties of closed physical systems. Such a framework is necessary if the process of prediction in quantum mechanics is to be genuinely quantum mechanical at every single step. That process consists of first logically reconstructing the past history of the universe from records existing in the classical domain at the present and then using the present records together with the deduced past history to make predictions about the future (strictly speaking, about correlations between records at a fixed moment of time in the future). A framework for reasoning may also lead to clarification of many of the conceptually troublesome aspects of quantum mechanics, such as the EPR paradox.

In more detail, the primary mathematical aim of the histories approach is to assign probabilities to histories of a closed system. The approach is a modest generalization of ordinary quantum mechanics, but relies on a far smaller list of axioms. These axioms are basically the statements that the closed system is described by the usual mathematical machinery of Hilbert together with a formula for the probabilities of histories and a rule of interpretation. The approach makes no distinction between microscopic and macroscopic, nor does it assume a "system-environment" split; in particular, a separate classical domain is not assumed. It makes no essential use of measurement or collapse of the wave function, although these notions may be discussed within the framework of the approach. What replaces measurement is the more general and objective notion of consistency (or the stronger notion of decoherence), determining which histories may be assigned probabilities. The approach also stresses classical (i.e., Boolean) logic, the conditions under which it may be applied, and thus the conditions under which ordinary reasoning may be applied to physical systems.

The decoherent histories approach is not designed to answer the question held by some to be the most important problem of quantum measurement theory: why one particular history for the universe "actually happens", whereas the other potential histories allowed by quantum mechanics fade away. Although some aspects of this problem are clarified by the decoherent histories approach, a satisfactory solution does not appear to be possible unless something external is added (see, for example, reference 27); nor is the approach intended to meet some philosophical prejudice about the way the world appears to be. Its aims, for the large part, are of a rather pragmatic nature, namely, answering the very physical question of why the world is described so well by classical mechanics and ordinary logic when its atomic constituents are described by quantum mechanics.

Why Histories?

The basic building blocks in the decoherent histories approach are the histories of a closed system—sequences of alternatives at a succession of times. Why are these objects of particular interest?

(a) Histories are the most general class of situations that one might be interested in. In a typical experiment, for example, a particle is emitted from a decaying nucleus at time t_1; then, it passes through a magnetic field at time t_2; finally, it is absorbed by a detector at time t_3.

(b) We would like to understand how classical behavior can emerge from the quantum mechanics of closed systems. This involves showing, amongst other things, that successive positions in time of a particle are approximately correlated according to classical laws. This involves the probabilities for approximate positions at different times.

(c) The basic pragmatic aim of theoretical physics is to find patterns in presently existing data. In cosmology, for example, one tries to explain the connections between observed data about the microwave background, the expansion of the universe, the distribution of matter in the universe, the spectrum of gravitational waves, etc. Why, then, should we not attempt to formulate our theories in the terms of the density matrix of the entire universe at the present moment? There are at least two reasons of why not. First, present records are stored in a wide variety of different ways—in computer memories, on photographic plates, on paper, in our own personal memories, in measuring devices, etc. The dynamical variables describing those records could be very hard to identify. The correlations between present records are far easier to understand in terms of histories. The patterns in current cosmological data, for example, are explained most economically by appealing to the big bang model of the history of the universe. Second, the correlation between present records and past events can never be perfect. In order to discuss the approximate nature of correlations between the past and the present, it becomes necessary to talk about the histories of a system.

THE FORMALISM OF DECOHERENT HISTORIES

I now briefly outline the mathematical formalism of the decoherent histories approach. Further details may be found in the original papers cited above.

Probabilities for Histories

In quantum mechanics, propositions about the attributes of a system at a fixed moment of time are represented by sets of projection operators. The projection operators P_α effect a partition of the possible alternatives α that a system may exhibit at each moment of time. They are exhaustive and exclusive:

$$\sum_\alpha P_\alpha = 1, \qquad P_\alpha P_\beta = \delta_{\alpha\beta} P_\alpha. \tag{2.1}$$

A projector is said to be fine-grained if it is of the form $|\alpha\rangle\langle\alpha|$, where $\{|\alpha\rangle\}$ comprises a complete set of states; otherwise, it is coarse-grained. A quantum mechanical history is characterized by a string of time-dependent projections, $P^1_{\alpha_1}(t_1), \ldots, P^n_{\alpha_n}(t_n)$, together with an initial state ρ. The time-dependent projections are related to the time-independent ones by

$$P^k_{\alpha_k}(t_k) = e^{iH(t_k - t_0)} P^k_{\alpha_k} e^{-iH(t_k - t_0)}, \tag{2.2}$$

where H is the Hamiltonian. The candidate probability for such histories is

$$p(\alpha_1, \alpha_2, \ldots, \alpha_n) = \mathrm{Tr}[P^n_{\alpha_n}(t_n) \ldots P^1_{\alpha_1}(t_1) \rho P^1_{\alpha_1}(t_1) \ldots P^n_{\alpha_n}(t_n)]. \tag{2.3}$$

This expression is a familiar one from quantum measurement theory, but the interpretation is different. Here, it is the probability for a sequence of alternatives for a closed system. The alternatives at each moment of time are characterized by projectors. The projectors are not generally associated with measurements, as they would be in the Copenhagen view of the formula (equation 2.3). They cannot because the system is closed.

It is straightforward to show that equation 2.3 is both nonnegative and normalized to unity when summed over $\alpha_1, \ldots, \alpha_n$. However, equation 2.3 does not satisfy all the axioms of probability theory and, for that reason, it is referred to as a candidate probability. It does not satisfy the requirement of additivity on disjoint regions of sample space. More precisely, for each set of histories, one may construct coarser-grained histories by grouping the histories together. This may be achieved, for example, by summing over the projections at each moment of time:

$$\bar{P}_{\bar{\alpha}} = \sum_{\alpha \in \bar{\alpha}} P_\alpha \tag{2.4}$$

(although this is not the most general type of coarse-graining). The additivity requirement is then that the probabilities for each coarser-grained history should be the sum of the probabilities of the finer-grained histories of which each comprises. Quantum mechanical interference generally prevents this requirement from being satisfied; thus, in general, histories of closed quantum systems cannot be assigned probabilities.

The standard illustrative example is the double-slit experiment. The histories consist of projections at two moments of time: projections determining which slit the particle went through at time t_1 and projections determining the point at which the particle hit the screen at time t_2. As is well known, the probability distribution for the interference pattern on the screen cannot be written as a sum of the probabilities for going through each slit; hence, the candidate probabilities do not satisfy the additivity requirement.

There are, however, certain types of histories for which interference is negligible and the candidate probabilities for histories do satisfy the sum rules. These histories may be found using the decoherence functional:

$$D(\underline{\alpha}, \underline{\alpha}') = \mathrm{Tr}[P^n_{\alpha_n}(t_n) \ldots P^1_{\alpha_1}(t_1)\rho P^1_{\alpha_1'}(t_1) \ldots P^n_{\alpha_n'}(t_n)]. \tag{2.5}$$

Here, $\underline{\alpha}$ denotes the string $\alpha_1, \alpha_2, \ldots, \alpha_n$. Intuitively, the decoherence functional measures the amount of interference between pairs of histories. It may be shown that the additivity requirement is satisfied for all coarse-grainings if and only if

$$\mathrm{Re}\, D(\underline{\alpha}, \underline{\alpha}') = 0 \tag{2.6}$$

for all distinct pairs of histories $\underline{\alpha}, \underline{\alpha}'$.[5] Such sets of histories are said to be consistent, or weakly decoherent. (Note that this definition of consistency is stronger than that originally introduced by Griffiths.[5] See reference 12 for a discussion of this point.)

Consistency and Classical Logic

Why are sets of consistent histories of interest? As stated, propositions about the attributes of a quantum system may be represented by projection operators. The set of all projections has the mathematical structure of a lattice. This lattice is nondistributive and this means that the corresponding propositions may not be submitted to Boolean logic. Similar remarks hold for the more complex propositions expressed by general sets of quantum mechanical histories.

The reason why consistent sets of histories are of interest is that they can be submitted to Boolean logic. Indeed, a theorem of Omnès states that a set of histories forms a consistent representation of Boolean logic if and only if it is a consistent set;[20,25,26] that is, in a consistent set of histories, each history corresponds to a proposition about the properties of a physical system and we can meaningfully manipulate these propositions without contradiction using ordinary classical logic. It is in this sense that the decoherent histories approach supplies a foundation for reasoning about closed physical systems.

An important example is the case of retrodiction of the past from present data. Suppose we have a consistent set of histories. We would say that the alternative α_n (present data) implies the alternatives $\alpha_{n-1} \ldots \alpha_1$ (past events) if

$$p(\alpha_1, \ldots, \alpha_{n-1} | \alpha_n) \equiv \frac{p(\alpha_1, \ldots, \alpha_n)}{p(\alpha_n)} = 1. \tag{2.7}$$

In this way, we can in quantum mechanics build a picture of the history of the universe, given the present data and the initial state, using only logic and the consistency of the histories. We can talk meaningfully about the past properties of the universe even though there was no measuring device there to record them.

There is, however, a caveat. It is very frequently the case that the same initial state and present data will admit two or more inequivalent sets of consistent histories, the union of which is not a consistent set. Then, there often exist propositions about the past properties of the system that are logically implied by the present data in some sets of histories, but not in others. Omnès refers to such propositions as "reliable", whereas propositions that are implied by the present data in every consistent set of histories are labeled "true"[28] (see also reference 29). The existence of these so-called multiple logics means that one cannot say that past properties corresponding to reliable propositions "actually happened" because they depend on a particular choice of consistent histories. In the histories approach, the reconstruction of history from present records is therefore not unique. This means that the approach does not in general allow one to talk about the past history of the universe as "the way it really is".

Is this a problem? Some feel that it is.[30] For the immediate practical purposes of quantum cosmology, however, it does not appear to be a difficulty. Recall that what quantum mechanics must ultimately explain is the correlation between records at a fixed moment of time. As stated earlier, it is easiest to understand those correlations in terms of histories, but histories enter as an intermediate step. The correlations between two records at a fixed moment of time predicted by quantum mechanics are unambiguous, even though the histories corresponding to these records may not be unique.

DECOHERENCE, CORRELATION, AND RECORDS

How may the consistency condition of equation 2.6 come to be satisfied? First of all, it is straightforward to show that, with some exceptions, histories of completely fine-grained projection operators will not generally lead to consistency. The consistency condition is generally satisfied only by sets of histories that are coarse-grained. When sets of histories satisfy the consistency condition of equation 2.6 as a result of coarse-graining, they typically satisfy, in addition, the stronger condition that both the real and imaginary parts of the off-diagonal terms of the decoherence functional vanish:

$$D(\underline{\alpha}, \underline{\alpha}') = 0 \qquad \text{for} \qquad \underline{\alpha} \neq \underline{\alpha}'. \tag{3.1}$$

This I shall refer to quite simply as decoherence. (It is sometimes referred to more specifically as medium decoherence,[12] but we shall not do so here.)

Physically, decoherence is intimately related to the existence of records about the system somewhere in the universe. In this sense, decoherence replaces and generalizes the notion of measurement in ordinary quantum mechanics. Sets of histories decohere and hence the system "acquires definite properties", not necessarily through measurement, but through the interactions and correlations of the variables that are followed with the variables that are ignored as a result of the coarse-graining.

Decoherence is typically only approximate, so measures of approximate decoherence are required. First, note that the decoherence functional obeys the simple inequality,[31]

$$|D(\underline{\alpha}, \underline{\alpha}')|^2 \leq D(\underline{\alpha}, \underline{\alpha}) D(\underline{\alpha}', \underline{\alpha}'). \tag{3.2}$$

Intuitively, this result indicates that there can be no interference with a history that has candidate probability zero. It also suggests a possible measure of approximate decoherence: we say that a system decoheres to order ϵ if the decoherence functional satisfies equation 3.2 with a factor of ϵ^2 multiplying the right-hand side. This condition may be shown to imply that most (but not all) probability sum rules will then be satisfied to order ϵ.[31]

Approximate decoherence to order ϵ means that the probabilities are defined only up to that order. In typical cases, ϵ is substantially smaller than any other effect that could conceivably modify the probabilities and, hence, they may be thought of as precisely defined for all practical purposes. Alternatively, it has been conjectured that a generic approximately decoherent set of histories may be turned into an exactly decoherent set by modifying to order ϵ the operators that are projected onto at each moment of time.[30]

Records Imply Decoherence

I now exemplify the connection between records and decoherence. Consider a closed system S that consists of two weakly interacting subsystems A and B. The Hilbert space $\mathcal{H}$ of S is therefore of the form $\mathcal{H}_A \otimes \mathcal{H}_B$. For simplicity, let $\mathcal{H}_A$ and $\mathcal{H}_B$ have the same dimension. Suppose we are interested in the histories characterized solely by properties of system A; thus, B is regarded as the environment. The system is analyzed using the decoherence functional (equation 2.5), where we take the P_α to denote a projection on $\mathcal{H}_A$ (hence, the projections in the decoherence functional are of the form $P_\alpha \otimes I^B$, where I^B denotes the identity on $\mathcal{H}_B$). I also introduce projections R_β on the Hilbert space $\mathcal{H}_B$.

I shall show that histories of A satisfy the decoherence condition (equation 3.1) if the sequences of alternatives that make up the histories exhibit exact and persistent correlations with sequences of alternatives of B. To be precise, suppose that the alternatives of A characterized by $P^k_{\alpha_k}$ at each moment of time t_k are perfectly recorded in B as a result of their interaction. Suppose also that this record in B is perfectly persistent (i.e., permanent). This means that at any time t_f after the time t_n of the last projection on A there will exist a sequence of alternatives of B, namely, $\beta_1 \ldots \beta_n$, in perfect correlation with the alternatives of A, namely, $\alpha_1 \ldots \alpha_n$, at times $t_1 \ldots t_n$.

For each moment of time t_k, the decoherence functional (equation 2.5) may be written

$$D(\underline{\alpha}, \underline{\alpha}') = \sum_{\beta_k} \mathrm{Tr}(I^A \otimes R^k_{\beta_k} \ldots P^k_{\alpha_k} \otimes I^B \ldots \rho \ldots P^k_{\alpha_k'} \otimes I^B \ldots) \qquad \textbf{(3.3)}$$

using the exhaustivity of the projections $R^k_{\beta_k}$, where the dots denote the projections at times other than t_k and the unitary evolution operators between them. Now, because $R^k_{\beta_k}$ is a projector, it may be replaced by $(R^k_{\beta_k})^2$. Furthermore, the assumption of persistence then allows us to move the projector $R^k_{\beta_k}$ through all the unitary evolution operators occurring after time t_k on each side of the decoherence functional, with the result

$$D(\underline{\alpha}, \underline{\alpha}') = \sum_{\beta_k} \mathrm{Tr}(\ldots P^k_{\alpha_k} \otimes R^k_{\beta_k} \ldots \rho \ldots P^k_{\alpha_k'} \otimes R^k_{\beta_k} \ldots). \qquad \textbf{(3.4)}$$

Finally, the assumed correlation between the alternative α_k in A and β_k in B means that the terms of the form $P^k_{\alpha_k} \otimes R^k_{\beta_k}$ on each side will yield zero when operating on everything that came earlier in the chain, unless $\alpha_k = \beta_k$. Equation 3.4 will therefore be diagonal in α_k. Repeating the argument for all other values of k, we thus find that, as advertised, a perfect and persistent correlation of alternatives of A with those of B leads to exact decoherence of the histories of A. It is not just the consistency condition of equation 2.6 that is satisfied through persistent correlation with another subsystem, but the stronger condition of decoherence (equation 3.1). This argument was inspired by an argument given by Hartle[14] in his discussion of the recovery of the Copenhagen interpretation from the decoherent histories approach. A more detailed version of it is given in reference 32.

Decoherence Implies Generalized Records

There is a converse to the above result: namely, that equation 3.1, in a certain sense, implies the existence of records.[12] Consider the decoherence functional (equation 2.5) for any system (not just the special one discussed above). Introduce the convenient notation

$$C_{\underline{\alpha}} = P_{\alpha_n}(t_n) \ldots P_{\alpha_1}(t_1). \tag{3.5}$$

Let the initial state be pure, $\rho = |\Psi\rangle\langle\Psi|$. In this case, the decoherence condition (equation 3.1) is referred tc as medium decoherence. It implies that the states $C_{\underline{\alpha}}|\Psi\rangle$ are an orthogonal (but in general incomplete) set. Thus, there will exist a set of projection operators $R_{\underline{\beta}}$ (not in general unique) of which these states are eigenstates:

$$R_{\underline{\beta}} C_{\underline{\alpha}} |\Psi\rangle = \delta_{\underline{\alpha\beta}} C_{\underline{\alpha}} |\Psi\rangle. \tag{3.6}$$

Note that the $C_{\underline{\alpha}}$ terms are not themselves projectors in general. One may then consider histories consisting of the string of projections (equation 3.5) adjoined by the projections $R_{\underline{\beta}}$ at any time after the final time. The decoherence functional for such histories is

$$D(\underline{\alpha}, \underline{\beta} | \underline{\alpha}', \underline{\beta}') = \mathrm{Tr}(R_{\underline{\beta}} C_{\underline{\alpha}} |\Psi\rangle\langle\Psi| C^{\dagger}_{\underline{\alpha}'} R_{\underline{\beta}'}). \tag{3.7}$$

These extended histories decohere exactly by virtue of equations 3.1 and 3.6 and, thus, the diagonal elements of equation 3.7, which we denote $p(\underline{\alpha}, \underline{\beta})$, are true probabilities. The correlations contained in these probabilities may therefore be discussed. Indeed, equation 3.6 implies that $p(\underline{\alpha}, \underline{\beta}) = \delta_{\underline{\alpha\beta}} p(\underline{\alpha})$ and thus $\underline{\alpha}$ and $\underline{\beta}$ are perfectly correlated.

Medium decoherence thus implies the existence of a string of alternatives $\beta_1 \ldots \beta_n$ at some fixed moment of time after t_n perfectly correlated with the string $\alpha_1 \ldots \alpha_n$ at the sequence of times $t_1 \ldots t_n$. For this reason, the projection operators $R_{\underline{\alpha}}$ are referred to as generalized records: information about the histories characterized by alternatives $\alpha_1 \ldots \alpha_n$ is recorded somewhere. However, it is not possible to say that the information resides in a particular subsystem because we have not specified the form of the system S; indeed, it is generally not possible to divide it into subsystems.

TOWARDS A QUASI-CLASSICAL DOMAIN

Given the framework sketched above, one of the principal aims of the decoherent histories approach is to demonstrate the emergence of an approximately classical world from an underlying quantum one, together with the quantum fluctuations about it described by the familiar Copenhagen quantum mechanics of measured subsystems. Such a state of affairs is referred to as a quasi-classical domain.[11–13] In more technical terms, a quasi-classical domain consists of a decoherent set of histories, characterized largely by the same types of variables at different times, and whose probabilities are peaked about deterministic evolution equations for the variables characterizing the histories.

Moreover, the histories should be maximally refined with respect to a specified degree of approximate decoherence; that is, one specifies a decoherence factor ϵ in the approximate decoherence condition discussed above. This should, for example, be chosen so that the probabilities are defined to a precision far beyond any conceivable test. Then, the histories should be fine-grained (e.g., by reducing the widths of the projections) to the point that further fine-graining would lead to violation of the specified degree of approximate decoherence. The resulting set of histories will then be called maximally refined. The reason for maximally refining the histories is to reduce as much as possible any apparent subjective element in the choice of coarse-graining.

Given the Hamiltonian and initial state of the system, one's task is to compute the decoherence functional for various different choices of histories and to see which ones lead to quasi-classical behavior. As suggested by the discussion at the end of the second section, there could be—and probably are—many such sets of variables leading to quasi-classical behavior. An important problem is to find as many such sets as possible and to develop criteria to distinguish between them. One useful criterion is whether a quasi-classical domain can support the existence of an information gathering and utilizing system (IGUS). This is a complex adaptive system that exploits the regularities in its environment in such a way as to ensure its own survival. This particular criterion may rule out domains described by particularly bizarre decoherent sets of histories, such as ones described by completely different variables at each moment of time, because the IGUS may not have sufficient information-processing capabilities to assimilate its environment. Also, criteria such as the existence of IGUSs alleviate to some degree the multiplicity of consistent sets of histories discussed in the second section. These issues are discussed further in references 11–13, 30, and 33–35.

Histories of Hydrodynamic Variables

What are the sets of variables that can lead to quasi-classical behavior? One particular set of variables that are strong candidates for it are the integrals over small volumes of locally conserved densities. A generic system will usually not have a natural separation into "system" and "environment", and it is one of the strengths of the decoherent histories approach that it does not rely on such a separation. Certain variables, however, will be distinguished by the existence of conservation laws for total energy, momentum, charge, particle number, etc. Associated with such conservation laws are local conservation laws of the form,

$$\frac{\partial \rho}{\partial t} + \nabla \cdot \mathbf{j} = 0. \tag{4.1}$$

The candidate quasi-classical variables are then

$$Q_V = \int_V d^3x \rho(\mathbf{x}). \tag{4.2}$$

If the volume V over which the local densities are smeared is infinite, Q_V will be an exactly conserved quantity. In quantum mechanics, it will commute with the Hamiltonian and, as is easily seen, histories characterized by projections onto Q_V will decohere exactly. If the volume is finite, but large compared to the microscopic scale, Q_V will be slowly varying compared to all other dynamical variables. This is because the local conservation law (equation 4.1) permits Q_V to change only by redistribution, which is limited by the rate at which the locally conserved quantity can flow out of the volume. Because these quantities are slowly varying, histories of them should approximately decohere. Furthermore, the fact that the Q_V terms are slowly varying may also be used, at least classically, to derive an approximately closed set of equations involving only those quantities singled out by the conservation laws. These equations are, for example, the Navier-Stokes equations and the derivation of them is a standard (although generally nontrivial) exercise in nonequilibrium statistical mechanics.[36] One of the current goals of the decoherent histories approach is to reexpress this derivation in the language of histories.[37]

Quantum Brownian Motion Models

Many concrete investigations of the mechanics of decoherence have actually concerned quantum Brownian motion models, primarily because calculations can be carried out with comparative ease.[12,31] These have proved to be quite instructive. Very briefly, such models consist of a particle of mass M in a potential $V(x)$ linearly coupled to an environment consisting of a large bath of harmonic oscillators in a thermal state at temperature T and characterized by a dissipation coefficient γ. The types of histories commonly considered are sequences of approximate positions of the Brownian particle, specified up to some width σ, whereas the environment of oscillators is traced over.

The results may be summarized briefly as follows. Decoherence through interaction with the environment is an extremely effective process. For example, for a particle whose macroscopic parameters (mass, time scale, etc.) are of order 1 in cgs units and for an environment at room temperature, the degree of approximate decoherence is of order $\exp(-10^{40})$, a very small number. The probabilities for histories of positions are then strongly peaked about the classical equations of motion, but modified by a dissipation term,

$$M\ddot{x} + M\gamma\dot{x} + V'(x) = 0. \tag{4.3}$$

There are fluctuations about classical predictability, consisting of the ubiquitous quantum fluctuations, adjoined by thermal fluctuations from the interaction with the environment. There is generally a tension between the demands of decoherence and classical predictability due to the fact that the degree of decoherence improves with increasing environment temperature, but predictability deteriorates because the fluctuations about equation 4.3 grow. However, if the particle is sufficiently massive, it can resist the thermal fluctuations and a compromise regime can be found in which there is a reasonable degree of both decoherence and classical predictability.

DECOHERENT HISTORIES AND QUANTUM STATE DIFFUSION

The decoherent histories approach is closely connected to the quantum state diffusion (QSD) approach to open systems. In that approach, the master equation for the reduced density operator of an open system (essentially, a closed system in which one focuses on a particular subsystem) is solved by exploiting a purely mathematical connection with a certain nonlinear stochastic Schrödinger equation (Ito equation).[38] Solutions to the Ito equation turn out to correspond rather closely to the results of actual laboratory experiments (e.g., in quantum optics) and are therefore held to describe individual systems and processes. For example, in a quantum Brownian motion model, the solutions to the Ito equation become localized about points in phase space following the classical equations of motion. The connection with the decoherent histories approach is that, loosely speaking, the solutions to the Ito equation may be thought of as the individual histories belonging to a decoherent set.[39] More precisely, the variables that localize in the QSD approach also define a decoherent set of histories in the decoherent histories approach. The degrees of localization and of decoherence are related and the probabilities assigned to the histories in each case are essentially the same. This connection could be a very useful one, both conceptually and computationally, and efforts to exploit it are being made.

WHAT HAVE WE GAINED?

In this contribution, I have tried to give a brief overview of the decoherent histories approach to quantum theory. What has the decoherent histories approach taught us?

At the level of ordinary quantum mechanics applied to laboratory situations, two things have been gained. First of all, a minimal view of the decoherent histories approach is that, in a sense, it is a more refined version of the Copenhagen interpretation. It rests on a considerable smaller number of axioms and, in particular, it is a predictive formulation of quantum mechanics that does not rely on any kind of assumptions referring to measurement or to a classical domain. It is internally consistent and reproduces all the experimental predictions of the Copenhagen approach. Second, it provides a clear set of criteria for the application of ordinary logic in quantum mechanics. Because many of the conceptual difficulties of quantum mechanics are essentially logical ones (e.g., the EPR paradox), a clarification of the applicability of logic has been argued to lead to their resolution.[7,21,24] Such a resolution is not strictly possible in Copenhagen quantum mechanics because it does not offer clear guidelines for the application of ordinary logic.

Of course, there will always be some who claim that they can finesse their way through any difficulty of quantum mechanics without having to worry about the somewhat cumbersome machinery of the histories approach described here. In this connection, Omnès has to say the following:[26]

> It may be true, as some people say, that everything is in Bohr, but this has been a matter for hermeneutics, with the endless disputes any scripture will lead to. It may also happen that he guessed the right answers, but the pedagogical means and the necessary technique details were not yet available to him. Science cannot, however, proceed by quotations, however elevated the source. It proceeds by elucidation so that feats of genius can become ordinary learning for beginners.

Intuition alone may be sufficient to see some through the difficulties of nonrelativistic quantum mechanics, but if we are to extend quantum theory to the entire universe, a reliable vehicle for travel beyond the domain of our intuition is required. For quantum cosmology, the development of the decoherent histories approach has been a considerable bonus. The decoherent histories approach supplies an unambiguous, workable, and predictive scheme for actually applying quantum mechanics to genuinely closed systems. Furthermore, as discussed at some length in this report, it supplies a conceptually clear method of discussing the emergence of classicality in closed quantum systems, and this is perhaps its greatest success.

Still outstanding are the largely technical difficulties of quantum cosmology connected with quantizing gravity. However, it is possible that the histories approach might be of use there also. The focus on histories may circumvent the "problem of time" encountered in most canonical approaches to quantum gravity. Isham and collaborators, for example, are currently exploring the possibility of histories-based formulations of quantum theory that do not rely on the conventional Hilbert space structure or on the existence of a preferred time coordinate,[40–42] building on an earlier suggestion of Hartle.[16] Much remains to be done, but the histories approach to quantum cosmology appears on both conceptual and technical grounds to be a particularly promising avenue for future research.

Further aspects of the decoherent histories approach are discussed in references 43–63.

ACKNOWLEDGMENTS

I am grateful to the organizers for the opportunity to take part in such an interesting meeting. I would also like to thank numerous colleagues for useful conversations, especially Lajos Diósi, Fay Dowker, Murray Gell-Mann, Nicolas Gisin, Jim Hartle, Chris Isham, Adrian Kent, Seth Lloyd, Roland Omnès, Ian Percival, Trevor Samols, Dieter Zeh, and Wojciech Zurek.

REFERENCES

1. A useful source of literature on the Copenhagen interpretation is: WHEELER, J. A. & W. ZUREK, Eds. 1983. Quantum Theory and Measurement. Princeton University Press. Princeton, New Jersey.

2. LEGGETT, A. J. 1980. Macroscopic quantum systems and the quantum theory of measurement. Suppl. Prog. Theor. Phys. **69:** 80.
3. HALLIWELL, J. J. 1993. The interpretation of quantum cosmological models. *In* General Relativity and Gravitation 1992. R. J. Gleiser, C. N. Kozameh & O. M. Moreschi, Eds. IOP Pub. Bristol.
4. WHEELER, J. A. 1963. *In* Relativity, Groups, and Topology. C. DeWitt & B. DeWitt, Eds. Gordon & Breach. New York.
5. GRIFFITHS, R. B. 1984. Consistent histories and the interpretation of quantum mechanics. J. Stat. Phys. **36:** 219.
6. GRIFFITHS, R. B. 1993. Consistent interpretations of quantum mechanics using quantum trajectories. Phys. Rev. Lett. **70:** 2201.
7. GRIFFITHS, R. B. 1987. Correlations in separated quantum systems: a consistent history analysis of the EPR problem. Am. J. Phys. **55:** 11.
8. GRIFFITHS, R. B. 1994. A consistent history approach to the logic of quantum mechanics. Pittsburgh preprint.
9. GRIFFITHS, R. B. 1994. Consistent quantum counterfactuals. Pittsburgh preprint.
10. GELL-MANN, M. 1994. The Quark and the Jaguar. Little, Brown. Boston.
11. GELL-MANN, M. & J. B. HARTLE. 1990. *In* Complexity, Entropy, and the Physics of Information: SFI Studies in the Sciences of Complexity. Vol. VIII. W. Zurek, Ed. Addison–Wesley. Reading, Massachusetts; 1990. Quantum mechanics in the light of quantum cosmology. *In* Proceedings of the Third International Symposium on the Foundations of Quantum Mechanics in the Light of New Technology. S. Kobayashi, H. Ezawa, Y. Murayama & S. Nomura, Eds. Phys. Soc. Japan. Kyoto.
12. GELL-MANN, M. & J. B. HARTLE. 1993. Classical equations for quantum systems. Phys. Rev. **D47:** 3345.
13. GELL-MANN, M. & J. B. HARTLE. 1994. Equivalent sets of histories and multiple quasiclassical domains. Santa Barbara preprint.
14. HARTLE, J. B. 1991. The quantum mechanics of cosmology. *In* Quantum Cosmology and Baby Universes. S. Coleman, J. Hartle, T. Piran & S. Weinberg, Eds. World Scientific. Singapore.
15. HARTLE, J. B. 1991. Spacetime coarse-grainings in nonrelativistic quantum mechanics. Phys. Rev. **D44:** 3173.
16. HARTLE, J. B. 1992. Spacetime quantum mechanics and the quantum mechanics of spacetime. *In* Proceedings of the 1992 Les Houches Summer School: Gravitation and Quantifications. B. Julia, Ed.
17. HARTLE, J. B. 1993. The quantum mechanics of closed systems. *In* The Festschrift for C. Misner. B. L. Hu, M. P. Ryan & C. V. Vishveshwara, Eds. Cambridge University Press. London/New York.
18. HARTLE, J. B. 1993. The reduction of the state vector and limitations on measurement in the quantum mechanics of closed systems. *In* The Festschrift for D. Brill. B. L. Hu & T. Jacobson, Eds. Cambridge University Press. London/New York.
19. HARTLE, J. B. 1994. Quasiclassical domains in a quantum universe. *In* Proceedings of the Lanczos Centenary Meeting. In press.
20. OMNÈS, R. 1988. Logical reformulation of quantum mechanics. I. Foundations. J. Stat. Phys. **53:** 893.
21. OMNÈS, R. 1988. Logical reformulation of quantum mechanics. II. Interferences and the Einstein-Podolsky-Rosen experiment. J. Stat. Phys. **53:** 933.
22. OMNÈS, R. 1988. Logical reformulation of quantum mechanics. III. Classical limit and irreversibility. J. Stat. Phys. **53:** 957.
23. OMNÈS, R. 1989. Logical reformulation of quantum mechanics. IV. Projectors in semiclassical physics. J. Stat. Phys. **57:** 357.
24. OMNÈS, R. 1989. The Einstein-Podolsky-Rosen problem: a new solution. Phys. Lett. **A138:** 157.
25. OMNÈS, R. 1990. From Hilbert space to common sense: a synthesis of recent progress in the interpretation of quantum mechanics. Ann. Phys. **201:** 354.
26. OMNÈS, R. 1992. Consistent interpretations of quantum mechanics. Rev. Mod. Phys. **64:** 339.

27. OMNÈS, R. 1994. A model for the uniqueness of data and decoherent histories. Phys. Lett. **A187:** 26.
28. OMNÈS, R. 1991. About the notion of truth in quantum mechanics. J. Stat. Phys. **62:** 841.
29. D'ESPAGNAT, B. 1989. Are there realistically interpretable local theories? J. Stat. Phys. **56:** 747.
30. DOWKER, H. F. & A. KENT. 1994. On the consistent histories formulation of quantum mechanics. Newton Institute preprint no. NI-94006.
31. DOWKER, H. F. & J. J. HALLIWELL. 1992. Quantum mechanics of history: the decoherence functional in quantum mechanics. Phys. Rev. **D46:** 1580.
32. HALLIWELL, J. J. 1994. Aspects of the decoherent histories approach to quantum mechanics. *In* Stochastic Evolution of Quantum States in Open Systems and Measurement Processes. L. Diósi, Ed. World Scientific. Singapore.
33. SAUNDERS, S. 1992. The quantum block universe. Harvard preprint.
34. SAUNDERS, S. 1993. Decoherence and evolutionary adaptation. Phys. Lett. **A184:** 1.
35. SAUNDERS, S. 1993. Decoherence, relative states, and evolutionary adaptation. Harvard preprint.
36. FORSTER, D. 1975. Hydrodynamic Fluctuations, Broken Symmetry, and Correlation Functions. Benjamin. New York.
37. HALLIWELL, J. J. & J. B. HARTLE. Work in progress.
38. GISIN, N. & I. C. PERCIVAL. 1992. The quantum state diffusion model applied to open systems. J. Phys. **A25:** 5677.
39. DIÓSI, L., N. GISIN, J. J. HALLIWELL & I. C. PERCIVAL. 1994. Decoherent histories and quantum state diffusion. Phys. Rev. Lett. To appear.
40. ISHAM, C. 1993. Quantum logic and the histories approach to quantum theory. Imperial College preprint no. IC/92-93/39.
41. ISHAM, C. & N. LINDEN. 1994. Quantum temporal logic and decoherence functionals in the histories approach to generalized quantum theory. Imperial College preprint no. IC/TP/93-94/35.
42. ISHAM, C., N. LINDEN & S. SCHRECKENBERG. 1994. The classification of decoherence functionals: an analogue of Gleason's theorem. Imperial College preprint no. IC/TP/93-94/42.
43. ALBRECHT, A. 1992. Investigating decoherence in a simple system. Phys. Rev. **D46:** 5504.
44. ALBRECHT, A. 1993. Following a collapsing wave function. Imperial College preprint no. IC/92-93/03.
45. BLENCOWE, M. 1991. The consistent histories interpretation of quantum fields in curved spacetime. Ann. Phys. **211:** 87.
46. BRUN, T. 1993. Quasiclassical equations of motion for nonlinear Brownian systems. Phys. Rev. **D47:** 3383.
47. BRUN, T. 1994. The decoherence of phase space histories. Cal-Tech preprint.
48. CALZETTA, E. & B. L. HU. 1993. Decoherence of correlation histories. *In* Directions in General Relativity. B. L. Hu & T. A. Jacobson, Eds. Cambridge University Press. London/New York.
49. DIÓSI, L. 1992. Coarse-graining and decoherence translated into von Neumann language. Budapest preprint no. KFKI-1992-03/A.
50. DIÓSI, L. 1993. Unique family of consistent histories in the Caldeira-Leggett model. Budapest preprint no. KFKI-RMKI-23.
51. DIÓSI, L. 1993. Comment on "consistent interpretations of quantum mechanics using quantum trajectories". Budapest preprint no. KFKI-RMKI-25.
52. DIÓSI, L. 1993. Unique quantum paths by continuous diagonalization of the density operator. Budapest preprint no. KFKI-RMKI-28.
53. FINKELSTEIN, J. 1993. On the definition of decoherence. San Jose preprint no. SJSU/TP-93-10.
54. GOLDSTEIN, S. & D. N. PAGE. 1993. Linearly positive histories. University of Alberta preprint no. Alberta-Thy-43-93/gr-qc 9403055.
55. HALLIWELL, J. J. 1993. Quantum mechanical histories and the uncertainty principle. I. Information-theoretic inequalities. Phys. Rev. **D48:** 2739.

56. HALLIWELL, J. J. 1993. Quantum mechanical histories and the uncertainty principle. II. Fluctuations about classical predictability. Phys. Rev. **D48:** 4785.
57. LAFLAMME, R. & A. MATACZ. 1993. Decoherence functional and inhomogeneities in the early universe. Los Alamos preprint.
58. OHKUWA, Y. 1992. Decoherence functional and probability interpretation. Santa Barbara preprint no. UCSBTH-92-40.
59. PAZ, J. P. & W. H. ZUREK. 1993. Environment-induced decoherence, classicality, and consistency of quantum histories. Phys. Rev. **D48:** 2728.
60. SORKIN, R. 1993. Quantum mechanics as quantum measure theory. Syracuse preprint no. SU-GP-93-12-1.
61. TWAMLEY, J. 1993. Inconsistency between alternative approaches to quantum decoherence in special systems. Adelaide preprint no. ADP-93-202/M16.
62. TWAMLEY, J. 1993. Phase space decoherence: a comparison between consistent histories and environment-induced superselection. Adelaide preprint no. ADP-93-208/M19.
63. ZUREK, W. 1994. Preferred sets of states, predictability, classicality, and environment-induced decoherence. *In* Physical Origins of Time Asymmetry. J. J. Halliwell, J. Perez-Mercader & W. Zurek, Eds. Cambridge University Press. London/New York.

Decoherence and Vacuum Fluctuations[a]

L. H. FORD

Department of Physics and Astronomy
Institute of Cosmology
Tufts University
Medford, Massachusetts 02155

VACUUM FLUCTUATIONS AND PHOTON EMISSION

It is well known that the interaction of a quantum system with its environment can destroy quantum coherence. In this report, a particular example of this phenomenon will be discussed: the coupling of coherent electrons to the quantized radiation field. This coupling gives rise to the possibility of photon emission and of interaction of the electrons with the electromagnetic vacuum fluctuations. Consider an electron interference experiment in which coherent electrons may travel from x_i to x_f along either of two classical paths, C_1 or C_2. First, let us recall the analysis of this experiment when the effects of the electromagnetic field are ignored. Let ψ_1 and ψ_2 be the amplitudes for an electron to travel along C_1 and C_2, respectively. Then, the superposed amplitude is $\psi = \psi_1 + \psi_2$ and the number density of electrons detected at x_f is

$$n_0(x_f) = |\psi|^2 = |\psi_1|^2 + |\psi_2|^2 + 2\,\mathrm{Re}(\psi_1\psi_2^*), \tag{1}$$

with the last term being responsible for the interference pattern. Note that C_1 and C_2 are space-time paths, as the events of emission and detection of the electrons occur at different times as well as at different points in space.

We now wish to (i) couple the electrons to the quantized electromagnetic field and (ii) examine the effect upon the interference pattern. This problem has been analyzed in detail in reference 1. Here, we will quote and discuss some of the main results. When both photon emission and vacuum fluctuation effects are present, the number density of electrons detected at x_f becomes

$$n(x_f) = |\psi_1|^2 + |\psi_2|^2 + 2e^W\mathrm{Re}(e^{i\phi}\psi_1\psi_2^*). \tag{2}$$

Here, ϕ is a phase shift introduced by the interaction and W is a function that describes the change in the amplitude of the interference oscillations (the contrast). The phase shift ϕ includes the Aharonov-Bohm shift due to any classical electromagnetic fields generated by the electrons. However, in this report, we will be primarily concerned with the amplitude of the interference oscillations, as this quantity carries the information about the vacuum fluctuations. The explicit form for W is

$$W = -2\pi\alpha \oint_C dx_\mu \oint_C dx'_\nu D^{\mu\nu}(x, x'), \tag{3}$$

where α is the fine-structure constant and $C = C_1 - C_2$ is the closed space-time path

[a] This work was supported in part by National Science Foundation Grant No. PHY-9208805.

obtained by traversing C_1 in the forward direction and C_2 in the backward direction. Note that equation 3 is independent of the direction in which the integrals around C are taken. The photon Hadamard (anticommutator) function, $D^{\mu\nu}(x, x')$, is defined by

$$D^{\mu\nu}(x, x') = (½)\langle 0|\{A^{\mu}(x), A^{\nu}(x')\}|0\rangle. \tag{4}$$

By means of the four-dimensional Stokes theorem, we may write

$$W = -2\pi\alpha \int da_{\mu\nu} \int da'_{\rho\sigma} D^{\mu\nu;\rho\sigma}(x, x'), \tag{5}$$

where $da_{\mu\nu}$ is the area element of the timelike two-surface enclosed by C and where

$$D^{\mu\nu;\rho\sigma}(x, x') = (½)\langle 0|\{F^{\mu\nu}(x), F^{\rho\sigma}(x')\}|0\rangle \tag{6}$$

is the Hadamard function for the field strengths. Equation 5 has the remarkable interpretation that the electrons are sensitive to vacuum fluctuations in regions from which they are excluded. This is analogous to the situation in the Aharonov-Bohm effect,[2] where the phase shift can depend upon classical electromagnetic fields in regions which the electrons cannot penetrate. The space-time geometry of the paths C_1 and C_2 encircling a region is illustrated in FIGURE 1.

In equation 2, it was assumed that both photon emission and vacuum fluctuation effects are present at the same time. This is the usual case in an interference experiment in which no attempt is made to detect the emitted photons. However, in principle, it is possible to distinguish the two by means of a *veto* experiment. Suppose that we arrange for the flux of electrons to be sufficiently low that only one electron is in the apparatus at any one time and that any photons emitted be detected. Whenever a photon is in fact detected, the electron counters are switched off for a sufficient time to insure that the associated electron is not counted. In this way, we guarantee that the interference pattern comprises only those electrons that have not emitted photons. In this case, the relevant contrast factor is no longer e^W, but rather e^{W_F}, where

$$W_F = -2\pi\alpha\left\{\int_{C_1} dx_\mu \int_{C_1} dx'_\nu + \int_{C_2} dx_\mu \int_{C_2} dx'_\nu\right\} D^{\mu\nu}(x, x'). \tag{7}$$

This function describes the effect of the vacuum fluctuations. Similarly, the effects of photon emission are described by the function,

$$W_\gamma = W - W_F = 2\pi\alpha\left\{\int_{C_1} dx_\mu \int_{C_2} dx'_\nu + \int_{C_2} dx_\mu \int_{C_1} dx'_\nu\right\} D^{\mu\nu}(x, x'). \tag{8}$$

Note that the vacuum fluctuation effect involves a double line integral over each path separately, whereas the photon emission contribution is a cross term involving a line integral over each path.

EFFECT OF A CONDUCTING PLATE

The line integrals in equations 3 and 7 are divergent due to the short distance singularity of $D^{\mu\nu}(x, x')$ when $x \to x'$. This is one of the familiar ultraviolet

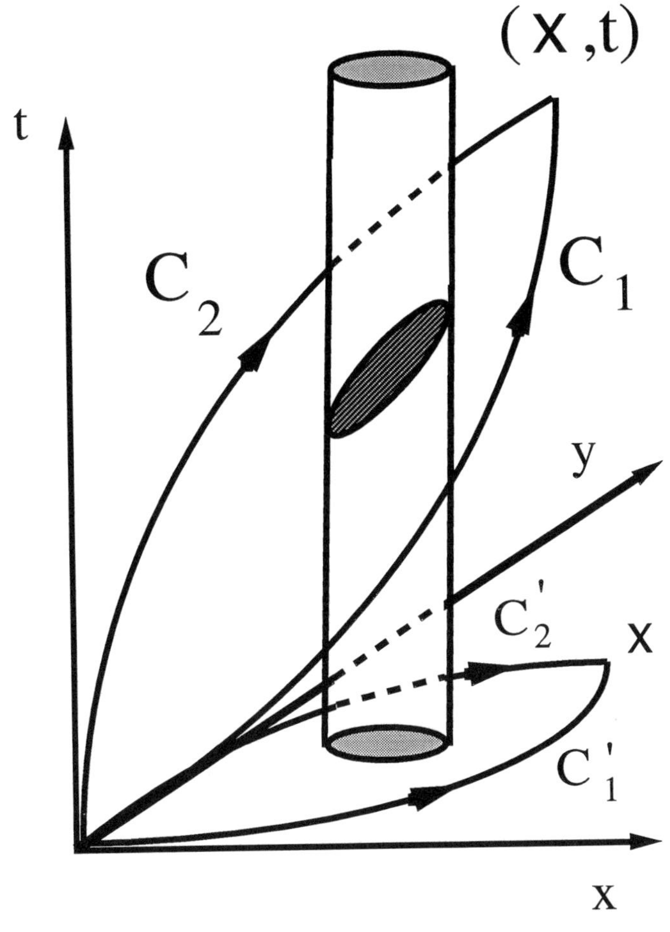

FIGURE 1. Electrons traverse space-time paths C_1 and C_2 to reach point $(\mathbf{x},t)$. The interference pattern depends upon vacuum fluctuations in the crosshatched region in the interior of these paths, which is the domain of integration in equation 5. The cylinder is the world history of an object contained within the electron paths. The spatial projections of the space-time paths C_1 and C_2 onto the xy plane are C_1' and C_2', respectively. The shaded region in this plane is the cross section in the xy plane of the object around which the electrons travel.

divergences in quantum field theory. A simple way to avoid dealing with this divergence is to consider only changes in W or W_F due to an external influence, such as the presence of a conducting boundary. In this case, we replace $D^{\mu\nu}(x,x')$ by $D_R^{\mu\nu}(x,x')$, that is, the renormalized Hadamard function obtained by subtracting the empty space function. Let W_R be the function obtained by making this replacement in equation 3. As compared to the empty space case, the amplitude of the interference oscillations is multiplied by the factor e^{W_R}. The interference pattern can now become sensitive to shifts in the vacuum fluctuations, including shifts occurring in excluded regions. In this case, one has an effect that combines aspects of both the Aharonov-Bohm and the Casimir effects.[3] The simplest geometry in which the effects of vacuum fluctuations may be investigated is where one of the electron paths skims above a perfectly conducting plate. Suppose that path C_1 travels for a distance L at a height z above the conducting plate and later recombines with path C_2, which travels far away from any conductors. Let v be the speed of the electrons. In the limit that $Lc/v \gg z$, that is, the electron's flight time over the plate is long compared to the light travel time to the plate, we have

$$W_R \approx \left(-\frac{\alpha}{\pi}\right)\left[1 + \log\left(\frac{Lc}{2vz}\right)\right]. \tag{9}$$

Note that this is the small z approximation to a function that vanishes in the limit that $z \to \infty$, as required by the fact that $W_R = 0$ in empty space. We first observe that

$$W_R < 0. \tag{10}$$

This means that the shift in the vacuum fluctuations due to the plate causes a decrease in the contrast and, hence, there has been a loss of quantum coherence.

The fact that photon emission can cause decoherence is no surprise. The detection of sufficiently short wavelength photons could reveal which path the electron has taken. It is perhaps less obvious that vacuum fluctuations are also capable of causing decoherence. One might intuitively think of the electrons as being subjected to random force fluctuations that eventually lead to decoherence. Although in the above example $W_R < 0$, there is in principle no reason why one could not have $W_R > 0$, in which case the presence of the conducting boundary would suppress the decohering effects of the vacuum fluctuations.

SQUEEZED STATES OF THE RADIATION FIELD

Such a situation may be displayed explicitly when one replaces the boundary by photons in a squeezed vacuum state. Let the photon field be in a squeezed vacuum state for a single mode, which can be defined as[4]

$$|\zeta\rangle = S(\zeta)|0\rangle = \exp\,[(\tfrac{1}{2})\zeta^* a^2 - (\tfrac{1}{2})\zeta(a^\dagger)^2]|0\rangle, \tag{11}$$

where $\zeta = re^{i\delta}$ is an arbitrary complex number. It may be shown that, in this state,

$$\langle a^\dagger a\rangle = \sinh^2 r \tag{12}$$

and

$$\langle a^2\rangle = -e^{i\delta}\sinh r\cosh r. \tag{13}$$

We wish to calculate the shift in W from the actual vacuum state. This shift, W_R, is given by equations 3 and 4 with the expectation value taken in the state $|\zeta\rangle$. The photon operator product is now understood to be normal-ordered with respect to the vacuum state. The result is

$$W_R = -4\pi\alpha|\eta|^2 \sinh r[\sinh r + \sin(2\theta + \delta)\cosh r], \tag{14}$$

where

$$\eta = |\eta| e^{i\theta} = \oint_C dx_\mu f^\mu(x) \tag{15}$$

and $f^\mu(x)$ is the mode function of the excited mode. The key point is that one can arrange for W_R to have either sign by an appropriate choice of the state parameter δ. Thus, the effect of the squeezed state can be to either enhance or suppress decoherence. In the above example, both vacuum fluctuation and photon emission contributions were included. However, one can treat them separately and show that the shifts in each of W_V and W_γ can have either sign. The fact that squeezed states can suppress the effects of quantum fluctuations below the vacuum level is well known. A squeezed vacuum state necessarily has the expectation value of the energy density negative in certain regions as a result of this suppression. A phenomenon somewhat analogous to the present situation arises when one considers the coupling of a spin system in a classical magnetic field to the electromagnetic vacuum fluctuations. These fluctuations tend to cause a depolarization of the system. Photons in a squeezed state can temporarily reduce this effect and cause the mean magnetic moment of the system to *increase* relative to the vacuum value.[5]

PROBING THE INTERIOR OF MATTER

Let us now return to the issue of nonlocality, that is, the ability of electrons to serve as remote probes. A simple illustration of this would arise in the following experiment: Send electrons around either side of a cylinder of radius R filled with a material of finite electrical conductivity, as illustrated in FIGURE 2. Also arrange that the outer wall of the cylinder is made of a material of very high conductivity. This wall both excludes the electrons and insures that the renormalized-field-strength Hadamard function, $D_R^{\mu\nu;\rho\sigma}(x,x')$, outside the cylinder is independent of the material on the interior. However, the change in the interference pattern contrast due to the presence of the cylinder also depends upon $D_R^{\mu\nu;\rho\sigma}(x,x')$ inside the cylinder and hence upon the conductivity of the material on the interior. An explicit calculation of $D_R^{\mu\nu;\rho\sigma}(x,x')$ in the interior requires a knowledge of the dielectric function $\epsilon(\omega)$ of the metal in question. However, we may make an order-of-magnitude estimate without this detailed information. We may adapt equation 9 to obtain an estimate of W_R for the present situation. Let λ_P be the wavelength associated with the plasma frequency in the metal in question. It is essentially the cutoff wavelength, in that only modes whose wavelengths are of this order or longer are affected by the interior material. It plays a role analogous to z in equation 9. Our estimate will be based upon the assumption that a result of the form of equation 9 also holds in the present situation. If we set $z \approx \lambda_P$ and $L \approx R$ (the radius of the rod), our estimate for W_R may be written as

$$W_R \approx (-10^{-3})\ln\left(\frac{cR}{v\lambda_P}\right), \qquad R \gg \lambda_P, \qquad v \ll c. \tag{16}$$

The result is only weakly (logarithmically) dependent upon the cutoff, λ_P. For example, if $R = 1$ cm, $\lambda_P = 810$ Å (the approximate plasma wavelength of alumi-

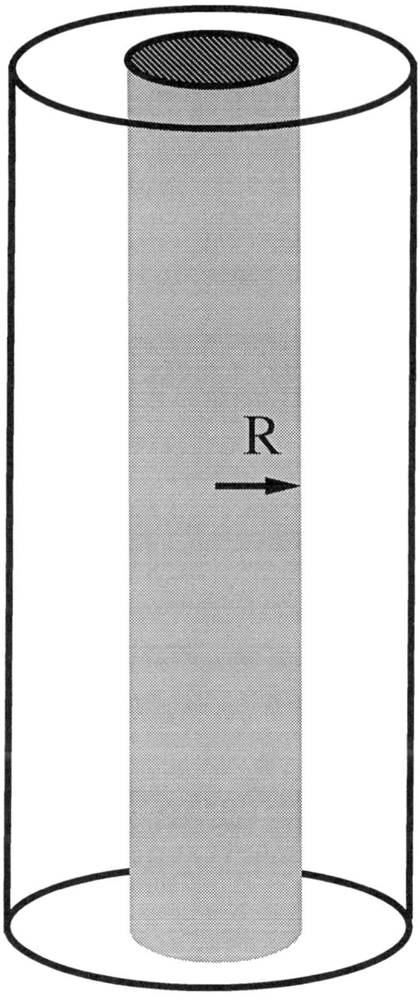

FIGURE 2. A "coaxial cable" consisting of an outer cylinder, which prevents electrons from penetrating to the interior, and an inner rod, with radius R, of material of finite conductivity.

num), and $v = 0.1c$ (corresponding to 2.5-keV electrons), we find that $W_R \approx -10^{-2}$. Because the amplitude of the interference oscillations is proportional to $e^{W_R} \approx 1 + W_R$, we see that the effect of the presence of the rod is to decrease this amplitude by about 1% in this example. If one were to use a material with a different plasma wavelength, λ'_P, the amplitude shift would now be $W'_R = W_R + \Delta W_R$, where

$$\Delta W_R \approx (10^{-3})\ln(\lambda'_P/\lambda_P). \tag{17}$$

This change will typically be of the order of a few times 10^{-4} of the interference oscillation amplitude. Thus, measuring the electron interference pattern to this accuracy would enable one to ascertain whether the material in the rod is, for example, aluminum or magnesium ($\lambda'_P = 1170$ Å). The effect that we have estimated is primarily due to the effects of the vacuum fluctuations, rather than due to photon emission, as the magnitudes of the integrals in equation 8 are determined by the separation of the paths, which is at least as large as R.

EXPERIMENTAL PROSPECTS

Let us conclude with some remarks on the possibility of experimentally verifying the effects discussed in this report. The effect discussed in the previous paragraph seems to be large enough to be possibly measurable. However, in this case, the theoretical prediction (equation 16) is only an order-of-magnitude estimate. For the case of the electrons moving parallel to a conducting plate, the theory is more clear-cut. In principle, W_R as given by equation 9 could be made arbitrarily large if one could make L large and z and v small. In an actual experiment, the image charge force experienced by an electron will limit the possible range of parameters. In particular, the image effect is negligible so long as

$$L \ll (6\ \text{cm})\left(\frac{v}{c}\right)\left(\frac{z}{1\,\mu}\right)^{3/2}. \tag{18}$$

It is not difficult to satisfy equation 18 with choices of parameters that yield a contrast shift, $|W_R|$, of the order of 1%. For example, if we again let $v = 0.1c$ and take $L = 1$ m and $z = 1$ mm, then equation 18 is satisfied and we have $|W_R| \approx 2.2\%$. Thus, detection of the vacuum decoherence effect seems to be a possibility.

REFERENCES

1. FORD, L. H. 1993. Phys. Rev. D **47:** 5571–5580.
2. AHARONOV, Y. & D. BOHM. 1959. Phys. Rev. **115:** 485–491.
3. CASIMIR, H. B. G. 1948. Proc. K. Ned. Akad. Wet. **51:** 793–795.
4. CAVES, C. M. 1981. Phys. Rev. D **23:** 1693–1708.
5. FORD, L. H., P. G. GROVE & A. C. OTTEWILL. 1992. Phys. Rev. D **46:** 4566–4573.

Testing the Randomness of Quantum Mechanics

Nature's Ultimate Cryptogram?[a]

T. ERBER

Departments of Physics and Mathematics
Illinois Institute of Technology
Chicago, Illinois 60616

Anybody who thinks that computers can generate random numbers is living in a state of sin!

—J. von Neumann

RANDOMNESS IN QUANTUM MECHANICS

Many of the basic concepts of Boltzmann's statistical mechanics[1]—including combinatorial optimization, the discretization of energy, detailed balancing, and ergodicity—were incorporated into theories of blackbody radiation, specific heats, level transitions (the A and B coefficients), the Sackur-Tetrode entropy formula, Bose-Einstein statistics, etc., during the era of the "old" quantum mechanics: this spanned the 25 years between Planck's initial lectures on blackbody radiation on 19 October[2] and 16 December 1900[3] and the advent of Heisenberg's matrix mechanics.[4] However, despite the wealth of fruitful physical applications, the foundations of statistical mechanics remained in uneasy disequilibrium with the evolution of rigorous mathematics. For instance, Boltzmann's "H"-theorem, with its implications of irreversibility, was vigorously disputed by Zermelo[5] on the basis of Poincaré's recurrence theorem,[6,7] and Plancherel[8] and Rosenthal[9] showed that Boltzmann's original formulation of the ergodic hypothesis was untenable. These polemics were exacerbated by the chronology of events: a measure theoretic proof of the recurrence theorem did not appear until 1919[10] and the ergodic hypothesis was finally proved in 1931, but with an altered meaning—Boltzmann's intuitive idealizations of sets of trajectories were replaced by the abstract notion of metric transitivity.[11] Some sharp-eyed critics have warned that in this case the resort to abstraction is actually " . . . a glittering deception . . . " that promises more than it delivers.[12]

Probability and wave mechanics were first combined by Schrödinger in his fourth paper[13] on "Quantization as an Eigenvalue Problem" (excerpted from p. 135):

> $\psi\,\psi^*$ is a kind of *weight function* in the configuration space of the system. The *wave-mechanical* configuration of the system is a *superposition* of many, in a strict sense *all*, kinematically possible point-mechanical configurations. Thereby every point-mechanical configuration contributes with a definite *weight* to the true wave-mechanical configuration; this weight being given by $\psi\,\psi^*$. *If one loves paradoxes*, one could say that

[a]This work was supported in part by the Research Corporation.

> the system finds itself simultaneously in all conceivable kinematic states, but is not "equally strong" in all of them.

Similar ideas were expressed concurrently by Born[14] (excerpted from p. 804):

> The guiding field, represented through a scalar function ψ of the coordinates of all participating particles, and the time, evolves according to Schrödinger's differential equation. Momentum and energy are exchanged as if particles (electrons) actually are flying around. The trajectories of these particles are however only determined to the extent that they satisfy the constraints of energy and momentum conservation; the selection of a particular trajectory can only be assigned a probability which is determined by the distribution of the values of the ψ function. One could summarize this *in the following somewhat paradoxical fashion*: The motion of the particles follows probability laws; the probabilities themselves however evolve in a causal manner; i.e., knowledge of a state in all points at a certain instant determines the distribution of states at all later times.

Schrödinger cautiously refers to weight functions (Gewichtsfunktionen), which do not necessarily entail random behavior. It is Born who says straightway that ψ is linked to a probability (Wahrscheinlichkeit). From this point on, applied probability became part of the standard tool-kit of many physicists. However, the link with the foundations remained tenuous and compounded the controversies associated with the problems of measurement.

Finally, in 1933, Kolmogorov found a concise and fruitful way of axiomatizing the foundations of probability within the framework of measure theory.[15] The essential idea is to define a random variable as a measurable function on a probability space, where the probability space is simply a set endowed with a normalized measure. The axiomatic development is deliberately silent concerning any requirements that these random variables be nondeterminate; nor is there any implication that the elements of the probability spaces correspond to inherently unpredictable events. Indeed, as emphasized by Kolmogorov, the axiomatics of random variables " . . . has applications in fields of science that have no relation to the concepts of random events and of probability in the precise measuring of these words." Plainly speaking, then, although a set of basic mathematical issues has been brilliantly resolved, the price is that a new axiomatic notion of randomness ($\equiv$ randomness$_k$) has been introduced. Obviously, randomness$_k$ does not carry any of the implications of unpredictable or "lawless" behavior that are often presumed in physical situations. If one traces the precise meaning of randomness—with or without the k—through the standard mathematical foundations of quantum statistics,[16] one is finally led to the following assertion: the inherent unpredictability of quantum phenomena, such as jumps between levels, is a *physical assumption* that is independent of the other foundations of quantum mechanics. A physical assumption of this kind is, in principle, subject to experimental test.

RANDOM AND PSEUDORANDOM PROCESSES—I

During the last few decades, advances in cryptography, complexity theory, and "chaos" have tended to blur the distinctions between random and pseudorandom processes. For example, a deterministic procedure that is effective in simulating many aspects of random behavior is the iteration of mixing transformations. Probably, the best-known and simplest transformation of this type is the logistic map[17] on the interval [0, 1]:

$$x_n \rightarrow 4x_n(1 - x_n) = x_{n+1}, \tag{2.1}$$

which is conjugate to the second-order Cebysev polynomial. There is a constructive procedure for associating a mixing transformation T with any strictly increasing probability measure P. This is based on the following:

LEMMA: Let P_1 be a continuous, strictly increasing function from an interval I_1 onto [0, 1]. If f is a continuous, strictly increasing function from I_1 onto the interval I_2 and if we define P_2 by the composition $P_2 = P_1 \circ f^{-1}$, then P_2 is a continuous, strictly increasing function from I_2 onto [0, 1]. Conversely, if P_2 is a continuous, strictly increasing function from I_2 onto [0, 1] and if f is defined by $f = P_2^{-1} \circ P_1$, then f is a continuous, strictly increasing function from I_1 onto I_2.

The functions P_1 and P_2 determine Lebesgue-Stieltjes measures on I_1 and I_2, respectively; P_1 and P_2 may also be interpreted as cumulative distribution functions. We now construct the mixing transformation T by means of the following:

CONJUGACY THEOREM:[18] Let I_1, I_2, P_1, P_2, and f satisfy the definitions and conditions of the lemma and suppose that T_1 is mixing on I_1 with respect to P_1. Then, the function T defined by

$$T = f \circ T_1 \circ f^{-1} = P_2^{-1} \circ P_1 \circ T_1 \circ P_1^{-1} \circ P_2 \tag{2.2}$$

is mixing on the interval I_2 with respect to the measure P_2.

What all this machinery effectively does is to mix anything "to order"; in particular, given any probability distribution function, it yields an explicit prescription for constructing a set of deterministic iterations that converge to this distribution. These results are sufficiently general to apply to quantum mechanical distributions. As an illustration, consider the ground state of a one-dimensional infinite symmetric well:

$$\psi(x) = \begin{cases} 2^{-1/2} \cos\left[\left(\frac{\pi}{4}\right)x\right] & \text{for} \quad -2 \le x \le 2, \\ 0 & \text{otherwise.} \end{cases} \tag{2.3}$$

In this case, it can be shown that the appropriate mapping is given by[18]

$$f^{-1}(x) = 2 \sin\left\{\left(\frac{\pi}{4}\right)x + \left(\frac{1}{2}\right)\sin\left[\left(\frac{\pi}{2}\right)x\right]\right\}. \tag{2.4}$$

Then, equation 2.2 leads to the composite function,

$$T(x) = f \circ C_m \circ f^{-1}(x), \tag{2.5}$$

where C_m $(m \ge 2)$ may be any of the Cebysev polynomials. The iterations of $T(x)$ generate the probability density corresponding to equation 2.3. Detailed discussions of these procedures and additional examples are given in reference 18. An important point is that the simulation of quantum mechanical distributions by mixing processes does not clash with von Neumann's "hidden variables" theorem.[19]

In summary, these mixing simulations are "demonstrations by construction" that a wide class of nominally random phenomena—including quantum mechanical distributions—can be reproduced by completely deterministic means.

This situation naturally leads to the converse question: given a nominally random string of symbols or numbers, is there a general procedure for distinguishing between inherently unpredictable or "lawless" sequences and sequences generated by deterministic means, even if the underlying stratagems are artfully concealed? In its rawest form, this problem appears with cryptographic adversaries: Group A sends encrypted messages interspersed among long strings of random gibberish. Group B intercepts the signals, but does not know which is which; this group will therefore be compelled to squander enormous effort in separating the encrypted information from the encrypted noise. However, one can make headway even in this situation. Specifically, suppose that one compares two long strings of ciphertext, each composed of the 26 letters of the alphabet. Then, as first shown by Friedman,[20] speech and gibberish could be distinguished by looking at the statistics of matching letters in the two strings. In the case of encryption by simple transposition, combinatorial and empirical results showed that the "index of coincidence" had the following values:

$$\begin{aligned} \text{encrypted English} &\rightarrow \frac{6.67 \text{ matches}}{100 \text{ pairs of ciphertext letters}} \\ \text{random letters} &\rightarrow 0.0385 \\ \text{encrypted German} &\rightarrow 0.0762 \\ \text{encrypted Japanese (romanized)} &\rightarrow 0.0819. \end{aligned} \tag{2.6}$$

Evidently, despite ignorance of the precise rule of encryption, it is even possible to infer the language of the concealed message. Of course, all of this becomes enormously more complicated when polyalphabetic memory-dependent encryption schemes are utilized, but sophisticated statistical analyses still manage to break intricate computer coding algorithms.[21]

RANDOM AND PSEUDORANDOM PROCESSES—II

In parallel with the cryptographic developments, general limitations on the behavior of "chaotic" processes were discovered. For instance, the simple quadratic map (equation 2.1) generates discrete ergodic flows when iterated and has many other "chaotic" properties, but its structure is far too trivial to qualify as a useful pseudorandom number generator. In analogy with cryptography, one can then attempt to improve matters by searching for "more chaotic" functions. Because the Cebysev polynomials form a kind of generic hierarchy of complex functions, their iterates exemplify this progressive disorder; it would, for instance, be quite tedious to reconstruct $C_{10,000}$ from a string of iterates. Nevertheless, there are ultimate constraints of order.

Consider a set S containing N distinct elements, $x_1, \ldots, x_N$, and let F denote the associated set of N^N functions that map S into itself; that is, F is the full transformation semigroup of S. Furthermore, suppose that h is a function selected at random from F and that x_i is an arbitrary element of S. Then, it is an obvious consequence of the "pigeonhole" principle that the sequence of iterates,

$$h(x_i), h^2(x_i), \ldots, h^m(x_i), \tag{3.1}$$

must contain at least two identical elements whenever $m > N$. In other words, equation 3.1 consists of a "free running" sequence containing no repetitions and a contiguous sequence that constitutes a terminal cycle. If h is a pseudorandom number generator, then the sequence of values in equation 3.1 will appear to vary erratically. However, it is precisely this requirement of pseudorandom behavior that imposes strong constraints on the orbit structure of h. For instance, in this pseudorandom evolution, the sequence in equation 3.1 must imitate the combinatorics of sampling from an urn with N distinct elements. It is then an elementary consequence of the *single birthday lemma* that the average length of the period of free running and the length of the terminal loop are both given by $(\pi N/8)^{1/2}$. Much stronger constraints emerge from the *double birthday lemma*:[22,23] From an urn containing N distinguishable objects, an observer R draws one object at a time, at random with replacement, until a repetition occurs. A similar set of draws is made by another observer B. Then, it can be shown that the probability P that R and B have selected at least one object in common is

$$P = \left(\frac{2}{3}\right) + \left(\frac{1}{12}\right)\left(\frac{2\pi}{N}\right)^{1/2} + \left(\frac{1}{9N}\right) + \cdots. \tag{3.2}$$

If this large coincidence probability is to be simulated by pseudorandom sequences such as equation 3.1, then arbitrary pairs of iteration trajectories, starting at distinct initial points x_i and x_j, will have to exhibit a strong tendency to merge into a common track whenever $m \sim O(N^{1/2})$.

The birthday lemmas are merely the simplest examples of a hierarchy of coincidence and recurrence criteria that must be satisfied by single and/or multiple strings of random elements. Every single one of these randomness tests further narrows the possible structure of the orbits of pseudorandom functions: thus, one is effectively left with a single generic type of orbit. The amount of quantitative information that can be extracted from the "statistical topology" of these orbits—without having to specify the precise structure of h—is astonishing.[23,24] For instance, 80% of all of the trajectories (equation 3.1) merge into a *single* terminal loop irrespective of the starting point x_i and 60% of these trajectories enter the terminal loop at *a single point.* Effectively, metric chaos results in topological order. A nontrivial practical application of the birthday lemmas occurs in breaking the United States Data Encryption Standard (promulgated by NIST), which is in widespread use for "secure" commercial communications.[25]

THE SINGLE-ATOM QUANTUM TELEGRAPH—I

During the Second World War, American code-breakers succeeded in physically reconstructing the Japanese "purple" enciphering machine purely on the basis of the ordered patterns discovered by cryptanalysis of diplomatic and military communications. Searching for ordered patterns in quantum phenomena appears to be far more speculative, but actually also has a practical purpose: if quantum mechanical events are inherently unpredictable, then it is technically feasible to construct devices that are inexhaustible sources of random numbers. The initial proposals to combine cryptography and quantum mechanics involved a number of basic issues:[26]

1. *The Necessity of Identifying Strictly Consecutive Sequences of Events*—Even the most ingenious cryptanalyst will be helpless if presented with arbitrary fragments haphazardly extracted from a message stream: random sampling obliterates information. It is this self-evident requirement that points in the direction of resonance fluorescence from single trapped atoms. As emphasized by Dehmelt in connection with his "shelved" electron scheme, this experimental arrangement for the first time permits a reliable discrimination between the emission of *a photon* (sampling) and the emission of *the next photon*.[27] The succession of bright and dark intervals in the resonance fluorescence is the quantum mechanical analogue of a message stream.
2. *Finite State Spaces*—The cryptanalysts' holy grail is a very, very long message stream because the underlying assumption of decryption is that the adversary's enciphering device has access to only a finite number of possible configurations (no infinite Turing tapes). The complexity then lies in the rules, which may be history-dependent, but the multiplicity of states is merely a complication of scale and enumeration. Similarly, all of the general criteria of statistical topology, or "ordered chaos", discussed in the previous section rest on the fundamental presumption that S is a finite set.

Thus, is the resonance fluorescence of a single trapped (mercury) ion equivalent to a process in a finite state space? Opinions are divided on this point. What can be said with certainty is that all cooling schemes result in a contraction of the phase space available to the ion. Whether this phase space includes all the degrees of freedom that are relevant to the generation of the resonance fluorescence telegraph remains to be seen. In any event, it is easy to estimate the number of phase-space cells accessible to the ion:

$$\left[\frac{1000\ \text{Å} \times 0.5\ \text{MeV}/c}{\hbar}\right]^3 \sim 10^{16}. \tag{4.1}$$

This small [*sic*] number is the result of the extraordinary cooling and localization of ions that is attained with current techniques. The momentum transfer is borrowed from relativistic cutoffs in quantum electrodynamics. If the estimate in equation 4.1 is identified with the cardinality of the set S in the previous section, then one would crudely estimate that about 10^8 consecutive signals would be necessary to find traces of long range order. This is roughly the recurrence period of some "good" pseudorandom number generators.

Another basic issue is the following:

3. *Complexity and Randomness*—The basic connection between randomness and information can be illustrated by a simple example. Consider a long string of

random numbers: if a single number is removed, there is no contextual way of estimating the magnitude of the missing object; information compression is impossible with random strings. In this sense, Kolmogorov[28] and Chaitin[29,30] showed that random processes are necessarily complex. This result, of course, is also consistent with cryptographic experience: there are no algorithms that are simultaneously simple and unbreakable. The single-atom situation provides an interesting contrast. From a physical point of view, this is a system stripped down to utter simplicity and yet it apparently provides an inexhaustible stream of random events.

THE SINGLE-ATOM QUANTUM TELEGRAPH—II

The technical details of generating and analyzing single-atom telegraph data are described elsewhere.[31] In the original work, the durations of 640 successive bright and dark periods of a double-beam resonance fluorescence telegraph were extracted from data furnished by NIST-Boulder and were subjected to 39 statistical tests of random behavior. Subsequently, we received more data from a single-beam telegraph experiment.[32] This file listed the number of 194-nm photons detected in 10^5 consecutive 1-ms time intervals. Typical "on" counts were 40/ms and typical "off" counts were 1/ms. The total yield of on and off pulses from this file was 1450. The series of 725 dark periods were subject to 20 randomness tests. Again, the results were consistent with random behavior, but the statistics are not really meaningful because this is such a small sample. Finally, we received 15 additional files containing data obtained in connection with the experiment described in reference 32. From this information, we were able to reconstruct about 10^4 on and off pulses, although, unfortunately, not all in strictly consecutive order. Nevertheless, the entire exercise was very fruitful because all the computer routines are now in place to process really long runs of data. A detailed description of these analyses will appear elsewhere.[33]

The long range recurrences that we are interested in locating are somewhat unusual. A typical example is the repeat test:[34] Consider a data set consisting of N elements, of which m are distinct. The repeat test compares the number of distinct orderings of every ℓ consecutive elements to the number of distinct orderings that are expected to appear with random drawings. The following sequence illustrates the rules used to count the number of repeated elements, element pairs, etc.:

$$a, b, c, d, e, a, b, d, b, c, e, e, b, a, b. \tag{5.1}$$

In this sequence, there are 5 distinct elements, a–e. Element a occurs 3 times; we count 2 repeats of a. Similarly, there are 4 repeats of element b; 1 repeat each of the elements c and d; and 2 repeats of element e. Therefore, in this 15-element sequence, there are $2 + 4 + 1 + 1 + 2 = 10$ repeated singles. There are 5^2 possible pairs of these 5 elements. In equation 5.1, there are in fact 14 consecutive pairs of elements: the pair a,b occurs 3 times and the pair b,c occurs twice; this yields 3 repeated doubles. In equation 5.1, there are no repeated subsequences of three or more elements in length. With this nomenclature, the following display should be intelligible:

Length of Sequence	Doubles		Triples		Quadruples	
500	400	400	102	101	11	13
1000	900	900	365	369	49	48
2500	2400	2400	1574	1580	280	271
5000	4900	4900	4007	4005	1050	1048

(5.2)

In every double column, the first set of entries is derived from the mercury telegraph signals, whereas the second set is from a pseudorandom number generator.[31] This test illustrates a crucial point in pattern recognition: one does not need high precision in order to identify the onset of nonrandom recurrences.

OUTLOOK

There are many classical versions of the problems discussed in the preceding sections. For example, given 10^6 strictly consecutive outcomes of a game of roulette, deduce that the motion of the ball is governed by deterministic mechanics. If one cannot solve this problem, why bother with quantum mechanics?

The roulette example emphasizes two features: (i) repetitive processes cannot necessarily be idealized as iterations and (ii) "chaotic" or mixing processes are sensitive to initial conditions. The first point already appears in the simplest instance when the logistic iteration (equation 2.1) is implemented on a computer. Despite formal "shadowing" theorems, when Cebysev polynomials are iterated on computers, the resulting orbit structures have little resemblance to analytical predictions.[35] The deviations, of course, are due to round-off and truncations. One might think that this could be taken into account by replacing the simple Cebysev iterations, $C_2 \circ C_2 \circ C_2 \circ \ldots$, by a composite sequence,

$$C_2 \circ g \circ C_2 \circ g \circ C_2 \circ \cdots, \tag{6.1}$$

where the g function represents the stepwise modifications introduced by the computer. However, it is easy to show that this scheme is inadequate because the computer effects, although completely determinate, are also history-dependent, like cipher machines.

In the case of roulette, the role of the "g" is played by the croupier, who picks up the ball and puts it into play again. In this obvious sense, repetition is not equivalent to iteration. The croupier of resonance fluorescence is the laser illumination that "resets" the atom after the shelf state has decayed.

Some reflection will show what is necessary in the roulette case: one has to intervene in order to eliminate the history-dependence (feature i) and to reduce the extent of the instabilities (feature ii). Replacing the croupier by a precisely controlled mechanical robot will make the underlying determinism more explicit, but it will take the fun out of the game. Whether a corresponding set of manipulations (Zener pulses?) is feasible in the quantum situation is an open question.

ACKNOWLEDGMENTS

It is a pleasure to acknowledge numerous informative conversations with P. Hammerling, S. Putterman, and A. Sklar. All experimental data were kindly furnished by D. Wineland and W. Itano. The randomness analyses were carried out in collaboration with D. R. Gavelek.

REFERENCES

1. Boltzmann, L. 1872. Wien. Sitzungsber. **66:** 275–370.
2. Planck, M. 1900. Verh. Dtsch. Phys. Ges. **2:** 202.
3. Planck, M. 1900. Verh. Dtsch. Phys. Ges. **2:** 237.
4. Heisenberg, W. 1925. Z. Phys. **33:** 879–893.
5. Zermelo, E. 1896. Ann. Phys. **57:** 485–494; **59:** 791–801.
6. Poincaré, H. 1889. C. R. Acad. Sci. **108:** 550–553; 1893. C. R. Acad. Sci. **116:** 1017–1021.
7. Olson, E. T. 1993. Found. Phys. Lett. **6**(no. 4): 327–337.
8. Plancherel, M. 1913. Ann. Phys. **42:** 1061–1063.
9. Rosenthal, A. 1913. Ann. Phys. **42:** 796–806; 1914. Ann. Phys. **43:** 894–904.
10. Carathéodory, C. 1919. Sitzungsber. Preuss. Akad. Wiss. Phys. Math. Kl., p. 580–584.
11. Birkhoff, G. D. 1931. Proc. Natl. Acad. Sci. U.S.A. **17:** 650–660.
12. Schwartz, J. 1960. The pernicious influence of mathematics on science. *In* Logic, Methodology, and Philosophy of Science. E. Nagel, P. Suppes & A. Tarski, Eds.: 356–360. Stanford University Press. Stanford, California.
13. Schrödinger, E. 1926. Ann. Phys. **81:** 109–139.
14. Born, M. 1926. Z. Phys. **38:** 803–827.
15. Kolmogorov, A. N. 1933. Grundbegriffe der Wahrscheinlichkeitsrechnung. Springer-Verlag. Berlin.
16. Khinchin, A. A. 1960. Mathematical Foundations of Quantum Statistics. Graylock Press. Baltimore.
17. May, R. M. 1976. Nature **261:** 459–467.
18. Erber, T., T. M. Rynne & A. Sklar. 1981. Acta Phys. Austriaca **53:** 145–155.
19. von Neumann, J. 1932. Mathematische Grundlagen der Quantenmechanik, p. 160 & 256 and notes 160–163. Springer-Verlag. Berlin.
20. Friedman, W. F. 1922. The Index of Coincidence and Its Applications in Cryptography. Riverbank Laboratories Publication No. 22. Geneva, Illinois.
21. Deavours, C. A. & L. Kruh. 1985. Machine Cryptography and Modern Cryptanalysis. Artech House. Dedham, Massachusetts.
22. Erber, T., T. M. Rynne, W. F. Darsow & M. J. Frank. 1983. J. Comput. Phys. **49:** 394–419.
23. Erber, T. & D. Gavelek. 1991. Physica **A177:** 394–400.
24. Harris, B. 1960. Ann. Math. Stat. **31:** 1045–1062.
25. Biham, E. & A. Shamir. 1993. Differential Cryptanalysis of the Data Encryption Standard. Springer-Verlag. New York/Berlin.
26. Erber, T. & S. Putterman. 1985. Nature **318**(no. 6041): 41–43.
27. Dehmelt, H. G. 1975. Bull. Am. Phys. Soc. **20:** 60.
28. Kolmogorov, A. N. 1968. IEEE Trans. **IT-14:** 662–664.
29. Chaitin, G. 1974. J. Assoc. Comput. Math. **21:** 403–424.
30. Chaitin, G. 1987. Adv. Appl. Math. **8:** 119–146.
31. Erber, T., P. Hammerling, G. Hockney, M. Porrati & S. Putterman. 1989. Ann. Phys. **190:** 254–309.
32. Itano, W., J. C. Bergquist, R. G. Hulet & D. J. Wineland. 1987. Phys. Rev. Lett. **59**(no. 24): 2732–2735.
33. Erber, T. & D. R. Gavelek. In preparation.
34. Gavelek, D. R. 1990. Computer orbits of Chebyshev mixing transformations. Ph.D. thesis. Illinois Institute of Technology. Chicago.
35. Gavelek, D. R. & T. Erber. 1992. J. Comput. Phys. **101:** 25–50.

Quantum Chaos and Statistical Mechanics[a]

MARK SREDNICKI

Department of Physics
University of California
Santa Barbara, California 93106

Consider a dilute gas of hard spheres in a box with hard walls. Give the spheres some arbitrary initial distribution of momenta (and positions). Classically, after a few mean free times have passed, we expect that the distribution of momenta will be given by the Maxwell-Boltzmann (MB) formula,

$$f_{\mathrm{MB}}(\mathbf{p}) = (2\pi mkT)^{-3/2}\exp(-\mathbf{p}^2/2mkT), \tag{1}$$

where the temperature T is given in terms of the conserved total energy U by the ideal-gas relation, $U = (3/2)NkT$.

To see why this should be so, first note that the Hamiltonian is simply

$$H = \frac{1}{2m}\sum_{i=1}^{N}\mathbf{p}_i^2 = \frac{1}{2m}\mathbf{P}^2, \tag{2}$$

where $\mathbf{P}$ is a vector with $3N$ components. Because H takes on the constant value U, the allowed values of $\mathbf{P}$ form a sphere that we will call the P-sphere. Suppose we now choose $\mathbf{P}$ "at random". For this to be a meaningful statement, we must have a measure that tells us which sets of $\mathbf{P}$'s are equally likely *a priori.* The obvious choice is to assign equal *a priori* probabilities to equal areas on the P-sphere. Then, if we choose $\mathbf{P}$ at random with respect to this measure, the probability that our choice makes an angle between θ and $\theta + d\theta$ with respect to any particular axis is simply

$$\begin{aligned} f(\theta)d\theta &\sim (\sin\theta)^{3N-2}d\theta \\ &\sim (\sin\theta)^{3N-3}d\cos\theta \\ &\sim (1-\cos^2\theta)^{(3N-3)/2}d\cos\theta. \end{aligned} \tag{3}$$

If we now identify $(2mU)^{1/2}\cos\theta$ as, say, the value of p_{1z} (the z component of the first particle's momentum), we find

$$\begin{aligned} f(p_{1z})dp_{1z} &\sim (1 - p_{1z}^2/2mU)^{(3N-3)/2}dp_{1z} \\ &\sim \exp(-p_{1z}^2/2mkT)dp_{1z}, \end{aligned} \tag{4}$$

where in the second line we have set $U = (3/2)NkT$ and have taken the large-N limit. Thus, we have recovered the MB distribution for p_{1z}. Now, consider the probability distribution for p_{1y} when p_{1z} is fixed; it is given by the first line of equation 4 with $3N$

[a]This work was supported by National Science Foundation Grant No. PHY-91-16964.

replaced by $3N - 1$ (because there is one less coordinate when p_{1z} is fixed) and with $2mU$ replaced by $2mU - p_{1z}^2$. In the large-N limit, we can neglect p_{1z}^2 compared to $2mU$ and we find the MB distribution for p_{1y}. In similar fashion, we get the MB distribution for any n components of $\mathbf{P}$ as long as $n \ll N$.

Now, our task is to justify the assumption that equal areas on the P-sphere are equally likely *a priori.* That such a justification is needed can be seen by considering how we would go about filling a real box with a real gas (say, helium). If the box did not already have some sort of valve on it, we would install one and would pump the air out through it. Then, we would close the valve, attach it to a tank of helium with a hose, and open the valve. The helium atoms would rush in, moving preferentially in the direction parallel to the hose. Thus, their initial distribution of momenta would be strongly anisotropic. This is in sharp contrast to the prediction of the equal-area measure, which tells us that we will find a thermal, isotropic distribution. Clearly, then, the equal-area measure has nothing to do with how we put real gases in real boxes and thus we must seek its justification elsewhere.

That justification comes from Sinai's theorem,[1] which states that a box of hard spheres is a *chaotic* system. The meaning of this statement in the present context is simple. Start off with arbitrary initial momenta and positions; the momenta can be as nonthermal as you like. (Actually, we must exclude a set of measure zero; for example, it is possible to set up initial conditions such that no two hard spheres ever collide, in which case the following discussion obviously does not apply.) Wait a few mean free times and then note the current location of $\mathbf{P}$ on the P-sphere. Continue this procedure, keeping track of the location of $\mathbf{P}$ each time. Chaos implies that this sequence of $\mathbf{P}$'s appears to be chosen at random with respect to the equal-area measure.

We are done. Even if we started off with a $\mathbf{P}$ representing a strongly anisotropic distribution, the next $\mathbf{P}$ will appear to be chosen "at random" and thus predicts a thermal distribution for the individual momenta.

So much for classical mechanics. What about quantum mechanics?

Now, we have a completely different problem.[2] The N-particle Schrödinger equation can always be solved by going to the energy eigenstate basis: $H|\alpha\rangle = U_\alpha|\alpha\rangle$. The Hamiltonian is given by equation 2, supplemented by the boundary condition that the energy eigenfunctions $\psi_\alpha(\mathbf{X})$ vanish whenever one of the hard spheres touches a wall of the box or whenever two hard spheres touch each other. The wave function in momentum space at time t is then

$$\tilde{\psi}(\mathbf{P}, t) = \sum_\alpha C_\alpha \exp(-iU_\alpha t/\hbar)\tilde{\psi}_\alpha(\mathbf{P}), \tag{5}$$

where the C_α's specify the initial state. The probability that the first particle has momentum $\mathbf{p}_1$ at time t is found by squaring the wave function and integrating over all momenta, except the first:

$$\begin{aligned} f(\mathbf{p}_1, t) &= \int d^3p_2 \ldots d^3p_N |\tilde{\psi}(\mathbf{P}, t)|^2 \\ &= \sum_{\alpha\beta} C_\alpha^* C_\beta e^{i(U_\alpha - U_\beta)t/\hbar} \int d^3p_2 \ldots d^3p_N \tilde{\psi}_\alpha^*(\mathbf{P})\tilde{\psi}_\beta(\mathbf{P}) \\ &= \sum_{\alpha\beta} C_\alpha^* C_\beta e^{i(U_\alpha - U_\beta)t/\hbar}\Phi_{\alpha\beta}(\mathbf{p}_1). \end{aligned} \tag{6}$$

In the last line, we have introduced

$$\Phi_{\alpha\beta}(\mathbf{p}_1) \equiv \int d^3p_2 \ldots d^3p_N \tilde{\psi}^*_\alpha(\mathbf{P})\tilde{\psi}_\beta(\mathbf{P}), \tag{7}$$

which obeys the normalization condition,

$$\int d^3p_1 \Phi_{\alpha\beta}(\mathbf{p}_1) = \delta_{\alpha\beta}. \tag{8}$$

If we symmetrize or antisymmetrize each $\tilde{\psi}_\alpha(\mathbf{P})$ on exchange of any two $\mathbf{p}_i$'s to reflect Bose-Einstein (BE) or Fermi-Dirac (FD) statistics, then $f(\mathbf{p}_1, t)$ is independent of whichever $\mathbf{p}_i$ we choose.

On physical grounds, we expect that $f(\mathbf{p}_1, t)$ should be the MB (or BE or FD) distribution for any time t greater than a few mean free times. It is not obvious how this can occur. Consider, for example, what happens if we take the infinite time average of $f(\mathbf{p}_1, t)$:

$$\lim_{\tau\to\infty} \frac{1}{\tau} \int_0^\tau dt\, f(\mathbf{p}_1, t) = \sum_\alpha |C_\alpha|^2 \Phi_{\alpha\alpha}(\mathbf{p}_1). \tag{9}$$

The infinite time average is obviously not something we can actually observe, but theoretically, if anything is going to be thermal, this is it. The problem is that the C_α's are essentially arbitrary; hence, how can we possibly get the MB distribution?

There is only one way: each $\Phi_{\alpha\alpha}(\mathbf{p}_1)$ must *individually* be equal to the MB (or BE or FD) distribution at a temperature T_α that is given (at least approximately) by the ideal-gas relation, $U_\alpha = (3/2)NkT_\alpha$. We call this hypothesis *eigenstate thermalization.* If eigenstate thermalization is valid, then equation 9 will indeed be a thermal distribution as long as the uncertainty in the total energy is much less than its expectation value.

Furthermore, if $\Phi_{\alpha\beta}(\mathbf{p}_1)$ is always sufficiently small whenever $\alpha \neq \beta$, then the $\alpha \neq \beta$ terms in equation 6 will usually make a negligible contribution and $f(\mathbf{p}_1, t)$ will be a thermal distribution at most times t, without any time averaging at all. However, if the magnitudes and phases of the C_α's are carefully chosen, then we can "line up" the $\Phi_{\alpha\beta}(\mathbf{p}_1)$'s so as to get any $f(\mathbf{p}_1, t)$ that we might want at any one particular time (say, $t = 0$). Afterwards, however, as we see in equation 6, the phases will change in the usual manner; the carefully contrived coherence among the various $\Phi_{\alpha\beta}(\mathbf{p}_1)$'s will be destroyed and we will again find a thermal distribution for $\mathbf{p}_1$.

I find this to be a clear and satisfying explanation for the validity of quantum statistical mechanics, at least in this particular problem, and even without any further evidence in favor of it. However, there is more to be said: a very strong case can be made for the two necessary ingredients—the thermal nature of $\Phi_{\alpha\alpha}(\mathbf{p}_1)$ and the smallness of $\Phi_{\alpha\beta}(\mathbf{p}_1)$—based on the theory of quantum chaos.

Quantum chaos is the study of quantum systems whose classical counterparts are chaotic. The result we will need is known as Berry's conjecture.[3–5] As its name implies, Berry's conjecture is as yet unproved, but there is significant numerical evidence (reviewed in reference 2) in support of it.

Berry's conjecture has two parts. Part one says that the energy eigenfunctions of a bounded, isolated quantum system that is classically chaotic appear to be Gaussian random variables, in the sense that

$$\lim_{\alpha\to\infty} \int d\mathbf{X}\psi_\alpha(\mathbf{X}+\mathbf{X}_1)\ldots\psi_\alpha(\mathbf{X}+\mathbf{X}_n) = \sum_{\text{pairs}} J(\mathbf{X}_{i_1}-\mathbf{X}_{i_2})\ldots J(\mathbf{X}_{i_{n-1}}-\mathbf{X}_{i_n}). \quad (10)$$

Here, the integration measure is normalized so that $\int d\mathbf{X} = 1$, and the sum is over all possible ways to pair up the $\mathbf{X}_i$'s; if n is odd, the result is zero. Part two says that the correlation function $J(\mathbf{X})$ is given by

$$J(\mathbf{X}) \sim \int d\mathbf{P} \exp(i\mathbf{P}\cdot\mathbf{X}/\hbar)\delta[H(\mathbf{P},\mathbf{X}) - U_\alpha], \quad (11)$$

where $\delta(x)$ is the Dirac delta function and $J(\mathbf{0}) = 1$.

It is straightforward to show that Berry's conjecture gives us the two necessary ingredients for quantum statistical mechanics. First, when $\alpha \neq \beta$, we find that $\Phi_{\alpha\beta}(\mathbf{p}_1)$ is exponentially small in the number of particles N. More importantly, we find eigenstate thermalization: $\Phi_{\alpha\alpha}(\mathbf{p}_1)$ is given by the MB or BE or FD distribution (depending on whether we use nonsymmetric, completely symmetric, or completely antisymmetric energy eigenfunctions) plus corrections that depend on the specific energy eigenfunction, but that are exponentially small in N. To derive this, the gas must be dilute; there are also other, hard-to-compute corrections to $\Phi_{\alpha\alpha}(\mathbf{p}_1)$ due to the finite radii of the hard spheres. We expect these to reproduce the usual hard-sphere corrections to ideal-gas behavior, but this remains to be demonstrated. Another important unsettled issue is to determine how high the energy needs to be before equation 10 is sufficiently accurate. A naive estimate is $\lambda_\alpha \lesssim a$, where $\lambda_\alpha = (2\pi\hbar^2/mkT_\alpha)^{1/2}$ is a typical particle's de Broglie wavelength and a is the hard-sphere radius. With a in angstroms and m in amu, this condition becomes $T_\alpha \gtrsim (300/ma^2)$ kelvin.

To summarize, the appearance of a thermal distribution of momenta in an isolated, bounded quantum system of many particles can only be understood if each energy eigenfunction individually predicts a thermal probability for the momentum of each constituent particle and if overlaps of different energy eigenfunctions are sufficiently small when one particle's momentum is left unintegrated. Both these statements can be derived as consequences of Berry's conjecture, which is expected to hold only for quantum systems whose classical counterparts are chaotic. Thus, the well-known connection between classical statistical mechanics and classical chaos is now seen to be mirrored by an analogous connection between quantum statistical mechanics and quantum chaos.

ACKNOWLEDGMENTS

I would like to thank Edouard Brezin, John Cardy, Marty Halpern, and Walter Kohn for helpful discussions.

REFERENCES

1. SINAI, YA. G. 1970. Usp. Mat. Nauk **25:** 137 (Russ. Math. Surv. **25:** 137).
2. SREDNICKI, M. 1994. Phys. Rev. E **50:** 888.
3. BERRY, M. V. 1977. J. Phys. A **10:** 2083.
4. BERRY, M. V. 1981. *In* Les Houches XXXVI: Chaotic Behavior of Deterministic Systems. G. Iooss, R. H. G. Helleman & R. Stora, Eds. North-Holland. Amsterdam.
5. BERRY, M. V. 1991. *In* Les Houches LII: Chaos and Quantum Physics. M-J. Giannoni, A. Voros & J. Zinn-Justin, Eds. North-Holland. Amsterdam.

Why Not Take All Observables as Beables?

JEFFREY BUB

Department of Philosophy
University of Maryland
College Park, Maryland 20742-7615

Why not take all quantum mechanical observables as beables? Jeroen Vink[1] proposed just this and explicitly formulated a dynamics for beable trajectories that reproduces the quantum statistics. At first sight, the idea seems preposterous: surely, the phenomenon of interference, induced by the noncommutativity of the algebra of observables or the non-Booleanity of the algebra of idempotent ("yes-no") observables, excludes this possibility. In fact, we could take all observables as beables and could account for the Born probabilities as the probabilities of measurement outcomes, in terms of an appropriate theory of measurement, but I shall argue that nothing is gained by this radical interpretation of Vink's theory. Instead, the theory should be interpreted as a type of Bohmian mechanics—a generalization of Bohm's hidden variable theory[2] in which some privileged discrete observable, rather than position in configuration space, is selected as always having a determinate value.

The term "beable" was introduced by Bell:[3]

> It would be foolish to expect that the next basic development in theoretical physics will yield an accurate and final theory. But it is interesting to speculate on the possibility that a future theory will not be *intrinsically* ambiguous and approximate. Such a theory could not be fundamentally about "measurements", for that would again imply incompleteness of the system and unanalyzed interventions from outside. Rather it should again become possible to say of a system not that such and such may be *observed* to be so, but that such and such *be* so. The theory would not be about "*observ*ables", but about "*be*ables".

In an article entitled "Beables for Quantum Field Theory", Bell sketched a generalization of Bohm's theory for fermion number density as the privileged beable. As Bell saw it:[4]

> Not all "observables" can be given beable status, for they do not all have simultaneous eigenvalues, i.e., do not all commute. It is important to realize therefore that most of these "observables" are entirely redundant. What is essential is to be able to define the positions of things, including the positions of instrument pointers or (the modern equivalent) of ink on computer output. . . . The distribution of fermion number in the world certainly includes the positions of instruments, instrument pointers, ink on paper, . . . , and much more.

Bell proposed a stochastic dynamics for the distribution of fermion number that replaces the deterministic equations of motion of Bohm's theory. He remarked that he found the introduction of a stochastic element for beables with discrete spectra "unwelcome", but he suspected that this would go away in some sense in the continuum limit.[5]

Vink begins by reformulating and extending Bell's stochastic dynamics and shows explicitly in what sense the theory reduces to Bohm's theory in the continuum limit.

Recall that Bohm extracts two real equations from Schrödinger's time-dependent complex equation of motion for the wave function in configuration space,

$$i\hbar\partial\psi/\partial t = -(\hbar^2/2m)\nabla^2\psi + V\psi,$$

by substituting $\psi = R\exp(iS/\hbar)$:

$$\partial S/\partial t + (\nabla S)^2/2m + V - (\hbar^2/2m)\nabla^2 R/R = 0,$$

$$\partial R^2/\partial t + \nabla\cdot(R^2\nabla S/m) = 0.$$

The first equation (derived from the real part of the Schrödinger equation) can be interpreted as a Hamilton-Jacobi equation for the motion of particles under the influence of a potential function V and an additional "quantum potential", specifically, $-(\hbar^2/2m)\nabla^2 R/R$. The trajectories of these particles are given by the solutions to the equation

$$d\boldsymbol{x}/dt = \boldsymbol{j}/\rho = \nabla S/m = (\hbar/m)\,\mathrm{Im}(\nabla\psi/\psi) = (\hbar/2im)(\psi^*\nabla\psi - \psi\nabla\psi^*)/|\psi|^2,$$

where $\rho = R^2 = |\psi|^2$ and $\boldsymbol{j} = R^2\nabla S/m = (\hbar/m)\,\mathrm{Im}(\psi^*\nabla\psi)$, that is, by the integral curves of a velocity field defined by the gradient of the phase S. Thus, the trajectories $\boldsymbol{x}(t)$ depend on the wave function ψ. The second equation (derived from the imaginary part of the Schrödinger equation) can be written as a continuity equation for an ensemble density $\rho = |\psi|^2$ and a probability current $\boldsymbol{j}$:

$$\partial\rho/\partial t + \nabla\cdot\boldsymbol{j} = 0.$$

The continuity equation guarantees that ρ will remain equal to $|\psi|^2$ at all times if $\rho = |\psi|^2$ initially.

Vink considers an arbitrary complete set of commuting observables Q^i $(i = 1, 2, \ldots, I)$, with eigenvectors $|q^1_{n^1}, q^2_{n^2}, \ldots, q^I_{n^I}\rangle$, where the subscripts $n^i = 1, 2, \ldots, N^i$ label the finite and discrete eigenvalues of Q^i. Suppressing the index i, these are written as $|q_n\rangle$. The time evolution of the state vector is given by the equation of motion,

$$i\hbar d|\psi(t)\rangle/dt = H|\psi(t)\rangle,$$

or

$$i\hbar d\langle q_n|\psi\rangle/dt = \langle q_n|H\psi\rangle = \sum_m \langle q_n|H|q_m\rangle\langle q_m|\psi\rangle$$

in the Q-representation.

The imaginary part of this equation yields the continuity equation:

$$dP_n/dt = (1/\hbar)\sum_m J_{nm},$$

where the probability density P_n and the current matrix J_{nm} are defined by

$$P_n(t) = |\langle q_n|\psi(t)\rangle|^2,$$

$$J_{nm}(t) = 2\,\mathrm{Im}[\langle\psi(t)|q_n\rangle\langle q_n|H|q_m\rangle\langle q_m|\psi(t)\rangle].$$

For the nonmaximal (degenerate) observables Q^i, the probability density and current matrices are defined by summing over the remaining indices, for example,

$$P^i_n = \sum_r |\langle q^i_{n^i}, r|\psi\rangle^2,$$

where r denotes $q^j_{m^j}$ ($j \neq i$), and similarly for J^i_{nm}.

We want a stochastic dynamics for the discrete observable Q consistent with the continuity equation. Suppose that the jumps in Q-values are governed by transition probabilities $T_{mn}dt$, where $T_{mn}dt$ denotes the probability of a jump from value q_n to value q_m in time dt. The transition matrix gives rise to time-dependent probability distributions of Q-values, $P_n(t)$, which must satisfy the master equation,

$$dP_n(t)/dt = \sum_m T_{nm}P_m - T_{mn}P_n,$$

and this equation must be consistent with the continuity equation,

$$dP_n(t)/dt = (1/\hbar) \sum_m J_{nm};$$

that is, we require

$$J_{nm}/\hbar = T_{nm}P_m - T_{mn}P_n.$$

We want solutions for T, given P and J, with $T_{mn} \geq 0$. Because $J_{mn} = -J_{nm}$ (hence, $J_{nn} = 0$), the above equation yields $n(n-1)/2$ equations for the n^2 elements of T. Thus, there are many solutions. Bell's choice was[6]

$$T_{nm} = J_{nm}/\hbar P_m, \qquad J_{nm} \geq 0,$$

$$T_{nm} = 0, \qquad J_{nm} \leq 0.$$

For $n = m$, T_{nn} is fixed by the normalization $\Sigma_m T_{mn}dt = 1$.

Vink shows that Bell's choice leads to Bohm's theory in the continuum limit, when Q is the position in configuration space. For example, consider a single particle on a one-dimensional lattice. Let $x = nd$, with $n = 1, 2, \ldots, N$ and d being the lattice distance. Writing $\psi = R\exp(iS/\hbar)$ and taking the derivative, F', of a function F on the lattice as defined by $F'(x) = [F(x+d) - F(x)]/d$, Vink shows that

$$J_{mn} = (\hbar/md)[S'(nd)P_n\delta_{n,m-1} - S'(nd)P_n\delta_{n,m+1}]$$

and

$$T_{mn} = [S'(nd)/md]\delta_{n,m-1}, \qquad S'(nd) \geq 0,$$

$$T_{mn} = [S'(nd)/md]\delta_{n,m+1}, \qquad S'(nd) \leq 0.$$

For positive $S'(nd)$, the particle can jump from site n to site $n+1$ with probability $S'(nd)dt/md$; for negative $S'(nd)$, the particle can jump from site n to $n-1$ with the same probability. Because each jump is over a distance d, the average displacement in a time interval dt is

$$dx = S'(x)dt/m,$$

that is,

$$dx/dt = S'(x)/m.$$

As $d \to 0$, $S' \to \partial_x S$; thus, in the continuum limit,

$$dx/dt = \partial_x S/m$$

as for the continuous trajectories in Bohm's theory. Vink shows that the dispersion vanishes in the limit as $d \to 0$ and, hence, the trajectories become smooth and identical to the trajectories in Bohm's theory as $d \to 0$.

Having shown how to reformulate Bohm's equations of motion for any discrete observable, Vink considers two options: either we select a suitable privileged observable that always has a determinate value (suitable in the sense that all measurements can be reduced to determinations of the value of this observable) or we regard all observables as having determinate values simultaneously, at all times. Vink advocates the second option, although recognizing that the Kochen and Specker theorem[7] imposes severe restrictions on the simultaneous attribution of values to observables. If the functional relation

$$f(Q^1, Q^2, \ldots, Q^n) = 0$$

holds as an operator identity among the observables of a mutually commuting set, the same relation

$$f[v(Q^1), v(Q^2), \ldots, v(Q^n)] = 0$$

cannot in general be satisfied for the values $v(Q^i)$ of the observables.

Vink's insight (following Bell[8]) is that functional relations between commuting observables need only be recovered for the measured values of observables. If the state is an eigenstate of the observables $Q^1, Q^2, \ldots, Q^n$, then the functional relation holds for the corresponding eigenvalues. A simultaneous measurement of Q^i on a system S involves considering the state of a composite system $S + M$, where M is some measuring instrument, with a large number of degrees of freedom. Decoherence arguments show that the state function of $S + M$ develops sharp peaks in the value space of the "pointer" observable R, with negligible overlap and interference, where each peak is associated with the product of a simultaneous eigenstate of the Q^i and a correlated state of the measuring instrument indicating the measurement result.

Just as in Bohm's theory, where the decoherence occurs in configuration space, the trajectory of R follows one particular peak of the state function, with negligible probability (after a very short time) of a jump to a value associated with a different peak. Essentially, this is because (i) the transition matrix induces transitions only between neighboring values, so the probability of a jump between values of R associated with macroscopically separated peaks is negligible, and (ii) the state function is effectively zero between peaks, so the trajectory of R would have to move through an effectively zero probability region between peaks (more precisely, the transition matrix would be effectively zero between values associated with the region between different peaks). Because different simultaneous eigenstates of the Q^i are correlated with values of R associated with different peaks in the state function, the trajectories of the Q^i will follow the trajectory of R. In effect, then, it is as if the state evolves into a proper mixture of eigenstates of the Q^i: all but one peak can be neglected for the evolution of the trajectories and, for this peak, the functional relation holds for the values of the Q^i.

Vink remarks:[9]

> In general, the [functional relationship constraint] will fail to hold, even if the operators involved commute. However, it was discussed above that during a measurement the wave function of the quantum system effectively evolves into an eigenstate of the observable being measured, and then the [functional relationship constraint] holds among any set of operators that commute with the one being measured.

Note that decoherence in the value space of the "pointer" observable R is crucial here. Suppose that the measured system S is not initially in an eigenstate of the commuting observables A and B, where $B = g(A)$. [Hence, the relation $f(A, B) = 0$ is $B - g(A) = 0$.] For definiteness, suppose that $B = A^2$, where A is the spin of a spin-1 particle in some direction. Then, the values of A and B will not in general satisfy the functional relation $v(B) = v(A)^2$; that is, A might take the value $+1$, whereas B takes the value 0.

As the state function of the composite system $S + M$ begins to separate into three sharp peaks with negligible overlap and interference in R-space—each peak corresponding to one of the eigenvalues -1, 0, $+1$ of A—the values $v(A)$ will evolve stochastically in time so as to eventually (after a very short time) follow one of the separating peaks of the state function, with negligible probability of jumping to a value associated with a different peak. Suppose that the value of A eventually evolves to the value associated with a particular peak—say, the peak associated with the eigenvalue $+1$. The particular peak that the trajectory of A follows depends effectively on the initial value *of R*, not of A. Similarly, the trajectory of B depends effectively on the initial value of R, not of B. If the initial state of S is not an eigenstate of spin, the initial value of B will not in general be equal to the square of the initial value of A. However, the trajectory of B will evolve to a value associated with the same peak as the value of A, with negligible probability of evolving to a different value. If A ends up with the value ± 1, then B will end up (effectively, with probability 1) with the value 1, not 0; if A ends up with the value 0, then B will end up (effectively, with probability 1) with the value 0, not 1.

Thus, the fact that functional relations between commuting observables are recovered on measurement hardly supports the interpretation of all observables as beables because "measurement" in this theory is not a process for ascertaining the values of beables. Rather, it is a process that transforms the values of beables in a certain way. The situation is entirely analogous to measurement in Bohm's theory, where R is the position in configuration space. As Bell remarks on Bohm's theory:[10]

> Note that from the present point of view the description of the experiment as "measurement" of "spin observable" [A] is an unfortunate one. Our particle has no internal degrees of freedom. It is guided however by a multicomponent field and, when this suffers the analogue of optical multiple refraction, the particle is dragged one way or another depending only on its initial position.

Consider the application of Vink's theory to a model quantum mechanical universe consisting of systems $S_1, S_2, \ldots, S_n$ that undergo measurements by other systems $M_1, M_2, \ldots, M_n$ that function as measuring instruments. Each M_i is associated with a "pointer" observable R^i. The complete state of this universe at a particular time t is given by the quantum state at t and the values of all the observables as beables at t. The trajectory of any beable A of S_i depends on the quantum state and the initial value of A. However, because of decoherence in a measurement of A by an instrument M_j with "pointer" R^j, the trajectory of A during measurement depends effectively on the quantum state and the value of R^j at the start of the measurement interaction.

Therefore, during and immediately after measurement, the value of A is effectively given by the value of R^j. Between measurements, the value of A and the values of other beables represented by operators that commute with A do not in general satisfy functional relations that hold between these beables. However, these values play no role in the dynamical evolution of the trajectories of the "pointer" beables $\{R^i\}$ because the trajectory of each R^i depends entirely on the initial value of R^i and the quantum state. One might say that the values of beables other than the $\{R^i\}$ supervene on the values of the $\{R^i\}$—they are quite redundant in the theory. In this quantum world, all phenomena can be completely described by the evolution of the commuting set of "pointer" beables $\{R^i\}$ or by a single "pointer" beable R of which each R^i is a function. Events described in terms of the changes to the values of beables other than the $\{R^i\}$ have the status of epiphenomena. Thus, nothing is gained by taking all observables as beables, beyond the interpretative advantage achieved by taking R alone as the only beable.

What we want for a consistent interpretation of quantum mechanics is a single suitable "pointer" observable, R, that can be taken as determinate at all times and in terms of which the measurement of any observable, Q, can be represented "internally" in the theory as the correlation of R-values with Q-values. Although Bohm's choice of position in configuration space might be appropriate, it would not be appropriate to take R as momentum rather than position. A momentum "pointer" would behave rather strangely, largely because of the nature of the potentials that we encounter in our universe. Even with a suitable choice of a privileged beable R in this sense, there will always be some probability of apparently anomalous behavior, for example, stochastic transitions between different values of R associated with different values of a measured observable in repeated measurements as a consequence of interference effects. Decoherence renders such probabilities negligible, as in Bohm's theory. In this interpretation of quantum mechanics, decoherence explains why certain events that we would regard as anomalous occur very rarely. The events in question are always determinate and do not depend for their determinateness on the smallness of certain probabilities, as in orthodox quantum mechanics.

Vink's theory should be interpreted as a generalized Bohmian mechanics for any discrete observable as a beable rather than a theory of all observables as beables. We might suppose with Vink that, on a sufficiently small scale, all dynamical variables take discrete values. For a suitable choice of a privileged observable R as a beable, the picture of a quantum mechanical universe is this: The quantum state evolves dynamically in time and can be understood as representing an objectively real field in R-space that influences the evolution of R-values. The always determinate values of R evolve stochastically under the influence of a transition matrix T_{mn} that depends on the quantum state. The only real change in a quantum mechanical universe is the change in the quantum state and the change in R, and this suffices to account for all quantum phenomena. All other observables are ersatz observables whose role in the theory is fully accounted for by the evolution of the quantum state. Insofar as they can be said to have determinate values, these values supervene on the values of R. The virtue of the Bell-Vink analysis is that we see how adopting a suitable discrete observable as a beable yields a consistent interpretation of quantum mechanics, in which measurements can be represented internally as dynamical interactions between quantum systems.

REFERENCES

1. Vink, J. 1993. Quantum mechanics in terms of discrete beables. Phys. Rev. A **48:** 1808–1818.
2. Bohm, D. 1952. A suggested interpretation of the quantum theory in terms of "hidden" variables: I and II. Phys. Rev. **85:** 166–179 and 180–193.
3. Bell, J. S. 1987. Speakable and Unspeakable in Quantum Mechanics, p. 41. Cambridge University Press. London/New York.
4. Ibid., p. 175.
5. Ibid., p. 177.
6. Ibid., p. 176.
7. Kochen, S. & E. P. Specker. 1967. On the problem of hidden variables in quantum mechanics. J. Math. Mech. **17:** 59–87.
8. Bell, J. S. Op. cit., p. 8–9 and 164–165.
9. Vink, J. Op. cit., p. 1811.
10. Bell, J. S. Op. cit., p. 163.

Vacuum Correlations and Outcome Dependence in Algebraic Quantum Field Theory

JEREMY BUTTERFIELD

Jesus College
Cambridge CB5 8BL, United Kingdom

INTRODUCTION

This report has three aims, one for each of the following sections. The first is to advertise some of the work of authors such as Landau, Summers, and Werner.[1–7] Through their work, it is now well established that the Bell inequality is generically violated in rigorous quantum field theories, especially algebraic (relativistic) quantum field theories (AQFT). The second section summarizes some of their technical notions and results.

My second aim is to make a conceptual point about this violation: in effect, to relate it to the usual assumptions of Bell's theorem, especially for local stochastic "hidden variable" models. The above authors briefly discuss such assumptions, but it is worth considering the case of stochastic models in more detail. Here, my main point will be that AQFT violates what Shimony[8] dubbed "outcome independence". More precisely, I will show that we can present the usual three assumptions of a stochastic Bell's theorem within the framework of AQFT. Of these assumptions, AQFT violates just outcome independence.

Of course, that verdict is not surprising: elementary quantum theory also violates just outcome independence. However, the verdict yields two subsidiary points about a recent result of Redhead[9] about strict correlations in AQFT. First, we see the result as directly proving outcome dependence in AQFT. Second, we see the result as reducing stochastic models to deterministic models, in a way analogous to a theorem of Suppes and Zanotti.[10]

More important, AQFT's violation of outcome independence also leads to my third aim, namely, to connect outcome independence with stochastic Einstein locality (SEL) or, rather, to connect outcome dependence with the violation of SEL. SEL was first formulated by Hellman.[11] It is a natural locality condition and has been cited often in the philosophical literature about Bell's theorem. However, its relation to the usual roster of locality conditions discussed in AQFT has not been investigated much; so far as I know, Redei was the first to do so.[12,13]

I will argue that SEL has two natural formulations. On the one hand, it can be formulated in terms of the expectation value of a single quantity. So formulated, it holds. (This is Redei's own verdict;[12] here, I shall report an amended proof, from joint work with Muller.[14]) On the other hand, SEL can also be formulated in terms of a quantity's expectation value, conditional on another quantity taking a certain value. So formulated, SEL is a particularly plausible version of outcome independence.

However, by the results of the second section, it fails: therein, say I, lies the strangeness of the Bell-type correlations.

AQFT VIOLATES THE BELL INEQUALITY

In this section, I will for the most part take notation and terminology from the review by Summers.[15] I begin by summarizing the situation in algebraic quantum theory without regard to space-time structure. First, briefly recall that a C*-algebra (with a unit 1), $\boldsymbol{C}$, is a Banach *-algebra (i.e., a normed algebra with a unit and an involution, $*$, that is complete in its norm $\|\ldots\|$), with the norm satisfying the C*-condition that $\|A^*A\| = \|A\|^2$ for all $A \in \boldsymbol{C}$. However, in this report, we can think of C*-algebras as norm-closed *-algebras of bounded operators on a Hilbert space, containing the identity operator.

The outcomes of an experiment are represented by projectors of a C*-algebra. All that follows would be unaffected if instead we took outcomes to be represented by effects, that is, by positive self-adjoint elements of the C*-algebra with the norm less than or equal to 1, of which projectors are a special case. However, for the sake of conformity with most authors considered, I concentrate on projectors. The projectors representing the possible outcomes for a given measurement apparatus sum to the identity. For convenience later on (in the next section), I label outcomes by a superscript x: $\Sigma_x E^x = 1$.

States on $\boldsymbol{C}$ are expectation functionals: that is, linear functionals ϕ sending elements of $\boldsymbol{C}$ to their expectation values (complex for nonself-adjoint elements of $\boldsymbol{C}$), which are also required to be (i) continuous [and so bounded in the sup-norm defined by $\|\phi\| := \sup\{|\phi(A)|: A \in \boldsymbol{C}, \|A\| = 1\}$], (ii) positive [i.e., for all $A \in \boldsymbol{C}$, $\phi(A^*A) \geq 0$], and (iii) normalized [i.e., $\phi(1) = 1$]. A state ϕ is called normal, if, for every monotone-increasing bounded $\{A_k\}$ with least upper bound A, $\phi(A_k) \uparrow \phi(A)$. These correspond in the Hilbert space formalism to density matrices.

Accordingly, the setting for discussing Bell's inequality is as follows: the observables of the system are a C*-algebra $\boldsymbol{C}$ (with a unit 1), containing a pair $(\boldsymbol{A}, \boldsymbol{B})$ of commuting subalgebras, with the outcomes of the Bell experiment represented by projectors $E \in \boldsymbol{A}$ and $F \in \boldsymbol{B}$. Because the subalgebras commute, joint probabilities are given by $\phi(EF)$. Furthermore, the linearity of ϕ along with the existence of a *common* unit to which the (projectors representing) outcomes of each experiment sum directly implies that single probabilities are independent of what is measured in the other wing—the condition often called "locality". Thus, for the "left" single probabilities being independent of what is measured on the right, we have the following:

$$\text{Because } \sum_x F^x = 1, \qquad \sum_x \phi(EF^x) = \phi(E). \tag{2.1}$$

I should also report another way in which, in the Hilbert space formalism, having $\boldsymbol{A}$ and $\boldsymbol{B}$ commute captures locality. Thus, Schlieder[16] proved a version of the so-called "no-signaling theorem": that is, that communication is equivalent to the preservation of one observable's statistics under nonselective Luders-rule-like measurement of the other observable. To be precise, he proved that two bounded self-adjoint operators, A and B, on a Hilbert space commute if and only if, for any normal state ϕ and any partition $\{I_j\}$ of the spectrum of B, with $B = \int \lambda dE(\lambda)$ (that is, the spectral resolution of B),

$$\phi(A) = \sum_j \phi\left[\int_{I_j} dE(\lambda) A \int_{I_j} dE(\lambda)\right]. \tag{2.2}$$

It is convenient to work with a Bell inequality expressed not in terms of projectors, but in terms of self-adjoint contractions. Thus, given a projector $E \in \boldsymbol{A}$, we define $A \in \boldsymbol{A}$ by the following: $A := 2E - 1$ so that $-1 \leq A = A^* \leq 1$. The Bell inequality (in the most familiar form[17]) is then that for self-adjoint contractions $A_i \in \boldsymbol{A}$, $B_j \in \boldsymbol{B}$ $(i, j = 1, 2)$:

$$|\phi[A_1(B_1 + B_2) + A_2(B_1 - B_2)]| \leq 2. \tag{2.3}$$

We then define the maximal Bell correlation of the pair $(\boldsymbol{A}, \boldsymbol{B})$ of commuting subalgebras of the C*-algebra $\boldsymbol{C}$ in the state $\phi \in \boldsymbol{C}^*$ to be

$$\beta(\phi, \boldsymbol{A}, \boldsymbol{B}) := \sup(\tfrac{1}{2})\phi[A_1(B_1 + B_2) + A_2(B_1 - B_2)], \tag{2.4}$$

where the supremum is taken over all self-adjoint contractions $A_i \in \boldsymbol{A}$, $B_j \in \boldsymbol{B}$. Hence, the Bell inequality (equation 2.3) is $\beta(\phi, \boldsymbol{A}, \boldsymbol{B}) \leq 1$.

In the next section, I will further discuss the traditional "classical" sufficient conditions for the Bell inequality to hold, namely, local "hidden variable" models of the Bell experiment. For the moment, note that even in the general C*-framework (and so within any quantum theory) there are some conditions in which the Bell inequality must be satisfied—and it is *not* necessary that both $\boldsymbol{A}$ and $\boldsymbol{B}$ be Abelian. Thus, we have (theorem 2.1.3 of reference 4 and theorem 7 of reference 18) the following proposition:

PROPOSITION 0. Let $(\boldsymbol{A}, \boldsymbol{B})$ be a pair of commuting subalgebras of a C*-algebra $\boldsymbol{C}$.

(i) If either $\boldsymbol{A}$ or $\boldsymbol{B}$ is Abelian, then $\beta(\phi, \boldsymbol{A}, \boldsymbol{B}) = 1$ for all states ϕ.
(ii) If $\phi|_{\boldsymbol{A} \vee \boldsymbol{B}}$, the restriction of ϕ to the C*-algebra generated by A and B, is a convex sum of product states over $(\boldsymbol{A}, \boldsymbol{B})$, then $\beta(\phi, \boldsymbol{A}, \boldsymbol{B}) = 1$.
(iii) If $\phi|_{\boldsymbol{A}}$ or $\phi|_{\boldsymbol{B}}$ is a pure state, then $\beta(\phi, \boldsymbol{A}, \boldsymbol{B}) = 1$.

Of course, the Bell inequality is generally violated by quantum theories. Less well known is the fact that quantum theories have their own bound (proposition 1 of reference 1; see also reference 19):

PROPOSITION 1. For any C*-algebra $\boldsymbol{C}$, any commuting subalgebras $\boldsymbol{A}$ and $\boldsymbol{B}$, and any state $\phi \in \boldsymbol{C}^*$,

$$\beta(\phi, \boldsymbol{A}, \boldsymbol{B}) \leq \sqrt{2}. \tag{2.5}$$

Hence, we define the maximal violation of the Bell inequality by $\beta = \sqrt{2}$. As originally discovered by Bell,[20] quantum theory attains this bound, provided that $\boldsymbol{A}$ and $\boldsymbol{B}$ each contain (copies of) the 2×2 complex matrices and thus the Pauli spin matrices; one takes ϕ as the familiar singlet state.

Landau (propositions 3 and 5 of reference 1) showed that this maximal violation was generic in quantum theories using von Neumann algebras (i.e., C*-algebras closed in the weak topology) in the sense of the following:

PROPOSITION 2. Let $\boldsymbol{A}$ and $\boldsymbol{B}$ be commuting von Neumann algebras on a Hilbert space H that satisfy the Schlieder property, namely, that $0 \neq \mathrm{A} \in \boldsymbol{A}$ and $0 \neq \mathrm{B} \in \boldsymbol{B}$ imply that $\mathrm{AB} \neq 0$. Then, if neither $\boldsymbol{A}$ nor $\boldsymbol{B}$ is Abelian, there is a normal state ϕ on the set of all bounded operators $B(H)$ such that $\beta(\phi, \boldsymbol{A}, \boldsymbol{B}) = \sqrt{2}$.

Sketch of Proof: First, for any projectors, $\mathrm{P}_i \in \boldsymbol{A}$ and $\mathrm{Q}_i \in \boldsymbol{B}$ $(i = 1, 2)$, and corresponding self-adjoint contractions, $\mathrm{A}_i = 2\mathrm{P}_i - 1$ and $\mathrm{B}_i = 2\mathrm{Q}_i - 1$, we have

$$\|\mathrm{A}_1(\mathrm{B}_1 + \mathrm{B}_2) + \mathrm{A}_2(\mathrm{B}_1 - \mathrm{B}_2)\| = 2\sqrt{\{1 + 4\|[\mathrm{P}_1, \mathrm{P}_2][\mathrm{Q}_1, \mathrm{Q}_2]\|\}}. \tag{2.6}$$

Then, by theorem 2 of Roos,[21] the Schlieder condition entails that the norm on the right factorizes into the product of the norms in the two algebras $\boldsymbol{A}$ and $\boldsymbol{B}$. Finally, one uses the fact that any non-Abelian von Neumann algebra contains projectors whose commutator norm equals ½.

Remarks: The fact that the norm of the commutator of any two projectors is less than or equal to ½ has two consequences. First, equation 2.6 corresponds to the limit on β given by proposition 1. Second, the proof also establishes that, if we are content with nonmaximal violation of the Bell inequality, we can choose *any* projectors P_i, Q_i $(i = 1, 2)$ with the P terms not commuting and the Q terms not commuting (in algebras $\boldsymbol{A}$ and $\boldsymbol{B}$ that are commuting, non-Abelian, with Schlieder property): that is, there is a state violating the Bell inequality for these projectors.

On the other hand, Summers and Werner (p. 2442 of reference 4; proposition 5.7 of reference 15) show that Bell's original violation using the singlet state on the Pauli matrices was archetypal in the sense of the following:

PROPOSITION 3. Let $\boldsymbol{A}$ and $\boldsymbol{B}$ be commuting subalgebras of a C*-algebra $\boldsymbol{C}$ and let A_i and B_i be self-adjoint contractions. Then, for a state ϕ on $\boldsymbol{C}$, with $\phi|_{\boldsymbol{A}}$ faithful (i.e., if $\mathrm{A} \in \boldsymbol{A}$, then both $\mathrm{A} \geq 0$ and $\phi(\mathrm{A}) = 0$ imply that $\mathrm{A} = 0$) and $\phi|_{\boldsymbol{B}}$ faithful, we have the maximal violation, that is,

$$\phi[\mathrm{A}_1(\mathrm{B}_1 + \mathrm{B}_2) + \mathrm{A}_2(\mathrm{B}_1 - \mathrm{B}_2)] = 2\sqrt{2}. \tag{2.7}$$

Then, A_1, A_2, and $\mathrm{A}_3 := (-i/2)[\mathrm{A}_1, \mathrm{A}_2]$ are a realization of the Pauli matrices in $\boldsymbol{A}$ and similarly for the B_i in $\boldsymbol{B}$.

I now specialize the discussion to algebraic quantum field theory.[22,23] The basic framework is that each open bounded region O of Minkowski space-time is associated with a C*-algebra $\boldsymbol{A}(O)$ of observables, subject to various axioms that are motivated by the idea that $\boldsymbol{A}(O)$ is the algebra generated by all observables measurable in the region O. However, we will specialize this framework, assuming the $\boldsymbol{A}(O)$ to be concrete von Neumann algebras acting on a separable Hilbert space H. The system's overall C*-algebra of observables, $\boldsymbol{C}$ as above, is then the so-called quasi-local algebra generated by all the $\boldsymbol{A}(O)$. We require that $\boldsymbol{C}$ is irreducible, that is, that the von Neumann algebra, $\boldsymbol{C}''$, that is generated by $\boldsymbol{C}$ equals $B(H)$.

With this specialization, I can state the axioms needed for my discussion as follows:

(i) Isotony: If $O_1 \subseteq O_2$, then $A(O_1) \subseteq A(O_2)$.

(ii) Spacelike Commutativity: If all points in O_1 are spacelike from all points of O_2, written $O_1 \times O_2$, then $A(O_1) \subseteq [A(O_2)]'$, the commutant of $A(O_2)$ (this commutativity, of course, represents the joint measurability, and the existence of joint distributions, of mutually spacelike observables).

(iii) Poincaré Invariance: There exists on H a strongly continuous unitary representation $U(P)$ of (the covering group of) the Poincaré group, which meshes with the automorphisms of the quasi-local algebra C, in the sense of the following: to each (active) Poincaré transformation g acting on Minkowski space-time [$g := (\Lambda, a)$, where Λ is a Lorentz transformation and $a \in \mathrm{R}^4$ is a space-time translation], there is associated an automorphism α_g of the quasi-local algebra C, with

$$\alpha_g[A(O)] = A[g(O)]$$

and such that $\mathrm{U}(g)\mathrm{A}\mathrm{U}(g)^{-1} = \alpha_g(\mathrm{A})$ for all $\mathrm{A} \in A(O)$.

(iv) Diamond: First, the domain of dependence, $D(O)$, of a region O is defined to be $D^+(O) \cup D^-(O)$, where the future domain of dependence, $D^+(O)$, is the set of points through which any smooth past-inextendable nonspacelike curve intersects O; similarly for the past domain of dependence. Then, Diamond requires that $A(O) = A[D(O)]$.

(v) Weak Additivity: For any nonempty region O, C'' equals the von Neumann algebra generated by the set of all translates of $A(O)$: that is, $C'' = \{\cup_{a\in\mathrm{R}^4} A(O_a)\}''$.

(vi) Irreducibility: H has no nontrivial subspace invariant under all elements of all the $A(O)$: that is, $C'' = B(H)$.

(vii) Spectrum: The four mutually commuting self-adjoint generators $\{\mathrm{P}_0, \mathrm{P}_1, \mathrm{P}_2, \mathrm{P}_3\}$ of the translation subgroup $U(\mathrm{R}^4)$ obey $\mathrm{P}_0 \geq 0$ and $\mathrm{P} \cdot \mathrm{P} \geq 0$.

(viii) Vacuum: There is a unique state Ω (called the vacuum) that is invariant under all translations.

These axioms are for the most part independent. I remark that, intuitively, this is reasonable as regards the three "locality" axioms—Spacelike Commutativity, Poincaré Invariance, and Spectrum—because they clearly make different statements. Thus, it is widely accepted that some kinds of violations of Spectrum ("tachyons") are logically compatible with Poincaré Invariance, even if physically unreasonable. Similarly, violation of Spacelike Commutativity seems compatible with Poincaré Invariance and need not involve tachyons.

So much by way of summarizing the framework. Returning to the Bell inequality, proposition 2 obviously makes us expect maximal violation of the inequality to be endemic in the local algebras of spacelike-related regions. Our first result in this direction is again by Landau. Regions O_1 and O_2 are said to be strictly spacelike if there is a neighborhood N of the origin in R^4 with $\{O_1 + \mathrm{N}\} \times O_2$ (the relation is symmetric). Schlieder[24] proved, using Spectrum and Vacuum, that the algebras of strictly spacelike regions satisfy the Schlieder property. Hence, we get the following proposition directly from proposition 2 (p. 56 of reference 1; cf. also p. 231–233 of reference 15):

PROPOSITION 4. Under the stated axioms [(i)–(viii)], if O_1 and O_2 are strictly spacelike, there is a normal state ϕ on $B(H)$ such that $\beta[\phi, A(O_1), A(O_2)] = \sqrt{2}$.

Remarks: First, this result establishes the remarkable fact that there is maximal violation of the Bell inequality *independent of spatial separation.* If, however, we fix the state considered, the degree of violation falls off rapidly with separation. For the vacuum, it is known that, roughly speaking, the degree of correlation (i.e., the difference of the expectation of a product observable from the product of individual expectations) in the massless case decays like R^{-2}, where R is the spacelike distance between regions; in the massive case, it decays like $\exp(-mR)$, where m is the mass gap. (See theorem 4.1 of reference 4; also see references 25 and 26.)

Second, as Landau remarks, under the stated assumptions, C is simple and this means that there are vector states violating the Bell inequality arbitrarily closely to $\sqrt{2}$.

Third, the requirement that the regions be *strictly* spacelike can be dropped if the O_i are simple in the sense that $O''_1 = O_1$, with O''_1 standing for the interior of the causal complement of O_1. (The causal complement of O is the set of points spacelike to O.) For such regions, if O_1 is merely spacelike to O_2, then theorem 3.5 of Driessler[27] implies the Schlieder property.

Proposition 4 is just the beginning. Much of Summers and Werner's work concerns special pairs of regions in which there is maximal violation of the Bell inequality in *all* normal states. (Of course, the observables concerned vary from one state to another.) Accordingly, they say that, in such a case, the algebras of the pair of regions are maximally correlated. They show, very remarkably, that such a maximal correlation of algebras is endemic for most rigorous quantum field theories: for all pairs of regions that are spacelike, but tangent (intersecting closures), and of prescribed shapes—namely, a pair is to consist of a wedge and its causal complement or of two tangent double cones. They also relate this maximal correlation to AQFT's usual roster of locality conditions. I can only give a taste of these results (for details, cf. theorems 5.8–5.15 and 6.9–6.13 of reference 15).

First, recall that a pair (A, B) of subalgebras of a C*-algebra C are called C*-independent if and only if any pair of states, one on each subalgebra, have a common extension to a state on C. Roos[21] showed that (A, B) are C*-independent if and only if they have the Schlieder property and if and only if the common extension is a product state across $A \vee B$.

The results of Roos,[21] Schlieder,[24] and Driessler[27] entail that C*-independence is typical in AQFT in the sense that, in an irreducible vacuum representation, the following results hold: for any strictly spacelike O_1 and O_2, $[A(O_1), A(O_2)]$ is C*-independent; for any tangent double cones, O_1 and O_2, $[A(O_1), A(O_2)]$ is C*-independent; and, for any wedge W, $[A(W), A(W')]$ is C*-independent.

Recall that a W*-algebra is a C*-algebra that is the dual of some Banach space (but we can think of them as von Neumann algebras because every W*-algebra is W*-isomorphic to a von Neumann algebra acting on a Hilbert space). A pair (A, B) of subalgebras of a W*-algebra C can be called W*-independent if and only if any pair of *normal* states, one on each subalgebra, have a common *normal* extension to a state on $\boldsymbol{C}$. Thus, W*-independence implies C*-independence.

It turns out that under the assumptions (i) to (viii), for any spacelike double cones or wedges O_1 and O_2, $[A(O_1), A(O_2)]$ is W*-independent if and only if any pair of normal states, one on each subalgebra, have a common normal extension that is a *product* state across $\boldsymbol{A} \vee \boldsymbol{B}$.

However, we know from part (ii) of proposition 0 above that convex sums of product states obey the Bell inequality. Thus, using Summers and Werner's results showing that algebras of wedges and their complements, and of two tangent double cones, are maximally correlated, we can infer that such pairs of algebras are *not* W*-independent.

For my purposes in the following two sections, though, I do not need details about (the occurrence and properties of) maximally correlated algebras and their consequences for AQFT's locality conditions. Rather, I need to report two theorems about correlations in the vacuum: a second theorem by Landau and one by Redhead. They both depend on the Reeh-Schlieder theorem[28] (e.g., cf. p. 101 of reference 22), which states that, under our assumptions (i)–(viii), the vacuum Ω is cyclic for any local algebra; that is, the following proposition:

PROPOSITION 5. For any open region O, the set of vectors $A(O)\Omega$ is dense in H.

Remarks: In the proof, the vacuum can be replaced by any vector state with bounded energy. Also, the vacuum can be similarly replaced in propositions 6 and 7 below.

The idea of Landau's second theorem is to use proposition 5 to hit the vacuum so as to produce one of the states giving the (arbitrarily close to) maximal violations of the Bell inequality that proposition 4 provides and then to argue that this implies that, in the vacuum state itself, some *conditional* expectations violate the inequality. The hitting is done by an operator, say A, in the local algebra of a third region O_3 that is spacelike to the given two regions. Landau then uses the fact that one can approximate the bounded self-adjoint operator A†A arbitrarily closely by a finite sum of its spectral projectors $\{E_j\}$ to argue that, for one of these projectors, say E, some expectation values on $A(O_1), A(O_2)$ according to the state EΩ violate the Bell inequality. (These values can be interpreted in the usual way as expectations conditional on getting a result of +1 in a measurement of E performed in the vacuum.) More precisely, for any projector $E \in A(O_3)$, Landau defines the vacuum conditional expectation, R_E, of the usual quantity, $A_1(B_1 + B_2) + A_2(B_1 - B_2) \in A(O_1) \vee A(O_2)$, by

$$R_E := \Omega([A_1(B_1 + B_2) + A_2(B_1 - B_2)]E)/\Omega(E) \qquad \mathbf{(2.8)}$$

and then proves the following proposition (proposition 1 of reference 2):

PROPOSITION 6. For any three strictly spacelike regions O_1, O_2, and O_3 and for any $\epsilon > 0$, there are self-adjoint contractions $A_i \in A(O_1)$ and $B_i \in A(O_2)$ and a projector $E \in A(O_3)$ such that the vacuum conditional expectation R_E statisfies $|R_E| > 2\sqrt{2} - \epsilon$.

Remarks: First, as in the second remark on proposition 2, equation 2.6 means that, if we are content with nonmaximal violation of the Bell inequality, we can get violation with (the contractions corresponding to) *any* projectors $P_i \in A(O_1)$, $Q_i \in A(O_2)$ $(i = 1, 2)$, with the P terms not commuting and the Q terms not commuting. Second, the rapid decay of correlations with spacelike separation, mentioned in the first remark on proposition 4, means that, for macroscopic distances, it is very rare to get a result of +1 in a measurement of E in the vacuum; that is, $\Omega(E)$ is very small.

Finally, there is another sense (needed in the next section) in which the correlations do not decay. This concerns not the vacuum's violation of the Bell inequality, but its strict correlations. Using a similar method of proof (approximation of a bounded self-adjoint operator $A^\dagger A$ [with A given by the Reeh-Schlieder theorem] by a finite sum of its spectral projectors), Redhead[9] has proved the following:

PROPOSITION 7. For any two spacelike regions O_1 and O_2, for any $\epsilon > 0$, and for any projector $P_1 \in A(O_1)$, there is a projector $P_2 \in A(O_2)$ such that $\Omega(P_1P_2) > (1 - \epsilon)\Omega(P_2)$.

Remarks: First, this is compatible with the rapid decay of correlations with spacelike separation for the reason given in the second remark on proposition 6. In more detail, the rapid decay means that the difference, $\Omega(P_1P_2) - [\Omega(P_2) \cdot \Omega(P_1)]$, falls off rapidly; proposition 7, that is, $\Omega(P_1P_2) \approx \Omega(P_2)$, then implies simply that $\Omega(P_2)$ falls off rapidly for given $\Omega(P_1)$. Second, I again emphasize that the vacuum is replaceable by any vector state with bounded energy.

OUTCOME DEPENDENCE IN AQFT

I now turn to relating the results of the previous section to analyses of exactly which "classical" assumptions imply a Bell inequality (and so are refuted by the inequality's violation). This is a big topic and I cannot do full justice to all the alternatives, even to those discussed by the authors mentioned in the previous section (Summers and Werner[4,29] give the fullest discussion). I shall confine myself (as they do) to the traditional two-particle Bell experiment. Moreover, I aim only to relate the results of the previous section to a distinction that is not articulated by these authors, but is traditional in the literature: namely, the distinction between deterministic and stochastic local models of the Bell experiment. (I drop the usual phrase, "hidden variable".) As we shall see, this distinction has been criticized (sympathy with those criticisms may be the reason why these authors do not articulate the distinction). However, I consider it a valid distinction and (as promised in the INTRODUCTION) deploying it will in any case yield a novel conclusion, in the next section, about Hellman's condition of stochastic Einstein locality.

Therefore, I first recall the relevant aspects of such deterministic and stochastic models (e.g., cf. references 17 and 30 for more details). A deterministic local model of the Bell experiment consists of a space Λ (of complete states of the pairs of particles; of "hidden variables"), on which the four quantities, A_1, A_2, B_1, B_2, to be measured in the two wings (A terms on the left, B terms on the right) are represented as real-valued functions. Λ is a probability space, with measure ρ, for example, so that the A_i and B_i are random variables. The model assumes locality in two ways. First, ρ is independent of which quantities are measured. Second, the model has a single function, A_1, to represent the results obtained in a left-wing measurement, irrespective of which quantity, if any, is measured on the right. These two conditions imply that the predicted observable probabilities for results of ± 1, obtained by averaging over Λ using ρ, satisfy locality in the sense of equation 2.1. Thus, defining the preimage set for a joint result by, for example,

$$[1, +1; 2, -1] := \{\lambda \in \Lambda: A_1(\lambda) = +1 \text{ and } B_2(\lambda) = -1\}, \quad \mathbf{(3.1)}$$

the observable probability is given by the ρ-measure of the preimage,

$$p(A_1 = +1 \text{ and } B_2 = -1) := \int_{[1,+1;2,-1]} d\rho, \quad \mathbf{(3.2)}$$

and locality is trivial. Bell's theorem states that any such deterministic local model is committed to a Bell inequality.[20]

On the other hand, in a stochastic local model, specifying $\lambda \in \Lambda$ does not specify measurement results (by applying a random variable); it specifies probabilities—in general, nontrivial (neither 1 nor 0)—for the various possible joint and single measurements:

$$\mathrm{pr}_{\lambda A_i B_j}(\pm 1, \pm 1),\ \mathrm{pr}_{\lambda A_i}(\pm 1),\ \mathrm{pr}_{\lambda B_j}(\pm 1).$$

The intuitive idea is that either the state λ of a pair of particles evolves indeterministically in flight or the measurement result is also influenced by features of the apparatus that vary from one run to another; or both. Furthermore, such a model need not use a single probability space, as a deterministic model does. As the subscripts suggest, each choice of a single quantity or a pair of comeasurable quantities can label a separate probability space, thus avoiding joint probabilities for noncomeasurable quantities. In these respects, stochastic local models generalize deterministic local models (but cf. the discussion below).

However, these models assume locality in a very similar form to the deterministic models. Again, there are two assumptions. First, ρ is independent of which quantities are measured. Second, all the joint probabilities are assumed to factorize into single probabilities; that is, for all λ, A_i, and B_j and results $x, y = \pm 1$,

$$\mathrm{pr}_{\lambda A_i B_j}(x, y) = \mathrm{pr}_{\lambda A_i}(x) \cdot \mathrm{pr}_{\lambda B_j}(y). \quad \mathbf{(3.3)}$$

I shall adopt a common terminology and shall use the term "factorizability" to describe equation 3.3. (Thus, what I have called "stochastic local models" are often called "factorizable stochastic models".) The predicted observable probabilities for results of ± 1 are obtained by averaging over Λ using ρ. Bell's theorem is now that any such stochastic local model is committed to a Bell inequality.

In assessing these models, attention has focused on "factorizability". For us, the crucial point is that it is the conjunction of two separate independence conditions on a single (marginal) probability. Roughly speaking, one is independence from the choice of quantity to be measured in the other wing; the other is probabilistic independence from the measurement result obtained in the other wing. The fullest presentation of this point is by Jarrett.[31] However, I shall follow Shimony,[8] who calls the conditions "parameter independence" and "outcome independence", respectively. Thus, parameter independence says that, for all λ, A_i, and B_j and results $x, y = \pm 1$,

$$\mathrm{pr}_{\lambda A_i}(x) = \mathrm{pr}_{\lambda A_i B_j}(x) := \mathrm{pr}_{\lambda A_i B_j}(x, y) + \mathrm{pr}_{\lambda A_i B_j}(x, -y) \qquad \textbf{(3.4)}$$

(and similarly for R-probabilities), whereas outcome independence says that

$$\mathrm{pr}_{\lambda A_i B_j}(x, y) = \mathrm{pr}_{\lambda A_i B_j}(x) \cdot \mathrm{pr}_{\lambda A_i B_j}(y). \qquad \textbf{(3.5)}$$

It is easy to see that only parameter independence (equation 3.4) corresponds to equation 2.1 and that deterministic local models automatically satisfy outcome independence (equation 3.5) (because each factor in equation 3.5 will be 0 or 1). These points suggest the well-known verdict: that the "culprit", that is, the assumption that we should give up so as to avoid a Bell inequality, is outcome independence. By doing so, we can reject both deterministic and stochastic local models.

Furthermore, this verdict is supported by quantum theory itself. In effect, it obeys equation 3.4, but not equation 3.5. More precisely, suppose that we fill in sufficient details about the experiment so as to be able to write down a quantum theoretic model of it (choosing, say, photon pairs and polarization measurements). In our model, either a single pure state ψ or an ignorance-interpreted mixture will replace all the λ's, and the probabilities will be given by the Born rule. Our model will then obey equation 3.4: that is just the no-signaling theorem, based on the commutation of the L- and R-quantities. On the other hand, our model will violate equation 3.5, except in some special cases, such as the model's quantum state being a product state. [Thus, reserving "locality" for equation 3.4, we might say that quantum theory itself supplies local stochastic models of the experiment, albeit nonfactorizable (more specifically, outcome-dependent) models.]

Although this verdict may be compelling, note that it is by no means trivial. Agreed, it is trivial that stochastic theories describe correlations and that quantum states typically prescribe correlations between commuting quantities. However, in everyday life and classical physics (i.e., outside quantum theory), correlations between events that are not causally related (say, because they are mutually spacelike) are eliminated once we consider probabilities conditional on a sufficiently rich specification of the state prior to the correlated events (in Reichenbach's terminology, a sufficiently rich specification of the common cause; cf. p. 163 of reference 32). Bell's theorem (understood with this verdict) tells us that the Bell correlations cannot be thus eliminated. In this sense, they are indeed strange. (The following section develops this point, especially as regards the specification of the prior state; see also, for example, reference 33.)

So much by way of recalling deterministic and stochastic local models, and the verdict of outcome dependence, in the usual context of the Bell experiment. We can now very easily lift these considerations to the previous section's context, AQFT, so as to produce a Bell's theorem for AQFT, in which the culprit (i.e., false assumption) is obviously factorizability, specifically outcome independence.

We can work with a fixed state, say ϕ, so that the locality assumption that ρ is independent of which quantity is measured falls away. I write A_i^x for the effect representing outcome x for the contraction A_i and, similarly, I write B_j^y $(i, j = 1, 2)$. Then, a marginal probability, for example, the probability of x for A_i in the context of measuring B_j, is given by summing out the B_j outcome. Hence, parameter independence says that this marginal is independent of the choice of $j = 1$ or $j = 2$, as follows:

$$\phi(A_i^x) = \phi[A_i^x(\Sigma_y B_1^y)] = \phi[A_i^x(\Sigma_y B_2^y)] \quad \textbf{(3.6)}$$

and similarly for $\phi(B_j^y)$; this condition holds, just as in the general setting at the start of the previous section, because of the common unit in the algebra. On the other hand, outcome independence says that joint probabilities factorize into their marginals:

$$\phi(A_i^x B_j^y) = \phi[A_i^x(\Sigma_y B_j^y)] \cdot \phi[(\Sigma_x A_i^x)B_j^y]. \quad \textbf{(3.7)}$$

Just as before, the conjunction of parameter and outcome independence is factorizability, namely,

$$\phi(A_i^x B_j^y) = \phi(A_i^x) \cdot \phi(B_j^y). \quad \textbf{(3.8)}$$

From these assumptions, we can exactly mimic the usual proof for a factorizable stochastic model (e.g., see reference 17) to get the Bell inequality (equation 2.3) governing the expectation values of the contractions.

Of course, outcome independence (equation 3.7) and so factorizability (equation 3.8) are manifestly false in AQFT, just as the corresponding conditions in elementary quantum theory were; again, except in some special states, for example, those that when restricted to $\boldsymbol{A} \vee \boldsymbol{B}$ are product states. However, the familiarity and triviality of the mathematics should not blind us to the nontriviality of the verdict. Just as in the above discussion concerning elementary quantum theory, we naturally expect correlations between spacelike events to be eliminated, conditional on sufficiently rich information about the prior state. Again, see the next section for development.

However, before turning to that, I should briefly state two criticisms that have been made of the distinction between factorizable stochastic models and deterministic local models, report my replies to them, and then relate these criticisms to algebraic quantum theory. I shall set more store by the second criticism.

The first is based on the fact that a given collection of single probabilities and comeasurable-pair joint probabilities can be modeled by a factorizable stochastic model if and only if it can also be modeled by a deterministic local model. The reason lies in the fact that the existence of each kind of model is equivalent to the existence of a total joint probability function (i.e., for all the quantities). (See, for example, proposition 3 of reference 34 and theorem 4 of reference 35.) Thus, says the critic, the move to factorizable stochastic models brings no generality. In short, my reply is that this equivalence is "merely mathematical". Because the physical ideas motivating a factorizable stochastic model, such as indeterminism during flight and avoidance of joint probabilities for noncomeasurable pairs, are clearly coherent, one can regard the total joint probability that also delivers the given probabilities as marginal probabilities, as not physically real. (For more details of this reply, see section 8 of reference 30 and section 6 of reference 36.)

The second criticism is also based on a result that factorizable stochastic models reduce to deterministic local models, but it cannot be dismissed as "merely mathematical". It assumes another condition, famously satisfied in the singlet state usually considered: namely, strict anticorrelation. The result depends on each of the quantities A_i and B_j having just two possible results, x and y; however, this allows more than the two quantities per wing usually considered. Thus, if for every A_i and x there is a B_j and y such that the predicted observable probabilities obtained by averaging equation 3.3 over ρ are 0, that is,

$$\int \mathrm{pr}_{\lambda A_i}(x) \cdot \mathrm{pr}_{\lambda B_j}(y)\, d\rho = 0, \tag{3.9}$$

we can conclude that, for almost all λ, one of the integrand's factors is 0; hence, by setting aside measure-zero sets, the model is deterministic. (The first statement of this seems to be theorem 2 of reference 10.) In short, my reply is that it remains valuable to study factorizable stochastic models or, rather, to study how they are motivated by our experience, outside quantum theory, of correlations being eliminated by a sufficiently rich specification of the "common cause": because doing so pinpoints the way in which this experiment's correlations are strange. (Again, for more details, see section 8 of reference 30, as well as the next section.)

Each of these two criticisms prompts a remark in the previous section's context of algebraic quantum theory. The first is mathematical: namely, that the equivalence of a deterministic local model with a total probability function is a trivially simple case of the famous theorem of Gelfand about the representation of commutative C*-algebras as a C*-algebra of functions on a locally compact Hausdorff space; cf. section 4.4 of reference 37. [To prove the existence of such a model, given a total probability function p for N quantities, each with a finite number of values, we just define Λ to be the set of all N-tuples, $\langle x_1, \ldots, x_N\rangle$, of possible values of the quantities; we define the random variables representing quantities to be projection functions; and, for each $\lambda = \langle x_1, \ldots, x_N\rangle$, we define $\rho(\lambda) := p(\langle x_1, \ldots, x_N\rangle)$. Similarly, Gelfand's theorem builds the representation space from simultaneous spectral values of elements of the abstract C*-algebra.]

The second remark is that Redhead's strict correlation result, that is, proposition 7 of the previous section, secures a reduction to determinism (for the vacuum state Ω) once we assume outcome as well as parameter independence, just as in the second criticism above. Thus, proposition 7 tells us that, for any two spacelike regions O_1 and O_2, any $\epsilon > 0$, and any projector $P_1 \in A(O_1)$, there is a projector $P_2 \in A(O_2)$ such that $\Omega(P_1P_2) > (1-\epsilon)\Omega(P_1)$. Assuming factorizability, this yields $\Omega(P_1)\Omega(P_2) > (1-\epsilon)\Omega(P_1)$. Hence, because ϵ is arbitrary, $\Omega(P_2) = 1$.

STOCHASTIC EINSTEIN NONLOCALITY IN AQFT

Having admitted that outcome independence is clearly false in quantum theory and, in particular, in AQFT, I will try in this section to make it look plausible; that is, I will develop the point made in the previous section: that outcome dependence is strange because we naturally expect correlations between spacelike events to be eliminated, conditional on sufficiently rich information about the prior state. To do so, I will adapt an intuitively plausible locality condition for Minkowski space-time, namely, Hellman's "stochastic Einstein locality" (SEL),[11] which gives some precision to the idea of "sufficiently rich information about the prior state" in terms of the physical state of a space-time region lying in the causal past of the events concerned. (Of course, Hellman is not the only author in the literature to appeal to such a state of a region: cf. also, preeminently, the exchange on local Beables between Bell and Shimony *et al.*[38])

Thus, I will exhibit how outcome dependence involves violation of SEL, a violation that is strong (and thus strange) because the space-time region, on which the "prior state" is defined, is *very* extensive. I will confine myself to making this point for AQFT. (Sections 5–7 of an earlier work of mine[39] make the point for elementary quantum theory. There, I also discuss how Hellman proposed SEL as expressing the prohibition of spacelike causation for any stochastic theory and how he, being keen to avoid the verdict that the Bell experiment involved spacelike causation, hedged his SEL with provisos to prevent its violation. For the reasons given there, I drop the provisos.)

The idea of SEL, roughly speaking, is that the probability at a time t of an event E occurring future to t is determined by the part of history up to t lying within the causal past of E. [Here, the causal past of E, written C^-(E), is the set of space-time points from which a signal, at most as fast as light, can reach E.] To make this precise, we need to be precise about the ideas of probability, event, history, and determination. How we do so will depend, in general, on both our general philosophical views and the specific context of discussion. Thus, Hellman's general views prompted him to make SEL precise in terms of formalized physical theories, with vocabulary for physical probability and isomorphisms of parts of their models. At this level of generality, I myself am less linguistic and more metaphysical. I instead talk of possible worlds, that is, total possible courses of history, where history consists of events (contingent, localized matters of particular fact) that have chances (time-dependent objective single-case probabilities). I also express determination in terms of worlds matching in their history on a region. This metaphysical framework is inspired by Lewis,[40] but neither here nor in reference 39 do I need the details of his views. Furthermore, this difference between Hellman and myself makes no difference in application to quantum theory (e.g., in our dispute about whether SEL needs provisos). Besides, when I discuss AQFT below, the context will be specific enough that I can talk of models of the theory instead of worlds, projectors instead of events, and states (expectation functionals) instead of chance functions.

Before discussing AQFT, though, I should exhibit two general formulations of SEL, for three reasons. First, it is striking that there are two inequivalent formulations of the basic idea stated above [even though, as reference 39 shows, (i) they are both violated by the Bell experiment when we make the obvious identifications and (ii) they are logically equivalent under some simplifying assumptions]. Second, the existence of inequivalent formulations is not just an artifact of my metaphysical framework: my formulations correspond to two in Hellman, which he recognizes as inequivalent. Third and most important, these formulations will prompt two more specific formulations for AQFT and these latter are more than just inequivalent. The first provably holds in AQFT, but the second is endemically false because of the outcome dependence shown in the previous section.

First, there is the direct expression, in terms of worlds, chances, etc., of the basic idea stated above:

(SEL1): for any two worlds w and w' and any hypersurface t earlier than the region for event E, if w and w' match in their history in $C^-(\mathrm{E}) \cap C^-(t)$, then they match in their chance at t of E, that is, with pr_{tw} for chance at t in w:

$$pr_{tw}(E) = pr_{tw'}(E). \tag{4.1}$$

Second, suppose that we assume that, within a world, history up to t and within $C^-(E)$ prescribes a chance function, whose domain contains possible events F occurring outside $C^-(E)$, but earlier than t. This is a contentious assumption because F can be influenced by subluminal signals from spatial infinity that never register on this history. Thus, assuming that this history prescribes a chance of F amounts to assuming that there are no such signals or that the chance function averages over them. However, once we make this assumption, we can take SEL to require that, according to this chance function, E is stochastically independent of any such event F. Hence, writing H for this history and pr_{Hw} for this chance function, we get the following:

(SEL2): for any world w, for any hypersurface t earlier than the region for event E, and for any possible event F in the difference, $C^-(t) - C^-(E)$,

$$pr_{Hw}(E/F) = pr_{Hw}(E), \tag{4.2}$$

where H is the history of w within $C^-(E) \cap C^-(t)$.

These two formulations of SEL correspond to Hellman's inequivalent (4) and (5), respectively (p. 446 and 495–497 of reference 11); his (5) is based on a condition of Bell [(2) in his contribution to reference 38]. They are both violated in the Bell experiment as a result of outcome dependence; cf. sections 5–7 of reference 39.

To formulate SEL for AQFT, I follow the lead of Redei[12,13] and Muller (in joint work with myself).[14] We replace worlds by models M = $\langle M, A, \phi \rangle$, where M is the Minkowski space-time, equipped with its open-bounded regions O_i; A is the map, characteristic of the system of interest, associating to each O_i its algebra $A(O_i)$; and ϕ is the system's state (expectation functional). We replace the event E by a projector, say A, associated with some region O, say A $\in A(O)$. (As mentioned at the start of the second section, we could use effects instead of projectors, but for the sake of uniformity with other authors we forego this.) We replace E's chance as prescribed by history up to time t by A's expectation in ϕ. (We shall see in a moment how SEL captures "as prescribed by history up to t", despite the global nature of ϕ.) We think of these models as satisfying (making true) various sentences of AQFT, which is cast in a putative formal language. These sentences will in general be about certain regions; thus, we write, for example, M $|=$ R$(O_1, \ldots, O_k)$, where R stands for some open sentence of the language. Also, among these sentences will be ascriptions of expectation values, say $\phi(A) = x$.

Finally, we can take the region that SEL asserts to determine the chance of A to be bounded rather than the whole of $C^-(O) \cap C^-(t)$ and, in doing so, we can drop reference to the hypersurface t. This will make our formulations of SEL rather stronger; hence, whereas AQFT's satisfying the first formulation is more interesting than otherwise, its violating the second formulation is apparently less interesting. However, this is only appearance: indeed, it will be clear that the second formulation is violated even if the determining region is *all of M.*

Thus, we take the determining region to be a bounded region lying fully across the causal past, $C^-(O)$. To be precise, recall for any region O' that the future domain of dependence, $D^+(O')$, of O' is the set of all points through which every smooth past-inextendable nonspacelike curve intersects O'. We then define a region O' to be a *past slab* of region O if and only if $O' \subset C^-(O)$ and $O \subset D^+(O')$.

Now, we can write our first formulation of SEL, corresponding to SEL1, which I will call SELS. Here, S stands for "single" because the consequent is like equation 4.1 in considering only a single quantity A. It is, however, harmless to consider expectations of all elements of the algebra $\boldsymbol{A}(O)$ rather than just projectors. Thus, we get the following:

(SELS): for any two models $M_j = \langle M, \boldsymbol{A}_j, \phi_j \rangle$ $(j = 1, 2)$; for any region $O \subset M$; and for any past slab O' of O:

IF: the models match in O' in the sense that:

$$\text{for all } k \in N, \text{ for all } O_1, \ldots, O_k \subseteq O', \text{ for all R:}$$

$$\mathrm{M}_1 \models \mathrm{R}(O_1, \ldots, O_k) \text{ iff } \mathrm{M}_2 \models \mathrm{R}(O_1, \ldots, O_k),$$

THEN: they match in their expectations on O in the sense that:

for all $\mathrm{A} \in \boldsymbol{A}(O)$, $\phi_1(\mathrm{A}) = \phi_2(\mathrm{A})$; that is, for all complex numbers z:

$$\mathrm{M}_1 \models \phi_1(\mathrm{A}) = z \text{ iff } \mathrm{M}_2 \models \phi_2(\mathrm{A}) = z.$$

Here, S also stands for "satisfied". It is easy to prove, regardless of the details of the putative formal language, the following proposition:

PROPOSITION 8. AQFT satisfies SELS.

Remarks: The proposition follows directly from AQFT's Isotony and Diamond axioms; see section 5 of reference 14. The proof amends a previous one by Redei.[12]

However, suppose that we want a formulation of SEL that, like SEL2 above, considers expectations of the projector (or a harmless generality: observable), say A, conditional on the occurrence of a spacelike event (projector), say B. Thus, like SEL2, SEL says that this conditional expectation of A equals A's unconditional expectation. Of course, it should not say merely that, for all states ϕ, for all $\mathrm{A} \in \boldsymbol{A}(O_1)$, and for all projectors B (associated to any region, say O_2) spacelike to O_1, $\phi(\mathrm{A}/\mathrm{B}) = \phi(\mathrm{A})$ because that would be a direct transcription of "naive" outcome independence (cf. equation 3.7 or 3.8), making no allowance for correlations.

We need some analogue of SEL2's pr_{Hw} expressing the idea of probabilities conditional on sufficiently rich information about the prior state, probabilities for which we naturally expect such correlations to be eliminated. The most obvious analogue, namely, conditioning ϕ on many projectors associated with the region $C^-(O_1) \cap C^-(t)$, is clearly fraught with difficulties, both mathematical and interpretative (i.e., about value ascriptions in quantum theory): for example, are we to condition ϕ on one of every pair of a projector and its orthocomplement?

However, fortunately, we can avoid these difficulties by returning to the idea of matching between models: that is, we again take SEL to be a conditional, with an antecedent supposing that a pair of models match on a region that is sufficiently extensive to incorporate all pairs of causal influences, one on each of (the expectations of) A and B, which have a common origin. [Again, such a pair could set up a correlation between A and B, violating the equality of conditional and unconditional expectation, and could give an utterly straightforward ("joint effects of a common cause"), and so spurious, violation of SEL.] Thus, this matching on the region expresses the restriction to probabilities conditional on sufficiently rich information about the prior state, and the consequent of SEL is then an equality of conditional with unconditional expectations of A.

How should we choose this region? We cannot just choose a slab, say S, lying across the "apex" ("peak") of $C^-(O_1) \cap C^-(O_2)$. Admittedly, if we also require that this apex lie in the future domain of dependence of S, then (by the Diamond axiom) pairs of influences that originate in the apex will in effect register on S; that is, they will be determined by the state and algebra in S. However, there could still be a pair of influences, one on each of A and B, that have a common origin sufficiently far in the past [sufficiently far back in $C^-(O_1) \cap C^-(O_2)$] and that then diverge spatially sufficiently rapidly so as to not register on S (nor in its domain of dependence).

There are various possible choices for this region. (The choice will not affect my punch line: namely, that the second section's results show that this second formulation of SEL is endemically violated.) One is the intersection of the causal pasts of (the regions for) A and B: that is, all of $C^-(O_1) \cap C^-(O_2)$. With this choice, the two models will in general differ as regards the expectation of A. Because A can be influenced by the "infinite strip" region, $C^-(O_1) - C^-(O_2)$, and because this infinite strip lies outside the region of matching, the influences on A can be different in the two models. (Similarly, of course, for B, but we are not concerned with its unconditional expectation.) Hence, with this choice, SEL's consequent will say merely that, within each of the two models M_1 and M_2, A's conditional and unconditional expectations are equal: $\phi_1(A/B) = \phi_1(A)$ and $\phi_2(A/B) = \phi_2(A)$.

On the other hand, we can choose a bounded region that is extensive enough to register all the influences on A that have, or might have, a common origin with an influence on B. The obvious choice is a past slab, as defined above, of O_1: call it O'. Because $O_1 \subset D^+(O')$, all influences on A whatsoever (and so all those that have a common origin with an influence on B) register on O'. [We could require, although we do not need to, that this past slab has the apex of $C^-(O_1) \cap C^-(O_2)$ in its future domain of dependence; then, the past slab will register, among the "last-minute" influences on both A and B, the influence on B as well as the influence on A.] Another obvious choice is symmetrical between A and B: we choose a past slab, as defined above, of the union $O_1 \cup O_2$. [Again, we could require, but do not need to, that this past slab has the apex of $C^-(O_1) \cap C^-(O_2)$ in its future domain of dependence.]

With both these choices—a past slab of O_1 and a past slab of $O_1 \cup O_2$—the two models will agree as regards the expectation of A just because SELS holds; that is, we will have $\phi_1(A) = \phi_2(A)$.

Finally, there is a minor point about the projector B. Above, I just said it is associated to O_2. However, after our discussion of matching, we can see that this assertion is vague because all the above choices for a matching region allow that $A_1(O_2) \neq A_2(O_2)$. Thus, should we strengthen SEL's antecedent to assume a matching, or at least a matching of algebras, on O_2, that is, $A_1(O_2) = A_2(O_2)$? Alternatively, should we allow different algebras and just require that B lie in the intersection? Again, the choice will not affect my punch line: namely, that this second formulation of SEL is endemically violated. I will arbitrarily take the second choice: $B \in A_1(O_2) \cap A_2(O_2)$.

We can at last write our second formulation of SEL, corresponding to SEL2, which I will call SELD. Of course, D stands for "double" because we now consider B. As my choice for the matching region, I will take the first bounded choice above, namely, a past slab of A's region O_1. (Again, this choice is arbitrary, but perhaps natural because we are not concerned with B's unconditional expectation; thus, we can let it vary between matching models.) Therefore, we get the following:

(SELD): for any two models $M_j = \langle M, A_j, \phi_j \rangle$ $(j = 1, 2)$;
for any region $O_1 \subset M$; for any O_2 that is spacelike to O_1;
for any projector $B \in A_1(O_2) \cap A_2(O_2)$; and for any past slab O' of O_1:
IF: the models match in O' in the sense that:

$$\text{for all } k \in N, \text{ for all } O_1, \ldots, O_k \subseteq O', \text{ for all R:}$$

$$M_1 \models R(O_1, \ldots, O_k) \text{ iff } M_2 \models R(O_1, \ldots, O_k)$$

THEN: they match in all their expectations on O in the sense that:

$$\text{(1) for all } A \in A(O),\ \phi_1(A) = \phi_2(A) =: \Phi(A);$$
$$\text{(2) } \phi_j(A/B) = \Phi(A)\ (j = 1, 2).$$

Here, D also stands for "denied". As mentioned, SELS implies that, given the antecedent, the first conjunct of the consequent, (1), holds, thus defining $\Phi(A)$. However, the second section's results directly imply that, given the antecedent, (2) is endemically false. Consider in particular propositions 6 and 7, recalling that they hold not just for the vacuum, but for any vector state with bounded energy. These propositions combine matching aplenty—throughout space-time—and therefore satisfaction of (1) with failure of (2).

ACKNOWLEDGMENTS

I thank L. Landau, S. Summers, and D. Malament for their helpfulness and inspiration; G. Bacciagaluppi, G. Fleming, N. Landsman, and M. Redhead for discussions of the material in the second and third sections; and F. Muller and M. Redei for discussions and correspondence about the material in the fourth section. I also thank H. Stapp and A. Shimony for remarks at the conference, as well as the organizers for such generous and enjoyable conference arrangements.

REFERENCES

1. LANDAU, L. 1987. Phys. Lett. **120A:** 54–56.
2. LANDAU, L. 1987. Phys. Lett. **123A:** 115–118.
3. SUMMERS, S. & R. WERNER. 1985. Phys. Lett. **110A:** 257–259.

4. SUMMERS, S. & R. WERNER. 1987. J. Math. Phys. **28:** 2440–2447.
5. SUMMERS, S. & R. WERNER. 1987. J. Math. Phys. **28:** 2448–2456.
6. SUMMERS, S. & R. WERNER. 1987. Commun. Math. Phys. **110:** 247–259.
7. SUMMERS, S. & R. WERNER. 1988. Ann. Inst. Henri Poincaré **49:** 215–243.
8. SHIMONY, A. 1986. Events and processes in the quantum world. *In* Quantum Concepts in Space and Time. C. Isham & R. Penrose, Eds. Oxford University Press. London/New York. *Reprinted in*: SHIMONY, A. 1993. Search for a Naturalistic Worldview, Vol. II. Cambridge University Press. London/New York.
9. REDHEAD, M. 1995. More ado about nothing. Found. Phys. In press.
10. SUPPES, P. & M. ZANOTTI. 1976. On the determinism of hidden variable theories with strict correlation and conditional stochastic independence of observables. *In* Logic and Probability in Quantum Mechanics. P. Suppes, Ed. Reidel. Dordrecht.
11. HELLMAN, G. 1982. Synthese **53:** 461–504.
12. REDEI, M. 1991. Philos. Sci. **58:** 628–638.
13. REDEI, M. 1993. Philos. Sci. **60:** 608–618.
14. MULLER, F. & J. BUTTERFIELD. 1994. Philos. Sci. In press.
15. SUMMERS, S. 1990. Rev. Math. Phys. **2:** 201–247.
16. SCHLIEDER, S. 1968. Commun. Math. Phys. **7:** 305–331.
17. SHIMONY, A. 1990. An exposition of Bell's theorem. *In* Sixty-two Years of Uncertainty. A. Miller, Ed. Plenum. New York. *Reprinted in*: SHIMONY, A. 1993. Search for a Naturalistic Worldview, Vol. II. Cambridge University Press. London/New York.
18. BACCIAGALUPPI, G. 1993. Separation theorems and Bell inequalities in algebraic quantum mechanics. *In* Symposium on the Foundations of Modern Physics—1993. P. Busch, P. Lahti & P. Mittelstaedt, Eds. World Scientific. Singapore.
19. CIREL'SON, B. 1980. Lett. Math. Phys. **4:** 93–100.
20. BELL, J. 1964. Physics **1:** 95–100. *Reprinted in*: BELL, J. 1987. Speakable and Unspeakable in Quantum Mechanics. Cambridge University Press. London/New York.
21. ROOS, H. 1970. Commun. Math. Phys. **16:** 238–246.
22. HAAG, R. 1992. Local Quantum Physics. Springer-Verlag. New York/Berlin.
23. HORUZHY, S. 1990. Introduction to Algebraic Quantum Field Theory. Kluwer. Dordrecht.
24. SCHLIEDER, S. 1969. Commun. Math. Phys. **13:** 216–225.
25. ARAKI, H., K. HEPP & D. RUELLE. 1962. Helv. Phys. Acta **35:** 164–175.
26. FREDENHAGEN, K. 1985. Commun. Math. Phys. **97:** 461–463.
27. DRIESSLER, W. 1975. Commun. Math. Phys. **44:** 133–141.
28. REEH, H. & S. SCHLIEDER. 1961. Nuovo Cimento **22:** 1051–1068.
29. SUMMERS, S. & R. WERNER. 1987. Preprint. Unabridged version of reference 4.
30. BUTTERFIELD, J. 1992. Br. J. Philos. Sci. **43:** 41–83.
31. JARRETT, J. 1984. Nous **18:** 569–589.
32. REICHENBACH, H. 1971. The Direction of Time. University of California Press. Berkeley.
33. BUTTERFIELD, J. 1989. A space-time approach to the Bell inequality. *In* Philosophical Consequences of Quantum Theory. Notre Dame University Press. South Bend, Indiana.
34. FINE, A. 1982. Phys. Rev. Lett. **48:** 291–295.
35. FINE, A. 1982. J. Math. Phys. **23:** 1306–1310.
36. SHIMONY, A. 1984. Br. J. Philos. Sci. **35:** 25–45. *Reprinted in*: SHIMONY, A. 1993. Search for a Naturalistic Worldview, Vol. II. Cambridge University Press. London/New York.
37. KADISON, R. & J. RINGROSE. 1983/1986. Fundamentals of the Theory of Operator Algebras. Volumes 1 & 2. Academic Press. New York.
38. BELL, J., A. SHIMONY, M. HORNE & J. CLAUSER. 1985. An exchange on local Beables. Dialectica **39:** 86–109. Respective contributions reprinted in the collections given in references 17 and 20.
39. BUTTERFIELD, J. 1994. Outcome dependence and stochastic Einstein nonlocality. *In* Logic and Philosophy of Science in Uppsala. D. Prawitz & D. Westerståhl, Eds. Kluwer. Dordrecht.
40. LEWIS, D. 1986. Philosophical Papers, Vol. II. Oxford University Press. London/New York.

Geometry of Quantum States[a]

SAMUEL L. BRAUNSTEIN[b] AND CARLTON M. CAVES[c]

Center for Advanced Studies
Department of Physics and Astronomy
University of New Mexico
Albuquerque, New Mexico 87131-1156

INTRODUCTION

There have been two threads running through John A. Wheeler's distinguished career: geometry as central to physics; the puzzle of the quantum. What could be more appropriate than that, in this volume dedicated to Wheeler, we pursue these threads in our own small way by formulating a natural Riemannian geometry on the space of quantum states—a geometry built on a concept, statistical distinguishability, that can be traced to Wheeler, who encouraged Bill Wootters' investigation of it. Another Wheeler theme is also important: communicate ideas clearly, as in his elegant summary of the key ideas of general relativity—space-time tells matter how to move; matter tells space-time how to curve. We can do no better than to characterize statistical distinguishability as Wootters did—a distance between quantum states, a statistical distance, quantified by how well measurements distinguish the states. It is a pleasure and an honor for us to dedicate this report to John A. Wheeler—a consummate researcher and extraordinary teacher and always a gentleman.

GEOMETRY ON THE PROBABILITY SIMPLEX

In this section, we review Wootters' derivation[1] of the distinguishability metric on the space of probability distributions, the probability simplex. Let the n-tuple $\vec{p} \equiv (p^1, \ldots, p^n)$ denote a probability distribution for discrete alternatives $j = 1, \ldots, n$. After N samplings from this distribution, we have a frequency f^j for each alternative. These frequencies are distributed according to the multinomial distribution,

$$\text{prob}(f^1, \ldots, f^N) = \left[\frac{N!}{\prod_{j=1}^{n} (Nf^j)!}\right] \prod_{k=1}^{n} (p^k)^{Nf^k}, \tag{2.1}$$

[a] This work was supported in part by the Office of Naval Research (Grant No. N00014-93-1-0116).

[b] Current address: Department of Chemical Physics, Weizmann Institute of Science, 76100 Rehovot, Israel.

[c] To whom all correspondence should be addressed.

which, as N tends towards infinity, becomes proportional to a Gaussian

$$\exp\left[\left(-\frac{N}{2}\right)\sum_j \frac{(f^j - p^j)^2}{p^j}\right]. \tag{2.2}$$

Thus, if $\vec{p} + \Delta\vec{p}$ is a nearby distribution, it can be distinguished reliably from $\vec{p}$ so long as $N\Sigma_j(\Delta p^j)^2/p^j$ is large. We can summarize this conclusion by defining a Riemannian line element

$$ds^2 = \sum_j \frac{(dp^j)^2}{p^j} \tag{2.3}$$

based on asymptotic distinguishability. If we use "probability amplitudes" $\sqrt{p^j}$ as coordinates, the line element becomes that of a sphere of radius 2:

$$ds^2 = 4\sum_j (d\sqrt{p^j})^2. \tag{2.4}$$

Notice that the use of probability amplitudes removes the coordinate singularity on the boundary of the simplex, where one or more alternatives have zero probability. The statistical distance s along the shortest path (great circles on the sphere) between distributions $\vec{p}$ and $\vec{q}$,

$$s = 2\cos^{-1}\left(\sum_j \sqrt{p^j q^j}\right), \tag{2.5}$$

is twice the angle separating the distributions on the sphere.[1]

Just how natural is this distinguishability metric? Founded on distinguishability in Wootters' derivation, it is also the metric associated with the bilinear structure of statistical correlations, as we now show. To get to the metric, however, requires a brief digression into the differential geometry of the probability simplex. Introduce a set of basis vectors $\vec{e}_j$ ($j = 1, \ldots, n$) so that the n-tuple for a probability distribution can be written as a vector

$$\vec{p} = \sum_j p^j \vec{e}_j \tag{2.6}$$

in an n-dimensional real vector space. The probability simplex is a manifold with boundary: normalization,

$$\sum_j p^j = 1, \tag{2.7}$$

defines an $(n - 1)$-dimensional "plane" within the vector space and nonnegativity of the probabilities,

$$p^j \geq 0 \qquad \text{for} \qquad j = 1, \ldots, n, \tag{2.8}$$

gives the plane a boundary. A path $\vec{p}(\lambda)$ through the probability simplex has tangent vector

$$\frac{d\vec{p}(\lambda)}{d\lambda} = \sum_j \left(\frac{dp^j}{d\lambda}\right)\vec{e}_j, \tag{2.9}$$

which, due to normalization, satisfies $\Sigma_j dp^j/d\lambda = 0$. Thus, the tangent space to the probability simplex at $\vec{p}$ is an $(n-1)$-dimensional vector space consisting of vectors $\vec{v} = \Sigma_j v^j \vec{e}_j$ that satisfy

$$\sum_j v^j = 0. \tag{2.10}$$

Affixing $\vec{p}$ to the tangent space at $\vec{p}$ recovers the original n-dimensional vector space.

Consider now the dual vector space of linear mappings that map vectors to real numbers. An element of the dual space, denoted $\tilde{A}$, is called a 1-form and acts on a vector $\vec{m}$ according to

$$\tilde{A}(\vec{m}) \equiv \langle \tilde{A}, \vec{m} \rangle = \sum_j A_j m^j, \tag{2.11}$$

where $A_j \equiv \langle \tilde{A}, \vec{e}_j \rangle$. Choosing a dual basis of 1-forms $\tilde{\omega}^j$, satisfying

$$\langle \tilde{\omega}^j, \vec{e}_k \rangle = \delta^j_k, \tag{2.12}$$

one can write an arbitrary 1-form as

$$\tilde{A} = \sum_j A_j \tilde{\omega}^j. \tag{2.13}$$

One sees that a 1-form $\tilde{A}$ corresponds, in statistical parlance, to a random variable, with A_j being the value of the random variable for alternative j and

$$\langle \tilde{A}, \vec{p} \rangle = \sum_j A_j p^j \equiv \langle \tilde{A} \rangle \tag{2.14}$$

being the mean value of $\tilde{A}$ with respect to the distribution $\vec{p}$. Notice that the product of random variables defines a natural product of 1-forms,

$$\tilde{A}\tilde{B} = \sum_j A_j B_j \tilde{\omega}^j, \tag{2.15}$$

which is the component-by-component product in the basis $\tilde{\omega}^j$.

There is a special 1-form,

$$\tilde{\mathbb{1}} \equiv \sum_j \tilde{\omega}^j. \tag{2.16}$$

This 1-form defines the plane of the probability simplex because $\langle \tilde{\mathbb{1}}, \vec{p} \rangle = 1$ for any probability distribution $\vec{p}$, and any tangent vector $\vec{v}$ lies in the surfaces of $\tilde{\mathbb{1}}$ because $\langle \tilde{\mathbb{1}}, \vec{v} \rangle = 0$. Any random variable can be decomposed uniquely as

$$\tilde{A} = \langle \tilde{A}, \vec{p} \rangle \tilde{\mathbb{1}} + \Delta\tilde{A}, \tag{2.17}$$

where $\Delta\tilde{A}$ has zero mean with respect to $\vec{p}$. The cotangent space to the probability simplex at $\vec{p}$ is an $(n-1)$-dimensional vector space made up of the zero-mean random variables, $\Delta\tilde{A}$; affixing $\tilde{\mathbb{1}}$ to the cotangent space recovers the n-dimensional vector space of all 1-forms. Because the tangent and cotangent spaces can be extended by affixing $\vec{p}$ and $\tilde{\mathbb{1}}$, we are somewhat sloppy about distinguishing these spaces from the spaces of all vectors and all 1-forms.

The linear structure of vectors and forms suffices to build the statistical structure of probabilities and random variables. Where then is the bilinear structure required for a metric? A metric tensor $\mathbf{g}_{\vec{p}}$ at point $\vec{p}$ is a symmetric bilinear "machine" that maps pairs of vectors to the reals (like an inner product). Such a machine is equivalent to an (invertible) "lowering" operator $\mathscr{L}_{\vec{p}}$ that maps vectors linearly to 1-forms according to

$$\mathbf{g}_{\vec{p}}(\vec{m}, \vec{n}) \equiv \langle \mathscr{L}_{\vec{p}}(\vec{m}), \vec{n} \rangle = \langle \mathscr{L}_{\vec{p}}(\vec{n}), \vec{m} \rangle \equiv \mathbf{g}_{\vec{p}}(\vec{n}, \vec{m}). \tag{2.18}$$

The adjective "lowering" is used because $\mathscr{L}_{\vec{p}}$ maps vectors, which have upper (contravariant) components, to 1-forms, which have lower (covariant) components. The inverse linear operator $\mathscr{R}_{\vec{p}}$, called the "raising" operator, maps 1-forms to vectors; it extends the action of the metric to pairs of 1-forms,

$$\mathbf{g}_{\vec{p}}(\tilde{A}, \tilde{B}) \equiv \mathbf{g}_{\vec{p}}[\mathscr{R}_{\vec{p}}(\tilde{A}), \mathscr{R}_{\vec{p}}(\tilde{B})] = \langle \tilde{A}, \mathscr{R}_{\vec{p}}(\tilde{B}) \rangle. \tag{2.19}$$

To get a bilinear metric structure, one must consider pairs of random variables. It is natural to define the metric's action on a pair of random variables to be the statistical correlation between the two variables,

$$\mathbf{g}_{\vec{p}}(\tilde{A}, \tilde{B}) \equiv \langle \tilde{A}\tilde{B} \rangle = \langle \tilde{A}\tilde{B}, \vec{p} \rangle = \sum_j A_j B_j p^j, \tag{2.20}$$

which is equivalent to the raising and lowering operators,

$$\mathscr{R}_{\vec{p}}(\tilde{B}) = \sum_j p^j B_j \vec{e}_j \tag{2.21}$$

and

$$\mathscr{L}_{\vec{p}}(\vec{m}) = \sum_j \left(\frac{m^j}{p^j} \right) \tilde{\omega}^j. \tag{2.22}$$

Notice that $\mathscr{L}_{\vec{p}}(\vec{p}) = \tilde{\mathbb{1}}$ and that the tangent space is mapped to the cotangent space—that is, a tangent vector $\vec{v}$, satisfying equation 2.10, lowers to a zero-mean random variable.

The metric's action on vectors,

$$\mathbf{g}_{\vec{p}}(\vec{m}, \vec{n}) = \langle \mathscr{L}_{\vec{p}}(\vec{m}), \vec{n} \rangle = \sum_j \frac{m^j n^j}{p^j}, \tag{2.23}$$

leads to a line element

$$ds^2 = \mathbf{g}_{\vec{p}}(d\vec{p}, d\vec{p}) = \langle \mathscr{L}_{\vec{p}}(d\vec{p}), d\vec{p} \rangle = \sum_j \frac{(dp^j)^2}{p^j}, \tag{2.24}$$

which is identical to the line element (equation 2.3) obtained from Wootters' distinguishability argument. Campbell[2] has reviewed the history of the correlation metric (equation 2.23) in the statistics literature.

GEOMETRY OF DENSITY OPERATORS

The geometric structure formulated for the probability simplex can be transferred effortlessly to the space of quantum states (density operators). We begin with an arbitrary linear operator on an n-dimensional Hilbert space, written as a vector

$$\vec{O} = \sum_{j,k} O^{jk}\vec{e}_{jk}, \tag{3.1}$$

where the operator vectors

$$\vec{e}_{jk} = |j\rangle\langle k|, \qquad j, k = 1, \ldots, n, \tag{3.2}$$

formed from vectors $|j\rangle$ that make up an orthonormal basis for Hilbert space, are a basis for the n^2-dimensional complex vector space of linear operators, and $O^{jk} = \langle j|\vec{O}|k\rangle$ is a matrix element of the operator $\vec{O}$ in this basis.

The space of density operators

$$\vec{\rho} = \sum_{j,k} \rho^{jk}\vec{e}_{jk} \tag{3.3}$$

is a manifold with boundary: Hermiticity,

$$\vec{\rho} = \vec{\rho}^{\dagger} \Leftrightarrow \rho^{jk} = (\rho^{kj})^*, \tag{3.4}$$

reduces the operator vector space to n^2 real dimensions; normalization,

$$\mathrm{tr}(\vec{\rho}) = 1 \Leftrightarrow \sum_{j} \rho^{jj} = 1, \tag{3.5}$$

defines an $(n^2 - 1)$-dimensional "plane" within this real vector space; and nonnegativity,

$$\vec{\rho} \geq 0 \Leftrightarrow \vec{\rho} = \vec{\rho}^{\dagger} \text{ has nonnegative eigenvalues}, \tag{3.6}$$

gives the plane a complicated boundary. By considering paths through the space of density operators, one sees that the tangent space at $\vec{\rho}$ is an $(n^2 - 1)$-dimensional real vector space consisting of vectors $\vec{v} = \Sigma_{j,k} v^{jk}\vec{e}_{jk}$ that satisfy

$$\mathrm{tr}(\vec{v}) = 0 \Leftrightarrow \sum_{j} v^{jj} = 0. \tag{3.7}$$

Affixing $\vec{\rho}$ to the tangent space at $\vec{\rho}$ recovers the entire n^2-dimensional vector space of linear operators.

Consider now the dual space of 1-forms. An arbitrary 1-form $\tilde{Q}$ acts on a vector $\vec{O}$ according to

$$\tilde{Q}(\vec{O}) \equiv \langle \tilde{Q}, \vec{O}\rangle = \sum_{j,k} Q_{kj}O^{jk} = \mathrm{tr}(\tilde{Q}\vec{O}), \tag{3.8}$$

where $Q_{kj} \equiv \langle \tilde{Q}, \vec{e}_{jk} \rangle$. Thus,

$$\tilde{Q} = \sum_{j,k} Q_{jk} \tilde{\omega}^{kj} \tag{3.9}$$

is an arbitrary linear operator expanded in terms of the dual basis

$$\tilde{\omega}^{kj} = |\, j \rangle \langle k \,|, \tag{3.10}$$

which satisfies

$$\langle \tilde{\omega}^{jk}, \vec{e}_{lm} \rangle = \mathrm{tr}(\tilde{\omega}^{jk} \vec{e}_{lm}) = \delta^j_l \delta^k_m. \tag{3.11}$$

The dual space is a copy of the original vector space of linear operators. We find it useful, however, to distinguish the two spaces because a Hermitian 1-form $\tilde{A} = \tilde{A}^\dagger$ is an ordinary quantum observable, with

$$\langle \tilde{A}, \vec{\rho} \rangle = \sum_{j,\,k} A^*_{jk} \rho^{jk} = \mathrm{tr}(\tilde{A}\vec{\rho}) \equiv \langle \tilde{A} \rangle \tag{3.12}$$

being the expectation value of $\tilde{A}$ with respect to the density operator $\vec{\rho}$. The matrix product of two operators defines a natural product of 1-forms,

$$\tilde{Q}\tilde{P} = \sum_{j,k,\,l} Q_{jl} P_{lk} \tilde{\omega}^{kj}. \tag{3.13}$$

The unit operator

$$\tilde{\mathbb{1}} \equiv \sum_j \tilde{\omega}^{jj} \tag{3.14}$$

is a special 1-form: it defines the density-operator plane because $\langle \tilde{\mathbb{1}}, \vec{\rho} \rangle = 1$ for any density operator $\vec{\rho}$, and any tangent vector $\vec{v}$ lies in the surfaces of $\tilde{\mathbb{1}}$ because $\langle \tilde{\mathbb{1}}, \vec{v} \rangle = 0$. Any observable can be decomposed uniquely as

$$\tilde{A} = \langle \tilde{A}, \vec{\rho} \rangle \tilde{\mathbb{1}} + \Delta\tilde{A}, \tag{3.15}$$

where $\Delta\tilde{A}$ has zero expectation value with respect to $\vec{\rho}$. The cotangent space at $\vec{\rho}$ is an $(n^2 - 1)$-dimensional real vector space made up of the zero-mean observables, $\Delta\tilde{A}$; affixing $\tilde{\mathbb{1}}$ to the cotangent space recovers the n^2-dimensional vector space of all 1-forms.

Balian, Alhassid, and Reinhardt[3] have stressed how quantum expectation values arise from the linear structure of observables as 1-forms acting on density operators as vectors. They go on to define a metric on density operators by taking the second variation of the von Neumann entropy. In contrast, we continue the development begun in the previous section by formulating a metric in terms of the statistical correlations of quantum observables.

We define the metric's action on a pair of observables to be the statistical correlation of the two observables,

$$\begin{aligned} \mathbf{g}_{\vec{\rho}}(\tilde{A}, \tilde{B}) &\equiv \langle (\tfrac{1}{2})(\tilde{A}\tilde{B} + \tilde{B}\tilde{A}) \rangle = \mathrm{tr}[(\tfrac{1}{2})(\tilde{A}\tilde{B} + \tilde{B}\tilde{A})\vec{\rho}] \\ &= \mathrm{tr}[\tilde{A}(\tfrac{1}{2})(\vec{\rho}\tilde{B} + \tilde{B}\vec{\rho})] = \langle \tilde{A}, \mathscr{R}_{\vec{\rho}}(\tilde{B}) \rangle, \end{aligned} \tag{3.16}$$

which is equivalent to a raising operator,

$$\mathscr{R}_{\vec{\rho}}(\tilde{Q}) = (½)(\vec{\rho}\tilde{Q} + \tilde{Q}\vec{\rho}). \tag{3.17}$$

Because the raising operator is symmetric multiplication by $\vec{\rho}$, the lowering operator $\mathscr{L}_{\vec{\rho}}$ is a sort of "symmetric division" by $\vec{\rho}$. In the orthonormal basis that diagonalizes $\vec{\rho} = \Sigma_j p^j |j\rangle\langle j|$, the actions of the raising and lowering operators become

$$\mathscr{R}_{\vec{\rho}}(\tilde{Q}) = \sum_{j,k}\left[\frac{p^j + p^k}{2}\right] Q_{jk}\vec{e}_{jk}, \tag{3.18}$$

$$\mathscr{L}_{\vec{\rho}}(\vec{O}) = \sum_{j,k}\left[\frac{2}{p^j + p^k}\right] O^{jk}\tilde{\omega}^{kj}. \tag{3.19}$$

Notice that $\mathscr{L}_{\vec{\rho}}(\vec{\rho}) = \tilde{\mathbb{1}}$ and that a tangent vector $\vec{v}$, satisfying equation 3.7, lowers to a zero-mean observable, that is, to a 1-form in the cotangent space.

The metric's action on vectors,

$$\mathbf{g}_{\vec{\rho}}(\vec{O}, \vec{N}) = \langle \mathscr{L}_{\vec{\rho}}(\vec{O}), \vec{N}\rangle = \mathrm{tr}[\vec{N}\mathscr{L}_{\vec{\rho}}(\vec{O})] = \sum_{j,k}\left[\frac{2}{(p^j + p^k)}\right] O^{kj}N^{jk}, \tag{3.20}$$

leads to a line element

$$ds^2 = \mathbf{g}_{\vec{\rho}}(d\vec{\rho}, d\vec{\rho}) = \langle \mathscr{L}_{\vec{\rho}}(d\vec{\rho}), d\vec{\rho}\rangle = \mathrm{tr}[d\vec{\rho}\mathscr{L}_{\vec{\rho}}(d\vec{\rho})] = \sum_{j,k}\left[\frac{2}{(p^j + p^k)}\right] |d\rho^{jk}|^2. \tag{3.21}$$

The final forms in equations 3.20 and 3.21 are written in the orthonormal basis that diagonalizes $\vec{\rho}$. In this basis, any tangent vector can be written as

$$d\vec{\rho} = \sum_j dp^j |j\rangle\langle j| - i[\widetilde{dh}, \vec{\rho}] = \sum_j dp^j |j\rangle\langle j| + i(p^j - p^k)dh_{jk}|j\rangle\langle k|, \tag{3.22}$$

where $\widetilde{dh}$ is an infinitesimal Hermitian operator. Plugging this form of the tangent vector into the line element (equation 3.21) gives

$$ds^2 = \sum_j \left[\frac{(dp^j)^2}{p^j}\right] + 2\sum_{j\neq k}\left[\frac{(p^j - p^k)^2}{(p^j + p^k)}\right] |dh_{jk}|^2. \tag{3.23}$$

This form of the metric is particularly useful for discussing what happens on the boundary, where one or more of the eigenvalues p^j vanish. The second term in equation 3.23 has no singularity or other misbehavior on the boundary and the first term has only the coordinate singularity of the classical distinguishability metric (equation 2.3), which can be removed by using coordinates $\sqrt{p^j}$.

The quantum line element (equation 3.21) is precisely the distinguishability metric for density operators obtained recently by us in reference 4 by optimizing over all generalized quantum measurements that can be used to distinguish neighboring quantum states $\vec{\rho}$ and $\vec{\rho} + d\vec{\rho}$. This metric has also arisen in another context: it is the infinitesimal form of a distance function introduced by Bures,[5] which Uhlmann[6] interpreted as a generalization of transition probabilities to mixed states and which has been investigated further by Hübner[7,8] and by Jozsa.[9] We return briefly to the theme of distinguishability in the fifth section, but first illustrate the density-operator geometry in two Hilbert-space dimensions.

GEOMETRY OF THE INTERIOR OF THE BLOCH SPHERE

As a simple example, we study the geometry of density operators on a two-dimensional Hilbert space. A convenient basis for the four-dimensional space of linear operators (2 × 2 matrices) is provided by the unit matrix and the three Pauli matrices, which we write as either vectors or 1-forms as appropriate. An arbitrary density operator can be written as

$$\vec{\rho} = (1/2)(\vec{\mathbb{1}} + \boldsymbol{x} \cdot \vec{\boldsymbol{\sigma}}) = (1/2)(\vec{\mathbb{1}} + r\boldsymbol{n} \cdot \vec{\boldsymbol{\sigma}}). \tag{4.1}$$

Here,

$$\boldsymbol{x} = \sum_{i=1}^{3} x^i \boldsymbol{e}_i = r\boldsymbol{n} \tag{4.2}$$

is the three-dimensional Bloch vector ($\boldsymbol{n}$ is the corresponding unit vector, with $r = |\boldsymbol{x}|$ being the magnitude of $\boldsymbol{x}$) and

$$\vec{\boldsymbol{\sigma}} = \sum_{i=1}^{3} \vec{\sigma}^i \boldsymbol{e}_i \tag{4.3}$$

is the 3-vector of Pauli matrices. In equations 4.1–4.3, vectors in the space of linear operators are distinguished by an arrow and 3-vectors in Bloch space are written in boldface, with the 3-vector of Pauli matrices being both. The fact that states satisfy $\mathrm{tr}(\vec{\rho}^2) \leq 1$ implies that $r \leq 1$: pure states, with $r = 1$, lie on the surface of a unit 2-sphere, whereas mixed states, with $r < 1$, lie in the interior.

To find the line element on the interior of the Bloch sphere, we need to calculate the action of the lowering operator. The raising operator acts according to the following rules:

$$\mathscr{R}_{\vec{\rho}}(\vec{\mathbb{1}}) = \vec{\rho} = (1/2)(\vec{\mathbb{1}} + r\boldsymbol{n} \cdot \vec{\boldsymbol{\sigma}}), \tag{4.4}$$

$$\mathscr{R}_{\vec{\rho}}(\boldsymbol{n} \cdot \vec{\boldsymbol{\sigma}}) = (1/2)(r\vec{\mathbb{1}} + \boldsymbol{n} \cdot \vec{\boldsymbol{\sigma}}), \tag{4.5}$$

$$\mathscr{R}_{\vec{\rho}}(\boldsymbol{n}_\perp \cdot \vec{\boldsymbol{\sigma}}) = (1/2)\boldsymbol{n}_\perp \cdot \vec{\boldsymbol{\sigma}}, \tag{4.6}$$

where $\boldsymbol{n}_\perp$ is any unit vector orthogonal to $\boldsymbol{n}$. Inverting these rules gives the action of the lowering operator:

$$\mathscr{L}_{\vec{\rho}}(\vec{\mathbb{1}}) = \left[\frac{2}{(1 - r^2)}\right](\tilde{\mathbb{1}} - r\boldsymbol{n} \cdot \tilde{\boldsymbol{\sigma}}), \tag{4.7}$$

$$\mathscr{L}_{\vec{\rho}}(\boldsymbol{n} \cdot \vec{\boldsymbol{\sigma}}) = \left[\frac{2}{(1 - r^2)}\right](-r\tilde{\mathbb{1}} + \boldsymbol{n} \cdot \tilde{\boldsymbol{\sigma}}), \tag{4.8}$$

$$\mathscr{L}_{\vec{\rho}}(\boldsymbol{n}_\perp \cdot \vec{\boldsymbol{\sigma}}) = 2\boldsymbol{n}_\perp \cdot \tilde{\boldsymbol{\sigma}}. \tag{4.9}$$

Applying the lowering operator to the tangent vector

$$d\vec{\rho} = (\tfrac{1}{2})d\boldsymbol{x} \cdot \vec{\boldsymbol{\sigma}} = (\tfrac{1}{2})(dr\,\boldsymbol{n} + rd\boldsymbol{n}) \cdot \vec{\boldsymbol{\sigma}}, \qquad \textbf{(4.10)}$$

where $\boldsymbol{n} \cdot d\boldsymbol{n} = 0$, yields

$$\begin{aligned}\mathscr{L}_{\vec{\rho}}(d\vec{\rho}) &= \left[\frac{1}{(1 - r^2)}\right]\{-rdr\,\vec{\mathbb{1}} + [dr\,\boldsymbol{n} + r(1 - r^2)d\boldsymbol{n}] \cdot \vec{\boldsymbol{\sigma}}\} \\ &= \left[\frac{1}{(1 - r^2)}\right]\{-(\boldsymbol{x} \cdot d\boldsymbol{x})\,\vec{\mathbb{1}} + [(1 - r^2)d\boldsymbol{x} + (\boldsymbol{x} \cdot d\boldsymbol{x})\boldsymbol{x}] \cdot \vec{\boldsymbol{\sigma}}\}. \qquad \textbf{(4.11)}\end{aligned}$$

The resulting line element on the interior of the Bloch sphere is given by

$$ds^2 = \mathrm{tr}[d\vec{\rho}\mathscr{L}_{\vec{\rho}}(d\vec{\rho})] = \left[\frac{dr^2}{(1 - r^2)}\right] + r^2d\boldsymbol{n} \cdot d\boldsymbol{n} = \left[\frac{(\boldsymbol{x} \cdot d\boldsymbol{x})^2}{(1 - r^2)}\right] + d\boldsymbol{x} \cdot d\boldsymbol{x}. \qquad \textbf{(4.12)}$$

The term $d\boldsymbol{n} \cdot d\boldsymbol{n}$ is the standard line element $d\Omega^2 = d\theta^2 + \sin^2\theta d\phi^2$ on a unit 2-sphere, so a surface of constant r within the Bloch sphere has the geometry of a 2-sphere of area $4\pi r^2$. The term $dr^2/(1 - r^2)$ shows that the interior of the Bloch sphere is curved because, as one moves away from the center, the circumference of the 2-spheres grows more slowly than distance from the center.

Indeed, the form of the line element (equation 4.12) suggests the introduction of a fourth coordinate:

$$x^0 \equiv \sqrt{1 - r^2} \geq 0 \Rightarrow (dx^0)^2 = \frac{r^2dr^2}{(1 - r^2)} = \frac{(\boldsymbol{x} \cdot d\boldsymbol{x})^2}{(1 - r^2)}. \qquad \textbf{(4.13)}$$

In terms of the four coordinates x^μ ($\mu = 0, 1, 2, 3$), the interior of the Bloch sphere is a three-dimensional surface defined by

$$\sum_{\mu=0}^{3} (x^\mu)^2 = 1, \qquad \textbf{(4.14)}$$

whose geometry is induced by the four-dimensional flat Euclidean line element,

$$ds^2 = \sum_{\mu=0}^{3} (dx^\mu)^2, \qquad \textbf{(4.15)}$$

a result obtained by Hübner.[7] Thus, the interior of the Bloch sphere is the northern hemisphere of a unit 3-sphere. The north pole of the 3-sphere is the unit density operator $(\tfrac{1}{2})\vec{\mathbb{1}}$ and the equator is the unit 2-sphere of pure states. Another useful coordinate system uses the hyperspherical angle χ defined by

$$r = \sin\chi, \qquad \textbf{(4.16)}$$

in terms of which the line element becomes

$$ds^2 = d\chi^2 + \sin^2\chi d\boldsymbol{n} \cdot d\boldsymbol{n}. \qquad \textbf{(4.17)}$$

Geodesics of this metric (equation 4.15) are great circles on the unit 3-sphere. Given any two orthogonal unit 4-vectors u^μ and v^μ, a great circle, parameterized by arc-length, can be written as

$$x^\mu(s) = u^\mu \cos s + v^\mu \sin s. \qquad \textbf{(4.18)}$$

Writing $u^i = n^i \sin\chi$ and $v^i = m^i \cos\xi$ (where $\boldsymbol{n}$ and $\boldsymbol{m}$ are unit 3-vectors), $u^0 = \cos\chi$, and $v^0 = \sin\xi$, one can write the geodesic in terms of 3-vectors on Bloch space:

$$\boldsymbol{x}(s) = \boldsymbol{n} \sin\chi \cos s + \boldsymbol{m} \cos\xi \sin s, \qquad \textbf{(4.19)}$$

where $\tan\xi = -(\boldsymbol{n}\cdot\boldsymbol{m})\tan\chi$.

STATISTICAL DISTINGUISHABILITY AND UNCERTAINTY RELATIONS

At the end of the third section, we promised to return to the theme of statistical distinguishability and to apply it to density operators. To formulate precisely the notion of statistical distinguishability requires that we change the language slightly and thereby consider an apparently different problem. In addition to precision, this new language leads to a class of uncertainty relations that limit our ability to determine parameters from the results of quantum measurements. A fuller account can be found in references 4 and 10. Suppose that we wish to distinguish two neighboring density operators. The new language imagines that these two density operators lie on a smooth path through density-operator space, parameterized by X. Instead of concentrating on distinguishing two states $\vec{\rho}(X)$ and $\vec{\rho}(X+dX) = \vec{\rho}(X) + (d\vec{\rho}/dX)dX$, we ask the following: for state $\vec{\rho}(X)$ or, in general, for N copies of it, how well can we determine the value of its parameter X, that is, its location on the path?

After measurements on the N copies, we must use some estimator function X_{est} to convert the measurement results into a number representing the estimated value of the parameter. It is natural to measure statistical distance along the path in units of the statistical noise in the estimator. If one lets δX_{est} denote the size of the statistical noise, the natural measure of statistical distance is

$$ds \equiv \frac{dX}{\min(\sqrt{N}\delta X_{\text{est}})}. \qquad \textbf{(5.1)}$$

The $\sqrt{N}$ is included to account for the expected $\sqrt{N}$ improvement in statistical noise with increasing N, and the minimum, taken over all estimators and over all quantum measurements, ensures that statistical distance is measured in terms of the most discriminating procedure for determining X. A suitable definition of the statistical deviation of the estimator away from the parameter value X is[4,10]

$$\delta X_{\text{est}} \equiv \left\langle \left[\left(\frac{X_{\text{est}}}{|d\langle X_{\text{est}}\rangle_X/dX|} \right) - X \right]^2 \right\rangle_X^{1/2}, \qquad \textbf{(5.2)}$$

where the derivative is included to remove any local difference in the "units" of the estimator and the parameter.

The most general quantum measurement is described by a positive-operator-valued measure (POVM), $\tilde{E}(\xi)d\xi$;[11] the probability density to obtain an outcome ξ, given density operator $\vec{\rho}(X)$, is given by

$$p(\xi|X) = \mathrm{tr}[\tilde{E}(\xi)\vec{\rho}(X)]. \tag{5.3}$$

We can use the Cramér-Rao bound[4,12] of classical estimation theory to perform the minimization over all estimator functions, that is,

$$\frac{1}{\min\limits_{X_{\mathrm{est}}}[N(\delta X_{\mathrm{est}})^2]} = F(X) \equiv \int d\xi\, \frac{[\partial p(\xi|X)/\partial X]^2}{p(\xi|X)} \equiv \left(\frac{ds}{dX}\right)^2_{\mathrm{classical}}, \tag{5.4}$$

where $F(X)$, called the Fisher information, is a continuum version of the line element (equation 2.3) for classical distinguishability. Generally, the Fisher-information minimum in equation 5.4 can be achieved only asymptotically for large N. Further optimization over all quantum measurements can be accomplished by appropriate use of the Schwarz inequality, giving[4]

$$\frac{1}{\min\limits_{X_{\mathrm{est}},\{\tilde{E}(\xi)\}}[N(\delta X_{\mathrm{est}})^2]} = \max_{\{\tilde{E}(\xi)\}} F(X) = \mathrm{tr}\left[\left(\frac{d\vec{\rho}}{dX}\right)\mathscr{L}_{\vec{\rho}}\left(\frac{d\vec{\rho}}{dX}\right)\right] \equiv \left(\frac{ds}{dX}\right)^2_{\mathrm{quantum}}, \tag{5.5}$$

with optimal measurements described by the eigenvectors of $\mathscr{L}_{\vec{\rho}}(d\vec{\rho}/dX)$. By formulating statistical distinguishability in terms of the asymptotic ability of measurements to determine the location of a density operator on a one-parameter curve, we are able to show that the density-operator metric based on statistical distinguishability is identical to the metric (equation 3.21) based on statistical correlations; moreover, the optimal measurement for locating a density operator is obtained by lowering the tangent vector $d\vec{\rho}$ to be an observable.

This change of language gives us for free a set of uncertainty relations[4,10] that limit the precision determination of parameters. In particular, we have

$$(\delta X_{\mathrm{est}})^2\left(\frac{ds}{dX}\right)^2_{\mathrm{quantum}} \geq \frac{1}{N}. \tag{5.6}$$

For paths generated by a single-parameter unitary operator,

$$\vec{\rho}(X) = e^{-i\tilde{h}X}\vec{\rho}(0)e^{i\tilde{h}X}, \tag{5.7}$$

there is a simple bound to the quantum metric (cf. equation 3.23),

$$\left(\frac{ds}{dX}\right)^2_{\mathrm{quantum}} = \mathrm{tr}\left[\left(\frac{d\vec{\rho}}{dX}\right)\mathscr{L}_{\vec{\rho}}\left(\frac{d\vec{\rho}}{dX}\right)\right] \leq 4\langle(\Delta\tilde{h})^2\rangle. \tag{5.8}$$

This gives an uncertainty relation

$$(\delta X_{\mathrm{est}})^2\langle(\Delta\tilde{h})^2\rangle \geq \frac{1}{4N}, \tag{5.9}$$

which, although less strict than equation 5.6, has the form of a standard uncertainty relation.

CONCLUSIONS

As Anandan[13] has discussed, geometric concepts have found little application in quantum mechanics. What little application has been found has tended to deal with the geometry of pure states, where the Hilbert-space inner product provides a natural and unavoidable notion of distinguishability, which leads to a metric called the Fubini-Study metric. By generalizing the distinguishability metric to mixed states, where no natural inner product fixes the notion of distinguishability, we aim to tackle old problems with new tools. The real test is whether geometric concepts can be used to make global, rather than just local, statements about quantum distinguishability. The fact that the distinguishability metric on the probability simplex can be used to make global statements[14] is cause for optimism. The report by Fuchs and Caves in this volume provides a hint that global distinguishability problems in quantum mechanics are also susceptible to geometric attack.

REFERENCES

1. WOOTTERS, W. K. 1981. Phys. Rev. D **23:** 357. Wootters' distance is half the distance that we use here.
2. CAMPBELL, L. L. 1985. Inf. Sci. **35:** 199.
3. BALIAN, R., Y. ALHASSID & H. REINHARDT. 1986. Phys. Rep. **131:** 1.
4. BRAUNSTEIN, S. L. & C. M. CAVES. 1994. Phys. Rev. Lett. **72:** 3439.
5. BURES, D. J. C. 1969. Trans. Am. Math. Soc. **135:** 199.
6. UHLMANN, A. 1976. Rep. Math. Phys. **9:** 273.
7. HÜBNER, M. 1992. Phys. Lett. A **163:** 239.
8. HÜBNER, M. 1993. Phys. Lett. A **179:** 226.
9. JOZSA, R. 1994. Fidelity for mixed quantum states. J. Mod. Opt. (special issue on quantum communication). To be published.
10. BRAUNSTEIN, S. L., C. M. CAVES & G. J. MILBURN. 1994. Generalized uncertainty relations: theory, examples, and Lorentz invariance. Phys. Rev. A. Submitted.
11. KRAUS, K. 1983. States, Effects, and Operations: Fundamental Notions of Quantum Theory. Springer-Verlag. Berlin/New York.
12. CRAMÉR, H. 1946. Mathematical Methods of Statistics, p. 500. Princeton University Press. Princeton, New Jersey.
13. ANANDAN, J. 1991. Found. Phys. **21:** 1265.
14. DEMBO, A., T. M. COVER & J. A. THOMAS. 1991. IEEE Trans. Inf. Theory **37:** 1501.

Local Propensities and Quantum Theory

F. SELLERI

Dipartimento di Fisica
Università di Bari
Istituto Nazionale di Fisica Nucleare
Sezione di Bari
I-70126 Bari, Italy

INTRODUCTION

Three different approaches have been considered for the theoretical description of quantum phenomena:

(1) The first one, based on standard quantum theory, is accepted by the large majority of theoretical physicists, who are impressed by the great predictive power of the theory. It is, however, a fact that several researchers had the feeling that the presently available theory cannot be considered a reasonable representation of the atomic reality, in spite of its undeniable successes. The rather long list of dissenters includes Planck, Einstein, Ehrenfest, Schrödinger, de Broglie, and Dirac.[1] There are also some contemporary physicists who feel the same way and it is rather clear that this line of dissent is not going to disappear soon.

(2) The second approach that has been considered is a modified quantum theory in which nonfactorizable state vectors for two correlated systems spontaneously collapse to mixtures of factorizable state vectors when the distance between the two systems increases beyond a certain value. The idea was first studied by Furry[2] and by Schrödinger[3] and could have come to an end when Bohm and Aharonov[4] showed that it was in disagreement with the experiment carried out by Wu and Shaknov.[5] Actually, it was later rediscovered by several other investigators, including Jauch,[6] de Broglie,[7] and Piccioni,[8] who probably did not know about the experimental situation. Today, the idea of spontaneous collapse has been falsified in many other experiments, including those performed with atomic photon pairs.[9] This is because "reduced quantum theory", as this approach is sometimes called, predicts not only the validity of the "weak" Bell-type inequalities,[10] but of the "strong" ones as well (for the distinction between weak and strong inequalities, see the paper by Lepore and myself[11]). Therefore, reduced quantum theory is completely ruled out.

(3) The third approach is based on space-time causal models of the atomic reality. If a new starting point is to be found after the discovery of Bell's theorem, it cannot be a simple rephrasing of the old theory, but must contain radically new ideas. Looking for a reformulation of quantum theory is therefore a bet about future developments: if Bell's inequalities were to be found violated in a new generation of experiments using nearly perfect detectors,[12] there would be no point in seeking for a better theory because

the highly counterintuitive feature of action at a distance[13] would then have been found to be present in nature. An "unreasonable" nature can be well represented, of course, by an "unreasonable" theory. Local realistic models have been developed for the explaining of the violations of the strong inequalities,[14] for the prediction of new effects in experiments with three polarizers,[15] for the interpretation of the Bohm-Aharonov effect,[16] and for the understanding of two-photon interference experiments.[17] These models have the common feature of explaining the violations of strong inequalities, but of predicting the validity of the weak ones.

In the present report, it will be assumed that there is a concretely real propensity behind every probability that can be predicted correctly before it is measured: the previous "reality criterion" will be shown to be compatible with Popper's idea of propensities. A suitable formalism will be introduced by starting from standard probability calculus and its frequency interpretation. A set of similar systems described as an eigenstate in quantum theory will be shown to be necessarily nonhomogeneous, if propensities are actually at work in nature. Therefore, variable probability is the key to the quantum world. Local propensities lead to a new proof of Bell's theorem and, at the same time, explain the well-known experimental results concerning it. Naturally, for nearly ideal detectors, the variable probability approach remains incompatible with the quantum mechanical predictions.

NEED FOR PROPENSITIES

The alternative starting point that we are looking for should look so strong and natural that all physicists should find it easy to accept. We will show that probabilistic local realism can be formulated in such general terms that it is a very natural candidate for such a role. In fact, (i) if there is no reality of any type behind even a stable and predictable probability, then no material reality is left at all and (ii) even the quantum mechanical probabilities are local in the sense advocated in the present report.

A new description of the physical reality in probabilistic terms, based on the conception of propensity, was proposed by Popper.[18] The main task to be achieved in this way is an objectivistic and realistic reformulation of quantum theory and of its probability calculus without relying heavily on idealistic elements. The same program was discussed in a paper by J. B. Bastos Filho and myself,[19] and it was shown that any objectivistic and local formulation of probabilities leads to the validity of Bell's theorem and thus to the impossibility of a full agreement with existing quantum theory. In spite of this, it is interesting to work on the idea of locality because the experimental evidence concerning the quantum mechanical predictions violating Bell's inequality is far from conclusive.[11] We assume, therefore, that the reality behind every stable and predictable probability is always a local propensity (the meaning of "locality" is discussed in the next section).

The idea of propensities is best introduced by observing that the classical theory of probability is based upon the following definition: "The probability of an event is the number of favorable possibilities divided by the number of all the (equal) possibilities." Thus, the probability of the event "tails turning up" would be 1 divided by 2 because there are altogether two equal possibilities and only one is "favorable". Similarly, the probability of throwing an odd number smaller than five with a perfect die is 2 divided by 6 because there are six equal possibilities and only two of them (the sides marked "1" and "3") are favorable to the event "an odd number smaller than five".

However, what happens if the die is loaded? Then, according to the classical theory of Laplace, we can no longer say that its six sides represent equal possibilities and we cannot speak anymore of probabilities in the classical numerical sense. If we insert in a wooden die a small piece of lead near, say, the face bearing the number "3", then this number will turn up less frequently than it does in throws with a fair die and the number of the opposite face (which is "4") will turn up more frequently.

A good theory of probability must, of course, include weighted possibilities, where the statistical weight can be interpreted as a measure of the propensity or tendency of the figure "3" to turn up upon repetition of the throws, always in the example of the loaded die. The weight, or frequency, of the event "three turning up" can be measured by repeating many throws with the loaded die and counting the number of times the face "3" shows up. If we find this frequency to be 0.05, instead of $1/6 = 0.1667$, we can conclude that the die is not perfect and can take this frequency as a quantitative measure of the propensity that the die has to produce the event "three turning up".

There is an important difference between the classical situation and the one met in atomic physics. When dealing with photons, electrons, or atoms, one cannot perform more than one measurement on a single individual object because its quantum state is modified. A statistical experiment in the atomic domain is performed with a large set of objects of the same type prepared in the same quantum state. Obviously, this is the same as performing many experiments with one object only if the atomic systems of the same type (photons, electrons, . . .) are identical.

Quantum theory is usually assumed to describe identical systems, for example, in connection with the Bose-Einstein and the Fermi-Dirac statistics, but, realistically speaking, the perfect identity is difficult to conceive and the explanation of the quantum statistics could, for example, be along the lines proposed by Tersoff and Bayer,[20] that is, by introducing variable statistical weights of the allowed states. One can also observe that unstable "identical" neutrons live different individual lives before disintegrating and that such a concrete behavior is not quite compatible with their presumed identity.

It turns out in atomic physics that one deals with situations similar to a set of dice loaded differently from one another, wherein each of which only one throw can be made. From a single act of measurement, one cannot deduce a probability and the individual propensities are thus not measurable. One can, however, introduce a propensity of a set of atomic systems to give different results with different frequencies. The latter propensity is instead easily measurable and can be considered as some kind of physical average of (undetectable) individual propensities.

LOCAL PROPENSITIES AND PROBABILITIES

The words "probability" and "frequency" are essentially identifiable, in the sense that the probability of an event can be considered as nothing but its measured (or predicted) frequency in a given set of similar objects. A probabilistic reality criterion will now be given stating essentially that there is a real propensity behind every probability that can be predicted correctly before it is measured.[21] An example can be the case of the loaded die considered in the previous section: if one can predict that the side marked "3" will turn up in many throws with frequency 0.05, one can infer the existence of a certain objectively real propensity of the die to give the result "3" with this particular frequency.

Let S and T be two sets of N physical objects of the same type (photons, neutrons, kaons, . . .):

$$S = \{\alpha_1, \alpha_2, \ldots, \alpha_N\},$$
$$T = \{\beta_1, \beta_2, \ldots, \beta_N\}. \tag{1}$$

Let the objects be produced in pairs, α_1 with β_1, α_2 with β_2, . . . , α_N with β_N, and let different pairs be totally independent from one another.

Suppose that one measures a dichotomic physical quantity $A(a) = \pm 1$ on the α's of S and that one can predict the frequencies $f(a+)$ of $A(a) = +1$ and $f(a-)$ of $A(a) = -1$ in a subset S' of S identified by performing some operation on the β's of T, when S and T are physically separated from one another. It is natural to conclude that these predicted frequencies $f(a\pm)$ reflect some reality of S', in the sense that they are necessary consequences of some concrete physical property ("propensity") of S' and, possibly, of the physical apparatus used. We can therefore assume the following probabilistic reality criterion (PRC):

> Given a set S of N α particles, if it is possible [1] to predict the existence of a subset S' of S such that future measurements of $A(a)$ on the α's of S' will give the results $+1$ and -1 with the probabilities (frequencies) $f(a+)$ and $f(a-)$, respectively, [2] to predict the population N' of S' $(0 < N' \leq N)$, and [3] to make the previous predictions without disturbing in any way the α objects of S and S', then a propensity λ_a belonging to S' is assumed to fix the probabilities:

$$f(a+) = f(a+, \lambda_a), \qquad f(a-) = f(a-, \lambda_a). \tag{2}$$

Actually, Popper also considered competing propensities for different outcomes of a process (e.g., of a measurement).[22] In fact, in the present connection, he would probably have introduced two different propensities $\lambda_{a\pm}$ for the two outcomes $A(a) = \pm 1$, but our symbol λ_a actually includes this because it can be written as a set of two propensities:

$$\lambda_a = \{\lambda_{a+}, \lambda_{a-}\}. \tag{3}$$

We can thus say that the frequency $f(a+)$ of the outcome $A(a) = +1$ is the synthetic result of all the propensities actively present in S' and can write it as in equation 2; similarly for $f(a-)$. The above PRC is not an attempt to "define" a physical reality, but only a criterion useful for recognizing the existence of an objective physical property (propensity) once a realistic attitude has been adopted. The previous PRC provides a natural generalization of the famous deterministic reality criterion of Einstein, Podolsky, and Rosen (EPR).[23] In this connection, the PRC can be applied to the subsets $S(b\pm)$ defined below that will thus play the role of S'.

In a typical EPR experiment, there are two observers: O_α, who measures the dichotomic observable $A(a)$ on the α particles of S, and O_β, who measures a second dichotomic observable $B(b)$ on the β particles of T. Assuming that O_β performs his/her measurements before O_α (in the laboratory frame), his/her results split T into two subsets, $T(b+)$ [all cases $B(b) = +1$] and $T(b-)$ [all cases $B(b) = -1$]. Corresponding to this splitting of T, one can also consider the following:

(1) a splitting of S into $S(b+)$ and $S(b-)$ composed of the α particles correlated individually with the β particles contained in $T(b+)$ and $T(b-)$, respectively;
(2) a splitting of the set E of (α, β) pairs,

$$E = \{(\alpha_1, \beta_1), (\alpha_2, \beta_2), \ldots, (\alpha_N, \beta_N)\}, \tag{4}$$

into two parts, $E(b+)$ and $E(b-)$, such that $S(b+)$ and $T(b+)$ compose physically $E(b+)$, and $S(b-)$ and $T(b-)$ compose physically $E(b-)$.

Let $N(b+)$ be the population of $E(b+)$, $S(b+)$, and $T(b+)$ and let $N(b-)$ be that of $E(b-)$, $S(b-)$, and $T(b-)$, with

$$N(b+) + N(b-) = N. \tag{5}$$

These populations are obviously related to the probabilities $g(b\pm)$ of measuring $B(b)$ in T and finding ± 1, respectively. In fact,

$$g(b\pm) = \left(\frac{1}{N}\right)N(b\pm). \tag{6}$$

Given enough previous experimental activity with repetitions of the sets E, S, and T, it is clear that O_β himself/herself will be able to *predict* that (later) O_α will find $A(a) = \pm 1$ with the frequencies $f(a\pm|b+)$ in $S(b+)$, and $A(a) = \pm 1$ with the frequencies $f(a\pm|b-)$ in $S(b-)$. However, $f(a\pm|b+)$ in $S(b+)$ and $f(a\pm|b-)$ in $S(b-)$ are in general different from the probabilities $f(a\pm)$ of finding $A(a) = \pm 1$ in the whole set S. Therefore, if O_α does find experimentally the frequencies earlier predicted by O_β, he/she can conclude that, in the physical reality of $S(b\pm)$, there is something (namely, a propensity) that generates these frequencies. By applying equation 2 to $S(b\pm)$, one has

$$f(a\pm|b+) = f[a\pm, \lambda_a(b+)],$$
$$f(a\pm|b-) = f[a\pm, \lambda_a(b-)]. \tag{7}$$

The b dependence of $\lambda_a(b\pm)$ is established at the moment of the birth of the pairs. The notation $\lambda_a(b\pm)$ only means that the sets $S(b\pm)$ and their propensities are defined by the measurement of $B(b)$ and is not meant to imply the existence of an action at a distance from T to S capable of modifying λ_a. In fact, to the PRC, one can naturally add the following locality assumption:

> Measurement performed on the β objects of T do not in any way generate or modify the propensities of the α objects of S, and vice versa.

The propensities $\lambda_a(b\pm)$ are therefore assumed to be real properties of existing subjects $S(b\pm)$ of S even if no measurement of $B(b)$ is performed. Not to admit this would imply that the propensities for the α particles are created via some action-at-a-distance mechanism by the measurements on the β particles, which is precisely what the locality assumption excludes. Actually, there should be many equivalent splittings of S in $S(b\pm)$ because, in a truly probabilistic situation, it cannot be fixed a priori which particular β particles give the results +1 and −1 upon measurement of $B(b)$. This complication is irrelevant for our purpose: knowing that at least one splitting of the previous type exists is all we need.

LIMITATION ON CONDITIONAL PROBABILITIES

Other probabilities can similarly be introduced for a different splitting of E into $E(b'+)$ and $E(b'-)$ arising from the actual (or the possible) measurements of $B(b')$ on the β particles. Naturally, S also splits into $S(b'+)$ and $S(b'-)$. Considering again the observable $A(a)$ of the α particles, from the application of the PRC to the present case one has

$$f(a\pm|b'+) = f[a\pm, \lambda_a(b'+)],$$
$$f(a\pm|b'-) = f[a\pm, \lambda_a(b'-)], \tag{8}$$

where the first frequency holds for $S(b'+)$ and the second one for $S(b'-)$. In general, different subsets imply different probabilities and $\lambda_a(b'+)$ will thus be different from $\lambda_a(b+)$ because different effects (probabilities) imply different causes (propensities). Clearly,

$$S = S(b+) \cup S(b-) \tag{9}$$

and therefore $S(b'+)$, which is a subset of S, is necessarily composed partly of objects of $S(b+)$ and partly of objects of $S(b-)$. In fact, by equation 9, one has

$$S(b'+) = S(b'+) \cap S = S(b'+) \cap [S(b+) \cup S(b-)]$$
$$= [S(b'+) \cap S(b+)] \cup [S(b'+) \cap S(b-)], \tag{10}$$

with the last step being a consequence of the fact that $S(b+)$ and $S(b-)$ do not overlap.

Consider now four of the eight probabilities (equations 7 and 8), namely, those referring to $A(a) = +1$. If $S(b+)$ were homogeneous, in the sense that every subset of it had a propensity $\lambda_a(b+)$ for the measurements of $A(a)$, and similarly for $S(b-)$, it would follow from equation 10 that $f[a+, \lambda_a(b+)]$ would apply to $S(b'+) \cap S(b+)$, containing a fraction γ of the population of $S(b'+)$, and that $f[a+, \lambda_a(b-)]$ would apply to $S(b'+) \cap S(b-)$, containing a fraction $(1 - \gamma)$ of the population of $S(b'+)$ (with $0 \leq \gamma \leq 1$), so one would necessarily have

$$f[a+, \lambda_a(b'+)] = \gamma f[a+, \lambda_a(b+)] + (1 - \gamma) f[a+, \lambda_a(b-)]. \tag{11}$$

This is a consequence of locality: if probabilities were instead created via action at a distance, they would exist literally only after a measurement on T is performed. However, $B(b)$ and $B(b')$ are incompatible and the previous equation could not be deduced. Equation 11 makes full sense in the local realistic approach and it gives the left-hand side as a weighted average of the other two probabilities. Therefore, the left-hand side must be *internal* to the interval of positive numbers in the right-hand side and one has

$$f[a+, \lambda_a(b+)] \geq f[a+, \lambda_a(b'+)] \geq f[a+, \lambda_a(b-)] \tag{12}$$

if, for instance, $f[a+, \lambda_a(b+)] \geq f[a+, \lambda_a(b-)]$. One can obviously repeat the previous argument for $f[a+, \lambda_a(b'-)]$ by considering that an expression similar to equation 10 can be deduced also for $S(b'-)$ and therefore that $S(b'-)$ is composed of a fraction γ' of elements of $S(b+)$ and of a fraction $(1 - \gamma')$ of elements of $S(b-)$, with $0 \leq \gamma' \leq 1$. Under the hypothesis that the sets $S(b\pm)$ are homogeneous, one now gets

$$f[a+, \lambda_a(b+)] \geq f[a+, \lambda_a(b'-)] \geq f[a+, \lambda_a(b-)]. \tag{13}$$

We thus see from equations 12 and 13 that the probabilities $f[a+, \lambda_a(b'\pm)]$ must be internal to the interval defined by the other two probabilities $f[a+, \lambda_a(b\pm)]$. However, b' is arbitrary and the previous conclusion must hold for all possible b'. On the other hand, b is also arbitrary and it cannot be true that it is always that value for which one gets the largest interval of probabilities, unless of course the probabilities are actually constant. The latter condition, however, is far too restrictive for theoretical and practical reasons, in particular because correlation experiments of the EPR type have shown that probabilities of the previous type do actually vary with a and b. Therefore, the sets $S(b\pm)$ in general cannot be homogeneous.

"SINGLET" STATE AS AN EXAMPLE

The considerations carried out up to now are totally general and apply to sets S, T, and E of any nature. It is, however, interesting to see how far the idea of local propensities applies to quantum systems. Of course, it is out of the question to reproduce totally the predictions of the quantum singlet state because of Bell's theorem. Nevertheless, it is useful to learn directly from quantum theory some needed features of the propensity approach. Consider the quantum observables

$$A(a) = \sigma \cdot \boldsymbol{a}, \qquad B(b) = \tau \cdot \boldsymbol{b}, \tag{14}$$

where σ_i and τ_j ($i, j = 1, 2, 3$) are the Pauli matrices for two spin-½ particles, α and β, respectively. They are dichotomic observables with ± 1 as eigenvalues and, apart from a trivial factor, they represent the spin components of particles α and β along the directions $\boldsymbol{a}$ and $\boldsymbol{b}$. The well-known singlet state is given by

$$\eta_s = \left(\frac{1}{\sqrt{2}}\right)[u(b+)v(b-) - u(b-)v(b+)], \tag{15}$$

where

$$(\sigma \cdot \boldsymbol{b})u(b\pm) = \pm u(b\pm), \qquad (\tau \cdot \boldsymbol{b})v(b\pm) = \pm v(b\pm), \tag{16}$$

so η_s is an eigenstate of $\sigma \cdot \boldsymbol{b} + \tau \cdot \boldsymbol{b}$ with eigenvalue zero. Because the singlet state is rotationally invariant, the previous conclusion holds for an arbitrary direction $\boldsymbol{b}$. Suppose next that $\tau \cdot \boldsymbol{b}$ is measured on the set T of β particles: according to quantum theory, the results ± 1 will be found with the same frequency ½. With the notation of equation 5, one has

$$N(b+) = N(b-) = N/2.$$

These populations $N(b\pm)$, of course, refer not only to the subsets $T(b\pm)$, but also to $E(b\pm)$ and $S(b\pm)$. According to quantum theory, the "reduction of the state vector" takes place during the measurement and one must consider the new states:

$$u(b-)v(b+) \text{ in } E(b+), \qquad u(b+)v(b-) \text{ in } E(b-). \tag{17}$$

It is now a simple matter to calculate the quantum mechanically predicted probabilities $f(a\pm\,|\,b\pm)$ of the previous section. One has, for example,

$$f_{\rm qm}(a+\,|\,b+) = \langle u(b-)v(b+)\,|\,(1 + \sigma\cdot\boldsymbol{a})/2\,|\,u(b-)v(b+)\rangle = \cos^2(a - b).$$

More generally, one has

$$f_{\rm qm}(a+\,|\,b+) = f_{\rm qm}(a-\,|\,b-) = \cos^2(a - b),$$

$$f_{\rm qm}(a+\,|\,b-) = f_{\rm qm}(a-\,|\,b-) = \sin^2(a - b), \tag{18}$$

where $a - b$ is the angle between the directions $\boldsymbol{a}$ and $\boldsymbol{b}$. The main point deduced in the previous section from probabilistic local realism is that the conditional probabilities must emerge as necessary consequences of objectively existing propensities. Of course, propensities are always there, but probabilities become concrete frequencies only when they are measured. From equation 7, one has

$$f_{\rm lr}(a+\,|\,b+) = f[a+, \lambda_a(b+)] = F(a - b),$$

$$f_{\rm lr}(a+\,|\,b-) = f[a+, \lambda_a(b-)] = 1 - F(a - b), \tag{19}$$

where F is the (presently unknown) prediction of probabilistic local realism and the dependence on $a - b$ can be justified on the basis of rotational invariance. Of course, Bell's theorem does not allow F to equal the quantum mechanical prediction. From equation 8, it follows that

$$f_{\rm lr}(a+\,|\,b'+) = f[a+, \lambda_a(b'+)] = F(a - b'),$$

$$f_{\rm lr}(a+\,|\,b'-) = f[a+, \lambda_a(b'-)] = 1 - F(a - b'). \tag{20}$$

Equation 10 shows that $S(b'+)$ splits into two subsets. By assuming the homogeneity of $S(b\pm)$, it would now follow, as a consequence of equation 12 or 13, that

$$F(a - b) \geq F(a - b') \geq 1 - F(a - b) \tag{21}$$

if, for instance, $F(a - b) \geq 1 - F(a - b)$. However, equation 21 cannot be valid for arbitrary values of a, b, b' if the predictions (equations 19 and 20), although respecting Bell's inequality, are even vaguely similar to the quantum theoretical expressions. Therefore, the subsets $S(b\pm)$ cannot be homogeneous. Because to these subsets we have attributed the quantum mechanical states $u(b\pm)$ (see equation 17), we are forced to the conclusion that, in an EPR situation, the eigenvectors of the spin observables cannot represent homogeneous ensembles. This conclusion follows rigorously by starting from the singlet state, but one can guess that it could be valid for all states of spin-½ particles independently of EPR correlations.

We have so discovered a new and more general sense in which quantum theory is incomplete. J. von Neumann showed[24] that the assumption of quantum mechanical completeness implies that a set described as an eigenstate is homogeneous (that is, every single one of its subsets must be described by the same eigenstate). We have instead come to the conclusion that, at least in some cases, quantum mechanical eigenstates actually describe inhomogeneous ensembles, if our assumptions concerning realism (propensities) and locality are true in nature. In conclusion:

> If probabilities are consequences of local propensities belonging to the sets $S(b\pm)$ for which they are predicted, they must result from averages of other probabilities, in general different for different subsets of $S(b\pm)$, but constant within every such subset.

These probabilities of homogeneous subsets of S and T will be dependent on local propensities and will therefore be totally independent of what is eventually measured on the other set of particles (T and S, respectively). This is a consequence of the locality assumption.

HOMOGENEOUS SETS OF SIMILAR SYSTEMS

The conclusion of a previous section concerning the existence of a splitting of S into $S(b+)$ and $S(b-)$ even if no measurement of $B(b)$ is made is now our starting point; the same splitting can be attributed to E, which consists of S and T.

Previously, we have also been led to the conclusion that $S(b+)$ and $S(b-)$—and therefore S that represents their union—cannot be homogeneous as far as the probabilities of finding $A(a) = \pm 1$ are concerned. A symmetrical conclusion is obtained for the probabilities of finding $B(b) = \pm 1$ in T by repeating the proof with the roles of particles α and β interchanged. Recalling the results of the previous section, we consider now a splitting of E (arising from a similar splitting of S) into the subsets

$$\sigma_1(a), \sigma_2(a), \ldots, \sigma_r(a), \tag{22}$$

where $\sigma_i(a)$ $(i = 1, 2, \ldots, r)$ is by definition homogeneous for the local probabilities $f_i(a\pm, \lambda_a)$ of obtaining $A(a) = \pm 1$, respectively, and λ_a is the local propensity for the same results, which is now totally independent of what is done with the β's.

We consider also a different splitting of E (arising from a similar splitting of T) into

$$\tau_1(b), \tau_2(b), \ldots, \tau_s(b), \tag{23}$$

where $\tau_j(b)$ is by definition homogeneous for the local probabilities $g_j(b\pm, \lambda_b)$ of obtaining $B(b) = \pm 1$ $(j = 1, 2, \ldots, s)$. Also, λ_b is a local propensity, independent of what is done with the α's.

Obviously, the union of these subsets must give E in both cases, that is,

$$\sigma_1(a) \cup \sigma_2(a) \cup \ldots \cup \sigma_r(a) = E,$$

$$\tau_1(b) \cup \tau_2(b) \cup \ldots \cup \tau_s(b) = E. \tag{24}$$

The dependence of σ_i on a and of τ_j on b reflects the fact that a subset that is homogeneous for an observable [e.g., $A(a)$] is not expected in general to be homogeneous for a different observable [e.g., $B(b)$].

One can, however, easily find smaller subsets that are homogeneous for two observables. For $A(a)$ and $B(b)$, they are

$$E_k(a, b) = \sigma_i(a) \cap \tau_j(b), \tag{25}$$

where the single index k has been chosen, for simplicity, to correspond in a one-to-one way to the pair of indices (i, j). Because i can take r values and j can take s values, the index k can take rs different values. Notice that $\sigma_i(a)$ [$\tau_j(b)$] remains homogeneous for $A(a)$ [$B(b)$], no matter what is measured on the β [α] objects. This is a consequence of our assumption of separability.

The previous approach can easily be generalized to an arbitrary number of observables of α and β. Considering m values $(a_1, a_2, \ldots, a_m)$ of the argument of $A(a)$ and n values $(b_1, b_2, \ldots, b_n)$ of the argument of $B(b)$, the following subsets of E can be introduced that are homogeneous for the probabilities of the values ± 1 of the indicated observable:

$$\begin{array}{l}
\sigma_1(a_1), \sigma_2(a_1), \ldots, \sigma_{r_1}(a_1) \text{ homogeneous for } A(a_1), \\
\sigma_1(a_2), \sigma_2(a_2), \ldots, \sigma_{r_2}(a_2) \text{ homogeneous for } A(a_2), \\
\ldots\ldots\ldots\ldots\ldots\ldots\ldots\ldots\ldots\ldots, \\
\sigma_1(a_m), \sigma_2(a_m), \ldots, \sigma_{r_m}(a_m) \text{ homogeneous for } A(a_m)
\end{array} \tag{26}$$

and

$$\begin{array}{l}
\tau_1(b_1), \tau_2(b_1), \ldots, \tau_{s_1}(b_1) \text{ homogeneous for } B(b_1), \\
\tau_1(b_2), \tau_2(b_2), \ldots, \tau_{s_2}(b_2) \text{ homogeneous for } B(b_2), \\
\ldots\ldots\ldots\ldots\ldots\ldots\ldots\ldots\ldots\ldots, \\
\tau_1(b_n), \tau_2(b_n), \ldots, \tau_{s_n}(b_n) \text{ homogeneous for } B(b_n).
\end{array} \tag{27}$$

The homogeneous nature of the subsets (equations 26 and 27) has the usual meaning: all the subsets of $\sigma_1(a_1)$ have the same probabilities $f_1(a_1\pm, \lambda_{a_1})$ for the results $A(a_1) = \pm 1$, where λ_{a_1} is a local propensity; all the subsets of $\sigma_2(a_1)$ have the same probabilities $f_2(a_1\pm, \lambda_{a_1})$ for the results $A(a_1) = \pm 1$; and so on.

Obviously, the union of the sets of every line of equations 26 and 27 always gives E, just as in equation 24.

By means of suitable intersections, one can introduce smaller subsets in which all the considered observables have constant probability. One can write

$$E_k(a_1, \ldots, a_m, b_1, \ldots, b_n) = \sigma_{i_1}(a_1) \cap \ldots \cap \sigma_{i_m}(a_m) \cap \tau_{j_1}(b_1) \cap \ldots \cap \tau_{j_n}(b_n) \tag{28}$$

for a typical subset homogeneous for all the probabilities of the ± 1 results of the $(m + n)$ considered observables. In equation 28, the single index k has been chosen, for simplicity, to correspond in a one-to-one way to the set of indices

$$\{i_1, i_2, \ldots, i_m, j_1, j_2, \ldots, j_n\}. \tag{29}$$

The number $t(m, n)$ of such sets can easily be calculated if one notices that i_1 can be chosen in r_1 different ways, . . . , i_m in r_m different ways, j_1 in s_1 different ways, . . . , j_n in s_n different ways, all choices being independent. Therefore,

$$t(m, n) = r_1 r_2 \dots r_m s_1 s_2 \dots s_n. \tag{30}$$

Notice, however, that $t(m, n)$ is in general expected to change if some arguments $a_1, \dots, b_n$ are modified, as it is also clear from the notation of equations 26 and 27, where s_1 [which is the number of subsets of E homogeneous for $B(b_1)$] has the same index as b_1 precisely because they depend on one another and a modification of b_1 can give rise to a change of s_1. Therefore, the set I of indices k [set of integers from 1 to $t(m, n)$] depends on the arguments of the observables:

$$I = I(a_1, \dots, a_m, b_1, \dots, b_n). \tag{31}$$

The notation can be simplified if one introduces a "vector" V having, as $(m + n)$ components, the arguments of the observables and given by

$$V = \{a_1, \dots, a_m, b_1, \dots, b_n\} \tag{32}$$

because the homogeneous subsets and their populations can then be written as

$$E_k(V) = E_k(a_1, \dots, a_m, \; b_1, \dots, b_n) \tag{33}$$

and

$$N_k(V) = N_k(a_1, \dots, a_m, \; b_1, \dots, b_n). \tag{34}$$

This simplified notation will soon turn out to be useful.

PROBABILITIES OF HOMOGENEOUS SETS

The basic probabilities, which are all constant for pairs belonging to a subset $E_k(V)$, are

$$\begin{aligned} &f_k(a_1\pm, \lambda_{a_1}), \text{ probabilities for } A(a_1) = \pm 1,\\ &f_k(a_2\pm, \lambda_{a_2}), \text{ probabilities for } A(a_2) = \pm 1,\\ &\dots\dots\dots\dots\dots\dots\dots\dots\dots,\\ &f_k(a_m\pm, \lambda_{a_m}), \text{ probabilities for } A(a_m) = \pm 1 \end{aligned} \tag{35}$$

and

$$\begin{aligned} &g_k(b_1\pm, \lambda_{b_1}), \text{ probabilities for } B(b_1) = \pm 1,\\ &g_k(b_2\pm, \lambda_{b_2}), \text{ probabilities for } B(b_2) = \pm 1,\\ &\dots\dots\dots\dots\dots\dots\dots\dots\dots,\\ &g_k(b_n\pm, \lambda_{b_n}), \text{ probabilities for } B(b_n) = \pm 1. \end{aligned} \tag{36}$$

Of course, the probabilities f_k belong to the α particles and the g_k to the β particles. Each of them depends only on the propensity of the measured observable. These propensities are now assumed to be strictly local. Thus, λ_{a_1} does not depend in any way on the particular b_i that is eventually being measured on the β particles, not even on the fact that a b_i is being measured or not. This property of the propensities, of course, is a consequence of the homogeneous nature of the sets $E_k(V)$ once the locality assumption has been made.

It is important to stress that the present formulation of locality is strictly analogous to the quantum mechanical locality of probabilities (which is well known to hold in spite of the overall nonlocal nature of the theory). The present formulation of locality is only applied to a wider set of probabilities.

An important feature of quantum probabilities is fully expected to hold also for equations 35 and 36: if $f(a_\mu\pm)$ is measured in a set S, this destroys the possibility to measure $f(a_\nu\pm)$ with $\nu \neq \mu$ in the same S because the observables $A(a_\mu)$ and $A(a_\nu)$ are in general incompatible; similarly for $g(b_\sigma\pm)$ and $g(b_\tau\pm)$ in a set T, if $\sigma \neq \tau$.

Obviously, if one writes

$$\rho_k(V) = \left(\frac{1}{N}\right)N_k(V), \tag{37}$$

then

$$\sum_{k\in I} \rho_k(V) = 1 \tag{38}$$

and $\rho_k(V)$ is the "statistical weight" of the subset $E_k(V)$ in the full set E.

JOINT PROBABILITIES AND BELL'S THEOREM

The definition of equation 37 is useful for calculating all types of probabilities in the set E, for example, those needed for a new proof of Bell's theorem. These are typically joint probabilities for observations carried out on correlated systems. Considering the joint probability for $A(a_\mu) = \pm 1$ and $B(b_\nu) = +1$, one has

$$\Omega(a_\mu\pm, b_\nu+) = \sum_{k\in I} \rho_k(V)\Omega_k(a_\mu\pm, b_\nu+), \tag{39}$$

where the index k indicates, as before, that a probability is calculated in the homogeneous set $E_k(V)$. One can obviously apply Bayes' formula and write

$$\Omega_k(a_\mu\pm, b_\nu+) = \omega_k(a_\mu\pm, \lambda_{a_\mu}/b_\nu+)g_k(b_\nu+, \lambda_{b_\nu}), \tag{40}$$

where $g_k(b_\nu+, \lambda_{b_\nu})$ has been introduced in equation 36 and $\omega_k(a_\mu\pm, \lambda_{a_\mu}/b_\nu+)$ is the conditional probability of $A(a_\mu) = \pm 1$, given that $B(b_\nu) = +1$. One can also say that $\omega_k(a_\mu\pm, \lambda_{a_\mu}/b_\nu+)$ is the probability of $A(a_\mu) = \pm 1$ in

$$E_k(V) \cap E(b_\nu+) \tag{41}$$

because it has been concluded that a set $E(b_\nu+)$ exists with the right properties even if $B(b_\nu)$ is not measured. Remember now that $E_k(V)$ is homogeneous, meaning that a probability valid for the whole of $E_k(V)$ applies as well to every conceivable part of it, for example, to the intersection (equation 41). Remembering equation 35, this gives immediately

$$\omega_k(a_\mu\pm, \lambda_{a_\mu}/b_\nu+) = f_k(a_\mu\pm, \lambda_{a_\mu}). \quad \textbf{(42)}$$

Inserting equation 42 in equation 40, one gets

$$\Omega_k(a_\mu\pm, b_\nu+) = f_k(a_\mu\pm, \lambda_{a_\mu})g_k(b_\nu+, \lambda_{b_\nu}) \quad \textbf{(43)}$$

and consequently, from equation 39, it follows that

$$\Omega(a_\mu\pm, b_\nu+) = \sum_{k\in I} \rho_k(V) f_k(a_\mu\pm, \lambda_{a_\mu})g_k(b_\nu+, \lambda_{b_\nu}). \quad \textbf{(44)}$$

With a trivial generalization of the reasoning leading from equation 39 to equation 44, one gets

$$\Omega(a_\mu\pm, b_\nu-) = \sum_{k\in I} \rho_k(V) f_k(a_\mu\pm, \lambda_{a_\mu})g_k(b_\nu-, \lambda_{b_\nu}). \quad \textbf{(45)}$$

The joint probabilities (equations 44 and 45) look like Clauser-Horne factorizability formulae,[25] with the role of the hidden variable λ being played by the much more transparent index k. There are, however, important differences because now the parameters of the considered observables enter in the probability density $\rho_k(V)$. This does not imply the introduction of any nonlocality because, in principle, all the conceivable observables should enter in $\rho_k(V)$. In practice, however, one can take into account only the observables relevant to the set of correlation functions measured in the considered experiment. The other ones are averaged away. A similar situation has been described by Wigner[26] with his deterministically based proof of Bell's theorem.

In spite of this dependence, it is very easy to deduce from equations 44 and 45 the usual inequalities of the Bell type.[27] It is amusing to notice that we never needed to introduce joint probabilities for two incompatible observables, described by quantum mechanics with two noncommuting operators. Observables like $A(a_\mu)$ and $B(b_\nu)$ are instead relative to two different objects and are therefore always compatible, so the previous obstacle does not exist and the joint probabilities (equations 44 and 45) are fully meaningful.

REFERENCES AND NOTES

1. For an account of the ideas of these authors, see: SELLERI, F. 1990. Quantum Paradoxes and Physical Reality. Kluwer. Dordrecht.
2. FURRY, W. H. 1936. Phys. Rev. **49:** 393 & 476.
3. SCHRÖDINGER, E. 1935. Proc. Cambridge Philos. Soc. **31:** 555.
4. BOHM, D. & Y. AHARONOV. 1957. Phys. Rev. **108:** 1070.
5. WU, C. S. & I. SHAKNOV. 1950. Phys. Rev. **77:** 136.
6. JAUCH, J. M. 1971. *In* Foundations of Quantum Mechanics. B. d'Espagnat, Ed. Academic Press. New York.
7. DE BROGLIE, L. 1974. C. R. Acad. Sci. (Paris) **278:** 721.
8. PICCIONI, O. & W. MEHLOP. 1987. *In* Microphysical Reality and Quantum Formalism. A. van der Merwe *et al.*, Eds. Reidel. Dordrecht.
9. The most important experiments performed during the two previous decades are: FREEDMAN, S. J. & J. F. CLAUSER. 1972. Phys. Rev. Lett. **28:** 938; FRY, E. S. & R. C.

THOMPSON. 1976. Phys. Rev. Lett. **37:** 465; ASPECT, A., P. GRANGIER & G. ROGER. 1981. Phys. Rev. Lett. **47:** 460; ASPECT, A., P. GRANGIER & G. ROGER. 1982. Phys. Rev. Lett. **49:** 91; ASPECT, A., J. DALIBARD & G. ROGER. 1982. Phys. Rev. Lett. **49:** 1804; PERRIE, W., A. J. DUNCAN, H. J. BEYER & H. KLEINPOPPEN. 1985. Phys. Rev. Lett. **54:** 1790.

10. BELL, J. S. 1965. Physics **1:** 195.
11. The limited detector efficiency of the previous experiments was discussed by: LEPORE, V. L. & F. SELLERI. 1990. Found. Phys. Lett. **3:** 203.
12. As an example of experiments that can be carried out with nearly perfect detectors, see: PRIVITERA, P. & F. SELLERI. 1992. Phys. Lett. **B296:** 261.
13. Any violation of Bell's theorem is called "action-at-a-distance" in the present report.
14. MARSHALL, T. W., E. SANTOS & F. SELLERI. 1983. Phys. Lett. **A98:** 5; MARSHALL, T. W. 1983. Phys. Lett. **A99:** 163; CASER, S. 1984. Phys. Lett. **A102:** 152; FERRERO, M. & E. SANTOS. 1986. Phys. Lett. **A116:** 356.
15. GARUCCIO, A. & F. SELLERI. 1984. Phys. Lett. **A103:** 99; SELLERI, F. 1985. Phys. Lett. **A108:** 197.
16. HOME, D. & F. SELLERI. 1992. The Aharonov-Bohm effect from the point of view of local realism. *In* Wave-Particle Duality. F. Selleri, Ed.: 127–137. Plenum. New York.
17. SELLERI, F. 1992. Two-photon interference and the question of empty waves. *In* Wave-Particle Duality. F. Selleri, Ed.: 277–298. Plenum. New York; Einstein–de Broglie waves and two-photon detection. *In* Bell's Theorem and the Foundations of Modern Physics. A. van der Merwe *et al.,* Eds.: 422–427. World Scientific. Singapore.
18. POPPER, K. R. 1988. Quantum Theory and the Schism in Physics. Hutchinson. London.
19. BASTOS FILHO, J. B. & F. SELLERI. 1994. Propensity, probability, and quantum physics. Found. Phys. In press.
20. TERSOFF, J. & D. BAYER. 1983. Phys. Rev. Lett. **50:** 553.
21. The same "reality criterion" was presented at the 1986 Conference of the New York Academy of Sciences, but was not related to propensities. See: SELLERI, F. 1986. Ann. N.Y. Acad. Sci. **480:** 518.
22. POPPER, K. R. 1990. A World of Propensities. Thoemmes. Bristol, United Kingdom.
23. EINSTEIN, A., B. PODOLSKY & N. ROSEN. 1935. Phys. Rev. **47:** 777.
24. VON NEUMANN, J. 1955. The Mathematical Foundations of Quantum Mechanics. Princeton University Press. Princeton, New Jersey.
25. CLAUSER, J. F. & M. A. HORNE. 1974. Phys. Rev. **D10:** 526.
26. WIGNER, E. P. 1970. Am. J. Phys. **38:** 1005.
27. See chapter 6 of reference 1.

A Quantum Theory of Angle[a]

SCOTT R. SHEPARD

Department of Physics
Baylor University
Waco, Texas 76798-7316

The significant role that the quantum theory of angular momentum plays in virtually every aspect of modern physics can hardly be overemphasized. Until recently,[1] however, there was no acceptable quantum theory of the complementary quantity, that is, the angles themselves. Traditionally, in quantum mechanics, angles have been treated as parameters, with no adequate means of describing their direct *measurement*, so angular distributions had to be *inferred* from the measurement of some other quantity. Herein, a complete quantum theory of angle measurement is presented and shown to provide a variety of novel insights when applied to spin dynamics, atomic structure, and polarized quantum fields. Previous attempts to formulate an Hermitian angle operator (or, equivalently, a collapsible angle wave function) have been plagued by mathematical obstacles, resulting in either a nonobservable operator[2] or a loss of complementarity with respect to angular momentum.[3] These difficulties can only be overcome by working in a Hilbert space of suitably large dimension. The emphasis herein, however, will be on the implications of the general theory, with the degree of formalism increasing only as the application requires it. Nevertheless, it should be noted that the methods used in the first two sections can only be formally justified via the complete formalism, which is presented and utilized in the third section.

QUANTUM ANGLE FOR A SYSTEM OF UNIQUE *J*

Just as $\hat{J}_z$ is the generator of translations in the azimuthal angle ϕ (about the z-axis), complementarity implies that the angle operator, $\hat{\phi}$, generates translations in m (the eigenspectra of $\hat{J}_z/\hbar$). For a system of known total angular momentum j, where $j(j + 1)$ is the eigenvalue of $\hat{J}^2$,[4] these translations can be written in the form,

$$\hat{T}(\delta m = 1) = \sum_{m=-j}^{j-1} |j, m\rangle\langle j, m + 1|, \tag{1}$$

but this cannot be expressed as $e^{i\hat{\phi}\delta m}$ (with $\hat{\phi}$ Hermitian) because $\hat{T}$ cannot be unitary due to the fact the m is bounded for a system of finite j. This difficulty is similar to the one encountered in the quantum phase problem and its solution[5] is also similar—to completely describe the measurement of an angle, we must work in a Hilbert space large enough for m to range from $-\infty$ to $+\infty$. The formalism of this complete description will be presented when it becomes necessary to do so in order to discuss its application to quantum polarization states of an electromagnetic field (this, in

[a] This work was financially supported by the Welch Foundation.

turn, will help to illustrate the meaning of the complete formalism). For the moment, we need only note that such spaces exist and that we can easily define complete translations through them as diagramed in FIGURE 1 for $j = 1$ (techniques for excitations of m that are not unique in j are presented via the discussion on photons in the third section). FIGURE 1 also depicts the technique of Vourdas,[3] which consists of adding a "wraparound" term, $|j, m = j\rangle\langle j, m = -j|$, to $\hat{T}$ so that unitarity can be achieved on a single j subspace. However, this results in a "discrete-angle" measurement that is not complementary to the $\hat{J}_z$ measurement.

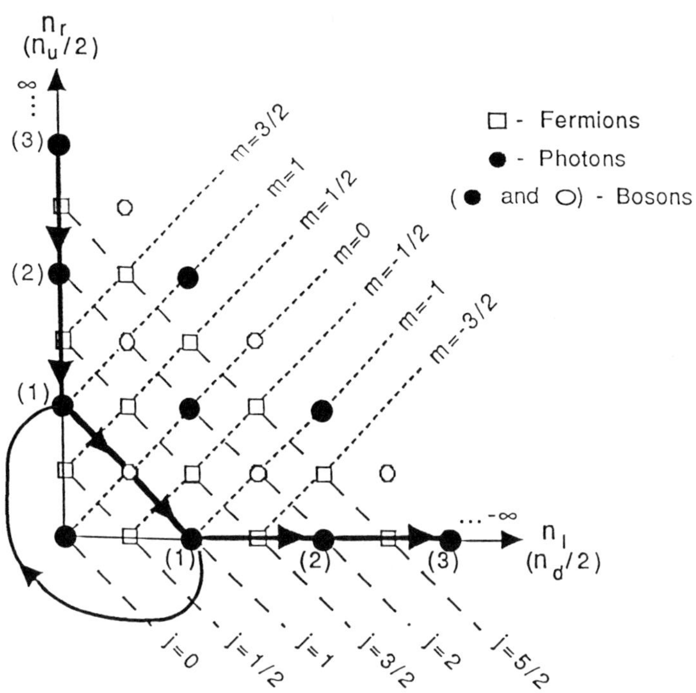

FIGURE 1. Translations through an infinite m-space versus wraparound.

Quantum measurements can be described in terms of wave functions, as well as operators, and the wave function perspective is particularly expedient here because complementary wave functions are related via a Fourier transform (even for the cases of bounded or discrete spectra[1]). Thus, because the m-ket expansion coefficients, $\psi_m \equiv \langle j, m | \psi \rangle$, can be considered to be a wave function in discrete m-space, these must also be the Fourier-series coefficients for the complementary angle wave function:

$$\psi^{(j)}(\phi) = \sum_{m=-j}^{j} \psi_m \, e^{-im\phi}. \tag{2}$$

Thus, for a state of fixed j, the angle statistics can then be readily calculated from the probability distribution,

$$P^{(j)}(\phi) = |\psi^{(j)}(\phi)|^2/2\pi. \tag{3}$$

It must be emphasized, however, that this "obvious" approach might also be considered to be "utter nonsense" unless we go to a larger Hilbert space. The necessity of this larger space can be seen from the wave function perspective by virtue of the fact that a bounded Fourier series, such as equation 2, cannot yield a distribution that in any sense approaches a delta-function. Thus, laying aside the issue of *how* wave functions collapse, this level of description of the angle measurement is incomplete because the single j m-space is not large enough to even *permit* collapse of the angle wave function.[1,5] As an example, the underlying angle kets,

$$\psi^{(j)}(\phi) = {}^{(j)}\langle\phi|\psi\rangle \qquad \text{so that} \qquad |\phi\rangle^{(j)} = \sum_{m=-j}^{j} e^{im\phi}|j, m\rangle, \tag{4}$$

are not *orthogonal*: ${}^{(j)}\langle\phi'|\phi\rangle^{(j)} \neq \delta(\phi - \phi')$, which indicates that we do not yet have the complete description of this measurement. They are, however, *complete*:

$$\int_{-\pi}^{\pi}\left(\frac{d\phi}{2\pi}\right)|\phi\rangle^{(j)(j)}\langle\phi| = \sum_{m=-j}^{j}|j, m\rangle\langle j, m| = \hat{I}_j, \tag{5}$$

which guarantees that equation 3 is a perfectly valid probability distribution function; thus, it must somehow correspond to a realizable quantum measurement (exactly how will be shown in the third section). The "discrete-angle" measurement (obtained via an orthogonal subset of the angle kets) would only provide a sample of the actual (continuous) angle distribution as depicted in FIGURE 2. Moreover, due to the wraparound term, these "discrete-angle" statistics are complementary to a periodically replicated version of the $\hat{J}_z$ spectrum rather than to the $\hat{J}_z$ spectrum itself.

IMPLICATIONS FOR SYSTEMS OF UNIQUE *J*

Prior to the solution of the "angle problem", one had to infer an angular distribution from the measurement of some other quantity. The "vector model", for example, attempts to *infer* the longitudinal angle θ (about the x-axis) from the *measurement* of $\hat{J}_z$. If angular momentum were a *classical* vector of length $j(j + 1)$, then one would infer that discrete valued projections of that vector onto the z-axis must imply that the angle θ is also discrete (simply because m is, as illustrated in FIGURES 3a and 3c for the case of spin ½). In quantum mechanics, however, it is imperative to use a formalism that describes the measurement of quantity rather than attempting to *infer* it from some other measurement. The quantum theory of angle attributes the discreteness of m to the periodicity of the angle about the ϕ axis (as an immediate consequence of the Fourier transform, although this well-known result has been proved in other ways). Moreover, the θ distribution can be obtained from its *measurement*, given the $\hat{J}_z$ representation, via a change of basis, followed by a

Fourier transform, that is,

$$\psi_{m'}^{(x)} = \sum_{m=-j}^{j} \psi_m d_{m',m}^{(j)} \qquad (\beta = -\pi/2) \tag{6}$$

and

$$\psi^{(j)}(\theta) = \sum_{m=-j}^{j} \psi_m^{(x)} e^{-im\theta} \qquad \text{so that} \qquad p^{(j)}(\theta) = |\psi^{(j)}(\theta)|^2/2\pi, \tag{7}$$

where $d_{m',m}^{(j)}$ $(\beta = -\pi/2)$ is the Wigner D matrix for rotating the z-axis into the x-axis.

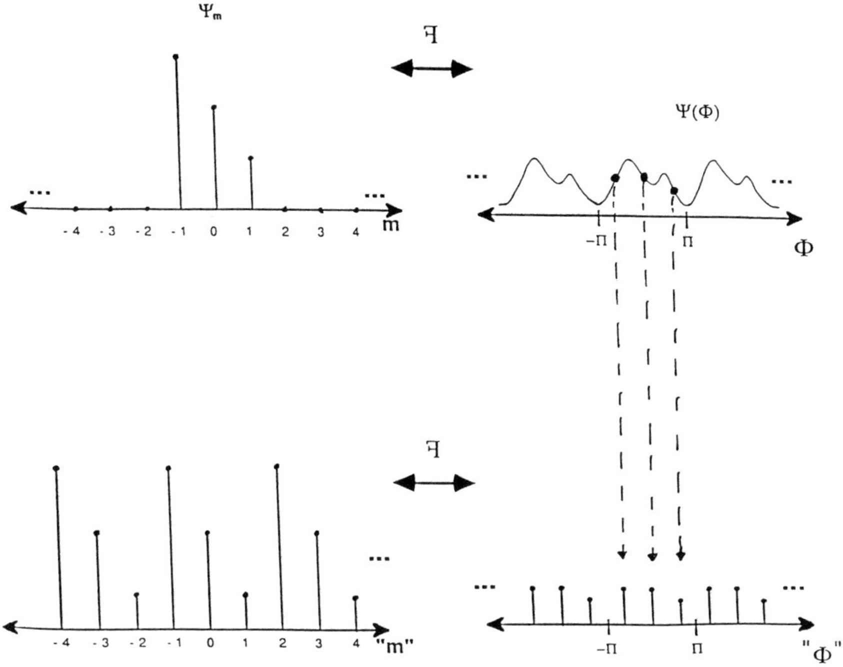

FIGURE 2. Continuous- versus discrete-angle representations.

The results for a spin-½ particle in the "up" state, shown in FIGURES 3b and 3d, demonstrate that *spin up really does point up* in the sense that the measurement of the angle between the spin vector and the z-axis yields $\theta = 0$ as its most likely value—in dramatic contrast with the "vector model", which erroneously predicts that $\theta = 0$ can never occur. Thus, the angle representation should prove quite useful in visualizing and calculating spin dynamics in magnetic resonance experiments and analogous phenomena.

Similarly, the ability to correctly describe the angular distribution of orbital angular momentum yields dramatic new insights on atomic/molecular structure and the nature of electronic "orbits". FIGURE 4 presents polar plots of the associated Legendre functions for $l = 3$ with $m = 0, 1, 2,$ and 3 in parts a, b, c, and d, respectively. Parts e, f, g, and h show the θ distribution (as calculated via equations 6 and 7) for each of these same states. The radial coordinate indicates the probability (compressed via a logarithmic scale so as to show the finer details) associated with each θ value (all distributions are uniform in ϕ). The spherical harmonics correspond

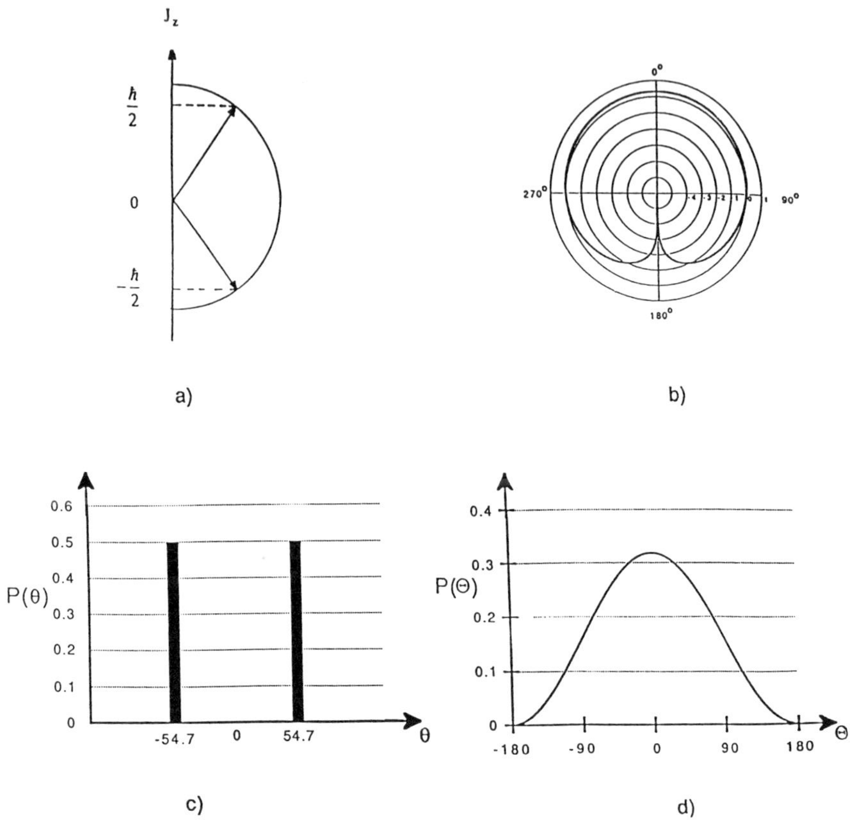

FIGURE 3. (a) Vector model for spin ½. (b) Polar plot of $P(\theta)$ for $|s = ½, m = +½\rangle$. (c) Inferred θ for $|s = ½, m = +½\rangle$. (d) Cartesian plot of $P(\theta)$ for $|s = ½, m = +½\rangle$.

to the following: (1) first measure the three commuting, Cartesian spatial components $\hat{x}$, $\hat{y}$, and $\hat{z}$; then (2) rather than recording the measured result as (x, y, z), these data are transformed into spherical coordinates (r, θ, ϕ) and the radial dependence is ignored. Thus, parts a through d plot the probability that when we perform this *spatial* measurement we will find the electron (or whatever system is in this $|l, m\rangle$ state) anywhere along a radial line that is at the angle θ with respect to the z-axis. In

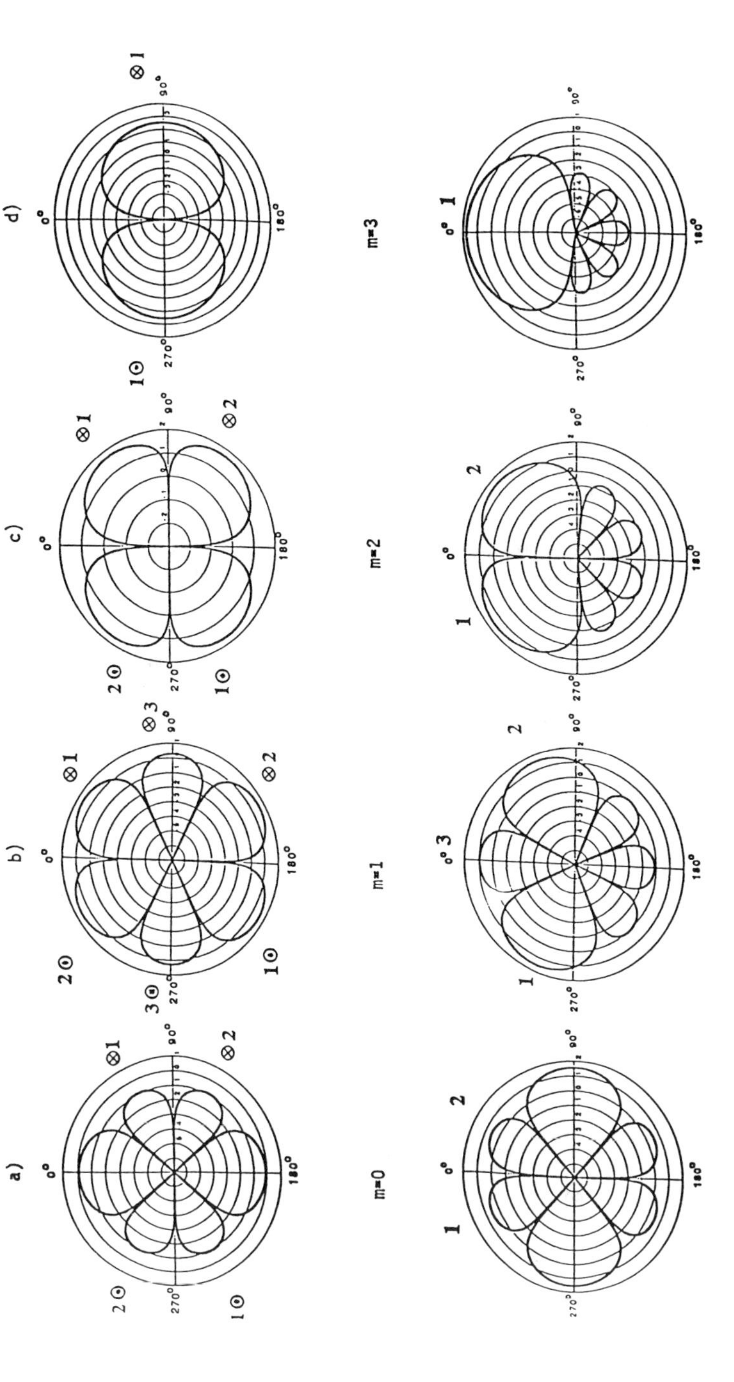

FIGURE 4. Legendre functions for $l = 3$ with (a) $m = 0$, (b) $m = 1$, (c) $m = 2$, (d) $m = 3$. $P(\theta)$ for $l = 3$ with (e) $m = 0$, (f) $m = 1$, (g) $m = 2$, (h) $m = 3$.

contrast, parts e through h plot the probability that the angle between the orbital angular momentum vector and the z-axis will be found to be θ when that angle is measured.

The spherical harmonics by themselves do not give an indication that anything is moving or "orbiting". The probability flux, however, can be readily calculated for these cases and is shown to be only ϕ-directed with an $m/\sin\theta$ dependence. Thus, we can visualize each of these "electron clouds" as undergoing a pure and constant rotation about the z-axis. Classically, if we rotate a mass density (given by the lobes of the Legendre polynomials) about the z-axis, we would obtain an angular momentum vector at the angles θ that correspond to the main peaks given by the θ distribution. The lobes of the spherical harmonics (or spatial lobes) are labeled in correspondence with these θ peaks to facilitate the visualization (but recall that all of these distributions are uniform in θ so that, in the $m = 2$ case, for example, lobes 1 and 2 are really the same—just rotate θ by π—as are the corresponding θ peaks 1 and 2). The width of these θ peaks (in part) reflects the width of the spatial lobes because we are actually rotating a "*probability* mass distribution" in this semiclassical picture of an orbit. This correspondence with previous pictures is, no doubt, comforting (and indeed the ability to infer an "orbit" without resorting to probability flux—a nonmeasurable entity—may prove to be useful), yet the most interesting results are always the unanticipated ones. Such is the case with the unlabeled θ peaks, which form part of what will be referred to as "substructure".

These substructure features are not predicted by the semiclassical notion of an orbit that we formed by supplementing the spherical harmonics with probability flux. Indeed, if we are to associate an orbit with the orbital angular momentum, then any time the θ measurement yields $|\theta| > \pi/2$ the associated orbit is in a direction opposite to that of the probability flux. It should be emphasized that we are discussing the behavior of a pure $|l, m\rangle$ state; there is no amplitude to be in the standard sense of its time-reversed version, $|l, -m\rangle$ (which would have a probability flux opposite to that of the states considered here). Portions of the labeled peaks can be thought of as substructure because they are time-reversed in this new sense ($|\theta| < \pi/2$), yet not all of the substructure (any peak or feature without a semiclassical counterpart) is time-reversed in this way. The probability flux for the $j = 3, m = 0$ case was labeled as if $m > 0$ to aid in the visualization, even though for $m = 0$ the flux is identically zero. The corresponding θ distribution, however, reveals that this comprises equal amounts of "forward and backward orbits". The application of the angle representation to Rydberg states should prove interesting and, in general, we expect that a variety of atomic and molecular processes can be further understood via this new window.

QUANTUM ANGLE FOR AN ARBITRARY SYSTEM

Mathematically, there are many similarities between the quantum theory of an angle in physical space and that of an angle in a phase-plane. A more physical connection between the angle and phase problems is obtained through the use of Schwinger's harmonic oscillator model of angular momentum,[6] resulting in an identification of the equivalence of the angle measurement and the measurement of the differential phase between the two oscillators.[1,7] Schwinger's scheme ascribes z-component angular momenta to a difference in the number of "up- and down-type" oscillator quanta via

$$\hat{J}_z \equiv \left(\frac{\hbar}{2}\right)(n_u - n_d) \qquad \text{and} \qquad \hat{J}_- \equiv \hbar \hat{a}_d^\dagger \hat{a}_u. \tag{8}$$

From the algebra of these two uncoupled oscillators, one can recover the algebra of angular momenta. However, these quanta are bosons that behave like spin-½ primitives (via equation 8) and hence they are deemed unphysical. The quantum associated with an electromagnetic plane wave, though, is a boson that resembles a fermion in the sense that its spin space is two-dimensional; that is, "the photon is spin 1 with $m = 0$ missing".[8] Indeed, we can reconstruct the algebra of angular momenta from these physically significant photonic primitives via

$$\hat{J}_z \equiv \hbar(n_u - n_d) \qquad \text{and} \qquad \hat{J}_- \equiv 2\hbar \hat{a}_d^\dagger \hat{a}_u. \tag{9}$$

We now present the complete formalism of the angle measurement and identify its connection with the measurement of differential phase between two oscillators. Complementarity suggests that for an arbitrary two-mode (two-oscillator) excitation, with number representation $\psi_{n_1,n_2} \equiv {}_1\langle n_1|{}_2\langle n_2\|\psi\rangle_{1\otimes 2}$, we take a two-dimensional Fourier transform,

$$\Psi(\phi_1, \phi_2) \equiv {}_1\langle\phi_1|{}_2\langle\phi_2\|\psi\rangle_{1\otimes 2} = \sum_{n_1=0}^{\infty}\sum_{n_2=0}^{\infty} \psi_{n_1,n_2} e^{-in_1\phi_1} e^{-in_2\phi_2}. \tag{10}$$

Rewriting this in terms of $\phi_\Sigma \equiv (\phi_1 + \phi_2)/2$ and $\phi_\Delta \equiv (\phi_1 - \phi_2)/2$, we have

$$\psi(\phi_\Delta, \phi_\Sigma) \equiv \langle\phi_\Delta, \phi_\Sigma\|\psi\rangle_{1\otimes 2} = \sum_{j=0}^{\infty}\sum_{m=-j}^{+j} \psi_{j,m} e^{-im\phi_\Delta} e^{-ij\phi_\Sigma}, \tag{11}$$

where

$$|\phi_\Delta, \phi_\Sigma\rangle \equiv \sum_{j=0}^{\infty}\sum_{m=-j}^{+j} |j, m\rangle e^{im\phi_\Delta} e^{ij\phi_\Sigma} \quad \text{and} \quad |j, m\rangle \equiv |n_1\rangle_1 |n_2\rangle_2 |_{j=n_1+n_2, m=n_1-n_2}. \tag{12}$$

We see that the ϕ_Σ part of $\psi(\phi_\Delta, \phi_\Sigma)$ cannot collapse due to the boundedness of its complement, namely, the number sum ($j \geq 0$).

We can eliminate ϕ_Σ to obtain a complete description of the "marginal measurement of ϕ_Δ" on $\mathscr{H}_{1\otimes 2}$ by applying an "absolute time average" to $|\phi_\Delta, \phi_\Sigma\rangle\langle\phi_\Delta, \phi_\Sigma|$, resulting in

$$(2\pi) d\hat{\Pi}_M(\phi_\Delta) \equiv \int_{-\pi}^{+\pi} \left(\frac{d\phi_\Sigma}{2\pi}\right) |\phi_\Delta, \phi_\Sigma\rangle\langle\phi_\Delta, \phi_\Sigma|$$

$$= \sum_{j=0}^{\infty} \left[\left(\sum_{m=-j}^{+j} |j, m\rangle e^{im\phi_\Delta}\right)\left(\sum_{m'=-j}^{+j} \langle j, m'| e^{-im'\phi_\Delta}\right)\right]. \tag{13}$$

Because both of the inner sums use the same value of j, interference among the different j states is excluded and we have (for pure states) the following probability distribution function:

$$P_M(\phi_\Delta) = \mathrm{Tr}[\hat{\rho}\, d\,\hat{\Pi}_M(\phi_\Delta)] = \left(\frac{1}{2\pi}\right)\sum_{j=0}^{\infty} |\psi^{(j)}(\phi_\Delta)|^2, \tag{14}$$

where Tr denotes trace; $\hat{\rho}$ is the density matrix; and

$$\psi^{(j)}(\phi_\Delta) \equiv \sum_{m=-j}^{+j} \psi_{j,m} e^{-im\phi_\Delta}. \tag{15}$$

Thus, when we take the two oscillators to be the right and left circularly polarized modes of an electromagnetic plane wave, the number sum and difference are the j and m of the angular momentum and $\psi^{(j)}(\phi_\Delta)$ is the single j angular wave function with $\phi_\Delta = \phi$. Hence, the "marginal measurement of ϕ_Δ" corresponds to an angle measurement in which the ϕ results from each of the individual j states are (in principle) distinguishable. Note that it is the distinguishability of the final (rather than initial) state that is of issue and that equation 14 describes the adding of the noninterfering probabilities that each j state contributes to the measurement.

We can also eliminate ϕ_Σ by conditioning $|\phi_\Delta, \phi_\Sigma\rangle\langle\phi_\Delta, \phi_\Sigma|$ to $\phi_\Sigma = 0$, resulting in

$$\begin{aligned}(2\pi)d\,\hat{\Pi}_C(\phi_\Delta) &\equiv \left[\frac{1}{P(\phi_\Sigma = 0)}\right] |\phi_\Delta, \phi_\Sigma = 0\rangle\langle\phi_\Delta, \phi_\Sigma = 0| \\ &= \left[\frac{1}{P(\phi_\Sigma = 0)}\right]\left(\sum_{j=0}^{\infty}\sum_{m=-j}^{+j} |j, m\rangle e^{im\phi_\Delta}\right)\left(\sum_{j'=0}^{\infty}\sum_{m'=-j'}^{+j'} \langle j', m'| e^{-im'\phi_\Delta}\right),\end{aligned} \tag{16}$$

where the renormalization constant is

$$P(\phi_\Sigma = 0) = \sum_{m=-\infty}^{\infty} \left\{\left|\left(\sum_{j=0}^{\infty} \psi_{j,m}\right)\right|^2\right\}. \tag{17}$$

In this "conditional measurement of ϕ_Δ", we are taking a "snapshot" in absolute time. Thus, the inner sums use different values of j, thereby permitting interference among the different j states. We therefore have (for pure states) the probability distribution function,

$$P_C(\phi_\Delta) = \mathrm{Tr}[\hat{\rho}\, d\,\hat{\Pi}_C(\phi_\Delta)] = \left[\frac{1}{2\pi P(\phi_\Sigma = 0)}\right]\left|\sum_{j=0}^{\infty} \psi^{(j)}(\phi_\Delta)\right|^2. \tag{18}$$

Hence, if we take the oscillators to be the photonic primitives, then this "conditional measurement of ϕ_Δ" corresponds to an angle measurement in which the ϕ results from each of the individual j states are (in principle) indistinguishable (because equation 18 is adding the interfering amplitudes that each contributes). Note that either procedure (marginal or conditional) will reduce to equations 2 and 3 for an initial excitation of unique j.

The connection between distinguishability and absolute time is somewhat surprising, but can be made palatable via example. Consider the case of the left-hand polarization being in the vacuum state ($n_l = 0$) so that each $\Psi_{n_r,0}$ corresponds to a different value of j. If we add interfering amplitudes, then we will get the nonuniform angular distribution of the field vectors (and their quantum fluctuations) at a point in absolute time. In contrast, if we add the noninterfering probabilities, then we will get a uniform angular distribution [because each of the $p^{(j)}(\phi)$ will be uniform in this case]. This uniform distribution corresponds to letting the nonuniform angular "snapshot" evolve in absolute time (which we average over), thereby tracing out a quantum version of the polarization ellipse (which is uniform, that is, a circle, for this case of pure right-handed polarization).

ACKNOWLEDGMENTS

I would like to thank Laura Bringol, Debbie Jones, and Tom Pilcher for their assistance in the preparation of the figures.

REFERENCES AND NOTES

1. SHEPARD, S. R. 1990. Ph.D. thesis proposal. Massachusetts Institute of Technology, Cambridge, Massachusetts; 1991. Workshop on Squeezed States and Uncertainty Relations (NASA CP-3135, Washington, District of Columbia); 1992. Workshop on Harmonic Oscillators (NASA CP-3197, Washington, District of Columbia).
2. CARRUTHERS, P. & N. NIETO. 1968. Rev. Mod. Phys. **40:** 411.
3. VOURDAS, A. 1991. Phys. Rev. **A43:** 1594.
4. For example, in the case of no orbital angular momentum, a system of this type could be a particle of spin $s = j$.
5. SHEPARD, S. R. & J. H. SHAPIRO. 1988. Optical Society of America Technical Digest (Washington, District of Columbia); SHEPARD, S. R. 1992. Ph.D. thesis. Massachusetts Institute of Technology, Cambridge, Massachusetts; 1993. Third International Workshop on Squeezed States and Uncertainty Relations (Washington, District of Columbia).
6. SCHWINGER, J. 1952. United States Atomic Energy Report No. NYO-3071. U.S. Govt. Printing Office. Washington, District of Columbia.
7. SHEPARD, S. R. 1993. Quantum Electronics and Laser Science Conference Technical Digest (Washington, District of Columbia).
8. BERESTETSKII, V. B. *et al.* 1982. Quantum Electrodynamics. Pergamon. Elmsford, New York.

The Integration of Mind into Physics

HENRY P. STAPP

Lawrence Berkeley Laboratory
University of California
Berkeley, California 94720

A THEORY OF MIND AND MATTER

John Wheeler has a knack for asking the right question. At the beginning of this conference, he directed our attention to what he deemed to be a key question in the foundations of quantum theory:

> Høffding asked, "Where can the photon be said to be?"
> Bohr replied, "To be. To be. What does it mean 'to be'?"

This report proposes to answer this question in a way that yields a parsimonious theory of mind and matter that reconciles the opposing views of Bohr and Einstein. Bohr held that quantum theory describes relationships between aspects of our knowledge, whereas Einstein insisted that our basic theory should describe what *could* be reality itself, not merely our knowledge of it.

Wheeler provided guidance in our search for an answer by offering several further quotations:

- "The concept of the physical object is a convenient myth."
- "Observations are the iron posts upon which everything is based; all else is papier-mâché."
- "No phenomenon is a phenomenon until it is an observed phenomenon."

Modern science grew out of Descartes' disjunction of mind and matter.[1] However, that separation led to classical physics, not to ultimate science. Mind is injected back into physics by quantum mechanics because the basic problem in quantum mechanics is to reconcile the nonclassical character of the quantum world with the classical character of our perceptions of it.

Bohr confronted this problem by adopting an epistemological approach based on "our knowledge": he regarded the quantum formalism as merely a set of rules that give statistical predictions relating our classically describable perceptions. Yet, Bohr's theory is limited in scope by its exclusion of biological systems. This exclusion entails that the physical carriers of our knowledge, namely, our brains, are not represented within the quantum system described by the theory. This omission is the basic cause of the difficulties that beset Bohr's version of quantum theory. An adequate dynamical theory should contain representations of the things that need to be related, and this means, for basic theory, both the quantum microrealities and the classically described experienced facts. Ideally, these two things should be represented within basic theory as two aspects of a seamless dynamical unity.

Von Neumann developed an alternative approach[2] that extends quantum theory to the entire universe, including our brains. However, he needed *two* dynamical processes:

- process I, consisting of abrupt "quantum jumps", called events;
- process II, consisting of continuous deterministic evolution according to the Schrödinger equation.

As regards these jumps, von Neumann's main result was that each one could be placed at any one of a sequence of alternative locations along the chain of causal connections that lead from the microworld of elementary particles to the macroscopic level of brain activity that corresponds to our conscious thoughts: differences in the placement of the jump have virtually no effect on the predictions of quantum theory. However, von Neumann noted that there is one placement of the jumps that is naturally singled out from all others, namely, the placement at the level of brain functioning where our conscious thoughts enter. All other placements are ad hoc and artificial, and they disrupt the natural linkage between the physical world and the quantum principles.

Wigner affirmed and reinforced the idea that the quantum jumps be placed exclusively at the level of brain action where conscious thoughts enter.[3] This placement of the jumps reduces the *triality* consisting of the mental world, the classical world, and the quantum world to a *duality* composed of the mental world and the quantum world: the observer-independent classical level of description is eliminated, and hence the world of classically describable thoughts rides directly on the world of quantum potentialities. This brings von Neumann's theory into alignment with Bohr's, in the sense that in both theories the classically described perceived facts are linked to each other via a purely quantum mechanical system, without the introduction of an observer-independent classical level of being.

Wigner later recanted, claiming that the disruptive effects of the environment make quantum theory inapplicable to macroscopic systems. However, his argument is not conclusive. The effects of the environment on macroscopic systems have been studied in great detail in recent years. Interactions with the environment certainly produce a great loss of effective phase coherence, but the overall practical effect of these interactions is to convert the quantum state at the level of macrovariables to an approximate statistical mixture of states. Macrovariables are collective variables that carry a large mass compared to the energies of the disturbances coming from the environment. Thus, interaction with the environment converts the quantum state, effectively and approximately, to a statistical mixture of a thin veneer of sluggish macrovariables riding on an ocean of tumultuous microvariable activity. This conversion justifies in practice, in many cases, the use of classical statistical mechanics at the macrolevel, but it leaves unresolved the central problem: where do we place the jump that reduces the approximate statistical mixtures of classically describable macrostates to the individual state that we perceive?

There is no empirical evidence for the occurrence of jumps at any place other than the mind-brain interface. Hence, there is no scientific basis for introducing other jumps. Certainly, the goal of bringing our mental image of the macroworld into concordance with our notoriously fallible classical intuition is not a sufficient reason.

Jumps are definitely needed inside our brains.[4] Thus, the law of parsimony, and the lack of a natural criterion that picks out any other location, enjoins us to place the quantum jumps exclusively at the mind-brain interface.

The following question then arises: Which brains?

I "know" only that I, myself, am conscious. However, conversations with other human beings, and the writings of psychologists and philosophers, have convinced me that some other human beings are, in all probability, also conscious. In this connection, it is important to recognize that the goal in science is not certainty: certainty is unattainable. We create general theories, test them, and use them if they work; we never verify them. Thus, in view of the great similarities, both structurally and behaviorally, of myself and other human beings, it is reasonable to posit, as the foundation of a tentative theory of nature, that all normal and alert human beings have thoughts similar to my own.

According to Bohr, "The task of science is to expand our experience and reduce it to order." Here, the "our" is, in the first instance, the human race. Science is an ongoing human endeavor, and the facts that it must coordinate are the facts defined by our collective experience. By taking only human brains to be associated with the emergence of the classically describable facts, we obtain a theory that is maximally parsimonious with respect to human-based science: the facts that are defined within the theory are precisely the facts that need to be explained by the theory.

The theory specified in this way describes what could be reality itself, including our knowledge of it. Thus, it meets Einstein's demand that basic theory must describe a possible reality. The virtue of theories of this kind is that they must conform to the strong condition that they be able, in principle, to describe *all* of nature in a completely consistent way. This rules out pretenders and retains theories that have a greater promise to carry us beyond what we presently know.

One can consider theories that differ from this "standard" human-based one by having a larger set of brains that harbor quantum jumps: theories in which the set of human brains is augmented to include some nonhuman ones, such as dog brains or computer brains. In this connection, I note the following:

(1) Theories based on larger sets of brains lead to consequences that differ *within the set of facts defined by human experience* from those of the standard human-based theory. This is because the additional brains produce additional quantum jumps, and these jumps, occurring outside human brains, will generally lead to eventual consequences also within human brains. This situation differs from the one in classical physics, where the occurrence or nonoccurrences of thoughts, per se, in other brains have no empirical consequences for me, because the laws of classical physics make no reference to subjective experiences. Thus, according to the ideas of classical physics, the occurrence or nonoccurrence of "experiences" in conjunction with the activities of a brain/body make no difference at all in the physical world and hence no difference in the consciousness of anyone else.

(2) Theories that allow jumps in nonhuman brains, although different *in principle* from the standard theory, are virtually identical to it *in practice*, within the realm of human experiences. This is a corollary of von Neumann's analysis. Thus, for all practical scientific purposes, we, as human beings, can, without introducing any significant error, use the standard human-based theory, even if quantum jumps do actually occur in nonhuman brains: inclusion of those

other jumps would make virtually no changes in the predictions pertaining to human experiences.

(3) By virtue of point 2, any extension of the standard human-based theory to a theory with jump-possessing nonhuman brains has no secure scientific justification. The standard human-based theory is the most parsimonious theory of nature consistent with my knowledge that I have thoughts, and that other human beings are structurally and behaviorally very similar to me, and therefore ought to be treated on a par with me in a general theory of nature.
(4) "Science is a tiny island of knowledge in a vast ocean of nescience." Let us not pretend to know more than we do. There exists a huge collection of theories that are different from one another in principle, but that, by presently available techniques, are indistinguishable in practice. In full awareness of this fact, we can choose, tentatively, the one that is best adapted to the human scientific enterprise, namely, the one that describes a possible reality in which the dynamically generated facts are exactly the facts specified by the experiences of the community of communicating human observers.
(5) After we have developed a satisfactory detailed understanding of the connection between human brains and human thoughts, we may be in a position to make a reasonable extrapolation to nonhuman brains. We may then wish to shift to what might seem by then to be a more reasonable theory.
(6) Within this theory, each train of thought is dynamically connected to the process going on *contemporaneously* in the associated brain, rather than, as in certain AI theories, to a computer program, which might be instantiated by a variety of alternative and different dynamical processes.
(7) In contrast to behavioristic approaches, the primary scientific data here are the facts specified by *our collective experience.*
(8) Within this theory, the history of the universe is defined only insofar as it is defined by the facts specified by accumulated human knowledge.

This formulation of quantum theory reduces the problem of quantum measurement to the problem of the dynamical connection of mind to brain. Twenty years ago, such a "reduction" might have been tantamount to casting the problem out of science. Today, that is not true. Scientific pursuit of the question of the relationship between brain process and conscious process has become an important confluence of interest among increasing numbers of brain scientists, psychologists, neuropsychologists, philosophers of science, and quantum physicists. The topic is rightly a "hot" subject because it bears directly on the core issue of our conception of ourselves: the question of the relationship of our thoughts to our bodies and brains, and to the universe around us.

An adequate quantum theory of the relationship between brain and mind must be:

(1) concordant with all results from neurophysiology and neuroanatomy;
(2) concordant with all results from psychology;
(3) concordant with all results from neuropsychology;
(4) concordant with the demand that the full structure of each thought (i.e., each consciously experienced event) be fully represented in the brain state actualized by the corresponding quantum event;

(5) concordant with the demand that the brain state associated with a thought have the functional character appropriate to that thought: for example, my thought of "I will now raise my arm" must actualize a brain state that, under normal conditions, will cause my arm to rise.

An outline of a theory that appears to meet these requirements is given in reference 4. In that book, I followed Heisenberg and allowed the quantum jumps to occur *also* at the level of quantum measuring devices, such as Geiger counters. This raises, though, the question of the rule that fixes the exact placement of these external jumps. In the present work, I have shifted to the position described above. However, that shift makes no change in the theory of the mind-brain interaction given in reference 4, which I now briefly summarize.

QUANTUM THEORY OF BRAIN EVENTS

The motivation within psychology and brain science for this theory of brain events is described in reference 4 and will not be repeated here. The essential postulate is that each human thought is an event that actualizes a particular pattern of activity in a human brain. This thought is said to belong to that brain and it is represented in the physicist's description of nature by the action of a projection operator that projects the prior (Heisenberg-picture) state vector onto its successor. This successor contains the pattern of brain activity that is actualized by the thought and it contains none of the alternative possible thought-related patterns that, according to the quantum analogue of Newton's laws of motion, could have occurred, but in fact did not occur. The pattern of brain activity actualized by the thought is called the brain correlate of the thought and it must contain within its structure all of the information and structure that is present in the felt content of the thought.

Each rudimentary thought always represents an image of the self in its environment and it either updates or adds to the latest image in the case of a thought that "attends", or it creates a projected (into the future) image of the self in its environment in the case of a thought that "intends". The latter sort of image forms a template for subsequent motor action. Each pattern of brain activity that is actualized by such an event persists for a certain time and is thereby "facilitated": it is strengthened in such a way that subsequent excitations of portions of this pattern tend to excite the whole, leading to associative recall. Most such patterns are largely prefabricated, in the sense that they are formed from earlier patterns, or their parts, joined together in new configurations. By virtue of the architecture of the brain, in conjunction with learning, the only allowed configurations are those that correspond to an image of a physically possible self in a physically possible environment, or to some generalization of this basic form. Details can be found in reference 4.

However, what is the physical structure of the patterns of brain activity that are actualized by thoughts? In the first place, each one must be an enduring, and presumably oscillatory, pattern of activity that, through its composition in terms of subpatterns, represents all of the felt structure of the associated thought. Hence, it must evidently cover a macroscopic portion of the brain. These spatially separated parts of the pattern are bound together dynamically: the entire structure hangs together as a resonating system by virtue of the mechanical feedback and feed-forward linkages in the brain. As a nonlinear (at the classical level) system with an energy supply and feedbacks, the system is nonstable in the sense that, like a system of microphone and amplifier, once the energy in the system passes a certain critical value, it evolves rapidly into an oscillatory mode that soaks up all of the available power. Fatigue properties of neurons eventually cause the pattern to fade out, and hence the conscious brain advances, step-by-step, from one of these resonance states to the next.

No human being can predict the exact progression of these states. Even in a classical idealization, such predictions are rendered impossible by our lack of knowledge of the unknown and uncontrollable effects of thermal noise and interactions with the environment. Hence, our knowledge about the "next" state can be represented only by a probability function, even though, according to classical ideas, this "next" state is completely fixed and predetermined. In a quantum world, this lack of knowledge about the "next" state is elevated, through Heisenberg's principle of indeterminacy, to a matter of principle, and hence the form of the next state remains undetermined and indeterminate, even in principle, until it is actualized by a thought.

A SIMPLIFIED MODEL OF THE MIND/BRAIN

In order to make the ideas outlined above more concrete, I shall describe a simplified model of the mind/brain. What I shall give is very much a toy model. It should be understood as a simplification of what was described in more detail in reference 4 (chapter 6 and appendix). Nevertheless, it may be useful, by starting at a still simpler stage, to bring out certain rudimentary features.

The "brain" will be taken to consist of (1) a source of power, consisting, in this idealization, of a very massive simple harmonic oscillator, and (2) a set of simple harmonic oscillators that represent the different patterns of brain activity that are the alternative possibilities for what the next thought can actualize. Thus, the classical unperturbed Hamiltonian is[5]

$$H_0 = (p^2 + M^2\omega^2x^2)/2M + \sum_{i=1}^{n} (p_i^2 + m^2\omega^2x_i^2)/2m.$$

Note that I have taken the frequency of the power source, ω, to be the same as the frequency of the modes i that are the brain correlates of the possible thoughts. Introducing[5]

$$a_i = p_i/(2m\omega)^{1/2} - i(m\omega/2)^{1/2}x_i$$

and its complex conjugate a_i^* and the analogous a_0 and a_0^*, one may rewrite H_0 in the simpler form,

$$H_0 = \sum_{i=0}^{n} \omega a_i^* a_i.$$

The interaction Hamiltonian has the form,

$$H_1 = i \sum_{i=1}^{n} (a_i^* a_0 - a_0^* a_i)(f_i - g_i),$$

where f_i and g_i are positive real functions of the variables of the problem. Thus, the Poisson bracket (i.e., classical) equation of motion[6]

$$du/dt = \{u, H\} \equiv -i \sum_i (\partial u/\partial a_i \partial H/\partial a_i^* - \partial H/\partial a_i \partial u/\partial a_i^*) \equiv -i[u, H],$$

where $[a_i, a_i^*] = 1$, etc., gives

$$da_i/dt = -i\omega a_i + a_0(f_i - g_i) + (a_i^* a_0 - a_0^* a_i)[a_i, (f_i - g_i)],$$

which is also the Heisenberg (i.e., quantum) equation of motion if the final bracket is interpreted as the commutator, $a_i(f_i - g_i) - (f_i - g_i)a_i$. The source/sink mode is supposed to carry a very large amount of energy. Thus, we assume that $a_0 = A\exp(-i\omega t)$, with A very large, positive, and essentially constant. Then, the ansatz $a_i = A_i \exp(-i\omega t)$, with A_i real and positive, and the definitions $f_i = f'_i/A$ and $g_i = g'_i/A$ convert this equation of motion to

$$dA_i/dt = f'_i - g'_i.$$

The term proportional to f'_i feeds energy from the power source into the mode i, whereas the term proportional to g'_i provides for dissipation: it gives the flow of energy back into the source (and sink) mode described by the variable a_0.

This power-supply term is required to have two main features. The first is that the coupling be nonlinear and lead to a very rapid buildup of the energy in a mode i if a certain critical value of that energy is reached. The energy will build up to a point where an equilibrium with dissipation is reached. The second feature is that the coupling should tend to divert, eventually, virtually all of the power flowing into this set of modes i into a single one of them. The rationale behind this second property is that the purpose of conscious thinking is to construct, as soon as possible, some single coordinated plan of action and to initiate it. Thus, at the classical level, the conscious brain process should produce one single plan, not several conflicting plans. Hence, the coupling should be such that it will lead fairly quickly, at the classical level, to a steady state where all of the available power is passing through just one of the oscillator modes i. At the quantum level, the upcoming thought belonging to this brain will be an event (i.e., a quantum jump) that actualizes such a state. In this state, one, and only one, of these modes i will be excited and, hence, there will be a clear distinction, at the level of the brain, between the various alternative possibilities among which the conscious thought will decide. Moreover, the selected mode will have the energy to initiate the chosen plan of action.

Several points are worth emphasizing right away. The first is that the oscillator coordinates x_i are collective coordinates: each one represents the amplitude of an entire organized pattern of activity, not just the position of an individual particle, nor even an individual neuron pulse. Each variable x_i can be likened to the angle of rotation of a wheel or the displacement of the center of mass of a large pendulum. It is this whole pattern of activity that is the brain correlate of a possible thought, namely, the brain activity that can, at the quantum level, be actualized by the next thought.

In the framework provided by classical physics, it is hard to understand how such an extended pattern of activity could be one single thought. The basic idea in classical physics is to reduce things at the fundamental level to tiny localized objects or to localized values of fields: each extended thing is regarded as fundamentally an aggregate of tiny fundamental localized parts. However, a thought is, psychologically, one single unified entity. In the words of William James: "Your acquaintance with reality grows literally by buds or drops of perception. Intellectually and upon reflection you can divide these into components, but as immediately given, they come totally or not at all."[7] The point here is that a thought is, at the level of its ultimate essence, exactly what is given, namely, one single entity. However, it is represented in terms of the motions of billions of particles that are scattered all over the brain. In classical physics, an aggregation of localized interacting particles can certainly *act* as a whole, but it is nevertheless conceived to *be* an aggregation of localized parts: at the level of its ultimate essence, it *is* an aggregation.

In quantum thinking, this "wholeness", or binding, problem vanishes: the quantum event is one single thing, which, however, can actualize an extended pattern of brain activity. Indeed, in quantum theory, a quantum event is allowed to actualize *only* a pattern that is sufficiently "macroscopic"; otherwise, the successful predictions of the theory will be lost.

Note also that the quantum thought can be regarded as playing an essentially creative role: by choosing to actualize *together* patterns that have not previously occurred together, the thought creates a newly composed pattern of brain activity that is a new thought, or idea. In fact, every thought is in this sense a new invention, brought into being by the quantum event that is, or corresponds to, that thought itself. Of course, some thoughts are more radically inventive than others.

Possible forms for f_i and g_i are

$$f_i = (C/A)(a_i^* a_i)^2 \Big/ \sum_j (a_j^* a_j)^2$$

and

$$g_i = (D/A) a_i^* a_i,$$

with C and D being positive constants satisfying $C \gg D$. Then, the equation of motion reduces to

$$dA_i/dt = CA_i^4 \Big/ \sum_j A_j^4 - DA_i^2.$$

Let the largest of the A_i be $A_1 > 1$. If A_1 is not too large, then it will grow. Some other A_i's may also grow, but they cannot overtake A_1. When A_1 gets close to its upper limit at

$$CA_1^4 \Big/ \sum_j A_j^4 = DA_1^2,$$

all of the other A_i with $A_i > 1$ will be decreasing. Hence, A_1 will tend to its upper limit, whereas all of the other A_i's will tend to the neighborhood of zero.

The emergence of this particular mode was, in this classical description, a consequence of the particular initial conditions: a different initial condition would have led, in general, to a different final state. Due to the initial uncertainties coming from our lack of knowledge, arising perhaps from thermal fluctuations, our knowledge about the final state will be represented by a probability function, even though, in the classical framework, the actual final state will be fixed and definite.

In the quantum generalization of this classical model, the quantum indeterminacy will lead to a quantum state that is roughly a *superposition* of states ψ_i, one for each of the alternative possible final states i occurring in the classical description. The state ψ_i has one of the modes, mode i, highly excited and all the others unexcited. These different quantum states correspond to distinct and well-separated possibilities for the brain state and also to different classically describable perceptions of the self in its environment. A quantum jump will then occur: it will actualize one of these possibilities and eradicate all the others.

REASONS FOR BELIEVING THAT THE MIND-BRAIN CONNECTION IS QUANTUM MECHANICAL

The association of our thoughts with quantum jumps was postulated, above, in order to resolve the quantum measurement problem, that is, to provide a coherent conception of nature that parsimoniously accommodates the quantum character of the microworld. However, there is a long causal chain from the microscopic world of elementary particles to the macroscopic patterns of activity that correspond to our thoughts, and other ways of forming a coherent conception of nature might be entertained. There are, however, at least three reasons to favor the linkage of the mind-brain problem to quantum theory proposed here:

(1) the parallel *dual* structures of mind/matter and the quantum world;
(2) the occurrence of consciousness where *choices* are needed;
(3) the *unity* of thoughts, in contrast to the local reductionistic character of classical physics.

As regards *duality*, it should be noted, first of all, that the proposed dualistic quantum view of the mind/brain does not contradict the mind/brain identity theory. Thoughts are not only *represented* as aspects of the physicist's description of brain activity, but can reasonably be imagined to *be* aspects of brain activity. Indeed, with the classical image of the nature of physical systems eliminated, we are invited to form a new image of what brains are actually made of. Part of what they are made of would seem to be a sort of objective tendency for a thought to occur. However, the other part of what a brain is made of could be the actually occurring thoughts themselves. Indeed, how else can we make an aspect of brain activity "actual" other than by identifying it with something that is truly real, and thoughts (including feelings, etc.) are the only sorts of things that we know to be real. In a naturalistic scientific approach, one will want a thought to be an *actual constituent* of the physical system described by the physical theory, that is, by quantum theory, and not some mysterious disconnected thing that hovers around outside the physical system described by the physical theory. This line of argument leads one to *identify* the thought as the actual physical counterpart of the associated quantum-theoretic event.

In the literature supporting the mind-brain identity theory, the usual position seems to be that mind-brain identity entails a monistic rather than dualistic ontology.

This bias evidently stems from the notion that the correct physical theory is classical physics. However, whereas the physical world of classical physics is monistic, the physical world as described by quantum theory is basically dualistic. It has two kinds of entities, operators and states, that evolve according to two different laws of motion (in the Heisenberg picture): one continuous and deterministic; the other abrupt and stochastic. It also has two types of beingness: that of potentiality, represented by the wave function; and that of factuality, fixed by the events. This dual character matches that of matter and thought: each mind/body has its own private subjective thoughts and also a tendency to produce images of its public or objective aspects in the thoughts, associated with other minds/bodies; that is, each mind/body has two different aspects—the private/subjective/mental aspect and the public/objective/material aspect. This duality is not destroyed by admitting the *identity* of thoughts with certain dynamical aspects of the mind/brain system, if that dynamical system itself is essentially dualistic.

As regards *choice*, it is a fact that thoughts occur when choices are apparently needed. However, in classical physics, there are no choices: everything is fixed at the birth of the universe. Hence, thoughts play no role in the unfolding of nature; they are superfluous. In the quantum theory of the mind/brain, our thoughts occur in conjunction with choices between bona fide alternatives.

The issue of *unity* was discussed above.

SUMMARY AND CONCLUSIONS

Good science introduces no superfluous entities. Bohr followed this dictum: recognizing that the basic problem in the interpretation of quantum theory was the incompatibility of the formal quantum principles with the classical character of our experiences, Bohr made our classically describable experiences, and the knowledge derived from them, the basis of his interpretation. He interpreted the quantum formalism as a procedure acting on and within this knowledge, and thereby avoided the need to draw any line in the external world between its quantum and classical parts.

The scope of Bohr's version of quantum theory was, however, limited by its exclusion of biological systems. Universalization of the quantum principles brings human beings and their brains into the quantum mechanically described system and this converts the problem of the interpretation of quantum theory to the problem of the dynamical connection of conscious process to brain process. As in Bohr's approach, there is no need to introduce into the external world any classicalization process or any exact classical variables. Such things are alien to the quantum principles and are superfluous: they have no empirical ramifications or, if they do have such ramifications, then these represent deviations from the "pure" quantum predictions. Of course, certain macroscopic variables have, due to their local interactions with the microscopic degrees of freedom of their environment, a strong tendency to become approximately classical in various ways, which can be specified, and this allows us to imagine that various "classical" macroscopic variables in the universe have reasonably well-defined values even if they are not being observed by anyone, and have never been observed by anyone. However, this approximate effective classicalization of macrovariables is an automatic consequence of quantum theory, and there is no need within science to make it exact merely to satisfy our classical intuition. On the other hand, we do need to introduce quantum jumps at the level of our thoughts because the occurrences of these thoughts are, empirically, precisely the events that the quantum probabilities are the probabilities of.

This formulation of quantum theory is a composite of elements coming from Bohr, Einstein, von Neumann, Wigner, Heisenberg, and James. In particular, it integrates:

(1) Bohr's recognition that physical science rests on the empirical foundation of human thoughts that have classically describable content;
(2) Einstein's demand that, to ensure sufficient scope, logical coherence, and an adequate foundation for future developments, our basic physical theory should describe something that at least *could* be reality itself;
(3) von Neumann's demonstration that the quantum jumps can be placed exclusively at that level of brain activity where our thoughts enter;
(4) Wigner's interactionist view that the quantum jumps occur at the brain-mind interface, and that mind and matter interact there;
(5) Heisenberg's linkage of quantum process to the idea that nature's process proceeds by deterministic continuous evolution of potentialities, punctuated by abrupt actual events, where each actual event constitutes a choice between the various possibilities generated by the prior deterministic evolution;
(6) James' emphasis on the wholeness of our thoughts and their close association with our choices.

The cited parts of the ideas of the above-named physicists differ in important ways from the full ideas of these scientists:

(1) Bohr's own formulation of the idea that our classically describable knowledge is the foundation of physical theory was expressed in ways that emphasized the intersubjective agreement of the classically described aspects of our experience. Hence it provided an intimation, though no explicit claim, that there may be classically describable properties existing outside of our thoughts, even though no such things are brought into Bohr's formulation of quantum theory.
(2) Einstein's view that basic physical theory should describe "reality" referred, in fact, to a reality that did *not* include our thoughts.
(3) Von Neumann's main point was the practical equivalence of various possible placements of the quantum/classical divide. He did mention the special role of consciousness, but the significance of this remark was obscure. It was his close friend and colleague, Wigner,[3] who put the clear "mind-matter interaction" interpretation on von Neumann's words.
(4) Wigner initially espoused this "interactionist" view, but later[8] argued that quantum theory did not apply to the macroscopic systems (hence, to brains) because of the large effects of noise. However, the effects of noise on macroscopic variables are primarily to *effectively* decompose the pure quantum state into an approximate statistical mixture at the level of appropriate macrovariables. There is no reason to claim that the quantum principles,

viewed as the rules that govern our model of reality, must break down at the points where, due to decoherence effects, their empirical verification becomes difficult in practice.

(5) Heisenberg, although he used the concepts of potentia and actual events to describe reality, did not institute the tight connection proposed here between these "real" things and the quantum formalism: he continued to view the latter, in accordance with the Copenhagen interpretation, as a tool for making predictions about our observations.

Thus, the approach to quantum theory being proposed here is not an amalgamation of the *complete* views of the above-named scientists.

ACKNOWLEDGMENTS

I acknowledge very useful correspondence with A. Jadczyk.

REFERENCES

1. DESCARTES, R. 1974. Discourse on Method; The Meditations. Translated by F. E. Sutcliffe. Harmondsworth. Middlesex, United Kingdom.
2. VON NEUMANN, J. 1955. Mathematical Foundations of Quantum Theory. Princeton University Press. Princeton, New Jersey.
3. WIGNER, E. 1962. *In* The Scientist Speculates. I. J. Good, Ed. Heinemann. London.
4. STAPP, H. P. 1993. Mind, Matter, and Quantum Mechanics. Springer-Verlag. Berlin/New York.
5. DIRAC, P. A. M. 1947. The Principles of Quantum Mechanics. Chapter VI. Third edition. Oxford University Press (Clarendon). London/New York.
6. DIRAC, P. A. M. 1947. The Principles of Quantum Mechanics. Chapter IV. Third edition. Oxford University Press (Clarendon). London/New York.
7. JAMES, W. 1987. *In* William James—Writings 1902–1910, p. 1016. The Library of America. New York.
8. WIGNER, E. 1983. Review of the quantum mechanical measurement problem. *In* Quantum Optics, Experimental Gravity, and Measurement Theory. NATO ASI Series: Physics, Series B, Volume 94. P. Meystre & M. O. Scully, Eds.: 58.

On the Computational Power of Physical Systems, Undecidability, the Consistency of Phenomena, and the Practical Uses of Paradoxes

K. SVOZIL

Institut für Theoretische Physik
Technische Universität Wien
A-1040 Vienna, Austria

Undecidability in mathematics comes in different varieties; so does undecidability in physics. In physics, we have to make sure that the theory is a suitable formal representation of the phenomenology. For example, if the outcome of an experiment cannot be predicted, does that mean that "God plays dice"? Alternatively, does it mean that we are unable to compute a prediction, although the causes are in principle known, or does it simply mean that there are causes, but these are unknown to us? These questions may never be fully settled,[1] but, since Gödel's[2,3] and Turing's[4] centennial findings, remarkable advances have been made in the formal perception of undecidability. In addition to this, today's computers not only serve as number crunchers, but are becoming a medium to "virtual" realities. This greatly promotes the interaction between theoretical computer sciences, formal logic, and the physics of "real" reality.

Let us briefly consider the physical correspondents of two forms of mathematical undecidability, the first being associated with the assumption of the continuum (oracle computation) and the second arising in the context of finite computation. If one wishes to order theories with respect to the computational power necessary to implement them, continuum theories require more resources than theories based on universal computation (e.g., Cellular Automata[5–7]), which in turn are more powerful than finite models.

Continuum theories require the generation, storage, and processing of numbers that are uncomputable in the sense of Turing. More precisely (and worse), "almost all" (with probability one) elements of the continuum "urn" must be represented by random sequences; stated pointedly, any bit in its binary expansion is as uncorrelated to the previous and the following bits as is any toss of a fair coin from other tosses.[8,9] In continuum theories, "God plays dice". In such theories, undecidability, as not caused otherwise, is implemented by absolute randomness. How come then, one may ask, that classical mechanics has been long considered as the prototype for a "deterministic" model? The reason for this is twofold: First, one may conjecture that it is possible to keep all the nice features of continuum mechanics (e.g., calculus) while getting rid of the nasty aspects (e.g., nonconstructive randomness) at the same time; indeed, there are indications that this might be possible.[10] Second, there are "nonchaotic" dynamical systems, in which arbitrary initial conditions yield solutions that converge rapidly toward periodic behavior or at least converge toward a computable function (attractor).

Continuum theory (any dense set) allows the construction of "infinity machines", which could serve as oracles for the halting problem. Their construction closely follows Zeno's paradox of Achilles and the Tortoise (Hector) by squeezing the time it takes for successive steps of computation τ with geometric progression:

that is, the time necessary for the n-th step becomes $\tau(n) = k^n$, $k < 0$. The limit of infinite computation is then reached in finite physical time: $\lim_{N\to\infty} \Sigma_{n=1}^{N} \tau(n) = \lim_{N\to\infty} \Sigma_{n=1}^{N} k^n = 1/(1-k)$. It can be shown by a diagonalization argument that the application of such oracle subroutines would result in a paradox in classical physics (cf. p. 24 and p. 114 of reference 11). Therefore, at least in this example, too-powerful physical models (of computation) are inconsistent.

A second type of undecidability that occurs in finite systems is computational complementarity, which is realizable already at a very elementary prediagonalization level,[12] that is, without the requirement of computational universality or its arithmetic equivalent. The resulting "static" automaton logic has great similarities to quantum logic.[11,13]

Our major concern here shall be a third type of undecidability. It will be demonstrated how diagonalization techniques lead to the exclusion of time paradoxes and how quantum physics implements causality.

Classical information theory (e.g., see reference 14) is based on the bit as the fundamental atom. This classical bit, henceforth called *cbit*, is physically represented by one of two classical states. It is customary to use the symbols "0" and "1" as the names of these states. The corresponding classical computational basis is $\{0, 1\} = \mathscr{Z}_2$.

In quantum information theory (cf. references 15–22), the most elementary unit of information, henceforth called *qbit*, may be physically represented by a coherent superposition of the two states that correspond to the symbols 0 and 1. The corresponding quantum computational basis is the undenumerable set $\{|a, b\rangle \,|\, |a, b\rangle = a|0\rangle + b|1\rangle, |a|^2 + |b|^2 = 1, a, b \in \mathscr{C}\}$.

In what follows, we shall consider the hypothetical transmission of information backward in time. To be more specific, we shall use an EPR-type telegraph that uses entangled particles in a singlet state (i.e., the total angular momentum of the two particles is zero) as drawn in FIGURE 1. The apparatus is tuned to convey perfect correlations of the direction of angular momentum labeled by "+" and "−"; that is, the outcomes are either $++$ or $--$. Perfect correlations can be achieved by choosing a relative angle of measurement of π. The (unphysical) assumption necessary for signaling backward in time is that on one side, say for particles in path 1, the *outcome can be controlled.* This means that it will be possible to produce a particle with, say, direction of angular momentum "+" ("−") in path 1 at t_A, thereby transmitting a signal "+" ("−") via its perfectly correlated entangled partner in path 2 to a second observer back in time at t_B; hence, $t_A > t_B > t_S$, but otherwise arbitrary.

An alternative setup for information transmission backward in time by an EPR-type *quantum* telegraph would use the stronger-than-classical correlations for relative measurement angles not equal to 0, $\pi/2$, and π. In this case, the (unphysical)

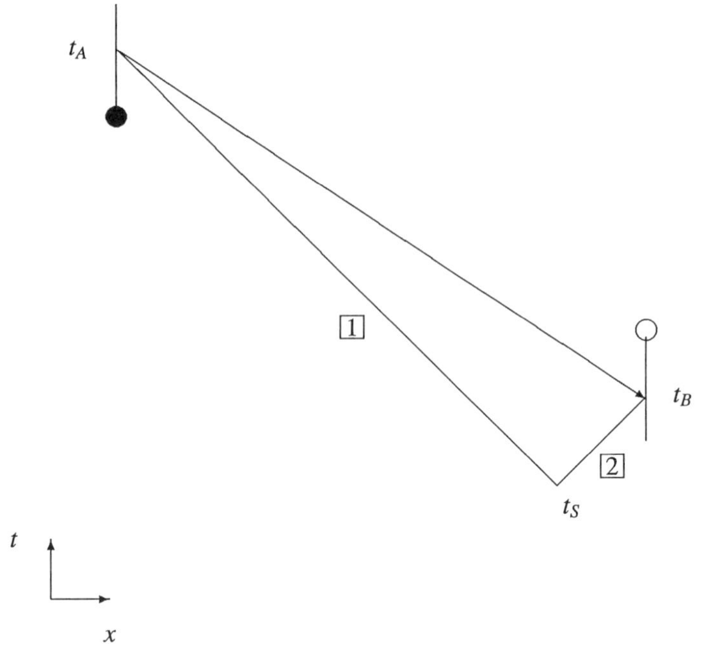

FIGURE 1. Scheme of backward-in-time signaling by an EPR-type telegraph. The postulated controllability of outcomes in 1, mediated via 2, is used to transmit information. The flow of information is indicated by the arrow. Symbols: "●" stands for the *active* mode, that is, controllable outcome (preparation); "○" stands for the *passive* mode, that is, measurement. The two signs are drawn on top and at bottom to indicate the orientation (relative angle π).

assumption necessary for signaling backward in time is the *parameter dependence* on one side, say for particles in path 2, on the angle chosen for measurements on beam 1 (e.g., by "cloning"; cf. references 23–27).

Of course, this kind of *outcome control* or *parameter dependence* would not be allowed either in relativistic mechanics or in quantum mechanics. The stronger-than-classical quantum expectation functions are often considered manifestations of "nonlocality"[28] (or, alternatively, of failure of classical probability theory[29]), but they only effect parameter dependence, not outcome dependence, of single events.[30,31]

We shall make use of the EPR-type telegraph to construct a time paradox and to argue against outcome predictability and outcome controllability in any form. In a similar manner, the liar paradox[32] was translated by Gödel into arithmetic[2] to argue against a complete description of a formal system within that very system.[33] For instance, the Gödelian sentence[34] claiming its own unprovability in a particular system appears undecidable within that very system. In physical terms, undecidability must be translated onto the level of *phenomena* and, only in a secondary step, into their theoretic description. On the phenomenological level, there is no such thing as an inconsistent phenomenon. In a typical *yes-no* experiment that can have two possible outcomes, only one of these outcomes will actually be measured. However, this uniqueness of phenomenology does not guarantee that a theory exists that predicts it completely. There might even be a "meta-physical" (extrinsic,[35] exo-[36]) arena in which this particular outcome could be deterministically accounted for. Yet, for an intrinsic observer who is embedded in the system,[37] this "meta-physical" level might be permanently inaccessible.[11,38] As will be shown below, quantum mechanics implements this phenomenological undecidability both by the postulate of randomness of certain outcomes and by the superposition principle. Related arguments have been put forward in references 34 and 39–44.

Consider two backward-in-time signaling EPR-type telegraphs of the above type arranged as drawn in FIGURE 2. Physically, the flow of information is mediated via the two entangled pairs in paths 1–2 and 3–4. An information in 2 is mirrored by M in 3. By this instrument, some mechanistic agent A (e.g., computer, deterministic observer) that is given the power of outcome control can exchange information with

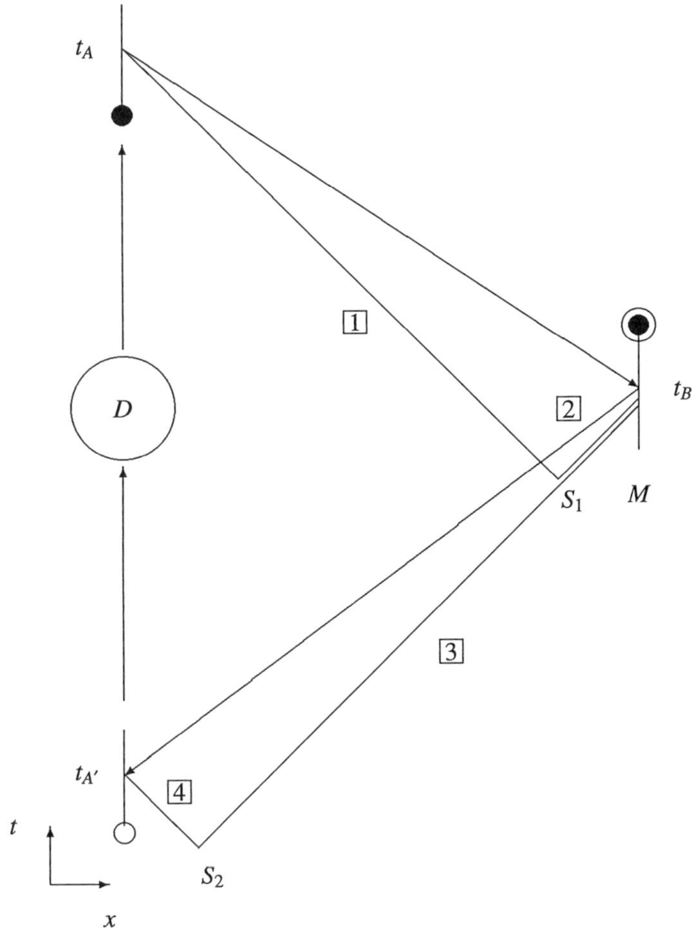

FIGURE 2. Time paradox (see caption to FIGURE 1 and the text).

itself on closed timelike lines.[45,46] Agent A shall be confronted with the following paradoxical task. Whenever A registers the information "+" ("−") at time $t_{A'}$, A must stimulate the opposite outcome "−" ("+") at the later time t_A.

Before discussing the paradox, let us consider the two states $|0\rangle \equiv$ "−" and $|1\rangle \equiv$ "+" that are accessible to A. These states can be the basis of a cbit with the identification of the symbols "0" and "1" for $|0\rangle$ and $|1\rangle$, respectively. Quantum mechanically, any coherent superposition of them is allowed. Agent A's paradoxical task can be formalized by a unitary evolution operator $\hat{D}$ as follows:

$$\hat{D}|0\rangle = |1\rangle, \qquad \hat{D}|1\rangle = |0\rangle. \tag{1}$$

In the state basis $\{|0\rangle, |1\rangle\}$ (τ_1 stands for the Pauli spin operator), $\hat{D}$ is just equivalent to the NOT-gate, that is,

$$\hat{D} = \tau_1 = \begin{pmatrix} 0 & 1 \\ 1 & 0 \end{pmatrix} = |1\rangle\langle 0| + |0\rangle\langle 1|. \tag{2}$$

The syntactic structure of the paradox closely resembles Cantor's diagonalization method, which has been applied by Gödel, Turing, and others for undecidability proofs in a recursion theoretic setup.[11,47–49] Therefore, $\hat{D}$ will be called the *diagonalization* operator, despite the fact that its only nonvanishing components are off-diagonal. (Notice that A's task would be perfectly consistent if there were no "bit switch" and if thus $\hat{D} = \mathcal{I}$.)

The paradoxical feature of the construction reveals itself in the following question: what happens to agent A? In particular, what does A register and send?

Let us first consider these questions from a classical perspective. Classically, the particles with which A operates can only be in one of two possible states, namely, in $|0\rangle$ or in $|1\rangle$, corresponding to the classical computational basis $\mathcal{Z}_2$. By measuring the particle in beam 4, A gets either the outcome "+" or "−". In both cases, agent A is led to a complete contradiction.

If A receives "+", corresponding to cbit state 1, then A is obliged to send out "−", corresponding to cbit state 0 (A has been assumed to be able to control the outcomes in beam 1). Due to the perfect EPR correlations, the partner particle in beam 2 is registered as "−" at the mirror at time t_B. By controlling the outcome in beam 3, this mirrored cbit can again be sent backward in time, where "−" is received by A via a measurement of the particle in beam 4. This, however, contradicts the initial assumption that the outcome in beam 4 is "+".

On the other hand, if A receives "−", corresponding to cbit state 0, then A is obliged to send out "+", corresponding to cbit state 1; yet, because at t_B the cbit is just reflected as described above, A should have received "+". Thus, classically, agent A is in an inescapable dilemma.

The defense strategy in formal logic and classical recursion theory against such inconsistencies is to avoid the appearance of a paradox by claiming (stronger: requiring) overall consistency, resulting in no-go theorems, that is, in the postulate of the impossibility of any operational method, procedure, or device that would have the potentiality to cause a paradox. (Among the many impossible objects giving rise to paradoxes are such seemingly innocent devices as a "halting algorithm" computing whether or not another arbitrary computable algorithm produces a particular output or an algorithm identifying another arbitrary algorithm by input-output experiments.)

In the above case, the defense strategy would result in the postulate of the impossibility of any backward-in-time information flow or, more general, of closed timelike lines. Because the only nontrivial feature of the backward-in-time information flow has been outcome controllability or parameter dependence, the diagonalization argument can be used against outcome controllability and parameter dependence, resulting in an intrinsic randomness of the individual outcomes.

Quantum mechanics implements exactly this kind of recursion theoretic argument, yet in a form that is not common in recursion theory. Observe that the paradox is resolved when A is allowed a nonclassical qbit of information. In particular, agent A's task can be consistently performed if it inputs a qbit corresponding to the *fixed point* state of $\hat{D}$; that is,

$$\hat{D}|*\rangle = |*\rangle. \tag{3}$$

The fixed point state $|*\rangle$ is just the eigenstate of the diagonalization operator $\hat{D}$ with eigenvalue 1. Notice that the eigenstates of $\hat{D}$ are

$$|I\rangle, |II\rangle = \left(\frac{1}{\sqrt{2}}\right)\left[\begin{pmatrix}1\\0\end{pmatrix} \pm \begin{pmatrix}0\\1\end{pmatrix}\right] = \left(\frac{1}{\sqrt{2}}\right)(|0\rangle \pm |1\rangle) \tag{4}$$

with the eigenvalues $+1$ and -1, respectively. Thus, the nonparadoxical, fixed point qbit in the basis of $|0\rangle$ and $|1\rangle$ is given by

$$|*\rangle = |I\rangle. \tag{5}$$

This qbit solution corresponds to the statement that it is impossible for the agent to control the outcome, a situation actually encountered in quantum mechanics.[50]

We close this short discussion on the consistent use of paradoxes in physics with a few comments. First, it is important to recognize that the above considerations have *no* immediate bearing on quantum complementarity. We believe that complementarity is a general feature of the intrinsic perception of computer-generated universes, which is realizable already at a very elementary prediagonalization level,[11–13] that is, without the requirement of computational universality or its arithmetic equivalent.

Second, the above argument remains valid for any conceivable (local or nonlocal[29,51]) hidden-variable theory. The consistency of the physical phenomenology requires that hidden variables remain inaccessible to an intrinsic observer. From an intrinsic, operational point of view, a paradox marks the appearance of uncertainty and uncontrollability.

Third and last, an application for quantum information theory is the handling of inconsistent information in data bases. Hence, two contradicting cbits of information $|a\rangle$ and $|b\rangle$ are resolved by the qbit $|1/\sqrt{2}, 1/\sqrt{2}\rangle = (1/\sqrt{2})(|a\rangle + |b\rangle)$. Throughout the rest of the computation, the coherence is maintained. After the processing, the result is obtained by a measurement. The processing of qbits requires an exponential space overhead on classical computers in cbit base.[7] Thus, in order to remain tractable, the corresponding qbits should be implemented on truly quantum universal computers.

REFERENCES AND NOTES

1. FRANK, PH. 1932. Das Kausalgesetz und seine Grenzen. Springer-Verlag. Berlin.
2. GÖDEL, K. 1931. Monatsh. Math. Phys. **38:** 173. English translation in reference 3.
3. GÖDEL, K. 1986. Collected Works, Volume I, Publications 1929–1936. S. Feferman, J. W. Dawson, Jr., St. C. Kleene, G. H. Moore, R. M. Solovay & J. van Heijenoort, Eds. Oxford University Press. London/New York.
4. TURING, A. M. 1936/37. Proc. London Math. Soc. **42**(2): 230.
5. FREDKIN, E. 1982. Int. J. Theor. Phys. **21:** 219; 1990. Physica **D45:** 254.
6. TOFFOLI, T. & N. MARGOLUS. 1987. Cellular Automata Machines. MIT Press. Cambridge, Massachusetts.
7. FEYNMAN, R. P. 1982. Int. J. Theor. Phys. **21:** 467.
8. CHAITIN, G. J. 1982. Int. J. Theor. Phys. **21:** 941. Reprinted in: CHAITIN, G. J. 1990. Information, Randomness, and Incompleteness. Second edition. World Scientific. Singapore.
9. CALUDE, C. 1994. Information and Randomness—An Algorithmic Perspective. Springer-Verlag. Berlin/New York.
10. BISHOP, E. & D. S. BRIDGES. 1985. Constructive Analysis. Springer-Verlag. Berlin/New York; BRIDGES, D. & R. RICHMAN. 1987. Varieties of Constructive Mathematics. Cambridge University Press. London/New York.
11. SVOZIL, K. 1993. Randomness and Undecidability in Physics. World Scientific. Singapore.
12. MOORE, E. F. 1956. Gedanken-experiments on sequential machines. *In* Automata Studies. C. E. Shannon & J. McCarthy, Eds. Princeton University Press. Princeton, New Jersey.
13. FINKELSTEIN, D. & S. R. FINKELSTEIN. 1983. Int. J. Theor. Phys. **22:** 753.
14. HAMMING, R. W. 1980. Coding and Information Theory. Second edition. Prentice–Hall. Englewood Cliffs, New Jersey.
15. ALBERT, D. Z. 1983. Phys. Lett. **94A:** 249.
16. DEUTSCH, D. 1985. Proc. R. Soc. London Ser. A **400:** 97.
17. FEYNMAN, R. P. 1985. Opt. News **11:** 11.
18. PERES, A. 1985. Phys. Rev. **A32:** 3266.
19. BENIOFF, P. 1986. Ann. N.Y. Acad. Sci. **480:** 475.
20. MARGOLUS, N. 1986. Ann. N.Y. Acad. Sci. **480:** 487.
21. DEUTSCH, D. 1989. Proc. R. Soc. London Ser. A **425:** 73.
22. DEUTSCH, D. & R. JOZSA. 1992. Proc. R. Soc. London Ser. A **439:** 553.
23. HERBERT, N. 1982. Found. Phys. **12:** 1171.
24. WOOTTERS, W. K. & W. H. ZUREK. 1982. Nature **299:** 802.
25. MILONNI, P. W. & M. L. HARDIES. 1982. Phys. Lett. **92A:** 321.
26. MANDEL, L. 1983. Nature **304:** 188.
27. GLAUBER, R. J. 1986. Amplifiers, attenuators, and the quantum theory of measurement. *In* Frontiers in Quantum Optics. E. R. Pikes & S. Sarkar, Eds. Adam Hilger. Bristol.
28. BELL, J. S. 1964. Physics **1:** 195; 1987. Speakable and Unspeakable in Quantum Mechanics. Cambridge University Press. London/New York.
29. PITOWSKY, I. 1989. Quantum Probability—Quantum Logic. Springer-Verlag. Berlin/New York.
30. SHIMONY, A. 1984. Controllable and uncontrollable non-locality. *In* Proceedings of the International Symposium on Foundations of Quantum Mechanics. S. Kamefuchi *et al.*, Eds. Physical Society of Japan. Kyoto.
31. SHIMONY, A. 1986. Events and processes in the quantum world. *In* Quantum Concepts in Space and Time. R. Penrose & C. I. Isham, Eds. Oxford University Press (Clarendon). London/New York.
32. The Bible contains the following passage, which refers to Epimenides, a Cretan living in the capital city of Knossos: "One of themselves, a prophet of their own, said, 'Cretans are always liars, evil beasts, lazy gluttons' "—St. Paul, Epistle to Titus I (12–13). For more details, see: ANDERSON, A. R. 1970. St. Paul's epistle to Titus. *In* The Paradox of the Liar. R. L. Martin, Ed. Yale University Press. New Haven.
33. The following citation is taken from K. Gödel's reply to a letter by A. W. Burks. Reprinted in: VON NEUMANN, J. 1966. Theory of Self-reproducing Automata. A. W. Burks, Ed.: 55.

University of Illinois Press. Urbana, Illinois (see also: FEFERMAN, S. 1984. Philos. Nat. **21:** 546): "... the fact that a complete epistemological description of a language *A* cannot be given in the same language *A* because the concept of truth of sentences of *A* cannot be defined in *A*. It is this theorem which is the true reason for the existence of undecidable propositions in the formal systems containing arithmetic." Nonetheless, it is possible for a system to contain its own "blueprint description" within itself. This "blueprint description" can be thought of as a string of symbols encoding an algorithm for a reconstruction of the original system. A. Shimony deserves acknowledgment for pointing to the importance of a distinction between a complete epistemological description and "statements of principle"; cf. SHIMONY, A. 1993. Search for a Naturalistic Worldview. Cambridge University Press. London/New York.

34. POPPER, K. R. 1950. Br. J. Philos. Sci. **1:** 117 & 173.
35. SVOZIL, K. 1986. Europhys. Lett. **2:** 83; 1986. Nuovo Cimento **96B:** 127; also see chapter 6 of reference 11.
36. RÖSSLER, O. E. 1987. Endophysics. *In* Real Brains, Artificial Minds. J. L. Casti & A. Karlquist, Eds.: 25. North-Holland. Amsterdam; 1992. Endophysics, Die Welt des inneren Beobachters. P. Weibel, Ed. Merwe Verlag. Berlin.
37. TOFFOLI, T. 1978. The role of the observer in uniform systems. *In* Applied General Systems Research. G. Klir, Ed. Plenum. New York.
38. The observer is situated, in O. E. Rössler's dictum, in a "Cartesian prison"; cf. ATMANSPACHER, H. & G. DALENOORT, Eds. 1994. Inside versus Outside, p. 156. Springer-Verlag. Berlin/New York.
39. ROTHSTEIN, J. 1982. Int. J. Theor. Phys. **21:** 327.
40. PERES, A. 1985. Found. Phys. **15:** 201.
41. WOLFRAM, ST. 1985. Phys. Rev. Lett. **54:** 735; 1984. Physica **D10:** 1; 1983. Rev. Mod. Phys. **55:** 601.
42. MOORE, CH. D. 1990. Phys. Rev. Lett. **64:** 2354; cf. BENNETT, CH. 1990. Nature **346:** 606.
43. ELITZUR, A. C. 1992. Phys. Lett. **A167:** 335; also see: POPESCU, S. & D. ROHRLICH. 1994. Found. Phys. **24:** 379. These sources refer to a talk of Y. Aharonov delivered at the Einstein centennial.
44. POSIEWNIK, A. 1993. On inconsistency of the language of physics. University of Gdansk preprint.
45. GÖDEL, K. 1949. Rev. Mod. Phys. **21:** 447; 1949. A remark about the relationship between relativity theory and idealistic philosophy. *In* Albert Einstein, Philosopher-Scientist. P. A. Schilpp, Ed.: 555. Tudor. New York.
46. DEUTSCH, D. 1991. Phys. Rev. **D44:** 3197.
47. DAVIS, M. 1965. The Undecidable. Raven Press. New York.
48. ROGERS, H. 1967. Theory of Recursive Functions and Effective Computability. McGraw–Hill. New York.
49. ODIFREDDI, P. 1989. Classical Recursion Theory. North-Holland. Amsterdam.
50. In an abstract form, the above argument is the analogon to the fixed point or paradoxical combinator in combinatory logic; cf. CURRY, H. B. & R. FEYS. 1958. Combinatory Logic, p. 151 & 178. North-Holland. Amsterdam; BARENDREGT, H. P. 1984. The Lambda Calculus (revised edition). North-Holland. Amsterdam.
51. PITOWSKY, I. 1982. Phys. Rev. Lett. **48:** 1299; 1983. Phys. Rev. **D27:** 2316; MERMIN, N. D. 1982. Phys. Rev. Lett. **49:** 1214; MACDONALD, A. L. 1982. Phys. Rev. Lett. **49:** 1215; PITOWSKY, I. 1982. Phys. Rev. Lett. **49:** 1216.

Pump-coupled Micromasers[a]

PÁL BOGÁR, JÁNOS A. BERGOU, AND MARK HILLERY

Department of Physics and Astronomy
Hunter College of the City University of New York
New York, New York 10021

Two micromasers coupled in series by the common pumping atomic beam of two-level atoms are studied. Both lossless and lossy cavities are investigated at zero as well as at finite temperature. Quantum correlation between the fields builds up due to the two atomic paths that an atom can follow to reach the same final state, reminiscent of Young's double-slit experiment.

Let us first assume no dissipation and consider the system by conditionally measuring the final state of the atoms emerging from the second micromaser. In the case of energy-preserving measurement schemes, when atoms are detected in the same final states as their initial ones, the steady state of the fields is an ensemble of distinct trapping number states. They correspond to Rabi angles of integer multiples of π and are located under the envelope of the initial fields. Starting from uncorrelated coherent states of the fields, the two micromasers are *uncorrelated* at steady state in this scheme. We can generate a *transient* entanglement of the two fields from initial coherent states via energy-transferring schemes when atoms are detected in final states different than their initial ones. Assuming initially excited atoms, correlation builds up due to the two possible atomic paths along which the atom can drop its photon in one or the other cavity. The photon number of the fields continuously increases until reaching a value where the measurement scheme becomes difficult to follow due to the small detection probabilities of the lower state of the atoms. This is a result of the so-called "single-" and "two-cavity" trapping effects. However, a combination of the two kinds of schemes can provide us with pure *steady-state entanglement* of the fields from initial coherent states by switching from an energy-transferring scheme to an energy-preserving scheme. The steady-state trapping number states produced by the latter scheme will be located in this case under the envelope of the correlated fields generated by the former scheme at the moment of the switching. In other words, the transient correlation produced by the energy-transferring scheme is frozen into a steady state by the trapping states of the energy-preserving scheme. An example illustrating this effect is depicted in FIGURE 1. Arbitrary entanglement of the fields can be engineered using this method in the absence of losses by choosing the proper interaction parameters, $g\tau$, in the cavities (g, atom-field coupling constant; τ, interaction time).

Starting from number states, $|N, N\rangle$, the system can be driven into trapping states of entangled fields of the form, $|N, N+M\rangle \pm |N+M, N\rangle$, without performing conditional measurements of the final atomic states. The procedure consists of two steps. First, we introduce some correlation by generating the above state of $M = 1$ and then amplify it to a larger M corresponding to the trapping state above. Both

[a]This work was supported by Grant No. N00014-92-J-1233 of the Office of Naval Research.

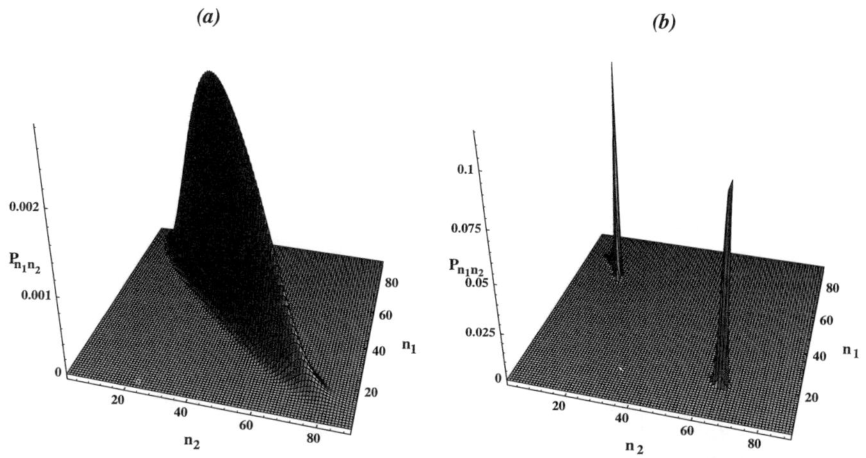

FIGURE 1. Photon statistics of the fields: (*a*) at the 60th atom, starting from coherent fields of the average of 20 photons using an energy-transferring scheme at $g\tau = 0.176$; (*b*) at the 200th atom, after starting from the field produced in part *a* using an energy-preserving scheme at $g\tau = \pi/4$.

steps are based on conditions regarding the interaction parameters, $g't'$ and $g''t''$, of the two cavities. They ensure that the fields evolve along statistical mixtures of state vectors of the above form only. An inclusion of other quantum states or an asymmetrical amplification of the state vector (when $g't' \neq g''t''$) would result in a

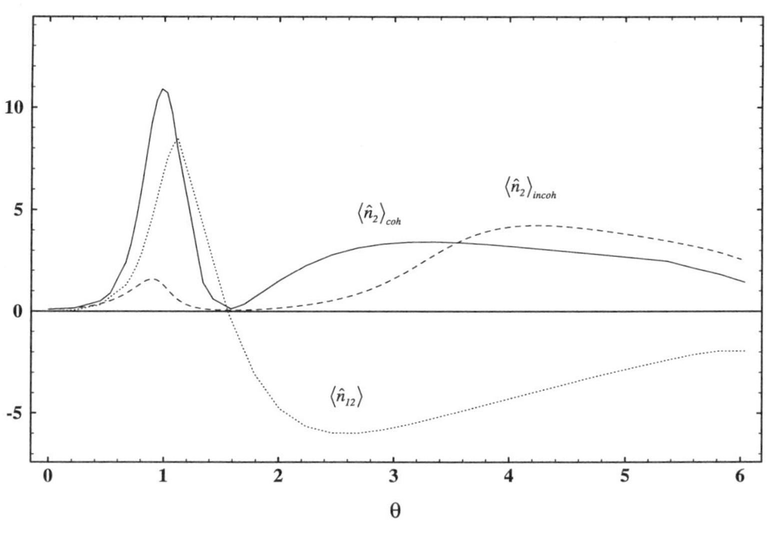

FIGURE 2. Steady-state average photon numbers in the second cavity, $\langle\hat{n}_2\rangle_{\text{coh}}$ and $\langle\hat{n}_2\rangle_{\text{incoh}}$, and the correlation between the two micromaser fields, $\langle\hat{n}_{12}\rangle$, as functions of the pumping parameter, $\theta = g\tau\sqrt{N_{\text{ex}}}$, depicted by the solid, dashed, and dotted lines, respectively. Here, $N_{\text{ex}} = 20$ and the average number of thermal photons is $\overline{n}_b = 0.1$.

loss of purity and mixed quantum states of no correlation at steady state. According to our numerical results, the entanglement exhibiting the above form does not survive at steady state in the presence of dissipation, but it still can be produced in the short-time transient regime when the pumping, $N_{ex} = r/\gamma$ (r, injection rate of atoms; γ, cavity decay rate), exceeds a certain threshold established by the losses. Although finite temperature allows for other trapping states to contribute and at the same time enhances the decay of the correlation, it does not significantly affect the transient behavior. Our estimate suggests that these entangled trapping states are experimentally feasible by the presently available facilities.

The steady-state behavior of the system in the presence of significant dissipations is shown in FIGURE 2, obtained from numerical simulations on computer. The average photon number of the second micromaser, $\langle \hat{n}_2 \rangle_{coh}$ (solid line), exhibits a second threshold below which the first-order correlation given by $\langle \hat{n}_{12} \rangle = (\frac{1}{2})(\langle \hat{a}_1 \hat{a}_2^\dagger \rangle + \langle \hat{a}_1^\dagger \hat{a}_2 \rangle)$ (dotted line) is positive and above which it is negative, corresponding to a switch in the locking of the relative phase of the fields from zero to π. The correlation is small far above threshold. The reason for the second threshold is that, in the vicinity of this point, all atoms enter the second cavity in their lower state prepared by the first micromaser; that is, it is an incoherent effect. This is why it can also be found in the incoherent case, $\langle \hat{n}_2 \rangle_{incoh}$ (dashed line), where the state of the atoms is measured between the cavities. Here, the loss of atomic coherence results in a loss of correlation between the fields due to the "which-path" information. Thus, $\langle \hat{n}_2 \rangle_{coh}$ significantly differs from $\langle \hat{n}_2 \rangle_{incoh}$ in those regions where the correlation is large. It can be seen in FIGURE 2 that coherence contributes to the photon number of the second micromaser and, on the other hand, results in an interference effect between the two fields when superposed.

MQC

An Experiment for Detecting Macroscopic Quantum Coherence with a System of SQUIDs[a]

P. CARELLI,[b] M. G. CASTELLANO,[c] F. CHIARELLO,[d]
L. CHIATTI,[e] M. CIRILLO,[f] C. COSMELLI,[d]
G. DIAMBRINI PALAZZI,[d] D. FARGION,[d] R. LEONI,[c]
G. ROTOLI,[f] F. SCARAMUZZI,[g] AND G. TORRIOLI[c]

[b]*Dipartimento di Energetica*
Università de L'Aquila
L'Aquila, Italy

[c]*Istituto di Elettronica dello Stato Solido*
CNR
Roma, Italy

[d]*Dipartimento di Fisica*
Università "La Sapienza"
Roma, Italy

[e]*CRS*
Istituto Regina Elena
Roma, Italy

[f]*Dipartimento di Fisica*
Università di Tor Vergata
Roma, Italy

[g]*Laboratori ENEA*
Frascati, Roma, Italy

The quantum mechanics (QM) behavior of microscopic quantum states has been successfully established after about 70 years of experimental results obtained mainly in the field of elementary particle physics. However, problems concerning consistency of QM predictions with causality, relativity, and macroscopic behavior have been raised and discussed since the very beginning of QM.

In the view of realizing experiments to detect the quantum behavior of macroscopic coherent states, as investigated in several papers,[1–3] the quantum behavior of an rf-SQUID has been analyzed. This device, consisting of a superconducting loop interrupted by a Josephson junction, is described by a single macroscopic degree of freedom (the phase difference of the Cooper pairs across the junction) moving, under proper bias conditions, in a double-well potential. Each well corresponds to a definite macroscopic state, with a definite valuc of magnetic flux (say, Φ_- and Φ_+). In

[a]This work was supported by the Istituto Nazionale di Fisica Nucleare (INFN), Italy.

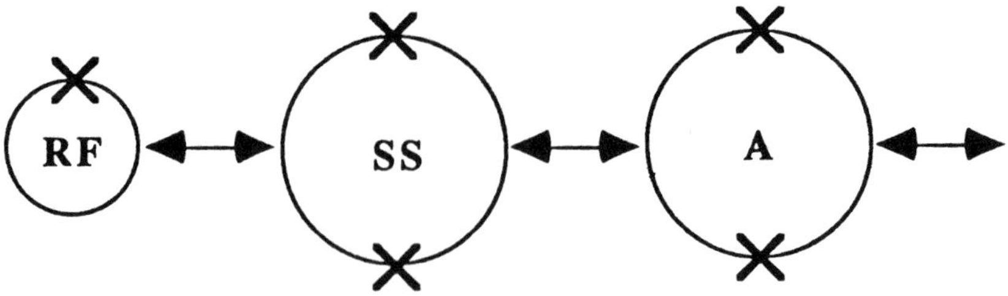

FIGURE 1. Scheme of the experimental apparatus.

principle, it is possible to perform an experiment to detect coherent tunneling oscillations from one well to the other, provided that we are able to fulfill all the experimental constraints necessary to get a coherent oscillation for a time interval of the order of a few microseconds.

The scheme of the experiment is shown in FIGURE 1, where RF represents the rf-SQUID under study. To read the status of the rf-SQUID, a superconducting switch (SS) is used. This consists of an hysteretic dc-SQUID that can either make a transition in the normal state or remain superconducting, depending on the sign of the circulating current in the rf-SQUID. A linear amplifier (A), in turn, reads the status of the switch.

The experiment that we present here is a modification[4,5] of the original one, discussed by Leggett and Tesche. Here, we want to present a very short description of the proposed measurements (see FIGURE 2). Suppose that the rf-SQUID has been

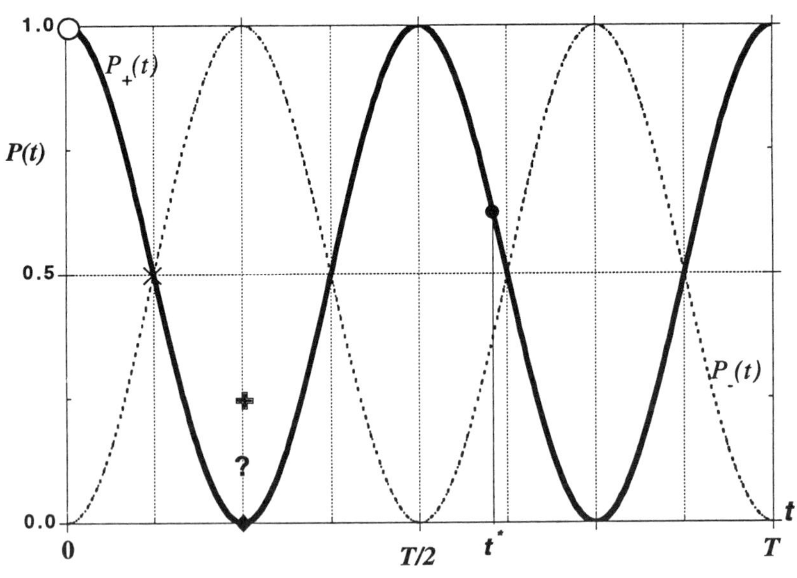

FIGURE 2. Quantum behavior of the probabilities to find the system in the plus or minus flux state.

prepared in the state Φ_+ at $t = t_0$ and is oscillating back and forth between Φ_+ and Φ_- with frequency ω_τ. We want to perform the following flux measurements on the system:

(A) Make a measurement at a generic time t^*. Repeat the measurement for different times t^*. The expected probability of finding the system in the state Φ_+ is

$$P_+(t^*) = \cos^2[\omega_\tau(t^* - t_0)].$$

This test should check the existence of the oscillations according to the QM formalism. If not observed, it would mean that QM predictions are violated at the macroscopic level.

(B) Make a measurement at a time $t = T_\tau/2$ and another at $t^* \geq T_\tau/2$. If the measured $P_+(t^*)$ is equal to that found in the absence of the measurement at $t = T_\tau/2$, the noninvasivity of the measurement (NIM) in the strict quantum definition is therefore proven.

(C) Make a measurement (with the same technique used in part B to perform an NIM) at $t = T_\tau/8$. Retain only the measurements where $\Phi_{rf} = \Phi_+$. Make a new measurement at $t = T_\tau/4$. If the system had a restart due to the first measurement, then one expects to find

$$P^+\left(\frac{T_\tau}{4}\right) = \frac{1}{4}.$$

In contrast, the above probability should be 0 if no restart has taken place. The $P = \frac{1}{4}$ result should then be evidence for the nonclassical behavior of our macroscopic system.

The realization of such an ambitious experiment is related, of course, to the fulfillment of all the experimental constraints necessary to maintain the coherence of the oscillation. The major problem at the moment seems to be the correct interpretation of the "quantum dissipation" that, in increasing the entropy of the system, should bring the oscillation from the coherent into the incoherent state. Work is in progress to evaluate the quantum dissipation associated with the rf-SQUID from both the experimental and theoretical point of view.

REFERENCES

1. LEGGETT, A. J. 1980. Macroscopic quantum systems and the quantum theory of measurement. Suppl. Prog. Theor. Phys. **69:** 80.
2. LEGGETT, A. J. & A. GARG. 1985. Quantum mechanics versus macroscopic realism: is the flux there when nobody looks? Phys. Rev. Lett. **54:** 857.
3. TESCHE, C. D. 1990. Can a noninvasive measurement of the magnetic flux be performed with superconducting circuits? Phys. Rev. Lett. **64:** 2358.
4. BENATTI, F., G. C. GHIRARDI & R. GRASSI. 1994. On some recent proposals of testing macrorealism versus quantum mechanics. Found. Phys. Lett. In press.
5. COSMELLI, C., G. DIAMBRINI PALAZZI *et al.* 1993. Proposal for an experiment for detecting macroscopic quantum coherence with a system of SQUIDs. Unpublished.

A Failure of Einsteinian Reality in an Unequal-Arm Interferometer

HYUNG SUP CHOI

Department of Physics
Center for Theology and the Natural Sciences
Berkeley, California 94709

Bernstein, Greenberger, Horne, and Zeilinger (BGHZ) extended the results of the many-particle Greenberger-Horne-Zeilinger theorem back to a two-particle system.[1] Unlike usual spin models, the BGHZ system uses spatial degrees of freedom in the form of a two-particle interferometer and identifies them as the elements of reality in Einsteinian sense. It was claimed that the restriction given by the spatial degree of freedom is powerful enough to provide a conclusive proof against the Einstein-Podolsky-Rosen-type local realism.

However, the BGHZ theorem as it was presented (the theorem used the usual equal-arm interferometer such as the one devised by Rarity and Tapster[2]) was vulnerable to some classical models such as de Broglie's pilot wave that has succeeded in producing a double-slit interference pattern.[3] The theorem would thus require an *unequal*-arm interferometer such as the one devised by Franson (see figure 1 of reference 4). In an unequal-arm experiment, although quantum theory predicts interference fringes due to the indistinguishability of two possible paths (i.e., either both photons take the shorter path or both photons take the longer path), any local realistic theory that tries to exploit the possibility of the interference between the two paths would violate the special relativity due to the path difference between the shorter path and the longer path. The other possibility that one photon takes a shorter path and the other takes a longer path is a distinguishable process and one can block this possibility by making the path difference long enough so that the coincidence of this type would violate the special relativity on which the local realism is based.

In this revised version of the BGHZ experiment, one may establish Einsteinian elements of reality by placing very fast shutters at the openings and by measuring the time taken for a pair of photons to travel to the coincidence counters. If the photon 1 is found to have taken the shorter path, we know that the photon 2 has also taken the shorter path, with certainty and without disturbing in any way the path of the particle. Similarly, the same thing holds for the longer paths. Therefore, this securely establishes the paths as elements of reality.

The kind of separation of integral that occurs in local realistic theories in the proof of BGHZ was not possible in the spin model because each particle takes only one path. Because this separation of hidden variables by means of separation of paths imposes such a tremendous restriction on local realism, the loopholes that were exploited most by the advocates of the local realism can be effectively blocked. The proof goes through for a general case of an *im*perfectly correlated system with low counter efficiency. Local hidden-variable theories would produce *one* flat straight line. This two-particle GHZ theorem, therefore, can be regarded as an altogether more powerful theorem than the other theorems that preceded it against local realism.

The photon versions of the BGHZ gedankenexperiment may readily be implemented by a parametric downconversion of light technique. The experimental works done for Franson's model by Ou *et al.*[5] and Kwiat *et al.*,[6] which were originally performed for the case of Bell's theorem, can be used to implement the two-particle GHZ theorem, with slight modifications.[7]

REFERENCES

1. BERNSTEIN, H., D. GREENBERGER, M. HORNE & A. ZEILINGER. 1991. Catastrophic failure of EPR realism for two spinless particles (preprint).
2. RARITY, J. & P. TAPSTER. 1990. Experimental violation of Bell's inequality based on phase and momentum. Phys. Rev. Lett. **64:** 2495.
3. See, for example: CROCA, J. R. 1987. Neutron interferometry can prove (or refute) the existence of de Broglie's waves. Found. Phys. **17**(no. 10): 971; MANDEL, L. *et al.* 1991. Experimental test of the de Broglie guided-wave theory for photons. Phys. Rev. Lett. **66:** 1111.
4. FRANSON, J. 1989. Bell inequality for position and time. Phys. Rev. Lett. **62:** 2205.
5. OU, Z., X. ZOU, L. WANG & L. MANDEL. 1990. Observation of nonlocal interference in separated photon channels. Phys. Rev. Lett. **65:** 321.
6. KWIAT, P., W. VAREKA, C. HONG, H. NATHEL & R. CHIAO. 1990. Correlated two-photon interference in a dual-beam Michelson interferometer. Phys. Rev. A **41:** 2910.
7. Detailed proofs and implementations are described in: CHOI, H. S. 1991. Ph.D. thesis (chapters IX and X). City University of New York.

"Which-Path" Experiments

CHARLES E. ENGELKE[a] AND CHARLES W. ENGELKE[b]

[a]*Department of Physics*
Lehman College
The City University of New York
Bronx, New York 10468

[b]*Department of Physics*
Northeastern University
Boston, Massachusetts 02115

INTRODUCTION

The Bell inequality now provides the key link in a chain of reasoning that is accepted by a majority of physicists as a proof that it is impossible to construct any theory consistent with experimental results that does not violate locality. A substantial minority of physicists, however, suspect that this impossibility proof is flawed, as was von Neumann's famous 1932 proof of the impossibility of constructing a more complete theory such as de Broglie's and Bohm's, consistent with the then-known experimental results. This minority included J. S. Bell himself, who refuted von Neumann's proof in 1966 and who often expressed his continuing interest in the de Broglie pilot-wave model. Bell presented his doubts in these words:[1]

> Of the various impossibility proofs, only those concerned with local causality seem now to retain some significance. . . . Let us hope that these analyses also may one day be illuminated, perhaps harshly, by some simple constructive model. However that may be, long may Louis de Broglie continue to inspire those who suspect that what is proved by impossibility proofs is lack of imagination.

It is therefore appropriate to follow up new lines of attack, independent of the Bell inequality, to further exploit correlated microsystems in shedding light on the entwined issues of locality violation and the completeness and correctness of quantum theory. One such line of attack is provided by "which-path" experiments utilizing correlations.

Von Neumann's treatment of correlated systems avoided predicting overt, directly observable locality violations by requiring that the expectation value of an observable such as a phototube singles count rate, which refers to one subsystem alone, must be the same whether calculated from the generally nonfactorable representation of the composite system or for the subsystem alone.[2] This requirement forced von Neumann to postulate incoherent addition of amplitudes for paths from the same microscopic source that could be distinguished by observations on a correlated system. It is remarkable that von Neumann's 1932 textbook thus anticipated some possible conflicts with locality in dealing with correlated systems and built a resolution of these conflicts into the mathematical formalism of quantum theory.

ESSENTIAL FEATURES OF A DEFINITIVE "WHICH-PATH" EXPERIMENT

Scully, Englert, and Walther[3] have recently returned attention to the fact that the loss of coherence predicted by von Neumann's formalism is separate and additional to any uncertainty principle effects that may occur. For example, in the experiment of Scully *et al.*, when the initial maser cavity states do not have precisely known photon numbers, the final cavity states are then not strongly correlated with the two amplitudes for arrival of the atom and the ordinarily observed coherent addition is predicted. However, if the cavity is cooled to produce initial cavity states of zero photon number, the final states are perfectly correlated so that a one-photon final cavity state would indicate that the atom has passed through that cavity and the disappearance of interference is predicted. Scully *et al.* designate experiments like theirs that can demonstrate the presence of von Neumann's postulated loss of coherence due solely to correlations as "which-path" experiments. The unique and essential features of a definitive "which-path" experiment such as theirs are as follows:

(1) two amplitudes for the arrival of a single quantum via different paths from a single radiating source are put into strong correlation with orthogonal states of another system that can be experimentally distinguished;
(2) the apparatus change (i.e., cooling a maser cavity or widening a slit traversed by an atomic beam) producing this strong correlation should have no other known physical influence on the coherence of the amplitudes.

The prediction to be tested in a definitive experiment is that, because the correlation makes the paths distinguishable, for this reason only and without any other physical cause, the amplitudes will combine incoherently, no longer producing the interference previously observed when the correlations were weak or absent.

PARAMETRIC DOWNCONVERSION EXPERIMENTS

Recent work performed with pairs of correlated photons produced by parametric downconversion in laser-pumped crystals has found the postulate that amplitudes from distinguishable paths never interfere to be sufficient to explain the experimental results without the mechanism of uncertainty principle effects. However, these important experiments do not bear strongly on the question of whether the postulate is necessary as well as sufficient and cannot be considered as definitive "which-path" experiments as defined here.

A SIMPLE OPTICAL "WHICH-PATH" EXPERIMENT

FIGURES 1 and 2 show the experimental setup and the optical geometry for a "which-path" experiment designed by the authors.[4] The optical source volume is defined by the intersection of the atomic beam and laser beam and thus has a thickness equal to the slit width w, which determines the atomic beam dimension. A photomultiplier tube (fitted with a filter passing only the desired 0.6164-micron

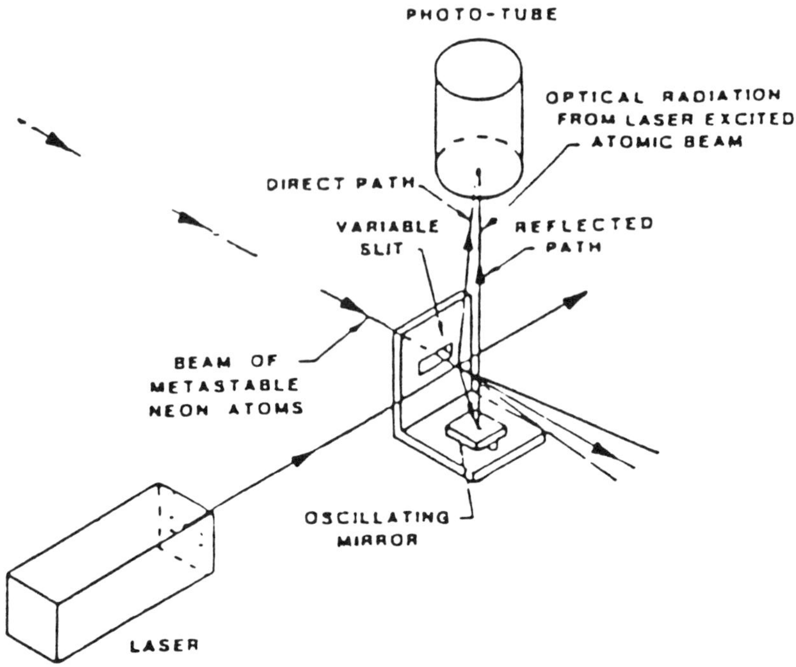

FIGURE 1. Experimental setup.

radiation) receives both direct radiation coming upward from the source and also radiation that starts downward, but is reflected from an oscillating mirror at a small distance d below the lower slit edge.

In contrast to Scully's "which-path" experiment, the experiment proposed here must live with an indirect uncertainty principle limitation. The width of the slit defining the atomic beam—and, with it, the important optical source dimension—must be comparable to the optical wavelength to obtain the small uncertainty in initial total momentum needed for significant localization. Thus, the near-optimum choice of width at the $w = (5/4)\lambda$ recurrence of fringe visibility results in classical and quantum fringe visibilities of 13% and 3.6%, respectively, compared to the values of 100% and 0% expected by Scully *et al.*

A detailed presentation of this experiment and the quantum mechanical and classical calculations of the interference term can be found in reference 4.

For a one-quarter-wavelength slit, the initial atom momentum is poorly known because of diffraction of the atomic beam and the final state of the recoiling atom is not strongly correlated with the photon arrival amplitudes, so the ordinarily observed coherent addition is predicted. The resulting 64% interference term due to the coherent amplitude addition then predicted by both classical electromagnetic theory and quantum mechanics would show up in a single one-hour run, at the projected signal and background rates, as a sinusoidal variation with time whose amplitude is 24 times the standard deviation of the data points. Adjusting piezoelectric transducers to increase the slit width to five-quarters of the wavelength, the 13% interference term predicted by coherent amplitude addition would have an amplitude of about 18 standard deviations, whereas the 3.6% interference term predicted by the quantum calculation for the 98%-distinguishable paths would have an amplitude of 5 standard deviations. The two possible results of widening the slit to five-quarters of a wavelength are thus unambiguously recognizable in even a one-hour data run, once the adequacy of the experimental geometry has been demonstrated by the run at one-quarter wavelength. However, the achievement of adequate experimental geometry is expected to take several years of continuous effort.

CONCLUSIONS

In both experiments discussed, a change in the apparatus (cooling a maser cavity or widening a slit) varies only the initial state of the correlated system and has no known direct physical influence on the coherence of the amplitudes involved. It is precisely this fact that would make observations of the predicted reduced coherence following the apparatus change convincing evidence that distinguishability alone eliminates coherence. It should also be noted that the de Broglie concept of a particlelike photon, guided by an associated real wave, requires coherent addition of the real wave amplitudes emitted by a single source and would be disproved by the observation of reduced interference in such a definitive "which-path" experiment. On the other hand, such experiments test a hitherto unverified prediction introduced by von Neumann in what now appears to have been an only partially successful

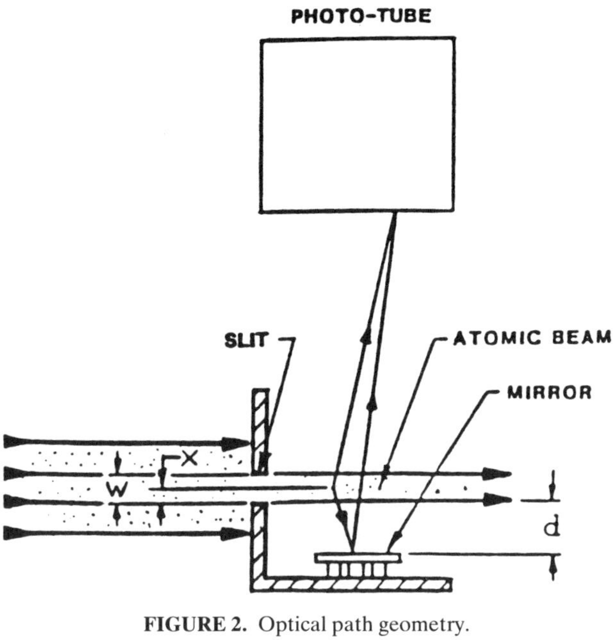

FIGURE 2. Optical path geometry.

attempt to impose the external attributes of locality on a fundamentally nonlocal theory.

REFERENCES

1. BELL, J. S. 1982. Found. Phys. **12:** 989.
2. VON NEUMANN, J. 1955. Mathematical Foundations of Quantum Mechanics. Chapter VI, section 2, p. 424–425. Princeton University Press. Princeton, New Jersey.
3. SCULLY, M. O., B. ENGLERT & H. WALTHER. 1991. Nature **351:** 111.
4. ENGELKE, C. E. & C. W. ENGELKE. 1986. Found. Phys. **16:** 905.

Quantization of the Wave Function

EDUARDO V. FLORES

Department of Chemistry and Physics
Rowan College of New Jersey
Glassboro, New Jersey 08028

At each point in space, the wave function can take on a continuous range of values. In this work, the wave function is allowed to take discrete values only. This has important consequences, such as the possibility of giving a *clearer* interpretation of quantum mechanics. The technique used in these calculations is the principle of least action. The results for the free particle case reveal that the approach is sensible: the wave effects are limited to a smaller region around the particle; outside this region, no wave effect is present. A wave packet will not spread beyond certain limits.

In real life, we cannot measure infinitesimal distances, times, or masses. It is possible to write theories that take into consideration this limitation, but a calculation is simpler if we use continuous instead of discrete calculus. Whenever we use continuous calculus, we extrapolate our measuring abilities to unknown regions with the possibility of error. Today, computers can help us handle the difficulties of discrete calculus and our theories do not require continuity.

In this work, the wave function is discrete or quantized. This approach does not discard the mathematical framework of quantum mechanics, but treats it as the continuous limit of a discrete theory. To calculate, I use the principle of least action.[1] For the time-independent case with no magnetic field, only the real part of the wave function matters. The discrete action A in one dimension is

$$A = \sum_n a\left[\left(\frac{\hbar^2}{2m}\right)\left(\frac{\psi_{n+1} - \psi_n}{a}\right)^2 + (V_n - E)\psi^2\right], \tag{1}$$

with constraint

$$\sum_n \psi_n^2 = \text{constant}. \tag{2}$$

The wave function ψ_n takes values proportional to 0, ± 1, ± 2, ± 3, ... A minimum for the action represents a stable ground state configuration for the particle. For the free particle and the infinite square well potential, the term $\Sigma(V_n - E)\psi^2$ is a constant and does not play any role in minimizing the action. Only the kinetic energy term need be considered. Thus, the action A will be a minimum when $B = \Sigma_n(\psi_{n+1} - \psi_n)^2$ is a minimum. For the free particle case, the minimum for B is found when all ψ_n's are equal to 1 (or -1) in the largest connected region with the smallest boundary. Hence, the ground state wave function is a rectangle of height 1 and spatial extent l. This is what we would expect if the wave function ψ_n represented an actual particle. FIGURE 1 shows the case when the range a of an infinite square well potential is larger than l. Notice that the ground state of the particle in the well is the same as the ground state of a free particle. In FIGURE 2, a is smaller than l and the

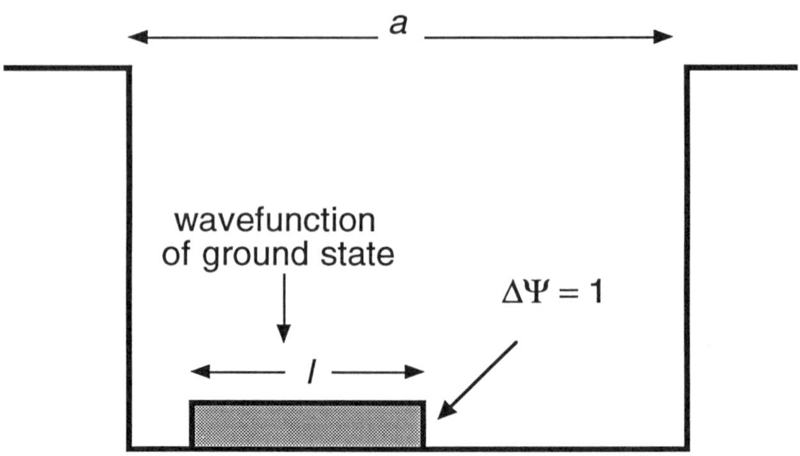

FIGURE 1. When l is smaller than a, the ground state of a particle in an infinite square well potential is similar to the ground state of a free particle.

wave function is compressed and should approach the sine function of the continuous case.

The examples above show that the wave function for a free particle at the lowest energy level is zero in a large region of space and nonzero in a small connected region. The small nonzero region could be thought of as the region where the particle experiences wave effects. The large region would be the outside where no wave effects are present. Thus, the wave function of a macroscopic object will always be zero beyond a relatively small distance from that object and classical physics could describe its dynamics.

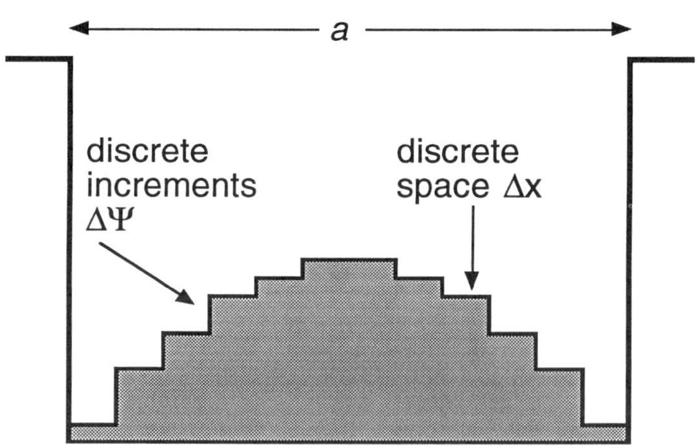

FIGURE 2. When l is greater than a, the particle wave function approaches the ground state of the continuous case. A smaller Δx will give a closer approximation.

The wave function corresponding to higher energies can also be obtained using the principle of least-action technique. For instance, the wave function for the free particle of nonzero momentum is the product of a constant amplitude, a rectangular box, times the phase $e^{i2\pi x/\lambda}$. This wave function has the expected continuous limit.

A discrete wave function also implies that, at a microscopic level, space and time are discrete too. Consider the kinetic term $\Delta\psi/\Delta x$ found in the Lagrangian. If the wave function changes in discrete steps ($\Delta\psi$), then the space interval Δx cannot be zero; otherwise, the ratio $\Delta\psi/\Delta x$ would be infinite. The discrete distances and times may be very small, on the order of Planck's length. Our concept of space and time may suffer another change. Quantum field theory presents many divergences at the short-distance scale. It is widely felt that the divergences are symptomatic of a chronic disorder in the small-distance behavior of the theory.[2] Thus, a discrete space may be a very helpful concept in the study of quantum field theory.

REFERENCES

1. LANDAU, L. D. & E. M. LIFSHITZ. 1977. Quantum Mechanics. Volume 3. Pergamon. Elmsford, New York.
2. BJORKEN, J. D. & S. D. DRELL. 1965. Relativistic Quantum Fields. McGraw–Hill. New York.

Neutron Multiray Reflection[a]

A. I. FRANK AND D. B. AMANDZHOLOVA

Frank Laboratory of Neutron Physics
Joint Institute for Nuclear Research
141980 Dubna, Moscow Region, Russia

In this report, the problem of neutron reflection from a medium with an oscillating potential is considered. The present report is closely related with previous work.[1,2]

Let us consider the reflection of neutrons from a mirror located at $x = 0$ and characterized by a periodical variable potential, $V(t) = U + u(t)$. Its constant part is just the optical potential, U_{opt}. It is natural to assume the time-dependent part of the potential, $u(t) = u(t + T)$, to be determined by the magnetic interaction.[1] Because the left-hand half-space is free, the state of the reflected waves may be represented in the form,[3]

$$\psi_r(x, t) = \sum_{n=-\infty}^{\infty} C_n e^{i(k_n x - \omega_n t)}, \qquad C_n = \left(\frac{1}{T}\right) \int_0^T r(t) e^{in\Omega t} dt, \tag{1}$$

where

$$\omega_n = \omega + n\Omega, \qquad k_n = k(1 + n\gamma)^{1/2}, \qquad \gamma = \Omega/\omega \ll 1, \qquad \Omega = \frac{2\pi}{T}. \tag{2}$$

Therefore, the aim is to establish the form of the time-dependent amplitude, $r(t)$. It is obvious that, in a low-frequency limit, it is possible to use the usual Fresnel formulas, which are deduced from the stationary continuity equation by a formal substitution in it, namely, the dependence on time. However, the quasi-stationary approximation is also valid in the nearby region $x \ll 1/k\gamma$, $x \ll 1/k\gamma_2$ [$\gamma_2 = \hbar\Omega/|E - V|$]. Consequently, we may use the quasi-stationary continuity equation for $x = 0$ and the Fresnel formula will follow from it, even when we do not find the exact solution in the right-hand half-space. As a result, we have

$$r(t) = \begin{cases} \dfrac{\sqrt{E} - \sqrt{E - V(t)}}{\sqrt{E} + \sqrt{E - V(t)}} & \text{if } [E - V(t)] > 0 \\[2ex] \dfrac{\sqrt{E} - i\sqrt{|E - V(t)|}}{\sqrt{E} + i\sqrt{|E - V(t)|}} & \text{if } [E - V(t)] < 0 \end{cases} \tag{3}$$

where E is the neutron energy. Naturally, the above arguments are also valid in the case of thermal or cold neutrons exhibiting grazing incidence on the mirror and not only the normal incidence upon the medium. In this case, the normal component, $k_\perp$, and the quantity, $E_p = \hbar^2 k_\perp^2/2m$, must be respectively substituted for the wave

[a]This work was partially supported by the International Science Foundation (Grant No. RFH 000).

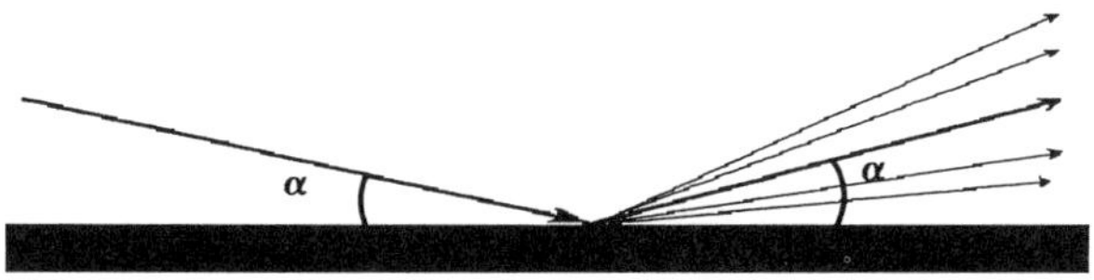

FIGURE 1. Quantum reflection.

number, k, and the energy, E, in all the above expressions. Thus, the reflection results in a set of reflected waves with differing normal wave number components:

$$k_{\perp n} = k_{\perp}\left[1 + n\left(\frac{2m\Omega}{\hbar k_{\perp}^{2}}\right)\right]^{1/2}, \tag{4}$$

where $k_{\perp}$ corresponds to the wave number of the incident wave.

Because the longitudinal component of the wave vector is not changing when a reflection occurs, the reflected waves corresponding to various satellites will exhibit different reflection angles and energies. The picture that arises can be illustrated by FIGURE 1.

It seems extremely similar to the diffraction on a plane grating. This similarity, however, is purely superficial. In the case considered, the adjoined variables of the Fourier transformation are the time and frequency (and, consequently, energy), but not the coordinates and wave numbers. We shall term this phenomenon as "quantum multiray reflection".

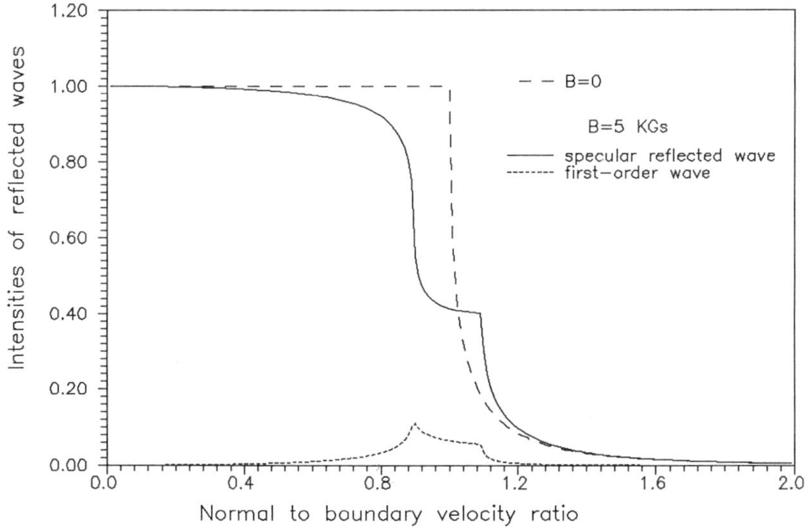

FIGURE 2. The intensities of reflected waves. A typical value of U_{opt} = 150 neV is assumed for the optical potential.

The actual form of the dependent $u(t)$—and, consequently, $r(t)$—is determined by the experimental conditions. For simplifying the calculations, we shall deal with the case of a ferromagnetic mirror magnetized up to saturation and assume the orientation of magnetization to change after a time $T/2$. Then, the quantity $r(t)$ is constant during a semiperiod and the calculation of the Fourier coefficients, C_n, in equation 1 becomes trivial.

The intensities of reflected waves of the zeroth (specular reflection) and first orders are plotted in FIGURE 2 versus the normal component of the incident neutron velocity. The angular distribution of reflected waves is given by equation 4.

Experimental observation of a new quantum effect, such as, for instance, the multiray reflection considered above, is evidently of great interest for its own virtue. A simple estimation shows that an experiment on the demonstration of neutron multiray reflection is quite within the range of resolutions exhibited by ordinary reflectometers.

One may try, however, to foresee the possible applications of the said effect. Note that the time-dependence may be related not only to the remagnetization of the material of the mirror, but also to rapid changes of the spin orientation.

Moreover, we note that quantum reflection results in coherent separation of the beams. In essence, this gives rise to the possibility of creating a neutron interferometer based on a new principle. The arrangement of this device may be quite similar to that of an interferometer based on diffraction gratings[4] or on reflection from a vibrating surface.[5]

Quantum reflection may turn out to be useful for the development of new techniques for magnetic studies in neutron reflectometry.

REFERENCES

1. GERASIMOV, A. S. & M. V. KAZARNOVSKY. 1976. JETP **71:** 1700.
2. GÄHLER, R. & J. FELBER. 1993. *In* Quantum Interferometry. F. De Martini & A. Zeilinger, Eds. World Scientific. Singapore.
3. FRANK, A. I. & V. G. NOSOV. 1993. *In* Quantum Interferometry. F. De Martini & A. Zeilinger, Eds. World Scientific. Singapore.
4. IOFFE, A. I. 1988. Physica **B151:** 50.
5. GOLUB, R. & S. K. LAMOREAX. 1992. Phys. Lett. A **162:** 122.

Resonant Expansions and the Time Evolution of Decay

GASTÓN GARCÍA-CALDERÓN

Instituto de Física
Universidad Nacional Autónoma de México
México 01000, D.F., México

Quantum decay refers to the description of the time evolution of an initially confined arbitrary state, $\psi(r, t = 0)$. Consider a single-channel problem characterized by a potential of arbitrary shape $V(r)$ that vanishes after a distance; that is, $V(r) = 0$ for $r > R$. One may solve the corresponding time-dependent Schrödinger equation in the radial variable r for s-waves as an initial value problem. The solution $\psi(r, t)$ at time t may be written in terms of the retarded time-dependent Green's function $g(r, r'; t)$ as

$$\psi(r, t) = \int_0^R g(r, r'; t)\psi(r', 0)dr'. \tag{1}$$

Two notions that have been proposed to describe the time evolution of decay are the survival probability $S(t)$ of the initial state at a time t after the start of the process and the nonescape probability $P(t)$ to find the particle still confined at time t. The survival probability is defined as

$$S(t) = \left| \int_0^R \psi^*(r, 0)\psi(r, t)dr \right|^2 \tag{2}$$

and, similarly, the nonescape probability reads

$$P(t) = \int_0^R \psi^*(r, t)\psi(r, t)dr. \tag{3}$$

One sees that the relevant quantity to describe the above notions is $g(r, r'; t)$. A convenient way to determine it is by a Laplace transform into the complex momentum k-plane to exploit the analytical properties of the corresponding outgoing Green's function $G^+(r, r'; k)$:

$$g(r, r'; t) = \left(\frac{i}{2\pi}\right) \int_{C_0} G^+(r, r'; k)e^{-ik^2t}2kdk, \tag{4}$$

where C_0 represents a hyperbolic contour along the first quadrant on the k-plane.[1] Because $V(r) = 0$ for $r > R$, $G^+(r, r'; k)$ may be extended analytically to the whole complex k-plane where it has an infinite number of complex poles distributed in a well-known manner.[2] Moreover, it may be proved that[3]

$$G^+(r, r'; k) = \sum_{n=-\infty}^{\infty} \frac{u_n(r)u_n(r')}{2k_n(k - k_n)}, \qquad (r, r') < R. \tag{5}$$

The validity of equation 5 is based on the proof that $G^+(r, r'; k) \to 0$ as $|k| \to \infty$ along all directions of the complex k-plane, provided $(r, r') < R$.[4] This implies that the contour C_0 in equation 4 may be deformed along the real k-axis. Then, using equation 5 into equation 4 leads to an expansion over the full set of resonant states that may be written as[4]

$$g(r, r'; t) = \sum_{n=-\infty}^{\infty} u_n(r)u_n(r')M(k_n, t), \qquad (r, r') < R, \tag{6}$$

where the functions $M(k_n, t)$ are defined as[4]

$$M(k_n, t) \equiv \left(\frac{i}{2\pi}\right) \int_{-\infty}^{\infty} \left(\frac{e^{-ik^2t}}{k - k_n}\right) dk = \left(\frac{1}{2}\right) e^{u^2} \operatorname{erfc}(u), \tag{7}$$

with $u = -\exp[(-i\pi/4)k_n t^{1/2}]$. One may use the symmetry relation, $M(k_p, t) = e^{-ik_p^2 t} - M(-k_p, t)$, with k_p being a complex pole on the fourth quadrant, to exhibit the exponential and nonexponential contributions to equation 5.

Equation 6 may be used to obtain resonant expansions of both the survival $S(t)$ and the nonescape $P(t)$ probabilities and to study their corresponding nonexponential contributions involving the full resonant spectra. For example, it may be proved that at long times $S(t)$ behaves according to the well-known result $\sim t^{-3}$, whereas, surprisingly, $P(t) \sim t^{-1}$.[5]

REFERENCES

1. García-Calderón, G. & R. E. Peierls. 1976. Resonant states and their uses. Nucl. Phys. **A265:** 443–460.
2. Newton, R. G. 1982. Scattering Theory of Waves and Particles. Second edition. Springer-Verlag. New York/Berlin.
3. More, R. 1973. Theory of decaying states. Phys. Rev. **A4:** 1782–1790.
4. García-Calderón, G. 1992. Resonant states and the decay process. *In* Symmetries in Physics. A. Frank & K. B. Wolf, Eds. Springer-Verlag. New York/Berlin.
5. García-Calderón, G., J. L. Mateos & M. Moshinsky. 1995. Resonant spectra and the time evolution of the survival and nonescape probabilities. Phys. Rev. Lett. **74:** 337–340.

Role of the Schrödinger Cat State in Quantum Information

OSAMU HIROTA

Research Center for Quantum Communications
Tamagawa University
Tokyo 194, Japan

INTRODUCTION

This report presents the fact that the standard quantum limit (SQL) in the statistical decision problem can be broken by a class of Schrödinger cat states (SC states). Here, the SQL in the decision problem is defined as the minimum error probability based on the projection valued measure (PVM) of the observable as the signal. The problem of finding the best quantum measurement process in order to distinguish the quantum states is called the quantum decision theory. Then, this theory predicts a possibility of overcoming the SQL. However, it is not clear what it means in the physical sense.

We showed that there does not exist a distinction between classical and quantum probability when studying the SQL.[1] In the decision based on classical probability, if one wishes to break the limit, one must predict signals before observation. As a result, to break the SQL requires a noncausal process in the sense of information theory. If the SQL is broken, this is by a special feature of the quantum probability.

In this work, it is shown that a quantum interference effect provides an interpretation to break the SQL. Then, it is demonstrated that a class of SC states, being a representative example of such a quantum interference, may provide a performance over the SQL.

According to the quantum decision theory,[1] the best decision operator cannot be constructed by the PVM of the signal observable. On the other hand, the SQL is achieved by the following decision operator, which is called the standard decision operator:

$$\Pi_1 = \int_{-\infty}^{\infty} f(x_d)\,|x\rangle\langle x|\,dx, \qquad \Pi_2 = \int_{-\infty}^{\infty} [1 - f(x_d)]\,|x\rangle\langle x|\,dx, \tag{1}$$

where $f(x_d)$ is Wald's decision function and $|x\rangle$ is the eigenstate of the signal observable, respectively. The generalized decision operator by which the SQL may

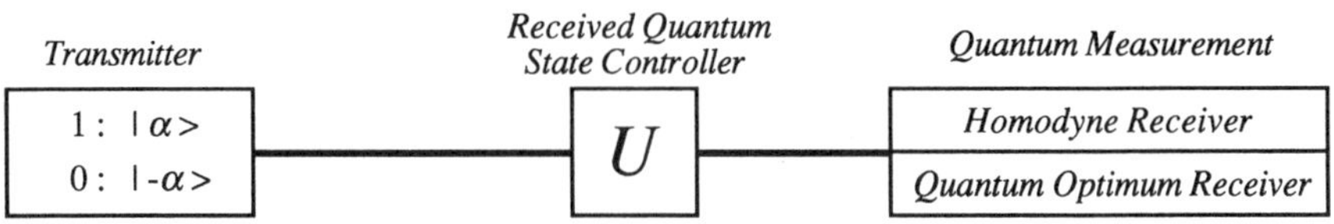

FIGURE 1. Schematic diagram of the received quantum state control system. U produces the Schrödinger cat state.

be broken can be constructed by a certain evolution operator[1] as follows:

$$\Pi_i^* = U^\dagger \Pi_i U. \quad \textbf{(2)}$$

As an example, $U = \exp[ig(\hat{n})] = \Sigma_{n=0}^{\infty} K_n |n\rangle\langle n|$ and $g(\hat{n})$ is a function of the number operator. In this form, the problem is to find an evolution operator that provides

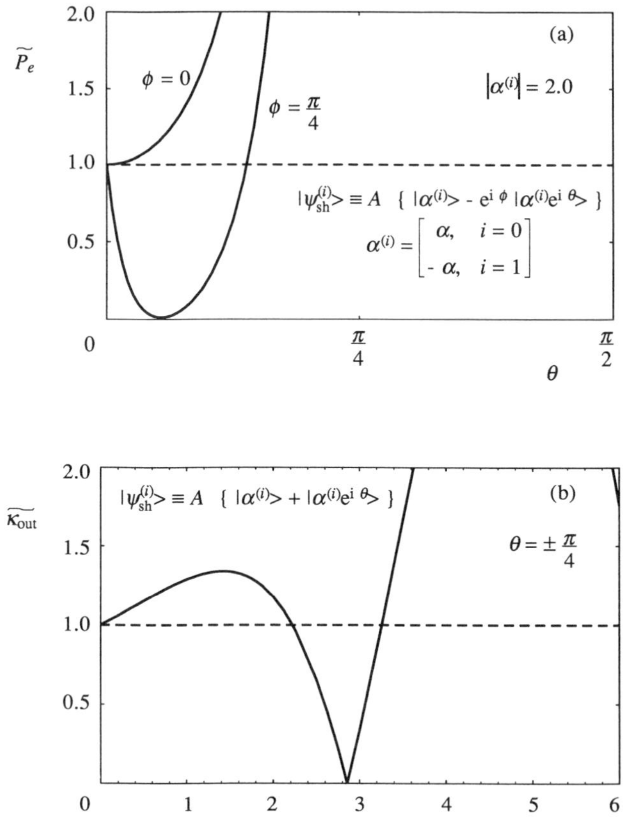

FIGURE 2. Numerical examples of system performance in quantum decision—(a) normalized error probabilities with respect to the phase shift of the cat state for a homodyne system: $\tilde{P}_e = P_e/P_e(\mathrm{SQL})$; (b) property of the normalized inner product of the Schrödinger cat state with respect to the average photon number for the quantum optimum receiver: $\tilde{\kappa}_{out} = (\kappa_{out}/\kappa_{in})$.

below the SQL. The criterion of the decision problem becomes as follows:

$$T_r\rho_2\Pi_1^* = \int_{-\infty}^{\infty} f(x_d)\left|\sum_{n=0}^{\infty} K_n^* h^*(x,n)\left(\frac{\alpha_{in}^n}{\sqrt{n!}}\right)e^{-(1/2)|\alpha_{in}|^2}\right|^2 dx$$

$$< \int_{-\infty}^{\infty} f(x_d)\left|\sum_{n=0}^{\infty} h^*(x,n)\left(\frac{\alpha_{in}^n}{\sqrt{n!}}\right)e^{-(1/2)|\alpha_{in}|^2}\right|^2 dx = \mathrm{SQL}, \quad (3)$$

where $h(x, n) = \langle x|n\rangle$. Thus, we can see that a quantum interference effect must be generated by an effect of K_n if one wishes to overcome the SQL.

ROLE OF THE SCHRÖDINGER CAT STATE

We have shown by numerical analysis that the SQL can be broken by the evolution of the coherent state in a Kerr medium based on equation 3.[2] Here, we give an explicit example that the SQL is also broken by the SC state as a result of certain evolution. The SC state as discussed by Yurke and Stoler[3] gives a typical example of the quantum interference phenomenon. This is related with our problem because it is generated by the anharmonic oscillation interaction and the interference occurs when we employ the homodyne process as the quantum measurement. Thus, it is easy to understand that this class is one of the candidates for satisfying equation 3. In general, the SC state is represented by $|\psi\rangle_{sh} = \Sigma_{m=0}^{N-1} C_m|\alpha_m\rangle$. This is an eigenstate of a^N for all values of C_m, which can be generated by unitary or nonunitary evolution from the coherent state. In many cases, the SQL is broken and the Helstrom bound given by the inner product of signal states[4] is also broken because the inner product may be reduced. Examples are shown in FIGURES 1 and 2. Hence, the received quantum state control[4] may be realizable. Thus, it has been shown in this work that the breaking of the SQL may be related with the quantum interference effect in the quantum measurement process.

REFERENCES

1. HIROTA, O. 1994. A study of signal detection based on quantum probability. Tech. Rep. IEICE Jpn. **IT93-124:** 25–30.
2. USUDA, S. T. & O. HIROTA. 1994. An example of a received quantum state controller by optical Kerr effect. *In* Technical Digest of QCM '94. Nottingham University.
3. YURKE, B. & D. STOLER. 1986. Generating quantum mechanical superpositions of macroscopically distinguishable states. Phys. Rev. Lett. **57:** 13–16.
4. HIROTA, O. 1991. Some remarks on conditional unitary operators. Phys. Lett. **A155:** 343–347.

Geometrical Phase Effects and Bohm's Quantum Potential

R. E. KASTNER

Department of Philosophy
University of Maryland
College Park, Maryland 20742

There has recently been renewed interest in David Bohm's 40-year-old alternative interpretation of quantum mechanics.[1] A key concept is Bohm's nonlocal "quantum potential" Q, which arises directly from the Schrödinger equation when the wave function is expressed in polar form ($\psi = Re^{iS/\hbar}$). We present here a simple application of Q to a gedankenexperiment proposed by J-M. Lévy-LeBlond[2] and further developed by M. Razavy.[3]

Lévy-LeBlond imagines a particle sent through a tube with square cross section of side a and length L and finds via energy conservation that its momentum in the tube will be reduced by an amount $\Delta p = \pi^2\hbar^2/pa^2$. When the particle exits the tube, its momentum will return to the original value; however, the particle's wave function will have undergone a phase shift $L\Delta p$, which is in principle measurable by techniques such as neutron interferometry.

Razavy gives a detailed analysis of a particle propagating through a constriction and shows that a spatial boundary in one of the (separable) coordinates gives rise to a nonlocal potential in the wave equation of the other coordinate. He interprets this effect as the result of what he terms a "fictitious force law" proportional to $\hbar^2$. As it turns out, if one simply applies Bohm's Q definition to either of these problems, one obtains exactly the same nonlocal effect discussed by both authors.

For the Lévy-LeBlond case, we have the two wave functions,

$$\psi_{\mathrm{f}} \propto e^{ipz/\hbar}$$

and

$$\psi_{\mathrm{t}} \propto \sin\frac{\pi x}{a}\sin\frac{\pi y}{a}e^{ip'z/\hbar},$$

for the free particle and the particle in the tube, respectively. Then, Q in the tube is calculated from the definition (see reference 1) $Q = (-\hbar^2/2m)(\nabla^2 R/R)$:

$$Q = \frac{\pi^2\hbar^2}{ma^2}.$$

This represents a nonlocal, but physically real potential, and we can calculate the particle's momentum in the tube as in the usual method for a step potential:

$$p' = [2m(E - Q)]^{1/2} = \left[p^2 - \frac{2\pi^2\hbar^2}{a^2}\right]^{1/2} \simeq p - \frac{\pi^2\hbar^2}{pa^2}.$$

Thus, we obtain the same Δp and the same phase shift obtained by Lévy-LeBlond.

From an historical standpoint, it is striking that these two authors discovered an interesting nonlocal effect, but did not recognize it as attributable to Bohm's quantum potential, although Bohm had explicitly predicted just this kind of effect in his original papers.[1] This perhaps points up how neglected Bohm's approach has been over the past four decades.

REFERENCES

1. BOHM, D. 1952. Phys. Rev. **85:** 166–193.
2. LÉVY-LEBLOND, J-M. 1987. Phys. Lett. A **125:** 441–442.
3. RAZAVY, M. 1989. Phys. Rev. A **40:** 1–4.

Noise Limits for Nonlinear, Phase-invariant Amplifiers[a]

DMITRI KOUZNETSOV,[b,c] ROBERTO ORTEGA,[b] AND DANIEL ROHRLICH[d]

[b]*Centro de Instrumentos*
Universidad Nacional Autónoma de México
04510 D.F., México

[c]*Lebedev Physics Institute*
Moscow, 117924 Russia

[d]*School of Physics and Astronomy*
Tel Aviv University
Ramat Aviv 69978, Tel Aviv, Israel

INTRODUCTION

Two basic values characterize a phase-invariant amplifier: the amplification coefficient $G = \langle A\rangle/\langle a\rangle$ and the noise $D = \langle A^\dagger A\rangle - \langle A^\dagger\rangle\langle A\rangle$. Here, we focus on a single field mode; a and A denote the field before and after amplification. We take expectation values in a coherent initial state $|\alpha\rangle$. For a *linear* quantum amplifier, G is a constant and the minimal noise is known:[1–3] $D \geq |G|^2 - 1$ for $|G| \geq 1$. Here, we derive lower limits for a nonlinear amplifier, in which G depends on $x \equiv \alpha^*\alpha$ (but not on the phase of α).

Mode operators before and after amplification are related by a unitary transformation, $A = U^\dagger a U$. To produce amplification, U must depend on the amplifier degrees of freedom as well as on a and $a^\dagger$. The simplest U is $U = e^{-iHt}$, with $H = ia^\dagger b^\dagger - iab$; it defines a *linear* amplifier. (The amplifier degree of freedom obeys $[b, b^\dagger] = 1$; t is real.) We find $A = a\cosh t + b^\dagger \sinh t$; for an amplifier prepared in the ground state, $G = \cosh t$; the noise $D = \sinh^2 t$ saturates the bound $|G|^2 - 1$. For *nonlinear* amplifiers, we have the following theorem:

THEOREM: Let $D = \langle\alpha|A^\dagger A|\alpha\rangle - \langle\alpha|U^\dagger A^\dagger U|\alpha\rangle\langle\alpha|U^\dagger A U|\alpha\rangle$ and

$$E(x) = \sum_{n=1}^{\infty} \frac{x^{n+1}}{n!}\left|\frac{d^n}{dx^n}G(x)\right|^2, \qquad F(x) = \sum_{n=1}^{\infty} \frac{x^{n-1}}{n!}\left|\frac{d^n}{dx^n}[xG(x)]\right|^2 - 1. \tag{1}$$

Then $D \geq E(x)$ and $D \geq F(x)$.

The proof is given in reference 4. If G is constant, then $F(x)$ reduces to the linear bound $|G|^2 - 1$.

[a]This work was supported by the SNI of México (to D. Kouznetsov) and the Ticho Fund (to D. Rohrlich).

EXAMPLES

A class of amplifiers that do realize the bound F have a linear amplification followed by a nonlinear refraction: $A = e^{itH}[G_0 a + (|G_0|^2 - 1)^{1/2} b^\dagger] e^{-itH}$, where H depends only on $a^\dagger a$. For $H = a^\dagger a(a^\dagger a - 1)/2$, we have $G(x) = G_0 \exp(-qx)$ and $D = |G_0|^2[1 + x - x \exp(-|q|^2 x)] - 1$, where $q = 1 - e^{-it}$. The amplifier realizes the lower bound $F(x) = D$, whereas $E(x) = D - 1 + |G_0|^2$. Here, the linear bound $D \geq |G(x)|^2 - 1$ also holds. It is broken in the next example: the resonant interaction of N identical two-level atoms with a single-mode field can be described[5] by the Hamiltonian $H = iab^\dagger - ia^\dagger b$, where $b = \Sigma_{k=1}^N b_k$, $b_k b_j = (1 - \delta_{jk}) b_j b_k$, $b_k^\dagger b_j = b_j b_k^\dagger$ for $j \neq k$, and $b_k^\dagger b_k + b_k b_k^\dagger = 1$. (Note that $b_k^\dagger$ is the operator for exciting the k-th atom.) Let all the atoms be initially excited and let $U = \exp(-iHt)$. For $N = 10$, we compute the evolution of G and D numerically, at input intensity $x = 1$. FIGURE 1 represents

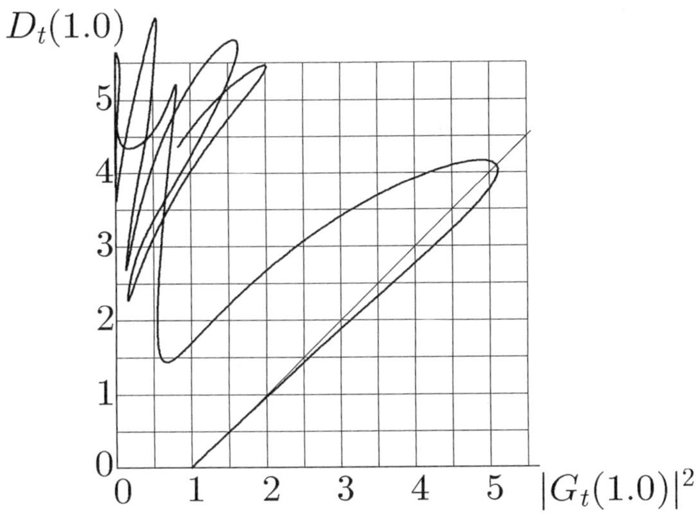

FIGURE 1. The noise D versus $|G|^2$ for the Cummings-Tavis amplifier with $N = 10$ atoms at $x = 1$ and $0 \leq t \leq 10$.

D_t versus $|G_t|^2$. The noise D_t oscillates while G_t has "revivals", and $D \geq |G|^2 - 1$ is valid only at small t (for $|G_t|^2 \lesssim 2$) and is then violated. FIGURE 2 shows all the bounds versus $|G(x)|^2$, for $t = 0.5$ and $0 \leq x \leq 20$. The nonlinear amplifier beats the linear limit, whereas the lower limits of the theorem hold.

CONCLUSIONS

We have obtained lower bounds (equation 1) on the irreducible noise of a nonlinear, phase-preserving quantum amplifier, in terms of the amplification factor G and its derivatives. For G constant, the bound F reproduces the linear bound. For

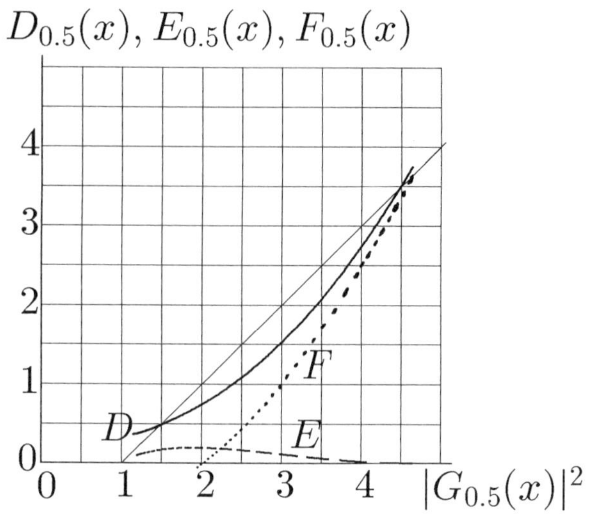

FIGURE 2. The noise D (solid curve) and lower bounds E (dashed curve) and F (dotted curve) versus $|G|^2$ for the Cummings-Tavis amplifier with $N = 10$ atoms at $t = 0.5$ and $0 \leq x \leq 20$.

G not constant, the noise can undercut the linear bound, as examples confirm. However, we have not determined in general which functions $G(x)$ correspond to nonlinear amplifiers.

REFERENCES

1. HAUS, H. A. & J. A. MULLEN. 1962. Phys. Rev. **128:** 2407–2413.
2. CAVES, C. M. 1982. Phys. Rev. **D26:** 1817–1839.
3. YAMAMOTO, Y. & H. A. HAUS. 1986. Rev. Mod. Phys. **58:** 1001–1020.
4. KOUZNETSOV, D., R. ORTEGA & D. ROHRLICH. 1994. Tel Aviv University report no. TAUP 2171-94. Phys. Rev. A. Submitted.
5. TAVIS, M. & F. M. CUMMINGS. 1969. Phys. Rev. **188:** 692–695.

Non-Abelian Quantum Kinematic Foundations of Quantum Dynamics

J. KRAUSE

Facultad de Física
Pontificia Universidad Católica de Chile
Santiago 22, Chile

Here, we present an attempt to interpret the quantum theory of symmetries as the cornerstone of quantum dynamics.[1] This program of "direct quantization through the symmetry" is conceivable (at least) because Lie groups can be "quantized", quite generally, by means of well-defined geometric methods. We refer to this generalized approach to quantization as non-Abelian quantum kinematics. The important point to remark is that the standard quantization procedure can be enlarged as a purely group-theoretic procedure. We next present this procedure in a rather sketchy manner.

Within the rigged Hilbert space $\tilde{\mathscr{H}}(G)$ that carries both (left and right) regular representations of a Lie group G, we introduce a set of commuting generalized position operators Q^a, which have the parameters for spectra. After some standard manipulations, we obtain

$$[Q^a, L_b] = i\hbar R^a_b(Q), \qquad [Q^a, R_b] = i\hbar L^a_b(Q). \tag{1}$$

These are the generalized Heisenberg commutation relations obeyed by the generalized position operators and the generalized non-Abelian momenta represented by the generators. Such a result is of potential interest for physics. In this fashion, a new set of invariant operators appear as linear combinations of the generators, whose coefficients are functions of the generalized position operators. In fact, in the left regular representation (for instance), the invariant Hermitian operators are given by[2]

$$R_a(Q; L) = R^\dagger_a(Q; L) = \overline{A}^{\,b}_a(Q)L_b - (1/2)i\hbar f^{\,b}_{ab} \tag{2}$$

for it can be shown that $U^\dagger_L(q)R_a(Q; L)U_L(q) = R_a(Q; L)$. [$\overline{A}^{\,b}_a(q)$ denotes the matrix of the adjoint representation.] As a matter of fact, these operators are the generators of the right regular representation, acting as invariant operators within the left regular representation of G.

Now, any reasonable physical interpretation of a quantum kinematic model of a Lie group G must be based on the following postulate: The only allowable physical states of a system, having G as symmetry group, correspond to simultaneous eigenkets in $\tilde{\mathscr{H}}(G)$ of a maximal set of compatible quantum-kinematic Hermitian invariant operators of G. Indeed, every maximal set of compatible invariant operators yields a set of superselection rules, by means of which the incoherent Hilbert space $\tilde{\mathscr{H}}(G)$ can be "diagonalized" into invariant subspaces, each carrying a physical model of a G-invariant system. On the other hand, concerning the permanent properties characterizing the system, one also considers the following second heuristic postulate: The eigenvalues of the operators pertaining to the superselection rules correspond to permanent intrinsic physical properties characterizing the system. These postulates enhance some features of the right regular representation with the character of an isotopic structure, describing the internal nature of a system, within the left regular representation of the group used as the working frame.

Next, one builds a quantum model of the system over its configuration space X that preserves all the symmetry transformations $x' = f(x; q)$ of X described by G; that is, we find kets $|x\rangle \in \tilde{\mathcal{H}}(G)$ that are in one-to-one correspondence with the points $x \in X$ and that transform under the action of G in a covariant manner: $U_L(q)|x\rangle = |f(x; q)\rangle$. Such kets carry a configuration representation of G on X. They have the general form,

$$|x; \xi\rangle = \int d\mu_L(q)\xi^*[f(x; \bar{q})]|q\rangle_L, \tag{3}$$

where the generating function $\xi(x)$ is arbitrary. Accordingly, one then solves the simultaneous eigenvalue problem for a maximal set of compatible invariant operators $S_\alpha(R)$. Thus, one searches for generating functions ξ such that they satisfy a system of coupled eigenvalue wave equations of the form,

$$S_\alpha\left\{-i\hbar\left[Z_a(x) - \left(\frac{1}{2}\right)f^b_{ab}\right]\right\}\xi(x; \epsilon) = \epsilon_\alpha\xi(x; \epsilon), \tag{4}$$

for each compatible superselection rule ($\alpha = 1, \ldots, s$). These are generalized Schrödinger wave equations in configuration space-time. Once the physical configuration kets of a model have been found, one calculates the transition amplitudes,

$$\langle x_2; \epsilon_{(2)}; \xi|x_1; \epsilon_{(1)}; \xi\rangle = \int d\mu_L(q)\xi[f(x_2; \bar{q}); \epsilon_{(2)}]\xi^*[f(x_1; \bar{q}); \epsilon_{(1)}], \tag{5}$$

for finding the system at configuration event x_2 given its configuration at x_1. This invariant integral corresponds to the propagator kernel associated with the system of generalized Schrödinger equations.[3] Thus, concerning the general mathematical traits of the formalism, we deem it as already complete. As it stands, however, quantum kinematics is a general framework rather than a specific physical theory.

REFERENCES

1. KRAUSE, J. 1986. J. Math. Phys. **27:** 2922.
2. KRAUSE, J. 1991. J. Math. Phys. **32:** 348.
3. KRAUSE, J. 1994. Configuration ray representations in non-Abelian quantum kinematics and dynamics. Int. J. Theor. Phys. In press.

Entangled Entanglement[a]

GÜNTHER KRENN[b] AND ANTON ZEILINGER[c]

[b]*Atominstitut der Österreichischen Universitäten*
A-1020 Wien, Austria

[c]*Institut für Experimentalphysik*
Universität Innsbruck
A-6020 Innsbruck, Austria

It is a well-known fact that, in general, values cannot be ascribed to quantum mechanical observables independent of the measurement context.[1] In entangled systems, contextuality gives rise to the astonishing property that we cannot think of the results of measurements on one subsystem as being specified independent of the parameters of the measurements on the other subsystems. Although the different measurements can be spacelike separated, the results of each measurement depend on the complete experimental context of the whole system. An interesting consequence of this fact is that even entanglement itself can be an entangled property, which will be discussed in the following.

Let us consider a three-particle system described by the GHZ state[2,3] proposed by Mermin:[4]

$$|\Psi\rangle = \left(\frac{1}{\sqrt{2}}\right)(|z+\rangle|z+\rangle|z+\rangle + |z-\rangle|z-\rangle|z-\rangle). \tag{1}$$

The three spin-½ particles are emitted by a common source into distinct directions. By adequately oriented Stern-Gerlach magnets, spin measurements along arbitrary directions are performed in spacelike separated regions by three observers. The three directions $\vec{e}_1$, $\vec{e}_2$, and $\vec{e}_3$ are denoted by the spherical coordinates (ϑ^1, φ^1), (ϑ^2, φ^2), and (ϑ^3, φ^3). For the results of the measurements on particles 1 and 2, the correlation function (expectation value of the product of the results) is given by $E_{12}(\vartheta^1, \vartheta^2) = \cos(\vartheta^1)\cos(\vartheta^2)$. This function does not violate Bell's inequality because it is factorized with respect to the parameters ϑ^1 and ϑ^2. Therefore, here, the results of observers 1 and 2 are always correlated in a classical way. This implies that we can think of a hidden local-realistic arena that accounts for the measured results. Nevertheless, as we will show, this interpretation is inconsistent with possible further observations.

Let us now turn to observer 3. Independent of observers 1 and 2, he/she decides to measure the spin of particle 3 along some direction $\vec{e}_3$. With probability ½, he/she gets the result +1 and −1, respectively. His/her results can be used to separate the results of observers 1 and 2 into two subensembles. Whenever the result +1 (−1) occurs in a spin measurement of particle 3, the corresponding particles 1 and 2 are assigned to subensemble + (−) (cf. FIGURE 1).

[a]This work has been supported by the Austrian FWF under Grant Nos. S6502 and P-8781-PHY and by the NSF under Grant No. PHY92-13964.

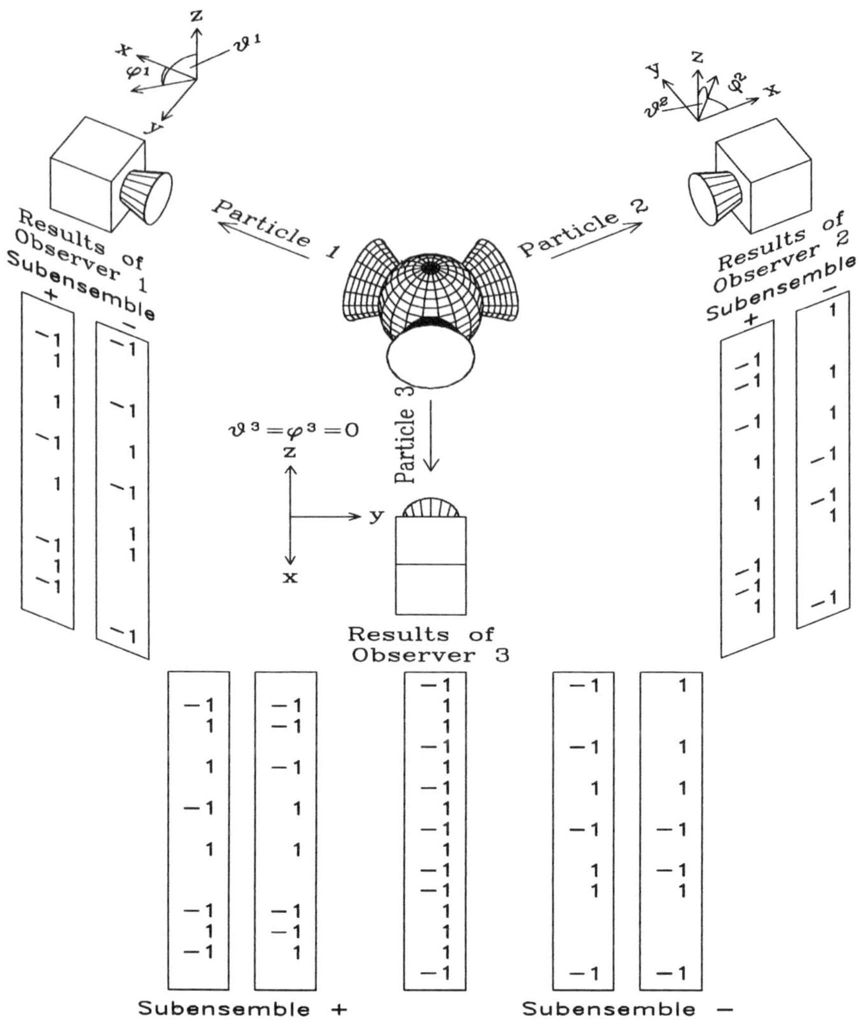

FIGURE 1. After a spin measurement on particle 3 along the *z*-direction, the results of observers 1 and 2 are separated into subensembles + and −. The data within each subensemble are maximally random and do not contradict local-realism.

We now consider the special case that observers 1 and 2 perform spin measurements within the *x*-*y* plane ($\vartheta^1 = \vartheta^2 = \pi/2$). Then, $E_{12} = 0$, which means that the results of observers 1 and 2 are maximally random. If we now separate these data into two subensembles by using the results of observer 3, we get the correlation function E^+_{12} (E^-_{12}) for subensemble + (−):

$$E^{\pm}_{12} = \pm \sin(\vartheta^3) \cos(\varphi^1 + \varphi^2 + \varphi^3). \tag{2}$$

In the following, these functions are discussed for two specific measurement directions $\vec{e}_3$ (ϑ^3, φ^3). In particular, we investigate if the correlation functions $E^{\pm}_{12}(\varphi^1, \varphi^2)$ (equation 2) violate Bell's inequality:[5]

$$-2 \leq E^{\pm}_{12}\left(\frac{\pi}{4}, 0\right) + E^{\pm}_{12}\left(\frac{\pi}{4}, \frac{\pi}{2}\right) + E^{\pm}_{12}\left(\frac{3\pi}{4}, 0\right) - E^{\pm}_{12}\left(\frac{3\pi}{4}, \frac{\pi}{2}\right) \leq 2. \qquad \textbf{(3)}$$

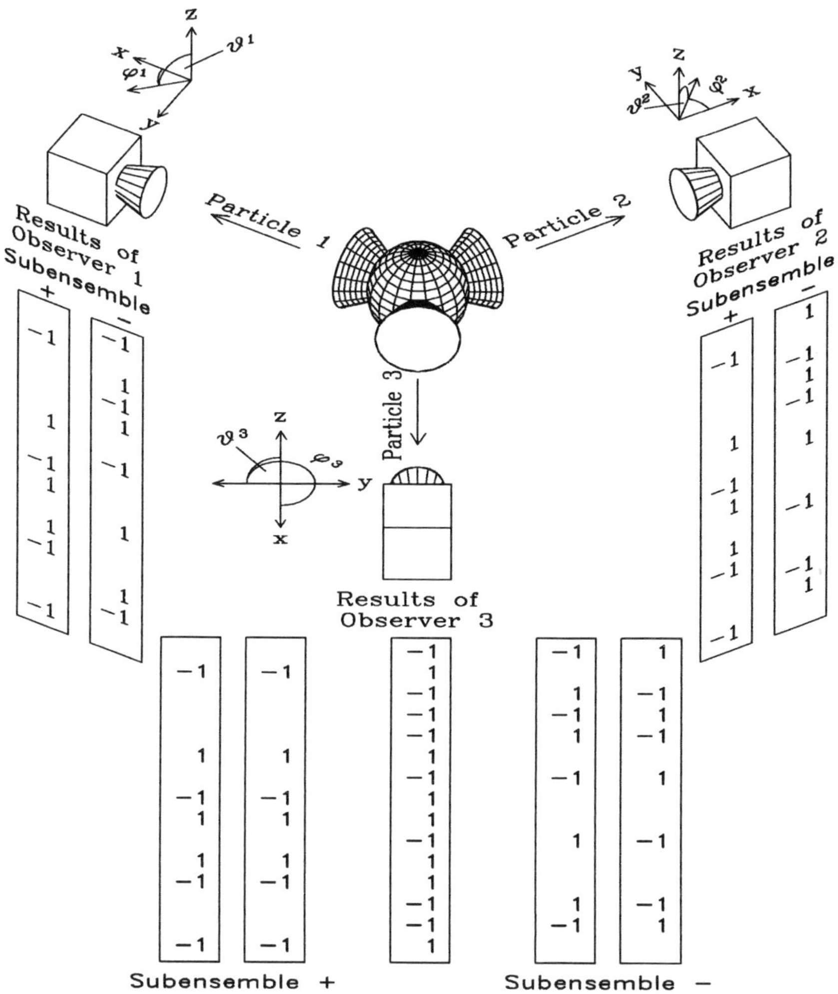

FIGURE 2. After a spin measurement on particle 3 along the negative y–axis, the results of observers 1 and 2 are separated into subensembles + and −. For specific directions $\vec{e}_1$ and $\vec{e}_2$, the data within each subensemble are perfectly (anti-)correlated. The correlation function (equation 2) violates Bell's inequality by the maximum amount $(2\sqrt{2})$ permitted by quantum mechanics. This implies that a local-realistic interpretation of the classically correlated results of observers 1 and 2 is not consistent in this case. Particles 1 and 2 have to be considered as entangled for certain measurement directions $\vec{e}_3$.

In the case that observer 3 measures along the z-direction ($\vartheta^3 = 0$) (cf. FIGURE 1), there is no violation of Bell's inequality (equation 3) because $E_{12}^{+} = E_{12}^{-} = 0$. Hence, the assumption that the results of observers 1 and 2 are local-realistic is confirmed.

In contrast, we now consider the case that observer 3 measures along the negative y–axis ($\vartheta^3 = \pi/2$, $\varphi^3 = 3\pi/2$) (cf. FIGURE 2) for the same measurements on particles 1 and 2. Then, inequality 3 is violated by the maximum amount ($2\sqrt{2}$) permitted by quantum mechanics. Therefore, in this case, the results of observers 1 and 2 cannot be understood in a local realistic way. It can be shown that, for general directions $\vec{e}_1$, $\vec{e}_2$, and $\vec{e}_3$, a separation into nonlocal subensembles occurs as soon as $\sin(\vartheta^1)\sin(\vartheta^2)\sin(\vartheta^3) > 1/\sqrt{2}$. With $\vartheta^1 = \vartheta^2 = \pi/2$, this inequality simplifies to $\sin\vartheta^3 > 1/\sqrt{2}$ or $\pi/4 < \vartheta^3 < 3\pi/4$.

In spite of the fact that the data of observers 1 and 2 are correlated classically, this demonstrates that a local-realistic interpretation is not always consistent with further observations. For certain measurement directions $\vec{e}_3$, particles 1 and 2 have to be considered as entangled. Clearly, as shown explicitly by Popescu and Rohrlich,[6] such directions exist for any entangled n-particle state ($n \geq 3$). Therefore, we conclude that in this case the interpretation itself is a contextual property and thus cannot be given consistently without taking into account the whole experimental situation.

REFERENCES

1. BELL, J. S. 1966. Rev. Mod. Phys. **38:** 447.
2. GREENBERGER, D. M., M. A. HORNE & A. ZEILINGER. 1989. Going beyond Bell's theorem. *In* Bell's Theorem, Quantum Theory, and Conceptions of the Universe. M. Kafatos, Ed.: 69–72. Kluwer. Dordrecht.
3. GREENBERGER, D. M., M. A. HORNE, A. SHIMONY & A. ZEILINGER. 1990. Bell's theorem without inequalities. Am. J. Phys. **58**(12): 1131–1143.
4. MERMIN, N. D. 1990. Quantum mysteries revisited. Am. J. Phys. **58**(8): 731–734.
5. BELL, J. S. 1964. On the Einstein-Podolsky-Rosen paradox. Physics **1:** 195–200.
6. POPESCU, S. & D. ROHRLICH. 1992. Generic quantum nonlocality. Phys. Lett. **A166:** 293–297.

Experiments on the Influence of Electro-Magnetic and Gravito-Inertial Potentials and Fields on the Quantum Mechanical Phase of Matter Waves

MARC NICKLAUS[a] AND FRANZ HASSELBACH

Institut für Angewandte Physik
Universität Tübingen
D-72076 Tübingen, Germany

Since the first experiments of Chambers[1] and Bayh[2] on the (magnetic) Aharonov-Bohm (A-B)[3] phase shift, a number of experiments with electrons, neutrons, and atoms have been performed that demonstrate the coupling of the phase of matter waves not only to electric and magnetic, but also to gravitational and inertial potentials and fields.

The formal identity between the forces in electro-magnetic fields on the one hand and gravito-inertial fields on the other hand,

$$\vec{F}_{\text{em}} = e\vec{E} + e(\vec{v} \times \vec{B}) \quad \text{(electro-magnetic)},$$
$$\vec{F}_{\text{gi}} = m\vec{g} + 2m(\vec{v} \times \vec{\omega}) \quad \text{(gravito-inertial)},$$

immediately leads to a list of corresponding phase shifts in the gravito-inertial case, which are compiled in TABLE 1. They are classified as type I and type II in TABLE 1 (right-hand side): Type I comprises those phase shifts caused exclusively by the action of potentials; their characteristic feature is that there is no deflection of the envelope of the entire pattern of interference fringes. In type II, the forces exerted on the particles by nonvanishing fields give rise to a lateral deflection of the whole fringe field.

When an electric or a magnetic field deflects laterally separated coherent electron wave packets, one always observes the same interference field irrespective of the deflection angle; that is, the fringes remain at constant positions within the interference field. This happens even though a path length difference between the separated wave packets is introduced by the deflection, which should give rise to a corresponding phase shift. The fact that this is not the case means that the phase shift caused by the change of the path length is exactly compensated by some counteracting effect. This compensating effect is the A-B phase shift. For an electric field, this is the scalar A-B phase shift; for a magnetic field, it is the "conventional" A-B phase shift of the vector potential $\vec{A}$. Ehrenberg and Siday[16] realized already in 1949 that these phase shifts exist and that they are, in fact, absolutely essential for the fact that electron optics works; that is, that electron optical components can be used for image formation in electron microscopy, lithography, etc.

[a] Present address: National Institutes of Health, Bethesda, Maryland 20892.

TABLE 1. Corresponding Electro-Magnetic and Gravito-Inertial Matter-Wave Phase Shift Effects and Experiments

Electro-Magnetic		Gravito-Inertial	Type
phase shift by vector potentials: enclosed flux $\int_A \vec{B} d\vec{\sigma} = \oint \vec{A} \cdot d\vec{s}$ (magnetic A-B effect with e^-);[1-3] "vector potential" $\oint \vec{\mu} \times \vec{E} \cdot d\vec{l}$ (Aharonov-Casher effect)[4]	⇔	phase shift by enclosed rotational ($\vec{\omega}$)-flux (Zimmerman & Mercereau experiment with electron Cooper pairs[5])	I
phase shift by scalar potentials (electric A-B effect with e^-,[3] proposed; scalar A-B effect with neutrons[6])			I
phase shift and deflection by a magnetic field (enclosed magnetic flux + magnetic field)	⇔	phase shift and deflection by enclosed $\vec{\omega}$-flux and $\vec{\omega}$-field (Sagnac experiments,[7-9] inertial acceleration[10])	II
phase shift and deflection by an electric field	⇔	gravitational acceleration (free-fall of neutrons and atoms[11,12])	II
mutually canceled deflection and phase shifts by combined electric and magnetic fields (Wien filter),[13] only longitudinal shift of the wave packets[14]	⇔	mutually canceled effects of rotation and gravitation, proposed[15]	

An example of a gravito-inertial phase shift is the Sagnac effect (already predicted by Heer in 1961[17]). This is shown by the expressions for both phase shifts:

Sagnac phase shift Aharonov-Bohm phase shift

$$\Delta\phi = \left(\frac{2m}{\hbar}\right) \int_A \vec{\omega} d\vec{\sigma} \qquad \Delta\phi = \left(\frac{e}{\hbar}\right) \oint \vec{A} \cdot d\vec{s} = \left(\frac{e}{\hbar}\right) \int_A \vec{B} d\vec{\sigma}$$

where the surface integrals are to be taken over the oriented area A enclosed between the two partial waves. In its "pure", that is, force-free, form, it is the inertial counterpart of the A-B effect, and this experiment was successfully conducted by Zimmerman and Mercereau in a rotating superconducting quantum interference device (SQUID) in 1965.[5]

It is of interest to note the following: (1) in fields produced by scalar potentials, energy is exchanged between the field and the particles (if no special precautions are taken to avoid field gradients, such as for a scalar A-B experiment), whereas this is not the case in fields produced by vector potentials; the latter therefore do not affect the temporal coherence of the matter waves;[14] (2) the Sagnac experiments with free electrons[9] and with Cooper pairs[5] prove that the presence of charge does not influence the electron's coupling to an accelerational field.

REFERENCES

1. Chambers, R. G. 1960. Phys. Rev. Lett. **5:** 3–5.
2. Bayh, W. 1962. Z. Phys. **169:** 492–510.
3. Aharonov, Y. & D. Bohm. 1959. Phys. Rev. **115:** 485–491.

4. CIMMINO, A., G. I. OPAT, A. G. KLEIN, H. KAISER, S. A. WERNER, M. ARIF & R. CLOTHIER. 1989. Phys. Rev. Lett. **63:** 380.
5. ZIMMERMAN, J. E. & J. E. MERCEREAU. 1965. Phys. Rev. Lett. **14:** 887.
6. ALLMANN, B. E., A. CIMMINO, A. G. KLEIN, G. I. OPAT, H. KAISER & S. A. WERNER. 1992. Phys. Rev. Lett. **68:** 2409–2412.
7. WERNER, S. A., J-L. STAUDENMANN & R. COLELLA. 1979. Phys. Rev. Lett. **42:** 1103–1106; ATWOOD, D. K., M. A. HORNE, C. G. SHULL & J. ARTHUR. 1984. Phys. Rev. Lett. **52:** 1673–1676.
8. RIEHLE, F., A. WITTE, T. KISTERS & J. HELMCKE. 1992. Appl. Phys. B **54:** 333–340.
9. HASSELBACH, F. & M. NICKLAUS. 1993. Phys. Rev. A **48:** 143–151.
10. BONSE, U. & T. WROBLEWSKI. 1983. Phys. Rev. Lett. **51:** 1401.
11. COLELLA, R., A. W. OVERHAUSER & S. A. WERNER. 1975. Phys. Rev. Lett. **34:** 1472–1474.
12. KASEVICH, M. & S. CHU. 1991. Phys. Rev. Lett. **67:** 181; SHIMIZU, F., K. SHIMIZU & H. TAKUMA. 1992. Phys. Rev. A **46:** R17.
13. MÖLLENSTEDT, G. & G. WOHLAND. 1980. *In* Electron Microscopy 1980. Volume 1 (Seventh European Congress on Electron Microscopy Foundation, Leiden, the Netherlands). P. Bredoro & G. Boom, Eds.: 28–29.
14. NICKLAUS, M. & F. HASSELBACH. 1993. Phys. Rev. A **48:** 152–160.
15. AHARONOV, Y. & G. CARMI. 1973. Found. Phys. **3:** 493–498.
16. EHRENBERG, W. E. & R. E. SIDAY. 1949. Proc. Phys. Soc. London Sect. B **62:** 8–21.
17. HEER, C. V. 1961. Bull. Am. Phys. Soc. **6:** 58.

The Einstein-Podolsky-Rosen Paradox and the Pauli Exclusion Principle

PAUL O'HARA

Department of Mathematics
Northeastern Illinois University
Chicago, Illinois 60625-4699

Let $\vec{\sigma}$ be a quantum spin operator of a particle and let $\vec{\sigma} \cdot \vec{a}$ be the observable measuring the spin component in direction $\vec{a}$. Let $\lambda_{\vec{a}}$ be an eigenvalue random variable with values in $\{-1, 1\}$ associated with the observable $\vec{\sigma} \cdot \vec{a}$. This induces a probability measure P_{QM} such that $P_{QM}(\lambda_{\vec{a}} = 1) = P_{QM}(\lambda_{\vec{a}} = -1) = ½$. In quantum mechanics, two particles can be in a singlet spin state. We generalize this notion by saying that two particles can be symmetrically linked with respect to the spin operator $\sigma^{(1)} \otimes \sigma^{(2)}$, where (1) and (2) refer to particles 1 and 2, respectively, if the probability measure P_{QM} can be extended to cover both particles by defining a conditional probability $P_{QM}(\lambda_{\vec{b}}^{(2)} = x \mid \lambda_{\vec{a}}^{(1)} = 1) = \cos^2(\theta_{12}/2)$ and $P_{QM}(\lambda_{\vec{b}}^{(2)} = -x \mid \lambda_{\vec{a}}^{(1)} = 1) = \sin^2(\theta_{12}/2)$, where $x = 1$ or $x = -1$ and θ_{12} is the angle between $\vec{a}$ and $\vec{b}$.

In the case of symmetrically linked particles, it is clear that if an observation is made on particle 1, in a direction $\vec{a}$, then with probability 1 the observed value in the same direction can be predicted for particle 2. It can now be asked if more than two particles can be symmetrically linked to the spin operators. This would mean that at least three simultaneous measurements could be made on the system. It is a remarkable fact that they cannot. For example, if we were to assume that three symmetrically linked particles existed, with θ_{12}, θ_{23}, and θ_{13} being the angles between $\vec{a}$ and $\vec{b}$, $\vec{b}$ and $\vec{c}$, and $\vec{c}$ and $\vec{a}$, a simple argument would lead to Bell's inequality:

$$\sin^2(\theta_{12}/2) + \sin^2(\theta_{23}/2) \geq \sin^2(\theta_{31}/2).$$

Taking $\theta_{12} = \theta_{23} = \pi/3$ and $\theta_{31} = 2\pi/3$ leads to a contradiction. Therefore, symmetrically linked particles occur only in pairs. This is called the coupling principle. Armed with this, it is easy to deduce that, in a system of n particles, if the eigenstates are indistinguishable with respect to the spin operator, then the joint eigenstates of the n-particle system will obey Fermi-Dirac statistics if the system contains coupled particles; otherwise, they will obey Bose-Einstein statistics. This suggests a natural definition of fermions and bosons. Moreover, because two photons can be in a singlet state, they exhibit Fermi-Dirac statistics with respect to the permutation group s_2 while coupled. This would suggest that neutrinos are perhaps photons. Parastatistics arises by partially relaxing the indistinguishability condition. Cooper pairing in solid-state physics might be a case in point.

Finally, the EPR paradox can be simply resolved by seeing coupled particles in a relativistic way. Once a frame of reference is specified, a specified value can be given to the spin relative to that frame. This means that, in the case of coupled particles, if a spin value in a certain direction is assigned to one particle, the spin value of the other particle can be determined without any need for action at a distance. Moreover, in this interpretation, once a frame of reference is fixed and everything is defined relative to that frame, Bell's inequality becomes a nonissue. An analogy might help. The surface of the particle can be envisioned as a string or a loop with two twists in it, obtained by combining two Möbius strips. It is a "two-faced" surface, so to speak. Depending on the direction that the face is hit by the interacting field, it will rotate in one way or another with respect to a reference frame. Finally, note that what might be considered +1 in one reference frame may be seen as −1 in another, although the particle is in a predetermined state.

A comparable situation occurs in a discussion of the spin of a black hole. Wheeler, in referring to work of Claudio Teitelboim, points out that the spin of a black hole "can be given one value or another depending on our choice of reference frame—except that now the frame of reference that comes into consideration is not the Lorentz frame, but the spinor reference frame."[1] If the EPR problem is viewed as a comparable phenomenon on the subatomic scale, then the paradox disappears.

To conclude, we can say that Bell's inequality gives rise to a paradox if the spinor reference frame is ignored. On the other hand, once this frame of reference is acknowledged, Bell's inequality can be used to directly prove that particles are coupled with probability 1 or that they are independent of each other. Moreover, the Pauli exclusion principle follows directly from this coupling, once the notion of indistinguishability is introduced.

REFERENCE

1. WHEELER, J. 1994. At Home in the Universe, p. 280f. Amer. Inst. Phys. New York.

Interplay of Aharonov-Bohm and Berry Phases for a Quantum Cloud of Charge[a]

SANDU POPESCU,[b] YAKIR AHARONOV,[b,c]
SIDNEY COLEMAN,[d] ALFRED S. GOLDHABER,[e]
SHMUEL NUSSINOV,[b] BENNI REZNIK,[b]
DANIEL ROHRLICH,[b] AND LEV VAIDMAN[b]

[b]*School of Physics and Astronomy*
Tel Aviv University
Ramat Aviv 69978, Israel

[c]*Department of Physics*
University of South Carolina
Columbia, South Carolina 29208

[d]*Lyman Laboratory*
Harvard University
Cambridge, Massachusetts 02138

[e]*Institute for Theoretical Physics*
State University of New York
Stony Brook, New York 11794-3840

The Aharonov-Bohm (AB) effect[1] is simple and topological: an electron encircling a solenoid containing a magnetic flux Φ acquires a geometrical phase equal to $ne\Phi/\hbar c$, where n is equal to the winding number of the electron around the solenoid. However, when a solenoid enters a quantum cloud of charge and there is no way to associate a well-defined path to the electron, the consequences of the AB effect might be complicated. For example, consider an electron bound in a potential well V, in an energy eigenstate. A solenoid crosses the well. How many times did the electron encircle the solenoid? There is no definite answer to this question. Of course, we can decompose the movement of the electron into a superposition of different Feynman paths, compute the phase acquired in each path, and resum, but no simple result will emerge. In general, the final state of the electron (once the solenoid has left the well) is different from the initial one (before the solenoid entered) and it depends on all the different parameters of the problem: the initial state, the potential V, the precise path of the solenoid and its velocity, and the value of the enclosed magnetic flux Φ. However, we have found a surprising topological effect for a solenoid containing exactly half a flux quantum $[\Phi = (\frac{1}{2})\Phi_0 = (\frac{1}{2})2\pi\hbar c/e]$ when it adiabatically crosses the quantum "cloud" of an electron in a nondegenerate energy eigenstate[f] (see

[a]This research was supported in part by Grant No. 425-91-1 of the Basic Research Foundation, administrated by the Israel Academy of Sciences and Humanities; Grant Nos. PHY-8807812 and PHY-9309888 of the National Science Foundation; and National Science Foundation Grant No. PHY-9321992 (to S. Popescu).

[f]Not counting the spin degeneracy.

reference 2). This topological effect is an interplay of the Aharonov-Bohm and Berry[3] phases.

Consider an electron in a nondegenerate energy eigenstate of an arbitrary potential well V and a semifluxon moving adiabatically on a closed path C that crosses the electron cloud. We take the solenoid to be linelike, that is, infinitely narrow and long. According to the adiabatic approximation, the solenoid does not induce transitions and the final state of the electron is identical to the initial one, up to a phase $e^{i\varphi}$. Now, φ contains a dynamical phase φ_d and a geometrical phase φ_g such that

$$\varphi = \varphi_d + \varphi_g. \tag{1}$$

We are interested in the geometrical part of the phase. For simplicity, we first consider the two-dimensional situation illustrated in FIGURE 1. Two limiting cases,

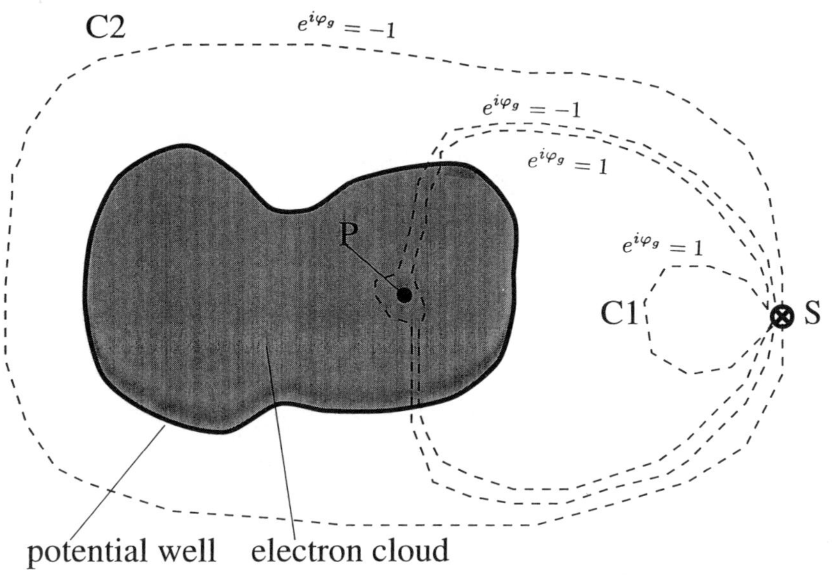

FIGURE 1. For simplicity, we consider an infinitely deep potential well so that the electron cloud vanishes outside it.

where we know how much charge the solenoid encircles, are easily computed.[g] When the solenoid moves along the path $C1$, it does not encircle any charge at all. The AB phase is 0. On the other hand, when the solenoid moves along $C2$, the electron is encircled with certainty and the AB phase is $e\Phi/\hbar c$, which in the case of a semifluxon (half a quantum of flux) yields π. However, how are we to interpolate the phase for intermediate paths that cross the electron cloud? Apparently, as the solenoid moves

[g] The AB effect is usually illustrated by an electron that encircles a solenoid. Here, a solenoid encircles an electron. However, whether we consider the movement in the reference frame of the solenoid or of the electron, the phase is the same.

on paths encircling larger and larger portions of the cloud, the effective charge it encircles gradually increases from 0 to e. Thus, the AB phase should gradually change from 0 to π. Closer inspection, though, leads to a different conclusion. The reason is that the physics manifests time-reversal symmetry. The initial wave function of the electron is nondegenerate and therefore unchanged under time reversal. (Assume that, initially, the semifluxon is infinitely far from the electron and no other vector fields act on it.) Under time reversal, the magnetic field inside the solenoid changes sign and thus also the magnetic flux ($\Phi \to -\Phi$); however, in the particular case of half a flux quantum, this change is not observable as long as the electron cannot penetrate into the solenoid because the difference between $(\frac{1}{2})\Phi_0$ and $-(\frac{1}{2})\Phi_0$ is exactly a flux quantum. Consequently, for any path C, the geometric phase $e^{i\varphi_g}$ must be the same in whatever direction the solenoid moves. On the other hand, φ_g must change sign when the solenoid changes direction (because it can be written as a line integral along the path C). Thus, we obtain

$$e^{i\varphi_g} = e^{-i\varphi_g}, \tag{2}$$

implying $e^{i\varphi_g} = \pm 1$. This result, corresponding to φ_g equal to an integer multiple of π, contradicts our naive expectation that the AB phase gradually changes from 0 to π.

What happened? Let us try to interpolate between the paths $C1$ and $C2$. We can gradually distort the path $C1$ into $C2$ by many steps that enlarge the loop by an infinitesimal region. In a certain region, the phase factor jumps from 1 to -1. We probe this particular infinitesimal region until we encounter a point P with the property that the geometrical phase jumps by π when P is encircled. However, what is the phase when the solenoid moves on a path crossing P? Our best guess is that the phase is not well defined. Our assumption that the solenoid moves adiabatically breaks down on this path. In other words, although initially the wave function of the electron was a nondegenerate energy eigenstate, it is no longer nondegenerate when the solenoid goes through the point P: at P, the solenoid induces a degeneracy.

Thus, we find a clue to the puzzle of abrupt phase changes. Indeed, we can say that our argument for a gradual change in the Aharonov-Bohm phase was correct. However, we neglected a second contribution to the geometric phase. As the solenoid crosses, it distorts the wave function of the electron and generates a Berry phase that adds to the AB phase:

$$\varphi_g = \varphi_{AB} + \varphi_{Berry}. \tag{3}$$

The Berry phase is responsible for maintaining the total geometric phase factor (1 or -1) despite gradual changes in the AB phase. The Berry phase is also responsible for the jump in the total geometric phase around the point P: as Berry[3] showed, isolated energy degeneracies can add π to the geometric phase.

Returning to our original problem, we can add a few details. First, the existence of points of degeneracy, for any arbitrary potential well V, can be proved by using time-reversal symmetry in the context of a Born-Oppenheimer approximation. Nevertheless, finding such points appears to be a difficult problem; the only explicit examples we know are for rotationally symmetric potential wells and for wells with even discrete rotational symmetries [$V(r, \theta) = V(r, \theta + 2\pi/2n)$], where the center of the well is such a point. Second, it is clear that there might be more than one such point. Any odd number of points is consistent with the phases of the extremal paths $C1$ and $C2$. Third, the adiabatic approximation might break down not only at some isolated point P, but in a whole region if the initial nondegenerate state becomes degenerate with states in the continuum. Last, but not least, a similar effect of phase jumps and energy level crossings arises even if the solenoid is not straight and also when several solenoids, each carrying half a flux quantum, enter the electron cloud.

We find a simple rule for the geometric phase of an atom with a heavy nucleus, initially in a spherically symmetric eigenstate, moving around semifluxons. (In this case, geometrical phases arise for both the electron and the nucleus, but the wave function of the nucleus is much more concentrated and semifluxons rarely penetrate it. Thus, its geometrical phase is simply the usual AB phase, which we neglect in the following.) The rule allows that we can replace the electron cloud with a point charge at the center of the atom and the semifluxons with "shadow" fluxons. A shadow fluxon is a point at which two electronic energy levels cross, if the center of the atom sits there. The winding number of the path of the point charge around the shadow fluxons gives the geometric phase accumulated by the atom.

To derive this rule, consider two straight and parallel solenoids situated a distance L apart. Two extreme cases are easily solved. When the distance between the solenoids is much larger than the size of the atom, we can move the atom in the vicinity of one of the solenoids without the electron cloud touching the other solenoid. In this case, the atom collects a phase of π each time its center encircles the solenoid, exactly as if the other solenoid were not present. There are thus two shadow fluxes, coinciding with the original solenoids. On the other hand, for $L = 0$, the two solenoids are at the same point, with their magnetic fluxes adding to an integer flux quantum. However, an integer flux quantum has no effect on an electron. There are therefore no energy level crossings and thus no shadow fluxons. When the solenoids are slightly separated, they do affect energy levels, but, by continuity, this effect is small and does not induce energy level crossing; rather, a minimal distance L^* is required. Thus, we conclude that, in adiabatic motion, the geometric phase accumulated by the atom due to two parallel semifluxons is zero once their separation is less than some critical distance L^*. We can now interpolate between these two extreme cases (large and small L) (FIGURE 2). When the distance between the solenoids is very large, the shadows coincide with the original solenoids. When the distance is still large, but comparable to the size of the atom, the shadow fluxes no longer coincide with the original solenoids. Instead, the shadow fluxon associated to each solenoid is shifted towards the other solenoid. When the two solenoids are at a critical distance L^*, their shadows overlap and therefore have no effect whatsoever on the atom. For separations smaller than L^*, the shadow fluxons disappear.

It is amusing to consider more general patterns of semifluxons carrying half a quantum of flux and the resulting shadows. Even in the case of a single solenoid, the shadow need not coincide with the original, if the solenoid is not straight. For example, a solenoid in the form of a ring should have a circular shadow, but of smaller radius. Just as in the case of two parallel solenoids, there is a critical radius for the ring (depending on the electron cloud) below which there will be no shadow at all. As a consequence, there will be no topological scattering of the atoms from small toroidal solenoids. For two intersecting straight solenoids, we expect hyperbolic shadows situated in the plane of the solenoids, in the acute angles. When the

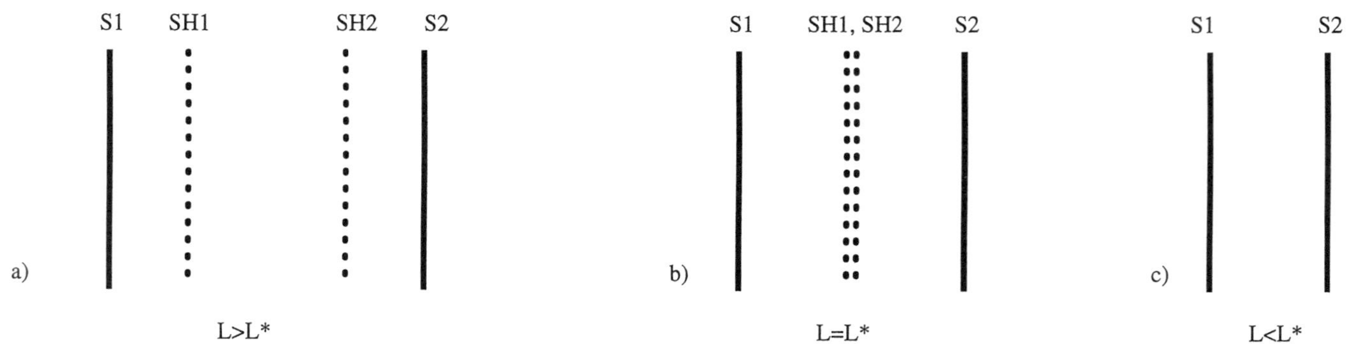

FIGURE 2. S1 and S2 represent the original solenoids and SH1 and SH2 represent their shadows (three cases—a, b, c—for varying lengths of L).

solenoids are perpendicular to each other, the shadows will coincide with the solenoids.

REFERENCES

1. AHARONOV, Y. & D. BOHM. 1959. Phys. Rev. **115:** 485.
2. AHARONOV, Y., S. COLEMAN, A. GOLDHABER, S. NUSSINOV, S. POPESCU, B. REZNIK, D. ROHRLICH & L. VAIDMAN. 1994. Phys. Rev. Lett. **73:** 918.
3. BERRY, M. V. 1984. Proc. R. Soc. London Ser. A **392:** 45.

Effective Mass–enhanced Deflection of Neutrons in Noninertial Frames[a]

K. RAUM,[b,c] M. KOELLNER,[c] A. ZEILINGER,[b] AND R. GÄHLER[c]

[b]*Institut für Experimentalphysik Universität Innsbruck A-6020 Innsbruck, Austria*

[c]*Department of Physics Technische Universität München 85747 Garching, Germany*

There are rather few experiments involving both quantum mechanics and the equivalence principle of relativity theory. We report two new experiments of this type. In the first one, we observed the deflection of neutrons in a slowly rotating perfect silicon crystal. In the second one, which is in preparation at present, we want to measure the deflection due to gravity. In earlier experiments, the phase shift in a neutron interferometer due to gravity[1] or due to a rotation of the interferometer was experimentally observed.[2,3] It is generally not realized that at present, for these interference experiments, a statistically significant discrepancy exists between the theoretical prediction and the experimental results.[4–6]

Our experiments rely on the greatly reduced effective mass m^* of neutrons under diffraction conditions in perfect crystals.[7] In our case, the ratio m^*/m is $\pm 4.7 \times 10^{-6}$. In the first experiment, we observed the deflection of the neutrons due to the Coriolis force in a rotating frame. If the crystal rotates with an angular velocity ω, the effective acceleration is approximately given by $2m/m^* \omega v_{\mathrm{d}}$, with v_{d} being the drift velocity[8] of the neutrons perpendicular to the rotating axis. For the gravity experiment, this has to be replaced by $m/m^* g \cos \Theta$, where Θ is the angle between the gravitational acceleration g and the reciprocal lattice vector G. In order to observe a deflection due to this effective acceleration, we use a specially shaped crystal consisting of two sections of a length L. Between them is a gap of width s. The whole structure is cut from one single crystal, which provides parallel-oriented lattice planes in the two sections. The first section of the crystal acts as a collimator by selecting only neutron trajectories with an initial slope that compensates the deflection on the way through this section (see FIGURE 1). In the second section, the two trajectories corresponding to the two signs of the effective mass are separated. A detailed calculation leads to the following expression for the separation of the outcoming neutron beams:

$$z(\omega) = 2\,\frac{L^2 \hbar G}{V_G} \sin(\Theta_{\mathrm{Bragg}}) \left(1 + \frac{s}{L}\right) \omega, \tag{1}$$

[a]This work was supported by the Austrian Science Foundation (FWF Project No. P8767), by the United States National Science Foundation, and by a travel grant from the Austrian Academy of Sciences.

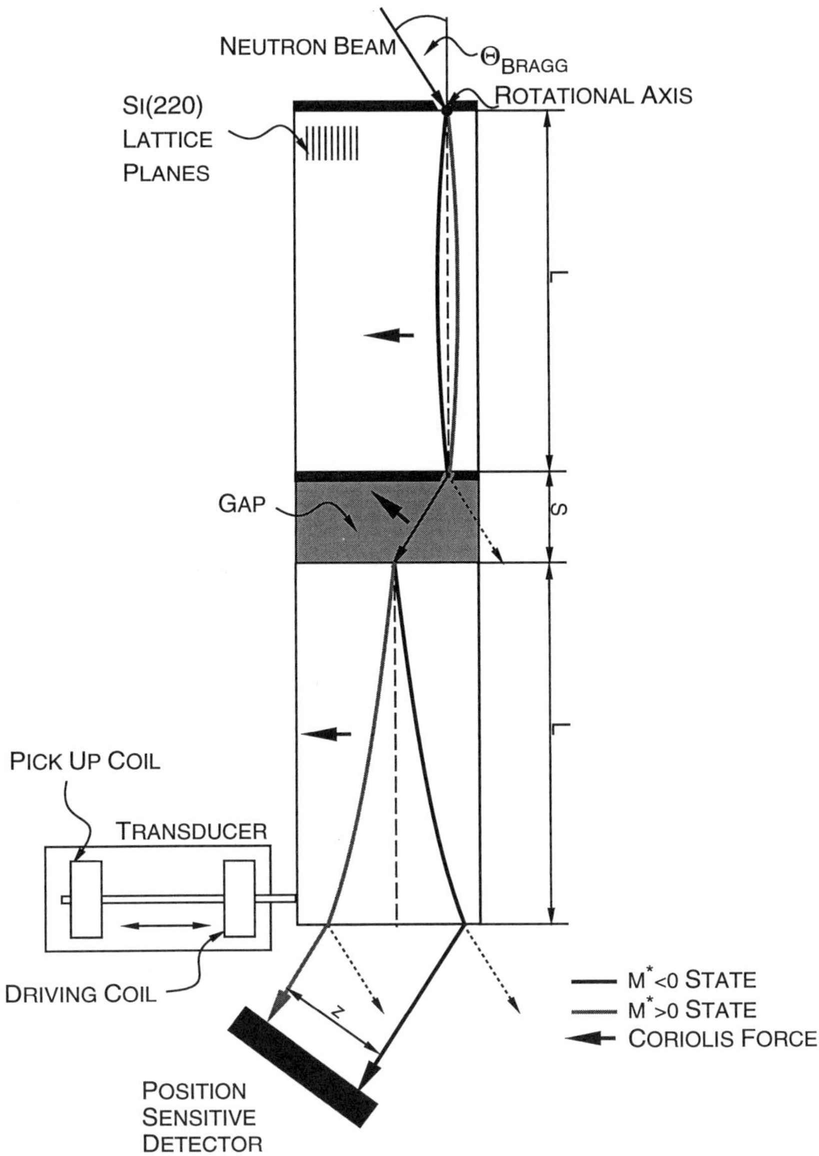

FIGURE 1. Principle of the setup viewed from above. The dashed lines are the neutron trajectories without external force and the continuous trajectories are shown for one sense of rotation.

where G is the reciprocal lattice vector and V_G is the crystal potential. Instead of a continuous rotation of the crystal, we used a rotating oscillation with a frequency of 13 or 20 Hz and a maximum amplitude between 2 and 7 μm. The neutrons came from a graphite monochromator at a mean wavelength of 2.35 Å. They passed through the crystal and were detected by a position-sensitive detector. The experiment was carried out at the NIST Gaithersburg reactor.

The maximum splitting that we observed was 9.5 mm, compared to a deflection of 42 nm for a free neutron on the same distance. FIGURE 2 shows the measured splitting of the two beams as a function of the angular velocity. The experimental results show a splitting for zero angular velocity not predicted by theory. Most likely, this effect was caused by an intrinsic bending of the crystal. If the additional deflection due to bending is small, this does not affect the slope of the $z(\omega)$ curve, but causes a constant offset independent of ω.

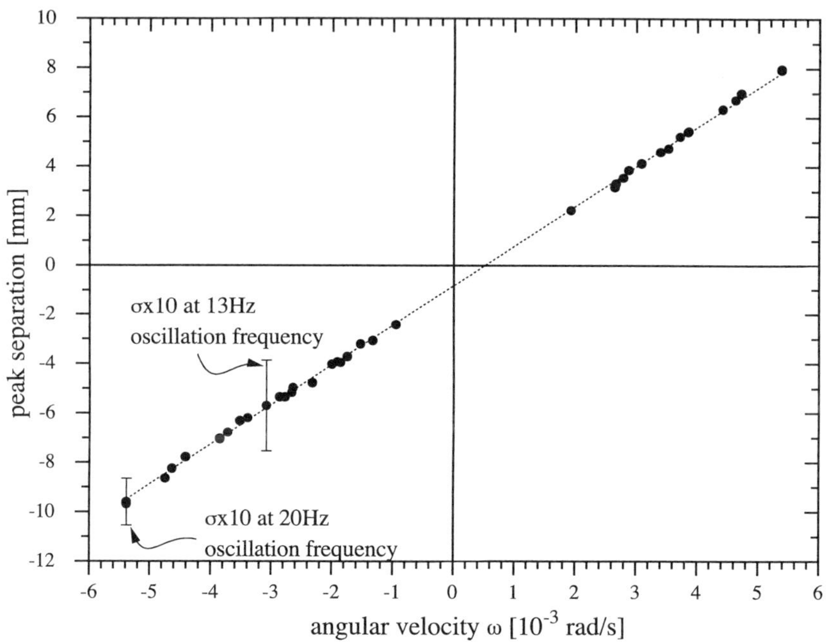

FIGURE 2. Measured distance of the peaks corresponding to the two effective mass states as a function of the angular velocity. The deflection for counterclockwise rotation of the crystal is defined as negative and is defined as positive for the other sense of rotation. The dashed line is the result of a linear fit to all shown data points.

Very good agreement between theory and experiment was obtained by comparing the predicted and the measured slope of the $z(\omega)$ curve. The predicted slope of (1.614 ± 0.008) m-s/rad and the value obtained from a linear fit to the measured data, namely, (1.609 ± 0.014) m-s/rad, agree within the errors. The error of the predicted value results mostly from uncertainties in the wavelength. The error in the measured value is determined by statistical errors, calibration errors, and environmental vibrations.

The slowest detectable rotation in the current experiment was of the order of the rotation of the earth. This can possibly be improved by more than one order magnitude by using a longer neutron wavelength, vibration isolation, better counting statistics, and smaller slit widths.

In experiments that are in preparation at the moment, we will measure the deflection due to gravity. These experiments will allow us to compare the inertial and heavy mass of a free neutron in the quantum mechanical limit.

ACKNOWLEDGMENTS

We wish to acknowledge the support of G. Greene, M. Arif, J. Felber, and C. Rausch.

REFERENCES

1. COLELLA, R., A. W. OVERHAUSER & S. A. WERNER. 1975. Phys. Rev. Lett. **34:** 1472.
2. ATWOOD, D. K., M. A. HORNE, C. G. SHULL & J. ARTHUR. 1984. Phys. Rev. Lett. **52:** 1673.
3. WERNER, S. A., J. L. STAUDENMANN & R. COLELLA. 1979. Phys. Rev. Lett. **42:** 1103.
4. BONSE, U. & T. WROBLEWSKI. 1984. Phys. Rev. D **30:** 1214.
5. HORNE, M. A. 1986. Physica B **137:** 260.
6. WERNER, S. A., H. KAISER, M. ARIF & R. CLOTHIER. 1988. Physica B **151:** 22.
7. ZEILINGER, A., C. G. SHULL, M. A. HORNE & K. D. FINKELSTEIN. 1986. Phys. Rev. Lett. **57:** 3092.
8. SHULL, C. G. *et al.* 1980. Phys. Rev. Lett. **44:** 1715.

Hidden-Variables Model and Nonlocality in the Bohm/EPR Experiment

KENNETH H. SCHATTEN

Sensors Development and Characterization Branch
Goddard Space Flight Center
National Aeronautics and Space Administration
Greenbelt, Maryland 20771

Einstein, Podolsky, and Rosen[1] (EPR) developed an experiment that questioned whether quantum mechanics (QM) was complete. The experiment was modified by Bohm[2] by considering two spin-½ particles, labeled E and P, that fly off in opposite directions from a system prepared in a "zero spin state". These particles, if attached to neutral atoms, may have their spin "responses" in any direction determined by Stern-Gerlach magnets. A QM probabilistic equation holds for the probability of finding the P particle in a spin UP—or, using the generic word, YES—state at an angle θ with respect to the same direction in which the E particle was observed to have spin UP, or YES, namely,

$$P(\text{YES}, \theta) = \tfrac{1}{2}(1 - \cos\theta) \text{ and } P(\text{NO}, \theta) = \tfrac{1}{2}(1 + \cos\theta). \quad (1)$$

Often, this is expressed in a correlation form:

$$c(\vec{z}_E, \vec{z}_P) = P(\text{YES}, \theta) - P(\text{NO}, \theta) = -\cos\theta, \quad (2)$$

with θ being the angle between $\vec{z}_E$ and $\vec{z}_P$, the orientation of the Stern-Gerlach magnets.

Bell[3] developed a local hidden-variables (HV) model (shown in FIGURE 1) that allows a 50%–50% YES-NO detector response to spin orientation and showed that it did not give the QM statistical behavior. It seems clear that the responses need to be divided differently than in Bell's model if one is to achieve QM statistics. FIGURE 2 shows the results of a form that allows any fractional division of an orientation sphere by dividing the sphere with a small circle to form two spherical caps, one occupying a fraction (to be determined), f, of the area. Bell's hidden-variable hemispherical caps have been shifted vertically by an offset, h, allowing an extra degree of flexibility. Thus, if we wish an arbitrary fraction, f, of YES responses, we choose a spherical cap at a height, $h = 1 - 2f$, and angle, $\theta = \arccos(h) = \arccos(1 - 2f)$. The hidden variable is again chosen to be a "random vector" that selects the response. This spherical cap correlation allows the response of the P detector, R_P, to be based upon two random vectors, $\vec{c}$ and $\vec{s}$, on the following basis: if $\{[(\vec{z}_E - \vec{c}) \cdot \vec{z}_P](\vec{s} \cdot \vec{z}_E) \geq 0\}$, then R_P = YES; else R_P = NO. This provides the quantum statistics of equations 1 and 2.

Now, we show how angular momentum conservation (AMC) can be used to justify a fractional division different from the 50%–50% YES/NO Bell HV model and

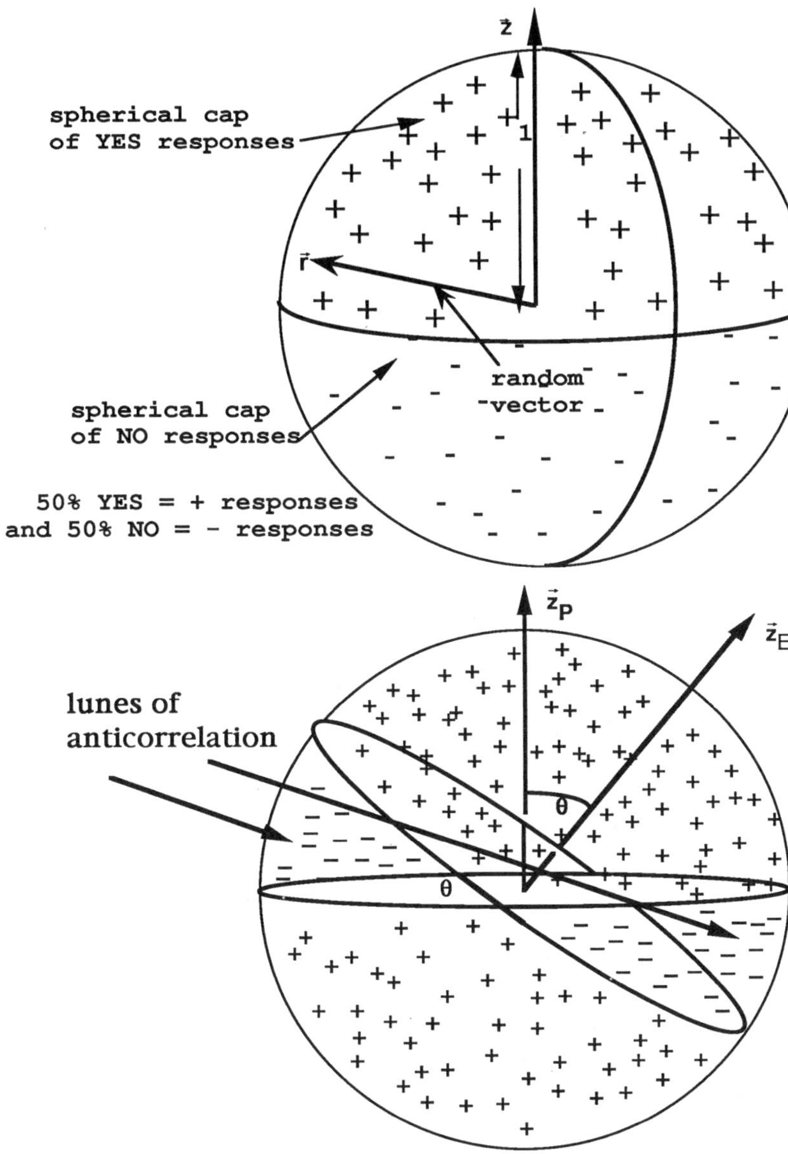

FIGURE 1. Bell's random vector model: view of the geometry considered by Bell to calculate the probability of obtaining a + correlation or a − correlation for spins oriented in space. Using Bell's reasoning, if a particle has a spin within either of the "lunes of anticorrelation" shown (for the two observing instruments directed towards $\vec{z}_E$ and $\vec{z}_P$), then the observations should be anticorrelated.

thereby provide the quantum correlations. We define an Angular Momentum Conservation Principle (AMCP) as follows:

> An observation, $s_L = \vec{L}_L \cdot \vec{z}_L$, in any "local" direction (e.g., $\vec{z}_L$) is constrained to provide, on average, angular momentum balance in this local observing direction, subject to any external "boundary conditions" imposed.

Let us express the AMCP in equation form for the current situation. The outside boundary condition includes any "constraint" associated with losses of angular momentum from the opposite, "nonlocal", particle. Then, the average amount of angular momentum in the local direction, $\vec{z}_L$, must add to zero:

$$\langle \vec{L}_{NL} \cdot \vec{z}_L \rangle + \langle \vec{L}_L \cdot \vec{z}_L \rangle = 0. \tag{3}$$

When the nonlocal detector observes a certain amount of angular momentum, s_{NL}, in a particular direction (e.g., $\vec{z}_{NL}$), then, in this observing direction, $s_{NL} = \vec{L}_{NL} \cdot \vec{z}_{NL}$

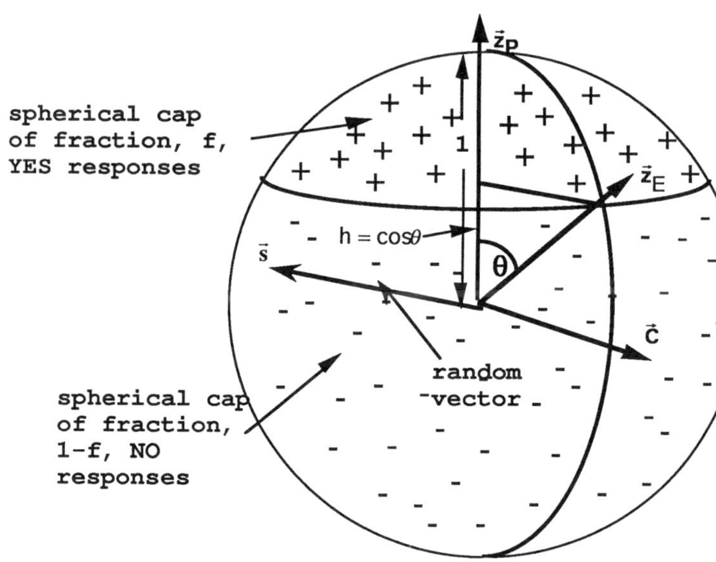

FIGURE 2. The geometry involved in calculating correlations in the nonlocal random vector model described by equation 3. The random vector $\vec{c}$ governs whether the correlation between the two detectors' observations oriented towards $\vec{z}_E$ and $\vec{z}_P$ will be positive or negative; this is indicated by the + and − signs. The overall probability of a positive correlation is related to the spherical cap area. The calculation in the text follows directly.

units of angular momentum has been lost. The average amount of angular momentum lost by the nonlocal particle in the local direction is $\langle \vec{L}_{NL} \cdot \vec{z}_L \rangle = \langle s_{NL} \vec{z}_{NL} \cdot \vec{z}_L \rangle = \langle s_{NL} \cos\theta \rangle$.

In QM, the detectors are restricted to quantized responses. Thus, our measuring device cannot observe a graduated amount of angular momentum; the responses must be quantized in units of angular momentum, $\pm C$ units, let us say $\pm C = \pm\hbar/2$

units, which we refer to as YES or NO. Let us now see how angular momentum can be conserved in the presence of this restricted binary response.

As an example, we examine the two terms in equation 3 for the subset of cases where our nonlocal detector, labeled the E detector, measures YES 100% of the time in the $\vec{z}_E$ direction (one can choose any subset of E responses and investigate the statistics). Let us evaluate the first term in equation 3. This quantized YES observation is taken to indicate that $+\hbar/2$ units of angular momentum, per particle, has been carried by the E particles in that direction. Thus, an amount of angular momentum in the $\vec{z}_P$ direction, at an angle θ from $\vec{z}_E$, equal to $\hbar/2 \cos\theta$ has been lost, per particle, to the YES E particles. This corresponds to a loss of angular momentum of $\vec{L}_{NL} \cdot \vec{z}_L = (\hbar/2)\cos\theta$ per particle, along the direction of our local P detector.

Now, let us evaluate the second term in equation 3. Let the local P detector respond YES or NO, with probabilities $P(\mathrm{Y}, \theta)$ and $P(\mathrm{N}, \theta)$, respectively, with each carrying $\pm\hbar/2$ units of angular momentum; then, the average amount of angular momentum per P particle carried in the local $\vec{z}_P$ direction is $\vec{L}_P = \vec{L}_L = \vec{z}_L[P(\mathrm{Y}, q) - P(\mathrm{N}, q)]\hbar/2$. Then, equation 3 requires $[P(\mathrm{Y}, \theta) - P(\mathrm{N}, \theta)] = -\cos\theta$, which is just the correlation of the two detectors, $c(\vec{z}_E, \vec{z}_P)$, of equation 2. With $P(\mathrm{Y}, \theta) + P(\mathrm{N}, \theta) = 1$, one arrives at equation 1. Thus, we see that AMC appears to require the offset or vertical shift, $h = 1 - 2f = \cos\theta$, discussed before. Hence, the QM "action at a distance" and nonlocal behavior emerge as consequences of AMC. We further view nonlocality as an indication of the inseparability of the whole system into separate components and thus as a consistency requirement, which spans space and time, rather than as a causal influencing agent affecting phenomena.

ACKNOWLEDGMENTS

I am indebted to Dave Rubincam for stimulating discussions and to Abner Shimony for excellent tutelage.

REFERENCES

1. EINSTEIN, A., B. PODOLSKY & N. ROSEN. 1935. Phys. Rev. **47:** 777.
2. BOHM, D. 1952. Phys. Rev. **85:** 166 & 180.
3. BELL, J. S. 1964. Physics (Long Island City, N.Y.) **1:** 195–200.

First-Order Dark Matter and the Effectiveness of Mathematics in the Natural Sciences[a]

ANTHONY ELLSWORTH SCOVILLE

Aerovest, Incorporated
Washington, District of Columbia 20008

On the one hand, Wigner has pondered the "unreasonable effectiveness of mathematics in the natural sciences."[2] On the other hand, standard big-bang cosmology has difficulty explaining the 90–99% nonbaryonic dark matter or "missing mass" required by observed galactic dynamics and required for $\Omega = \Omega_{crit} = 1$, which closes the universe at $t = \infty$. Here, it is proposed that dark-matter missing mass is a necessary consequence of the effectiveness (algorithmic locality) of mathematics in any complete cosmology, that is, a "Theory of Everything" (TOE).

Zurek and others have developed an information-theoretic formulation of quantum mechanics.[3] Omitting higher-order cosmic texture, I suggest that at the Planck scale the universe can be modeled as a homogenous vector field of Szilard engines with an average 8π cylinders (degrees of freedom). Strict algorithmic locality requires that this field of Planck engines have no memory of past states. As shown by Zurek, the energy of state erasure dissipated in each Planck-time cycle is $\Delta W^{-} = k_bT$ per degree of freedom.[4] The erased state energy constitutes the dark-matter missing mass of the Planck engine field. With suitable encoding of past states prior to erasure, it is shown that the proportion of dark-matter missing mass is $M_m \approx \{\log_2(t)\}/\{8\pi + \log_2(t)\}$. Assuming $H_0 = 37$, $M_0 = 1 - M_m = 0.1103\Omega_{crit}$.[5] This result closely agrees with $\Omega_{baryon} = 0.1096\Omega_{crit}$ suggested by elemental abundances, Jones' estimates of the Sunyaev-Zel'dovich effect, and the apparent gravitational lensing of MACHOs.[6] It appears, therefore, that the mathematical effectiveness of physical law is a good candidate for the cause of dark matter.

Three general implications deserve mention. First, employing the exchange of photons (and other B-E exchange force particles) between electrons and other Fermi particles as the quantum mechanism of state computation, it seems possible to model the universe as a computer. Mathematically effective computation generates dark matter. There should be a physical counterpart to every component of the computational process, including (for a complete system like the universe) undecidable states whose mathematical existence has been demonstrated by Gödel and others. Second, state computation (present observations of the universe) occurs on the surface of a cosmological-scale expanding black hole comprising the erased past states of the universe. The black hole is irreversibly expanding through time "into" a gas of undecidable states (the Gödel gas) comprising the future of the universe. Third, a cosmological Turing test (extended to include undecidable states) seems

[a] See reference 1, which develops the findings given here.

required to verify the rationality or nonteleological "intelligence" of the laws of physics.[7]

REFERENCES AND NOTES

1. SCOVILLE, A. E. 1993. Symmetry Breaking of the Gödel Gas [and sequels in progress]. Aerovest, Incorporated. Washington, District of Columbia.
2. WIGNER, E. P. 1960. The unreasonable effectiveness of mathematics in the natural sciences. Commun. Pure Appl. Math. **1:** 1.
3. See articles by Zurek, Bennett, and others in: ZUREK, W. H., Ed. 1990. Complexity, Entropy, and the Physics of Information. Addison–Wesley. Reading, Massachusetts.
4. Ibid., p. 82.
5. Ibid., p. 83–85. Also, see reference 1 (p. 49–65) for a detailed derivation of the exact equation as well as the approximation given here.
6. Numerous citations given in reference 1, p. 45–48.
7. See reference 1 for a discussion of these topics.

Geometric Phase Shifts in Pilot-Wave Theory

ERIK SJÖQVIST

Department of Quantum Chemistry
Uppsala University
S-751 20 Uppsala, Sweden

The foundation of Bohm's pilot-wave theory[1] consists of three elements: (i) the wave function, $\psi(\boldsymbol{q}, t) = R(\boldsymbol{q}, t) \exp [iS(\boldsymbol{q}, t)/\hbar]$, whose time evolution is given by the time-dependent Schrödinger equation; (ii) well-defined particle trajectories given by the guiding equation, $\boldsymbol{p} = \nabla S(\boldsymbol{q}, t)$; (iii) averaging over trajectories with the density $R^2(\boldsymbol{q}, t)$ as weight. Using these elements, I will show how one can obtain an anholonomic effect on the particle trajectories in this formulation of quantum mechanics.[2]

Consider a particle with unit mass on a ring of unit radius. The position of the particle is described by the polar angle θ. The normalized wave function, assumed to be cyclic with period T, can be written as $\psi(\theta, t) = R(\theta, t) \exp [iS(\theta, t)/\hbar]$ and the guiding equation becomes $\dot{\theta} = \partial S/\partial\theta \equiv \Omega[\theta, \boldsymbol{X}(t)]$, where $\boldsymbol{X}(t) = [X_1(t), \ldots, X_N(t)]$ are complex coordinates in the N-dimensional ray space $P(\mathscr{H})$.[3] Ω depends on t through $\boldsymbol{X}$ because the guiding equation is completely determined by the physical state; that is, it is invariant under a time-dependent phase change of the wave function.

The phase of the motion φ is defined by the equal-time parametrization, $d\varphi/dt\,|_X = 2\pi[\int_0^{2\pi} d\theta/\Omega(\theta, \boldsymbol{X})]^{-1} \equiv \omega(\boldsymbol{X})$, where $|_X$ means fixed $\boldsymbol{X}$.[4] X are now $2N$ real coordinates in $P(\mathscr{H})$. Using the guiding equation and averaging over the Bohm trajectories, the total phase shift after the cyclic change of the guiding wave function becomes

$$\Delta\varphi = \int_0^T \omega[\boldsymbol{X}(t)]\, dt + \oint_{\mathscr{C}} \mathscr{A}, \tag{1}$$

where the one-form $\mathscr{A}$ is given by

$$\mathscr{A} = \int_0^{2\pi} d\theta R^2(\theta, \boldsymbol{X}) \partial_\mu \left[\omega(\boldsymbol{X}) \int_0^{\theta} \frac{d\vartheta}{\Omega(\vartheta, \boldsymbol{X})} \right] dX^\mu \tag{2}$$

and $\mathscr{C}$ is the closed path in $P(\mathscr{H})$. In addition to the expected dynamical shift $\int_0^T \omega[\boldsymbol{X}(t)]\, dt$, we see that there is a geometric contribution given by a line integral over a one-form in the ray space. This extra part is the geometric phase shift $\Delta\varphi_g$.

Using Stokes' theorem, the geometric shift can be written as

$$\Delta\varphi_g = \int_S d\mathscr{A} \equiv \int_S \mathscr{F}, \tag{3}$$

where $\partial S = \mathscr{C}$. Explicitly, the two-form $\mathscr{F}$ is given by

$$\mathscr{F} = \int_0^{2\pi} d\theta\, \partial_\mu R^2(\theta, X) \partial_\nu \varphi(\theta, X)\, dX^\mu \wedge dX^\nu. \tag{4}$$

From this expression, one sees that the averaging procedure leads in general to a nonzero geometric phase shift.

Moreover, an X-dependent change of the origin described by the transformation

$$\varphi(\theta, X) \rightarrow \varphi(\theta, X) + \lambda(X) \tag{5}$$

does not change the geometric phase shift. The reason is that this transformation just adds an exact differential $d\lambda$ to the one-form $\mathscr{A}$, which clearly does not affect the line integral. In other words, the geometric phase shift is gauge-invariant.

Finally, I want to emphasize the close analogy with the classical Hannay angle.[5] The Hannay angle is generated by a closed excursion of slowly changing parameters and becomes nonzero because of the adiabatic averaging.[6] This slow change corresponds here to a closed excursion in $P(\mathscr{H})$ and the adiabatic averaging corresponds to averaging over Bohm trajectories.

REFERENCES

1. BOHM, D. 1952. Phys. Rev. **85:** 166 & 180.
2. SJÖQVIST, E. 1994. Phys. Rev. Lett. Submitted.
3. PAGE, D. N. 1987. Phys. Rev. A **36:** 3479.
4. KEPLER, T. B. & M. L. KAGAN. 1991. Phys. Rev. Lett. **66:** 847.
5. HANNAY, J. H. 1985. J. Phys. A **18:** 221.
6. ARNOLD, V. I. 1989. Mathematical Methods of Classical Mechanics. Second edition. Springer-Verlag. New York/Berlin.

Tunneling Times and Weak Measurements[a]

AEPHRAIM M. STEINBERG[b]

Department of Physics
University of California
Berkeley, California 94720

The question of how much time a tunneling particle spends in a barrier region before being transmitted has long been controversial.[1] One way to answer the question is to explicitly consider a von Neumann–style interaction with a measuring apparatus or "pointer" and examine the pointer position only on those occasions where the particle is found to have tunneled. I have recently done this[2] by applying the "weak measurement" formalism of Aharonov and Vaidman.[3] As explained in reference 2, this is essentially a way of defining Bayesian "conditional probabilities" in the framework of quantum mechanics. I find that the transmission time so defined is complex in general, but that the physical meaning of the real and imaginary parts is clear. The real part represents an actual shift in the pointer position, that is, the desired measurement outcome. The imaginary part (which vanishes in the "classical" limit where all incident particles reach the same final state) represents instead a shift in the *momentum* of the pointer. This reflects a back-action on the particle provoked by the measurement interaction rather than a physical property of the tunneling particle itself. To see this, one can consider a pointer whose momentum uncertainty is very small before the measurement. One finds that the position shift of the pointer is unchanged, whereas the amount of momentum it exchanges with the tunneling particle vanishes along with the square of the initial uncertainty. In other words, in the limit of a "weak" measurement, it is the real part of the calculated time that one will observe. For a rectangular barrier, I find that the real part of the transmission time is equal to the "dwell time".[1] The popular Larmor times[4] are simply one realization of such a measurement; in reference 2, I show how it is possible to make the Larmor clock "weaker" by using spin-squeezed states in order to confirm that only the portion corresponding to the dwell time should be thought of as part of the measurement outcome.

Instead of considering spin, one could imagine a scenario as in FIGURE 1. A heavy charged particle such as a proton is tunneling in one dimension. It passes through a series of grounded conducting plates (these plates themselves could even form the tunnel barrier). Between each pair of plates, one may have a distant electron, initially at rest, serve as a test particle. Each electron only feels a transverse Coulomb force while the proton is between the same pair of plates as that electron, so the electron will accumulate a momentum kick proportional to the amount of time spent by the proton between those plates (note that this momentum is the "pointer position" and

[a]This work was supported by the United States Office of Naval Research.

[b]Present address: National Institute of Standards and Technology, Gaithersburg, Maryland 20899.

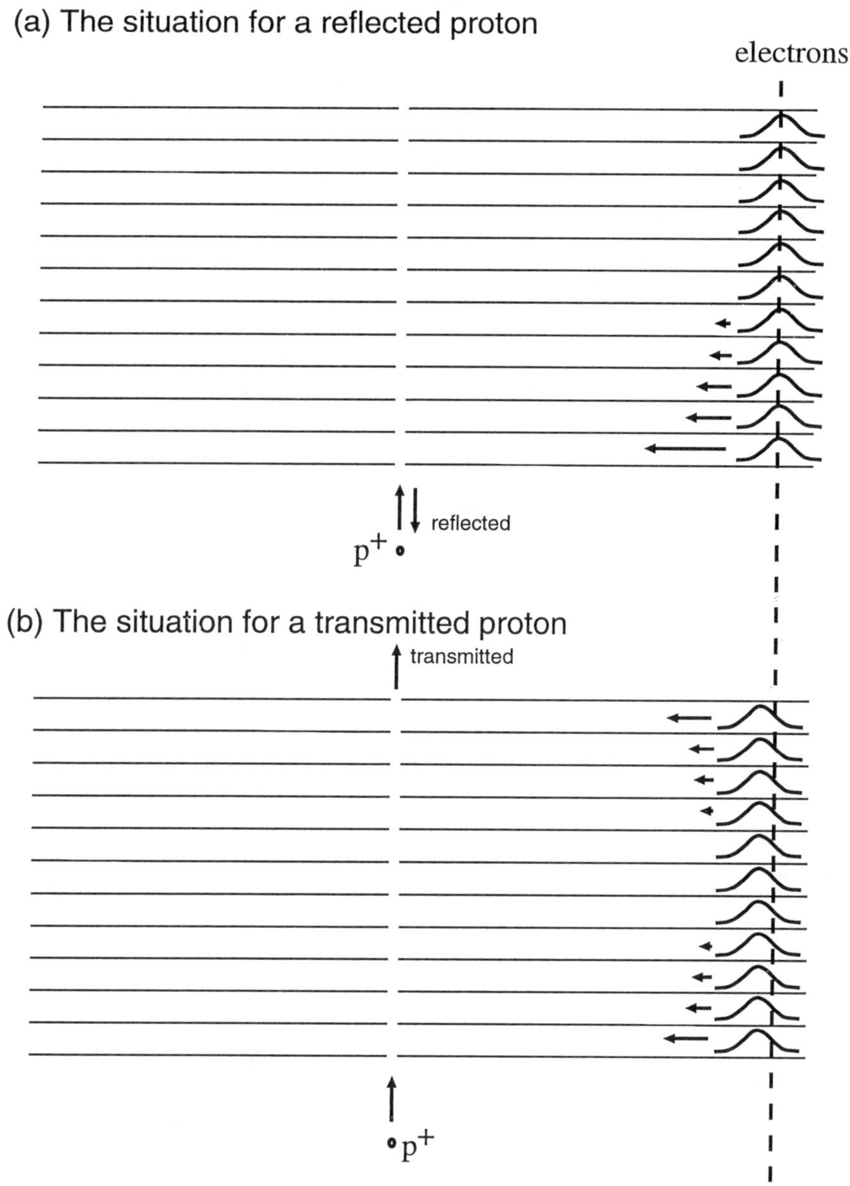

FIGURE 1. A gedankenexperiment using distant electrons to measure how much time a tunneling proton spends in each of several shielded regions of space. In (a), we see the final state of the electrons for cases where the proton is reflected: only those electrons in the first evanescent decay length of the tunnel barrier feel the proton's Coulomb potential and acquire a consequent momentum kick. In (b), we see what happens if the proton is transmitted: electrons near both edges of the barrier acquire a momentum kick, but those near the center do not. The tunneling proton does not seem to have spent any time in the center of the barrier. However, the *position* of the electrons gets shifted when the proton is transmitted. As explained in the text, this is a measure of the back-action of the measuring electrons on the tunneling particle.

the electron's position thus plays the role of "pointer momentum"). The meaning of the present result is as follows: when the proton is reflected, only the electrons near the front edge of the barrier will be set in motion; in contrast, when the proton is transmitted, the electrons toward *both* edges of the barrier will receive momentum kicks. The electrons near the center of the barrier will receive no momentum kick at all: it is as if the proton "hopped" across the barrier without spending any time in the middle. Like the group delay and the dwell time, the time spent by tunneling particles in the barrier regions saturates to a finite value as the barrier becomes infinitely thick. On the other hand, the back-action (the imaginary part of the complex time) grows linearly with barrier thickness: electrons near the middle of the barrier get no momentum kick, but instead are shifted in *position* toward the proton.

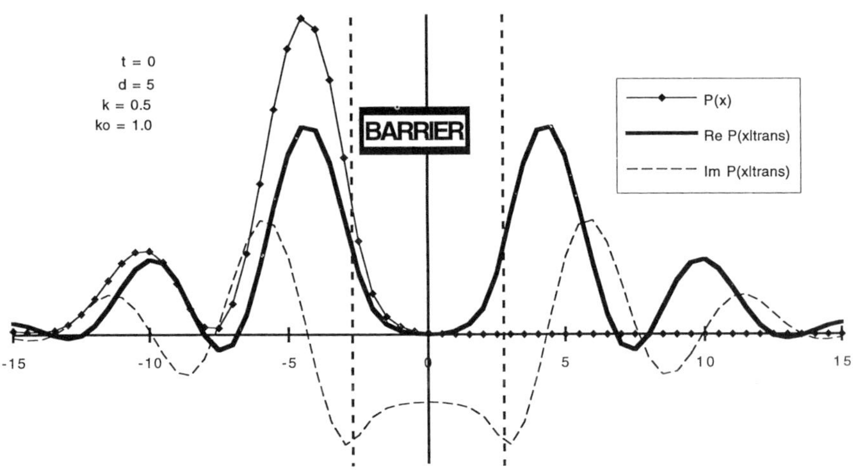

FIGURE 2. The thick solid curve shows the "conditional probability" distribution for a tunneling particle as a function of position, at an instant about halfway through the tunneling event. This represents the force that an electron at a given x would experience in the example of FIGURE 1. The dashed curve represents the back-action, that is, the position shift that an electron would experience. For comparison, the thin solid curve shows the probability distribution for the ensemble as a whole, without distinguishing between transmitted and reflected particles.

This is simple to understand; the electrons had some uncertainty in their position to begin with. Due to the attractive Coulomb interaction, the closer the electron was to the proton, the smaller the potential barrier that the proton had to traverse. Thus, by selecting protons that succeeded in tunneling, we are postselecting states where the electron was nearby to begin with. Unlike the momentum kick—the measurement outcome itself—this effect is entirely dependent on the initial uncertainty in the electron position.

In FIGURE 2, I extend this approach to calculate the conditional probability distribution for the tunneling particle's position at a given instant of time during the tunneling process. At early times, this distribution is real and mimics the incident wave packet; at late times, it is also real and mimics the transmitted wave packet.

(Consequently, this approach does not support the assertion[5] that particles that tunnel originated toward the front of the incident wave packet.) During the tunneling, as shown here, the real part displays a growing peak at the exit face and an ebbing peak at the entrance face, but little penetration into the center of the barrier. This conditional probability distribution may traverse the barrier superluminally in this fashion. Meanwhile, the imaginary part of the distribution measures the back-action and is essentially constant across the barrier region.

REFERENCES

1. LANDAUER, R. & TH. MARTIN. 1994. Rev. Mod. Phys. **66**(1): 217–228.
2. STEINBERG, A. M. 1995. How much time does a tunneling particle spend in the barrier region? Phys. Rev. Lett. To appear.
3. AHARONOV, Y. & L. VAIDMAN. 1990. Phys. Rev. A **41**(1): 11–20.
4. BÜTTIKER, M. 1983. Phys. Rev. B **27**(10): 6178–6188.
5. STEINBERG, A. M., P. G. KWIAT & R. Y. CHIAO. 1984. Found. Phys. Lett. **7**(3): 223–239.

Is Quantum Mechanics an Exotic Probability Theory?

SAUL YOUSSEF

Supercomputer Computations Research Institute
Florida State University
Tallahassee, Florida 32306-4052

Quantum mechanics is often introduced with a discussion of the two-slit experiment where the observed interference pattern forces us to conclude that the particle goes through both slits at once and is "both a particle and a wave". This most-basic argument has a loophole, however. The conclusion rests on probability theory and, in particular, on the fact that probabilities are nonnegative so that, when the second slit is opened,

$$P(x) = P(x \text{ via slit } 1) + P(x \text{ via slit } 2) \geq P(x \text{ via slit } 1), \qquad \textbf{(1)}$$

where $P(x)$ is the probability for the particle to arrive at position x on the screen. Perhaps, then, it is possible that the particle does go through either one slit or the other after all and the interference effects can be explained by modifying probability theory itself. Here, we give a brief overview of this program, which is put forward in references 1–3. For other approaches to exotic probability theory, see references 4–7.

Probability theory is most often presented with the "frequentist" approach where probabilities are defined as limits of experimental frequencies. For example, if n successes occur in N trials of an experiment, the large N limit of n/N is called the "probability of success". Such probabilities are then assumed to follow Kolmogorov's axioms.[8] From this point of view, probability theory is a theory of "random phenomena" that are well described by these axioms and, of course, probabilities are necessarily nonnegative. For our purposes, then, it is necessary to adopt the more general "Bayesian" view where probabilities are not a priori defined as frequencies, but where, instead, a frequency interpretation is derived as a consequence of the fundamental axioms.[9] With the Bayesian view, probability is introduced as a measure of "likelihood" where, if proposition a is known, the likelihood that proposition b is true is denoted[10] "$(a \to b)$". For "$\to$" to be a useful likelihood measure, one expects it to have a few properties such as (i) if $(a \to b)$ is known, this should determine $(a \to \neg b)$ and (ii) the procedure to get from $(a \to b)$ to $(a \to \neg b)$ should be independent of a and b. As shown by Cox,[11] such considerations are enough to entirely fix probability theory, which then takes the form

$$(a \to b \wedge c) = (a \to b)(a \wedge b \to c) \qquad \textbf{(2a)}$$

$$(a \to b) + (a \to \neg b) = 1 \qquad \textbf{(2b)}$$

$$(a \to \neg a) = 0 \qquad \textbf{(2c)}$$

for all propositions a, b, and c.

Adopting the Bayesian view, we introduce complex probabilities while keeping all of Cox's remaining conditions. Cox's analysis then leads to equations 2a–c again, except with complex $(a \rightarrow b \wedge c)$, $(a \rightarrow b)$, etc. In the Bayesian view of ordinary probabilities, the frequency interpretation can be derived from the axioms using an additional condition—namely, if $(a \rightarrow b) = 0$, then b never actually happens if a is known. In complex probability, however, a may be known and $(a \rightarrow b)$ may be zero, but b may still be found to be true. If we suppose, though, that a special set of propositions U exists where, for $x \in U$, $(a \rightarrow x) = 0$ does imply that x is never true if a is known, then a frequency interpretation can once again be derived. This motivates the introduction of the realistic "state space" axioms as described in references 1 and 2. The resulting complex probability theory is then shown to have a superposition principle, to include wave functions that are expansions in eigenfunctions of Hermitian operators, to have a path integral representation, and to describe both pure and mixed systems. A scalar particle in R^d is shown to obey the Schrödinger equation with mass, vector potential, and metric appearing as moments of a fundamental complex probability.

Although probability theory has been historically considered to be a branch of mathematics, its frequency meaning makes it a physical theory that may succeed or fail by experimental test. From this point of view, it is natural to attempt to explain quantum mechanical effects by modifying probability theory. We find that there is essentially one possible modification of probability theory that introduces complex probabilities while maintaining the Bayesian consistency conditions. Although this modified probability theory is both realistic and local,[3] it nevertheless reproduces the predictions of standard quantum mechanics while coexisting with Bell's theorem and other limitations on local realistic theories.

REFERENCES AND NOTES

1. YOUSSEF, S. 1991. Mod. Phys. Lett. **6:** 225.
2. YOUSSEF, S. 1994. Mod. Phys. Lett. **9:** 2571.
3. YOUSSEF, S. 1994. Is complex probability theory consistent with Bell's theorem? SCRI technical report no. FSU-SCRI-94-53. Phys. Lett. A. Submitted.
4. PITOWSKY, I. 1983. Phys. Rev. **D27:** 2316.
5. FEYNMAN, R. P. 1987. *In* Quantum Implications. B. J. Hiley & F. D. Peat, Eds. Routledge & Kegan Paul. London.
6. MÜCKENHEIM, W. *et al.* 1986. Phys. Rep. **133:** 339.
7. GUDDER, S. 1992. Int. J. Theor. Phys. **31:** 15.
8. KOLMOGOROV, A. N. 1956. Foundations of the Theory of Probability. Chelsea. New York.
9. For example, see: JAYNES, E. *et al.* 1989. *In* Maximum-Entropy and Bayesian Methods. J. Skilling, Ed. Kluwer. Dordrecht.
10. We use the standard notation "$\wedge$" for *and* and "$\neg$" for *not*. Probabilities that we denote by $(a \rightarrow b)$ might otherwise be written as $P(b|a)$. The arrow notation $(a \rightarrow b)$ was chosen to suggest a numerical version of logical implication and is meant to be read as "a goes to b".
11. COX, R. T. 1946. Am. J. Phys. **14:** 1.

Index of Contributors